DEDICATION

This work is respectfully dedicated

to the everlasting memories of

my father, mother and

Brothers Abrar Husain

and Ishtiaq Husain

PREFACE TO THE FIRST EDITION

This book is intended to serve as a text book for the undergraduate students of Electrical, Electronics, Computer, Instrumentation, Control and Communication Engineering. 'Networks and Systems' deals with a comprehensive treatment of fundamental topics in Circuit Theory/Network Analysis. It covers the syllabi of the core courses of almost all Universities and Engineering Institutions in the country. It will serve as a valuable reference source for candidates appearing for AMIE, IETE, GATE, UPSC Engineering services, IAS entrance examinations.

The book presents systematic presentation of basic concepts, principles and techniques involved in Circuit Analysis. The subject matter in each chapter has been developed systematically from first principles. Definitions statements of laws, network theorems have been dealt very clearly. Emphasis is placed on precise and logical presentation of principles, concepts and applications.

The book is written in a very simple language and therefore, easy to follow for self study. The subject matter has been presented in a graded manner. Basic concepts have been illustrated as much as possible through the use of resistive circuits. These concepts are then expanded to add other circuit elements for analysis in the frequency domain.

The illustration of principles and concepts has been done by means of a large number of carefully selected worked examples. Step-by-step procedures for solving problems are provided. Examples have been worked in sufficient details so that the reader can follow each example from beginning to end. Most simplified methods have been provided for solving problems to enable the reader to grasp the subject more conveniently and confidently.

At the end of each chapter a large number of unsolved problems with answers have been included for practice.

ASHFAQ HUSAIN

Networks and Systems

[For BE./B.Tech, AMIE, IETE, GATE, UPSE (IAS, IES)]

SECOND EDITION

ASHFAQ HUSAIN

Formerly Reader in Electrical Engineering, University Polytechnic,
Faculty of Engineering and Technology
Aligarh Muslim University, Aligarh - 202002
U.P. (India)

KHANNA BOOK PUBLISHING CO. (P) LTD.

Publisher of Engineering and Computer Books

4C/4344, Ansari Road, Darya Ganj, New Delhi-110002

Phone : 011-23244447-48 **Mobile:** +91-9910909320

E-mail : contact@khannabooks.com

Website : www.khannabooks.com

Dear Readers,

To prevent the piracy, this book is secured with HIGH SECURITY HOLOGRAM on the front title cover. In case you don't find the hologram on the front cover title, please write us to at contact@khannabooks.com or whatsapp us at +91-99109 09320 and avail special gift voucher for yourself.

Specimen of Hologram on front Cover title:

Moreover, there is a SPECIAL DISCOUNT COUPON for you with EVERY HOLOGRAM.

How to avail this SPECIAL DISCOUNT:

Step 1: Scratch the hologram
Step 2: Under the scratch area, your "coupon code" is available
Step 3: Logon to www.khannabooks.com
Step 4: Use your "coupon code" in the shopping cart and get your copy at a special discount
Step 5: Enjoy your reading!

Price: ₹ 525.00

ISBN: 978-81-87522-08-9

Networks and Systems *by*
Ashfaq Husain

First Edition: 2001
Second Edition: 2015
Reprint: 2021,2023, **2025**

To view complete list of books, Please scan the QR Code:

Published by:
Khanna Book Publishing Co. (P) Ltd.
CIN: U22110DL1998PTC095547

Visit us at: **www.khannabooks.com**
Write us at: **contact@khannabooks.com**

Printed in India.

PREFACE TO THE SECOND EDITION

In response to the suggestions for teachers and students the following new chapter is incorporated in the revised edition :

Passive Network Synthesis

An introductory topic on **Convolution** has been included. Some sections of the book have been rewritten.

Additional worked examples have been provided in some chapters.

I hope that this book will be useful to teachers and students alike.

I will be greatful to the readers for their suggestions for further improvement of the book.

ASHFAQ HUSAIN

ACKNOWLEDGEMENTS

I appreciate the patience, understanding and support of my wife Dr. Nigar Minhaj, Reader in Electronics, Womens Polytechnic, Aligarh Muslim University, Aligarh. Her valuable suggestions and comments have made it possible for me to complete the book.

I also appreciate the patience and concern of my sons Ahmad Ashfaq, M. Tech. (Environmental Engineering), Lecturer in Civil Engineering, University Polytechnic, Aligarh Muslim University, Aligarh and Haroon Ashfaq, M.Tech. (Electrical Engineering), Lecturer, Department of Electrical Engineering, Faculty of Engineering and Technology, Jamia Millia Islamia, New Delhi (India), who have helped me a lot in the completion of this book.

I am grateful to my publisher Khanna Book Publishing Co. (P) Ltd. for bringing out this book in a scheduled time.

ASHFAQ HUSAIN

CONTENTS

Chapter 3

SIGNAL WAVEFORMS 36–66

Chapter 4

KIRCHHHOFF'S LAWS AND EQUIVALENT NETWORKS 67–127

Chapter 9

Chapter 10

Chaper 12

NETWORK THEOREMS 425–526

Chapter 13

NETWORK GRAPH THEORY 527–593

Chapter 15

STATE VARIABLE ANALYSIS 690–718

Chapter 16

Chapter 17

COUPLED ELEMENTS AND CIRCUITS 829–887

Chapter 18

NETWORK FUNCTIONS 888–952

Chapter 19

Chapter 20

Chapter 21

1

INTRODUCTORY

1.1 INTRODUCTION

An electrical system (or a network) is an interconnection of electrical elements. In general, a stimulus (or excitation, or input) is applied to the system. The system then performs the desired functions and results in an effect (or response, or output). A block diagram showing this general situation is given in Fig. 1.1 where only one input and one output are shown. In general, the system may have more than one input and one output. The single-input, single-output system is one of the most important types of the system met in practice. Furthermore a multi-input system can often be treated as a combination of several single-input systems.

Figure 1.1 shows that the system operates on the input and transforms it into the output. A set of mathematical equations relating the output to the input is called as **mathematical model** (or simply a **model**) of the system. The excitation signal is denoted by $x(t)$ and the response by $y(t)$.

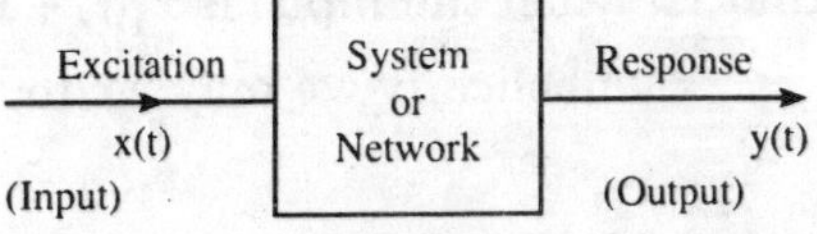

FIG. 1.1
Excitation-response relationship

In the study of electric networks we are mainly concerned with the following types of problems :

1. **Circuit analysis** is the process of determining the current that flows in one or more components of an electric circuit and/or the voltage that exists at various points in the circuit, or the power delivered.
2. **Circuit synthesis** is the process of relating components having certain specified properties, and interconnecting them in a way necessary to achieve a desired goal.
3. **Design** is a process wherein the results of synthesis are subjected to additional criteria, such as economics, safety, aesthetics, resource conservation, environmental factors, and the like.
4. **Circuit troubleshooting** is identifying a defective component, or other circuit fault, based on a knowledge of voltages and currents that are supposed to exist in a circuit and on the actual values measured in the defective circuit.

Although there are similarities between the problems of analysis and synthesis, it is a fact that while a problem in analysis usually has a unique solution, a problem in synthesis generally has several equivalent solutions and, in some cases, may have no solution at all.

In this book we shall deal only with the analysis of linear time invariant casual networks assumed to be made up of ideal elements. We shall discuss these properties :

1.2 LINEARITY

Mathematically linear systems are those whose behavior is governed by linear algebraic equations, linear difference equations, or linear differential equations. A system is said to be linear if, and only if, it satisfies the properties of superposition and homogeneity.

1.2.1. Superposition or Additive Property

The superposition property states that the response to a sum of independent excitations acting on a relaxed system is equal to the sum of the responses due to each of these excitations acting alone, assuming that the other excitations are zero. *A system is said to be* ***at rest*** *or* ***relaxed,*** *when there is zero energy stored in the system prior to the application of the input.*

According to this principle the presence of one excitation does not affect the response due to other excitations and there are no interactions among different responses. To determine the total effect of a number of excitations acting on a relaxed system we determine the effect of each individual excitation at a time assuming that the other excitations are not present. The total response (effect) is found by combining (adding or superposing) all such outputs (responses) computed by considering one excitation (input) at a time.

Suppose that $y_1(t)$ is the response of a relaxed system to the input $x_1(t)$ and $y_2(t)$ is the response of the same system to the input $x_2(t)$, then if both inputs are acting simultaneously, that is, when the input is $x_1(t) + x_2(t)$ then the response will be $y_1(t) + y_2(t)$.

Symbolically we may write

$$x_1(t) \rightarrow y_1(t)$$

$$x_2(t) \rightarrow y_2(t)$$

then $$x_1(t) + x_2(t) \rightarrow y_1(t) + y_2(t) \tag{1.2.1}$$

Here the symbol $\rightarrow$ means "gives rise to". This property is shown in Fig. 1.2.

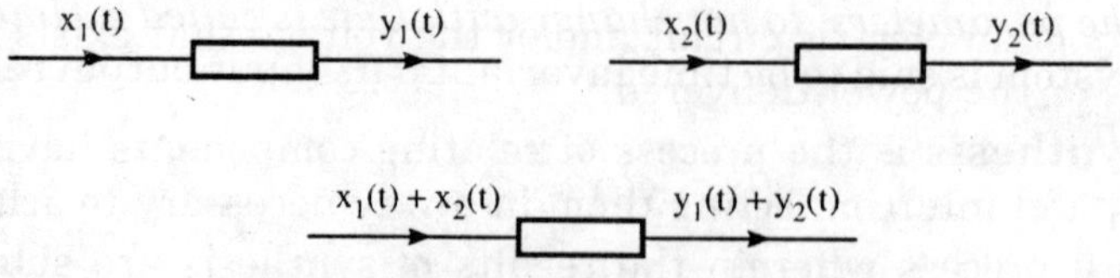

FIG. 1.2 Superposition property

1.2.2 Homogeneity or Proportionality Property

If there are k identical inputs applied to the same part of the system, that is, if

$$x_1(t) = x_2(t) = \ldots = x_k(t) = x(t)$$

and $$x_1(t) \rightarrow y_1(t)\ ,\ x_2(t) \rightarrow y_2(t)\ ,\ \ldots$$

$$x_k(t) \rightarrow y_k(t),$$

then by superposition theorem

$$x_1(t) + x_2(t) + \ldots + x_k(t)$$
$$\rightarrow y_1(t) + y_2(t) + \ldots + y_k(t)$$

$$k\, x_k(t) \rightarrow k\, y_k(t) \tag{1.2.2}$$

Thus if the excitation (input) is multiplied by a real constant, the response (output) is also multiplied by the same constant. This property is known as the **property of homogeneity** or **proportionality** and is shown in Fig. 1.3.

$x_k(t)$ $y_k(t)$ $k\,x_k(t)$ $k\,y_k(t)$

FIG. 1.3 Homogeneity property

The properties of superposition and homogeneity can be combined together to give the property of linearity.

If $\quad x_1(t) \rightarrow y_1(t)$

$\quad x_2(t) \rightarrow y_2(t)$

then $\quad k\,[x_1(t) + x_2(t)] \rightarrow k\,[y_1(t) + y_2(t)] \quad (1\ 2.3)$

Equation (1.2.3) represents the **linearity property** (**superposition property + homo-geneity property**). The property of linearity is shown in Fig. 1.4.

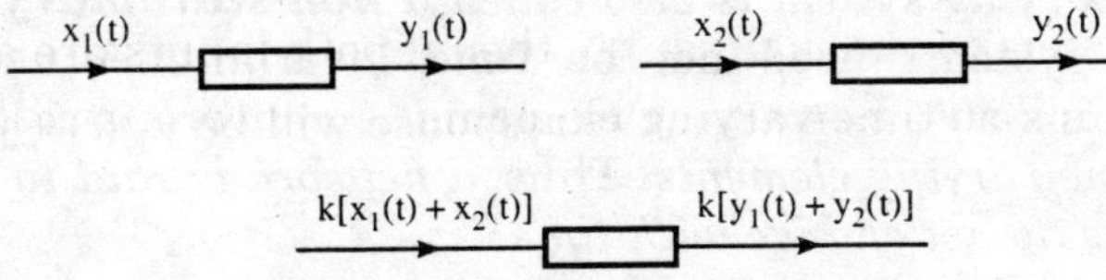

$x_1(t)$ $y_1(t)$ $x_2(t)$ $y_2(t)$

$k[x_1(t) + x_2(t)]$ $k[y_1(t) + y_2(t)]$

FIG. 1.4 Linearity Property

1.3 TIME INVARIANCE

A system in which the parameters do not change with time is called a ***time-invariant system.*** Mathematically, a system is said to be time invariant if its input-output relations do not change with time. That is, if

$$x(t) \rightarrow y(t) \tag{1.3.1}$$

then $\quad x\,(t - \tau) \rightarrow y\,(t - \tau) \quad (1.3.2a)$

or $\quad x\,(t + \tau) \rightarrow y\,(t + \tau) \quad (1.3.2b)$

for all t and τ. Here τ is any time shift. The response (output) of a time invariant system depends only on the shape of the excitation (input) signal and not on the instant of application of the input. Thus in a time-invariant system the relation between its response and excitation always remains the same, regardless at the time of application of the input.

The time invariance property is shown in Fig. 1.5.

In Fig. 1.5a, the excitation is applied at $t = 0$ and it gives rise to a response $y(t)$. If the excitation is applied at $t = \tau$ and if the shape of the response waveform remains the same as in the first case, but delayed by a time τ (Fig. 1.5b) then we say that the system is time invariant. Time-invariant systems are also called **constant-parameter systems** or **stationary systems.**

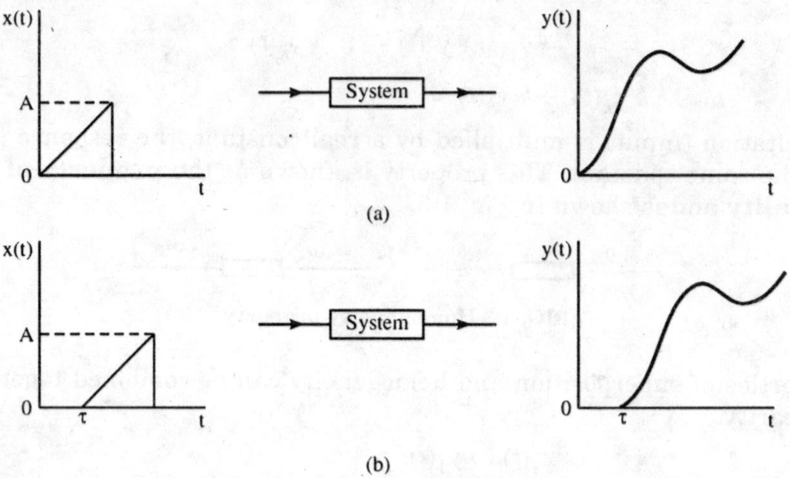

FIG. 1.5 Time invariance property

Whenever Eq. (1.3.2) is not satisfied or the parameters of the system are time dependent, the system is said to be *time varying*. In other words, the system is time varying if it is not time invariant. *A time varying system is also called a* ***non-stationary system.*** It should be noted that the linear systems need not be time invariant. A network is said to be *time invariant* if it contains no time-varying elements. *A network is called* ***time varying*** *if it contains one or more time-varying elements.* Thus, *a network is said to be* ***time invariant*** *if its input-output relations do not change with time.*

1.4 CASUALITY

A network is said to be ***casual*** *if its response to an excitation is zero until after the excitation is applied.* That is, response does not precede excitation. Thus, for a casual network, if

$$x(t) = 0 \qquad t < T$$

then

$$y(t) = 0 \qquad t < T$$

as shown in Fig. 1.6.

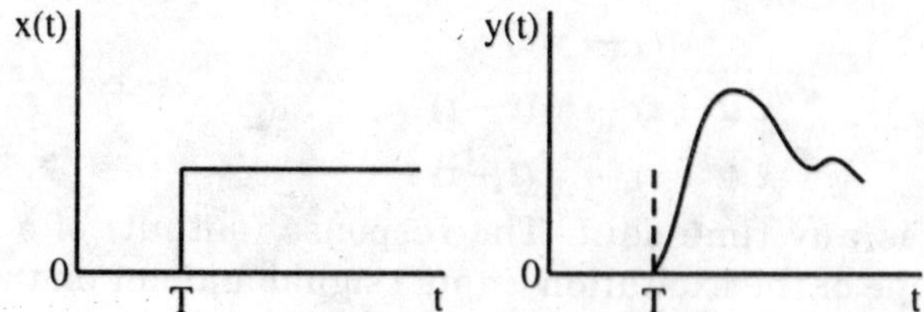

FIG. 1.6 Excitation and response of a casual system

A casual system is non-predictive. That is, a casual system's output at any particular time depends only upon the input prior to that time, and not upon any future values of the input.

Thus, $y(t_0)$ depends only upon $x(t)$ $(t \le t_0)$. The study of casuality is important since, in general, casual systems are physically realizable. It is not possible to synthesize networks which are noncasual.

1.5 PASSIVE AND ACTIVE ELEMENTS

A network element is defined to be passive if the energy delivered to it is non-negative for *all* times t and for *all* combinations of voltage $v(t)$ and current $i(t)$. A two-terminal (or 1-port) element is said to be *passive* if its terminal voltage $v(t)$ and current $i(t)$ satisfy the condition

$$W(t) \triangleq \int_{-\infty}^{t} v(\tau)\, i(\tau)\, d\tau \ge 0 \tag{1.5.1.}$$

for all $t \ge -\infty$ and for all possible combinations of $v(t)$ and $i(t)$. Equation (1.5.1) can also be written as

$$W(t) \triangleq \int_{t_0}^{t} v(\tau)\, i(\tau)\, d\tau + W(t_0) \ge 0 \tag{1.5.2}$$

for all $t \ge t_0$, where $W(t_0)$ is the energy stored in the network at time $t = t_0$. *A 1-port is said to be* ***intially relaxed*** *at t_0 if all initial conditions at $t = t_0$ are zero.* Consequently,

$$W(-\infty) = 0 \quad \text{and} \quad W(t_0) = 0$$

Hence, an initially relaxed one-port is passive if

$$\int_{t_0}^{t} v(\tau)\, i(\tau)\, d\tau \ge 0 \tag{1.5.3}$$

for all $t \ge t_0$ and for all possible values of $v(t)$, $i(t)$ pairs.

Eq. (1.5.2) simply means that the energy consumed by the one port is greater than or equal to zero.

A 2-terminal element (or 1-port) is said to be ***active*** *if it is not passive.* An active element generally supplies electric energy to the network to which it is connected. For an active element $W(t) < 0$.

A 1-port passive network is said to be *lossless* if $W(t) = 0$, other wise, the passive element is called *lossy*. In the next section we shall generalize the concept of passivity of n-port networks.

1.6 PASSIVE AND ACTIVE *n*-PORT NETWORKS

Consider an n-port network which is linear and time invariant. Let $v_k(t)$ and $i_k(t)$ denote the instantaneous voltage and current respectively, at port k–k' (say), as shown in Fig. 1.7.

An n-port network is said to be **passive** if for all real admissible voltage-current pairs, the total energy supplied to its ports is non-negative for all t, that is,

$$W(t) \triangleq \int_{-\infty}^{t} \sum_{k=1}^{n} [v_k(\tau)\, i_k(\tau)\, d\tau] \ge 0$$

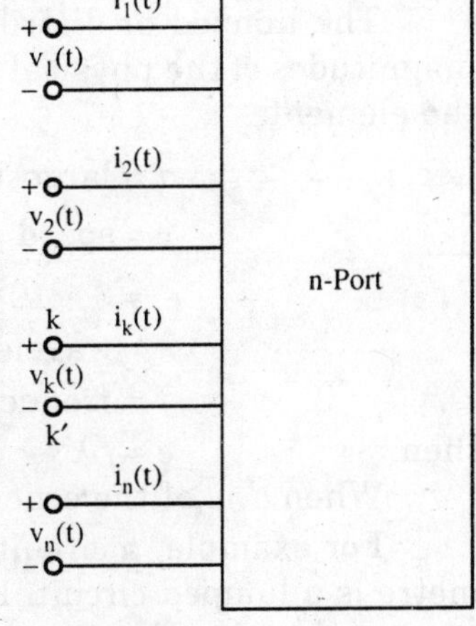

FIG. 1.7 ***n*-port**

$$\text{for all } t \geq -\infty \qquad (1.6.1)$$

assuming that at $t = -\infty$, the network is *relaxed*, that is,

$$v_k(-\infty) = 0 \quad \text{and} \quad i_k(-\infty) = 0$$

Thus a network is *passive*, if it cannot generate or amplify energy, that is, if the energy which it has supplied since the beginning of time cannot exceed the energy which was fed into it.

An n-port network is said to be lossless if the total energy absorbed or delivered by the network at its ports is zero. In Eq. (1.6.1) $W(t)$ is the difference between energies entering and leaving the network, If $W(t)$ is equal to the energy stored in the network then the network neither generates nor loses energy. Such a passive network is called a *lossless* network, otherwise, the passive network is lossy.

An n-port network is said to be *active* if $W(t) < 0$. An active network can supply excess energy to the outside world. In other words, an active network supplies energy to the network to which it is connected.

1.7 LUMPED AND DISTRIBUTED NETWORKS

An electric **network** consists of a number of elements connected together. The word **circuit** is synonymously used with the term **network**. The components, of a network have definite physical dimensions. The voltage and current are functions of time. A time varying field produces electromagnetic waves. *A circuit is said to be* ***lumped*** *if the physical dimensions of all its components are negligible compared with the wavelength of the electromagnetic signal inside the component.* In such a circuit Kirchhoff's current and voltage relations hold. The current within any branch is the same at all points of the branch. The voltage along any wire is constant and it changes only at the components.

If the physical dimensions of the elements are comparable to the signal wavelength, the voltage and current are functions of time as well as space. Such a circuit in which the voltage and current are functions of time and position is called a **distributed parameter circuit**. *A circuit in which the voltage and current are functions of time only is called a* ***lumped parameter circuit*** *or simply* ***lumped circuit***. In a distributed parameter circuit Kirchhoff's laws are no longer valid and the more general laws of Maxwell must be applied. In a distributed parameter system, it takes a finite amount of time for the electromagnetic wave at one point to be propagated to the other point.

The lumped or distributed nature of the network is decided by comparing the relative magnitudes of the physical dimensions of the elements and the wavelength of the signal inside the element.

Let d = largest dimension of the circuit

c = speed of propagation of electromagnetic waves

$= 3 \times 10^8$ m/s

λ = wavelength of the highest frequency of interest

f = frequency

then $c = f\lambda$

When d is of the order of or larger than λ it is a distributed circuit.

For example, a circuit at 50 Hz and having the physical dimensions of the order of one metre is a lumped circuit. For such a circuit

$$\lambda = \frac{c}{f} = \frac{3 \times 10^8}{50} = 6 \times 10^6 \text{ m}$$

For a circuit with $f = 10$ Ghz $= 10^{10}$ Hz

$$\lambda = \frac{c}{f} = \frac{3 \times 10^8}{10^{10}} = 0.03 \text{ m}$$

Such a circuit is a distributed circuit. A 100 km power line at 50 Hz is a distributed circuit. For a microwave circuit, λ is between 10 cm and 1 mm and KCl and KVL do not apply. Therefore a microwave circuit is a distributed circuit.

In this book, we shall restrict our discussion to lumped circuits, that is, circuits consisting of lumped elements only.

2

CIRCUIT ELEMENTS

2.1 TWO-TERMINAL CIRCUIT ELEMENTS

An electrical device may be divided into regions or volumes. Each region has at least two accessible points called *terminals*. A simple circuit element may be defined as a two-terminal electrical device which cannot be subdivided into other two terminal devices.

Fig. 2.1 shows the symbolic representation of a two-terminal circuit element. The end points marked A and B represent the two terminals which are available for interconnection with other circuit elements. Let the currents entering and leaving the element be $i_1(t)$ and $i_2(t)$ respectively. A two-terminal circuit element is said to be *lumped* if it possesses the following properties :

(1) The instantaneous current entering one terminal is equal to the instantaneous

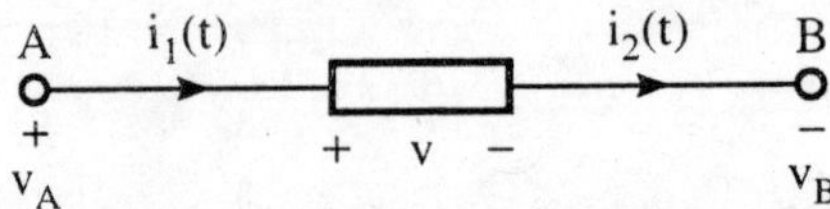

FIG. 2.1 Symbolic representation of a two-terminal lumped circuit element

current leaving the other terminal. For the element shown in Fig. 2.1,

$i_1(t) = i_2(t) = i(t)$ for all t where $i(t)$ is called the *element current*.

(2) Let v_A and v_B denote the potentials of A and B with respect to an arbitrary reference. Then to the lumped element connected to these terminals there corresponds a unique potential difference $(v_A - v_B)$ called the terminal potential difference, which is denoted by $v\ (t)$. That is

$$v\ (t) = v_A(t) - v_B(t)$$

2.2 VOLTAGE POLARITY

Voltage polarity on a circuit element is assigned arbitrarily by assuming that one terminal is at a higher potential than the other terminal. The terminal assumed at higher potential is marked with a plus (+) sign and the terminal assumed to be at the lower potential is marked with a minus (–) sign as shown in Fig. 2.2. Let v_A and v_B be the potentials at A and B respectively. In Fig. 2.2a the terminal A is assumed to be at higher potential than B, and if the potential difference between A and B is $v_1(t)$ then

$$v_1(t) \overset{\Delta}{=} v_A - v_B$$

In Fig. 2.2b the terminal B is assumed to be at higher potential than A, and $v_2(t)$ denotes the potential difference between B and A, then

$$v_2(t) \overset{\Delta}{=} v_B - v_A$$

From these two definitions it implies that

$$v_1(t) = -\,v_2(t) \qquad \text{for any } t$$

FIG. 2.2 Voltage sign convention
(a) *A* is at higher potential than *B*, (b) *B* is at higher potential than *A*

2.3 REFERENCE DIRECTION OF CURRENT

Based on the assumed voltage sign convention we can assign the reference direction of current through an element. The reference direction for the current in a circuit element is shown by an arrow through it. It should be noted that the reference directions are chosen arbitrarily and they need not represent the actual direction of current flowing in the element. If the actual current in the element is in the same direction as the assigned current reference, the current is said to be positive and equal to the actual current. If the actual current is flowing in the opposite direction to our arbitrarily assigned current reference, we say that the current is negative and equal to the negative (absolute value) of the actual current.

2.4 ASSOCIATED REFERENCE DIRECTIONS

The power delivered to a circuit element in terms of the sign conventions of Fig. 2.3, is given by the product of the voltage *across* and current *through* it. Thus

$$p(t) = v\,(t)\,i(t)$$

The energy delivered over an interval from $t = t_1$ to $t = t_2$ is given by

$$W\,(t_1\,,\,t_2) = \int_{t1}^{t2} p(t)\,dt = \int_{t1}^{t2} v\,(t)\,i(t)\,dt \qquad (2.4.1)$$

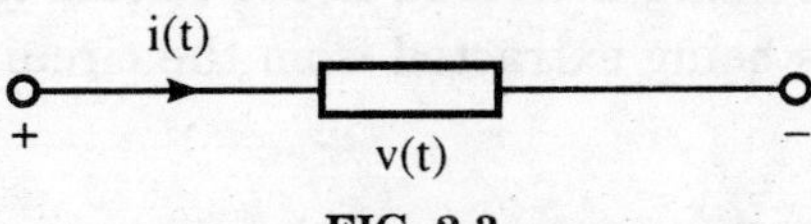

FIG. 2.3

When the power and energy are positive quantities, they are flowing *into* the circuit element. When the power and energy are negative quantities, the power and energy are flowing *out* of the circuit element. It is to be noted that the reference directions of voltage and current are assigned arbitrarily. In practice, *associated reference directions* are used. They are as follows :

1. For an *non-source element* (such as resistors, inductors, capacitors etc.) the positive current enters element at the terminal with plus (+) sign and leaves the element at the terminal with minus (–) sign as shown in Fig. 2.4a.

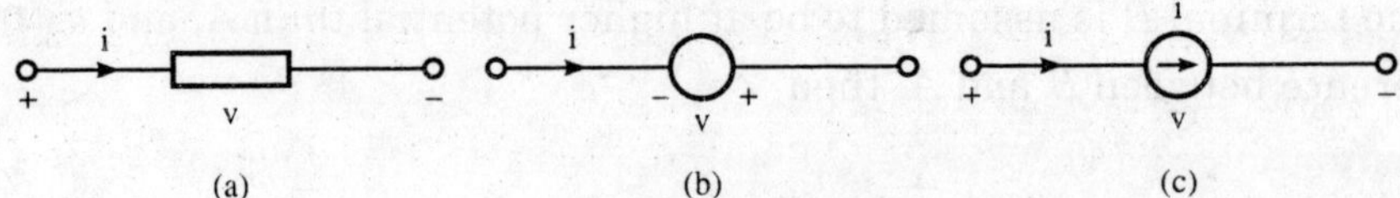

FIG. 2.4 Associated reference directions (a) Two-terminal nonsource element (b) Independent voltage source (c) Independent current source

2. *For source element* (voltage source or current source) the positive current enters the negative (–) terminal and leaves the terminal marked with plus (+) sign. Figs. 2.4b and 2.4c show the associated reference directions for an independent voltage source and an independent current source respectively.

Fig. 2.5 summarizes the relationship between the polarity references for voltage and current and the expression for power.

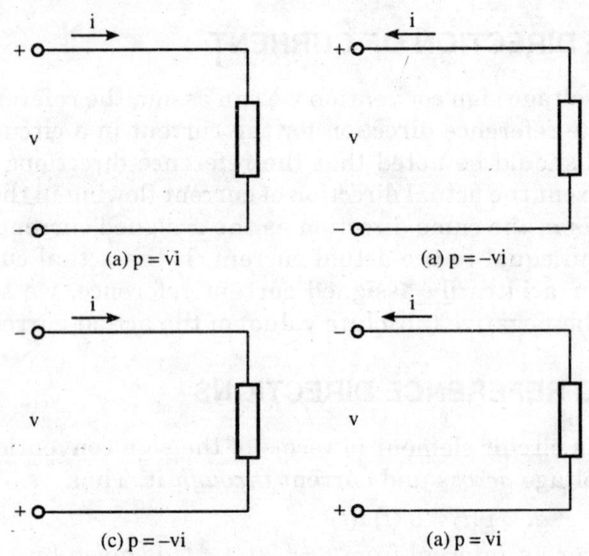

FIG. 2.5 Polarity references and the expressions for power

The algebraic sign of power is interpreted as follows :

(a) If $p(t) > 0$, the power is being delivered to the circuit or circuit component.

(b) If $p(t) < 0$, the power is being extracted from the circuit or circuit component.

2.5 SHORT CIRCUIT

Short circuit between any two points can be defined in terms of Fig. 2.6. If the circuit element is replaced by a short length of an ideal conductor (short and thick good conducting material), the terminals are said to be short circuited.

The voltage across the short circuit is always zero regardless of the current through it. The current in the short circuit depends upon the circuit connected to the terminals. Thus, a short circuit (SC) is a two terminal element for which the *voltage must be zero* when the *current is not zero*. This statement is mathematically written as

$$v\ (t) \equiv 0\,, \qquad i\ (t) \neq 0 \tag{2.5.1}$$

A short, heavy slab of good conducting material, such as copper or aluminium, is usually a good approximation to a short circuit. Figure 2.6b shows the schematic representation of a short circuit.

2.6 OPEN CIRCUIT

Open circuit between any two points can be defined in terms of Fig. 2.6.

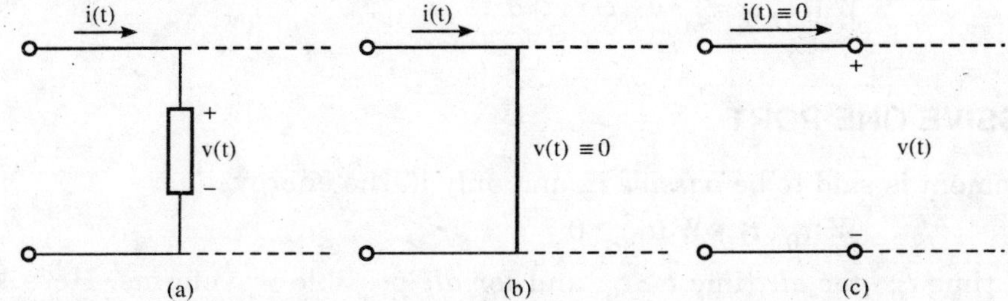

FIG. 2.6 (a) Two-terminal circuit element (b) Short circuit (c) An open circuit

The terminals in Fig. 2.6a are open circuited when the circuit element is removed. Thus, there is no electrical path between the terminals. *The current in an open circuit is always zero regardless of the voltage across it.* The voltage across an open circuit depends upon the circuit connected to the terminals. Thus, an open circuit is a 2-terminal element for which the *current must be zero* when the *voltage is not zero*. This statement is written mathematically as

$$i\ (t) \equiv 0\ ,\ v\ \ (t) \neq 0 \tag{2.6.1}$$

The air surrounding two terminals is usually a good approximation of an open circuit. This element can be realized in practice by cutting a length of conductor into two. The two ends constitute the element. Figure 2.6c shows the schematic representation of an open circuit.

2.7 ONE PORT

A **port** is a pair of terminals of a circuit in which, at all times, the instantaneous current entering one terminal is equal to the instantaneous current leaving the other terminal. That is,

$$i_{\text{in}}\ (t) = i_{\text{out}}\ (t) \quad \text{for all } t \tag{2.7.1}$$

A one-port is shown in Fig. 2.7. The reference directions of a 1-port are shown in Fig. 2.7*b*. The current $i\ (t)$ entering the port is called the **port current** and the voltage $\upsilon\ (t)$ across the port is called the **port voltage**. A 2-terminal circuit is a one-port, Thus, resistors, inductors, capacitors, and independent voltage sources are examples of one ports consisting of a single element. The *instantaneous power* entering the one port is equal to the product of the port voltage and the port current provided the reference directions of the port voltage and the port current are associated reference directions.

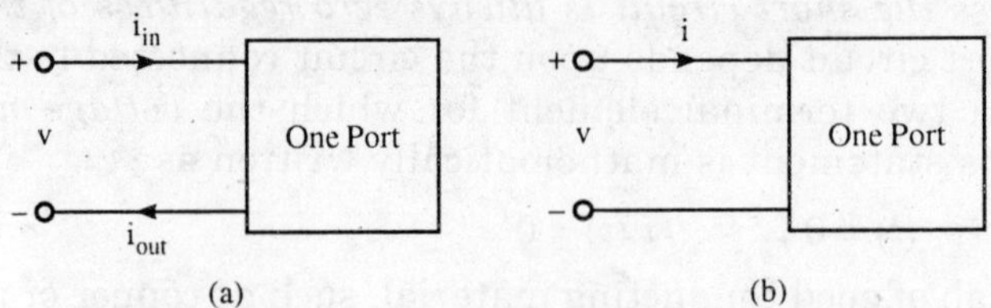

FIG. 2.7 (a) One port (b) Reference directions

The instantaneous power entering the port at time t is

$$p(t) = v(t)\, i(t) \tag{2.7.2}$$

The energy delivered to the one port from time t_0 to time t is

$$W(t_0, t) = \int_{t_0}^{t} v(\tau)\, i(\tau)\, d\tau \tag{2.7.3}$$

2.8 PASSIVE ONE PORT

A one-port element is said to be *passive* if, and only if, the energy

$$W(t_0, t) + W(t_0) \geq 0 \tag{2.8.1}$$

for all initial time t_0, for *all* time $t \geq t_0$, and for *all* possible waveforms. Here $W(t_0)$ is the energy stored in the one port at time t_0, and $W(t_0, t)$ is the energy supplied to the port from time t_0 to time t.

$$W(t_0, t) = \int_{t_0}^{t} v(\tau)\, i(\tau)\, d\tau \tag{2.8.2}$$

A one-port is said to be ***initially relaxed*** *at t_0 if all intial conditions at t_0 are zero.* Consequently, $W(t_0) = 0$. Hence, an initially relaxed one port is passive if

$$\int_{t_0}^{t} v(\tau)\, i(\tau)\, d\tau \geq 0 \tag{2.8.3}$$

for all $t \geq t_0$, for all initial times t_0, and for all possible $v(t)$, $i(t)$ pairs. Relation (2.8.3) simply means that the energy consumed by the one port is greater than or equal to zero.

A one port is said to be **active** if it is not passive. Accordingly, a one port is active if $W(t_0, t) < 0$ for *some* $t \geq t_0$, or for *some* $i(t)$ or $v(t)$.

2.9 ELECTRICAL ENERGY SOURCES

The two sources of electrical energy are the votage source and the current source. These sources may be classified as ideal or nonideal, dependent or independent.

2.10 IDEAL VOLTAGE SOURCE

An **ideal voltage source** is a 2-terminal circuit element that is capable of producing a specified voltage across its terminals regardless of what is connected to it. Such a source, therefore, can supply any amount of current while maintaining the specified voltage v_s (t) at its terminals. The amount of current supplied depends on the circuit connected to the voltage source. Only the voltage is constrained to be v_s (t) volts at the terminals. The magnitude and the waveform of the voltages source remain the same regardless of the network connected at its terminals. The voltage may be a constant, or it may be a function of time t. The voltage source for which the value of the source voltage is a constant (that is, it does not change with time), is represented by the symbol shown in Fig. 2.8.

FIG. 2.8 Constant voltage source of voltage V_0

FIG. 2.9 General symbol for an ideal voltage source

A constant voltage source is called a dc source or an *ideal battery.*

In general, the voltage v_s (t) produced by an ideal voltage source is a function of time t. Fig. 2.9 shows the general representation of an ideal voltage source.

Fig. 2.10 shows an ideal voltage source connected to an arbitrary network. Let

v_s (t) = source voltage

v (t) = terminal voltage

i (t) = source current

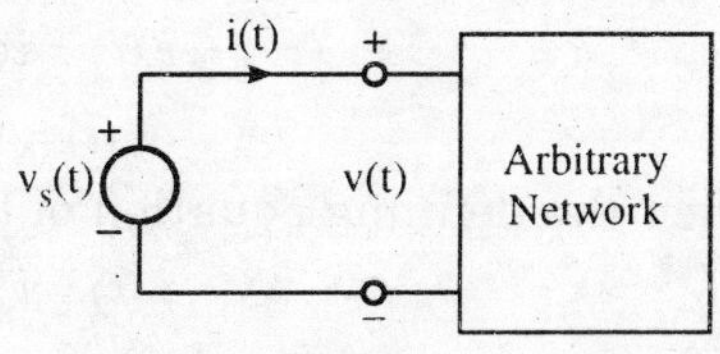

FIG. 2.10 Ideal voltage source connected to an arbitrary network

then the defining equation of an ideal voltage source may be written as

$$v\ (t) = v_s\ (t) \quad \text{for any } i\ (t) \tag{2.10.1}$$

A voltage source is completely specified by its voltage vu_s (t) for all t. Since the voltage of the source does not depend in any manner on the connected network, an ideal voltage source is also called an independent voltage source.

Some of the examples of independent voltage sources are :

$v_s = 10$

$v_s = 120 \sin 314\, t$

$v_s = 15\, t^2$ (t is time)

$v_s = 32\, (T - 273)$ (T = temperature in K)

In Eq. (2.10.1),if we put v_s $(t) = 0$, then v $(t) = 0$, independent of the current i (t) flowing through the source. Zero voltage across a pair of terminals with the current unconstrained, defines a *short circuit.* Thus, an ideal voltage source whose voltage v_s (t) is identically zero represents a short circuit, provided that the current through it is finite. The independent ideal voltage source is a *concept* and not a real-world device.

2.11 IDEAL CURRENT SOURCE

An **ideal current source** (Fig. 2.11) is a 2-terminal circuit element that is capable of producing a specifed current $i_s(t)$ through it in the direction of the arrow, regardless of what is connected to it. The current supplied by the source is indepedent of the voltage at the source terminals. The voltage across the terminals is determined by the circuit connected to the source terminals. Only the current is constrained to be $i_s(t)$ through the source. The magnitude and waveform of current at the terminals of a current source remain the same irrespective of the network connected at its terminals.

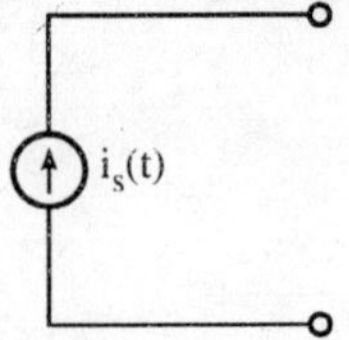

FIG. 2.11 Ideal current source

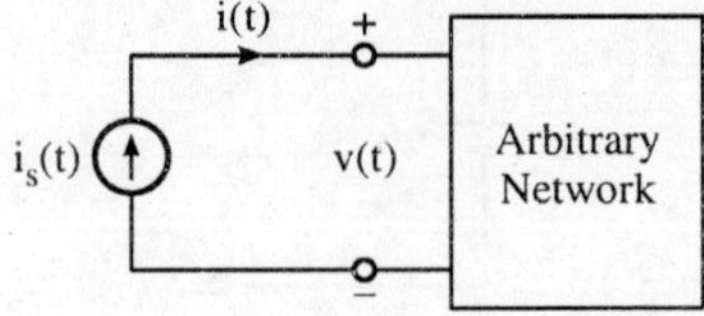

FIG. 2.12 Ideal current source connected to an arbitrary network

Fig. 2.12 shows an ideal current source connected to any arbitrary network. Let

$i_s(t)$ = source current

$i(t)$ = terminal current

$v(t)$ = terminal voltage

then the defining equation of the ideal current source may be written as

$$i(t) = i_s(t) \qquad \text{for any } v(t) \tag{2.11.1}$$

A current source is completely specified by its current $i_s(t)$ for all t. From the definition of the ideal current source it is clear that the ideal current sources shown in Fig. 2.13 are equivalent.

Again, the current produced by an ideal current source may be a constant or a function of time. Since the current through the ideal current source does not depend in any manner on the connected network, it is also called an *independent current source*.

FIG. 2.13 Equivalent current sources

Some of the examples of current sources are :

$$i_s = 5$$

$$i_s = 8 \sin 377\, t$$

$$i_s = a\,T^2 - b\,T + c \qquad (T \text{ is temperature in } K \text{ and } a\,,\,b\,,\,c \text{ are constants})$$

In Eq. (2.11.1), if we put $i_s(t) = 0$, then $i(t) = 0$, independent of voltage $v(t)$ across the source terminals. Zero current through an element, with the voltage unconstrained, defines an open circuit. Thus, an ideal current source whose current $i_s(t)$ is identically zero represents an open circuit.

2.12 REAL, PRACTICAL, OR NONIDEAL VOLTAGE SOURCE

An ideal voltage source is a source of energy capable of supplying any amount of electrical energy at constant voltage. This implies that the internal impedance of an ideal voltage source is zero. A real (practical) voltage source, however, has a small but finite internal impedance Z_{sv} , so that the terminal voltage $v(t)$ of the source depends on the current that flows through the source. The circuit symbol for a practical voltage source is shown in Fig. 2.14a. A practical voltage source is also called *nonideal voltage source.* If Z_{sv} becomes zero, the terminal voltage is no longer a function of the load conditions. Such a voltage source then becomes an ideal voltage source. It should be noted that a practical voltage source is *not* constructed from an ideal voltage source and an impedance.

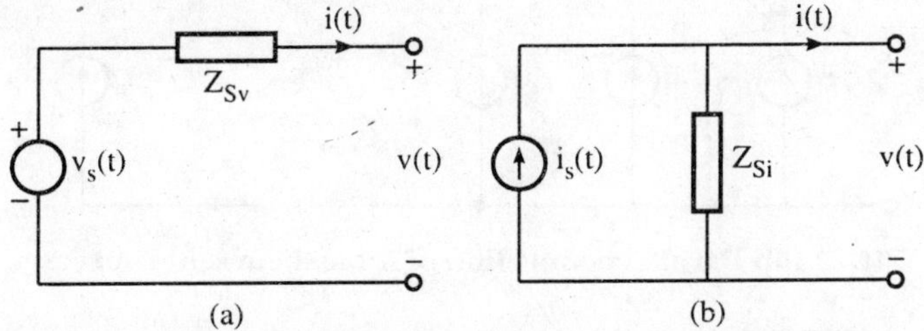

FIG. 2.14 (a) Practical voltage source (b) Practical current source

2.12.1 Practical or Nonideal Current Source

An ideal current source is a source of energy capable of delivering any amount of electrical energy at constant current. This implies that the internal impedance of an ideal current source is infinite, so that the current it supplies to an external impedance is independent of the magnitude of the load impedance. A practical current source, however, has a large but finite internal impedance Z_{si} , so that the current supplied to the external load is dependent on the load. The symbolic representation of a practical current source is shown in Fig. 2.14 b.

A practical current source is also called *nonideal current source..* If Z_{st} becomes infinite, the current, in the current source is no longer a function of the load conditions. Such a current then becomes an ideal current source.

2.13 SERIES AND PARALLEL CONNECTION OF SOURCES

(a) Series connection of Voltage Sources

Fig. 2.15a shows a series connection of n ideal voltage sources. By *KVL* the total voltage $v(t)$ is given by

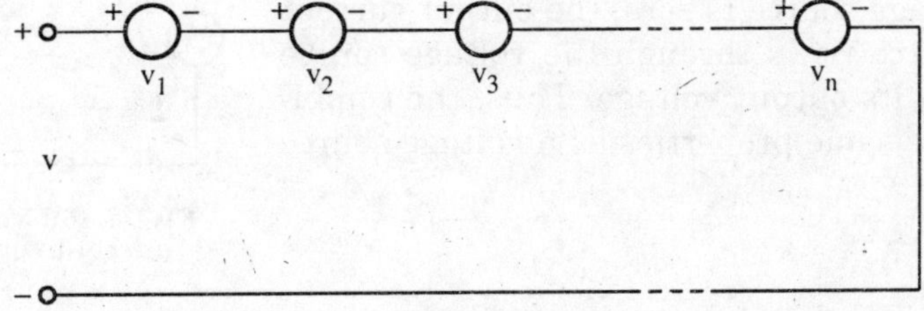

FIG. 2.15a Series connection of *n* ideal voltage sources

$$v(t) = v_1(t) + v_2(t) + v_3(t) + \ldots + v_n(t)$$

(b) Parallel Connection of Current Sources

Fig. 2.15b shows a parallel connection of n ideal current sources. By KCL the total current $i(t)$ is given by

$$i(t) = i_1(t) + i_2(t) + i_3(t) + \ldots + i_n(t)$$

Thus, the total current is the sum of currents from the individual sources.

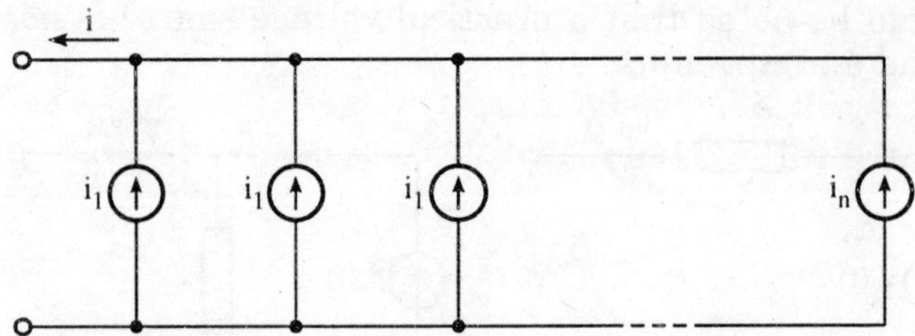

FIG. 2.15b Parallel connection of *n* ideal current sources

The current resulting from a series connection of current sources is not defined except for the case where all the sources have exactly the same output current for all values of time, since otherwise such a connection would violate KCL at any node connecting two sources.

(c) Parallel connection of Identical Voltage Sources

Fig. 2.15c shows the parallel connection of n ideal and identical voltage sources. The overall voltage is equal to the voltage of a single source.

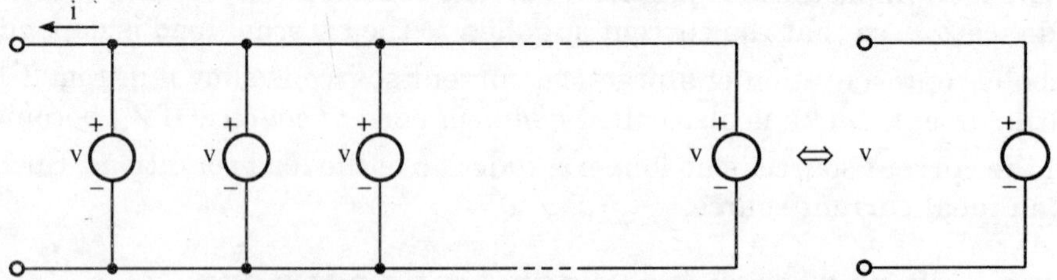

FIG. 2.15c Parallel connection of *n* ideal and identical sources

(d) Voltage Source and Current Source in Parallel

Fig. 2.15d shows an ideal voltage source and an ideal current source in parallel. In this case, since the internal resistance of the voltage source is zero, the output current from the current source flows through the voltage source without any effect on its output voltage. Thus, the combination has exactly the same properties as a voltage source alone.

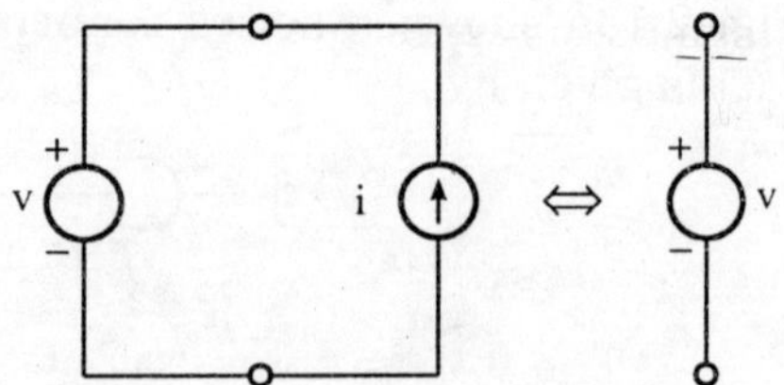

FIG. 2.15d Voltage source and current source in parallel

(e) Voltage Source and Current Source in Series

Fig. 2.15e shows a voltage source connected in series with a current source. In this case, since the voltage across the current source is undefined, the addition of the output voltage of the voltage source gives a quantity which is undefined. Hence the combination has the same properties as a current source alone.

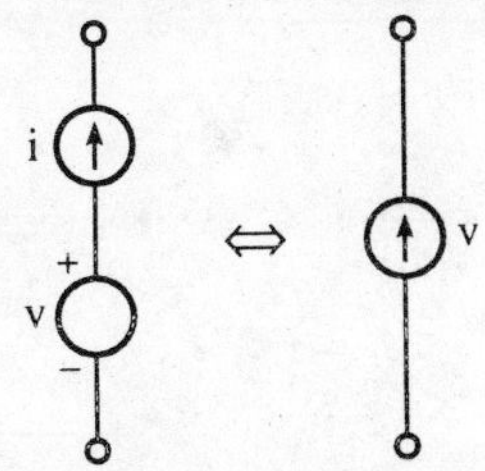

FIG. 2.15e Voltage source and current source in series

2.14 RESISTORS

A 2-terminal element is called a resistor if any time t the current $i(t)$ *through* it and the voltage $v(t)$ *across* it are related through an algebraic relation

$$f[v(t), i(t), t] = 0$$

A curve can be drawn on the v-i (or i-v) plane to satisfy this relationship. This curve is called the *characteristic of the resistor at time t*. This is a general definition classifying diodes, switches and most other 2-terminal elements without memory as resistors. Depending on the shape of their i-v characteristic, resistors can be classified in several categories. The most commonly used resistor is *linear time invariant*. A resistor is called time invariant if its characteristic does not vary with time. If the characteristic changes with time, the resistor is called a *time–varying resistor*. A resistor is called *linear* if its characteristic is at all times a straight line through the origin. Conversely, if at any time the characteristic is not a straight line through the origin of the i-v plane, the resistor is called *nonlinear*.

2.14.1 Linear Time-Invariant Resistor

A linear time-invariant resistor is a 2-terminal network element whose v-i relationship can be written in the form

$$v(t) = R\, i(t) \tag{2.14.1}$$

or

$$i(t) = G\, v(t) \tag{2.14.2}$$

where

$$R = \frac{1}{G} \tag{2.14.3}$$

R and G are constants independent of i, v and t. The quantity R is called the *resistance* and G is called the *conductance*. The relation given by Eq. (2.14.1) or Eq. (2.14.2) is called *Ohm's law*. If v is expressed in volts (V) and i in amperes (A), the resistance of the resistor is in ohms (symbolically written as Ω). The conductance G is then expressed in siemens (S). The symbolic representation for a linear time-invariant resistor is shown in Fig. 2.16.

i(t) R
+ v(t) −

FIG. 2.16. Symbolic representation for R or G

The v-i characteristic of a linear time-invariant resistor is shown in Fig. 2.17. It is a straight linear through the origin in the v-i plane. The slope of this is the resistance R of the resistor.

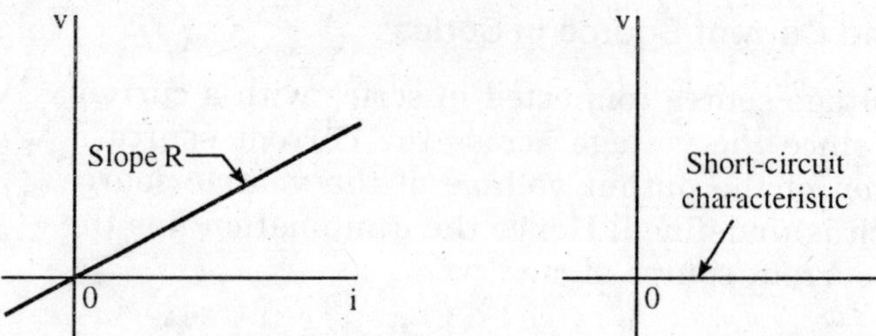

FIG. 2.17. Typical v-i characteristic of a linear time-invariant resistor

FIG. 2.18 Short-circuit (SC) characteristic

In Eq. (2.14.1), if we put $R = 0$, then $v(t) = 0$ whatever be the current through the element. In this case the 2-terminal resistor is said to be *short circuit*. In the *v-i* plane a short-circuit characteristic is the *i* axis as shown in Fig. 2.18, zero, that is, $R = 0$ and equivalently $G = \infty$.

In Eq. (2.14.2), if we put $G = 0$, then $i(t) = 0$ whatever be the voltage across the element. In this case, the two-terminal resistor is said to be an *open circuit*. In the *v-i* plane an open circuit characteristic is the *v* axis as shown in Fig. 2.19.

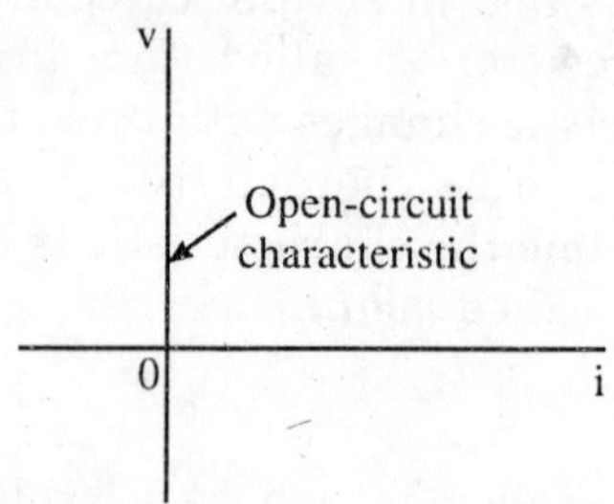

FIG. 2.19 Open-circuit (OC)

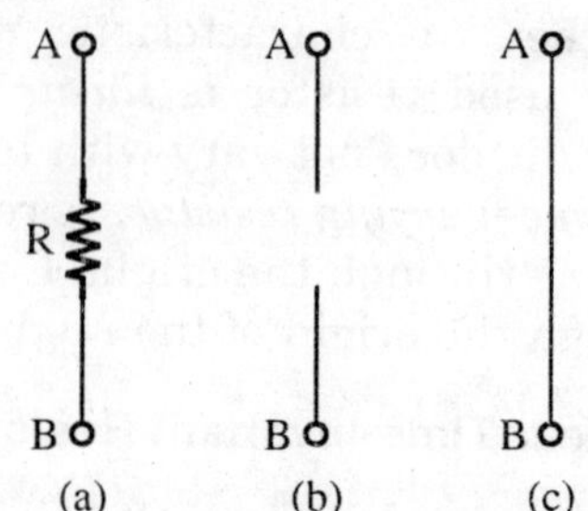

FIG. 2.20 Symbolic representations of characteristic (a) resistor, (b) open circuit (c) short circuit

The slope of this characteristic is infinite, that is, $R = \infty$ and equivalently $G = 0$. The symbolic representations of a resistor, an open circuit, and a short circuit are shown in Fig. 2.20.

As indicated, an open-circuit between the terminals *A* and *B* is shown by a broken line, while a short circuit is shown by a completed line between *A* and *B*.

2.14.2 Application of Ohm's Law

It should be noted that while writing Ohm's law for a linear, time-invariant resistor, directions of currents and polarities of voltages should be taken into account. The accepted sign convention is given in Fig. 2.21a and by Ohm's law $v(t) = R\,i(t)$. In Fig. 2.21b, the current's direction is opposite to that indicated by convention. Since a current of $i(t)$ flowing through the resistor to the left is the same as a current of $-i(t)$ flowing to the right, we can redraw Fig. 2.21b in the equivalent form shown in Fig. 2.21c. Now, the direction of current and polarity of the voltage conform to the convention, and therefore we can apply Ohm's law to write.

$$v(t) = R\,[-i(t)]$$

or

$$v(t) = -R\,i(t)$$

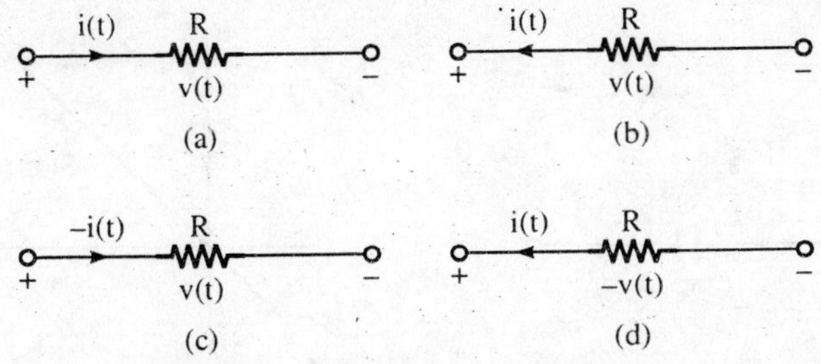

FIG. 2.21 Application of Ohm's law

Alternatively, we can redraw Fig. 2.21b in another equivalent form shown in Fig. 2.21d and apply Ohm's law to write

$$-v(t) = R\, i(t)$$

or
$$v(t) = -R\, i(t)$$

2.15 CAPACITORS

Capacitance is the property of a circuit element or device which is capable of storing charge or energy in n electric field. A 2-terminal circuit element is called a *capacitor* if at any time t its stored charge $q(t)$ and the voltage $v(t)$ across it are related through an algebraic relation

$$f\,[q(t)\,,\, v(t)\,,\, t] = 0$$

Hence a curve can be drawn on the q-v (or v-q) plane to satisfy the algebraic relation. This curve is called the characteristic of the capacitor at time t as shown in Fig 2.22. A capacitor whose characteristic is at all times a straight line through the origin of the v-q plane, is called a *linear capacitor*. Conversely, if at any time the characteristic is not a straight line through the origin of the v-q plane, the capacitor is called *nonlinear*. A capacitor whose characteristic does not change with time is called a *time-invariant capacitor*. If the chracteristic changes with time, the capacitor is called a *time-varying capacitor*.

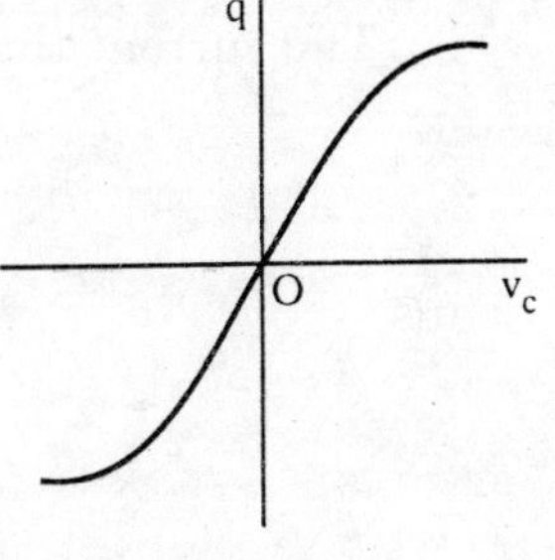

FIG. 2.22

2.15.1 Linear Time-Invariant Capacitor

A linear time-invariant capacitor is a 2-terminal element whose v -q relationship can be written in the form

$$q(t) = C\, v_c(t) \qquad (2.15.1)$$

where $q(t)$ = charge stored in the capacitor at time t

$v_c(t)$ = voltage across the terminals of the capacitor

C = capacitance of the element

The units of capacitance are expressd in farads (F) when the units of $q(t)$ are coulombs and the units of $v_c(t)$ are volts.

Eq. (2.15.1) represents a straight line through the origin on the v-q plane. The symbolic representation of a linear time-invariant capacitor and its q-v characteristic are shown in Figs. 2.23a and 2.23b respectively. The slope of this characteristic is the capacitance C.

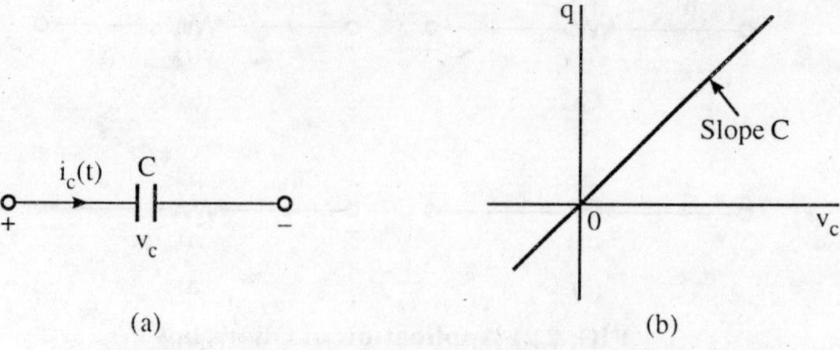

FIG. 2.23 Linear-time invariant capacitor (a) symbolic representation (b) q–v_c characteristic

Physically, a capacitor consists of two conducting plates or films separated by a nonconducting material such as air, paper, oil, ceramic etc. When the capacitor terminals are connected across a voltage $v_c(t)$ charge $q(t)$ flows into one plate and an equal amount of charge flows out of the other plate. Since the current is the rate of flow of charge, the current flowing into and out of the capacitor depends on the rate of change of charge. We thus refer to current *through* the capacitor even though, strictly speaking, current in the form of electron flow does not occur between the plates. The capacitor charge $q(t)$ is positive on the plate into which the reference arrow of the current $i(t)$ is directed.

The current through the capacitor can be obtained from the relation

$$i_c(t) = \frac{d}{dt} q(t) \tag{2.15.2}$$

Thus,

$$i_c(t) = \frac{d}{dt} [Cv_c(t)]$$

or

$$i_c(t) = C \frac{dv_c}{dt} \tag{2.15.3}$$

The voltage across the capacitor can be found as follows :
From Eq. (2.15.3)

$$dv_c = \frac{1}{C} i_c(t)\, dt \tag{2.15.4}$$

Integrating both the sides of Eq. (2.15.4) within the limits t_0 and t we get

$$\int_{t_0}^{t} d\, v_c(t) = \frac{1}{C} \int_{t_0}^{t} i\,(\tau)\, d\tau$$

where τ is the dummy integration variable

$$v_c(t) - v_c(t_0) = \frac{1}{C} \int_{t_0}^{t} i_c(\tau)\, d\tau$$

or

$$v_c(t) = v_c\,(t_0) + \frac{1}{C} \int_{t_0}^{t} i_c(\tau)\, d\tau \tag{2.15.5}$$

The quantity $v_c(t_0)$ is called the *initial voltage* across the capacitor at time $t = t_0$. If the initial voltage is zero, we say that the capacitor is initially uncharged. If the initial voltage is not zero, the capacitor is initially charged. Eq. (2.15.5) gives the relationship between the voltage across and current through the capacitor. This relation shows that the value of voltage $v_c(t)$ at any time, depends on its initial value $v_c(t_0)$ and all the values of the current between the interval from t_0 to t. This fact is commonly stated as : **Capacitors have memory.**

2.16 ENERGY STORED IN A CAPACITOR

Capacitors are energy-storing elements. The energy stored in a time-invariant capacitor at time t is given by

$$W_c(t) = \int_{-\infty}^{t} v_c(\tau)\, i_c(\tau)\, d\tau$$

For a linear time-invariant capacitor

$$i_c(\tau) = C\frac{dv_c(\tau)}{dt}$$

therefore,

$$W_c(t) = C\int_{-\infty}^{t} C\, v_c(\tau)\frac{dv_c(\tau)}{dt}\, dt$$

By change of variables we can write

$$W_c(t) = C\int_{v_c(-\infty)}^{v_c(t)} v_c(\tau)\, dv_c(t)$$

$$= C\left[\frac{v_c^2}{2}\right]_{v_c(-\infty)}^{v_c(t)}$$

or

$$W_c(t) = \frac{1}{2}C\,[v_c^2(t) - v_c^2(-\infty)] \tag{2.16.1}$$

It is assumed that at $t = -\infty$, the capacitor is uncharged and $q(-\infty) = 0$, that is, $v(-\infty) = 0$. It is logical to assume that the uncharged state of the capacitor is the state corresponding to zero energy stored in the capacitor. Thus, the energy stored in the capacitor at the time t is

$$W_c(t) = \frac{1}{2}C\,v_c^2(t) \tag{2.16.2}$$

Some of the important characteristics of a capacitor are as follows :

1. The current through a capacitor is zero, if the voltage across it is constant. Therefore a capacitor acts as an open circuit to *dc*.
2. The voltage across a capacitor cannot change instantaneously since it would require an infinite current.
3. The capacitor can store a finite amount of energy, even if the current through it zero.
4. A pure capacitor never disipates energy, but only stores it.

2.17 INDUCTORS

Inductance is the property of a device or circuit element which is capable of storing energy in a magnetic field by a current through it. This property is exhibited only when there is a changing current in the element. A circuit element that exhibits the property of inductance is called an *inductor*. It is a two-terminal circuit element. A 2-terminal circuit element is said to be an inductor if the current $i_L(t)$, and the magnetic flux $\phi(t)$, through it are related by an algebraic relation

$$\phi(t) = L\, i_L(t) \tag{2.17.1}$$

where the constant L is called the inductance of the inductor. If $i_L(t)$ is in amperes and $\phi(t)$ is in webers (Wb), then the unit of L is henry (H). The symbolic representation of a linear time-invariant inductor is shown in Fig. 2.24. The associated current and voltage are also shown in Fig. 2.24. The relationship between the voltage at the terminals and the current through it is given by

$$v_L(t) = L\frac{di_L}{dt} \tag{2.17.2}$$

FIG. 2.24 Symbolic representation of a linear time-invariant inductor

The current through the inductor can be found as follows :

$$di_L = \frac{1}{L} v_L(\tau)\, d\tau \tag{2.17.3}$$

Integration of both the sides of Eq. (2.17.3) gives

$$\int_{i(t_0)}^{i(t)} d\,i_L = \frac{1}{L}\int_{t_0}^{t} v_L(\tau)\, d\tau$$

$$i_L(t) - i_L(t_0) = \frac{1}{L}\int_{t_0}^{t} v_L(\tau)\, d\tau$$

Thus, the current through the inductor is given by

$$i_L(t) = i_L(t_0) + \frac{1}{L}\int_{t_0}^{t} v_L(\tau)\, d\tau \tag{2.17.4}$$

Eq. (2.17.4) shows that the current through an inductor at time t depends on the past value of i_L. This fact is commonly stated as : **An inductor has a memory of the past.**

2.18 ENERGY STORED IN AN INDUCTOR

The inductor has the ability to store electrical energy and, therefore, acts as a source of energy at a future time.

The energy stored in an inductor of constant value L during the interval (t_0, t_1) is given by

$$W_L(t_0, t_1) = \int_{t_0}^{t_1} v_L(\tau)\, i_L(\tau)\, d\tau$$

$$= \int_{t_0}^{t_1} L\frac{d\,i_L(\tau)}{d\tau} \cdot i_L(\tau)\, d\tau$$

$$= L \int_{i(t_0)}^{i(t_1)} i_L(\tau)\, d\, i_L\, \tau$$

$$= L \left[\frac{i_L^2(\tau)}{2} \right]_{i(t_0)}^{i(t_1)}$$

or
$$W_L(t_0, t_1) = \frac{1}{2} L[i_L^2(t_1) - i_L^2(t_0)] \quad (2.18.1)$$

Eq. (2.18.1) shows that the total energy flow over the interval (t_0, t_1) can be either into or out of the inductor depending upon the relative values of $i_L^2(t_1)$ and $i_L^2(t_0)$. A negative value of $W_L(t_0, t_1)$ shows that the energy flows out of the inductor. In other words, if $W_L(t_0, t_1)$ is positive the energy is stored in the inductor, and if $W_L(t_0, t_1)$ is negative the energy is supplied by the inductor. If the initial current is zero, that is $i_L(t_0) = 0$, then the energy stored in time t is given by

$$W_L(t) = \frac{1}{2} L\, i_L^2(t) \quad (2.18.2)$$

Some of the *important characteristics of an inductor* are as follows :

1. The induced voltage across an inductor is zero if the current through it is constant. Thus, an inductor acts as a short circuit to *dc*.
2. The current in an inductor cannot change instantaneously, (that is, in zero time), for this requires an infinite voltage across the inductor.
3. The inductor can store finite amount of energy, even if the voltage across the inductor is zero.
4. The inductor never dissipates energy, but only stores, it. Although this is true for the mathematical model, it is not true for a physical inductor.

2.19 TWO-PORT ELEMENTS

So far we have discussed two-terminal elements namely-independent voltage and current sources, resistors, capacitors, and inductors. These elements are also called *one-port elements*. There are other circuit elements of importance that have four terminals. Such elements are called two-port elements. These elements have two input terminals and two output terminals. Fig. 2.25 shows a two-port element. Here the terminals 1 and 1′ from the input port and the terminals 2 and 2′ from the output port. There are many important 2-port elements, for example, controlled or dependent sources, coupled coils, transformers, gyrators, negative impedance converters etc. In this chapter we shall discuss controlled sources. Other two-port elements will be discussed in Chapter 17.

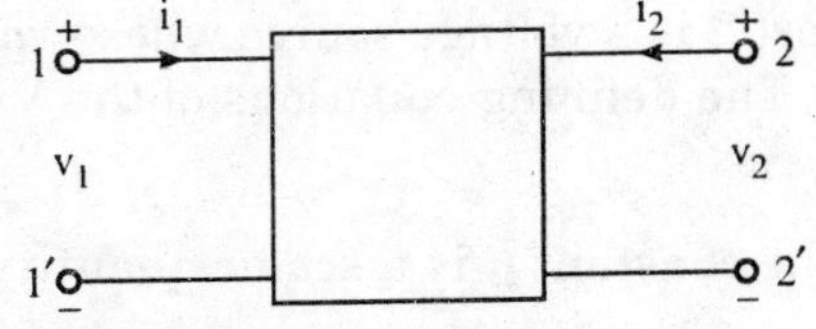

FIG. 2.25 Two-port element

2.20 DEPENDENT OR CONTROLLED SOURCES

A **dependent** or **controlled voltage** source is a voltage source whose terminal voltage depends on, or is controlled by, a voltage or a current defined at some other location in the circuit. Controlled voltage sources are categorized by the type of controlling variable. A *voltage-controlled voltage source* (VCVS) is controlled by a voltage and a *current-controlled voltage source* (CCVS) by a current.

A **dependent** or **controlled-current source** is a current source whose current depends on, or is controlled by, a voltage or a current defined at some other location in the circuit. A **voltage-controlled current source** (VCCS) is controlled by a voltage and a *current-controlled current source* (CCCS) by a current.

Dependent sources are represented by *diamond-shaped* symbols to distingiush them from the circular symbols for independent sources. The symbol for a dependent voltage source with value (source function) v_s is shown in Fig. 2.26a. The symbol for a dependent current source with value (source function) i_s is shown in Fig. 2.26b.

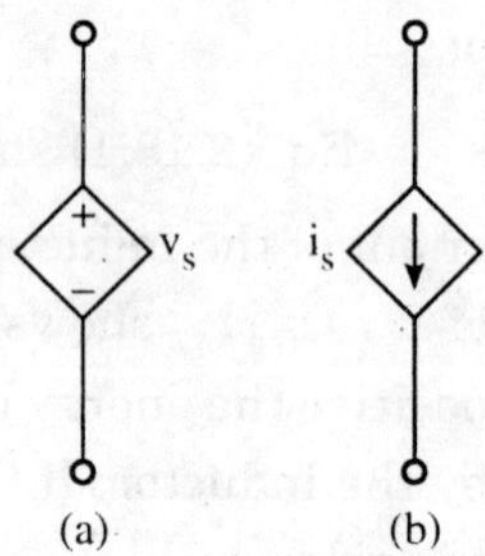

FIG. 2.26 (a) Dependent voltage source (b) Dependent current source

Controlled sources (or dependent sources) are 2-port unidirectional devices having an output (or controlled) variable dependent upon an input (or controlling) variable and independent of any other variable in the network. Port 2 is either a voltage source or a current source, and port 1 is either an open circuit or a short circuit. The source in port 2 is controlled by the voltage across the open circuit of port 1 or the current through the short circuit of port 1.

The four basic types of controlled sources are :

(a) Voltage-controlled voltage source (VCVS)

(b) Voltage-controlled current source (VCCS or VCIS)

(c) Current-controlled current source (CCCS or ICIS)

(d) Current-controlled voltage source (CCVS or ICVS)

The four types of dependent source, are shown in Fig. 2.27.

In the figure, dashed lines are used to indicate that although the controlling variables may be located anywhere in the network, frequently they are connected to the source so as to form a three-terminal element.

Voltage-Controlled Voltage Source (VCVS)

It is an element with output voltage controlled by input voltage as shown in Fig. 2.27a. Here port 2 is a voltage source whose voltage depends on the voltage across the open circuit of port 1. The defining equations of the VCVS are

$$i_1 = 0 \,, \; v_2 = \mu v_1$$

The constant μ is a scalar quantity and is called the **voltage gain**.

$$\mu = \left.\frac{v_2}{v_1}\right|_{i_1 = 0}$$

The constant μ is dimensionless.

Voltage-Controlled Current Source (VCCS)

It is an element with output current controlled by input voltage as shown in Fig. 2.27c. Here port 2 is a current source whose current depends on the voltage across the open circuit of port 1. The defining equations of VCCS are

$$i_1 = 0 \, , i_2 = g_m \, v_1$$

The constant g_m is scalar and is called **transconductance**.

$$g_m = \left.\frac{i_2}{v_1}\right|_{i_1 = 0}$$

The prefix *trans* is used to indicate that the current and voltage are not measured at the same location.

The transconductance g_m has the dimensions of siemens.

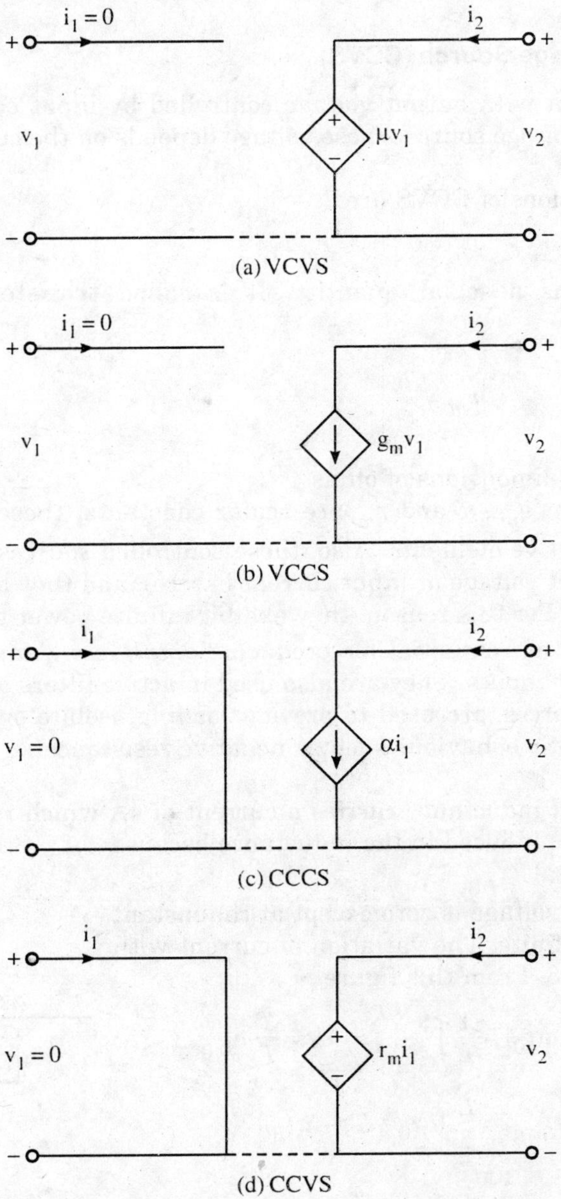

FIG. 2.27 (a) VCVS (b) VCCS (c) CCCS (d) CCVS

Current-Controlled Current Source (CCCS)

It is an element with output current controlled by input current as shown in Fig. 2.27c. Here port 2 is a current source whose current depends on the current through the short circuit of port 1. The defining equations of CCCS are

$$v_1 = 0 \, , \, i_2 = \alpha \, i_1$$

The quantity α is scalar and is known as **current gain**.

$$\alpha = \left.\frac{i_2}{i_1}\right|_{v_1 = 0}$$

Current-Controlled Voltage Source (CCVS)

The CCVS is an element with output voltage controlled by input current as shown in Fig. 2.27d. Here port 2 is a voltage source whose voltage depends on the current through the short circuit of port 1.

The defining equations of CCVS are

$$v_1 = 0 \, , \, v_2 = r_m \, i_1$$

The constant r_m is a scalar quantity. It is called **transfer resistance** or **trans resistance**.

$$r_m = \left.\frac{v_2}{i_1}\right|_{v_1 = 0}$$

Transresistance has the dimensions of ohms.

Since the factors μ , g_m , α and r_m are scalar constants, these controlled sources are linear, time-invariant active elements. Also, these controlled sources have zero input power requirement (either input voltage or input current is zero), and they are capable of delivering unlimited output power. For this reason, they exhibit infinite power gain.

Dependent sources are essential for producing *amplifiers*. Amplifiers produce outputs more powerful than their inputs. They are also used in active filters and electronic circuits of all kinds. Dependent sources, are used to prevent loading, isolate one part of a circuit from another, and provide exotic behaviour such as negative resistance.

■ **EXAMPLE 2.1** A 20mH inductance carries a current of 4A which reverses in 32 ms. What is the average voltage induced in the inductance because of this current reversal ?

SOLUTION : The induced voltage is zero except at the instant of reversal, when it is infinite. The variation of current with time is shown in Fig. 2.28. From this figure

$$i(t_0 + T) - i(t_0) = \frac{1}{L}\int_{t_0}^{t_0 + T} v_L \, dt = \frac{T}{L} v_{L\,(avg)}$$

therefore

$$v_{L\,(avg)} = \frac{L}{T}\left[i(t_0 + T) - i(t_0)\right]$$

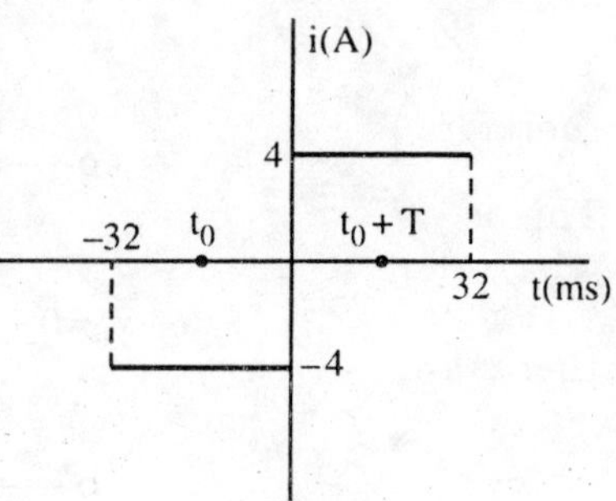

FIG. 2.28

$$= \frac{20 \times 10^{-3}}{32 \times 10^{-3}} [4 - (-4)] = 5\text{V}$$

EXAMPLE 2.2 The wavefrom of the current through an inductor of 10H is shown in Fig 2.29*a*. Sketch the waveform of the voltage across the inductor.

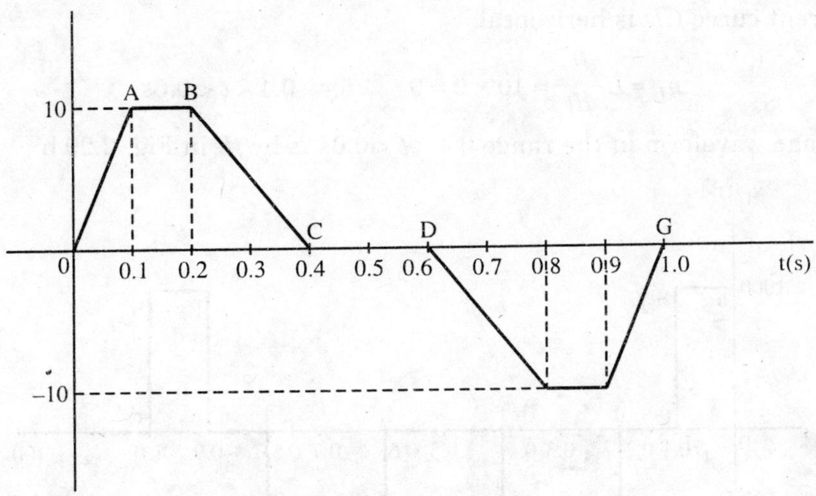

FIG. 2.29 (a) Waveform of i_L

SOLUTION : The voltage across an inductor is

$$v_L = L\frac{di_L}{dt}$$

(a) Between 0 and 0.1s (that is, for $0 < t < 0.1$), the derivative, $\frac{di_L}{dt}$, which is the slope of the i_L curve, is constant since $i_L(t)$ is linear.

$$\frac{di_L}{dt} = \text{slope of line } OA = \frac{OA}{AM} = \frac{10}{0.1} = 100 \text{ A/s (constant)}$$

therefore, $$v_L = L\frac{di_L}{dt} = 10 \times 100 = 1000 \text{ V (constant)} \quad \text{for } 0 < t < 0.1$$

This constant value is plotted as a horizontal line *ab* in Fig. 2.29b

(b) For $0.1 < t < 0.2s$, the current curve *AB* is horizontal, that is, its slope is zero.

therefore, $$v_L = L\frac{di_L}{dt} = 10 \times 0 = 0 \text{ V} \quad \text{for } 0.1 < t < 0.2s$$

This is plotted as horizontal line *cd* in Fig. 2.29b.

(c) For $0.2 < t < 0.4s$, the current curve *BC*. The slope of *BC* is negative.

therefore, $$\frac{di_L}{dt} = \text{slope of line } BC$$

$$= -\frac{BN}{NC} = \frac{10}{0.2} = -50 \text{ A/s}$$

therefore, $$v_L = L\frac{di_L}{dt} = 10 \times (-50) = -500 \text{ V} \qquad \text{for } 0.2 < t < 0.4\text{s}$$

This is shown by the horizontal line *eg* in Fig. 2.29b

(d) For $0.4 < t < 0.6$s, $\frac{di_L}{dt}$ is zero, not because the current is zero, but because the current curve *CD* is horizontal.

therefore, $$v_L = L\frac{di_L}{dt} = 10 \times 0 = 0 \qquad \text{for } 0.4 < t < 0.6\text{s}$$

The voltage waveform in the range $0.4 < t < 0.6$s is by *fh* in Fig. 2.29.b

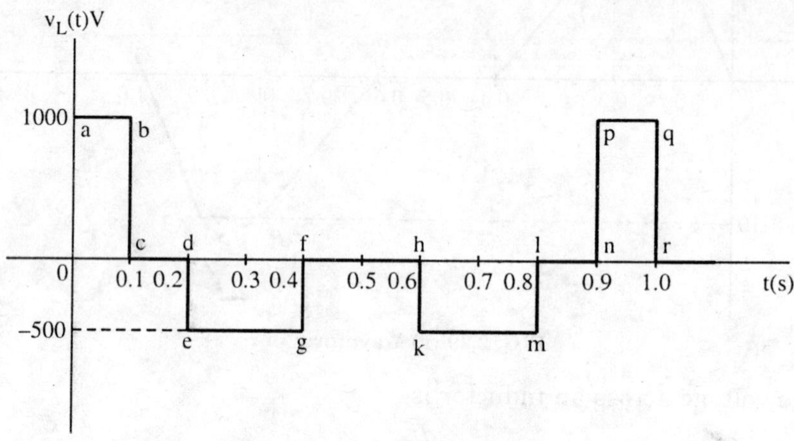

FIG. 2.29 (b) Waveform of v_L

(e) For $0.6 < t < 0.8$s, the current curve is *DE*. It has a negative slope given by

$$\frac{di_L}{dt} = -\frac{EP}{PD} = -\frac{10}{0.2} = -50\text{A/s}$$

therefore, $$v_L = L\frac{di_L}{dt} = 10 \times (-50) = -500\text{V} \quad \text{for } 0.6 < t < 0.8\text{s}$$

The voltage waveform in this range is shown by *km* in Fig. 2.29b.

(f) For $0.8 < t < 0.9$s, since the current curve *EF* is horizontal, $\frac{di_L}{dt} = 0$

therefore $$v_L = L\frac{di_L}{dt} = 10 \times 0 = 0\text{V}$$

This is shown by curve *ln* in Fig. 2.29b

(g) For $0.9 < t < 1.0$s, the current curve *FG* has a slope

$$\frac{di_L}{dt} = \frac{FQ}{QG} = \frac{10}{0.1} = 100 \text{ A/s}$$

therefore $$v_L = L\frac{di_L}{dt} = 10 \times 100 = 1000 \text{ V}$$

This is shown by curve *pq* in Fig. 2.29b.

EXAMPLE 2.3 The waveform of the voltage across a capacitor of 2F is shown in Fig. 2.30a. Sketch the waveform of the current i_c through the capacitor. Interpret the waveform of i_c.

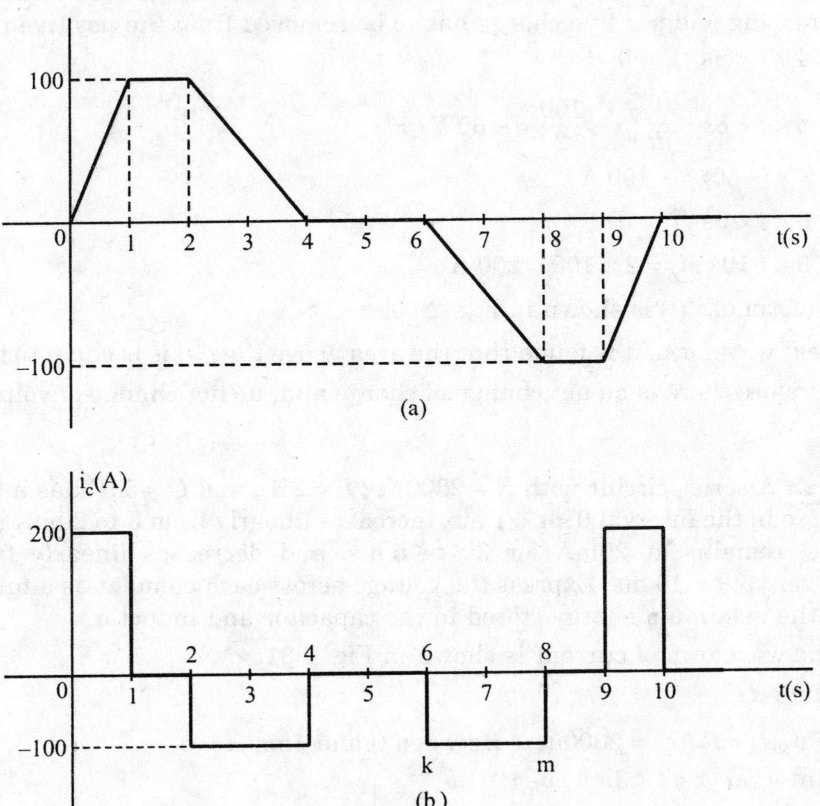

Fig. 2.30 (a) Waveform of υ (*t*) (b) Waveform of i_c (*t*)

SOLUTION : The current through a capacitor is given by

$$i_c = C\frac{dv_c}{dt}$$

The waveform of the current i_c through the capacitor is obtained in a manner identical to that applied to i_L in Example 2.2

(a) For $0 < t < 1\text{s}$, $\frac{dv_c}{dt} = \frac{100}{1} = 100\text{V/s}$, $i_c = C\frac{dv_c}{dt} = 2 \times 100 = 200$ V

(b) For $1 < t < 2\text{s}$, $\frac{dv_i}{dt} = 0$, $i_c = 2 \times 0 = 0\text{A}$

In this interval, $v_c = 100$ V , $i_c = 0$, since the voltage is *not changing* and charges do not move to, or from, the capacitor plates.

(c) For $2 < t < 4\text{s}$, $\dfrac{dv_c}{dt} = -\dfrac{100}{2} = -50$ V/s

$i_c = C\dfrac{dv_c}{dt} = 2 \times (-50) = -100$ A

In this interval, the voltage is positive but the current is *negative*. This is because the voltage is decreasing and positive charge has to be removed from the positive plate.

(d) For $4 < t < 6\text{s}$, $i_c = 0$

(e) For $6 < t < 8s$, $\dfrac{dv_c}{dt} = -\dfrac{100}{2} = -50$ V/s

$i_c = 2 \times (-50) = -100$ A

(f) For $8 < t < 9\text{s}$, $i_c = 0$

(g) For $9 < t\ 10\text{s}$, $i_c = 2 \times 100 = 200$ A

The waveform of $i_c(t)$ is shown in Fig. 2.30b

From the i_c waveform, it is found that the area above the t axis is equal to the area below the t axis. Therefore there is no net change of charge and, no net change of voltage across the capacitor.

EXAMPLE 2.4 A series circuit with $R = 2000\ \Omega$, $L = 2\text{H}$, and $C = 5\mu\text{F}$ has a current which remains zero in the interval $0 < t < 1$ ms, increases linearly from 0 to 20mA in the interval $1 < t < 3$ ms, remains at 20mA for $3 \le t \le 5$ ms, and decreases linearly from 20mA at $t = 5$ms to zero at $t = 10$ ms. Express the voltage across each element as a function of time. Find also the maximum energy stored in the capacitor and inductor.

SOLUTION : The waveform of current is shown in Fig. 2.31.

Waveform of $v_R(t)$

We have $v_R(t) = R\,i(t) = 2000i(t)$. Between 0 and 1ms, $i = 0$

In the interval $1 < t < 3\text{ms}$, $v_R(t)$ increases uniformly from 0 at $t = 1$ms to

$$2000 \times 20 \times 10^{-3} = 40 \text{ V at 3 ms.}$$

In the interval $3 < t < 5$,

$$v_R = 2000 \times 20 \times 10^{-3} = 40 \text{ V}$$

In the interval $5 < t < 10$, v_R decreases linearly from 40V at $t = 5$ms to zero volt at $t = 10$ms.

Thus, the waveform of $v_R(t)$ is identical to the waveform of $i(t)$ on a different scale.

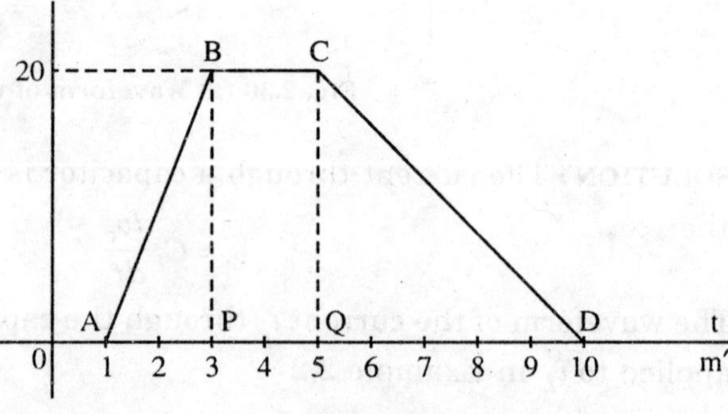

FIG. 2.31

Waveform of $v_c(t)$

$$i_c = C\frac{dv_c}{dt}$$

$$v_c = \frac{1}{C}\int i_c\,dt = \frac{1}{5 \times 10^{-6}}\int i_c\,dt$$

If i_c is in mA and t in ms, then

$$v_c = 0.2 \times 10^{+6} \int i_c \times 10^{-3} \times 10^{-3}\, dt$$

$$= 0.2 \int i_c\, dt$$

For $0 < t < 1\text{ms}$, $i_c = i = 0 \therefore v_c = 0$

For $1 < t < 3\text{ms}$, the current is a straight line AB.

$$\text{slope of } AB = m_1 = \frac{BP}{AP} = \frac{20 \text{ mA}}{2\text{ms}} = 10 \text{ A/s}$$

The equation of the line AB is

$$i(t) = m_1 (t - t_0)$$

$$i(t) = 10\,(t - 1)$$

Hence the voltage at any time t in this interval is

$$v_c(t) = 0.2 \int_0^t i_c\, dt$$

$$= 0.2 \int_0^1 i_c\, dt + 0.2 \int_1^t i_c\, dt$$

$$= 0 + 0.2 \int_1^t 10\,(t - 1)\, dt$$

$$= \frac{2\,(t-1)^2}{2} = (t - 1)^2 \qquad 1 < t < 3\text{ms}$$

At $t = 3\text{ms}$

$$v_c(t) = (3 - 1)^2 = 4\text{V}$$

For $3 < t < 5\text{ms}$, $i_c = 20\text{mA}$

$$v_c(t) = 4 + 0.2 \int_3^t 20\, dt$$

$$= 4 + [4t]_3^t = 4 + 4\,(t - 3) = 4t - 8$$

therefore, $$v_c(t) = 4t - 8 \qquad \text{for } 3 < t < 5\text{ms}$$

At $t = 5$ ms, $v_c(t) = 4 \times 5 - 8 = 12\text{V}$

For $5 < t < 10\text{ms}$, the current is straight line CD.

$$\text{slope of } CD,\ m_2 = -\frac{CQ}{QD} = -\frac{20\text{mA}}{(10 - 5)\text{ ms}}$$

$$= -4 \text{ A/s}$$

Therefore the equation of CD is

$$i_c(t) = m_2\,(t - 10)$$

$$i_c(t) = -4\,(t - 10) = 40 - 4t$$

therefore, $$v_c(t) = 12 + 0.2 \int_5^t (40 - 4t)\, dt$$

$$= 12 - 0.8 \int_5^t (t - 10)\, dt$$

$$= 12 - 0.8 \left[\frac{(t-10)^2}{2} \right]_5^t$$

$$= 12 - 0.4\, [(t - 10)^2 - (5 - 10)^2]$$

or $$v_c(t) = 22 - 0.4\,(t - 10)^2 \quad \text{for } 5 < t < 10 \text{ ms}$$

At t = 10ms,

$$v_c(t) = 22 - 0.4\,(10 - 10)^2 = 22\text{V}$$

This is the maximum value of the voltage. Therefore, maximum stored energy in the capacitor

$$W_{c\max} = \frac{1}{2} C V_m^2 = \frac{1}{2} \times 5 \times 10^{-6}\, (22)^2 \text{ J} = 1210\ \mu\text{J}$$

Waveform of $v_L(t)$

$$v_L = L \frac{di_L}{dt} = 2 \frac{di_L}{dt}$$

Between 0 and 1 ms, $i_L = 0$, and therefore $v_L = 0$.

Between t = 1 ms and t = 3ms,

$$\frac{di_L}{dt} = \frac{20 \text{ mA}}{(3-1) \text{ ms}} = 10 \text{ A/s}$$

therefore, $$v_L = 2 \frac{di_L}{dt} = 2 \times 10 = 20 \text{ V}$$

That is, the voltage remains constant at 20V from t = 1 ms to t = 3 ms.

For $3 < t < 5$ms , $i_L(t)$ is constant equal to 20mA. Therefore,

$$\frac{di_L}{dt} = \frac{d(20)}{dt} = 0$$

therefore, $$v_L = L \frac{di_L}{d} = 2 \times 0 = 0\text{V} \qquad 3 < t < 5 \text{ ms}$$

For $5 < t < 10$ ms, the current $i_L(t)$ follows the line CD.

Slope of CD = – 4 A/s

therefore, $$v_L = L \frac{di_L}{dt}$$

$$= 2 \times -4 = -8\text{V}$$

The maximum current is 20mA. Therefore, the maximum energy stored in the inductor is

$$W_{L\max} = \frac{1}{2} L I_m^2$$

$$= \frac{1}{2} \times 2\,(20 \times 10^{-3})^2 \text{J}$$

$$= 400\ \mu\text{J}$$

EXAMPLE 2.5 The waveform of the current $i_L(t)$ through a 2–H inductor is shown in Fig. 2.32a. Determine and sketch waveforms of (a) voltage $v_L(t)$, (b) instantaneous power $p(t)$ and (c) the energy stored in the inductor.

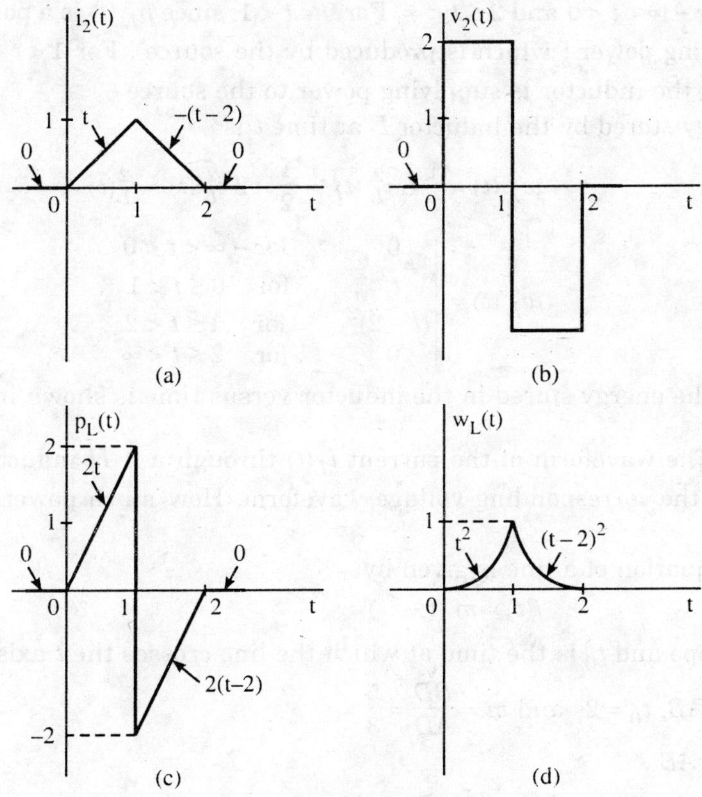

FIG. 2.32

SOLUTION : The current $i_L(t)$ can be written in analytical form as follows :

$$i_L(t) = \begin{cases} 0 & \text{for } -\infty < t < 0 \\ t & \text{for } \ 0 \le t < 1\text{s} \\ -(t-2) & \text{for } \ 1 \le t < 2\text{s} \\ 0 & \text{for } \ 2 \le t < \infty \end{cases}$$

(a) Since

$$v_L = L\frac{di_L}{dt} = 2\frac{di_L}{dt}$$

then
$$v_L(t) = \begin{cases} 0 & \text{for } -\infty < t < 0 \\ 2 & \text{for } 0 \le t < 1\text{s} \\ -2 & \text{for } 1 \le t < 2\text{s} \\ 0 & \text{for } 2 \le t < \infty \end{cases}$$

The waveform of $v_L(t)$ is shown in Fig 2.32b.

(b) Since $p_L(t) = v_L(t)\, i_L(t)$, we have

$$p_L(t) = \begin{cases} 0 & \text{for } -\infty < t < 0 \\ 2t & \text{for } 0 \le t < 1\text{s} \\ 2(t-2) & \text{for } 1 \le t < 2\text{s} \\ 0 & \text{for } 2 \le t < \infty \end{cases}$$

The waveform of $p_L(t)$ is shown in Fig. 2.32c. It is seen that the power absorbed by the inductor is zero for $-\infty < t < 0$ and $2 \le t < \infty$. For $0 < t < 1$, since $p_L(t)$ is a positive quantity, the inductor is absorbing power (which is produced by the source). For $1 < t < 2$, since $p_L(t)$ is a negative quantity, the inductor is supplying power to the source.

(c) The energy stored by the inductor L at time t is

$$w_L(t) = \frac{1}{2} L\, i_L^2(t) = \frac{1}{2} \times 2\, i_L^2(t) = i_L^2(t)$$

$$w_L(t) = \begin{cases} 0 & \text{for } -\infty < t < 0 \\ t^2 & \text{for } 0 \le t < 1 \\ (t-2)^2 & \text{for } 1 \le t < 2 \\ 0 & \text{for } 2 \le t < \infty \end{cases}$$

A sketch of the energy stored in the inductor versus time is shown in Fig 2.32d.

■ **EXAMPLE 2.6** The waveform of the current $i_L(t)$ through a 2–H inductor is shown in Fig 2.33a. Sketch the corresponding voltage waveform. How much power is associated with the inductor?

SOLUTION : The equation of a line is given by

$$f(t) = m\,(t - t_0)$$

where m is the slope and t_0 is the time at which the line crosses the t axis.

For the line AB, $t_0 = 2$s and $m = \dfrac{BD}{AD} = \dfrac{5}{3}$

therefore for AB ,

$$i_L(t) = \frac{5}{3}(t-2) = \frac{5}{3}t - \frac{10}{3} \qquad \text{for } 2 < t < 5\text{s}$$

$$\frac{di_L}{dt} = 5/3\ ,\ v_L = L\frac{di_L}{dt} = \frac{10}{3}\ \text{V}$$

For the line BC , $t_0 = 6$, $m = -\dfrac{BD}{DC} = \dfrac{-5}{1} = -5$

therefore $i_L(t) = -5\,(t-6)$

$$= -5t + 30 \qquad \text{for } 5 < t < 6s$$

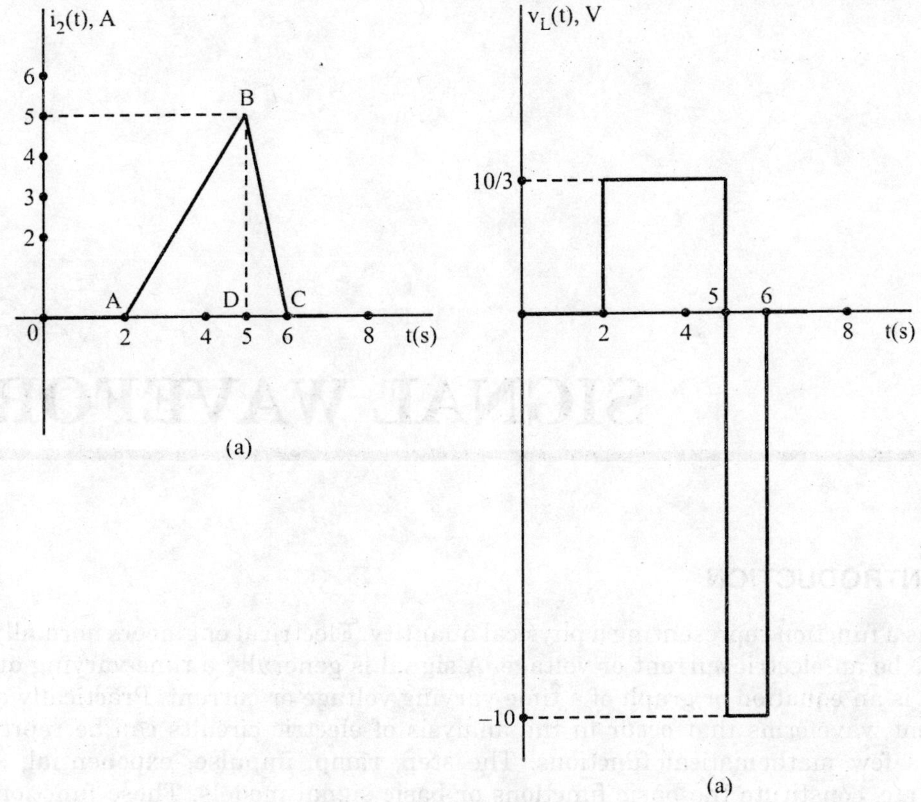

Fig. 2.33

$$\frac{di_L}{dt} = -5\ ,\ v_L = \frac{L\,di_L}{dt} = 2\,(-5) = -10\text{V}$$

Therefore, the voltage $v_L\,(t)$ across the inductor can be written in analytical form as follows :

$$v_L\,(t) = \begin{cases} 0 & \text{for} \quad 0 \le t < 2\text{s} \\ \dfrac{10}{3} & \text{for} \quad 2 < t < 5\text{s} \\ -10 & \text{for} \quad 5 < t < 6\text{s} \end{cases}$$

The waveform of $v_L\,(t)$ is shown in Fig 2.33b.

Since $\quad p_L\,(t) = v_L(t)\, i_L\,(t)$, we have

for $\quad 0 < t < 2\text{s}\ , \quad p_L\,(t) = 0$

for $\quad 2 < t < 5\text{s}$

$$p_L\,(t) = \frac{10}{3}\left(\frac{5}{3}t - \frac{10}{3}\right) = \left(\frac{50}{9}t - 100\right) W$$

and for $5 < t < 6\text{s}$

$$p_L\,(t) = -10\,(-5t + 30) = (50t - 300)\ W$$

3

SIGNAL WAVEFORMS

3.1 INTRODUCTION

A **signal** is a function representing a physical quantity. Electrical engineers normally consider a signal to be an electric current or voltage. A signal is generally a time-varying quantity. A waveform is an equation or graph of a time varying voltage or current. Practically all voltage and current waveforms that occur in the analysis of electric circuits can be represented in terms of a few mathematical functions. The step, ramp, impulse, exponential, sinusoidal functions etc. constitute the basic functions or basic signal models. These functions may be combined by addition or subtraction to build a variety of more general waveforms used in practice.

3.2 STEP FUNCTION

Suppose that a constant voltage or a constant current is switched into a circuit at time $t = 0$. The switching process can be described mathematically by defining a function called the step function. The unit step function is denoted by the symbol $u_{-1}(t)$ and is defined by equations

$$u_{-1}(t) \overset{\Delta}{=} \begin{cases} 0 & \text{for } t < 0 \\ 1 & \text{for } t > 0 \end{cases} \tag{3.2.1}$$

$u_{-1}(t)$ is not defined right at $t = 0$. It has a jump or discountinuity at $t = 0$. A graph of the function is shown in Fig. 3.1.

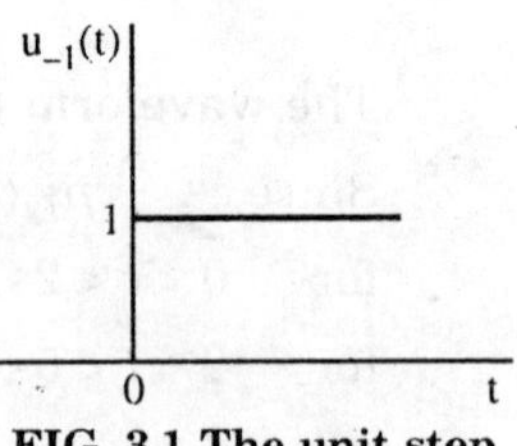

FIG. 3.1 The unit step function

The unit step function is zero for negative values of its argument t. It has a value of unity for any positive value of argument t. We have assumed that the switch is closed at $t = 0$. It is convenient to break this instant into three parts :

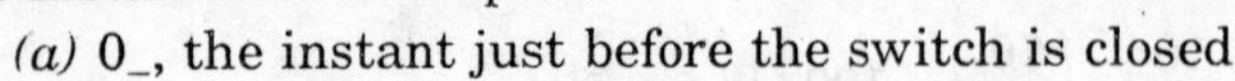

(*a*) 0_-, the instant just before the switch is closed,

(*b*) 0, the instant during which the switch is being closed,

and (*c*) 0_+, the instant just after the switch is closed.

These three instants are separated by negligibly small but finite intervals. Thus the switching operation occurs in the short interval between 0_-, when the function is zero, and 0_+ when it is unity. The value of the function precisely at $t = 0$ is undefined and is discontinuous at $t = 0$. Thus, we can write

$$\lim_{t \to 0_+} u_{-1}(t) = 1$$

and

$$\lim_{t \to 0_-} u_{-1}(t) = 0$$

One way we can construct an independent source that produce, a step function is the use of a switch. A **normally open ideal switch** can be represented as in Fig. 3.2a. The resistance between the two terminals of the switch is infinite for time $t < a$ seconds. During this time interval we say that the switch is **open.** For $t \geq a$ seconds, the resistance between the two terminals is zero, and we say that the switch is **closed** for this time interval.

t = a　　t = a　　t = a 2 1 3

(a)　　(b)　　(c)

FIG. 3.2

A **normally** closed ideal switch is shown in Fig. 3.2b. It is an ideal switch which is closed for $t < a$ seconds and open for $t \geq a$ seconds.

Both switches described above are examples of **single-pole, single-throw switches.** By **single-pole** we mean that only one pair of terminals is connected. By **single-throw** we mean that there is only one pair of terminals. A **single-pole, double-throw** switch is shown in Fig. 3.2a. Here the switch connects terminals 1 and 2 for time $t < a$ seconds, and connects terminals 1 and 3 for time $t \geq a$ seconds.

Consider the independent voltage source in series with a switch S as shown in Fig. 3.3a.

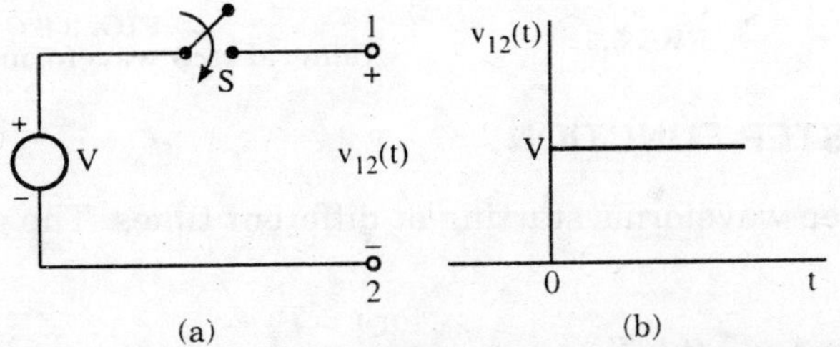

FIG. 3.3 (a) Independent voltage source (b) Step function

The terminals marked 1 and 2 are to be connected to a network. Suppose that the switch S is closed at $t = 0$. The voltage $v_{12}(t)$ to be applied to the network is zero when the switch is open. This statement can be expressed mathematically as

$$v_{12}(t) = 0 \qquad -\infty < t < 0 \tag{3.2.2}$$

When the switch is closed, the voltage $v_{12}(t)$ becomes V, that is,

$$v_{12}(t) = V \qquad 0 \leq t < \infty$$

Equations (3.2.2) and (3.2.3) can be combined as

$$v_{12}(t) = \begin{cases} 0 & -\infty \le t < 0 \\ V & 0 \le t < \infty \end{cases} \tag{3.2.4}$$

The unit step function concept allows us to mathematically describe the switching process given in Eq. (3.2.4). We can write

$$v_{12}(t) = V\, u_{-1}(t)$$

This is shown in Fig. 3.3b.

Thus, we can see that the function $u_{-1}(t)$ multiplies V by zero for negative values of t and by unity for positive values of t.

Consider the current source shown in Fig. 3.4. The current source switched into a circuit at $t = 0$ can also be expressed in terms of the unit step function. Before $t = 0$, the switch is closed and the current from the current source passes to the short circuit. At $t = 0$, the switch is opened, the current passes to the network connected between terminals 1 and 2. We can write $i(t)$ by two separate equations:

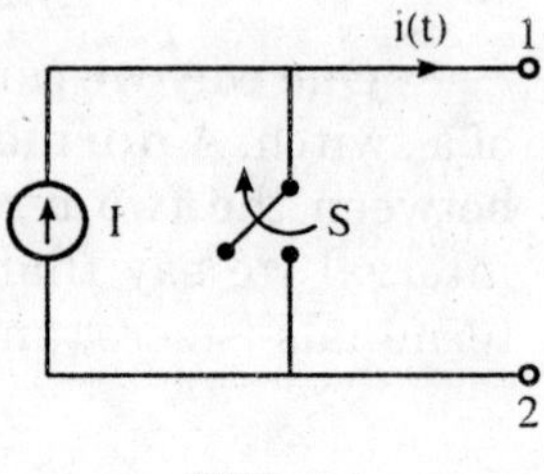

FIG. 3.4

$$i(t) = 0 \qquad -\infty < t < 0 \tag{3.2.6}$$

$$i(t) = I \qquad 0 \le t < 0 \tag{3.2.7}$$

These two relationships can be expressed in terms of the step function as

$$i\,(t) = I\, u_{-1}\,(t) \tag{3.2.8}$$

This is shown in Fig. 3.5

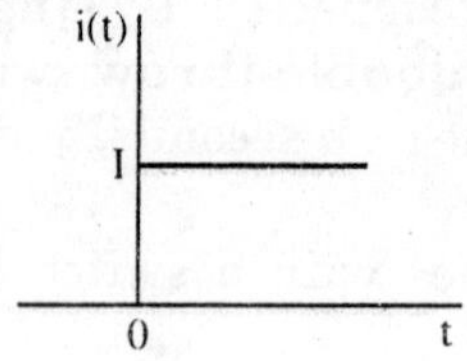

FIG. 3.5

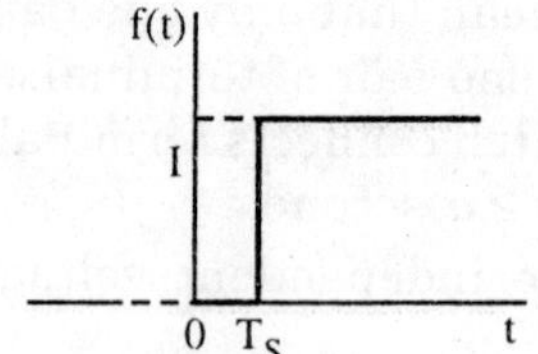

FIG. 3.6
General step waveform $A\, u_{-1}\,(t - T_S)$

3.3 GENERAL STEP FUNCTION

Let us consider the step waveforms starting at different times. The general step waveform is described by

$$A\, u_{-1}\,(t - T_s) = \begin{cases} A & \text{for } t > T_s \\ 0 & \text{for } t < T_s \end{cases} \tag{3.3.1}$$

and is shown in Fig. 3.6.

Here A is a real constant called the **amplitude** of the waveform. This waveform changes instantaneously from zero to A at the point in time at which its argument $(t - T_s)$ is zero. That is, the time T_s determines the time at which the step occurs. The constant T_s is called the **step time.** When the argument $(t - T_s)$ is positive, the step function is nonzero of value of A. When the argument $(t - T_s)$ is negative the step function is zero. It should be noted that either A, or T_s, or both can be negative. When $A = 1$, *a* unit step function results. Thus, we require two constants A and T_s to define a step signal. It should be noted clearly that we shall be concerned with the **sign** of the argument and not its actual value. If the sign of the argument is negative, the function will be zero. If the sign of the argument is positive, the value of the function will be equal to A. By changing A and T_s and adding several step functions, we can represent a number of important waveforms. The following examples illustrate the technique.

■ **EXAMPLE 3.1** Sketch the function $f(t) = 2u_{-1}(t - 4)$

SOLUTION : The step of amplitude $A = 2$ occurs at the point at which the argument $(t - 4)$ of the function is zero. That is, the step occurs at $t = 4$. For $t < 4$, the function $f(t)$ is zero, because the argument $(t - 4)$ has a negative value for $t < 4$. For any value of $t > 4$, the argument $(t - 4)$ has a positive value and therefore $f(t)$ is equal to 2. The sketch of the given function is shown in Fig. 3.7.

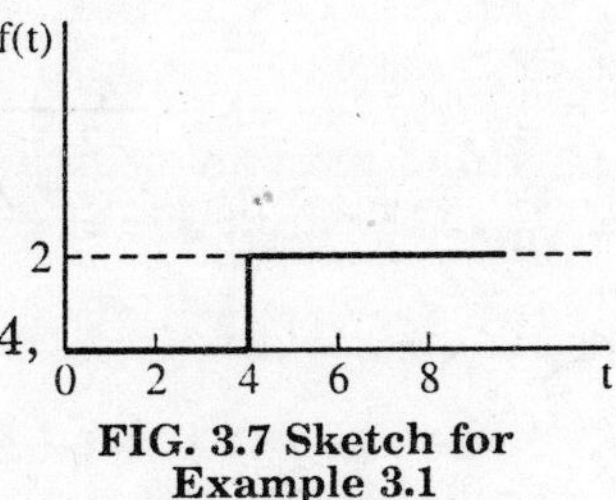

FIG. 3.7 Sketch for Example 3.1

■ **EXAMPLE 3.2** Sketch the function $v_1(t) = u_{-1}(t + 3)$. Find the value of the function $v_1(t)$ at $t = -5$ and $t = +5$.

SOLUTION : When $t < -3$, the argument $(t + 3)$ is negative and therefore $v_1(t)$ is zero. When $t > -3$, the argument $(t + 3)$ is positive and therefore v_1 is equal to one. The function $v_1(t)$ is shown in Fig. 3.8.

$$v_1(-5) = u_{-1}(-5 + 3) = u_{-1}(-2) = 0$$

$$v_1(+5) = u_{-1}(5 + 3) = u_{-1}(8) = 1$$

FIG. 3.8 Sketch for Example 3.2

3.4 MULTIPLICATION OF A FUNCTION BY A STEP FUNCTION

When any function of time is multiplied by the unit step function a new function of time is produced. Consider the function described by the relationship

$$f_2(t) = f_1(t)\, u_{-1}(t - 3)$$

where $$f_1(t) = 2t$$

The function $f_1(t)$ and $u_{-1}(t - 3)$ are shown in Figs. 3.9a and 3.9b.

Since $u_{-1}(t - 3)$ is zero for $t < 3$, the *product* of $u_{-1}(t - 3)$ and $2t$ is also zero for $t < 3$. When $t > 3$, $u_{-1}(t - 3)$ is equal to 1. Therefore the product of $2t$ and $u_{-1}(t - 3)$ for $t > 3$ is equal to $2t$. The function $f_2(t)$ which is the product of $f_1(t)$ and $u_{-1}(t - 3)$ is therefore shown in Fig. 3.9c.

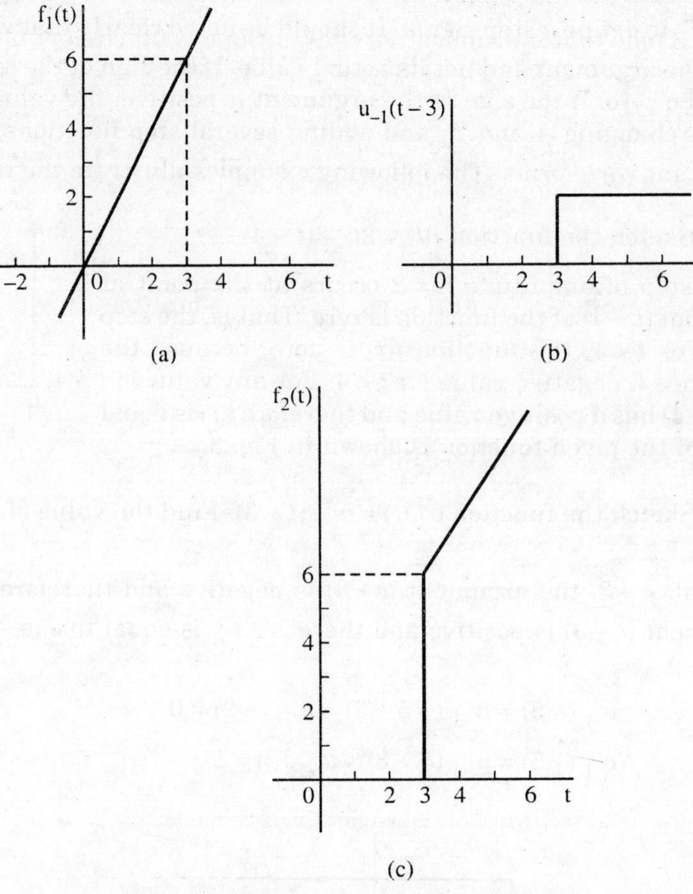

FIG. 3.9 Multiplication by a step function

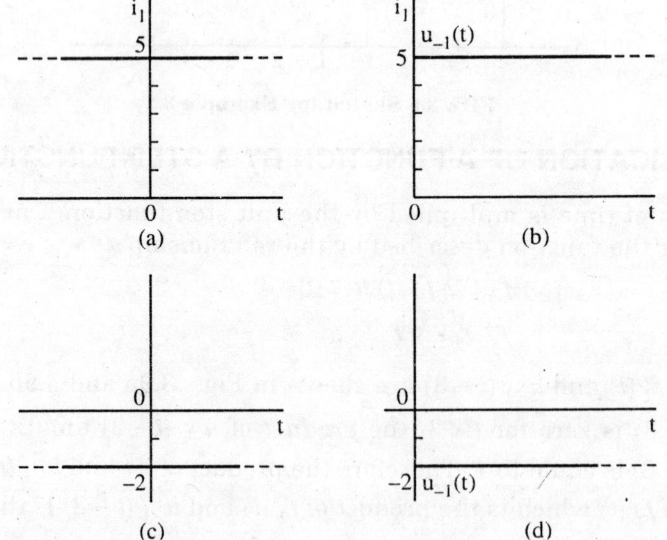

FIG. 3.10 Multiplication of constants by the unit step function

Hence we conclude that multiplication of a continuous function of time, $f_1(t)$, by the unit step function cuts off the function for negative values of the argument and maintains $f_1(t)$, for positive values of the argument, to be unchanged. Similarly, multiplication of a continuous function of time, $f_1(t)$, by the unit step function $u_{-1}(t)$, forces the resulting function to be zero for $t > 0$ and produces the original function $f_1(t)$ for $t > 0$.

Fig. 3.10 shows the effect of multiplying constants by the unit step function. The variable i_1 exists for *all* time. when i_1 is multiplied by a unit step function the waveform i_2 is generated (Fig. 3.10b). Similarlly, v_1 exists for *all* time. When v_1 is multiplied by a unit step function, a different function v_2 is generated (Fig. 3.10d).

■ **EXAMPLE 3.3** Sketch the functions $-u_{-1}(t-3)$, $u_{-1}(t+3)$ and $u_{-1}(3-t)$.

SOLUTION : The function $-u_{-1}(t-3)$ is shown in Fig. 3.11a. It is a negative unit step starting at $t = 3$. The function $u_{-1}(t+3)$ is shown in Fig. 3.11b. It is a positive unit step starting at $t = -3$, that is, the step changes from zero to unit value at $t = -3$.

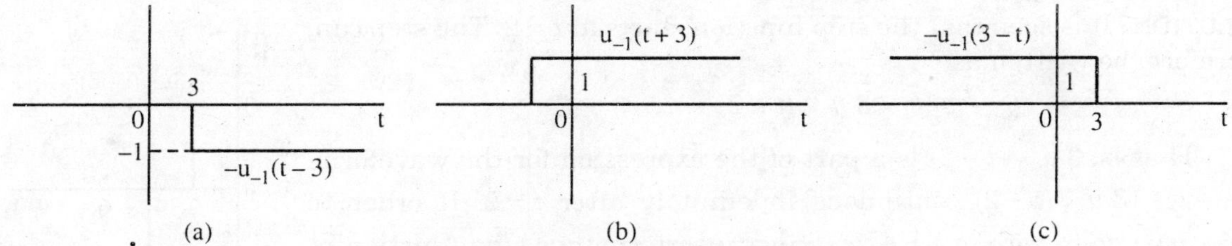

FIG. 3.11 Sketches for Example 3.3

Similarly, the function $u_{-1}(3-t)$, shown in Fig. 3.11c is a positive unit step that changes from unit to zero value (with increasing time) at $t = 3$. It is better to write the function $u_{-1}(3-t)$ as $u_{-1}[-(t-3)]$. Thus, a negative sign in front of a function denotes a negative step, whereas a negative sign in front of t in the argument denotes the termination of a step.

3.5 ADDITION OF STEP FUNCTIONS

When two or more step functions are added, a number of flat-line functions can be generated. The sum of two or more step functions is the algebraic sum of the individual functions evaluated in each range where a new function begins. Example 3.4 illustrates the addition of three step functions.

■ **EXAMPLE 3.4** Draw the waveform given by the equation

$$i(t) = 10\,u_{-1}(t-1) + 5\,u_{-1}(t-3) - 20\,u_{-1}(t\_5)$$

SOLUTION : The function $10\,u_{-1}(t-1)$ starts at $t = 1$ and continues for all time. The function $5\,u_{-1}(t-3)$

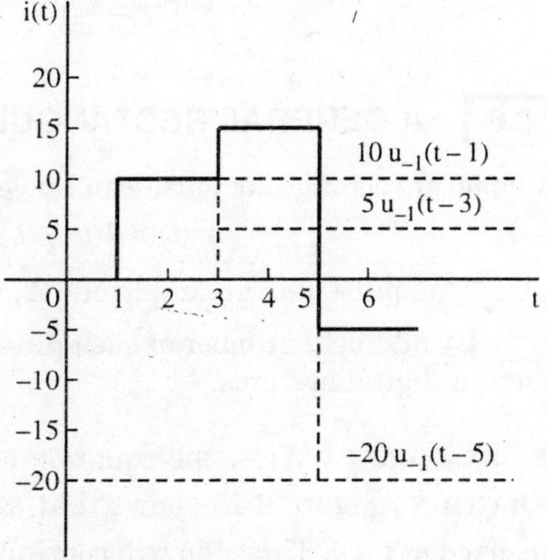

FIG. 3.12
The addition of three step functions

starts at $t = 3$ and continues for all time. At $t = 5$, the step function $-20\, u_{-1}(t-5)$ begins and continues for all time. Fig. 3.12 shows the resultant waveform $i(t)$ obtained by the graphical addition of the three step functions. The above procedure also explains the technique of writing the equation describing a given waveform.

EXAMPLE 3.5 Fig. 3.13 shows a rectangular pulse. Write the equation describing the pulse.

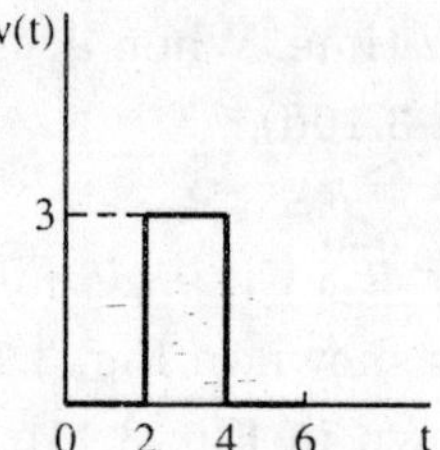

FIG. 3.13 Waveform of Example 3.5

SOLUTION : It is seen that the step function starts at $t = 2$. The step can, therefore, be written as

$$v(t) = 3\, u_{-1}(t-2) + \ldots..$$

That is, $3\, u_{-1}(t-2)$ is a part of the expression for the waveform. However, $3\, u_{-1}(t-2)$ continuous indefinitely after $t = 2$. If order to make $v(t)$ go to zero at $t = 4$, we must insert another step function at $t = 4$ such that when graphically added to $3\, u_{-1}(t-2)$ will cancel its effect. The inserted step should have the value $-3\, u_{-1}(t-4)$ as shown in Fig. 3.14. Therefore the required equation describing the rectangular pulse can be expressed as

$$v(t) = 3u_{-1}(t-2) - 3\, u_{-1}(t-4)$$

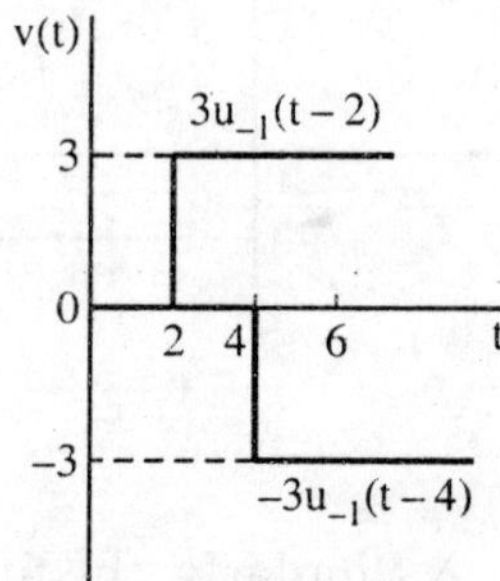

FIG. 3.14 Use of step functions to realize a rectangular pulse

3.6 A GENERAL RECTANGULAR PULSE

A general rectangular pulse can be written in the form

$$v(t) = A\, u_{-1}(t - T_1) - A\, u_{-1}(t - T_2)$$

This pulse has an amplitude A, which turns on at $t = T_1$ and off at $t = T_2$.

By adding a number of such pulses we can represent a long sequence of pulses commonly used in digital systems.

EXAMPLE 3.6 Write the equation of $v(t)$ given in Fig. 3.15.

SOLUTION : Figure 3.15 shows that a step of 6 starts at $t = 0$. Therefore a step of $6\, u_{-1}(t)$ is inserted at $t = 0$. This step will continue for all time. According to our requirement of Fig. 3.15, another step $-12\, u_{-1}(t-1)$ is introduced at $t = 1$, so that it adds to $6\, u_{-1}(t)$ to reduce it to

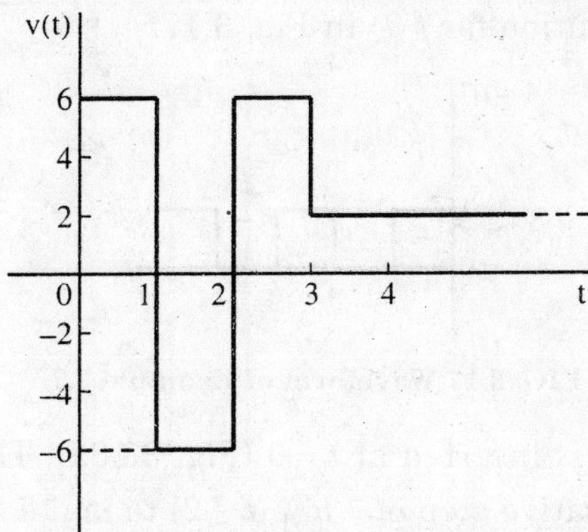

FIG. 3.15 Waveform of Example 3.6

– 6. Similarly, at $t = 2$, a step of $12\ u_{-1}\ (t - 2)$ is introduced to make the combined effect of the first two steps to + 6. At $t = 3$, we add a negative step of $-\ 4\ u_{-1}\ (t - 3)$ to get the required waveform for $t > 3$. The addition of the step functions is shown in Fig. 3.16.

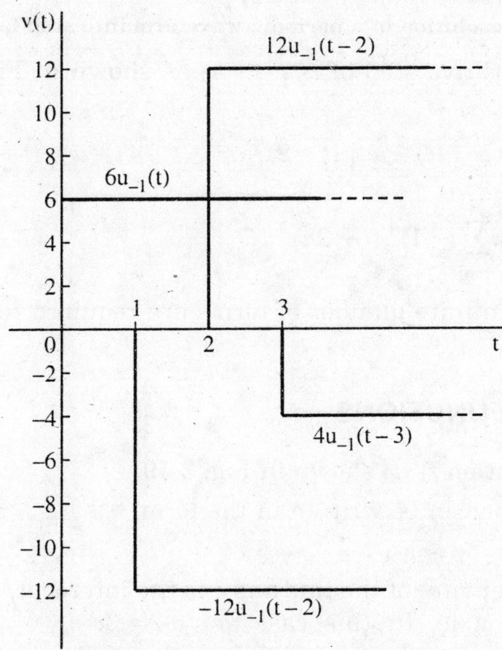

FIG. 3.16 Addition of step functions

Hence $$v\ (t) = 6\ u_{-1}\ (t) - 12\ u_{-1}\ (t - 1) + 12\ u_{-1}\ (t - 2) - 4\ u_{-1}\ (t - 3)$$

■ **EXAMPLE 3.7** Write the equation for $f(t)$ in Fig. 3.17.

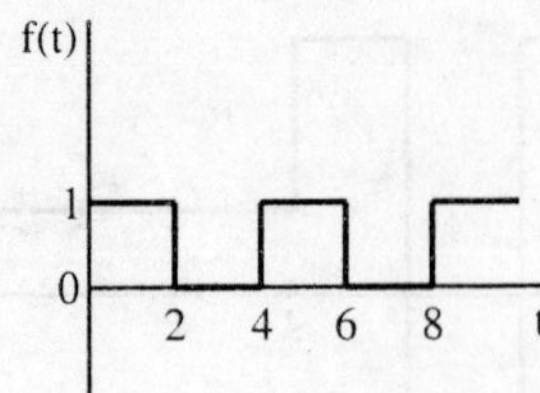

FIG. 3.17 Waveform of Example 3.7

SOLUTION : A unit step $u_{-1}(t)$ is inserted at $t = 0$ (Fig. 3.18a). This step will continue for all time. At $t = -2$ we insert a negative step of $-u_{-1}(t-2)$ to make the original function to zero. This is shown in Fig. 3.18b.

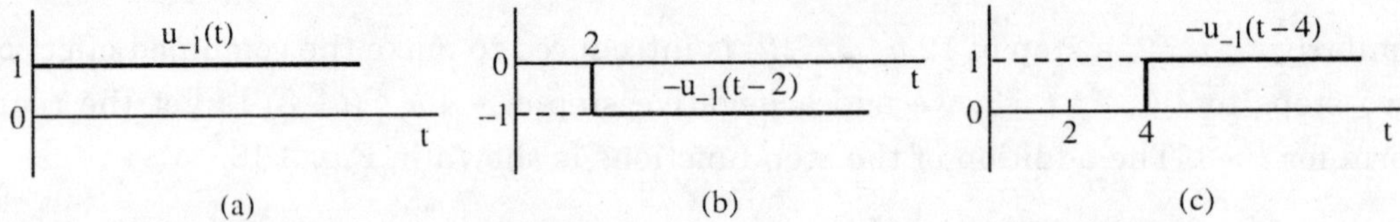

FIG. 3.18 Resolution of a periodic waveform into step functions

At $t = 4$ we insert a positive step of $u_{-1}(t-4)$ as shown in Fig. 3.18c. This process is repeated. Hence

$$f(t) = u_{-1}(t) - u_{-1}(t-2) + u_{-1}(t-4) - u_{-1}(t-6) + \ldots..$$

or

$$f(t) = \sum_{n=0}^{\infty} (-1)^n (t - 2n)$$

It is observed that an infinite number of terms are required to express the function for an infinite time range.

3.7 STRAIGHT-LINE FUNCTIONS

Consider a straight line function $f_1(t)$ shown in Fig. 3.19.

The equation of this line can be written in the form

$$y = m\,x + c$$

where m is the slope (rise over run) of the line and c is the intercept of the line on the vertical axis. In this case $m = 6/2 = 3,\; c = 6$. Therefore we can write

$$f_1(t) = 3t + 6$$

This equation can be written in the form

$$f_1(t) = 3(t + 2)$$

Here the constant *outside* the brackets is the slope of the line; the constant *within* the brackets is the *negative* of the time at which the line crosses the t-axis. Usually it is more convenient to write the equation of a line where it crosses the horizontal axis than where it

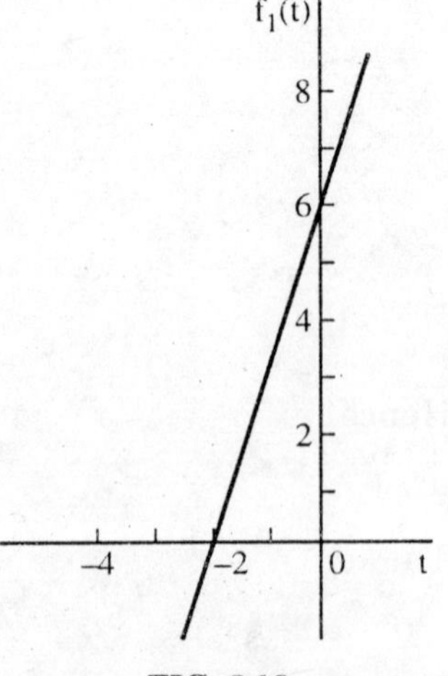

FIG. 3.19
A straight-line function

crosses the vertical axis. In general, a straight line can be written in the form

$$f(t) = m\ (t - t_0)$$

where m is the slope and t_0 is the time at which the line crosses the t-axis.

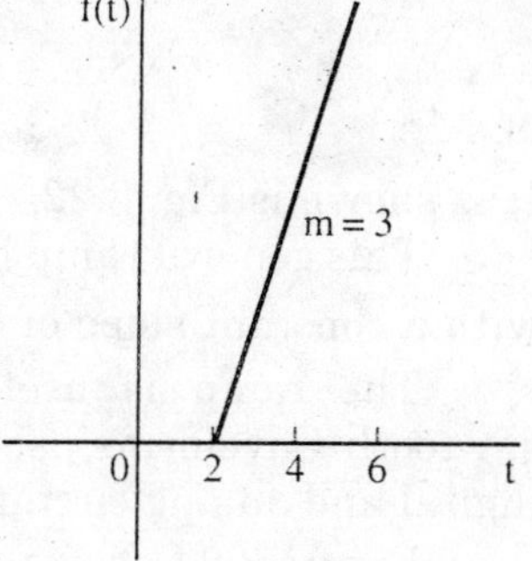

FIG. 3.20
Wave form of Example 3.8

EXAMPLE 3.8 Write the equation of the waveform shown in Fig. 3.20.

SOLUTION : The function $f(t)$ is zero for $t > 2$. It is equal to 3 $(t - 2)$ for $t > 2$. Therefore we can write

$$f(t) = 3\ (t - 2)\ u_{-1}\ (t - 2)$$

3.8 RAMP FUNCTION

The **basic ramp function** is a straight line beginning at the origin and increasing (or decreasing) linearly with time as shown in Fig. 3.21. Let $f(t)$ be the ramp function. The ramp can be defined as follows :

$$\left.\begin{aligned} f(t) &= 0 \quad \text{for } t < 0 \\ f(t) &= kt \quad \text{for } t > 0 \end{aligned}\right\} \tag{3.8.1}$$

Eq. (3.8.1) can also be written as

$$f(t) = k\ t\ u_{-1}\ (t) \tag{3.8.2}$$

The constant k represents the *slope* of the ramp. If k is positive, the slope is upward as shown in Fig. 3.21a. If k is negative, the slope is downward as shown in Fig. 3.21b.

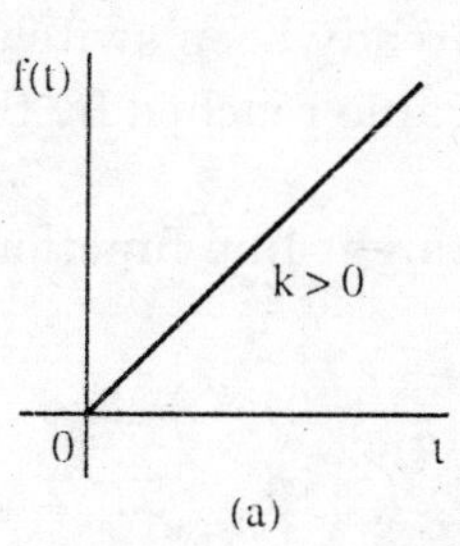

(a)

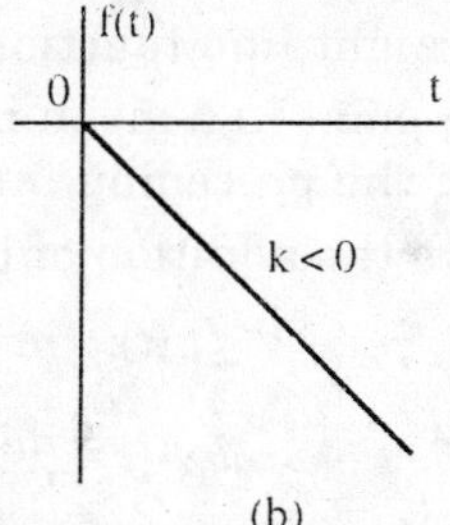

(b)

FIG. 3.21 Ramp functions

If $k = 1$, the ramp is called a **unit ramp.** The unit ramp is denoted by $r\ (t)$. Thus

$$r\ (t) = t\ u_{-1}\ (t) \tag{3.8.3}$$

or

$$\left.\begin{aligned} r\ (t) &= 0 \quad \text{for } t < 0 \\ r\ (t) &= t \quad \text{for } t > 0 \end{aligned}\right\} \tag{3.8.4}$$

The unit ramp function is also denoted by $u_{-2}\ (t)$. Thus

$$u_{-2}\ (t) = r\ (t) = t\ u_{-1}\ (t) \tag{3.8.5}$$

The **general ramp function** can be written as follows :

$$k\, r\,(t - T_s)$$

or $$k\, u_{-2}\,(t - T_s)$$

It is shown is Fig. 3.22.

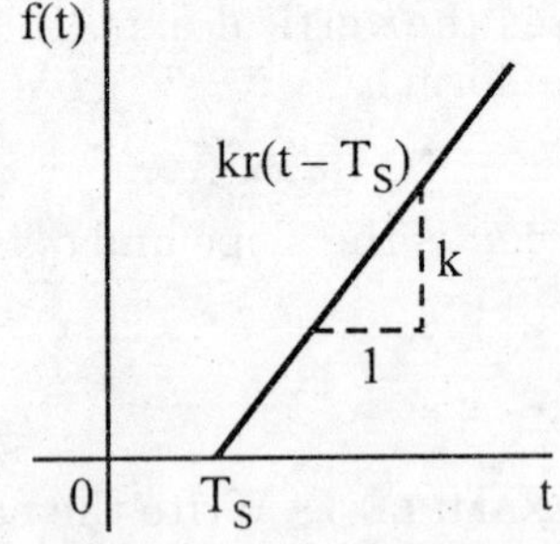

FIG. 3.22 General ramp waveform

This general ramp is zero until $t = T_s$ and then increases, with a constant slope of k.

The ramp is useful in representing triangular and sawtooth waveforms used as timing or sweep signals in both digital and analog circuits.

Figure 3.24 shows the addition of two positive straight-line functions

$$f_1\,(t) = m_1\, t\, u_{-1}\,(t)$$

and $$f_2\,(t) = m_2\,(t - a)\, u_{-1}\,(t - a)$$

$$f\,(t) = f_1\,(t) + f_2\,(t) = m_1\, t\, u_{-1}\,(t) + m_2\,(t - a)\, u_{-1}\,(t - a)$$

■ **EXAMPLE 3.9** A ramp voltage with a slope of 20 V/s is to start at $t = 4$s. Write an equation and sketch the voltage.

SOLUTION : The required equation is

$$v\,(t) = 20\,(t - 4)\, u_{-1}\,(t - 4)$$

The function $v\;(t)$ is shown in Fig. 3.23 by the solid line.

Fig. 3.23

3.9 ADDITION OF STRAIGHT-LINE FUNCTIONS

The following points are to be noted while adding straight-line functions:

1. The slope of the resulting function in a given interval is equal to the algebraic sum of all the straight-line functions that have already been switched on.
2. The starting point in a given interval is the value reached by the resulting function at the end of the preceding interval.

Figure 3.24 shows the addition of two positive straight-line functions

$$f_1\,(t) = m_1\, t\, u_{-1}\,(t)$$

and $$f_2\,(t) = m_2\,(t - a)\, u_{-1}\,(t - a)$$

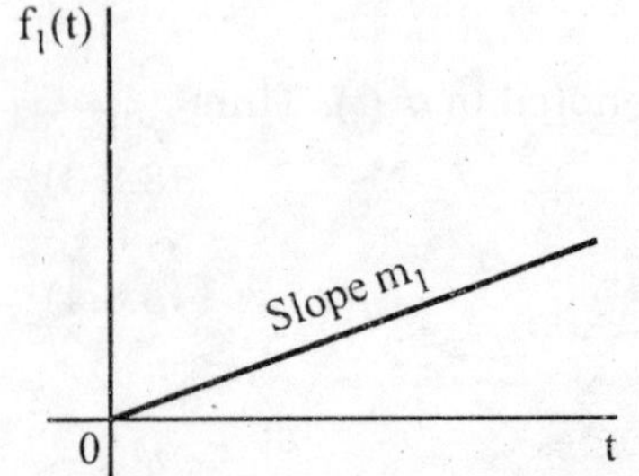

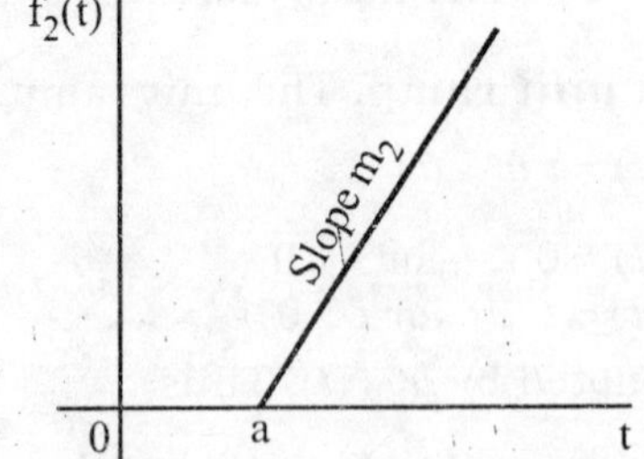

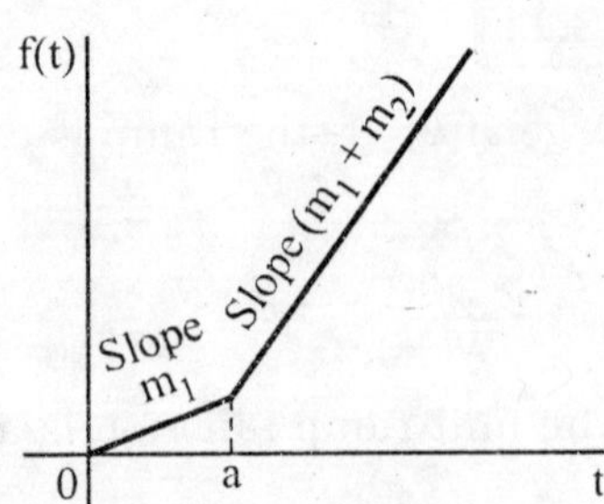

FIG. 3.24 Addition of two positive ramp functions

$$f(t) = f_1(t) + f_2(t)$$
$$= m_1\, t\, u_{-1}(t) + m_2\,(t-a)\, u_{-1}(t-a)$$

Fig. 3.25 shows the addition of a positive ramp function $m_2\, t\, u_{-1}(t)$ and a negative ramp function $-m_3\,(t-a)\, u_{-1}(t-a)$ where m_2 and m_3 are positive constants and $m_2 > m_3$.

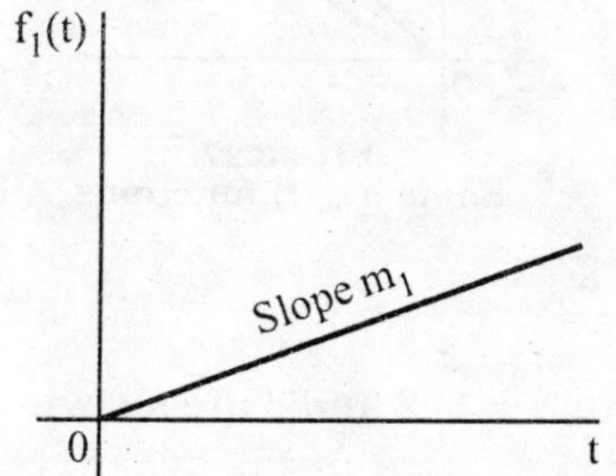

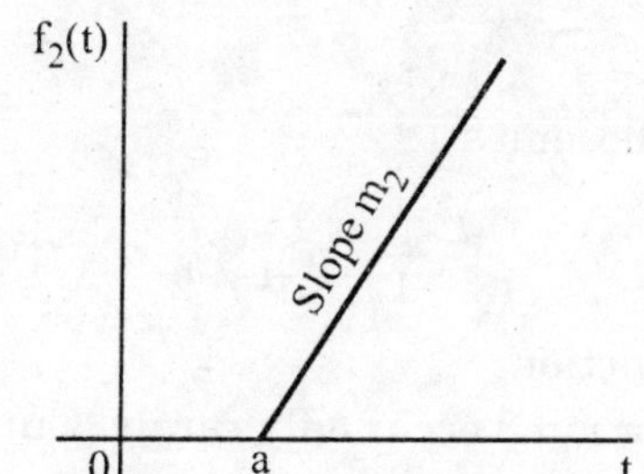

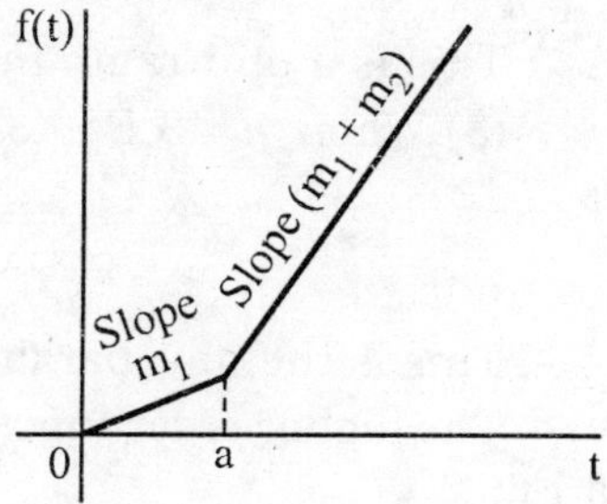

FIG. 3.25 Addition of positive and negative ramps

$$f_3(t) = m_2\, t\, u_{-1}(t)$$
$$f_4(t) = -\,m_3(t-a)\, u_{-1}(t-a)$$

therefore,
$$f(t) = f_3(t) + f_4(t)$$
$$= m_2\, t\, u_{-1}(t) - m_3\,(t-a)\, u_{-1}(t-a)$$

3.10 PARABOLIC FUNCTION

The **unit parabolic function,** $u_{-3}(t)$, is defined as follows :

$$u_{-3}(t) = \begin{cases} 0 & \text{for } t < 0 \\ t^2/2 & \text{for } t > 0 \end{cases} \tag{3.10.1}$$

It can also be written as

$$u_{-3}(t) = \frac{t^2}{2}\, u_{-1}(t) \tag{3.10.2}$$

This function is shown in Fig. 3.26. It is zero for negative values of t and a parabola for positive values of t.

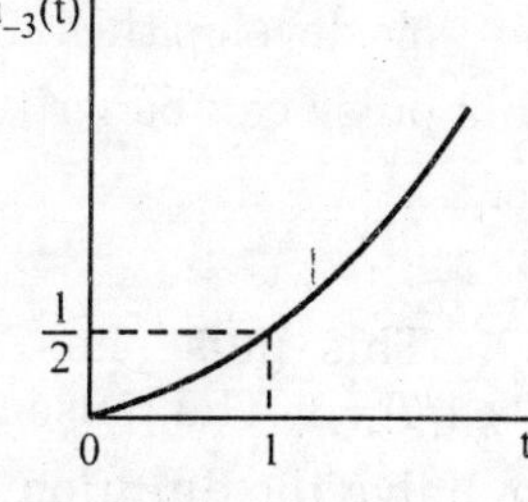

FIG. 3.26
Unit parabolic function

3.11 GENERAL $u_{-n}(t)$ FUNCTION

A general set of possible functions is given by

$$u_{-n}(t) = \frac{t^{n-1}}{(n-1)!}\, u_{-1}(t) \tag{3.11.1}$$

We shall discuss three cases.

(1) When $n = 1$, Eq. (3.11.1) becomes

$$u_{-1}(t) = \frac{t^0}{(t-1)!}\, u_{-1}(t) = \frac{1}{1}\, u_{-1}(t) = u_{-1}(t)$$

This is a unit step function.

(2) When $n = 2$, Eq. (3.11.1) becomes

$$u_{-2}(t) = \frac{t^{2-1}}{(2-1)!} u_{-1}(t)$$
$$= t\, u_{-1}(t)$$

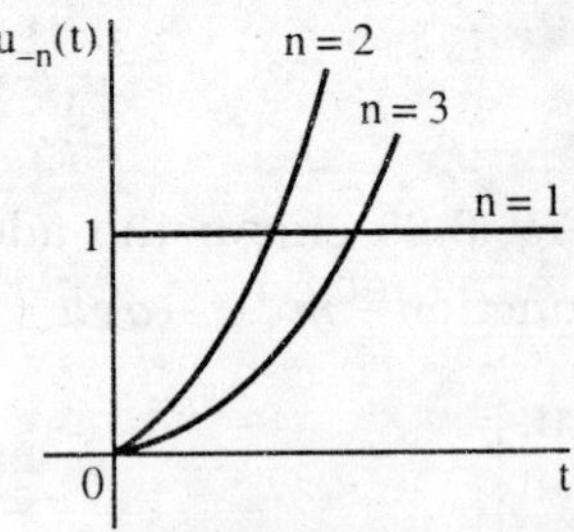

FIG. 3.27
Some $u_{-n}(t)$ functions

This is a unit ramp function.

(3) When, $n = 3$ Eq. (3.11.1) becomes

$$u_{-3}(t) = \frac{t^{3-1}}{(3-1)!} u_{-1} = \frac{t^2}{2} u_{-1}(t)$$

This is the unit parabolic function.

The higher order functions for $n > 3$ occur only rarely. Curves for $n = 1$, 2 and 3 are shown in Fig. 3.27.

3.12 IMPULSE FUNCTION OR DELTA FUNCTION

The **impulse function** or the **delta function** was first used by the physicist Dirac in 1930 and is quite often called the **Dirac-delta function.** The validity of the delta function and the manner of its use has been a matter of great deal of controversy for a long time. The concept called *The Theory of Distributions* developed by Schwartz provided a fully rigorous and satisfactory foundation for the impulse function. In this approach the function under study is considered a distribution rather than a point function. The impulse function or the delta function is denoted by $u_0(t)$ or $\delta(t)$.

To develop the impulse function, consider the rectangular pulse $v_1(t)$ shown in Fig. 3.28a. This pulse can be written as

$$v_1(t) = \frac{1}{T}[u_{-1}(t) - u_{-1}(t - T)]$$

This pulse has a width T and a height $1/T$ such that the area of the pulse is $T \times 1/T = 1$. The pulse $v_1(t)$ has an amplitude that is inversely proportional to its duration. If we halve the duration and double the amplitude we obtain the pulse v_2 shown in Fig. 3.28a. Let the duration be further halved and the amplitude doubled we obtain the pulse v_3 of Fig. 3.28a. It is seen that the areas under all the three waveforms v_1, v_2 and v_3 remain constant

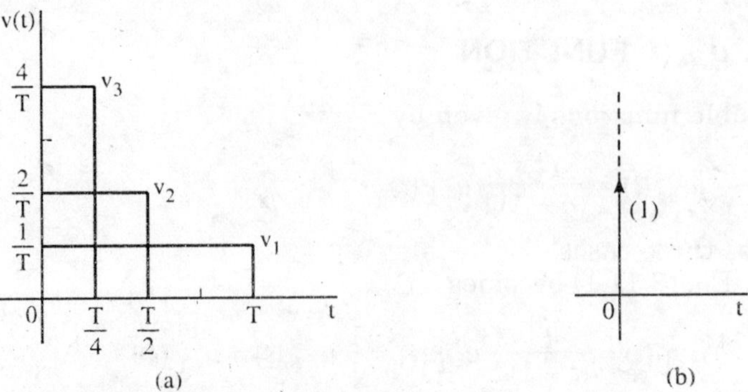

FIG. 3.28 Development of the impulse function

at unity. Now consider the special function obtained by carrying this process to the limit. In the limit when the duration (width) of the pulse approaches zero, the amplitude of the pulse approaches infinity, but the area of the pulse still remains unity. The pulse for which the duration tends to zero and the amplitude tends to infinity is called the *impulse function* or the delta function. Since it is impossible to show an infinite amplitude (height) and zero duration (width) the unit impulse occuring at $t = 0$ is represented graphically by a vertical arrow (spike) as shown in Fig. 3.28b. The arrow, pointing up, with the mark ∞ above, indicates that the function goes to infinity. The number with parentheses beside the impulse denotes the area under the impulse. The impulse function is specified by the area under it. Mathematically, the unit impulse function (also called unit delta function) occuring at the origin ($t = 0$) is denoted by $u_0(t)$ or $\delta(t)$. It is defined by

$$u_0(t) = \begin{cases} 0 & t \neq 0 \\ \infty & t = 0 \end{cases} \tag{3.12.1}$$

Since the area under the impulse is unity

$$\int_{-\infty}^{\infty} u_0(t)\, dt = 1 \tag{3.12.2}$$

Thus, the unit impulse has infinite amplitude and area equal to unity at $t = 0$. It has no value for any value of the argument, t, not equal to zero.

A unit impulse function which occurs at $t = t_0$ is $u_0\,(t - t_0)$. It is defined by the relations

$$u_0\,(t - t_0) = \begin{cases} 0 & t \neq t_0 \\ \infty & t = t_0 \end{cases} \tag{3.12.3}$$

$$\int_{-\infty}^{\infty} u_0\,(t - t_0)\, dt = 1 \tag{3.12.4}$$

The delayed unit impulse function is shown in Fig. 3.29. Thus $u_0\,(t - t_0)$ exists only at the time of occurence $t = t_0$. It takes an infinite value at $t = t_0$ and zero elsewhere with the

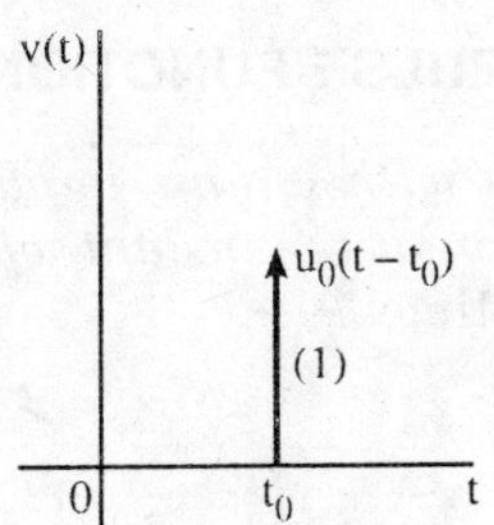

FIG. 3.29
Delayed unit impulse function

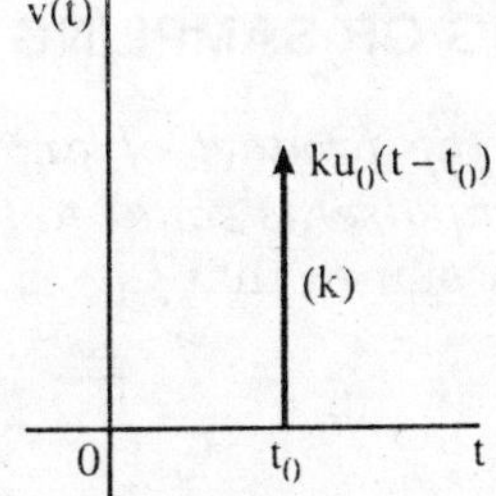

FIG. 3.30
General delayed impulse function of weight *k*

further property that the area under the function $u_0\,(t - t_0)$ is unity.

The impulse is characterized by the area under it. If the area is k, the result is called the impulse of value k, or of *weight k*. The total area under the impulse is sometimes called the *strength*. The impulse of weight k, occuring at time $t = t_0$ is represented by the function $k\,u_0\,(t - t_0)$. This general impulse function is shown in Fig. 3.30.

It is to be noted that the magnitude of the impulse is represented by its area and not by its amplitude which is infinite. Actually, we call $u_0(t)$ a function rather loosely, because the only place it has a nonzero value is not defined. However, this function is very useful.

A negative value of k is represented by negatively directed arrow. With this convention we can represent a number of impulses as shown in Fig. 3.31. The equation for $v(t)$ can be written as

$$v(t) = u_0(t) + 2u_0(t-1) - 3u_0(t-2)$$

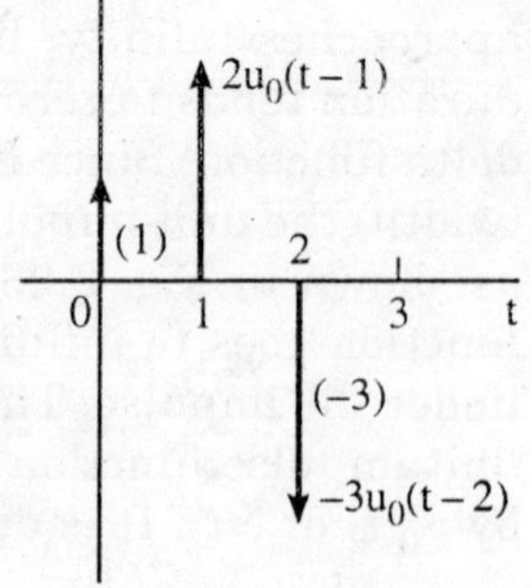

FIG. 3.31

In our earlier discussion we have introduced the concept of the impulse function by approximating a pulse of sufficiently small width Any waveform whose area remains unity as the base approaches zero can be used to generate the impulse function. The impulse function can be written as a limit

$$u_0(t) = \lim_{\in \to 0} Z_{\in}(t)$$

where $Z_{\in}(t)$ is a family of functions of unit area vanishing outside the interval $(0, \in)$. Thus,

$$\int_0^{\in} Z_{\in}(t)\, dt = 1$$

$$Z_{\in}(t) = 0 \text{ for } t < 0$$

and $t < \in$.

The function $Z_{\in}(t)$ is shown in Fig. 3.32.

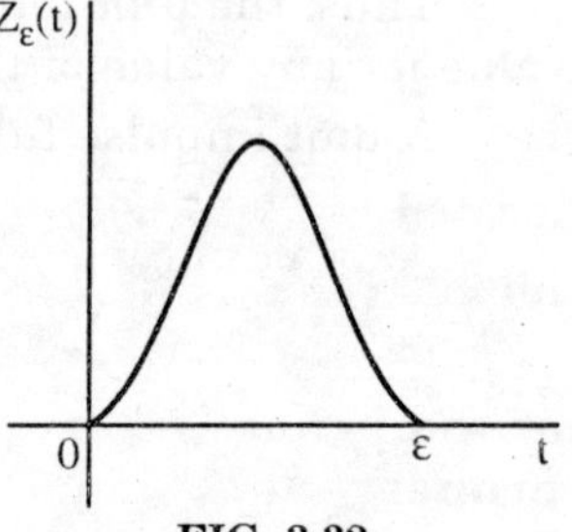

FIG. 3.32

The impulse function is an approximation of pulses of short duration of large magnitudes. Because of its zero width and infinite height it is not possible physically to obtain a unit impulse function as either a voltage or a current. However, we shall find the unit impulse function to be a convenient and very useful mathematical tool.

3.13 SIFTING OR SAMPLING PROPERTY OF IMPULSE FUNCTION

The integral of the product of any time of function f(t) with a unit impulse (over limits containing the impulse) is equal to the value of the function at the instant where the impulse is located. It is assumed that $f(t)$ is single valued at that time.

Thus
$$\int_{-\infty}^{\infty} f(t)\, u_0(t-t_0)\, dt = f(t_0) \qquad (3.13.1)$$

Proof

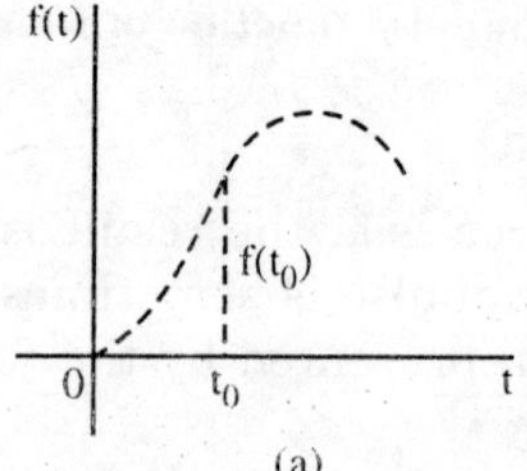

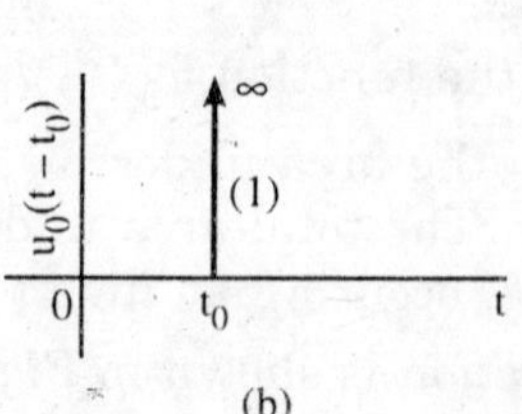

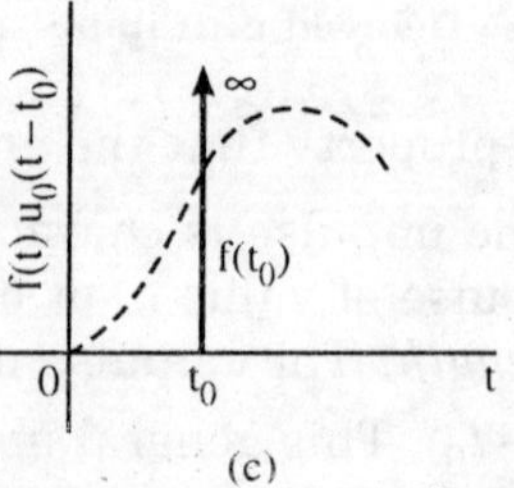

FIG. 3.33 Sifting property of unit impulse function
(a) An arbitrary function $f(t)$ (b) The unit impulse $u_0(t-t_0)$ (c) The product of (a) and (b)

Let $$I = \int_{-\infty}^{\infty} f(t)\, u_0(t - t_0)\, dt$$

The value of the product $f(t)\, u_0(t - t_0)$ is zero everywhere except at $t = t_0$, because $u_0(t)$ is zero everywhere except at $t = t_0$. Thus we can write

$$I = \int_{t_{0-}}^{t_{0+}} f(t)\, u_0(t - t_0)\, dt$$

Assuming that $f(t)$ does not change over the infinitesimally small interval $t_{0-} < t < t_{0+}$, we can treat it like a constant and take it outside the integral.

therefore, $$I = f(t_0) \int_{t_{0}}^{t_{0+}} u_0(t - t_0)\, dt$$

Since the area of the impulse is unity,

$$I = f(t_0)$$

therefore, $$\int_{-\infty}^{\infty} f(t)\, u_0(t - t_0)\, dt = f(t_0)$$

Fig. 3.33a shows an arbitrary function $f(t)$ and Fig. 3.33b shows the unit impulse $u_0(t - t_0)$ located at $t = t_0$. The product of $f(t)$ and $u_0(t - t_0)$ is shown in Fig. 3.33c. This product is an impulse of area $f(t_0)$.

The above property of impulse function to sample or to sift the value of a function $f(t)$ at the instant where the impulse exists is known as the **sampling property,** or **sifting property** of the impulse function. This is a very significant property of the impulse function. Thus, the impulse function may be used as a scanning function to pick out the value of $f(t)$ at some point $t = t_0$.

3.14 HIGHER-ORDER IMPULSE FUNCTIONS

The first derivative of $u_0(t)$ is called the **doublet.** It is represented by the symbol $u_{+1}(t)$.

$$u_{+1}(t) = \frac{d}{dt} u_0(t) \tag{3.14.1}$$

Higher-order impulse functions are defined as derivatives of $u_0(t)$

$$u_{+2}(t) = \frac{d}{dt} u_{+1}(t) = \frac{d^2}{dt^2} u_0(t) \tag{3.14.2}$$

$$u_{+3}(t) = \frac{d}{dt} u_{+2}(t) = \frac{d^2}{dt^3} u_0(t) \tag{3.14.3}$$

The nth derivative of $u_0(t)$ is denoted by $u_{+n}(1)$.

We have seen that any member of the impulse family of functions can be found by successive differenciations or integrations of any other member of the family.

Each of the members of the impulse family of functions is related to its neighbours through the expression

$$\frac{d}{dt}u_{n-1}(t) = u_n(t) = \int_{-\infty}^{t} u_{n+1}(t)\,dt \qquad (3.14.4)$$

Thus, each member of the family of impulse functions can be obtained by integrating the function just before it. Similarly, each member of this family can be obtained by differenciating the function just after it. That is,

$$u_0(t) = \frac{d}{dt}u_{-1}(t)$$

$$u_{-3}(t) = \frac{d}{dt}u_{-4}(t)$$

If we begin at $u_0(t)$ and integrate, the $u_{-n}(t)$ part of this infinite family of functions is generated. Similarly, if we begin at $u_0(t)$ and differenciate, we generate the $u_{+n}(t)$ part of this family. Thus, the first derivative of $u_0(t)$ is denoted by $u_{+1}(t)$, the second derivative by $u_{+2}(t)$. The nth derivative of $u_0(t)$ is denoted by $u_{+n}(t)$.

We have used the set of symbols $u_{+n}(t)$, $u_0(t)$ and $u_{-n}(t)$ to denote the impulse family of functions. The advantage of this notation is that the symbol itself describes the function being used. Many authors use different symbols for these functions. For example, the symbol $u(t)$ is used to have the same meaning as $u_{-1}(t)$. Similarly, the symbol $\delta(t)$ is used to have the same meaning as $u_0(t)$.

■ **EXAMPLE 3.10** Show that the derivative of a parabolic function is a ramp function.

SOLUTION : The parabolic function is given by

$$u_{-3}(t) = \begin{cases} 0 & \text{for} \quad t < 0 \\ \dfrac{t^2}{2} & \text{for} \quad t > 0 \end{cases}$$

therefore, $\qquad \dfrac{d}{dt}u_{-3}(t) = t \text{ for } t > 0$

Also, $\qquad u_{-2}(t) = t$

therefore $\qquad \dfrac{d}{dt}u_{-3}(t) = u_{-2}(t)$

■ **EXAMPLE 3.11** Show that the derivative of a ramp function is a step function.

SOLUTION : The unit ramp function is given by

$$u_{-2}(t) = \begin{cases} 0 & \text{for} \quad t < 0 \\ t & \text{for} \quad t > 0 \end{cases}$$

therefore, $\qquad \dfrac{d}{dt}u_{-2}(t) = \dfrac{d}{dt}t \text{ for } t > 0$

$$= 1 \text{ for } t > 0$$

This unit step function is given by

$$u_{-1}(t) = 1,\ t > 0$$

therefore, $$\frac{d}{dt}u_{-2}(t) = u_{-1}(t)$$

■ **EXAMPLE 3.12** Show that the derivative of unit step function is the unit impulse function.

SOLUTION : Since the derivative of a function is simply the slope of the function and because the slope of $u_{-1}(t)$ is zero except at $t = 0$ where a discontinuity exists (infinite slope), the unit impulse function accounts for the derivative of discontinuities.

therefore, $$\frac{d}{dt}u_{-1}(t) = u_0(t)$$

When a function has a discontinuity and is differenciated, a $ku_0(t)$ impulse results whose value k is equal to the magnitude of the discontinuity. When the discontinuity takes a positive direction, the area of the impulse is positive. When the discontinuity takes a negative direction, the area is negative.

■ **EXAMPLE 3.13** Show that the integral of the unit impulse is the unit step.

SOLUTION :

When $t > 0$

$$\int_{-\infty}^{t} u_0(t)\,dt = \int_{0-}^{0+} u_0(t)\,dt = 1$$

When $t < 0$

$$\int_{-\infty}^{t} u_0(t)\,dt = \int_{-\infty}^{t} 0\,dt = 0$$

Combining these two results we get

$$\int_{-\infty}^{t} u_0(t)\,dt = u_{-1}(t) \qquad \text{for all } t$$

This result is also true for any location of the impulse, that is,

$$\int_{-\infty}^{t} u_0(t - t_0)\,dt = u_{-1}(t - t_0) \qquad \text{for all } t$$

■ **EXAMPLE 3.14** Determine the integral of the signal shown in Fig. 3.34.

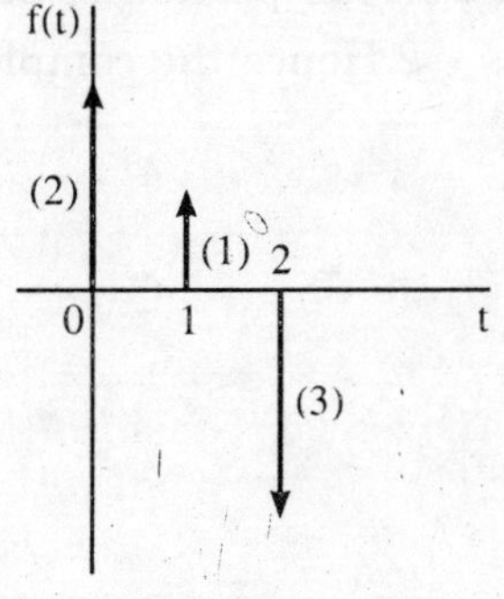

FIG. 3.34

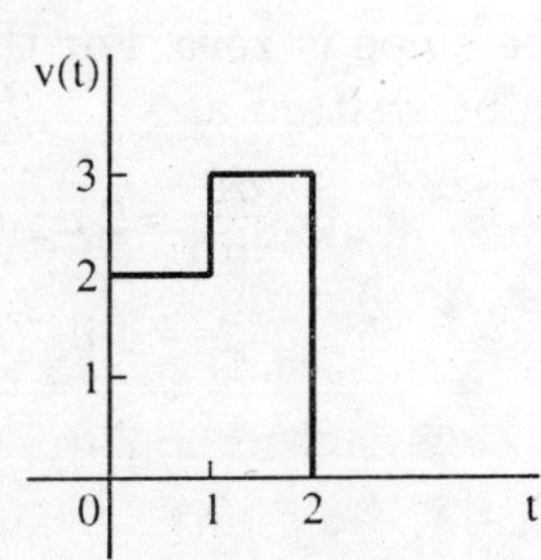

FIG. 3.35

SOLUTION :

$$f(t) = 2u_0(t) + u_0(t-1) - 3\,u_0\,(t-2)$$

Let the integral of $f(t)$ be denoted by $v(t)$

therefore, $v(t) = 2u_{-1}(t) + 2\,u_{-1}\,(t-1) - 3\,u_{-1}\,(t-2)$

This integral is shown in Fig. 3.35. There are jumps of height 2 at $t = 0$, height 1 at $t = 1$, and height -3 at $t = 2$. It is seen that at every jump discontinuity of $v(t)$, the function $f(t)$ has an impulse whose area is equal to the height of the jump of $v(t)$.

■ **EXAMPLE 3.15** For the voltage waveform shown in Fig. 3.36 sketch $\frac{dv}{dt}$ and write the equation for $\frac{dv}{dt}$.

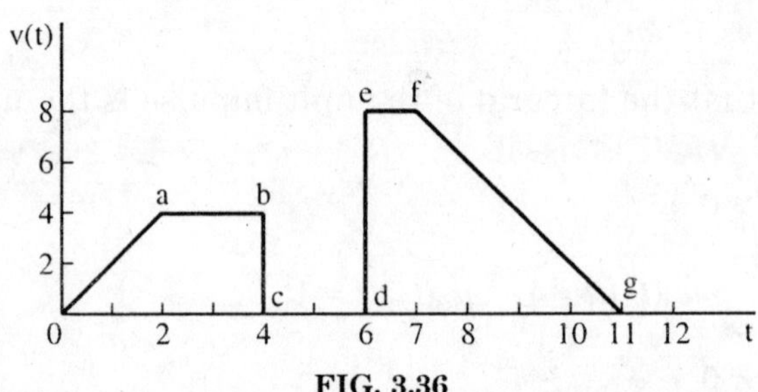

FIG. 3.36

SOLUTION : The derivative dv/dt is the slope of v . For the straight line Oa the slope is 4/2, that is, 2. Hence for this portion of the waveform $dv/dt = 2u_{-1}(t)$ which is a step function. From $t = 2$ to $t = 4$ for the portion ab, the slope is zero. At $t = 4$, there is a discontinuity in the voltage waveform. That is, at b the voltage drops in zero time from a value of 4 to zero. It is the derivative of this quantity that gives rise to an impulse at $t = 4$. The area of the impulse is -4 due to the discontinuity occuring in a negative direction. Since the impulse can exist only for an argument value of zero, the argument is written as $(t-4)$. Hence the expression for this impulse can be written as $-4\,u_0(t-4)$. For the portion cd the value of $v(t)$ is zero. At $t = 6$ there is again a discontinuity. Here $v(t)$ rises from zero to a positive value 8 in zero time. This discontinuity gives rise to an impulse of area 8 in positive direction. Since the impulse can exist only for an argument value of zero, the argument is written as $(t-6)$. Hence the expression for this impulse can be written as $+8u_0(t-6)$. For the portion ef of the voltage waveform, the slope is zero. For the portion fg, the slope is $-$ 2.Hence the complete equation for dv/dt can be written as

$$\frac{dv}{dt} = 2\,u_{-1}(t) - 2\,u_{-1}\,(t-2)$$

$$- 4\,u_0\,(t-4) + 8\,u_0\,(t-6)$$

$$- 2\,u_{-1}\,(t-7) + 2\,u_{-1}\,(t-11)$$

The waveform for $\frac{dv}{dt}$ is shown in Fig. 3.37.

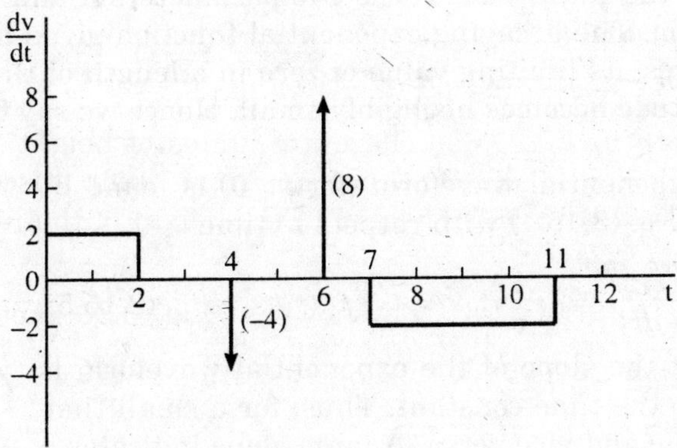

FIG. 3.37

3.15 EXPONENTIAL WAVEFORMS

A very commonly used exponential function is given by

$$f(t) = Ae^{-at} \tag{3.15.1}$$

A plot of this waveform is shown in Fig. 3.38. This is a decaying exponential function. At $t = 0, f(t) = Ae^0 = A$. For times less than zero, the exponential function given by Eq. (3.15.1) is greater than A. If we multiply the right-hand side of Eq. (3.15.1) by the unit step function $u_{-1}(t)$ we get

$$f_1(t) = Ae^{-at}\, u_{-1}(t) \tag{3.15.2}$$

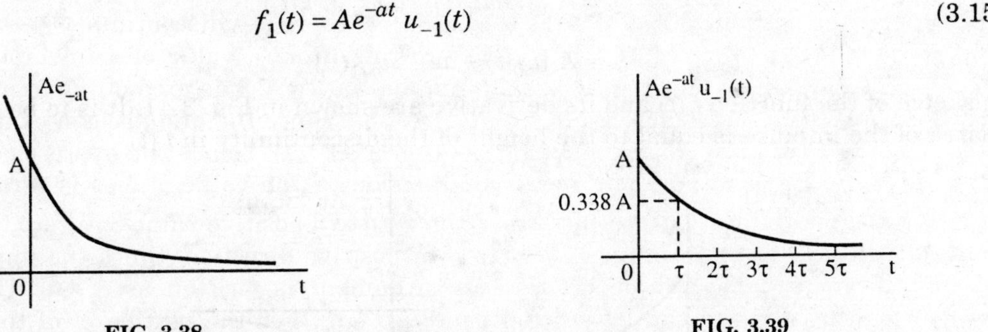

FIG. 3.38

FIG. 3.39

The waveform representing Eq. (3.15.2) is zero for $t < 0$ and jumps to an *amplitude* A at $t = 0$. Thereafter the waveform decreases towards zero as time approaches infinity. The relationship given by Eq. (3.15.2) is equivalent to

$$f_1(t) = \begin{cases} 0 & \text{for } t < 0 \\ Ae^{-at} & \text{for } t > 0 \end{cases} \tag{3.15.3}$$

A plot of this waveform is shown is Fig. 3.39.

The value of time t that makes the value of the exponent equal to – 1 is called the *time constant* τ. So the value of t for which $-at = -1$ is the time constant $t = \tau$. Therefore $\tau = 1/a$. Hence we can write

$$f_1(t) = Ae^{-t/\tau}\, u_{-1}(t) \tag{3.15.4}$$

The quantity A is called the *amplitude.* These two parameters A and τ are required to define the exponential waveform. The decaying exponential function given by Eq. (3.15.4), for most practical purposes reaches, its limiting value of zero in a length of time approximately equal to 5τ. After 5τ the amplitude becomes negligibly small. Hence we say that the *signal duration* is practically 5τ.

The slope of the exponential waveform (for $t > 0$) is found by taking the derivative of Eq. (3.15.4) with respect to time :

$$\frac{d f_1(t)}{dt} = -\frac{A}{\tau} e^{-t/\tau} = -\frac{1}{\tau} f_1(t) \tag{3.15.5}$$

This relation shows that the slope of the exponential waveform is inversely proportional to the time constant. Thus, for a small time constant the slope is large and vice versa. A large slope indicates a rapid decay. It is also seen that the derivative of an exponential signal is also an exponential with the same time constant but a different amplitude. The derivative of Ae^{-at} is shown in Fig. 3.40.

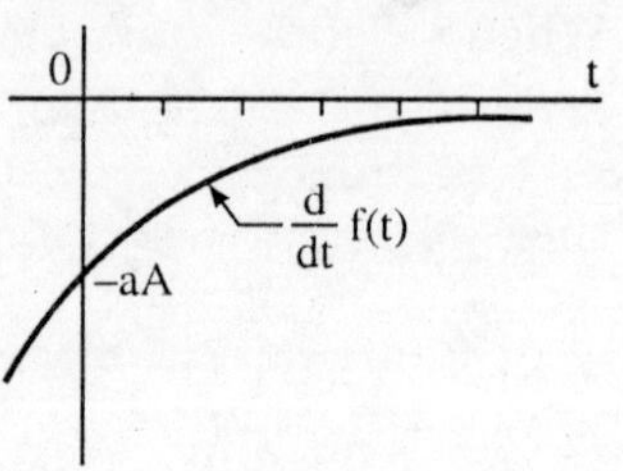

FIG. 3.40

■ **EXAMPLE 3.16** Determine and sketch the derivative of the function given by

$$f(t) = Ae^{-at}\, u_{-1}(t)$$

SOLUTION :

$$\frac{d}{dt} f(t) = A\left[e^{-at} \frac{d}{dt} u_{-1}(t) + u_{-1}(t) \frac{d}{dt} e^{-at}\right]$$

$$= A\,[e^{-at}\, u_0(t) - ae^{-at}\, u_{-1}(t)]$$

$$= A\,[u_0(t) - ae^{-at}\, u_{-1}(t)]$$

The sketch of the function $f(t)$ and its derivative are shown in Fig. 3.41. It is to be noted that the area of the impulse is equal to the height of the discontinuity in $f(t)$.

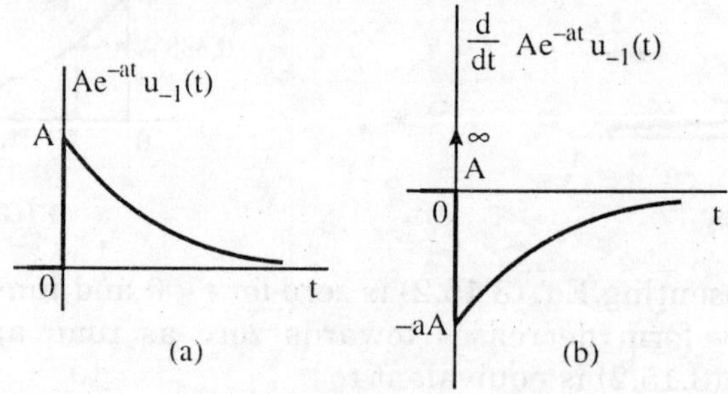

FIG. 3.41

■ **EXAMPLE 3.17** Find the function $g(t)$ given by

$$g(t) = \int_{-\infty}^{t} f(t)\, dt$$

where
$$f(t) = Ae^{-at}\, u_{-1}(t)$$

SOLUTION : When $t > 0$, the step function $u_{-1}(t) = 1$. Therefore for $t > 0$,

$$g(t) = \int_0^t Ae^{-at}\,dt = A\left[\frac{e^{-at}}{a}\right]_0^t$$

$$= -\frac{A}{a}(e^{-at} - e^0) = \frac{A}{a}(1 - e^{-at})$$

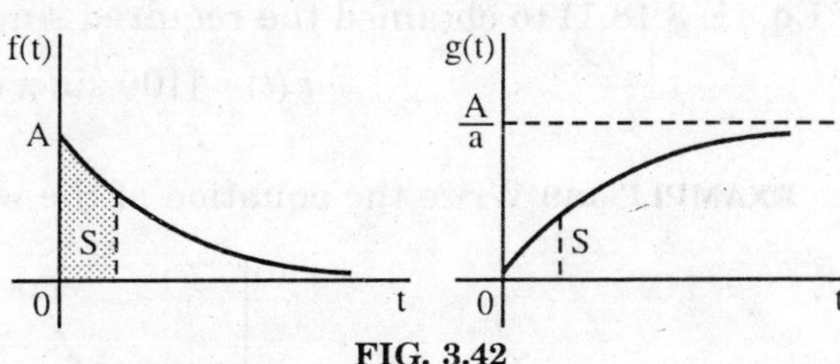

FIG. 3.42

When $t < 0$, the step function $u_{-1}(t)$ in $f(t)$ is zero.

therefore, $$g(t) = \int_{-\infty}^t (Ae^{-at}) \times 0\,dt = 0$$

Combination of these results gives

$$g(t) = \frac{A}{a}(1 - e^{-at})\,u_{-1}(t) \quad \text{for all } t$$

The waveforms of $f(t)$ and $g(t)$ are shown in Fig. 3.42. It is to be noted that the area S in Fig. 3.42a equals the height S in Fig. 3.42b.

3.16 SINUSOIDAL WAVEFORMS

The equation of a sinusoid depends upon three factors namely, its peak value, its period and where it crosses the t-axis going in a positive direction. In terms of these quantities, the equation of a sinusoid can be written as

$$f(t) = F_m \sin \omega (t - t_c)$$

where F_m = peak value

ω = radian frequency = $\frac{2\pi}{T}$

T = period

t_c = time where the waveform crosses the t-axis going in a positive direction.

■ **EXAMPLE 3.18** A sinusoidal current with a period of 2s and an amplitude of 100 mA begins at t = 4s. Write an equation for the current.

SOLUTION : A sinusoidal current with a period of 2s and an amplitude of 100 mA can be written as

$$i_1 = I_m \sin 2\pi ft$$

or $$i_1 = I_m \sin \frac{2\pi}{T} t$$

$$= 100 \sin \frac{2\pi}{2} t$$

$$i_1(t) = 100 \sin \pi t$$

For $t > 0$, we can write

$$i_1(t) = (100 \sin \pi t)\, u_{-1}(t) \qquad \text{(E 3.18.1)}$$

Since the given sinusoid is to start at $t = 4$s, we replace t by $(t - 4)$ in the right-hand side of Eq. (E 3.18.1) to obtained the required sinusoid $i(t)$.

$$i(t) = [100 \sin \pi (t - 4)]\, u_{-1} (t - 4)$$

■ **EXAMPLE 3.19** Write the equation of the waveform shown in Fig. 3.43.

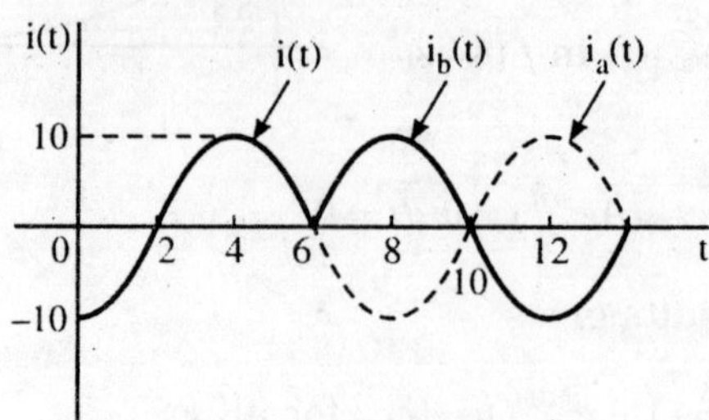

FIG. 3.43

SOLUTION : For the waveform shown in Fig. 3.43.

$$I_m = 10,\ T = 8,$$

$$\omega = \frac{2\pi}{T} = \frac{2\pi}{8} = \frac{\pi}{4} \text{ rad/s}$$

Let $t_c = 2$. It is seen that $i(t)$ is zero for $t < 0$ and $t > 6$. The sinusoid i_a can be written as

$$i_a = I_m \sin \omega (t - t_c)\, u_{-1}(t)$$

$$= 10 \sin \frac{\pi}{4} (t - 2)\, u_{-1}(t)$$

The multiplying unit function is necessary to that $i_a(t)$ is zero for $t < 0$. Since $i_a(t)$ continues forever, therefore, other sinusoid $i_b(t)$ must be inserted at $t = 6$ to cancel the remaining effect of $i_a(t)$. The sinusoid $i_b(t)$ can be written as

$$i_b(t) = \left[10 \sin \frac{\pi}{4} (t - 6)\right] u_{-1}(t - 6)$$

therefore,

$$i(t) = i_a(t) + i_b(t)$$

$$= \left[10 \sin \frac{\pi}{4} (t - 2)\right] u_{-1}(t) + \left[10 \sin \frac{\pi}{4} (t - 6)\right] u_{-1}(t - 6)$$

3.17 SYNTHESIS OF GENERAL WAVEFORMS

The basic waveforms can be combined to obtain a desired waveform. Waveform synthesis is the problem of determining the component waveforms of a desired waveform. The synthesis of some commonly used waveforms is given here.

3.17.1 Rectangular Pulse

Let us find an expression for the rectangular pulse $f(t)$ shown in Fig. 3.44a is terms of simpler functions.

To match the rectangular pulse from 0 to T we must have a step function with height k as shown by $f_1(t)$ in Fig. 3.44b. This step continues for all time, but according the requirement

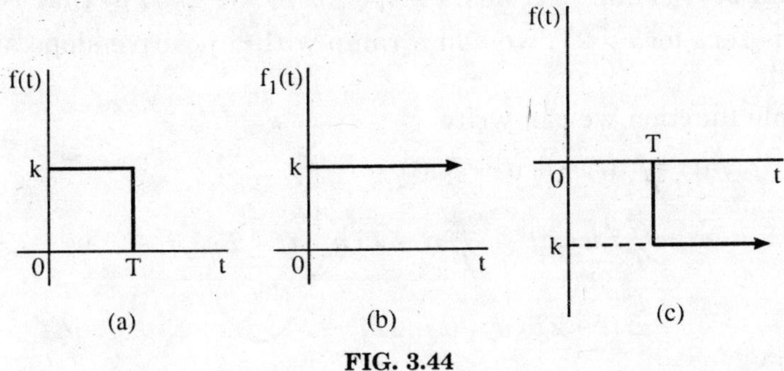

FIG. 3.44

of Fig. 3.44a the step of height k should become zero after $t = T$. Hence we add a negative step of magnitude k that is delayed by T seconds, as shown by $f_2(t)$ in Fig. 3.44c. In terms of time functions we can write

$$f(t) = f_1(t) + f_2(t)$$
$$= k\, u_{-1}(t) - k\, u_{-1}(t - T) \qquad (3.17.1)$$

3.17.2 Triangular Pulse

Consider the triangular pulse shown in Fig. 3.45a.To match the triangular pulse from 0 to V, we must have a ramp function with a slope of k/T as shown by $f_1(t)$ in Fig. 3.45b. According

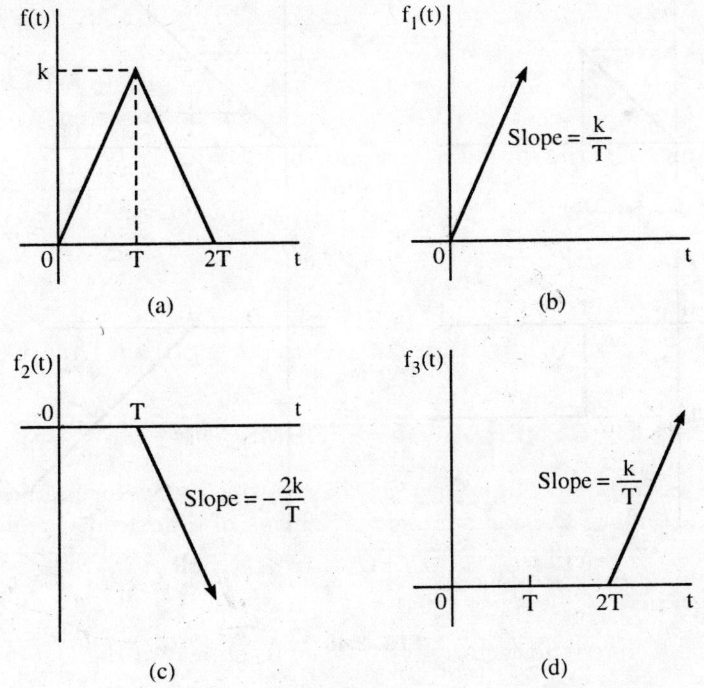

FIG. 3.45

to the requirement of Fig. 3.45a another ramp function with a slope of $-k/T$ begins at $t = T$. Hence we add a ramp with a slope of $-2k/T$, delayed by T seconds as shown by $f_2(t)$ in Fig. 3.45c. Thus, the sum of $f_1(t)$ and $f_2(t)$ has a slope $[(k/T) + (-2k/T)]$, that is, $-k/T$ for $t > 2T$. Since the function is zero for $t > 2T$, we add a ramp with a positive slope at $t = 2T$ as shown by $f_3(t)$ in Fig. 3.45d.

In terms of time function we can write

$$f(t) = f_1(t) + f_2(t) + f_3(t)$$

$$= \frac{k}{T} t\, u_{-1}(t) - \frac{2k}{T}(t - T)\, u_{-1}(t - T)$$

$$+ \frac{k}{T}(t - 2T)\, u_{-1}(t - 2T) \quad (3.17.2)$$

3.17.3 Sawtooth Pulse

Consider the sawtooth pulse shown in Fig. 3.46a. To match the sawtooth pulse from 0 to T we must have a ramp function with a slope k/T as shown by $f_1(t)$ in Fig. 3.46b. We add a negative step function of magnitude k at $t = T$ to reduce the sawtooth function to zero. This negative step is shown in Fig. 3.46c. The sum of $f_1(t)$ and $f_2(t)$ has a slope $(k/T + 0)$, that is, k/T for $t > T$. Since the function is zero for $t > T$, we add a ramp with a negative slope $-k/T$ at $t = T$ as shown by $f_3(t)$ in Fig. 3.46d.

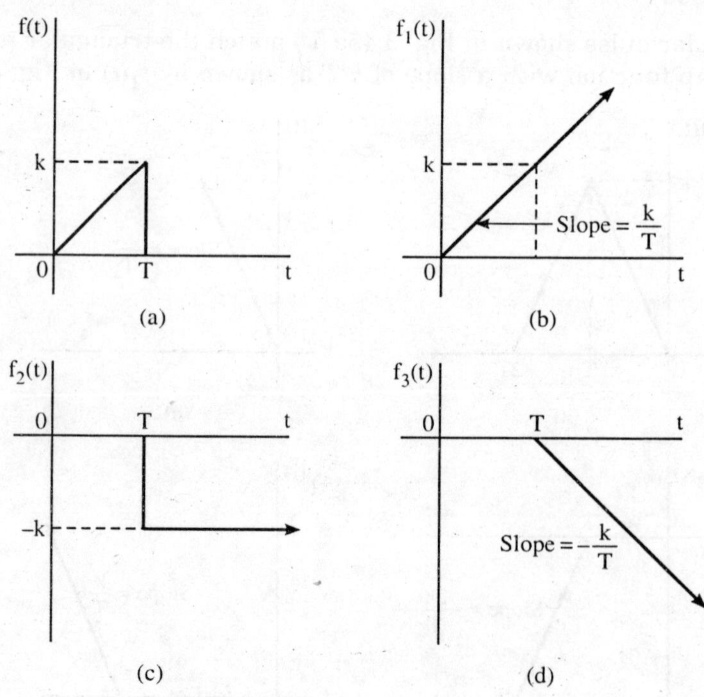

FIG. 3.46

Thus the sawtooth pulse function can be written as

$$f(t) = f_1(t) + f_2(t) + f_3(t)$$

$$= \frac{k}{T} t\, u_{-1}(t) - k\, u_{-1}(t - T)$$

$$- \frac{k}{T}(t - T)\, u_{-1}(t - T) \qquad (3.17.3)$$

3.17.4 Sinusoidal Pulse

Fig. 3.47 shows the pulse consisting of the first half cycle of a sine wave of amplitude A and periodic time T seconds. A sine wave of amplitude A and periodic time T seconds and defined for all positive values of time t given by

$$f_1(t) = A\left(\sin \frac{2\pi}{T} t\right) u_{-1}(t) \qquad (3.17.4)$$

This is shown in Fig. 3.47b. Figure 3.47c shows the same sine wave shifted by $T/2$ seconds. This shifted sine wave is defined by the relation

$$f_2(t) = A\left[\sin \frac{2\pi}{T}\left(t - \frac{T}{2}\right)\right] u_{-1}\left(t - \frac{T}{2}\right) \qquad (3.17.5)$$

It should be noted that the two sinusoids of Fig. 3.47b and Fig. 3.47c become out of phase by shifting one to a time $T/2$. The addition of these two waveforms results in the cancellation of 1 with 3 and 2 with 4 and so on, leaving only the desired half cycle of the sine wave as shown in Fig. 3.47a. We can write

$$f(t) = f_1(t) + f_2(t)$$

$$= A\left(\sin \frac{2\pi}{T} t\right) u_{-1}(t)$$

$$+ A\left[\sin 2\frac{\pi}{T}\left(t - \frac{T}{2}\right)\right] u_{-1}\left(t - \frac{T}{2}\right) \qquad (3.17.6)$$

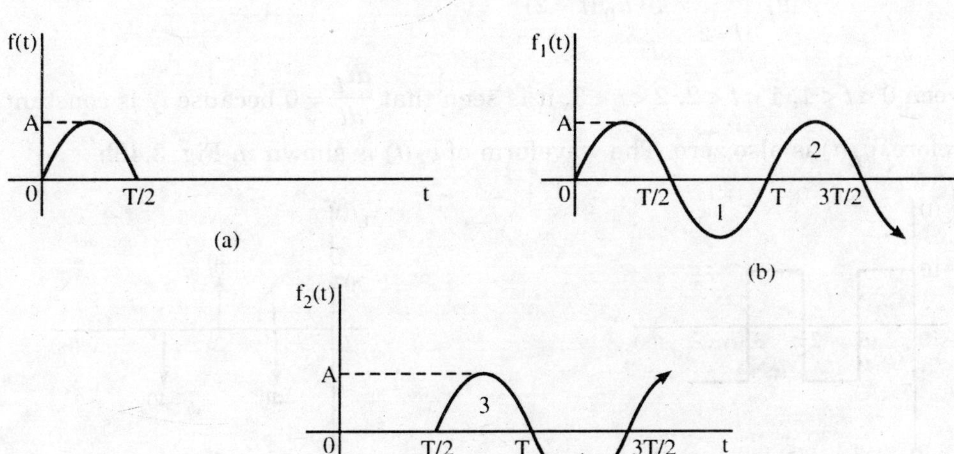

FIG. 3.47

■ **EXAMPLE 3.20** The current through an inductor of $2H$ is a square wave as shown in Fig. 3.48a. Determine its voltage and draw its waveshape.

SOLUTION :

$$v_L = L\frac{di_L}{dt}$$

(a) At $t = 0$, there is a sudden change of current from 0 to 10A. Therefore $\frac{di_L}{dt}$ is an impulse of magnitude 10, that is, $\frac{di_L}{dt} = 10\ u_0(t)$.

$v_L = L\frac{di_L}{dt} = 2 \times 10\ u_0(t) = 20\ u_0(t)$ That is, the voltage is also an impulse of magnitude 20.

(b) At $t = 1$ s, the current changes from 10A to –10A. Therefore, $\frac{di_L}{dt}$ is an impulse of magnitude – 20. That is,

$$\left.\frac{di_L}{dt}\right|_{t=1} = -20\ u_0\ (t-1)$$

and

$$\left.v_L\right|_{t=1} = \left.L\frac{di_L}{dt}\right|_{t=1} = 2 \times (-20)\ u_0(t-1)$$

$$= -40\ u_0(t-1)$$

Similarly, at $t = 2s$,

$$\left.\frac{di_L}{dt}\right|_{t=2} = 20\ u_0\ (t-2)$$

and

$$\left.v_L\right|_{t=2} = 40\ u_0\ (t-2)$$

Between $0 < t < 1$, $1 < t < 2$, $2 < t < 3$, it is seen that $\frac{di_L}{dt} = 0$ because i_L is constant. Therefore, $v_L(t)$ is also zero. The waveform of $v_L(t)$ is shown in Fig. 3.48b.

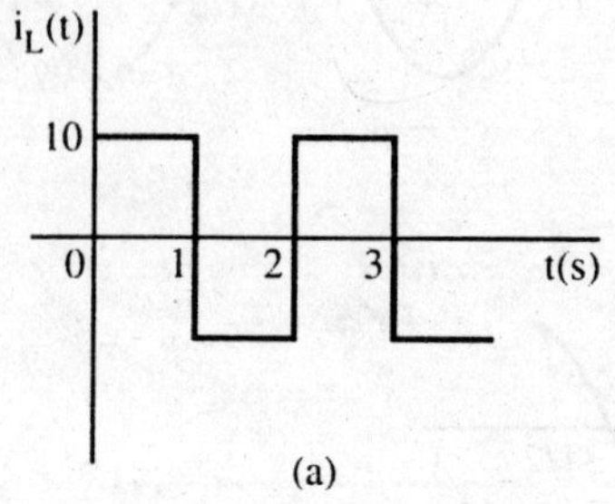

(a)

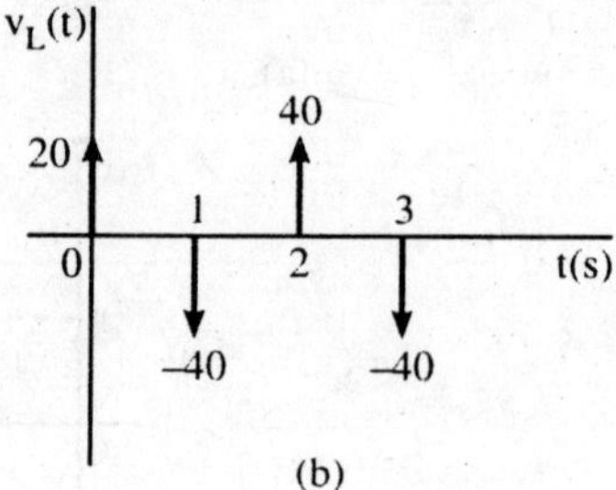

(b)

FIG. 3.48

EXERCISES

Write the mathematical expressions for the waveforms shown in the following figures : (P 3.1—P 3.14)

ANSWERS

1. $f(t) = u_{-1}(t-1) + u_{-1}(t-2) + u_{-1}(t-3) + u_{-1}(t-4) - 4\,u_{-1}(t-5)$

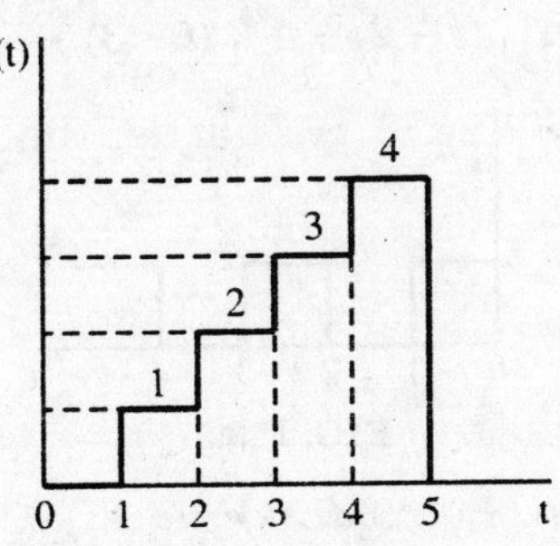

FIG. P 3.1

2. $v(t) = u_{-1}(t-a) - u_{-1}(t-b)$

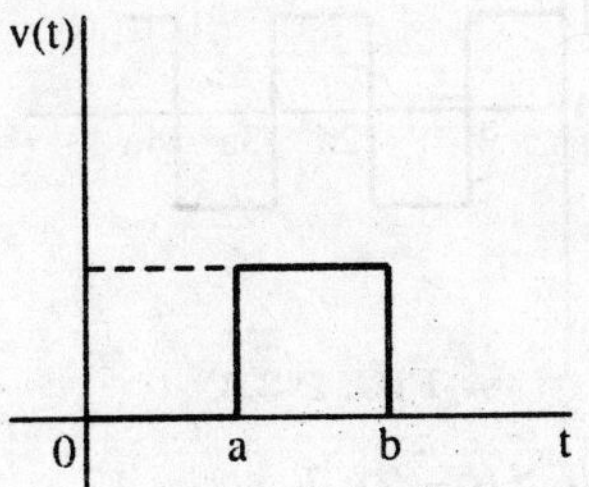

FIG. P 3.2

3. $f(t) = u_{-1}(t) - 2\,u_{-1}\left(t - \dfrac{T}{2}\right) + u_{-1}(t-T)$

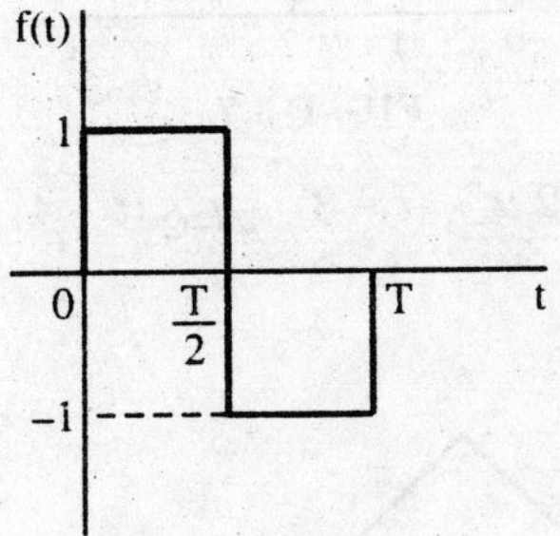

FIG. P 3.3

4. $i(t) = 10\,u_{-1}(t-2) - 10\,u_{-1}(t-4) + 10\,u_{-1}(t-6)$
$\quad - 10\,u_{-1}(t-8) + 10\,u_{-1}(t-10) - 10\,u_{-1}(t-12)$

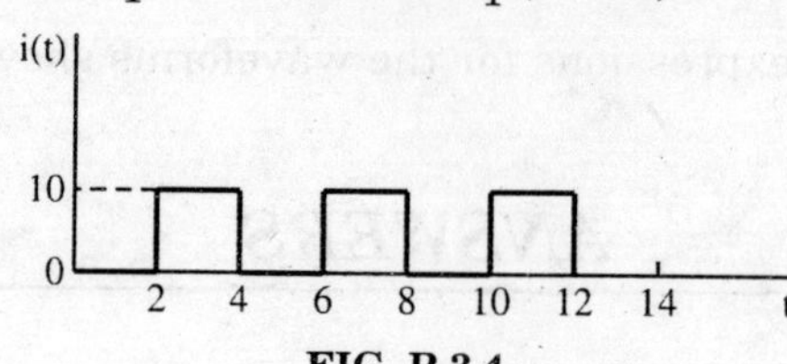

FIG. P 3.4

5. $f(t) = u_{-1}(t) - u_{-1}(t-1) + u_{-1}(t-2) - u_{-1}(t-3) + \ldots$

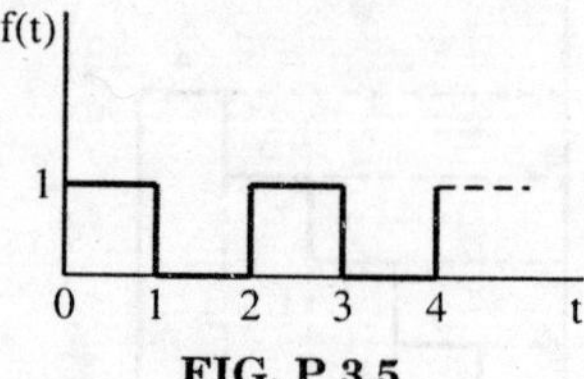

FIG. P 3.5

6. $f(t) = u_{-1}(t) + \sum_{k=1}^{\infty} (-1)^k\, 2\, u_{-1}(t-ka)$

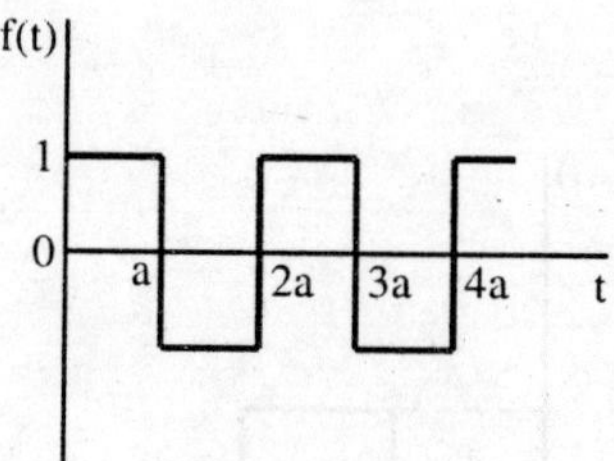

FIG. P 3.6

7. $f(t) = u_{-2}(t) - 2\,u_{-2}(t-1) + u_{-2}(t-2)$

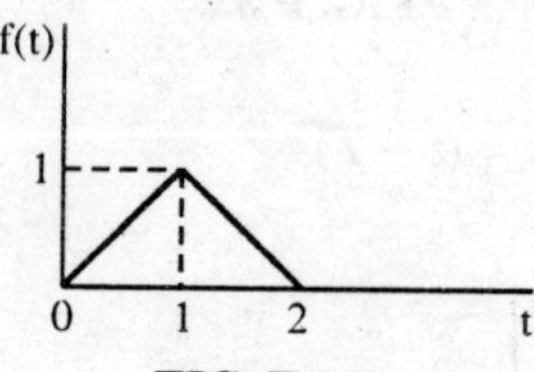

FIG. P 3.7

8. $v(t) = u_{-2}(t) - 2\,u_{-2}(t-1) + 2\,u_{-2}(t-3) - u_{-2}(t-4)$

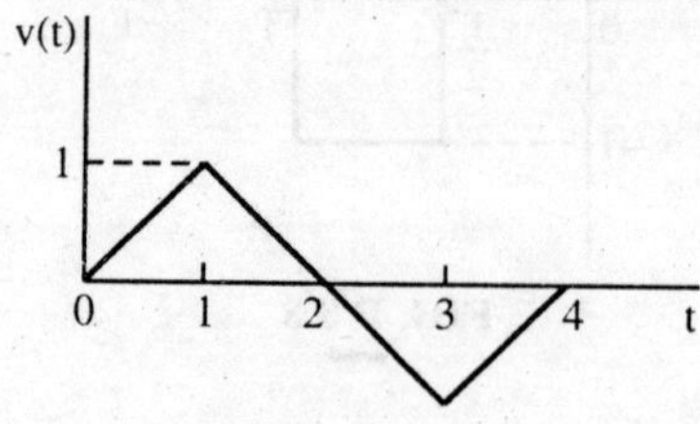

FIG. P 3.8

9. $f(t) = u_{-2}(t) - \frac{3}{2} u_{-2}(t-1) + \frac{1}{2} u_{-2}(t-3)$

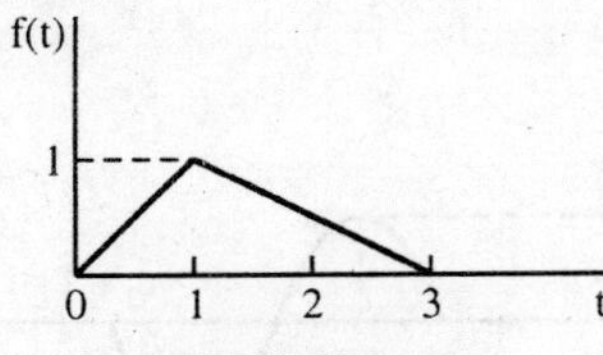

FIG. P 3.9

10. $f(t) = u_{-2}(t) - u_{-1}(t-1) - u_{-1}(t-2) - u_{-1}(t-3) - \ldots$

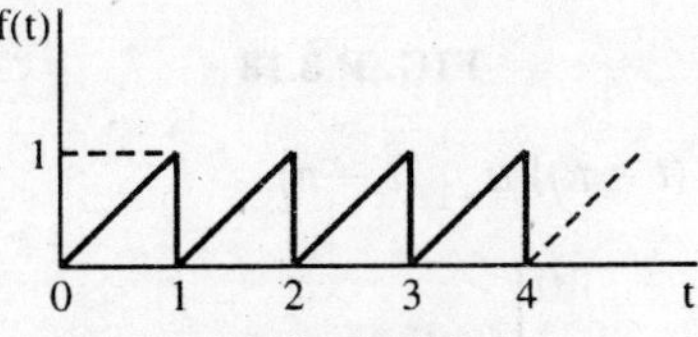

FIG. P 3.10

11. $f(t) = 2\,u_{-2}(t) - 2\,u_{-2}(t-1) - 2\,u_{-2}(t-3) + 2u_{-2}(t-4)$

$$= 2\,t\,u_{-1}(t) - 2\,(t-1)\,u_{-1}(t-1)$$

$$- 2\,(t-3)\,u_{-1}(t-3) + 2\,(t-4)\,u_{-1}(t-4)$$

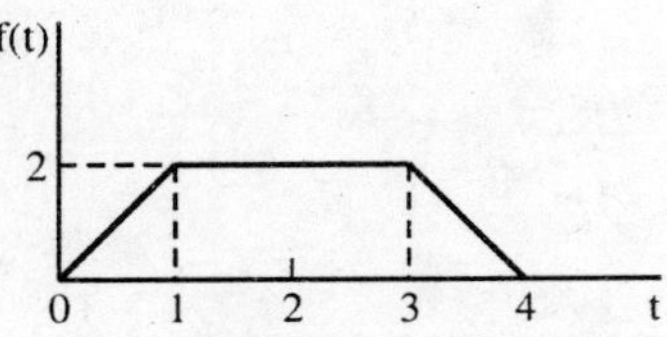

FIG. P 3.11

12. $f(t) = -\frac{2k}{T}\left[t\,u_{-1}(t) - tu_{-1}(t-T) - \frac{T}{2} u_{-1}(t) + \frac{T}{2} u_{-1}(t-T)\right]$

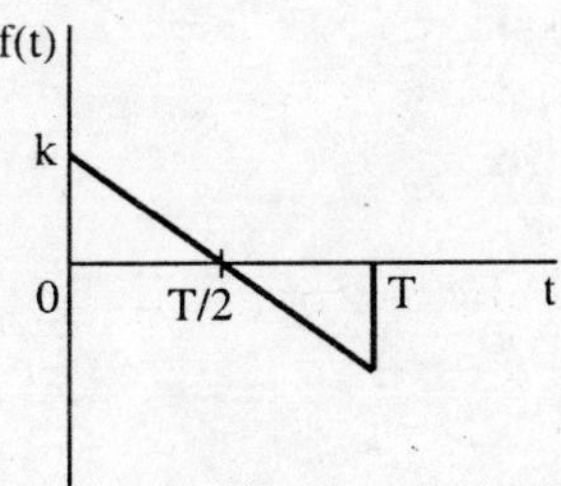

FIG. P 3.12

13. $v(t) = A\left[\sin\frac{\pi}{2}(t - t_0)\right]u_{-1}(t - t_0) - A\left[\sin\frac{\pi}{2}(t - t_0)\right]u_{-1}(t - t_0 - T)$

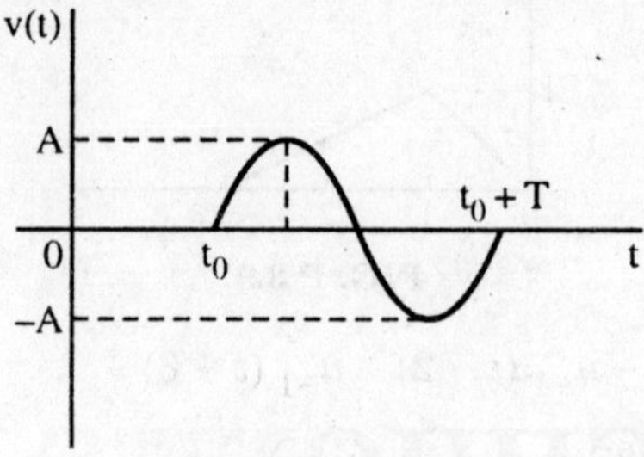

FIG. P 3.13

14. $f(t) = (\sin t)\, u_{-1}(t) + [\sin(t - \pi)]\, u_{-1}(t - \pi).$

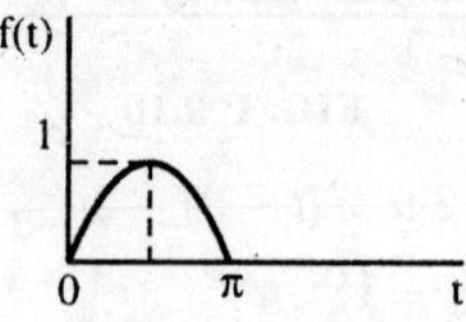

FIG. P 3.14

4

KIRCHHOFF'S LAWS AND EQUIVALENT NETWORKS

4.1 NETWORK TERMINOLOGY

An electrical device may be divided into regions and volumes. Each region has at least two accessible points called **terminals.** A simple circuit element may be defined as a two-terminal electrical device which cannot be sub-divided into other two-terminal devices. A circuit element is any circuit component such as a resistor, an inductor, a capacitor, a voltage source, a current source. An interconnection (or coupling) of circuit elements is called a **circuit** or a **network**. These two terms are used synonymously. In general, the term circuit is used to describe a simple configuration, while the term network more often refers to complicated element interconnections.

In order to describe the manner in which the elements are interconnected in a network, the following terms are used :

Node. We define a **node** to be any point in a circuit where the terminals of two or more elements are connected together. If three or more elements are connected together at a node, then that node is sometimes called a **junction** or **essential node** or **principal node**. All points connected together by resistanceless wires represent the same node. Thus, any two points which are short-circuited represent the same node.

Branch. A branch is a part of the circuit which extends from one node to another. A branch may contain one element or several elements in series. It has two terminals.

Essential Branch. A path which connects two essential nodes without passing through an essential node is called an **essential branch.**

Closed Path. A circuit may contain one or more closed paths. The term **closed path** means any path, along various branches of the circuit which finishes at the point where it started. It is not necessary that a closed path is made up of closed sequence of connected elements. The path can jump from any node to any other, whether or not these two nodes are directly connected by an element. Thus, starting at any node n in the circuit, we form a closed path by traversing through elements (open circuits included) and returning to the starting node n, never encountering any other node more than once.

Mesh. Any path which contains no other paths within it is called a **mesh.**

Loop. A path which contains more than two meshes is called a **loop**. Thus a loop contains meshes but a mesh does not contain a loop.

The circuit in Fig. 4.1 is used to explain the terms mentioned above. Here there are nine circuit elements, two voltage sources V_{S1} and V_{S2}, one current source I_{S1}, three resistors R_1, R_2 and R_3, two inductors L_1, L_2 and one capacitor C. The points a, b, c, d, g are the nodes. It is to be noted that the whole line forms one node g. The points b, c, d are essential nodes.

The branches in Fig. 4.1 are as follows :

Branch ag consisting of a voltage source V_{S1},

branch ab consisting of a resitor R_1,

branch bg consisting of a resistor R_2,

branch bc consisting of an inductor L_1,

branch cg consisting of a current source I_{S1},

branch cd consisting of an inductor L_2 and a capacitor C,

branch dg consisting of a voltage source V_{S2},

branch df consisting of a resistor R_3,

branches bc, cd are essential branches

The following are the meshes in Fig. 4.1 :

$$abga, bcgb, cdgc, dfgd$$

Some of the loops in Fig. 4.1 are

$$abcga, bcdgb, cdfgc.$$

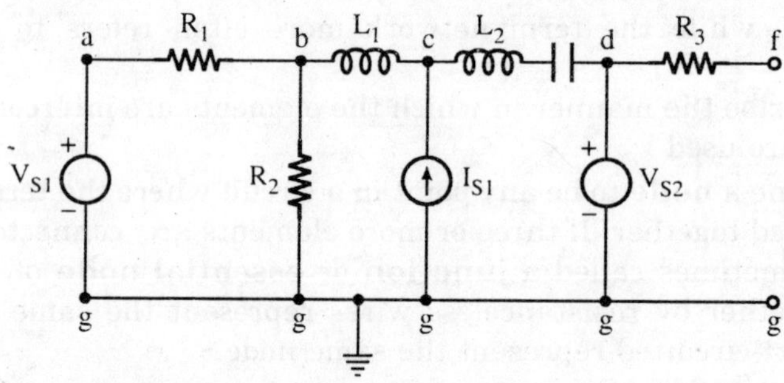

FIG. 4.1. Circuit to explain terminology.

4.2 KIRCHHOFF'S LAWS

The two basic laws which concern the voltage across and current through the elements of a network are Kirchhoff's voltage law (KVL) and Kirchhoff's current law (KCL) respectively. These laws form the basis of all network analysis and are applicable to any network whether the elements of the network are linear, nonlinear, time-varying or time invariant. The equations necessary to describe a network can be written with the help of KVL and KCL and the v-i relations of the circuit elements.

4.3 SIGN CONVENTIONS

4.3.1 Polarity of a Voltage Source

The polarity of a voltage source is fixed and is independent of the direction of current. The positive (+) terminal is at a higher potential and negative (–) terminal at a lower potential. Consider a voltage source shown in Fig. 4.2. When we move from a to b (positive terminal to negative terminal) there is a *voltage drop*. When we go from b to a (negative terminal to positive terminal) there is a *voltage rise*. A *voltage rise* indicates a *positive voltage* and a *voltage drop* indicates a *negative voltage*. Thus, in Fig. 4.2 A a voltage (equal to $- V_s$) results when we go from a to b and a voltage (equal to $+ V_s$) results when we move from b to a.

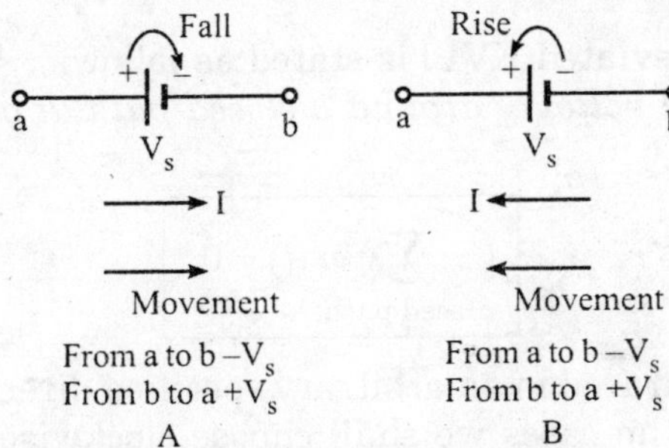

FIG. 4.2. Polarity of a voltage source.

4.3.2 Polarity of a Resistor

Current flows from a point of higher potential to another point on lower potential. Thus, in Fig. 4.3 since the current I flows from point c to point d, the point c is at a higher potential and the point d is at a lower potential. When we move from c to d, that is, along the direction of current, there is a voltage drop V_{cd} (= – RI). While going from d to c, that is, opposite to the flow of current, there is a voltage rise V_{dc} (= + RI). While traversing a closed path through a resistor the following rule can easily be remembered :

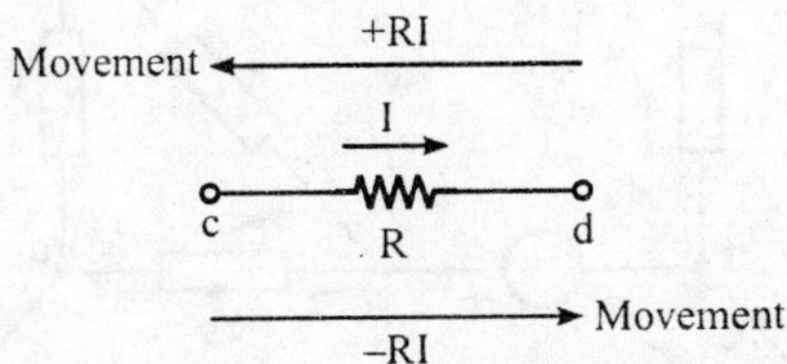

FIG. 4.3. Sign convention for voltage across a resistor.

Movement along the flow of current gives *N*egative voltage (*i.e.*, voltage drop). Movement opposite to the flow of current gives *P*ositive voltage (*i.e.*, voltage rise).

4.3.3 Another Convention

We assign a positive (+) sign for the voltage to the terminal of the element where the current enters and a negative (–) sign to the terminal of the element where the current leaves it. While traversing a closed path, if we move from + terminal to – terminal there is a voltage drop. If we move from – terminal to + terminal, there is a voltage rise. A **voltage rise** indicates **positive voltage** and a **voltage drop** indicates a **negative voltage**.

In Fig. 4.4 current enters the terminal a of the element. Hence point a is assigned + sign. Since the current leaves the element at point b, the point b is assigned a (–) sign. While traversing a closed path through the element between ab of Fig. 4.4, if we move from a to b there is a voltage drop equal to $-v$. Conversely, if we move from b to a, there is a voltage rise, equal to $+v$.

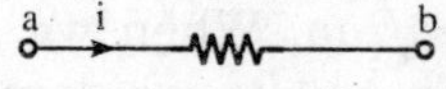

FIG. 4.4. Sign convention.

4.4 KIRCHHOFF'S VOLTAGE LAW (KVL)

Kirchhoff's voltage law (abbreviated KVL) is stated as follows :

The algebraic sum of all voltages around a closed path at any instant is zero.

Mathematically,

$$\sum_{\text{closed path}} v(t) = 0$$

The closed path is to be traced in an arbitrary specified direction, which may be clockwise or anticlockwise. In most of the cases we shall choose clockwise direction for traversing the closed path.

The voltage across elements that are traversed from – to + are taken positive and the voltages across elements that are traversed from + to – are taken negative.

An algebraic sum is the one in which the sign of the quantity is taken into account.

Fig. 4.5 shows a portion of a network. The voltages across the elements are shown in the network. By KVL around mesh *abcefa*,

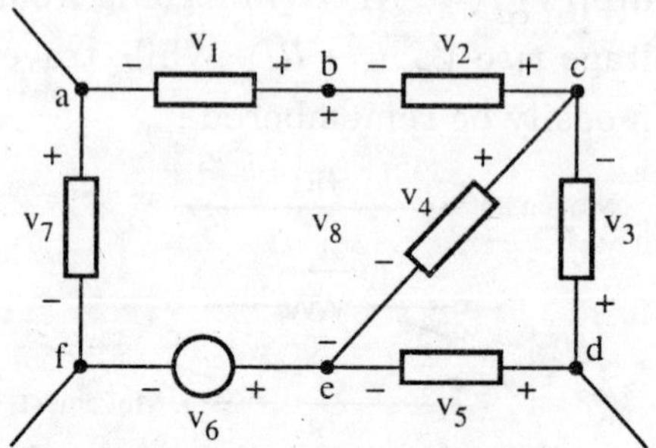

FIG. 4.5. Illustration of KVL

$$v_{ab} + v_{bc} + v_{ce} + v_{ef} + v_{fa} = 0$$

$$(-v_1) + (+v_2) + (-v_4) + (-v_6) + (+v_7) = 0$$

or $$-v_1 + v_2 - v_4 - v_6 + v_7 = 0$$

It is to be noted that while writing KVL equations, it is not necessary to have a physically closed connection of elements. It should be a *closed path.* It may not be closed circuit. For example, there is no element connected between b and e. We can write KVL equation for the closed path which has one of its elements between nodes b and e an open circuit.

$$v_{bc} + v_{cd} + v_{de} + v_{eb} = 0$$

$$v_2 + v_3 - v_5 + v_8 = 0$$

Since voltage is energy (or work) per unit charge, KVL is an alternative method of stating the *law of conservation of energy.*

4.5 KIRCHHOFF'S CURRENT LAW (KCL)

Kirchhoff's current (abbreviated KCL) is stated as follows :

At any instant of time, the algebraic sum of current at a node is zero.

Mathematically,

$$\sum_{\text{node } n} i(t) = 0$$

If the currents **entering** a node are assigned **positive** sign, then the currents leaving the node will be assigned **negative** sign or vice versa. The choice of the sign convention is arbitrary, but once a sign convention is chosen, it should not be changed with respect to any particular node while writing KCL equation. In applying KCL let us adopt the sign convention that the **incoming currents are positive and the outgoing currents are negative.**

Consider a portion of some network as shown in Fig. 4.6. Current $i_1(t)$, $i_2(t)$ and $i_4(t)$ are entering the node n. Hence they are assigned positive sign. Currents $i_3(t)$, $i_5(t)$ and $i_6(t)$ are leaving n and hence they are assigned negative sign. Applying KCL at node n we get

$$i_1(t) + i_2(t) - i_3(t) + i_4(t) - i_5(t) - i_6(t) = 0 \tag{4.5.1}$$

From Eq. (4.5.1)

$$i_1(t) + i_2(t) + i_4(t) = i_3(t) + i_5(t) + i_6(t)$$

or sum of the incoming current = sum of the outgoing currents

Thus, as **alternative form of KCL** can be stated as follows :

At any instant of time, the sum of all the currents flowing into a node is equal to the sum of all the currents leaving the same node.

It should be noted that KCL puts a constraint on the currents at a node. Nothing is said about the branch characteristics. For example, in Fig. 4.6 the element in which current $i_3(t)$ is flowing a short circuit and KCL still holds. In other words, KCL applies to currents at a node regardless of the nature of the elements in the circuit and how the currents come about.

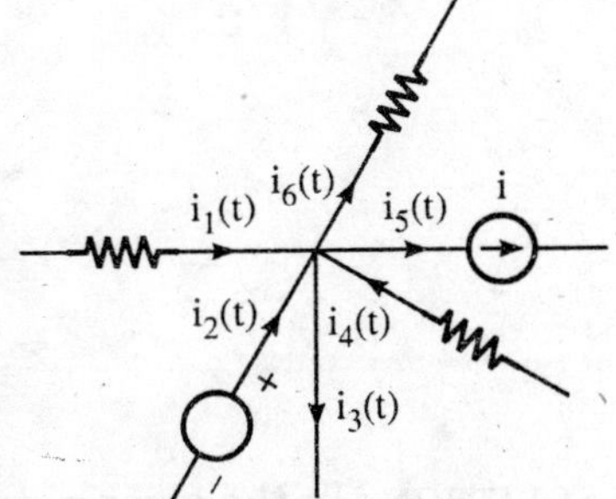

FIG. 4.6. Illustration of KCL.

Kirchhoff's current law (KCL) is based on the *conservation of charge* at a node. According to the law of conservation of charge, the charge flowing into a node is equal to the net charge flowing out of the node. The current law given by Eq. (4.5.1) then follows directly by differentiation with respect to time (since $i = dq/dt$).

4.6 APPLICATION OF KCL AND KVL

The procedure to apply KCL and KVL for resistive circuits is as follows :

(1) Draw the circuit diagram.

(2) Mark the values of resistance and emfs including their polarities.

(3) Assume the reference directions of currents in different branches arbitrarily.

(4) Apply KCL at each node directly on the circuit diagram to keep the minimum number of currents.

(5) For every resistor assign positive (+) sign for the voltage to the resistor terminal where the current enters, and negative (–) sign to the terminal of the resistor where the current leaves it.

(6) Apply KVL in various closed paths.
To apply KVL in a closed path, start any point in the path and proceed round the path in either clockwise or anticlockwise direction. While traversing a closed path if we move from positive terminal to negative terminal there is a voltage drop, which is assigned a negative sign. If we move from negative terminal to positive terminal, there is a voltage rise which is assigned a positive sign. Add all the voltages across the elements algebraically till the starting point is reached. The equation so obtained is called the *KVL equation.* The final KVL equation is the same whether the closed path is traversed in clockwise or anticlockwise direction.

(7) Solve the KVL equations and obtain the values of the assumed currents.

(8) If the calculated value of a particular current has positive sign, it means that its assumed direction is correct. If it has a negative value then the actual direction is opposite to that assumed. In such a case draw a fresh circuit diagram and make the correct directions of currents on this diagram.

The above procedure is illustrated by the following examples :

■ **EXAMPLE 4.1** Write the KVL equations for the network shown in Fig. 4.7.

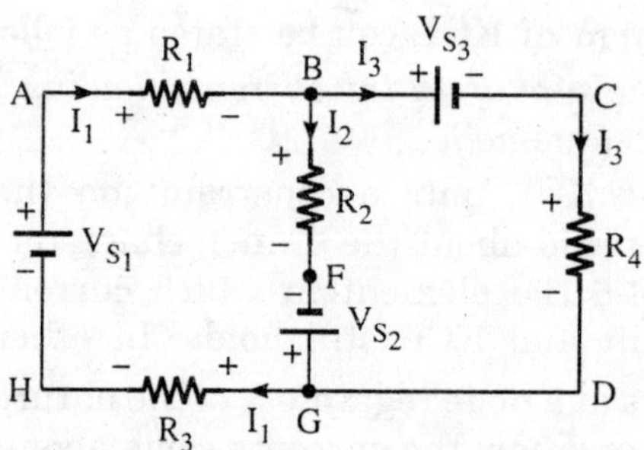

FIG. 4.7. Illustrating Example 4.1.

SOLUTION : In the given network the reference directions of currents in different branches are assumed. On each resistor the current is marked. A positive (+) sign is assigned for the voltage to the terminal of each resistor where the current enters and negative (–) sign to the terminal where the current leaves.

We shall write down the KVL equations in various meshes. It should be noted that traversing of a closed path either in clockwise or anticlockwise direction does not affect the final KVL equation provided that the same convention of signs is used throughout the closed path. In other words, the choice of direction round the mesh is immaterial. We shall prove this statement.

Traversing of meshes in clockwise direction

(a) *KVL in mesh* $A-B-F-G-H-A$

We have to traverse the mesh from A to B, B to F, F to G, G to H and H to A.

(1) From A to B, since we are moving along the current I_1, there is a voltage drop $-R_1I_1$.

(2) From B to F again we move along the current I_2, therefore, there is a voltage drop $-R_2I_2$.

(3) From F to G (– to +) there is a rise of voltage $+V_{S2}$.

(4) From G to H, because of the movement along the current I_1, there is a voltage drop $-R_3I_1$.

(5) From H to A (– to +) there is a voltage rise $+V_{S1}$.

By KVL the algebraic sum of all these voltages is zero.

$$V_{AB}+V_{BF}+V_{FG}+V_{GH}+V_{HA}=0 \tag{E4.1.1}$$

$$(-R_1I_1)+(-R_2I_2)+(+V_{S2})+(-R_3I_1)+(+V_{S1})=0 \tag{E4.1.2}$$

$$-R_1I_1-R_2I_2+V_{S2}-R_3I_1+V_{S1}=0 \tag{E4.1.3}$$

Alternative method

We have already assigned positive (+) and negative (–) signs on various resistors and voltage sources. While traversing a closed path, if we move from (+) terminal to (–) terminal there is a voltage drop, and if we move from (–) terminal to (+) terminal there is a voltage rise. A voltage rise is positive and a voltage drop is negative. With this convention of signs Eq. (E4.1.2) can directly be written from Eq. (E 4.1.1) ignoring the steps from (1) to (5) given above. This alternative method is easier to apply. Hence it will be used to write KVL equations in other meshes.

(*b*) KVL *in mesh* $B-C-D-G-F-B$

$$V_{BC}+V_{CD}+V_{DG}+V_{GF}+V_{FB}=0 \tag{E4.1.4}$$

$$(-V_{S3})+(-R_4I_3)+(0)+(-V_{S2})+(+R_2I_2)=0$$

$$-V_{S3}-R_4I_3-V_{S2}+R_2I_2=0 \tag{E4.1.5}$$

Traversing of meshes in anticlockwise direction

(c) KVL in mesh $A-H-G-F-B-A$

$$V_{AH}+V_{HG}+V_{GF}+V_{FB}+V_{BA}=0 \tag{E4.1.6}$$

$$(-V_{S1})+(+R_3I_1)+(-V_{S2})+(+R_2I_2)+(+R_1I_1)=0$$

$$-V_{S1}+R_3I_1-V_{S2}+R_2I_2+R_1I_1=0 \tag{E4.1.7}$$

Multiplying Eq. (E4.1.7) by – 1 on both the sides and rearranging the terms we get

$$-R_1I_1-R_2I_2+V_{S2}-R_3I_1+V_{S1}=0 \tag{E4.1.8}$$

It is seen that Eqs. (E4.1.3) and (E4.1.8) are identical. This shows that the choice of direction round a mesh is immaterial to write KVL equation.

(*d*) KVL in mesh $B-F-G-D-C-B$

$$V_{BF}+V_{FG}+V_{GD}+V_{DC}+V_{CB}=0 \quad \text{(E4.1.9)}$$

$$(-R_2I_2)+(+V_{S2})+(0)+(+R_4I_3)+(+V_{S3})=0$$

$$-R_2I_2+V_{S2}+R_4I_3+V_{S3}=0 \quad \text{(E4.1.10)}$$

Multiplying Eq. (E4.1.10) by – 1 on both the sides and rearranging the terms

$$-V_{S3}-R_4I_3-V_{S2}+R_2I_2=0 \quad \text{(E4.1.11)}$$

It is observed that Eqs. (E4.1.5) and (E4.1.11) are identical. Thus, the choice of direction round a mesh is immaterial to write KVL equation.

EXAMPLE 4.2 A battery of e.m.f. 40 V and internal resistance 2 Ω is connected in parallel with a second battery of 44 V and internal resistance 4 Ω. A load resistance of 6 Ω is connected across the ends of the parallel circuit. Calculate the current in each battery and in the load.

SOLUTION : The circuit is shown in Fig. 4.8. Let the currents supplied by the batteries be I_1 and I_2. By KCL at node D, the current in the load is (I_1+I_2).

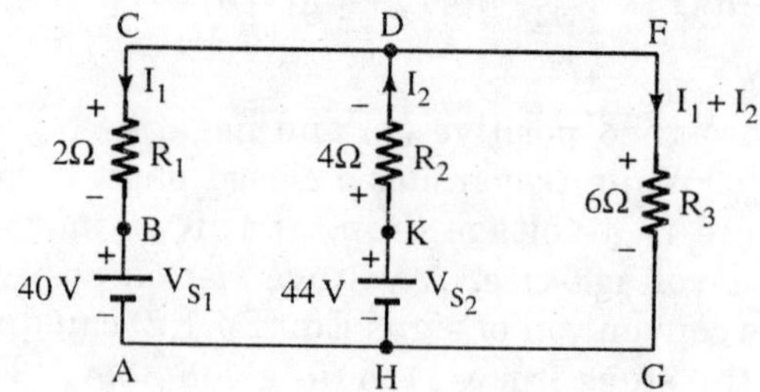

FIG. 4.8. Illustrating Example 4.2.

By KVL in mesh *ABCDFGHA*,

$$V_{AB}+V_{BC}+V_{CD}+V_{DF}+V_{FG}+V_{GH}+V_{HA}=0$$

$$V_{S1}-R_1I_1+0+0-R_3(I_1+I_2)+0+0=0$$

$$V_{S1}-R_1I_1-R_3(I_1+I_2)=0$$

$$40-2I_1-6(I_1+I_2)=0$$

$$8I_1+6I_2=40 \quad \text{(E4.2.1)}$$

By KVL in mesh *ABCDKHA*,

$$V_{AB}+V_{BC}+V_{CD}+V_{DK}+V_{KH}+V_{HA}=0$$

$$V_{S1}-R_1I_1+0+R_2I_2-V_{S2}+0=0$$

$$40-2I_1+4I_2-44=0$$

$$2I_1-4I_2=-4 \quad \text{(E4.2.2)}$$

Multiplying Eq. (E4.2.2) by 4 and subtracting from Eq. (E4.2.1)

$$22I_2 = 56$$

$$I_2 = \frac{56}{22} = 2.545 \text{ A}$$

Substituting the value of I_2 in Eq. (E4.2.2)

$$2I_1 = -4 + 4 \times 2.545 = 4 \times 1.545$$

$$I_1 = 3.090$$

$$\text{Load current} = I_1 + I_2 = 3.090 + 2.545 = 5.635 \text{ A}$$

■ **EXAMPLE 4.3** A battery having an e.m.f. of 105 V and an internal resistance of 1 Ω is connected in parallel wth a.d.c. generator of e.m.f. 110 V and internal resistance of 0.5 Ω to supply a load having a resistance of 8 Ω. Calculate (*a*) the currents in the battery, the generator and the load; (*b*) the p.d. across the load.

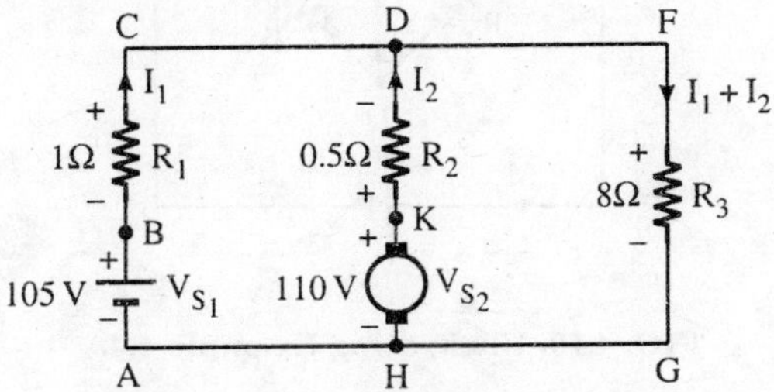

FIG. 4.9. Illustrating Example 4.3.

SOLUTION : The circuit is shown in Fig. 4.9. Let I_1 and I_2 be the current supplied by the battery and the generator respectively. By KCL at node *D*, the current in the load is $(I_1 + I_2)$.

(*a*) By KVL in mesh *ABCDFGHA*

$$V_{AB} + V_{BC} + V_{CD} + V_{DF} + V_{FG} + V_{GH} + V_{HA} = 0$$

$$V_{S1} - R_1 I_1 + 0 + 0 - R_3 (I_1 + I_2) + 0 + 0 = 0;$$

$$105 - 1 \times I_1 - 8 (I_1 + I_2) = 0$$

$$-9I_1 - 8I_2 = -105; \; 9I_1 + 8I_2 = 105 \quad \text{(E4.3.1)}$$

By KVL in mesh *ABCDKHA*

$$V_{AB} + V_{BC} + V_{CD} + V_{DK} + V_{KH} + V_{HA} = 0$$

$$V_{S1} - R_1 I_1 + 0 + R_2 I_2 - V_{S2} + 0 = 0$$

$$105 - 1 \times I_1 + 0.5I_2 - 110 = 0; \; -I_1 + 0.5I_2 = 5 \quad \text{(E4.3.2)}$$

Multiplying both the side of Eq. (E4.3.2) by 9 and adding to Eq. (E4.3.1)

$$12.5I_2 = 150;$$

$$I_2 = \frac{150}{12.5} = 12 \text{ A}$$

Substituting the value of I_2 in Eq. (E4.3.2)

$$-I_1 + 0.5 \times 12 = 5;$$

$$I_1 = 6 - 5 = 1 \text{ A}$$

$$\text{Load current} = I_1 + I_2 = 1 + 12 = 13 \text{ A}$$

(*b*) Voltage across the load $= 8\ (I_1 + I_2) = 8 \times 13 = 104$ V

■ **EXAMPLE 4.4** Fig. 4.10 shows a bridge network. With the currents as marked, (*a*) write KCL at the four nodes; (*b*) write KVL in meshes *ABDA*, *BCDB*, *ADCMA*

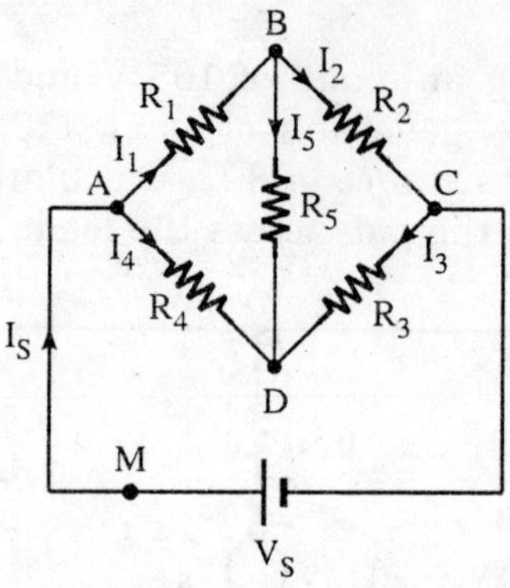

FIG. 4.10. Illustrating Example 4.4.

SOLUTION :

(*a*) Node *A* : $I_S = I_1 + I_4$

Node *B* : $I_1 = I_2 + I_5$

Node *C* : $I_2 = I_3 + I_S$

Node *D* : $I_3 + I_4 + I_5 = 0$

(*b*) By KVL in mesh *ABDA*,

$$-R_1I_1 - R_5I_5 + R_4I_4 = 0$$

By KVL in mesh *BCDB*,

$$-R_2I_2 - R_3I_3 + R_5I_5 = 0$$

By KVL in mesh *ADCMA*

$$-R_4I_4 + R_3I_3 + V_S = 0$$

■ **EXAMPLE 4.5** A bridge network *ABCD* has *AB* = 6 Ω, *BC* = 12 Ω, *CD* = 10 Ω and *DA* = 3 Ω. A galvanometer of resistance 20 Ω is connected between *B* and *D* and a battery of e.m.f. 2 V and negligible internal resistance is connected between *A* and *C*. Calculate the current in the galvanometer.

SOLUTION : The network is shown in Fig. 4.11. Let I_1, I_2 and I_3 be the currents in branches AB, AD and BD respectively.

By KCL,

current in branch $BC = I_1 - I_3$; and

current in branch $DC = I_2 + I_3$

By KL in mesh $ABDA$,

$$V_{AB} + V_{BD} + V_{DA} = 0;$$

$$-6I_1 - 20I_3 + 3I_2 = 0 \qquad \text{(E4.5.1)}$$

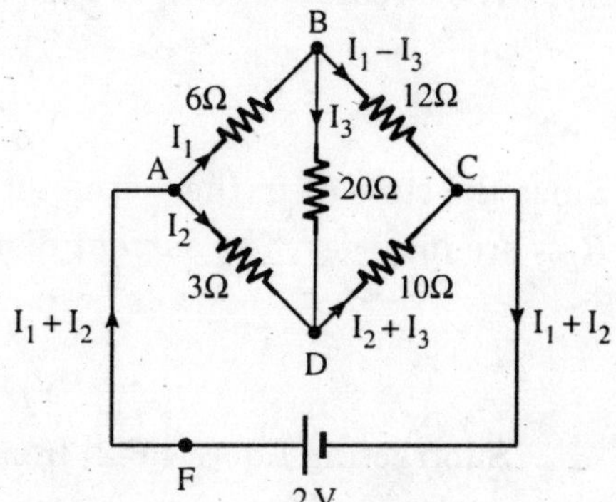

FIG. 4.11. Illustrating Example 4.5.

By KVL in mesh $ABCFA$,

$$V_{AB} + V_{BC} + V_{CF} + V_{FA} = 0$$

$$-6I_1 - 12(I_1 - I_3) + 2 + 0 = 0$$

$$-18I_1 + 12I_3 + 2 = 0$$

$$9I_1 - 6I_3 = 1 \qquad \text{(E4.5.2)}$$

By KVL in mesh $ADCFA$,

$$V_{AD} + V_{DC} + V_{CF} + V_{FA} = 0; \; -3I_2 - 10(I_2 + I_3) + 2 + 0 = 0$$

$$-13I_2 - 10I_3 = -2 \qquad \text{(E4.5.3)}$$

To eliminate I_1 from Eqs (E4.5.1) and (E4.5.2) multiplying Eq. (E4.5.1) by 3 and Eq. (E4.5.2) by 2 and adding

$$9I_2 - 60I_3 - 12I_3 = 2; \; 9I_2 - 72I_3 = 2$$

$$I_2 - 8I_3 = \frac{2}{9} \qquad \text{(E4.5.4)}$$

To eliminate I_2, multiplying Eq. (E4.5.4) by 13 and adding it to Eq. (E4.5.3)

$$-104I_3 - 10I_3 = -2 + \frac{26}{9}; \; -114I_3 = \frac{8}{9}$$

$$I_3 = \frac{-8}{9 \times 114} = -7.797 \times 10^{-3}\text{ A} = -7.797\text{ mA}$$

The minus sign indicates that the current is in the opposite direction to that shown in Fig. 4.11. Hence the current in the galvanometer is 7.797 mA from D to B.

■ **EXAMPLE 4.6** Four resistance AB, BC, AD and DC are connected together to form a closed square $ABCD$. The known resistance values are; AD = 12 ohms, AB = 35 ohms and DC = 12 ohms.

A d.c. supply of 120 volts is connected to A and C so that the current enters the combination at A leaves at C. A high-registance voltmeter is connected between B and D and whilst carrying negligible current, registers a voltage drop of 10 volts from B to D.

(*a*) Calculate the value of the resistance BC and the total current taken from the supply.

(*b*) Calculate also the value of BC, such that the potential difference between B and D is in the reverse direction, *i.e.*, from D to B (C and G).

SOLUTION :

The circuit is shown in Fig. 4.12.

(*a*) $$V_A - V_C = 120 \tag{E4.6.1}$$

$$V_B - V_D = 10 \tag{E4.6.2}$$

Since the current in the voltmeter is negligible and $R_{AD} = R_{DC} = 12\ \Omega$, the resistors R_{AD} and R_{DC} are in series. The current distributions are shown in the figure. Therefore,

$$V_A - V_D = 60 \tag{E4.6.3}$$

$$V_D - V_C = 60 \tag{E4.6.4}$$

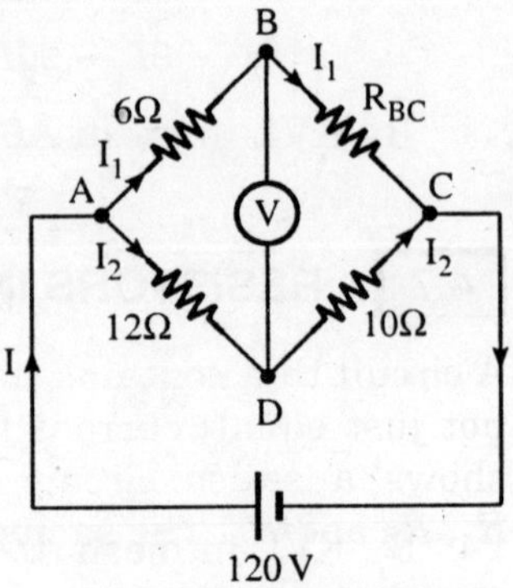

Subtracting Eq. (E4.6.2) from Eq. (E4.6.3),

$$(V_A - V_D) - (V_B - V_D) = 60 - 10$$

$$V_A - V_B = 50 \tag{E4.6.5}$$

FIG. 4.12. Illustrating Example 4.6.

By Ohm's law,

$$V_A - V_B = 35 I_1$$

therefore,

$$35 I_1 = 50$$

$$I_1 = \frac{50}{35} = 1.428 \text{ A}$$

By Ohm's law,

$$V_D - V_C = 12 I_2;\ 60 = 12 I_2$$

$$I_2 = \frac{60}{12} = 5 \text{ A}$$

Total current from the supply

$$I = I_1 + I_2 = 1.428 + 5 = 6.428 \text{ A}$$

Subtracting Eq. (E4.6.5) from Eq. (E4.6.1)

$$(V_A - V_C) - (V_A - V_B) = 120 - 50$$

$$V_B - V_C = 70$$

$$R_{BC} I_1 = 70$$

$$R_{BC} \times \frac{10}{7} = 70$$

$$R_{BC} = \frac{70 \times 7}{10} = 49\ \Omega$$

(*b*) If $V_D - V_B = 10$ (E4.6.6)

Adding Eqs. (E4.6.3) and (E4.6.6),

$$(V_A - V_D) + (V_D - V_B) = 70$$

$$V_A - V_B = 70$$

$$35 I_1 = 70$$

$$I_1 = \frac{70}{35} = 2 \text{ A}$$

Subtracting Eq. (E4.6.6) from Eq. (E4.6.4)

$$(V_D - V_C) - (V_D - V_B) = 50$$

$$V_B - V_C = 50$$

$$R_{BC} I_1 = 50$$

$$R_{BC} \times 2 = 50$$

$$R_{BC} = \frac{50}{2} = 25\ \Omega$$

4.7 RESISTORS IN SERIES

A circuit that contains two or more elements connected end to end, so that the *same* (identical, not just equal) current flows through all the elements, is called a *series circuit*. Fig. 4.13a shows a series circuit that contains three resistors R_1, R_2 and R_3. The same current flows through all three resistors. Thus,

$$I_1 = I_2 = I_3 = I$$

We have to find the equivalent resistance between the two terminals A and B of the series combination. That is, we have to determine a single equivalent resistance R_S which has the same volt-ampere relation as the combination. In general, an *equivalent resistance* is a single resistance which has the same volt-ampere relation as the network which it replaces.

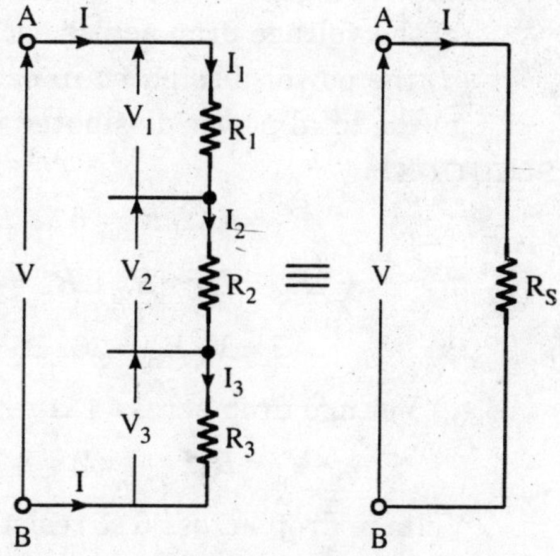

FIG. 4.13. Series circuit and its equivalent.

By KVL in mesh $A - C - D - B - A$

$$V_{AC} + V_{CD} + V_{DB} + V_{BA} = 0;$$

$$-V_1 - V_2 - V_3 + V = 0$$

$$V = V_1 + V_2 + V_3 \tag{4.7.1}$$

By Ohm's law

$$V_1 = R_1 I \tag{4.7.2}$$

$$V_2 = R_2 I \tag{4.7.3}$$

$$V_3 = R_3 I \tag{4.7.4}$$

From the equivalent circuit shown in Fig. 4.13b,

$$V = R_S I \tag{4.7.5}$$

Combining the Eqs. (4.7.1) to (4.7.5), we get

$$R_S I = R_1 I + R_2 I + R_3 I$$

$$R_S = R_1 + R_2 + R_3 \tag{4.7.6}$$

Thus, we see that the equivalent resistance of three series resistors is equal to the sum of the individual resistances. This result can be extended to several resistors in series. If there are n resistors in series, the equivalent resistance of the combination may be written as

$$R_S = R_1 + R_2 + R_3 + \ldots + R_n \tag{4.7.7}$$

In terms of conductance,

$$G_1 = \frac{1}{R_1}, G_2 = \frac{1}{R_2} \ldots G_n = \frac{1}{R_n} \text{ and } G_S = \frac{1}{R_S}$$

$$\frac{1}{G_S} = \frac{1}{G_1} + \frac{1}{G_2} + \frac{1}{G_3} + \ldots + \frac{1}{G_n} \tag{4.7.8}$$

■ **EXAMPLE 4.7** Three resistances of values 4, 6 and 8 Ω are connected in series across a 36 V d.c. supply. Calculate :

(*a*) the total resistance of the circuit,
(*b*) the total current in the circuit,
(*c*) the voltage drop across each resistance
(*d*) the power dissipated in each resistance,
(*e*) the total power dissipated in the circuit.

SOLUTION :

$$R_1 = 4\ \Omega, R_2 = 6\ \Omega, R_3 = 8\ \Omega$$

(*a*) $$R_S = R_1 + R_2 + R_3 = 4 + 6 + 8 = 18\ \Omega$$

(*b*) $$I = V/R_S = 36/18 = 2\text{ A}$$

(*c*) Voltage drop across 4 Ω resistor

$$V_1 = R_1 I = 4 \times 2 = 8\text{ V}$$

Voltage drop across 6 Ω resistor

$$V_2 = R_2 I = 6 \times 2 = 12\text{ V}$$

Voltage drop across 8 Ω resistor

$$V_3 = R_3 I = 8 \times 2 = 16\text{ V}$$

Check $V_1 + V_2 + V_3 = 8 + 12 + 16 = 36$ V

(*d*) Power dissipated in 4 Ω resistance

$$P_1 = I^2 R_1 = (2)^2 \times 4 = 16\text{ W}$$

Power dissipated in 6 Ω resistance

$$P_2 = I^2 R_2 = (2)^2 \times 6 = 24\text{ W}$$

Power dissipated in 8 Ω resistance

$$P_3 = I^2 R_3 = (2)^2 \times 8 = 32\text{ W}$$

(*e*) Total power dissipated

$$P = P_1 + P_2 + P_3 = 16 + 24 + 32 = 72 \text{ W}$$

The total power can also be found by

$$P = VI = 36 \times 2 = 72 \text{ W}$$

or $$P = I^2 R_S = (2)^2 \times 18 = 72 \text{ W}$$

4.8 VOLTAGE-DIVIDER EQUATIONS

Suppose that we are interested in determining V_1, V_2 and V_3 in terms of the total voltage V.

From Eq. (4.7.5)

$$I = \frac{V}{R_S}$$

From Eq. (4.7.2)

$$V_1 = R_1 I$$

$$V_1 = \frac{R_1}{R_S} V \tag{4.8.1}$$

Similarly,

$$V_2 = \frac{R_2}{R_S} V \tag{4.8.2}$$

$$V_3 = \frac{R_3}{R_S} V \tag{4.8.3}$$

These relations indicate how the voltage across a series combination in distributed between the resistors. The above equations are known as **voltage-divider equations.**

Voltage Division between Two Resistors

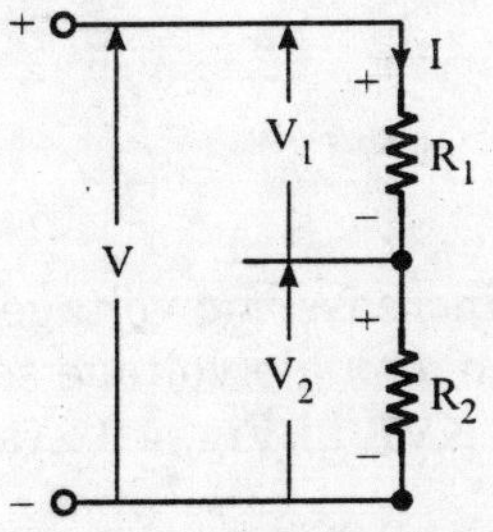

FIG. 4.14. Voltage division between two resistors in series.

When there are only two resistors in series (Fig. 4.14), the voltage division between them can be found as follows :

Total current in the circuit

$$I = \frac{V}{R_1 + R_2} \tag{4.8.4}$$

By Ohm's law

$$V_1 = R_1 I$$

$$V_1 = \frac{R_1}{R_1 + R_2} V \tag{4.8.5}$$

Similarly,

$$V_2 = R_2 I$$

$$V_2 = \frac{R_2}{R_1 + R_2} V \tag{4.8.6}$$

Thus, the voltage across either of the series resistors is equal to the applied voltage times the ratio of that resistance to the total resistance. The use of voltage-divider equation eliminates calculations for current when determining voltage across resistors. This can be seen from Eqs. (4.8.5) and (4.8.6) which are independent of current.

4.9 INTERNAL RESISTANCE OF A VOLTAGE SOURCE

The resistance offered to the flow of current by the internal parts of a voltage source is called its *internal resistance*. Because of the presence of internal resistance r, the terminal voltage V is different from the source e.m.f. V_S. The internal resistance is shown *external* to the source for circuit representation. Consider a cell with e.m.f. V_S and internal resistane r connected to an external resistance R as shown in Fig. 4.15.

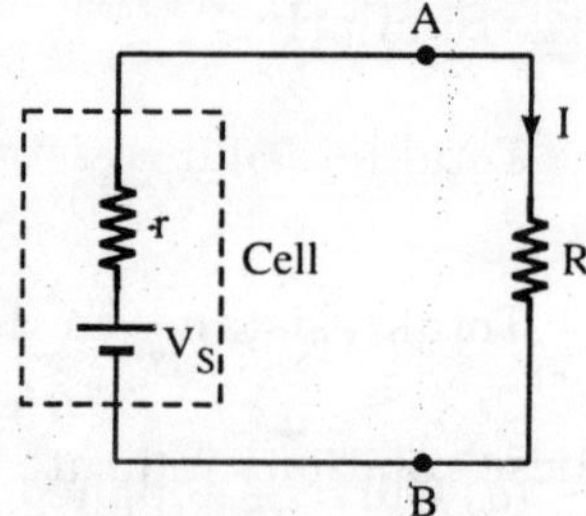

FIG. 4.15. Internal resistance of a voltage source.

The total circuit resistance $= R + r$

circuit current $I = \dfrac{V_S}{R + r}$

Voltage drop in the internal resistance of the cell $= rI$

Voltage available at the terminals of the cell $V_{AB} = V_S - rI$

4.10 CELLS IN SERIES

Cells are often connected in series to increase the voltage. When they are connected in series they can be considered as equivalent to a single voltage source. Three cells in series and their equivalent are shown in Fig. 4.16. By KVL in Fig. 4.16 (a).

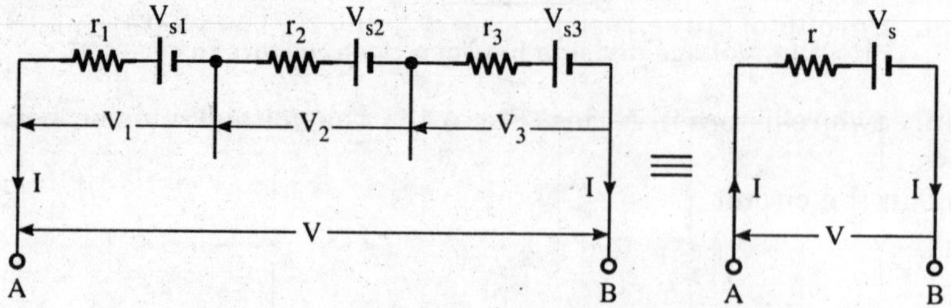

FIG. 4.16 Three cells in series and their equivalent.

$$V = V_1 + V_2 + V_3 = (V_{s1} - r_1 I) + (V_{s2} - r_2 I) + (V_{s3} - r_3 I)$$

$$= (V_{s1} + V_{s2} + V_{s3}) + (r_1 + r_2 + r_3)\, I \qquad (4.10.1)$$

From the equivalent circuit diagram (Fig. 4.15b)

$$V = V_S - rI \qquad (4.10.2)$$

Comparison of Eqs. (4.10.1) and (4.10.2) gives

$$V_S = V_{S1} + V_{S2} + V_{S3} \qquad (4.10.3)$$

$$r = r_1 + r_2 + r_3 \qquad (4.10.4)$$

Thus, the total e.m.f. of the cells is equal to the sum of the e.m.fs of each cell, and the total internal resistance is the sum of the internal resistances of each cell.

■ **EXAMPLE 4.8** A battery consists of five cells, each having an e.m.f. of 1.2 V and internal resistance of 0.4 Ω, joined in series. If this battery supplies current to a 6 Ω resistor, calculate:

(*a*) the current supplied by the battery,

(*b*) the p.d. across the battery terminals.

SOLUTION : Total e.m.f. of the battery

$$V_s = 5 \times 1.2 = 6 \text{ V}$$

Total internal resistance of the battery

$$r = 5 \times 0.4 = 2\ \Omega$$

Total resistance in the circuit

$$R = 6 + 2 = 8\ \Omega$$

(*a*) Current supplied by the battery

$$I = \frac{V_s}{R} = \frac{6}{8} = 0.75 \text{ A}$$

(*b*) Voltage across the battery terminals

$$V = V_s - rI = 6 - 2 \times 0.75 = 4.5 \text{ V}$$

4.11 PARALLEL RESISTORS

A group of elements is said to be connected in *parallel* when all are connected between the same two points. Elements connected in parallel will have the same (identical, not just equal) voltage across them. A group of three resistors R_1, R_2 and R_3 connected in parallel is shown in Fig. 4.17.

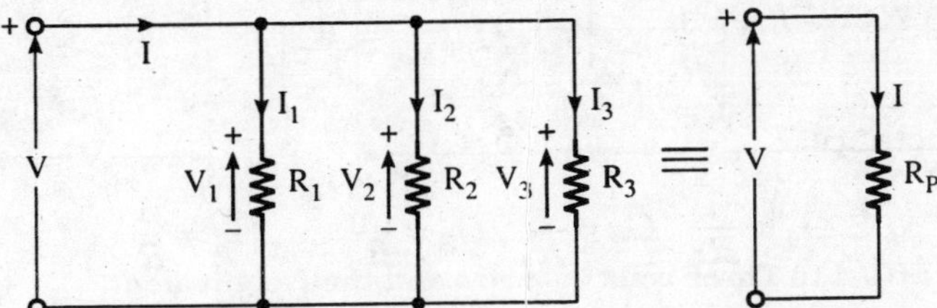

FIG. 4.17. Resistors in parallel and their equivalent.

Thus, $V_1 = V_2 = V_3 = V$

We are interested in finding a single resistance R_p which has the same volt-ampere relation as the combination.

By KCL,

$$I = I_1 + I_2 + I_3 \tag{4.11.1}$$

By Ohm's law

$$I_1 = \frac{V_1}{R_1} = \frac{V}{R_1}$$

$$I_2 = \frac{V_2}{R_2} = \frac{V}{R_2}$$

$$I_3 = \frac{V_3}{R_3} = \frac{V}{R_3}$$

$$I = \frac{V}{R_p}$$

Substituting these values in Eq. (4.11.1)

$$\frac{V}{R_p} = \frac{V}{R_1} + \frac{V}{R_2} + \frac{V}{R_3} = V\left(\frac{1}{R_1} + \frac{1}{R_2} + \frac{1}{R_3}\right)$$

$$\frac{1}{R_p} = \frac{1}{R_1} + \frac{1}{R_2} + \frac{1}{R_3} \tag{4.11.2}$$

When only *two* resistors are in parallel (Fig. 4.18)

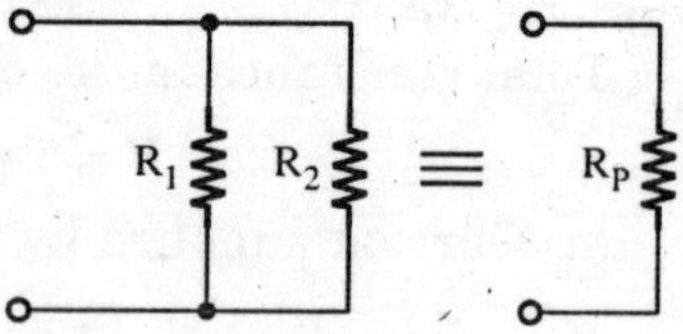

$$\frac{1}{R_p} = \frac{1}{R_1} + \frac{1}{R_2} = \frac{R_1 + R_2}{R_1 R_2}$$

$$R_p = \frac{R_1 R_2}{R_1 + R_2} = \frac{\text{product}}{\text{sum}} \tag{4.11.3}$$

R1 R2 ≡ Rp

FIG. 4.18.
Two resistors in parallel.

Eq. (4.11.3) is called the *product over the sum* equation for determining the total resistance of two resistors connected in parallel.

The result given in Eq. (4.11.2) can be extended to several resistors in parallel. If there are n resistors in parallel the equivalent resistance R_p of the combination may be written as

$$\frac{1}{R_p} = \frac{1}{R_1} + \frac{1}{R_2} + \frac{1}{R_3} + \ldots + \frac{1}{R_n} \tag{4.11.4}$$

$$R_p = \frac{1}{\left(\frac{1}{R_1} + \frac{1}{R_2} + \frac{1}{R_3} + \ldots + \frac{1}{R_n}\right)} \tag{4.11.5}$$

In terms of conductance

$$G_1 = \frac{1}{R_1}, G_2 = \frac{1}{R_2}, G_3 = \frac{1}{R_3} \ldots, G_p = \frac{1}{R_p}$$

$$G_p = G_1 + G_2 + G_3 + \ldots + G_n \tag{4.11.6}$$

Thus, the equivalent conductance of a parallel combination is equal to the sum of individual conductances.

4.12 CURRENT-DIVIDER EQUATIONS

Suppose we wish to determine the current in any one of the branches of a parallel circuit. Consider two resistances R_1 and R_2 connected in parallel as shown in Fig. 4.19.

Let $\quad R_p$ = total resistance of the circuit

$\quad I$ = total current in the circuit

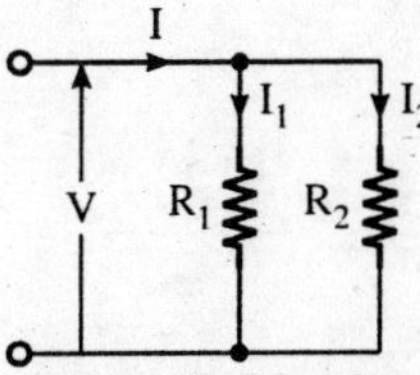

$$R_p = \frac{R_1 R_2}{R_1 + R_2}$$

$$I = \frac{V}{R_p}$$

$$I_1 = \frac{V}{R_1}$$

$$I_2 = \frac{V}{R_2}$$

FIG. 4.19. Division of current in a parallel

therefore,
$$\frac{I_1}{I} = \frac{V/R_1}{V/R_p} = \frac{R_p}{R_1} = \frac{1}{R_1} \cdot \frac{R_1 R_2}{R_1 + R_2} = \frac{R_2}{R_1 + R_2}$$

$$I_1 = \frac{R_2 I}{R_1 + R_2} \tag{4.12.1}$$

Also,
$$\frac{I_2}{I} = \frac{V/R_2}{V/R_p} = \frac{R_p}{R_2} = \frac{1}{R_2} \cdot \frac{R_1 R_2}{R_1 + R_2} = \frac{R_1}{R_1 + R_2}$$

$$I_2 = \frac{R_2 I}{R_1 + R_2} \tag{4.12.2}$$

Thus, the current in one of the two parallel resistors is found by multiplying the total current by the *other* resistance and dividing by the sum of the two resistances.

If there are n resistances in parallel, by Ohm's law

$$I_1 = VG_1$$

$$I_2 = VG_2$$

$$I_3 = VG_3$$

$$\vdots \qquad \vdots$$

$$\vdots \qquad \vdots$$

$$I_n = VG_n$$

$$I = VG_p$$

where

$$G_p = G_1 + G_2 + G_3 + \dots + G_n$$

therefore,

$$\frac{I_1}{I} = \frac{G_1}{G_2}$$

$$I_1 = \frac{G_1}{G_p} I$$

$$\frac{I_2}{I} = \frac{G_2}{G_p}$$

$$I_2 = \frac{G_2}{G_p} I$$

$$I_3 = \frac{G_3}{G_p} I$$

$$I_n = \frac{G_n}{G_p} I$$

Since

$$G_p = \frac{1}{R_p},\ G_n = \frac{1}{R_n}$$

$$I_n = \frac{R_p}{R_n} I$$

These equations are called the **current-divider equations**, since they show how the total current is divided in separate branches of a parallel combination of resistors.

■ **EXAMPLE 4.9** Three resistors of value 3, 8 and 24 Ω are connected in parallel across a 12 V d.c. supply. Calculate : (*a*) the total resistance of the combination; (*b*) the current in each branch; (*c*) the total current.

SOLUTION :

$$R_1 = 3\ \Omega,\ R_2 = 8\ \Omega,\ R_3 = 24\ \Omega$$

(*a*)

$$\frac{1}{R_p} = \frac{1}{R_1} + \frac{1}{R_2} + \frac{1}{R_3} = \frac{1}{3} + \frac{1}{8} + \frac{1}{24}$$

$$= \frac{8+3+1}{24} = \frac{12}{24} = \frac{1}{2}$$

$$R_p = 2\ \Omega$$

(*b*) Current in 3 Ω resistor, $I_1 = \frac{12}{3} = 4\ \text{A}$

Current in 6 Ω resistor, $I_2 = \frac{12}{8} = 1.5\ \text{A}$

Current in 24 Ω resistor, $I_3 = \frac{12}{24} = 0.5$ A

(*c*) Total current

$$I = I_1 + I_2 + I_3 = 4 + 1.5 + 0.5 = 6 \text{ A}$$

The total current can also be found from the relation

$$I = \frac{V}{R_p} = \frac{12}{2} = 6 \text{ A}$$

4.13 SOLUTION OF SERIES-PARALLEL RESISTIVE CIRCUITS

Circuits with a combination of series and parallel resistors may be called complex circuits. Such circuits can be solved completely by successive applications of the series and parallel equivalents, voltage, and current-divider relationships and KCL and KVL. A complete solution of a circuit means finding the current through and the voltage across each element of the circuit.

The following procedure is used in solving a circuit containing resistances in series and parallel connected to a single source of supply :

1. Draw the circuit diagram.
2. Label the circuit. Mark on the circuit diagram the direction of current through each resistor.
3. Combine all series resistor by using the formula
   $R_s = R_1 + R_2 + R_3 + \ldots + R_N$
4. Combine all parallel resistors by using the formula
   $\frac{1}{R_p} = \frac{1}{R_1} + \frac{1}{R_2} + \frac{1}{R_3} + \ldots + \frac{1}{R_N}$
   Replace all parallel resistors by their equivalent values.
5. Combine the equivalent resistance obtained in steps 3 and 4 and the remaining resistors to obtain the total equivalent resistance of the circuit.
6. Calculate the total circuit current by using Ohm's law.
7. Calculate the total voltage drop across and the current through each resistor keeping in mind the following rules :
   (*a*) The total voltage across a number of resistors connected in series is equal to the sum of the voltage drops across the individual resistors.
   (*b*) Each resistor in series carries the same current.
   (*c*) All sections connected in parallel have the same voltage drop.
   (*d*) The total current of a number of sections connected in parallel is equal to the sum of the individual branch currents.

   The procedure is illustrated by the following examples.

■ **EXAMPLE 4.10** For the network shown in Fig. 4.20 determine (*a*) the voltage drop in each resistor; (*b*) the current in each resistor.

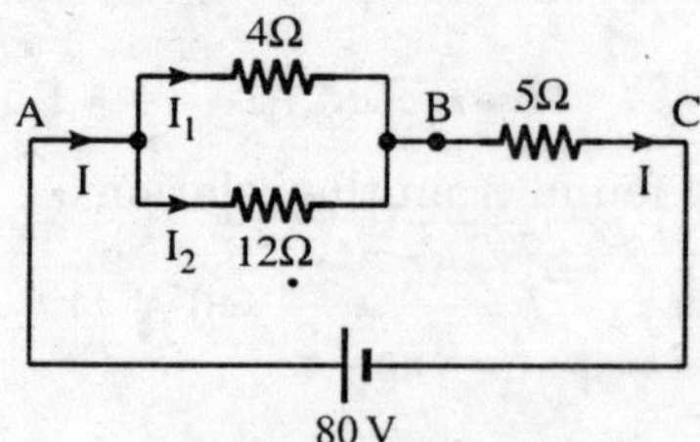

FIG. 4.20. Illustrating Example 4.10

SOLUTION :

$$\frac{1}{R_{AB}} = \frac{1}{4} + \frac{1}{12} = \frac{3+1}{12} = \frac{1}{3}$$

$$R_{AB} = 3\ \Omega$$

$$R_{AC} = R_{AB} + R_{BC} = 3 + 5 = 8\ \Omega$$

$$I = \frac{V}{R_{AC}} = \frac{80}{8} = 10\ \text{A}$$

Voltage drop across 5 Ω resistor

$$V_{BC} = R_{BC} I = 5 \times 10 = 50\ \text{V}$$

$$V_{AC} = V_{AB} + V_{BC}$$

$$80 = V_{AB} + 50$$

$$V_{AB} = 80 - 50 = 30\ \text{V}$$

Voltage drop across 4 Ω resistor = voltage drop across 12 Ω resistor = 30 V

$$I_1 = \frac{V_{AB}}{4} = \frac{30}{4} = 7.5\ \text{A}$$

$$I_2 = \frac{V_{AB}}{12} = \frac{30}{12} = 2.5\ \text{A}$$

■ **EXAMPLE 4.11** A 150 Ω resistance coil *AB* is connected across 240 V d.c. supply. Calculate the value of the resistance which, when connected between the midpoint of *AB* and end *A*, will carry a current of 0.8 A.

SOLUTION : The circuit is shown in Fig. 4.21. Let *M* be the midpoint of *AB*, and *R* be the resistance connected between *M* and *A*. The current in *R* is 0.8 A.

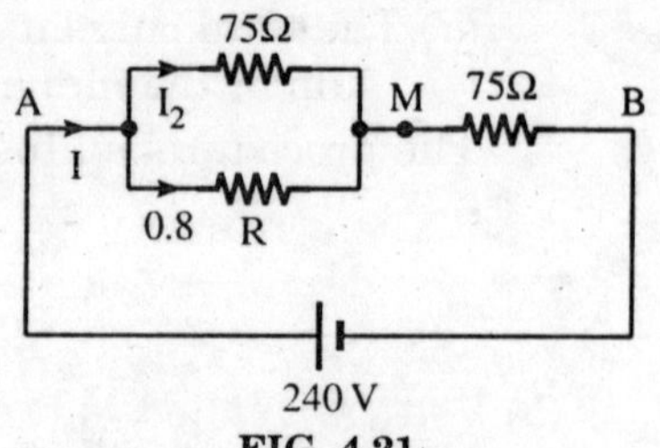

FIG. 4.21. Illustrating Example 4.11.

Voltage drop across

$$R_{MB} = R_{MB} I_1 = 75 I_1$$

$$V_{AM} + V_{MB} = V_{AB}$$

$$V_{AM} = V_{AB} - V_{MB} = 240 - 75I_1$$

By KCL at point A,

$$I_1 = I_2 + 0.8 = \frac{V_{AM}}{75} + 0.8$$

$$= \frac{240 - 75I_1}{75} + 0.8$$

$$75I_1 = 240 - 75I_1 + 75 \times 0.8$$

$$150I_1 = 240 + 60$$

$$I_1 = \frac{300}{150} = 2 \text{ A}$$

$$V_{AM} = 240 - 75 \times 2 = 90$$

$$R = \frac{V_{AM}}{0.8} = \frac{90}{0.8} = 112.5\ \Omega$$

■ **EXAMPLE 4.12** A direct-current curcuit comprises two resistors, A of value 25 ohms and B of unknown value, connected in parallel, together with a third resistor C of value 5 ohms connected in series with the parallel group. The potential difference across C is found to 90 V. If the total power in the circuit is 4320 W, calculate :

(*a*) the value of resistor B,

(*b*) the voltage applied to the ends of the whole circuit,

(*c*) the current in each resistors.

(*C & G*)

SOLUTION : The circuit is shown in Fig. 4.22.

(*a*) $I = \frac{90}{5} = 18$ A

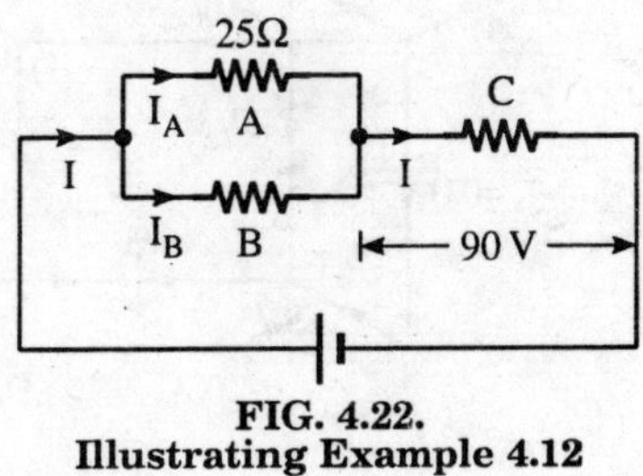

FIG. 4.22.
Illustrating Example 4.12

If R be the total resistance of the circuit,

$$I^2R = 4320$$

$$R = \frac{4320}{(18)^2} = \frac{40}{3}\ \Omega$$

$$\frac{25B}{B+25} + 5 = \frac{40}{3}$$

$$\frac{25B}{B+25} = \frac{40}{3} - 5$$

$$\frac{B}{B+25} = \frac{1}{3}$$

$$3B = B + 25$$

$$B = 12.5\ \Omega$$

(*b*) Voltage applied to the ends of the whole circuit

$$= RI = \frac{40}{3} \times 18 = 240 \text{ V}$$

(c) Current in $C = I = 18$ A

Voltage across the parallel branch

$$= 240 - 90 = 150 \text{ V}$$

Current in A, $$I_A = \frac{150}{25} = 6 \text{ A}$$

Current in B, $$I_B = \frac{150}{12.5} = 12 \text{ A}$$

EXAMPLE 4.13 Twelve identical wires of resistance 6 Ω each are arranged to form edges of a cube. A current of 40 mA is led into the cube at one corner and out at the other corner. Calculate the p.d. developed between these corners and the effective resistance of the network.

SOLUTION: The cube is shown in Fig. 4.23 (*a*). The distribution of current among various edges of the cube is found by KCL. The current of 40 mA enters the cube at corner A. It is then equally divided along AB, AD and AE. Hence

$$\text{current in } AB = \text{current in } AD = \text{current in } AE = \frac{40}{3} \text{ mA}$$

At B the current of 40/3 mA enters. This current is then equally divided along BC and BF. Hence current in BC = current in $BF = \frac{20}{3}$ mA.

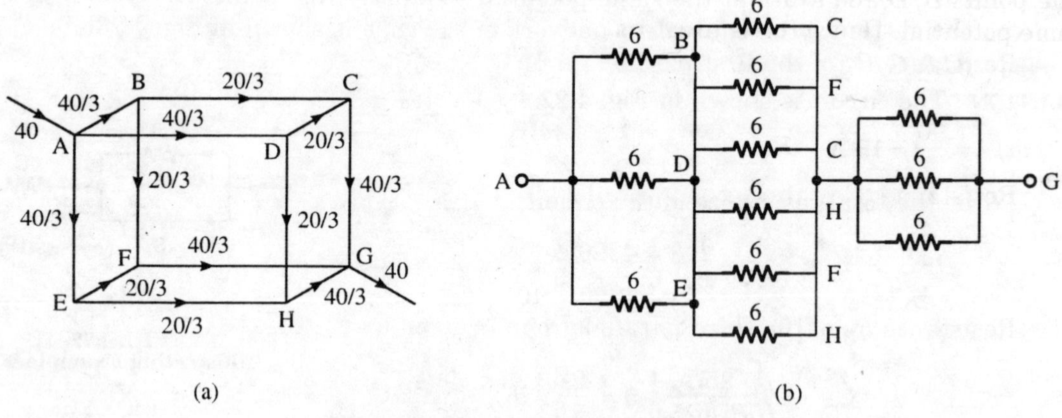

FIG. 4.23. Illustrating Example 4.13.

Similarly,

$$\text{current in } DC = \text{current in } DH = \frac{20}{3} \text{ mA}$$

$$\text{current in } EF = \text{current in } EH = \frac{20}{3} \text{ mA}$$

By applying KCL at the point F,

$$\text{current in } FG = \frac{40}{3} \text{ mA}$$

By applying KCL at the point H,

current in $HG = \frac{40}{3}$ mA

Current coming out from G

$$= I_{CG} + I_{FG} + I_{HG} = 40 \text{ mA}$$

One of the parallel paths for the current of 40 mA is along AD, DH and HG.

Potential difference between A and G is given by

$$V_{AG} = V_{AD} + V_{DH} + V_{HG}$$

$$= R_{AD}I_{AD} + R_{DH}I_{DH} + R_{HG}I_{HG}$$

$$= 6 \times \frac{40}{3} + 6 \times \frac{20}{3} + 6 \times \frac{40}{3}$$

$$= 200 \text{ mV} = 200 \times 10^{-3} \text{ V} = 0.2 \text{ V}$$

If R_{AG} is the resistance between A and G

$$V_{AG} = R_{AG}I_{AG}$$

$$0.2 = R_{AG} \times 40 \times 10^{-3}$$

$$R_{AG} = \frac{0.2}{40 \times 10^{-3}} = 5\ \Omega$$

Alternative method

The points B, D and E are at the same potential. Similarly, the points C, F and H are at the same potential. Hence the equivalent network of the cube is shown in Fig. 4.23b.

Resistance R_1 of the first parallel group is given by

$$\frac{1}{R_1} = \frac{1}{6} + \frac{1}{6} + \frac{1}{6} = \frac{3}{6}; \; R_1 = \frac{6}{3} = 2\ \Omega$$

Resistance R_2 of the second parallel group is given by

$$\frac{1}{R_2} = \frac{1}{6} + \frac{1}{6} + \frac{1}{6} + \frac{1}{6} + \frac{1}{6} + \frac{1}{6} = 1; \; R_2 = 1\ \Omega$$

Resistance R_3 of the third parallel group is given by

$$\frac{1}{R_3} = \frac{1}{6} + \frac{1}{6} + \frac{1}{6} = \frac{3}{6}; \; R_3 = \frac{6}{3} = 2\ \Omega$$

Resistance between A and G,

$$R_{AG} = R_1 + R_2 + R_3 = 2 + 1 + 2 = 5\ \Omega$$

Potential difference between A and G

$$V_{AG} = R_{AG}I_{AG} = 5 \times 40 \times 10^{-3} = 0.2 \text{ V}$$

■ **EXAMPLE 4.14** Determine the resistance between the terminals *a-b* in the network shown in Fig. 4.24.

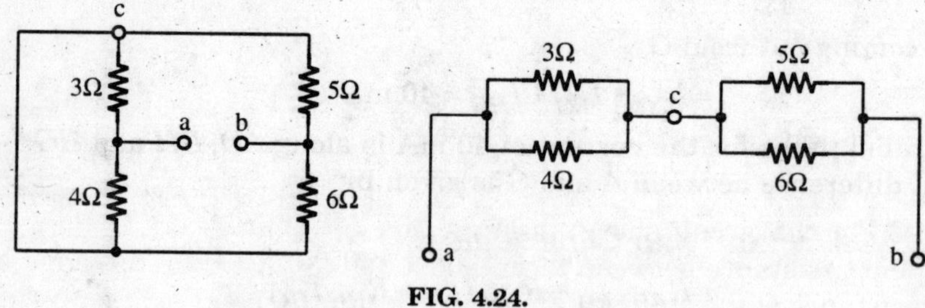

FIG. 4.24.

SOLUTION The network of Fig. 4.24a may be redrawn as Fig. 4.24b .

$$R_{ac} = 3 \mid\mid 4 = \frac{3 \times 4}{3+4} = \frac{132}{7}\ \Omega$$

$$R_{bc} = 5 \mid\mid 6 = \frac{5 \times 6}{5+6} = \frac{30}{11}\ \Omega$$

$$R_{ab} = R_{ac} + R_{bc} = \frac{12}{7} + \frac{30}{11} = 4.44\ \Omega$$

■ **EXAMPLE 4.15** Reduce the network shown in Fig. 4.25 to a single equivalent resistance placed across terminals *A-B*.

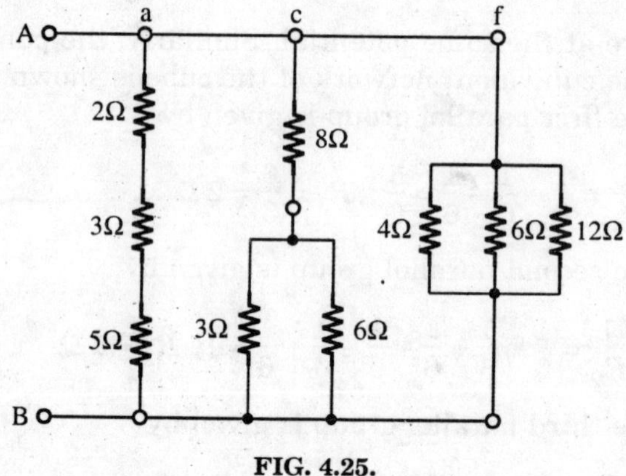

FIG. 4.25.

SOLUTION

$$R_{ab} = 2 + 3 + 5 = 10\ \Omega$$

$$R_{de} = 3 \mid\mid 6 = \frac{3 \times 6}{3+6} = 2\ \Omega$$

$$R_{ce} = R_{cd} + R_{de} = 8 + 2 = 10\ \Omega$$

$$R_{fg} = 4 \mid\mid 6 \mid\mid 12$$

threrefore, $$\frac{1}{R_{fg}} = \frac{1}{4} + \frac{1}{6} + \frac{1}{12} = \frac{3+2+1}{12} = \frac{1}{2}, \; R_{fg} = 2\ \Omega$$

$$\frac{1}{R_{AB}} = \frac{1}{R_{ab}} + \frac{1}{R_{ce}} + \frac{1}{R_{fg}} = \frac{1}{10} + \frac{1}{10} + \frac{1}{2} = \frac{7}{10}$$

therefore, $$R_{AB} = \frac{10}{7}\ \Omega$$

■ **EXAMPLE 4.16** For the circuit shown in Fig. 4.26a determine

(*a*) equivalent resistance between terminals *a* and *f*, (*b*) current supplied by the battery, (*c*) equivalent resistance of the circuit with an open circuit at point *b*, (*d*) equivalent resistance of the circuit when 2 Ω resistance is removed, (*e*) equivalent resistance of the circuit when the points *c* and *g* are short-circuited, (*f*) equivalent resistance of the circuit when the points *e* and *f* are short-circuited.

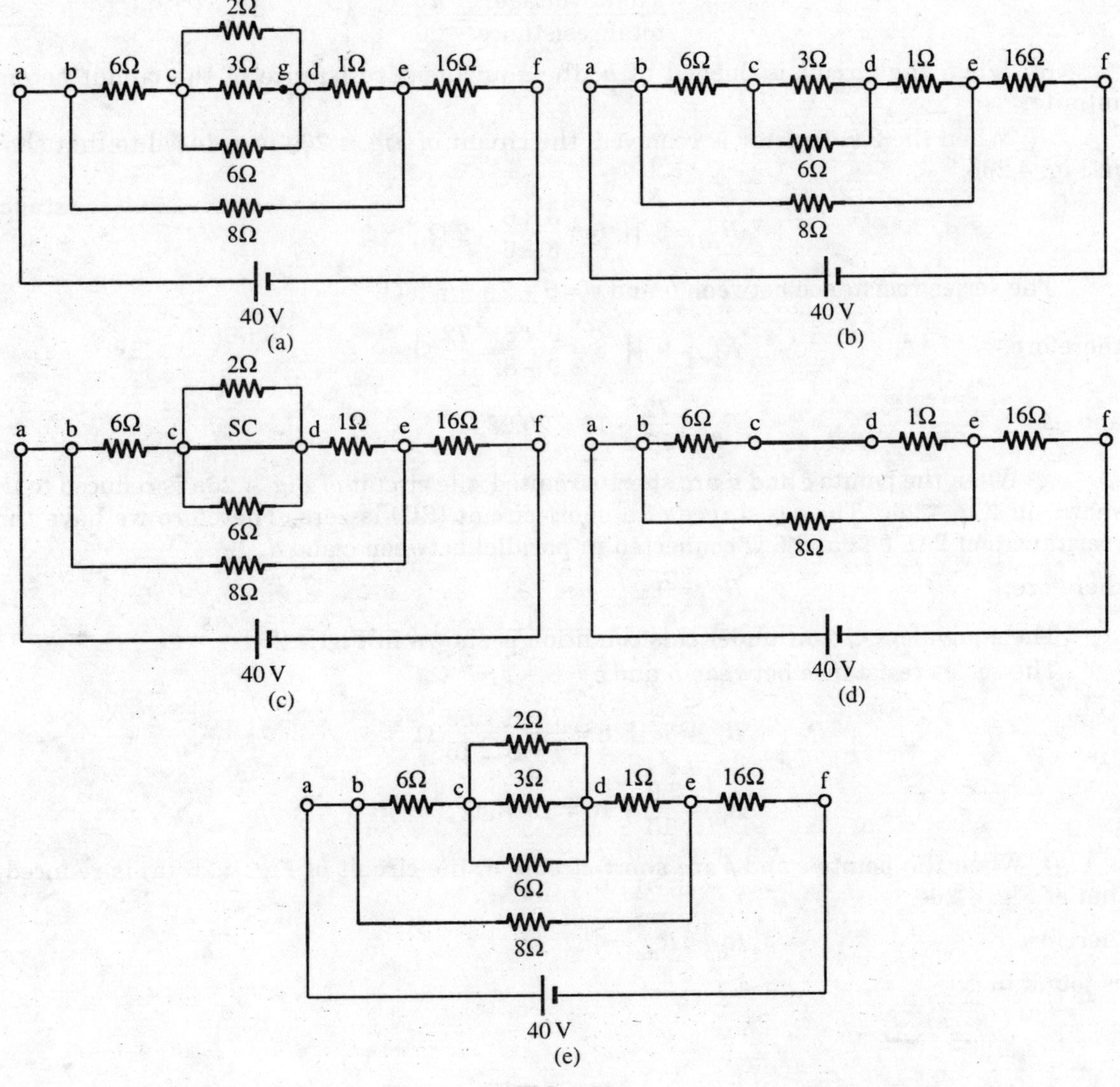

FIG. 4.26.

SOLUTION 1. Solve for the equivalent resistance of the parallel circuit.

(*a*) From the law of parallel resistors

$$\frac{1}{R_{cd}} = \frac{1}{2} + \frac{1}{3} + \frac{1}{6} = \frac{3+2+1}{6} = 1$$

therefore, $R_{cd} = 1\ \Omega$

The series resistance between the terminals *b* and *e*

$$= R_{bc} + R_{cd} + R_{de} = 6 + 1 + 1 = 8\ \Omega$$

2. Solve for the total resistance.

$$R_{af} = 4 + 16 = 20\ \Omega$$

3. Solve for the total current.

(*b*) Current supplied by the battery

$$I = \frac{\text{total voltage}}{\text{total resistance}} = \frac{40}{20} = 2\ \text{A}$$

(*c*) When the circuit is opened at *b*, the equivalent resistance of the circuit becomes infinite.

(*d*) When the 2 Ω resistor is removed, the circuit of Fig. 4.26a is reduced to that shown in Fig. 4.26b.

$$R_{cd} = 3 \ || \ 6 = \frac{3 \times 6}{3+6} = 2\ \Omega$$

The series resistance between *b* and *e* = 6 + 2 + 1 = 9 Ω

therefore, $R_{be} = 9 \ || \ 8 = \frac{9 \times 8}{9+8} = \frac{72}{17}\ \Omega$

and $R_{af} = \frac{72}{17} + 16 = 20.23\ \Omega$

(*e*) When the points *c* and *g* are short-circuited, the circuit of Fig. 4.26a is reduced to that shown in Fig. 4.26c. The resistance of a short circuit (SC) is zero. Therefore we have three resistances of 2 Ω, 0 Ω and 6 Ω connected in parallel between *c* and *d*.

therefore, $R_{cd} = 0$

The equivalent circuit under this condition is shown in Fig. 4.26d.

The series resistance between *b* and *e* = 6 + 1 = 7 Ω

$$R_{be} = 7 \ || \ 8 = \frac{7 \times 8}{7+8} = \frac{56}{15}\ \Omega$$

$$R_{af} = \frac{56}{15} + 16 = 19.73\ \Omega$$

(*f*) When the points *e* and *f* are short-circuited, the circuit of Fig. 4.26 (a) is reduced to that of Fig. 4.26e.

therefore, $R_{af} = R_{ae} = 4\ \Omega$

as found in (*a*).

■ **EXAMPLE 4.17** The voltage divider AB shown in Fig. 4.27a has a total resistance of 120 Ω across a 180-V supply. It is required connected to supply current to a 60 Ω load. Calculate the current in the load when the sliding contact S is set (a) at position A, (b) at a point midway between A and B. Also, determine the resistance between the position of the slider from B to supply a current of 2 A to the load.

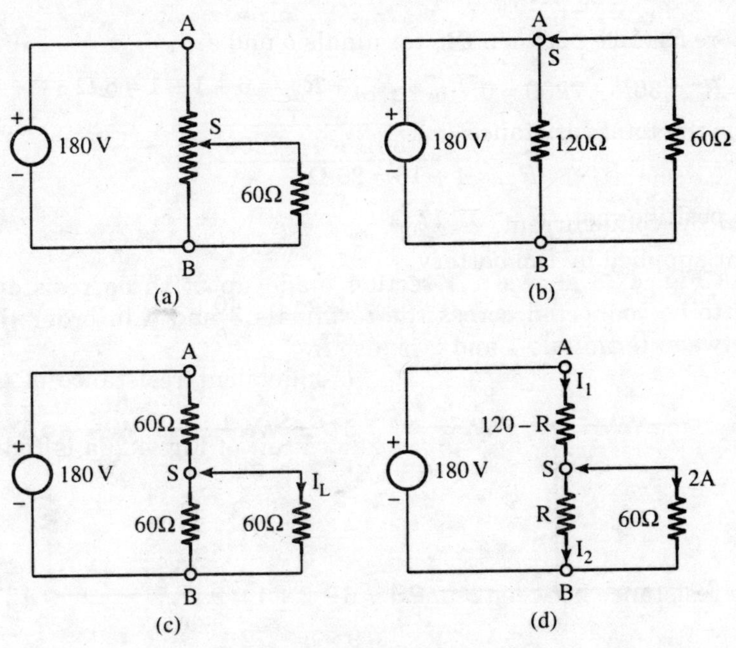

FIG. 4.27.

SOLUTION (a) When the sliding contact S is set at position A (Fig. 4.27b), the resistances 60 Ω and 120 Ω are in parallel.

$$\text{Current in the load resistor} = \frac{180}{60} = 3 \text{ A}$$

(b) When the sliding contact S is set at a point midway between A and B (Fig. 4.27c), total resistance between the terminals A and B is

$$R_{AB} = 60 + 60 \,||\, 60$$

$$= 60 + \frac{60 \times 60}{60 + 60} = 90 \ \Omega$$

$$\text{Total current} = \frac{180}{90} = 2 \text{ A}$$

therefore, load current $I_L = 1$ A

(c) The circuit to supply a current of 2 A is shown in Fig. 4.27d

Voltage across the load = 60 × 2 = 120 V

Voltage between A and S = 180 – 120 = 60 V

therefore, $$I_1 = \frac{60}{120 - R}, \quad I_2 = \frac{120}{R}$$

$$I_1 = I_2 + 2$$

$$\frac{60}{120 - R} = \frac{120}{R} + 2$$

$$\frac{60}{120 - R} = \frac{120 + 2R}{R}$$

$$(120 - R)(120 + 2R) = 60R$$

$$R^2 - 30R - 7200 = 0$$

therefore, $$R = \frac{30 \pm \sqrt{(30)^2 + 4 \times 7200}}{2}$$

Taking the positive root, $R = 101.17\ \Omega$

■ **EXAMPLE 4.18** Fig. 4.28 shows a T-section made up of three resistors. Determine the resistance R to be connected across the terminals 3 and 4 in order that the resistance measured between terminals 1 and 2 is also R.

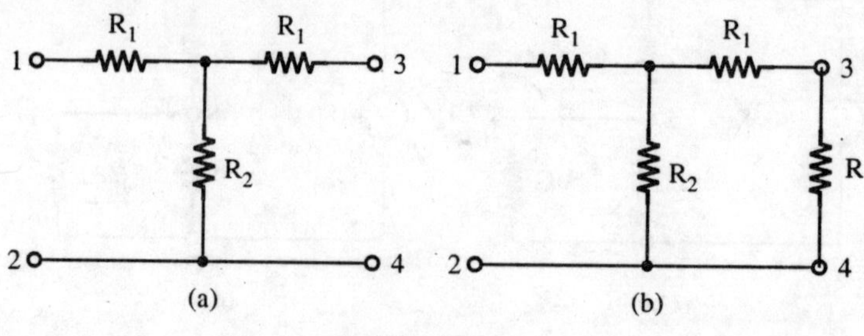

FIG. 4.28.

SOLUTION Fig. 4.28b shows the network when the resistance R is connected between terminals 3 and 4 of Fig. 4.28a.

From Fig. 4.28b

$$R_{ab} = R_2 \,||\, (R + R_1) = \frac{R_2 (R + R_1)}{R_2 + R + R_1}$$

$$R_{12} = R_1 + R_{ab} = R_1 + \frac{R_2 (R + R_1)}{R_2 + R + R_1}$$

Also, $R_{12} = R$

therefore, $$R_1 + \frac{R_2 (R + R_1)}{R_2 + R + R_1} = R$$

$$R_1 (R_2 + R + R_1) + R_2 (R + R_1) = R (R_2 + R + R_1)$$

$$R_1R_2 + R_1R + R_1^2 + R_2R + R_1R_2 = RR_2 + R^2 + RR_1$$

$$R^2 = R_1^2 + 2R_1R_2$$

$$R = \sqrt{R_1^2 + 2R_1R_2}$$

4.14 EQUIVALENT NETWORKS

Two networks are defined to be equivalent if the mathematical equations relating the terminal voltages and currents in one are the same as the mathematical equations relating the terminal voltages and currents in the other.

Equivalent circuit concept is used to simplify the analysis of the circuits. It is also used to replace a complicated network or a network whose internal structure is not known, by a network consisting of few elements only. Equivalent circuits are also used in modelling of physical elements or circuits. The term equivalent circuit of a physical device is also called its **model**.

It should be noted that the idea of equivalence is valid *only* at the specified terminals at which two devices have the same voltage-current (v-i) relationships. Nothing can be said about the equality of currents or voltages anywhere inside the device.

In this chapter several techniques using the concept of equivalent circuits are described. The simplest example of the equivalent circuit is the one in which two resistors connected in series are equivalent to a single resistor whose resistance is equal to the resistance of the sum of the series resistances.

EXAMPLE 4.19 Find the input resistance at AB for the lattice network shown in Fig. 4.29a, when terminals CD are (a) open-circuited; (b) short-circuited.

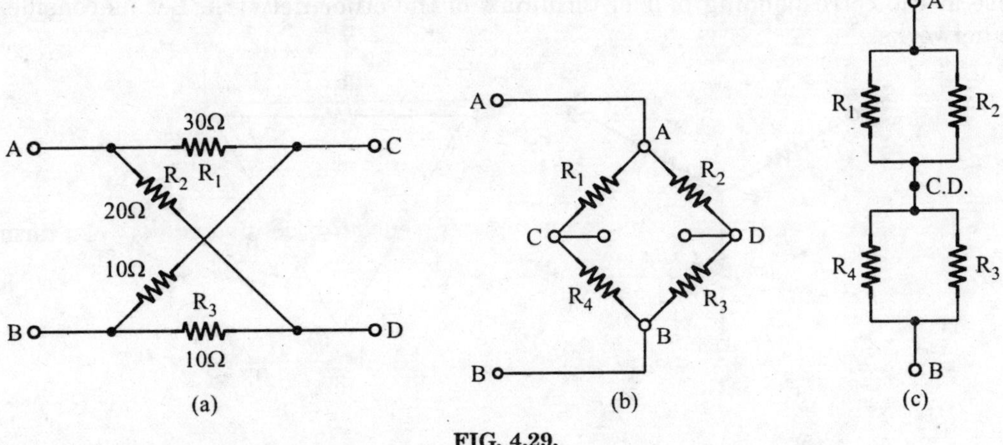

FIG. 4.29.

SOLUTION : (a) When the terminals C and D are open-circuited the network in Fig. 4.29a can be redrawn as that shown in Fig. 4.29b.

Hence R_1 and R_4 are in series. Similarly, R_2 and R_3 are in series.

$$R_{AB} = (R_2 + R_3) \;||\; (R_1 + R_4)$$

$$= (20 + 10) \;||\; (30 + 10) = 30 \;||\; 40$$

$$= \frac{30 \times 40}{30 + 40} = 17.143\ \Omega$$

(b) When the terminals C and D are short-circuited the network in Fig. 4.29a becomes the circuit in Fig. 4.29c

Here
$$R_{AB} = (R_1 \parallel R_2) + (R_3 \parallel R_4)$$
$$= \frac{R_1R_2}{R_1+R_2} + \frac{R_3R_4}{R_3+R_4}$$
$$= \frac{30\times 20}{30+20} + \frac{10\times 10}{10+10} = 17\ \Omega$$

4.15 STAR ↔ DELTA (OR $T \leftrightarrow \Pi$) TRANSFORMATION

Star-delta transformation leads to the simplification of the networks. By such a stranformation new elements are created which can sometimes be combined with elements in the rest of the network.

Consider star (Y) and delta (Δ) networks shown in Figs. 4.30a and 4.30b respectively. The star (or wye) network can be drawn as a T network as shown in Fig. 4.30c. A delta (or mesh) network shown in Fig. 4.30b can be redrawn as a Π-network shown in Fig. 4.30d. Although T and Π networks are three-terminals networks, but for the purpose of analysis they are often drawn as four-terminal or two-port networks (Figs. 4.30c and 4.30d), two of the terminals being connected to the same point. Tee (T) and pi (Π) networks are very commonly used networks. Hence $T \leftrightarrow \Pi$ conversion is of special significance. These networks are said to be *equivalent* if the impedance at any pair of terminals of one network is equal to the impedance at the corresponding pair of terminals of the other network. Let us consider the resistive networks.

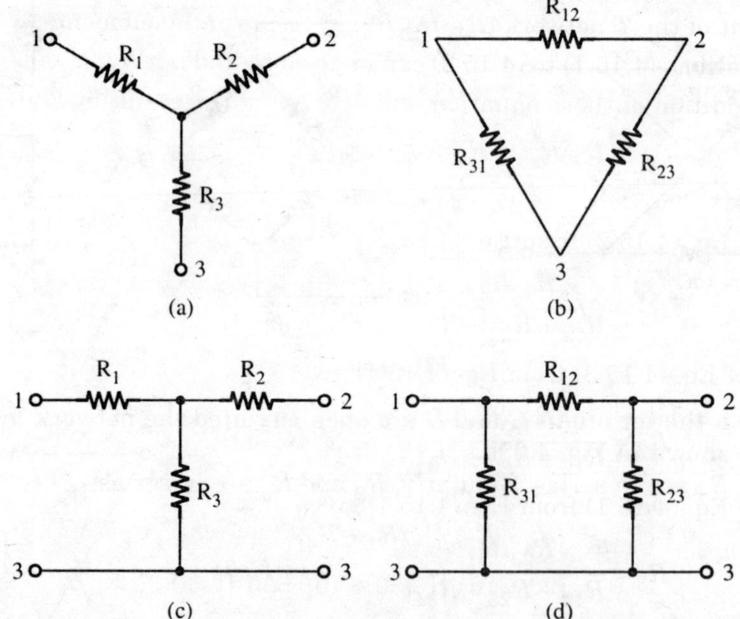

FIG. 4.30.(a) Wye (Y) network; (b) Delta (Δ) network; (c) Tee (T) network; (d) (Π) network.

From Figs. 4.30a and 4.30b resistance between terminals 1 and 2 for star network = resistance between terminals 1 and 2 for delta network

$$R_1 + R_2 = R_{12} \,||\, (R_{23} + R_{31})$$

$$= \frac{R_{12}\,(R_{23} + R_{31})}{R_{12} + R_{23} + R_{31}} \tag{4.15.1}$$

Similarly, resistance between terminals 2 and 3 for star network = resistance between terminals 2 and 3 for delta network

$$R_2 + R_3 = R_{23} \,||\, (R_{31} + R_{12})$$

$$= \frac{R_{23}\,(R_{31} + R_{12})}{R_{12} + R_{23} + R_{31}} \tag{4.15.2}$$

Similarly, resistance between terminals 3 and 1 for star network = resistance between terminals 3 and 1 for delta network

$$R_3 + R_1 = R_{31} \,||\, (R_{12} + R_{23})$$

$$= \frac{R_{31}\,(R_{12} + R_{23})}{R_{12} + R_{23} + R_{31}} \tag{4.15.3}$$

It is seen that Eqs. (4.15.1) to (4.15.3) possess symmetry because of the symmetry of the networks. Thus, Eq. (4.15.2) can be written directly from Eq. (4.15.1) by a cyclical change in subscripts.

Delta-to-star (or Π-to-*T*) Transformation

To convert the delta (Π) network in Fig. 4.30 (*b*) to its star (*T*) equivalent, it is necessary to express each element of the *T* network (R_1, R_2, R_3) in terms of the elements of the Π network (R_{12}, R_{23}, R_{31}). Equations (4.15.1) to (4.15.3) can be used to find out R_1, R_2 and R_3 in terms of R_{12}, R_{23} and R_{31}. Addition of these equation and division of the result by 2 gives

$$R_1 + R_2 + R_3 = \frac{R_{12}R_{23} + R_{23}R_{31} + R_{31}R_{12}}{R_{12} + R_{23} + R_{31}} \tag{4.15.4}$$

Subtraction of Eq. (4.15.2) from Eq. (4.15.4) gives

$$R_1 = \frac{R_{12}R_{31}}{R_{12} + R_{23} + R_{31}} \tag{4.15.5}$$

Substraction of Eq. (4.15.3) from Eq. (4.15.4) gives

$$R_2 = \frac{R_{12}R_{23}}{R_{12} + R_{23} + R_{31}} \tag{4.15.6}$$

Subtraction of Eq. (4.15.1) from Eq. (4.15.4) gives

$$R_3 = \frac{R_{23}R_{31}}{R_{12} + R_{23} + R_{31}} \tag{4.15.7}$$

Eqs. (4.15.5) to (4.15.7) are the formulae for converting from a delta connection to a star connection, that is, these are the results for Δ to *Y* (or Π to *T*) transformation. Here again the results possess symmetry. It is better to learn the transforms by configuration rather than by equations. For this

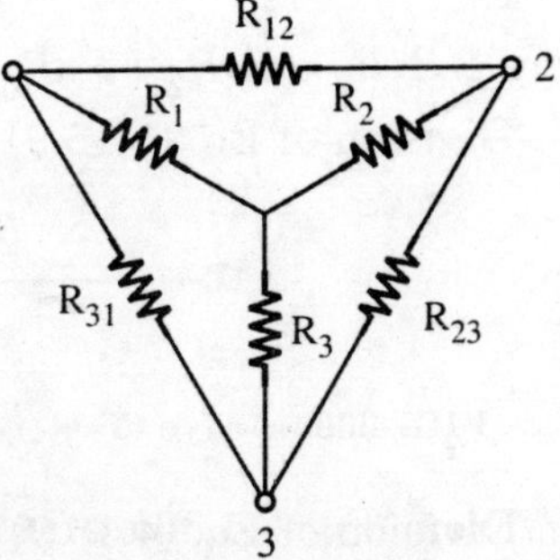

FIG. 4.31. Superimposed delta and star networks

purpose, the delta-connected circuit and star connected circuit are superimposed as in Fig. 4.31. It is seen that the resistance of any branch of the equivalent star connection is obtained by multiplying the adjacent delta resistances and dividing by the sum of the three delta resistances. That is, to replace a delta by an equivalent star, the star resistance is given by

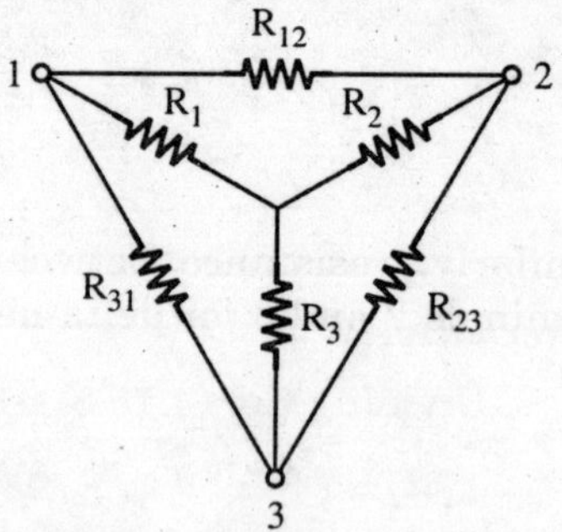

FIG. 4.31. Superimposed delta and

$$Y \text{ resistance} = \frac{\text{product of adjacent } \Delta \text{ resistances}}{\text{sum of } \Delta \text{ resistances}} \quad (4.15.8)$$

Star-to delta (or *T*-to-Π) Transformation

Delta-connected resistances can be converted in terms of star-connected resistances. Here we have to find R_{12}, R_{23} and R_{31} in terms of R_1, R_2 and R_3.

From Eqs. (4.15.5), (4.15.6) and (4.15.7)

$$R_1R_2 + R_2R_3 + R_3R_1 = \frac{R_{12}^2R_{23}R_{31} + R_{23}^2R_{31}R_{12} + R_{31}^2R_{12}R_{23}}{(R_{12}+R_{23}+R_{31})^2}$$

$$= \frac{(R_{12}+R_{23}+R_{31})\,R_{12}R_{23}R_{31}}{(R_{12}+R_{23}+R_{31})^2}$$

$$= \frac{R_{12}R_{23}R_{31}}{R_{12}+R_{23}+R_{31}} \quad (4.15.9)$$

Division of Eq. (4.15.9) by Eq. (4.15.7) gives

$$R_{12} = \frac{1}{R_3}(R_1R_2 + R_2R_3 + R_3R_1)$$

$$= \frac{\Sigma R_1R_2}{R_3} \quad (4.15.10)$$

where $\Sigma R_1R_2 = R_1R_2 + R_2R_3 + R_3R_1$

Division of Eq. (4.15.9) by Eq. (4.15.5) gives

$$R_{23} = \frac{1}{R_1}(R_1R_2 + R_2R_3 + R_3R_1)$$

$$= \frac{\Sigma R_1R_2}{R_1} \quad (4.15.11)$$

Division of Eq. (4.15.9) by Eq. (4.15.6) gives

$$R_{31} = \frac{1}{R_2}(R_1R_2 + R_2R_3 + R_3R_1)$$

$$= \frac{\Sigma R_1R_2}{R_2} \tag{4.15.12}$$

Alternatively

Dividing Eq. (4.15.5) by Eq. (4.15.6), we get

$$\frac{R_1}{R_2} = \frac{R_{31}}{R_{23}}$$

therefore,

$$R_{31} = R_{23}\frac{R_1}{R_2}$$

Similarly, dividing Eq. (4.15.5) by Eq. (4.15.7), we get

$$\frac{R_1}{R_3} = \frac{R_{12}}{R_{23}}$$

therefore,

$$R_{12} = R_{23}\frac{R_1}{R_3}$$

Substituting the values of R_{31} and R_{12} in Eq. (4.15.5), we get

$$R_1 = \frac{R_{23}\frac{R_1}{R_3} \times R_{23}\frac{R_1}{R_2}}{R_{23}\frac{R_1}{R_3} + R_{23} + R_{23}\frac{R_1}{R_2}}$$

$$R_1 = \frac{R_1^2 \times R_{23}}{R_1R_2 + R_2R_3 + R_3R_1}$$

therefore,

$$R_{23} = \frac{R_1R_2 + R_2R_3 + R_3R_1}{R_1}$$

$$= \frac{\Sigma R_1R_2}{R_1}$$

Similarly,

$$R_{31} = \frac{\Sigma R_1R_2}{R_2}$$

$$R_{12} = \frac{\Sigma R_1R_2}{R_3}$$

Eqs. (4.15.10) to (4.15.12) are the formulae for converting from a star connection to a delta connection. These results also possess symmetry. It is seen that the resistance of any branch of the equivalent delta is equal to the sum of the products of the star resistance taken two at a time divided by the *opposite* star resistance. That is, to replace a star by an equivalent delta, the delta resistances are given by

$$\Delta \text{ resistance} = \frac{\text{sum of the products of } Y \text{ resistances taken two at a time}}{\text{opposite } Y \text{ resistance}} \qquad (4.15.13)$$

The following examples illustrate star ↔ delta transformations.

■ **EXAMPLE 4.20** Convert the Π network in Fig. 4.32a to its T equivalent in Fig. 4.32b. As a check convert the star circuit back to its delta circuit.

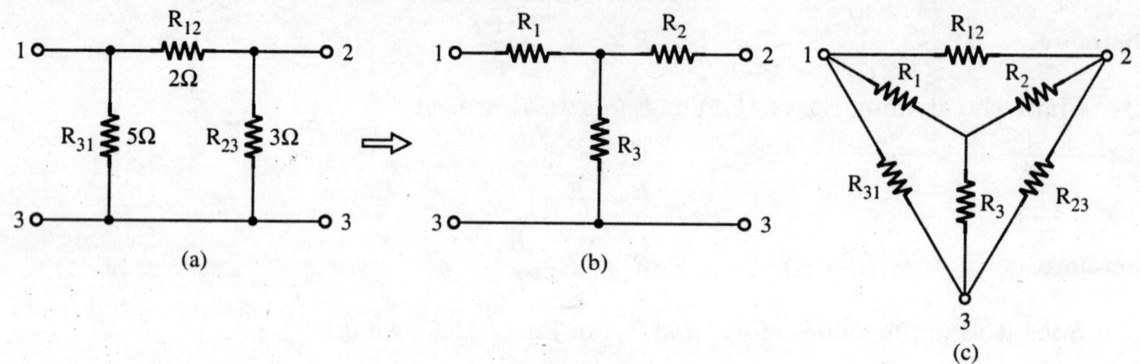

FIG. 4.32.

SOLUTION :

(*a*) **Delta to Star Transformation**

The superimposed delta and star networks are shown in Fig. 4.32c

$$\Sigma R_\Delta = R_{12} + R_{23} + R_{31} = 2 + 3 + 5 = 10\ \Omega$$

$$R_1 = \frac{R_{12}R_{31}}{\Sigma R_\Delta} = \frac{2 \times 5}{10} = 1\ \Omega$$

$$R_2 = \frac{R_{12}R_{23}}{\Sigma R_\Delta} = \frac{2 \times 3}{10} = 0.6\ \Omega$$

$$R_3 = \frac{R_{23}R_{31}}{\Sigma R_\Delta} = \frac{3 \times 5}{10} = 1.5\ \Omega$$

(*b*) **Star to Delta Transformation**

$$\Sigma R_1R_2 = R_1R_2 + R_2R_3 + R_3R_1 = 1 \times 0.6 + 0.6 \times 1.5 + 1.5 \times 1 = 3\ \Omega$$

$$R_{12} = \frac{\Sigma R_1R_2}{R_3} = \frac{3}{1.5} = 2\ \Omega$$

$$R_{23} = \Sigma \frac{R_1R_2}{R_1} = \frac{3}{1} = 3\ \Omega$$

$$R_{31} = \frac{\Sigma R_1 R_2}{R_2} = \frac{3}{0.6} = 5\ \Omega$$

■ **EXAMPLE 4.21** (*a*) Simplify the network given in Fig. 4.33 to a three-resistor star network between terminals 1, 1′, 2 and 2′.
(*b*) Convert it to be the equivalent delta network.

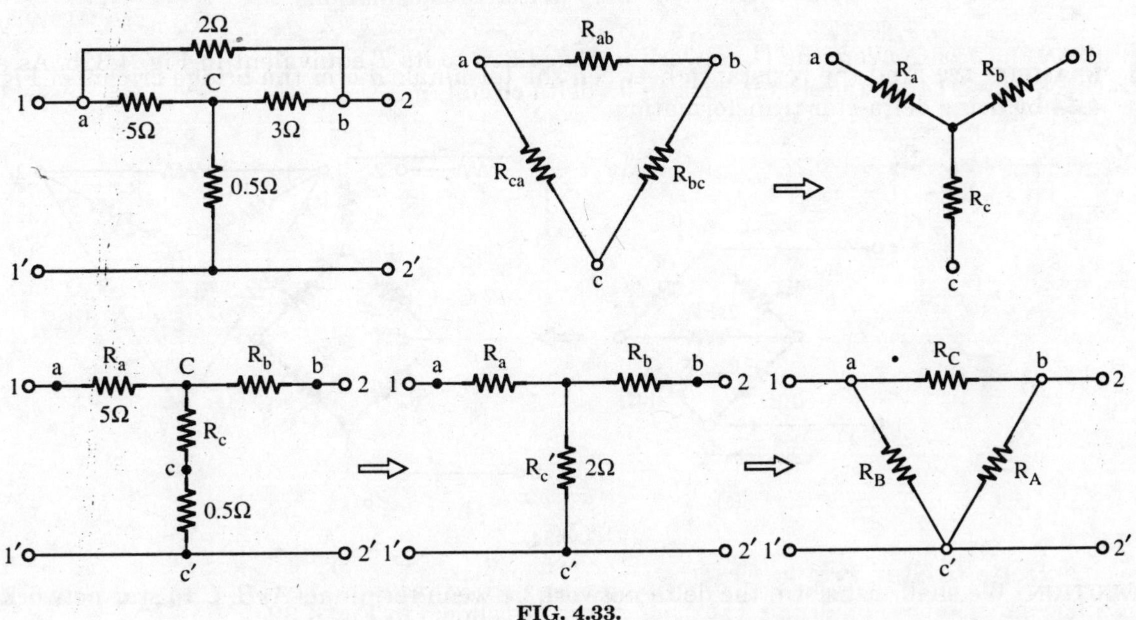

FIG. 4.33.

SOLUTION (*a*) We shall first transform the delta network between terminals *abc* to star network as shown in Fig. 4.33b.

$$\Sigma R_\Delta = R_{ab} + R_{bc} + R_{ca} = 2 + 3 + 5 = 10\ \Omega$$

$$R_a = \frac{R_{ab} R_{ca}}{\Sigma R_\Delta} = \frac{2 \times 5}{10} = 1\ \Omega$$

$$R_b = \frac{R_{ab} R_{bc}}{\Sigma R_\Delta} = \frac{2 \times 3}{10} = 0.6\ \Omega$$

$$R_c = \frac{R_{ca} R_{bc}}{\Sigma R_\Delta} = \frac{5 \times 3}{10} = 1.5\ \Omega$$

The network in Fig. 4.33a can now be simplified to the network in Fig. 4.33c. Here R_c and 0.5 Ω in series are combined into a single resistor $R_{c'} = R_c + 0.5 = 1.5 + 0.5 = 2\ \Omega$. Hence the network in Fig. 4.33c takes the form shown in Fig. 4.33d. This is the required three-resistor star network between terminals 1, 1′ and 2, 2′.

(*b*) The equivalent delta network of Fig. 4.33d is shown in Fig. 4.33e

$$\Sigma R_Y = R_a R_b + R_b R_{c'} + R_{c'} R_a = 1 \times 0.6 + 0.6 \times 2 + 2 \times 1 = 3.8\ \Omega$$

$$R_A = \frac{\Sigma R_Y}{R_a} = \frac{3.8}{1} = 3.8\ \Omega$$

$$R_B = \frac{\Sigma R_Y}{R_b} = \frac{3.8}{0.6} = 6.33\ \Omega$$

$$R_C = \frac{\Sigma R_Y}{R_c} = \frac{3.8}{2} = 1.9\ \Omega$$

■ **EXAMPLE 4.22** Find the resistance between the terminals *a-b* of the bridge cricuit in Fig. 4.34 by using delta-star transformation.

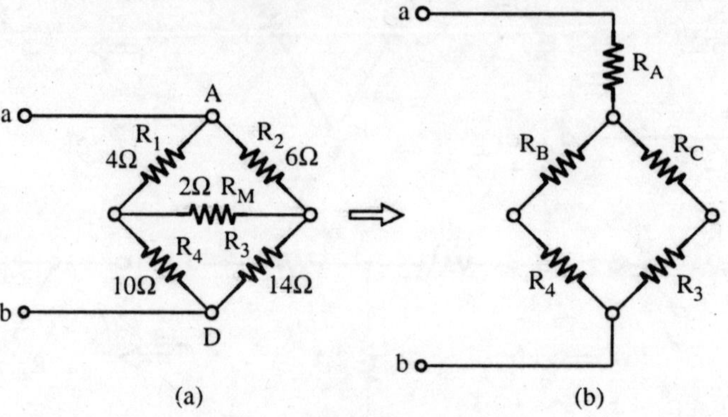

FIG. 4.34.

SOLUTION : We shall transform the delta network between terminals A, B, C to star network.

$$\Sigma R_\Delta = R_1 + R_2 + R_M = 4 + 6 + 2 = 12\ \Omega$$

$$R_A = \frac{R_1 R_2}{\Sigma R_\Delta} = \frac{4 \times 6}{12} = 2\ \Omega$$

$$R_B = \frac{R_1 R_M}{\Sigma R_\Delta} = \frac{4 \times 2}{12} = \frac{2}{3}\ \Omega$$

$$R_C = \frac{R_2 R_M}{\Sigma R_\Delta} = \frac{6 \times 2}{12} = 1\ \Omega$$

Fig. 4.34b shows that equivalent network of Fig. 4.34a after delta star transformation. The resistors R_C and R_3 are in series. Their total resistance $R_5 = R_C + R_3 = 1 + 14 = 15\ \Omega$. Similarly, R_B and R_4 are in series. Their total resistance

$$R_6 = R_B + R_5 = \frac{2}{3} + 10 = \frac{32}{3}\ \Omega$$

The parallel combination of R_5 and R_6 is in series with R_A. Therefore,

$$R_{ab} = R_A + (R_5 \mid\mid R_6)$$

$$= R_A + \frac{R_5 R_6}{R_5 + R_6}$$

$$= 2 + \frac{15 \times (32/3)}{15 + (32/3)} = 8.234\ \Omega$$

EXAMPLE 4.23 Determine the resistance R_{ab} in the network shown in Fig. 4.35a.

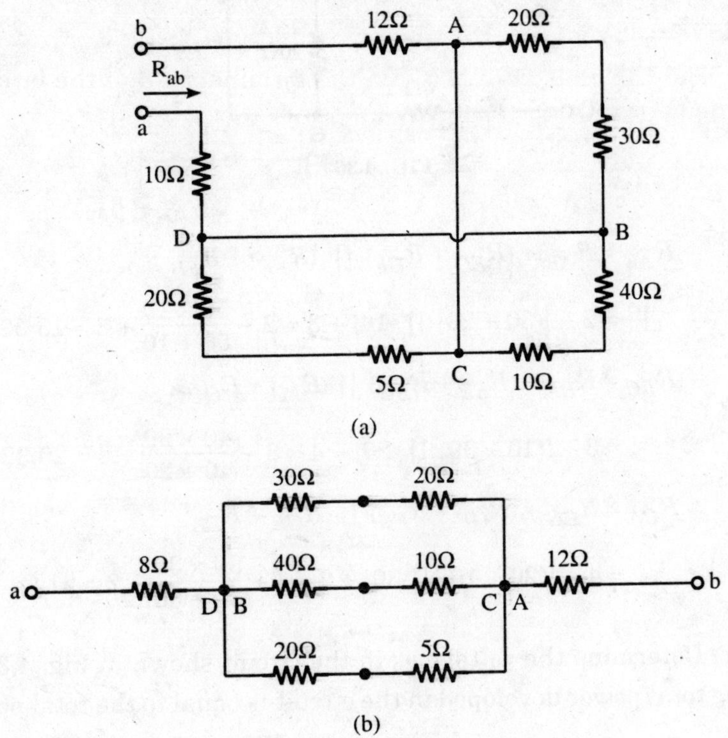

FIG. 4.35.

SOLUTION : The circuit given in Fig. 4.35 (a) is redrawn as shown in Fig. 4.35b. It is seen that points A and C are directly connected. Similarly, points B and D are directly connected. If R_p is the resistance of the parallel branch

$$\frac{1}{R_p} = \frac{1}{30+20} + \frac{1}{40+10} + \frac{1}{20+5} = \frac{4}{50}$$

therefore,

$$R_p = \frac{50}{4} = 12.5\ \Omega$$

$$R_{ab} = R_{aD} + R_p + R_{Ab}$$

$$= 8 + 12.5 + 12$$

$$= 32.5\ \Omega$$

■ **EXAMPLE 4.24** Determine the resistances between the terminals AB, BC and CA in the network shown in Fig. 4.36.

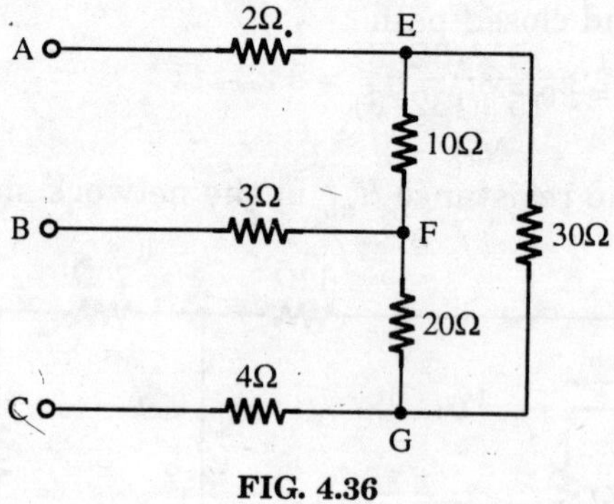

FIG. 4.36

SOLUTION :

$$R_{AB} = R_{AE} + (R_{EG} + R_{GF}) \;||\; (R_{EF}) + R_{FB}$$

$$= 2 + [(30 + 20) \;||\; 10] + 3 = 2 + \frac{50 \times 10}{50 + 10} + 3 = 13.33\ \Omega$$

$$R_{BC} = R_{BF} + (R_{FE} + R_{EG}) \;||\; R_{FG} + R_{GC}$$

$$= 3 + [(10 + 30) \;||\; 20] + 4 = 3 + \frac{40 \times 20}{40 + 20} + 4 = 20.33\ \Omega$$

$$R_{CA} = R_{CG} + (R_{GF} + R_{FE}) \;||\; R_{GE} + R_{EA}$$

$$= 4 + [(20 + 10) \;||\; 30] + 2 = 4 + \frac{30 \times 30}{30 + 30} + 2 = 21\ \Omega$$

■ **EXAMPLE 4.25** (*a*) Determine the voltage v_0 in the circuit shown in Fig. 4.37.
(*b*) Show that the total power developed in the circuit is equal to the total power dissipated.

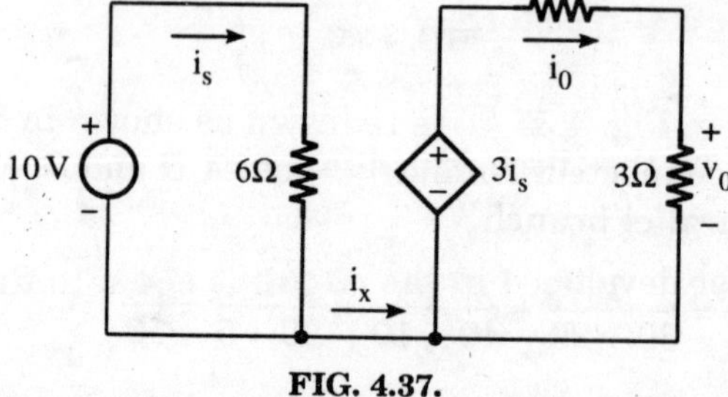

FIG. 4.37.

SOLUTION : (*a*) In the given circuit, there are two closed paths. One circuit is on the left carrying current i_0. The other circuit is on the right with current i_0. The current i_x shown in the circuit must be zero, because there is no return path for i_x to enter the left hand portion of the circuit. By KVL in the left-hand closed path

$$10 - 6i_s = 0$$

or $$i_s = \frac{10}{6} = \frac{5}{3}\text{ A}$$

By KVL in the right-hand closed path

$$3i_s - 2i_0 - 3i_0 = 0$$

$$5i_0 = 3i_s$$

$$i_0 = \frac{3}{5} i_s = \frac{3}{5} \times \frac{5}{3} = 1\text{ A}$$

By Ohm's law

$$v_0 = 3i_0 = 3 \times 1 = 3\text{ V}$$

(*b*) We have $p = vi$

Power developed by the independent 10 V source

$$= 10 \times i_s = 10 \times \frac{5}{3} = \frac{50}{3}\text{ W}$$

Power developed by the dependent $3i_s$ source

$$= 3i_s \times i_0 = 5 \times 1 = 5\text{ W}$$

The total power *developed* by the two voltage sources

$$= \frac{50}{3} + 5 = \frac{65}{3} = 21.7\text{ W}$$

Power dissipated in the 6 Ω resistor

$$= i_s^2 \times 6 = \left(\frac{5}{3}\right)^2 \times 6 = 16.7\text{ W}$$

Power dissipated in the 2 Ω resistor

$$= i_0^2 \times 2 = (1)^2 \times 2 = 2\text{ W}$$

Power dissipated in the 3 Ω resistor

$$= i_0^2 \times 3 = (1)^2 \times 3 = 3\text{ W}$$

Total power dissipated by the three resistors

$$16.7 + 2 + 3 = 21.7\text{ W}$$

It is seen that the total power developed in the circuit is equal to the total power dissipated.

4.16.1 Inductors In Series

Consider three inductors L_1, L_2 and L_3 connected in series as shown in Fig. 4.38. Let the current through each inductor be i. The rate of change of current in each conductor is the same. Let v be the supply voltage and the voltages across L_1, L_2 and L_3 be v_1, v_2 and v_3 respectively.

therefore, $$v_1 = L_1 \frac{di}{dt}, \quad v_2 = L_2 \frac{di}{dt}, \quad v_3 = L_3 \frac{di}{dt}$$

If the equivalent inductance of the series combination is L_s, then

$$v = L_s \frac{di}{dt}$$

Assuming that there is no coupling between the inductor, we have by KVL

$$v = v_1 + v_2 + v_3$$

$$L_s \frac{di}{dt} = L_1 \frac{di}{dt} + L_2 \frac{di}{dt} + L_3 \frac{di}{dt}$$

or
$$L_s = L_1 + L_2 + L_3 \tag{4.16.1}$$

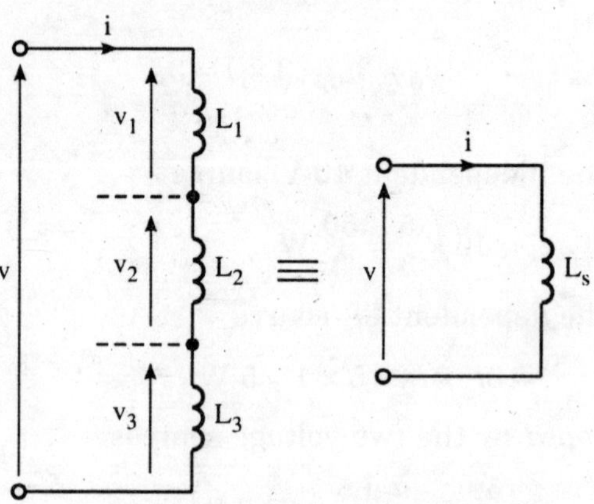

FIG. 4.38. Inductors in series

This result can be extended to several inductors in series. If there are n inductors in series, the equivalent inductance of the combination may be written as

$$L_s = L_1 + L_2 + L_3 + \ldots + L_n \tag{4.16.2}$$

4.16.2 Inductors In Parallel

Consider three inductors L_1, L_2 and L_3 in parallel as shown in Fig. 4.39. Let the currents through inductors L_1, L_2 and L_3 be i_1, i_2, i_3 respectively. The voltage across each inductor is the same all and it is equal to the supply voltage v. Let the total current be i.

therefore,
$$v = L_1 \frac{di_1}{dt}, \quad v = L_2 \frac{di_2}{dt}, \quad v = L_3 = \frac{di_3}{dt}$$

If the equivalent inductance of the parallel combination is L_p, then

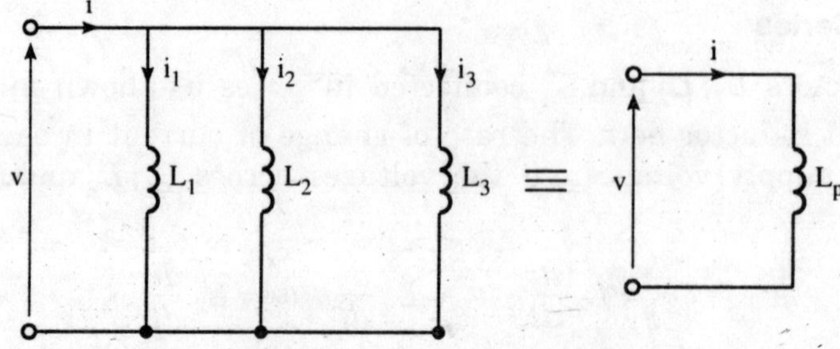

FIG. 4.39. Inductors in parallel

$$v = L_p \frac{di}{dt}$$

By KCL

$$i = i_1 + i_2 + i_3$$

Differentiating both the sides with respect to t

$$\frac{di}{dt} = \frac{di_1}{dt} + \frac{di_2}{dt} + \frac{di_3}{dt}$$

$$\frac{v}{L_p} = \frac{v}{L_1} + \frac{v}{L_2} + \frac{v}{L_3}$$

$$\frac{1}{L_p} = \frac{1}{L_1} + \frac{1}{L_2} + \frac{1}{L_3} \tag{4.16.3}$$

In general, for n inductors in parallel

$$\frac{1}{L_p} = \frac{1}{L_1} + \frac{1}{L_2} + \frac{1}{L_3} + \ldots + \frac{1}{L_n} \tag{4.16.4}$$

4.17.1 Capacitors In Series

Consider three capacitors connected in series as shown in Fig. 4.40. Let the current through each capacitor be i. It is assumed that there is no initial charge across any of the capacitors. Let v be the supply voltage and the voltages across C_1, C_2, C_3 be v_1, v_2 and v_3 respectively.

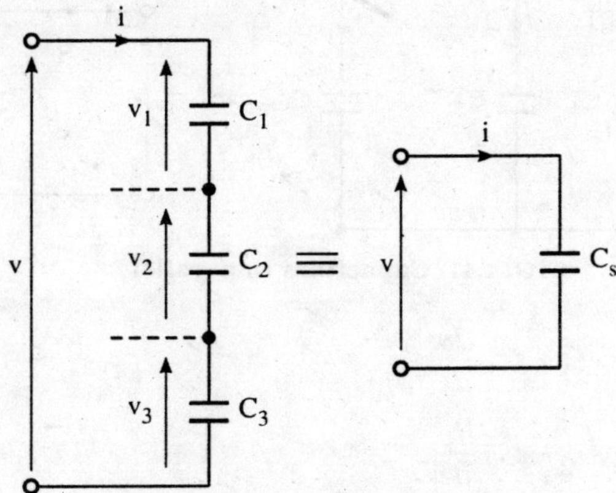

FIG. 4.40. Capacitors in series

therefore,
$$v_1 = \frac{1}{C_1}\int_0^t i\,dt$$

$$v_2 = \frac{1}{C_2}\int_0^t i\,dt$$

$$v_3 = \frac{1}{C_3}\int_0^t i\,dt$$

If the equivalent capacitance of the series combination is C_s, then

$$v = \frac{1}{C_s}\int_0^t i\,dt$$

By KVL,

$$v = v_1 + v_2 + v_3$$

$$\frac{1}{C_s}\int_0^t i\,dt = \frac{1}{C_1}\int_0^t i\,dt + \frac{1}{C_2}\int_0^t i\,dt + \frac{1}{C_3}\int_0^t i\,dt$$

or

$$\frac{1}{C_s} = \frac{1}{C_1} + \frac{1}{C_2} + \frac{1}{C_3} \tag{4.17.1}$$

If there are n capacitors in series, the equivalent capacitance C_s is given by

$$\frac{1}{C_s} = \frac{1}{C_1} + \frac{1}{C_2} + \frac{1}{C_3} + \dots + \frac{1}{C_n} \tag{4.17.2}$$

4.17.2 Capacitors In Parallel

Consider three capacitors C_1, C_2 and C_3 connected in parallel as shown in Fig. 4.41. Let the currents through the capacitors C_1, C_2 and C_3 be i_1, i_2 and i_3 respectively. The voltage across each capacitor is the same and it is equal to the supply voltage v. Let the total current be i.

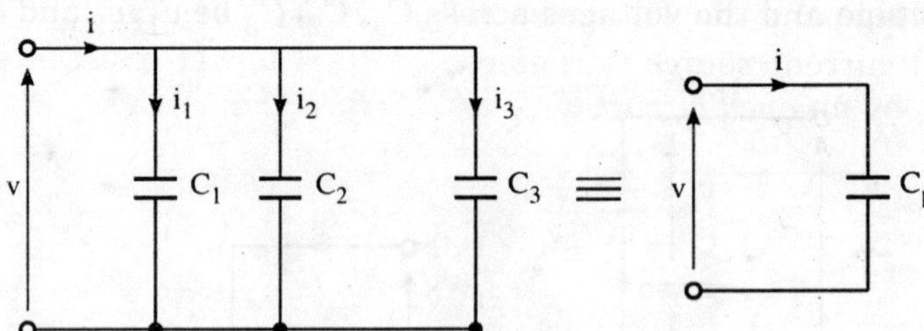

FIG. 4.41. Capacitors in parallel

therefore,

$$i_1 = C_1\frac{dv}{dt}$$

$$i_2 = C_2\frac{dv}{dt}$$

$$i_3 = C_3\frac{dv}{dt}$$

If the equivalent capacitance of the parallel combination is C_p, then

$$i = C_p\frac{dv}{dt}$$

By KCL,

$$i = i_1 + i_2 + i_3$$

$$C_p\frac{dv}{dt} = C_1\frac{dv}{dt} + C_2\frac{dv}{dt} + C_3\frac{dv}{dt} + C_3\frac{dv}{dt}$$

$$\therefore \qquad C_p = C_1 + C_2 + C_3 \tag{4.17.3}$$

If n capacitors $C_1, C_2, \ldots, C_n$ are connected in parallel, the equivalent capacitance C_p is given by

$$C_p = C_1 + C_2 + C_3 + \ldots + C_n \tag{4.17.4}$$

It is to be noted that *capacitors in parallel combine in the same way as resistors in series.*

4.18 REPLACEMENT OF PRACTICAL VOLTAGE AND CURRENT SOURCES BY THEIR INTERNAL RESISTANCES

A *practical voltage source* may be represented by an ideal voltage source of voltage v_S in series with a resistance R_{sv}, equal to the internal resistance of the source. Such a voltage source is shown in Fig. 4.42a. Some problems may involve the replacement of a practical (non-ideal) voltage source by its internal resistance. When a practical voltage source is replaced by its internal resistance we replace the ideal voltage source v_S by a *short circuit* (SC) as shown in Fig. 4.42b.

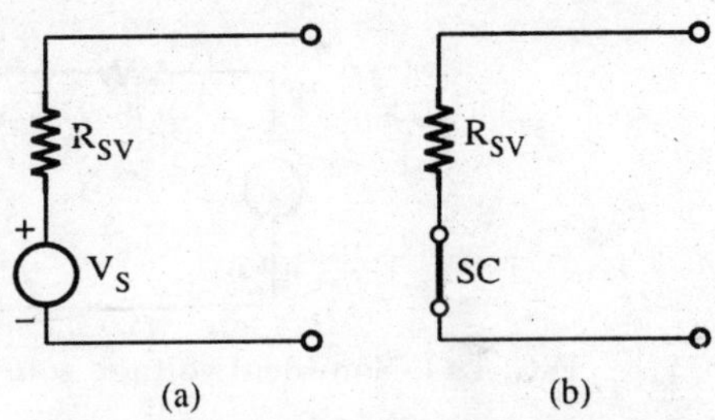

FIG. 4.42. A practical voltage source.

A *practical current source* may be represented by an ideal current source of current i_S in parallel with a resistance R_{Si}, equal to the internal resistance of the source. Such a current source is shown in Fig. 4.43a.

When a practical current source is replaced by its internal resistance, we replace the ideal current source i_S by an *open circuit* (OC) as shown in Fig. 4.43b.

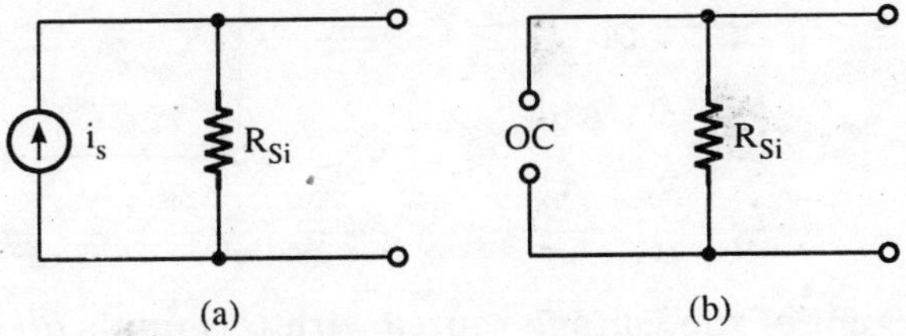

FIG. 4.43. A practical current source.

4.19 SOURCE TRANSFORMATION

A non-ideal (practical) voltage source can be transformed to an equivalent non-ideal current source. Similarly, a non-ideal current source can be transformed to an equivalent non-ideal voltage source. Replacing one source by an equivalent source is often called a *source transformation*. It should be noted that this is a fictional change in type of source. Actually, it is a mathematical equivalence in terms of equations in the circuits. This transformation serves as a powerful tool to simplify the analysis of circuits specially with mixed sources. Source transformation can be used for independent as well as dependant sources. In case of source transformations for dependent (controlled) sources, care should be taken to keep the dependent variables intact.

A non-ideal voltage source and a non-ideal current source are said to be equivalent if their voltampere (v-i) relations are the same for all terminal conditions. For equivalence of the two sources the following conditions should be satisfied :

1. The open-circuit voltages at their terminals are equal.
2. The short-circuit currents at their terminals are equal.

4.20 EQUIVALENCE OF RESISTIVE VOLTAGE AND CURRENT SOURCES

Fig. 4.44 shows a non-ideal voltage source with terminals ab. It has a resistane R between the ideal source and its terminals. Fig. 4.45 shows a non-ideal current source with terminals cd. It has a resistance R_1 in parallel with the ideal source. The open-circuit voltage across the terminals ab of the voltage source is

$$v_{OC} = v_S \tag{4.20.1}$$

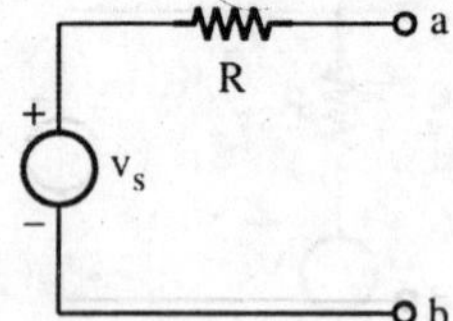

FIG. 4.44. Non-ideal voltage source.

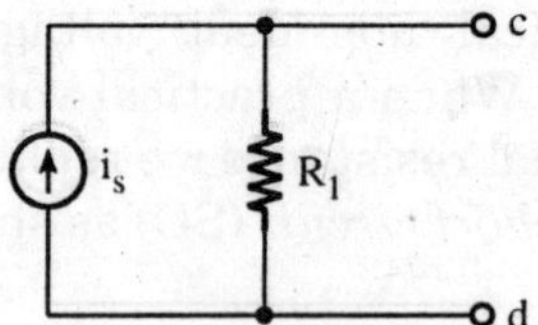

FIG. 4.45. Non-ideal current source.

The open-circuit voltage across the terminals cd of the current source is

$$v'_{OC} = R_1 i_S \tag{4.20.2}$$

For equivalence, the open-circuit voltage at ab must be equal to the open-circuit voltage at cd. Therefore,

$$v_{OC} = v'_{OC} \tag{4.20.3}$$

$$v_S = R_1 i_S \tag{4.20.4}$$

$$i_S = \frac{v_S}{R_1} \tag{4.20.4}$$

The short-circuit current of the voltage source with terminals ab short-circuited is

$$i_{SC} = \frac{v_S}{R} \tag{4.20.5}$$

The short-circuit current of the current source with terminals cd short-circuited is

$$i'_{SC} = i_S \tag{4.20.6}$$

For equivalence of the two sources

$$i'_{SC} = i_{SC}$$

$$i_S = \frac{v_S}{R} \tag{4.20.7}$$

Comparing Eqs. (4.20.4) and (4.20.7), we get

$$R_1 = R, \; i_S = \frac{v_S}{R}, \; v_S = R i_S$$

These are the conditions for the equivalence of non-ideal voltage and currrent sources. Thus, *a voltage source v_S in series with the resistance R is transformed into a current source i_S in parallel with the resistance R where $i_S = v_S/R$*. This equivalence is shown in Fig. 4.46.

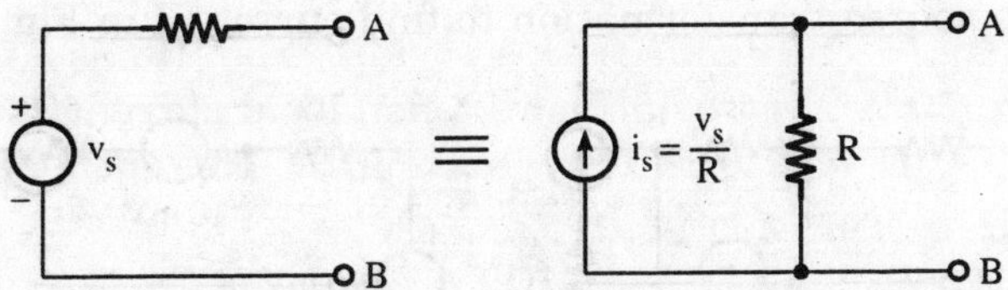

FIG. 4.46. Transformation of a resistive voltage source to a resistive current source.

Similarly, a *current source i_S in parallel with a resistance R is transformed into a voltage source v_S in series with a resistance R where $v_S = R i_S$*. This equivalence is shown in Fig. 4.47.

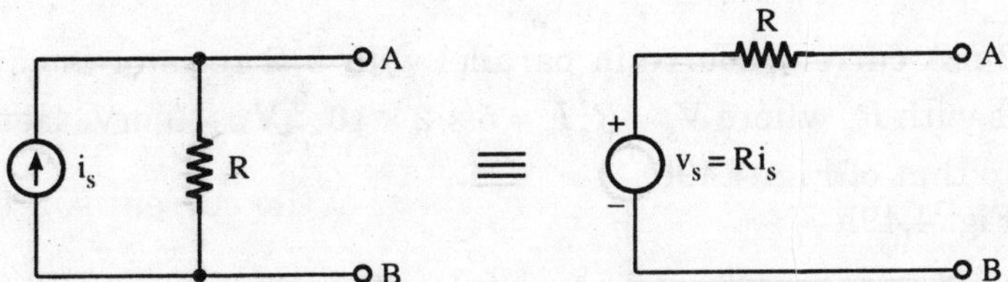

FIG. 4.47. Transformation of a resistive currents source to a resistive voltage source.

It is to be noted that the relationships $i_S = v_S/R$ and $v_S = R i_S$ have no meaning if $R = \infty$ or $R = 0$. The equivalence applies only to the load connected to the terminal external to the sources. There is no equivalence within the sources.

The polarity of the equivalent voltage source will always be in the same direction as the current through the current source. Similarly, current flows from the equivalent current source to the terminal originally connected to the positive side of the voltage source.

It should be noted that equivalence cannot be applied to ideal sources.

■ **EXAMPLE 4.26** Convert the voltage source in Fig. 4.48a to a current source.

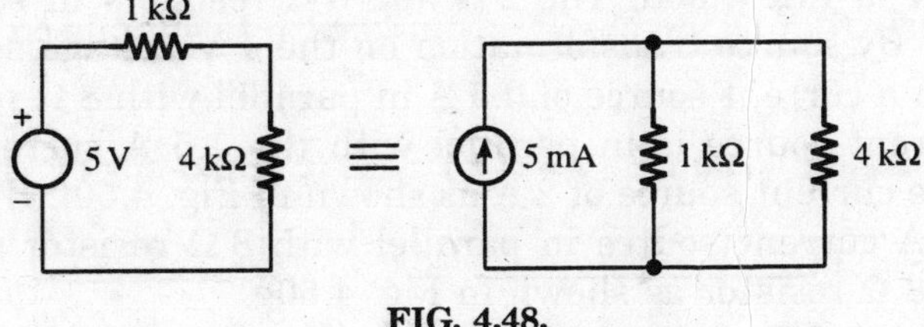

FIG. 4.48.

SOLUTION :

Here $\quad V_s = 5 \text{ V}, \; R_s = 1 \text{ k}\Omega = 1 \times 10^3 \; \Omega$

$$\therefore \quad I_s = \frac{V_s}{R_s} = \frac{5}{1 \times 10^3}$$

$$= 5 \times 10^{-3} \text{ A} = 5 \text{ mA}$$

Hence the equivalent current source is shown in Fig. 4.48b.

■ **EXAMPLE 4.27** Use source transformation to find current I in Fig. 4.49a.

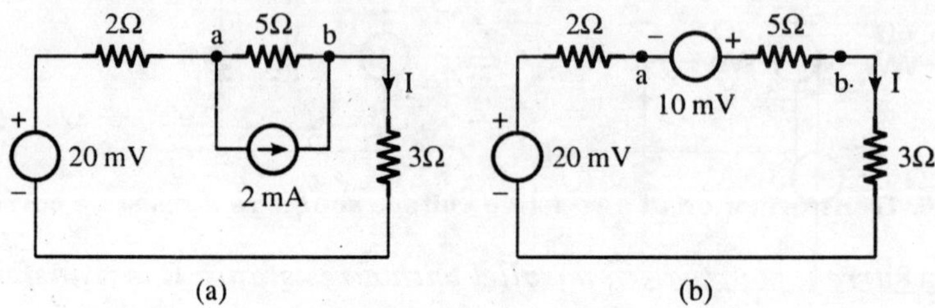

FIG. 4.49.

SOLUTION : The 2 mA current source in parallel with 5 Ω resistor is equivalent to a voltage source V_s in series with R_s where $V_s = R_s I_s = 5 \times 2 \times 10^{-3}$ V = 10 mV. Hence the circuit of Fig. 4.49a is reduced to that of Fig. 4.49b.

By KVL in Fig. 4.49b

$$20 \times 10^{-3} - 2I + 10 \times 10^{-3} - 5I - 3I = 0$$

$$10I = 30 \times 10^{-3}$$

$$I = 3 \times 10^{-3} \text{ A} = 3 \text{ mA}$$

■ **EXAMPLE 4.28** In the network shown in Fig. 4.50a, use source transformations to determine the current through and voltage across the 5 Ω resistor.

SOLUTION : The 6 V voltage source is in series with 3 Ω resistor. By source transformation we obtain the circuit shown in Fig. 4.50b. The 3 Ω and 6 Ω resistors in parallel can be combined into a single resistor $\frac{3 \times 6}{3 + 6}$, that is, 2 Ω resistor as shown in Fig. 4.50c. The 2A current source in parallel with the 2 Ω resistor can be transformed to a 4 V voltage source in series with the 2 Ω resistor as shown in Fig. 4.50d. The 2 Ω and 6 Ω resistors in series are combined to a single resistor of 8 Ω. By source transformation on the 4 V voltage source in series with the 8 Ω resistor we obtain a current source of 0.5 A in parallel with 8 Ω resistor as shown in Fig. 4.50e. The 0.5 A current source is in parallel with the 1.5 A current source. They can be combined into a single current source of 2 A as shown in Fig. 4.50f. By performing the source transformation on 2 A current source in parallel with 8 Ω resistor we obtain 16 V voltage source in series with 8 Ω resistor as shown in Fig. 4.50g.

The current through 5 Ω resistor

$$i = \frac{16}{8 + 3 + 5} = 1 \text{ A}$$

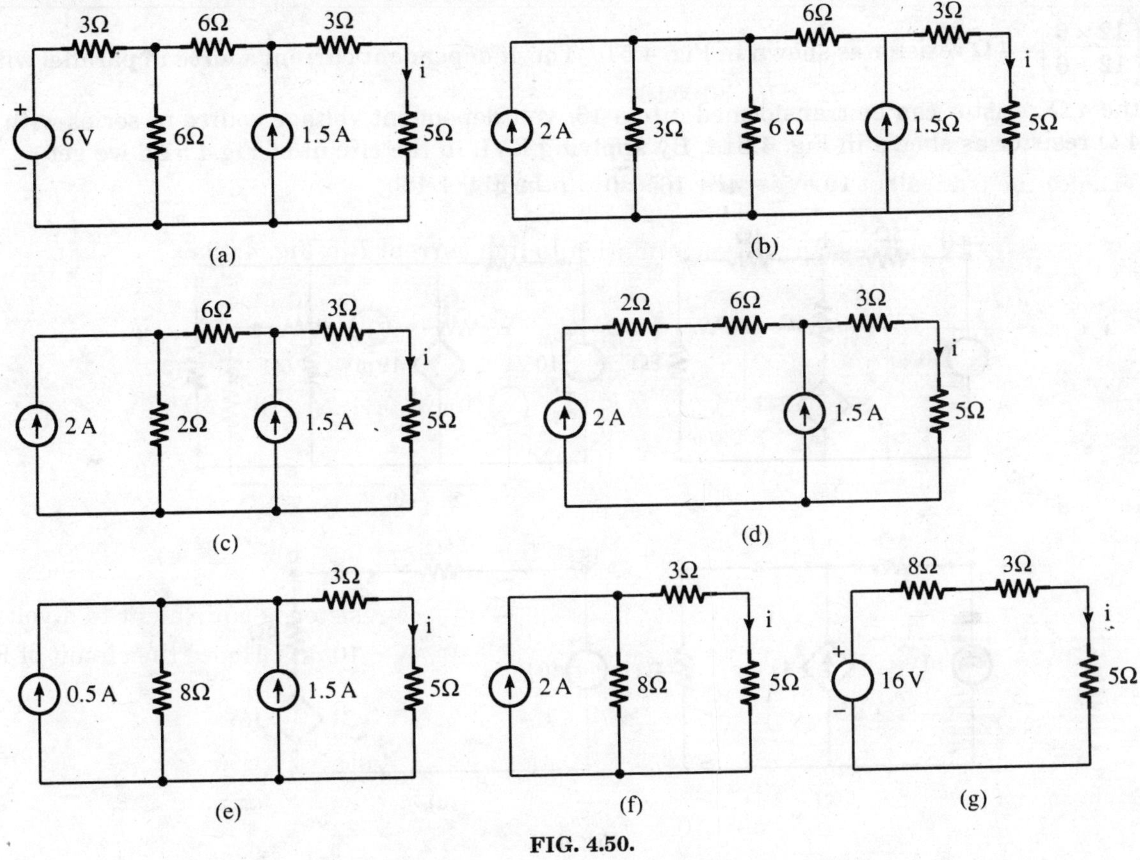

FIG. 4.50.

Therefore the voltage across 5 Ω resistor

$$= 5i = 5 \times 1 = 5 \text{ V}$$

4.21 SOURCE TRANSFORMATION FOR DEPENDENT SOURCES

Source transformation can also be applied to dependent (controlled) sources. However, the dependent variable should be kept intact, since, the operation of the dependent source depends on it. The following examples illustrate the transformation of a dependent source.

■ **EXAMPLE 4.29** By performing source transformation find the value of the voltage across the 5 Ω resistor of the network shown in Fig. 4.51a.

SOLUTION : In the network of Fig. 4.51a, the current-controlled voltage source S is controlled by the current in the 5 Ω resistor and, therefore, we should not change the circuit between a and b. The 5 Ω resistor must remain explicity in the circuit. The 4 Ω and 8 Ω resistors in series can be combined into a 12 Ω resistor. Performing the source transformation on the $24i$ current dependent voltage source in series with the 6 Ω resistor we obtain a dependent current source of value $(24i/6) = 4i$ amperes in parallel with a 6 Ω resistor as shown in Fig. 4.51 b. The 12 Ω and 6 Ω resistors in parallel in Fig. 4.51 b can be combined into a

$\left(\frac{12 \times 6}{12+6}\right) = 4\ \Omega$ resistor as shown in Fig. 4.51c. The $4i$ dependent current source in parallel with the 4 Ω resistor can be transformed into a $16i$ volt dependent voltage source in series with a 4 Ω resistor as shown in Fig. 4.51d. By applying KVL in the circuit of Fig.4.51.d we get

$$10 - 5i - 4i - 16i = 0, \quad i = 0.4 \text{ A}$$

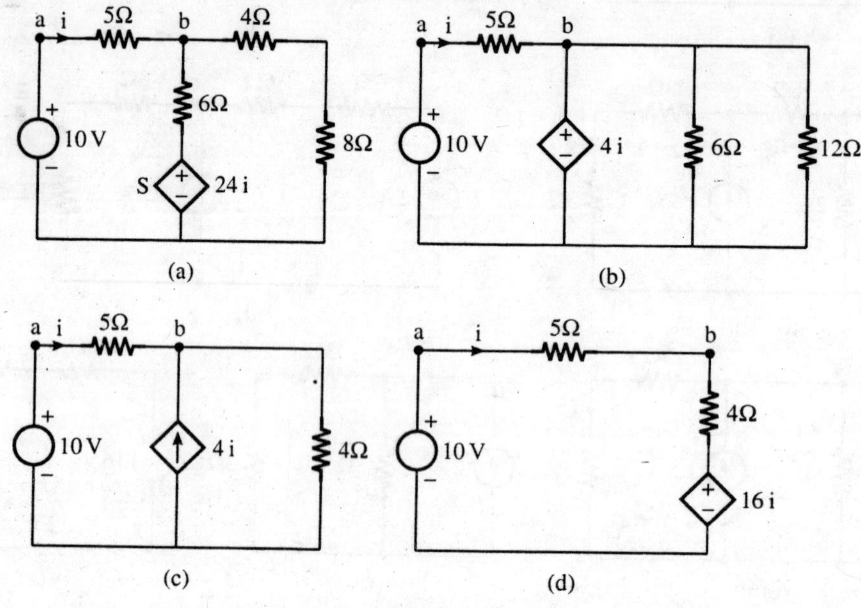

Fig. 4.51

The voltage drop V_{ab} across ab is given by

$$V_{ab} = 5i = 5 \times 0.4 = 2 \text{ V}$$

■ **EXAMPLE 4.30** In the circuit shown in Fig. 4.52a, use source transformation to determine the voltage v_2.

SOLUTION : Fig. 4.52b, 4.52c and 4.52d show various steps in source transformations.

By KVL in Fig. 4.52d,

$$\frac{2.5}{3} - \frac{4}{3}i - 1 \times i - \frac{1}{2}i + v = 0$$

$$\frac{17}{6}i - v = \frac{2.5}{3} \tag{E4.30.1}$$

Also, by KVL

$$\frac{2.5}{3}v - \frac{4}{3}i - v = 0$$

$$\frac{4}{3}i + v = \frac{2.5}{3} \tag{E4.30.2}$$

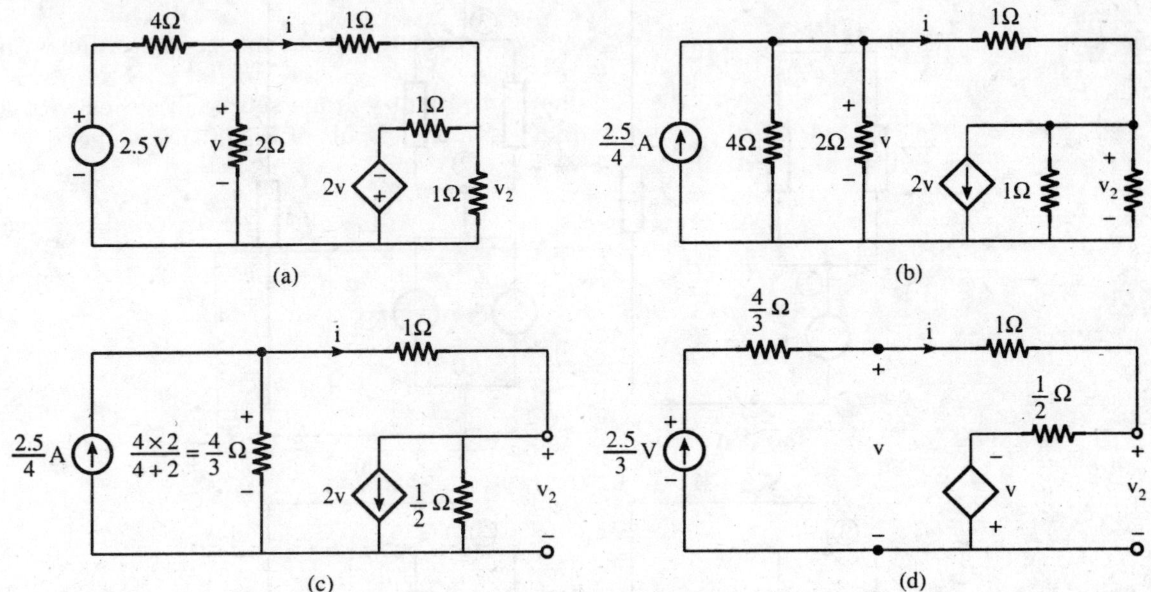

FIG. 4.52.

Addition of Eqs. (E4.30.1) and (E4.30.2) given

$$\frac{25}{6}i=\frac{5}{3},\quad i=\frac{2}{5}\text{ A}$$

Again, by KVL

$$\frac{2.5}{3}-\frac{4}{3}i-1\times i-v_2=0$$

$$v_2=\frac{2.5}{3}-\frac{4}{3}i-i$$

$$=\frac{2.5}{3}-\frac{4}{3}\times\frac{2}{5}-\frac{2}{5}=-0.1\text{ V}$$

4.22 SOURCE SHIFTING

A voltage source in series with an impedance can be transformed to a current source in parallel with an impedance. Similarly, a current source in parallel with an impedance can be transformed to a voltage source in series with an impedance. If, however, there is a voltage source without a series impedance or a current source connected across several branches, the source transformation cannot be directly applied. In such cases source shifting methods are found useful. The voltage source shift method is known as **E-shift method** and the current source shift method is called **I-shift method**. These methods allow us to shift the voltage and current sources without changing the analysis problem.

4.23 VOLTAGE SOURCE SHIFTING

Consider the circuit shown in Fig. 4.53a. The voltage source e_s has no single impedance in

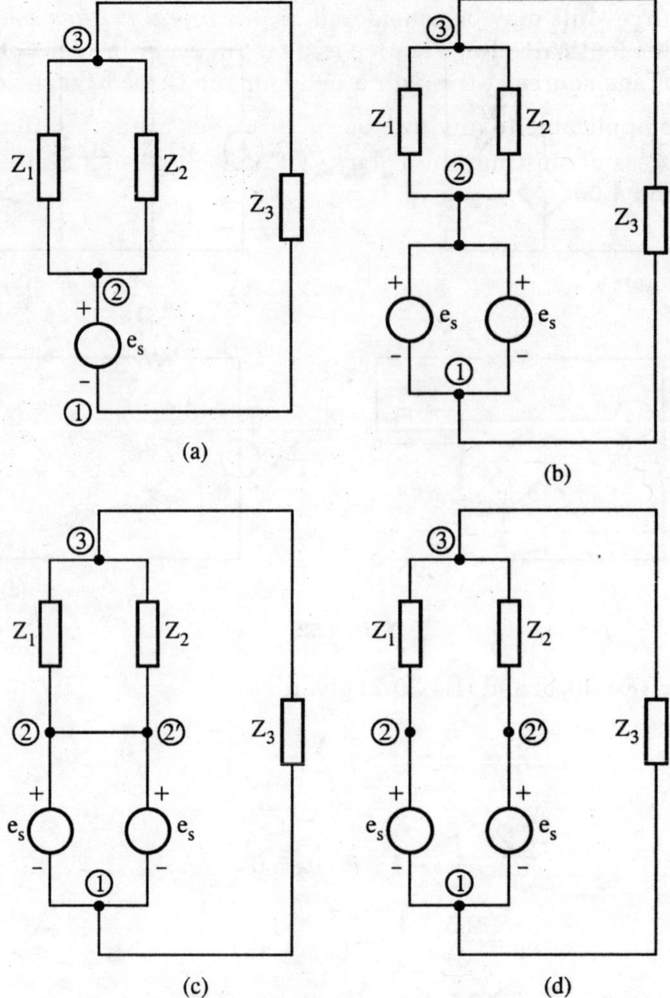

FIG. 4.53.

series with it. The single voltage source may be considered to be equivalent to two identical voltage sources as shown in Fig. 4.53b. The network shown in Fig. 4.53b is identical to that of Fig. 4.53b.

Node (b) of Fig. 4.53b is split as 2 and 2′ in Fig. 4.53c. Since nodes 2 and 2′ are maintained at the same potential by the voltage sources, no current can flow between them. The link wire between 2 and 2′ is therefore, unnecessary, and can be removed, giving the circuit of Fig. 4.53d. Here the two voltage sources now have impedances in series, and source transformation may be done if desired. If we write the KVL equations for the closed paths of Fig. 4.53a and Fig. 4.53d, the corresponding equations for both networks are the same showing that the two networks are equivalent.

The voltage source shift may be considered as *pushing a voltage source through a node.* In this shifting a new identical voltage source appears in every branch connected to the node. Fig. 4.54 shows a voltage source e_s through a node for the three branches.

The methods is applicable to any number of branches connected to the node.

During the process of shifting, the polarity of the source should be marked properly as shown in Fig. 4.54 and 4.55.

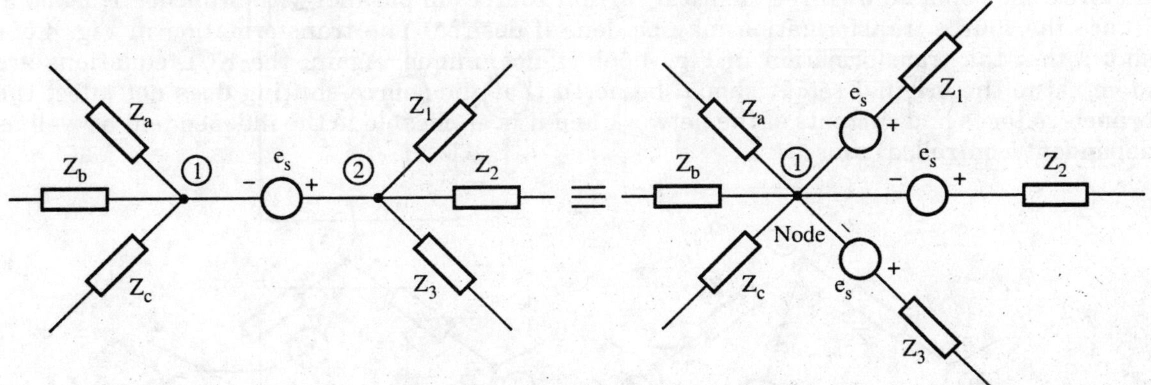

FIG. 4.54.

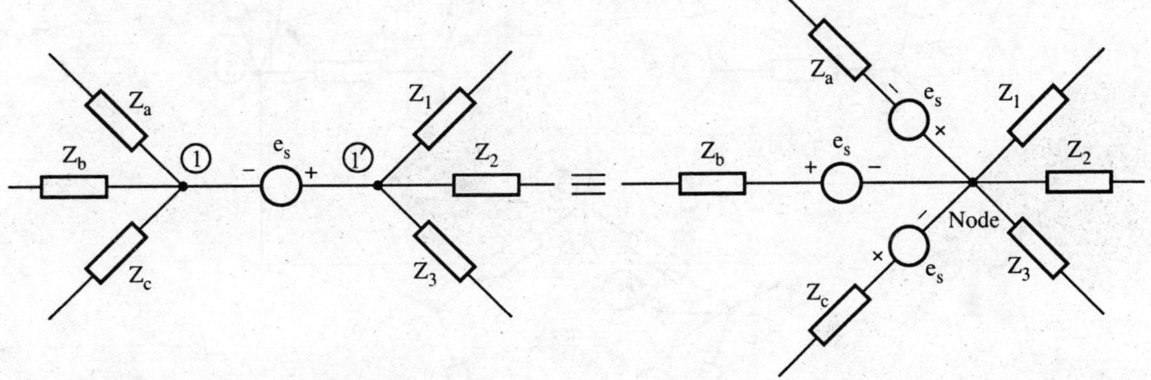

FIG.4.55

4.24 CURRENT SOURCE SHIFTING

The current source shifting method or the I-sift method is illustrated in Fig. 4.56 in which branch 6 is a current source i_s connected between nodes 5 and 3. Here the current source appears across more than one branch. The current i_s leaves node 5 and enters node 3. This current source may be shifted in the network by introducing additional identical current sources in such a way that currents at all nodes maintain their original values. Fig. 4.56b shows a circuit in which the current source in branch 6 has been removed and replaced by two identical current sources in parallel with branches 5 and 4. In Fig. 4.56a, the branch current

i_s leaves node 5 and enters node 3. In Fig. 4.56b , current at node 4 is not affected, since the same current i_s enters and leaves node 4. Also, the currents at nodes 5 and 3 remain the same as in Fig. 4.56a. Since the KCL equations at each nodes of the networks shown in Figs. 4.56a and 4.56b are identical, we conclude that these networks are equivalent and the current source shifting does not change the solution of the network. Fig. 4.56c shows another equivalent circuit of Fig. 4.56a. Here the original current source in branch 6 of Fig. 4.56a has been removed and replaced by three identical current sources in parallel with branches 1, 2 and 3. Hence the source transformation may be done if desired. The transformation in Fig. 4.56c shows that the transformation in Fig. 4.56b is not unique. Again, the KCL equations are identical to the original set. It should be noted that the source shifting does not affect the branch voltages and currents of the network and it is applicable to the independent as well as dependent (controlled) sources.

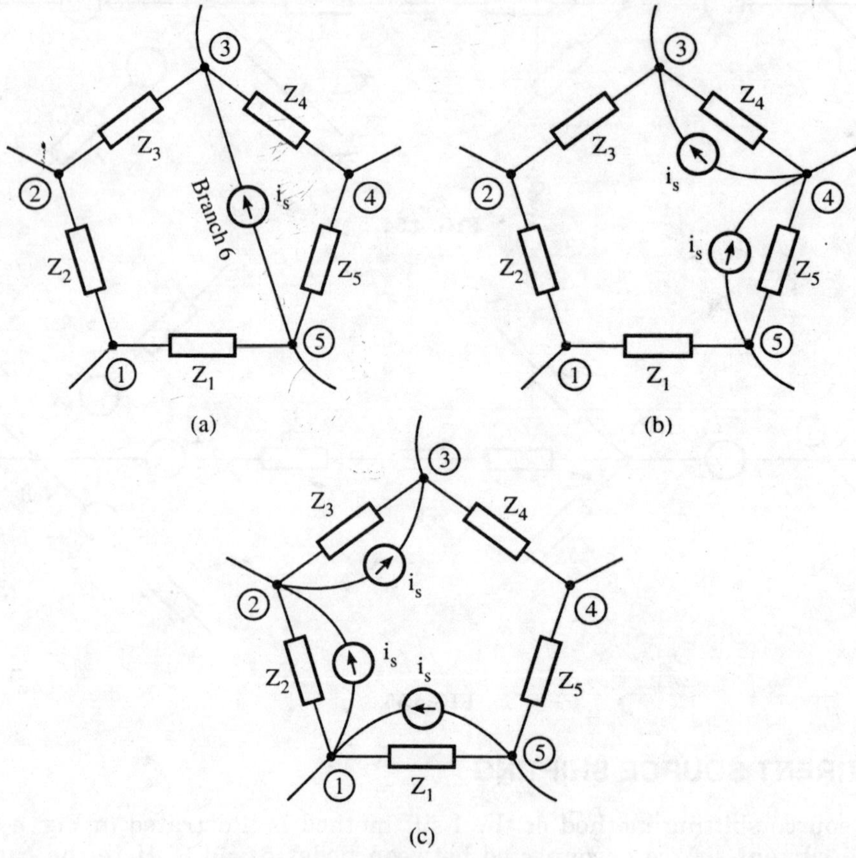

FIG. 4.56

■ **EXAMPLE 4.31** Find the voltage v_0 of the network shown in Fig. 4.57a by the method of source shifting.

SOLUTION : By the current-source shifting method the circuit of Fig. 4.57a can be redrawn as shown in Fig. 4.57b. The circuit of Fig. 4.57b can be shown to be equivalent to the circuit of Fig. 4.57c. From Fig. 4.57c, by currrent-division method

$$i_1 = \frac{R_s}{R_s + [R_1 + (\beta + 1) R_2]} i_s$$

therefore,
$$v_0 = -(\beta i_1 R_3) = \frac{-\beta R_s R_3}{R_s + R_1 + (\beta + 1) R_2} i_s$$

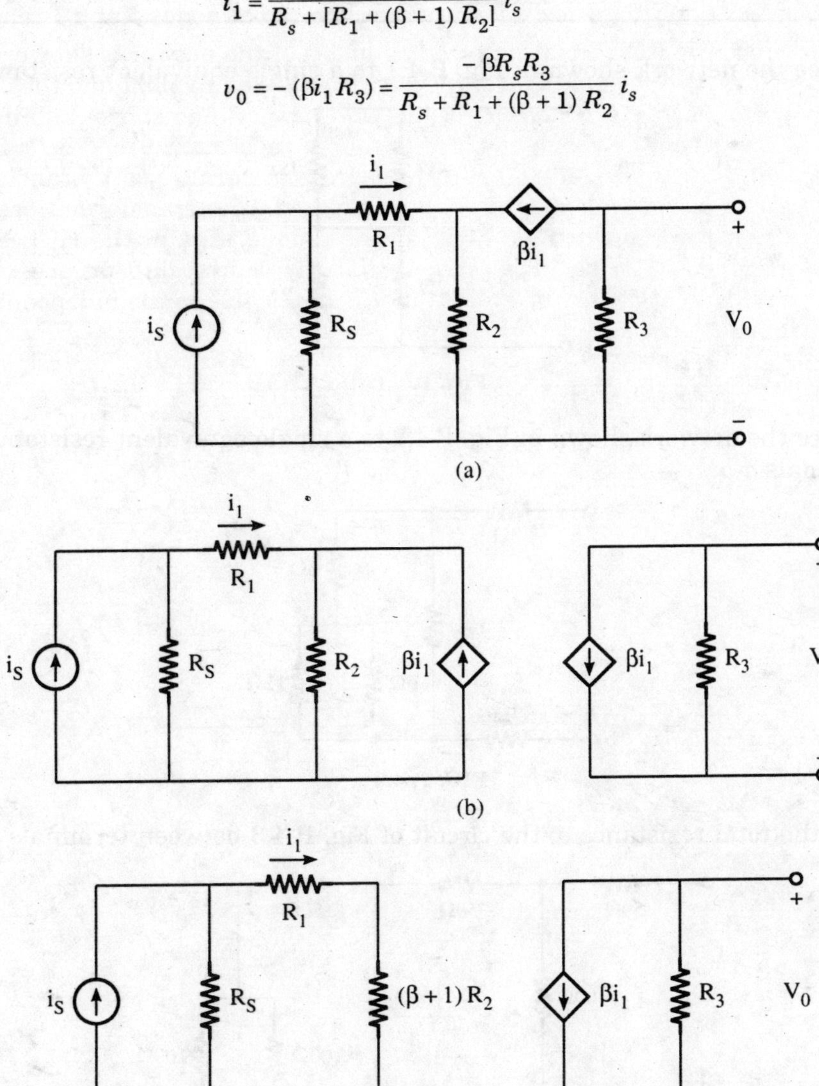

FIG. 4.57

EXERCISES

1. Reduce the network shown in Fig. P 4.1 to a single equivalent resistance.

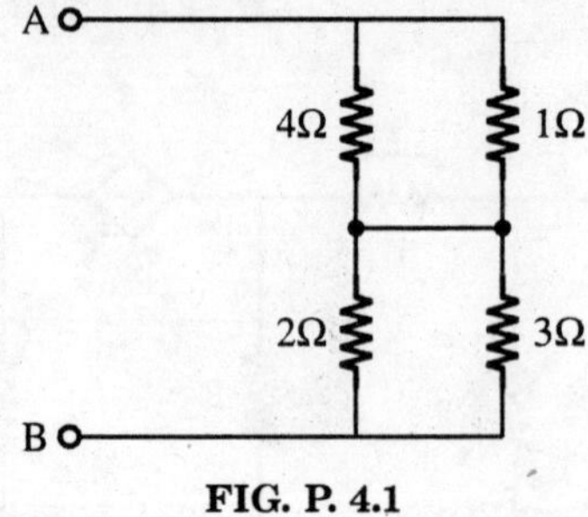

FIG. P. 4.1

2. Reduce the network shown in Fig. P 4.2 to a single equivalent resistance across the terminals *a-b*.

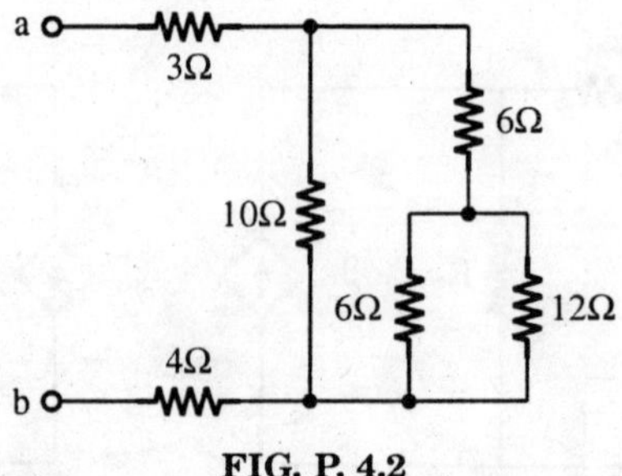

FIG. P. 4.2

3. Find the total resistance of the circuit of Fig. P 4.3 between terminals *a-b*.

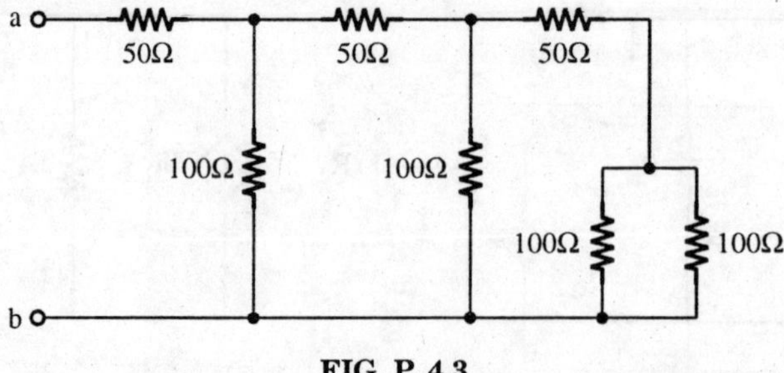

FIG. P. 4.3

4. Two batteries having emfs of 10 V and 8 V and internal resistances of 2 Ω and 3 Ω respectively, are connected in parallel across of a load of resistance 1 Ω. Calculate (*a*) the individual battery currents, (*b*) the current through the load, (*c*) the voltage across the load.

5. Determine I, I_1, I_2, V_{ab} and V_{bc} in the network shown in Fig. P 4.4.

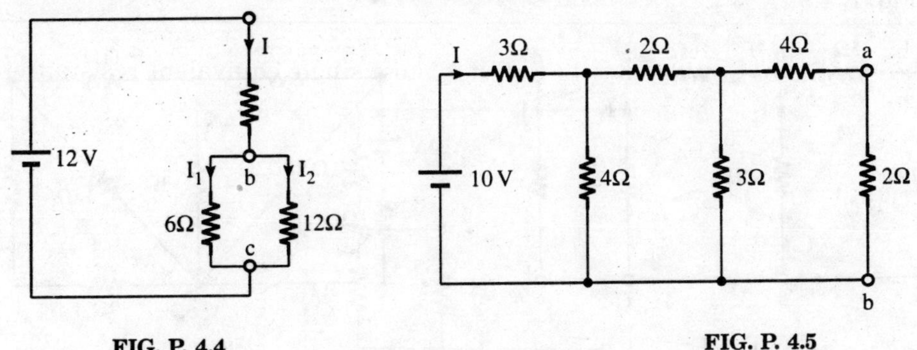

FIG. P. 4.4

FIG. P. 4.5

6. In the network shown in Fig. P 4.5, calculate V_{ab} and I.

7. Determine the current through each resistor in the network shown in Fig. P. 4.6.

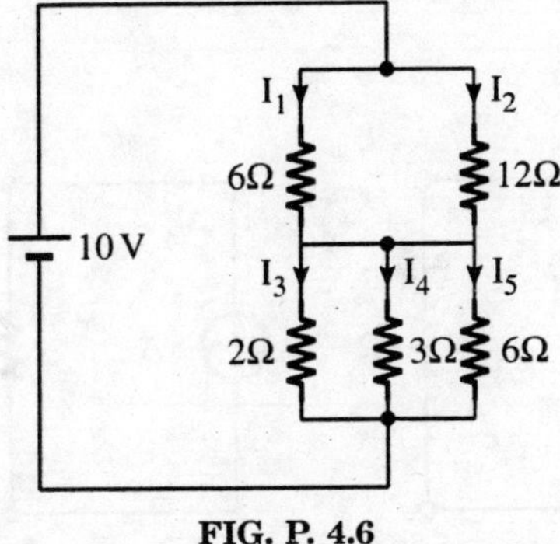

FIG. P. 4.6

8. Explain the difference between the emf and the terminal voltage of a battery. Calculate the current through 10 Ω resistor shown in Fig. P 4.7.

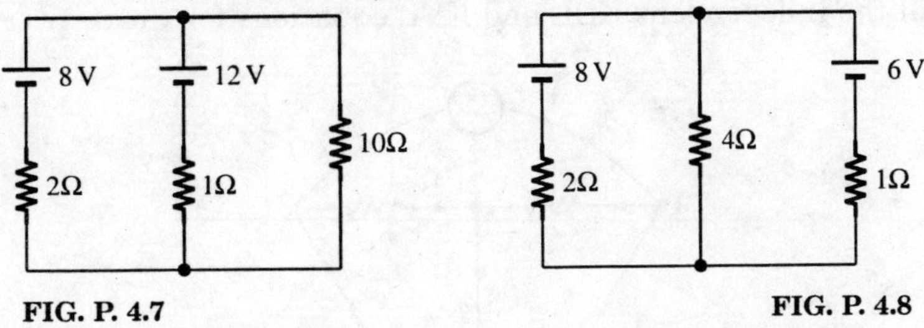

FIG. P. 4.7

FIG. P. 4.8

9. Calculate the power consumed in 4 Ω resistor shown in Fig. P 4.8.

10. Determine the current through and voltage across each resistor in the network of Fig. P 4.9.

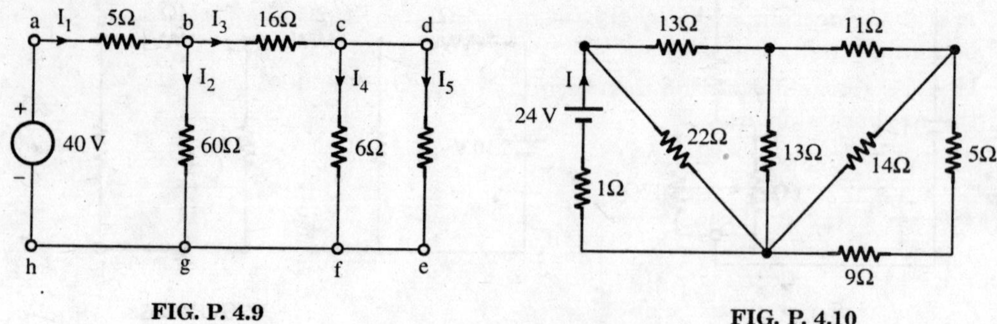

FIG. P. 4.9 FIG. P. 4.10

11. In the network shown in Fig. P. 4.10., determine (*a*) the current *I*, (*b*) voltage across 5 Ω resistor, (*c*) power loss in 18 Ω resistor.
12. Apply KCL and KVL to determine current I_L through 1 Ω resistor in the network of Fig. P 4.11.

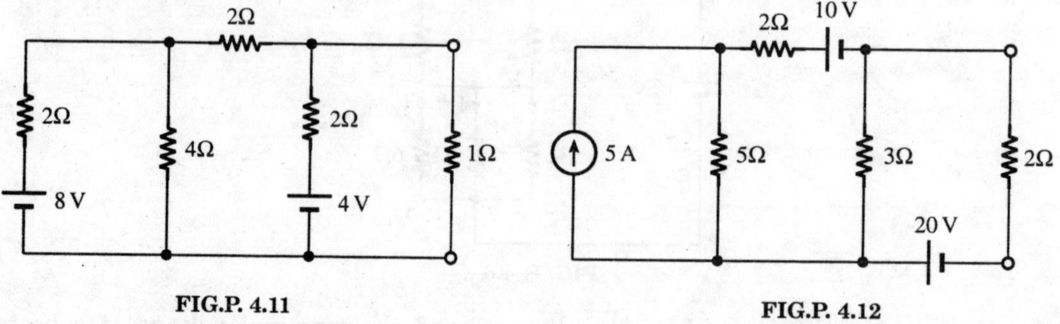

FIG.P. 4.11 FIG.P. 4.12

13. Determine the current through and voltage across the 2 Ω resistor in Fig. P. 4.12.
14. Write all the independent KCL and KVL equations for the network of Fig. P 4.13.

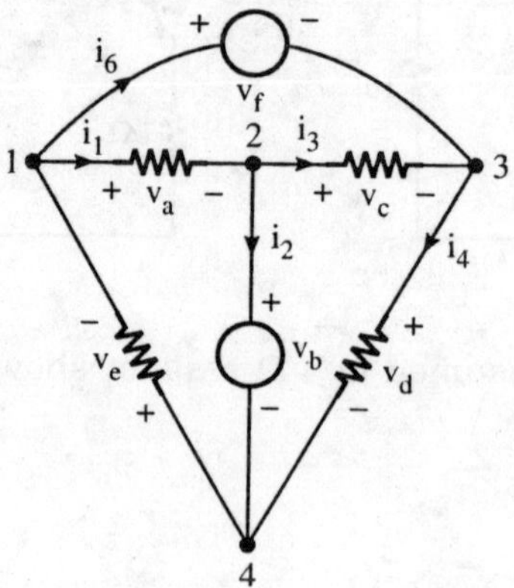

FIG.P. 4.13

15. A delta connection contains three equal resistances R. Find the resistances of the equivalent star connection.
16. A star connection contains three equal resistances R. Find the resistances of the equivalent delta connection.
17. Use $\Delta - Y$ transformation to determine the resistance between terminals a and b of the network shown in Fig. P 4.14.

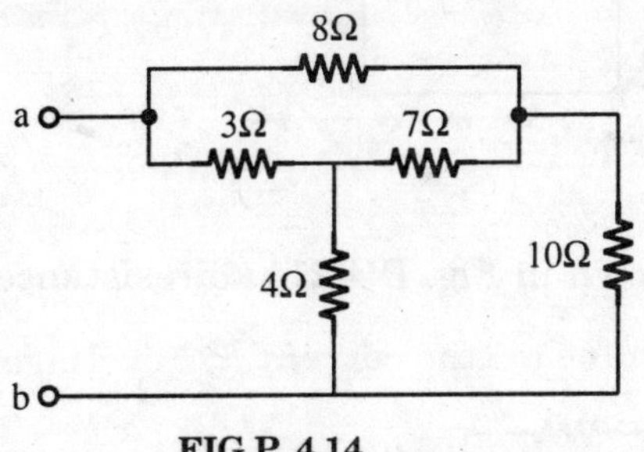

FIG.P. 4.14

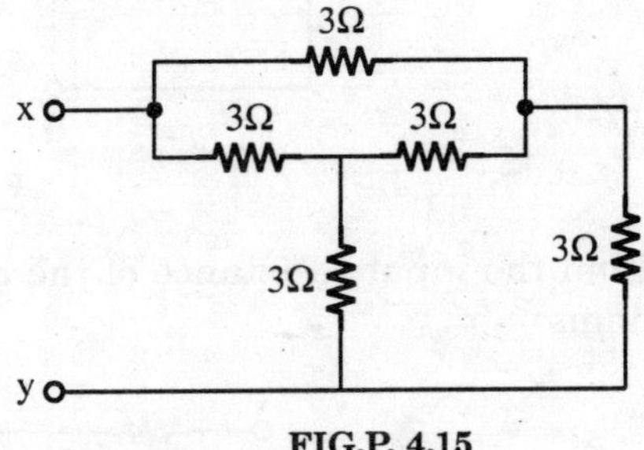

FIG.P. 4.15

18. Use $\Delta - Y$ transformation to determine the resistance between terminals x and y of the network shown in Fig. P 4.15.
19. Find the resistance R_{AB} at the input of the circuit in Fig. P. 4.16 by simplifying the circuit (a) with a Δ–Y transformation (b) with a Y–Δ transformation.

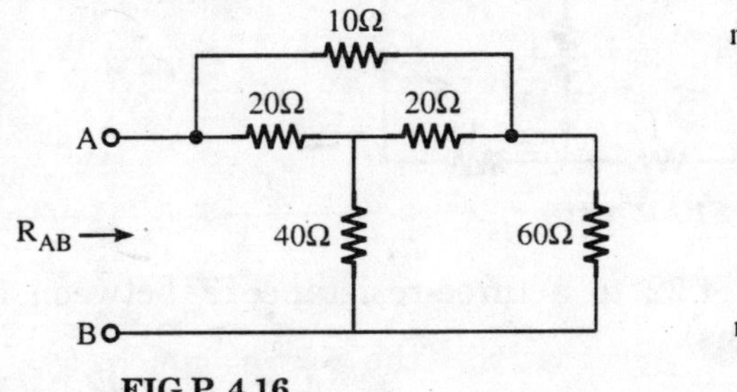

FIG.P. 4.16

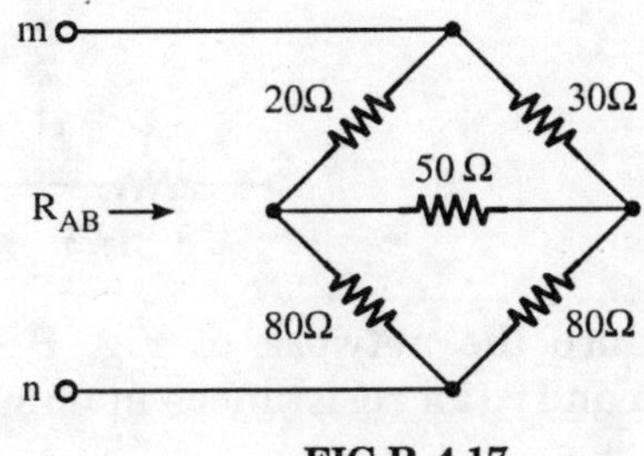

FIG.P. 4.17

20. Find the resistance R_{mn} at the input of the circuit in Fig. P 4.17 by simplifying the circuit (a) with Δ–Y transformation (b) with Y–Δ transformation.
21. Find the equivalent resistance at the terminals ab in Fig. P. 4.18.
22. Find the equivalent resistance at terminals ab in Fig. P 4.19.

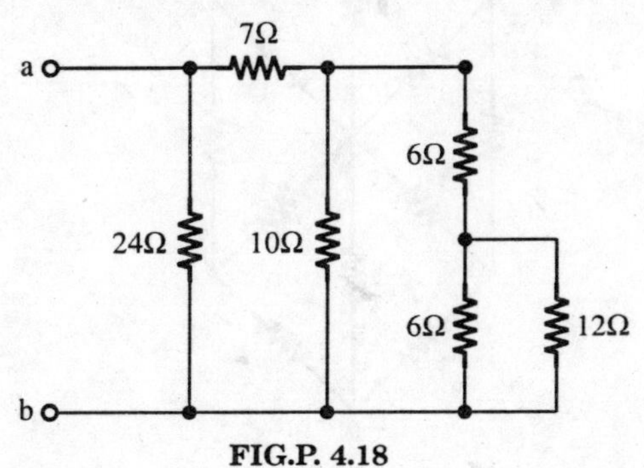

FIG.P. 4.18

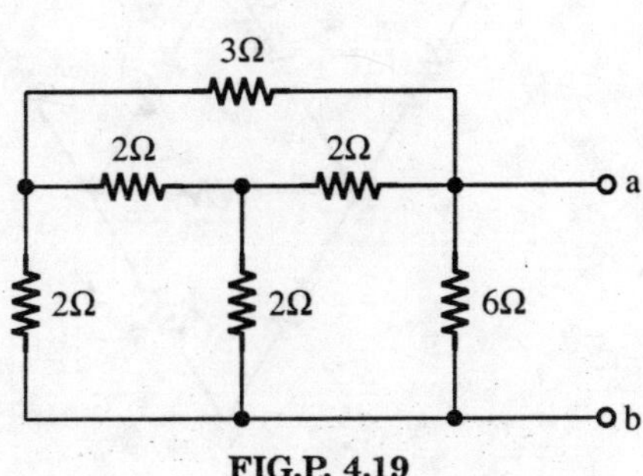

FIG.P. 4.19

23. Find the voltage between A and B in Fig. P 4.20.

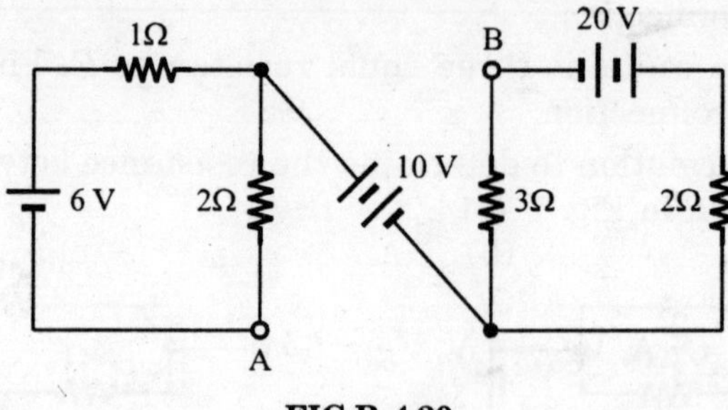

FIG.P. 4.20

24. Find the input resistance of the circuit shown in Fig. P 4.21 (all resistances are in ohms).

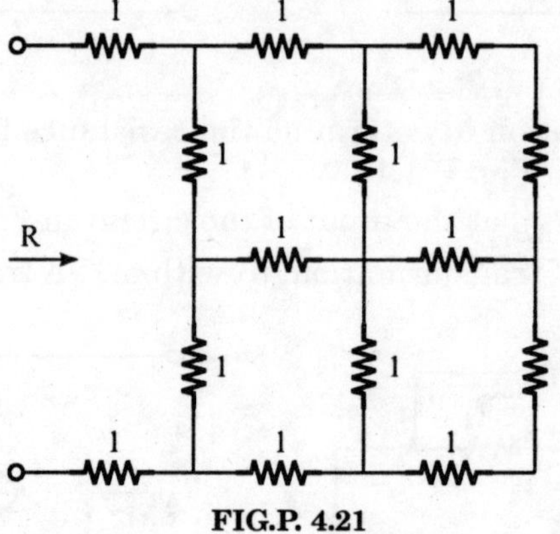

FIG.P. 4.21

25. Reduce the network of Fig. P 4.22 to a three-resistance Y between the points a, b and c (all resistances in ohms).

26. In Fig. P 4.23 show that the resistance between AB is $\frac{4}{5}$ Ω.

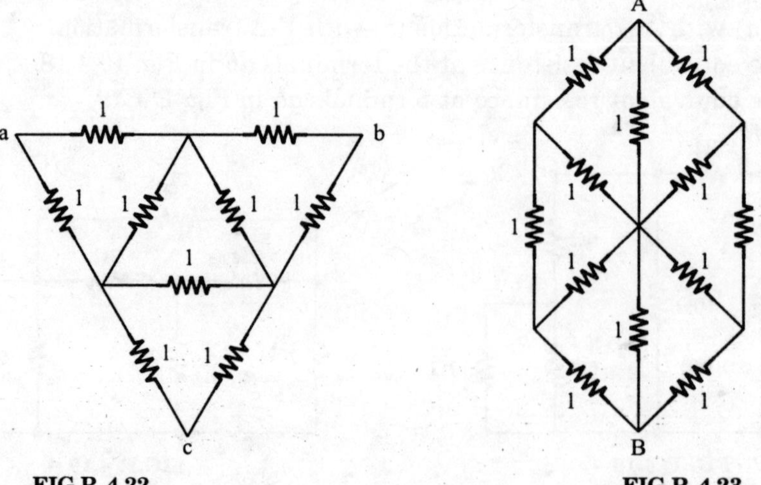

FIG.P. 4.22

FIG.P. 4.23

ANSWERS

1. $R_{AB} = 2\ \Omega$
2. $R_{ab} = 12\ \Omega$
3. $R_{ab} = 100\ \Omega$
4. (a) $\frac{14}{11}$ A, $\frac{32}{11}$ A (b) $\frac{46}{11}$ A (c) $\frac{46}{11}$ V
5. $I = 2$ A, $I_1 = \frac{4}{3}$ A, $I_2 = \frac{2}{3}$ A, $V_{ab} = 4$ V, $V_{bc} = 8$ V
6. $\frac{2}{3}$ V, 2 A
7. $I_1 = \frac{4}{3}$ A, $I_2 = \frac{2}{3}$ A, $I_3 = 1$ A, $I_4 = \frac{2}{3}$ A, $I_5 = \frac{1}{3}$ A
8. 0.5 A
9. 8.3 W
10. $I_1 = 2$ A, $I_2 = 0.5$ A, $I_3 = 1.5$ A, $I_4 = 1$ A, $I_5 = 0.5$ A,
    $V_{ab} = 10$ V, $V_{bg} = 30$ V, $V_{bc} = 24$ V, $V_{ce} = V_{de} = 6$ V
11. (a) 2 A (b) 1.25 V (c) 4.5 W
12. 2 A
13. 5 A , 10 V
14. KCL

    node 1 : $i_5 = i_1 + i_6$

    node 2 : $i_1 = i_2 + i_3$

    node 3 : $i_3 + i_6 = i_4$

    KVL

    Upper mesh

    $$v_f - v_c - v_a = 0$$

    Left-hand mesh

    $$v_a + v_b + v_e = 0$$

    Right-hand mesh

    $$-v_b + v_c + v_d = 0$$
15. $\frac{R}{3}, \frac{R}{3}, \frac{R}{3}$
16. $3R, 3R, 3R$
17. 5 Ω
18. 3 Ω
19. 31.4285 Ω
20. 46.9 Ω
21. 8 Ω
22. 2 Ω
23. A is 18 V above B
24. $\frac{13}{4}\ \Omega$
25. $\frac{5}{9}\ \Omega, \frac{5}{9}\ \Omega, \frac{5}{9}\ \Omega$

5

SINUSOIDS AND PHASORS

5.1 ALTERNATING QUANTITIES

An **alternating voltage** is any voltage that varies both in magnitude and polarity with respect to time. Similarly, **alternating current** is any current that varies both in magnitude and direction with respect to time. The reversal of polarity of voltage or direction of current occurs at regular intervals of time.

Alternating quantities may be represented graphically. The shape of the curve obtained by plotting the values of the alternating quantity at different instants on vertical axis and time (or angle) on horizontal axis is called the **waveform** or **waveshape** of the alternating quantity. An alternating quantity may have any waveform. It can be proved that any waveform can be considered to be made up of various combinations of sine waves. Thus, sine wave is basic to all alternating quantities. The quantities obeying sine law are called **sinusoidal quantities** or **sinusoids**. Sinusoid is the most important signal or waveform in electrical engineering.

5.2 SINUSOIDAL FUNCTIONS

A sinusoidal alternating current may be expressed in the form

$$i = I_m \sin \omega t \tag{5.2.1}$$

If $\omega t = \theta$, Eq. (5.2.1) may be written as

$$i = I_m \sin \theta \tag{5.2.2}$$

where i = value of current at time t

I_m = maximum or peak value of current

ω = radian or angular frequency in radians per second (rad/s)

By putting a particular value for t the corresponding value of i is obtained from Eq. (5.2.1). Hence Eq. (5.2.1) is called the **equation for instantaneous value of current**. Instantaneous values are denoted by small letters Thus, i denotes the instantaneous current and v denotes the instaneous voltage.

The graph of Eq. (5.2.2) is shown in Fig. 5.1. The shape of the curve is obtained by plotting the values of current at different times on vertical axis and angle θ on horizontal axis. It is noted from Eq. (5.2.2) or from the waveform that the value of current is continuously changing when θ lies between 0 and 2π radians. The maximum value of current is I_m. The maximum value is also called the **peak value, crest value** or **amplitude**. It is denoted by a capital letter and a subscript *m*. Thus, I_m denotes the maximum current and V_m denotes the maximum voltage.

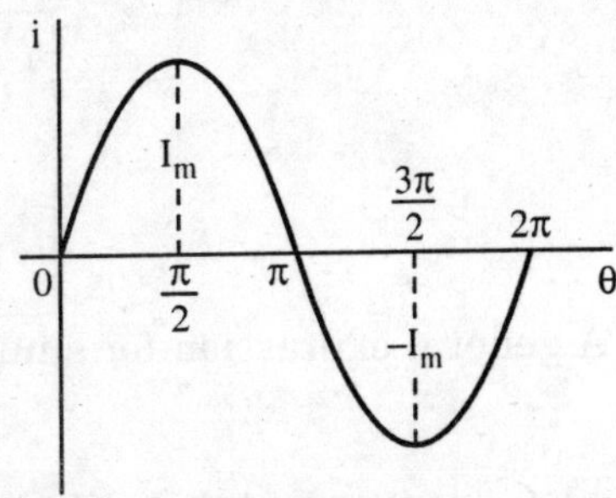

FIG. 5.1 Sinusoidal waveform

There is a repetition of the first set of values when θ lies between 2π and 4π. One complete alternation or repetition of a set of values of an alternating quantity is called a **cycle**. Thus, in Fig. 5.2 one cycle of the waveform is completed from point *o* to point *d*. The number of cycles completed in one second is called the **frequency** (*f*). The unit of frequency is hertz (Hz).

1 hertz = 1 cycle per second

$$\boxed{1 \text{ Hz} = 1 \text{ c/s}} \tag{5.2.3}$$

The time required to complete one cycle is called the **periodic time** or simply the **period** (*T*). The unit of period is the second (*s*). Period is measured between two points in a similar position on the waveform as shown in Fig. 5.2.

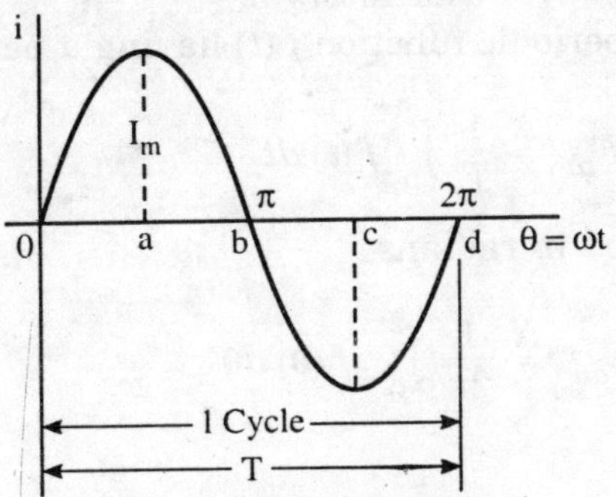

FIG. 5.2 Period of a sinusoid

Since the time required for one cycle is *T* seconds,

$$f = \frac{\text{number of cycles}}{\text{time in seconds}} = \frac{1}{T} \tag{5.2.4}$$

From Fig. 5.2, it is seen that the value of ωt at the end of one complete cycle is 2π. Since *T* is the time for one cycle, $\omega t = 2\pi$

$$\boxed{\omega = \frac{2\pi}{T} = 2\pi f} \quad \text{rad/s} \tag{5.2.5}$$

We call ω as the **radian frequency** or **angular frequency**. Thus Eq. (5.2.1) can be written in the following forms :

$$i = I_m \sin \omega t \quad (5.2.6)$$

$$i = I_m \sin 2\pi ft \quad (5.2.7)$$

$$i = I_m \sin \frac{2\pi}{T} t \quad (5.2.8)$$

A general expression for sinusoidal curent is given by

$$i = I_m \sin (\omega t + \phi_0) \quad (5.2.9)$$

where ϕ_0 is called the *initial phase angle*.

When $\phi_0 = \pm 90°$, Eq. (5.2.9) becomes

$$i = I_m \cos \omega t$$

This is a cosine function. When $\phi_0 = 0$, $i = I_m \sin \omega t$, which is a sine function. Hence sinusoid is a more general term which includes both sine and cosine functions. Equation (5.2.9) shows that a sinusoid is characterized by three quantities namely, the magnitude, phase angle and frequency.

5.3 AVERAGE VALUE

The **average or mean value** of an alternating quantity over a given interval is the sum of all instantaneous values taken over that interval.

The average value of a periodic function $f(t)$ having a period T seconds is defined as

$$F_{av} \triangleq \frac{1}{T} \int_0^T f(t)\, dt \quad (5.3.1)$$

If the time function is expessed in radians

$$F_{av} \triangleq \frac{1}{2\pi} \int_0^{2\pi} f(\theta)\, d\theta \quad (5.3.2)$$

For example,

$$I_{av} = \frac{1}{2\pi} \int_0^{2\pi} i\, d\theta \quad (5.3.3)$$

$$V_{av} = \frac{1}{2\pi} \int_0^{2\pi} v\, d\theta \quad (5.3.4)$$

$$P_{av} = \frac{1}{2\pi} \int_0^{2\pi} p\, d\theta \quad (5.3.5)$$

If the alternating quantity is represented by a curve, the average value is the average height of the curve. The average height of any curve is found by dividing the area under the curve by the length of the base or the interval of the curve over which the curve extends.

$$\text{Average value} = \frac{\text{area under the curve}}{\text{length of the base of the curve}} \quad (5.3.6)$$

5.4 AVERAGE VALUE OF A SINUSOID

(a) *Complete Cycle Average*

Let $$i = I_m \sin \theta$$

$$I_{av} = \frac{1}{2\pi}\int_0^{2\pi} i\, d\theta$$

$$= \frac{1}{2\pi}\int_0^{2\pi} I_m \sin \theta\, d\theta$$

$$= \frac{I_m}{2\pi}[-\cos\theta]_0^{2\pi} = \frac{-I_m}{2\pi}(1-1) = 0$$

Thus, **the average value of a sinusoid over a complete cycle is zero.**

(b) *Half-Cycle Average*

By convention, the average value of a sinusoid is defined as the average value over half a cycle.

$$I_{av} = \frac{1}{\pi}\int_0^{\pi} i\, d\theta = \frac{1}{\pi}\int_0^{\pi} I_m \sin\theta\, d\theta$$

$$= \frac{I_m}{\pi}[-\cos\theta]_0^{\pi} = \frac{2\,I_m}{\pi} = 0.636\, I_m$$

$$I_{av} = \frac{I_m}{\pi/2}$$

In general, for a sine wave half-cycle average is found by multiplying the maximum value by 0.636. Thus,

$$V_{av} = 0.636\ V$$

5.5 EFFECTIVE OR ROOT MEAN SQUARE VALUE

An alternting current varies from instant to instant. It is therefore, necessary to determine an equivalent direct current that will produce the same heat in the same time, in the same resistor as the alternating current. In other words we measure alternating current (A.C.) in terms of direct current (D.C.) which has a constant value. The particular value of alternating current is called its effective value. *The **effective value** of an alternating current is by definition, equal to that value of direct current which will produce the same heat in the same time in the same resistor.*

Heat produced by an alternating current of instantaneous value i in a resistor R in time dt is $i^2 R\, dt$. Total heat produced in one cycle (that is, in time T) is given by

$$H_{ac} = \int_0^T i^2 R\, dt \tag{5.5.1}$$

Heat produced by the equivalent direct current I_d in resistor R is time T is given by

$$H_{dc} = I_d^2 RT \tag{5.5.2}$$

If $$H_{dc} = H_{ac}$$

$$I_d^2 RT = \int_0^T i^2 R\, dt$$

$$I_d = \sqrt{\frac{1}{T}\int_0^T i^2\, dt} \tag{5.5.4}$$

But this particular value of I_d of direct current is the effective value I_{eff} of alternating current.

$$I_{eff} = I_d = \sqrt{\left[\frac{1}{T}\int_0^T i^2 R\, dt\right]} \tag{5.5.5}$$

The *effective value of an alternating voltage* is defined in a similar manner. The relation for effective value of voltage can be deduced from Eq. (5.5.3) by substituting

$$I_d = \frac{V_d}{R} \text{ and } i = \frac{v}{R}$$

$$\left(\frac{V_d}{R}\right)^2 RT = \int_0^T \left(\frac{v}{R}\right)^2 R\, dt$$

$$V_d = V_{eff} = \sqrt{\left[\frac{1}{T}\int_0^T v^2\, dt\right]} \tag{5.5.6}$$

Eq. (5.5.6) gives the effective value of voltage. *The* ***effective value of alternating voltage*** *is that value of direct voltage which will produce the same heat in the same resistor in the same time.* Eq. (5.5.5) can be written as

I_{ejf} = square *root* of the *mean* of the *squares* of the currents

= root-mean-square (or r.m.s.) value of the current

= I_{rms}

The effective value of current is, therefore, called its *root mean square value* or *r.m.s. value*. Sometimes it is called its *virtual value*. Similarly, the effective value of voltage is also called its r.m.s. value.

If the instantaneous values of current are $i_1, i_2, \ldots, i_n$ then

$$I_{rms} = \sqrt{\left(\frac{i_1^2 + i_2^2 + \ldots + i_n^2}{n}\right)} \tag{5.5.7}$$

The r.m.s. value are denoted by capital letters without any suffix. For example,

$$I_{rms} = I, \quad V_{rms} = V$$

The r.m.s. values are commonly used in practice. **Unless stated otherwise, the values of the given voltage and current are always taken as the r.m.s. values.**

5.6 R.M.S. VALUE OF A SINUSOID

Let
$$i = I_m \sin\theta$$

$$I^2 = \frac{1}{2\pi}\int_0^{2\pi} i^2\, d\theta$$

$$= \frac{1}{2\pi} \int_0^{2\pi} I_m^2 \sin^2 \theta \, d\theta$$

$$= \frac{I_m^2}{2\pi} \int_0^{2\pi} \frac{1}{2} (1 - \cos 2\theta) \, d\theta$$

$$= \frac{I_m^2}{4\pi} \left[\theta - \frac{1}{2} \sin 2\theta \right]_0^{2\pi}$$

$$= \frac{I_m^2}{4\pi} \left[\theta \right]_0^{2\pi} - \frac{I_m^2}{8\pi} \left[\sin 2\theta \right]_0^{2\pi}$$

$$= \frac{I_m^2}{4\pi} (2\pi - 0) - \frac{I_m^2}{8\pi} (\sin 4\pi - \sin 0^\circ)$$

Since $\sin 4\pi = 0$ and $\sin 0^\circ = 0$

$$I^2 = \frac{I_m^2}{4\pi} \times 2\pi$$

$$I = \frac{I_m}{\sqrt{2}} = 0.707\, I_m \qquad (5.6.1)$$

The rm.s. value of a sinusoidal voltage may be found in a similar manner.

$$V = \frac{V_m}{\sqrt{2}} = 0.707\, V_m \qquad (5.6.2)$$

In general, for a sine wave

$$\text{r.m.s. value} = \frac{1}{\sqrt{2}} \times \text{maximum value} = 0.707 \text{ maximum value}$$

It is to be noted that the r.m.s. value calculated for one complete cycle is equal to that calculated for half a cycle.

5.7 FORM FACTOR AND PEAK FACTOR

The **form factor** *for a particular waveform is defined as the ratio of the r.m.s. value and the average value. It is denoted by* k_f.

$$k_f \overset{\Delta}{=} \frac{\text{r.m.s. value}}{\text{average value}} \qquad (5.7.1)$$

For a sine wave

$$k_f = \frac{0.707 \times \text{maximum value}}{0.636 \times \text{maximum value}} = 1.10 \qquad (5.7.2)$$

The **peak factor** *for a given waveform is defined as the ratio of the peak (or maximum) value and the r.m.s. value. It is denoted by* k_p.

$$k_p \overset{\Delta}{=} \frac{\text{peak value}}{\text{r.m.s. value}} \qquad (5.7.3)$$

Peak factor is also called the **crest factor** or the **amplitude factor**. For a sine wave,

$$k_p = \frac{\text{maximum value}}{0.707 \times \text{maximum value}} = 1.414 \tag{5.7.4}$$

The form factor and peak factor give an indication of the shape of the waveform. The more pointed the peak of the wave, the higher are the values of these factors.

5.8 PHASOR REPRESENTATION OF SINUSOIDS

A sinusoidal quantity may be represented by a line fixed at one end and rotating counterclockwise at a constant angular velocity ω radians per second. The length of the line is equal to the peak value of the sinusoid. The rotating line is called the **phasor**. Phasors are denoted by bold face upper case letter symbols. Suppose that the phasor $OA = I_m$ is rotated with an angular velocity ω rad/s. After t second it occupies a position OA′ as shown in Fig. 5.3. Let OA take t seconds to rotate through an angle θ then $\theta = \omega t$.

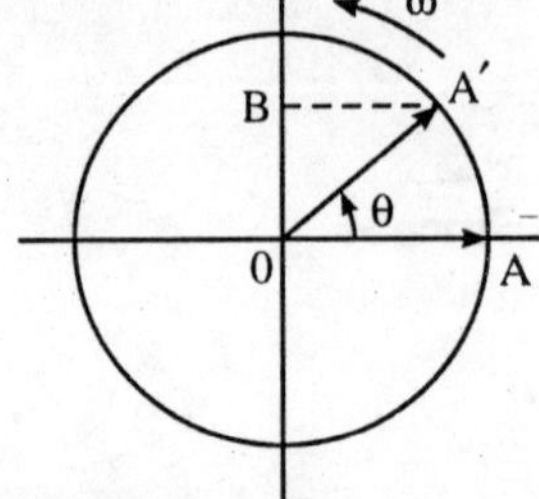

FIG. 5.3 Phasor representation of a sinusoid

The vertical component of OA'

$$= OB = OA' \sin \omega t = I_m \sin \omega t = i$$

But i is the value of current at instant t. Hence the vertical component of the phasor represents to scale the instantaneous value of current.

The **phase** of an alternating quantity at any time t is defined by the angle which the phasor makes with the reference line. The phase angle is taken positive when measured anticlockwise and negative in the clockwise direction. The angular position of the phasor represents a position in time, not space.

A general expression for sinusoidal current is given by

$$i = I_m \sin (\omega t + \phi_0) \tag{5.8.1}$$

Here ϕ_0 is called the *initial phase angle*

Consider the following sinusoids

$$i_1 = I_{m1} \sin \omega t \tag{5.8.2}$$

$$i_2 = I_{m2} \sin (\omega t + \alpha) \tag{5.8.3}$$

$$v_3 = V_{m3} \sin (\omega t - \beta) \tag{5.8.4}$$

$$v_4 = V_{m4} \sin \omega t \tag{5.8.5}$$

These sinusoids may be represented by phasors. Since all the phasors are rotating with the same angular velocity ω, they are all stationary with respect to each other. By convention, the phasors are drawn for the instant $t = 0$. Since $\sin \omega t = 0$ when $t = 0$ the phasor represented by Eq. (5.8.2) is drawn horizontal. The phasor $\mathbf{I}_{m2}$ makes an angle $+\alpha$ with the horizonal and the phasor V_{m4} is also horizonal. *The graphic representation of sinusoidal quantities taken all at the same frequency and with proper phase relationships with respect to each other is called a **phasor diagram**.* The phasor diagram showing the phasors $\mathbf{I}_{m1}$, $\mathbf{I}_{m2}$, $\mathbf{V}_{m3}$ and $\mathbf{V}_{m4}$ is shown in Fig. 5.4.

Voltage and current phasors are drawn on the same diagram. It is not necessary to use the same scale for voltage and current phasors. However, if several current phasors are drawn on the same diagram, they should all be drawn to the same scale. Similarly, all voltage phasors in the same diagram should be drawn to the same scale.

In view of greater importance of r.m.s. values, it is a common practice to draw the phasor diagrams in terms of the r.m.s values rather than maximum values. The shape of the phasor diagram with r.m.s. values is similar to that with maximum values, but to a different scale.

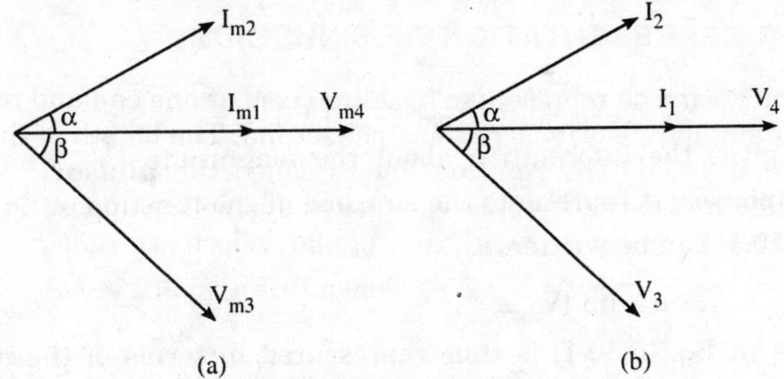

FIG. 5.4 Phasor diagram

Fig. 5.4a shows the phasor diagram in terms of the maximum values and Fig. 5.4b shows the phasor diagram in terms of the r.m.s. values.

The representation of sinusoidal voltages and currents as phasors eliminates time as a variable in circuit equations, but the phase differences between them are retained.

5.9 INPHASE, LAGGING AND LEADNG QUANTITIES

When two alternating quantities have the same phase angle they are said to be *inphase* with each other. Two waveforms are said to be inphase when they begin simultaneously and end simultaneously. They also reach their maximum values at the same time. Similarly, they pass through zero at the same time.

The terms *lead* or *lag* are used to describe the relative positions in time of two sinusoids that are not in phase. The one that is *ahead* in time is said to *lead*, while the one *behind lags*. Fig. 5.4b shows that the phasors $\mathbf{I}_1$ and $\mathbf{V}_4$ are in phase. The phasor $\mathbf{I}_2$ leads $\mathbf{I}_1$ by an angle α and the phasor $\mathbf{V}_3$ lags behind $\mathbf{I}_1$ or $\mathbf{V}_4$ by and angle β.

5.10 SINUSOIDS AND EXPONENTIALS

Consider a sinusoidal voltage

$$v = V_m \sin(\omega t + \phi) \tag{5.10.1}$$

By the Euler's identity

$$e^{j\omega t} = \cos \omega t + j \sin \omega t \tag{5.10.2}$$

Eq. (5.10.2) shows that a complex exponential function can be represented as a complex sum of sinusoids.

Also,

$$V e^{j(\omega t + \phi)} = V_m \cos(\omega t + \phi) + j V_m \sin(\omega t + \phi) \tag{5.10.3}$$

Therefore,
$$v = V_m \sin(\omega t + \phi)$$
$$= Im\,[V_m e^{j(\omega t + \phi)}]$$
$$= Im\,[V_m e^{j\phi} \cdot e^{j\omega t}] \qquad (5.10.4)$$

where "*Im*" means the imaginary part of the quantity. The only term which varies with time is $e^{j\omega t}$. Let us define

$$\mathbf{V}_m \overset{\Delta}{=} V_m e^{j\phi} = |\mathbf{V}_m|\, e^{j\phi} = |\mathbf{V}_m| \underline{/\phi} \qquad (5.10.5)$$

The quantity $\mathbf{V}_m$ gives the information about the magnitude $|\mathbf{V}_m|$ and the phase angle ϕ. $\mathbf{V}_m$ is called the phasor. It represents the sinusoid of the function of Eq. (5.10.1).

Thus, Eq. (5.10.4) can be written as

$$v = \text{Im}\,[\mathbf{V}_m e^{j\omega t}] \qquad (5.10.6)$$

The sinusoid given in Eq. (5.10.1) is thus represented in terms of the phasor $\mathbf{V}_m$ and the complex exponential

We have
$$|e^{j\omega t}| = \sqrt{(\cos^2 \omega t + \sin^2 \omega t)} = 1$$

Thus, the magnitude of $e^{j\omega t}$ is always unity.

At $t = 0$, $\omega t = 0$, $e^{j\omega t} = e^0 = 1 + j0$

At $\omega t = \pi/2$, $e^{j\omega t} = e^{j\pi/2} = 0 + j1$

At $\omega t = \pi$, $e^{j\omega t} = e^{j\pi} = -1 + j0$

At $\omega t = 2\pi$, $e^{j\omega t} = e^{j2\pi} = 1 + j0$

As the angle ωt increases with time, the function represents an operator of unit magnitude which rotates around the origin uniformly with time. Thus, $e^{j\omega t}$ may be called a rotating phasor of unit magnitude, the unit phasor (rotating counter clockwise with angular frequency ω) can be interpreted as the geometrical representation of the exponential $e^{j\omega t}$. It should be noted that the exponential $e^{-j\omega t}$ represents a unit phasor rotating in the clockwise direction. By convention, counter-clockwise direction is taken as the positive direction of rotation and clockwise directions the negative direction. Fig. 5.5 shows the rotating phasors in the complex plane.

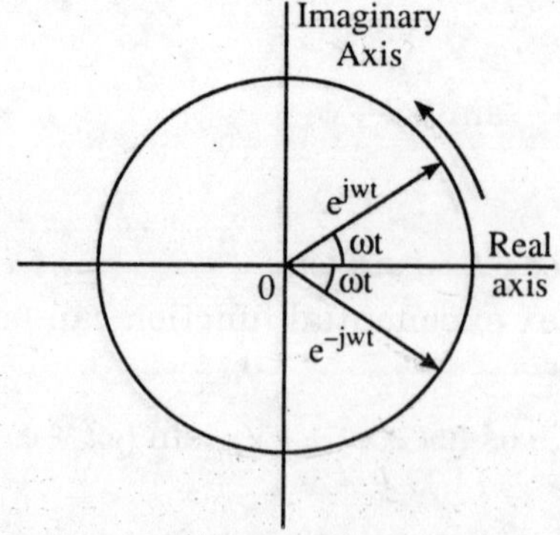

FIG. 5.5 Rotating phasors in the complex plane

Eqs. (5.10.5) and (5.10.6) form the basis for the phasor representation of sinusoids of section 5.8. It is generally easier to work with the exponentials than with the sinusoids. Eq. (5.10.6) is the relationship between the sinusoidal function and the exponential function.

For the purpose of computation with sinusoids represented as complex exponentials the imaginary part designation "*Im*" is dropped and unless derivatives and integrals are involved the exponential $e^{j\omega t}$ is also dropped. Let us prove these statements. We shall find the sum of two sinusoids of the same frequency given by

$$v_1 = V_{m1} \sin (\omega t + \phi_1)$$

and

$$v_2 = V_{m2} \sin (\omega t + \phi_2)$$

Now

$$v_1 = V_{m1} \sin (\omega t + \phi_1) = Im\ [Vm_1\, e^{j(\omega t + \phi_1)}]$$

$$= Im\ [V_{m1}\, e^{j\phi_1} \,.\, e^{j\omega t}] = Im\ [\mathbf{V}_{m1}\, e^{j\omega t}] \tag{5.10.7}$$

where

$$\mathbf{V}_{m1} = V_{m1}\, e^{j\phi_1} \tag{5.10.8}$$

Similarly,

$$v_2 = V_{m2} \sin (\omega t + \phi_2) = Im\ [\mathbf{V}_{m2}\, e^{j\omega t}] \tag{5.10.9}$$

where

$$\mathbf{V}_{m2} = V_{m2}\, e^{j\phi_2} \tag{5.10.10}$$

The algebraic sum of any number of sinusoids of the same angular frequency, say, ω, and of number of their derivatives, of any order is also a sinusoid of the same angular frequency ω.

This theorem can be applied to find the sum of v_1 and v_2. The sum will also be sinusoid of the same angular frequency. It may be written as

$$v = V_m \sin (\omega t + \phi) = Im\ [\mathbf{V}_m\, e^{j\omega t}] \tag{5.10.11}$$

where

$$\mathbf{V}_m = V_m\, e^{j\phi} \tag{5.10.12}$$

$$v = v_1 + v_2 \tag{5.10.13}$$

$$Im\ [\mathbf{V}_m\, e^{j\omega t}] = Im\ [\mathbf{V}_{m1}\, e^{j\omega t}] + Im\ [\mathbf{V}_{m2}\, e^{j\omega t}]$$

$$= Im\ [(\mathbf{V}_{m1} + \mathbf{V}_{m2})\, e^{j\omega t}]$$

The imaginary parts on both sides will be equal if the complex quantities within brackets are equal, Therefore

$$\mathbf{V}_m\, e^{j\omega t} = \mathbf{V}_{m1}\, e^{j\omega t} + \mathbf{V}_{m2}\, e^{j\omega t} \tag{5.10.14}$$

If we cancel $e^{j\omega t}$ from both the sides of Eq. (5.10.14) we get

$$\mathbf{V}_m = \mathbf{V}_{m1} + \mathbf{V}_{m2} \tag{5.10.15}$$

Eq. (5.10.15) shows that the sum of two sinusoids of the same frequency can be found by adding the corresponding phasors.

When **V** is obtained, the corresponding time expression for voltage can be found as follows :

$$v = Im\ [(\mathbf{V}_{mt} + \mathbf{V}_{m2})\, e^{j\omega t}]$$

$$= Im\,[\mathbf{V}_m\, e^{j\omega t}]$$

$$= Im\,[V_m\, e^{j\phi}\, e^{j\omega t}]$$

$$= Im\,[V_m\, e^{j(\omega t + \phi)}]$$

$$= Im\,[V_m \cos(\omega t + \phi) + j\, V_m \sin(\omega t + \phi)]$$

$$\therefore \qquad v = V_m \sin(\omega t + \phi)$$

If we specify a sinusoid by a cosine function rather than by a sine function, we have

$$v = V_m \cos(\omega t + \phi)$$

Then the phasor representation

$$\mathbf{V}_m = V_m\, e^{j\phi} \quad \text{is applicable again.}$$

However the sinusoid should be recovered from the following equation :

$v = Re\,[V_m\, e^{j\phi}]$ where "Re" means the real part of the quantity.

In this text the imaginary part representation is used throughout.

5.11 COMPLEX NUMBER REPRESENTATION OF SINUSOIDS

We have seen that sinusoids can be represented by phasors. Since phasors can be expressed in terms of complex numbers, sinusoids can be expressed in terms of complex numbers.
Consider the voltage phasor **V** shown in Fig. 5.6. It can be expressed in rectangular, polar, exponential, and trigonometric forms as follows :

Rectangular form

$$\mathbf{V} = V_1 + j\, V_2$$

Polar form

$$\mathbf{V} = V\, \underline{/\theta}$$

Exponential form

$$\mathbf{V} = V e^{j\theta}$$

Trigonometric form

$$\mathbf{V} = V \cos\theta + j\, V \sin\theta$$

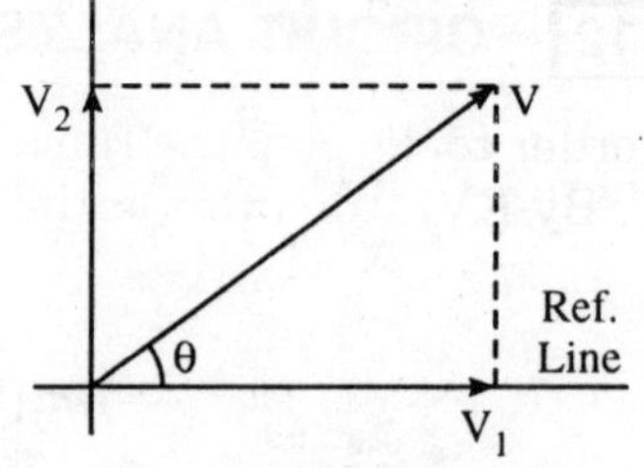

FIG. 5.6 Voltage phasor

The complex number representation of sinusoids simplifies the solution of a.c. circuit problems.

■ **EXAMPLE 5.1** In a parallel circuit with three branches the instantaneous branch currents are represented by

$$i_1 = 10 \sin(\theta - 45^\circ)$$

$$i_2 = 6 \cos\theta$$

$$i_3 = 5 \sin(\theta + 30^\circ)$$

Calculate the total instantaneous current and express the result in the same form.

SOLUTION :

Here $\qquad i_2 = 6 \cos\theta = 6 \sin(\theta + 90^\circ)$

The phasor expressions for the currents are

$$\mathbf{I}_1 = \frac{10}{\sqrt{2}} \underline{/-45^\circ} = \frac{10}{\sqrt{2}} [\cos(-45^\circ) + j\sin(-45^\circ)]$$

$$= 5 - j5$$

$$\mathbf{I}_2 = \frac{6}{\sqrt{2}} \underline{/90^\circ} = \frac{6}{\sqrt{2}} (\cos 90^\circ + j\sin 90^\circ)$$

$$= 0 + j\,4.24$$

$$\mathbf{I}_3 = \frac{5}{\sqrt{2}} \underline{/30^\circ} = \frac{5}{\sqrt{2}} (\cos 30^\circ + j\sin 30^\circ)$$

$$= 3.06 + j\,1.77$$

Resultant phasor current

$$\mathbf{I} = \mathbf{I}_1 + \mathbf{I}_2 + \mathbf{I}_3$$

$$= 5 - j5 + 0 + j\,4.24 + 3.06 + j\,1.77$$

$$= 8.06 + j\,1.01$$

$$= \sqrt{[(8.06)^2 + (1.01)^2]} \;\underline{/\tan^{-1}\frac{1.01}{8.06}} = 8.12 \underline{/7.14^\circ}$$

In terms of time, the resultant current is

$$i = (8.12 \times \sqrt{2}) \sin(\theta + 7.14^\circ)$$

$$i = 11.5 \sin(\theta + 7.14^\circ)$$

5.12 CIRCUIT ANALYSIS BY PHASOR TRANSFORMS

In order to develop the concept of phasor transform let us consider a series RLC circuit of Fig. 5.7. By KVL the differential equation of the circuit can be written as

$$v = Ri + L\frac{di}{dt} + \frac{1}{C}\int i\,dt \qquad (5.12.1)$$

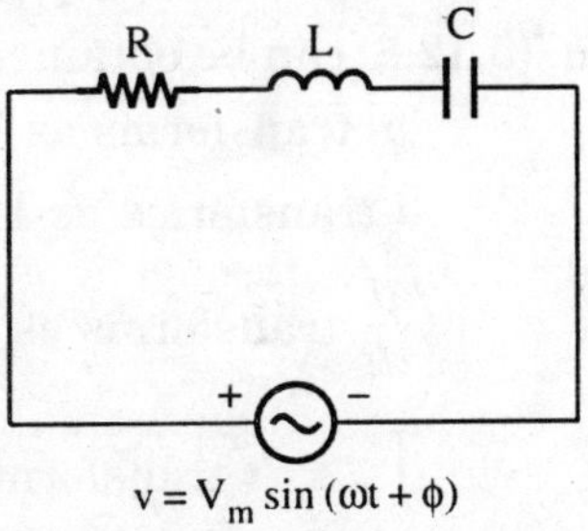

This equation is the time-domain equation because it contains functions of time. Let us define the phasors $\mathbf{V}_m$ and $\mathbf{I}_m$ as follows :

$$v \stackrel{\Delta}{=} Im\,[\mathbf{V}_m\, e^{j\omega t}] \qquad (5.12.2)$$

$$v \stackrel{\Delta}{=} Im\,[\mathbf{I}_m\, e^{j\omega t}] \qquad (5.12.3)$$

FIG. 5.7 Series RLC circuit

where '*Im*' means the imaginary part of the quantity. Eq. (5.12.1) can now be written as

$$Im\,[\mathbf{V}_m\, e^{j\omega t}] = R\, Im\,[\mathbf{I}_m\, e^{j\omega t}]$$

$$+ L\frac{d}{dt}[Im\,(\mathbf{I}_m\, e^{j\omega t})] + \frac{1}{C}\int Im\,[\mathbf{I}_m\, e^{j\omega t}]\,dt \qquad (5.12.4)$$

Since $\mathbf{V}_m$ and $\mathbf{I}_m$ are not functions of time, they are treated as constants in the differenciation and integration operations. Thus Eq. (5.12.4) becomes

$$Im\ [\mathbf{V}_m\, e^{j\omega t}] = Im\ [R\,\mathbf{I}_m\, e^{j\omega t}] + Im\ [j\omega L\,\mathbf{I}_m\, e^{j\omega t}] + Im\left[\frac{1}{j\omega C}\mathbf{I}_m\, e^{j\omega t}\right]$$

$$= Im\left[(R\,\mathbf{I}_m + j\omega\, L\,\mathbf{I}_m + \frac{1}{j\omega\, C}\mathbf{I}_m)\, e^{j\omega t}\right] \qquad (5.12.5)$$

The *Im* (imaginary) components on both sides of Eq. (5.12.5) will be equal if the complex quantities themselves are equal. Therefore

$$\mathbf{V}_m\, e^{j\omega t} = (R\mathbf{I}_m + j\omega L\,\mathbf{I}_m + \frac{1}{j\omega\, C}\mathbf{I}_m)\, e^{j\omega t} \qquad (5.12.6)$$

Dividing both sides of Eq. (5.12.6) by $e^{j\omega t}$ we obtain

$$\mathbf{V}_m = R\,\mathbf{I}_m + j\omega L\,\mathbf{I}_m + \frac{1}{j\omega\, C}\mathbf{I}_m$$

$$\mathbf{V}_m = \left(R + j\omega L + \frac{1}{j\omega\, L}\right)\mathbf{I}_m \qquad (5.12.7)$$

$$\mathbf{V} = \left(R + j\omega L + \frac{1}{j\omega C}\right)\mathbf{I} \qquad (5.12.8)$$

where **V** and **I** are r.m.s phasor voltage and current respectively. Eq. (5.12.8) is independent of time functions and involves constants only. It is an algebraic equation, that is, it does not contain any derivatives. It is the **frequency-domain equation** because it contains functions of radian frequency ω. We speak of Eq. (5.12.8) as being the **phasor transform** of Eq. (5.12.1) and we say that Eq. (5.12.1) in the time domain is transformed to Eq. (5.12.8) in the frequency domain. The inherent advantage of the phasor transform is that differential equation in the time domain is transformed into an algebraic equation in the frequency domain. It is clear that algebraic equations are much easier to solve than differential equations.

5.13 RULES FOR PHASOR TRANSFORMS

Eq. (5.12.8) can be obtained directly from Eq. (5.12.1) by noting the following transformations:

v transforms as **V** $\qquad (v \Rightarrow \mathbf{V}) \qquad$ (5.13.1)

i transforms as **I** $\qquad (i \Rightarrow \mathbf{I}) \qquad$ (5.13.2)

$\dfrac{d}{dt}$ transforms as $j\omega$ $\qquad \left(\dfrac{d}{dt} \Rightarrow j\omega\right) \qquad$ (5.13.3)

$\int \ldots dt$ transforms as $\dfrac{1}{j\omega}$ $\qquad \left(\int dt \Rightarrow \dfrac{1}{j\omega}\right) \qquad$ (5.13.4)

Further it can be proved that

$\dfrac{d^n i}{d t^n}$ transforms as $(j\omega)^n\,\mathbf{I}$ $\qquad \left[\dfrac{d^n i}{d t^n} \Rightarrow (j\omega)^n\,\mathbf{I}\right] \qquad$ (5.13.5)

and $\sin(\omega t + \phi)$ transforms as $e^{j\phi}$ $\qquad [\sin(\omega t + \phi) \Rightarrow e^{j\phi}] \qquad$ (5.13.6)

We can write these results in more convenient forms as

$$\wp\,[f] = \mathbf{F} \qquad (5.13.7)$$

$$\wp\,[af] = a\mathbf{F} \qquad (5.13.8)$$

$$\wp\left[\frac{d^n f}{d\,t^n}\right] = (j\omega)^n \tag{5.13.9}$$

$$\wp\,[\sin(\omega t + \phi)] = e^{j\phi} \tag{5.13.10}$$

where the script $\wp$ means **"the phasor transform of "**. Eq. (5.13.7) is the basic definition of the phasor transform. Eq. (5.13.8) states that the transform of a constant times a function is equal to the constant times the transform of the function. Eq. (5.13.9) shows a relation for the transform of a derivative. Eq. (5.13.10) gives a relation for the transform of a sinusoid.

■ **EXAMPLE 5.2.** Apply the method of phasor transform to the differential equation of the circuit of Fig. 5.8.

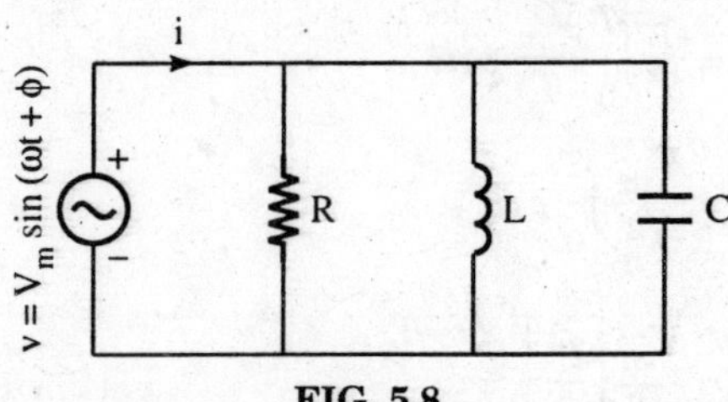

FIG. 5.8

SOLUTION : By KVL, the circuit differential equation is

$$i = \frac{1}{R}v + C\frac{dv}{dt} + \frac{1}{L}\int v\,dt \tag{E 5.2.1}$$

By the rule of phasor transforms

$$i \Rightarrow \mathbf{I},\; v \Rightarrow \mathbf{V},\; \frac{d}{dt} \Rightarrow j\omega,\; \int dt \Rightarrow \frac{1}{j\omega}$$

Therefore Eq. (E 5.2.1) can be written as

$$\mathbf{I} = \frac{1}{R}\mathbf{V} + j\omega\,C\mathbf{V} + \frac{1}{j\,\omega L}\mathbf{V} \tag{E 5.2.2}$$

Eq. (E 5.2.2) is the desired transform equation.

■ **EXAMPLE 5.3** Calculate the average and rms values of a triangular wave in which voltage rises uniformly from zero to V_m and completes the cycle by falling instantly back to zero.

SOLUTION :

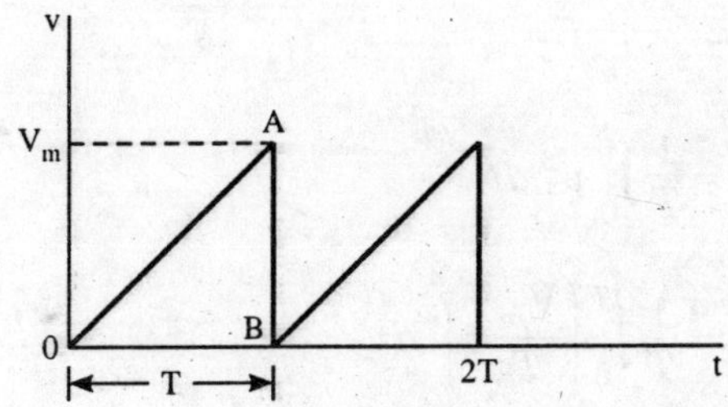

FIG. 5.9. Triangular wave.

The triangular wave is shown in Fig. 5.9.

Here T is the time required to complete one cycle. The voltage v at time t is given by

$$\frac{v}{t} = \frac{V_m}{T}$$

$$v = \frac{V_m}{T} t$$

The average value is given by

$$V_{av} = \frac{1}{T}\int_0^T v\,dt$$

$$= \frac{1}{T}\int_0^T \frac{V_m}{T} t\,dt$$

$$= \frac{V_m}{T^2}\int_0^T t\,dt$$

$$= \frac{V_m}{T^2}\left[\frac{t^2}{2}\right]_0^T$$

$$= \frac{V_m}{t^2} \cdot \frac{T^2}{2} = \frac{V_m}{2}$$

therefore, $$V_{av} = \frac{V_m}{2}$$

Alternatively

$$V_{av} = \frac{\text{area under one cycle}}{\text{base}}$$

$$= \frac{\text{area of triangle OAB}}{\text{OB}}$$

$$= \frac{1}{OB} \cdot \left(\frac{1}{2} V_m \times OB\right)$$

therefore, $$V_{av} = \frac{V_m}{2}$$

RMS value

$$V_{rms}^2 = \frac{1}{T}\int_0^T v^2\,dt$$

$$= \frac{1}{T}\int_0^T \left(\frac{V_m}{T} t\right)^2 dt$$

$$= \frac{V_m^2}{T^3}\left[\frac{t^3}{3}\right]_0^T$$

$$= \frac{V_m^2}{T^3}\left(\frac{T^3}{3} - 0\right)$$

$$= \frac{V_m^2}{3}$$

therefore, $$V_{rms} = \frac{V_m}{\sqrt{3}}$$

■ **EXAMPLE 5.4** Find the form factor and peak factor of the waveform of Example 5.3.

SOLUTION :

$$\text{Form factor} = \frac{\text{rms value}}{\text{average value}}$$

$$= \frac{V_m/\sqrt{3}}{V_m/2}$$

therefore, $$\text{form factor} = 1{\cdot}1547$$

$$\text{Peak factor} = \frac{\text{peak value}}{\text{rms value}}$$

$$= \frac{V_m}{(V_m/\sqrt{3})}$$

$$= 1{\cdot}732$$

■ **EXAMPLE 5.5** A full rectified sine wave is clipped as shown in Fig. 5.10. Find the average and rms values of the waveform.

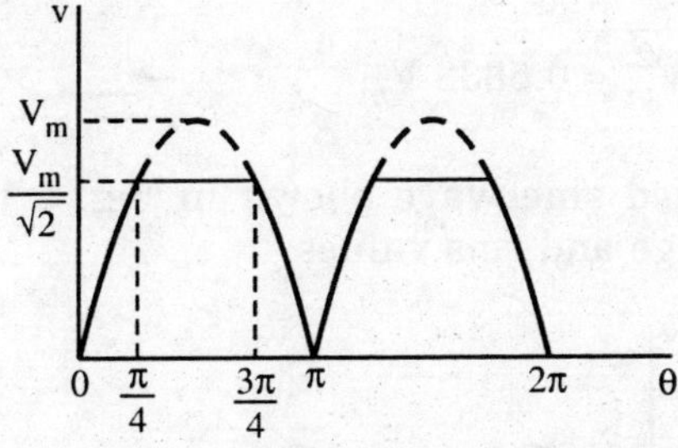

FIG. 5.10

SOLUTION :

$$\text{Average value} = \frac{1}{\pi}\int_0^{\pi} v \; d\theta$$

$$V_{av} = \frac{1}{\pi}\int_0^{\pi/4} V_m \sin\theta\, d\theta + \frac{1}{\pi}\int_{\pi/4}^{3\pi/4} \frac{V_m}{\sqrt{2}}\, d\theta + \frac{1}{\pi}\int_{3\pi/4}^{\pi} V_m \sin\theta\, d\theta$$

$$= \frac{V_m}{\pi}\Big[-\cos\theta\Big]_0^{\pi/4} + \frac{V_m}{\pi\sqrt{2}}\Big[\theta\Big]_{\pi/4}^{3\pi/4} + \frac{V_m}{\pi}\Big[-\cos\theta\Big]_{3\pi/4}^{\pi}$$

$$= -\frac{V_m}{\pi}\left(\cos\frac{\pi}{4} - \cos 0\right) + \frac{V_m}{\pi\sqrt{2}}\left(\frac{3\pi}{4} - \frac{\pi}{4}\right) - \frac{V_m}{\pi}\left(\cos\pi - \cos\frac{3\pi}{4}\right)$$

$$\therefore \quad V_{av} = 0.54\ V_m$$

Mean square value of voltage

$$V^2 = \frac{1}{\pi}\int_0^{\pi} v^2\, d\theta$$

$$= \frac{1}{\pi}\left[2 \times \int_0^{\pi/4} V_m^2 \sin^2\theta\, d\theta + \int_{\pi/4}^{3\pi/4}\left(\frac{V_m}{\sqrt{2}}\right)^2 d\theta\right]$$

$$= \frac{V_m^2}{\pi}\left[\int_0^{\pi/4}(1 - \cos 2\theta)\, d\theta\right] + \frac{V_m^2}{2\pi}\Big[\theta\Big]_{\pi/4}^{3\pi/4}$$

$$= \frac{V_m^2}{\pi}\left[\theta - \frac{\sin 2\theta}{2}\right]_0^{\pi/4} + \frac{V_m^2}{2\pi}\left(\frac{3\pi}{4} - \frac{\pi}{4}\right)$$

$$= \frac{V_m^2}{4}\left[\left(\frac{\pi}{4} - 0\right) - \frac{1}{2}\left(\sin\frac{\pi}{2} - 0\right)\right] + \frac{V_m^2}{4}$$

$$= \frac{V_m^2}{4} - \frac{V_m^2}{2\pi} + \frac{V_m^2}{4} = \frac{V_m^2}{2}\left(1 - \frac{1}{\pi}\right)$$

$$= 0.30845\ V_m^2$$

therefore, the rms value of voltage

$$V = \sqrt{0.30845\ V_m^2} = 0.5838\ V_m$$

■ **EXAMPLE 5.6** The full rectified sine wave shown in Fig. 5.11 has a delay angle of $\pi/4$ radians. Calculate the average and rms values.

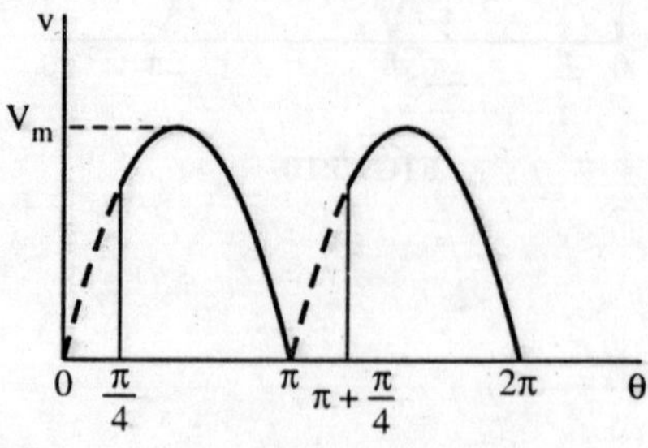

FIG. 5.11

SOLUTION : $v = 0$ from 0 to $\pi/4$

$v = V_m \sin\theta$ from $\pi/4$ to π

Area under half cycle

$$= \int_{\pi/4}^{\pi} V_m \sin\theta \, d\theta = V_m \Big[-\cos\theta \Big]_{\pi/4}^{\pi}$$

$$= -V_m \left(\cos\pi - \cos\frac{\pi}{4} \right)$$

$$= 1.707 \text{ V}_m$$

Average value of voltage

$$V_{av} = \frac{\text{area under half cycle}}{\text{base}} = \frac{1.707 V_m}{\pi} = 0.5434 V_m$$

Mean square value of voltage

$$V^2 = \frac{1}{\pi} \int_{\pi/4}^{\pi} V_m^2 \sin^2\theta \, d\theta$$

$$= \frac{V_m^2}{2\pi} \int_{\pi/4}^{\pi} (1 - \cos 2\theta) \, d\theta$$

$$= \frac{V_m^2}{2\pi} \left[\theta - \frac{\sin 2\theta}{2} \right]_{\pi/4}^{\pi}$$

$$= \frac{V_m^2}{2\pi} \left[\left(\pi - \frac{\pi}{4} \right) - \frac{1}{2} \left(\sin 2\pi - \sin\frac{\pi}{2} \right) \right]$$

$$= \frac{V_m^2}{2\pi} \left(\frac{3\pi}{4} + \frac{1}{2} \right) = 0.45457 \, V_m^2$$

Therefore the root mean square value of the voltage

$$V = \sqrt{0.45457 \, V_m^2} = 0.6742 \, \mathbf{V}_m$$

■ **EXAMPLE 5.7** Determine the average and rms values of the trapezoidal waveform shown in Fig. 5.12.

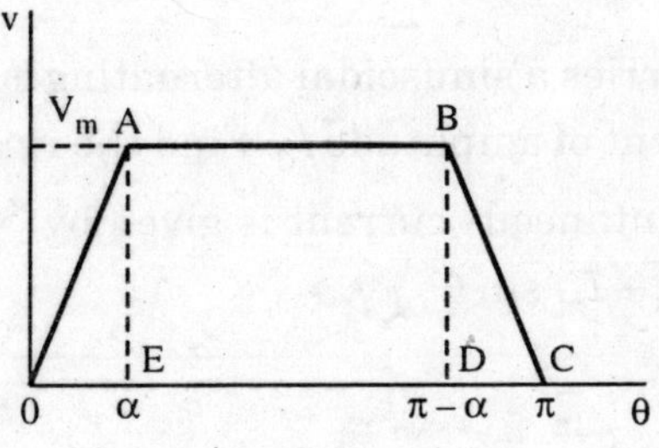

FIG. 5.12

SOLUTION : The given waveform shows that the voltage varies with θ as follows :

$$v = \frac{V_m}{\alpha}\theta \qquad \text{from } \theta = 0 \text{ to } \theta = \alpha$$

$$v = V_m \qquad \text{from } \theta = \alpha \text{ to } \theta = \pi - \alpha$$

$$v = \frac{V_m}{\alpha}(\pi - \theta) \qquad \text{from } \theta = \pi - \alpha \text{ to } \theta = \pi$$

Area over half cycle

$$= \text{area } OAE + \text{area } ABDE + \text{area } BCD$$

$$= \frac{1}{2}\alpha V_m + (\pi - 2\alpha) V_m + \frac{1}{2}\alpha V_m$$

Average value over half cycle

$$= \frac{\text{area over half cycle}}{\text{base}} = \frac{(\pi - \alpha) V_m}{\pi}$$

therefore, $$V_{av} = \left(1 - \frac{\alpha}{\pi}\right) V_m$$

Mean square value of voltage

$$V^2 = \frac{1}{\pi}\int_0^{\pi} v^2\, d\theta$$

$$= \frac{1}{\pi}\left[\int_0^{\alpha}\left(\frac{V_m}{\alpha}\theta\right)^2 d\theta + \int_{\alpha}^{\pi-\alpha} V_m^2\, d\theta + \int_{\pi-\alpha}^{\pi}\frac{V_m^2}{\alpha^2}(\pi - \theta)^2\, d\theta\right]$$

$$\pi V^2 = \frac{V_m^2}{\alpha^2}\left[\frac{\theta^3}{3}\right]_0^{\pi} + V_m^2\left[\theta\right]_{\alpha}^{\pi-\alpha} - \frac{V_m^2}{\alpha^2}\left[\frac{(\pi - \theta)^3}{3}\right]_{\pi - x}^{\pi}$$

$$= \frac{V_m^2}{3\alpha^2}(\alpha^3 - 0) + V_m^2(\pi - \alpha - \alpha) - \frac{V_m^2}{\alpha^2}\left(-\frac{\alpha^3}{3}\right)$$

$$= V_m^2\left(\pi - \frac{4\alpha}{3}\right)$$

therefore, $$V = V_m\sqrt{\left(1 - \frac{4\alpha}{3\pi}\right)}$$

■ **EXAMPLE 5.8** A conductor carries a sinusoidal alternating current of peak value I_m amperes superposed on a direct current of amplitude I_0. Find the rms value of the resultant current.

SOLUTION : The resultant instantaneous current is given by

$$i = I_0 + I_m \sin\theta$$

Mean square value of current

$$I^2 = \frac{1}{2\pi}\int_0^{2\pi} i^2\, d\theta$$

$$= \frac{1}{2\pi}\int_0^{2\pi} (I_0 + I_m \sin\theta)^2\, d\theta$$

$$= \frac{1}{2\pi}\int_0^{2\pi} (I_0^2 + I_m^2 \sin^2\theta + 2I_0 I_m \sin\theta)\, d\theta$$

$$= \frac{I_0^2}{2\pi}\int_0^{2\pi} 1\, d\theta + \frac{I_m^2}{2\pi}\int_0^{2\pi} \sin^2\theta\, d\theta + \frac{2I_0 I_m}{2\pi}\int_0^{2\pi} \sin\theta\, d\theta$$

$$= \frac{I_0^2}{2\pi}\Big[\theta\Big]_0^{2\pi} + \frac{I_m^2}{2\pi}\int_0^{2\pi}\left(\frac{1-\cos 2\theta}{2}\right) d\theta + \frac{2I_0 I_m}{2\pi}\Big[-\cos\theta\Big]_0^{2\pi}$$

$$= \frac{I_0^2}{2\pi}(2\pi - 0) + \frac{I_m^2}{4\pi}\Big[\theta\Big]_0^{2\pi} - \frac{I_m^2}{4\pi}\left[\frac{\sin 2\theta}{2}\right]_0^{2\pi} - \frac{2I_0 I_m}{4\pi}(\cos 4\pi - \cos 0°)$$

$$= I_0^2 + \frac{I_m^2}{4\pi}(2\pi - 0) - \frac{I_m^2}{8\pi}(\sin 4\pi - \sin 0°) - \frac{2 I_0 I_m}{4\pi}(1 - 1)$$

$$= I_0^2 + \frac{I_m^2}{2} - \frac{I_m^2}{8\pi}(0 - 0) - 0$$

$$= I_0^2 + \frac{I_m^2}{2}$$

Therefore the rms value of resultant current

$$I_0 = \sqrt{I_0^2 + \frac{I_m^2}{2}}\ A$$

■ **EXAMPLE 5.9** Find the rms value of a resultant current in a wire which carries simultaneously a direct current of $5A$ and a sinusoidal alternating current with a peak value of $5A$.

SOLUTION : The rms value of *ac* with peak value of $5A$ is $\frac{5}{\sqrt{2}} A$.

Let I be the rms value of resultant current flowing through a resistance R for t seconds.

Energy consumed due to resultant current I = energy consumed due to *dc* + energy consumed due to *ac* .

$$I^2 Rt = 5^2 Rt + \left(\frac{5}{\sqrt{2}}\right)^2 Rt$$

$$I^2 = 25 + \frac{25}{2} = 37{\cdot}5$$

therefore,

$$I = \sqrt{37{\cdot}5} = 6{\cdot}12\ \text{A}$$

EXERCISES

Find the average and root mean square values of the waveforms shown in Figs P 5.1a to P 5.1f.

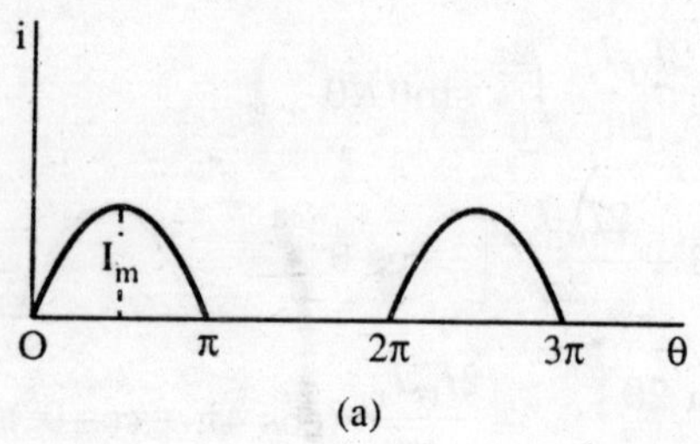

(a)

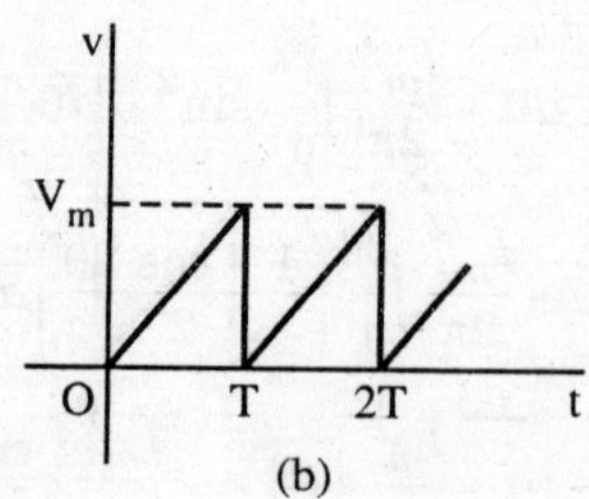

(b)

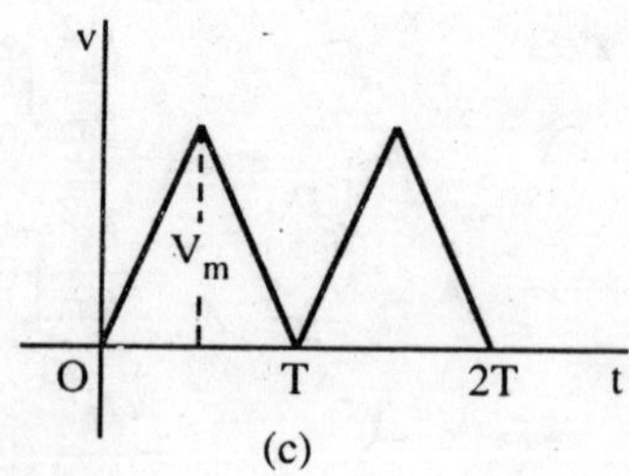

(c)

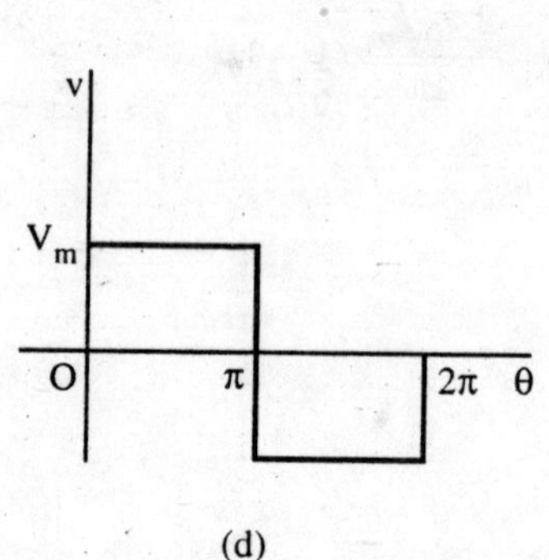

(d)

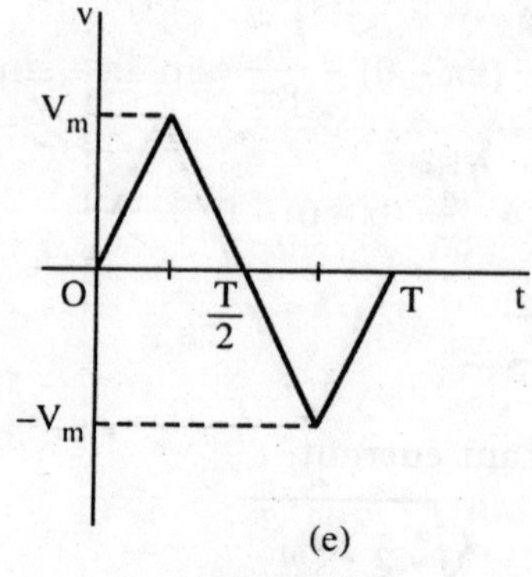

(e)

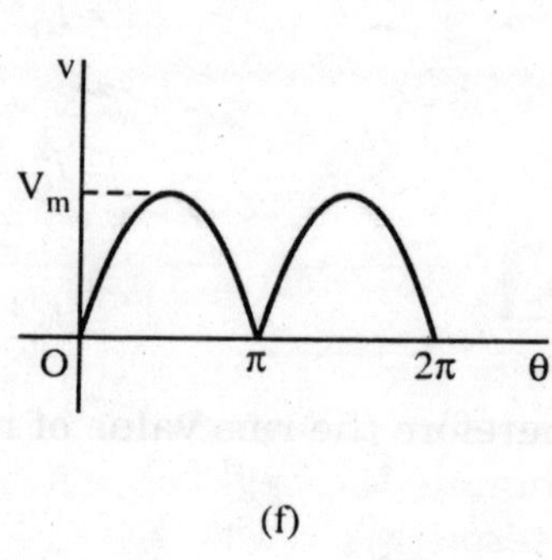

(f)

FIG. P 5.1

ANSWERS

1. (a) $\frac{I_m}{\pi}, \frac{I_m}{2}$, (b) $\frac{V_m}{2}, \frac{V_m}{\sqrt{3}}$, (c) $\frac{V_m}{2}, \frac{V_m}{\sqrt{3}}$,

   (d) V_m, V_m, (e) $0, \frac{V_m}{\sqrt{3}}$, (f) $0.637\ V_m$, $0.707\ V_m$

6

SINGLE-PHASE SERIES A.C. CIRCUITS

6.1 INTRODUCTION

Circuits in which currents and voltages vary sinusoidally are called **alternating current circuits or a.c. circuits.** In this chapter we shall consider a.c. circuits that are made up of combinations of resistance R, inductance L and capacitance C. The parameters R, L and C are two-terminal circuit elements. In order to study a general a.c. circuit, it is necessary to consider the effects of R, L and C separately. We are insterested in determining the relationships between phasors **V** and **I** for these elements.

The ratio of the voltage phasor to the current phasor is defined as **impedance Z** *in the frequency domain.*

$$\text{Impedance} \overset{\Delta}{=} \frac{\text{voltage phasor}}{\text{current phasor}}$$

$$\boxed{\mathbf{Z} \overset{\Delta}{=} \frac{\mathbf{V}}{\mathbf{I}} \text{ ohms}} \qquad (6.1.1)$$

where **V** and **I** are the r.m.s. values of voltage and current respectively.

6.2 PURELY RESISTIVE CIRCUIT

Fig. 6.1 a shows a circuit containing only a resistance R connected across a sinusoidal voltage given by

$$v_R = V_{Rm} \sin \omega t \qquad (6.2.1)$$

By Ohm's law, the instantaneous current in the circuit is given by

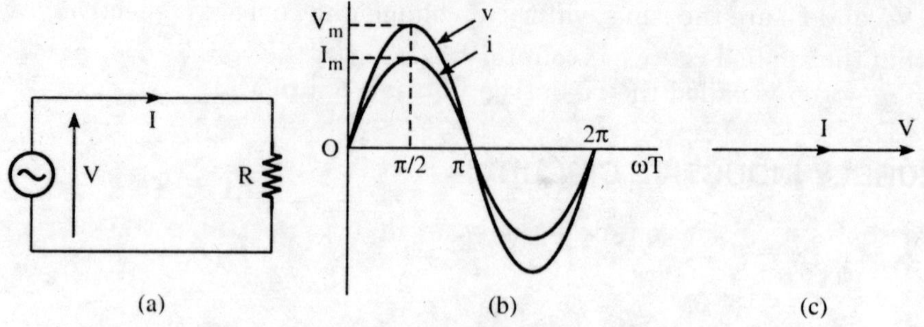

Fig. 6.1. (a) circuit diagram ; (b) waveforms ; (c) phasor diagram

$$i_R = \frac{v_R}{R} = \frac{V_{Rm}}{R} \sin \omega t$$

If $$\frac{V_{Rm}}{R} \stackrel{\Delta}{=} I_{Rm} \tag{6.2.2}$$

then $$i_R = I_{Rm} \sin \omega t \tag{6.2.3}$$

Comparison of the voltage relation (6.2.1) and the current relation (6.2.3) shows that the phase difference ϕ between the voltage and current is zero. Hence, *in a circuit with resistance only the voltage and current are in the same phase.*

From Eq. (6.2.1.), the voltage phasor $\mathbf{V}_R$ can be written as

$$\mathbf{V}_R = V_R \underline{/ 0^\circ} \tag{6.2.4}$$

where V_R = r.m.s. value of voltage $= \frac{1}{\sqrt{2}} \times$ maximum value of voltage $= \frac{V_{Rm}}{\sqrt{2}}$

The current phasor $\mathbf{I}_R$ can be written from Eq. (6.2.3) as

$$\mathbf{I}_R = I_R \underline{/ 0^\circ} = \frac{V_R}{R} \underline{/ 0^\circ} \tag{6.2.5}$$

where I_R = r.m.s. value of current $= \frac{1}{\sqrt{2}} \times$ maximum value of current $= \frac{I_{Rm}}{\sqrt{2}}$

The complex value of circuit impedance is

$$\mathbf{Z}_R = \frac{\mathbf{V}_R}{\mathbf{I}_R} = \frac{V_R \underline{/ 0^\circ}}{I_R \underline{/ 0^\circ}} = \frac{R\, I_R \underline{/ 0^\circ}}{I_R \underline{/ 0^\circ}} = R \underline{/ 0^\circ} \tag{6.2.6}$$

therefore, $$\mathbf{Z}_R = R + j0$$

Thus, the impedance of a resistor in complex notation is R.

Figs. 6.1b and 6.1c show the waveforms and phasor diagram respectively of the voltage and current in a circuit containing only a resistance.

From Eq. (6.2.6)

$$\boxed{\mathbf{V}_R = R\, \mathbf{I}_R} \tag{6.2.8}$$

Eq. (6.2.8) is called **Ohm's law for a resistor in phasor form.**

Here $\mathbf{V}_R$ and $\mathbf{I}_R$ are the r.m.s. values of voltage and current respectively. From Fig. 6.1.*a* it is noted that the applied voltage is counter-balanced by the voltage drop across the resistance R. This voltage drop is called the **resistive voltage drop** ($\mathbf{V}_R$).

6.3 PURELY INDUCTIVE CIRCUIT

Figure 6.2a shows a circuit containing only an inductance L. Let the current taken by the circuit be

$$i_L = I_{Lm} \sin \omega t \tag{6.3.1}$$

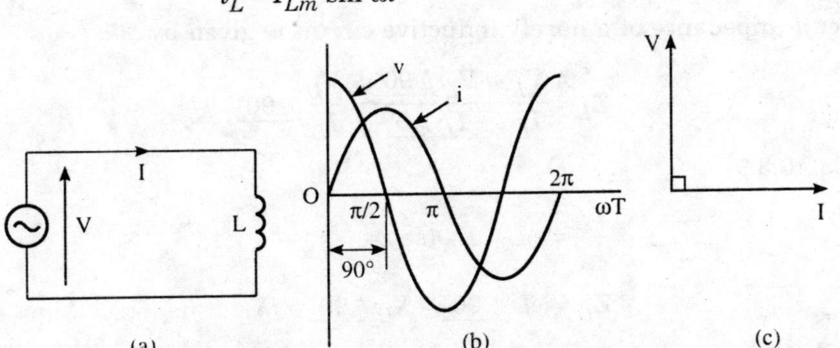

FIG. 6.2 Purely inductive circuit
(a) circuit diagram ; (b) waveforms ; (c) phasor diagram

This current produces a self induced e.m.f., e_L given by

$$e_L = -L \frac{di_L}{dt} \tag{6.3.2}$$

At any instant two voltages are present in the circuit, one is the applied voltage v_L and the other is the induced voltage e_L. By KVL

$$v_L + e_L = 0 \tag{6.3.3}$$

$$v_L = -e_L = L \frac{di_L}{dt} = L \frac{d}{dt}(I_{Lm} \sin \omega t)$$

$$= \omega L\, I_{Lm} \cos \omega t = \omega L\, I_{Lm} \sin(\omega t + 90^\circ)$$

If
$$\omega L\, I_{Lm} \overset{\Delta}{=} V_{Lm} \tag{6.3.4}$$

then
$$v_L = V_{Lm} \sin(\omega t + 90^\circ) \tag{6.3.5}$$

The current phasor $\mathbf{I}_L$ can be written from Eq. (6.3.2) as

$$\mathbf{I}_L = I_L \underline{/ 0^\circ} \tag{6.3.6}$$

where
$$I_L = \frac{I_{Lm}}{\sqrt{2}} \tag{6.3.7}$$

The voltage phasor $\mathbf{V}_L$ can be written from Eq. (6.3.5) as

$$\mathbf{V}_L = V_L \angle 90^\circ \tag{6.3.8}$$

where $$V_L = \frac{V_{Lm}}{\sqrt{2}} = \omega L \frac{I_{Lm}}{\sqrt{2}} = \omega L\, I_L \tag{6.3.9}$$

Comparison of Eqs. (6.3.6) and (6.3.8) shows that *for a purely inductive circuit the voltage leads the current by 90° or equivalently, the current lags behind the voltage by 90°.*

This is shown in Fig. 6.2c. The waveforms are shown in Fig. 6.2b.

If ϕ is the phase difference between $\mathbf{V}_L$ and $\mathbf{I}_L$, then $\phi = 90^\circ$ or $\frac{\pi}{2}$ radians.

The circuit impedance of a purely inductive circuit is given by

$$\mathbf{Z}_L \overset{\Delta}{=} \frac{\mathbf{V}_L}{\mathbf{I}_L} = \frac{V_L \angle 90^\circ}{I_L \angle 0^\circ} = \frac{V_L}{I_L} \angle 90^\circ \tag{6.3.10}$$

From Eq. (6.3.9)

$$\frac{V_L}{I_L} = \omega L = X_L \text{ (say)} \tag{6.3.11}$$

therefore, $$\mathbf{Z}_L = \omega L \angle 90^\circ = X_L \angle 90^\circ = jX_L \tag{6.3.12}$$

The quantity X_L (= ωL) is called the **inductive reactance** of the circuit. It is measured in ohms.

therefore, $$\boxed{X_L = \omega L = 2\pi fL \quad \Omega} \tag{6.3.13}$$

and $$\boxed{\mathbf{Z}_L = jX_L = j\omega L = j2\pi fL \quad \Omega} \tag{6.3.14}$$

where L is in henrys.

From Eqs. (6.3.10) and (6.3.14)

$$\boxed{\mathbf{V}_L = \mathbf{Z}_L\, \mathbf{I}_L = jX_L\, \mathbf{I}_L} \tag{6.3.15}$$

Eq. (6.3.15) is called the **Ohm's law for a purely inductive circuit.**

The voltage drop across the inductance L is called the inductive voltage drop $\mathbf{V_L}$.

6.4 PURELY CAPACITIVE CIRCUIT

Figure 6.3a shows a purely capacitive circuit containing only a capacitor C connected to an a.c. supply voltage given by

$$v_C = V_{Cm} \sin \omega t \tag{6.4.1}$$

The current through the circuit at any instant is

$$i_C = \frac{dq}{dt} = \frac{d}{dt}(v_C C) = C\frac{dv_C}{dt}$$

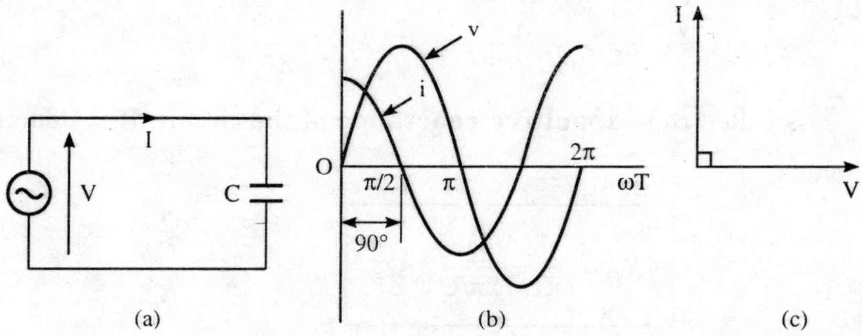

FIG. 6.3 Purely capacitive circuit

$$= C\frac{d}{dt}(V_{Cm}\sin\omega t) = \omega C\,V_{Cm}\cos\omega t$$

or
$$i_C = \omega C\,V_{Cm}\sin(\omega t + 90^\circ) \tag{6.4.2}$$

If
$$\omega C V_{Cm} \overset{\Delta}{=} I_{Cm} \tag{6.4.3}$$

then
$$i_C = I_{Cm}\sin(\omega t + 90^\circ) \tag{6.4.4}$$

The current phasor $\mathbf{I}_C$ can be written from Eq. (6.4.4) as

$$\mathbf{I}_C = I_C \underline{/\ 90^\circ} \tag{6.4.5}$$

where
$$I_C = \frac{I_{Cm}}{\sqrt{2}} \tag{6.4.6}$$

The voltage phasor $\mathbf{V}_C$ can be written from Eq. (6.4.1) as

$$\mathbf{V}_C = V_C \underline{/\ 0^\circ} \tag{6.4.7}$$

From Eq. (6.4.3)

$$\omega C\,\frac{V_{Cm}}{\sqrt{2}} = \frac{I_{Cm}}{\sqrt{2}}$$

$$\omega C\,V_C = I_C \tag{6.4.8}$$

Combination of Eqs. (6.4.5) and (6.4.8) gives

$$\mathbf{I}_C = \omega C\,V_C \underline{/\ 90^\circ} \tag{6.4.9}$$

Comparison of Eqs. (6.4.5) and (6.4.7) shows that *for a purely capacitive circuit current leads the voltage by 90° or equivalently, the voltage lags behind the current by 90°*. Figure 6.3b and 6.3c show the waveforms and phasor diagram respectively for a circuit containing capacitance alone.

If ϕ is the phase difference between $\mathbf{V}_C$ and $\mathbf{I}_C$, then $\phi = 90^\circ$ or $\frac{\pi}{2}$ radians.

The circuit impedance of a purely capacitive circuit is given by

$$\mathbf{Z}_C \overset{\Delta}{=} \frac{\mathbf{V}_C}{\mathbf{I}_C} = \frac{V_C \underline{/\ 0^\circ}}{I_C \underline{/\ 90^\circ}} = \frac{V_C \underline{/\ 0^\circ}}{\omega C\,V_C \underline{/\ 90^\circ}} \tag{6.4.10}$$

$$\therefore \qquad \mathbf{Z}_C = \frac{1}{\omega C \angle 90^\circ} = \frac{1}{j\omega C} = -j\frac{1}{\omega C}$$

The quantity $\frac{1}{\omega C}$ is called the **capacitive reactance** of the circuit. It is denoted by X_C and is measured in ohms.

$$\therefore \qquad \boxed{X_C = \frac{1}{\omega C} = \frac{1}{2\pi f C} \quad \Omega} \tag{6.4.11}$$

$$\text{and} \qquad \boxed{\mathbf{Z}_C = -jX_C = -j\frac{1}{\omega C} \quad \Omega} \tag{6.4.12}$$

where C is in farads (F) and f in Hz. From Eqs. (6.4.10) and (6.4.12)

$$\boxed{\mathbf{V}_C = \mathbf{Z}_C \, \mathbf{I}_C = -jX_C \, I_C} \tag{6.4.13}$$

Eq. (6.4.13) is called the **Ohm's law for a purely capacitive circuit.**

The part of the supply voltage which charges the capacitor C is called the **capacitive voltage drop $\mathbf{V}_C$**.

$$V_C = X_C \, I_C \tag{6.4.14}$$

6.5 PHASE RELATIONSHIP BETWEEN V AND I IN L AND C

In an inductance, the current *lags* the voltage by 90°, and in a capacitance the current *leads* the voltage by 90°. These relationships are very important in drawing phasor diagrams. it is very convenient to remember these relationships with the help of the word **CIVIL** as follows:

In C, **I** leads **V**

C I V I L

V leads **I** in L

6.6 CIRCUIT LAWS IN THE FREQUENCY DOMAIN

Ohm's law, KCL and KVL can be applied to a.c. circuits provided that currents, voltages and impedances are expressed in complex form.

Kirchhoff's current law (KCL) in the frequency domain states that the phasor sum of currents at a node is zero. Thus,

$$\sum_{\text{node}} \mathbf{I} = 0 \qquad (6.6.1)$$

Kirchhoff 's voltage law (KVL) in the frequency domain states that the phasor sum of voltage in a closed path is zero. Thus,

$$\sum_{\text{closed path}} \mathbf{V} = 0 \qquad (6.6.2)$$

6.7 SERIES AC CIRCUIT

The general rules of series ac circuits are the same as for dc circuits. However, we work with phasors.

1. Current

Current is the *same* throughout in any series circuit.

$$\text{Total current} = \frac{\text{total voltage}}{\text{total impedance}} = \frac{\mathbf{V}_T}{\mathbf{Z}_T}$$

Also,
$$\mathbf{I}_T = \frac{\mathbf{V}_1}{\mathbf{Z}_1} = \frac{\mathbf{V}_2}{\mathbf{Z}_2} = \frac{\mathbf{V}_n}{\mathbf{Z}_n}$$

where $\mathbf{V}_1$, $\mathbf{V}_2$, $\mathbf{V}_n$ are the voltage drops in the impedances $\mathbf{Z}_1$, $\mathbf{Z}_2$ and $\mathbf{Z}_n$ connected in series.

2. Impedance

The total impedance of a series circuit is the phasor sum of the impedances of the circuit.

$$\mathbf{Z}_T = \mathbf{Z}_1 + \mathbf{Z}_2 + \ldots + \mathbf{Z}_n$$

Total impedance may also be found by using Ohm's law.

$$\mathbf{Z}_T = \frac{\mathbf{V}_T}{\mathbf{I}_T}$$

Similarly,

$$\mathbf{Z}_1 = \frac{\mathbf{V}_1}{\mathbf{I}_T}, \; \mathbf{Z}_2 = \frac{\mathbf{V}_2}{\mathbf{I}_T}, \; \mathbf{Z}_n = \frac{\mathbf{V}_n}{\mathbf{I}_T}$$

Also
$$\mathbf{Z} = R + j\,(X_L - X_C)$$

3. Voltage

The phasor sum of the voltage drops is equal to the applied voltage.

$$\mathbf{V}_T = \mathbf{V}_1 + \mathbf{V}_2 + \ldots + \mathbf{V}_n$$

By Ohm's law

$$\mathbf{V}_T = \mathbf{I}_T\,\mathbf{Z}_T$$

$$\mathbf{V}_1 = \mathbf{I}_T\,\mathbf{Z}_1, \; \mathbf{V}_2 = \mathbf{I}_T\,\mathbf{Z}_2, \; \mathbf{V}_n = \mathbf{I}_n\,\mathbf{Z}_n$$

6.8 SERIES RLC CIRCUIT

A circuit having R , L and C in series is called a *general series circuit.* Such a circuit is shown in Fig. 6.4.

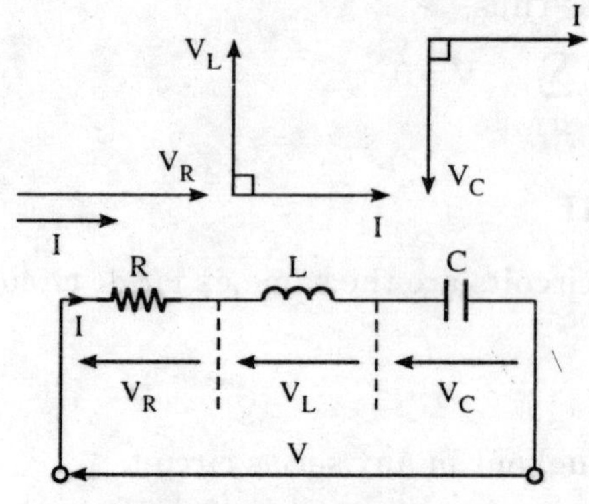

FIG. 6.4 Series RLC circuit

Current **I** is used as reference phasor in series circuit since it is common to all the elements of the circuit. There are four voltages in the circuit, namely

$\mathbf{V}_R$ in phase with **I**,

$\mathbf{V}_L$ leading **I** by 90°,

$\mathbf{V}_C$ lagging **I** by 90°,

By KVL, the applied voltage **V** is the phasor sum of $\mathbf{V}_R$, $\mathbf{V}_L$ and $\mathbf{V}_C$. That is,

$$\mathbf{V} = \mathbf{V}_R + \mathbf{V}_L + \mathbf{V}_C \tag{6.8.1}$$

But $\mathbf{V}_R = \mathbf{Z}_R\,\mathbf{I} = R\,\mathbf{I}$

$\mathbf{V}_L = \mathbf{Z}_L\,\mathbf{I} = jX_L\,\mathbf{I}$

$\mathbf{V}_C = \mathbf{Z}_C\,\mathbf{I} = -jX_C\,\mathbf{I}$

therefore, $\mathbf{V} = R\,\mathbf{I} + jX_L\,\mathbf{I} - jX_C\,\mathbf{I}$

$$\mathbf{V} = [R + j(X_L - X_C)]\,\mathbf{I}$$

If Z is the total impedance of the series RLC circuit

$$\boxed{\mathbf{Z} = \frac{\mathbf{V}}{\mathbf{I}} = R + j(X_L - X_C)} \tag{6.8.2}$$

Let $$X \overset{\Delta}{=} X_L - X_C \tag{6.8.3}$$

The quantity X is called the **net reactance** or **total reactance** of the circuit.

therefore, $$\boxed{\mathbf{Z} = R + jX} \tag{6.8.4}$$

From Eq. (6.8.2),

$$\mathbf{Z} = \frac{\mathbf{V}}{\mathbf{I}},\ \mathbf{V} = \mathbf{ZI},\ \mathbf{I} = \frac{\mathbf{V}}{\mathbf{Z}}$$

The relation

$$\boxed{\mathbf{V} = \mathbf{ZI}} \tag{6.8.5}$$

is called the **generalized Ohm's law for a.c. circuits.**

We have

$$\mathbf{Z} = \frac{\mathbf{V}}{\mathbf{I}} = R + j\,(X_L - X_C) = Z\ \underline{/\ \phi} \tag{6.8.6}$$

$$\boxed{Z = \sqrt{R^2 + (X_L - X_C)^2}} \tag{6.8.7}$$

The circuit phase angle ϕ is given by

$$\tan\phi = \frac{X_L - X_C}{R} \tag{6.8.8}$$

$$\phi = \tan^{-1}\frac{X_L - X_C}{R} \tag{6.8.9}$$

The angle of the impedance ϕ is the angle by which **I** lags behind **V**. Although impedance is a complex quantity, *it is not a phasor.*

6.9 PHASOR DIAGRAM FOR A SERIES RLC CIRCUIT

Since current **I** is common to all the elements of the circuit, it is taken as reference phasor. The voltage $\mathbf{V}_R$ is in phase with **I**, $\mathbf{V}_L$ is leading **I** by 90° and $\mathbf{V}_C$ is lagging behind **I** by 90°. It is seen that $\mathbf{V}_L$ and $\mathbf{V}_C$ are is opposite directions.

There are three possible cases of the series *RLC* circuit.

(a) $V_L > V_C$ that is, $X_L > X_C$

(b) $V_L < V_C$ that is, $X_L < X_C$

(c) $V_L = V_C$ that is, $X_L = X_C$

Case (c) will be dealt with in detail in Chapter 8.

(a) $V_L > V_C$ **that is,** $X_L > X_C$

When $X_L > X_C$, the circuit is predominantly inductive. The phasor diagram for this condition is shown in Fig. 6.5. The applied voltage **V** is equal to the phasor sum of $\mathbf{V}_R$, $\mathbf{V}_L$ and $\mathbf{V}_C$. Since $\mathbf{V}_L$ and $\mathbf{V}_C$ are in opposite directions, the resultant of $\mathbf{V}_L$ and $\mathbf{V}_C$ is therefore, the *arithmetical difference* between them. The phasor sum of $(\mathbf{V}_L - \mathbf{V}_C)$ and $\mathbf{V}_R$ is found by parallelogram law. This phasor sum is the applied voltage **V**.

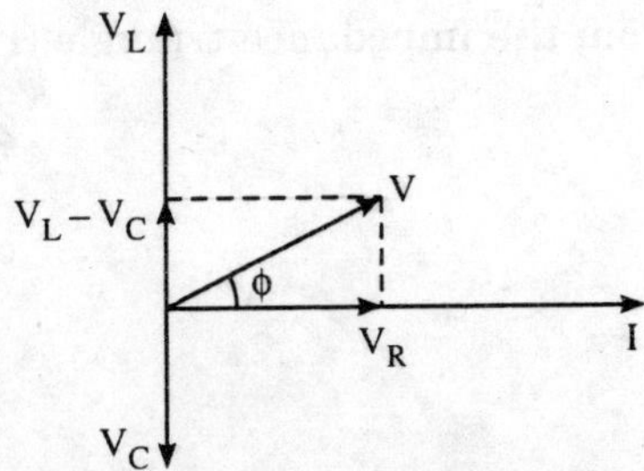

FIG. 6.5. Phasor diagram for a series RLC circuit when $X_L > X_C$

From the phasor diagram of Fig. 6.5,

$$V^2 = V_R^2 + (V_L - V_C)^2$$

$$= (RI)^2 + (X_L I - X_C I)^2$$

therefore, $$V = I\sqrt{R^2 + (X_L - X_C)^2} \quad (6.9.1)$$

Voltage Triangle for a series RLC Circuit

The triangle having V_R, $(V_L - V_C)$ and V as its sides is called the voltage triangle for a series RLC circuit. It is shown in Fig. 6.6.

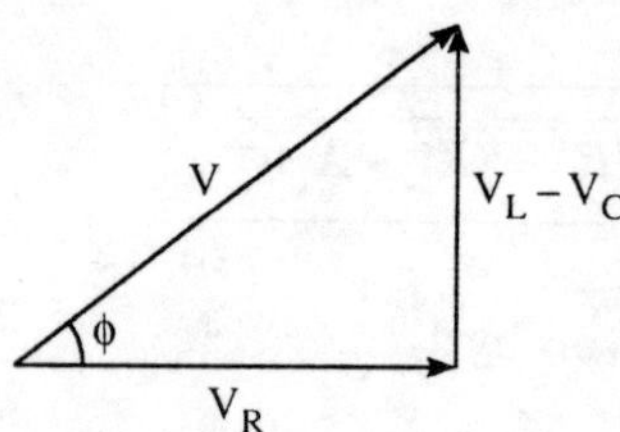

FIG. 6.6. Voltage triangle for a series RLC circuit

Impedance Triangle for a series RLC Circuit

If the length of each side of a voltage triangle is divided by current I, the **impedance triangle** is obtained. The impedance triangle for a series RLC circuit is shown in Fig. 6.7. from the phasor diagram of Fig. 6.5, it is is seen that **I** lags **V** by an angle ϕ.

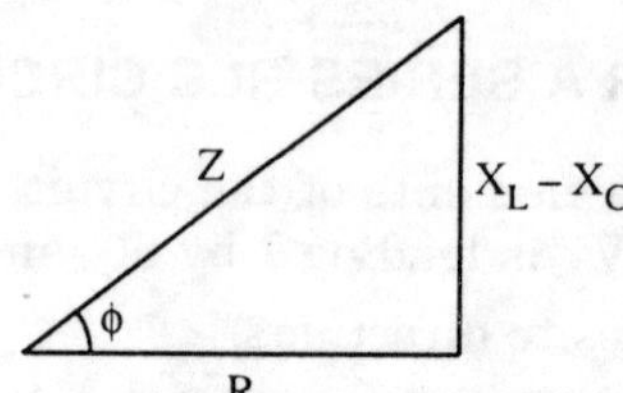

FIG. 6.7. Impedance triangle for a series RLC circuit when $X_L > X_C$

The phase angle ϕ can be found either from the voltage triangle or impedance triangle. From the impedance triangle (Fig. 6.7)

$$Z = \sqrt{R^2 + (X_L - X_L)^2}$$

$$\cos\phi = \frac{R}{Z}$$

$$\sin\phi = \frac{X_L - X_C}{Z}$$

$$\tan\phi = \frac{X_L - X_C}{R}$$

These expressions should not be memorized, but derived from the impedance triangle when required. This is the advantage of the impedance triangle. It should be noted that the sides of the impedance triangle are *not* phasors.

(b) $V_L < V_C$ that is $X_L < X_C$

When $X_L < X_C$ the circuit is predominantly capacitive. The phasor diagram and the impedance triangle for this case are shown in Figs. 6.8 and 6.9 respectively

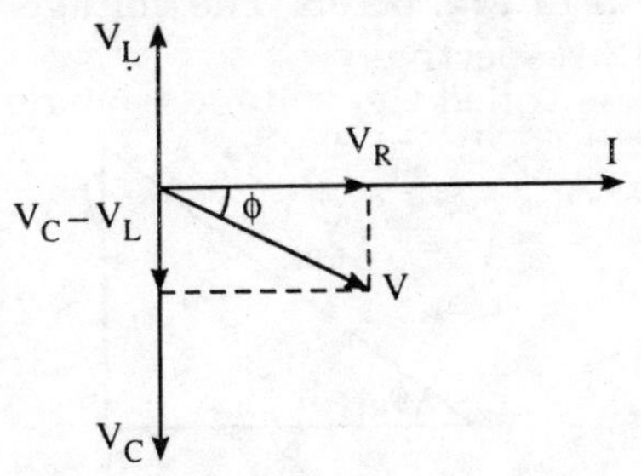

FIG. 6.8 Phasor diagram for series RLC circuit when $X_L < X_C$.

FIG. 6.9 Impedance triangle for series RLC circuit when $X_L < X_C$

6.10 SERIES RL CIRCUIT

A series *RL* circuit is shown in Fig. 6.10a. It is a special case of a general series RLC circuit when $X_C = 0$.

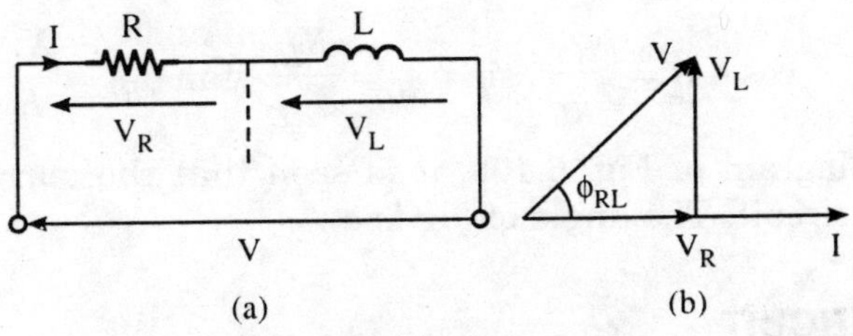

FIG. 6.10. Series RL circuit
(a) Circuit diagram (b) Phasor diagram

By KVL

$$\mathbf{V} = \mathbf{V}_R + \mathbf{V}_L \tag{6.10.1}$$

$$= R\mathbf{I} + jX_L\mathbf{I} = (R + jX_L)\,\mathbf{I}$$

Let

$$\mathbf{Z}_{RL} \overset{\Delta}{=} \frac{\mathbf{V}}{\mathbf{I}} = R + jX_L \tag{6.10.2}$$

The quantity Z_{RL} is called the *impedance of a series RL circuuit.*

From Eq. (6.10.2)

$$\boxed{\mathbf{V} = \mathbf{Z}_{RL}\,\mathbf{I}} \tag{6.10.3}$$

Relationship given by Eq. (6.10.3) is called **Ohm's law for a series RL circuit.**

$$\mathbf{Z}_{RL} = \frac{\mathbf{V}}{\mathbf{I}} = R + jX_L = Z_{RL} \angle \phi_{RL} \tag{6.10.4}$$

$$Z_{RL} = \sqrt{R^2 + X_L^2}$$

$$\tan \phi_{RL} = \frac{X_L}{R}$$

The phasor diagram for a series RL circuit is shown in Fig. 6.10b. The voltage triangle and mpedance triangle are shown in Figs. 6.11 and 6.12 respectively.

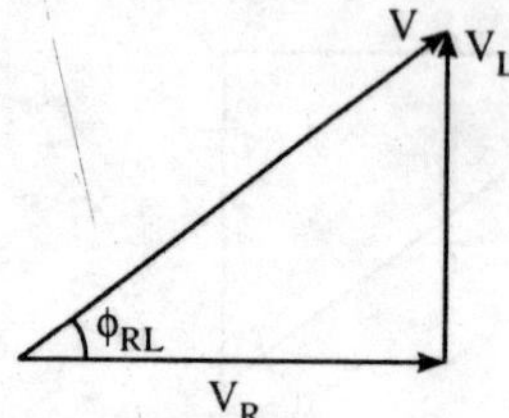

FIG. 6.11 Voltage triangle for a series RL circuit

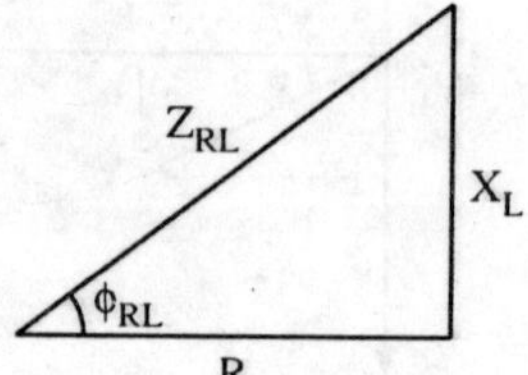

FIG. 6.12 Impedance triangle for a series RL circuit

The circuit phase angle ϕ_{RL} between the supply voltage **V** and the circuit current **I** can be found either from the voltage triangle or impedance triangle. From the impedance triangle of Fig. 6.12,

$$Z_{RL} = \sqrt{R^2 + X_L^2}$$

$$\cos \phi_{RL} = \frac{R}{Z_{RL}}, \ \sin \phi_{RL} = \frac{X_L}{Z_{RL}}, \ \tan \phi_{RL} = \frac{X_L}{R}$$

From the phasor diagram of Fig. 6.10b, it is seen that the current I lags behind the voltage V in a series RL circuit. The angle of lag is ϕ_{RL}.

6.11 SERIES RC CIRCUIT

A series RC circuit is shown in Fig. 6.13a. It is also a special case of a general series RLC circuit when $X_L = 0$.

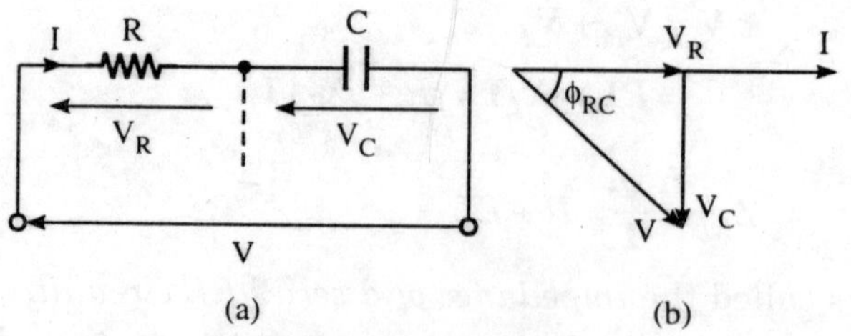

FIG. 6.13 Series RC circuit
(a) circuit diagram (b) phasor diagram

By KVL

$$\mathbf{V} = \mathbf{V}_R + \mathbf{V}_C \tag{6.11.1}$$

$$= R\mathbf{I} - jX_C\,\mathbf{I} = (R - jX_C)\,\mathbf{I}$$

Let $$\mathbf{Z}_{RC} \stackrel{\Delta}{=} \frac{\mathbf{V}}{\mathbf{I}} = R - jX_C \tag{6.11.2}$$

The quantity Z_{RC} is called the *impedance of a series RC circuit.* From Eqs. (6.11.2)

$$\boxed{\mathbf{V} = \mathbf{Z}_{RC}\,\mathbf{I}} \tag{6.11.3}$$

Relationship given by Eq. (6.11.3) is called **Ohm's law for a series RC circuit.**

$$\mathbf{Z}_{RC} = \frac{\mathbf{V}}{\mathbf{I}} = R - jX_C = Z_{RC} \angle \phi_{RC}$$

$$Z_{RC} = \sqrt{R^2 + X_C^2}\ ,\ \tan \phi_{RC} = \frac{-X_C}{R}$$

The phasor diagram is shown in Fig. 6.13b. The voltage triangle and impedance triangle for a series RC circuit are shown in Figs. 6.14 and 6.15 respectively.

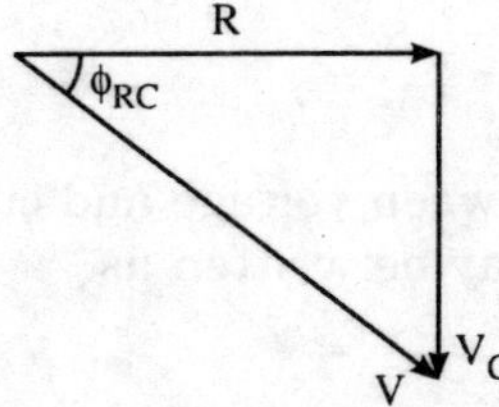

FIG 6.14 Voltage triangle for a series RC circuit

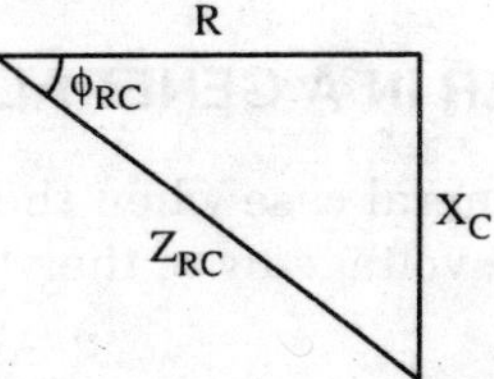

FIG. 6.15 Impedance triangle for a series RC circuit

The circuit phase angle ϕ_{RC} between the supply voltage **V** and circuit current **I** can be found either from the voltage triangle or impedance triangle. From impedance triangle of Fig. 6.15,

$$Z_{RC} = \sqrt{R^2 + X_C^2}$$

$$\cos \phi_{RC} = \frac{R}{Z_{RC}}\ ,\ \sin \phi_{RC} = \frac{-X_C}{Z_{RC}}\ ,\ \tan \phi_{RC} = \frac{-X_C}{R}$$

From the phasor diagram of Fig 6.13b it is seen that the current **I** leads the voltage **V** by an angle ϕ_{RC}.

6.12 POWER IN A.C. CIRCUITS

Power in a *dc* circuit is given by

$$p_{dc} = VI = I^2 R = \frac{V^2}{R}$$

In an *ac* circuit the power at any instant is called **instantaneous power.** It is equal to the product of the values of voltage and current at that instant. It is given by

$$p = vi$$

Since the voltage and current are continuously changing in an *ac* circuit, the power values will also change with time. By definition, the average value of a varying power p is the steady value of power p_{av} which in period T would transfer the same energy W.

$$P_{av}\,T = W = \int_0^T p\,dt$$

$$P_{av} = \frac{1}{T}\int_0^T p\,dt \tag{6.12.1}$$

By convention, P always means average power and no subscript is used.

therefore,
$$P = \frac{1}{T}\int_0^T p\,dt \tag{6.12.2}$$

Also,
$$P = \frac{1}{2\pi}\int_0^{2\pi} p\,d\theta = \frac{1}{2\pi}\int_0^{2\pi} vi\,d\theta \tag{6.12.3}$$

Average power is also called **active power** or **true power**. It is measured in watts (W). Kilowatts (kW) and megawatts (MW) are its bigger units.

$$1\text{ kW} = 10^3\text{ W},\ 1\text{ MW} = 10^6\text{ W}$$

6.13 POWER IN A GENERAL SERIES CIRCUIT

Consider the general case when the phase difference between voltage and current is ϕ. If the current lags the voltage by ϕ, the voltage and current may be written as

$$v = V_m \sin\theta \tag{6.13.1}$$

$$i = I_m \sin(\theta - \phi) \tag{6.13.2}$$

Instantaneous power

$$\begin{aligned} p &= vi \\ &= V_m I_m \sin\theta \sin(\theta - \phi) \\ &= (\sqrt{2}\,V)(\sqrt{2}\,I)\sin\theta(\sin\theta\cos\phi - \cos\theta\sin\phi) \\ &= (VI\cos\phi)(2\sin^2\theta) - (VI\sin\phi)(2\sin\theta\cos\theta) \\ &= (VI\cos\phi)(1 - \cos 2\theta) - (VI\sin\phi)\sin 2\theta \end{aligned}$$

The first term $(VI\cos\phi)(1 - \cos 2\theta)$ is always positive. It represents the instantaneous power transferred from the source to the load. It is called instantaneous active power.

Average power suppled to the circuit is

$$\begin{aligned} P &= \frac{1}{2\pi}\int_0^{2\pi} p\,d\theta \\ &= \frac{1}{2\pi}\int_0^{2\pi}(VI\cos\phi)(1-\cos 2\theta)\,d\theta - \frac{1}{2\pi}\int_0^{2\pi}(VI\cos\phi)\sin 2\theta\,d\theta \\ &= (VI\cos\phi)\left[\frac{1}{2\pi}\int_0^{2\pi} 1\,d\theta - \frac{1}{2\pi}\int_0^{2\pi}\cos 2\theta\,d\theta\right] \\ &\qquad - (VI\sin\phi)\frac{1}{2\pi}\int_0^{2\pi}\sin 2\theta\,d\theta \\ &= \frac{VI\cos\phi}{2\pi}\Big[\theta\Big]_0^{2\pi} - 0 - (VI\sin\phi)\times 0 \end{aligned}$$

or $$P = VI \cos \phi \tag{6.13.3}$$

The second term $(VI \sin \phi) \sin 2\theta$ represents the continual interchange of power between the source and the reactive elements in the circuit. This term is called instantaneous reactive voltamperes. The average value of this term over a complete cycle is zero. The quantity $VI \sin \phi$ is called **reactive voltamperes**. It is denoted by Q and is measured in vars (V Ar). Sometimes this is called **reactive power.** Thus,

$$\boxed{Q = VI \sin \phi} \tag{6.13.4}$$

Kilovars (kVAr) and megavars (MVAr) are bigger units of reactive voltamperes. The product of r.m.s. values of voltage and current in a circuit is called the circuit **voltamperes.** Sometimes it is known as apparent power. It is denoted by S and is measured in voltamperes (VA). Thus,

$$\boxed{\mathbf{S = VI}} \tag{6.13.5}$$

Since $$\mathbf{V = ZI}$$

$$\mathbf{S = ZI^2}$$

6.14 REACTIVE VOLTAMPERES

Reactive voltampers in a circuit

$$Q = VI \sin \phi$$

By Ohm's law,

$$V = ZI$$

and from impedance triangle

$$X = Z \sin \phi$$

therefore, $$Q = I^2 Z \sin \phi = I^2 X = I^2 (X_L - X_c)$$

$$= I^2 X_L - I^2 X_c$$

or $$\boxed{Q = Q_L - Q_c}$$

where Q_L = reactive voltamperes for inductive element

Q_c = reactive voltamperes for capacitive element.

By convention, Q_L is taken positive and Q_c negative.

6.15 POWER FACTOR

The ratio of active power to the voltamperes in an a.c. circuit is defined as the power factor (p.f) of the circuit. Thus,

$$\text{power factor} \overset{\Delta}{=} \frac{\text{active power}}{\text{voltamperes}} = \frac{P}{S} \tag{6.15.1}$$

For sinusoidal waveforms

$$\text{power factor} = \frac{P}{VI} = \frac{VI \cos \phi}{VI} = \cos \phi \tag{6.15.2}$$

The general relationship given by Eq. (6.15.1) is valid for nonsinusoidal waveforms also. It is to be noted that phase angle ϕ has no significance for nonsinusoidal waveforms hence in such cases $\cos \phi$ is not the power factor.

An a.c. circuit in which the current lags behind the voltage is said to have a ***lagging power factor.*** A predominantly inductive circuit has a *lagging power factor. A circuit in which the current leads the voltage is said to have a* ***leading power factor.*** A predominantly capacitive circuit has a leading power factor. Power factor can also be found in terms of circuit constants. From impedance triangle

$$\cos \phi = \frac{R}{Z} \tag{6.15.3}$$

Power factor for a purely resistive circuit = $\cos 0° = 1$

Power factor for a purely inductive circuit = $\cos 90° = 0$

Power factor for a purely capacitive circuit = $\cos (-90°) = 0$

The power factor (p.f.) can never be greater than unity since the maximum value of $\cos \phi$ is unity.

Power consumed in a circuit

$$P = VI \cos \phi$$

Power consumed in a purely resistive circuit

$$P_R = VI \times I = VI = I^2R$$

Power consumed in a purely inductive circuit

$$P_L = VI \times 0 = 0$$

Power consumed in a purely capacitive circuit

$$P_c = VI \times 0 = 0$$

Hence it is found that in a circuit the power is consumed only in the resistor and there is no power consumption in either pure inductor or pure capacitor.

Total power in a circuit

$$P = VI \cos \phi = I^2R \tag{6.15.4}$$

6.16 COMPONENTS OF CURRENT

When the current is not in phase with the voltage, it may lag or lead the voltage by an angle ϕ as shown in Fig. 6.19. In both the cases the current has two components namely, $I \cos \phi$ which is in phase with voltage and $I \sin \phi$ which is at 90° with the voltage. These components are shown in Fig. 5.19. Since active power is $VI \cos \phi$, and the component $I \cos \phi$ multiplied by the voltage gives the active power, the inphase component $I \cos \phi$ is called the **power component, active component** or **wattful component of current.**

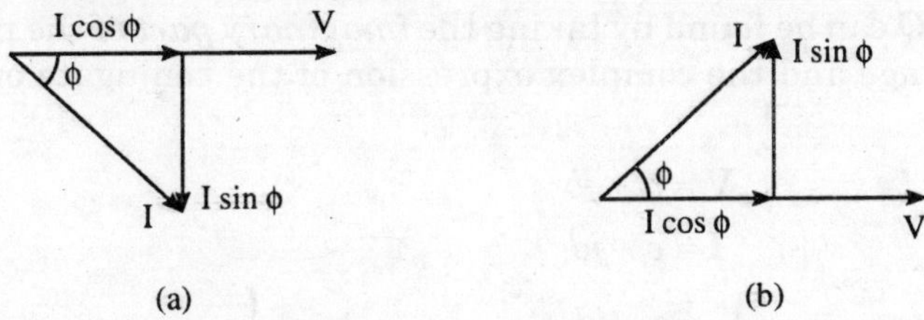

FIG. 6.16 Components of current (a) Lagging current (b) Leading current

Also, reactive voltamperes is $Q = VI \sin \phi$ and the component $I \sin \phi$ multiplied by the voltage gives the reactive voltamperes, the component $I \sin \phi$ is called the **reactive component** or **wattless component of current.**

6.17 COMPLEX POWER

Consider a circuit in which the voltage leads the current by an angle ϕ. Let

$$\mathbf{I} = I \,\underline{/\ \theta} \quad \text{and} \quad \mathbf{V} = V \,\underline{/\ \theta + \phi}$$

We know that

$$P = VI \cos \phi, \; Q = VI \sin \phi$$

We shall define complex power **S** as

$$\mathbf{S} \overset{\Delta}{=} P + jQ \qquad (6.14.1)$$

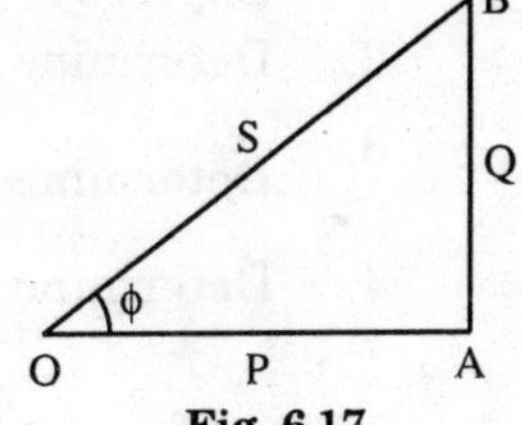

Fig. 6.17 Power triangle

P, Q and S are shown in Fig. 6.17.

The right angled triangle OAB formed by P, Q and S is called the **power triangle.** From this triangle

$$P = S \cos \phi \,, \; Q = S \sin \phi = P \tan \phi$$

and

$$S^2 = P^2 + Q^2$$

$$\mathbf{S} = P + jQ = S \cos \phi + jS \sin \phi$$

$$= VI \cos \phi + j\, VI \sin \phi$$

$$= VI\,(\cos \phi + j \sin \phi)$$

$$= VI\, e^{j\phi} = VI \,\underline{/\ \phi} \qquad (6.17.2)$$

Since

$$\mathbf{I} = I \angle \theta$$

$$\mathbf{I}^* = I \angle -\theta$$

$$\mathbf{VI}^* = (V \,\underline{/\ \theta + \phi})\,(I \angle -\theta)$$

therefore, $$= VI \angle \phi = S \tag{6.17.3}$$

Thus $$\mathbf{S} = P + jQ = \mathbf{VI}^* \tag{6.17.4}$$

$$P = \text{Re}\,[\mathbf{VI}^*] = VI\cos\phi = I^2 R \tag{6.17.5}$$

$$Q = \text{Im}\,[\mathbf{VI}^*] = VI\sin\phi = I^2 X \tag{6.17.6}$$

Eq. (6.17.5) shows that active power P can be obtained by taking the *real part* of the product of the complex expressions of voltage and the *conjugate* of current. Eq. (6.17.6) shows that reactive voltamperes Q can be found by taking the *imaginary part* of the product of the complex expression of the voltage and the complex expression of the conjugate of current.

For example, if

$$\mathbf{V} = a + jb$$

$$\mathbf{I} = c + jd$$

then $$\mathbf{I}^* = c - jd$$

$$\mathbf{S} = \mathbf{VI}^*$$

$$= (a + jb)(c - jd) = (ac + bd) + j(bc - ad) = P + jQ$$

therefore, $$P = \text{Re}\,[\mathbf{VI}^*] = (ac + bd) \quad \text{watts} \tag{6.17.7}$$

$$Q = \text{Im}\,[\mathbf{VI}^*] = (bc - ad) \quad \text{vars} \tag{6.17.8}$$

6.18 STEP-BY-STEP CALCULATION PROCEDURE FOR SERIES AC CIRCUITS

1. Express R in ohms, L in henrys, C in farads and f in hertz.
2. Determine $X_L = 2\pi f L\ \Omega$
3. Determine $X_C = \dfrac{1}{2\pi f C}\ \Omega$
4. Determine the impedance in rectangular form by the formula

   $$\mathbf{Z} = R + j(X_L - X_C)\ \Omega$$

5. Determine the impedance in polar form also by the formula

   $$\mathbf{Z} = |\,\mathbf{Z}\,| \angle \phi$$

   where $|\,\mathbf{Z}\,| = \sqrt{R^2 + (X_L - X_C)^2}\ ;\ \phi = \tan^{-1}\dfrac{X_L - X_C}{R}$

6. Take the supply voltage $\mathbf{V}$ as reference phasor

   $$\mathbf{V} = V \angle 0° = V + j0$$

7. Determine the circuit current by Ohm's law

   $$\mathbf{I} = \frac{\mathbf{V}}{\mathbf{Z}} = \frac{V \angle 0°}{Z \angle \phi} = \frac{V}{Z} \angle -\phi\ \Omega$$

8. Express $\mathbf{I}$ in rectangular form also.
9. Determine the power factor $\cos\phi$ to four places of decimal. Specify whether the power factor is *lagging* or *leading*. If ϕ is negative in step 7, the power factor is lagging. If ϕ is positive in step 7, the power factor is leading.

10. Determine $\mathbf{I}^*$, the complex conjugate of $\mathbf{I}$.
11. Determine $\mathbf{S} = \mathbf{VI}^* = P + jQ$. The real part of $\mathbf{VI}^*$ gives the active power P in watts and the imaginary part gives the reactive voltamperes in VAr. The magnitude of $\mathbf{VI}^*$ gives voltamperes (VA).

therefore,
$$P = \text{Re}\,[\mathbf{VI}^*] \quad \text{W}$$
$$Q = \text{Im}\,[\mathbf{VI}^*] \quad \text{VAr}$$
$$S = |\mathbf{VI}^*| \quad \text{VA}$$

Alternatively

$$P = VI\cos\phi = I^2R$$
$$S = VI$$
$$Q = VI\sin\phi$$

The lagging current gives a positive of Q, and the leading current gives a negative value of Q. By convention, lagging VAr is taken positive and leading VAr is taken negative.

12. Determine the voltage drop each element.

$$\mathbf{V}_R = \mathbf{IZ}_R = \mathbf{I}R$$
$$\mathbf{V}_L = \mathbf{IZ}_L = \mathbf{I}\,(jX_L)$$
$$\mathbf{V}_C = \mathbf{IZ}_C = \mathbf{I}\,(-jX_C)$$

13. Draw the phasor diagram.

■ **EXAMPLE 6.1** A resistance of 60 Ω is connected in series with a 120 μF capacitor across a 230 V, 50 Hz supply. Determine
(a) the current through the circuit,
(b) the phase difference between voltage and current,
(c) the voltage across the resistor,
(d) the voltage across the capacitor,
(e) the power consumed in the circuit.

SOLUTION :

$$R = 60\ \Omega,\ C = 120\ \mu\text{F} = 120 \times 10^{-6}\ \text{F}$$

$$X_C = \frac{1}{2\pi fC} = \frac{1}{2\pi \times 50 \times 120 \times 10^{-6}} = 26.526\ \Omega$$

$$\mathbf{Z} = R - jX_C = 60 - j\,26.526 = 65.6\angle -23.85^\circ\ \Omega$$

Taking the supply voltage $\mathbf{V}$ as reference phasor we have

$$\mathbf{V} = V\angle 0^\circ = 230\angle 0^\circ\ \text{V}$$

(a) By Ohm's law

$$\mathbf{I} = \frac{\mathbf{V}}{\mathbf{Z}} = \frac{230\angle 0^\circ}{65.6\angle -23.85^\circ} = 3.506\angle +23.85^\circ\ \text{A}$$

(b) Phase difference between **V** and **I**

$$\phi = \phi_i - \phi_v = 23.85^\circ - 0^\circ = 23.85^\circ$$

The current leads the voltage by 23.85°.

(c) Voltage across the resistor

$$V_R = IR = 3.506 \times 60 = 210.36 \text{ V}$$

(d) Voltage across the capacitor

$$V_C = IX_C = 3.506 \times 26.526 = 93 \text{ V}$$

(e) Power consumed in the circuit

$$P = I^2R = (3.506)^2 \times 60 = 737.5 \text{ W}$$

■ **EXAMPLE 6.2** The voltage across and current through a circuit are given by

$$v = 250 \sin(314\,t - 10^\circ) \text{ V}$$

$$i = 10 \sin(314\,t + 50^\circ) \text{ A}$$

Calculate the impedance, resistance, and reactance of the circuit.

SOLUTION : Converting the time-domain forms of voltage and current into polar forms we get

$$\mathbf{V} = \frac{250}{\sqrt{2}} \angle -10^\circ \text{ V}$$

$$\mathbf{I} = \frac{10}{\sqrt{2}} \angle +50^\circ \text{ A}$$

Impedance of the circuit

$$\mathbf{Z} = \frac{\mathbf{V}}{\mathbf{I}} = \frac{(250/\sqrt{2}) \angle -10^\circ}{(10/\sqrt{2}) \angle +50^\circ}$$

$$= 25 \angle -60^\circ = (12.5 - j\,21.65)\ \Omega$$

This is of the form

$$\mathbf{Z} = R - jX_C$$

therefore, resistance $R = 12.5\ \Omega$

and Capacitive reactance $X_C = 21.65\ \Omega$

■ **EXAMPLE 6.3** A voltage of $250 \angle 0^\circ$ V is applied to an inductive circuit of impedance $(5 + j10)\ \Omega$. Calculate

(a) the circuit current, (b) the power factor,

(c) the power consumption, (d) the apparent power,

(e) the reactive power.

SOLUTION :

$$\mathbf{Z} = 5 + j10 = 11.18 \angle 63.43^\circ\ \Omega$$

(a)
$$\mathbf{I} = \frac{\mathbf{V}}{\mathbf{Z}} = \frac{250 \angle 0^\circ}{11.18 \angle 63.43^\circ} = 22.36 \angle -63.43^\circ \text{ A}$$

(b) $\cos\phi = \cos(-63.43^\circ) = 0.4473$ (lagging)

(c) Power consumption

$$P = I^2R = (22.36)^2 \times 5 = 2500 \text{ W}$$

or $$P = VI\cos\phi = 250 \times 22.36 \times 0.4473 = 2500 \text{ W}$$

(d) Apparent power

$$S = VI = 250 \times 22.36 = 5590 \text{ VA}$$

(e) Reactive power

$$Q = VI\sin\phi = 250 \times 22.36 \times \sin 63.43^\circ = 5000 \text{ VAr (lagging)}$$

Alternative method

The complex conjugate of current is

$$\mathbf{I}^* = 22.36 \angle +63.43^\circ \text{ A}$$

therefore, $$\mathbf{VI}^* = (250 \angle 0^\circ)(22.36 \angle 63.43^\circ) = 5590 \angle 63.43^\circ \text{ VA}$$

Also $$\mathbf{S} = \mathbf{VI}^* = P + jQ$$

therefore, $$S = 5590 \text{ VA}$$

$$5590 \underline{/ 63.43^\circ} = P + jQ$$

$$2500 + j5000 = P + jQ$$

therefore, $$P = 2500 \text{ W}, Q = 5000 \text{ VAr (lagging)}$$

■ **EXAMPLE 6.4** An alternating voltage $(80 + j60)$ V is applied to a circuit and the current flowing is $(-4 + j10)$ A. Find (*a*) the impedance of the circuit, (*b*) the power consumed, and (*c*) the phase angle.

SOLUTION :

$$\mathbf{V} = 80 + j60 = 100 \underline{/ 36.87^\circ} \text{ V}$$

$$\mathbf{I} = -4 + j10 = 10.77 \underline{/ 111.8^\circ} \text{ A}$$

The phase difference between the current **I** and voltage **V** is

$$\phi = 111.8 - 36.87 = 74.93^\circ$$

The current leads the voltage by an angle 74.93°. Therefore, the power factor is leading and is given by

$$\cos\phi = \cos 74.93^\circ = 0.26$$

The impedance of the circuit

$$\mathbf{Z} = \frac{\mathbf{V}}{\mathbf{I}} = \frac{100 \angle 36.87^\circ}{10.77 \angle 111.8^\circ} = 9.285 \angle -74.9^\circ \ \Omega$$

$$= (2.418 - j8.96)\ \Omega = R - jX_C$$

Power consumed in the circuit

$$P = I^2 R = (10.77)^2 \times 2.418 = 280.47 \text{ W}$$

or $$P = VI\cos\phi = 100 \times 100.77 \times 0.26 = 280.5 \text{ W}$$

or $$P = \text{real part of } \mathbf{VI}^*$$
$$= \text{real part of } (80 + j60)(-4 - j10)$$
$$= -320 + 600 = 280 \text{ W}$$

■ **EXAMPLE 6.5** A resistance of 5 Ω is connected in series with an inductive reactance of 8 Ω. The current in the circuit is 4 A. Calculate the supply voltage. Write expression for the instantaneous values of current and supply voltage.

SOLUTION :

$$R = 5\ \Omega\ , X_L = 8\ \Omega\ , I = 4 \text{ A}$$
$$\mathbf{Z} = R + jX_L = 5 + j8\ \Omega = 9.43 \angle 58^\circ\ \Omega$$

Taking the current **I** as reference phasor.

$$\mathbf{I} = I \angle 0^\circ = 4 \angle 0^\circ \text{ A}$$

By Ohm's law $$\mathbf{V} = \mathbf{IZ} = 4 \angle 0^\circ \times 9.43 \angle 58^\circ$$
$$= 37.72 \angle 58^\circ \text{ V}$$
$$V_m = \sqrt{2}\ V = \sqrt{2} \times 37.72 = 53.34 \text{ V}$$

therefore, $$i = I_m \sin \omega t = \sqrt{2} \times 4 \sin \omega t$$
$$i = 5.66 \sin \omega t$$
$$v = V_m \sin(\omega t + \phi)\ ;\ v = 53.34 \sin(\omega t + 58^\circ)$$

■ **EXAMPLE 6.6** A coil of resistance 10 Ω and inductance 0.02 H is connected in series with another coil of resistance 6 Ω and inductance 15 mH across a 230 V, 50 Hz supply. Calculate (a) the impedance of the circuit, (b) the voltage drop across each coil, (c) the total power consumed by the circuit.

SOLUTION :

$$R_1 = 10\ \Omega,\ L_1 = 0.02 \text{ H}$$
$$R_2 = 6\ \Omega\ ,\ L_2 = 15 \text{ mH} = 15 \times 10^{-3} \text{ H}$$
$$X_{L_1} = 2\pi f L_1 = 2\pi \times 50 \times 0.02 = 6.28\ \Omega$$
$$X_{L_2} = 2\pi f L_2 = 2\pi \times 50 \times 15 \times 10^{-3} = 4.71\ \Omega$$
$$\mathbf{Z}_1 = R_1 + jX_{L_1} = 10 + j6.28 = 11.81 \underline{/\ 32.13^\circ}\ \Omega$$
$$\mathbf{Z}_2 = R_2 + jX_{L_2} = 6 + j4.71 = 7.63 \underline{/\ 38.13^\circ}\ \Omega$$

Total impedance of the circuit

$$\mathbf{Z} = \mathbf{Z}_1 + \mathbf{Z}_2 = 10 + j6.28 + 6 + j4.71$$
$$= 16 + j10.99 = 19.41 \underline{/\ 34.48^\circ}\ \Omega$$

Taking the voltage phasor **V** as reference.

$$\mathbf{V} = V \angle 0^\circ = 230 \angle 0^\circ \text{ V}$$

therefore, total current $\mathbf{I} = \dfrac{\mathbf{V}}{\mathbf{Z}} = \dfrac{230 \angle 0^\circ}{19.41 \angle 34.48^\circ}$

$= 11.85 \angle -34.48^\circ$ A

Voltage across coil 1 $= \mathbf{IZ}_1 = (11.85 \angle -34.48^\circ)(11.81 \angle 32.13^\circ)$

therefore, $\mathbf{V}_1 = 139.95 \angle -2.35^\circ$ V

Voltage across coil 2 $= \mathbf{IZ}_2 = (11.85 \angle -34.48^\circ)(7.63 \angle 38.13^\circ)$

therefore, $\mathbf{V}_2 = 90.42 \angle 3.65.$ V

Total power consumed by the circuit

$$P = I^2 R_1 + I^2 R_2 = I^2 (R_1 + R_2)$$

$$= (11.85)^2 (10 + 6) = 2246.76 \text{ W}$$

■ **EXAMPLE 6.7** When a supply voltage of $v = 400 \sin\left(\omega t + \dfrac{\pi}{4}\right)$ V is applied to three impedance $\mathbf{Z}_1$, $\mathbf{Z}_2$ and $\mathbf{Z}_3$ in series, the voltages across $\mathbf{Z}_1$ and $\mathbf{Z}_2$ are 100 sin ωt and 50 $\sin\left(\omega t - \dfrac{\pi}{4}\right)$ volts respectively. Determine the voltage across $\mathbf{Z}_3$. If the current is $i = 30 \sin \omega t$ A, calculate the circuit impedance.

SOLUTION :

Given $v = 400 \sin\left(\omega t + \dfrac{\pi}{4}\right)$ V

$v_1 = 100 \sin \omega t$; $v_2 = 50 \sin\left(\omega t - \dfrac{\pi}{4}\right)$

$i = 300 \sin \omega t$

therefore, $V = \dfrac{400}{\sqrt{2}} \angle 45^\circ \text{ V} = 200 + j200$ V

$$V_1 = \frac{100}{\sqrt{2}} \angle 0^\circ \text{ V} = 70.71 + j0$$

$$\mathbf{V}_2 = \frac{50}{\sqrt{2}} \angle -45^\circ = 25 - j25$$

$$\mathbf{I} = \frac{30}{\sqrt{2}} \angle 0^\circ$$

Since the impedances are in series

$$\mathbf{V} = \mathbf{V}_1 + \mathbf{V}_2 + \mathbf{V}_3$$

$$\mathbf{V}_3 = \mathbf{V} - (\mathbf{V}_1 + \mathbf{V}_2)$$

$$= 200 + j200 - (70.71 + j0 + 25 - j25)$$

$$= 104.29 + j225 = 248 \angle 65.13^\circ \text{ V}$$

therefore, $v_3 = 248 \times \sqrt{2} \sin(\omega t + 65.13^\circ)$

$$v_3 = 350.72 \sin(\omega t + 65.13^\circ)$$

Circuit impedance

$$\mathbf{Z} = \frac{\mathbf{V}}{\mathbf{I}} = \frac{(400/\sqrt{2}) \angle 45^\circ}{(30/\sqrt{2}) \angle 0^\circ} = 13.33 \angle 45^\circ \ \Omega$$

■ **EXAMPLE 6.8** An inductive circuit takes 5 A from a 200 V supply at a power factor of 0.8 lagging. Calculate (a) the impedance, (b) the resistance, (c) the reactance, (d) the phase angle, (e) the voltamperes, (f) the reactive voltamperes, (g) the active power.

SOLUTION :

$$V = 200 \text{ V}, I = 5 \text{ A}, \cos\phi = 0.8 \text{ lagging}$$

$$\mathbf{V} = 200 \angle 0^\circ$$

$$\mathbf{I} = 5 \,\underline{/ -\cos^{-1} 0.8} = 5 \,\underline{/ -36.87^\circ}$$

$$\mathbf{Z} = \frac{\mathbf{V}}{\mathbf{I}} = \frac{200 \angle 0^\circ}{5 \,/ -36.87^\circ} = 40 \,\underline{/ +36.87^\circ} \ \Omega$$

$$= 32 + j24$$

Also $\mathbf{Z} = R + jX_L$

therefore, $R = 32 \ \Omega, X_L = 24 \ \Omega$

$$\mathbf{S} = \mathbf{V}\mathbf{I}^*$$

$$= (200 \angle 0^\circ)(5 \,\underline{/ +36.87^\circ}) = 1000 \,\underline{/ 36.87^\circ} \text{ VA}$$

$$\mathbf{S} = 800 + j600$$

Also $\mathbf{S} = P + jQ$

therefore, $S = 1000$ VA

$$P = 800 \text{ W}, Q = 600 \text{ VAr}$$

■ **EXAMPLE 6.9** An impedance of $20 \angle 50^\circ \ \Omega$ is connected in series with another impedance of $15 \angle -45^\circ \ \Omega$ and the combination takes a current of 2 A. Calculate the voltage across each impedance.

SOLUTION :

$$\mathbf{Z}_1 = 20 \angle 50^\circ = 12.86 + j15.32 \ \Omega$$

$$\mathbf{Z}_2 = 15 \angle -45^\circ = 10.61 - j10.61 \ \Omega$$

Total impedance of the series circuit

$$\mathbf{Z} = \mathbf{Z}_1 + \mathbf{Z}_2 = (12.86 + j15.32) + (10.61 - j10.61)$$

$$= 23.47 + j4.71 = 23.94 \,\underline{/ 11.35^\circ} \ \Omega$$

Taking **I** as reference phasor.

$$\mathbf{I} = 2 \angle 0^\circ$$

By Ohm's law

$$\mathbf{V} = \mathbf{IZ} = (2 \angle 0^\circ)(23.94 \underline{/ 11.35^\circ})$$

$$= 47.88 \underline{/ 11.35^\circ} \text{ V}$$

$$\mathbf{V}_1 = \mathbf{I}_1 \mathbf{Z}_1 = (2 \angle 0^\circ)(20 \angle 50^\circ) = 40 \angle 50^\circ \text{ V}$$

$$\mathbf{V}_2 = \mathbf{I}_2 \mathbf{Z}_2 = (2 \angle 0^\circ)(15 \angle -45^\circ) = 30 \angle -45^\circ \text{ V}$$

■ **EXAMPLE 6.10** **Three impedances $\mathbf{Z}_1$, $\mathbf{Z}_2$ and $\mathbf{Z}_3$ are connected in series across a 200 V, 50 Hz supply. If $\mathbf{Z}_1 = 20\ \Omega$, $\mathbf{Z}_2 = (8 + j10)\ \Omega$, $\mathbf{Z}_3 = (15 - j15)\ \Omega$, determine (a) the current through the circuit, (b) the power factor of the circuit, (c) the voltage across each impedance.**

SOLUTION :

$$\mathbf{Z}_1 = 20\ \Omega = 20 + j0\ \Omega = 20 \angle 0^\circ\ \Omega$$

$$\mathbf{Z}_2 = 8 + j10 = 12.81 \angle 51.34^\circ\ \Omega$$

$$\mathbf{Z}_3 = 15 - j15 = 21.21 \angle -45^\circ\ \Omega$$

Total impedance of the series circuit

$$\mathbf{Z} = \mathbf{Z}_1 + \mathbf{Z}_2 + \mathbf{Z}_3$$

$$= 20 + 8 + j10 + 15 - j15$$

$$= 43 - j5\ \Omega = 43.29 \underline{/ -6.63^\circ}\ \Omega$$

Let **V** be taken as reference phasor.

$$\mathbf{V} = 200 \angle 0^\circ \text{ V}$$

By Ohm's law

$$\text{(a) } \mathbf{I} = \frac{\mathbf{V}}{\mathbf{Z}} = \frac{200 \angle 0^\circ}{43.29 \angle -6.63} = 4.62 \angle 6.63^\circ \text{ A}$$

(b) Phase angle between current and voltage

$$\phi = 6.63^\circ$$

therefore, $\cos\phi = \cos 6.63^\circ = 0.9933$ (leading)

(c) Voltage across $\mathbf{Z}_1$ is

$$\mathbf{V}_1 = \mathbf{I}_1\mathbf{Z}_1 = \mathbf{IZ}_1 = (4.62 \angle 6.63^\circ)(20 \angle 0^\circ) = 92.2 \angle 6.63^\circ \text{ V}$$

Voltage across $\mathbf{Z}_2$ is

$$\mathbf{V}_2 = \mathbf{I}_1\mathbf{Z}_2 = \mathbf{IZ}_2 = (4.62 \angle 6.63^\circ)(12.81 \angle 51.34^\circ)$$

$$= 59.81 \underline{/ 57.97^\circ} \text{ V}$$

Voltage across $\mathbf{Z}_3$ is

$$\mathbf{V}_3 = \mathbf{I}_3\mathbf{Z}_3 = \mathbf{IZ}_3 = (4.62 \angle 6.63^\circ)(21.21 \angle -45^\circ)$$

$$= 97.99 \underline{/ 38.37^\circ} \text{ V}$$

■ **EXAMPLE 6.11** A voltage $v = 100 \sin(314t + 5°)$ V is applied to a circuit. The resulting current is $i = 5 \sin(314t - 40°)$ A. Determine power factor and the average power delivered to the circuit.

SOLUTION : The phase difference between current and votage phasors is given by

$$\phi = \phi_i - \phi_v$$

$$= -40° - 5 = -45°$$

Therefore the power factor

$$\cos\phi = \cos(-45°) = 0.707 \text{ (lagging)}$$

The power factor is lagging because the current lags behind the voltage. Power delivered to the circuit

$$P = VI\cos\phi$$

$$= \frac{V_m}{\sqrt{2}} \times \frac{I_m}{\sqrt{2}} \cos\phi$$

$$= \frac{100}{\sqrt{2}} \times \frac{5}{\sqrt{2}} \cos 45° = 176.78 \text{ W}$$

■ **EXAMPLE 6.12** A 100 V, 60 W lamp is to be operated on a 250 V 50 Hz supply. Calculate the value of (a) noninductive resistor, (b) pure inductance, and (c) capacitor, that would be required to be placed in series with the lamp in order that it may be used at its rated voltage.

SOLUTION : Current taken by the lamp

$$I = \frac{P}{V_1} = \frac{60}{100} = 0.6 \text{ A}$$

If R_1 is the resistance of the lamp,

$$P = I^2 R_1$$

$$R_1 = \frac{P}{I^2} = \frac{60}{(0.6)^2} = 166.66\ \Omega$$

(a) *Noninductive Resistor R*

When a noninductive resistance R is placed in series with the lamp, the total resistance of the circuit becomes R_T (say) where

$$R_T = R_1 + R = 166.66 + R$$

Since the circuit is purely resistive

$$V = R_T I$$

$$250 = (166.66 + R) \times 0.6$$

$$R = \frac{250}{0.6} - 166.66 = 250\ \Omega$$

(b) *Pure Inductance L*

When a pure inductance L is placed in sseries with the lamp the total impedance of the circuit is given by

$$\mathbf{Z}_2 = R_1 + jX_L = 166.66 + j\,2\pi \times 50\,L$$

By Ohm's law

$$\mathbf{V} = \mathbf{Z}_2\mathbf{I}$$

$$250 = (166.66 + j\,2\pi \times 50\,L) \times 0.6$$

$$\frac{250}{0.6} = \sqrt{(166.66)^2 + (2\pi \times 50\,L)^2}$$

$$\left(\frac{250}{0.6}\right)^2 - (166.66)^2 = (2\pi \times 50\,L)^2$$

$$\sqrt{(173611 - 27775)} = 2\pi \times 50\,L$$

$$L = 1.2155 \text{ H}$$

(c) When a capactior C is placed in series with the lamp, the total impedance of the circuit is given by

$$\mathbf{Z}_3 = R_1 - jX_c$$

$$|\,\mathbf{Z}_3\,| = \sqrt{R_1^2 + X_c^2}$$

$$\mathbf{V} = \mathbf{Z}_3\mathbf{I}$$

$$Z_3 = \frac{V}{I}$$

$$\sqrt{R_1^2 + X_c^2} = \frac{V}{I}$$

$$\sqrt{(166.66)^2 + \left(\frac{1}{2\pi fC}\right)^2} = \frac{250}{0.6}$$

therefore, $$C = 8.339 \times 10^{-6} \text{ F} = 8.339\ \mu\text{F}$$

■ **EXAMPLE 6.13** A coil, when connected to a 12 V dc supply, takes a current of 3 A, but when connected to a 100 Hz ac supply of 50 V takes a current of 10 A. Calculate the resistance, reactance and inductance of the coil. What is the power consumed by the coil when connected to the ac supply ?

SOLUTION : Under steady-state conditions, the inductance has no effect on direct current.

Resistance of the coil

$$R = \frac{V_{dc}}{I_{dc}} = \frac{12}{3} = 4\ \Omega$$

The impedance of the coil

$$\mathbf{Z} = \frac{\mathbf{V}_{ac}}{\mathbf{I}_{ac}} = \frac{50}{10} = 5\ \Omega$$

The reactance of the coil

$$X_L = \sqrt{Z^2 - R^2} = \sqrt{5^2 - 4^4} = 3\ \Omega$$

Therefore the inductance of the coil

$$L = \frac{X_L}{2\pi f} = \frac{3}{2\pi \times 100} = 4.774 \times 10^{-3}\text{ H}$$

$$= 4.774 \text{ mH}$$

Power consumed by the coil

$$P = I^2 R = (10)^2 \times 4 = 400 \text{ W}$$

■ **EXAMPLE 6.14** A series circuit consists of a 300 Ω noninductive resistor, a 7.95 μF capactior and a 2.06 H inductor of negligible resistance.
If the suppply voltage is

$$v(t) = 250\sqrt{2}\sin(314\,t + 30°)\text{, calculate}$$

(a) the circuit current,
(b) the voltage drop across each component in the circuit,
(c) the power consumed in the circuit.

SOLUTION : Resistance of the circuit, $R = 300\ \Omega$

Inductive reactance of the circuit,

$$X_L = 2\pi f L = 314 \times 2.06 = 646.8\ \Omega$$

Capacitive reactance of the circuit,

$$X_c = \frac{1}{2\pi f C} = \frac{1}{314 \times 7.95 \times 10^{-6}} = 400\ \Omega$$

Impedance of the circuit,

$$\mathbf{Z} = R + j(X_L - X_c)$$

$$= 300 + j(646.8 - 400)$$

$$= 300 + j\,246.8 = 388.5\ \underline{/\ 39.44°}\ \Omega$$

Since $v(t) = 250\sqrt{2}\sin(314\,t + 30°)$

$$\mathbf{V} = \frac{250\sqrt{2}}{\sqrt{2}}\ \underline{/\ 30°} = 250\ \underline{/\ 30°}\text{ V}$$

(a) The circuit current,

$$\mathbf{I} = \frac{\mathbf{V}}{\mathbf{Z}} = \frac{250\ \underline{/\ 30°}}{388.5\ \underline{/\ 39.44°}}$$

$$= 0.6435\ \underline{/ -9.44°}\text{ A}$$

(b) Voltage drop across the noninductive resistor,

$$\mathbf{V}_R = R\mathbf{I} = 300 \times 0.6435\ \underline{/ -9.44°}\text{ V}$$

$$= 193\ \underline{/ -9.44°}\ V$$

Also, $$v_R = \sqrt{2} \times 193\sin(314\,t - 9.44°)\text{ V}$$

Voltage drop across the inductor,

$$V_L = (jX_L)\,\mathbf{I} = (646.8\ \underline{/\ 90°})(0.6435\ \underline{/ -9.44°})$$

$$= 416.2 \underline{/\ 80.56^\circ}\ \text{V}$$

Also, $v_L = \sqrt{2} \times 416.2 \sin(314\,t + 88.56^\circ)$ V

Voltage drop across the capacitor,

$$\mathbf{V}_c = (-jX_c)\,\mathbf{I} = (400 \underline{/\ -90^\circ})\ (0.6435 \underline{/\ -9.44^\circ})$$

$$= 257.4 \underline{/\ -99.44^\circ}\ \text{V}$$

Also, $v_c = \sqrt{2} \times 257.4 \sin(314\,t - 99.44^\circ)$ V

(c) Power consumed in the circuit

$$P = I^2R = (0.6435)^2 \times 300 = 124.2\ \text{W}$$

Alternatively,

$$P = VI\cos\phi = 250 \times 0.6435 \cos[30^\circ - (-9.44^\circ)]$$

$$= 124.2\ \text{W}$$

■ **EXAMPLE 6.15** A two-element series circuit has voltage $\mathbf{V} = 200 \underline{/\ 0^\circ}$ V and circuit $\mathbf{I} = 10 \underline{/\ -30^\circ}$ A. Determine the current which results when the resistance is reduced to (a) 25%, (b) 50% of its former value.

SOLUTION :

$$\mathbf{Z} = \frac{\mathbf{V}}{\mathbf{I}} = \frac{200 \underline{/\ 0^\circ}}{10 \underline{/\ -30^\circ}} = 20 \underline{/\ 30^\circ} = (17.32 + j\,10)\ \Omega$$

(a) 25% of 17.32 = 4.33 Ω

therefore, $\mathbf{Z}_1 = 4.33 + j\,10 = 10.897 \underline{/\ 66.58^\circ}\ \Omega$

$$\mathbf{I}_1 = \frac{\mathbf{V}}{\mathbf{Z}_1} = \frac{200 \underline{/\ 0^\circ}}{10.897 \underline{/\ 66.58^\circ}}$$

$$= 18.35 \underline{/\ -66.58^\circ}\ \text{A}$$

(b) 50% of 17.32 = 8.66

therefore, $\mathbf{Z}_2 = 8.66 + j\,10 = 13.23 \underline{/\ 49.1^\circ}\ \Omega$

$$\mathbf{I}_2 = \frac{\mathbf{V}}{\mathbf{Z}_2} = \frac{200 \underline{/\ 0^\circ}}{13.23 \underline{/\ 49.1}} = 15.1 \underline{/\ -49.1^\circ}\ \text{A}$$

■ **EXAMPLE 6.16** Two coils A and B are connected in series to a supply of 230 V, 50 Hz. Coil A has an inductance of 0.2 H and a resistance of 20 Ω, and coil B has an inductance of 0.05 H and a resistance of 60 Ω. Calculate the current and its phase angle relative to the supply voltage. Also, determine the voltage across each coil.

SOLUTION :

$$\mathbf{Z}_A = R_A + jX_{LA} = R_A + j\,2\pi f L_A$$

$$= 20 + j\,2\pi \times 50 \times 0.2 = 20 + j\,62.8$$

$$Z_B = R_B + jX_{LB} = R_B + j\,2\pi f L_B$$

$$= 60 + j\,2\pi \times 50 \times 0.05 = 60 + j\,15.7$$

Total impedance of the circuit

$$\mathbf{Z} = \mathbf{Z}_A + \mathbf{Z}_B$$

$$= 20 + j\,62.8 + 60 + j\,15.7$$

$$= 80 + j\,78.5 = 112.08 \underline{/ 44.45^\circ}\ \Omega$$

Let the supply voltage **V** be taken as the reference. Therefore

$$\mathbf{V} = V \underline{/ 0^\circ} = 230 \underline{/ 0^\circ}$$

Current in the circuit

$$\mathbf{I} = \frac{\mathbf{V}}{\mathbf{Z}} = \frac{230 \underline{/ 0^\circ}}{112.08 \underline{/ 44.45^\circ}}$$

$$= 2.052 \underline{/ -44.45^\circ}\ \text{A}$$

Voltage drop across coil A

$$\mathbf{V}_A = \mathbf{Z}_A\mathbf{I} = (20 + j\,62.8)\,(2.052 \underline{/ -44.45^\circ})$$

$$= (65.908 \underline{/ 72.33^\circ})\,(2.052 \underline{/ -44.45})$$

$$= 135.24 \underline{/ 27.88^\circ}\ \text{V}$$

Voltage drop across coil B

$$\mathbf{V}_B = \mathbf{Z}_B\,\mathbf{I} = (60 + j\,15.7)\,(2.052 \underline{/ -44.45^\circ})$$

$$= (65.908 \underline{/ 14.66^\circ})\,(2.052 \underline{/ -44.45^\circ})$$

$$= 127.26 \underline{/ -29.79^\circ}\ \text{V}$$

■ **EXAMPLE 6.17** An iron cored coil A connected to a 100 V, 50 Hz supply is found to take a current of 5 A and to dissipate a power of 200 W. When the same supply is connected to coil B, the current is 8 A and the power dissipated is 450 W. Calculate the current and power taken when the two coils are connected in series across the same 100 V, 50 Hz supply.

SOLUTION : ***Coil A***

$$P_A = I_A^2\,R_A$$

$$200 = (5)^2\,R_A, \quad R_A = \frac{200}{25} = 8\ \Omega$$

$$Z_A = \frac{V_A}{I_A} = \frac{100}{5} = 20\ \Omega$$

Inductive reactance

$$X_A = \sqrt{Z_A^2 - R_A^2}$$

$$= \sqrt{20^2 - 8^2} = 18.33\ \Omega$$

therefore, $\mathbf{Z}_A = R_A + jX_A = 8 + j\,18.33\ \Omega$

Coil B

$$P_B = I_B^2\,R_B$$

$$450 = (8)^2 R_B, \quad R_B = \frac{450}{64} = 7.03\ \Omega$$

$$Z_B = \frac{V_B}{I_B} = \frac{100}{8} = 12.5\ \Omega$$

Inductive reactance,

$$X_B = \sqrt{Z_B^2 - R_B^2}$$

$$= \sqrt{12.5^2 - 7.03^2} = 10.36\ \Omega$$

$$\mathbf{Z}_B = R_B + jX_B = 7.03 + j\,10.36$$

When the two coils are connected in series, the total impedance

$$\mathbf{Z} = \mathbf{Z}_A + \mathbf{Z}_B = 8 + j\,18.33 + 7.03 + j\,10.36$$

$$= 15.03 + j\,28.69 = 32.388\ \underline{/\ 62.35^\circ}\ \Omega$$

Current taken by the combination

$$\mathbf{I} = \frac{\mathbf{V}}{\mathbf{Z}} = \frac{100\ \underline{/\ 0^\circ}}{32.388\ \underline{/\ 62.35^\circ}} = 3.087\ \underline{/\ -62.35^\circ}\ \text{A}$$

Power dissipated by the combination

$$P = VI\cos\phi = 100 \times 3.087 \cos 62.35^\circ = 143.26\ \text{W}$$

Alternatively

$$P = I^2R = (3.087)^2 \times 15.03 = 143.23\ \text{W}$$

EXERCISES

1. A series combination of resistance and inductance when connected across a 100 V dc supply dissipates 500 W. When connected across a 100 V, 50 Hz supply dissipates 200 V. Calculate the values of resistance and inductance.
2. Find the two elements in a series circuit, given that the current and voltage are

   $$i = 5 \sin(314\,t - 30^\circ)\ \text{A}$$

   $$v = 20 \sin(314\,t + 15^\circ)\ \text{V}$$
3. A circuit takes a current

   $$i = 20 \sin\left(314\,t - \frac{\pi}{6}\right)\ \text{A}$$

   when the supply voltage is

   $$v = 100 \sin 314\,t\ \text{V}.$$

   Calculate the impedance, phase angle, resistance and inductance of the circuit
4. (a) When a certain inductive coil is connected to a d.c. supply at 240 V, the current in the coil is 16 A. When the same coil is connected to an a.c. supply at 240 V, 50 Hz, the current is 12.27 A. Calculate the resistance, impedance, reactance and inductance of the coil.

(b) If the supply frequency were to be altered to 60 Hz at 240 V, would the current in the coil be greater or lesser than the value given above ? Give reasons for your answer.

5. A 240 V, single-phase induction motor delivers 1.64 kW at full load. The efficiency of the motor at this load is 0.86 p.u. and the power factor is 0.75 lagging. Calculate the input current, input power, and input kVA to the motor.
6. A single-phase ac motor takes, 8 A at 250 V. Assuming a power factor of 0.75 lagging and an efficiency of 0.80 p.u., determine the power input and output.
7. A noninductive resistor R is connected in series with a coil across a 230 V, 50 Hz supply. The voltage across the resistor and coil are 100 V and 180 V respectively. The circuit current is 5A. Calculate : (a) the resistance R ; (b) the impedance of the coil ; (c) the impedance of the circuit ; (d) the resistance of the coil ; (e) the power factor of the circuit ; (f) the power loss in the circuit.
8. An iron-cored coil connected to a 100 V, 50 Hz supply is found to take a current of 5A and to dissipate a power of 200 W. Find : (a) the impedance ; (b) the effective resistance ; (c) the inductance ; (d) the circuit power factor.
9. The voltage across a circuit is given by $(300 + j60)$ V and the current through it by $(10 - j5)$ A. Determine (a) the active power ; (b) the reactive voltampers ; (c) the apparent power.
10. An impedance $80 \underline{/ 60^\circ}$ Ω is connected in series with another of $50 \underline{/ -45^\circ}$ Ω, and the combination takes a current of 1.5 A. Calculate the total voltage and its phase angle relative to the current.

ANSWERS

1. $R = 20\ \Omega,\ L = 0.078\ H$
2. $R = 2.828\ \Omega,\ L = 9\ mH$
3. $Z = 5\Omega,\ \phi = 30^\circ$ (lag.)
   $R = 4.33\ \Omega,\ L = 7.96\ mH$
4. (a) $R = 15\ \Omega,\ Z = 19.6\ \Omega,\ X_L = 12.6\ \Omega,\ L = 0.04\ H$
   (b) The current decreases.
5. $I = 10.59\ A,\ P_i = 1.91\ kW,\ S_i = 2.54\ kVA$
6. $1500\ W;\ 1200\ W$
7. (a) $R = 20\ \Omega$; (b) $Z_{coil} = 36\ \Omega$; (c) $Z = 46\ \Omega$;
   (d) $R_{coil} = 10.5\ \Omega$; (e) $\cos\phi = 0.663$ (lag) ; (f) $P = 76.25\ W$
8. $20\ \Omega$; 8Ω , $57.3\ mH$; 0.4 (lag)
9. (a) $2.7\ kW$; (b) $2.1\ kVAr$ (lagging) ; (c) $3.42\ kVA$
10. $\mathbf{V} = 123.95 \underline{/ 24.24^\circ}$ V

7

SINGLE-PHASE PARALLEL AND SERIES-PARALLEL A.C. CIRCUITS

7.1 PARALLEL CIRCUITS

A parallel circuit consists of two or more branches connected in parallel. The voltage is the same for each branch, but the branch currents may differ in magnitude and phase depending on the branch impedances. Each branch of the circuit is solved separately. The voltage is taken as the reference phasor. The magnitude and phase of each current is determined. The total current in the circuit is equal to the phasor sum of the branch currents.

7.2 STEP-BY-STEP CALCULATION PROCEDURE FOR SINGLE-PHASE PARALLEL CIRCUITS

1. Solve each branch by the procedure given for single-phase series circuits.
2. Determine the current through each branch in rectangular form.
3. Add the branch currents in rectangular form to determine the total supply current $\mathbf{I}_T$.
4. Determine the total current in polar form. This gives the magnitude I_T and phase difference ϕ between $\mathbf{V}$ and $\mathbf{I}_T$.
5. Determine the overall power factor $\cos \phi$.
6. Determine the total impedance $\mathbf{Z}_T$ by Ohm's law

$$\mathbf{Z}_T = \frac{\mathbf{V}}{\mathbf{I}_T}$$

The total impedance of the parallel circuit may also be found by the relation

$$\frac{1}{\mathbf{Z}_T} = \frac{1}{\mathbf{Z}_1} + \frac{1}{\mathbf{Z}_2} + \ldots + \frac{1}{\mathbf{Z}_n}$$

For two impedance in parallel

$$\frac{1}{\mathbf{Z}_T} = \frac{1}{\mathbf{Z}_1} + \frac{1}{\mathbf{Z}_2}$$

$$\mathbf{Z}_T = \frac{\mathbf{Z}_1 \mathbf{Z}_2}{\mathbf{Z}_1 + \mathbf{Z}_2} = \frac{\text{product of impedances}}{\text{sum of impedances}}$$

When we solve for $\mathbf{Z}_T$, we should keep $\mathbf{Z}_1$ and $\mathbf{Z}_2$ in polar form in the numerator to make multiplication easier. Magnitudes are multiplied and angles are added algebraically. In the denominator, $\mathbf{Z}_1$ and $\mathbf{Z}_2$ are changed to rectangular form in order to add them. The resultant sum is then changed to polar form. Now divide the numerator by the denominator and subtract the angles algebraically.

7.3 ADMITTANCE, CONDUCTANCE AND SUSCEPTANCE

The admittance **Y** is defined as the reciprocal of impedance

$$\mathbf{Y} \triangleq \frac{1}{\mathbf{Z}} = \frac{\mathbf{I}}{\mathbf{V}} \tag{7.3.1}$$

The unit of admittance is siemen (S). Like impedance admittance may also be expressed in complex form.

Pure resistance

For a circuit containing pure resistance R, the admittance is

$$\mathbf{Y}_R = \frac{1}{\mathbf{Z}_R} = \frac{1}{R} = G \tag{7.3.2}$$

where G is the *conductance.* it is the reciprocal of resistance.

Pure inductance

For a circuit containing pure inductance L, the admittance is

$$\mathbf{Y} = \frac{1}{\mathbf{Z}_L} = \frac{1}{jX_L} = -j\frac{1}{X_L} = -jB_L \tag{7.3.3}$$

where B_L is called the *inductive susceptance.* It is the reciprocal of the inductive reactance X_L.

Inductive susceptance

$$B_L = \frac{1}{X_L} = \frac{1}{\omega L} \tag{7.3.4}$$

Pure capacitance

For a circuit containing pure capacitance C, the admittance is

$$\mathbf{Y}_C = \frac{1}{\mathbf{Z}_C} = \frac{1}{-jX_C} = j\frac{1}{X_C} = j\omega C = jBC \tag{7.3.5}$$

where B_C is called the *capacitive susceptance.* It is the reciprocal of capacitor reactance X_C.

Capacitive susceptance

$$B_C = \frac{1}{X_C} = \omega C \tag{7.3.6}$$

A circuit in which $X_L > X_C$ is called an *inductive circuit*, similarly, a circuit in which $X_C > X_L$ is called a *capacitive circuit.*

7.4 ADMITTANCES IN PARALLEL

The circuit in Fig. 7.1 shows three admittances in *parallel*. Each admittance could be a single resistor, inductor, or capacitor, or a combination of these three elements. Since the admittance are all connected in parallel, the *voltage* must be the *same* across each of them.

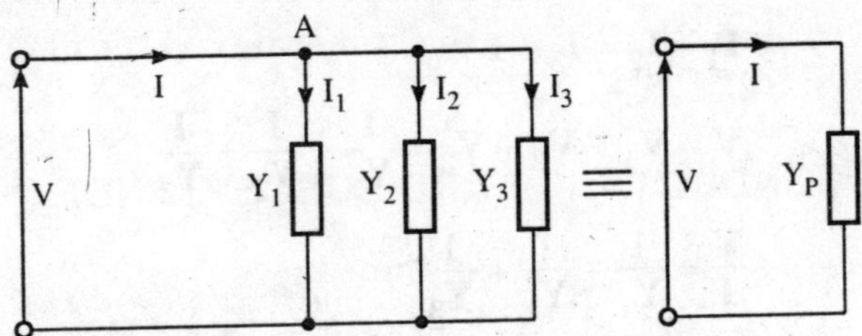

FIG. 7.1 (a) Admittances in parallel , (b) Equivalent circuit of (*a*)

The currents in impedances are

$$\mathbf{I}_1 = \mathbf{V}\mathbf{Y}_1,\ \ \mathbf{I}_2 = \mathbf{V}\mathbf{Y}_2,\ \ \mathbf{I}_3 = \mathbf{V}\mathbf{Y}_3$$

By KCL at node *A*

$$\mathbf{I} = \mathbf{I}_1 + \mathbf{I}_2 + \mathbf{I}_3 \tag{7.4.1}$$

Substituting the values of current we get

$$\mathbf{I} = \mathbf{V}\mathbf{Y}_1 + \mathbf{V}\mathbf{Y}_2 + \mathbf{V}\mathbf{Y}_3$$

$$\frac{\mathbf{I}}{\mathbf{V}} = \mathbf{Y}_1 + \mathbf{Y}_2 + \mathbf{Y}_3 \tag{7.4.2}$$

But **I/V** is the total admittance $\mathbf{Y_p}$ of the circuit shown in Fig. 7.1b.

therefore,

$$\mathbf{Y}_p = \mathbf{Y}_1 + \mathbf{Y}_2 + \mathbf{Y}_3 \tag{7.4.3}$$

This result can be extended to a parallel circuit consisting of any number of admittances. Thus, the total admittance of a parallel circuit is

$$\mathbf{Y}_p = \mathbf{Y}_1 + \mathbf{Y}_2 + \mathbf{Y}_3 + \ldots + \mathbf{Y}_n \tag{7.4.4}$$

where n is the total number of admittances.

The admittance concept is particularly useful in dealing with parallel circuits since admittances of parallel elements add directly.

7.5 ADMITTANCES IN SERIES

The circuit in Fig. 7.2 shows three admittances $\mathbf{Y_1}$, $\mathbf{Y_2}$ and $\mathbf{Y_3}$ in series. The current I is the same through each admittance.

The voltage across the admittances are

$$\mathbf{V}_1 = \frac{\mathbf{I}_1}{\mathbf{Y}_1},\ \mathbf{V}_2 = \frac{\mathbf{I}_2}{\mathbf{Y}_2},\ \mathbf{V}_3 = \frac{\mathbf{I}_3}{\mathbf{Y}_3}$$

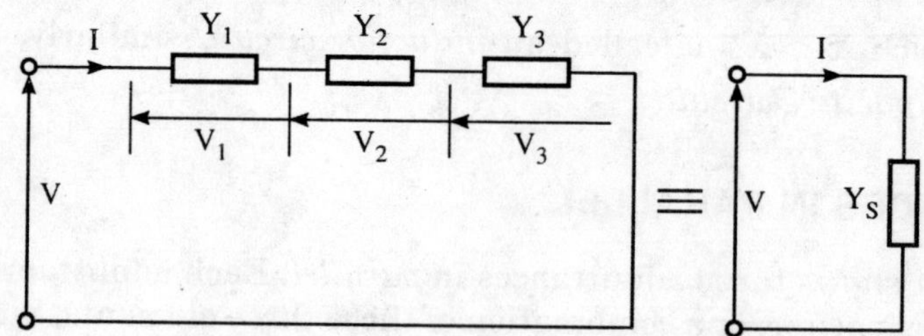

FIG. 7.2 (a) Admittances in series, (b) Equivalent circuit of (*a*)

For a series circuit $\mathbf{I}_1 = \mathbf{I}_2 = \mathbf{I}_3 = \mathbf{I}$

By KVL $\mathbf{V} = \mathbf{V}_1 + \mathbf{V}_2 + \mathbf{V}_3 = \dfrac{\mathbf{I}}{\mathbf{Y}_1} + \dfrac{\mathbf{I}}{\mathbf{Y}_2} + \dfrac{\mathbf{I}}{\mathbf{Y}_3}$

$$\frac{\mathbf{V}}{\mathbf{I}} = \frac{1}{\mathbf{Y}_1} + \frac{1}{\mathbf{Y}_2} + \frac{1}{\mathbf{Y}_3}$$

But **I/V** is the total admittance $\mathbf{Y}_S$ of the circuit shown in Fig. 7.2b.

therefore,

$$\frac{1}{\mathbf{Y}_S} = \frac{1}{\mathbf{Y}_1} + \frac{1}{\mathbf{Y}_2} + \frac{1}{\mathbf{Y}_3} \tag{7.5.1}$$

If the circuit has n admittances in series.

$$\frac{1}{\mathbf{Y}_S} = \frac{1}{\mathbf{Y}_1} + \frac{1}{\mathbf{Y}_2} + \frac{1}{\mathbf{Y}_3} + \ldots + \frac{1}{\mathbf{Y}_n} \tag{7.5.2}$$

and

$$\mathbf{Y}_S = \frac{1}{\dfrac{1}{\mathbf{Y}_1} + \dfrac{1}{\mathbf{Y}_2} + \dfrac{1}{\mathbf{Y}_3} + \ldots + \dfrac{1}{\mathbf{Y}_n}} \tag{7.5.3}$$

This result is generally quite tedious to use.

7.6 ADMITTANCE OF A PARALLEL RLC CIRCUIT

Fig. 17.3 shows a parallel RLC circuit.

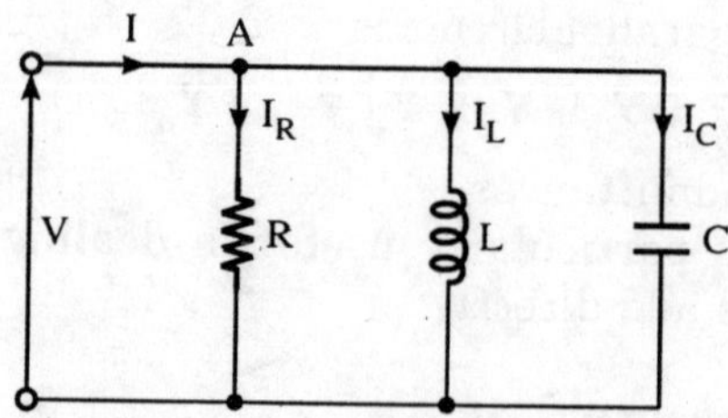

FIG. 7.3 A parallel RLC circuit

Since the elements R, L and C are in parallel, the voltage is the same across each element. The currents in R, L and C are

$$\mathbf{I}_R = \frac{\mathbf{V}}{R},\ \mathbf{I}_L = \frac{\mathbf{V}}{jX_L},\ \mathbf{I}_C = \frac{\mathbf{V}}{-jX_C}$$

By KCL at node A

$$\mathbf{I} = \mathbf{I}_R + \mathbf{I}_L + \mathbf{I}_C \quad (7.6.1)$$

$$= \frac{\mathbf{V}}{R} + \frac{\mathbf{V}}{jX_L} + \frac{\mathbf{V}}{-jX_C}$$

$$= \mathbf{V}\left(\frac{1}{R} + \frac{1}{jX_L} + \frac{1}{-jX_C}\right)$$

Multiplying the numerator and denominator of the two j-terms by j, we get

$$\frac{\mathbf{I}}{\mathbf{V}} = \frac{1}{R} + \frac{j}{j^2X_L} + \frac{j}{-j^2X_C}$$

$$\mathbf{Y} = \frac{1}{R} + j\frac{1}{X_C} - j\frac{1}{X_L} \qquad [\because j^2 = -1]$$

or

$$\mathbf{Y} = G + jB_C - jB_L$$

$$\mathbf{Y} = G + j(B_C - B_L) \quad (7.6.2)$$

or

$$\mathbf{Y} = G + jB \quad (7.6.3)$$

where

$$G = \frac{1}{R}$$

$$B = B_C - B_L$$

$$B_C = \frac{1}{X_C}$$

$$B_L = \frac{1}{X_L}$$

The quantity G is called the *conductance* of the circuit. It is the real part of **Y**. The quantity B is called the *total susceptance or net susceptance* of the circuit. It is the imaginary part of **Y**. The susceptance of an inductance is specifically called inductive susceptance, B_L. The susceptance of a capacitance is called *capacitive susceptance*, B_C. Thus

conductance $\quad G = \frac{1}{R}$

inductive susceptance $\quad B_L = \frac{1}{X_L} = \frac{1}{\omega L}$

capacitive susceptance $\quad B_C = \frac{1}{X_C} = \omega C$

Total susceptance $\quad B = B_C - B_L = \frac{1}{X_C} - \frac{1}{X_L}$

The quantities G, B_L, B_C and B are all the reciprocals of quantities expressed in ohms and therefore G, B_L, B_C and B are expressed in *siemens* (S).

7.7 ADMITTANCE DIAGRAM

The admittance of a circuit is a complex quantity. It can be expressed either in rectangular form or polar form. The rectangular form is

$$\mathbf{Y} = G + j\,(B_C - B_L) \tag{7.7.1}$$

where

$$Y = \sqrt{G^2 + (B_C - B_L)^2}$$

$$\tan \phi_Y = \frac{B_C - B_L}{G},\ \phi_Y = \tan^{-1}\frac{B_C - B_L}{G}$$

The polar form of admittance is

$$\mathbf{Y} = Y \angle \phi_Y \tag{7.7.2}$$

The admittance of a circuit expressed in either the rectangular form or the polar form can be plotted on the complex plane as shown in Fig. 7.4.

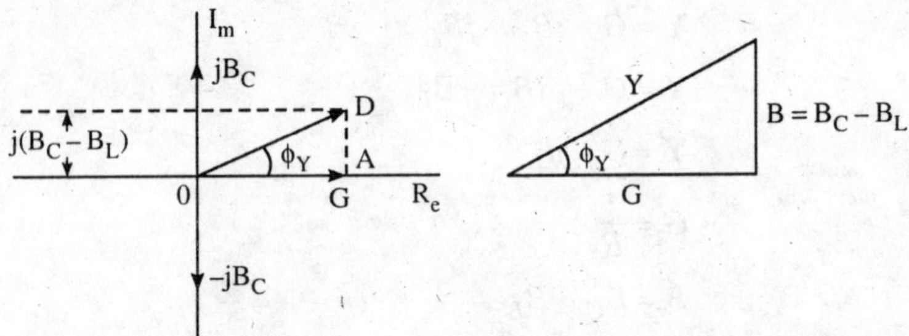

FIG. 7.4 Admittance diagram. **FIG. 7.5 Admittance triangle**

The triangle OAD is called the *admittance triangle.* It is shown separately in Fig. 7.5.

From admittance triangle $Y = \sqrt{G^2 + (B_C - B_L)^2}$ (7.7.3)

$$\phi_Y = \tan^{-1}\frac{B_C - B_L}{G} \tag{7.7.4}$$

$$B_C - B_L = Y \sin \phi_Y \tag{7.7.5}$$

$$G = Y \cos \phi_Y \tag{7.7.6}$$

where ϕ_Y is the phase angle between voltage and current.

If net susceptance is B, then $B = B_C - B_L$

$$Y = G + jB$$

$$Y^2 = G^2 + B^2$$

$$\cos \phi_Y = G/Y,\ \sin \phi_Y = \frac{B}{Y},\ \tan \phi_Y = \frac{B}{G}$$

■ **EXAMPLE 7.1** A parallel circuit consists of two branches, one consisting of a noninductive resistor of 150 Ω in series with an inductor of inductance 318 μH, the other a noninductive resistance of 300 Ω in series with a capacitor of 0.01 μF capacitance. The circuit is connected across a 25V, 100 kHz supply, Determine (*a*) the current in each branch; (*b*) the total current; (*c*) the circuit phase angle; (*d*) the circuit impedance; (*e*) the components of an equivalent circuit consisting of a resistance and reactance.

SOLUTION :

$$R_1 = 150\ \Omega,\ L_1 = 318\ \mu H = 318 \times 10^{-6}\ H$$

$$R_2 = 300\ \Omega,\ C_2 = 0.01\ \mu F = 0.01 \times 10^{-6}\ F$$

$$f = 100\ kHz = 100 \times 10^3\ Hz$$

(a) Branch 1

$$R_1 = 150\ \Omega,\ L_1 = 318\ \mu H = 318 \times 10^{-6} H$$

Inductive reactive

$$X_{L_1} = 2\pi f L_1 = 2\pi \times 100 \times 10^3 \times 318 \times 10^{-6} \Omega$$

2 × [π] × 100 × 3 [10^x] × 318 × 6 [+/–] [10^x] [=]
↓
199.8

therefore, $X_{L_1} = 199.8\ \Omega$

Impedance of branch 1

$$\mathbf{Z}_1 = R_1 + jX_L = 150 + j199.8$$

For changing from rectangular to polar

R→P
150 [INV] [+] 199.8 [=] [X↔Y]
↓ ↓
249.84 53.1

therefore, $\mathbf{Z}_1 = 150 + j199.8 = 249.84 \angle 53.1°\ \Omega$

Let the supply voltage be taken as reference

$$\mathbf{V} = V \angle 0° = 25 \angle 0°\ V$$

Current through branch 1

$$\mathbf{I}_1 = \frac{\mathbf{V}}{\mathbf{Z}_1} = \frac{25 \angle 0°}{249.84 \underline{/53.1°}} = 0.1 \underline{/-53.1°}\ A$$

This current is changed to rectangular form as follows :

R→P
0.1 [INV] [–] 53.1 [+/–] [=] [X↔Y]
↓ ↓
0.06 – 0.07996

therefore, $0.1 \angle -53.1° = (0.06 - j0.08)\ A$

Branch 2

$$R_2 = 300\ \Omega,\ C_2 = 0.01\ \mu F = 0.01 \times 10^{-6}\ F$$

Capacitive reactance $X_{C_2} = \dfrac{1}{2\pi f C_2} = \dfrac{1}{2\pi \times 100 \times 10^3 \times 0.01 \times 10^{-6}}$

$2 \times \pi \times 100 \times 3$ [10^x] $\times 0.01 \times 6$ [+/–] [10^x] [=] [$1/x$]
↓
159.15

therefore, $X_{C_2} = 159\ \Omega$

Impedance $\mathbf{Z}_2 = R_2 - jX_{C_2}$

$= 300 - j159\Omega$

For changing from rectangular to polar

$R \to P$
300 [INV] [+] 159 [+/–] [=] [$X \leftrightarrow Y$]
↓ ↓
339.5 – 27.92

therefore, $\mathbf{Z}_2 = 300 - j159 = 339.5 \angle -27.92°\ \Omega$

By Ohm's law, the current through branch 2 is

$$\mathbf{I}_2 = \frac{\mathbf{V}}{\mathbf{Z}_2}$$

$$= \frac{25 \angle 0°}{339.5 \angle -27.92°} = 0.0736 \angle 27.92°\ A$$

Polar to rectangular

$R \to P$
0.0736 [INV] [–] 27.92 [=] [$X \leftrightarrow Y$]
↓ ↓
0.065 0.0344

therefore, $\mathbf{I}_2 = 0.0736 \angle 27.92° = 0.065 + j0.0344$ A

(b) Total circuit current I_T

By KCL, the total current is equal to the phasor sum of branch currents.

$\mathbf{I}_T = \mathbf{I}_1 + \mathbf{I}_2$

$= 0.06 - j0.08$

$+ 0.065 + j0.0344$

$\mathbf{I}_T = 0.125 - j0.0456$ A

For changing this value to polar form

$R \to P$
0.125 [INV] [+] 0.0456 [+/–] [=] [$X \leftrightarrow Y$]
↓ ↓
0.133 – 20.04

therefore, $\mathbf{I}_T = 0.125 - j0.0456 = 0.133 \angle -20°\ A$

(c) The current $\mathbf{I}_T$ lags behind the voltage by 20°. That is, $\phi = -20°$.

(d) Impedance of the circuit

$$\mathbf{Z}_T = \frac{\mathbf{V}}{\mathbf{I}_T} = \frac{25 \angle 0°}{0.133 \angle -20°} = 187.97 \angle 20$$

(e) Polar → rectangular

$$R \to P$$

187.97 [INV] [–] 20 [=] [X↔Y]

176.63 64.28

therefore, $\mathbf{Z}_T = 187.97 \angle 20° = 176.63 + j64.28\ \Omega$

This is of the form $\mathbf{Z} = R + jX_L$

Thus, the circuit is equivalent to 176.63 Ω resistor is series with a 64.28 Ω *inductive* reactance.

■ **EXAMPLE 7.2** Two circuits, the impedances of which are given by $Z_1 = 10 + j15\ \Omega$ and $Z_2 = 6 - j8\ \Omega$ are connected in parallel. If the total current supplied is 15 A, what is the power taken by each branch ?

SOLUTION :

$$\mathbf{Z}_1 = 10 + j15 = 18.027 \angle 56.31°\ \Omega$$

$$\mathbf{Z}_2 = 6 - j8 = 10 \angle -53.13°\ \Omega$$

$$\mathbf{Z}_p = \mathbf{Z}_1 \,||\, \mathbf{Z}_2 = \frac{\mathbf{Z}_1\,\mathbf{Z}_2}{\mathbf{Z}_1 + \mathbf{Z}_2}$$

$$= 18.027 \times 10 \frac{/56.31° - 53.13°}{10 + j15 + 6 - j8}$$

$$= \frac{180.27 \angle 3.18°}{16 + j7} = \frac{180.27 \angle 3.18°}{17.46 \angle 23.63°}$$

$$= 10.325 \angle -20.45°\ \Omega$$

Total supply voltage

$$V = IZ_p = 15 \times 10.325 = 154.9 \text{ V}$$

$$I_1 = \frac{V}{Z_1} = \frac{154.9}{18.027} = 8.59 \text{ A}$$

$$P_1 = I_1^2 R_1 = (8.59)^2 \times 10 = 738 \text{ W}$$

$$I_2 = \frac{V}{Z_2} = \frac{154.9}{10} = 15.49 \text{ A}$$

$$P_2 = I_2^2 R_2 = (15.49)^2 \times 6 = 1439.6 \text{ W}$$

■ **EXAMPLE 7.3** Two inductive coils *A* and *B* are in parallel across a 100V, 50 Hz supply. Coil

A takes 12 A at 0.9 power factor and the total current for both the coils is 20 A at 0.8 power factor. Determine (*a*) the equivalent resistance and equivalent reactance, (*b*) the individual resistances and reactances of both coils, (*c*) the power factor of coil *B*.

SOLUTION : Let the supply voltage *V* be taken as the reference phasor :

therefore, $\mathbf{V} = 100 \angle 0° = 100 + j0\text{V}$

$$I_A = 12\text{A}, \cos \phi_A = 0.9 \text{ lagging}$$

The power factor is lagging since the coil is inductive.

therefore, $\phi_A = -\cos^{-1} 0.9 = -25.84°$

$$\mathbf{I}_A = 12 \angle -25.84° = 10.8 - j5.23 \text{ A}$$

Total current $I = 20$ A

Overall power factor $\cos \phi = 0.8$ lagging

therefore, $\mathbf{I} = I \angle -\phi = 20 \angle -\cos^{-1} 0.8 = 20 \angle -36.87° \text{ A}$

$$= 16 - j12 \text{ A}$$

(a) Equivalent impedance of the circuit

$$\mathbf{Z}_{eq} = \frac{\mathbf{V}}{\mathbf{I}} = \frac{100 \angle 0°}{20 \angle -36.87°} = 5 \angle +36.87° \text{ A}$$

$$= 4 + j3 = R_{eq} + jX_{eq}$$

therefore, $R_{eq} = 4\ \Omega,\ X_{eq} = 3\ \Omega$

Impedance of coil *A*

$$\mathbf{Z}_A = \frac{\mathbf{V}}{\mathbf{I}_A} = \frac{100 \angle 0°}{12 \angle 25.84°} = 8.33 \angle 25.84°\ \Omega$$

$$= 7.5 + j3.63 = R_A + jX_A$$

therefore, $R_A = 7.5\ \Omega,\quad X_A = 3.63\ \Omega$

By *KCL* $\mathbf{I}_A + \mathbf{I}_B = \mathbf{I}$

$$\mathbf{I}_B = \mathbf{I} - \mathbf{I}_A = 16 - j12 - (10.8 - j5.23)$$

$$= 5.2 - j6.77 = 8.536 \angle -52.47° \text{ A}$$

Impedance of coil *B*

$$\mathbf{Z}_B = \frac{\mathbf{V}}{\mathbf{I}_B} = \frac{100 \angle 0°}{8.536 \angle -52.47°}$$

$$= 11.71 \angle 52.47° = 7.133 + j9.286\ \Omega$$

Also, $\mathbf{Z}_B = R_B + jX_B$

therefore, $R_B = 7.133\ \Omega$

and $X_B = 9.286\ \Omega$

$$\cos \phi_B = \cos(-52.47°) = 0.6091 \text{ (lagging)}$$

■ **EXAMPLE 7.4** Two circuits having the same numerical ohmic impedance are joined in parallel. The power factor of one circuit is 0.8 and the other 0.6. What is the power factor of the combination ?

SOLUTION : Let Z be the impedance of each circuit

therefore,
$$\mathbf{Z}_1 = Z \angle \cos^{-1} 0.8 = Z \angle 36.87^\circ = Z\,(0.8 + j0.6)$$
$$\mathbf{Z}_2 = Z \angle \cos^{-1} 0.6 = Z\angle 53.13^\circ = Z\,(0.6 + j0.8)$$

Since the two impedances are connected in parallel, the equivalent impedance of the combination is given by

$$\mathbf{Z}_p = \mathbf{Z}_1 \,||\, \mathbf{Z}_2 = \frac{\mathbf{Z}_1\mathbf{Z}_2}{\mathbf{Z}_1 + \mathbf{Z}_2}$$
$$= \frac{(Z \angle 36.87^\circ)\,(Z \angle 53.13^\circ)}{Z\,(0.8 + j0.6 + 0.6 + j0.8)} = \frac{Z^2 \angle 90^\circ}{Z\,(1.4 + j1.4)}$$
$$= \frac{Z^2 \angle 90^\circ}{Z\,(1.98 \angle 45^\circ)} = \frac{Z \angle 45^\circ}{1.98}$$

Therefore, the power factor of the combination is

$$\cos\phi = \cos 45^\circ = 0.707$$

■ **EXAMPLE 7.5** A voltage of $200 \angle 30^\circ$ V is applied to two circuits connected in parallel. The currents in the respective branches are $20 \angle 60^\circ A$ and $40 \angle -30^\circ$ A. Find the kVA and kW in each branch circuit and in the main circuit.

SOLUTION :

$$\mathbf{V} = 200 \angle 30^\circ \text{ V} = 173.2 + j100 \text{ V}$$
$$\mathbf{I}_1 = 20 \angle 60^\circ \text{ V} = 10 + j17.32 \text{ A}$$
$$\mathbf{I}_2 = 40 \angle -30^\circ \text{ V} = 34.64 - j20 \text{ V}$$

$$\mathbf{S}_1 = \mathbf{V}I_1^* = (200 \angle 30^\circ)\,(20 \,\underline{/-60^\circ})$$
$$= 4000 \angle -30^\circ \text{ VA} = 3464 - j2000$$
$$S_1 = 4000 \text{ VA} = 4 \text{ kVA}$$
$$P_1 = \text{real part of } \mathbf{VI}_1^* = 3464 \text{ W} = 3.464 \text{ kW}$$

$$\mathbf{S}_2 = \mathbf{VI}_2^* = (200 \angle 30^\circ)\,(40 \angle +30^\circ)$$
$$= 8000 \angle 60^\circ \text{ VA} = 4000 + j6928$$
$$S_2 = 8000 \text{ VA} = 8 \text{ kVA}$$
$$P_2 = \text{Re}\,[\mathbf{VI}^*] = 4000 \text{ W} = 4 \text{ kW}$$

Total current of the circuit

$$\mathbf{I} = \mathbf{I}_1 + \mathbf{I}_2 = 10 + j17.32 + 34.64 - j20$$
$$= 44.64 - j2.68 = 44.72 \angle -3.43^\circ \text{ A}$$

$$\mathbf{S} = \mathbf{VI}^* = (200 \angle 30^\circ)(44.72 \angle +3.43^\circ)$$
$$= 8944 \angle 33.43^\circ \text{VA} = 7464 + j4927 = P + jQ$$

therefore,
$$S = 8944 \text{ VA} = 8.944 \text{ kVA}$$
$$P = 7464 \text{ W} = 7.464 \text{ kW}$$

■ **EXAMPLE 7.6** Three impedances $(6 + j5)\ \Omega$, $(8 - j6)\ \Omega$, and $(8 + j10)\ \Omega$ are connected in parallel. Calculate the current in each branch when the total current is 20 A.

SOLUTION : Let the total current **I** be taken as reference phasor.

$$\mathbf{I} = 20 \angle 0^\circ = 20 + j0 \text{ A}$$
$$\mathbf{Z}_1 = 8 + j5 = 7.81 \angle 39.8^\circ\ \Omega$$
$$\mathbf{Y}_1 = \frac{1}{\mathbf{Z}_1} = \frac{1}{6 + j5} = \frac{6 - j5}{(6 + j5)(6 - j5)}$$
$$= \frac{6 - j5}{6^2 + 5^2} = 0.09836 - j0.08196 \text{ S}$$
$$\mathbf{Z}_2 = 8 - j6 = 10 \angle -36.87^\circ\ \Omega$$
$$\mathbf{Y}_2 = \frac{1}{\mathbf{Z}_2} = \frac{1}{8 - j6} = \frac{8 - j6}{(8 + j6)(8 - j6)} = \frac{8 - j6}{8^2 + 6^2}$$
$$= 0.08 + j0.06 \text{ S}$$
$$\mathbf{Z}_3 = 8 + j10 = 12.8 \angle 51.34^\circ\ \Omega$$
$$\mathbf{Y}_3 = \frac{1}{\mathbf{Z}_3} = \frac{1}{8 + j10} = \frac{8 - j10}{(8 + j10)(8 - j10)} = \frac{8 - j10}{8^2 + 10^2}$$
$$= 0.04878 - j0.06097 \text{ S}$$

Total admittance of the circuit

$$\mathbf{Y} = \mathbf{Y}_1 + \mathbf{Y}_2 + \mathbf{Y}_3$$
$$= 0.09836 - j0.08196 + 0.08 - j0.06 + 0.04878 - j0.06097$$
$$= 0.22714 - j0.08293 = 0.2418 \angle -20.06^\circ \text{ S}$$

Total circuit voltage
$$\mathbf{V} = \mathbf{IZ} = \frac{\mathbf{I}}{\mathbf{Y}} = \frac{20 < 0^\circ}{0.2418 \angle -20.06^\circ} = 82.71 \angle 20.06^\circ \text{ V}$$

$$\mathbf{I}_1 = \frac{\mathbf{V}}{\mathbf{Z}_1} = \frac{82.71 \angle 20.06^\circ}{7.81 \angle 39.8^\circ}$$
$$= 10.59 \angle -19.74^\circ \text{ A} = 9.967 - j3.576 \text{ A}$$
$$\mathbf{I}_2 = \frac{\mathbf{V}}{\mathbf{Z}_2} = \frac{82.71 \angle 20.06^\circ}{10 \angle -36.87^\circ}$$
$$= 8.271 \angle 56.93^\circ \text{ A} = 4.513 + j6.930 \text{ A}$$
$$\mathbf{I}_3 = \frac{\mathbf{V}}{\mathbf{Z}_3} = \frac{82.71 \angle 20.06^\circ}{12.8 \angle -51.34^\circ}$$

$$= 6.46 \angle -31.28^\circ \text{ A} = 5.52 - j3.35 \text{ A}$$

Check

$$\mathbf{I}_1 + \mathbf{I}_2 + \mathbf{I}_3 = 9.967 - j3.576$$
$$+ 4.513 + j6.930$$
$$+ 5.52 - j3.354$$
$$\mathbf{I}_1 + \mathbf{I}_2 + \mathbf{I}_3 = 20 + j0 = \mathbf{I}$$

■ **EXAMPLE 7.7** **A single phase circuit consists of three parallel branches, the admittances of the branches are**

$$\mathbf{Y}_1 = 0.4 + j0.6$$
$$\mathbf{Y}_2 = 0.1 + j0.4$$
$$\mathbf{Y}_3 = 0.06 + j0.23$$

Determine the total admittance and impedance of the circuit.

SOLUTION :

Since the admittances are in parallel

$$\mathbf{Y} = \mathbf{Y}_1 + \mathbf{Y}_2 + \mathbf{Y}_3$$
$$= (0.4 + j0.6) + (0.1 + j0.42) + (0.06 + j0.23)$$
$$= (0.4 + 0.1 + 0.06) + j\,(0.6 + 0.42 + 0.23)$$
$$= 0.56 + j\,1.25$$
$$= \sqrt{(0.56)^2 + (1.25)^2} \angle \tan^{-1}\frac{1.25}{0.56}$$
$$= 1.3697 \angle 65.87^\circ \text{ S}$$

The impedance is

$$\mathbf{Z} = \frac{1}{\mathbf{Y}} = \frac{1}{1.3697 \angle 65.87^\circ} = 0.73 \angle -65.87^\circ\ \Omega$$

■ **EXAMPLE 7.8** **An alternating supply at 200 V is connected to a circuit having a conductance** $G = 0.03$ **S and a capacitive susceptance, at frequency of the supply,** $B = 0.04$ **S. Calculate the current, power factor, and power consumed in the circuit.**

SOLUTION :

$$Y = \sqrt{G^2 + B^2} = \sqrt{0.03^2 + 0.04^2} = 0.05 \text{ S}$$

Current $\quad I = YV = 0.05 \times 200 = 10\text{A}$

$$\cos\phi = \frac{G}{Y} = \frac{0.03}{0.05} = 0.6$$

Power consumed, $P = VI \cos\phi = 200 \times 10 \times 0.6 = 1200$ W

Alternative method

$$P = V^2 G = (200)^2 \times 0.03 = 1200 \text{ W}$$

EXAMPLE 7.9 A circuit comprises a conductance G in parallel with a susceptance B. Calculate the admittance $G+jB$ if the impedance is $(10+j5)\ \Omega$.

SOLUTION :

$$\mathbf{Y}=\frac{1}{\mathbf{Z}}=\frac{1}{10+j5}=\frac{10-j5}{(10+j5)\,(10-j5)}$$

$$=\frac{10-j5}{10^2+5^2}=\frac{2}{25}-j\frac{1}{25}=(0.08-j0.004)\ \text{S}$$

EXAMPLE 7.10 Two impedance, Z_1 and Z_2, are connected in parallel. The first branch takes a leading current of 16 A and has a resistance of 5 Ω, while the second branch takes a lagging current at power factor 0.8. The total power supplied is 5 kW, the applied voltage being $(100+j200)$ V. Determine the complex expressions for branch and total currents, and for the circuit constants.

SOLUTION :

$$\mathbf{V}=100+j200\ \text{V}=223.6\angle 63.43^\circ\ \text{V}$$

Branch 1

Since the first branch takes a leading current, it contains a capacitive reactance X_{C_1}.

Impedance $$Z_1=\frac{V}{I_1}=\frac{223.6}{16}=13.975\ \Omega$$

$$Z_1{}^2=R_1{}^2+X_{C_1}^2$$

$$(13.975)^2=5^2\times X_{C_1}^2$$

$$X_{C_1}^2=(13.975)^2-(5)^2=170.3,\ X_{C_1}=13.04\ \Omega$$

therefore, $$\mathbf{Z}_1=R_1-jX_{C_1}=5-j13.04=13.975\angle-69^\circ\ \Omega$$

By Ohm's law $$\mathbf{I}_1=\frac{\mathbf{V}}{\mathbf{Z}_1}=\frac{223.6\angle 63.43^\circ}{13.975\angle-69^\circ}=16\angle 132.43^\circ\ \text{A}=-10.79+j11.8\ \text{A}$$

Branch 2

Total power consumed in both the branches

$$P=I_1{}^2R_1+I_2{}^2R_2$$

$$5000=(16)^2\times 5+I_2{}^2R_2,\qquad I_2{}^2R_2=3720\ \text{W}$$

Since the power factor is lagging for branch 2, X_2 is inductive.

$$\cos\phi_2=\frac{R_2}{Z_2},\ Z_2=\frac{R_2}{\cos\phi_2}=\frac{R_2}{0.8}=1.25\,R_2$$

$$I_2=\frac{V}{Z_2}=\frac{223.6}{1.25\,R_2}=\frac{178.88}{R_2}$$

therefore, $$I_2R_2=178.88\ \text{V}$$

$$\frac{I_2^2 R_2}{I_2 R_2} = \frac{3720}{178.88}, \; I_2 = 20.796 \text{ A}$$

$$R_2 = \frac{I_2 R_2}{I_2} = \frac{178.88}{20.796} = 8.60 \text{ A}$$

$$Z_2 = 1.25\, R_2 = 1.25 \times 8.6 = 10.75\ \Omega$$

$$Z_2^2 = R_2^2 + X_2^2$$

$$X_2 = \sqrt{Z_2^2 - R_2^2} = \sqrt{(10.75)^2 - (8.6)^2} = 6.45\ \Omega$$

therefore,

$$\mathbf{Z}_2 = R_2 + jX_2 = 8.6 + j6.45 = 10.75 \angle 36.87^\circ\ \Omega$$

By Ohm's law

$$\mathbf{I}_2 = \frac{\mathbf{V}}{\mathbf{Z}_2} = \frac{223.6 \angle 63.43^\circ}{10.75 \angle 36.87^\circ}$$

$$= 20.8 \angle 26.56^\circ \text{ A} = 18.6 + j9.3 \text{ A}$$

By KCL

$$\mathbf{I} = \mathbf{I}_2 + \mathbf{I}_2$$

$$= -\,10.79 + j11.8 + 18.6 + j9.3$$

$$= 7.81 + j21.1 = 22.5 \angle 69.69^\circ \text{ A}$$

Total impedance of the circuit

$$\mathbf{Z} = \frac{\mathbf{V}}{\mathbf{I}} = \frac{223.6 \angle 63.43^\circ}{22.5 \angle 69.69^\circ}$$

$$= 9.938 \angle -6.26^\circ\ \Omega = 9.878 - j1.083\ \Omega$$

■ **EXAMPLE 7.11** An industrial load consists of the following :

(*a*) a load of 200 kVA at 0.8 power factor lagging.

(*b*) a load of 50 kW at unity power factor,

(*c*) a load of 48 kW at 0.6 power factor leading.

Calculate the total kW, total kVAr, total kVA and the overall power factor.

SOLUTION : By convention, Q_L is taken positive and Q_C negative.

Load a

$$\cos \phi_a = 0.8, \; \sin \phi_a = 0.6$$

$$\mathbf{S}_a = S_a \angle \phi_a = S_a (\cos \phi_a + j \sin \phi_a)$$

$$= 200\,(0.8 + j0.6) = (160 + j120) \text{ kVA}$$

Load b

$$\cos \phi_b = 1, \; \sin \phi_b = 0$$

$$P_b = S_b \cos \phi_b$$

$$50 = S_b \times 1, \; S_b = 50 \text{ kVA}$$

$$\mathbf{S}_b = S_b (\cos \phi_b + j \sin \phi_b) = 50 (1 + j0) = 50 + j\,0 \text{ kVA}$$

Load c

$$\cos \phi_C = 0.6, \; \sin \phi_C = 0.8$$

$$P_c = S_c \cos \phi_c$$

$$48 = S_c \times 0.6, \; S_c = \frac{48}{0.6} = 80 \text{ kVA}$$

Since the power factor $\cos \phi_c$ is leading, Q_C will be negative

therefore,
$$\mathbf{S}_c = S_c \angle -\phi = S_c (\cos \phi_c - j \sin \phi_C)$$
$$= 80 (0.6 - j0.8) = (48 - j64) \text{ kVA}$$

Total complex power

$$\mathbf{S} = \mathbf{S}_a + \mathbf{S}_b + \mathbf{S}_c$$
$$= 160 + j120$$
$$+ 50 + j0$$
$$+ 48 + j64$$
$$= 258 + j56 = 264 \angle 12.246° \text{ kVA}$$

Also,

$$\mathbf{S} = S \angle \phi = P + jQ$$

therefore,
$$S = 264 \text{ kVA}$$
$$P = 258 \text{ kW}$$
$$Q = 56 \text{ kVAr (lagging)}$$
$$\phi = 12.246°$$

Overall power factor, $\cos \phi = \cos 12.246° = 0.9772$ (lagging)

It is to be noted that the overall power factor is lagging because the overall Q is positive and by convention lagging Q is taken positive.

■ **EXAMPLE 7.12** Four loads are connected across a 230 V, 50 Hz line :

(*a*) lights 10 kVA at unity power factor,

(*b*) a motor 4 kW at power factor 0.8 lagging,

(*c*) a rectifier 3.6 kW at power factor 0.6 leading,

(*d*) a capacitor 8 kVA

Determine the total kW, the total kVAr, the total kVA, the overall power factor, and the total line current.

SOLUTION : By convention, Q_L is taken positive and Q_C negative.

Load a

$$\cos \phi_a = 1, \; \sin \phi_a = 0$$

$$\mathbf{S}_a = S_a \angle \phi_a = S_a (\cos \phi_a + j \sin \phi_a)$$

$$= 10\,(1 + j0) = (10 + j0)\text{ kVA}$$

Load b

$$\cos\phi_b = 0.8,\ \sin\phi_b = 0.6$$

$$P_b = S_b \cos\phi_b$$

$$4 = S_b \times 0.8,\ S_b = \frac{4}{0.8} = 5\text{ kVA}$$

$$\mathbf{S}_b = S_b\,(\cos\phi_b + j\sin\phi_b)$$

$$= 5\,(0.8 + j0.6) = (4 + j3)\text{ kVA}$$

Load c

$$\cos\phi_c = 0.6,\ \sin\phi_c = 0.8$$

$$P_c = S_c \cos\phi_c$$

$$3.6 = S_c \times 0.6,\ S_c = \frac{3.6}{0.6} = 6\text{ kVA}$$

$$\mathbf{S}_c = S_c \angle -\phi_c = S_c\,(\cos\phi_c - j\sin\phi_c)$$

$$= 6\,(0.6 - j0.8) = (3.6 - j4.8)\text{ kVA}$$

Load d

For a capacitor, the current leads the voltage by 90°. Therefore, the power factor $\cos\phi_d$ is leading.

$$\cos\phi_d = \cos 90^\circ = 0\text{ (leading)}$$

$$\sin\phi_d = \sin 90^\circ = 1$$

$$P_d = S_d \cos\phi_d$$

$$P_d = 8 \times 0,\ P_d = 0\text{ kW}$$

$$\mathbf{S}_d = S_d \angle -90^\circ = 8 \angle -90^\circ = (\cos 90^\circ - j\sin 90^\circ)$$

$$= 8\,(0 - j1) = 0 - j8\text{ kVA}$$

Total complex power

$$\mathbf{S} = \mathbf{S}_a + \mathbf{S}_b + \mathbf{S}_c + \mathbf{S}_d$$

$$= 10 + j0 + 4 + j3 + 3.6 - j4.8 + 0 - j8$$

$$= 17.6 - j9.8 = 20.144 \angle -29.1^\circ\text{kVA}$$

Also, $\mathbf{S} = P + jQ = S \angle \phi$

therefore, $S = 20.144$ kVA

$P = 17.6$ kW

$Q = 9.8$ kVAr (leading)

Overall power factor

$$\cos\phi = \cos(-29.1^\circ) = 0.8737\text{ (leading)}$$

Total current

$$I = \frac{\text{kVA} \times 1000}{V} = \frac{20.144 \times 1000}{230} = 87.58 \text{ A}$$

7.8 SERIES-PARALLEL A.C. CIRCUITS

Series parallel circuits are those which have a series group of elements connected to a parallel group so that the two groups are in series. It is to be noted that all laws, techniques and network theorems used for dc circuits are also applicable to solve ac circuits provided we use phasors throughout.

7.9 SOLUTION OF SERIES - PARALLEL A.C. CIRCUITS

The solution of problems in series-parallel ac circuits is done in the same way that we solved series-parallel dc circuits.

1. Reduce all parallel circuits to series equivalent impedances. Calculate these impedances in rectangular and polar forms because both forms are used in calculations.
2. The total impedance $\mathbf{Z}_T$ of the circuit is the phasor sum of all the series impedances. This sum is obtained by using the rectangular forms of the impedances. The final result is then changed from rectangular to polar form.
3. Total circuit current is found by Ohm's law.

$$\mathbf{I}_T = \frac{\mathbf{V}}{\mathbf{Z}_T}$$

4. Each voltage drop is calculated using Ohm's law.

$$\mathbf{V}_1 = \mathbf{I}_T \mathbf{Z}_1$$

$$\mathbf{V}_2 = \mathbf{I}_T \mathbf{Z}_2$$

$$\mathbf{V}_n = \mathbf{I}_T \mathbf{Z}_n$$

5. Individual branch currents are found by dividing parallel circuit voltage drops by branch impedances.

■ **EXAMPLE 7.13** For the network shown in Fig. 7.6, determine (*a*) the voltage V_{BC}, and (*b*) Z_1 and the element it contains.

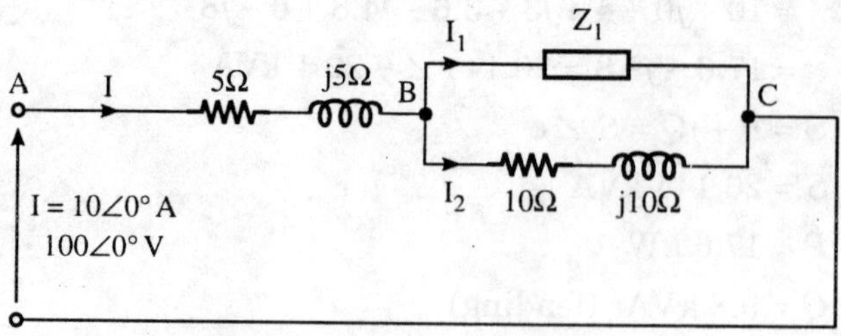

FIG. 7.6 Network of Example 7.13

SOLUTION : Let **Z** be the impedance of the complete circuit.

$$\mathbf{Z} = \frac{\mathbf{V}}{\mathbf{I}} = \frac{100 \angle 0^\circ}{10 \angle 0^\circ} = 10 \angle 0^\circ = 10 + j0\ \Omega$$

$$\mathbf{Z}_{AB} = 5 + j5$$

$$\mathbf{V}_{AB} = \mathbf{I}\,\mathbf{Z}_{AB} = (10 \angle 0^\circ)(5 + j5) = 50 + j50\ \text{V}$$

$$\mathbf{V}_{AC} = \mathbf{V}_{AB} + \mathbf{V}_{BC}$$

therefore, $\mathbf{V}_{BC} = \mathbf{V}_{AC} - \mathbf{V}_{AB}$

$$= 100 \angle 0^\circ - (50 + j50) = 100 + j0 - (50 + j50) = 50 - j50\ \text{V} = 70.7 \angle -45^\circ\ \text{V}$$

$$\mathbf{I}_2 = \frac{\mathbf{V}_{BC}}{10 + j10} = \frac{70.7 \angle -45^\circ}{14.14 \angle 45^\circ} = 5\ \underline{/(-45^\circ - 45^\circ)} = 5 \angle -90^\circ = -j5\ \text{A}$$

By KCL,

$$\mathbf{I} = \mathbf{I}_1 + \mathbf{I}_2$$

therefore, $\mathbf{I}_1 = \mathbf{I} - \mathbf{I}_2 = 10 + j0 - (-j5) = 10 + j5\ \text{A} = 11.18 \angle 26.57^\circ\ \text{A}$

$$\mathbf{Z}_1 = \frac{\mathbf{V}_{BC}}{\mathbf{I}_1} = \frac{70.7 \angle -45^\circ}{11.18 \angle 26.57^\circ} = 6.324 \angle -71.57^\circ\ \Omega$$

therefore, $\mathbf{Z}_1 = 2 - j6$

This is of the form $\mathbf{Z} = R - jX_c$

Thus, $\mathbf{Z}_1$ consists of a resistor of 2 Ω in series with a capacitor of reactance 6 Ω at the operating supply frequency.

■ **EXAMPLE 7.14** In the network shown in Fig. 7.7, determine (*a*) the total impedance, (*b*) the total current, (*c*) the current in each branch, (*d*) the overall power factor, (*e*) voltamperes, (*f*) active power, and (*g*) reactive voltamperes.

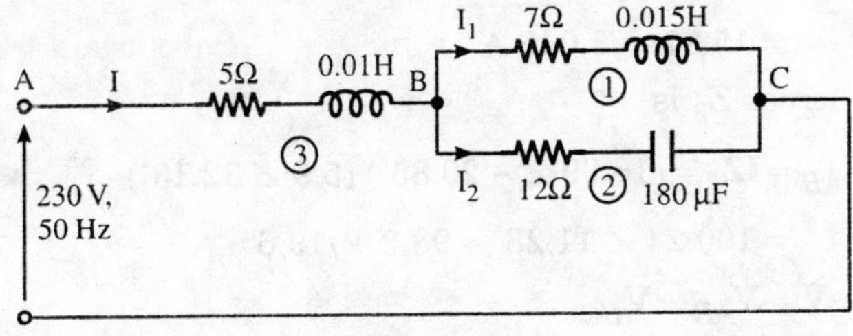

FIG. 7.7

SOLUTION :

(*a*) **Branch 1.**

$$R_1 = 7\Omega\ L_1 = 0.015\ \text{H}$$

$$X_{L_1} = 2\pi f L_1 = 2\pi \times 50 \times 0.015 = 4.71\ \Omega$$

$$\mathbf{Z}_1 = R_1 + jX_{L_1} = 7 + j4.71 = 8.437\ \underline{/33.93^\circ}\ \Omega$$

Branch 2. $R_2 = 12\ \Omega,\ C_2 = 180\ \mu\text{F} = 180 \times 10^{-6}\ \text{F}$

$$X_{c_2} = \frac{1}{2\pi f C_2} = \frac{1}{2\pi \times 50 \times 180 \times 10^{-6}} = 17.68\ \Omega$$

$$\mathbf{Z}_2 = R_2 - jX_{c_2} = 12 - j17.68 = 21.37 \angle -55.83^\circ\ \Omega$$

Since $\mathbf{Z}_1$ and $\mathbf{Z}_2$ are connected in parallel, their equivalent impedance $\mathbf{Z}_p$ is given by

$$\mathbf{Z}_p = \mathbf{Z}_1 \,||\, \mathbf{Z}_2 = \frac{\mathbf{Z}_1 \mathbf{Z}_2}{\mathbf{Z}_1 + \mathbf{Z}_2}$$

$$= \frac{(8.437 \angle 33.93^\circ)(21.37 \angle -55.83^\circ)}{(7 + j4.71) + (12 - j17.68)}$$

$$= \frac{180.3 \angle -21.9^\circ}{19 - j12.97} = \frac{180.3 \angle -21.9^\circ}{23 \angle -34.3^\circ}$$

$$= 7.839 \angle +12.4^\circ = 7.656 + j1.68\ \Omega$$

Branch 3. $R_3 = 5\ \Omega,\ L_3 = 0.01\ \text{H}$

$$X_{L_3} = 2\pi f L_3 = 2\pi \times 50 \times 0.01 = 3.14\ \Omega$$

$$\mathbf{Z}_3 = R_3 + jX_{L_3} = 5 + j3.14 = 5.9 \angle 32.13^\circ\ \Omega$$

Since $\mathbf{Z}_3$ and $\mathbf{Z}_p$ are connected in series, the total impedance of the circuit is

$$\mathbf{Z} = \mathbf{Z}_3 + \mathbf{Z}_p = 5 + j3.14 + 7.656 + j1.68 = 12.656 + j4.82 = 13.54 \angle 20.85^\circ\ \Omega$$

(*b*) Let the supply voltage **V** be taken as reference phasor.

$$\mathbf{V} = 230 \angle 0^\circ\ \text{V} = 230 + j0\ \text{V}$$

By Ohm's law the total circuit current is

$$\mathbf{I} = \frac{\mathbf{V}}{\mathbf{Z}} = \frac{230 \angle 0^\circ}{13.54 \angle 20.85^\circ} = 16.99 \angle -20.85^\circ\ \text{A}$$

$$= 15.87 - j6.046\ \text{A}$$

(*c*) Voltage drop across $\mathbf{Z}_3$ is

$$\mathbf{V}_{AB} = \mathbf{I}\mathbf{Z}_3 = (16.99 \angle -20.85^\circ)(5.9 \angle 32.13^\circ)$$

$$= 100.24 \angle 11.28^\circ = 98.3 + j19.6$$

By KVL $\quad \mathbf{V} = \mathbf{V}_{AB} + \mathbf{V}_{BC}$

therefore, $\quad \mathbf{V}_{BC} = \mathbf{V} - \mathbf{V}_{AB}$

$$= 230 + j0 - (98.3 + j19.6) = 131.7 - j19.6 = 133.15 \angle -8.46^\circ\ \text{V}$$

Alternatively,

$$\mathbf{V}_p = \mathbf{I}\mathbf{Z}_p = (16.99 \angle -20.85^\circ)(7.839 \angle 12.4^\circ)$$

$$= 133.18 \angle -8.45^\circ\ \text{V}$$

$$= 131.73 - j19.57\ \text{V}$$

Since $\mathbf{Z}_p$ is the parallel combination of $\mathbf{Z}_1$ and $\mathbf{Z}_2$

$$\mathbf{V}_{BC} = \mathbf{V}_p = \mathbf{V}_1 = \mathbf{V}_2 = 133.15 \angle -8.46° \text{ V}$$

By Ohm's law

$$\mathbf{I}_1 = \frac{\mathbf{V}_1}{\mathbf{Z}_1} = \frac{133.15 \angle -8.46°}{8.437 \angle 33.93°} = 15.78 \angle -42.39° \text{ A}$$

or $\quad \mathbf{I}_1 = 11.65 - j10.64 \text{ A}$

$$\mathbf{I}_2 = \frac{\mathbf{V}_2}{\mathbf{Z}_2} = \frac{133.15 \angle -8.46°}{21.37 \angle -55.83°}$$

$$= 6.23 \underline{/47.37°} \text{ A} = 4.22 + j4.58 \text{ A}$$

Check $\mathbf{I}_1 + \mathbf{I}_2 = 11.65 - j10.64 + 4.22 + j4.58 = 15.87 - j6.06 \text{ A} = \mathbf{I}$

(*d*) Total current

$$\mathbf{I} = 16.99 \angle -20.85° \text{ A}$$

therefore, $\quad \phi = -20.85°$

Overall power factor, $\cos\phi = \cos(-20.85°) = 0.9345$ (lagging)

(*e*) Voltamperes $S = VI = 230 \times 16.99 = 3907.7$ VA

(*f*) Active power $P = VI\cos\phi = 230 \times 16.99 \times 0.9345 = 3651.7$ W

(*g*) Reactive voltamperes

$$Q = VI\sin\phi = 230 \times 16.99 \times \sin 20.85° = 1390.8 \text{ VAr}$$

Alternative method

$$\mathbf{S} = \mathbf{VI}^* = (230 \angle 0°)(16.99 \angle +20.85°)$$

$$= 3907.7 \underline{/20.85°} = 3651.8 + j1390.8$$

Also $\quad \mathbf{S} = P + jQ = S \angle \phi$

therefore, $\quad S = 3907.7$ VA

$P = 3651.8$ W

$Q = 1390.8$ VAr

EXERCISES

7.1 A choke coil of inductance 0.08 H and resistance 12 Ω is connected in parallel with a 120 μF capacitor to a 240 V, 50 Hz supply. Determine the current taken from the supply and its phase angle.

7.2 A coil of resistance 15 Ω and reactance 25 Ω is connected in parallel with a capacitor of reactance 10 Ω and series resistance of 12 Ω to a 100 V, 50 Hz supply. Calculate the supply current and the circuit phase angle.

7.3 A circuit consists of three branches connected in parallel across a 100 V, 50 Hz supply. Branch 1 consists of an 8 Ω resistor in series with a 200 μF capacitor, branch 2 consists of an inductor of 4 Ω resistance and 20 mH inductance and branch 3 is a

10 Ω resistor. Calculate (*a*) the current in each branch and its phase relationship to the applied voltage; (*b*) the total current taken from the supply; (*c*) the overall power factor; (*d*) the power consumed in each branch; (*e*) the total power consumed in the circuit.

7.4 A 60 Ω resistor and a capacitor are connected in parallel across a 240 V, 50 Hz supply. The current taken from the supply is 6 A. Calculate the capacitance of the capacitor. Sketch the phasor diagram representing the supply voltage, the currents through the resistor and capacitor, and the current taken from the supply.

7.5 A 60 Ω resistor is connected in parallel with an inductive reactance of 80 Ω to a 240 V, 50 Hz supply. Calculate : (*a*) the current through the resistor and inductive reactance; (*b*) the supply current; (*c*) the circuit phase angle. Draw the phasor diagram.

7.6 A circuit consists of a 100 Ω resistor in parallel with a 60 μF capacitor, and is connected to a 200 V, 50 Hz supply. Calculate : (*a*) the branch currents and the supply current; (*b*) the circuit phase angle; (*c*) the circuit impedance.

7.7 A parallel circuit consists of two branches one containing a coil of resistance 20 Ω and inductance 0.07 H, the other a noninductive resistance 50 Ω in series a capacitor of 60 μF capacitance. The circuit is connected to a 200 V, 50 Hz supply. Determine (*a*) the current in each branch; (*b*) the supply current.

7.8 A parallel circuit consists of two branches, one containing a coil of resistance 5 Ω and inductance 38.2 mH, the other a noninductive resistance 16 Ω in series with a capacitor of 300 μF capacitance. The circuit is connected to a 240 V, 50 Hz supply. Determine : (*a*) the current in each branch; (*b*) the total current; (*c*) the circuit phase angle; (*d*) the circuit impedance; (*e*) the components of an equivalent circuit consisting of a resistance and a reactance.

7.9 A single-phase load consists of heating and lighting (10 kW at unity power factor) and induction motors (8 kW at a power factor of 0.8 lagging). Calculate : (*a*) the total kW; (*b*) the total kVAr; (*c*) the total kVA; (*d*) the overall power factor; (*e*) the total line current at 240 V.

7.10 The circuit shown in Fig. P7.1 is connected to a 50 Hz sinusoidal supply. The total current is 12 A leading the supply voltage by a phase angle of 30°. The current in the top branch is 8 A. Calculate : (*a*) the supply voltage; (*b*) the unknown branch currents; (*c*) the values of *R* and *C*; (*d*) the resistance and reactance of the equivalent series circuit.

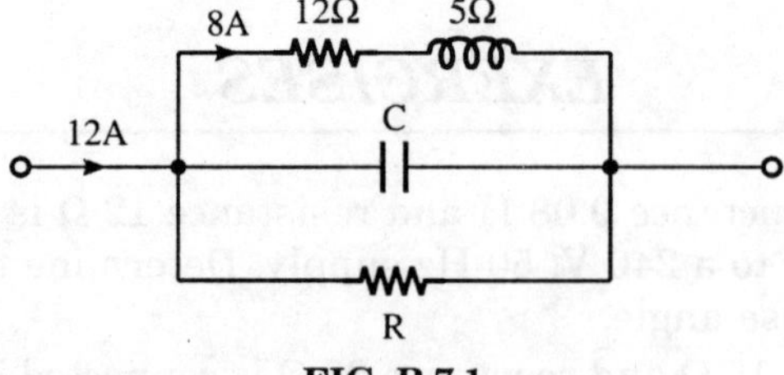

FIG. P 7.1

7.11 The following loads are connected in parallel :

(*a*) 100 kVA at 0.8 power factor lagging,

(*b*) 50 kW at unity power factor,

(*c*) 250 kVA at 0.8 power factor leading,

(*d*) 200 kVA at 0.6 power factor lagging

Determine : (*i*) the total kVA; (*ii*) the total kW; (*iii*) the total kVAr; (*iv*) the overall power factor.

7.12 Calculate the admittance, conductance, and susceptance of a circuit consisting of a resistance of 30 Ω in series with a capacitance of 25 μF at angular frequency ω = 1000 rad/s.

7.13 Three impedances $(4 - j6)$ Ω, $(6 + j8)$ Ω, and $(5 - j3)$ Ω are connected in parallel. Calculate the current in each branch when the total supply current is 20 A.

7.14 A circuit consists of parallel impedances $(3 + j4)$ Ω and $(5 - j3)$ Ω and is connected in series with an impedance of $(1 + j2)$ Ω. If a 200 V ac supply is connected across the combination, calculate the supply current, overall power factor of the circuit, voltage drop across each impedance, total voltamperes, total active power and total reactive voltamperes.

7.15 A resistance of 5 Ω is in parallel with an inductance (of negligible resistance) having a reactance of 8 Ω. This circuit is joined in series with another consisting of a resistance of 2 Ω in series with a capacitive reactance of 4 Ω. The combination is connected to 100 V, 50 Hz mains. Find :

(*a*) the current taken from the supply and its power factor ;

(*b*) potential difference across each circuit.

7.16 For the network shown in Fig. P7.2, determine the supply voltage *V* ?

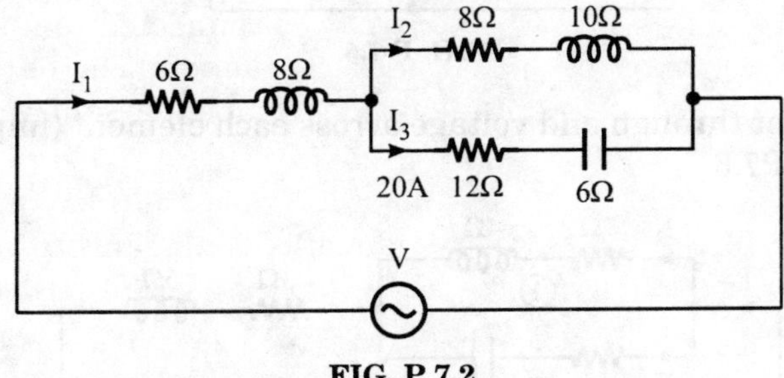

FIG. P 7.2

7.17 In the circuit of Fig. P7.3, determine

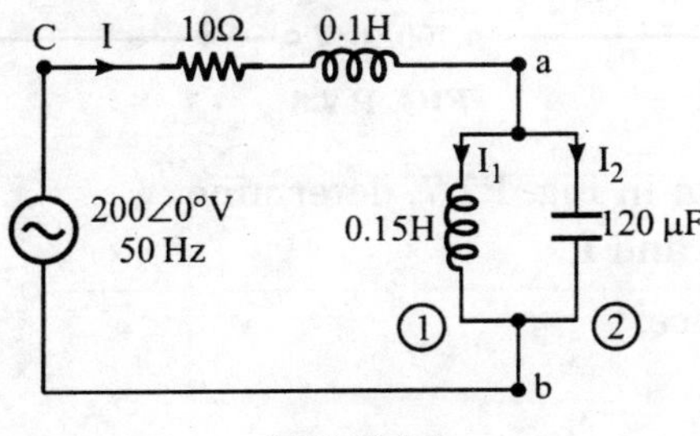

FIG. P 7.3

(*a*) the current **I**, $\mathbf{I}_1$ and $\mathbf{I}_2$;

(*b*) the voltage $\mathbf{V}_{ab}$ across the parallel circuit;

(*c*) the overall power factor;

(*d*) the total power consumed in the circuit.

7.18 Determine the total current, power factor and power consumed by the circuit shown in Fig. P7.4.

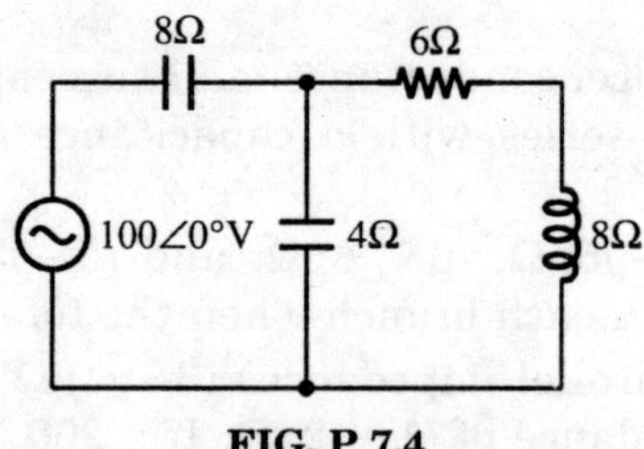

FIG. P 7.4

7.19 For the Fig. P7.5, determine currents **I**, $\mathbf{I}_1$ and $\mathbf{I}_2$, the total impedance, and the total power consumed.

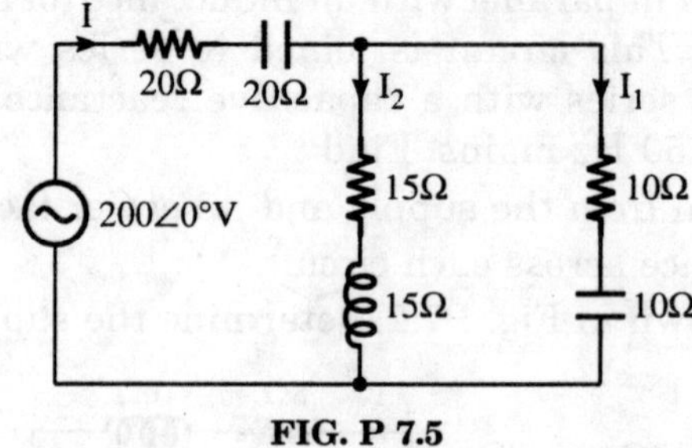

FIG. P 7.5

7.20 Find the current through and voltage across each element (impedance) in the circuit shown in Fig. P7.6.

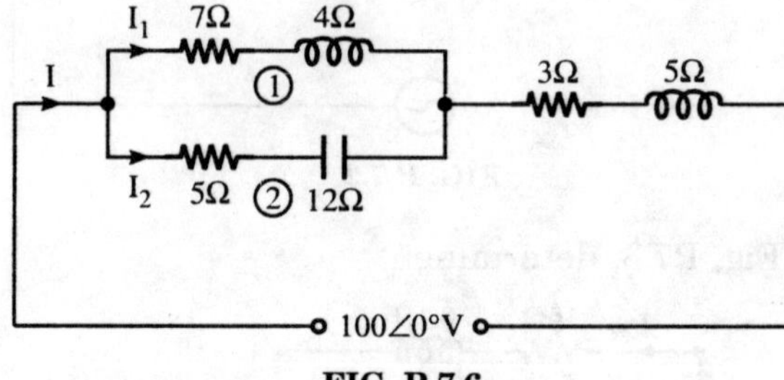

FIG. P 7.6

7.21 In the network shown in Fig. P7.7, determine :

(*a*) currents $\mathbf{I}_1$, $\mathbf{I}_2$, $\mathbf{I}_3$ and **I**;

(*b*) the total impedance;

(*c*) the total power.

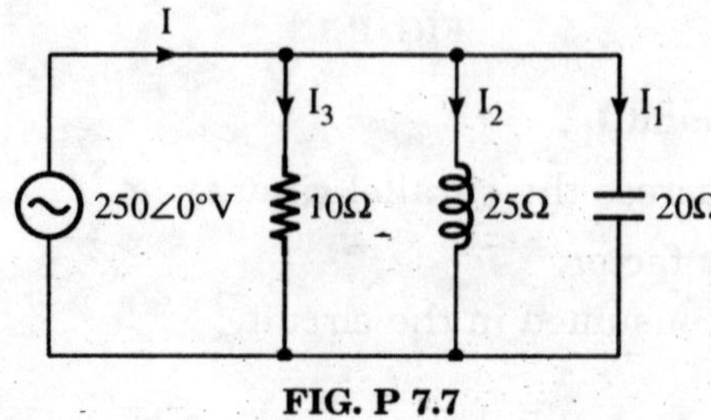

FIG. P 7.7

7.22 Find the current and power supplied by the 100 V source for the circuit of Fig. P7.8.

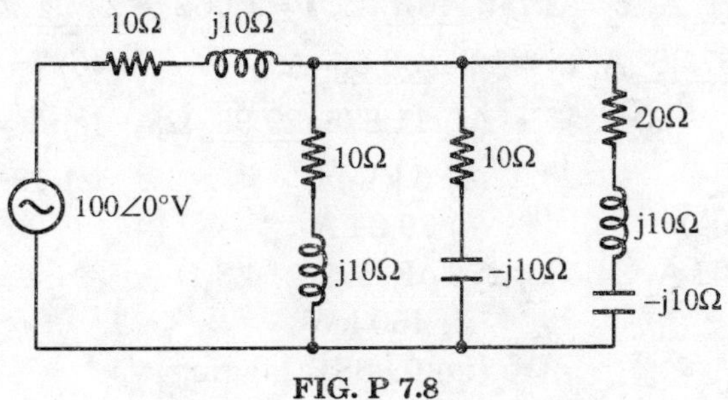

FIG. P 7.8

7.23 For the Fig. P7.9, determine :

(*a*) **I**, $\mathbf{I}_1$ and $\mathbf{I}_2$; (*b*) total impedance;

(*c*) total power.

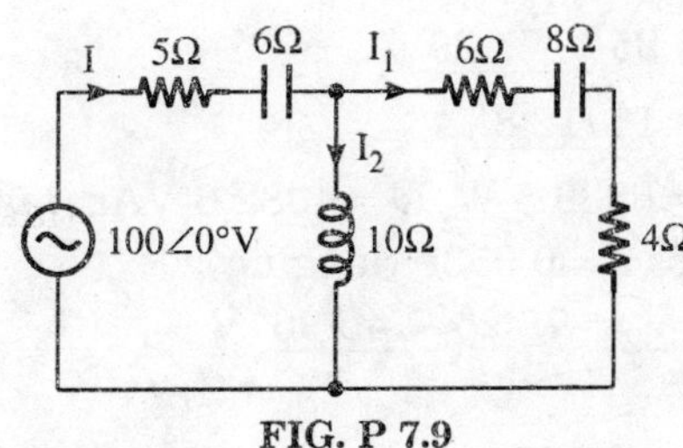

FIG. P 7.9

7.24 For the circuit of Fig. P7.10, determine the total impedance, total current, overall power factor and total power.

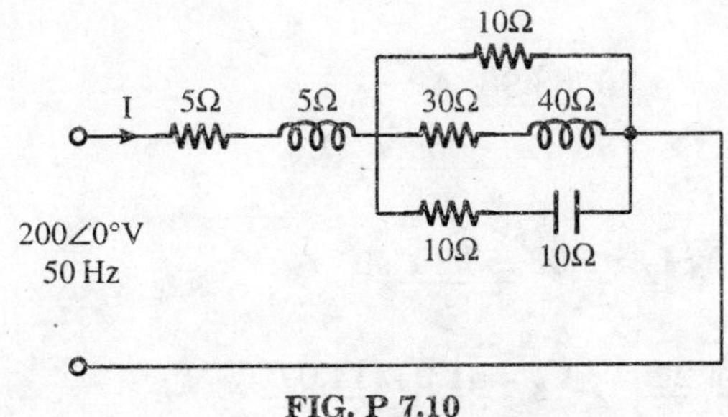

FIG. P 7.10

ANSWERS

1. $3.923 \underline{/18.9^\circ}$ A
2. $6.782 \underline{/9.82^\circ}$ A
3. (*a*) $\mathbf{I}_1 = 5.61 \underline{/63.3^\circ}$ A, $\mathbf{I}_2 = 13.426 \underline{/-57.52^\circ}$ A, $\mathbf{I}_3 = 10 \underline{/0^\circ}$ A;
   (*b*) 20.71 A; (*c*) 0.9526 (lagging);
   (*d*) $P_1 = 252$ W, $P_2 = 721$ W, $P_3 = 1000$ W; (*e*) 1973 W
4. 59.5 μF
5. (*a*) $\mathbf{I}_R = 4 \underline{/0^\circ}$ A, $\mathbf{I}_L = 3 \underline{/-90^\circ}$ A; (*b*) 5 A; (*c*) $\phi = 36.87^\circ$ (lagging)

6. (*a*) $\mathbf{I}_R = 2 \underline{/0^\circ}$ A; $\mathbf{I}_C = 3.78 \underline{/90^\circ}$ A; $\mathbf{I} = 4.277 \underline{/62.1^\circ}$ A
7. $\mathbf{I}_1 = 6.728 \underline{/-47.71^\circ}$ A, $\mathbf{I}_2 = 2.743 \underline{/46.7^\circ}$ A, $\mathbf{I} = 7.068 \underline{/-24.94^\circ}$ A
8. (*a*) $\mathbf{I}_1 = 18.46 \underline{/-67.38^\circ}$ A, $\mathbf{I}_2 = 12.5 \underline{/33.55^\circ}$ A; (*b*) $\mathbf{I} = 20.21 \underline{/-29.98^\circ}$ A;
(*c*) $\phi = -29.98^\circ$; (*d*) $11.875 \underline{/29.98^\circ}$ Ω; (*e*) $R = 10.287$ Ω, $X_L = 5.932$ Ω
9. (*a*) 18 kW; (*b*) 6 kVAr; (*c*) 18.97 kVA;
(*d*) 0.9487 (lagging); (*e*) 79.04 A
10. 104 V, 3.02 A, 9.08 A, 34.5 Ω, 278 μF, 7.5 Ω, 4.3 Ω
11. (*i*) 455.4 kVA; (*ii*) 450 kW;
(*iii*) 70 kVAr; (*iv*) 0.9881 (lagging)
12. $\mathbf{Y} = 0.012 + j0.016$ S, $G = 0.012$ S, $B = 0.016$ S
13. $\mathbf{I}_1 = 8.96 \underline{/32.79^\circ}$ A, $\mathbf{I}_2 = 6.46 \underline{/-76.65^\circ}$ A, $\mathbf{I}_3 = 11.08 \underline{/7.44^\circ}$ A
14. $\mathbf{I} = 37.27 \underline{/-33.21^\circ}$ A, $\cos\phi = 0.8367$ (lagging)
Voltage across $\mathbf{Z}_1$ = voltage across $\mathbf{Z}_2 = 134.75 \underline{/-18.16^\circ}$ V
Current through $\mathbf{Z}_1 = 26.95 \underline{/-71.29^\circ}$ A
Current through $\mathbf{Z}_2 = 23.11 \underline{/12.8^\circ}$ A
$\mathbf{S} = 7454 \underline{/32.21^\circ}$ VA, $P = 6236.5$ W, $Q = 4082.6$ VAr (lag)
15. (*a*) $\mathbf{I} = 17.06 \underline{/17.47^\circ}$ A, $\cos\phi = 0.9538$ (lagging);
(*b*) $\mathbf{V}_1 = 72.33 \underline{/49.48^\circ}$ V, $\mathbf{V}_2 = 76.25 \underline{/-45.96^\circ}$ V
16. 305 V
17. (*a*) $\mathbf{I} = 6.46 \underline{/71.14^\circ}$ A, $\mathbf{I}_1 = 8.338 \underline{/-108.8^\circ}$ A, $\mathbf{I}_2 = 14.796 \underline{/71.2^\circ}$ A;
(*b*) $392.7 \underline{/-18.8^\circ}$ V; (*c*) 0.3232 leading; (*d*) 417.57 W
18. $\mathbf{I} = 30.03 \underline{/123.5^\circ}$ A, $\cos\phi = 0.5519$ (leading),
$P = 1657.4$ W
19. $\mathbf{I} = 5.18 \underline{/35.27^\circ}$ A, $\mathbf{I}_1 = 4.30 \underline{/68.98^\circ}$ A,
$\mathbf{I}_2 = 2.86 \underline{/-20.83^\circ}$ A, $\mathbf{Z} = 38.61 \underline{/-35.27^\circ}$ Ω,
$P = 845.8$ W
20. $\mathbf{I}_1 = 10.23 \underline{/-67.37^\circ}$ A, $\mathbf{I}_2 = 3.93 \underline{/53.14^\circ}$ A,
$\mathbf{I} = 8.91 \underline{/-44.96^\circ}$ A,
$\mathbf{V}_1 = \mathbf{V}_2 = 51.19 \underline{/-14.24^\circ}$ V, $\mathbf{V}_3 = 51.94 \underline{/14.07^\circ}$ V
21. (*a*) $\mathbf{I}_1 + 12.5 \underline{/90^\circ}$ A, $\mathbf{I}_2 = 10 \underline{/-90^\circ}$ A, $\mathbf{I}_3 = 25 \underline{/0^\circ}$ A
$\mathbf{I} = 25.125 \underline{/5.71^\circ}$ A;
(*b*) $\mathbf{Z} = 9.951 \underline{/-5.71^\circ}$ Ω; (*c*) $P = 6250$ W
22. $\mathbf{I} = 5.14 \underline{/-30.97^\circ}$ A, $P = 440.7$ W
23. (*a*) $\mathbf{I} = 6.17 \underline{/0.77^\circ}$ A, $\mathbf{I}_1 = 8.46 \underline{/73.21^\circ}$ A, $\mathbf{I}_2 = 9 \underline{/-65.6^\circ}$ A;
(*b*) $\mathbf{Z} = 16.2 \underline{/0.77^\circ}$ Ω; (*c*) $P = 616.93$ W
24. $\mathbf{Z} = 60.8 \underline{/-72.4^\circ}$ Ω, $\mathbf{I} = 3.28 \underline{/72.4^\circ}$ A,
$\cos\phi = 0.3023$ (leading), $P = 198.3$ W

8

RESONANCE

8.1 INTRODUCTION

Resonance *is defined as the condition in a circuit containing at least one inductor and one capacitor, when the supply voltage and the supply current are in phase.* Thus, at resonance the equivalent impedance of the circuit is purely resistive. Since the supply voltage and supply current are in phase, the power factor of a resonant circuit is unity. At resonance the circuit impedance **Z** and the admittance **Y** are real quantities. Thus, if **Z** and **Y** are expressed in rectangular form their j terms are zero while if **Z** and **Y** are expressed in polar form their angles are zero.

For example,

(*a*) If $\mathbf{Z} = R + jX$ and $\mathbf{Y} = G + jB$

then at resonance

$$\text{Im}(\mathbf{Z}) = X = 0 \text{ or } \text{Im}(\mathbf{Y}) = B = 0$$

(*b*) If $\mathbf{Z} = Z\,\underline{/\phi}$ and $\mathbf{Y} = Y\,\underline{/\theta}$

then at resonance

$$\phi = 0 \text{ and } \theta = 0$$

(*c*) If $\mathbf{Z} = \dfrac{a + jb}{c + jd}$

then at resonance

$$\frac{b}{a} = \frac{d}{c}$$

Similarly,

if $\mathbf{Y} = \dfrac{l + jm}{p + jq}$

then at resonance

$$\frac{m}{l} = \frac{q}{p}$$

8.2 RESONANCE IN SERIES CIRCUITS

In a series circuit if resonance occurs, it is called **series resonance or voltage resonance.** Consider a general series circuit having resistance R, inductance L and capacitance C connected across a sinusoidal voltage source of constant-voltage amplitude and variable frequency as shown in Fig. 8.1.

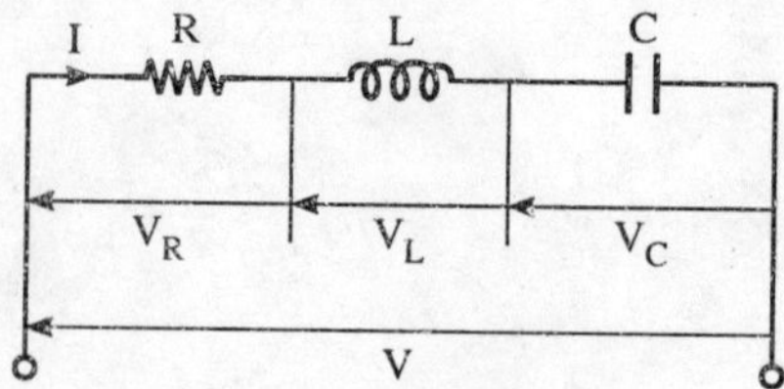

FIG. 8.1 General series circuit

Inductive reactance $\chi_L = 2\pi fL$

Capacitive reactance $\chi_C = \dfrac{1}{2\pi fC}$

Total reactance of the circuit

$$\chi = \chi_L - \chi_C \tag{8.2.1}$$

Total impedance of the circuit

$$Z = \sqrt{R^2 + (\chi_L - \chi_C)^2} \tag{8.2.2}$$

Let χ_{L_0} = inductive reactance at resonance

χ_{C_0} = capacitive reactance at resonance

Z_0 = impedance at resonance

From the definition of resonance, the applied voltage and current are in phase and the impedance of the circuit becomes purely resistive. For a circuit to behave as purely resistive, the total reactance χ is zero. That is,

$$\chi_{L_0} - \chi_{C_0} = 0$$

$$\chi_{L_0} = \chi_{C_0} \tag{8.2.3}$$

From Eqs. (8.2.2) and (8.2.3)

$$Z_0 = R \tag{8.2.4}$$

Thus, at resonance the circuit impedance Z is equal to the circuit resistance R. Since χ_{L_0} and χ_{C_0} cancel each other, Z_0 is a minimum and equal to R. The circuit impedance at resonance is called the **dynamic impedance Z_0.**

$$Z_0 = R \tag{8.2.5}$$

Since at resonance

$$\chi_{L_0} = \chi_{C_0}$$

$$I_0\chi_{L_0} = I_0\chi_{C_0}$$

$$V_{L_0} = V_{C_0}$$

Equation (8.2.6) shows that at resonance the voltage across the inductance is equal in magnitude and opposite in phase to the voltage across the capacitance.

Also, $$V = \sqrt{V_R^2 + (V_{L_0} - V_{C_0})^2} = \sqrt{(V_R^2 + 0)}$$

$$V = V_R \tag{8.2.7}$$

and $$V_R = RI_0$$

Thus, at resonance the supply voltage V is equal to V_R. It should be noted that although the total reactive voltage $(V_L - V_C)$ is zero at resonance, V_{L_0} and V_{C_0} may reach values greater than the supply voltage with the result that the circuit insulation may be damaged.

Let ϕ_0 be the phase difference between V and I at resonance.

Since V and I are in phase, the phase difference between V and I is zero, that is, $\phi_0 = 0$ and

$$\cos \phi_0 = 1 \tag{8.2.8}$$

That is, the power factor is a maximum at resonance. Current at resonance

$$I_0 = \frac{V}{Z_0} = \frac{V}{R} \tag{8.2.9}$$

Since the impedance has a minimum value, the current has a maximum value at resonance. *The series resonant circuit is also called an* **acceptor circuit** *since it accepts frequencies close to the resonant frequency when its impedance is a minimum.*

Also, the power factor has a maximum value (that is $\cos \phi_0 = 1$) the power absorbed by the circuit is also a maximum at resonance. The power at resonance is given by

$$P_0 = VI_0 \cos \phi_0 = V \times \frac{V}{R} \times 1$$

$$P_0 = \frac{V^2}{R} \tag{8.2.10}$$

In summary, for a series resonant circuit the main properties are :

(1) The supply voltage and supply current are in phase so that $\phi_0 = 0$, $\cos \phi_0 = 1$

(2) The circuit impedance Z_0 is a minimum and $Z_0 = R$

(3) The supply current I_0 is a maximum and is given by $I_0 = V/R$

(4) $\chi_{L_0} = \chi_{C_0}$

(5) $V_{L_0} = V_{C_0}$

(6) The power P_0 absorbed by the circuit is a maximum and $P_0 = V^2/R$.

8.3 PHASOR DIAGRAM AT RESONANCE

The phasor diagram at resonance is shown in Fig. 8.2. Here $\mathbf{V}_{L_0} = (= \chi_L \mathbf{I}_0)$ leads the current $\mathbf{I}_0$ by 90° and $\mathbf{V}_{C_0} (= \chi_{C_0} \mathbf{I}_0)$ lags behind the current by 90°. The total supply voltage $\mathbf{V} = \mathbf{V}_{R_0}$ and is in phase with the current.

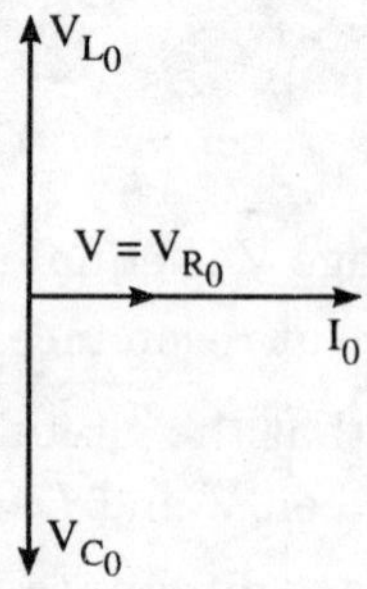

FIG. 8.2 Phasor diagram at resonance

8.4 RESONANT FREQUENCY

The frequency at which the resonance occurs is called the *resonant frequency* or frequency of resonance. it is denoted by f_0. At resonance

$$\chi_{L_0} = \chi_{C_0}$$

$$2\pi f_0 L = \frac{1}{2\pi f_0 C}$$

$$f_0^2 = \frac{1}{(2\pi)^2 LC}$$

$$\boxed{f_0 = \frac{1}{2\pi\sqrt{LC}}\ \text{Hz}} \qquad (8.4.1)$$

If ω_0 = resonant frequency in rad/s

$$\boxed{\omega_0 = 2\pi f_0 = \frac{1}{\sqrt{LC}}\ \text{rad/s}} \qquad (8.4.2)$$

Equation (8.4.1) shows that the series resonance effect may be produced by the following methods :

(*a*) varying frequency keeping L and C constant ;

(*b*) varying either L or C (or both) for a given frequency.

In practice, it is convenient to obtain resonance by varying the capacitance.

8.5 VARIATIONS OF DIFFERENT QUANTITIES WITH FREQUENCY

The variations of reactance with frequency are shown in Fig. 8.3

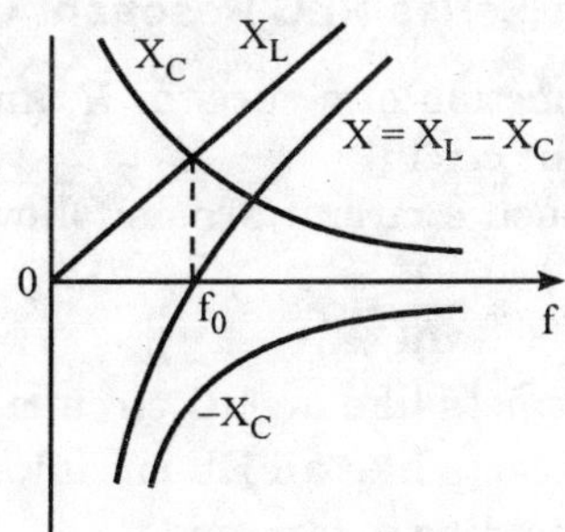

FIG. 8.3 Variations of reactances with frequency

Variation of inductive reactance χ_L

The inductive reactance χ_L (= $2\pi fL$) is directly proportional to the frequency f, hence its graph is a straight line through the origin.

Variation of capacitive reactance χ_C

The capacitive reactance χ_C [= $1/(2\pi fC)$] is inversely proportional to the frequency, hence its graph is a reactangular hyperbola. χ_L versus f and χ_C versus f curves cut at a point where $f = f_0$.

Variation of net reactance χ

The graph of net reactance χ (= $\chi_L - \chi_C$) is obtained from χ_L versus f and $-\chi_C$ versus f curves. It is a hyperbola (but not rectangular hyperbola). This curve crosses the frequency axis at a point where $f = f_0$.

The variations of resistance, impedance, current and power factor with frequency are shown in Fig. 8.4.

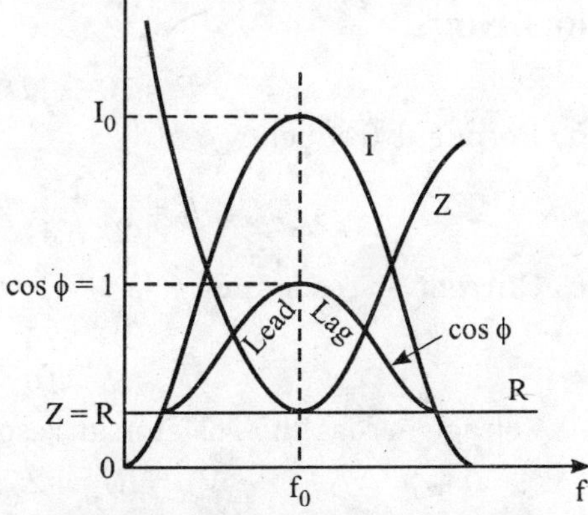

FIG. 8.4 Variations of resistance, impedance, current and power factor with frequency

Variation of resistance R

Since the resistance R is independent of frequency, the graph of R versus f is a horizontal straight line.

Variations of impedance and power factor

The impedance is a minimum at resonance given by $Z = R$. The value of Z increases on either side of resonance. At frequencies below resonance $\chi_C > \chi_L$ and the impedance is capacitive. Therefore the power factor is leading. At frequencies above resonance $\chi_L > \chi_C$ and the impedance is inductive. Therefore, the power factor is lagging. At resonant frequency the power factor is unity.

Variation of current

The current versus frequency curve is known as *resonance curve* or *response curve*. The current has a maximum value at resonance given by $I_0 = V/R$. The value of I decreases on either side of the resonance.

8.5.1 Properties of a Second-Order Series RLC Resonant Circuit

A circuit consisting of a *series* connection of a resistor R, an inductor L, and a capacitor C is called a second-order series resonant circuit.

The important properties of such a circuit are as follows :

- The resonant frequency is $\omega_0 = \dfrac{1}{\sqrt{LC}}$ rad/s.
- Below resonance, the circuit acts like an RC circuit.
- Above resonance, the circuit acts like an RL circuit.
- At resonance, the circuit acts, like a resistor.
- If the resistance is zero, at resonance the circuit acts like a short circuit.
- The bandwidth is $\dfrac{R}{L}$.
- The quality factor Q_{os} is $\dfrac{1}{R}\sqrt{\dfrac{L}{C}}$.

EXAMPLE 8.1 A circuit consists of a noninductive resistor of 10 Ω, an inductor of 0.1 H and a capacitor of 8 μF is series. Calculate : **(a)** the resonant frequency ; **(b)** the current at resonant frequency ; **(c)** the voltage across each component when a voltage of 25 V at resonant frequency is applied to the whole circuit.

SOLUTION :

$$R = 10\ \Omega\ , L = 0.1\text{ H}, C = 8\ \mu\text{F} = 8 \times 10^{-6}\text{ F}$$

(*a*) Resonant frequency

$$f_0 = \frac{1}{2\pi\sqrt{LC}} = \frac{1}{2\pi\sqrt{0.1 \times 8 \times 10^{-6}}} = 178\text{ Hz}$$

(*b*) Current at resonance

$$I_0 = \frac{V}{R} = \frac{25}{10} = 2.5\text{ A}$$

(*c*) Voltage across the resistor at resonance

$$V_R = RI_0 = 25\text{ volts}$$

Voltage across the inductor at resonance

$$V_L = \chi_L I_0 = 2\pi f_0 LI_0 = 2\pi \times 178 \times 0.1 \times 2.5 = 279.6\text{ V}$$

Voltage across the capacitor at resonance

$$V_C = V_L = 279.6\text{ V}$$

EXAMPLE 8.2 A large coil of inductance 1.405 H and resistance 40 Ω is conected in series with a capacitor of capacitance 20 μF. Calculate the frequency at which the circuit resonates. If a voltage of 100 V is applied to the circuit at resonant condition, calculate the current drawn from the supply and the voltage across the coil and the capacitor.

SOLUTION :

$$R = 40\ \Omega\ ,\ L = 1.405\ \text{H}\ ,\ C = 20\ \mu\text{F} = 20 \times 10^{-6}\ \text{F}$$

$$f_0 = \frac{1}{2\pi\sqrt{LC}} = \frac{1}{2\pi\ \sqrt{1.405 \times 20 \times 10^{-6}}} = 30\ \text{Hz}$$

$$I_0 = \frac{V}{R} = \frac{100}{40} = 2.5\ \text{A}$$

Reactance of the coil at resonance

$$X_{L_0} = 2\pi f_0 L = 2\pi \times 30 \times 1.405 = 264.8\ \Omega$$

Impedance of the coil at resonance

$$Z_{coil} = \sqrt{R^2 + \chi_{L_0}^2} = \sqrt{40^2 + 264.8^2}$$

$$= 267.8\ \Omega$$

Volate across the coil

$$= Z_{coil}\ I_0 = 267.8 \times 2.5 = 669.5\ \text{V}$$

Reactance of the capacitor at resonance

$$X_{C_0} = \frac{1}{2\pi f_0 C} = 1/(2\pi \times 30 \times 20 \times 10^{-6}) = 265.2\ \Omega$$

Voltage across the capacitor

$$V_{C_0} = X_{C_0}\ I_0 = 265.2 \times 2.5 = 663\ \text{V}$$

■ **EXAMPLE 8.3** A choke coil is connected in series with a 20 μF capacitor. With a constant supply voltage of 200 V the circuit takes its maximum current of 50 A when the supply frequency is 50 Hz. Calculate (*a*) the resistance and inductance of the choke coil ; (*b*) the voltage across the capacitor.

SOLUTION :

The current is a maximum at resonant frequency.

$$f_0 = 50\ \text{Hz}\ ,\ \ V = 200\ \text{volts},\ I = 50\ \text{A},\ C = 20 \times 10^{-6}\ \text{F}$$

(*a*) At resonance

$$I_0 = \frac{V}{R}$$

$$R = \frac{V}{I_0} = \frac{200}{50} = 4\ \Omega$$

$$f_0 = \frac{1}{2\pi\ \sqrt{LC}}$$

$$f_0^2 = \frac{1}{4\pi^2\ LC}$$

$$L = \frac{1}{(4\pi^2 f_0^2 C)} = \frac{1}{(4\pi^2 \times 50^2 \times 20 \times 10^{-6})} = 0.507\ \text{H}$$

(*b*) Reactance of the capacitor

$$X_{C_0} = \frac{1}{2\pi f_0 C} = \frac{1}{(2\pi \times 50 \times 20 \times 10^{-6})} = 159.2\ \Omega$$

Voltage across the capacitor $= X_{C_0} I_0 = 159.2 \times 50 = 7960$ V

8.6 QUALITY FACTOR Q

In general, the **quality factor** Q of coils, capacitors and circuits is defined by

$$Q \triangleq 2\pi \times \frac{\text{maximum stored energy per cycle}}{\text{energy dissipated per cycle}} \tag{8.6.1}$$

The expression for Q in a series resonant circuit can be derived as follows :
The total energy stored in a series RLC circuit is

$$W_S = \frac{1}{2} L i^2 + \frac{1}{2} C v_C^2 \tag{8.6.2}$$

If $i = I_m \sin \omega t$, the voltage across the capacitor is

$$v_C = \frac{1}{C}\int i\, dt = \frac{-I_m}{\omega C} \cos \omega t$$

$$\therefore \quad W_S = \frac{1}{2} L I_m^2 \sin^2 \omega t + \frac{1}{2}\frac{C I_m^2}{\omega^2 C^2} \cos^2 \omega t$$

At the resonant frequency ω_0 let the maximum value of current be I_{m0}. At resonance

$$\omega = \omega_0 = \frac{1}{\sqrt{LC}}$$

$$\therefore \quad W_{S_0} = \frac{1}{2} L I_{m0}^2 \sin^2 \omega_0 t + \frac{1}{2}\frac{I_{m0}^2}{2\omega_0^2 C} \cos^2 \omega_0 t$$

$$= \frac{1}{2} I_{m0}^2 L (\sin^2 \omega_0 t + \cos^2 \omega t)$$

or

$$W_{S0} = \frac{1}{2} L I_{m0}^2 \tag{8.6.3}$$

If I_0 = rms current at resonance

$$I_0 = \frac{I_{m0}}{\sqrt{2}}, \quad I_0^2 = \frac{I_{m0}^2}{2}$$

and

$$W_{S0} = L I_0^2 \tag{8.6.4}$$

Thus, at resonance the total energy stored is a constant.
The power dissipated in the circuit at resonance is

$$P_{d_0} = I_0^2 R \tag{8.6.5}$$

This is the value of energy dissipated per second. Hence the energy dissipated per cycle at resonance is

$$W_{d_0} = \frac{P_{d_0}}{f_0} = \frac{I_0^2 R}{(\omega_0/2\pi)} = \frac{2\pi I_0^2 R}{\omega_0}$$

- By definition the quality factor of a series RL circuit at resonance is

$$Q_{0S} = 2\pi \frac{W_{S_0}}{W_{d_0}} = \frac{2\pi L I_0^2}{(2\pi I_0^2 R/\omega_0)}$$

therefore

$$\boxed{Q_{0S} = \frac{\omega_0 L}{R} = \frac{1}{\omega_0 CR}} \qquad (8.6.6)$$

8.7 VOLTAGE MAGNIFICATION

Since the current at resonance is limited only by the resistance, it has maximum value. The voltages across L and C may therefore be very much higher than the supply voltage V. Thus, there is a voltage magnification at series resonance. The ratio of voltage across L or C and the supply voltage is called the **voltage magnification**. That is,

$$\text{Voltage magnification} = \frac{\text{voltage across } L \text{ at resonance}}{\text{supply voltage at resonance}} = \frac{V_{L_0}}{V}$$

Also,

$$\text{Voltage magnification} = \frac{\text{voltage across } C \text{ at resonance}}{\text{supply voltage at resonance}} = \frac{V_{C_0}}{V}$$

At resonance, $V = V_{R_0}$

therefore

$$\frac{V_{L_0}}{V} = \frac{V_{L_0}}{V_{R_0}} = \frac{X_{L_0} I_0}{R I_0} = \frac{X_{L_0}}{R} = \frac{\omega_0 L}{R} = Q_0 \qquad (8.7.1)$$

Also,

$$\frac{V_{C_0}}{V} = \frac{V_{C_0}}{V_{R_0}} = \frac{X_{C_0} I_0}{R I_0} = \frac{X_{C_0}}{R} = \frac{X_{L_0}}{R} = \frac{\omega_0 L}{R} = Q_0 \qquad (8.7.2)$$

Thus, for a series resonant circuit the factor Q_0 is a measure of the voltage magnification. For a series RLC circuit Q_{0S} can be found in terms of R, L and C as follow :

$$Q_{0S} = \frac{\omega_0 L}{R} = \frac{1}{\sqrt{LC}} \times \frac{L}{R} = \frac{1}{R}\sqrt{\frac{L}{C}} \qquad (8.7.3)$$

The effect of voltage magnification in series resonant circuit must be considered in the selection of circuit components.

8.8 BANDWIDTH OF SERIES RESONANT CIRCUIT

The **bandwidth of series resonant circuit** *is given by the band of frequencies which lie between two points on either side of resonant frequency where the current falls to* $1/\sqrt{2}$ *(or 0.707) of its maximum value at resonance.* On the frequency response curve shown in Fig. 8.5,

A and B are the points at which current falls to $1/\sqrt{2}$ or 0.707 of its value at resonance. The bandwidth is given by

$$BW = \Delta f = (f_2 - f_1) \text{ Hz} \tag{8.8.1}$$

or

$$BW = \Delta\omega = (\omega_2 - \omega_1) \text{ rad/s} \tag{8.8.2}$$

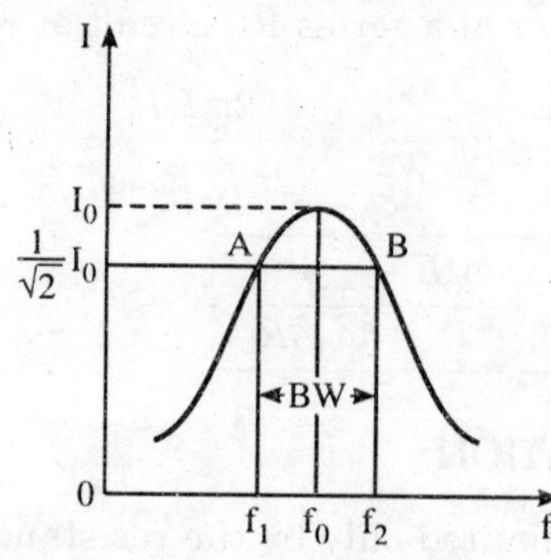

FIG. 8.5 Bandwidth of a series reasonant circuit

where f_1 and f_2 are the frequencies at the points A and B respectively. Point A is called the **lower cut-off frequency point** and point B is called the **upper cut-off frequency point**.

Current at half-power points $= \dfrac{I_0}{\sqrt{2}}$

Power dissipated in the circuit at frequency f_1

$$P_A = I_1^2 R = \left(\frac{I_0}{\sqrt{2}}\right)^2 R = \frac{1}{2} I_0^2 R$$

Power dissipated in the circuit at frequency f_2

$$P_B = I_2^2 R = \left(\frac{I_0}{\sqrt{2}}\right)^2 R = \frac{1}{2} I_0^2 R$$

Power dissipated in the circuit at resonant frequency

$$P_0 = I_0^2 R$$

Therefore,

$$P_A = P_B = \frac{1}{2} P_0$$

For this reason, the two points A and B on the response curve are known as half-power points.

The decibel power response at these points in terms of the maximum power at resonance is given by

$$\text{power ratio in decibels} = 10 \log_{10} \frac{\text{power at half-power points}}{\text{power at resonance}}$$

$$= 10 \log_{10} \frac{P_0/2}{P_0} = 10 \log_{10} \frac{1}{2} = -\,10 \log_{10} 2 = -\,10 \times 0.301 = -\,3 \text{ dB}$$

Hence, the half-power points are also called 3 dB points. At these points the power is 3 dB less than at resonance.

From the above discussion, it is seen that *the* **bandwidth of a series tuned circuit** *may also be defined as the band of frequencies which lie between lower and upper cut-off frequency or the band of frequencies which lie between half-power points on the frequency response curve.*

8.9 BANDWIDTH IN TERMS OF CIRCUIT PARAMETERS

The current in a series RLC circuit is

$$\mathbf{I} = \frac{\mathbf{V}}{\mathbf{Z}} = \frac{\mathbf{V}}{R + j(X_L - X_C)}$$

$$\mathbf{I} = \frac{\mathbf{V}}{R + j\left(\omega L - \frac{1}{\omega C}\right)}$$

$$|\mathbf{I}| = \frac{V}{\sqrt{R^2 + \left(\omega L - \frac{1}{\omega C}\right)^2}} \tag{8.9.1}$$

At half-power points,

$$|\mathbf{I}| = \frac{I_0}{\sqrt{2}} \tag{8.9.2}$$

Since $$|\mathbf{I}_0| = \frac{V}{R}$$

$$|\mathbf{I}| = \frac{V}{\sqrt{2}\,R} \tag{8.9.3}$$

Comparison of Eqs. (8.9.1) and (8.9.3) gives

$$\frac{V}{\sqrt{2}\,R} = \frac{V}{\sqrt{R^2 + \left(\omega L - \frac{1}{\omega C}\right)^2}}$$

$$\sqrt{R^2 + \left(\omega L - \frac{1}{\omega C}\right)^2} = \sqrt{2}\,R \tag{8.9.4}$$

Squaring both sides

$$R^2 + \left(\omega L - \frac{1}{\omega C}\right)^2 = 2R^2$$

$$\omega L - \frac{1}{\omega C} = \pm R \tag{8.9.5}$$

This relation shows that the magnitude of reactive component of **Z** is equal to the resistive component of **Z**, that is

$$\mathbf{Z} = R \pm jR = R\,(1 \pm j1) \tag{8.9.6}$$

Let ω_1 and ω_2 be the lower half-power and upper half-power frequencies respectively. Reactance at lower half-power frequency from Eq. (8.9.6) is

$$\omega_1 L - \frac{1}{\omega_1 C} = -R \tag{8.9.7}$$

The presence of negative sign of right side of Eq. (8.9.7) indicates that below resonance the capacitive reactance $1/(\omega_1 C)$ is greater than the inductive reactance $\omega_1 L$. Eq. (8.9.7) can be written as

$$\omega_1^2 + \frac{R}{L}\omega_1 - \frac{1}{LC} = 0$$

This is a quadratic equation in ω_1. The two values of ω_1 are given by

$$\omega_1 = -\frac{R}{2L} \pm \sqrt{\left(\frac{R}{2L}\right)^2 + \frac{1}{LC}}$$

The negative value of ω_1 is meaningless.

Hence
$$\omega_1 = -\frac{R}{2L} + \sqrt{\left(\frac{R}{2L}\right)^2 + \frac{1}{LC}}$$

Let
$$\alpha = \frac{R}{2L}.$$

Since
$$\omega_0^2 = \frac{1}{LC}$$

$$\omega_1 = -\alpha + \sqrt{\alpha^2 + \omega_0^2} \tag{8.9.8}$$

Reactance at upper half-power frequency can be written from Eq. (8.9.5) as

$$\omega_2 L - \frac{1}{\omega_2 C} = R$$

Rearrangement of this equation gives

$$\omega_2^2 - \frac{R}{L}\omega_2 - \frac{1}{LC} = 0$$

The two values of ω_2 are

$$\omega_2 = \frac{R}{2L} \pm \sqrt{\left(\frac{R}{2L}\right)^2 + \frac{1}{LC}}$$

Again rejecting the negative value of frequency we get

$$\omega_2 = \frac{R}{2L} + \sqrt{\left(\frac{R}{2L}\right)^2 + \frac{1}{LC}}$$

$$\omega_2 = \alpha + \sqrt{\alpha^2 + \omega_0^2} \tag{8.9.10}$$

Bandwidth
$$\Delta\omega = \omega_2 - \omega_1 \text{ rad/s}$$

Subtracting Eq. (8.9.8) from Eq. (8.9.10)

$$\omega_2 - \omega_1 = 2\alpha$$

therefore,
$$\Delta\omega = \omega_2 - \omega_1 = 2\alpha = -\frac{R}{L} \tag{8.9.11}$$

Equation (8.9.11) shows that the bandwidth of a series RLC circuit depends upon the R/L ratio and independent of C. But the value of ω_0 is dependent on C.

From Eqs. (8.9.8) and (8.9.10)

$$\omega_1 \omega_2 = (-\alpha + \sqrt{\alpha^2 + \omega_0^2})(\alpha + \sqrt{\alpha^2 + \omega_0^2}) = \omega_0^2$$

therefore,

$$\boxed{\omega_0 = \sqrt{\omega_1 \omega_2}} \tag{8.9.12}$$

This relation shows that *the resonant frequency is the geometric mean of the lower and upper half-power frequencies.*

■ **EXAMPLE 8.4** A series RLC circuit has $R = 10\ \Omega$, $L = 0.1$ H and $c = 8\ \mu$F. Determine (*a*) the resonant requency, (*b*) Q-factor of the circuit at resonance, and (*c*) the half-power frequencies.

SOLUTION :

(*a*) Resonant frequency $f_0 = \dfrac{1}{2\pi\sqrt{LC}} = \dfrac{1}{2\pi\sqrt{0.1 \times 8 \times 10^{-6}}} = 178$ Hz

(*b*) Q-factor at resonance $Q_0 = \dfrac{1}{R}\sqrt{\dfrac{L}{C}} = \dfrac{1}{10}\sqrt{\dfrac{0.1}{8 \times 10^{-6}}} = 11.2$

(*c*)

$$\omega_1 = -\frac{R}{2L} + \sqrt{\left(\frac{R}{2L}\right)^2 + \frac{1}{LC}}$$

$$= -\frac{10}{2 \times 0.1} + \sqrt{\left(\frac{10}{2 \times 0.1}\right)^2 + \frac{1}{0.1 \times 8 \times 10^{-6}}} = 1069 \text{ rad/s}$$

$$f_1 = \frac{\omega_1}{2\pi} = \frac{1069}{2\pi} = 170 \text{ Hz}$$

$$\omega_2 = +\frac{R}{2L} + \sqrt{\left(\frac{R}{2L}\right)^2 + \frac{1}{LC}} = \frac{10}{2 \times 0.1} + \sqrt{\left(\frac{10}{2 \times 0.1}\right)^2 + \frac{1}{0.1 \times 8 \times 10^{-6}}} = \mathbf{1169 \text{ rad/s}}$$

$$f_2 = \frac{\omega_2}{2\pi} = \frac{1169}{2\pi} = 186 \text{ Hz}$$

8.10 RELATION BETWEEN BANDWIDTH, RESONANT FREQUENCY AND QUALITY FACTOR FOR A SERIES RLC CIRCUIT

For a series RLC circuit

$$\mathbf{Z} = R + j\left(\omega L - \frac{1}{\omega C}\right)$$

$$= R\left[1 + j\left(\frac{\omega L}{R} - \frac{1}{\omega CR}\right)\right] = R\left[1 + j\left(\frac{\omega_0 L}{R} \times \frac{\omega}{\omega_0} - \frac{1}{\omega_0 CR} \times \frac{\omega_0}{\omega}\right)\right]$$

But

$$Q_0 = \frac{\omega_0 L}{R} = \frac{1}{\omega_0 CR}$$

therefore,

$$\mathbf{Z} = R\left[1 + jQ_0\left(\frac{\omega}{\omega_0} - \frac{\omega_0}{\omega}\right)\right] = R\left[1 + jQ_0\left(\frac{\omega^2 - \omega_0^2}{\omega_0 \omega}\right)\right]$$

$$= R\left[1 + jQ_0 \frac{(\omega - \omega_0)(\omega + \omega_0)}{\omega_0 \omega}\right]$$

A frequencies which are near resonance $\omega + \omega_0 = 2\omega$.

Therefore

$$\mathbf{Z} = R\left[1 + jQ_0 \times 2\left(\frac{\omega - \omega_0}{\omega_0}\right)\right]$$

$$\mathbf{Z} = R\,(1 + j2Q_0\,\delta) \tag{8.10.1}$$

where

$$\boxed{\delta \overset{\Delta}{=} \frac{\omega - \omega_0}{\omega_0} = \frac{f - f_0}{f_0}} \tag{8.10.2}$$

The quantity δ is called the **per unit frequency deviation**. It is the difference between the circuit frequency f and the resonant frequency f_0 expressed as fraction of f_0. If a $f < f_0$, δ is negative. If $f > f_0$, δ is positive. At resonance $\delta = 0$.

At half-power frequencies the circuit impedance is given by Eq. (8.9.6) as

$$\mathbf{Z} = R\,(1 \pm j\,1) \tag{8.10.3}$$

The circuit impedance at lower-half-power frequency is

$$\mathbf{Z}_{\omega_1} = R - jR = \sqrt{2}\,R\,\angle -45^\circ \tag{8.10.4}$$

The circuit impedance at upper half-power frequency is

$$\mathbf{Z}_{\omega_2} = R + jR = \sqrt{2}\,R\,\angle 45^\circ \tag{8.10.5}$$

Comparison of Eqs. (8.10.1) and (8.10.3) gives

$$2Q_0\,\delta = \pm 1 \tag{8.10.6}$$

$$\delta = \pm \frac{1}{2Q_0} \tag{8.10.7}$$

The positive sign is for f_2 and negative sign for f_1. Thus,

$$\frac{f_2 - f_0}{f_0} = +\frac{1}{2Q_0} \tag{8.10.8}$$

and

$$\frac{f_1 - f_0}{f_0} = -\frac{1}{2Q_0} \tag{8.10.9}$$

Subtraction of Eq. (8.10.9) from Eq. (8.10.8) gives

$$\frac{f_2 - f_1}{f_0} = \frac{1}{Q_0}$$

$$f_2 - f_1 = \frac{f_0}{Q_0} \quad \text{or} \quad \omega_2 - \omega_1 = \frac{\omega_0}{Q_0}$$

That is,

$$Q_0 = \frac{\omega_0}{\omega_2 - \omega_1} = \frac{\text{resonant frequency}}{\text{bandwidth}} \tag{8.10.10}$$

Thus, for a given resonant frequency, the higher the Q_0 the smaller the bandwidth and sharper the amplitude response curve. Such a circuit is said to have a greater frequency selectively. **Selectivity** of a resonant circuit is its ability to discriminate between signals of desired and undesired frequencies. **Selectivity** S *is defined as the ratio of resonant frequency and bandwidth.* That is

$$S \triangleq \frac{\text{resonant frequency}}{\text{bandwidth}} = \frac{\omega_0}{\Delta\omega} \tag{8.10.11}$$

For a series RLC circuit, since $\Delta\omega = R/L$

$$S = \frac{\omega_0 L}{R} = Q_{0S} \tag{8.10.12}$$

Thus, for a series RLC circuit the selectivity is equal to the quality factor. In general, Q_0 and S are different.

From the relation $Q_0 = \frac{1}{R}\sqrt{\frac{L}{C}}$, it is seen that if L and C of a series RLC circuit remain constant, Q_0 is inversely proportional to R. For low values of R, the factor Q_0 is high, and for high values of R the factor Q_0 is low. Fig. 8.6 shows the shape of frequency response curves for low and high values of R. The curve with high Q_0 has a much sharper peak and, therefore, has a greater frequency selectivity. That is, the current magnitude is high over a small range of frequencies. The relation $Q_0 = \frac{1}{R}\sqrt{\frac{L}{C}}$ also shows that Q_0 and hence the selectivity of a

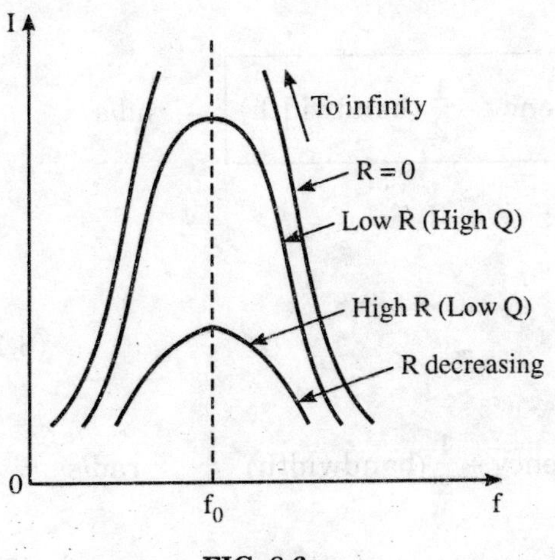

FIG. 8.6

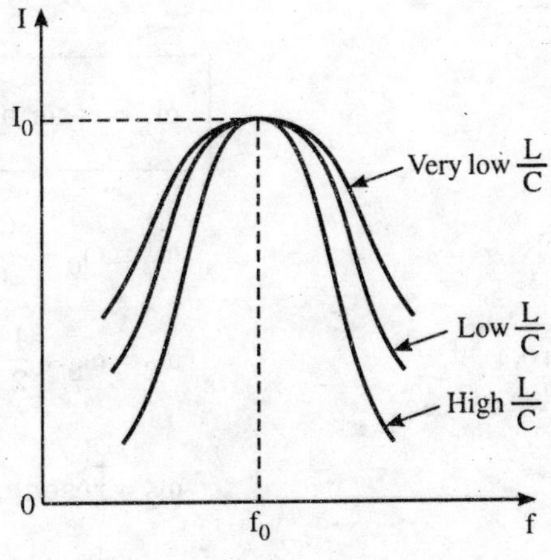

FIG. 8.7

series RLC resonant circuit also depends on L/C ratio. It increases with the increase in L/C ratio as shown in Fig. 8.7.

8.11 APPROXIMATE RELATIONSHIP BETWEEN ω_1, ω_2, ω_0 AND BW

From Eq. (8.9.8)

$$\omega_1 = -\frac{R}{2L} + \sqrt{\left(\frac{R}{2L}\right)^2 + \omega_0^2}$$

But $$Q_0 = \frac{\omega_0 L}{R}, \frac{R}{L} = \frac{\omega_0}{Q_0}$$

therefore, $$\omega_1 = -\frac{\omega_0}{2Q_0} + \sqrt{\frac{\omega_0^2}{4Q_0^2} + \omega_0^2} = -\frac{\omega_0}{2Q_0} + \omega_0\sqrt{\frac{1}{4Q_0^2} + 1}$$

Similarly, $$\omega_2 = \frac{\omega_0}{2Q_0} + \omega_0\sqrt{\frac{1}{4Q_0^2} + 1}$$

$$\boxed{BW = \omega_2 - \omega_1 = \frac{\omega_0}{Q_0} = 2\alpha}$$

Approximations

When $$\boxed{Q_0 \geq 5, \left(1 + \frac{1}{4Q_0^2}\right) \simeq 1 \text{ (approximately)}}$$

and $$\omega_1 = \omega_0 - \frac{\omega_0}{2Q_0}$$

$$\omega_1 = \omega_0 - \frac{1}{2}\Delta\omega \tag{8.11.1}$$

$$\boxed{\omega_1 = \text{resonant frequency} - \frac{1}{2}\text{(bandwidth)}} \quad \text{rad/s}$$

Similarly, $$\omega_2 = \omega_0 + \frac{\omega_0}{2Q_0}$$

therefore, $$\omega_2 = \omega_0 + \frac{1}{2}\Delta\omega \tag{8.11.12}$$

$$\omega_2 = \text{resonant frequency} + \frac{1}{2}\text{(bandwidth)} \quad \text{rad/s}$$

Also, $$\omega_0 = \frac{\omega_1 + \omega_2}{2} \tag{8.11.3}$$

■ **EXAMPLE 8.5** In a series RLC circuit, the power consumed at resonance is 0.1 W. The circuit is connected across 1 V supply. If $L = 0.05$ H and the voltage across L at resonance is 70.71 V, determine the values of R and C. The bandwidth of the circuit is 200 rad/s. Determine also the resonant frequency of the circiut.

SOLUTION : Power at resonance

$$P_0 = \frac{V^2}{R}$$

therefore,
$$R = \frac{V^2}{P_0} = \frac{(1)^2}{0.1} = 10\ \Omega$$

Current at resonance

$$I_0 = \frac{V}{R} = \frac{1}{10.} = 0.1\text{ A}$$

Voltage across the inductor at resonance

$$V_{L0} = I_0 X_{L0} = I_0 \omega_0 L$$

therefore,
$$\omega_0 = \frac{V_{L0}}{I_0 L} = \frac{70.71}{0.1 \times 0.05} = 14142\text{ rad/s}$$

Q-factor at resonance

$$Q_0 = \frac{\omega_0}{BW} = \frac{14142}{200} = 70.71$$

Also,
$$Q_0 = \frac{1}{\omega_0 RC}$$

therefore,
$$C = \frac{1}{\omega_0 Q_0 R} = \frac{1}{14142 \times 70.71 \times 10} = 10^{-7}\text{ F} = 0.1\ \mu\text{F}$$

■ **EXAMPLE 8.6** A series RLC circuit has $R = 10\ \Omega$, $L = 0.01$ H and $C = 0.01\ \mu$F. It is connected across a 10 mV supply. Calculate (*a*) the resonant frequency of the circuit, (*b*) the Q-factor at resonance, (*c*) the bandwidth, (*d*) the upper and lower cut-off frequencies, (*e*) the current at resonance.
Determine also the voltage across each element in the circuit. Calculate the current in the circuit at 90 per cent of the resonant frequency.

SOLUTION :

(*a*) Resonant-frequency

$$f_0 = \frac{1}{2\pi\sqrt{LC}} = \frac{1}{2\pi\sqrt{(0.01) \times 0.01 \times 10^{-6}}}$$

$$= 15915\text{ Hz}$$

$$\omega_0 = 2\pi f_0 = 100\ 000\text{ rad/s}$$

(b) Q-factor at resonance

$$Q_0 = \frac{1}{R}\sqrt{\frac{L}{C}} = \frac{1}{10}\sqrt{\frac{0.01}{0.01 \times 10^{-6}}} = 100$$

(c) Bandwidth of the resonant circuit

$$BW = \frac{\omega_0}{Q_0} = \frac{100\,000}{100} = 1000 \text{ rad/s}$$

or

$$BW = \frac{f_0}{Q_0} = \frac{15\,915}{100} = 159.15 \text{ Hz}$$

(d) Lower cut-off frequency

$$\omega_1 = \omega_0 - \frac{BW}{2} = 100\,000 - \frac{1000}{2} = 99\,500 \text{ rad/s}$$

Lower cut-off frquency in hertz

$$f_1 = f_0 - \frac{BW \text{ in Hz}}{2} = \frac{\omega_1}{2\pi} = \frac{99\,500}{2\pi} = 15\,836$$

(e) Current at resonance

$$I_0 = \frac{V}{R} = \frac{10 \times 10^{-3}}{10} = 10^{-3} \text{ A} \quad \text{or} \quad 1 \text{ mA}$$

Voltage across the inductance at resonance

$$V_{L_0} = j\,X_{L_0} I_0 = (2\pi f_0\, L)\, I_0 = j\omega_0 L\, I_0$$

$$= j\,100\,000 \times 0.01 \times 10^{-3} = j\,1 \text{ V}$$

Voltage across the capacitor at resonance

$$V_{C_0} = -j\,X_{C_0}\, I_0 = \frac{-j\,I_0}{2\pi f_0 C} = \frac{-j\,I_0}{\omega_0 C}$$

$$= \frac{-j\,10^{-3}}{100\,000 \times 0.01 \times 10^{-6}} = -j\,1 \text{ V}$$

At 90 per cent of the resonant frequency, the frequency is

$$\omega = 90\% \text{ of } 100\,000 = 90\,000 \text{ rad/s}$$

The circuit impedance is

$$\mathbf{Z} = R + j\,(X_L - X_C)$$

$$= R + j\left(\omega L - \frac{1}{\omega C}\right)$$

$$= 10 + j\left(90\,000 \times 0.01 - \frac{1}{90\,000 \times 0.01 \times 10^{-6}}\right)$$

$$= 10 - j\,211.1 = 211.34 \angle -87.29^\circ\ \Omega$$

The current at 90 000 rad/s is

$$\mathbf{I} = \frac{\mathbf{V}}{\mathbf{Z}} = \frac{10 \times 10^{-3} \angle 0^\circ}{211.34 \angle -87.29^\circ} = 0.0473 \angle 87.29^\circ \text{ A}$$

■ **EXAPLE 8.7** A 10 mH coil is connected in series with a loss-free capacitor to a variable-frequency source which supplies a constant voltage of 10 V. The circuit current has a maximum value of 0.1 A at a frequency of 80 kHz. Calculate *(a)* the capacitance of the capacitor, *(b)* the Q factor of the coil, and *(c)* the half-power frequencies.

SOLUTION : *(a)* The current is a maximum at resonance when $X_{L_0} = X_{C_0}$, that is,

$$\omega_0 L = \frac{1}{\omega_0 C},$$

$$C = \frac{1}{\omega_0^2 L} = \frac{1}{(2\pi f_0)^2 L}$$

$$L = 10 \text{ mH} = 10 \times 10^{-3} \text{ H}$$

$$f_0 = 80 \text{ kHz} = 80 \times 10^3 \text{ Hz}$$

therefore,

$$C = \frac{1}{(2\pi \times 80 \times 10^3)^2 \times 10 \times 10^{-3}} \text{ F} = 400 \text{ pF}$$

(*b*) At resonance

$$R = \frac{V}{I} = \frac{10}{0.1} = 100 \ \Omega$$

$$Q_{0S} = \frac{\omega_0 L}{R} = \frac{2\pi \times 80 \times 10^3 \times 10 \times 10^{-3}}{100} = 50$$

(*c*) Bandwidth

$$\Delta f = f_2 - f_1 = \frac{f_0}{Q_0} = \frac{80 \times 10^3}{50} = 1600 \text{ Hz}$$

$$f_1 = f_0 - \frac{1}{2}\Delta f$$

$$= 80 \times 10^3 - \frac{1}{2} \times 1600 = 79.2 \times 10^3 \text{ Hz} = 79.2 \text{ kHz}$$

$$f_2 = f_0 + \frac{1}{2}\Delta f$$

$$= 80 \times 10^3 + \frac{1}{2} \times 1600 = 80.8 \times 10^3 \text{ Hz} = 80.8 \text{ kHz}$$

8.12 RESONANCE IN PARALLEL RLC CIRCUIT

Consider the parallel RLC circuit shown in Fig. 8.8 connected to sinusoidal current source I_S.

The admittance 'seen' by the current source is given by

$$\mathbf{Y} = \mathbf{Y}_R + \mathbf{Y}_L + \mathbf{Y}_C$$

$$= \frac{1}{R} + \frac{1}{j\omega L} + j\omega C = \frac{1}{R} + j\left(\omega C - \frac{1}{\omega L}\right)$$

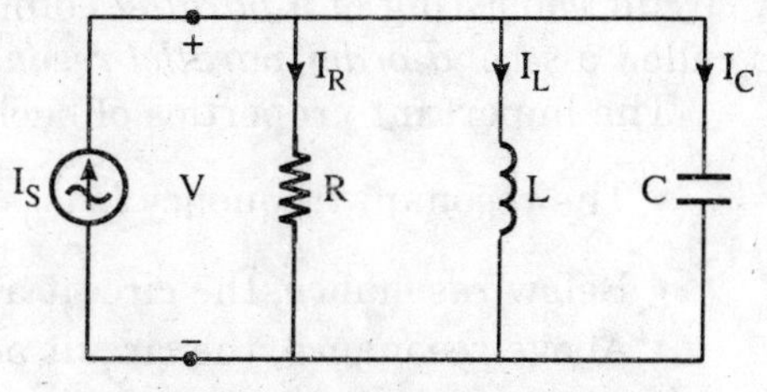

FIG. 8.8 Parallel RLC circuit

At resonance the imaginary part of **Y** is zero. That is, at $\omega = \omega_0$, Im (**Y**) = 0

$$\omega_0 C - \frac{1}{\omega_0 L} = 0$$

$$\omega_0 C = \frac{1}{\omega_0 L}$$

$\therefore$ resonant frequency,
$$\omega_0 = \sqrt{\frac{1}{LC}} \text{ rad/s}$$

$$2\pi f_0 = \sqrt{\frac{1}{LC}}$$

$$\boxed{f_0 = \frac{1}{2\pi}\sqrt{\frac{1}{LC}} \text{ Hz}} \qquad (8.12.1)$$

This is the same relation as was determined for the series resonant circuit.

The admittance at resonance is $Y_0 = \frac{1}{R}$. Thus the admittance is minimum at resonance,

Since
$$|\mathbf{V}| = \left|\frac{\mathbf{I}}{\mathbf{Y}}\right| = \frac{|\mathbf{I}|}{|\mathbf{Y}|}$$

we see that $|\mathbf{V}|$ is maximum at resonance. The voltage at resonance

$$|\mathbf{V}_0| = \frac{|\mathbf{I}|}{1/R} = RI$$

When $\omega = 0$, $Z_L = 0$ and therefore $\mathbf{V} = 0$. As $\omega \to \infty$, $Z_C = \frac{1}{j\omega C} \to 0$, and again $V \to 0$. Fig. 8.9 shows that variation of $|\mathbf{V}|$ with ω.

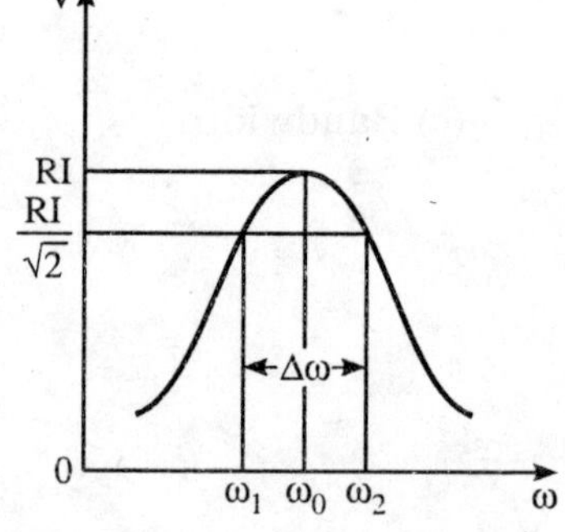

FIG. 8.9 $|\mathbf{V}|$ **versus** ω **curve**

Here ω_1 and ω_2 are the *lower* and *upper* half-power frequencies, respectively.

We have seen that at resonance the parallel RLC circuit behaves simply as the resistance, the parallel LC combination known as a tank circuit behaves as an open circuit. But it does not mean that there is no current in the inductor and capacitor. Acutally, the current circulates around the loop formed by the inductor and capacitor.

8.12.1 Properties of Second-Order Parallel RLC Resonant Circuit

A circuit consisting of a *parallel* connection of a resistor R, an inductor L, and a capacitor C is called a *second-order parallel resonant circuit.*

The important properties of such a circuit are as follows :

- The resonant frequency is $\omega_0 = \frac{1}{\sqrt{LC}}$ rad/s.
- Below resonance, the circuit acts like an RL circuit.
- Above resonance, the circuit acts like an RC circuit.
- At resonance, the circuit acts like a resistor.

- If the conductance is zero (the resistance is infinite), at resonance the circuit acts like an open circuit.
- The bandwidth is $\dfrac{1}{RC}$
- The quantity factor Q_{op} is $R\sqrt{\dfrac{C}{L}}$

EXAMPLE 8.8 Determine the resonant frequency for the network shown in Fig. 8.10.

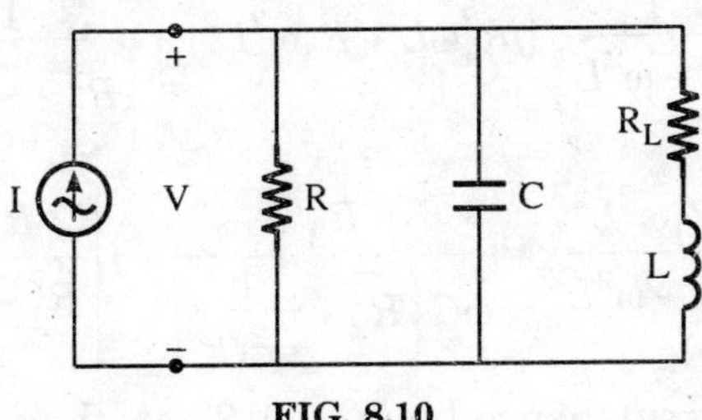

FIG. 8.10

SOLUTION : For the given network,

$$\mathbf{Y} = \frac{1}{R} + j\omega C + \frac{1}{R_L + j\omega L} = \frac{1}{R} + j\omega C + \frac{R_L - j\omega L}{R_L^2 + \omega^2 L^2}$$

$$= \frac{1}{R} + \frac{R_L}{R_L^2 + \omega^2 L^2} + j\left(\omega C - \frac{\omega L}{R_L^2 + \omega^2 L^2}\right)$$

At resonance $\omega = \omega_0$ and Im $(\mathbf{Y}) = 0$

therefore, $$\omega_0 C - \frac{\omega_0 L}{R_L^2 + \omega_0^2 L^2} = 0$$

$$R_L^2 + \omega_0^2 L^2 = \frac{L}{C}$$

$$\omega_0^2 L^2 = \frac{L}{C} - R_L^2$$

$$\omega_0 = \sqrt{\frac{1}{LC} - \frac{R_L^2}{L^2}}$$

EXAMPLE 8.9 Determine the resonant frequency of the network shown in Fig. 8.11.

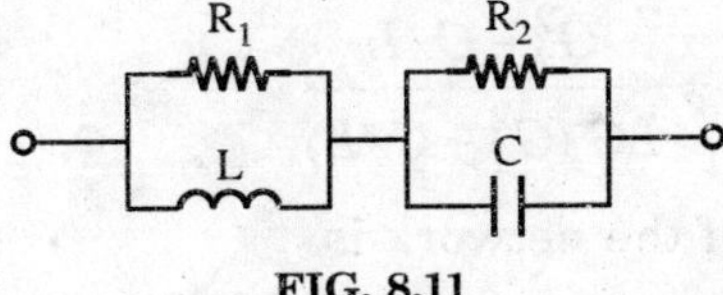

FIG. 8.11

SOLUTION : The total impedance of the network is

$$\mathbf{Z} = (R_1 \,||\, jX_L) + [R_2 \,||\, (-jX_C)]$$

$$= \frac{R_1 jX_L}{R_1 + jX_L} + \frac{R_2(-jX_C)}{R_2 - jX_C} = \frac{jR_1\omega L}{R_1 + j\omega L} - \frac{jR_2/(\omega C)}{R_2 - j/(\omega C)}$$

$$= \frac{jR_1\omega R\,(R_1 - j\omega L)}{R_1^2 + \omega^2 L^2} - j\frac{R_2/(\omega C)\,[R_2 + j/\omega C]}{R_2^2 + \dfrac{1}{\omega^2 C^2}}$$

$$= \frac{1}{R_1^2 + \omega^2 L^2}[jR_1^2\omega L + R_1\omega^2 L^2] + \frac{1}{\left(R_2^2 + \dfrac{1}{\omega^2 C^2}\right)}\left[-j\frac{R_2^2}{\omega C} + \frac{R_2}{\omega C}\right]$$

$$= \frac{R_1\omega^2 L^2}{R_1^2 + \omega^2 L^2} + \frac{R_2}{\omega C\left(R_2^2 + \dfrac{1}{\omega^2 C^2}\right)} + j\left[\frac{R_1^2\omega L}{R_1^2 + \omega^2 L^2} - \frac{R_2^2}{\omega C\left(R_2^2 + \dfrac{1}{\omega^2 C^2}\right)}\right]$$

At resonance, **Z** must be real, hence Im (**Z**) = 0. That, the j term of **Z** is zero and $\omega = \omega_0$

$$\frac{R_1^2\,\omega_0 L}{R_1^2 + \omega_0^2 L^2} - \frac{R_2^2}{\omega_0 C\left(R_2^2 + \dfrac{1}{\omega_0^2 C^2}\right)} = 0$$

$$\frac{R_1^2\,\omega_0 L}{R_1^2 + \omega_0^2 L^2} = \frac{R_2^2\,\omega_0^2\,C}{R_2^2\,\omega_0^2 C^2 + 1}$$

$$\frac{L}{1 + \omega_0^2\dfrac{L^2}{R_1^2}} = \frac{C}{\omega_0^2\,C^2 + \dfrac{1}{R_2^2}}$$

$$\omega_0^2\,C^2 + \frac{1}{R_2^2} = \frac{C}{L}\left(1 + \omega_0^2\frac{L^2}{R_1^2}\right)$$

$$\omega_0^2\,C^2 + G_2^2 = \frac{C}{L}(1 + \omega_0^2\,L^2\,G_1^2)$$

$$\omega_0^2\,C^2 - \frac{C}{L}(\omega_0^2\,L^2\,G_1^2) = \frac{C}{L} - G_2^2$$

$$\omega_0^2\,(C^2 - CL\,G_1^2) = \frac{C}{L} - G_2^2$$

$$\omega_0^2 = \frac{G_2^2 - C/L}{LC\,(G_1^2 - C/L)}$$

Hence resonant frequency of the network is

$$f_0 = \frac{\omega_0}{2\pi} = \frac{1}{2\pi}\sqrt{\left[\frac{G^2 - C/L}{LC\,(G_1^2 - C/L)}\right]}\ Hz$$

8.13 RESONANCE IN A PRACTICAL PARALLEL CIRCUIT

The most common form of parallel resonant circuit in practical use is shown in Fig. 8.12. It is called a **tank circuit**.

A coil of inductance L and effective series resistance R is connected in parallel with a capacitor.

The admittance of the circuit is $\mathbf{Y} = \frac{1}{\mathbf{Z}} = \frac{1}{\mathbf{Z}_C} + \frac{1}{\mathbf{Z}_L} = \frac{1}{-jX_C} + \frac{1}{R + jX_L}$

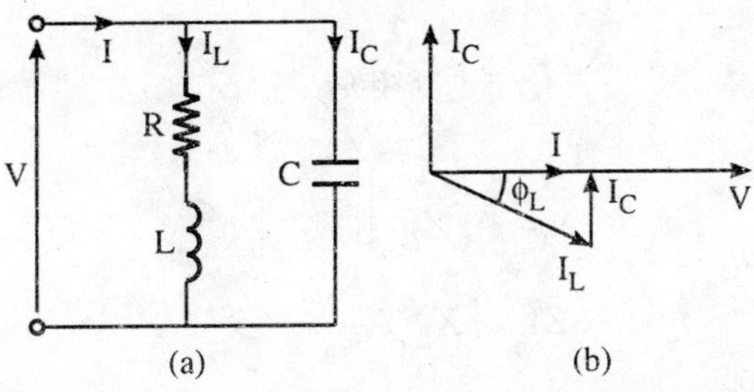

FIG. 8.12 (a) Parallel resonant circuit (b) Phasor diagram

$$\mathbf{Y} = j\omega C + \frac{1}{R + j\omega L} \tag{8.13.1}$$

$$= j\omega C + \frac{R - j\omega L}{R^2 + \omega^2 L^2}$$

$$= \frac{R}{R^2 + \omega^2 L^2} + j\omega\left(C - \frac{L}{R^2 + \omega^2 L^2}\right) \tag{8.13.2}$$

At resonance the j term is zero and $\omega = \omega_0$, so that

$$C - \frac{L}{R^2 + \omega_0^2 L^2} = 0$$

$$R^2 + \omega_0^2 L^2 = \frac{L}{C}$$

$$\omega_0^2 L^2 = \frac{L}{C} - R^2 \tag{8.13.3}$$

$$\omega_0 = \sqrt{\frac{1}{LC} - \frac{R^2}{L^2}} \tag{8.13.4}$$

Resonant frequency in hertz

$$f_0 = \frac{\omega_0}{2\pi} = \frac{1}{2\pi}\sqrt{\frac{1}{LC} - \frac{R^2}{L^2}} \tag{8.13.5}$$

If R is small so that (R^2/L^2) may be neglected in comparison with $1/(LC)$, Eq. (8.13.5) becomes

$$f_0 = \frac{1}{2\pi}\sqrt{\frac{1}{LC}}$$

which is the same as resonant frequency of a series RLC circuit.

Phasor diagram

The phasor diagram for the circuit of Fig. 8.12 a is shown in Fig. 8.12 b. The capacitor current $\mathbf{I}_c$ leads the supply voltage **V** by 90° and the coil current $\mathbf{I}_L$ lags behind the voltage **V** by the angle $\phi_L = \tan^{-1}\frac{X_L}{R}$. The supply current **I** is the phasor sum of I_C and I_L. At resonance I is in phase with **V** and

$$I_C = I_L \sin\phi_L$$

$$\frac{V}{X_{C_0}} = \left(\frac{V}{Z_{L_0}}\right)\left(\frac{X_{L_0}}{Z_{L_0}}\right)$$

$$Z_{L_0}^2 = X_{L_0} X_{C_0}$$

At resonance,

$$Z_{L_0}^2 = R^2 + (\omega_0 L)^2$$

$$X_{L_0} = \omega_0 L,\ X_{C_0} = \frac{1}{\omega_0 C}$$

therefore,

$$R^2 + (\omega_0 L)^2 = (\omega_0 L)\left(\frac{1}{\omega_0 C}\right)$$

$$\omega_0^2 L^2 = \frac{L}{C} - R^2$$

$$\omega_0 = \sqrt{\frac{1}{LC} - \frac{R^2}{L^2}} \quad \text{as before.}$$

Input admittance at resonance ($\mathbf{Y}_0$)

The input admittance $\mathbf{Y}_0$ at resonance is found by putting the j term equal to zero and $\omega = \omega_0$ in Eq. (8.13.2)

$$\mathbf{Y}_0 = \frac{R}{R^2 + \omega_0^2 L^2}$$

The impedance at resonance is

$$\mathbf{Z}_0 = \frac{1}{\mathbf{Y}_0} = \frac{R^2 + \omega_0^2 L^2}{R} = R + \frac{\omega_0^2 L^2}{R}$$

Substitution of the value of $\omega_0^2 L^2$ from Eq. (8.13.3) gives

$$\mathbf{Z}_0 = R + \frac{(L/C) - R^2}{R} = R + \frac{L}{CR} - R$$

therefore, $$Z_0 = \frac{L}{CR} \tag{8.13.6}$$

Z_0 is called the **dynamic impedance** of the circuit. This impedance is a pure resistance because it is independent of frequency. It is seen that lower the resistance of the coil, the higher the value of Z_0. Hence the value of impedance at resonance is a maximum and the resultant current is a minimum. A parallel resonant circuit is also called a **rejector circuit** since the current at resonance is a minimum. In other words, a tank circuit almost rejects the current at resonance. The current at resonance is given by

$$I_0 = \frac{V}{Z_0} = \frac{VCR}{L} \tag{8.13.7}$$

If the resistance of the coil $R = 0$, the circuit will draw no current at resonance.

The current versus frequency and impedance versus frequency curves for a tank circuit are shown in Fig. 8.13. If R is very small the current taken from the supply will also be very small. In an ideal case when R is zero, the supply current is zero. Under this condition there will be a large current circulating in the parallel circuit. This current is of oscillating nature. This is due to the transfer of energy from the inductor to the capacitor and then from the capacitor to the inductor The oscillations can be maintained provided that the losses in the coil are supplied by the source.

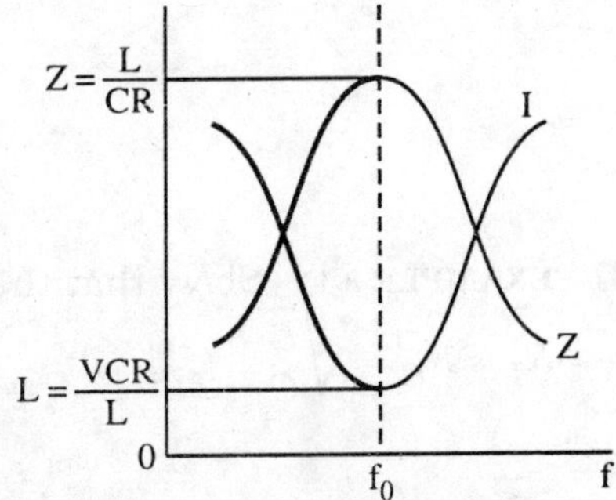

FIG. 8.13 *I* versus *f* and *Z* versus *f* curves for a tank circuit

8.14 CURRENT MAGNIFICATION

At resonance in a parallel circuit, the branch current may be many times greater than the supply current. Thus, by means of a parallel resonant circuit, the current taken from the supply can be greatly magnified. Hence this type of resonance is called *current resonance*.

$$\text{Current magnification} = \frac{I_C}{I}$$

$$= \frac{I_L \sin \phi_L}{I_L \cos \phi_L} = \tan \phi_L = \frac{\omega_0 L}{R}$$

$$= Q\text{-factor of the circuit at resonance} = Q_0 \tag{8.14.1}$$

Thus, *Q-factor is a measure of curent magnification in a parallel resonant circuit.*

■ **EXAMPLE 8.10** One branch of a parallel circuit consists of a coil of inductance 0.05 H and effective resistance 5 Ω and the other is a 0.1 μF capacitor. Calculate *(a)* the resonant frequency, *(b)* the dynamic impedance, and *(c)* the Q-factor at resonance.

SOLUTION : *(a)* Resonant frequency

$$f_0 = \frac{1}{2\pi}\sqrt{\frac{1}{LC} - \frac{R^2}{L^2}}$$

$$= \frac{1}{2\pi}\sqrt{\frac{1}{0.05 \times 0.1 \times 10^{-6}} - \frac{5^2}{0.05^2}} = 2250 \text{ Hz}$$

(b) Dymanic impedance = impedance at resonance

(b) $$Z_0 = \frac{L}{CR}$$

$$= \frac{0.05}{0.1 \times 10^{-6} \times 5} = 100 \times 10^3\ \Omega = 100\ \text{k}\Omega$$

(c) $$Q_0 = \frac{\omega_0 L}{R}$$

$$= \frac{2\pi f_0 L}{R} = \frac{2\pi \times 2250 \times 0.05}{5} = 141.4$$

■ **EXAMPLE 8.11** Show that the network shown in Fig. 8.14 is resonant at all frequencies if $R_L = R_C = \sqrt{\frac{L}{C}}$

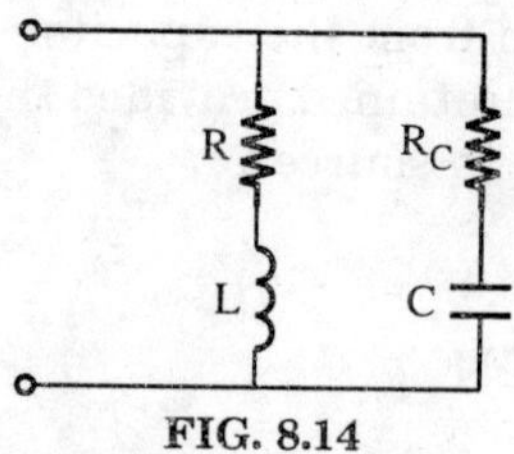

FIG. 8.14

SOLUTION : The admittance of the network is

$$\mathbf{Y} = \mathbf{Y}_L + \mathbf{Y}_C$$

$$= \frac{1}{R_L + jX_L} + \frac{1}{R_C - jX_C}$$

$$= \frac{R_C - jX_C + R_L + jX_L}{(R_L + jX_L)(R_C - jX_C)}$$

$$= \frac{(R_L + R_C) + j(X_L - X_C)}{R_L R_C + X_L X_C + j(R_C X_L - R_L X_C)}$$

$$X_L = 2\pi f L,\ X_C = \frac{1}{2\pi f C},\ X_L X_C = \frac{L}{C}$$

If we put $$R_L = R_C = \sqrt{\frac{L}{C}}, \text{ then}$$

$$\mathbf{Y} = \frac{\sqrt{\frac{L}{C}} + \sqrt{\frac{L}{C}} + j\,(X_L - X_C)}{\frac{L}{C} + \frac{L}{C} + j\left(\sqrt{\frac{L}{C}}\right)(X_L - X_C)}$$

$$= \frac{2 \times \sqrt{\frac{L}{C}} + j\,(X_L - X_C)}{\left(\sqrt{\frac{L}{C}}\right)\left[2\sqrt{\frac{L}{C}} + j\,(X_L - X_C)\right]} = \sqrt{\frac{C}{L}}$$

Thus, if $R_L = R_C = \sqrt{\frac{L}{C}}$ it is found that **Y** is real and independent of frequency. Hence the network is resonant at all frequencies.

■ **EXAMPLE 8.12** A circuit comprises a lossy inductor connected in series with a lossy capacitor. The inductor is represented by an inductance L in series with a resistance R_L, and the capacitor is represented by a capacitance C in parallel with a resistance R_c. Determine the resonant frequency and the impedance of the circuit.

Show that at resonance the maximum instantaneous energy stored in the inductor is equal to the maximum instantaneous energy stored in the capacitor.

If Q_L and Q_C are the Q-factors of the inductor and capacitor respectively at resonance show that Q value of the circuit at resonance is given by

$$Q_0 = \frac{Q_L\, Q_C}{Q_L + Q_C}$$

SOLUTION : The network is shown in Fig. 8.15.

The impedance of the network is

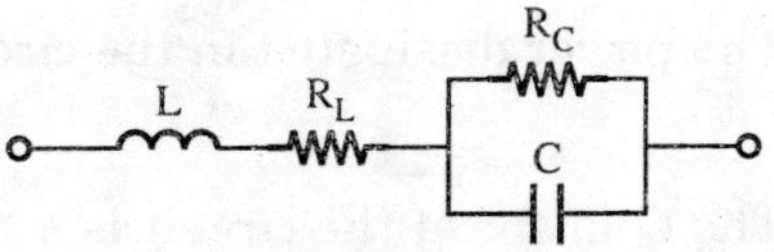

$$\mathbf{Z} = R_L + j\omega L + \frac{1}{G_c + j\omega C}$$

$$= R_L + j\omega L + \frac{G_c - j\omega C}{G_c^2 + \omega^2 C^2}$$

$$= R_L + \frac{G_c}{G_c^2 + \omega^2 C^2} + j\omega\left(L - \frac{C}{G_c^2 + \omega^2 C^2}\right)$$

FIG. 8.15

At resonance the j term is zero and $\omega = \omega_0$. Thus,

$$L - \frac{C}{G_c^2 + \omega_0^2 C^2} = 0$$

$$G_c^2 + \omega_0^2 C^2 = \frac{C}{L}$$

$$\omega_0^2 C^2 = \frac{C}{L} - G_c^2$$

$$\omega_0 = \sqrt{\frac{1}{LC} - \frac{G_c^2}{C^2}}$$

$$f_0 = \frac{\omega_0}{2\pi} = \frac{1}{2\pi}\sqrt{\frac{1}{LC} - \frac{G_2^2}{C^2}}$$

The impedance at resonance is equal to the real part of **Z** and $\omega = \omega_0$ that is,

$$\mathbf{Z}_0 = R_L + \frac{G_c}{G_c^2 + \omega_0^2 C^2}$$

$$= R_L + \frac{C_c}{(C/L)} = R_L + \frac{G_c L}{C}$$

Let I_0 = r.m.s. current in the circuit at resonance

V_{c0} = r.m.s. voltage across the capacitor at resonance

Maximum instantaneous energy stored in L is

$$W_{mL} = \tfrac{1}{2} L\,(\sqrt{2}\,I_0)^2 = L I_0^2$$

Maximum instantaneous energy stored in C is

$$W_{mC} = \tfrac{1}{2} C\,(\sqrt{2}\,V_{c0})^2 = C V_{c0}^2$$

$$V_{c0} = \frac{I_0}{|\,G_c + j\omega_0 C\,|} = \frac{I_0}{\sqrt{G_c^2 + \omega_0^2 C^2}} = \frac{I_0}{\sqrt{C/L}}$$

$$W_{mC} = C \times \frac{I_0^2}{C/L} = L I_0^2 = W_{mL}$$

The power dissipated in the circuit at responance is

$$P_0 = I_0^2 Z_0 = I_0^2 \left(R_L + \frac{G_c L}{C} \right)$$

The Q factor of the circuit is

$$Q_0 = \frac{\omega_0 W m_L}{P_0}$$

$$= \frac{\omega_0 (L I_0^2)}{I_0^2 \left(R_L + \frac{G_c L}{C} \right)} = \frac{1}{\left(\frac{R_L}{\omega_0 L} + \frac{G_c}{\omega_0 C} \right)}$$

Q-factor of the inductor

$$Q_L = \frac{\omega_0 L}{R_L}$$

Q-factor of the capacitor

$$Q_c = \omega_0 C R_c = \frac{\omega_0 C}{G_c}$$

Substitution of the value of Q_L and Q_c in the expression for Q_c gives

$$Q_0 = \frac{1}{\frac{1}{Q_L} + \frac{1}{Q_c}} = \frac{Q_L\, Q_c}{Q_L + Q_c}$$

■ **EXAMPLE 8.13** A series RLC has resonant frequency of 10^4 rad/s and bandwidth of 500 rad/s. If the voltage $V_s = 0.5$ V at resonant frequency produces a current of 50 mA, determine the element values and the voltage across L at resonance.

SOLUTION :

$$Z_0 = \frac{V}{I} = \frac{0.5}{50 \times 10^{-3}} = 10\ \Omega$$

In a series RLC circuit, $Z_0 = R$

therefore, $R = 10\ \Omega$

$$BW = \frac{R}{L}$$

$$500 = \frac{10}{L},\ L = \frac{10}{500} = 0.02\ \text{H}$$

$$\omega_0^2 = \frac{1}{LC}$$

$$C = \frac{1}{\omega_0^2 L} = \frac{1}{(10^4)^2 \times 0.02} = 0.5 \times 10^{-6}\ \text{F} = 0.5\ \mu\text{F}$$

$$V_{L_0} = (\omega_0 L)\, I_0 = 10^4 \times 0.02 \times 50 \times 10^{-3} = 10\ \text{V}$$

■ **EXAMPLE 8.14** A series RLC circuit is excited by a variable frequency sinusoidal voltage source having an r.m.s. value of 2 V. The maximum r.m.s. current in the circuit is to be 10 mA at a frequency of 10 kHz. At frequencies between 9.6 kHz and 10.4 kHz, the magnitude of the current is to be at least 7.07 mA. Determine the values of R, L and C and the voltage across the capacitor at resonance.

SOLUTION : The current at resonance is maximum and $Z_0 = R$

$$R = Z_0 = \frac{V_0}{I_0} = \frac{2}{10 \times 10^{-3}} = 200\ \Omega$$

$$BW = f_2 - f_1 = 10.4 - 9.6 = 0.8\ \text{kHz} = 800\ \text{Hz}$$

$$f_0 = 10\ \text{kHz} = 10 \times 10^3\ \text{Hz}$$

$$Q_0 = \frac{f_0}{BW} = \frac{10 \times 10^3}{800} = 12.5$$

$$Q_0 = \frac{\omega_0 L}{R}$$

$$L = \frac{Q_0 R}{\omega_0} = \frac{Q_0 R}{2\pi f_0} = \frac{12.5 \times 200}{2\pi \times 10 \times 10^3} \text{ H} = 39.79 \text{ mH}$$

$$Q_0 = \frac{1}{\omega_0 CR}$$

$$C = \frac{1}{\omega_0 Q_0 R} = \frac{1}{2\pi \times 10 \times 10^3 \times 12.5 \times 200} \text{ F} = 0.00637 \text{ μF}$$

Voltage across the capacitor at resonance

$$V_{c0} = X_{c0} I_0 = \frac{I_0}{\omega_0 C} = Q_0 R I_0 = 12.5 \times 200 \times 10 \times 10^{-3} = 25 \text{ V}$$

EXAMPLE 8.15 A circuit consists of a 4 μF capacitor in parallel with a coil of resistance 40 Ω and inductance 0.25 H. If the voltage applied to the circuit at this frequency is 230 V, calculate the current in each branch, supply current and the current magnification.

SOLUTION :

$$R = 40\ \Omega,\ L = 0.25 \text{ H},\ C = 4 \text{ μF} = 4 \times 10^{-6} \text{ F}$$

Resonance frequency

$$f_0 = \frac{1}{2\pi}\sqrt{\frac{1}{LC} - \frac{R^2}{L^2}}$$

$$= \frac{1}{2\pi}\sqrt{\frac{1}{0.25 \times 4 \times 10^{-6}} - \frac{40^2}{(0.25)^2}} = 157 \text{ Hz}$$

Current through the capacitor at resonance

$$\mathbf{I}_{c0} = \frac{\mathbf{V}}{\mathbf{Z}_{c0}} = \frac{V\angle 0^\circ}{-jX_{c0}} = jV(2\pi f_0 C)$$

$$= j\,230 \times 2\pi \times 157 \times 4 \times 10^{-6} = j\,0.908 \text{ A}$$

Current through the coil at resonance

$$\mathbf{I}_{coil} = \frac{\mathbf{V}}{\mathbf{Z}_{coil}} = \frac{V\angle 0^\circ}{R + jX_{L0}} = \frac{V\angle 0^\circ}{R + j\,2\pi f_0 L}$$

$$= \frac{230\angle 0^\circ}{40 + j\,2\pi \times 157 \times 0.25} = \frac{230\angle 0^\circ}{40 + j\,246.6}$$

$$= \frac{230\angle 0^\circ}{249.8\angle 80.78^\circ} = 0.9207\angle -80.78^\circ \text{ A}$$

$$= (0.1475 - j\,0.9088) \text{ A}$$

Supply current at resonance

$$\mathbf{I} = \mathbf{I}_{c0} + \mathbf{I}_{coil} = j\,0.908 + 0.1475 - j\,0.908$$

$$= 0.1475 \text{ A}$$

Alternatively

$$\mathbf{I} = \frac{VCR}{L} = \frac{230 \times 4 \times 10^{-6} \times 40}{0.25} = 0.1472 \text{ A}$$

8.14.1 Quality Factor for a Parallel RLC Resonant Circuit

Let $\quad v = V_m \sin \omega t$

At resonance, $\quad v = V_m \sin \omega_0 t$

Current through the inductor

$$i_L = \frac{V_m}{X_{L_0}} \sin (\omega_0 t - 90^\circ) = -\frac{V_m}{\omega_0 L} \cos \omega_0 t$$

Resonant frequency, $\quad \omega_0 = \frac{1}{\sqrt{LC}}$

Instantaneous energy in the circuit is

$$w(t) = \frac{1}{2} L i_L^2 + \frac{1}{2} C v^2$$

$$= \frac{1}{2} \frac{L V_m^2 \cos^2 \omega_0 t}{\omega_0^2 L^2} + \frac{1}{2} C V_m^2 \sin^2 \omega_0 t$$

$$= \frac{1}{2} C V_m^2 \cos^2 \omega_0 t + \frac{1}{2} C V_m^2 \sin^2 \omega_0 t$$

$$= \frac{1}{2} C V_m^2 (\cos^2 \omega_0 t + \sin^2 \omega_0 t) = \frac{1}{2} C V_m^2 = \text{constant}$$

Therefore, maximum energy stored in the circuit $= \frac{1}{2} C V_m^2$

Average power loss in the circuit is

$$P = \frac{V^2}{R} = \frac{(V_m/\sqrt{2})^2}{R} = \frac{V_m^2}{2R}$$

Therefore, energy lost per cycle

$$= P \times \frac{2\pi}{\omega_0} = \frac{\pi V_m^2}{\omega_0 R}$$

Quality factor of the parallel RLC circuit at resonance is

$$Q_{0p} = \frac{2\pi \text{ (maximum energy stored in the circuit)}}{\text{total energy lost per cycle}} = \frac{2\pi \left(\frac{1}{2} C V_m^2\right)}{(\pi V_m^2)/(\omega_0 R)}$$

therefore, $\quad \boxed{Q_{0p} = \omega_0 CR}$

The quality factors for a series resonant circuit and a parallel resonant circuit are

$$Q_{0s} = \frac{\omega_0 L}{R} = \frac{1}{\omega_0 CR}$$

$$Q_{0p} = \omega_0 CR = \frac{R}{\omega_0 L}$$

These expression correspond under the duality transformation

$$R \rightarrow \frac{1}{R}$$

$$L \rightarrow C$$

$$C \rightarrow L$$

For either circuit bandwith $= \dfrac{\omega_0}{Q_0}$

8.15 RELATION BETWEEN RESONANT FREQUENCY, BANDWIDTH AND QUALITY FACTOR IN A PARALLEL RLC RESONANT CIRCUIT

For the circuit shown in Fig. 8.8,

$$\mathbf{Y} = \frac{1}{R} + j\left(\omega C - \frac{1}{\omega L}\right) \tag{8.15.1}$$

therefore,

$$\mathbf{Z} = \frac{1}{\dfrac{1}{R} + j\left(\omega C - \dfrac{1}{\omega L}\right)}$$

$$= \frac{R}{1 + j\left(\omega CR - \dfrac{R}{\omega L}\right)}$$

$$\frac{\mathbf{Z}}{R} = \frac{1}{1 + j\left(\omega CR - \dfrac{R}{\omega L}\right)} \tag{8.15.2}$$

The quality factor for a parallel RLC resonant circuit is given by

$$Q_{0p} = \omega_0 CR = \frac{R}{\omega_0 L}$$

therefore, $\quad CR = \dfrac{Q_{0p}}{\omega_0}, \quad \dfrac{R}{L} = \omega_0 \, Q_{0p}$

Also, at resonance $\quad \mathbf{Z}_0 = R,$

Substituting the values of R, CR and $\dfrac{R}{L}$ in Eq. (8.15.2) we get

$$\frac{\mathbf{Z}}{\mathbf{Z}_0} = \frac{1}{1 + j\left(\dfrac{\omega}{\omega_0} Q_0 - \dfrac{\omega_0}{\omega} Q_0\right)}$$

$$\frac{\mathbf{Z}}{\mathbf{Z}_0} = \frac{1}{1 + jQ_0\left(\dfrac{\omega}{\omega_0} - \dfrac{\omega_0}{\omega}\right)} \tag{8.15.3}$$

The bandwidth of a parallel resonant circuit is defined as the difference between two frequencies at which the circuit impedance has fallen to $1/\sqrt{2}$ of its value at resonance. Thus, the half-power frequencies are obtained when the impedance **Z** becomes equal to $(\mathbf{Z}_0/\sqrt{2})$. This is shown in Fig. 8.16.

$$\left|\frac{\mathbf{Z}}{\mathbf{Z}_0}\right| = \frac{1}{\sqrt{2}} \tag{8.15.4}$$

At half-power points

$$Q_0\left(\frac{\omega}{\omega_0} - \frac{\omega_0}{\omega}\right) = \pm 1 \tag{8.15.5}$$

Let ω_1 be the lower value, and ω_2 the higher value. Then

$$\frac{\omega_1}{\omega_0} - \frac{\omega_0}{\omega_1} = -\frac{1}{Q_0} \tag{8.15.6}$$

$$\frac{\omega_2}{\omega_0} - \frac{\omega_0}{\omega_2} = +\frac{1}{Q_0} \tag{8.15.7}$$

Eq. (8.15.6) can be written as

$$\omega_1^2 + \frac{\omega_0}{Q_0}\,\omega_1 - \omega_0^2 = 0 \tag{*8.15.8}$$

This is a quadratic equation in ω_1. The two values of ω_1 are given by

$$\omega_1 = \frac{1}{2}\left(-\frac{\omega_0}{Q_0} \pm \sqrt{\frac{\omega_0^2}{Q_0^2} + 4\omega_0^2}\right)$$

$$= -\frac{1}{2}\frac{\omega_0}{Q_0} \pm \sqrt{\omega_0^2 + \frac{\omega_0^2}{4Q_0^2}}$$

The negative value of ω_1 is meaningless.

therefore,

$$\omega_1 = -\frac{1}{2}\frac{\omega_0}{Q_0} + \omega_0\sqrt{1 + \frac{1}{4Q_0^2}} \tag{8.15.9}$$

Similarly, from Eq. (8.15.7) we get

$$\omega_2 = +\frac{1}{2}\frac{\omega_0}{Q_0} + \omega_0\sqrt{1 + \frac{1}{4Q_0^2}} \tag{8.15.10}$$

Eqs. (8.15.9) and (8.15.10) give the *exact expression* for ω_1 and ω_2.

Bandwidth $\quad BW = \omega_2 - \omega_1 = \dfrac{\omega_0}{Q_0}$

therefore,
$$Q_0 = \frac{\omega_0}{BW} \tag{8.15.11}$$

Also, from Eqs. (8.15.9) and (8.15.10)

$$\omega_1\omega_2 = \left(\omega_0\sqrt{1+\frac{1}{4Q_0^2}}\right)^2 - \left(\frac{1}{2}\frac{\omega_0}{Q_0}\right)^2$$

$$= \omega_0^2\left(1+\frac{1}{4Q_0^2}\right) - \frac{\omega_0^2}{4Q_0^2}$$

$$\omega_1\omega_2 = \omega_0^2$$

therefore,
$$\boxed{\omega_0 = \sqrt{\omega_1\omega_2}} \tag{8.15.12}$$

Eq. (8.15.12) shows that *the resonant frequency is the geometric mean of the lower and upper half-power frequencies.*

Approximate Expressions

When $Q_0 \geq 5, \dfrac{1}{4Q_0^2} \leq \dfrac{1}{100}$

and $\left(1+\dfrac{1}{4Q_0^2}\right) = 1$ (approximately)

Since
$$\omega_1 = \omega_0\left[\sqrt{\left(1+\frac{1}{4Q_0^2}\right)} - \frac{1}{2Q_0}\right]$$

and
$$\omega_2 = \omega_0\left[\sqrt{1+\frac{1}{4Q_0^2}} + \frac{1}{2Q_0}\right]$$

for $Q_0 \geq 5$,

$$\omega_1 = \omega_0 - \frac{\omega_0}{2Q_0} \tag{8.15.13}$$

and
$$\omega_2 = \omega_0 + \frac{\omega_0}{2Q_0} \tag{8.15.14}$$

$$\omega_2 - \omega_1 = \frac{\omega_0}{Q_0} = BW \tag{8.15.15}$$

$\therefore$
$$\omega_1 = \omega_0 - \frac{1}{2}BW \tag{8.15.16}$$

$$\omega_2 = \omega_0 + \frac{1}{2}BW \tag{8.15.17}$$

$$\omega_0 = \frac{\omega_1 + \omega_2}{2} \tag{8.15.18}$$

Now,
$$\frac{\omega}{\omega_0} - \frac{\omega_0}{\omega} = \frac{\omega^2 - \omega_0^2}{\omega_0 \omega}$$

$$= \frac{(\omega - \omega_0)(\omega + \omega_0)}{\omega_0 \omega} = 2\left(\frac{\omega - \omega_0}{\omega_0}\right)$$

At frequencies which are near resonance

$$\omega + \omega_0 = 2\omega$$

Therefore, eq. (8.15.3) can be written as

$$\frac{\mathbf{Z}}{\mathbf{Z}_0} = \frac{1}{1 + j\,Q_0\left(\frac{\omega}{\omega_0} - \frac{\omega_0}{\omega}\right)} = \frac{1}{1 + j\,Q_0 \times 2\left(\frac{\omega - \omega_0}{\omega_0}\right)}$$

therefore,
$$\frac{\mathbf{Z}}{\mathbf{Z}_0} = \frac{1}{1 + j\,2Q_0\,\delta} \tag{8.15.19}$$

where
$$\delta \stackrel{\Delta}{=} \frac{\omega - \omega_0}{\omega_0} = \frac{f - f_0}{f_0}$$

= per unit frequency deviation

$$\left|\frac{\mathbf{Z}}{\mathbf{Z}_0}\right| = \frac{1}{\sqrt{1 + 4Q_0^2\,\delta^2}} \tag{8.15.20}$$

The half power frequencies are obtained when the impedance Z becomes equal to $\frac{Z_0}{\sqrt{2}}$.

therefore,
$$\left|\frac{\mathbf{Z}}{\mathbf{Z}_0}\right| = \frac{1}{\sqrt{2}} \tag{8.15.21}$$

$$1 + 4Q_0^2\,\delta^2 = 2$$

$$2Q_0\,\delta = \pm 1$$

$$\delta = \frac{1}{2Q_0} \tag{8.15.22}$$

The circuit impedance at lower-half power frequency

$$\mathbf{Z}(\omega_1) = \frac{\mathbf{Z}_0}{1 - j1} = \frac{Z_0}{\sqrt{2}} \underline{/ \; 45^\circ}$$

The circuit impedance at upper half-power frequency

$$\mathbf{Z}(\omega_2) = \frac{\mathbf{Z}_0}{1 + j1} = \frac{Z_0}{\sqrt{2}} \underline{/ - 45^\circ}$$

It should be noted that the impedance is predominantly inductive at frequencies below resonant frequency and the impedance is predominantly capacitive at frequencies above the resonant frequency.

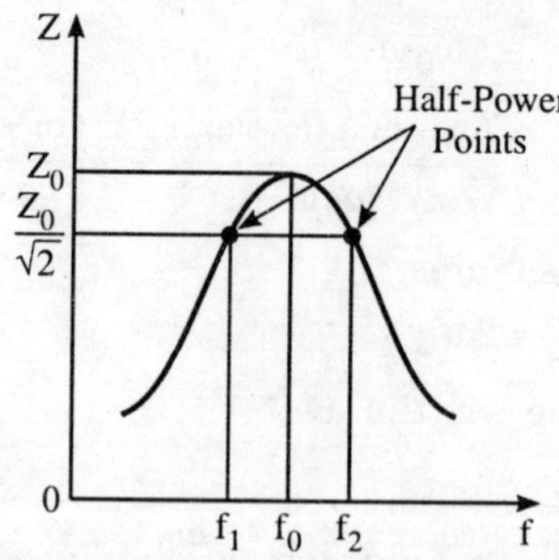

FIG. 8.16 Bandwidth in a parallel resonant circuit

EXAMPLE 8.16 A series resonant circuit is tuned to 1 MHz. The Q factor of the coil is 100. What is the ratio for the current at a frequency 20 kHz below resonance to that of the maximum current ? At what frequency above resonance is the current reduced to 90%.

SOLUTION :

(a)

$$f_0 = 1 \text{ MHz} = 1000 \text{ kHz} , Q_0 = 100$$

$$f = 1000 - 20 = 980 \text{ kHz}$$

$$\delta = \frac{f - f_0}{f_0} = \frac{980 - 1000}{1000} = -0.02$$

$$\left|\frac{\mathbf{I}}{\mathbf{I}_0}\right| = \left|\frac{1}{1 + jQ_0\left(\frac{f}{f_0} - \frac{f_0}{f}\right)}\right|$$

$$= \left|\frac{1}{1 + j + 100\left(\frac{980}{1000} - \frac{1000}{980}\right)}\right| = \left|\frac{1}{1 - j\,4.04}\right|$$

$$\frac{1}{\sqrt{1 + (4.04)^2}} = 0.24$$

(b)

$$\frac{I}{I_0} = \frac{1}{1 + j2Q_0\,\delta}$$

$$\left|\frac{\mathbf{I}}{\mathbf{I}_0}\right| = \sqrt{\frac{1}{1 + 4Q_0^2\,\delta^2}}$$

$$1 + 4Q_0^2\,\delta^2 = \left(\frac{I_0}{I}\right)^2$$

$$\delta = \frac{1}{2Q_0}\sqrt{\left(\frac{I_0}{I}\right)^2 - 1} = \frac{1}{2 \times 100}\sqrt{\left(\frac{1}{0.9}\right)^2 - 1}$$

$$= 2.4216 \times 10^{-3}$$

$$\frac{f - f_0}{f_0} = \delta$$

$$f = f_0 (1 + \delta) = 1000\,(1 + 2.4216 \times 10^{-3})$$

$$= 1002.4216 \text{ kHz}$$

■ **EXAMPLE 8.17** It is desired to design a series resonant circuit with the following specifications:

$$C = 250 \times 10^{-12} \text{ F}, \; f_0 = 600 \text{ kHz}$$

$$BW = 20 \text{ kHz}$$

Calculate Q_0, R and L of the circuit. Also calculate the current at 500 kHz as a fraction of the current at resonance.

SOLUTION :

$$f_0 = \frac{1}{2\pi\sqrt{LC}}$$

$$L = \frac{1}{4\pi^2 f_0^2 C} = \frac{1}{4\pi^2 (600 \times 10^3)^2 \times 250 \times 10^{-12}} \text{ H}$$

$$= 0.2814 \times 10^{-3} \text{ H} = 0.2814 \text{ mH}$$

$$Q_0 = \frac{f_0}{BW} = \frac{600 \times 10^3}{20 \times 10^3} = 30$$

Also,

$$Q_0 = \frac{\omega_0 L}{R}$$

$$R = \frac{\omega_0 L}{Q_0} = \frac{2\pi f_0 L}{Q_0}$$

$$= \frac{2\pi \times 600 \times 10^3 \times 0.2814 \times 10^{-3}}{30} = 35.36\ \Omega$$

$$\delta = \frac{f - f_0}{f_0} = \frac{500 - 600}{600} = -\frac{1}{6}$$

$$\frac{\mathbf{I_0}}{\mathbf{I}} = \frac{1}{1 + j\,2Q_0\delta} = \frac{1}{1 - j2 \times 30 \times \frac{1}{6}}$$

$$= \frac{1}{1 - j10}$$

$$\frac{\mathbf{I}_0}{\mathbf{I}} = \left| \frac{1}{1 - j10} \right| = \frac{1}{\sqrt{1^2 + 10^2}} = 0.0995$$

8.16 LOCUS DIAGRAMS

The magnitude and phase of current in a circuit depends upon the values of R, L, C and frequency. If one of these parameters is varied with constant applied voltage, the magnitude and phase of the circuit current will also vary. The variation of current **I** with the variation of any one of the parameters can be shown graphically. This graphical representation is called the **current locus**. In general, the path traced by the tip of the arrow of a phasor is called its locus. By locus of the current phasor is meant the path of the tip (end point) of the arrow of the phasor **I**. Graphical methods sometimes provide quick and useful informations than analytical methods specially under varying conditions. We shall draw the current locus diagrams for few general cases only. The technique can be extended to cover other cases with slight modifications.

8.17 SERIES RL CIRCUIT (CONSTANT X_L, VARIABLE R)

Consider the series RL circuit of Fig. 8.17 containing a constant reactance X_L and a variable resistance R. Suppose that the applied voltage has a constant r.m.s. value V.

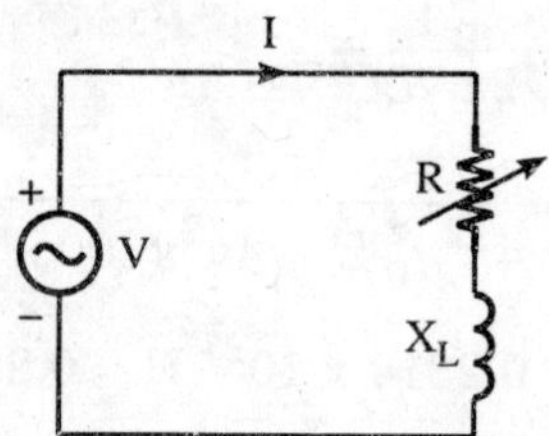

FIG. 8.17 Series RL circuit with constant X_L and variable R

Let V be taken as reference. Thus,

$$\mathbf{V} = V \angle 0°$$

Impedance of the circuit, $\mathbf{Z} = R + jX_L$. Current in the circuit

$$\mathbf{I} = \frac{\mathbf{V}}{\mathbf{Z}} = \frac{V}{R + jX_L} = \frac{V(R - jX_L)}{R^2 + X_L^2}$$

therefore,
$$\mathbf{I} = \frac{VR}{R^2 + X_L^2} - j\frac{VX_L}{R^2 + X_L^2} \tag{8.17.1}$$

Let I_x and I_y be the real and imaginary parts of **I** respectively. Therefore

$$\mathbf{I} = I_x + jI_y \tag{8.17.2}$$

Comparison of Eqs. (8.17.1) and (8.17.2) gives

$$I_x = \frac{VR}{R^2 + X_L^2} \tag{8.17.3}$$

$$I_y = -\frac{VX_L}{R^2 + X_L^2} \tag{8.17.4}$$

These equations are the parametric equations of the locus of **I**. Here R is the parameter. We shall eliminate this parameter by squaring and adding Eqs. (8.17.3) and (8.17.4)

$$I_x^2 + I_y^2 = \frac{V^2 (R^2 + X_L^2)}{(R^2 + X_L^2)^2}$$

therefore,

$$I_x^2 + I_y^2 = \frac{V^2}{R^2 + X_L^2} \tag{8.17.5}$$

Combination of Eqs. (8.17.4) and (8.17.5) gives

$$I_x^2 + I_y^2 = -\frac{VI_y}{X_L}$$

$$I_x^2 + I_y^2 + \frac{V}{X_L} I_y = 0$$

$$(I_x - 0)^2 + \left(I_y + \frac{V}{2X_L}\right)^2 = \left(\frac{V}{2X_L}\right)^2 \tag{8.17.6}$$

Eq. (7.15.6) represents a circle whose radius is $\frac{V}{2X_L}$ and the co-ordinates of its centre are $\left(0, -\frac{V}{2X_L}\right)$. It should be noted that Eq. (7.15.6) represents a full circle. Since the current **I** cannot lag behind **V** by an angle greater than 90°, the locus of **I** is a semicircle. Thus, the locus of **I** in an RL series circuit with R as a parameter is a semicircle as shown in Fig. 8.18.

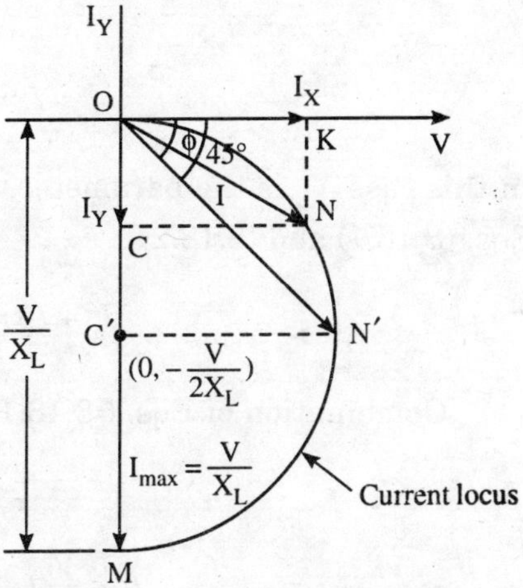

FIG. 8.18 Locus of I in a series *RL* circuit with *R* as a parameter

When

$$R = 0,\ \mathbf{I} = \frac{\mathbf{V}}{jX_L} = \frac{V}{X_L} \angle -90^\circ$$

In this case the current lags behind the voltage by 90°. It is shown by OM in Fig. 8.18. It is the maximum value of current.

$$OM = I_{max} = \frac{V}{X_L}$$

Under this condition, the power factor is cos (– 90°), that is, zero. The power supplied to the circuit, $P = VI \cos \phi$. Since V is constant the power supplied to the circuit will be proportional to $I \cos \phi$, that is proportional to length NC in Fig. 8.18. Thus, the power will be maximum when the length NC attains a maximum value. This is possible when $\phi = -45°$. Under this condition, power factor = $\cos \phi = \cos(-45°) = 1/\sqrt{2}$. Hence the maximum value of power supplied to the circuit is

$$P = V \times N'C' = V \times \frac{V}{2X_L} = \frac{V^2}{2X_L}$$

When the power input is a maximum,

$$R = X_L, \text{ and } I = \frac{V}{\sqrt{2}\, X_L}$$

8.18 SERIES RL CIRCUIT (CONSTANT *R*, VARIABLE X_L)

Consider the series RL circuit of Fig. 8.19 containing a constant R and a variable X_L. The applied voltage has a constant r.m.s. value V. We shall rewrite Eqs. (8.17.3) and (8.17.4)

$$I_x = \frac{VR}{R^2 + X_L^2} \qquad (8.18.1)$$

$$I_y = -\frac{VX_L}{R^2 + X_L^2} \qquad (8.18.2)$$

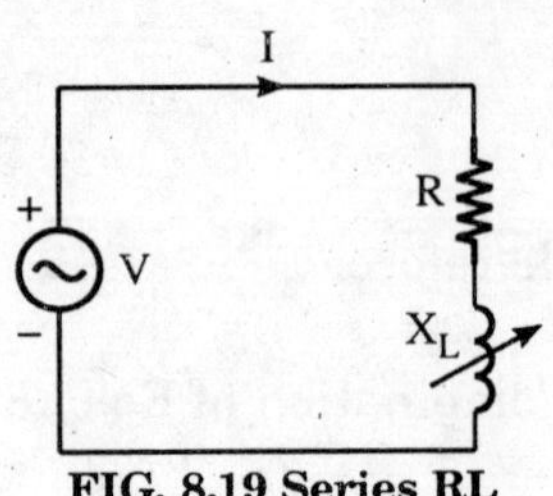

FIG. 8.19 Series RL circuit with constant *R* and variable X_L

In this case X_L is the parameter. We shall eliminate this parameter by squaring and adding Eqs. (8.18.1) and (8.18.2)

$$I_x^2 + I_y^2 = \frac{V^2}{R^2 + X_L^2} \qquad (8.18.3)$$

Combination of Eqs. (8.18.1) and (8.18.3) gives

$$I_x^2 + I_y^2 = \frac{V}{R} I_x$$

$$I_x^2 - \frac{V}{R} I_x + I_y^2 = 0$$

$$\left(I_x - \frac{V}{2R}\right)^2 + (I_y - 0)^2 = \left(\frac{V}{2R}\right)^2 \qquad (8.18.4)$$

Eq. (8.18.4) represents a circle whose radius is $(V/2R)$ and the coordinates of its centre are $(V/2R, 0)$.

Eq. (8.18.4) shows that the locus of **I** is a circle. Since the current **I** cannot lag behind **V** by an angle greater than 90°, the locus of **I** is a semicircle as shown in Fig. 8.20.

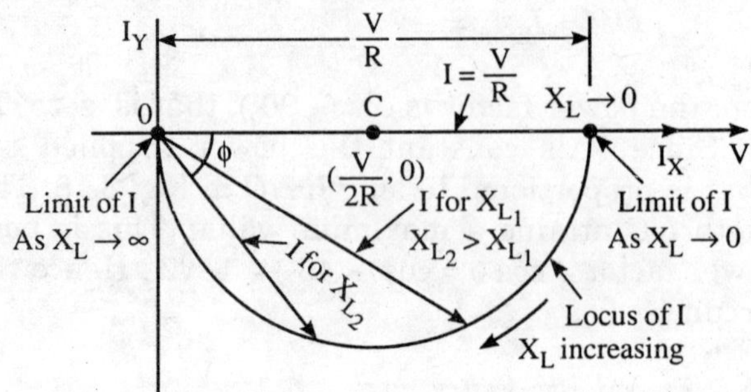

FIG. 8.20 Locus of I in a series *RL* circuit with X_L as parameter

The current is maximum when $X_L = 0$. $I_{max} = V/R$. Under this condition, $\phi = 0$. Maximum value of power input

$$= V \times \text{maximum value of } I \cos\phi = VI_{max} = \frac{V^2}{R}$$

8.19 SERIES RC CIRCUIT (CONSTANT *R*, VARIABLE X_C)

Consider the series RC circuit of Fig. 8.21 containing a constant R and a variable X_C. The applied voltage has a constant r.m.s value V.

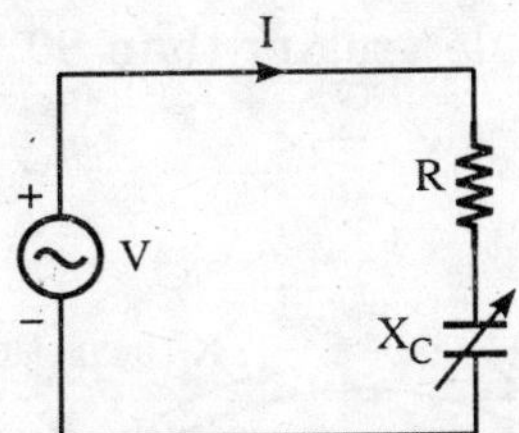

FIG. 8.21 Series RC circuit with constant *R* and variable X_C

Let **V** be taken as reference. Thus, $\mathbf{V} = V\angle 0°$. Impedance of the circuit $\mathbf{Z} = R - jX_C$. Current in the circuit

$$\mathbf{I} = \frac{\mathbf{V}}{\mathbf{Z}} = \frac{\mathbf{V}}{R - jX_C} = \frac{V(R + jX_C)}{R^2 + X_C^2}$$

$$= \frac{VR}{R^2 + X_C^2} + j\frac{VX_C}{R^2 + X_C^2}$$

Let I_x and I_y be the real and imaginary parts of **I**. Therefore

$$I_x = \frac{VR}{R^2 + X_C^2} \tag{8.19.1}$$

$$I_y = \frac{VX_C}{R^2 + X_C^2} \tag{8.19.2}$$

We shall eliminate the variable X_C by squaring and adding Eqs. (8.19.1) and (8.19.2)

$$I_x^2 + I_y^2 = \frac{V^2(R^2 + X_C^2)}{(R^2 + X_C^2)^2}$$

$$= \frac{V^2}{R^2 + X_C^2} \tag{8.19.3}$$

Combination of Eqs. (8.19.1) and (8.19.3) gives

$$I_x^2 + I_y^2 = \frac{VI_x}{R}$$

$$I_x^2 - \frac{V}{R} I_x + I_y^2 = 0$$

$$\left(I_x - \frac{V}{2R}\right)^2 + (I_y - 0)^2 = \left(\frac{V}{2R}\right)^2 \quad (8.19.4)$$

Eq. (8.19.4) shows that the locus of **I** is a circle.

Centre of the circle $\left(\frac{V}{2R}, 0\right)$

Radius of the circle $= \frac{V}{2R}$

Since **I** cannot lead **V** by an angle greater than 90° the locus of **I** is a semicircle. This semicircle is shown in Fig. 8.22.

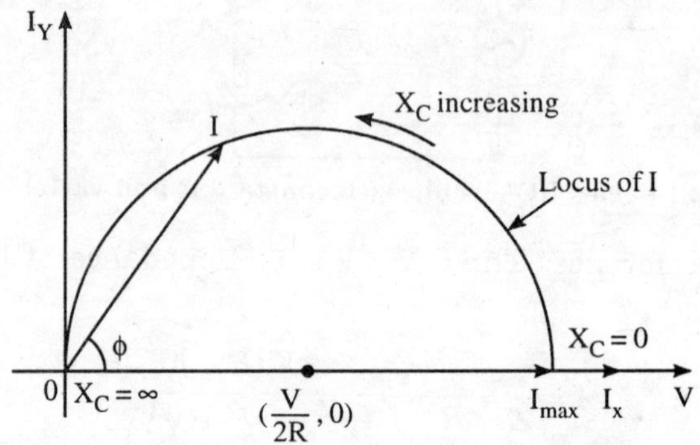

FIG. 8.22 Locus of I in a series RC circuit with X_C as a parameter

Here maximum value of current, $I_{max} = \frac{V}{R}$

when $$X_C = 0 \text{ and } \phi = 0$$

Minimum value of current = 0

when $$X_C = \infty \text{ and } \phi, = \frac{\pi}{2}$$

8.20 SERIES RC CIRCUIT (CONSTANT X_C, VARIABLE R)

Consider the series RC circuit of Fig. 8.23 containing a constant X_C and a variable R. The applied voltage has a constant r.m.s. value V.

Let us rewrite (8.19.3)

$$I_x^2 + I_y^2 = \frac{V^2}{R^2 + X_C^2} \quad (8.20.1)$$

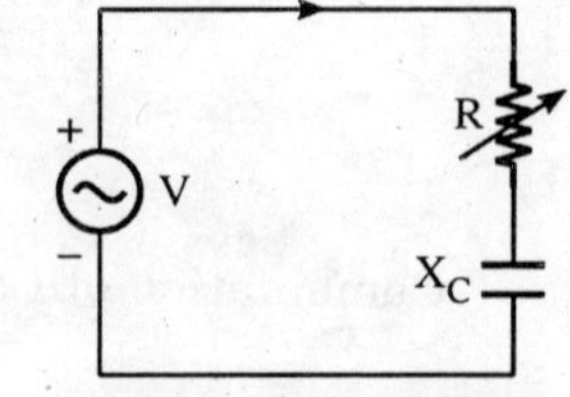

FIG. 8.23 Series RC circuit with constant X_C and variable R

Combination of Eqs. (8.19.2) and (8.20.1) gives

$$I_x^2 + I_y^2 = \frac{V}{X_C} I_y$$

$$I_x^2 + I_y^2 - \frac{V}{X_C} I_y = 0$$

$$(I_x - 0)^2 + \left(I_y - \frac{V}{2X_C}\right)^2 = \left(\frac{V}{2X_C}\right)^2 \tag{8.20.2}$$

Eq. (8.20.2) shows that the locus of I is a semicircle as shown in Fig. 8.24.

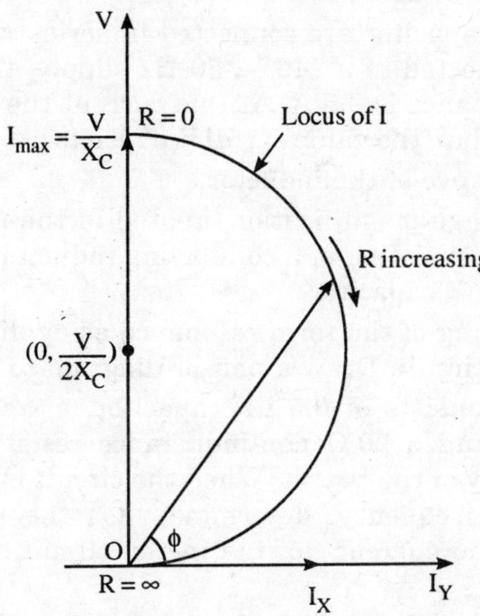

FIG. 8.24 Locus of I in a series RC circuit with *R* as parameter

Centre of the circle $\left(0, \frac{V}{2X_C}\right)$

Radius of the circle $= \frac{V}{2X_C}$

Here maximum current $= \frac{V}{X_C}$, **when** $R = 0$ **and** $\phi = 90°$

Minimum current $= 0$, $R = \infty$ **and** $\phi = 0°$,

EXERCISES

1. **Explain with the aid of a phasor diagram the phenomenon of resonance in a circuit containing an inductor, a capacitor and a resistor in series.**
   A resistor of 15 Ω, an inductance of 4 H and a capacitance of 25 μF are connected in series across a 240 V a.c. supply. Calculate : (*a*) the frequency at which the current shall be maximum; (*b*) the current at this frequency; (*c*) the p.d. across the inductance.
2. **A coil having a resistance of 5 Ω and an inductance of 0.1 H is connected in series with a 50 μF capacitor across a 200 V, variable frequency supply. Derive an expression for the frequency at which the current will be a maximum and hence**

determine the frequency. Calculate the voltages across the coil and the capacitor at this frequency.

3. Deduce an expression for the resonant frequency of a sereis a.c. circuit.

   A series circuit consisting of a 10 Ω resistor, a 30 mH inductor and a 1 μF capacitor is supplied from a 10 V variable frequency supply. Determine the frequency for which the voltage developed across the capacitor is a maximum. Calculate the magnitude of this voltage.

4. A resistor and a capacitor are connected in series with a variable inductor. When the circuit is connected to a 240 V, 50 Hz supply, the maximum current given by varying the inductance is 0.5 A. At this current the voltage across the capacitor is 250 V, hence calculate the value of : *(a)* the resistor; *(b)* the capacitor; *(c)* the inductor. Neglect the resistance of the inductor.

5. Calculate the voltage magnification created in the resonant circuit connected to a 240 V a.c. supply consisting of a coil having inductance 0.1 H and resistance 2 Ω in series with a 100 μF capacitor.

6. Explain the meaning of the term resonance as applied to : (*a*) a series a.c. circuit; (*b*) a parallel a.c. circuit. Draw a phasor diagram to illustrate each condition.

   A series circuit consists of 0.5 μF capacitor, a coil of inductance 0.32 H and a resistance 40 Ω and a 20 Ω non-inductance resistor. Calculate the value of the resonant frequency of the circuit. When the circuit is connected to a 30 V a.c. supply at this resonant frequency, determine : (*a*) the p.d. across each of the three components; (*b*) the current flowing in the circuit; (*c*) the power absorbed by the circuit.

7. Derive an expression for the resonant frequency of a parallel circuit, one branch consisting of a coil of inductance L henry and resistance R Ω and other branch of capacitor C farad.

   A coil of resistance 50 Ω and inductance 0.5 H is connected in parallel with a capacitor of 100 μF. Calculate the frequency at which the circuit acts as a non-inductive resistance. Calculate the value of this non-inductive resistance.

8. What is meant by the Q-factor of a tuned circuit ?

   Derive an expression for the resonant frequency and impedance at resonance of a circuit consisting of an inductor L having a resistance R connected in parallel with a capacitor C.

   A coil of resistance 10 Ω and inductance 10 mH is connected in parallel with a 25 μF capacitor. Calculate the frequency at resonance and the effective impedance of the circuit.

9. In a resonant series RLC circuit show that the total energy is a constant. Derive the quality factor of the series RLC circuit at resonance.

10. Show that the resonant frequency of a series RLC circuit is the geometric mean of the lower and upper half-power frequencies.

11. For either the series or parallel RLC circuit, determine the half-power frequencies ω_1 and ω_2 in terms of the quality factor Q_0 and the resonant frequency ω_0.

12. Define resonance. Determine the resonant frequency, the impedance at resonance, and the Q-factor at resonance for each of the networks shown in (a) Fig. 8.8, (b) Fig. P8.1

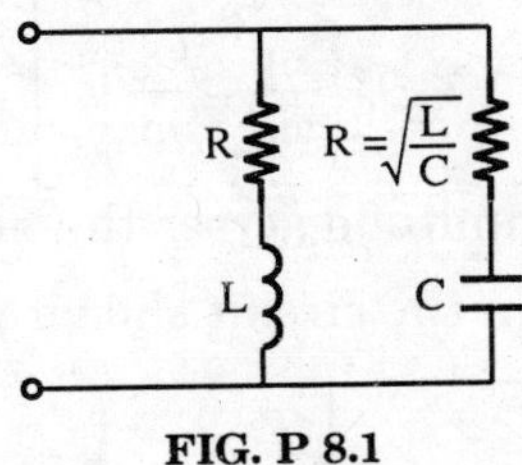

FIG. P 8.1

13. A series RLC curcuit is connected across a sinusoidal voltage of constant amplitude and variable frequency. Determine the frequency at which the voltage across the inductor is a maximum.
14. Show that the network Fig. P8.2 is resonant at all frequencies if $Z_1 Z_2 = R^2$.

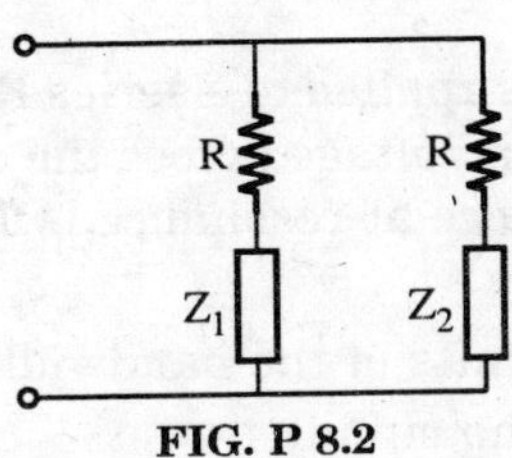

FIG. P 8.2

15. Show that for a series RLC circuit

$$\frac{\mathbf{Y}}{\mathbf{Y}_0} = \frac{1}{1 + j\,2Q_0\,\delta}$$

and for a parallel GCL circuit

$$\frac{\mathbf{Z}}{\mathbf{Z}_0} = \frac{1}{1 + j\,2Q_0\,\delta}$$

16. Determine the resonant frequency of the network shown in Fig. 8.14.
17. Define the Q-factor of a tuned coil and deduce the relationship between Q and bandwidth about the frequency of resonance.
A tuned coil of inductance 1 mH resonates at 500 kHz with a bandwidth of 50 kHz. What is its dynamic impedance at resonance?
18. A coil of inductance 100 μH has a Q-factor of 60 at a frequency of 1 MHz. Calculate the required value of shunting capacitor to cause resonance at this frequency and determine the dynamic impedance of this circuit.
19. Derive an expression for the impedance of a series RLC network in terms of the resistance R, the Q-factor at resonance Q_0 and the per unit frequency deviation δ. Hence show that if $\delta \ll 1$ the ratio of the current amplitude I at any frequency to that at resonance is given by

$$\frac{I}{I_0} = \frac{1}{\sqrt{1 + 4Q_2^2\,\delta^2}}$$

20. Show that the voltage amplitude V at any frequency across a parallel GCL network supplied by a constant current sinusoidal source is given by

$$\frac{V}{V_0} = \left[1 + Q_0^2\left(\frac{\omega}{\omega_0} - \frac{\omega_0}{\omega}\right)^2\right]^{-1/2}$$

where V_0 is the voltage amplitude across the network at resonance.

21. Determine the value of R in the circuit shown in Fig. P. 8.3 to give resonance.

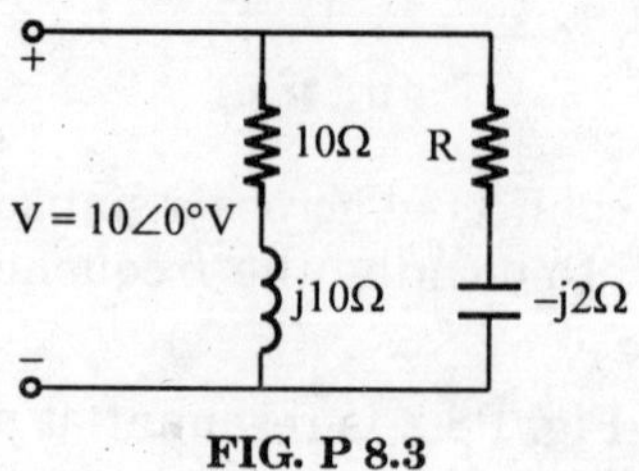

FIG. P 8.3

22. A voltage of $v = 10 \sin \omega t$ is applied to a series RLC circuit. At the resonant frequency of the circuit the maximum voltage across the capacitor is 500 V and the bandwidth is 400 rad/s.If the impedance at resonance is 100 Ω determine
   (a) resonant frequency
   (b) the upper and lower limits of the bandwidth
   (c) values of L and C for the circuit.

ANSWERS

1. (a) 15.9 Hz; (b) 16 A (c) 6402 V
2. 71.2 Hz ; 1800 V ; 1788 V
3. 918.9 Hz, 173 V
4. (a) 480 Ω (b) 6.38 μF (c) 1.59 H
5. 15.81
6. 398 Hz (a) 399.8 V, 400.6 V, 10 V, (b) 0.5 A, (c) 15 W
7. 15.93 Hz, 100 Ω
8. 275·7 Hz, 40Ω
9. $W_{s0} = \frac{1}{2} L I_0^2$, $Q_{0s} = \frac{\omega_0 L}{R}$
11. $\omega_1 = \omega_0 \sqrt{1 + \left(\frac{1}{2Q_0}\right)^2} - \frac{\omega_0}{2Q_0}$

    $\omega_2 = \omega_0 \sqrt{1 + \left(\frac{1}{2Q_0}\right)^2} + \frac{\omega_0}{2Q_0}$
12. (a) $\omega_0 = \frac{1}{\sqrt{LC}}$, $Z_0 = R$

$$Q_0 = \frac{R}{\omega_0 L} = \omega_0 CR = \frac{R}{\sqrt{L/C}}$$

(b) Network is resonant at all frequencies with a constant resistive impedance of

$$\sqrt{\frac{L}{C}}\ \Omega,\ Q_0 = 0$$

13. $f = \dfrac{1}{2\pi \sqrt{LC - \frac{1}{2} R^2 C^2}}$

16. $\omega_0 = \dfrac{1}{\sqrt{LC}} = \sqrt{\left(\dfrac{\frac{L}{C} - R_L^2}{\frac{L}{C} - R_C^2}\right)}$

17. 31.4 kΩ
18. 253 pF, 37.7 kΩ
21. 6 Ω
22. (a) 159.1 Hz (b) 191 Hz, 127.3 Hz (c) $L = 0.25$ H, $C = 4$ μF

9

THREE-PHASE CIRCUITS

9.1 POLYPHASE SYSTEM

Systems with more than one phase are called **polyphase systems**. A polyphase system contains two or more a.c. voltage sources of the same frequency. These source voltages have a fixed phase angle difference between them. The most widely used polyphase system is the three-phase system. Three-phase systems are in common use for generation, transmission, distribution and utilisation of electric energy.

The **main advantages of a 3–phase system** over single-phase system are :

1. A 3-phase machine has a smaller size than a single-phase machine of the same power output.
2. The conductor material required to transmit a given power at a given voltage over a given distance by a 3-phase system is lesser than that of an equivalent single-phase system.
3. The voltage drop in the conductors from the source to the load is less in a 3-phase system.
4. Since 3-phase supply produces a uniform rotating magnetic field, 3-phase motors are simpler in construction, smaller in size, start more conveniently, run more smoothly and are more efficient than equivalent single-phase motors.

For these reasons, 3-phase systems are commonly used in practice for generation, transmission, distribution and utilisation of electric energy.

9.2 GENERATION OF A THREE-PHASE SUPPLY

When three identical coils placed with their axes at 120° (elec.) apart from each other are rotated in a uniform magnetic field, a sinusoidal voltage is generated across each coil. This is the basis of a simple 3-phase a.c. generator (alternator). Fig 9.1 a shows a 3-phase, 2-pole alternator. It has three sets of coils aa' , bb' and cc' symmetrically mounted on a rotor such that their axes are 120° from each other. When the rotor is rotated in anticlockwise direction at a constant angular velocity ω radians per second, a sinusoidal voltage is generated across each coil. Since the three coils rotate at the same velocity ($\omega = 2\pi f$), the generated voltages have the same frequency. Also, since the coils are identical, the generated voltages have the same magnitudes, but there is a phase difference of 120 electrical degrees between these voltages. The generated voltages in the coils are given by

$$v_{aa'} = V_m \sin \omega t \tag{9.2.1}$$

$$v_{bb'} = V_m \sin (\omega t - 120^\circ) \tag{9.2.2}$$

$$v_{cc'} = V_m \sin (\omega t - 240^\circ) \tag{9.2.3}$$

or

$$v_{cc'} = V_m \sin (\omega t + 120^\circ) \tag{9.2.4}$$

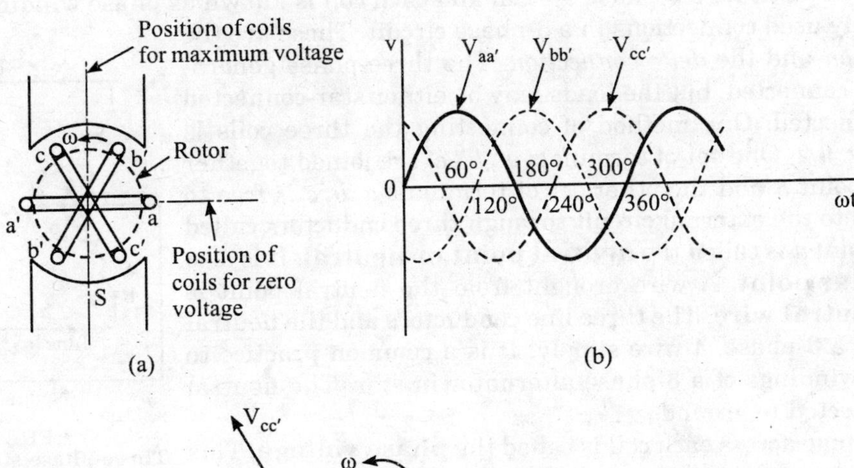

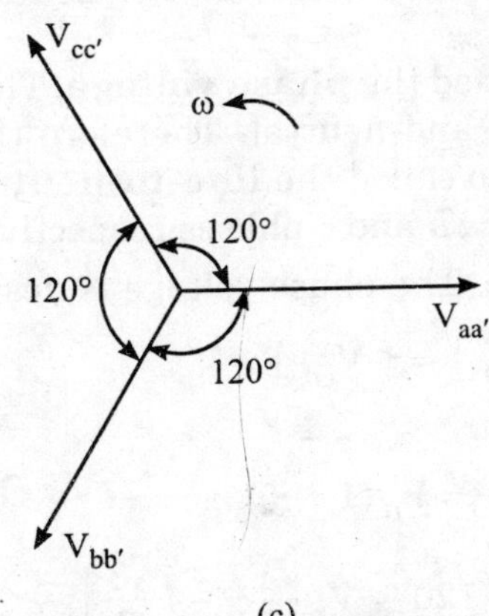

FIG. 9.1 Generation of 3-phase supply

It is to be noted that a phase angle of – 240° is the same as + 120°. In polar form,

$$\mathbf{V}_{aa'} = V \underline{/\ 0^\circ}$$

$$\mathbf{V}_{bb'} = V \underline{/ -120^\circ}$$

$$\mathbf{V}_{cc'} = V \underline{/ -240^\circ} = V \underline{/ +120^\circ}$$

The voltage waveforms are shown in Fig. 9.1 b These are sine waves. The phasor diagram is shown in Fig. 9.1. c.

It is seen that $\mathbf{V}_{aa'}$ leads $\mathbf{V}_{bb'}$ by 120° and $\mathbf{V}_{bb'}$ leads $\mathbf{V}_{cc'}$ by 120°. Also, the three voltages reach their positive maximum values in the order $\mathbf{V}_{aa'}$, $\mathbf{V}_{bb'}$, $\mathbf{V}_{cc'}$. The order in which the phase voltages reach their maximum values is called the **phase sequence**. For the arrangement shown in Fig. 9.1 a in which the coils are rotating in anticlockwise direction, the *phase sequence* is *abc*. If the rotor is rotated in clockwise direction the voltages reach their positive maximum values in the order $\mathbf{V}_{cc'}$, $\mathbf{V}_{bb'}$, $\mathbf{V}_{aa'}$ In this case the phase sequence is *cba* or *acb*. Thus, the phase sequence determines the direction of rotation.

9.3 STAR OR WYE (Y) CONNECTION

Since a voltage is generated in each coil, it may be considered as a source of voltage. The three coils together constitute a 3-phase system and each coil is known as phase winding. There are two commonly used connections in a 3-phase circuit. These are the *star connection* and the *delta connection*. The three-phase generators are star connected, but the loads may be either star-connected or delta connected. One method of connecting the three coils is shown in Fig. 9.2. One set of terminals a', b', c' are joined together at common point n and the other set of terminals a, b, c is free to be connected to the external circuit through three cnductors called **lines**. The point n is called the **neutral point or neutral**. It is also called the **star point**. A wire brought from the neutral point is called the **neutral wire**. The three line conductors and the neutral wire provide a 3-phase, 4-wire supply. It is a common practice to connect the windings of a 3-phase alternator in star. The neutral point is connected to ground.

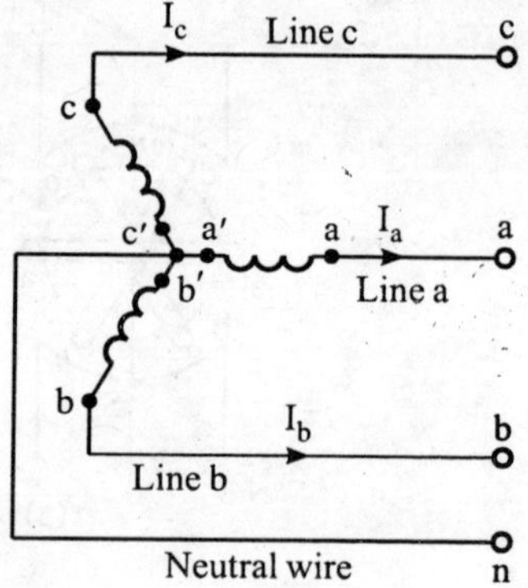

FIG. 9.2
Three-phase star connection

The voltage across each coil is called the **phase voltage**. This is also the voltage between each line and neutral. Therefore the phase voltage in star connection is also called the **line-to-neutral voltage**. The three phase voltages are $\mathbf{V}_{an}$, $\mathbf{V}_{bn}$ and $\mathbf{V}_{cn}$ in the a , b and c phases respectively. In a symmetrical system these voltages are equal in magnitude. The phase voltage is designated as $\mathbf{V}_p$.

$$|\mathbf{V}_{an}| = |\mathbf{V}_{bn}| = |\mathbf{V}_{cn}| = V_p \tag{9.3.1}$$

If $\mathbf{V}_{an}$ is taken as the reference

$$\mathbf{V}_{an} = V_p \underline{/0^\circ} = V_p(1 + j0) \tag{9.3.2}$$

$$\mathbf{V}_{bn} = V_p \underline{/-120^\circ} = V_p\left(-\frac{1}{2} - j\frac{\sqrt{3}}{2}\right) \tag{9.3.3}$$

$$\mathbf{V}_{cn} = V_p \underline{/120^\circ} = V_p\left(-\frac{1}{2} + j\frac{\sqrt{3}}{2}\right) \tag{9.3.4}$$

The voltage between two lines is called the **line-to-line voltage** or simply the **line voltage.** The line voltages are $\mathbf{V}_{ab}$, $\mathbf{V}_{bc}$ and $\mathbf{V}_{ca}$ between the pairs of line a–b, b–c and c–a respectively. In a symmetrical system these voltages are equal in magnitude and each is designated as V_L. The phase and line voltages in star connection are shown in Fig. 9.3. The phasor diagram is shown in Fig. 9.4.

The relationship between the phase and line voltages in a balanced 3-phase star connection can be found as follows

By KVL,

$$\mathbf{V}_{ab} = \mathbf{V}_{an} + \mathbf{V}_{nb}$$

$$= \mathbf{V}_{an} - \mathbf{V}_{bn}$$

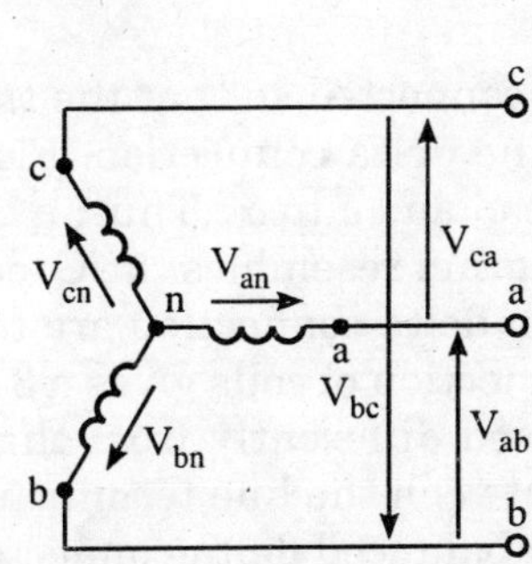

FIG. 9.3 Phase and line voltages in star connection

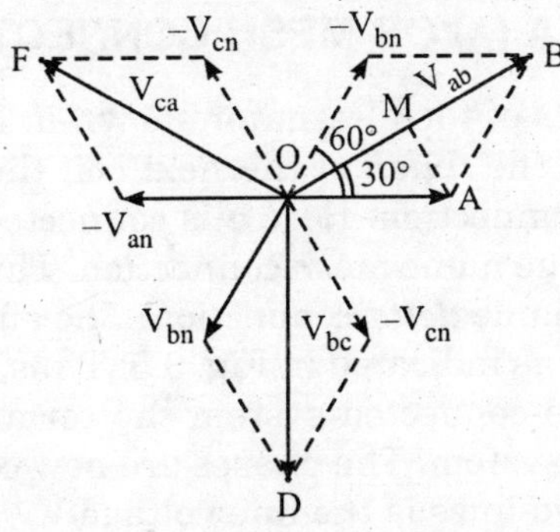

FIG. 9.4 Phasor diagram

$$= V_p\,(1 + j0) - V_p\left(-\frac{1}{2} - j\,\frac{\sqrt{3}}{2}\right)$$

$$= V_p\left(\frac{3}{2} + j\,\frac{\sqrt{3}}{2}\right)$$

$$= \sqrt{3}\ V_p\ \underline{/\ 30^\circ} \qquad (9.3.5)$$

Again by KVL,

$$\mathbf{V}_{bc} = \mathbf{V}_{bn} + \mathbf{V}_{nc}$$

$$= \mathbf{V}_{bn} - \mathbf{V}_{cn}$$

$$= V_p\left(-\frac{1}{2} - j\,\frac{\sqrt{3}}{2}\right) - V_p\left(-\frac{1}{2} + j\,\frac{\sqrt{3}}{2}\right)$$

$$= V_p\,(0 - j\,\sqrt{3})$$

$$= \sqrt{3}\ V_p\ \underline{/ -90^\circ} \qquad (9.3.6)$$

By KVL,

$$\mathbf{V}_{ca} = \mathbf{V}_{cn} + \mathbf{V}_{na}$$

$$= \mathbf{V}_{cn} - \mathbf{V}_{an} = V_p\left(-\frac{1}{2} + j\,\frac{\sqrt{3}}{2}\right) - V_p\,(1 + j0)$$

$$= V_p\left(-\frac{3}{2} + j\,\frac{\sqrt{3}}{2}\right)$$

$$= \sqrt{3}\ V_p\ \underline{/\ 150^\circ} \qquad (9.3.7)$$

$$|\,\mathbf{V}_{ab}\,| = |\,\mathbf{V}_{bc}\,| = |\,\mathbf{V}_{ca}\,| = V_L = \sqrt{3}\ V_P \qquad (9.3.8)$$

Thus in the star-connected alternator, the line voltage is equal to $\sqrt{3}$ times the phase voltage. Also, the phase difference between the line voltages is 120°.

The phasor sum of the phase voltages is zero. that is,

$$\mathbf{V}_{an} + \mathbf{V}_{bn} + \mathbf{V}_{cn} = 0 \qquad (9.3.9)$$

Similarly with the help of Eqs. (9.3.5) – (9.3.7) it can be shown that the phasor sum of the line voltages is zero. That is,

$$\mathbf{V}_{ab} + \mathbf{V}_{bc} + \mathbf{V}_{ca} = 0 \qquad (9.3.10)$$

9.4 DELTA (Δ) OR MESH CONNECTION

If the three coils of the alternator shown in Fig. 9.1 are connected so that the *start* of one coil is connected to the *finish* of the next coil the result is the delta connection. Fig. 9.5 shows a 3-phase delta connection. Here a is connected to b', b to c' and c to a'. Thus, a closed mesh is formed, hence the name mesh connection. The circuit diagram resembles the Greek letter delta (Δ), hence the name *delta* connection. The supply lines in delta connection are taken from the three junctions as indicated in Fig. 9.5. Thus, a delta connection of coils gives a 3-phase, 3-wire supply. In delta-connected system the term phase is used differently from that used in the star-connected system. The phases are now connected between the line terminals. The voltage between any two lines is the line voltage V_L. Hence, in a balanced delta-connected system the magnitude of the line voltage is equal to the magnitude of the phase voltage.

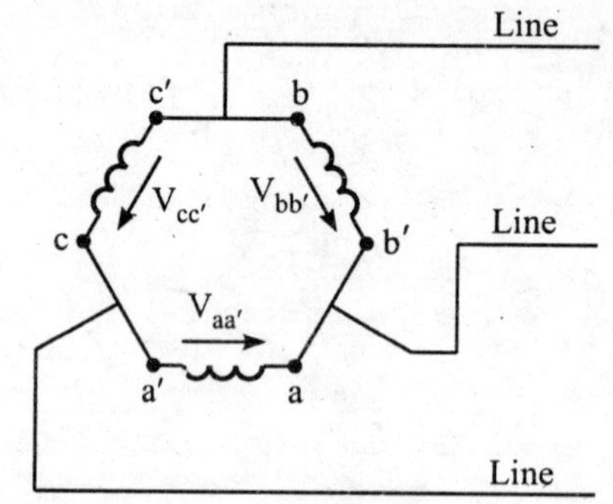

FIG. 9.5 Delta connection

FIG. 9.6 Phasor diagram

$$|\mathbf{V}_{ab}| = |\mathbf{V}_{bc}| = |\mathbf{V}_{ca}| = V_L = V_p \tag{9.4.1}$$

The voltages can be expressed in polar and rectangular coordinates

$$\mathbf{V}_{ab} = V_p \underline{/ 0^\circ} = V_p\,(1 + j0) \tag{9.4.2}$$

$$\mathbf{V}_{bc} = V_p \underline{/ -120^\circ} = V_p \left(-\frac{1}{2} - j\frac{\sqrt{3}}{2} \right) \tag{9.4.3}$$

$$\mathbf{V}_{ca} = V_p \underline{/ +120^\circ} = V_p \left(-\frac{1}{2} + j\frac{\sqrt{3}}{2} \right) \tag{9.4.4}$$

The phasor diagram is shown in Fig. 9.6.

Now,
$$\mathbf{V}_{ab} + \mathbf{V}_{bc} + \mathbf{V}_{ca} = 0 \tag{9.4.5}$$

This relation shows that the phasor sum of the line voltage is zero.

9.5 BALANCED STAR-CONNECTED LOAD

The three-phase generators supply energy to the loads which may be connected either in star or in delta. When the impedances in all three phases are identical the load is said to be *balanced*.

Fig. 9.7 shows a balanced star-connected load are connected of impedance **Z** in each phase where

$$\mathbf{Z} = Z \underline{/ \phi} \tag{9.5.1}$$

The common point s to which the impedances forming the load are connected is called the *star point* of the load. In a 3-phase 4-wire system the neutral point N of the supply is connected to the star point s of the load. The wire $N s$ is called the *neutral wire*. The current in the neutral wire is represented by I_n. From Fig. 9.7, it is seen that

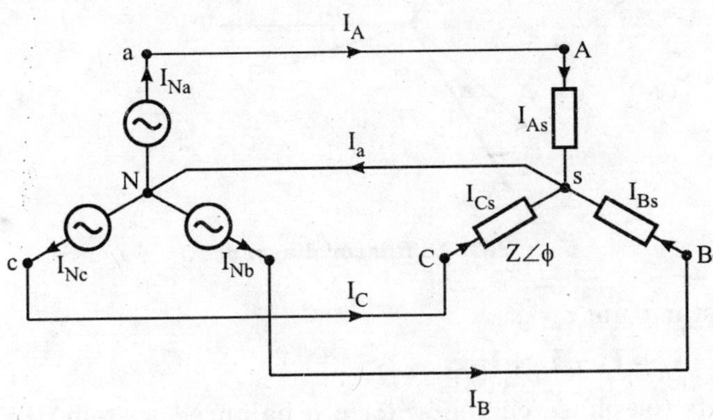

FIG. 9.7 Star-connected load

$$\mathbf{I}_{NA} = \mathbf{I}_A = \mathbf{I}_{As} \tag{9.5.2}$$

that is, current in phase a of the alternator

= current in line A = current in phase A of the load

Also, $$\mathbf{I}_{Nb} = \mathbf{I}_B = \mathbf{I}_{Bs} \tag{9.5.3}$$

and $$\mathbf{I}_{Nc} = \mathbf{I}_C = \mathbf{I}_{Cs} \tag{9.5.4}$$

Thus, the currents flowing through the voltage sources flow through the line conductors and through the load elements. Hence, the line current is same as phase current both in magnitude and phase that is,

$$\mathbf{I}_L = \mathbf{I}_p \text{ for a star connection} \tag{9.5.5}$$

In Fig. 9.7, each impedance is connected between a line and star point s. This is equivalent to three single-phase circuits. The line currents are given by

$$\mathbf{I}_A = \frac{\mathbf{V}_{aN}}{\mathbf{Z}} = \frac{V_p \underline{/0^\circ}}{Z \underline{/\phi}} = \frac{V_p}{Z} \underline{/-\phi} \tag{9.5.6}$$

$$\mathbf{I}_B = \frac{\mathbf{V}_{bN}}{\mathbf{Z}} = \frac{V_p \underline{/-120^\circ}}{Z \underline{/\phi}} = \frac{V_p}{Z} \underline{/-120^\circ - \phi} \tag{9.5.7}$$

$$\mathbf{I}_C = \frac{\mathbf{V}_{cN}}{\mathbf{Z}} = \frac{V_p \underline{/120^\circ}}{Z \underline{/\phi}} = \frac{V_p}{Z} \underline{/120^\circ - \phi} \tag{9.5.8}$$

It is seen that the magnitude of each line current is V_p/Z and the line currents and displaced by 120° phase angle from each other. The phasor diagram is shown in Fig. 9.8.

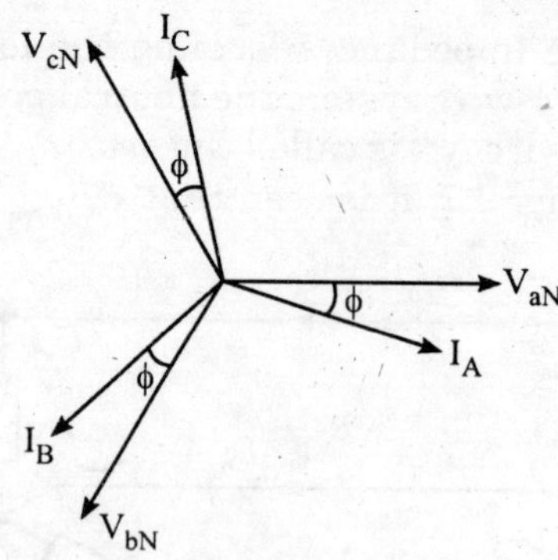

FIG. 9.8 Phasor diagram

By KCL at the star point s

$$\mathbf{I}_n = \mathbf{I}_A + \mathbf{I}_B + \mathbf{I}_C \tag{9.5.9}$$

Since the line currents (or phase currents) form a balanced system, the phasor sum of the these currents must be zero. That

$$\mathbf{I}_A + \mathbf{I}_B + \mathbf{I}_C \equiv 0 \tag{9.5.10}$$

Therefore, $\mathbf{I}_n = 0$

That is, in a balanced 3-phase star-connected load, the current in the neutral conductor is **zero.** The neutral wire is unnecessary in a balanced system and therefore it may be omitted without affecting the system. This gives a 3-phase, 3-wire, star-connected supply system.

9.6 BALANCED DELTA-CONNECTED LOAD

Figure 9.9 shows a balanced delta-connected load of impedance $\mathbf{Z}$ in each phase where

$$\mathbf{Z} = Z \angle \phi$$

The phasor diagram for this load is shown in Fig. 9.10. Here $\mathbf{V}_{AB}$ is taken as the reference phasor. The phase currents are given by

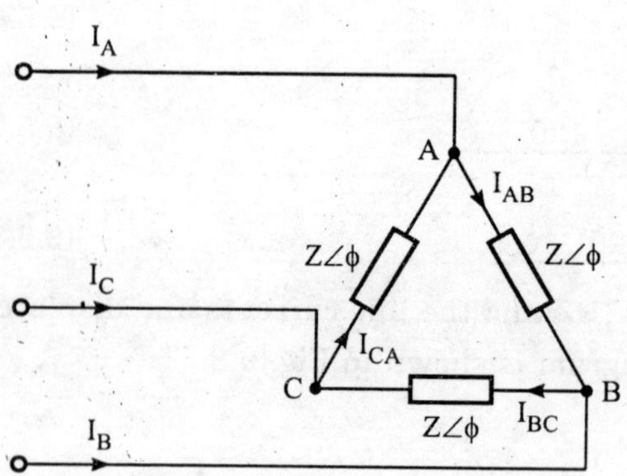

FIG. 9.9 Balanced delta-connected load

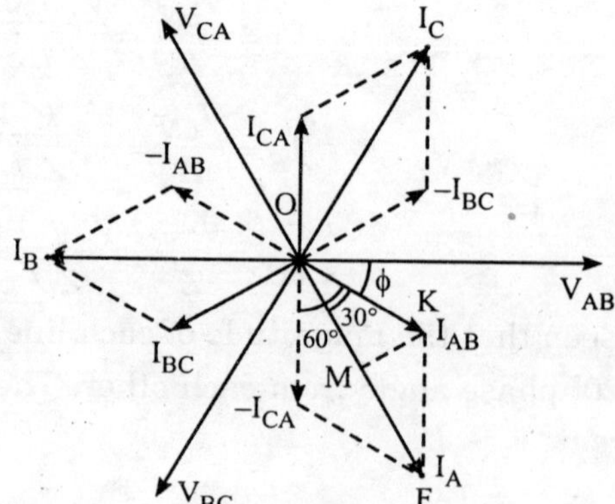

FIG. 9.10 Phasor diagram

$$\mathbf{I}_{AB} = \frac{\mathbf{V}_{AB}}{\mathbf{Z}} = \frac{V_L \angle 0^\circ}{Z \angle \phi} = \frac{V_L}{Z} \angle -\phi \tag{9.6.1}$$

$$\mathbf{I}_{BC} = \frac{\mathbf{V}_{BC}}{\mathbf{Z}} = \frac{V_L \angle -120^\circ}{Z \angle \phi} = \frac{V_L}{Z} \angle -120^\circ - \phi \tag{9.6.2}$$

$$\mathbf{I}_{CA} = \frac{\mathbf{V}_{CA}}{\mathbf{Z}} = \frac{V_L \angle 120^\circ}{Z \angle \phi} = \frac{V_L}{Z} \angle 120^\circ - \phi \tag{9.6.3}$$

The line currents $\mathbf{I}_A$, $\mathbf{I}_B$, and $\mathbf{I}_C$ can be found by applying KCL at the point A, B and C as follows :

$$\mathbf{I}_A = \mathbf{I}_{AB} - \mathbf{I}_{CA} \tag{9.6.4}$$

$$\mathbf{I}_B = \mathbf{I}_{BC} - \mathbf{I}_{AB} \tag{9.6.5}$$

$$\mathbf{I}_C = \mathbf{I}_{CA} - \mathbf{I}_{BC} \tag{9.6.6}$$

The magnitudes and the phase angles of the line currents can also be found from Fig. 9.10.

$$\mathbf{I}_A = \text{phasor sum of } \mathbf{I}_{AB} \text{ and } \mathbf{I}_{CA} \text{ reversed}$$

$$= \text{phasor OF} = \text{OF} \angle -\phi - 30^\circ$$

Draw a perpendicular KM on OF. In triangle OKF.

$$\text{OK} = \text{KF} = \text{phase current} = I_p$$

$$\text{OF} = I_A = \text{current in line } A = I_L$$

Now, $\quad \text{OF} = 20M = 2 \times OK \cos 30^\circ = 2\, I_p \times \frac{\sqrt{3}}{2}$

therefore, $\quad I_L = \sqrt{3}\, I_p$

Thus, we see that in a balanced delta-connected load the magnitude of the line current is equal to $\sqrt{3}$ times the magnitude of the phase current.

$$|\mathbf{I}_A| = |\mathbf{I}_B| = |\mathbf{I}_C| = I_L = \sqrt{3}\, I_p \tag{9.6.7}$$

From the geometry of Fig. 9.10 it is seen that $\mathbf{I}_A$, $\mathbf{I}_B$ and $\mathbf{I}_C$ are 120° apart. In polar form,

$$\mathbf{I}_A = \sqrt{3}\, I_p \angle -\phi - 30^\circ \tag{9.6.8}$$

$$\mathbf{I}_B = \sqrt{3}\, I_p \angle -\phi - 150^\circ \tag{9.6.9}$$

$$\mathbf{I}_C = \sqrt{3}\, I_p \angle -\phi + 90^\circ \tag{9.6.10}$$

9.7 UNBALANCED DELTA-CONNECTED LOAD

Fig. 9.11a shows an unbalanced delta-connected load with symmetrical line voltages. It is seen from the figure that the line voltages are identical with the phase voltages. They are equal in magnitude and displaced by 120°. Let the phase sequence be ABC and $\mathbf{V}_{AB}$ be taken as the reference phasor, so that

$$\mathbf{V}_{AB} = V_L \angle 0^\circ \tag{9.7.1}$$

$$\mathbf{V}_{BC} = V_L \angle -120^\circ \tag{9.7.2}$$

$$\mathbf{V}_{CA} = V_L \angle +120^\circ \tag{9.7.3}$$

Let $\mathbf{Z}_{AB} = Z_1 \angle \phi_1,\ \mathbf{Z}_{BC} = Z_2 \angle \phi_2,\ \mathbf{Z}_{CA} = Z_3 \angle \phi_3$

The phase currents are given by

$$\mathbf{I}_{AB} = \frac{\mathbf{V}_{AB}}{\mathbf{Z}} = \frac{V_L \angle 0^\circ}{Z_1 \angle \phi_1} \tag{9.7.4}$$

$$\mathbf{I}_{BC} = \frac{\mathbf{V}_{BC}}{\mathbf{Z}_{BC}} = \frac{V_L \angle -120^\circ}{Z_2 \angle \phi_2} \tag{9.7.5}$$

$$\mathbf{I}_{CA} = \frac{\mathbf{V}_{CA}}{\mathbf{Z}_{CA}} = \frac{V_L \angle 120^\circ}{Z_3 \angle \phi_3} \tag{9.7.6}$$

The line currents are found by applying KCL at the nodes A, B and C of Fig. 9.11a respectively,

$$\mathbf{I}_A = \mathbf{I}_{AB} - \mathbf{I}_{CA} \tag{9.7.7}$$

$$\mathbf{I}_B = \mathbf{I}_{BC} - \mathbf{I}_{AB} \tag{9.7.8}$$

$$\mathbf{I}_C = \mathbf{I}_{CA} - \mathbf{I}_{BC} \tag{9.7.9}$$

Because the load is unbalanced the currents in the three phases, and therefore the line currents, are not equal in magnitude. Even though the load is unbalanced it can be shown from the above relations that the phasor sum of the line currents is zero, that is,

$$\mathbf{I}_A + \mathbf{I}_B + \mathbf{I}_C = 0 \tag{9.7.10}$$

The phasor diagram is shown in Fig. 9.11 b.

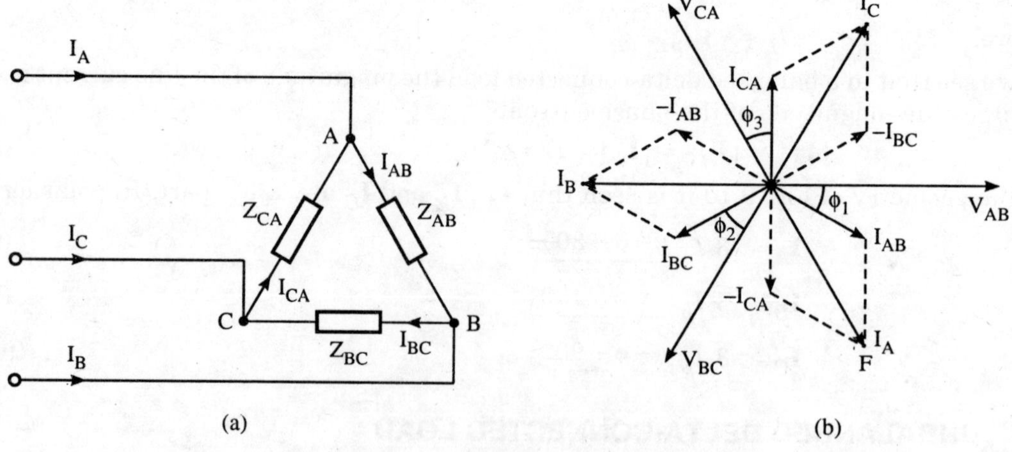

FIG. 9.11

■ **EXAMPLE 9.1** Three impedances $\mathbf{Z}_1$, $\mathbf{Z}_2$ and $\mathbf{Z}_3$ are delta connected to a symmetrical 3-phase 400 V 50 Hz supply of phase sequence ABC.

$\mathbf{Z}_1 = (10 + j\,0)\ \Omega$ connected between lines A and B

$\mathbf{Z}_2 = (8 + j\,6)\ \Omega$ connected between lines B and C

$\mathbf{Z}_3 = (5 - j\,5)\ \Omega$ connected between lines C and A

Calculate the phase and the line currents and the total active power consumed. Draw the phasor diagram.

SOLUTION : The circuit diagram is similar to that of Fig. 9.11a. Let V_{AB} be taken as the reference phasor, so that

$$\mathbf{V}_{AB} = 400 \underline{/ 0^\circ},$$

$$\mathbf{V}_{BC} = 400 \underline{/ -120^\circ},$$

$$\mathbf{V}_{CA} = 400 \underline{/ 120^\circ}$$

$$\mathbf{I}_{AB} = \frac{\mathbf{V}_{AB}}{\mathbf{Z}_{AB}} = \frac{400 \underline{/ 0^\circ}}{10 + j\,0} = \frac{400 \underline{/ 0^\circ}}{10 \underline{/ 0^\circ}}$$

$$= 40 \underline{/ 0^\circ} \text{ A} = (40 + j\,0) \text{ A}$$

$$\mathbf{I}_{BC} = \frac{\mathbf{V}_{BC}}{\mathbf{Z}_{BC}} = \frac{400 \underline{/ -120^\circ}}{8 + j\,6} = \frac{400 \underline{/ -120^\circ}}{10 \underline{/ 36.87^\circ}}$$

$$= 40 \underline{/ -156.87^\circ} \text{ A} = (-36.78 - j\,15.71) \text{ A}$$

$$\mathbf{I}_{CA} = \frac{\mathbf{V}_{CA}}{\mathbf{Z}_{CA}} = \frac{400 \underline{/ 120^\circ}}{5 - j\,5} = \frac{400 \underline{/ 120^\circ}}{7.07 \underline{/ -45^\circ}}$$

$$= 56.56 \underline{/ -165^\circ} \text{ A} = (-54.63 + j\,14.638) \text{ A}$$

The magnitudes of the phase currents are

$$|\mathbf{I}_{AB}| = 40 \text{ A}, \ |\mathbf{I}_{BC}| = 40 \text{ A}, \ |\mathbf{I}_{CA}| = 56.56 \text{ A}$$

The line currents are

$$\mathbf{I}_A = \mathbf{I}_{AB} - \mathbf{I}_{CA}$$

$$= (40 + j\,0) - (-54.53 + j\,14.638)$$

$$= (94.63 - j\,14.638) \text{ A} = 95.75 \underline{/ -8.79^\circ} \text{ A}$$

$$\mathbf{I}_B = \mathbf{I}_{BC} - \mathbf{I}_{AB}$$

$$= -36.78 - j\,15.71 - 40$$

$$= (-76.78 - j\,15.71) \text{ A} = 78.37 \underline{/ -168.4^\circ} \text{ A}$$

$$\mathbf{I}_C = \mathbf{I}_{CA} - \mathbf{I}_{BC}$$

$$= -54.63 + j\,14.638 - (-36.78 - j\,15.71)$$

$$= -17.85 + j\,30.348 = 35.2 \underline{/ 120.46^\circ} \text{ A}$$

Let R_{AB}, R_{BC} and R_{CA} be the resistive components of $\mathbf{Z}_{AB}$, $\mathbf{Z}_{BC}$ and $\mathbf{Z}_{CA}$.

Total power consumed

$$P = P_{AB} + P_{BC} + P_{CA}$$

$$= I_{AB}^2 R_{AB} + I_{BC}^2 R_{BC} + I_{CA}^2 R_{CA}$$

$$= (40)^2 \times 10 + (40)^2 \times 8 + (56.56)^2 \times 5$$

$$= 44795 \text{ W} = 44.795 \text{ kW}$$

The phasor diagram is shown in Fig. 9.11b.

9.8 UNBALANCED 4-WIRE STAR CONNECTED LOAD

Consider an unbalanced star-connected load connected to a 3-phase 4-wire supply as shown in Fig. 9.12. Suppose that the supply is balanced and the neutral point S of the load is connected to the source neutral. Thus, the phase voltage of the load would be balanced. Let the voltage across phase A be taken as reference, so that when the phase sequence is ABC,

$$\mathbf{V}_{AS} = V_p \underline{/ 0^\circ},$$

$$\mathbf{V}_{BS} = V_p \underline{/ -120^\circ},$$

$$\mathbf{V}_{CS} = V_p \underline{/ 120^\circ}$$

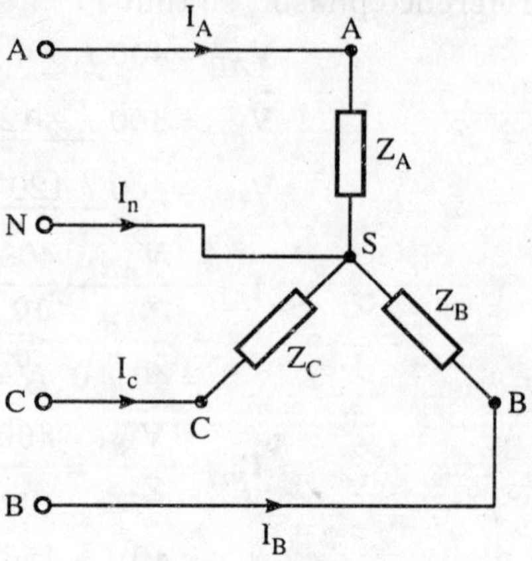

FIG. 9.12 Unbalanced 4 wire star-connected load

The phase currents are

$$\mathbf{I}_A = \frac{\mathbf{V}_{AS}}{\mathbf{Z}_A}, \ \mathbf{I}_B = \frac{\mathbf{V}_{BS}}{\mathbf{Z}_B}, \ \mathbf{I}_C = \frac{\mathbf{V}_{CS}}{\mathbf{Z}_C}$$

The line currents are equal to the phase currents.

EXAMPLE 9.2 An unbalanced 4 wire star connected load has balanced line voltages of 440 V. The load impedances are

$$\mathbf{Z}_A = 10\ \Omega, \ \mathbf{Z}_B = (5 + j\,10)\ \Omega,$$

$$\mathbf{Z}_C = (15 - j\,5)\ \Omega$$

Calculate the line currents, neutral current, power in each phase and the total power. The phase sequence is ABC.

SOLUTION : The phase voltage

$$V_p = \frac{V_L}{\sqrt{3}} = \frac{440}{\sqrt{3}} = 254 \text{ V}$$

Let S be the star point of the load.

Taking the voltage across phase A as reference

$$\mathbf{V}_{AS} = 254 \underline{/ 0^\circ} \text{ V}, \ \mathbf{V}_{BS} = 254 \underline{/ -120^\circ} \text{ V},$$

$$\mathbf{V}_{CS} = 254 \underline{/ 120^\circ} \text{ V},$$

The phase currents are

$$\mathbf{I}_A = \frac{\mathbf{V}_{AS}}{\mathbf{Z}_A} = \frac{254 \underline{/ 0^\circ}}{10} = 25.4 \underline{/ 0^\circ} \text{ A} = (25.4 - j\,0) \text{ A}$$

$$\mathbf{I}_B = \frac{\mathbf{V}_{BS}}{\mathbf{Z}_B} = \frac{254 \underline{/ -120^\circ}}{5 + j\,10} = \frac{254 \underline{/ -120^\circ}}{11.2 \underline{/ 63.4^\circ}}$$

$$= 22.678 \underline{/ -183.4^\circ} \text{ A} = (-22.63 + j\,1.345) \text{ A}$$

$$\mathbf{I}_C = \frac{\mathbf{V}_{CS}}{\mathbf{Z}_C} = \frac{254\ \angle 120^\circ}{15 - j\,5} = \frac{254\ \angle 120^\circ}{15.8\ \angle -18.4^\circ}$$

$$= 16.07\ \angle 138.4^\circ\ \text{A}$$

$$= (-12.02 + j\,10.67)\ \text{A}$$

The line currents are the same as phase currents. By KCL, the current in the neutral

$$\mathbf{I}_n = \mathbf{I}_A + \mathbf{I}_B + \mathbf{I}_C$$

$$= 25.4 - 22.63 + j\,1.345 - 12.02 + j\,10.67$$

$$= -9.25 + j\,12.015$$

$$= 15.16\ \angle 127.59^\circ\ \text{A}$$

Power consumed in different phases are

$$P_A = I_A^2 R_A = (25.4)^2 \times 10 = 6451\ \text{W}$$

$$P_B = I_B^2 R_B = (22.678)^2 \times 5 = 2571\ \text{W}$$

$$P_C = I_C^2 R_C = (16.07)^2 \times 15 = 3873\ \text{W}$$

where R_A, R_B and R_C are the resistive components of the impedances $\mathbf{Z}_A$, $\mathbf{Z}_B$, and $\mathbf{Z}_C$.

Total power consumed

$$P = P_A + P_B + P_C$$

$$= 6451 + 2571 + 3873$$

$$= 12\,895\ \text{W} = 12.895\ \text{kW}$$

9.9 STAR ⟷ DELTA (OR TEE-PI) TRANSFORMATION FOR AC CIRCUITS

The expressions derived in section 4.17 for converting a star network into a delta network or vice versa are valid for ac circuits also if we replace the resistances by *complex impedances.*

Consider star (Y) and delta (Δ) networks shown in Fig. 9.13 a and 9.13 b respectively. The star network can be drawn as a T network (Fig. 9.13 c) and the delta network can be drawn as a Π network (Fig. 9.13 e)

For converting *delta impedances to star impedances* we have

$$\mathbf{Z}_1 = \frac{\mathbf{Z}_{12}\,\mathbf{Z}_{31}}{\mathbf{Z}_{12} + \mathbf{Z}_{23} + \mathbf{Z}_{31}} \qquad (9.9.1)$$

$$\mathbf{Z}_2 = \frac{\mathbf{Z}_{12}\,\mathbf{Z}_{23}}{\mathbf{Z}_{12} + \mathbf{Z}_{23} + \mathbf{Z}_{31}} \qquad (9.9.2)$$

$$\mathbf{Z}_3 = \frac{\mathbf{Z}_{23}\,\mathbf{Z}_{31}}{\mathbf{Z}_{12} + \mathbf{Z}_{23} + \mathbf{Z}_{31}} \qquad (9.9.3)$$

For convenience, the delta and star connected circuits are superimposed as shown in Fig. 9.13e. It is seen that the impedance of any branch of the equivalent star connection is obtained by multiplying the adjacent delta impedances and dividing by the sum of the three delta impedances.

$$Y \text{ impedance} = \frac{\text{product of adjacent } \Delta \text{ impedances}}{\text{sum of delta impedances}} \tag{9.9.4}$$

By *adjacent* here, we mean "on each side of and terminating on the same node as."

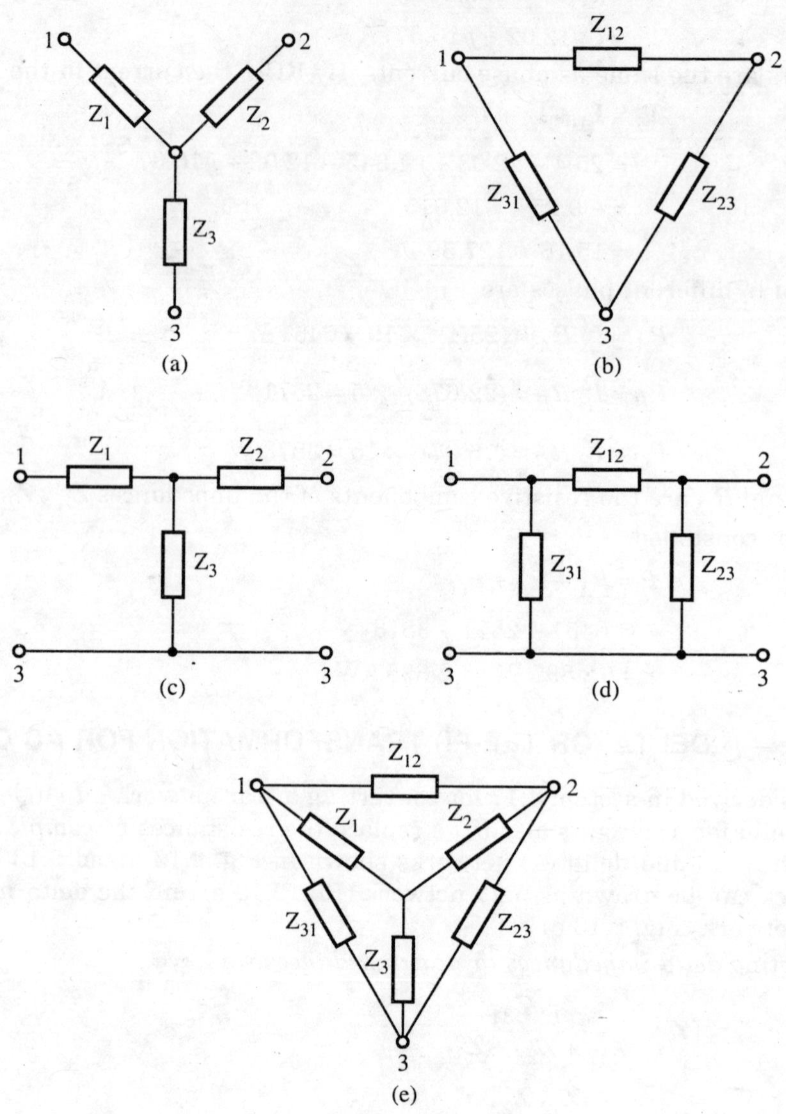

FIG. 9.13

Star-to-delta (or T to Π) Transformation

Here we find $\mathbf{Z}_{12}$, $\mathbf{Z}_{23}$ and $\mathbf{Z}_{31}$ in terms of $\mathbf{Z}_1$, $\mathbf{Z}_2$ and $\mathbf{Z}_3$.

$$\mathbf{Z}_{12} = \frac{\mathbf{Z}_1\,\mathbf{Z}_2 + \mathbf{Z}_2\,\mathbf{Z}_3 + \mathbf{Z}_3\,\mathbf{Z}_1}{\mathbf{Z}_3} \tag{9.9.5}$$

$$\mathbf{Z}_{23} = \frac{\mathbf{Z}_1\,\mathbf{Z}_2 + \mathbf{Z}_2\,\mathbf{Z}_3 + \mathbf{Z}_3\,\mathbf{Z}_1}{\mathbf{Z}_1} \tag{9.9.6}$$

$$\mathbf{Z}_{31} = \frac{\mathbf{Z}_1\,\mathbf{Z}_2 + \mathbf{Z}_2\,\mathbf{Z}_3 + \mathbf{Z}_3\,\mathbf{Z}_1}{\mathbf{Z}_2} \tag{9.9.7}$$

From Fig. 9.13 e

$$\Delta \text{ impedance} = \frac{\text{(sum of the products of Y impedances, taken two at a time)}}{\text{opposite Y impedance}} \tag{9.9.8}$$

These formulae can be applied even if the loads are unbalanced.

If a delta connection contains three equal impedances **Z** then any impedance of the equivalent star connection is **Z**/3.

If a star connection contains three equal impedances **Z** then any impedances of the equivalent delta connection is 3**Z**.

The $Y - \Delta$ transformation formulae do not require balanced loads or even three-phase circuits. They are used considerably in simplifying general circuits that are neither in series nor in parallel.

■ **EXAMPLE 9.3** In Fig. 9.14 a, $\mathbf{Z}_1 = 20\ \underline{/\ 45^\circ}\ \Omega$ and $\mathbf{Z}_2 = 10\ \underline{/\ -30^\circ}\ \Omega$. Determine the delta-connected equivalent circuit.

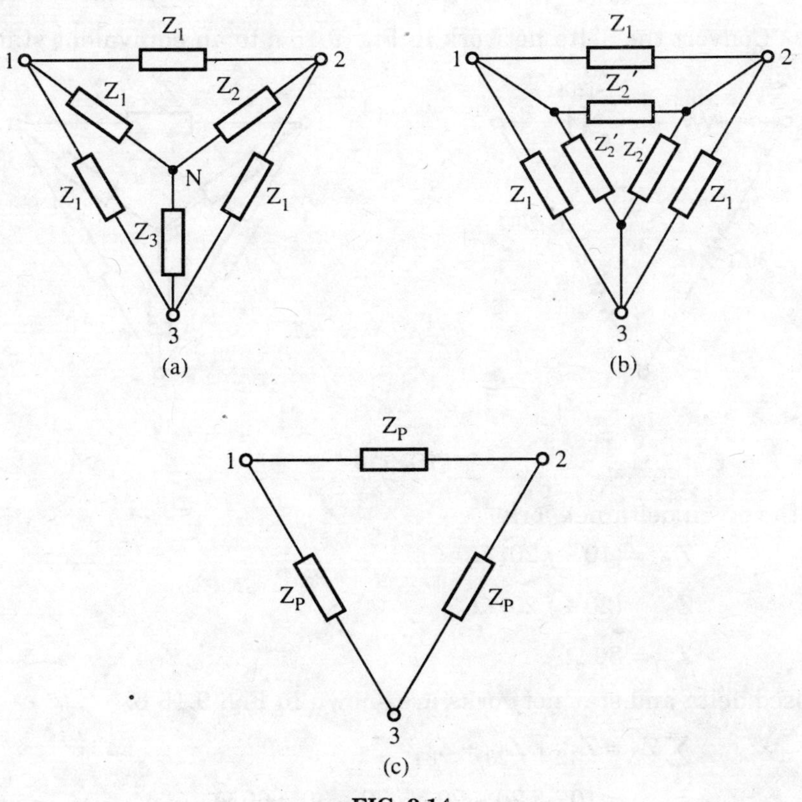

FIG. 9.14

SOLUTION : The inner star comprising three $\mathbf{Z}_2$ elements is converted into delta-connected impedances $\mathbf{Z}_2'$.

$$\mathbf{Z}_2' = 3\mathbf{Z}_2 = 3 \times 10 \angle -30^\circ = 30 \angle -30^\circ\ \Omega$$

Each $\mathbf{Z}_2'$ impedance is connected in parallel with a $\mathbf{Z}_1$ impedance to give a total parallel impedance given by

$$\mathbf{Z}_p = \frac{\mathbf{Z}_1\,\mathbf{Z}_2'}{\mathbf{Z}_1 + \mathbf{Z}_2'}$$

$$= \frac{(20 \angle 45^\circ) \times (30 \angle -30^\circ)}{20 \angle 45^\circ + 30 \angle -30^\circ}$$

$$= \frac{600 \angle 45^\circ - 30^\circ}{14.14 + j\,14.14 + 25.98 - j\,15}$$

$$= \frac{600 \angle 15^\circ}{40.12 - j\,0.86} = \frac{600 \angle 15^\circ}{40.129 \angle -1.23^\circ}$$

$$= 14.95 \angle 16.23^\circ = 14.35 + j\,4.18\ \Omega$$

Therefore the network in Fig. 9.14 a can be replaced by three delta-connected $\mathbf{Z}_p$ impedances each of $14.95 \angle 16.23^\circ = 14.35 + j\,4.18\ \Omega$ as shown in Fig. 9.14 c.

EXAMPLE 9.4 Convert the delta network in Fig. 9.15 a to an equivalent star.

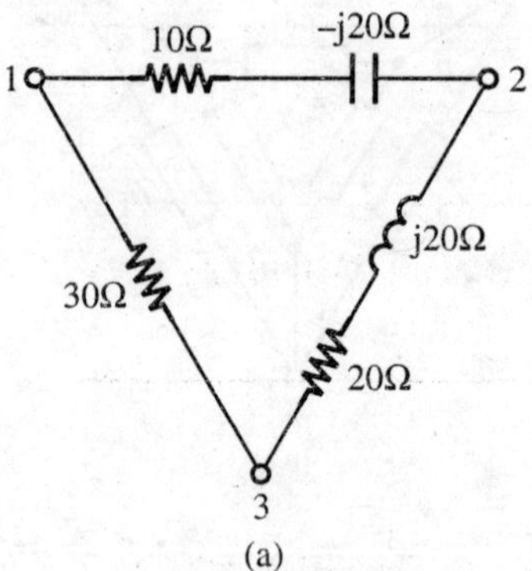

(a)

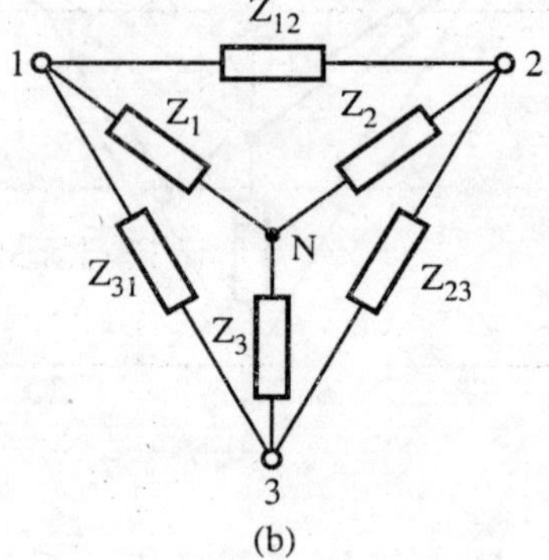

(b)

FIG. 9.15

SOLUTION : In the given delta network

$$\mathbf{Z}_{12} = (10 - j\,20)\ \Omega$$

$$\mathbf{Z}_{23} = (20 + j\,20)\ \Omega$$

$$\mathbf{Z}_{31} = 30\ \Omega$$

The superimposed delta and star networks are shown in Fig. 9.15 b.

$$\sum \mathbf{Z}_\Delta = \mathbf{Z}_{12} + \mathbf{Z}_{23} + \mathbf{Z}_{31}$$

$$= 10 - j\,20 + 20 + j\,20 + 30 = 60\ \Omega$$

$$\mathbf{Z}_1 = \frac{\mathbf{Z}_{12}\,\mathbf{Z}_{31}}{\sum \mathbf{Z}_\Delta} = \frac{(10 - j\,20)\,(30)}{60} = (5 - j\,10)\ \Omega$$

$$\mathbf{Z}_2 = \frac{\mathbf{Z}_{12}\,\mathbf{Z}_{23}}{\sum \mathbf{Z}_\Delta} = \frac{(10 - j\,20)\,(20 + j\,20)}{60} = (10 - j\,3.33)\ \Omega$$

$$\mathbf{Z}_3 = \frac{\mathbf{Z}_{23}\,\mathbf{Z}_{31}}{\sum \mathbf{Z}_\Delta} = \frac{(20 + j\,20)\,30}{60} = (10 + j\,10)\ \Omega$$

9.10 UNBALANCED STAR-CONNECTED LOAD

Figure 9.16 shows a star-connected unbalanced load supplied from a symmetrical star-connected source. The point *N* is the neutral point of the supply and normally it is at zero potential. The star point *S* of the load is not at the same potential as the neutral point *N*. Let us assume that there are no voltage drops in the *A* , *B* and *C* lines.

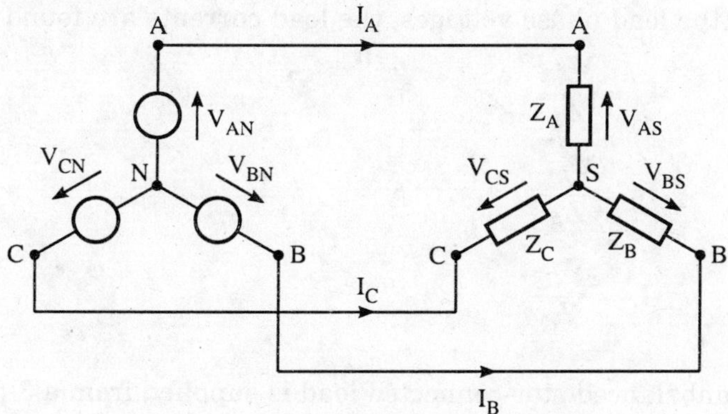

FIG. 9.16 Unbalanced star-connected load fed from a symmetrical star-connected supply.

Unbalanced star-connected loads having isolated star point may be solved by the following methods :

1. Star-delta transformation method
2. Maxwell's mesh current method
3. By applying Millman's theorem.

These methods are explained as follows :

Star-delta transformation method

The equivalent delta impedances of the load are calculated. The problem is then solved as a delta connected system, and the line currents are calculated.

Maxwell's mesh-current method

Maxwell's mesh equations are obtained by using complex values of impedances, voltages and currents. These equations are solved for mesh currents. The line currents are found with the help of the mesh currents.

Millman's theorem application method

For the supply we have

$$\mathbf{V}_{AN} + \mathbf{V}_{BN} + \mathbf{V}_{CN} = 0 \qquad (9.10.1)$$

These supply phase voltages are shown on the phasor diagram of Fig. 9.17.

According to Millman's theorem, the voltages of the load-star point S with respect to the neutral point N of the supply is given by

$$\mathbf{V}_{SN} = \frac{\mathbf{Y}_A\,\mathbf{V}_{AN} + \mathbf{Y}_B\,\mathbf{V}_{BN} + \mathbf{Y}_C\,\mathbf{V}_{CN}}{\mathbf{Y}_A + \mathbf{Y}_B + \mathbf{Y}_C} \qquad (9.10.2)$$

The load phase voltages are given by

$$\mathbf{V}_{AS} = \mathbf{V}_{AN} - \mathbf{V}_{SN} \qquad (9.10.3)$$

$$\mathbf{V}_{BS} = \mathbf{V}_{BN} - \mathbf{V}_{SN} \qquad (9.10.4)$$

$$\mathbf{V}_{CS} = \mathbf{V}_{CN} - \mathbf{V}_{SN} \qquad (9.10.5)$$

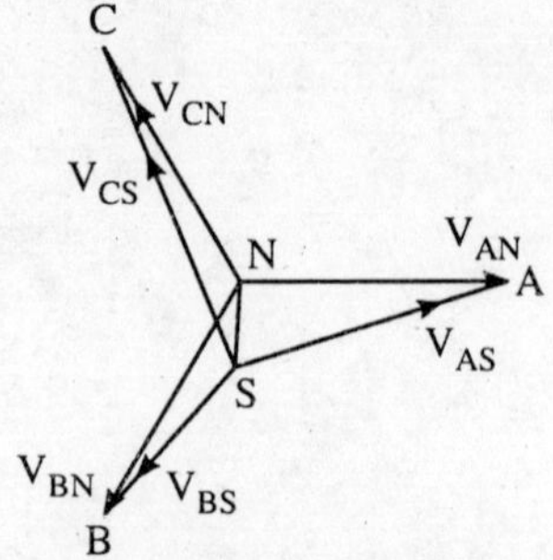

FIG. 9.17 Phasor diagram for unbalanced 3-wire star-connected load

The load phase voltages are shown in the phasor diagram of Fig. 9.17 Since there are no voltage drops in the A , B and C lines the extremeties of the supply and load voltages are coincident.

After obtaining the load phase voltages, the load currents are found from the relations

$$\mathbf{I}_A = \frac{\mathbf{V}_{AS}}{\mathbf{Z}_A} \qquad (9.10.6)$$

$$\mathbf{I}_B = \frac{\mathbf{V}_{BS}}{\mathbf{Z}_B} \qquad (9.10.7)$$

$$\mathbf{I}_C = \frac{\mathbf{V}_{CS}}{\mathbf{Z}_C} \qquad (9.10.8)$$

■ **EXAMPLE 9.5** An unbalanced star-connected load is supplied from a 3-phase, 440 V 3-wire system. The branch impedances of the load are

$$\mathbf{Z}_1 = 5\ \underline{/\ 30^\circ}\ \Omega\ ,\ \mathbf{Z}_2 = 10\ \underline{/\ 45^\circ}\ \Omega\ ,$$

and
$$\mathbf{Z}_3 = 10\ \underline{/\ 60^\circ}\ \Omega\ ,$$

The phase sequence is *ABC*. Calculate the line currents (*a*) by star-delta transformation method (*b*) by applying Millman's theorem.

SOLUTION :

(*a*) Star-delta conversion method

The given star-connected load is converted into the equivalent delta. The equivalent delta impedances are $\mathbf{Z}_{AB}$, $\mathbf{Z}_{BC}$ and $\mathbf{Z}_{CA}$.

$$\sum \mathbf{Z}_1\,\mathbf{Z}_2 = \mathbf{Z}_1\,\mathbf{Z}_2 + \mathbf{Z}_2\,\mathbf{Z}_3 + \mathbf{Z}_3\,\mathbf{Z}_1$$

$$= (5\ \underline{/\ 30^\circ})\,(10\ \underline{/\ 45^\circ}) + (10\ \underline{/\ 45^\circ})\,(10\ \underline{/\ 60^\circ}) + (10\ \underline{/\ 60^\circ})\,(5\ \underline{/\ 30^\circ})$$

$$= 50\ \underline{/\ 75^\circ} + 100\ \underline{/\ 105^\circ} + 50\ \underline{/\ 90^\circ}$$

$$= 12.94 + j\,48.3 - 25.88 + j\,96.59 + j\,50$$

$$= -12.94 + j\,194.89 = 195.3\ \underline{/\ 93.8^\circ}$$

$$Z_{AB} = \frac{\sum Z_1 Z_2}{Z_3} = \frac{195.3\ \underline{/\ 93.8^\circ}}{10\ \underline{/\ 60^\circ}} = 19.53\ \underline{/\ 33.8^\circ}\ \Omega$$

$$Z_{BC} = \frac{\sum Z_1 Z_2}{Z_1} = \frac{195.3\ \underline{/\ 93.8^\circ}}{5\ \underline{/\ 30^\circ}} = 39.06\ \underline{/\ 63.8^\circ}\ \Omega$$

$$Z_{CA} = \frac{\sum Z_1 Z_2}{Z_2} = \frac{195.3\ \underline{/\ 93.8^\circ}}{10\ \underline{/\ 45^\circ}} = 19.53\ \underline{/\ 48.8^\circ}\ \Omega$$

Let $\quad V_{AB} = 440\ \underline{/\ 0^\circ}\ ,\ V_{BC} = 440\ \underline{/\ -120^\circ}$

$$V_{CA} = 440\ \underline{/\ 120^\circ}$$

$$I_{AB} = \frac{V_{AB}}{Z_{AB}} = \frac{440\ \underline{/\ 0^\circ}}{19.53\ \underline{/\ 33.8^\circ}} = 22.529\ \underline{/\ -33.8^\circ}\ \text{A}$$

$$= (18.72 - j\,12.53)\ \text{A}$$

$$I_{BC} = \frac{V_{BC}}{Z_{BC}} = \frac{440\ \underline{/\ -120^\circ}}{39.06\ \underline{/\ 63.8^\circ}} = 11.26\ \underline{/\ -183.8^\circ}\ \text{A}$$

$$= (-11.24 + j\,0.746)\ \text{A}$$

$$I_{CA} = \frac{V_{CA}}{Z_{CA}} = \frac{440\ \underline{/\ 0^\circ}}{19.53\ \underline{/\ 48.8^\circ}} = 22.529\ \underline{/\ 71.2^\circ}\ \text{A}$$

$$= (7.26 + j\,21.32)\ \text{A}$$

The current I_{AB} . I_{BC} and I_{CA} are the phase currents of the equivalent-delta circuit. The line currents are I_A , I_B and I_C.

$$I_A = I_{AB} - I_{CA}$$

$$= (18.72 - j\,12.53) - (7.26 + j\,21.32)$$

$$= 11.46 - j\,33.85 = 35.73\ \underline{/\ -71.3^\circ}\ \text{A}$$

$$I_B = I_{BC} - I_{AB}$$

$$= (-11.24 + j\,0.746) - (18.72 - j\,12.53)$$

$$= -29.96 + j\,13.276 = 32.77\ \underline{/\ -156.1^\circ}\ \text{A}$$

$$I_C = I_{CA} - I_{BC}$$

$$= (7.26 + j\,21.32) - (-11.24 + j\,0.746)$$

$$= 18.5 + j\,20.574 = 27.66\ \underline{/\ 48^\circ}\ \text{A}$$

(*b*) By applying Millman's theorem

Let the supply phase voltage V_{AN} be chosen as the reference phasor.

Then $\quad V_{AN} = \frac{440}{\sqrt{3}}\ \underline{/\ 0^\circ} = 254\ \underline{/\ 0^\circ}\ \text{V} = (254 + j\,0)\ \text{V}$

$$\mathbf{V}_{BN} = \frac{440}{\sqrt{3}} \underline{/-120^\circ} = (-127 - j\,220)\text{ V}$$

$$\mathbf{V}_{CN} = \frac{440}{\sqrt{3}} \underline{/120^\circ} = (-127 + j\,220)\text{V}$$

The load admittances are

$$\mathbf{Y}_1 = \frac{1}{\mathbf{Z}_1} = \frac{1}{5\,\underline{/30^\circ}} = 0.2\,\underline{/-30^\circ}\text{ S} = (0.1732 - j\,0.1)\text{ S}$$

$$\mathbf{Y}_2 = \frac{1}{\mathbf{Z}_2} = \frac{1}{10\,\underline{/45^\circ}} = 0.1\,\underline{/-45^\circ}\text{ S} = (0.0707 - j\,0.0707)\text{ S}$$

$$\mathbf{Y}_3 = \frac{1}{\mathbf{Z}_3} = \frac{1}{10\,\underline{/60^\circ}} = 0.1\,\underline{/-60^\circ}\text{ S} = (0.05 - j\,0.0866)\text{ S}$$

$$\sum \mathbf{Y} = \mathbf{Y}_1 + \mathbf{Y}_2 + \mathbf{Y}_3 = 0.2939 - j\,0.2573\text{ S}$$

$$= 0.3906\,\underline{/-41.2^\circ}\text{ S}$$

By Millman's theorem

$$\mathbf{V}_{SN} = \frac{\mathbf{Y}_1\mathbf{V}_{AN} + \mathbf{Y}_2\mathbf{V}_{BN} + \mathbf{Y}_3\mathbf{V}_{CN}}{\mathbf{Y}_1 + \mathbf{Y}_2 + \mathbf{Y}_3}$$

$$\mathbf{Y}_1\mathbf{V}_{AN} = (0.2\,\underline{/-30^\circ})(254\,\underline{/0^\circ}) = 50.8\,\underline{/-30^\circ}$$

$$= 43.99 - j\,25.4$$

$$\mathbf{Y}_2\mathbf{V}_{BN} = (0.1\,\underline{/-45^\circ})(254\,\underline{/-120^\circ}) = 25.4\,\underline{/-165^\circ}$$

$$= 24.53 - j\,6.57$$

$$\mathbf{Y}_3\mathbf{V}_{CN} = (0.1\,\underline{/-60^\circ})(254\,\underline{/120^\circ}) = 25.4\,\underline{/-60^\circ}$$

$$= 12.7 - j\,21.99$$

$$\sum \mathbf{YV} = \mathbf{Y}_1\mathbf{V}_{AN} + \mathbf{Y}_2\mathbf{V}_{BN} + \mathbf{Y}_3\mathbf{V}_{CN}$$

$$= 32.16 - j\,9.98 = 33.67\,\underline{/-17.24^\circ}$$

Therefore

$$\mathbf{V}_{SN} = \frac{\sum \mathbf{YV}}{\sum \mathbf{Y}} = \frac{33.67\,\underline{/-17.24^\circ}}{0.3906\,\underline{/-41.20^\circ}} = 8.62\,\underline{/23\,96^\circ}$$

$$= (78.77 + j\,35)\text{ V}$$

The load phase voltages are given by

$$\mathbf{V}_{AS} = \mathbf{V}_{AN} - \mathbf{V}_{SN} = (254 + j\,0) - (78.77 + j\,35)$$

$$= 175.23 - j\,35 = 178.7\,\underline{/-11.29^\circ}\text{ V}$$

$$\mathbf{V}_{BS} = \mathbf{V}_{BN} - \mathbf{V}_{SN} = (-127 - j\,220) - (78.77 + j\,35)$$

$$= -205.77 - j\,255 = 327.6\,\underline{/-128.9^\circ}\text{ V}$$

$$\mathbf{V}_{CS} = \mathbf{V}_{CN} - \mathbf{V}_{SN} = (-127 + j\,220) - (78.77 + j\,35)$$

$$= -205.77 + j\,185 = 276.7 \underline{/\ 138^\circ}\ \text{V}$$

The load currents are given by

$$\mathbf{I}_A = \frac{\mathbf{V}_{AS}}{\mathbf{Z}_1} = \frac{178.7 \underline{/\ -11.29^\circ}}{5 \underline{/\ 30^\circ}} = 35.74 \underline{/\ -41.29^\circ}\ \text{A}$$

$$\mathbf{I}_B = \frac{\mathbf{V}_{BS}}{\mathbf{Z}_2} = \frac{327.6 \underline{/\ -128.9^\circ}}{10 \underline{/\ 45^\circ}} = 32.76 \underline{/\ -173.9^\circ}\ \text{A}$$

$$\mathbf{I}_C = \frac{\mathbf{V}_{CS}}{\mathbf{Z}_3} = \frac{276.7 \underline{/\ 138^\circ}}{10 \underline{/\ 60^\circ}} = 27.67 \underline{/\ 78^\circ}\ \text{A}$$

9.11 VOLTAMPERES, POWER AND REACTIVE VOLTAMPERES

Voltamperes

Voltamperes in a single-phase circuit is given by

$$S_{1\phi} = VI \quad \text{V A (voltamperes)}$$

The voltamperes of a three-phase system, whether balanced or un-balanced, is equal to the sum of the voltamperes in each phase. For a balanced system, the total voltamperes is equal to three times the voltamperes per phase. That is,

$$S_{3\phi} = 3\,S_{1\phi} = 3\,V_p\,I_p \tag{9.11.1}$$

where V_p and I_p are the r.m.s. phase voltage and r.m.s. phase current. The total voltamperes can also be calculated in terms of line values.

For a star-connected system

$$V_p = \frac{V_L}{\sqrt{3}},\ I_p = I_L$$

$$S_{3\phi} = 3\,V_p\,I_p = \sqrt{3}\,V_L\,I_L \quad \text{VA}$$

For a delta-connected system,

$$V_p = V_L,\ I_p = \frac{I_L}{\sqrt{3}}$$

$$S_{3\phi} = 3\,V_p\,I_p = \sqrt{3}\,V_L\,I_L \quad \text{VA} \tag{9.11.2}$$

Power *P*

Power in a single-phase circuit is given by

$$P_{1\phi} = VI\cos\phi \quad \text{W}$$

The power of a three-phase system, whether balanced or unbalanced, is equal to the sum of the power of each phase. For a balanced system, the total power is equal to three times the power per phase. That is,

$$P_{3\phi} = 3\,P_{1\phi} = 3\,V_p\,I_p\cos\phi \quad \text{W} \tag{9.11.3}$$

In terms of line values, for both star and delta connections

$$P_{3\phi} = \sqrt{3}\,V_L\,I_L\cos\phi \quad \text{W} \tag{9.11.4}$$

Reactive Voltamperes *Q*

Reactive voltamperes in a single-phase circuit is given by

$$Q_{1\phi} = VI \sin \phi \quad \text{VAr}$$

The reactive voltamperes of a 3-phase system, whether balanced or unbalanced, is equal to the sum of the reactive voltamperes in each phase. For a balanced system the total reactive voltamperes is equal to three times the reactive voltamperes per phase. That is,

$$Q_{3\phi} = 3\,Q_{1\phi} = 3\,V_p\,I_p \sin \phi \quad \text{VAr} \tag{9.11.5}$$

It terms of line values for both star and delta connections.

$$Q_{3\phi} = \sqrt{3}\,V_L\,I_L \sin \phi \quad \text{VAr} \tag{9.11.6}$$

Relationship Between *P*, *Q* and *S*

$$P_{3\phi}^2 + Q_{3\phi}^2 = (\sqrt{3}\,V_L\,I_L \cos \phi)^2 + (\sqrt{3}\,V_L\,I_L \sin \phi)^2 = (\sqrt{3}\,V_L\,I_L)^2 = S_{3\phi}^2$$

$$S_{3\phi} = \sqrt{(P_{3\phi}^2 + Q_{3\phi}^2)} \tag{9.11.7}$$

In a 3-phase system, balanced conditions are assumed unless stated otherwise; and the term voltage, current and power, unless described otherwise, are understood to mean a line-to-line voltage, line current, and total power of all three phases.

9.12 MEASUREMENT OF THREE-PHASE POWER

The power in a 3-phase circuit is equal to the sum of the powers in three phases. For a balanced system,

total power = 3 × power in one phase

The power in any one phase is measured by a wattmeter. The total power is obtained by multiplying the wattmeter reading by three.

9.11.1 One-Wattmeter Method

Figure 9.18 shows a circuit for the measurement of power in one phase of a 3-phase balanced circuit where the star point is available. The current coil is connected on one phase and the

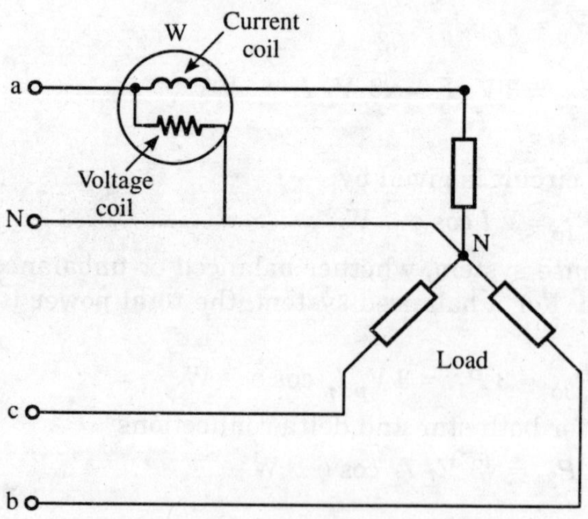

FIG. 9.18 One wattmeter method of power measurement where star point is accessible

voltage coil is connected between the same phase and star point. The reading on the wattmeter gives the power per phase.

The total 3-phase power is given by threee times the wattmeter reading.

If the neutral point is not available an artificial neutral point N' is created by connecting two resistors R , R as shown in Fig. 9.19.

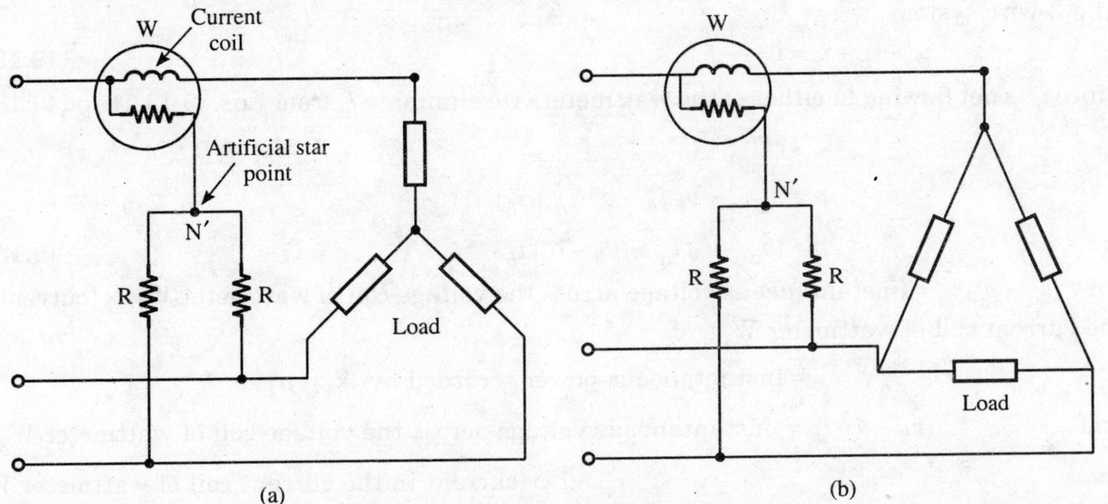

FIG. 9.19 One wattmeter method of power measurement in a 3-phase balanced system where neutral point is not available

The total power is given by three times the wattmeter reading.

9.13 TWO-WATTMETER METHOD

For a 3-phase 3-wire system, whether balanced or unbalanced, star or delta connected, only two wattmeters are required to measure the three-phase power. This is the most popular method of measuring power. Figure 9.20 shows the connections of the two wattmeters for a

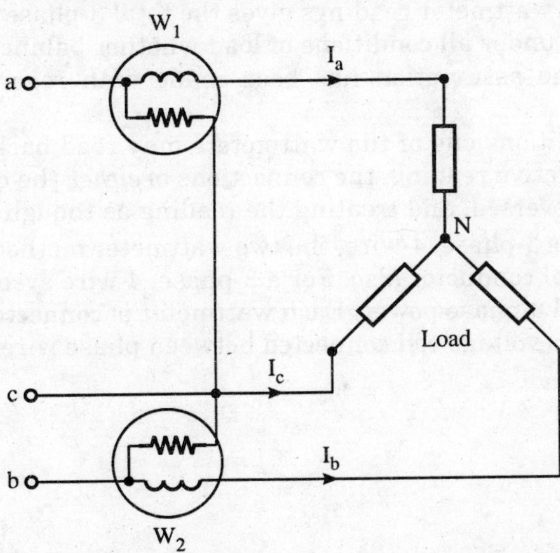

FIG. 9.20 Two-wattmeter method of 3-phase power measurement.

star-connected load. The wattmeter connections are the same for a delta connected load also. The current coils of the two wattmeters are connected to any two of the lines while their voltage coils are connected between the corresponding lines and the third line.

Total instantaneous power supplied to the load is

$$p = v_a i_a + v_b i_b + v_c i_c \tag{9.13.1}$$

In a 3-wire system

$$i_a + i_b + i_c = 0 \tag{9.13.2}$$

Since i_c is not flowing in either of the wattmeters we eliminate i_c from Eqs. (9.13.1) and (9.13.2)

$$i_c = -(i_a + i_b)$$

$$p = v_a i_a + v_b i_b - v_c (i_a + i_b)$$

or

$$p = (v_a - v_c) i_a + (v_b - v_c) i_b \tag{9.13.3}$$

Now $(v_a - v_c) i_a$ = (instantaneous voltage across the voltage coil of wattmeter W_1) × (current in the current coil of wattmeter W_1)

= instantaneous power recorded by $W_1 = p_1$

and $(v_b - v_c) i_b$ = (instantaneous voltage across the voltage coil of wattmeter W_2) × (current in the current coil of wattmeter W_2)

= instataneous power recorded by $W_2 = p_2$

Therefore

$$p = p_1 + p_2 \tag{9.13.4}$$

Average value of p

= average value of p_1 + average value of p_2

P = reading of W_1 + reading of W_2

$$P = P_1 + P_2 \tag{9.13.5}$$

Thus, the sum of the two wattmeter readings gives the total 3-phase power in a 3-wire system (star or delta connected) under all conditions of load whether balanced or unbalanced, star or delta connected. Also, no assumption has been made with regard to phase sequence or waveform.

Under certain conditions one of the wattmeters may read backwards giving a negative reading. To take this negative reading, the connections of *either* the current coil *or* the voltage coils (but *not* both) are reversed, and treating the reading as though it were a negative value.

In case the supply is 3-phase, 4-wire, the two wattmeter method cannot be used as there is a current in the neutral conductor also. For a 3-phase, 4-wire system three wattmeters are used to measure the total 3-phase power. Each wattmeter is connected with its current coil in series with the line and is voltage coil connected between phase wire and neutral as shown in Fig 9.21.

The total power consumed by the load

$$P_{3\phi} = P_A + P_B + P_C$$

where P_A, P_B and P_C are the readings of wattmeters W_A, W_B, and W_C.

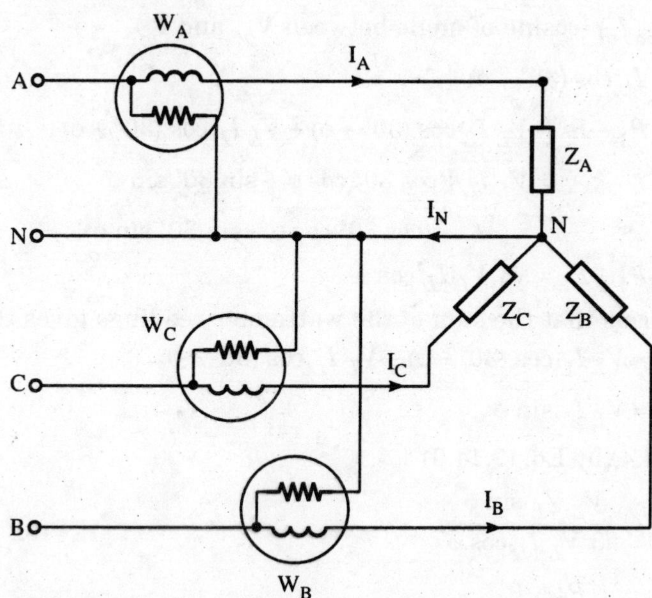

FIG. 9.21 Measurement of power in 3-phase, 4-wire system

9.14 POWER FACTOR OF A BALANCED LOAD

If the load is inductive, the power factor cos ϕ will be lagging. In case the load is capacitive the power factor cos ϕ will be leading.

9.14.1 Lagging Power Factor Load

In case of lagging power factor the phase currents $\mathbf{I}_a$, $\mathbf{I}_b$ and $\mathbf{I}_c$ will be lagging behind their respective phase voltages by an angle ϕ. The phasor diagram under balanced load conditions for lagging power factor cos ϕ is shown in Fig. 9.22.

The power indicated by a wattmeter in an ac circuit is equal to the product

(Voltage across the voltage coil) × (current through the current coil) × (cosine of the angle between the voltage and current in the coils)

In the phasor diagram of Fig. 9.22.

$$|\mathbf{I}_a| = |\mathbf{I}_b| = \mathbf{I}_c| = I_L$$

$$|\mathbf{V}_{ab}| = |\mathbf{V}_{bc}| = |\mathbf{V}_{ca}| = V_L$$

Reading of wattmeter W_1 is given by

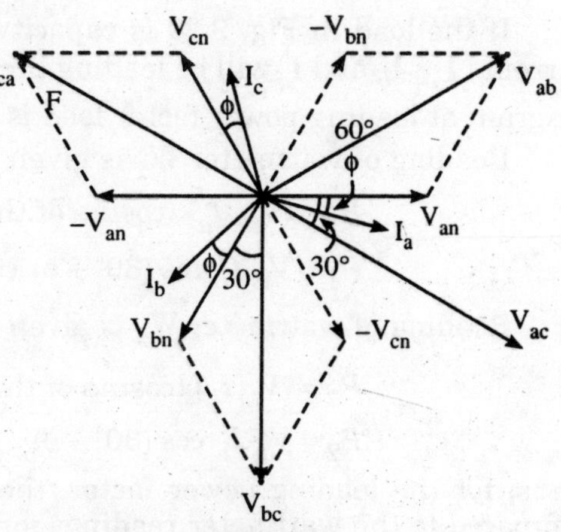

FIG. 9.22
Phasor diagram for lagging p.f. cos ϕ

$$P_1 = (V_{ac}\, I_a) \quad \text{(cosine of the angle between } \mathbf{V}_{ac} \text{ and } \mathbf{I}_a)$$

$$P_1 = V_L\, I_L \cos(30^\circ - \phi) \tag{9.14.1}$$

Reading of wattmeter W_2 is given by

$$P_2 = (V_{bc}\, I_b) \text{ (cosine of angle between } \mathbf{V}_{bc} \text{ and } \mathbf{I}_b)$$

$$P_2 = V_L\, I_L \cos(30^\circ + \phi) \tag{9.14.2}$$

$$P_1 + P_2 = V_L\, I_L \cos(30^\circ - \phi) + V_L\, I_L \cos(30^\circ + \phi)$$

$$= V_L\, I_L\, [(\cos 30^\circ \cos\phi + \sin 30^\circ \sin\phi)$$

$$+ (\cos 30^\circ \cos\phi - \sin 30^\circ \sin\phi)]$$

$$P_1 + P_2 = \sqrt{3}\, V_L\, I_L \cos\phi \tag{9.14.3}$$

This relation also confirms that the sum of the wattmeter readings gives the total power.

Also, $$P_1 - P_2 = V_L\, I_L \cos(30^\circ - \phi) - V_L\, I_L \cos(30^\circ + \phi)$$

$$P_1 - P_2 = V_L\, I_L \sin\phi \tag{9.14.4}$$

Dividing Eq. (9.14.4) by Eq. (9.14.3)

$$\frac{P_1 - P_2}{P_1 + P_2} = \frac{V_L\, V_L \sin\phi}{\sqrt{3}\, V_L\, I_L \cos\phi}$$

$$\tan\phi = \sqrt{3}\,\frac{P_1 - P_2}{P_1 + P_2} \tag{9.14.5}$$

or $$\tan\phi = \sqrt{3}\,\frac{\text{difference of wattmeter readings}}{\text{sum of wattmeter readings}}$$

Eq. (9.14.5) gives the value of $\tan\phi$. Hence the value of power factor $\cos\phi$ can be calculated. In making calculations it is usual to take P_1 as the greater value of the two readings of the wattmeter.

9.14.2 Leading Power Factor Load

If the load in Fig. 9.20 is capacitve, the power factor $\cos\phi$ will be leading. The phase currents $\mathbf{I}_a$, $\mathbf{I}_b$ and $\mathbf{I}_c$ will be leading their respective phase voltages by an angle ϕ. The phasor diagram at leading power factor load is shown in Fig. 9.23.

Reading of wattmeter W_1 is given by

$$P_1 = (V_{ac}\, I_a) \text{ (cosine of the angle between } \mathbf{V}_{ac} \text{ and } \mathbf{I}_a)$$

or $$P_1 = V_L\, I_L \cos(30^\circ + \phi) \quad (9.14.6)$$

Reading of wattmeter W_2 is given by

$$P_2 = (V_{bc} I_b) \text{ (cosine of the angle between } \mathbf{V}_{bc} \text{ and } \mathbf{I}_b)$$

or $$P_2 = V_L\, I_L \cos(30^\circ - \phi) \tag{9.14.7}$$

Thus, for the leading power factor, the readings of the wattmeters are interchanged when compared to the wattmeter readings for the lagging power factor.

Since $\cos(30° + \phi) < \cos(30° - \phi)$

$$P_1 < P_2$$

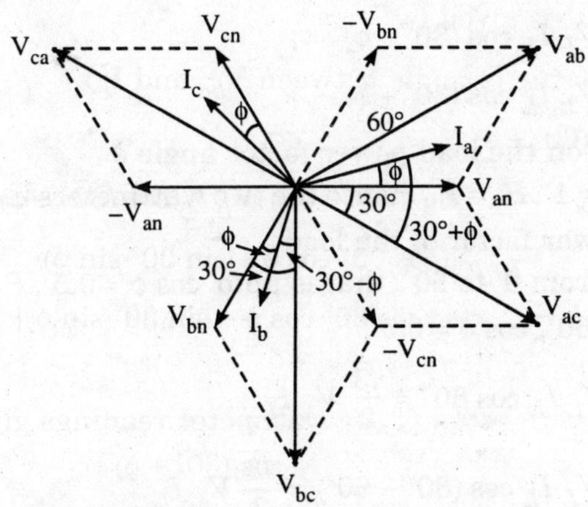

Fig. 9.23 Phasor diagram at leading p.f.

In case of lagging power factor, $P_1 > P_2$.

Also,
$$P_{3\phi} = P_1 + P_2$$
$$= V_L I_L \cos(30° + \phi) + V_L I_L \cos(30° - \phi)$$

Therefore,
$$P_1 + P_2 = \sqrt{3}\, V_L I_L \cos\phi \tag{9.14.8}$$
$$P_1 - P_2 = V_L I_L \cos(30° + \phi) - V_L I_L \cos(30° - \phi)$$
$$= V_L I_L\, [(\cos 30° \cos\phi - \sin 30° \sin\phi)$$
$$- (\cos 30° \cos\phi + \sin 30° \sin\phi)]$$
$$= -2\, V_L I_L \sin 30° \sin\phi$$
$$= -V_L I_L \sin\phi$$

Therefore,
$$P_2 - P_1 = V_L I_L \sin\phi \tag{9.14.9}$$

Dividing Eq. (9.14.9) by Eq. (9.14.8)

$$\frac{P_2 - P_1}{P_2 + P_1} = \frac{V_L I_L \sin\phi}{\sqrt{3}\, V_L I_L \cos\phi}$$

or
$$\tan\phi = \sqrt{3}\, \frac{P_2 - P_1}{P_2 + P_1} \tag{9.14.10}$$

Thus, it is seen that in both the cases of lagging and leading power factors, the smaller reading is to be subtracted from the greater reading in the numerator of Eq. (9.14.10). That is, for both lagging and leading power factor cos φ,

$$\tan\phi = \sqrt{3} \times \frac{\text{greater wattmeter reading} - \text{smaller wattmeter reading}}{\text{sum of wattmeter readings}} \tag{9.14.11}$$

9.15 VARIATION OF WATTMETER READINGS WITH LOAD POWER FACTOR

For lagging power factor load (inductive load) the readings of the two wattmeters are given by

$$P_1 = V_L I_L \cos(30° - \phi)$$

$$P_2 = V_L I_L \cos(30° + \phi)$$

These readings depend on the load power factor angle ϕ.

(*a*) When $\phi = 0°$, $\cos\phi = 1$, $P_1 = P_2$ Hence the two wattmeters indicate equal and positive values at unity power factor of the load.

(*b*) When ϕ increases from 0° to 60°, that is upto $\cos\phi = 0.5$, P_1 and P_2 are positive but $P_1 > P_2$. When $\phi = 60°$, $\cos\phi = 0.5$,

$$\text{Total power} = \sqrt{3}\, V_L I_L \cos 60° = \frac{\sqrt{3}}{2} V_L I_L$$

$$P_1 = V_L I_L \cos(30° - 60°) = \frac{\sqrt{3}}{2} V_L I_L$$

$$P_2 = V_L I_L \cos(30° + 60°) = 0$$

Hence one wattmeter will read zero and the other will read the total power.

(*c*) When $\cos\phi > 0.5$, $\phi < 60°$ In this case both the wattmeters indicate positive readings.

(*d*) When $\cos\phi < 0.5$, $\phi > 60°$. In this case wattmeter W_1 gives a positive reading while wattmeter W_2 gives a negative reading. The pointer of this wattmeter tries to go on the left side of the zero point of the scale. To take this reading on the wattmeter, it is necessary to reverse the connections of either current coil or voltage coil. The reading thus obtained must be given a negative sign. Then the total power in the circuit

$$P = P_1 + (-P_2) = P_1 - P_2$$

Wattmeter W_2 reads downscale for the phase angle ϕ between 60° and 90°.

(*e*) When $\phi = 90°$, $\cos\phi = 0$

$$P_1 = V_L I_L \cos(30° - 90°) = 0.5\, V_L I_L$$

$$P_2 = V_L I_L \cos(30° + 90°) = -\,0.5\, V_L I_L$$

Thus, at zero power factor (lag or lead), the two wattmeters read equal and opposite values.

The variation in the wattmeter readings with the capacitive load (leading power factor) follows the same sequence as in inductive load (logging power factor), with a change in roles of the wattmeters.

9.16 MEASUREMENT OF REACTIVE VOLTAMPERES

Two-wattmeter method may also be used to measure the reactive voltamperes in all the three phases.

Total reactive voltamperes

$$Q = \sqrt{3}\, V_L I_L \sin\phi$$

From Eq. (9.14.4)

$$P_1 - P_2 = V_L I_L \sin\phi$$

Therefore $$Q = \sqrt{3}\,(P_1 - P_2) \qquad (9.16.1)$$

■ **EXAMPLE 9.6** **Two wattmeters connected to measure the total power in a 3-phase load read 5 kW and 2kW. Calculate the total power and the power factor**

SOLUTION : Total power

$$P = P_1 + P_2$$

$$= 5 + 2 = 7 \text{ kW}$$

$$\tan\phi = \sqrt{3}\,\frac{P_1 - P_2}{P_1 + P_2} = \sqrt{3}\,\frac{5-2}{5+2} = 0.7423$$

$$\phi = 36.58^\circ\,,\ \cos\phi = 0.803$$

■ **EXAMPLE 9.7** **The input power to a 3-phase motor was measured by the 2-wattmeter method. The readings were 5.2 kW and 1.7 kW, the latter reading was obtained after reversal of current-coil connections. The line voltage was 400 V. Calculate (*a*) the total power, (*b*) the power factor, and (*c*) the line current.**

SOLUTION :

$$P_1 = 5.2 \text{ kW}\,,\ P_2 = -1.7 \text{ kW}$$

(*a*) Total power

$$P = P_1 + P_2$$

$$= 5.2 + (-1.7) = 3.5 \text{ kW}$$

(*b*) $$\tan\phi = \sqrt{3}\,\frac{P_1 - P_2}{P_1 + P_2} = \sqrt{3} \times \frac{5.2 - (-1.7)}{5.2 + (-1.7)}$$

$$= \frac{\sqrt{3} \times 6.9}{3.5} = 3.41$$

$$\phi = 73.67^\circ\,,\quad \cos\phi = 0.281$$

(*a*) $$P = \sqrt{3}\, V_L I_L \cos\phi$$

$$I_L = \frac{P}{\sqrt{3}\, V_L \cos\phi} = \frac{3.5 \times 10^3}{\sqrt{3} \times 400 \times 0.281}$$

$$= 17.98 \text{ A}$$

■ **EXAMPLE 9.8** **A three phase 400 V generator supplies a load of 1.8 kW at a line current of 10 A when three identical impedances are connected is star across the line terminals of the generator. Calculate the resistive and reactive components of each phase impedance.**

SOLUTION : Let R_p and X_p be the resistive and reactive components of each phase impedance. Since the load is connected in star, phase and line currents are equal.

$$P_{3\phi} = 3I_L^2 R_p = 3I_L^2 R_p$$

$$1.8 \times 10^3 = 3 \times 10^2 R_p \text{ , } R_p = 6\ \Omega$$

$$\text{Phase voltage} = \frac{1}{\sqrt{3}} \times \text{line voltage}$$

$$V_p = \frac{1}{\sqrt{3}} \times 400 = 230.9 \text{ V}$$

Phase impedance

$$Z_p = \frac{V_p}{I_p} = \frac{230.9}{10}$$

$$= 23.09\Omega$$

We have

$$R_p^2 + X_p^2 = Z_p^2$$

therefore,

$$X_p = \sqrt{Z_p^2 - R_p^2} = \sqrt{23.09^2 - 6^2} = 22.29\ \Omega$$

■ **EXAMPLE 9.9** A three phase 400 V generator supplies power to both delta- and star-connected load as shown in Fig. 9.24. All phase impedances are identical and equal to $(12 + j\ 16)\ \Omega$. Calculate the generator current flowing in line a.

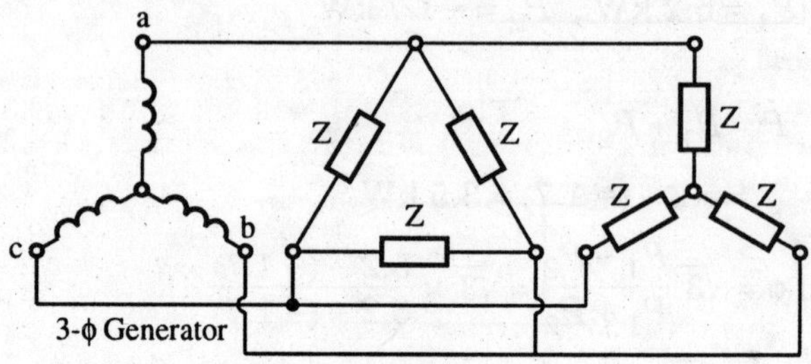

Fig. 9.24.

SOLUTION : Using delta-star transformation, the delta connected load is transformed to an equivalent star-connected load as shown in Fig. 9.25 a. The combined delta- and star-connected load is then reduced to a single star-connected load as shown in Fig. 9.25 b. The modified form of Fig. 9.24 is shown in Fig. 9.25 c.

Load impedance per phase

$$Z_a = Z \,||\, \frac{Z}{3} = \frac{Z\,(Z/3)}{Z + (Z/3)} = Z/4$$

Load voltage per phase

$$= \frac{V_L}{\sqrt{3}} = \frac{400}{\sqrt{43}} = 230 \text{ V}$$

Taking the voltage of phase a as the reference phasor $\mathbf{V}_a = 230 \angle 0°$. Load impedance of phase a is given by

$$\mathbf{Z}_a = \frac{\mathbf{Z}}{4} = \frac{1}{4}(12 + j16) = 3 + j4 = 5 \angle 53.13° \ \Omega$$

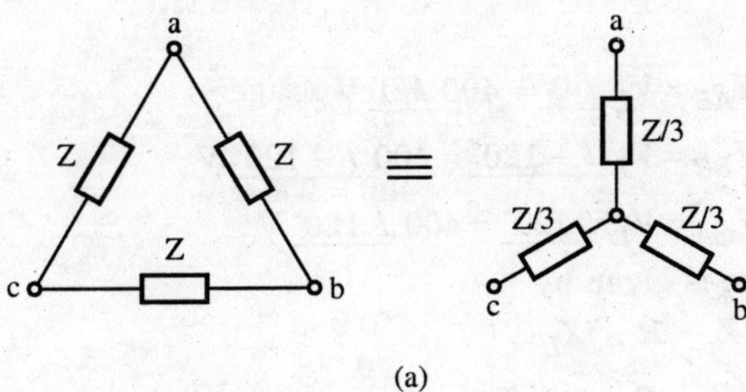

(a)

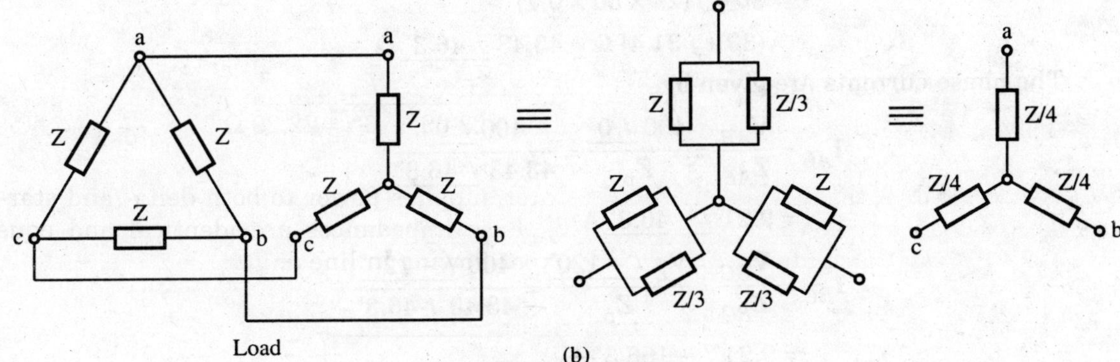

(b)

Fig. 9.25.(a) and (b)

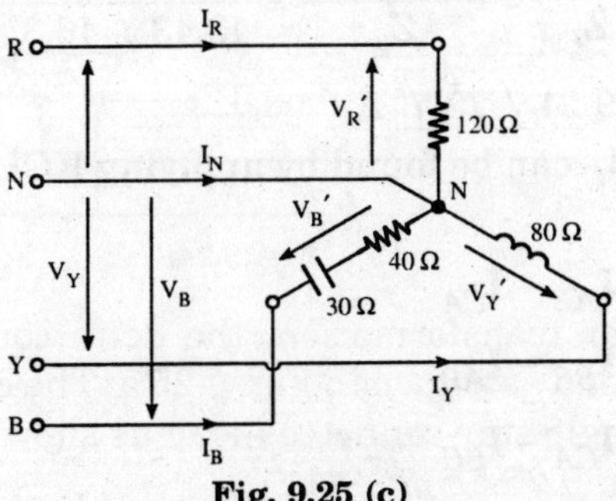

Fig. 9.25 (c)

Current in phase a

$$I_{pa} = \frac{\mathbf{V}_a}{\mathbf{Z}_a} = \frac{230 \angle 0°}{5 \angle 53.13°} = 46 \angle -53.13° \text{ A}$$

For star connection, the line and phase currents are the same. Hence the current flowing in line a is given by

$$\mathbf{I}_{La} = \mathbf{I}_{pa} = 46 \angle -53.13° \text{ A}$$

EXAMPLE 9.10 A three-phase delta load consists of three similar inductive coils each of resistance 30 Ω and inductance 0.1 H. The load is connected across a 400 V, three-phase 50 Hz balanced supply. Calculate the line currents and total power consumed by the load. Assume phase sequence to be *ABC*.

SOLUTION : Let the line voltage $\mathbf{V}_{AB}$ be taken as the reference voltage phasor.

Therefore

$$\mathbf{V}_{AB} = V_L \underline{/0^\circ} = 400 \underline{/0^\circ}\text{ V}$$

$$\mathbf{V}_{BC} = V_L \underline{/-120^\circ} = 400 \underline{/-120^\circ}\text{ V}$$

$$\mathbf{V}_{CA} = V_L \underline{/120^\circ} = 400 \underline{/120^\circ}\text{ V}$$

The impedance per phase is given by

$$\mathbf{Z}_p = R + jX_L$$

$$= R + j\,(2\pi f L)$$

$$= 30 + j\,(2\pi \times 50 \times 0.1)$$

$$= (30 + j\,31.4)\ \Omega = 43.43 \underline{/46.3^\circ}\ \Omega$$

The phase currents are given by

$$\mathbf{I}_{AB} = \frac{\mathbf{V}_{AB}}{\mathbf{Z}_{AB}} = \frac{400 \underline{/0^\circ}}{Z_p} = \frac{400 \underline{/0^\circ}}{43.43 \underline{/46.3^\circ}}$$

$$= 9.21 \underline{/-46.3^\circ}\text{ A}$$

$$\mathbf{I}_{BC} = \frac{\mathbf{V}_{BC}}{\mathbf{Z}_{BC}} = \frac{V_L \underline{/-120^\circ}}{Z_p} = \frac{400 \underline{/-120^\circ}}{43.43 \underline{/46.3^\circ}}$$

$$= 9.21 \underline{/-166.3^\circ}\text{ A}$$

$$\mathbf{I}_{CA} = \frac{\mathbf{V}_{CA}}{\mathbf{Z}_{CA}} = \frac{V_L \underline{/120^\circ}}{Z_p} = \frac{400 \underline{/120^\circ}}{43.43 \underline{/46.3^\circ}}$$

$$= 9.21 \underline{/73.7^\circ}\text{ A}$$

The line currents $\mathbf{I}_A$, $\mathbf{I}_B$ and $\mathbf{I}_C$ can be found by applying KCL at points *A* , *B* and *C* in Fig. 9.9 as follows :

$$\mathbf{I}_A = \mathbf{I}_{AB} - \mathbf{I}_{CA}$$

$$\mathbf{I}_B = \mathbf{I}_{BC} - \mathbf{I}_{AB}$$

$$\mathbf{I}_C = \mathbf{I}_{CA} - \mathbf{I}_{BC}$$

$$\mathbf{I}_A = 9.21 \underline{/-46.3^\circ} - 9.21 \underline{/73.7^\circ}$$

$$= 9.21\,(0.691 - j0.723)$$

$$- 9.21\,(0.2807 + j0.9598)$$

$$= 9.21\,(0.4103 - j1.6828)$$

$$= 9.21 \times 1.732 \underline{/-763^\circ}$$

$$= 15.952 \underline{/-76.3^\circ}\text{ A}$$

$$\mathbf{I}_B = 9.21 \underline{/ -166.3^\circ} - 9.21 \underline{/ -46.3^\circ}$$

$$= 15.952 \underline{/ -196.3^\circ} \text{ A}$$

$$\mathbf{I}_C = 9.21 \underline{/ 73.7^\circ} - 9.21 \underline{/ -166.3^\circ}$$

$$= 15.952 \underline{/ 43.7^\circ} \text{ A}$$

Alternatively

It is given that the load is balanced and the phase sequence is *ABC*. Therefore

$$|\mathbf{I}_A| = |\mathbf{I}_B| = |\mathbf{I}_C| = 15.952 \text{ A}$$

$\mathbf{I}_B$ lags behind $\mathbf{I}_A$ by 120° and $\mathbf{I}_C$ leads $\mathbf{I}_A$ by 120°. Hence the currents $\mathbf{I}_B$ and $\mathbf{I}_C$ can be determined by inspection after determining $\mathbf{I}_A$.

$$\mathbf{I}_B = 15.952 \underline{/ -76.3^\circ - 120^\circ}$$

$$= 15.952 \underline{/ -196.3^\circ} \text{ A}$$

$$\mathbf{I}_C = 15.952 \underline{/ -76.3^\circ + 120^\circ}$$

$$= 15.952 \underline{/ 73.7^\circ} \text{ A}$$

■ **EXAMPLE 9.11 A balanced load of 20 k VA is connected to a three phase three-wire system. Two wattmeters are connected in the usual manner to measure power. Determine the readings of the two wattmeters if the power factor of the load is (*a*) unity, (*b*) 0.866 lagging, (*c*) 0.5 leading, and (*d*) zero lagging. What is the maximum possible reading of either wattmeter.**

SOLUTION : The total power consumed by the load

$$= \text{kVA} \cos\phi = 20 \cos\phi \text{ kW}$$

therefore, $\quad P_1 + P_2 = 20 \cos\phi$

Also, $\quad \tan\phi = \dfrac{\sqrt{3}\,(P_1 - P_2)}{P_1 + P_2}$

therefore, $\quad \tan\phi = \dfrac{\sqrt{3}\,(P_1 - P_2)}{20 \cos\phi}$

$$P_1 - P_2 = \frac{1}{\sqrt{3}}(20 \cos\phi \tan\phi) = \frac{20}{\sqrt{3}} \sin\phi$$

(*a*) $\quad \cos\phi = 1,\ (\phi = 0^\circ)$

$$P_1 - P_2 = 20 \cos\phi = 20 \times 1 = 20$$

$$P_1 - P_2 = \frac{20}{\sqrt{3}} \sin\phi = 20 \times 0 = 0$$

therefore, $\quad P_1 = P_2 = 10 \text{ kW}$

(*b*) $\quad \cos\phi = 0.866$ (lagging), $\phi = 30^\circ$ lagging

$$P_1 + P_2 = 20 \cos\phi = 20 \times 0.866 = 17.32$$

$$P_1 - P_2 = \frac{20}{\sqrt{3}} \sin\phi = \frac{20}{\sqrt{3}} \sin 30° = 5.773$$

therefore, $P_1 = 11.546$ kW , $P_2 = 5.773$ kW

(*c*) $\cos\phi = 0.5$ leading, ($\phi = 60°$ leading) since ϕ is leading, we shall take $\phi = -60°$.

$$P_1 + P_2 = 20 \cos\phi = 20 \cos(-60°) = 10 \text{ kW}$$

$$P_1 - P_2 = \frac{20}{\sqrt{3}} \sin\phi = \frac{20}{\sqrt{3}} \sin(-60°) = -10 \text{ kW}$$

therefore, $P_1 = 0, \quad P_2 = 10$ kW

(*d*) $\cos\phi = 0. \ (\phi = 90°)$

$$P_1 + P_2 = 20 \cos\phi = 20 \cos 90° = 0$$

$$P_1 - P_2 = \frac{20}{\sqrt{3}} \sin\phi = \frac{20}{\sqrt{3}} \sin 90° = 11.547$$

therefore, $P_1 = 5.7736$ kW ,

$P_2 = -5.7736$ kW

The maximum reading on either wattmeter occurs when the current through the current coil of the wattmeter is in phase with the voltage across the voltage coil. This occurs for wattmeter W_1 at a phase angle of 30° lagging, and for W_2 at a phase angle of 30° leading. In either case the maximum reading is 11.546 kW as seen in solution (*b*).

■ **EXAMPLE 9.12** Two wattmeters in a three-phase, three-wire system with line voltage of 400 V read 1200 W and 400 W. What is the impedance per phase of the delta connected load.

SOLUTION :

$$V_L = 400 \text{ V}, P_1 = 1200 \text{ W}, P_2 = 400 \text{ W}$$

$$\tan\phi = \sqrt{3}\,\frac{P_1 - P_2}{P_1 + P_2}$$

$$= \sqrt{3} \times \frac{1200 - 400}{1200 + 400} = \frac{\sqrt{3}}{2}$$

therefore, $\phi = 40.89°, \cos\phi = 0.7559$

Total power

$$P_{3\phi} = 3\,V_p I_p \cos\phi$$

$$I_p = \frac{V_p}{Z_p}$$

$$P_{3\phi} = 3V_p \frac{V_p}{Z_p} \cos\phi$$

$$Z_p = \frac{3\,V_p^2 \cos\phi}{P_{3\phi}}$$

For a delta-connected load

$$V_p = V_L = 400 \text{ V}$$

$$P_{3\phi} = P_1 + P_2 = 1200 + 400 = 1600 \text{ W}$$

$$Z_p = \frac{3 \times (400)^2 \times 0.7559}{1600} = 226.77$$

$$\mathbf{Z}_p = Z_p \underline{/\ \phi} = 226.77 \underline{/\ 40.89^\circ}\ \Omega$$

■ **EXAMPLE 9.13** A three-phase, 400 V has a delta-connected load with

$$\mathbf{Z}_{AB} = 20 \underline{/\ 0^\circ}\ \Omega,\ \mathbf{Z}_{BC} = 20 \underline{/\ 30^\circ}\ \Omega,$$

$$\mathbf{Z}_{CA} = 20 \underline{/\ -30^\circ}\ \Omega$$

Two wattmeters are connected in the circuit to measure the power. Determine the wattmeter readings if the current coils are in the lines (*a*) *A* and *B*, (*b*) *B* and *C*, and (*c*) *C* and *A*. The phase sequence is *ABC*. Draw the connections of the wattmeters for the three cases. Check the sum of two wattmeter indications against total power.

SOLUTION : Taking $\mathbf{V}_{AB}$ as the reference phasor, we have

$$\mathbf{V}_{AB} = 400 \underline{/\ 0^\circ} \text{ V},\ \mathbf{V}_{BC} = 400 \underline{/\ -120^\circ} \text{ A},$$

$$\mathbf{V}_{CA} = 400 \underline{/\ 120^\circ} \text{ V}$$

The phase currents $\mathbf{I}_{AB}$, $\mathbf{I}_{BC}$ and $\mathbf{I}_{CA}$ are found as follows :

$$\mathbf{I}_{AB} = \frac{\mathbf{V}_{AB}}{\mathbf{Z}_{AB}} = \frac{400 \underline{/\ 0^\circ}}{20 \underline{/\ 0^\circ}} = 20 \underline{/\ 0^\circ} \text{ A} = (20 + j0) \text{ A}$$

$$\mathbf{I}_{BC} = \frac{\mathbf{V}_{BC}}{\mathbf{Z}_{BC}} = \frac{400 \underline{/\ -120^\circ}}{20 \underline{/\ 30^\circ}} = 20 \underline{/\ -150^\circ} \text{ A}$$

$$= (-17.32 - j\,10) \text{ A}$$

$$\mathbf{I}_{CA} = \frac{\mathbf{V}_{CA}}{\mathbf{Z}_{CA}} = \frac{400 \underline{/\ 120^\circ}}{20 \underline{/\ -30^\circ}} = 20 \underline{/\ 150^\circ} \text{ A}$$

$$= (-17.32 + j\,10) \text{ A}$$

The line currents $\mathbf{I}_A$, $\mathbf{I}_B$ and $\mathbf{I}_C$ are calculated by applying KCL at nodes *A*, *B* and *C* of the load.

$$\mathbf{I}_A = \mathbf{I}_{AB} - \mathbf{I}_{CA} = 37.2 - j10 = 38.63 \underline{/\ -15^\circ} \text{ A}$$

$$\mathbf{I}_B = \mathbf{I}_{BA} - \mathbf{I}_{AB} = -37.2 - j10 = 38.63 \underline{/\ -165^\circ} \text{ A}$$

$$\mathbf{I}_C = \mathbf{I}_{CA} - \mathbf{I}_{BC} = 0 + j20 = 20 \underline{/\ 90^\circ} \text{ A}$$

(*a*) When the current coils of the wattmeters are in the lines *A* and *B* (Fig. 9.26*a*), the voltage coil of wattmeter W_1 is across the lines *A* and *C* and the voltage coil of wattmeter W_2 is across the lines *B* and *C*.

$$P_1 = |\mathbf{V}_{AC}| \cdot |\mathbf{I}_A| \text{ (cosine of the angle between } \mathbf{V}_{AC} \text{ and } \mathbf{I}_A)$$

$$= 400 \times 38.63 \cos [-60° - (-15°)] = 10926 \text{ W}$$

Alternatively

$$P_1 = Re\ [\mathbf{V}_{AC}\ \mathbf{I}_A^*] = Re\ [(400\ \underline{/-60°})\ (38.63\ \underline{/+15°})]$$

$$= Re\ [400 \times 38.63\ \underline{/-45°}]$$

$$= 10\ 926 \text{ W}$$

P_1 can be also be found as follows,

$$P_1 = Re\ [\mathbf{V}_{AC}^*\ \mathbf{I}_A] = Re\ [(400\ \underline{/+60°})\ (38.63\ \underline{/-15°})]$$

$$= Re\ [400 \times 38.63\ \underline{/45°}]$$

$$= Re\ [400 \times 38.63 \cos 45°] = 10\ 926 \text{ W}$$

P_2 can also be calculated in a similar manner.

$$P_2 = Re\ [\mathbf{V}_{BC}^*\ \mathbf{I}_B]$$

$$= Re\ [(400\ \underline{/120°})\ (38.63\ \underline{/-165°})]$$

$$= 400 \times 38.63 \cos (-45°) = 10\ 926 \text{ W}$$

$$P_1 + P_2 = 10\ 926 + 10\ 926 = 21\ 852 \text{ W}$$

(*b*) When the current coils of the wattmeters are in the lines B and C (Fig. 9.26b), the voltage coil of wattmeter W_1 is across the lines B and A and the voltage coil of wattmeter W_2 is across the lines C and A.

$$P_1 = Re\ [\mathbf{V}_{BA}^*\ \mathbf{I}_B]$$

$$= Re\ [(400\ \underline{/180°})\ (38.63\ \underline{/-165°})]$$

$$= Re\ [400 \times 38.63\ \underline{/15°}] = 14925 \text{ W}$$

$$P_2 = Re\ [\mathbf{V}_{CA}^*\ \mathbf{I}_C]$$

$$= Re\ [(400\ \underline{/-120°})\ (20\ \underline{/90°})]$$

$$= 400 \times 20 \cos 30° = 6928 \text{ W}$$

$$P_1 + P_2 = 14\ 925 + 6928 = 21\ 835 \text{ W}$$

(*c*) When the current coils of the wattmeters are in the lines C and A (Fig. 9.26c), the voltage coil of wattmeter W_1 is across the lines C and B, and the voltage coil of wattmeter W_2 is across the lines A and B.

$$P_1 = Re\ [\mathbf{V}_{CB}^*\ \mathbf{I}_C]$$

$$= Re\ [(-400\ \underline{/120°})\ (20\ \underline{/90°})]$$

$$= -400 \times 20 \cos 210°$$

$$= 6928 \text{ W}$$

$$P_2 = Re\,[\mathbf{V}_{AB}^*\,\mathbf{I}_A]$$

$$= Re\,[(400\ \underline{/\ 0^\circ})\,(38.63\ \underline{/\ -15^\circ})]$$

$$= 400 \times 38.63 \times \cos(-15^\circ) = 14\ 925\ \text{W}$$

$$P_1 + P_2 = 6928 + 14\ 925 = 21\ 853\ \text{W}$$

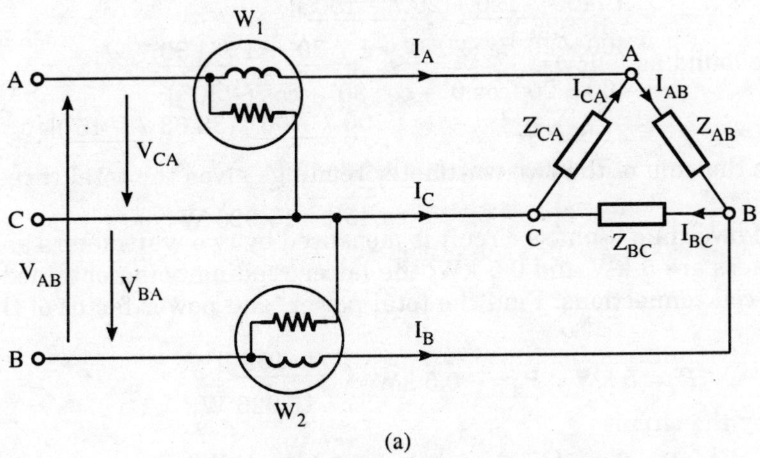

(a)

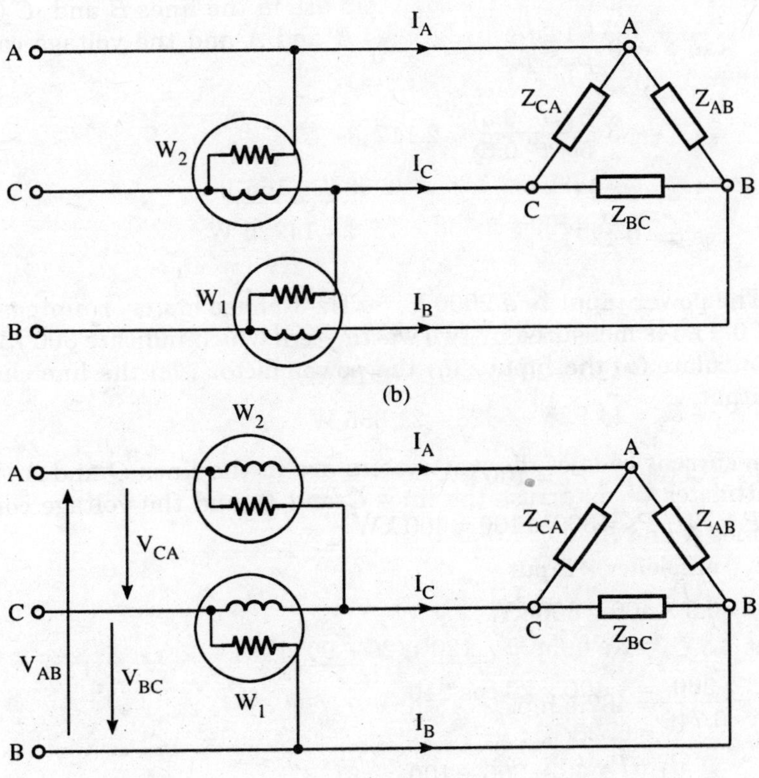

(b)

(c)

Fig. 9.26

It is to be noted that for all the three connections (*a*), (*b*) and (*c*), $P_1 + P_2$ remains same.

Total power consumed by the three-phase load

$$P_{3\phi} = Re\,[\mathbf{V}_{AB}\,\mathbf{I}^*_{AB} + \mathbf{V}_{BC}\,\mathbf{I}^*_{BC} + \mathbf{V}_{CA}\,\mathbf{I}^*_{CA}]$$

$$= Re\,[(400\,\underline{/\,0^\circ})\,(20\,\underline{/-0^\circ}) + (400\,\underline{/-120^\circ})\,(20\,\underline{/\,150^\circ}) + (400\,\underline{/\,120^\circ})\,(20\,\underline{/-150^\circ})]$$

$$= 400 \times 20\,Re\,(1\,\underline{/\,0^\circ} + 1\,\underline{/\,30^\circ} + 1\,\underline{/-30^\circ})$$

$$= 400 \times 20\,[\cos 0^\circ + \cos 30^\circ + \cos(-30^\circ)]$$

$$= 21\,856 \text{ W}$$

It is observed that the sum of the two wattmeter readings gives the total three-phase power.

■ **EXAMPLE 9.14** Power in a 3-phase circuit is measured by two wattmeters and the readings of the wattmeters are 5 kW and 0.5 kW, the latter reading being obtained after reversal of the current-coil connections. Find the total power, and power factor of the circuit.

SOLUTION :

$$P_1 = 5 \text{ kW}, \quad P_2 = -0.5 \text{ kW}$$

Power consumed by the circuit

$$P = P_1 + P_2 = 5 + (-0.5) = 4.5 \text{ kW}$$

$$\tan\phi = \sqrt{3}\,\frac{P_1 - P_2}{P_1 + P_2}$$

$$= \sqrt{3}\,\frac{5 - (-0.5)}{5 + (-0.5)} = 2.117$$

$$\phi = 64.71^\circ$$

$$\cos\phi = 0.4271$$

■ **EXAMPLE 9.15** The power input to a 2000 V, 50 Hz 3-phase motor, running on full load at an efficiency of 0.9 pu is measured by two wattmeters which indicate 300 kW and 100 kW respectively. Calculate (*a*) the input ; (*b*) the power factor ; (*c*) the line current ; (*d*) the horse-power output.

SOLUTION :

$$P_1 = 300 \text{ kW}, \; P_2 = 100 \text{ kW}$$

Power input $P = P_1 + P_2 = 300 + 100 = 400$ kW

Power output = efficiency × input

$$= 0.9 \times 400 = 360 \text{ kW}$$

Horse-power output

$$= \frac{360}{0.746} = 482.5 \text{ h.p.}$$

$$\tan\phi = \sqrt{3}\,\frac{P_1 - P_2}{P_1 + P_2} = \sqrt{3}\,\frac{300 - 100}{300 + 100} = \frac{\sqrt{3}}{2}$$

$$\cos\phi = 0.756$$

$$P = \sqrt{3}\, V_l I_l \cos\phi$$

therefore, $$I_l = \frac{P}{\sqrt{3}\, V_l \cos\phi} = \frac{400 \times 10^3}{\sqrt{3} \times 2000 \times 0.756} = 152.7 \text{ A}$$

■ **EXAMPLE 9.16** Each phase of a 3-phase, delta-connected load consists of an impedances $\mathbf{Z}_p = 40 \angle 30^\circ\ \Omega$. The line voltage is 400 V at 50 Hz. Determine

(*a*) the power factor of the load

(*b*) the phase and line currents,

(*c*) the power consumed by each phase impedance,

(*d*) the readings of the the two wattmeters used to measure total load power.

SOLUTION : Impedance per phase

$$\mathbf{Z}_p = 40 \angle 30^\circ\ \Omega$$

$$= 40 \cos 30^\circ + j\,40 \sin 30^\circ$$

Also $$\mathbf{Z}_p = R_p + j X_p$$

$\therefore$ resistance per phase , $R_p = 40 \cos 30^\circ = 34.64\ \Omega$

$$\phi = 30^\circ$$

(*a*) Power factor of the load

$$\cos\phi = \cos 30^\circ = 0.866 \text{ (lag)}$$

(*b*) For delta-connected system

$$V_p = V_L = 400 \text{ V}$$

Phase current, $$I_p = \frac{V_p}{Z_p} = \frac{400}{40} = 10 \text{ A}$$

Line current $$I_L = \sqrt{3}\, I_p = \sqrt{3} \times 10 = 17.32 \text{ A}$$

(*c*) Power consumed by each phase impedance

$$P_p = I_p^2 R_p = (10)^2 \times 34.64 = 3464 \text{ W}$$

(*d*) Total power consumed by the load

$$P = 3\, P_p = 3 \times 3464 = 10\,392 \text{ W}$$

Alternatively

$$P = \sqrt{3}\, V_l I_L \cos\phi$$

$$= \sqrt{3} \times 400 \times 17.32 \times 0.866 = 10\,392 \text{ W}$$

Now $$P_1 + P_2 = P = 10\,392 \text{ W}$$

$$\tan\phi = \sqrt{3}\, \frac{P_1 - P_2}{P_1 + P_2}$$

$$\tan 30^\circ = \sqrt{3}\,\frac{P_1 - P_2}{10\ 392}$$

$$P_1 - P_2 = \frac{10\ 392}{3} = 3464$$

$$P_1 + P_2 = 10\ 392$$

therefore, $P_1 = 6928$ W , $P_2 = 3464$ W

■ **EXAMPLE 9.17** The input power to a 3-phase induction motor running on a 400 V, 50 Hz supply was measured by two-wattmeter method. The readings were 10 kW and – 2 kW. Calculate (*a*) the total input power, (*b*) the power factor, and (*c*) the line current.

What is the load power factor when both wattmeters read identical ?

SOLUTION :

(*a*) $$P = P_1 + P_2 = 10 + (-2) = 8 \text{ kW}$$

(*b*) $$\tan \phi = \sqrt{3}\,\frac{P_1 - P_2}{P_1 + P_2}$$

$$= \sqrt{3}\,\frac{10 - (-2)}{10 + (-2)} = 2.598$$

therefore, $$\cos \phi = \cos(\tan^{-1} 2.598)$$

$$= 0.3592 \text{ (lagging)}$$

(*c*) $$P = \sqrt{3}\, V_L I_L \cos \phi$$

$$I_L = \frac{P}{\sqrt{3}\, V_L \cos \phi} = \frac{8 \times 10^3}{\sqrt{3} \times 400 \times 0.3592} = 32.15 \text{ A}$$

When $$P_1 = P_2$$

$$\tan \phi = \sqrt{3}\,\frac{P_1 - P_2}{P_1 + P_2} = 0$$

therefore, $$\phi = 0^\circ , \cos \phi = 1$$

■ **EXAMPLE 9.18** The power in a 3-phase circuit is measured by two wattmeters. If the input power is 15 kW and the power factor is 0.72 (lagging), (*a*) what will be the reading of each wattmeter? (*b*) For what power factor will one of the wattmeters read zero?

SOLUTION :

$$\cos \phi = 0.72 , \tan \phi = 0.9638$$

$$\tan \phi = \sqrt{3}\,\frac{P_1 - P_2}{P_1 + P_2}$$

Here $$P_1 + P_2 = 15 \qquad \text{(E 9.18.1)}$$

therefore, $$0.9638 = \sqrt{3}\,\frac{P_1 - P_2}{15}$$

$$P_1 - P_2 = \frac{15 \times 0.9638}{\sqrt{3}} = 8.113 \qquad \text{(E 9.18.2)}$$

Adding Eqs. (E 9.18.1) and (E 9.18.2)

$$2\,P_1 = 23.11\,,\; P_1 = 11.55 \text{ kW}$$

$$P_2 = 15 - P_1 = 15 - 11.55 = 3.45 \text{ kW}$$

(*b*) If $P_2 = 0$,

$$\tan\phi = \sqrt{3}\,\frac{P_1 - 0}{P_1 + 0} = \sqrt{3} = \tan 60^\circ$$

therefore, $\phi = 60^\circ\,,\; \cos\phi = \cos 60^\circ = 0.5$

■ **EXAMPLE 9.19** The readings of the two wattmeters to measure power in a capacitive load are – 3 kW and 8 kW respectively. Calculate the input power and the power factor of the load. The phase sequence is ABC.

SOLUTION : As the load is capacitive (leading p.f.), the wattmeter connected in the leading phase *A* will give less value.

therefore, $P_1 = -3 \text{ kW}\,,\; P_2 = +8 \text{ kW}$

$$\tan\phi = \sqrt{3}\,\frac{\text{greater reading} - \text{smaller reading}}{\text{sum of the wattmeter readings}}$$

$$= \sqrt{3}\,\frac{8-(-3)}{8+(-3)} = 3.81$$

$$\cos\phi = \cos(\tan^{-1} 3.81) = 0.2538 \text{ (leading)}$$

■ **EXAMPLE 9.20** In a balanced 3-phase system, power is measured by two wattmeters and the ratio of two wattmeter readings is 2 : 1. Determine the power factor of the system.

SOLUTION :

Here $P_1 = 2\,P_2$

$$\tan\phi = \sqrt{3}\,\frac{P_1 - P_2}{P_1 + P_2} = \sqrt{3}\,\frac{2\,P_2 - P_2}{2\,P_2 + P_2}$$

$$= \frac{1}{\sqrt{3}} = \tan 30^\circ$$

therefore, $\phi = 30^\circ\,,\; \cos\phi = \cos 30^\circ = 0.866 \text{ (lag)}$

■ **EXAMPLE 9.21** Three indentical impedances, each having a resistance *R* and a capacitance *C* are connected in star across a 3-phase, 400 V, 50 Hz supply. The power input to the load is measured by the 2-wattmeter method. The readings of the two wattmeters are 800 W and 2000 W. Determine

(*a*) the power factor of the circuit,

(*b*) the line current and

(*c*) the resistance and capacitance of each phase.

SOLUTION : Since each phase has a resistance and a capacitance, the power factor will be leading.

For leading *pf*

$$\tan\phi = \sqrt{3}\,\frac{\text{greater reading} - \text{smaller reading}}{\text{sum of the two readings}}$$

$$= \sqrt{3}\,\frac{2000 - 800}{2000 + 800} = 0.723$$

$$\phi = 36.59°$$

$$\cos\phi = 0.8029 \text{ (leading)}$$

$$P_{3\phi} = P_1 + P_2 = 800 + 2000 = 2800 \text{ W}$$

For a star-connected system,

$$I_l = I_p = \frac{V_p}{Z_p} \text{ and } V_p = \frac{V_l}{\sqrt{3}}$$

$$P_{3\phi} = 3V_p \,.\, I_p \cos\phi = 3\,V_p \frac{V_p}{Z_p} \cos\phi$$

$$Z_p = \frac{3\,V_p^2 \cos\phi}{P_{3\phi}}$$

$$= \frac{3\left(\frac{V_l}{\sqrt{3}}\right)^2 \cos\phi}{P_{3\phi}}$$

$$= \frac{V_l^2 \cos\phi}{P_{3\phi}}$$

$$= \frac{(400)^2 \times 0.8029}{2800} = 45.88\ \Omega$$

$$\mathbf{Z}_p = Z_p \underline{/\ \phi}$$

$$= 45.88 \underline{/\ 36.59°}\ \Omega$$

$$R_p + jX_{cp} = 36.84 + j\,27.35$$

Therefore, resistance per phase $R_p = 36.84\ \Omega$. Capacitive reactance per phase

$$X_{cp} = 27.35\ \Omega$$

$$\frac{1}{2\pi f C_p} = 27.35$$

$$C_p = \frac{1}{2\pi \times 50 \times 27.35} = 116.38 \times 10^{-6} \text{ F}$$

$$= 116.38\ \mu F$$

■ **EXAMPLE 9.22** Two wattmeters are used to measure the input to a 3-phase, 400 V balanced load taking 25 A at a power of 0.8 (lag). Calculate

(*a*) the total power consumed,

(*b*) the readings of the wattmeters when 2-wattmeter method is used to measure the power input to the load.

SOLUTION :

(*a*)
$$P_{3\phi} = \sqrt{3}\, V_L I_L \cos\phi$$
$$= \sqrt{3} \times 400 \times 25 \times 0.8 = 13\,856 \text{ W}$$

(*b*)
$$\cos\phi = 0.8\,,\ \phi = 36.87^\circ$$

Reading of wattmeter 1

$$P_1 = V_L I_L \cos(30^\circ - \phi)$$
$$= 400 \times 25 \cos(30^\circ - 36.87^\circ) = 9928 \text{ W}$$

Reading of wattmeter 2

$$P_2 = V_L I_L \cos(30^\circ + \phi)$$
$$= 400 \times 25 \cos(30^\circ + 36.87^\circ) = 3928 \text{ W}$$

Check $\quad P_1 + P_2 = 9928 + 3928 = 13\,856$ W

Alternatively

$$P_1 + P_2 = P_{3\phi} = \sqrt{3}\, V_L I_L \cos\phi = 13\,856$$

$$\tan\phi = \sqrt{3}\,\frac{P_1 - P_2}{P_1 + P_2}$$

$$\tan 36.87^\circ = \sqrt{3}\,\frac{P_1 - P_2}{13\,856}$$

$$P_1 - P_2 = \frac{13\,856\,(\tan 36.87^\circ)}{\sqrt{3}} = 6000$$

$$P_1 + P_2 = 13\,856$$

therefore, $\quad P_1 = 9928$ W , $P_2 = 3928$ W

■ **EXAMPLE 9.23** A 3-phase, 400 V, 50 Hz supply is connected to a 3-phase star-connected induction motor and 3-phase star-connected synchronous motor. The impedance of each phase of induction motor is $(3 + j\,4)\Omega$. The synchronous motor is over excited and it draws a current of 100 A at 0.85 leading power factor. Two wattmeters are connected in the usual manner to measure power drawn by the two motors. Determine

(*a*) the over all power factor.

and (*b*) the readings of the two wattmeters

SOLUTION :

Phase voltage $V_p = \dfrac{V_l}{\sqrt{3}} = \dfrac{400}{\sqrt{3}} = 230.9$ V

Impedance per phase of induction motor

$$\mathbf{Z}_{ip} = 3 + j\,4 = 5 \underline{/\ 53.13^\circ}\ \Omega$$

Phase current drawn by induction motor

$$\mathbf{I}_{ip} = \frac{\mathbf{V}_p}{\mathbf{Z}_{ip}} = \frac{230.9 \angle 0^\circ}{5 \angle 53.13^\circ} = 46.18 \angle -53.15^\circ \text{ A}$$

Line current of the induction motor

$$\mathbf{I}_{il} = \mathbf{I}_{ip} = 46.18 \angle -53.13^\circ \text{ A}$$

Line current drawn by the synchronous motor

$$\mathbf{I}_{sl} = 100 \angle \cos^{-1} 0.85 = 100 \angle 31.79^\circ \text{ A}$$

Total line current

$$\mathbf{I}_l = \mathbf{I}_{il} + \mathbf{I}_{sl}$$

$$= 46.18 \angle -53.13^\circ + 100 \angle 31.79^\circ$$

$$= 27.70 - j\,36.94 + 85 + j\,52.68$$

$$= 112.7 + j\,15.74 = 113.79 \angle 7.95^\circ \text{ A}$$

Here $\phi = +\,7.95^\circ$

Over all p.f. $= \cos\phi = 0.9904$ (leading)

For leading power factor of the load

$$P_1 = V_l\, I_l \cos(30^\circ + \phi)$$

$$= 400 \times 113.79 \cos(30^\circ + 7.95^\circ) = 35\,891 \text{ W}$$

$$P_2 = V_L\, I_L \cos(30^\circ - \phi)$$

$$= 400 \times 113.79 \cos(30^\circ - 7.95^\circ) = 42\,186 \text{ W}$$

■ **EXAMPLE 9.24** A 400 V, 3-phase motor has an output of 50 h.p. and operates at a power factor of 0.86 with an efficiency of 90%. Calculate the reading on each of two wattmeters connected to measure the input. What is the full-load line current ?

SOLUTION : Output of the motor

$$= 50 \text{ h.p.} = 50 \times 746 = 37\,300 \text{ W}$$

Input to motor

$$P = \frac{\text{output of the motor}}{\text{efficiency}} = \frac{37\,300}{0.9} = 41\,444 \text{ W}$$

therefore, $P = P_1 + P_2 = 41\,444$ (E 9.24.1)

Since $\cos\phi = 0.86$, $\tan\phi = 0.5934$

But
$$\tan\phi = \sqrt{3}\,\frac{P_1 - P_2}{P_1 + P_2}$$

therefore,
$$0.5934 = \sqrt{3}\,\frac{P_1 - P_2}{41\,444}$$

$$P_1 - P_2 = \frac{41\,444 \times 0.5934}{\sqrt{3}} = 18\,630 \qquad \text{(E 9.24.2)}$$

Adding Eqs. (E 9.24.1) and (E 9.24.2)

$$2P_1 = 41\,444 + 18\,630\ ,\ P_1 = 30\,037\ \text{W}$$

therefore, $$P_2 = 41\,444 - 30\,037 = 11\,407\ \text{W}$$

$$P = \sqrt{3}\ V_l\, I_l \cos\phi$$

$$I_l = \frac{P}{\sqrt{3}\ V_l \cos\phi} = \frac{41\,444}{\sqrt{3} \times 400 \times 0.86} = 69.56\ \text{A}$$

■ **EXAMPLE 9.25** Three equal impedances, each $(20 + j\,30)\ \Omega$ are connected in star across a 3-phase, 400 V, 50–Hz supply. Determine

(*a*) the phase and line currents,

(*b*) the power factor of the load,

(*c*) the readings of the wattmeters

when the 2-wattmeter method is used to measure power input to the load.

SOLUTION :

$$\mathbf{Z}_p = 20 + j\,30 = 36.05\ \underline{/\ 56.31^\circ}\ \Omega$$

$$\phi = 56.31^\circ\ ,\ \cos\phi = 0.5547\ \text{(lagging)}$$

$$\tan\phi = 1.5$$

$$\mathbf{I}_p = \frac{\mathbf{V}_p}{\mathbf{Z}_p} = \frac{(400/\sqrt{3})\ \underline{/\ 0^\circ}}{36.05\ \underline{/\ 56.31^\circ}} = 6.406\ \underline{/\ -56.31^\circ}\ \text{A}$$

$$I_L = I_p = 6.406\ \text{A}$$

Total power consumed

$$= 3I_p^2 R_p = 3 \times (6.406)^2 \times 20 = 2462\ \text{W}$$

$$\tan\phi = \sqrt{3}\ \frac{P_1 - P_2}{P_1 + P_2}$$

$$1.5 = \sqrt{3}\ \frac{P_1 - P_2}{2462}$$

$$P_1 - P_2 = \frac{1.5 \times 2462}{\sqrt{3}} = 2132\ \text{W}$$

$$P_1 + P_2 = 2462\ \text{W}$$

therefore, $$P_1 = 2297\ \text{W}\ ,\ P_2 = 165\ \text{W}$$

■ **EXAMPLE 9.26** Three equal impedances, each having a resistance R and inductance L in series are connected in star across a 3-phase, 400 V , 50-Hz supply. The power input to the load is measured by two-wattmeter method. The readings of the two wattmeters are 5kW and 2kW. Determine the resistance and inductance of each impedance.

SOLUTION : Total input power $= P_1 + P_2 = 5 + 2 = 7\text{kW}$

$$\tan\phi = \sqrt{3}\ \frac{P_1 - P_2}{P_1 + P_2} = \sqrt{3}\ \frac{5 - 2}{5 + 2} = 0.7423$$

$$\cos\phi = 0.8029 \text{ (lagging)}$$

$$P_{3\phi} = 3\,V_p\,I_p \cos\phi$$

$$7\times10^3 = 3\,V_p\,.\,\frac{V_p}{Z_p}\cos\phi$$

$$Z_p = \frac{3\,V_p^2\cos\phi}{7\times10^3} = \frac{3\left(\frac{400}{\sqrt{3}}\right)^2\times0.8029}{7\times10^3}$$

$$= 18.35\ \Omega$$

$$R_p = Z_p\cos\phi = 18.35\times0.8029 = 14.73\ \Omega$$

$$X_{Lp} = \sqrt{Z_p^2 - R_p^2} = \sqrt{(18.35)^2 - (14.73)^2}$$

$$= 10.94\ \Omega$$

$$X_{Lp} = 2\pi f L_p$$

Inductance per phase $L_p = \dfrac{X_{Lp}}{2\pi f} = \dfrac{10.94}{2\pi\times50} = 0.0348$ H

■ **EXAMPLE 9.27** A 3-phase, 20-kW, delta-connected induction motor operates at a rated load in parallel with a star-connected load having impedance of $(10 + j15)\ \Omega$ per phase. The motor power factor is 0.82 (lagging). If the supply is 3-phase, 400 V, 50 Hz determine

(*a*) the line and phase currents in motor,

(*b*) the line and phase currents in star-connected load ,

(*c*) total line current,

(*d*) total power consumed.

SOLUTION :

(*a*)

$$P = \sqrt{3}\,V_l\,I_{lm}\cos\phi$$

$$20\times10^3 = \sqrt{3}\times400\,I_{lm}\times0.82$$

$$I_{lm} = \frac{20\times10^3}{\sqrt{3}\times400\times0.82} = 35.2\text{ A}$$

Power factor of the motor = 0.82 lagging

Line current of the motor in phasor form

$$\mathbf{I}_{lm} = |\mathbf{I}_{lm}|\ \underline{/-\cos^{-1}0.82}$$

$$= 35.2\ \underline{/-34.9^\circ} = (28.869 - j\,20.139)\text{ A}$$

Phase current of the delta-connected motor,

$$I_{pm} = \frac{I_{lm}}{\sqrt{3}} = \frac{35.2}{\sqrt{3}} = 20.32\text{ A}$$

(*b*) Impedance of the star-connected load,

$$\mathbf{Z}_{Lp} = 10 + j\,15 = 18.03\ \underline{/\,56.3^\circ}\ \Omega$$

Since the load is connected in parallel with the delta-connected motor, the phase voltage of the load = $400/\sqrt{3}$ V phase current of the star-connected load

$$= \text{line-current of the } Y - \text{connected load}$$

$$\mathbf{I}_p = \frac{\mathbf{V}_p}{\mathbf{Z}_{Lp}} = \frac{400/\sqrt{3}}{18.03 \underline{/ \ 56.3^\circ}}$$

$$= 12.8 \underline{/ - 56.3^\circ} = 7.10 - j\, 10.65 \text{ A}$$

therefore, line current of the load = $I_{lL} = 7.10 - j\, 10.65$ A

(*c*) Total line current

$$\mathbf{I}_{lT} = \mathbf{I}_{lm} + \mathbf{I}_{lL}$$

$$= 28.869 - j\, 20.139 + 7.10 - j\, 10.65$$

$$= 35.969 - j\, 30.789 = 47.35 \underline{/ - 40.56^\circ} \text{ A}$$

(*d*) Total power consumed

$$= \sqrt{3}\, \text{V}_l\, \text{I}_{lT} \cos \phi_T$$

$$= \sqrt{3} \times 400 \times 47.35 \cos (- 40.56^\circ)$$

$$= 24\,922 \text{ W} = 24.922 \text{ kW}$$

■ **EXAMPLE 9.28** The unbalanced star-connected load of Fig. 9.27 is fed from a 3-phase, 4-wire, 240V, 50 Hz supply, the phase sequence of which is RYB. Determine the currents in each of the four wires.

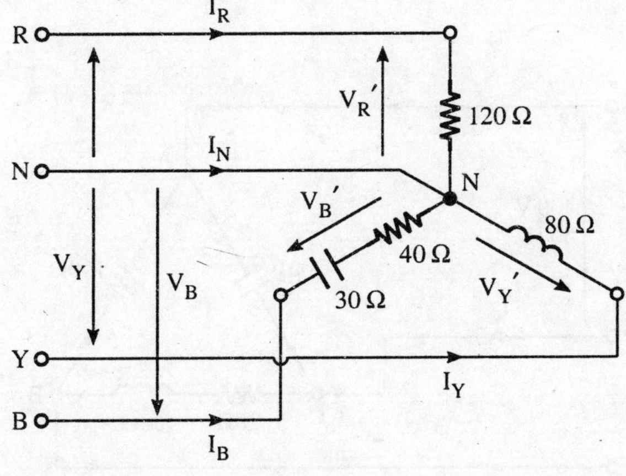

Fig. 9.27

SOLUTION :

By KVL, $\mathbf{V}_R{}' - \mathbf{V}_R$, $\mathbf{V}_Y{}' = \mathbf{V}_Y$, $\mathbf{V}_B{}' = \mathbf{V}_B$

Let $\mathbf{V}_{RN} = 240 \underline{/ \ 0^\circ}$ V

therefore, $\mathbf{V}_{YN} = 240 \underline{/ - 120^\circ}$ V , $\mathbf{V}_{BN} = 240 \underline{/ + 120^\circ}$ V

$$\mathbf{I}_R = \frac{\mathbf{V}_R'}{120} = \frac{240 \,\underline{/\, 0^\circ}}{120} = 2 \,\underline{/\, 0^\circ} \text{ A}$$

$$\mathbf{I}_Y = \frac{\mathbf{V}_Y'}{j\,80} = \frac{240 \,\underline{/\, -120^\circ}}{80 \,\underline{/\, 90^\circ}} = 3 \,\underline{/\, -210^\circ} \text{ A}$$

$$\mathbf{I}_B = \frac{\mathbf{V}_B'}{40 - j\,30} = \frac{240 \,\underline{/\, +120^\circ}}{50 \,\underline{/\, -36.87^\circ}} = 4.8 \,\underline{/\, 156.87^\circ} \text{ A}$$

By KCL

$$\mathbf{I}_N = \mathbf{I}_R + \mathbf{I}_Y = \mathbf{I}_B$$

$$= 2 \,\underline{/\, 0^\circ} + 3 \,\underline{/\, -210^\circ} + 4.8 \,\underline{/\, 156.87^\circ}$$

$$= 2 + j\,0 - 2.598 + j\,1.5 - 4.414 + j\,1.886$$

$$= -\,5.012 + j\,3.386 = 6.049 \,\underline{/\, 145.95^\circ} \text{ A}$$

■ **EXAMPLE 9.29** The unbalanced delta-connected load of Fig. 9.28 is fed from a 3-phase, 400 V, 50-Hz supply. Calculate the phase and line currents. Assume the phase sequence to be *A-B-C*.

SOLUTION :

$$\mathbf{V}_{AB} = 400 \,\underline{/\, 0^\circ} \text{ V}, \; \mathbf{V}_{BC} = 400 \,\underline{/\, -120^\circ} \text{ V}, \; \mathbf{V}_{CA} = 400 \,\underline{/\, +120^\circ} \text{ V}$$

$$\mathbf{Z}_{AB} = 100 + j\,0 = 100 \,\underline{/\, 0^\circ} \;\Omega$$

$$\mathbf{Z}_{BC} = 20 + j\,60 = 63.246 \,\underline{/\, 71.56^\circ} \;\Omega$$

$$\mathbf{Z}_{CA} = 0 - j\frac{1}{2\pi f C} = 0 - \frac{1}{j\,2\pi \times 50 \times 30 \times 10^{-6}} = -\,j\,106.1 = 106.1 \,\underline{/\, -90^\circ} \;\Omega$$

Phase Currents

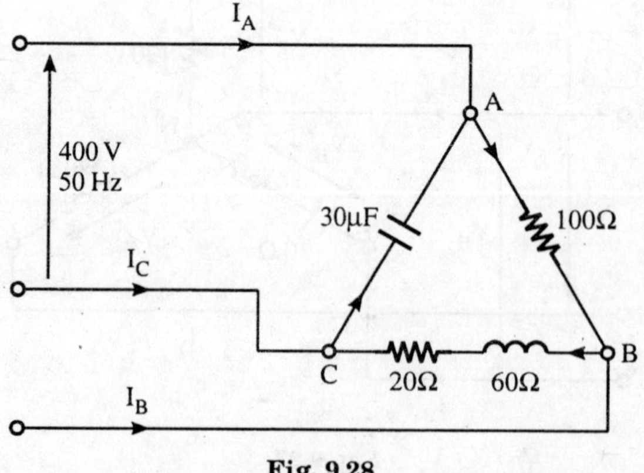

Fig. 9.28

$$\mathbf{I}_{AB} = \frac{\mathbf{V}_{AB}}{\mathbf{Z}_{AB}} = \frac{400 \,\underline{/\, 0^\circ}}{100 \,\underline{/\, 0^\circ}} = 4 \,\underline{/\, 0^\circ} \text{ A} = 4 + j\,0 \text{ A}$$

$$\mathbf{I}_{BC} = \frac{\mathbf{V}_{BC}}{\mathbf{Z}_{BC}} = \frac{400 \angle -120^\circ}{63.246 \angle 71.56^\circ} = 6.3245 \angle -191.56^\circ \text{ A}$$

$$= -6.196 + j\,1.267 \text{ A}$$

$$\mathbf{I}_{CA} = \frac{\mathbf{V}_{CA}}{\mathbf{Z}_{CA}} = \frac{400 \angle +120^\circ}{106.1 \angle -90^\circ} = 3.77 \angle 210^\circ = -3.265 - j\,1.885 \text{ A}$$

Line currents

$$\mathbf{I}_A = \mathbf{I}_{AB} - \mathbf{I}_{CA} = (4 + j\,0) - (-3.265 - j\,1.885)$$

$$= 7.265 + j\,1.885 = 7.505 \angle 14.54^\circ \text{ A}$$

$$\mathbf{I}_B = \mathbf{I}_{BC} - \mathbf{I}_{AB} = -6.196 + j\,1.267 - (4 + j\,0)$$

$$= -10.196 + j\,1.267 = 10.275 \angle 172.9^\circ \text{ A}$$

$$\mathbf{I}_C = \mathbf{I}_{CA} - \mathbf{I}_{BC} = (-3.265 - j\,1.885) - (-6.196 + j\,1.267)$$

$$= 2.931 - j\,3.152 = 4.304 \angle -47^\circ \text{ A}$$

EXERCISES

1. Show that the line-to-line voltage V_L in a three-phase star-connected system is $\sqrt{3}$ times the line-to-neutral voltage V_p.

   A balanced *Y*-connected load with impedances $16 \angle 30^\circ\ \Omega$, is connected to a three-phase, 4 –wire ABC system. The supply voltage is 400 V at 50 Hz. Determine (*a*) the line currents, (*b*) the current in the neutral wire and (*c*) the power drawn by the load.
2. A 400 V, three-phase 50 Hz supply is connected to a star-connected load. The three arms of the three-phase load each comprise an inductor of resistance 25 Ω and of inductance 0.15 H in series with a 120 μF capacitor. Calculate the line currents and the total power consumed. Assume phase sequence to be ABC.
3. Three similar impedances are connected in delta across a three-phase 400 V, 50 Hz supply. The line current is 12 A at a power factor of 0.8 lagging. Calculate the inductance and resistance of each impedance.
4. Three identical impedances of $20 \angle 60^\circ\ \Omega$ make up a balanced delta conected load. The load is connected across a 400 V, 50 Hz 3-ϕ balanced supply. Determine the line currents. Take the phase sequence as *ABC*.
5. A three-phase, 440 V, 50 Hz supply has delta-connected load, with

$$\mathbf{Z}_{AB} = 15 \angle 0^\circ\ \Omega,$$

$$\mathbf{Z}_{BC} = 25 \angle 60^\circ\ \Omega$$

$$\mathbf{Z}_{CA} = 10 \angle -30^\circ\ \Omega$$

Determine phase and line currents and draw the phasor diagram. Take the phase sequence as *ABC*.

6. A three-phase, 4-wire 400 V, 50 Hz, ABC system has a Y-connected load with

$$\mathbf{Z}_A = 8 \underline{/\ 0^\circ}\ \Omega, \quad \mathbf{Z}_B = 10 \underline{/ -45^\circ}\ \Omega$$

$$\mathbf{Z}_C = 12 \underline{/\ 30^\circ}\ \Omega$$

Determine
(a) phase currents
(b) line currents
(c) currents in the neutral.
Draw the phasor diagram

7. A balanced three-phase Y-connected source whose phase voltages are 115 V has an unbalanced Y-connected load with $Z_A = (3 + j4)\ \Omega$, $Z_B = 10\ \Omega$ and $Z_C = (5 + j12)\ \Omega$, determine the line currents and the total power taken by the load.

8. Two wattmeters are connected in a three-phase balanced circuit for measuring the total power. If the ratio of the two wattmeter readings is r, show that the power factor of the load is given by

$$\cos\phi = \frac{r+1}{2\,(r^2 - r + 1)^{1/2}}$$

9. Two wattmeters in a three-phase, three-wire system with line voltage of 480 V read 1000 W and 300 W. What is the impedance of the balanced delta-connected load.

10. A Y-connected load with

$$\mathbf{Z}_{AN} = \mathbf{Z}_{BN} = \mathbf{Z}_{CN} = 12 + j12\ \Omega$$

is connected to a 3-phase, 3-wire ABC system. Two wattmeters are connected in the usual manner to measure three-phase power. Determine the readings of the wattmeters if the line voltage is 230 V.

11. A delta connection contains three equal impedances $\mathbf{Z}_\Delta = 30 \underline{/\ 45^\circ}\ \Omega$. Find the impedances of the equivalent star connection.

12. Obtain the star-connected equivalent of the circuit shown in Fig. P 9.1.

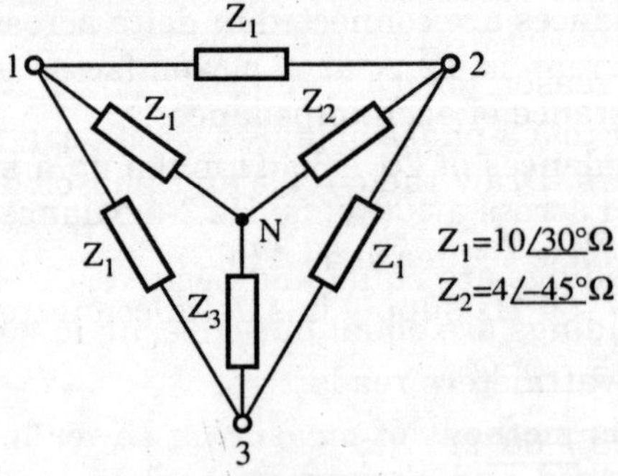

Fig. P. 9.1

13. Calculate the value of the impedance that would be measured across terminals *bc* of the network shown in Fig. P 9.2.

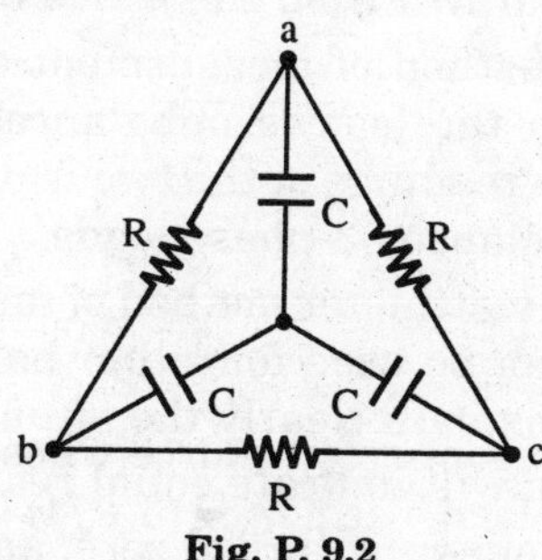

Fig. P. 9.2

14. Reduce the network shown in Fig. P 9.3 to a single equivalent impedance across terminals AB.

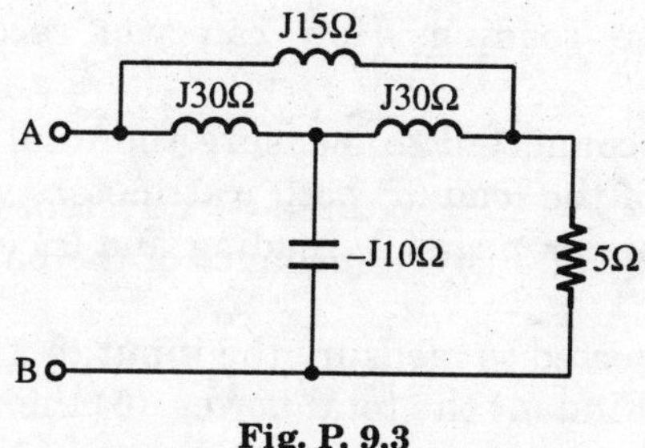

Fig. P. 9.3

15. Show that the total power and power factor in a 3-phase balanced system can be determined using two wattmeters. Draw the circuit diagram and phasor diagram.
16. Draw a circuit diagram for the measurement of power in a 3-phase balanced system by 2-wattmeter method. For a balanced load derive the expressions for the wattmeter readings.
17. The power in a 3-phase balanced system is measured by using two wattmeters. It is observed that one of the wattmeters shows zero reading. Explain with the help of phasor diagram the reason for it.
18. Prove that the power in a 3-phase balanced circuit can be deduced from the readings of the two wattmeters. Draw the circuit and phasor diagrams. Discuss the nature of power factor
    (*a*) when the two readings are equal and positive,
    (*b*) when the two readings are equal but opposite in sign,
    (*c*) when one of the wattmeters reads zero.
19. What are the various methods of measuring power in a 3-phase circuit ? Explain the 2-wattmeter method and show that these two wattmeters can read the true power in a 3-phase system.
20. In case of power measurement of a 3-phase circuit by 2-wattmeters method, the following readings were obtained in three different cases :
    (*a*) equal readings of both the wattmeters,
    (*b*) zero reading of one wattmeter, and

(*a*) negative kick in one wattmeter and after reversing the pressure coil, almost equal readings were observed in both the wattmeters.

What conclusions can be drawn from the above readings and why ?

21. Show that 2-wattmeter method of measurment of power can be used for 3-phase, 3-wire system even when the load is unbalanced. Draw the phasor diagram and obtain expressions for the readings of the two wattmeters and the total power when this method is used for balanced 3-phase loads.
22. Explain the principle of 2-wattmeter method of measuring power in 3-phase circuits. Show that this method can be used for either balanced or unbalanced load. Under balanced load condition, explain clearly the situations :
    (*a*) when both wattmeters will indicate equal readings,
    (*b*) when one of the wattmeters will read zero, and
    (*c*) when the other wattmeter will read zero.
23. In a 2-wattmeter method of power measurement in 3-phase circuits explain with relevant phasor diagrams the situations when one of the wattmeters indicates zero reading and negative reading. How can you record negative reading of the wattmeters?
24. Two wattmeters are connected to measure the total power in a balanced 3-phase circuit. At what p.f. of the load (*a*) both wattmeters will give the same reading, (*b*) one wattmeter indicates a negative reading and (*c*) one wattmeter reads twice the other.
25. Two wattmeters connected to measure the input to a 3-phase, 400V circuit indicate 3000W and 5000 W. Find (*a*) the total power, (*b*) the power factor of the circuit and (*c*) the line current.
26. Power in a 3-phase balanced circuit is measured by 2 wattmeters, and the readings of the wattmeters are 5 kW and – 0.5 kW. Find the total power and power factor of the circuit.
27. In a 3-phase circuit two wattmeters used to measure power indicate 1000 W and 500 W respectively. Find the power factor of the circuit
    (*a*) when both wattmeter readings are positive,
    (*b*) when the latter reading is obtained by reversing the current-coil connections.
28. A 3-phase, 400 V circuit has a power factor of 0.6 lagging. Two wattmeters connected to measure the power show the input to be 15 kW. Find the reading on each instrument.
29. In a 3-phase circuit, the power is measured by two wattmeters. If the total power is 25 kW and the power factor is 0.7 leading, what will be reading of each wattmeter? For what power factor will one of the wattmeters read zero.
30. Each phase of a delta-connected load consisting of 15 Ω resistance in series with a 38 mH inductance, is connected across a 3-phase, 400 V, 50 Hz supply. Determine
    (*a*) the power and line currents,
    (*b*) the power factor of the load,
    (*c*) the readings of the two wattmeters used to measure total load power.
31. Two wattmeters are connected to measure power in a 3-phase circuit. The reading of one of the wattmeters is 8 kW when the load power factor is unity. If the power

factor of the load is changed to 0.707 lagging, without changing the total input power, calculate the readings of the two wattmeters.

32. Each phase of a balanced star-connected load, consisting of a resistance of 10Ω in series with a reactance of 20 Ω, is connected across a 400 V, 50 Hz, 3-phase supply. Calculate the readings of the two wattmeters used to measure total load power.
33. The total power input to a 3-phase synchronous motor is 10 kW at 0.4 leading power factor. Determine the readings of the two wattmeters used to measure power input to the motor.
34. Two wattmeters are used to measure power input to a synchronous motor. Each wattmeter reads 30 kW. If the power factor of the motor is changed to 0.866 leading, determine the readings of the two wattmeters, the total input power remaining the same. Draw the phasor diagram for the second condition of the load.

ANSWERS

1. (a) $I_A = 14.43 \underline{/ -30^\circ}$ A
   $I_B = 14.43 \underline{/ -150^\circ}$ A
   $I_C = 14.43 \underline{/ 90^\circ}$ A
   (b) $I_N = 0$ (a) $P_{3\phi} = 8660$ W
2. $I_A = 7.132 \underline{/ -39.46^\circ}$ A
   $I_B = 7.132 \underline{/ -159.46^\circ}$ A
   $I_C = 7.132 \underline{/ 80.54^\circ}$ A
   $P_{3\phi} = 3815$ W
3. 0.11 H, 46.2 Ω
4. $I_A = 34.64 \underline{/ -90^\circ}$ A
   $I_B = 34.64 \underline{/ 150^\circ}$ A
   $I_C = 34.64 \underline{/ 30^\circ}$ A
5. $I_{AB} = 26.67 \underline{/ 0^\circ}$ A
   $I_{BC} = 16 \underline{/ -180^\circ}$ A
   $I_{CA} = 40 \underline{/ 150^\circ}$ A
   $I_A = 64.49 \underline{/ -18.06^\circ}$ A
   $I_B = 42.67 \underline{/ 180^\circ}$ A
   $I_C = 27.34 \underline{/ 132.98^\circ}$ A
6. $I_{pA} = I_{LA} = 28.87 \underline{/0^\circ}$ A
   $I_{pB} = I_{LB} = 23.09 \underline{/ -75^\circ}$ A
   $I_{pC} = I_{LC} = 19.24 \underline{/ 90^\circ}$ A
   $I_n = 34.97 \underline{/ -5.02^\circ}$ A

7. $I_A = 23 \underline{/ -53.13^\circ}$ A

   $I_B = 11.5 \underline{/ -120^\circ}$ A

   $I_C = 8.846 \underline{/ 52.62^\circ}$ A

   $P_{3\phi} = 3300.7$ W

9. $388.8 \underline{/ 43^\circ}\ \Omega$

10. $P_1 = 1738$ W, $P_2 = 466$ W

11. $\mathbf{Z}_1 = \mathbf{Z}_2 = \mathbf{Z}_3 = 10 \underline{/ 45^\circ}\ \Omega$

12. $\mathbf{Z} = 2.287 \underline{/ -3.52^\circ}\ \Omega$

13. $\mathbf{Z}_{bc} = \dfrac{2R\,(3 - j\omega CR)}{(9 + \omega^2 C^2 R^2)}$

14. $\mathbf{Z}_{AB} = (4.435 + j\,6.878)\ \Omega$

24. (*a*) 1.0, (*b*) $\phi > 60^\circ$, (*c*) $\cos\phi = 0.866$

25. (*a*) 8000 W, (*b*) 0.9177 (lagging), (*c*) 12.58 A

26. 4.5 kW, 0.4271 (lagging)

27. (*a*) 0.866 lagging, (*b*) 0.189 (lagging)

28. 13.27 kW, 1.73 kW

29. 19.863 kW, 5.137 kW, $\cos\phi = 0.5$

30. (*a*) $I_p = 20.86$ A, $I_l = 36.14$ A, (*b*) $\cos\phi = 0.7824$ (lag), (*c*) 14289 W, 2646 W.

31. $P_1 = 12.62$ kW, $P_2 = 3.38$ kW

32. $P_1 = -247$ W, $P_2 = 3448$ W

33. $P_1 = -1.61$ kW, $P_2 = 11.61$ kW

34. 20 kW, 40 kW

10

LAPLACE TRANSFORM ANALYSIS

10.1 EXPONENTIAL FUNCTION

The exponential function e^{st} is the most important function because a variety of important functions or waveforms in electrical engineering can be derived from it. This function has the property that it can be differentiated and integrated any number of times to again give an exponential function. Thus.

$$\frac{d}{dt} e^{st} = se^{st}$$

$$\int_{-\infty}^{t} e^{st}\, dt = \frac{1}{s} e^{st}$$

More generally,

$$\frac{d^n}{dt^n} (e^{st}) = s^n e^{st}$$

and

$$\underbrace{\int \ldots \int}_{n \text{ times}} e^{st}\, dt = \frac{1}{s^n} e^{st}$$

A **generalized exponential function** may be defined as

$$x(t) \stackrel{\Delta}{=} X e^{st} \qquad \text{for} \qquad -\infty < t < \infty \tag{10.1.1}$$

where X may be real or complex. The exponent **s** is complex in general and is given by

$$\mathbf{s} = \sigma + j\omega \tag{10.1.2}$$

where σ is the real part of **s**, and ω is the imaginary part. Since **s** has the dimensions of 1/time, which is frequency, it is called the **complex frequency.**

In general

$$e^{st} = e^{(\sigma + j\omega)t}$$

$$= e^{\sigma t} \, . \, e^{j\omega t} \tag{10.1.3}$$

The waveform of e^{st} depends upon the nature of s. We shall consider the following special cases :

1. *s* real

If s is real. that is, if $\omega = 0$ then the exponential function of Eq. (10.1.1) becomes

$$x(t) = Xe^{\sigma t} \tag{10.1.4}$$

The function increases exponentially for $\sigma > 0$ and decreases (decays) exponentially for $\sigma < 0$.

When $\sigma = 0$ so that $s = 0 + j0$

$$x(t) = Xe^{0} = X \tag{10.1.5}$$

This implies that $x(t)$ has a constant magnitude X. The plots of $e^{\sigma t}$ for positive, zero and negative σ are shown in Fig. 10.1

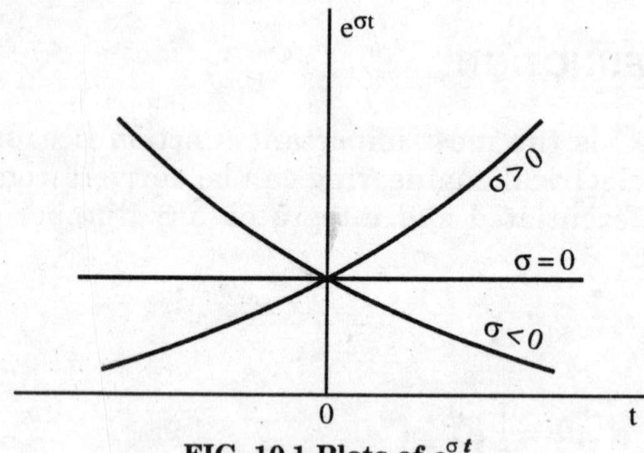

FIG. 10.1 Plots of $e^{\sigma t}$

2. *s* imaginary

If s is imaginary, that is, $\sigma = 0$, then the exponential function of Eq. (10.1.1) is given by

$$x(t) = Xe^{j\omega t} \tag{10.1.6}$$

The exponential $e^{j\omega t}$ may be considered as a unit phasor rotating with angular frequency ω as shown in Fig. 10.2

Sine $\quad e^{j\omega t} = \cos \omega t + j \sin \omega t$

$$\text{Re}\,(e^{j\omega t}) = \cos \omega t$$

and $\quad \text{Im}\,(e^{j\omega t}) = \sin \omega t$

Thus, the projection of this unit phasor onto the plane perpendicular to the real axis is $\sin \omega t$, while its projection onto the plane perpendicular to the imaginary axis is $\cos \omega t$. This is shown in Fig. 10.2.

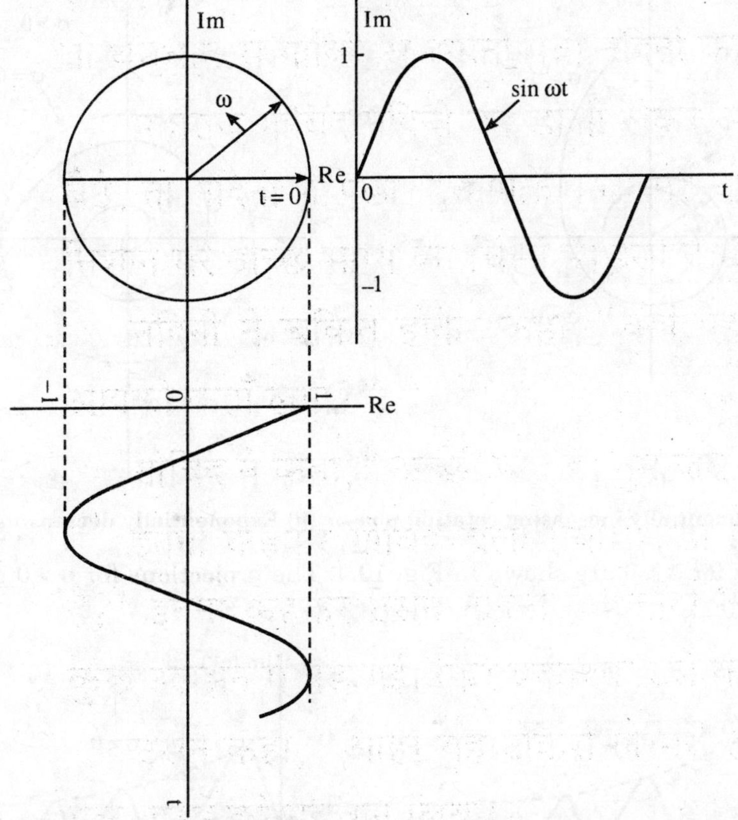

FIG. 10.2 A unit rotating phasor $e^{j\omega t}$ with its imaginary and real projections

The quantity ω is called the **real frequency** variable and $e^{j\omega t}$ is called **real frequency phasor**. It is to be noted that each of the sin ωt and cos ωt functions is a sinusoidal function of frequency ω and has a constant amplitude.

3. *s* complex

If s is complex, that is, $s = \sigma + j\omega$

$$e^{st} = e^{(\sigma + j\omega)t}$$

or

$$e^{st} = e^{\sigma t}\, e^{j\omega t} \tag{10.1.7}$$

Thus, e^{st} is the product of two functions $e^{\sigma t}$ and $e^{j\omega t}$. The function $e^{\sigma t}$ is an exponentially increasing function when $\sigma > 0$. It is an exponentially decreasing (decaying) function when $\sigma < 0$. The function $e^{j\omega t}$ is a unit rotating phasor. Therefore, the function $e^{(\sigma + j\omega)t}$ may be considered as a rotating phasor of length $e^{\sigma t}$ and angular frequency ω. The magnitude of the phasor changes with time. These phasors are shown in Fig. 10.3.

The real and imaginary projections of this phasor are

$$\text{Re}\,(e^{st}) = e^{\sigma t} \cos \omega t \tag{10.1.8}$$

$$\text{Im}\,(e^{st}) = e^{\sigma t} \sin \omega t \tag{10.1.9}$$

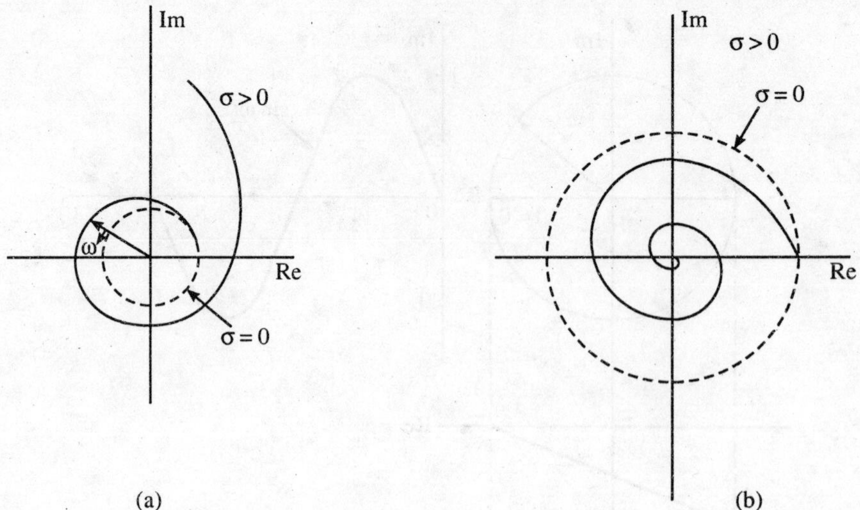

FIG. 10.3 (a) Exponentially increasing rotating phasor (b) Exponentially decreasing rotating phasor

These projections for $\sigma < 0$ are shown in Fig. 10.4. The projections for $\sigma > 0$ are shown in Fig. 10.5.

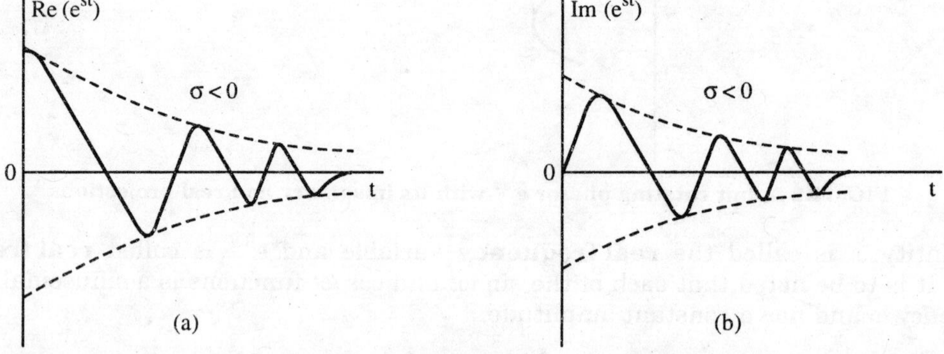

FIG. 10.4 (a) Real projection of decreasing rotating phasor. (b) Imaginary projection of decreasing rotating phasor

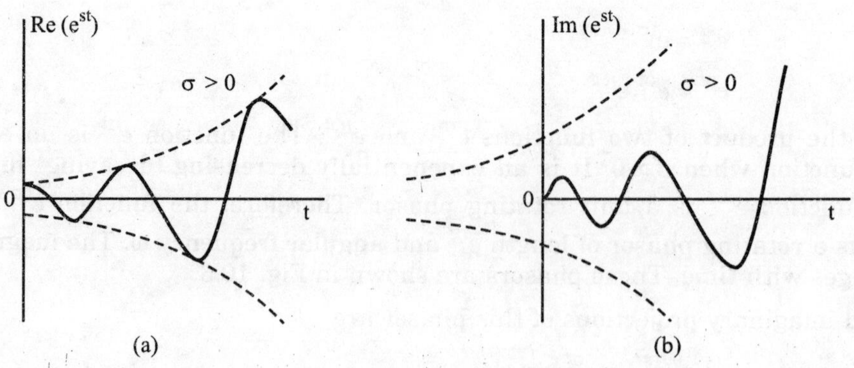

FIG. 10.5

The functions $e^{\sigma t}$ sin ωt and $e^{\sigma t}$ cos ωt oscillate at angular frequency ω. They vary in amplitude. The amplitude increases or decreases exponentially depending upon whether σ is positive ($\sigma > 0$) or negative ($\sigma < 0$). For $\sigma < 0$, the waveform is known as **damped sinusoid**. It may be represented as

$$x(t) = e^{-\sigma t} \sin \omega t \qquad (\sigma < 0) \tag{10.1.10}$$

It should be noted that the sinusoid simply oscillates between the bounds established by exponential $e^{\sigma t}$

Eq. (10.1.10) shows that, at any time t, the value of the damped sinusoidal function is the product of the decaying exponential function and the sinusoidal function, both evaluated at time t. The exponential function is shown in Fig 10.6a, and the sinusoidal function is shown in Fig. 10.6b. The damped sinusoidal function is shown in Fig. 10.6c.

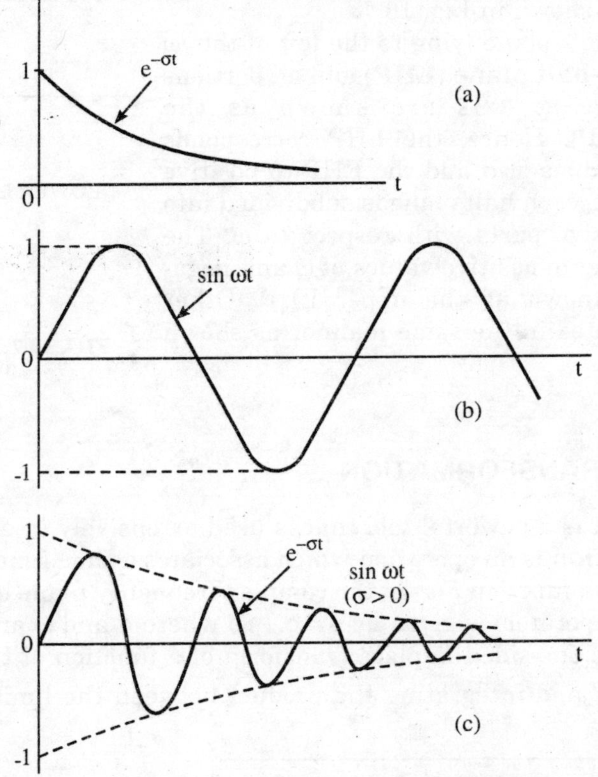

FIG. 10.6 (a) Exponential function (s) Sinsuoidal function (c) Damped sinusoidal function

The quantity e^{st} is called the **generalized frequency phasor** and **s** is known as the **generalized** or **complex frequency variable**. The imaginary part of the complex frequency is called **radian (or real) frequency**. It is denoted by ω and corresponds to oscillations. The real part of the complex frequency is σ. It is called **neper frequency**. If $\sigma > 0$, the amplitude of the generalized phasor increases without bound. If the $\sigma < 0$, the amplitude exponentially decays to zero. If $\sigma = 0$, then $s = j\omega$ and the amplitude remains constant and the function becomes $e^{j\omega t}$. Thus, it is found that $e^{j\omega t}$ is a special case of the generalized frequency phasor

e^{st}. The case $s = j\omega$ corresponds to the sinusoidal steady state. The radian frequency is expressed in radians per second (rad/s).

The neper frequency is expressed in nepers per second (Np/s).

10.1.1 Complex-frequency Plane

A **complex-frequency plane** is a rectangular coordinate system in which the horizontal and vertical axes represent the real and imaginary parts of the complex variable **s**. Since $\mathbf{s} = \sigma + j\omega$, the horizontal (or real) axis designated as σ axis and the vertical (or imaginary) axis is designated as $j\omega$ axis. The complex-frequency plane is also called the **s-plane**. To each pair of values (σ, ω) there corresponds a complex value of **s**, which can be represented as a point on a complex plane as shown in Fig. 10.7.

All portions of the s-plane lying to the left of the $j\omega$ axis are called the **left-half plane** (**LHP**) and all portions to the right of the $j\omega$ axis are known as the **right-half plane** (**RHP**). Hence, the LHP corresponds entirely to negative values of σ and the RHP to positive values of σ. Sometimes, each half plane is subdivided into its positive and negative parts with respect to ω. The quadrant corresponding to positive values of ω and negative values of σ is known as the upper LHP. Other quadrants are designated in the same manner as shown Fig. 10.7.

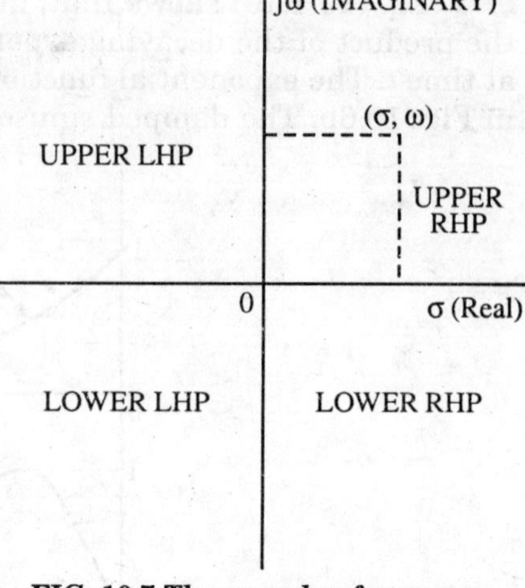

FIG. 10.7 The complex-frequency plane or *s*-plane.

10.2 LAPLACE TRANSFORMATION

The Laplace transform is a powerful tool that is used extensively in electric circuit analysis. **Laplace transformation** is an operation which associates a time function $f(t)$, defined in the interval $0 \le t \le \infty$, with a function $F(s)$ in the complex-frequency plane or **s**-plane. The complex frequency variable (**s**-operator) is given by $\mathbf{s} = \sigma + j\omega$ where σ and ω are both real quantities.

By definition, the one-sided Laplace transform of a function of time $f(t)$ is obtained by multiplying $f(t)$ by e^{-st},and integrating the product between the limits of zero and infinity, that is,

$$\pounds\,[f(t)] \overset{\Delta}{=} \int_0^{\infty} f(t)\, e^{-st}\, dt = F(s) \tag{10.2.1}$$

The symbol $\pounds\,[f(t)]$ is read as the *"Laplace transform of f(t)."* After integration, the time t is replaced with ∞ in the upper limit, and with 0 in the lower limit. Therefore, t does not appear in the result, which is a function of s as indicated by the symbol $F(s)$. Laplace transformation thus changes the time domain funcion $f(t)$ to frequency domain function $F(s)$. The function $f(t)$ and $F(s)$ constitute a **Laplace transform pair**.

The defining Eq. (10.2.1) is called **one-sided** or **unilateral Laplace transform**, because the limits in the integral are 0 and ∞. This means that all information, contained in $f(t)$ prior to $t = 0$, is ignored or considered to be zero.

In a **two-sided (bilateral) Laplace transformation,** the limits on the integration are from $-\infty$ to $+\infty$. In general, when we refer to the one-sided Laplace transform, we shall call it the Laplace transform.

The lower limit of integration in the defining Eq. (10.2.1.) is actually 0_-. This emphasizes the fact that if the time function $f(t)$ includes an impulse at the origin ($t = 0$), the defining integral in Eq. (10.2.1) includes the impulse. For functions not containing an impulse function, there is no difference between 0_- and 0. It is conventional to write the lower limit of integration as 0 rather then 0_-. Thus, the Laplace transform is written in the form of Eq. (10.2.1) although

$$F(s) = \int_{0-}^{\infty} f(t)\, e^{-st}\, dt \tag{10.2.2}$$

is understood.

The Laplace transform of a given function is denoted by a capital letter. Thus, $F(s)$, $G(s)$, and $H(s)$ denote the Laplace transforms of $f(t)$, $g(t)$, and $h(t)$, respectively.

Since Eq. (10.2.1.) is an improper integral, it exists only under certain conditions. The condition for a function $f(t)$ to be Laplace transformable is given by

$$\int_{0}^{\infty} |f(t)|\, e^{-\sigma_0 t}\, dt < \infty \tag{10.2.3}$$

for positive volues of σ_0.

Here $f(t)$ is defined in the interval $0 \le t \le \infty$. In order to satisfy the relationship given in Eq. (10.2.3), that is, to make the area under the curve finite, σ_0 is chosen sufficiently large. This condition is usually satisfied by functions of engineering interest.

10.2.1. Inverse Laplace Transformation

The operation of obtaining $f(t)$ from the Laplace transform $F(s)$ is called the *inverse Laplace transformation*. The inverse Laplace transform of $F(s)$ is given by

$$\pounds^{-1}\,[F(s)] = f(t) \tag{10.2.4}$$

The symbol $\pounds^{-1}\,[F(s)]$ is read as the *inverse Laplace transform* of $F(s)$. The inverse Laplace transform is said to transform from the frequency domain to the time domain. The inverse Laplace transform is given by

$$\pounds^{-1}\,[F(s)] = \frac{1}{2\pi j}\int_{c-j\infty}^{c+j\infty} F(s)\, e^{st}\, dt = f(t)\;,\; c < \sigma_0 \tag{10.2.5}$$

where c is a real constant that is greater than the real parts of all singularities of $F(s)$. Equation (10.2.5) represents a line integral that is to be evaluated in the complex s-plane. The integral involves complex integration, which is beyond the scope of this text. For most practical purposes Laplace transforms are obtained from Laplace transform tables. Table 10.1 gives basic Laplace transform properties and Table 10.2 shows some Laplace transform pairs.

10.3 SOME BASIC PROPERTIES OF THE LAPLACE TRANSFORMS

Property 1

The Laplace transform of a sum (or difference) of time functions is the sum (or difference) of the Laplace transforms of the individual time functions.

$$\pounds\,[f_1(t) + f_2(t) + \ldots + f_n(t)]$$
$$= \pounds\, f_1(t) + \pounds\, f_2(t) + \ldots + \pounds\, f_n(t)$$
$$= F_1(s) + F_2(s) + \ldots + F_n(s) \tag{10.3.1}$$

Proof

$$\pounds\,[f_1(t) + f_2(t) + \ldots + f_n(t)]$$
$$= \int_0^\infty [f_1(t) + f_2(t) + \ldots + f_n(t)]\, e^{-st}\, dt$$
$$= \int_0^\infty f_1(t)\, e^{-st}\, dt + \int_0^\infty f_2(t)\, e^{-st}\, dt + \ldots + \int_0^\infty f_n(t)\, e^{-st}\, dt$$
$$= \pounds\, f_1(t) + \pounds\, f_2(t) + \ldots + \pounds\, f_n(t)$$
$$= F_1(s) + F_2(s) + \ldots + F_n(s)$$

Property 2

The Laplace transform of a product of a constant and a time function is equal to the product of the constant and the Laplace transform of the time function.

If a is a constant independent of s and t, then

$$\boxed{\pounds\,[a\, f(t)] = a\, \pounds\,[f(t)] = a\, F(s)}$$

Proof,

$$\pounds\,[af(t)] = \int_0^\infty a\, f(t)\, e^{-st}\, dt$$
$$= a \int_0^\infty f(t)\, e^{-st}\, dt = a\, \pounds\, f(t) = a\, F(s)$$

Linearity Theorem

If $F_1(s)$ and $F_2(s)$ are the Laplace transforms of $f_1(t)$ and $f_2(t)$ and a_1 and a_2 are two arbitrary constants then

$$\pounds\,[a_1 f_1(t) + a_2 f_2(t)] = a_1 F_1(s) + a_2 F_2(s)$$

It should be noted that the Laplace transform of the product of two functions is not the product of the Laplace transforms, that is

$$\pounds\,[f_1(t) \,.\, f_2(t)] \neq F_1(s)\, F_2(s)$$

TABLE 10.1

Basic Laplace Transform Properties

Property	Time Domain	Frequency Domain
Independent variable	Time t	Complex frequency $s = \sigma + j\omega$
Signal representation	Waveform $f(t)$	Transform $F(s)$
Linearity property	$a_1 f_1(t) + a_2 f_2(t)$	$a_1 F_1(s) + a_2 F_2(s)$
Integration property	$\int_0^t f(t)\,dt$	$\frac{F(s)}{s}$
Differentiation property	$\frac{d\,f(t)}{dt}$	$s\,F(s) - f(0_-)$
	$\frac{d^2 f(t)}{dt^2}$	$s^2 F(s) - sf(0_-) - f'(0_-)$
	$\frac{d^3 f(t)}{dt^3}$	$s^3 F(s) - s^2 f(0_-) - sf'(0^-) - f''(0_-)$

10.4 LAPLACE TRANSFORMS OF SOME TYPICAL TIME FUNCTIONS

(1) Exponential Function

Let $f(t) = e^{at}$

By definition of Laplace transform, we have

$$£\,[e^{at}] = \int_0^\infty e^{at}\,e^{-st}\,dt$$

$$= \int_0^\infty e^{-(s-a)t}\,dt$$

$$= \frac{-1}{s-a}\left[e^{-(s-a)t}\right]_0^\infty = \frac{1}{s-a}$$

Similarly, $£\,[e^{-at}] = \dfrac{1}{s+a}$

and $£\,[A\,e^{-at}] = \dfrac{A}{s+a}$

(2) Unit Step Function

$$f(t) = u_{-1}(t) = \begin{cases} 0 & t < 0 \\ 1 & t > 1 \end{cases}$$

$$£\,u_{-1}(t) = \int_0^\infty u_{-1}(t)\,e^{-st}\,dt$$

$$= \int_0^\infty 1\, e^{-st}\, dt = -\frac{1}{s}\Big[\, e^{-st} \Big]_0^\infty$$

$$= -\frac{1}{s}\,(0 - 1) = \frac{1}{s}$$

Alternatively from the exponential function

$$£\, e^{-at} = \frac{1}{s + a}$$

Putting $a = 0$, we get

$$£\, u_{-1}(t) = \frac{1}{s}$$

(3) Sine Function

$$f(t) = \sin \omega t$$

$$£\,(\sin \omega t) = \int_0^\infty \sin \omega t\, e^{-st}\, dt$$

$$= \int_0^\infty \frac{1}{2j}\,(e^{j\omega t} - e^{-j\omega t})\, e^{-st}\, dt$$

$$= \frac{1}{2j}\left[\frac{e^{-(s - j\omega)t}}{-(s - j\omega)}\right]_0^\infty - \frac{1}{2j}\left[\frac{e^{-(s + j\omega)}}{-(s + j\omega)}\right]_0^\infty$$

$$= \frac{1}{2j}\left(\frac{1}{s - j\omega} - \frac{1}{s + j\omega}\right)$$

$$= \frac{\omega}{s^2 + \omega^2}$$

Alternative method

$$£\,(\sin \omega t) = £\left[\frac{1}{2j}\,(e^{j\omega t} - e^{-j\omega t})\right]$$

$$= \frac{1}{2j}\,(£\, e^{j\omega t} - £\, e^{-j\omega t})$$

$$= \frac{1}{2j}\left(\frac{1}{s - j\omega} - \frac{1}{s + j\omega}\right)$$

$$= \frac{\omega}{s^2 + \omega^2}$$

Similarly,

$$£\,(\cos \omega t) = £\,[\tfrac{1}{2}\,(e^{j\omega t} + e^{-j\omega t})]$$

$$= \tfrac{1}{2}\,(£\, e^{j\omega t} + £\, e^{-j\omega t})$$

$$= \frac{1}{2}\left(\frac{1}{s - j\omega} + \frac{1}{s + j\omega}\right) = \frac{s}{s^2 + \omega^2}$$

(4) Hyperbolic Sine and Cosine Functions

$$f_1(t) = \sin hbt$$

$$£\,(\sin hbt) = £\,[\tfrac{1}{2}\,(e^{bt} - e^{-bt})]$$

$$= \frac{1}{2}\left(\frac{1}{s - b} - \frac{1}{s + b}\right) = \frac{b}{s^2 - b^2}$$

$$f_2(t) = \cos hbt = \frac{1}{2}\,(e^{bt} + e^{-bt})$$

$$£\,(\cos hbt) = £\,[\tfrac{1}{2}\,(e^{bt} + e^{-bt})]$$

$$= \frac{1}{2}\left(\frac{1}{s - b} + \frac{1}{s + b}\right) = \frac{s}{s^2 - b^2}$$

(5) Damped Sine and Cosine Functions

(*a*)
$$f_1(t) = e^{-at}\sin bt = e^{-at}\left(\frac{e^{jbt} - e^{-jbt}}{2j}\right)$$

$$= \frac{1}{2j}\,[e^{(a - jb)t} - e^{-(a + jb)t}]$$

$$£\,e^{-at}\sin bt = \frac{1}{2j}\left[\frac{1}{s + (a - jb)} - \frac{1}{s + a + jb}\right]$$

$$= \frac{b}{(s + a)^2 + b^2}$$

(*b*)
$$f_2(t) = e^{-at}\cos bt = e^{-(at)}\left(\frac{e^{jbt} + e^{-jbt}}{2}\right)$$

$$= \frac{1}{2}\,[e^{-(a - jb)t} + e^{-(a + jb)t}]$$

$$£\,f_2(t) = £\,e^{-at}\cos bt$$

$$= \frac{1}{2}\left[\frac{1}{s + (a - jb)} + \frac{1}{s + (a + jb)}\right]$$

$$= \frac{s + a}{(s + a)^2 + b^2}$$

(6) Damped Hyperbolic Sine and Cosine Functions

$$£\,(e^{-at}\sin hbt) = £\left[e^{-at}\,\frac{e^{bt} - e^{-bt}}{2}\right]$$

$$= \frac{b}{(s+a)^2 - b^2}$$

$$£\,(e^{-at} \cos hbt) = £\left[e^{-at}\,\frac{e^{bt} + e^{-bt}}{2}\right]$$

$$= \frac{s+a}{(s+a)^2 - b^2}$$

10.5 LAPLACE TRANSFORM OF t^n

$$£\,t^n = \int_0^\infty t^n\, e^{-st}\, dt \tag{10.5.1.}$$

Let $st = x$, therefore, $dt = \frac{1}{s}\,dx,\ t^n = \left(\frac{x}{s}\right)^n$

$$£\,t^n = \int_0^\infty \left(\frac{x}{s}\right)^n e^{-x}\,\frac{1}{s}\,dx$$

$$= \frac{1}{s^{n+1}} \int_0^\infty x^n\, e^{-x}\, dx \tag{10.5.2}$$

The gamma function is defined as

$$\Gamma\,(n+1) = \int_0^\infty x^n\, e^{-x}\, dx \tag{10.5.3}$$

$\therefore$
$$£\,t^n = \frac{\Gamma\,(n+1)}{s^{n+1}} \tag{10.5.4}$$

Also, $\Gamma\,(n+1) = n\,\Gamma\,(n) = n\,!$

$\therefore$
$$£\,t^n = \frac{n\,!}{s^{n+1}} \tag{10.5.5}$$

If $n = 1$,

$$£\,(t) = \frac{1}{s^2} \tag{10.5.6}$$

If $n = 2$

$$£\,(t^2) = \frac{2\,!}{s^3} = \frac{2}{s^3} \tag{10.5.7}$$

■ **EXAMPLE 10.1** Find (*a*) $£\,A \sin(\omega_0\, t + \theta)$

(*b*) $£\,A \cos(\omega_0\, t + \theta)$

SOLUTION : (*a*) $£\,A \sin(\omega_0\, t + \theta) = £\,(A \sin \omega_0\, t \cos\theta + A \cos \omega_0\, t \sin\theta)$

$$= \frac{(A \cos \theta)\, \omega_0}{s^2 + \omega_0^2} + \frac{(A \sin \theta)\, s}{s^2 + \omega_0^2}$$

(b) $\pounds\, A \cos (\omega_0 t + \theta) = \pounds\, (A \cos \omega_0 t \cos \theta - A \sin \omega_0 t \sin \theta)$

$$= \frac{s\,(A \cos \theta)}{s^2 + \omega_0^2} - \frac{(A \sin \theta)\, \omega_0}{s^2 + \omega_0^2}$$

10.6 LAPLACE TRANSFORM OF THE UNIT IMPULSE FUNCTION

The unit impulse function is also called ***Dirac delta function****.* It is denoted by $u_0(t)$ or $\delta(t)$. We define the unit impulse function as follows :

$$u_0(t) = 0\,, \qquad t \neq 0$$

$$\int_{-\infty}^{\infty} u_0(t)\, dt = 1\,, \qquad \int_{-\infty}^{\infty} u_0(t)\, f(t)\, dt = f(0)$$

$$\pounds\, u_0(t) = \int_0^{\infty} u_0(t)\, e^{-st}\, dt$$

Let us analyse the product $u_0(t)\, e^{-st}$. This product is zero for any time, t, not equal to zero, because $u_0(t) = 0$ for $t \neq 0$. Therefore e^{-st} need be considered only at $t = 0$ at which time $u_0(t)$ is not zero. But $e^{-st} = 1$ at $t = 0$ so $u_0(t)\, e^{-st} = u_0(t)$

therefore, $$\pounds\, u_0(t) = \int_0^{\infty} u_0(t)\, e^{-st}\, dt = \int_0^{\infty} u_0(t)\, dt$$

This integration represents the area under the unit impulse function, which is unity. That is,

$$\int_0^{\infty} u_0(t)\, dt = 1$$

therefore, $$\pounds\, u_0(t) = 1$$

Thus *the Laplace transform of the unit imuplse is unity.*

10.7 LAPLACE TRANSFORM OF THE DERIVATIVE OF A FUNCTION

The time derivative of function $f(t)$ is $\frac{d}{dt} f(t)$ or $f'(t)$ and the Laplace transform is

$$\pounds \left[\frac{d}{dt} f(t) \right] = \pounds\, f'(t) = s\, F(s) - f(0_-)$$

where $f(0_-)$ is the value of $f(t)$ at $t = 0_-$.

Proof

$$\pounds\, f'(t) = \int_0^{\infty} f'(t)\, e^{-st}\, dt = \int_0^{\infty} e^{-st}\, f'(t)\, dt$$

Let $u = f(t)$ and $dv = e^{-st}\,dt$ and integrating by parts using

$$\int u\,dv = uv - \int v\,du$$

Since
$$du = \left[\frac{d}{dt} f(t)\right] dt,$$

integration by parts gives

$$£ f'(t) = \left[e^{-st} f(t)\right]_0^{\infty} - \int_0^{\infty} (-s\,e^{-st})\, f(t)\,dt$$

$$= \lim_{t \to \infty} [f(t)\,e^{-st}] - f(0)\,e^{-0} + \int_0^{\infty} se^{-st} f(t)\,dt$$

$$= 0 - f(0) + s\int_0^{\infty} f(t)\,e^{-st}\,dt$$

In the last term, the integration is with respect to t. therefore, s is a constant and is taken out of the integral sign.

therefore,
$$£ f'(t) = -f(0) + \mathbf{s}\,£ f(t)$$

$$£\left[\frac{d}{dt} f(t)\right] = \mathbf{s}\,F(\mathbf{s}) - f(0)$$

Alternative Method

$$£ f(t) = F(s) = \int_0^{\infty} f(t)\,e^{-st}\,dt$$

Integration by parts gives

$$F(s) = \left[f(t)\,\frac{e^{-st}}{-s}\right]_0^{\infty} + \int_0^{\infty} f'(t)\,\frac{e^{-st}}{s}\,dt$$

$$= \frac{f(0)}{s} + \frac{1}{s}\,£ f'(t)$$

therefore,
$$\boxed{£ f'(t) = \mathbf{s}\,F(\mathbf{s}) - f(0)}$$

Thus, *differentiation in the time domain transforms into algebraic operations in the s domain.*

10.8 LAPLACE TRANSFORMS OF HIGHER DERIVATIVES

The higher derivaties can be transformed as follows :

If
$$£ f(t) = F(s)$$

then
$$£ f'(t) = s\,F(s) - f(0) \tag{10.8.1}$$

Similarly, if

$$£ f'(t) = F'(s)$$

then
$$£ f''(t) = s\,F'(s) - f'(0) \tag{10.8.2}$$

where $$f''(t) \triangleq \frac{d}{dt} f'(t) = \frac{d^2}{dt^2} f(t)$$

Combination of Eqs. (10.8.1) and (10.8.2) gives

$$£ f''(t) = s\,[s\,F(s) - f(0)] - f'(0)$$

$$£\left[\frac{d^2}{dt^2} f(t)\right] = s^2 F(s) - s\,f(0) - f'(0)$$

In general, the transform of the nth derivative is given by

$$£\,[f^n(t)] = £\left[\frac{d^n}{dt^n} f(t)\right]$$

$$= s^n F(s) - s^{n-1} f(0) - s^{r-2} f(0) - \ldots - f^{n-1}(0) \qquad (10.8.3)$$

where $f^k(0)$ is the kth derivative of $f(t)$ with respect to t evaluated as $t \to 0$. It is assumed that $f^k(t)$, $0 \le k \le n$ are continuous Laplace transformable functions.

10.9 LAPLACE TRANSFORM OF THE INTEGRAL OF A FUNCTION

The Laplace transform of the integral of the function $f(t)$ with respect to time variable is given by

$$£\left[\int f(t)\,dt\right] = f^{-1}(t)$$

$$= \frac{F(s)}{s} + \frac{f^{-1}(0)}{s}$$

where $$f^{-1}(0) = \left[\int f(t)\,dt\right]_{t=0}$$

Proof We have

$$F(s) = \int_0^\infty f(t)\,e^{-st}\,dt \qquad (10.9.1)$$

Integrating the right hand side by parts taking e^{-st} as the first function and $f(t)$ as the second function

$$F(s) = \left[e^{-st}\int f(t)\,dt\right]_0^\infty + \int_0^\infty (se^{-st})\left[\int f(t)\,dt\right]dt$$

$$= 0 - \left[\int f(t)\,dt\right]_{t=0} + s\int_0^\infty \left[\int f(t)\,dt\right]e^{-st}\,dt$$

$$= -f^{-1}(0) + s\,£\left[\int f(t)\,dt\right]$$

$$£\left[\int f(t)\,dt\right] = \frac{F(s)}{s} + \frac{f^{-1}(0)}{s} \qquad (10.9.2)$$

Eq. (10.9.2) gives the Laplace transform of th indefinte integral. *Thus, the Laplace transform of the indefinite integral of a function is the Laplace transform of the function divided by s plus the initial value of the intergral divided by s.*

TABLE 10.2

Some Laplace Transform Pairs

$f(t)$	$F(s)$
$u_0 t$	1
$u_{-1}(t)$	$\frac{1}{s}$
$r(t) = tu_{-1}(t)$	$\frac{1}{s^2}$
$t^n u_{-1}(t)$	$\frac{n!}{s^{n+1}}$
$e^{-at} u_{-1}(t)$	$\frac{1}{s+a}$
$(\sin \omega t)\, u_{-1}(t)$	$\frac{\omega}{s^2+\omega^2}$
$(\cos \omega t)\, u_{-1}(t)$	$\frac{s}{s^2+\omega^2}$
$[\sin(\omega t+\alpha)]\, u_{-1}(t)$	$\frac{s\sin\alpha+\omega\cos\alpha}{s^2+\omega^2}$
$[\cos(\omega t+\alpha)]\, u_{-1}(t)$	$\frac{s\cos\alpha-\omega\sin\alpha}{s^2+\omega^2}$
$[(\sin h\, bt)]\, u_{-1}(t)$	$\frac{b}{s^2-b^2}$
$(\cos h\, bt)\, u_{-1}(t)$	$\frac{s}{s^2-b^2}$
$[e^{-at}\sin(\omega t+\alpha)]\, u_{-1}(t)$	$\frac{(s+a)\sin\alpha+\omega\cos\alpha}{(s+a)^2+\omega^2}$
$[e^{-at}\cos(\omega t+\alpha)]\, u_{-1}(t)$	$\frac{(s+a)\cos\alpha-\omega\sin\alpha}{(s+a)^2+\omega^2}$
$(e^{-at} t^n)\, u_{-1}(t)$	$\frac{n!}{(s+a)^{n+1}}$
$t \sin \omega t$	$\frac{2\omega s}{(s^2+\omega^2)^2}$

$t \cos \omega t$	$\dfrac{s^2 - \omega^2}{(s^2 + \omega^2)^2}$
$[t \sin \omega t]\, u_{-1}(t)$	$\dfrac{2\omega s}{(s^2 + \omega^2)^2}$
$[t \cos \omega t]\, u_{-1}(t)$	$\dfrac{s^2 - \omega^2}{(s^2 + \omega^2)^2}$
$[e^{-at} \sin \omega t]\, u_{-1}(t)$	$\dfrac{\omega}{(s+a)^2 + \omega^2}$
$[e^{-at} \cos \omega t]\, u_{-1}(t)$	$\dfrac{s+a}{(s+a)^2 + \omega^2}$

We make use of $u_{-1}(t)$ to make $f(t)$ casual.

It is seen that integration in time domain transforms into algebraic operations in the s domain.

For the definite integral $\int_0^t f(t)\, dt$, the Laplace transform is given by

$$\pounds\left[\int_0^t f(t)\, dt\right] = \frac{F(s)}{s} \qquad (10.9.3)$$

Alternative Method

$$f^{-1}(t) = \int f(t)\, dt \qquad (10.9.4)$$

Differenciating both sides of Eq. (10.9.4) we get

$$\frac{d}{dt}[f^{-1}(t)] - f(t) \qquad (10.9.5)$$

Taking the Laplace transform of both sides of Eq. (10.9.5) we obtain

$$s\,\pounds\,[f^{-1}(t)] - f^{-1}(0)] = F(s)$$

therefore,

$$\pounds\,[f^{-1}(t)] = \pounds\left[\int f(t)\, dt\right] = \frac{F(s)}{s} + \frac{f^{-1}(0)}{s}$$

10.10 LAPLACE TRANSFORM OF THE *n*th INTEGRAL OF A FUNCTION

We have $\quad \pounds f^{-1}(t) = \dfrac{F(s)}{s} + \dfrac{f^{-1}(0)}{s}$

Let $\quad f^{-n}(t) = n$th indefinite integral of $f(t)$, that is the nth integral of $f(t)$ between the limits $-\infty$ and t

$f^{-1}(0) =$ the value of the first indefinite inegral of $f(t)$ at $t = 0$

$f^{-n}(0) =$ the value of the nth indefinite intgral of $f(t)$ at $t = 0$

We have $$f^{-1}(t) = \int_{-\infty}^{t} f(t)\,dt \tag{10.10.1}$$

Let $$h(t) = \int_{-\infty}^{t} f(t)\,dt \tag{10.10.2}$$

$$f^{-2}(t) = \int_{-\infty}^{t}\left[\int_{-\infty}^{t} f(t)\,dt\,dt\right] \tag{10.10.3}$$

$$= \int_{-\infty}^{t} h(t)\,dt \tag{10.10.4}$$

$$£ f^{-2}(t) = £\left[\int_{-\infty}^{t} h(t)\,dt\right]$$

$$= \frac{£\,[h(t)]}{s} + \frac{(h^{-1})(0)}{s}$$

$$= \frac{1}{s}\,£\left[\int_{-\infty}^{t} f(t)\,dt\right] + \frac{1}{s}\,h^{-1}(0)$$

$$= \frac{1}{s}\left[\frac{£ f(t)}{s} + \frac{f^{-1}(0)}{s}\right] + \frac{1}{s}\,h^{-1}(0)$$

or $$£ f^{-2}(t) = \frac{1}{s^2}F(s) + \frac{1}{s^2}f^{-1}(0) + \frac{1}{s}f^{-2}(0) \tag{10.10.5}$$

since $$h^{-1}(0) = f^{-2}(0)$$

The result in Eq. (10.10.5) can be generalized to include repeated integration. The general result is given as

$$£ f^{-n}(t) = \frac{1}{s^n}F(s) + \frac{1}{s^{n-1}}f^{-1}(0) + \frac{1}{s^{n-2}}f^{-2}(0) + \ldots + \frac{1}{s}f^{-n}(0) \tag{10.10.6}$$

If we assume that

$$g(t) = \int_{0}^{t} f(t)\,dt \tag{10.10.7}$$

then the Laplace transform of the second *definite integral* of $f(t)$ can be written as

$$£\left[\int_{0}^{t}\left\{\int_{0}^{t} f(t)\,dt\right\}dt\right] = £\left[\int_{0}^{t} g(t)\,dt\right]$$

$$= \frac{£\,[g(t)]}{s} = \frac{1}{s}\,£\left[\int_{0}^{t} f(t)\,dt\right]$$

$$= \frac{1}{s}\left[\frac{1}{s}\,£ f(t)\right] = \frac{1}{s^2}\,£ f(t)$$

$$= \frac{F(s)}{s^2} \tag{10.10.8}$$

Similarly, the Laplace transform of the nth definite integral is given by

$$£\left[\int_0^t\int_0^t \ldots \int_0^t f(t)\,dt \ldots \underset{n}{dt}\right] = \frac{£ f(t)}{s^n} \tag{10.10.9}$$

10.11 COMPLEX-FREQUENCY (s-domain) DIFFERENTIATION

We have

$$£ f(t) = F(s) = \int_0^\infty f(t)\, e^{-st}\, dt \tag{10.11.1}$$

Differentiation of Eq. (10.11.1) gives

$$\frac{d}{ds} F(s) = \frac{d}{ds}\left[\int_0^\infty f(t)\, e^{-st}\, dt\right] \tag{10.11.2}$$

Since $f(t)$ is piecewise continuous and of exponential order, the differentiation can be performed under the integral sign.

$$\begin{aligned} \frac{d}{ds} F(s) &= \int_0^\infty f(t)\left(\frac{d}{ds} e^{-st}\right) dt \\ &= \int_0^\infty f(t)\,[-t\, e^{-st}]\, dt \\ &= \int_0^\infty [-t f(t)] e^{-st}\, dt \\ &= £\,[-t f(t)] \end{aligned} \tag{10.11.3}$$

Thus, *differentiation of the transform of a function with respect to s corresponds to the multiplication of the time function by* $-t$. Similarly, repeated differentiation of Eq. (10.11.1) gives

$$F^{2}(s) = \frac{d^2 F(s)}{ds^2} = £\,[t^2 f(t)] \tag{10.11.4}$$

In general,

$$F^{n}(s) = \frac{d^n F(s)}{ds^n} = £\,[(-t)^n f(t)] \qquad n = 1, 2, 3, \ldots \tag{10.11.5}$$

■ **EXAMPLE 10.2** Find the Laplace transform of $t \sin \omega t$

SOLUTION : Putting $f(t) = \sin \omega t$ and using the result

$$£\,[t f(t)] = -\frac{d}{ds} F(s)$$

We have

$$£\,[f(t)] = £ \sin \omega t = \frac{\omega}{s^2 + \omega^2}$$

$$£\,[t\,f(t)] = £\,[t \sin \omega t]$$

$$= -\frac{d}{ds}\left(\frac{\omega}{s^2+\omega^2}\right) = \frac{2\omega s}{(s^2+\omega^2)^2}$$

■ **EXAMPLE 10.3** Find the Laplace transform of $t^n\, u_{-1}(t)$

SOLUTION : We have $£\, u_{-1}(t) = \frac{1}{s}$

therefore, $$£\, t\, u_{-1}(t) = -\frac{d}{ds}\left(\frac{1}{s}\right) = \frac{1}{s^2}$$

$$£\, t^2 u_{-1}(t) = \frac{d^2}{ds^2}\left(\frac{1}{s}\right) = \frac{2}{s^3}$$

Applying multiplication time property

$$£\,[t^n\, u_{-1}(t)] = (-1)^n \frac{d^n}{ds^n}\left(\frac{1}{s}\right) = \frac{n\,!}{s^{n+1}}$$

■ **EXAMPLE 10.4** Find the Laplace transform of (*a*) $t\, e^{-at}\, u_{-1}(t)$, (*b*) $t^2 e^{-at}\, u_{-1}(t)$, (*c*) $t^n\, e^{-at}\, u_{-1}(t)$

SOLUTION : (*a*) We have $£e^{-at}\, u_{-1}(t) = \frac{1}{s+a}$

therefore, $$£\,(te^{-at})\, u_{-1}t = -\frac{d}{ds}\left(\frac{1}{s+a}\right) = \frac{1}{(s+a)^2}$$

(*b*) $$£(t^2 e^{-at})\, u_{-1}t = \frac{d^2}{ds^2}\left(\frac{1}{s+1}\right) = \frac{2}{(s+a)^3}$$

(*c*) From Example 10.3

$$£t^n\, u_{-1}(t) = \frac{n\,!}{s^{n+1}}$$

By *s*-domain displacement theorem

$$£\,[e^{-at} f(t)] = F(s+a)$$

Combination of these two results gives

$$£\,[t^n e^{-at}\, u_{-1}(t)] = \frac{n\,!}{(s+a)^{n+1}}$$

10.12 COMPLEX-FREQUENCY (*s*-domain) INTEGRATION

If $\frac{1}{t} f(t)$ is Laplace transformable, then

$$£\left[\frac{1}{t} f(t)\right] = \int_s^{\infty} F(s)\, ds$$

where $F(s)$ is the Laplace transform of $f(t)$.

Proof

$$\int_s^\infty F(s)\,ds = \int_s^\infty \left[\int_0^\infty f(t)\,e^{-st}\,ds\right] ds$$

since $f(t)$ and $\left(\frac{1}{t}\right) f(t)$ are Laplace transformable, we can interchange the orders of t and s-domain integrations. This gives

$$\int_s^\infty F(s)\,ds = \int_0^\infty f(t)\left[\int_s^\infty e^{-st}\,ds\right] dt$$

$$= \int_0^\infty f(t)\left[\frac{e^{-st}}{-t}\right]_s^\infty dt$$

$$= \int_0^\infty \frac{f(t)}{t}\,e^{-st}\,dt$$

$$= \pounds\left[\frac{1}{t} f(t)\right]$$

Thus, division by t in the t domain corresponds to integration with respect to s in the s domain. In general,

$$\pounds\,\frac{f(t)}{t^n} = \underbrace{\int_s^\infty ds \ldots \int_s^\infty F(s)\,ds}_{n \text{ integrals}}$$

10.13 TIME-SHIFT PROPERTY

If $\pounds\, f(t) = F(s)$ and $f(t) = 0$ when $t < 0$, then

$$\pounds\,[f(t-a)] = e^{-as}\,F(s)$$

provided that a is a constant which is positive and real.

Proof

If the function $f(t)$ is shifted by a units of time in the direction of increasing t, then the translated function can be written as $f(t-a)$. The function $f(t-a)$ is also written as $f(t-a)\,u_{-1}(t-a)$ to emphasize the fact that it is zero for $0 < t < a$.

$$\pounds\, f(t-a) = \int_0^\infty f(t-a)\,e^{-st}\,dt \tag{10.13.1}$$

Multiplication and division of the right hand side of Eq. (10.13.1) by e^{-as} gives

$$\pounds\, f(t-a) = \int_0^\infty e^{-as}\,e^{as}\,f(t-a)\,e^{-st}\,dt$$

since e^{-as} is a constant independent of time t, it can be moved outside the sign of integral as

$$\pounds\, f(t-a) = e^{-as}\int_0^\infty f(t-a)\,e^{-s(t-a)}\,dt \tag{10.13.2}$$

Let $t - a = \tau$, so that $dt = d\tau$.

The new limits of integration are found as follows :

As $t \to 0, \tau \to -a$

$t \to \infty, \tau \to \infty$

Eq. (10.13.2) now becomes

$$£\, f(t-a) = e^{-as} \int_{-a}^{\infty} f(\tau)\, e^{-s\tau}\, d\tau \tag{10.13.3}$$

Since $f(\tau) = 0$ for $\tau < 0$, the lower limit of integration can be replaced by zero. Hence Eq. (10.13.3) becomes

$$£\, f(t-a) = e^{-as} \int_{0}^{\infty} f(\tau)\, e^{-s\tau}\, d\tau = e^{-as}\, F(s) \tag{10.13.4}$$

Thus, *if the original time function is shifted to the right by a units of time, its Laplace transform is equal to the original Laplace transform multiplied by* e^{-as}. This theorem is called **time-shifting theorem.** It is very useful in obtaining the Laplace transforms of composite functions made up of standard functions shifted in time.

10.14 COMPLEX-FREQUENCY-SHIFT PROPERTY

If $£\, f(t) = F(s)$ and a is a real or complex number, then

$$£\, [e^{-at} f(t)] = F(s+a)$$

Proof

$$£\, [e^{-at} f(t)] = \int_0^{\infty} e^{-at} f(t)\, e^{-st}\, dt$$

$$= \int_0^{\infty} f(t)\, e^{-(s+a)}\, dt$$

therefore,

$$\boxed{£\, [e^{-at} f(t)] = F(s+a)} \tag{10.14.1}$$

Eq. (10.14.1) shows that multiplication be e^{-at} in the time domain is equivalent to translation by a in the s domain. Thus, if the original time function $f(t)$ multiplied by e^{-at}, the Laplace transform of $e^{-at} f(t)$ is obtained from $F(s)$ by replacing s by $(s+a)$

10.15 SCALING

Mathematically a change of scale in the time domain is represented by multiplying the time variable t by a positive scaling factor a. Hence the function $f(t)$ after rescaling of the time axis becomes $f(at)$. Let us determine the effect of time scale change on the Laplace transform.

$$£\, [f(at)] = \int_0^{\infty} f(at)]\, e^{-st}\, dt \tag{10.15.1}$$

Let us put $at = \tau$ with $a > 0$

therefore, $t = \dfrac{\tau}{a},\ dt = \dfrac{1}{a}\, d\tau$

$$£\,[f(at)] = \int_0^{\infty} f(\tau)\, e^{-s\tau/a} \,.\, \frac{1}{a}\, d\tau$$

or

$$\boxed{£\,[f(at)] = \frac{1}{a} F\left(\frac{s}{a}\right)} \qquad (10.15.2)$$

Eq. (10.15.2) shows that *scaling by the factor a in the time domain corresponds to scaling s by the factor* $\frac{1}{a}$.

■ **EXAMPLE 10.5** Find the Laplace transform of sin βt.

SOLUTION : We have

$$£\,\sin t = \frac{1}{s^2+1}$$

By scaling property,

therefore,

$$£\,\sin\beta t = \frac{1}{\beta} \frac{1}{\left(\frac{s}{\beta}\right)^2 + 1} = \frac{\beta}{s^2+\beta^2}$$

■ **EXAMPLE 10.6** Find the Laplace transform of $\frac{1}{t}$ sin$t$ and determine the value of $\int_0^{\infty} \frac{\sin t}{t}\, dt$.

SOLUTION : We have

$$£\,\sin t = \frac{1}{1+s^2}$$

therefore,

$$£\,\frac{\sin t}{t} = \int_s^{\infty} \frac{1}{1+s^2}\, ds = \left[\tan^{-1} s\right]_s^{\infty}$$

$$= \frac{\pi}{2} - \tan^{-1} s = \tan^{-1}\frac{1}{s}$$

Also

$$£\,\frac{\sin t}{t} = \int_0^{\infty} \frac{\sin t}{t}\, e^{-st}\, dt = \tan^{-1}\frac{1}{s}$$

If $\quad s \to 0$,

$$\int_0^{\infty} \frac{\sin t}{t}\,,\, dt = \frac{\pi}{2}$$

This is a well known definite integral.

■ **EXAMPLE 10.7** Use complex-frequency shift property to determine the Laplace transforms of the following functions :

(*a*) Damped sine function e^{-at} sinωt

(*b*) Damped cosine function e^{-at} cos ωt

(*c*) Damped hyperbolic sine function e^{-at} sin hbt

(*d*) Damped hyperbolic cosine function $e^{-at} \cos hbt$

SOLUTION : We know that by complex frequency shift peoperty the Laplace transform of e^{-at} times a function is equal to the Laplace transform of that function, with *s* replaced by $(s + a)$.

(*a*) Let $F(s) = \pounds(\sin \omega t) = \dfrac{\omega}{s^2 + \omega^2}$

therefore, $F(s + a) = \dfrac{\omega}{(s + a)^2 + \omega^2}$

and $\pounds(e^{-at} \sin \omega t) = \dfrac{\omega}{(s + a)^2 + \omega^2}$

(*b*) Let $F(s) = \pounds(\cos \omega t) = \dfrac{s}{s^2 + \omega^2}$

therefore, $F(s + a) = \dfrac{s + a}{(s + a)^2 + \omega^2}$

and $\pounds(e^{-at} \cos \omega t) = F(s + a) = \dfrac{s + a}{(s + a)^2 + \omega^2}$

(*c*) Let $F(s) = \pounds \sin hbt = \dfrac{b}{s^2 - b^2}$

therefore, $F(s + a) = \dfrac{b}{(s + a)^2 - b^2}$

and $\pounds(e^{-at} \sin hbt) = F(s + a) = \dfrac{b}{(s + a)^2 - b^2}$

(*d*) Let $F(s) = \pounds \cos hbt = \dfrac{s}{s^2 - b^2}$

therefore, $\pounds(e^{-at} \sin hbt) = F(s + a) = \dfrac{s + a}{(s + a)^2 - b^2}$

■ **EXAMPLE 10.8** Find the Laplace transform of $Ae^{-at} \sin(\omega_0 t + \theta)$.

SOLUTION : $A \sin(\omega_0 t + \theta) = A(\sin \omega_0 t \cos \theta + \cos \omega_0 t \sin \theta)$

therefore, $\pounds A \sin(\omega_0 t + \theta) = \dfrac{A\,\omega_0 \cos \theta}{s^2 + \omega_0^2} + \dfrac{A\,s \sin \theta}{s^2 + \omega_0^2}$

By complex-frequency shift property

$$\pounds\, e^{-at} f(t) = F(s + a)$$

therefore, $\pounds\,[Ae^{-at} \sin(\omega_0 t + \theta)]$

$$= A \frac{\omega_0 \cos\theta}{(s + a)^2 + \omega_0^2} + \frac{A(s + a) \sin \theta}{(s + a)^2 + \omega_0^2}$$

$$= A\left[\frac{\omega_0 \cos\theta + (s+a)\sin\theta}{(s+a)^2 + \omega_0^2}\right]$$

10.16 GATE FUNCTION

A function comprising two unit step functions of the form

$$G_{t_0}(T) = u_{-1}(t - t_0) - u_{-1}[t - (t_0 + T)] \tag{10.16.1}$$

is called a **gate function**. It is shown in Fig. 10.8. This function represents a rectangular pulse of unit height starting at $t = t_0$ and lasting for a time T. Such a function is very useful in the construction of the waveforms. Any function multiplied by the gate function $G_{t_0}(T)$ would have nonzero values only in the duration of the gate, $t_0 \le t \le (t_0 + T)$. The value of the function within the gate would be unaffected. Thus, the gate function provides a **gate** through which only those values of the function that exist for the range $t_0 \le t \le (t_0 + T)$ are permitted to pass.

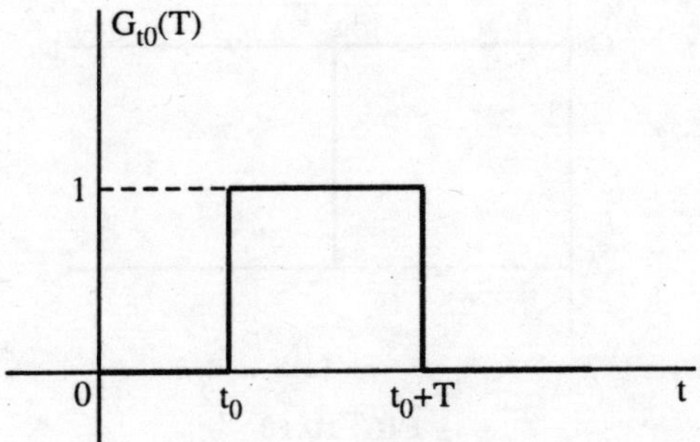

FIG. 10.8 A gate function

The usefulness of the gate lies in the ease with which the unwanted portions of a continuous function can be **erased** without tedious graphical composition by shifting and by addition and subtraction. Gate functions are frequently used in digital and computer circuits.

The equation of a gate function starting at origin and of duration T is obtained by putting $t_0 = 0$ in Eq. (10.16.1).

$$G_0(T) = u_{-1}(t) - u_{-1}(t - T) \tag{10.16.2}$$

This function is shown in Fig. 10.9

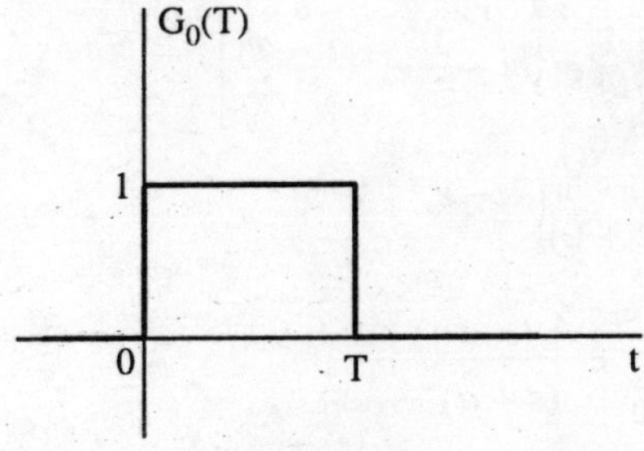

FIG. 10.9

The following examples will serve to illustrate its application.

■ **EXAMPLE 10.9** Express the pulse shown in Fig. 10.10a as a combination of step functions. Obtain its Laplace transform.

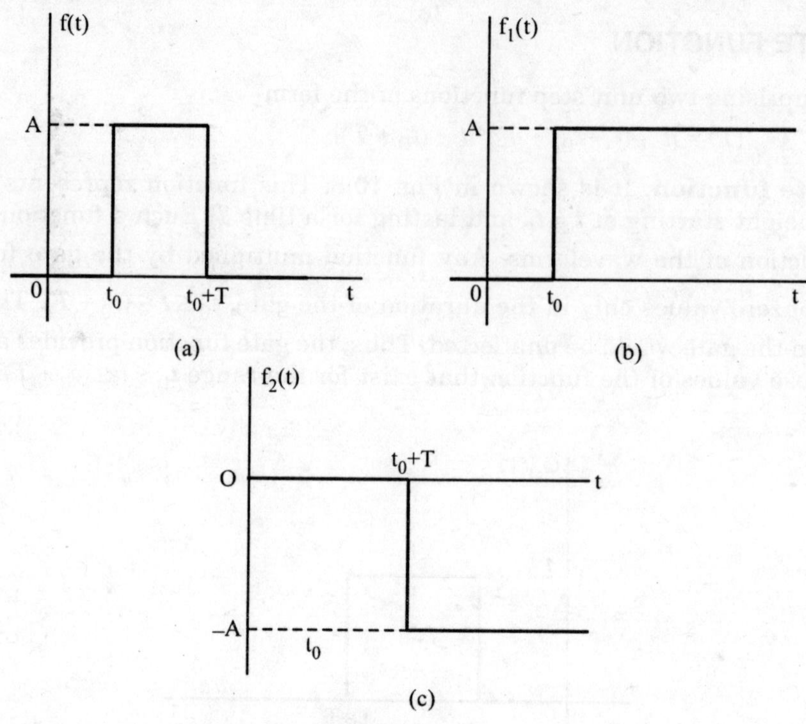

FIG. 10.10

SOLUTION : The pulse shown in Fig. 10.10a can be expressed as a combination of two step functions $f_1(t)$ and $f_2(t)$ shown in Fig. 10.10b and Fig 10.10c.

$$f_1(t) = A\, u_{-1}\,(t - t_0),$$

$$f_2(t) = -A\, u_{-1}\,[t - (t_0 + T)]$$

$$f(t) = f_1(t) + f_2(t)$$

$$= A\, u_{-1}\,(t - t_0) - A\, u_{-1}\,[t - (t_0 + T)]$$

$$£\, f(t) = F\,(s) = A\,[£\, u_{-1}\,(t - t_0) - £\, u_{-1}\,(t - \overline{t_0 + T})]$$

$$= A\left[\frac{1}{s}\, e^{-st_0} - \frac{1}{s}\, e^{-s\,(t_0 + T)}\right]$$

$$= \frac{A}{s}\, e^{-st_0}\left(1 - e^{-sT}\right)$$

EXAMPLE 10.10 Find the Laplace transform of the sawtooth function shown in Fig. 10.11 by making the use of a gate function.

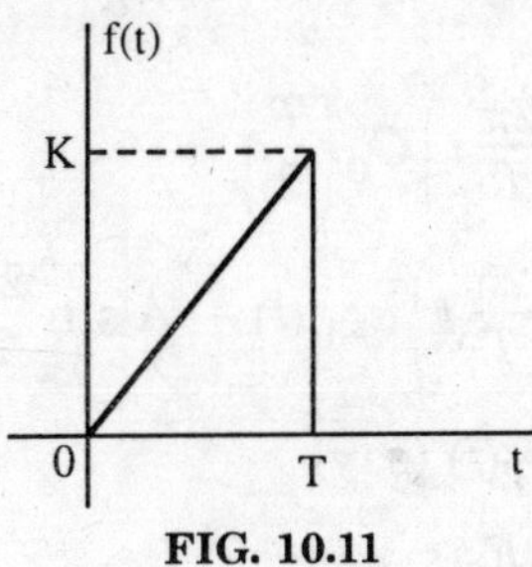

FIG. 10.11

SOLUTION :

$$f(t) = \frac{K}{T} t\, G_0(t)$$

$$= \frac{K}{T} t\, [u_{-1}(t) - u_{-1}(t - T)]$$

$$= \frac{K}{T} [tu_{-1}(t) - tu_{-1}(t - T)]$$

$$F(s) = \pounds\, f(t) = \frac{K}{T}\left[\pounds\, tu_{-1}(t) - \pounds\, tu_{-1}(t - T)\right]$$

$$= \frac{K}{T}\left[\frac{1}{s^2} - \pounds\, (\overline{t - T} + T)u_{-1}(t - T)\right]$$

It is to be noted that T has been added and subtracted to bring it to a standard form of shifted function $(t - T)\, u_{-1}(t - T)$.

therefore,
$$F(s) = \frac{K}{T}\left[\frac{1}{s^2} - \pounds\, (t - T)\, u_{-1}(t - T) - T\, \pounds\, u_{-1}(t - T)\right]$$

$$= \frac{K}{T}\left[\frac{1}{s^2} - \frac{1}{s^2} e^{-Ts} - T\frac{1}{s} e^{-Ts}\right]$$

$$= \frac{K}{Ts^2} [1 - (Ts + 1)e^{-Ts}]$$

EXAMPLE 10.11 Find the Laplace transform of the single half-sine cycle shown in Fig. 10.12 using a gate function.

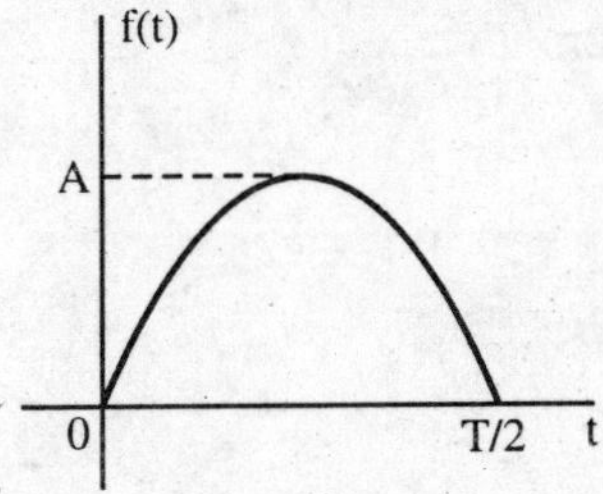

FIG. 10.12

SOLUTION : Let us use the gate function

$$G_0\left(\frac{T}{2}\right) = u_{-1}(t) - u_{-1}(t - T/2)$$

Now

$$f(t) = \left(A \sin \frac{2\pi}{T} t\right) G_0\left(\frac{T}{2}\right)$$

$$= \left(A \sin \frac{2\pi}{T} t\right) u_{-1}(t) - \left(A \sin \frac{2\pi}{T}\right) u_{-1}\left(t - \frac{T}{2}\right)$$

$$= f_1(t) - f_2(t) \text{ (say)}$$

$$F(s) = F_1(s) - F_2(s)$$

$$F_1(s) = \pounds f_1(t) = \pounds \left(A \sin \frac{2\pi}{T} t\right) u_{-1}(t)$$

$$= \frac{A(2\pi/T)}{s^2 + \left(\frac{2\pi}{T}\right)^2}$$

$$f_2(t) = \left(A \sin \frac{2\pi}{T} t\right) u_{-1}\left(t - \frac{T}{2}\right)$$

$$= A \sin \frac{2\pi}{T}\left[\left(t - \frac{T}{2}\right) + \frac{T}{2}\right] u_{-1}\left(t - \frac{T}{2}\right)$$

$$= A \sin\left[\frac{2\pi}{T}\left(t - \frac{T}{2}\right) + \pi\right] u_{-1}\left(t - \frac{T}{2}\right)$$

$$= -A \sin \frac{2\pi}{T}\left(t - \frac{T}{2}\right) u_{-1}\left(t - \frac{T}{2}\right)$$

$$F_2(s) = \pounds f_2(t)$$

$$= \frac{-A(2\pi/T)}{s^2 + \left(\frac{2\pi}{T}\right)^2} e^{-\frac{Ts}{2}}$$

$$F(s) = F_1(s) + F_2(s)$$

$$= \frac{A(2\pi/T)}{s^2 + (2\pi/T)^2}\left(1 + e^{-\frac{Ts}{2}}\right) \qquad \text{(E 10.10.1)}$$

10.17 LAPLACE TRANSFORM OF PERIODIC FUNCTIONS

Consider a periodic waveform with a period T as shown in Fig. 10.13. Let $f_1(t)$, $f_2(t)$, $f_3(t)$, . . . be the functions describing the first cycle, second cycle, third cycle, . . . Let $F_1(s)$, $F_2(s)$, $F_3(s)$, . . . be their Laplace transforms respectively. Then

$$f(t) = f_1(t) + f_2(t) + f_3(t) + \dots \qquad (10.17.1)$$

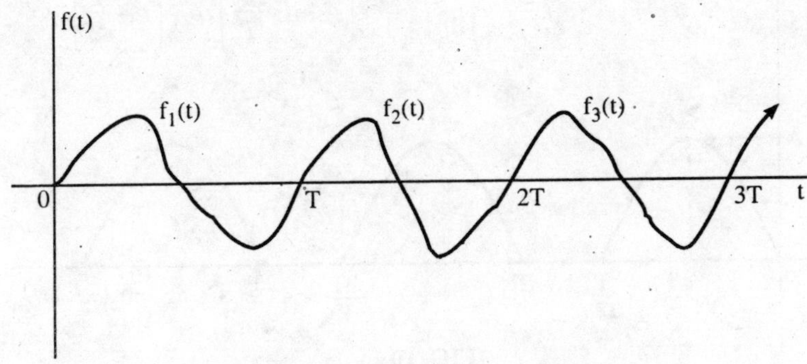

FIG. 10.13 Periodic waveform

It is to be noted that each cycle of the waveform is a repetition of the first cycle. Thus, $f_2(t)$ is $f_1(t)$ delayed by T seconds, $f_2(t)$ is $f_1(t)$ delayed by $2T$ seconds, and so on. Therefore Eq. (10.17.1) can be written as

$$f(t) = f_1(t) + f_1(t - T)\, u_{-1}(t - T) + f_1(t - 2T)\, u_{-1}(t - 2T) + \dots \qquad (10.17.2)$$

Taking Laplace transform of Eq. (10.17.1),

$$F(s) = F_1(s) + F_2(s) + F_3(s) + \dots \qquad (10.17.3)$$

Since $f_2(t)$ is $f_1(t)$ delayed by T seconds,

$$F_2(s) = F_1(s)\, e^{-Ts}$$

Also, $f_3(t)$ is $f_1(t)$ delayed by $2T$ seconds,

$$F_3(s) = F_1(s)\, e^{-2Ts}$$

Therefore, Eq. (10.17.3) may be written as

$$F(s) = F_1(s) + F_1(s)\, e^{-Ts} + F_1(s)\, e^{-2Ts} + \dots \qquad (10.17.4)$$

$$= F_1(s)\left[1 + e^{-Ts} + e^{-2Ts} + \dots + \infty\right]$$

But $(1 + e^{-Ts} + e^{-2Ts} + \dots + \infty)$ is a geometric series and its sum is $\dfrac{1}{1 - e^{-Ts}}$

$$\therefore \quad F(s) = \frac{F_1(s)}{1 - e^{-Ts}} \quad \text{for } |e^{-Ts}| < 1 \qquad (10.17.5)$$

Eq. (10.17.5) shows that the Laplace transform of a periodic function with period T is equal to $\dfrac{1}{1 - e^{-Ts}}$ times the Laplace transform of the first cycle.

■ **EXAMPLE 10.12** Obtain the Laplace transform of the periodic, rectified, half-sine wave.

SOLUTION: The period; rectified, half-sine wave is shown in Fig. 10.14. Let $F_1(s)$ be the Laplace transform of the first half sine wave. Then from Eq. (10.10.1) we have

$$F_1(s) = \frac{A\,(2\pi/T)}{s^2 + (2\pi/T)^2}\,(1 + e^{-\frac{T}{2}s})$$

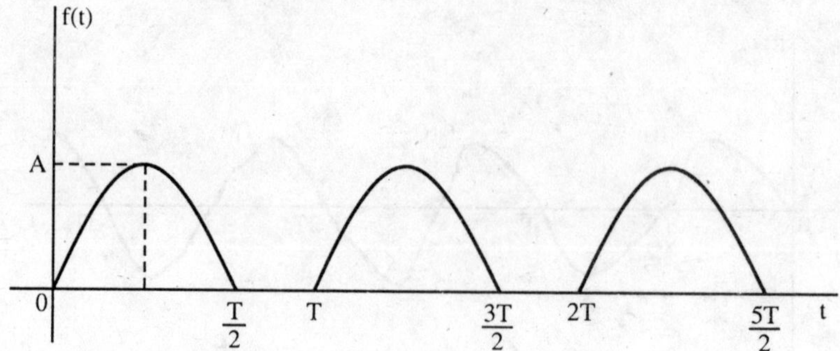

FIG. 10.14

Now, from Eq. (10.17.5) we have

$$F(s) = \frac{F_1(s)}{1 - e^{-Ts}}$$

$$= \frac{A\,(2\pi/T)}{s^2 + \left(\frac{2\pi}{T}\right)^2} - \frac{(1 + e^{-\frac{T}{2}s})}{(1 - e^{-Ts})}$$

$$= \frac{A\,(2\pi/T)}{s^2 + \left(\frac{2\pi}{T}\right)^2},\ \frac{1}{1 - e^{-\frac{T}{2}s}} \qquad \text{(E 10.12.1)}$$

■ **EXAMPLE 10.13** Obtain the Laplace transform of the periodic, full-wave rectified voltage shown in Fig. 10.15a.

SOLUTION: The periodic wave $f(t)$ shown in Fig. 10.15a can be expressed as a combination of two periodic waves $f_a(t)$ and $f_b(t)$ shown in Figs. 10.15b and 10.15c.

respectively.

$$f(t) = f_a(t) + f_b(t)$$

$$F(s) = F_a(s) + F_b(s)$$

But from Eq. (E 10.12.1)

$$F_a(s) = \frac{A\,(2\pi/T)}{\left[s^2 + \left(\frac{2\pi}{T}\right)^2\right](1 - e^{-\frac{T}{2}s})} \qquad \text{(E 10.13.1)}$$

Since $f_b(t)$ is $f_a(t)$ delayed by $\frac{T}{2}$ seconds,

$$F_b(s) = F_a(s)\, e^{-\frac{T}{2}s}$$

$$\therefore \quad F(s) = F_a(s) + F_b(s)$$

$$= F_a(s) + F_a(s)\, e^{-\frac{T}{2}s}$$

$$= F_a(s)\,.\,(1 + e^{-\frac{T}{2}s}) \qquad \text{(E 10.13.2)}$$

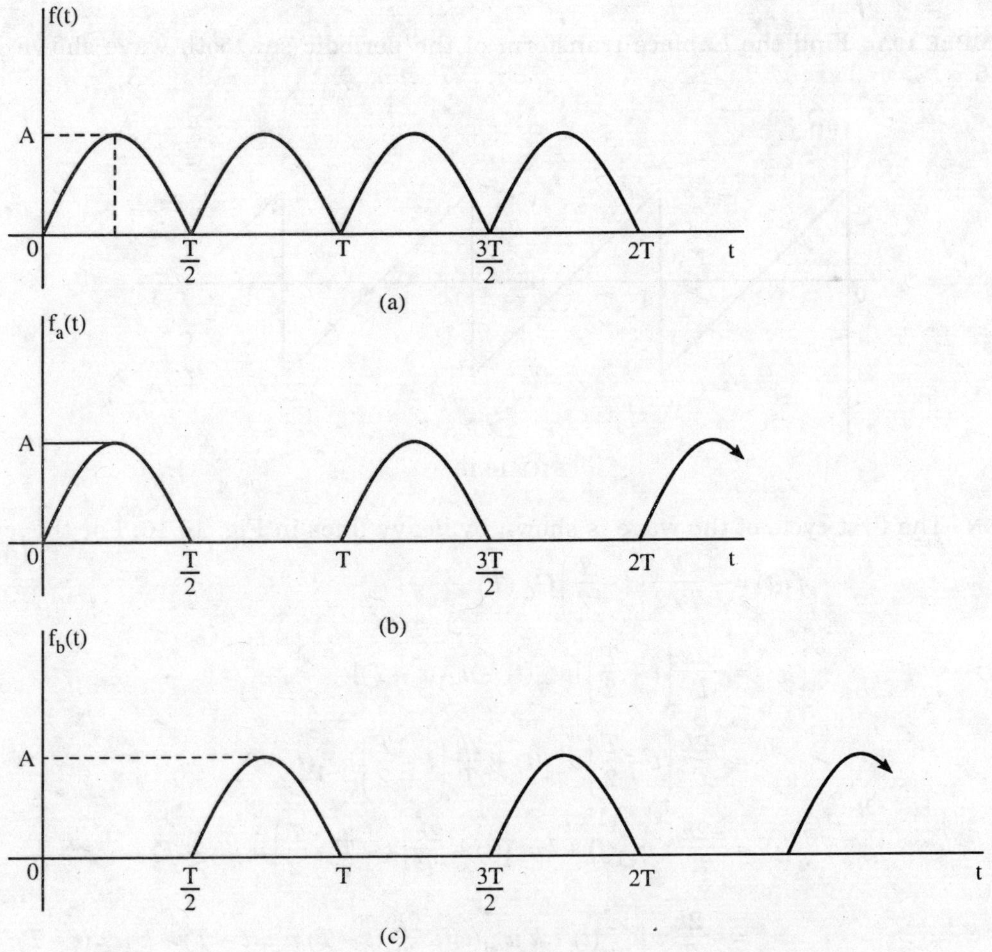

FIG. 10.15

Substituting the value of $F_a(s)$ from Eq (E 10.12.1) in Eq. (E 10.13.2)

$$F(s) = \frac{A\,(2\pi/T)\,(1 + e^{-\frac{T}{2}s})}{\left[s^2 + \left(\frac{2\pi}{T}\right)^2\right](1 - e^{-\frac{T}{2}s})}$$

$$= \frac{A\,(2\pi/T)}{\left[s^2 + \left(\frac{2\pi}{T}\right)^2\right]} \coth \frac{T}{4} s$$

EXAMPLE 10.14 Find the Laplace transform of the periodic sawtooth wave shown in Fig 10.16.

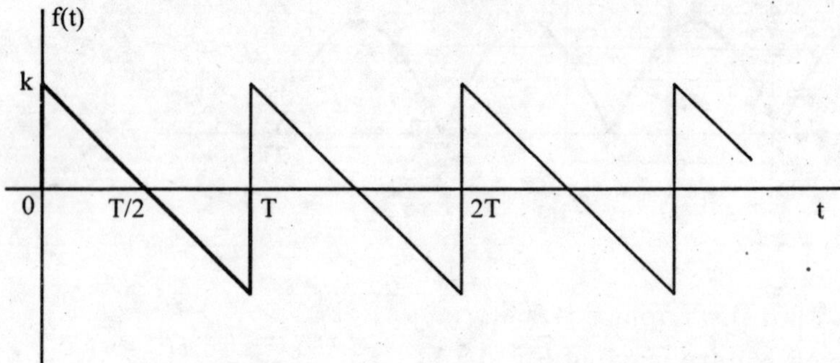

FIG. 10.16

SOLUTION : The first cycle of the wave is shown by heavy lines in Fig. 10.16. For this cycle

$$f_1(t) = -\frac{k}{T/2}\left(t - \frac{T}{2}\right) G_0\,(T)$$

$$= \frac{-2k}{T}\left(t - \frac{T}{2}\right) [u_{-1}(t) - u_{-1}(t - T)]$$

$$= \frac{-2k}{T}\left(t - \frac{T}{2}\right) u_{-1}(t) + \frac{2k}{T}\left(t - \frac{T}{2}\right) u_{-1}(t - T)$$

$$= \frac{-2k}{T}\, t\, u_{-1}(t) + k u_{-1}(t) + \frac{2k}{T}\left(t - T + \frac{T}{2}\right) u_{-1}(t - T)$$

$$= -\frac{2k}{T}\, t\, u_{-1}\,(t) + k\, u_{-1}(t) + \frac{2k}{T}\,(t - T)\, u_{-1}(t - T) + k\, u_{-1}(t - T)$$

$$F_1(s) = \pounds\, f_1(t)$$

$$= -\frac{2k}{T} \cdot \frac{1}{s^2} + \frac{k}{s} + \frac{2k}{Ts^2} e^{-Ts} + \frac{k}{s} e^{-Ts}$$

$$= \frac{2k}{Ts}\left[\frac{T}{2}(1+e^{-Ts}) - \frac{1}{s}(1-e^{-Ts})\right]$$

$$F(s) = \frac{F_1(s)}{1-e^{-Ts}}$$

$$= \frac{2k}{Ts}\left[\frac{T}{2}\left(\frac{1+e^{-Ts}}{1-e^{-Ts}}\right) - \frac{1}{s}\right]$$

$$= \frac{2k}{Ts}\left[\frac{T}{2}\coth\left(\frac{Ts}{2}\right) - \frac{1}{s}\right]$$

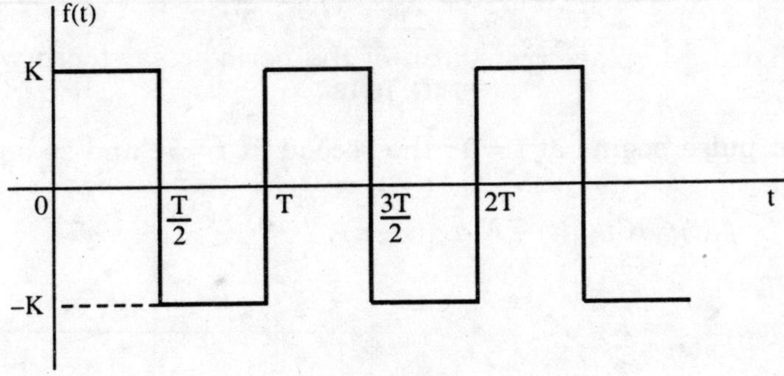

FIG. 10.17

■ **EXAMPLE 10.15** Find the Laplace transform of the square wave shown in Fig. 10.17.

SOLUTION : The given square wave of Fig. 10.17 can be expressed as

$$f(t) = K\,u_{-1}(t) - 2Ku_{-1}\left(t - \frac{T}{2}\right) + 2Ku_{-1}(t-T) - 2Ku_{-1}\left(t - \frac{3T}{2}\right) + \ldots \quad \text{(E 10.15.1)}$$

Laplace transform of Eq. (E 10.15.1) gives

$$F(s) = \frac{K}{s} - \frac{2K}{s}e^{-\frac{T}{2}s} + \frac{2K}{s}e^{-Ts} - \frac{2K}{s}e^{-\frac{3T}{2}s} + \ldots$$

$$= \frac{K}{s}\left[1 - 2e^{-\frac{T}{2}s}\left(1 - e^{-\frac{T}{2}s} + e^{-Ts} - e^{-\frac{3T}{2}s} + \ldots\right)\right]$$

But $\left(1 - e^{-\frac{T}{2}s} + e^{-Ts} - e^{-\frac{3}{2}Ts} + \ldots\right)$ is a geometric series and its sum is $\dfrac{1}{1+e^{-\frac{T}{2}s}}$

therefore,
$$F(s) = \frac{K}{s}\left(1 - \frac{2e^{-\frac{T}{2}s}}{1+e^{-\frac{T}{2}s}}\right)$$

$$= \frac{K}{s}\,\frac{1-e^{-\frac{T}{2}s}}{1+e^{-\frac{T}{2}s}} = \frac{K}{s}\tanh\frac{T}{4}s$$

EXAMPLE 10.16 Find the Laplace transform of a train of pulses of width a, amplitude k and periodic time T shown in Fig. 10.18.

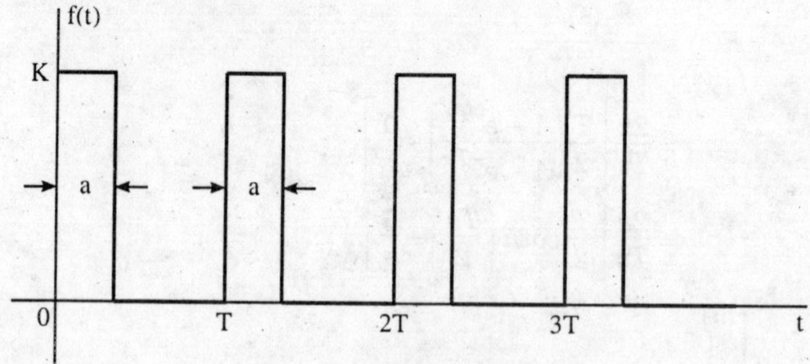

FIG. 10.18

SOLUTION : The first pulse begins at $t = 0$, the second at $t = T$, and so on. The first pulse is given by the relation

$$f_1(t) = K\,u_{-1}(t) - K\,u_{-1}(t-a)$$

Its Laplace transform is

$$F_1(s) = \frac{K}{s} - \frac{K}{s}e^{-as}$$

$$= \frac{K}{s}(1 - e^{-as})$$

For periodic waves

$$F(s) = \frac{F_1(s)}{1 - e^{-Ts}} = \frac{K}{s}\frac{1 - e^{-as}}{1 - e^{-Ts}}$$

EXAMPLE 10.17 Find the Laplace transform of the truncated ramp function shown in Fig. 10.19a.

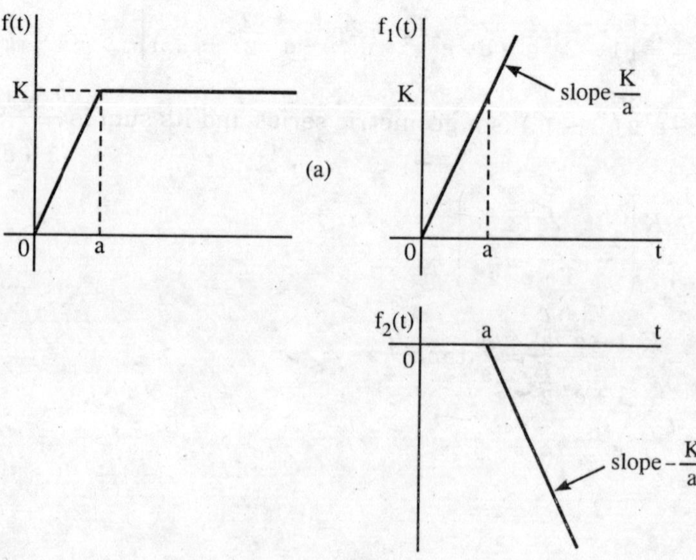

Fig. 10.19

SOLUTION : The truncated ramp may be considered to be constructed from the two ramps $f_1(t)$ and $f_2(t)$. The function $f_1(t)$ is a ramp starting at $t = 0$ and having a slope K/a. This is given by

$$f_1(t) = \frac{K}{a}\, t\, u_{-1}(t)$$

The function $f_2(t)$ is a ramp starting at $t = a$ and having a slope $-\frac{K}{a}$. This is given by

$$f_2(t) = -\frac{K}{a}(t-a)\, u_{-1}(t-a)$$

The function $f_1(t)$ and $f_2(t)$ are shown in Figs. 10.19b and 10.19c respectively.

The truncated function of Fig. 10.19a is given by

$$f(t) = f_1(t) + f_2(t)$$

$$= \frac{K}{a}\, t\, u_1(t) - \frac{K}{a}(t-a)\, u_{-1}(t-a)$$

therefore,

$$F(s) = F_1(s) + F_2(s)$$

$$= \frac{K}{a}\cdot\frac{1}{s^2} - \frac{K}{a}\cdot\frac{1}{s^2}\, e^{-as}$$

$$= \frac{K}{as^2}(1 - e^{-as})$$

10.18 TRANSFORM NETWORKS

*A **transform network** is the s-domain equivalent network of the time-domain network.* To obtain the transform network of a given time-domain network the R, L and C elements of the t-domain network are replaced by their transform impedances or admittances. The initial conditions of the storage elements are replaced by voltage and current sources in the transformed network. In addition, if there are independent voltage and current sources, they are replaced by their corresponding Laplace transformed functions.

For both the inductor and capacitor we have got the Thevenin's and Norton's equivalent models in the s-domain. When mesh or loop equations are to be written it is more convenient to use the Thevenin's equivalent model. If we are interested in writing node or cut-set equations, it is more conveneint to use the Norton's equivalent model.

10.19 OPERATIONAL IMPEDANCE

***Operational impedance** is defined as the ratio of the transformed voltage to the transformed current with all initial conditions neglected. it is also called the* **transform impedance** *or* **Laplacian impedance.** Thus

$$\mathbf{Z}(s) \triangleq \left.\frac{\mathbf{V}(s)}{\mathbf{I}(s)}\right|_{\text{zero initial conditions}} \qquad (10.19.1)$$

It is to be noted that in this definition, the network is in zero state that is, all its reactive initial conditions are zero. **Operational admittance** is defined in a similar manner.

$$\mathbf{Y}(s) \triangleq \frac{1}{\mathbf{Z}(s)} = \frac{\mathbf{I}(s)}{\mathbf{V}(s)}\bigg|_{\text{zero initial conditions}} \tag{10.19.2}$$

Operational admittance is also called the **transform admittance** or **Laplacian admittance.**

We shall now determine the operational impedances and admittances for the elements R, L and C.

10.20 RESISTOR IN THE s-DOMAIIN

In the time domain, the volt-ampere relationship of a resistor can be expressed in either of the two forms :

$$v_R(t) = R\, i_R(t) \tag{10.20.1}$$

$$i_R(t) = G\, v_R(t) \tag{10.20.2}$$

Laplace transforms of these equation give

$$\mathbf{V}_R(s) = R\, \mathbf{I}_R(s) \tag{10.20.3}$$

$$\mathbf{I}_R(s) = G\mathbf{V}_R(s) \tag{10.20.4}$$

Eq. (10.20.3) and (10.20.4) give the relationships between voltage and current transforms for a resistive element R. In other words, Eqs. (10.20.3) and (10.20.4) give the voltage-current relationships for R in the s-domain.

The operational impedence for a resistive element can be found from Eq. (10.20.3) as

$$\boxed{\mathbf{Z}_R(s) = \frac{\mathbf{V}_R(s)}{\mathbf{I}_R(s)} = \text{R}} \tag{10.20.5}$$

The operational admittance for a resistive element is found from Eq. (10.20.4) as

$$\boxed{\mathbf{Y}_R(s) = \frac{\mathbf{I}_R(s)}{\mathbf{V}_R(s)} = \text{G}} \tag{10.20.6}$$

The t-domain and s-domain representations of a resistive element R are shown in Fig 10.20a and 10.20b respectively. The s-domain equivalent network of the time-domain network is called the **transform network**.

It is seen that the resistive element is described by R in either the time or-s domain.

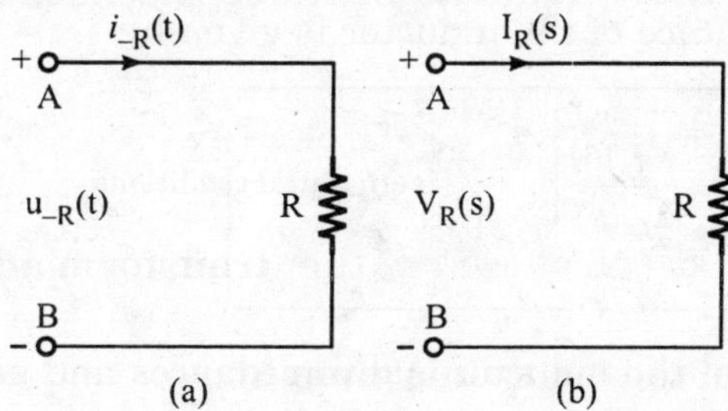

FIG. 10.20 (a) Time-domain representation of resistor *R*. (b) *s*-domain representation of resistor *R*.

10.21 INDUCTOR IN THE s-DOMAIN

Consider an inductor shown in Fig 10.21a with an initial current $i_L(0_-)$. In the time domain, the volt-ampere relationship of the inductor can be expressed in either of the two forms :

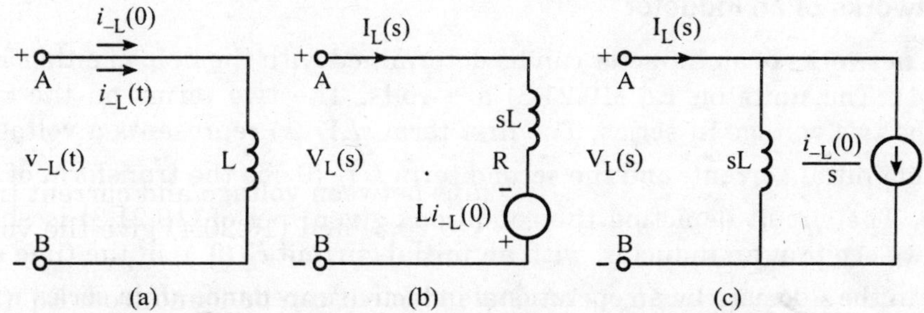

FIG. 10.21 (a) Representation of the inductor in the *t*-domain (b) Representation of the inductor in the *s*-domain by Thevenin's equivalent model (c) Representation of the inductor in the *s*-domain by Norton's equivalent model.

$$v_L(t) = L\frac{di_L}{dt} \tag{10.21.1}$$

$$i_L(t) = i_L(0_-) + \int_{0-}^{t} v_L(t)\,dt \tag{10.21.2}$$

The Laplace transform of Eq. (10.21.1) is given by

$$\mathbf{V}_L(s) = L\,[s\,\mathbf{I}_L(s) - i_L(0_-)]$$

or

$$\mathbf{V}_L(s) = s\,L\,\mathbf{I}_L(s) - L\,i_L(0_-) \tag{10.21.3}$$

The Laplace transform of Eq. (10.21.2) is given by

$$\mathbf{I}_L(s) = \frac{i_L(0_-)}{s} + \frac{1}{sL}\mathbf{V}_L(s) \tag{10.21.4}$$

Thus, an inductor which is described by either a derivative or an integral in the time domain is decribed by an algebraic relationship in the s-domain.

If the initial current $i_L(0_-)$ ia assumed to be zero, then Eq. (10.21.3) becomes

$$\mathbf{V}_L(s) = s\,L\mathbf{I}_L(s).$$

Hence the operational impedance of the inductor is given by

$$\mathbf{Z}_L(s) = \left.\frac{\mathbf{V}_L(s)}{\mathbf{I}_L(s)}\right|_{i_{L(0-)}=0} = sL\ \Omega \tag{10.21.5}$$

The operational admittance of the inductor is given by

$$\mathbf{Y}_L(s) = \frac{1}{\mathbf{Z}_L(s)} = \left.\frac{\mathbf{I}_L(s)}{\mathbf{V}_L(s)}\right|_{i_{L(0-)}=0} = \frac{1}{sL}\ \text{siemens} \tag{10.21.6}$$

The operational impedance and the operational admittance can also be found from Eq. (10.21.4).

Transform Networks of an Inductor

The transform networks of an inductor can be determined with the help of either Eq. (10.21.3) or Eq. (10.21.4). The units on Eq (10.21.3) are volts. The two terms on the right can be considered to be two voltage in series. The first term $sL\mathbf{I}_L(s)$ represents a voltage across an inductor with no initial current, and the second term $L\,i_L(0_-)$ is the transform of the impulse voltage source. The circuit depicting the conditions given in Eq. (10.21.3) is shown in Fig. 10.21b. Thus, we see that an inductor, with an initial current $i_L(0_-)$ in the time domain, can be represented in the *s* domain by an operational inductive impedance sL in series with a voltage source $L\,i_L(0_-)$ with polarity as shown in Fig. 20.21b. Such a transformed network of the inductor is called the **Thevenin's equivalent model.**

The units on Eq. (10.21.4) are amperes. The two terms on the right can be considered to be two currents in parallel. The first term $\frac{i_L(0_-)}{s}$ represents the transform of a constant current source due to initial current in the inductor. The second term $\frac{1}{sL}V_L(s)$ is the current in the inductor with no initial current. The circuit depicting the conditions given in Eq. (10.21.4) is shown in Fig. 10.21c. Thus, it is seen that an inductor with an initial current $i_L(0_-)$ in the time domain, can be represented in the *s*-domain by an operational inductive impedance sL in parallel with a transform ideal current source $\frac{i_L(0_-)}{s}$ as shown in Fig. 10.21c. This transformed network of the inductor is called the **Norton's equivalent model**. It should be noted that the independent current source $\frac{i_L(0_-)}{s}$ has the same direction as $i_L(0_-)$.

The choice between the Thevenin's and Norton's equivalent models in the *s* domain is decided by the nature of the problem and convenience.

It is to be noted that when the initial current is zero, the transformed network reduces to simply the impedance sL.

10.22 CAPACITOR IN THE *s*-DOMAIN

Consider a capacitor shown in Fig. 10.22a with the an initial voltage v_c (0_). In the time domair the volt-ampere relationship of a capacitor can be expressed in either of the two forms :

$$i_c(t) = C\frac{dv_c}{dt} \tag{10.22.1}$$

$$v_c(t) = v_c(0\_) + \frac{1}{C}\int_0^t i_c(t)\, dt \tag{10.22.2}$$

The Laplace transform of Eq. (10.22.1) is given by

$$\mathbf{I}_c(s) = C\,[s\mathbf{V}_c(s) - v_c(0\_)]$$

or
$$\mathbf{I}_c(s) = s\,C\mathbf{V}_c(s) - Cv_c(0\_) \tag{10.22.3}$$

The Laplace transform of Eq. (10.22.2) is given by

$$\mathbf{V}_c(s) = \frac{v_c(0\_)}{s} + \frac{1}{sC}\mathbf{I}_c(s) \tag{10.22.4}$$

Thus, a capacitor which is described by either a derivative or an integral in the time domain is described by an algebraic relationship in the *s*-domain.

If the initial charge of the capacitor is assumed to be zero, that is, if $q(0\_) = 0$, then $v_c(0\_) = 0$ and Eq. (10.22.3) becomes

$$\mathbf{I}_c(s) = s\,\mathbf{V}_c(s)$$

Hence the operational impedance of the capacitor is given by

$$\mathbf{Z}_c(s) = \left.\frac{\mathbf{V}_c(s)}{\mathbf{I}_c(s)}\right|_{v_c(0\_)=0} = \frac{1}{sC}\ \Omega \tag{10.22.5}$$

The operational admittance of the capacitor

$$\mathbf{Y}_c(s) = \frac{1}{\mathbf{Z}_c(s)} = \left.\frac{\mathbf{I}_c(s)}{\mathbf{V}_c(s)}\right|_{v_{c(0\_)}=0} = sC \text{ siemens} \tag{10.22.6}$$

Transform Networks of a Capacitor

Fig. 10.22a shows a capacitor with an initial voltage v_c (0_) in the time domain. This capacitor can be represented in the *s*-domain with the help of Eqs. (10.22.3) and (10.22.4). Eq. (10.22.4) shows that a capacitor, with an initial voltage v_c (0_) in the time domain is equivalent to a capacitive operational impedance $\frac{1}{sC}$ in series with an ideal voltage source $\frac{v_c(0\_)}{s}$ in the *s*-domain. It should be noted that the $\frac{v_c(0\_)}{s}$ independent voltage source has the same polarity as v_c (0_). The s-domain representation is shown in Fig. 10.22b. Such a transformed network of the capacitor is called the **Thevenin's equivalent model.**

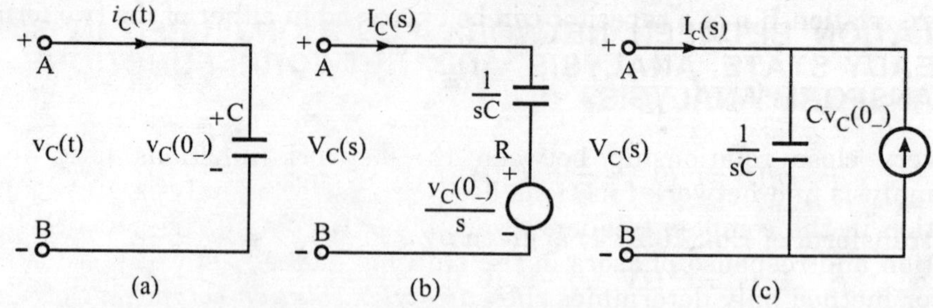

FIG. 10.22 (a) Representation of a capacitor in the *t* domain
(b) Representation of the capacitor in the *s*-domain by Thevenin's equivalent model
(c) Representation of the capacitor in the *s*-domain by Norton's equivalent model

Eq. (10.22.3) shows that a capacitor, with an initial voltage v_c (0_) in the time domain, can also be represented in the s-domain by a transform impedance $\frac{1}{sC}$ in parallel with a transform current source Cv_c (0_) due to the initial voltage across the capacitor. This representation is shown in Fig 10.22c. Such a transformed network of the capacitor is called the **Norton's equivalent model.**

It is to be noted that when the initial charge on the capacitor is zero, the representation reduces to simply the impedance $1/sC$. Again the choice between the two models depends upon the nature of the problem and convenience.

10.23 ANALYSIS OF TRANSFORM NETWORKS

The following results have been deduced in previous sections :

$$\mathbf{Z}_R(s) = R\ \Omega \quad , \quad \mathbf{Y}_R(s) = G \quad \text{siemens}$$

$$\mathbf{Z}_L(s) = sL\ \Omega \quad , \quad \mathbf{Y}_L(s) = \frac{1}{sL} \quad \text{siemens}$$

$$\mathbf{Z}_C(s) = \frac{1}{sC}\ \Omega \quad , \quad \mathbf{Y}_C(s) = sC \quad \text{siemens}$$

The concept of operational impedance/admittance is very useful in circuit theory. The transform network can be analysed in the same way as a resistive network. This is due to the fact that the relations between the transformed quantities in the transform network are purely algebraic in nature which is also the case with resistive networks. Hence all the techniques applicable to resistive networks apply to transform networks also. Thus, all the rules regarding series and parallel combination of impedances, Ohm's law, KCL, KVL, voltage division across impedances in series, current division among impedances in parallel are valid. Similarly, all the network theorems are applicable in transform networks. Mesh and node analyses are also applicable to transform networks. The frequency domain equivalent network is solved for the unknown transformed voltage or current with any of the above mentioned techniques. The unknown transformed voltage or current is then inverse transformed to find the more meaningful corresponding function of time.

10.24 RELATION BETWEEN NETWORK FUNCTIONS USED IN SINUSOIDAL STEADY-STATE ANALYSIS AND NETWORK FUNCTIONS USED IN TRANSFORM ANALYSIS

There is a very close relationship between the network functions used in sinusoidal steady-state analysis and network functions used in transform analysis. If we put $s = j\omega$ in a network function in the complex frequency variable s, we obtain the network function that relates excitation and response phasors in the sinusoidal steady state case. It is to be noted that the phasor method only determines the *particular solution component* of the zero state response, not the transient component. The transform network function method, however, determines, *both* components, of the zero-state response, namely, the *particular solution* and the *transient solution.* Also, the phasor method is applicable to sinusoidal input excitation while the transform network function method is applicable to any input excitation function.

10.25 INITIAL AND FINAL-VALUE THEOREMS

For a given transform $\boldsymbol{F}(s)$ it is often desired to determine the initial and final values of the corresponding time function $f(t)$. This may be done directly using the initial and final value theorems without finding the inverse transform. The initial value theorem is useful in determining the intial value of $f(t)$, and the final value theorem is used to determine the final value of $f(t)$. These theorems also serve as partial checks to the accuracy of the response transform from the known initial and final values before the determination of the inverse transform.

10.25.1 Initial-Value Theorem

The initial-value theorem states that if $f(t)$ and its derivative $f'(t)$ are Laplace transformable, then the initial value $f(0)$ of the function $f(t)$ is given by

$$\boxed{f(0) = \lim_{t \to 0} f(t) = \lim_{s \to \infty} sF(s)} \tag{10.25.1}$$

Proof The Laplace transform of $f'(t)$ is given by

$$\pounds\,[f'(t)] = \int_0^\infty f'(t)\, e^{-st}\, dt = sF(s) - f(0) \tag{10.25.2}$$

Taking the limit of both sides of Eq. (10.25.2) as $s \to \infty$ gives

$$\lim_{s \to \infty} \int_0^\infty f'(t)\, e^{-st}\, dt = \lim_{s \to \infty} [sF(s) - f(0)] \tag{10.25.3}$$

On the left hand side, the integration is with respect to time t and s is not a function of time. Hence the limit can be taken inside the integration. That is,

$$\lim_{s \to \infty} \int_0^\infty f'(t)\, e^{-st}\, dt - \int_0^\infty \lim_{s \to \infty} f'(t)\, e^{-st}\, dt = 0$$

This is due to the fact that as $s \to \infty$, $e^{-st} \to 0$. Therefore Eq. (10.25.3) becomes

$$0 = \lim_{s \to \infty} [sF(s) - f(0)]$$

Since $f(0)$ is independent of s, we have

$$\lim_{s \to \infty} sF(s) = f(0) \tag{10.25.4}$$

But

$$f(0) = \lim_{t \to 0} f(t) \tag{10.25.5}$$

and is independent of s. Therefore

$$\lim_{s \to \infty} sF(s) = f(0) = \text{initial value of } f(t)$$

It has been assumed that $\lim_{s \to \infty} sF(s)$ exists.

10.25.2 Final-Value Theorem

The final value theorem states that if the function $f(t)$ and its derivative $f'(t)$ are Laplace transformable then the final value $f(\infty)$ of the function $f(t)$ is given by

$$\boxed{f(\infty) = \lim_{t \to \infty} f(t) = \lim_{s \to 0} s\,F(s)}$$

Proof We have

$$\pounds\,[f'(t)] = \int_0^\infty f'(t)\,e^{-st}\,dt = sF(s) - f(0) \tag{10.25.6}$$

Taking the limit of both sides of Eq. (10.25.6) as $s \to 0$ gives

$$\lim_{s \to 0} \int_0^\infty f'(t)\,e^{-st}\,dt = \lim_{s \to 0} [sF(s) - f(0)] \tag{10.25.7}$$

As before, s is not a function of t, we allow $s \to 0$ before integrating so that e^{-st} tends to unity as $s \to 0$. Therefore, Eq. (10.25.7) becomes

$$\int_0^\infty f'(t)\,dt = \lim_{s \to 0} [sF(s) - f(0)] \tag{10.25.8}$$

But

$$\int_0^\infty f'(t)\,dt = \lim_{t \to \infty} \int_0^t f'(t)\,dt$$

$$= \lim_{t \to \infty} [f(t) - f(0)] \tag{10.25.9}$$

therefore,

$$\lim_{t \to \infty} f(t) = \lim_{s \to 0} sF(s) \tag{10.25.10}$$

The left-hand side of Eq. (10.25.10) is equal to the final value of the function $f(t)$ and so this is called the final-value theorem.

The final value can be obtained only for those functions that reach a final steady (constant) value. That is the final value theorem is valid if $\lim_{t \to \infty} f(t)$ exists. This requires that all the roots in the denominator of $sF(s)$ have negative real parts. This is equivalent to the statement that this theorem is not valid if $sF(s)$ contains any poles whose real part is zero or positive. For example, the final value theorem does not apply if $f(t)$ is sinusoid because $sF(s)$ will have poles on the imaginary axis.

■ **EXAMPLE 10.18** The response transform of a series *RL* circuit is given by

$$I(s) = \frac{V/L}{s(s + R/L)}$$

Find the initial value of the time response using initial-value theorem. Verify the result by solving it for $i(t)$.

SOLUTION : By the initial-value theorem

$$i(0) = \lim_{s \to \infty} sI(s) = \lim_{s \to \infty} \frac{V/L}{s + R/L} = 0$$

Check We have

$$I(s) = \frac{V/L}{s(s + R/L)}$$

$$= \frac{V}{L} \cdot \frac{L}{R}\left[\frac{1}{s} - \frac{1}{s + R/L}\right]$$

$$= \frac{V}{R}\left(\frac{1}{s} - \frac{1}{s + R/L}\right)$$

Taking the inverse Laplace transform we get

$$i(t) = \frac{V}{R}\left(1 - e^{-(R/L)\,t}\right) \tag{E 10.18.1}$$

If we put $t = 0$, then

$$i(0) = \frac{V}{R}(1 - e^{0}) = 0$$

Thus, it is seen that the initial-value theorem serves as a partial check on the accuracy of the inverse transform in Eq. (E 10.18.1).

■ **EXAMPLE 10.19** The response of a series *RC* circuit is given by

$$I(s) = \frac{V - q_0/C}{R\left[s + \dfrac{1}{RC}\right]}$$

where q_0 is the initial charge on the capacitor. Find the final value of the time response using final-value theorem. Verify the result by solving it for $i(t)$.

SOLUTION : By final-value theorem

$$i(\infty) = \lim_{t \to \infty} i(t) = \lim_{s \to 0} sI(s)$$

$$= \lim_{s \to 0} \frac{s(V - q_0/C)}{R\left(s + \dfrac{1}{RC}\right)} = 0$$

Thus the final value of the current is zero.

Check

$$I(s) = \frac{V - q_0/C}{R\left(s + \frac{1}{RC}\right)} \tag{E 10.19.1}$$

Taking the inverse Laplace transform of Eq. (E 10.19.1) we get

$$i(t) = \left(\frac{V - \frac{q_0}{C}}{R}\right) e^{-(t/RC)}$$

If $t \to \infty, \quad e^{-(t/RC)} \to 0$

$$\therefore \qquad i(\infty) = \frac{V - q_0/C}{R} \times 0 = 0$$

10.26 FINDING THE INVERSE LAPLACE TRANSFORM BY PARTIAL-FRACTION EXPANSION

Consider the network function

$$F(s) = \frac{a_m s^m + a_{m-1} s^{m-1} + \ldots + a_1 s + a_0}{b_n s^n + b_{n-1} s^{n-1} + \ldots + b_1 s + b_0} = \frac{N(s)}{D(s)}$$

$$= \frac{k(s - z_1)(s - z_2) \ldots (s - z_m)}{(s - p_1)(s - p_2) \ldots (s - p_n)}$$

where $N(s)$ is the numerator polynomial and $D(s)$ is the denominator polynomial. In case of the partial-fraction expansion we have to ensure that $F(s)$ is a proper fraction, that is, a fraction in which n is equal to or greater than m. If $F(s)$ is not a proper fraction, we shall divide $N(s)$ by $D(s)$ until the degree of the remainder numerator is one less than n. Thus, we write

$$F(s) = Q(s) + \frac{N_1(s)}{D(s)}$$

where $Q(s)$ is the quotient and $N_1(s)/D(s)$ is a proper fraction and $N_1(s)$ is of degree $(n - 1)$.

Suppose that we restrict ourselves to the case when $n > m$, that is, the degree of denominator polynomial is greater than the degree of numerator polynomial.

The first step is to factor $D(s)$ into linear and quadratic factors with real coefficients

$$F(s) = \frac{N(s)}{D(s)} = \frac{N(s)}{(s - p_1)(s - p_2) \ldots (s - p_n)} \qquad \ldots(10.26.1)$$

The values $p_1, p_2, \ldots p_n$ are called the **poles** of the function $F(s)$. The poles of $F(s)$ may be grouped into four categories.

1. Real poles of first order
2. Complex poles of first order
3. Real poles of multiple order

4. Complex poles of multiple order

In general, $F(s)$ may contain poles belonging to more than one of the above categories. With the help of the method of partial fractions, the given function of s is expanded as the sum of simple terms whose inverse Laplace transforms can be obtained directly from Laplace transform tables.

Simple Real Roots

If the poles of $F(s)$ are simple, that is, $p_1, p_2, \ldots, p_n$ are all different, so that the denominator of $F(s)$ does not have multiple roots, the partial fraction expansion of $F(s)$ takes the form

$$F(s) = \frac{N(s)}{D(s)} = \frac{A_1}{s - p_1} + \frac{A_2}{s - p_2} + \ldots + \frac{A_k}{s - p_k} + \ldots + \frac{A_n}{s - p_n} \qquad (10.26.2)$$

The coefficients of terms in a partial fraction with first degree denominators are called **residues**. Thus, $A_1, A_2, \ldots, A_n$ are the residues at $p_1, p_2, \ldots, p_n$ respectively.

To determine a particular coefficient A_k, multiply both the sides of the above equation by $(s - p_k)$, so that

$$(s - p_k) F(s) = A_1 \frac{s - p_k}{s - p_1} + A_2 \frac{s - p_k}{s - p_2} + \ldots + A_k + \ldots + A_n \frac{s - p_k}{s - p_n} \qquad (10.26.3)$$

By putting $s = p_k$ in Eq. (10.26.3), we get

$$A_k = [(s - p_k) F(s)]_{s = p_k} \qquad k = 1, 2, \ldots, n \qquad (10.26.4)$$

Alternatively, A_k can be found as follows :

$$A_k = \lim_{s \to p_k} (s - p_k) \frac{N(s)}{D(s)}$$

$$= \lim_{s \to p_k} \frac{(s - p_k) N'(s) + N(s)}{Q'(s)} \qquad \text{(By L'Hospital's rule)}$$

therefore,

$$A_k = \left[\frac{N(s)}{D'(s)}\right]_{s = p_k} \qquad (10.26.5)$$

where the primes indicate differenciation with respect to s. The result in Eq. (10.26.5) is useful when $D(s)$ is not in a factored form.

Once A_k is determined, its contribution in the expansion is found by determining the inverse Laplace transform of $\frac{A_k}{s - p_k}$ from the table of Laplace transform pairs. In this case it is found to be $A_k e^{p_k t}$. Hence the inverse Laplace transform of $F(s)$ is given by

$$f(t) = A_1 \pounds^{-1} \frac{1}{s - p_1} + A_2 \pounds^{-1} \frac{1}{s - p_2} + \ldots + A_n \pounds^{-1} \frac{1}{s - p_n}$$

$$= A_1 e^{p_1 t} + A_2 e^{p_2 t} + \ldots + A_n e^{p_n t}$$

$$= \sum_{k=1}^{n} A_k e^{p_k t}, \qquad t > 0 \qquad (10.26.6)$$

Special case

If one of the poles lies at the origin

$$F(s) = \frac{N(s)}{D(s)} = \frac{N(s)}{sD_1(s)}$$

where $D_1(s)$ is of the form

$$D_1(s) = (s - p_2)(s - p_3) \ldots (s - p_n)$$

Therefore,

$$F(s) = \frac{N(0)}{D(0)} \cdot \frac{1}{s} + \sum_{i=2}^{n} \frac{A_i}{s - p_i}$$

$$f(t) = \frac{N(0)}{D(0)} u_{-1}(t) + \sum_{i=2}^{n} A_i e^{p_i t}$$

■ **EXAMPLE 10.20** Find the time-domain expression for

$$F(s) = \frac{s^3 + 7s^2 + 14s + 7}{s^3 + 6s^2 + 11s + 1}$$

SOLUTION : The function $F(s)$ is not a proper fraction. The numerator is, therefore, divided by the denominator to obtain.

$$F(s) = 1 + \frac{s^2 + 3s + 1}{s^3 + 6s^2 + 11s + 6}$$

$$= 1 + \frac{s^2 + 3s + 1}{(s+1)(s+2)(s+3)}$$

$$= F_1(s) + F_2(s) \text{ (say)} \qquad \text{(E 10.20.1)}$$

The second term $F_2(s)$ is a new polynomial with the degree of the numerator less than the degree of the denominator. We can put it as follows :

$$F_2(s) = \frac{s^2 + 3s + 1}{(s+1)(s+2)(s+3)}$$

$$= \frac{A_1}{s+1} + \frac{A_2}{s+2} + \frac{A_3}{s+3} \qquad \text{(E 10.20.2)}$$

The result of Eq. (10.26.4) can be used to evaluate A_1, A_2 and A_3.

$$A_1 = [(s+1) F_2(s)]_{s=-1}$$

$$= \left[\frac{(s^2 + 3s + 1)}{(s+2)(s+3)}\right]_{s=-1} = -\frac{1}{2}$$

$$A_2 = [(s+2)\,F_2(s)]_{s=-2}$$

$$= \left[\frac{s^2+3s+1}{(s+1)(s+3)}\right]_{s=-2} = 1$$

$$A_3 = [(s+3)\,F_2(s)]_{s=-3}$$

$$= \left[\frac{s^2+3s+1}{(s+1)(s+2)}\right]_{s=-3} = \frac{1}{2}$$

Using the values of A_1, A_2 and A_3 in Eq. (E 10.19.2), we get

$$F_2(s) = -\frac{1/2}{s+1} + \frac{1}{s+2} + \frac{1/2}{s+3}$$

From the Laplace transform pair table, the time domain solution is

$$f_2(t) = \pounds^{-1} F_2(s) = -\frac{1}{2} e^{-t} + e^{-2t} + \frac{1}{2} e^{-3t}$$

Now $$\pounds^{-1} F_1(s) = \pounds^{-1}(1) = u_0(t)$$

Since $$F(s) = F_1(s) + F_2(s)$$

$$\pounds^{-1} F(s) = \pounds^{-1} F_1(s) + \pounds^{-1} F_2(s)$$

$$f(t) = u_0(t) - \frac{1}{2} e^{-t} + e^{-2t} + \frac{1}{2} e^{-3t}$$

The above expression gives the time-domain relationship for the given transform $F(s)$.

10.27 ALL ROOTS REPEATED

Let us assume that the denominator of $F(s)$ contains a factor of the form $(s-p_i)^r$, showing that $s = p_i$ is a real pole of the rth order. In other words, $D(s)$ is composed of a multiple root at $s = p_i$ of degree r. We may thus write $F(s)$ in the form

$$F(s) = \frac{N(s)}{D(s)} = \frac{N(s)}{(s-p_i)^r}$$

$$= \frac{A_{i1}}{s-p_i} + \frac{A_{i2}}{(s-p_i)^2} + \ldots + \frac{A_{ir}}{(s-p_i)^r} \qquad (10.27.1)$$

To designate the coefficients in the partial fractions, a single subscript is used for a first order pole. For multiple order poles a double subscript notation is used. The first subscript designates the pole and the second subscript designates the order of the pole in the partial fraction.

To evaluate A_{ir}, the coefficient of the highest-power term in $(s-p_i)$, we multiply both sides of Eq. (10.27.1) by $(s-p_i)^r$ to obtain

$$(s-p_i)^r F(s) = (s-p_i)^r \frac{N(s)}{D(s)}$$

$$= A_{i1}(s-p_i)^{r-1} + A_{i2}(s-p_i)^{r-2} + \ldots + A_{i(r-1)}(s-p_i) + A_{ir} \tag{10.27.2}$$

If we put $s = p_i$, all the terms of the right side of Eq. (10.27.2) become zero except A_{ir}.

$$A_{ir} = [(s-p_i)^r F(s)]_{s=p_i} \tag{10.27.3}$$

$A_{i(r-1)}$ cannot be evaluated in this manner. Multiplying both sides of Eq. (10.27.1) by $(s-p_i)^{r-1}$ and putting $s = p_i$ would make both sides infinite and $A_{i(r-1)}$ indeterminate. We need another procedure to evaluate $A_{i(r-1)}, A_{i(r-2)}, \ldots, A_{i1}$. To evaluate $A_{i(r-1)}$, the term A_{ir} is eliminated from Eq. (10.27.2). In order to do this, Eq. (10.27.2) is differentiated with respect to s. The coefficient A_{ir} disappears and the coefficient $A_{i(r-1)}$ is left alone without the multiplier $(s-p_i)$.

$$\frac{d}{ds}\left[(s-p_i)^r \frac{N(s)}{D(s)}\right]$$

$$= (r-1)A_{i1}(s-p_i)^{r-2} + (r-2)A_{i2}(s-p_i)^{r-3} + \ldots + 2A_{i(r-2)}(s-p_i) + A_{i(r-1)} \tag{10.27.4}$$

Putting $s = p_i$ gives

$$A_{i(r-1)} = \left[\frac{d}{ds}\left\{(s-p_i)^r \frac{N(s)}{D(s)}\right\}\right]_{s=p_i} \tag{10.27.5}$$

Differentiating Eq. (10.27.4) and then putting $s = p_i$ gives

$$A_{i(r-2)} = \frac{1}{2}\left[\frac{d^2}{ds^2}\left\{(s-p_i)^r \frac{N(s)}{D(s)}\right\}\right]_{s=p_i}$$

This differentiation process is continued to obtain the remaining coefficients $A_{i(r-3)}, A_{i(r-4)}, \ldots, A_{i2}$ and A_{i1}. The general formula for the repeated real pole of order r is

$$A_{i(r-k)} = \frac{1}{k!}\left[\frac{d^k}{ds^k}\{(s-p_i)^r F(s)\}\right]_{s=p_i} \tag{10.27.6}$$

$$A_{ik} = \frac{1}{(r-k)!}\left[\frac{d^{r-k}}{ds^{r-k}}\{(s-p_i)^r F(s)\}\right]_{s=p_i} \quad k \le r \tag{10.27.7}$$

If $r = k$, Eq. (10.27.3) is obtained and if $k = (r-1)$, Eq. (10.27.5) is obtained.

10.28 RESIDUE-SUM THEOREM

Residue-sum theorem is due to Hazony and Riley. It is presented here without proof.

$$\sum \text{Res} = \begin{cases} 0 & \text{if } n > m+1 \\ a_n/b_n & \text{if } n = m+1 \end{cases}$$

where Res stands for residue.

This theorem considerably reduces the labour involved in finding the coefficients of partial fraction expansion, specially when $D(s)$ has multiple roots.

10.29 BOTH SIMPLE AND REPEATED ROOTS

Let $F(s)$ be of the form

$$F(s) = \frac{N(s)}{D(s)} = \frac{N(s)}{(s-p_1)^r (s-p_2)(s-p_3)\dots(s-p_n)}$$

The partial fraction expansion is now of the form

$$F(s) = \frac{A_{11}}{s-p_1} + \frac{A_{12}}{(s-p_1)^2} + \dots + \frac{A_{1r}}{(s-p_1)^r} + \frac{A_2}{s-p_2} + \frac{A_3}{s-p_3} + \dots + \frac{A_n}{s-p_n}$$

Here $A_{11}, A_{12}, \dots, A_{1r}$ are found by using Eq. (10.27.7), and $A_2, A_3, \dots, A_n$ are found by using Eq. (10.26.4).

■ **EXAMPLE 10.21** Find the inverse transform of

$$F(s) = \frac{2s+1}{(s+1)^2 (s+2)}$$

SOLUTION : $F(s)$ can be written as

$$F(s) = \frac{2s+1}{(s+1)^2(s+2)} = \frac{A_{11}}{(s+1)} + \frac{A_{12}}{(s+1)^2} + \frac{A_2}{s+2}$$

By Eq. (10.27.7)

$$A_{11} = \frac{1}{1!}\left[\frac{d}{ds}\{(s+1)^2 F(s)\}\right]_{s=-1}$$

$$= \left[\frac{d}{ds}\left(\frac{2s+1}{s+2}\right)\right]_{s=-1} = \left[\frac{3}{(s+2)^2}\right]_{s=-1} = 3$$

$$A_{12} = \frac{1}{0!}[(s+1)^2 F(s)]_{s=-1} = \left[\frac{2s+1}{s+2}\right]_{s=-1} = -1$$

By Eq. (10.26.4)

$$A_2 = \left[(s+2)F(s)\right]_{s=-2} = \left[\frac{2s+1}{(s+1)^2}\right]_{s=-2} = -3$$

Therefore, the partial fraction expansion is

$$F(s) = \frac{3}{s+1} - \frac{1}{(s+1)^2} - \frac{3}{s+2}$$

The corresponding time function for $F(s)$ is

$$f(t) = \pounds^{-1} F(s) = \begin{cases} 3e^{-t} - t\,e^{-t} - 3e^{-2t} & \text{for } t \ge 0 \\ 0 & \text{for } t < 0 \end{cases}$$

10.30 SIMPLE COMPLEX CONJUGATE ROOTS

The two methods discussed above are applicable whether the poles of $F(s)$ are real or complex. If some of the roots of $D(s)$ are complex, some simplification is possible in the determination of the coefficients. This is due to the fact that if $F(s)$ is a rational function with *real* coefficients then the complex poles must appear in complex conjugate pairs. Let $F(s)$ have a simple pole at $s = p_1 = (\alpha + j\beta)$, then it has another simple pole at $s = p_1^* = (\alpha - j\beta)$. The portion of the partial fraction expansion of $F(s)$ relating to these poles must have the following two terms :

$$F_1(s) = \frac{A_1}{s - p_1} + \frac{A_2}{s - p_1^*} \tag{10.30.1}$$

$$= \frac{A_1}{s - \alpha - j\beta} + \frac{A_2}{s - \alpha + j\beta} \tag{10.30.2}$$

$$= \frac{(A_1 + A_2)\, s - (A_1 + A_2)\,\alpha + j\,(A_1 - A_2)\,\beta}{(s - \alpha)^2 + \beta^2} \tag{10.30.3}$$

Now, $F_1(s)$ has real coefficients. Thus, $(A_1 + A_2)$ must be real and $(A_1 - A_2)$ imaginary. It follows that $A_2 = A_1^*$. That is, the coefficients A_1 and A_2 in the partial fraction expansion must also be complex conjugates. Thus, we need only calculate one of the coefficients and the other follows by inspection.

Let
$$A_1 = Ae^{j\theta} \tag{10.30.4}$$

$\therefore$
$$A_2 = A_1^* = Ae^{-j\theta} \tag{10.30.5}$$

The Laplace inverse of Eq. (10.30.2) is given by

$$f_1(t) = A_1 e^{(\alpha + j\beta)t} + A_2 e^{(\alpha - j\beta)t}$$

$$= A\, e^{j\theta} e^{(\alpha + j\beta)t} + A\, e^{-j\theta} e^{(\alpha - j\beta)t}$$

$$= A\, e^{\alpha t}\, [e^{j(\beta t + \theta)} + e^{-j(\beta t + \theta)}]$$

$$= 2A\, e^{\alpha t} \cos(\beta t + \theta) \tag{10.30.6}$$

Then coefficient A_1 is found from the relationship

$$A_1 = [(s - \alpha - j\beta)\, F(s)]_{s = (\alpha + j\beta)} \tag{10.30.7}$$

Then A and θ are found from A_1.

If p_1 is imaginary, α is zero then the response is an undamped cosine wave given by

$$f(t) = 2A \cos(\beta t + \theta) \tag{10.30.8}$$

It is to be noted that it is only necessary to find A_1 by using Eq. (10.30.7) since the coerresponding time function for *both terms* due to complex conjugate poles in Eq. (10.30.2) can be written directly with the help of Eq. (10.30.6).

11

MESH AND NODAL ANALYSIS

11.1 INTRODUCTION

The branch currents and branch voltages in a network can be found by the application of Ohm's law, KCL, and KVL. For the solution of complicated networks the direct application of these basic methods is laborious and time consuming. Mesh-current and node-voltage methods reduce the number of unknown variables and the number of simultaneous equations needed to describe a network. This chapter is devoted to the formulation of network equations for linear time invariant networks.

11.2 MESH CURRENTS

A **mesh** is defined as a closed path around a circuit that does not contain any other closed path within it. A **closed path** is one that begins at a node, moving through successive nodes until the starting node is reached without going through any node more than once. A **mesh current** is a **hypothetical current** that is assumed to flow within the perimeter of that mesh. In other words, a mesh current flows around the boundary of that mesh. The reference direction of mesh currents is selected arbitrarily, but it must be consistent. That is, all mesh currents in a network must be selected either clockwise or counter-clockwise. But, once we have selected a particular direction we must stick to that direction until the end of the problem. For each mesh we assume **clockwise** currents to be **positive**. A mesh current is shown by a curved arrow drawn inside the mesh. The arrow is labelled with its identifying current symbol with a subscirpt.

Consider the network shown in Fig. 11.1. There are two meshes, namely, *abfga* and *bcdfb* in this network. For these meshes the mesh current are I_1 and I_2. Thus, I_1 flows around the boundary of the mesh *abfga* and I_2 flows around the boundary of the mesh *bcdfb*. In other words, the mesh current I_1 flows through *all* the elements in the mesh *abfga*, and the mesh current I_2 flows through *all* the elements in the mesh *bcdfb*.

The branch currents are I_A, I_B and I_C. We can write the following relations between branch and mesh currents :

$$I_A = I_1 , I_B = I_1 - I_2 , I_C = I_2$$

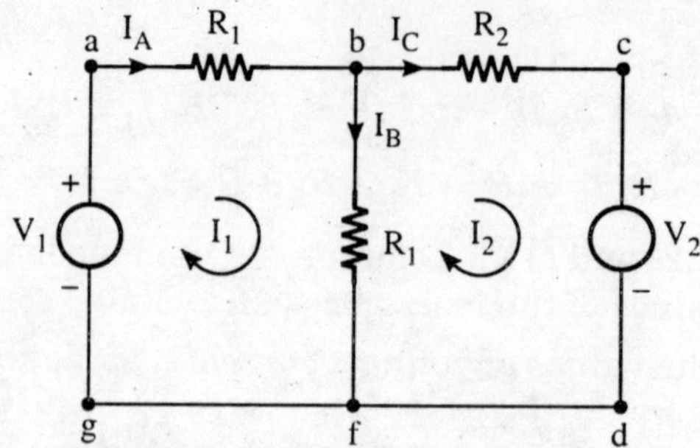

Fig 11.1 Mesh and branch currents

Thus, we see that each branch current is equal to the algebraic sum of the mesh currents in that branch. Hence, if the mesh currents are known, the branch currents may be determined.

11.3 MESH ANALYSIS

Mesh analysis is applicable to planar networks. A **planar network** *is a network that can be drawn on a plane surface such that no two branches intersect at a point which is not a node.*

The method of mesh analysis is based on assigning the mesh currents and then applying KVL around the meshes in the network. For obtaining symmetry in mesh equations, the mesh currents are all taken in the same direction. We shall assume that all the mesh currents flow in *clockwise* direction.

We describe the systematic procedure for writing mesh equations for resistive networks with independent voltage sources. The procedure is developed through the use of several examples.

Consider a planar network shown in Fig. 11.2. The network contains resistances and independent voltage sources. It has three meshes. We assign currents I_1, I_2 and I_3 to these meshes.

By KVL in mesh 1,

$$E_1 - R_1 I_1 - R_3 I_1 + R_3 I_3 - R_2 I_1 + R_2 I_2 = 0$$

or

$$(R_1 + R_2 + R_3)\, I_1 - R_2 I_2 - R_3 I_3 = E_1 \tag{11.3.1}$$

By KVL in mesh 2.

$$E_2 - R_2 I_2 + R_2 I_1 - R_5 I_2 + R_5 I_3 - R_4 I_2 = 0$$

or

$$-R_2 I_1 + (R_2 + R_4 + R_5)\, I_2 - R_5 I_3 = E_2 \tag{11.3.2}$$

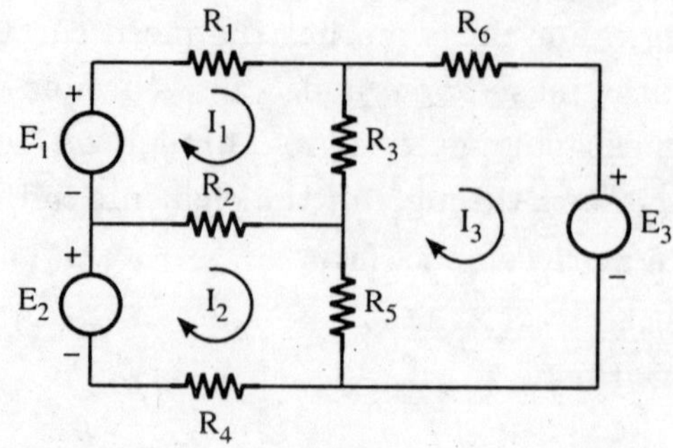

Fig 11.2

By KVL in mesh 3,

$$E_3 - R_7 I_3 + R_5 I_2 - R_5 I_3 - R_3 I_3 + R_3 I_1 - R_6 I_3 = 0$$

or

$$-R_3 I_1 - R_5 I_2 + (R_3 + R_5 + R_6 + R_7) I_3 = E_3 \tag{11.3.3}$$

Equations (11.3.1), (11.3.2) and (11.3.3) form a set of independent equations that can be solved for I_1, I_2 and I_3 if the values of the resistances and voltage sources are known. Once the mesh currents are found, all the values of voltage, currents, and power can be calculated from basic laws.

The matrix equivalent of Eqs. (11.3.1), (11.3.2) and (11.3.3) is

$$\begin{bmatrix} R_1 + R_2 + R_3 & -R_2 & -R_3 \\ -R_2 & R_2 + R_4 + R_5 & -R_5 \\ -R_3 & -R_5 & R_3 + R_5 + R_6 + R_7 \end{bmatrix} \begin{bmatrix} I_1 \\ I_2 \\ I_3 \end{bmatrix} = \begin{bmatrix} E_1 \\ E_2 \\ E_3 \end{bmatrix} \tag{11.3.4}$$

The general form for the set of Eqs. (11.3.4) will be

$$\begin{bmatrix} R_{11} & R_{12} & R_{13} \\ R_{21} & R_{22} & R_{23} \\ R_{31} & R_{32} & R_{33} \end{bmatrix} \begin{bmatrix} I_1 \\ I_2 \\ I_3 \end{bmatrix} = \begin{bmatrix} E_1 \\ E_2 \\ E_3 \end{bmatrix} \tag{11.3.5}$$

In expanded form these equations may be written as

$$\left.\begin{aligned} R_{11} I_1 + R_{12} I_2 + R_{13} I_3 = E_1 \\ R_{21} I_1 + R_{22} I_2 + R_{23} I_3 = E_2 \\ R_{31} I_1 + R_{32} I_2 + R_{33} I_3 = E_3 \end{aligned}\right\} \tag{11.3.6}$$

The results obtained in Eqs. (11.3.1) to Eq. (11.3.3) can be extended to any number of meshes. The **mesh equations** for a linear network having m independent meshes may be written in the **generalized form** as

$$\left.\begin{aligned} R_{11} I_1 + R_{12} I_2 + \ldots + R_{1m} I_m = E_1 \\ R_{21} I_1 + R_{22} I_2 + \ldots + R_{2m} I_m = E_2 \\ \ldots \quad \ldots \quad \ldots \quad \ldots \quad \ldots \\ R_{m1} I_1 + R_{m2} I_2 + \ldots + R_{mm} I_m = E_m \end{aligned}\right\} \tag{11.3.7}$$

These equations can be written in matrix form as

$$\begin{bmatrix} R_{11}\, R_{12} \ldots R_{1m} \\ R_{21}\, R_{22} \ldots R_{2m} \\ \ldots \quad \ldots \\ \ldots \quad \ldots \\ R_{m1}\, R_{m2} \ldots R_{mm} \end{bmatrix} \begin{bmatrix} I_1 \\ I_2 \\ \ldots \\ \ldots \\ I_m \end{bmatrix} = \begin{bmatrix} E_1 \\ E_2 \\ \ldots \\ \ldots \\ E_m \end{bmatrix} \tag{11.3.8}$$

Eq. (11.3.8) can be written in a more compact form as

$$[\mathbf{R}_m]\, [\mathbf{I}_m] = [\mathbf{E}_m] \tag{11.3.9}$$

which is **Ohm's law in matrix form**.

In Eq. (11.3.9)

$$[\mathbf{R}_m] = \begin{bmatrix} R_{11} \, R_{12} \, \dots \, R_{1m} \\ R_{21} \, R_{22} \, \dots \, R_{2m} \\ \dots \quad \dots \quad \dots \\ \dots \quad \dots \quad \dots \\ R_{m1} \, R_{m2} \, \dots \, R_{mm} \end{bmatrix}$$

$$[\mathbf{I}_m] = \begin{bmatrix} I_1 \\ I_2 \\ \dots \\ \dots \\ I_m \end{bmatrix}, \quad [\mathbf{E}_m] = \begin{bmatrix} E_1 \\ E_2 \\ \dots \\ \dots \\ E_m \end{bmatrix}$$

The matrix $[\mathbf{R}_m]$ is called the **mesh-resistance matrix**. The order of the subscripts among the elements of this matrix are to be noted. The first or left subscript attached to each element indicates the row that the element occupies and the second subscript identifies the column. For example, R_{11} is in row 1 and column 1 of the set of equations. R_{23} is located in row 2, column 3, etc. The elements of mth row are the R elements associated with the equation for mth mesh.

The values of the voltages and the resistances in mesh equations can be determined as follows :

E_1 = sum of all the voltages of the voltage sources around mesh 1

E_2 = sum of all the voltages of the voltages sources around mesh 2

E_m = sum of all the voltages of the voltages sources around mesh m

The sign of the voltage is taken positive if it acts in a direction to produce a current in the assumed positive direction of its mesh current.

The resistances with like subscripts (R_{11}, R_{22}, ... R_{mm}) are called the **self resistances of the meshes**. The self resistance of a mesh is equal to the sum of all the resistances around that mesh. Thus,

R_{11} = self resistance of mesh 1

= sum of all the resistances in mesh 1

R_{22} = self resistance of mesh 2

= sum of all the resistances in mesh 2

R_{mm} = self resistance of mesh m

= sum of all the resistances in mesh m

The resistances with unlike subscripts (R_{jk}, $j \neq k$) in the generalized mesh equations are called **mutual resistances** of the meshes. The mutual resistances are common to the meshes indicated by their subscripts. For example, R_{jk}, for which $j \neq k$, is equal to the sum of all the resistances common to both the jth and kth meshes. The sign of mutual resistance is positive or negative according as the directions of mesh currents I_j and I_k are the same or opposite in the mutual resistances. The directions of mesh currents may be selected arbitrarily. But if we assign them *clockwise* direction in every mesh, the mutual resistances will *always* be *negative.* This convention has a great advantage that in mesh equations **all the self resistance will always be positive and all the mutual resistances will always be negative.** We shall adopt this convention throughout. With this convention,

$$R_{jk} = R_{kj} = -\text{ (sum of all the resistances common to meshes } j \text{ and } k)$$

The advantage of the generalized form of the mesh equations is that these equations are easily remembered because of their symmetry. For a given network they can be written by inspection.

11.3.1 Procedure of Writing Mesh Equations

The procedure for writing mesh equations is as follows:

Step 1. Convert all practical current sources to equivalent voltage sources.

Step 2. The circuit is then redrawn.

Step 3. Select a mesh current for each mesh.

Step 4. Choose all the mesh currents in the same direction; a clockwise direction is usually convenient.

Step 5. Write a KVL equation for each mesh in direction of the mesh current.

The **mesh equations** for a linear network having m independent meshes with no dependent (controlled) sources may be written in the **generalized form** as

$$\begin{aligned} R_{11}I_1 + R_{12}I_2 + \dots + R_{1m}I_m &= E_1 \\ R_{21}I_1 + R_{22}I_2 + \dots + R_{2m}I_m &= E_2 \\ \dots \quad \dots \quad \dots \quad \dots \\ R_{m1}I_1 + R_{m2}I_2 + \dots + R_{mm}I_m &= E_m \end{aligned}$$

These equations can be written in matrix form as

$$\begin{bmatrix} R_{11} & R_{12} & \cdots & R_{1m} \\ R_{21} & R_{22} & \cdots & R_{2m} \\ \cdots & \cdots & \cdots & \cdots \\ R_{m1} & R_{m2} & \cdots & R_{mm} \end{bmatrix} \begin{bmatrix} I_1 \\ I_2 \\ \cdots \\ I_m \end{bmatrix} = \begin{bmatrix} E_1 \\ E_2 \\ \cdots \\ E_m \end{bmatrix}$$

Step 6. Solve the set of equations for the desired mesh currents. Other currents and voltage in the circuit are found by using ohm's law and KCL.

■ **EXAMPLE 11.1** Write the mesh equations for the network shown in Fig. 11.3.

SOLUTION : The given network is resistive. The mesh equations in the generalized form for the given network are

$$\left.\begin{aligned} R_{11}I_1 + R_{12}I_2 &= V_1 \\ R_{21}I_1 + R_{22}I_2 &= V_2 \end{aligned}\right\} \qquad \text{(E11.1.1)}$$

In matrix form, the mesh equations can be written as

$$\begin{bmatrix} R_{11} & R_{12} \\ R_{21} & R_{22} \end{bmatrix}\begin{bmatrix} I_1 \\ I_2 \end{bmatrix}=\begin{bmatrix} V_1 \\ V_2 \end{bmatrix} \tag{E11.1.2}$$

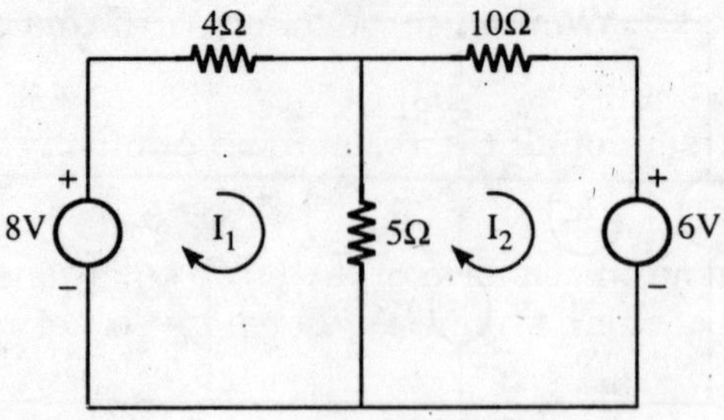

Fig 11.3

The elements of the resistance matrix [**R**] can be written directly by inspection of the given network as follows :

R_{11} = sum of all the resistances in mesh 1

$= 4 + 5 = 9\ \Omega$

R_{22} = sum of all the resistances in mesh 2

$= 10 + 5 = 15\ \Omega$

$R_{12} = R_{21} = -$ (sum of all the resistances common to meshes 1 and 2)

$= -5\ \Omega$

The mesh voltages V_1 and V_2 are written as follows :

V_1 = algebraic sum of all the voltages around mesh 1

Since there is only one voltage source of 8V and is acting in a direction to produce a current in the assumed positive direction of its mesh current I_1,

$V_1 = +8$ V.

V_2 = algebraic sum of all the voltages around mesh 2

$= -6$ V, since the voltage source in mesh 2 opposes the circulation of mesh current I_2 in it.

Substitution of the values of resistances and voltages in Eqs. (E11.1.1) and (E11.1.2) gives the mesh equations as

$$\left.\begin{aligned} 9I_1 - 5I_2 &= 8 \\ -5I_1 + 15I_2 &= -6 \end{aligned}\right\} \tag{E11.1.3}$$

or

$$\begin{bmatrix} 9 & -5 \\ -5 & 15 \end{bmatrix}\begin{bmatrix} I_1 \\ I_2 \end{bmatrix}=\begin{bmatrix} 8 \\ -6 \end{bmatrix} \tag{E11.1.4}$$

■ **EXAMPLE 11.2** Write the mesh equations for the network shown in Fig. 11.4.

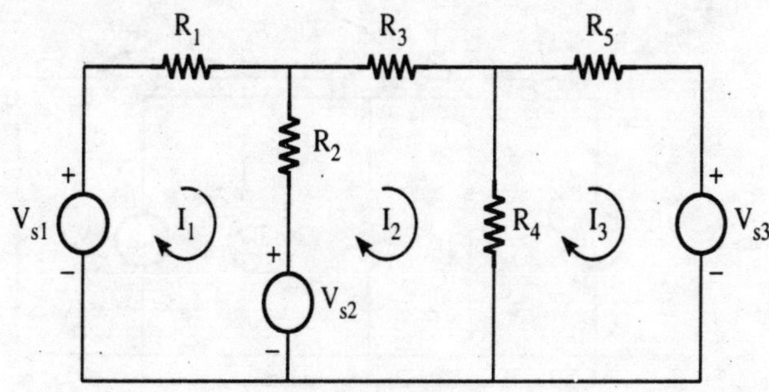

Fig 11.4

SOLUTION : The mesh equations in the generalized form for the given network are

$$R_{11} I_1 + R_{12} I_2 + R_{13} I_3 = V_1$$

$$R_{21} I_1 + R_{22} I_2 + R_{23} I_3 = V_2$$

$$R_{31} I_1 + R_{32} I_2 + R_{33} I_3 = V_3$$

Here $R_{11} = R_1 + R_2,\ R_{22} = R_2 + R_3 + R_4,\ R_{33} = R_4 + R_5$

$R_{12} = R_{21} = -R_2$

$R_{23} = R_{32} = -R_4$

$R_{13} = R_{31} = 0$, since there is no resistance common to meshes 1 and 3.

$V_1 = V_{s1} + V_{s2}$, since both the voltage sources, V_{s1} and V_{s2} produce clockwise currents in mesh 1

$V_2 = -V_{s2}$, since the voltage source V_{s2} in mesh 2 opposes the `circulation of mesh current in it.

Similarly, $V_3 = -V_{s3}$

Hence the mesh equations are

$$(R_1 + R_2) I_1 - R_2 I_2 - (0) I_3 = V_{s1} + V_{s2}$$

$$-R_2 I_1 + (R_2 + R_3 + R_4) I_2 - R_4 I_3 = -V_{s2}$$

$$(0) I_1 - R_4 I_2 + (R_4 + R_5) I_3 = -V_{s3}$$

11.4 MESH EQUATIONS FOR NETWORKS WITH INDEPENDENT PRACTICAL CURRENT SOURCES

We shall now consider mesh analysis for the case when a network contains one or more independent current sources. When practical current sources (those with parallel impedances) are present in a network we use source transformation to convert sources to equivalent voltage sources. Then we write KVL equations around each mesh as given previously.

EXAMPLE 11.3 Determine the mesh currents in the network shown in Fig. 11.5.

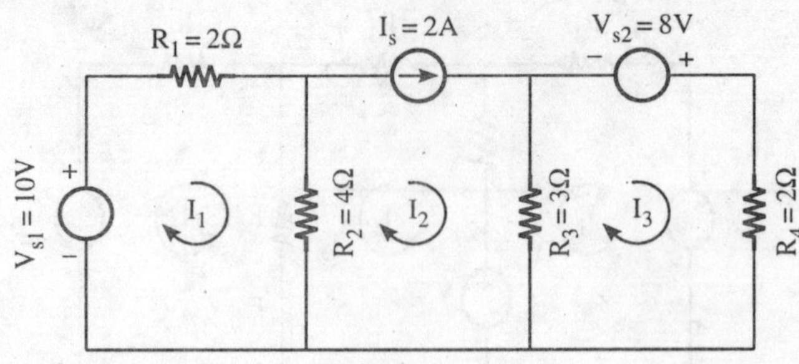

Fig 11.5

SOLUTION : The given network contains a current source also. The voltage across the current source is not known. However, the source current I_s is equal to the mesh current I_2. Thus

$$I_2 = I_s = 2\text{ A}$$

Since I_2 is known, it is sufficient to write mesh equations for meshes 1 and 3 only.

For mesh 1,

$$V_{s1} = (R_1 + R_2)\,I_1 - R_2\,I_2 - 0\,(I_3)$$

$$10 = (2 + 4)\,I_1 - 4 \times 2$$

$$10 = 6I_1 - 8$$

therefore, $$I_1 = \frac{1}{6}\,(10 + 8) = 3\text{ A}$$

For mesh 3,

$$V_{s2} = 0\,(I_1) - R_3\,I_2 + (R_3 + R_4)\,I_3$$

$$8 = -\,3 \times 2 + (3 + 2)\,I_3,\; I_3 = \frac{1}{5}\,(8 + 6) = 2.8\text{ A}$$

11.5 MESH EQUATIONS FOR NETWORKS WITH IDEAL CURRENT SOURCES

When ideal current sources are present in meshes, some of the mesh currents can be related to the source currents. If the current source is common to only *one* mesh, no KVL equation is written for that mesh. That mesh current is given by the source current. When a current source is common to *two* meshes, the source current can be expressed in terms of the mesh currents of these two meshes. A voltage is assumed across the current source common to the two meshes. Mesh equations are then written for both the meshes. The assumed *unknown* voltage appears in both the mesh equations. These two equations are then combined into one by eliminating the unknown voltage across the current source.

Another possible method of obtained mesh equation for networks containing current sources is the applcation of KVL to the supermeshes. If a current source is common to two meshes, a **supermesh** is obtained by combining the meshes containing the current source and considering the current source as an open circuit. It is to be noted that the supermesh equation cannot be written by inspection. The KVL equation around the supermesh is written in terms of the *original* mesh currents and resistances and the voltage sources.

■ **EXAMPLE 11.4** Write the mesh equations for the network in Fig. 11.6.

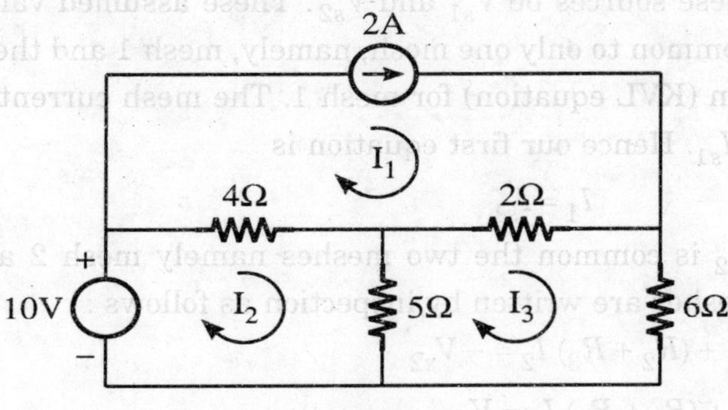

Fig 11.6

SOLUTION : The given network contains a 2A current source which is not directly in parallel with a resistance. Hence this current source cannot be transformed to its equivalent voltages source. Since the network has three meshes, we need three mesh equations. Here the source current is directly expressible in terms of mesh current I_1, that is

$$I_1 = 2\text{ A} \qquad \text{(E11.4.1)}$$

This is our first mesh equation. Only I_2 and I_3 are unkown. Hence we need to write only two additional independent equations.

By KVL in mesh 2,

$$R_{21} I_1 + R_{22} I_2 + R_{23} I_3 = V_2$$

$$-4I_1 + (4+5) I_2 - 5I_3 = 10$$

$$-4I_1 + 9I_2 - 5I_3 = 10 \qquad \text{(E11.4.2)}$$

By KVL in mesh 3,

$$R_{31} I_1 + R_{32} I_2 + R_{33} I_3 = V_3$$

$$-2I_1 - 5I_2 + (2+6+5) I_3 = 0$$

$$-2I_1 - 5I_2 + 13I_3 = 0 \qquad \text{(E11.4.3)}$$

Eqs. (E11.4.1), (E11.4.2) and (E11.4.3) are three equations with three unknowns. They can be solved to give I_1, I_2 and I_3.

■ **EXAMPLE 11.5** Write the mesh equations for the network shown in Fig. 11.7.

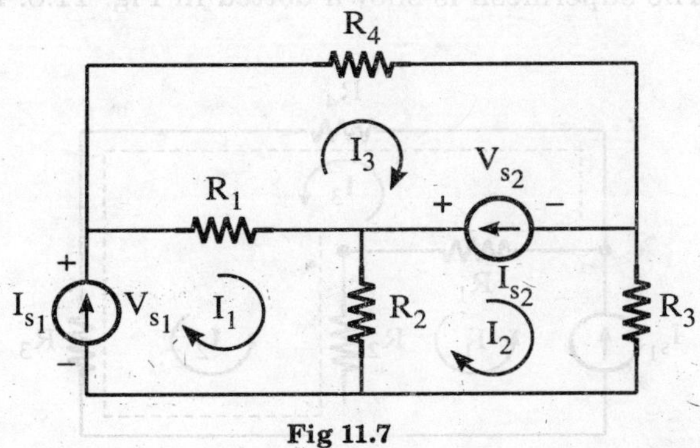

Fig 11.7

SOLUTION : The given network contains two ideal current sources with values I_{s1} and I_{s2}. Let the voltages across these sources be V_{s1} and V_{s2}. These assumed values are unknown. The current source I_{s1} is common to only one mesh, namely, mesh 1 and therefore there is no need to write mesh equation (KVL equation) for mesh 1. The mesh current I_1 for mesh 1 is equal to the source current I_{s1}. Hence our first equation is

$$I_1 = I_{s1} \tag{E11.5.1}$$

The current source I_{s2} is common the two meshes namely mesh 2 and mesh 3. The mesh equations for these meshes are written by inspection as follows :

Mesh 2 : $-R_2 I_1 + (R_2 + R_3) I_2 = -V_{s2}$ (E11.5.2)

Mesh 3 : $-R_1 I_1 + (R_1 + R_4) I_3 = V_{s2}$ (E11.5.3)

Since V_{s1} and V_{s2} are unknown and they cannot be expressed in terms of products of mesh currents and resistances by Ohm's law, we shall eliminate them by adding Eqs. (E11.5.1) and (E11.5.2) as follows :

$$-(R_1 + R_2) I_1 + (R_2 + R_3) I_2 + (R_1 + R_4) I_3 = 0 \tag{E11.5.4}$$

Meshes 2 and 3 contain the common current source I_{s2}. The current I_{s2} is related to the mesh currents I_2 and I_3 by the following relationship :

$$I_3 - I_2 = I_{s2}$$

or

$$I_2 = I_3 - I_{s2} \tag{E11.5.5}$$

Since the mesh current I_1 is already known from Eq. (E11.5.1), we can write Eq. (E11.5.4) as

$$-(R_1 + R_2) I_{s1} + (R_2 + R_3) I_2 + (R_1 + R_4) I_3 = 0 \tag{E11.5.6}$$

Substitution of the value of I_2 from Eq. (E11.5.5) in Eq. (E11.5.6) gives

$$-(R_1 + R_2) I_{s1} + (R_2 + R_3)(I_3 - I_{s2}) + (R_1 + R_4) I_3 = 0$$

or

$$(R_1 + R_2 + R_3 + R_4) I_3 = (R_1 + R_2) I_{s1} + (R_2 + R_3) I_{s2} \tag{E11.5.7}$$

Eq.(E11.5.7) can be solved to give I_3. When I_3 is known can find I_2 from Eq. (E11.5.5). The current I_1 is already known. Since, all the mesh currents are determined we can determine the behaviour of the network.

Alternative Method

Consider the supermesh formed by combining meshes 2 and 3 considering the current source I_{s2} as an open circuit. The supermesh is shown dotted in Fig. 11.8. By applying KVL in the supermesh we get

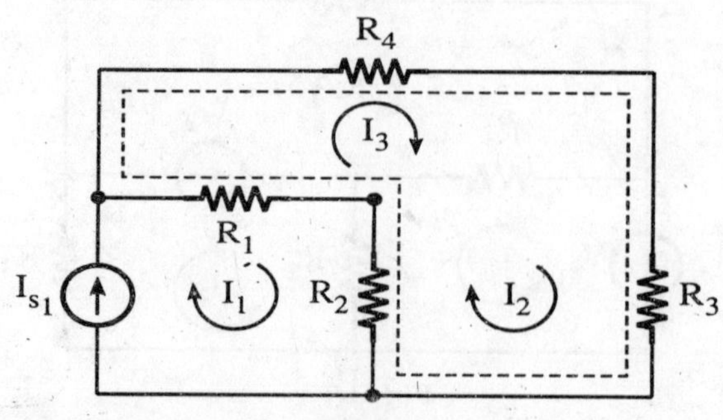

Fig 11.8

$$I_1R_2 - I_2R_2 + I_1R_1 - I_3R_1 - I_3R_4 - I_2R_3 = 0$$

$$(R_1 + R_2)\,I_1 - (R_2 + R_3)\,I_2 - (R_1 + R_4)\,I_3 = 0$$

$$-(R_1 + R_2)\,I_1 + (R_2 + R_3)\,I_2 + (R_1 + R_4)\,I_3 = 0 \qquad \text{(E11.5.8)}$$

It is seen that Eq. (E11.5.8) is the same as Eq. (E11.5.4).

■ **EXAMPLE 11.6** Determine the mesh currents in the network shown in Fig. 11.9 if $R_1 = 1\ \Omega$, $R_2 = 2\ \Omega$, $R_3 = 3\ \Omega$, $R_4 = 4\ \Omega$, $R_5 = 5\ \Omega$, $V_s = 12$ V, $I_s = 4$ A.

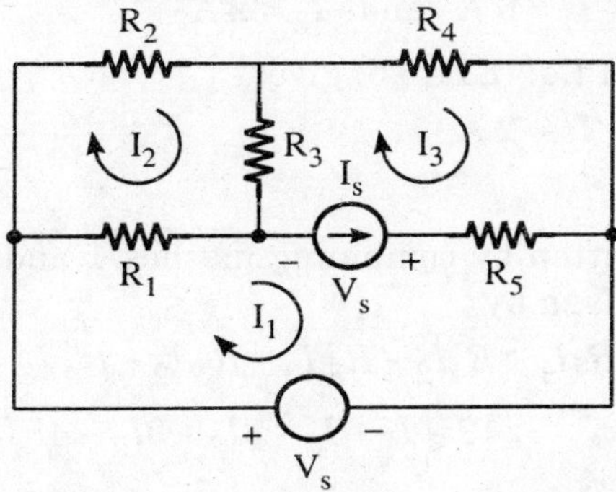

Fig 11.9

SOLUTION : The given network contains a current source common to two meshes 1 and 3. The voltage across the current source is not known. Let us assume the voltage across the current source be V_x. With this assumption the mesh equations can be written easily. An additional equation is written for the current source which relates the source current to the mesh currents passing through it.

The mesh equations can be written as follows :

Mesh 1 : $V_s + V_x = (R_1 + R_5)\,I_1 - R_1I_2 - R_5\,I_3$ (E11.6.1)

Mesh 2 : $0 = -R_1I_1 + (R_1 + R_2 + R_3)\,I_2 - R_3\,I_3$ (E11.6.2)

Mesh 3 : $-V_x = -R_5\,I_1 - R_3\,I_2 + (R_3 + R_4 + R_5)\,I_3$ (E11.6.3)

Substitution of the given values in the above equations gives

$$12 + V_x = 6I_1 - I_2 - 5I_3 \qquad \text{(E11.6.4)}$$

$$0 = -I_1 + 6I_2 - 3I_3 \qquad \text{(E11.6.5)}$$

$$-V_x = -5I_1 - 3I_2 + 12I_3 \qquad \text{(E11.6.6)}$$

Adding Eqs. (E11.6.4) and (E11.6.6) to eliminate V_x we get

$$12 = I_1 - 4I_2 + 7I_3 \qquad \text{(E11.6.7)}$$

The relation between the source current and mesh currents is given by

$$I_s = I_1 - I_3$$

or

$$4 = I_1 - I_3 \qquad \text{(E11.6.8)}$$

Thus we have got three Eq. (E11.6.5), (E11.6.7) and (E11.6.8) to determine the three unknowns I_1, I_2 and I_3.

Multiplying Eqs. (E11.6.5) by 2 and Eq. (E11.6.7) by 3 and adding the results to eliminate I_2 we get

$$I_1 + 15I_3 = 36 \qquad \text{(E11.6.9)}$$

Also $$I_1 - I_3 = 4 \qquad \text{(E11.6.10)}$$

Solution of these equations gives

$$I_1 = 6\text{ A} \quad \text{and} \quad I_3 = 2\text{ A}$$

Substitution of these values in Eq. (E11.6.5) gives

$$I_3 = 2\text{ A}$$

Alternative method

Equation (E11.6.7) can be written by combining meshes 1 and 3 to form a supermesh. The supermesh KVL equation is given by

$$+12 - R_1 I_1 + R_1 I_2 - R_3 I_3 + R_3 I_2 - R_4 I_3 = 0$$

$$12 - I_1 + I_2 - 3I_3 + 3I_2 - 4I_3 = 0$$

$$I_1 - 4I_2 + 7I_3 = 12$$

11.6 MESH EQUATIONS FOR NETWORKS WITH DEPENDENT SOURCES

When a network contains a dependent source, mesh equations are written in the same manner as networks without dependent sources. A dependent source is treated exactly as an independent source in obtaining the independent mesh equations. Then a dependent equation is written for each dependent source in terms of one or more of the mesh currents. These values are substituted into the main mesh equations from which the mesh currents can be determined. The method is illustrated by the following examples.

■ **EXAMPLE 11.7** Write the set of mesh equations that are required to solve the network given in Fig. 11.10.

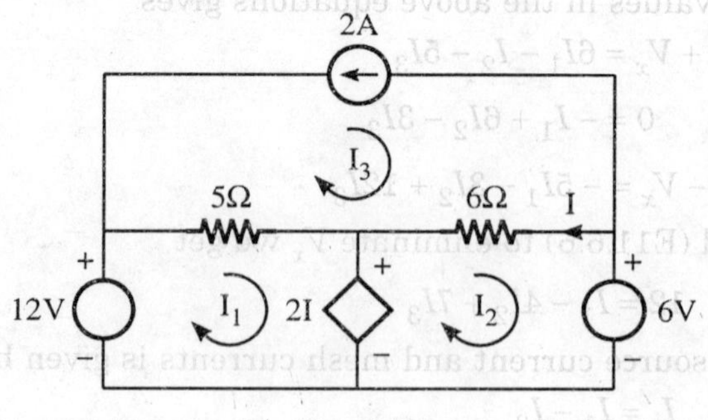

Fig 11.10

SOLUTION : The three independent mesh equations that are required for the network of Fig. 11.10 are two KVL equations for meshes 1 and 2 and a KCL equation that relates the 3 A source to the mesh current. Since there are no current sources in meshes 1 and 2, the two KVL equations can be written by inspection as follows :

$$\text{Mesh 1}: \; 5I_1 - 0 \times I_2 - 5I_3 = 12 - 2I$$

$$\text{Mesh 2}: \; -0 \times I_1 + 6I_2 - 6I_3 = 2I - 6$$

The KCL equation is

$$I_3 = -3$$

■ **EXAMPLE 11.8** Write the mesh equations for the network shown in Fig. 11.11.

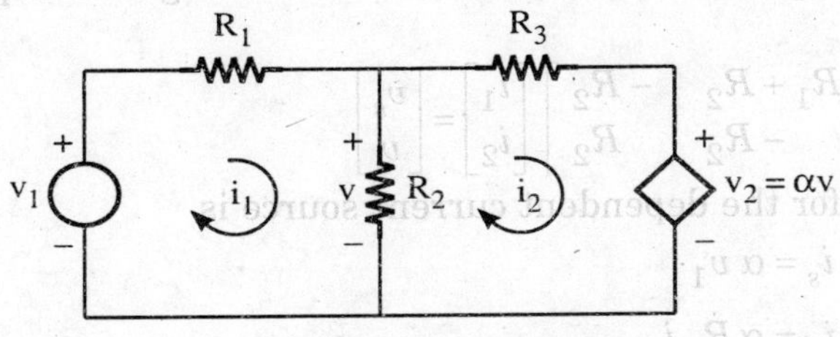

Fig 11.11

SOLUTION : The mesh equations of the network treating the dependent voltage source v_2 as an independent source are

$$\begin{bmatrix} R_1 + R_2 & -R_2 \\ -R_2 & R_2 + R_3 \end{bmatrix} \begin{bmatrix} i_1 \\ i_2 \end{bmatrix} = \begin{bmatrix} v_1 \\ -v_2 \end{bmatrix} \qquad \text{(E11.8.1)}$$

These equations can be written in the expanded form as

$$(R_1 + R_2)\, i_1 - R_2\, i_2 = v_1 \qquad \text{(E11.8.2)}$$

$$-R_2\, i_1 + (R_2 + R_3)\, i_2 = -v_2 \qquad \text{(E11.8.3)}$$

The constraint equation for the dependent voltage source is

$$v_2 = \alpha v = \alpha\,(i_1 - i_2)\, R_2 \qquad \text{(E11.8.4)}$$

Substituting the value of v_2 from Eq. (E11.8.4) in Eq. (E11.8.3) we get

$$-R_2 i_1 + (R_2 + R_3)\, i_2 = \alpha R_2\,(i_2 - i_1)$$

$$(-R_2 + \alpha R_2)\, i_1 + (R_2 + R_3 - \alpha R_2)\, i_2 = 0 \qquad \text{(E11.8.5)}$$

Eqs. (E11.8.3) and (E11.8.5) can be written in matrix form as

$$\begin{bmatrix} R_1 + R_2 & -R_2 \\ -R_2 + \alpha R_2 & R_2(1-\alpha) + R_3 \end{bmatrix} \begin{bmatrix} i_1 \\ i_2 \end{bmatrix} = \begin{bmatrix} v_1 \\ 0 \end{bmatrix}$$

■ **EXAMPLE 11.9** Write the mesh equations for the network of Fig. 11.12.

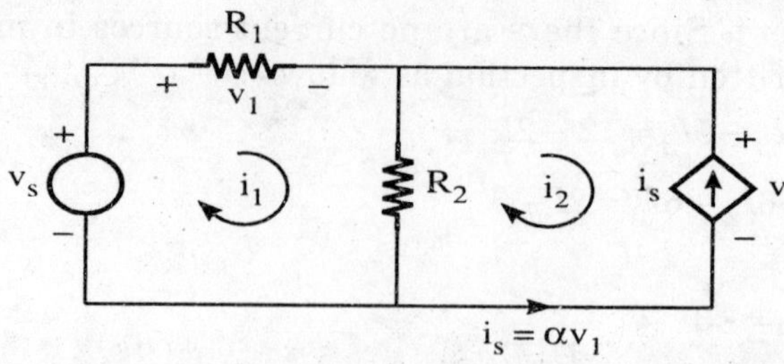

Fig 11.12

SOLUTION : The mesh equations for the network treating the dependent source as an independent source are

$$\begin{bmatrix} R_1+R_2 & -R_2 \\ -R_2 & R_2 \end{bmatrix}\begin{bmatrix} i_1 \\ i_2 \end{bmatrix}=\begin{bmatrix} v_s \\ v \end{bmatrix} \tag{E11.9.1}$$

The constraint equation for the dependent current source is

$$i_s = \alpha\, v_1$$

$$-\,i_2 = \alpha R_1\, i_1$$

$$i_2 = -\,\alpha R_1 i_1$$

Substituting the value of i_2 in Eq. (E11.9.1) we get

$$\begin{bmatrix} R_1+R_2 & -R_2 \\ -R_2 & R_2 \end{bmatrix}\begin{bmatrix} i_1 \\ -\,\alpha R_1 i_1 \end{bmatrix}=\begin{bmatrix} v_s \\ v \end{bmatrix}$$

11.7 NODAL ANALYSIS

A **node** is defined as a connection of two or more circuit elements. The first step in nodal analysis is that we identify all the *different* nodes in the network. A **different node** is one whose voltage is different from any other node. One of these nodes is arbitrarily chosen as the **reference node**. Generally, the node connecting the greatest number of elements is taken as the reference node. We assume that the reference node is at zero potential. The reference node is also called the **datum node** or the **ground node.** All the different nodes are assigned numbers. The datum node is assigned the label 0 and the remaining nodes are labelled 1, 2, 3, ... Let $v_1, v_2, ..., v_n$ be the node voltages of nodes 1, 2, ..., n measured with respect to the reference node. The voltages $v_1, v_2, ..., v_n$ are called **node-to-datum voltages** or simple **node voltages**. These node voltages are taken as the unknown variables in node analysis.

The voltage at a node is indicated by a voltage symbol written adjacent to that node. For example, the voltage at node 1 is identified by a voltage symbol v_1 written adjacent to node 1. The second step in nodal analysis is that we arbitrarily assign currents to the branches in which no current is designated. The third step is to write KCL equations at each *nonreference* node. In the fourth step we apply Ohm's law to write branch currents in terms of node voltages. These values of branch currents are now substituted in KCL equations obtained in the third step. Then we arrange these equations so that the unknown node voltages appear in the left-hand side and the known quantities on the right-hand side of equations. These equations are called **node equations.**

We shall develop the technique of nodal analysis by considering circuits containing only resistors and sources. The same technique can be applied to the networks containing other types of elements also.

11.8 NODE EQUATIONS FOR NETWORKS HAVING ONLY RESISTORS AND INDEPENDENT CURRENT SOURCES

Consider the network shown in Fig. 11.13. This network consists of resistors and independent current sources only. To apply nodal analysis we choose node 3 as the reference node indicated by the ground symbol . Let the voltages of nodes 1 and 2 with respect to the reference node be v_1 and v_2 respectively. These voltages are identified at nodes 1 and 2 in the network of Fig. 11.13. We will now arbitrarily assign currents to the elements in which no current is designated. In Fig. 11.13 the assigned currents are i_1, i_2 and i_3 while the currents I_{s1}, I_{s2} are known. Now we write KCL equations at each non-reference node.

By KCL at node 1,

sum of incoming currents = sum of outgoing currents

$$I_{s1} = i_1 + i_2 \tag{11.8.1}$$

By KCL at node 2,

$$i_2 + I_{s2} = i_3 \tag{11.8.2}$$

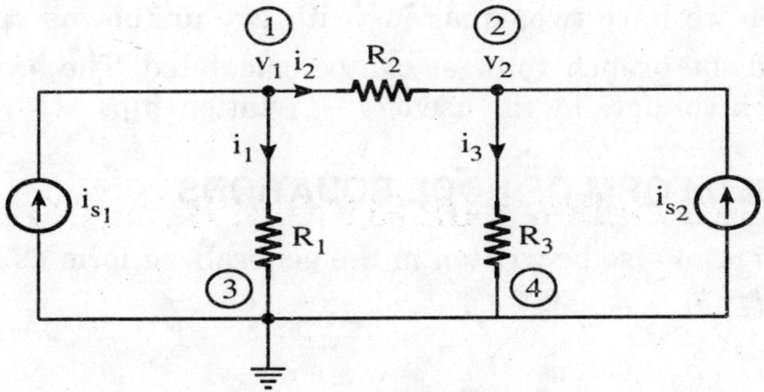

Fig 11.13

Now we apply Ohm's law to write branch currents in terms of node voltages as follows :

$$i_1 = \frac{\text{voltage between nodes 1 and 3}}{\text{resistance between nodes 1 and 3}}$$

$$= \frac{v_1 - 0}{R_1} = \frac{1}{R_1} v_1 = G_1 v_1 \tag{11.8.3}$$

$$i_2 = \frac{\text{voltage between nodes 1 and 2}}{\text{resistance between nodes 1 and 2}}$$

$$= \frac{v_1 - v_2}{R_2} = G_2 (v_1 - v_2) \tag{11.8.4}$$

$$i_3 = \frac{\text{voltage between nodes 2 and 3}}{\text{resistance between nodes 2 and 3}}$$

$$= \frac{v_2 - 0}{R_3} = G_3 v_2 \tag{11.8.5}$$

It is to be noted that the current flows from higher potential point to lower potential point. Since we have chosen the directions for i_1, i_2 and i_3 arbitrarily, the polarities of the voltage terms should be assigned properly in consistence with our assumption. For example, i_1 flows from node 1 to node 3, then the voltage between node 1 and node 3 should be taken as $(v_1 - 0)$ and not $(0 - v_1)$. Since i_2 is assumed to flow from node 1 to node 2, the voltage across R_2 is $(v_1 - v_2)$ and not $(v_2 - v_1)$.

These values of branch currents are now substituted in current Eqs. (11.8.1) and (11.8.2) to give

$$I_{s1} = G_1 v_1 + G_2 (v_1 - v_2)$$

and

$$G_2 (v_1 - v_2) + I_{s2} = G_3 v_2$$

We rearrange these equations so that the unknown node voltages appear in the left-hand side and the known quantities on the right-hand side of equations. We can write

$$(G_1 + G_2)\, v_1 - G_2 v_2 = I_{s1} \tag{11.8.6}$$

$$- G_2 v_1 + (G_2 + G_3)\, v_2 = I_{s2} \tag{11.8.7}$$

Eqs. (11.8.6) and (11.8.7) are called **node equations**. If the values of all conductances and sources are known, then we have two equations with two unknowns v_1 and v_2. Once these voltages are determined the branch voltages can be calculated. The branch currents can be obtained from the branch voltages by the use of $v - i$ relationships.

11.9 GENERALIZED FORM OF NODE EQUATIONS

Eqs. (11.8.6) and (11.8.7) can also be written in the generalized form as

$$G_{11}\, v_1 + G_{12}\, v_2 = I_{s1} \tag{11.9.1}$$

$$G_{21}\, v_1 + G_{22}\, v_2 = I_{s2} \tag{11.9.2}$$

Alternatively, expressing these two equations as a single matrix equation, we have

$$\begin{bmatrix} G_{11} & G_{12} \\ G_{21} & G_{22} \end{bmatrix} \begin{bmatrix} v_1 \\ v_2 \end{bmatrix} = \begin{bmatrix} I_{s1} \\ I_{s2} \end{bmatrix} \tag{11.9.3}$$

In a more compact form, we have

$$[\mathbf{G}]\, [\mathbf{v}] = [\mathbf{i}_s] \tag{11.9.4}$$

where $[\mathbf{G}] = \begin{bmatrix} G_{11} & G_{12} \\ G_{21} & G_{22} \end{bmatrix}$, $[\mathbf{v}] = \begin{bmatrix} v_1 \\ v_2 \end{bmatrix}$, $[\mathbf{i}_s] = \begin{bmatrix} I_{s1} \\ I_{s2} \end{bmatrix}$

Eq. (11.9.4) is just Ohm's law in matrix form. Here [**G**] is called the **node-conductance matrix**, [**v**] is the **node-voltage matrix** and [$\mathbf{i}_s$] is the **node-current source matrix**.

The values of the conductances in the above equations for network of Fig. 11.13 may be determined as follow :

$G_{11} = G_1 + G_2$ = sum of the conductances of all branches directly connected to node 1

$G_{22} = G_2 + G_3$ = sum of the conductances of all branches directly connected to node 2

$G_{12} = G_{21} = -G_2 = -$ (sum of the conductances of all branches directly connecting nodes 1 and 2)

The conductance G_{11} is called the **self conductance of node 1**. Similarly, the conductance G_{22} is called the **self conductance of node 2**. The conductance G_{12} or G_{21} is called the **mutual conductance between nodes 1 and 2.**

The results obtained for two nodes can be extended to any number of nodes. For $(n + 1)$ nodes we shall select $(n + 1)$th node as the reference node. For the remaining n non-reference nodes, the node equations may be written in the generalized form as

$$G_{11}v_1 + G_{12}v_2 + \dots + G_{1n}v_n = i_{s1}$$

$$G_{21}v_1 + G_{22}v_2 + \dots + G_{2n}v_n = i_{s2}$$

$$\vdots \qquad \vdots \qquad \vdots \qquad \vdots$$

$$G_{n1}v_1 + G_{n2}v_2 + \dots + G_{nn}v_n = i_{sn}$$

In matrix form

$$\begin{bmatrix} G_{11} & G_{12} & \cdots & G_{1n} \\ G_{21} & G_{22} & \cdots & G_{2n} \\ \cdot & \cdot & & \cdot \\ \cdot & \cdot & & \cdot \\ \cdot & \cdot & & \cdot \\ G_{n1} & G_{n2} & & G_{nn} \end{bmatrix} \begin{bmatrix} v_1 \\ v_2 \\ \cdot \\ \cdot \\ \cdot \\ v_n \end{bmatrix} = \begin{bmatrix} i_{s1} \\ i_{s2} \\ \cdot \\ \cdot \\ \cdot \\ i_{sn} \end{bmatrix}$$

In a more compact form

$$[\mathbf{G}]\,[\mathbf{v}] = [\mathbf{i}_s]$$

The values of the currents and conductances in the above equations may be determined as follows :

The current terms represent the sources that force current into the respective nodes.

i_{s1} = algebraic sum of all the source currents at node 1

i_{s2} = algebraic sum of all the source currents at node 2

i_{sn} = algebraic sum of all the source currents at node n

Generally, currents *entering* a node are taken positive and currents *leaving* the node are taken *negative*.

The conductances with like subscripts (G_{11}, G_{22}, ..., G_{nn}) are called the **self conductances** of the nodes. Each self conductance is equal to the sum of all the conductances of all branches directly connected to a given node. Thus,

G_{11} = sum of the conductances of all branches directly connected to node 1

G_{22} = sum of the conductances of all branches directly connected to node 2

G_{nn} = sum of the conductances of all branches directly connected to node n

The conductance G_{nn} is the self conductance of node n.

The conductances with unlike subscripts (G_{12}, G_{23}, ..., G_{ij} $i \neq j$) are known as **mutual node conductances**. Thus,

$G_{ij} = G_{ji}$ = mutual conductance between the node i and node j.

= – (sum of the conductances of all branches *directly* connecting node i and node j)

For networks containing resistors and independent current sources the procedure of writing nodal equations in matrix form is very systematic. The matrix equation can be written directly by inspection of the circuit diagram. This method avoids the error-prone steps of assuming branch currents, writing KCL equations and then applying v-i relationships.

It is to be noted that all diagonal terms in the conductance matrix are positive, whereas all off-diagonal terms are negative.

11.9.1 Procedure for Writing Node Voltage Equations

The procedure for writing node-voltage equations is as follows :

Step 1. Convert all voltage sources in series with resistance to convert sources in parallel with conductances. Convert all resistance values to conductances. The circuit in then redrawn.

Step 2. Identify all the different nodes in the network. A **different node** is one whose voltage is different from any other node.

Step 3. Choose one of these nodes as datum (reference) node. The datum node is assumed to be at zero potential. All the different nodes are assigned numbers. The datum node is assigned the label 0 and the remaining nodes are labelled 1, 2, 3, ..., n. Let $v_1, v_2, v_3, ..., v_n$ measured with respect to the reference node.

Step 4. Arbitrarily assign currents to the branches in which no current is designated.

Step 5. Write KCL equation at non reference node.

Step 6. Apply ohm's law to write branch currents in terms of node voltages. These values of branch currents are now substituted in KCL equations obtained in step 5.

Step 7. Arrange these equations so that the unknown node voltages appear in the left-hand side and the known quantities on the right-hand side of equations. These equations are called the **node voltage equations**. For networks which contain no dependent sources with $(n + 1)$ nodes the set of node equations may be written in the **generalized form** as

$$G_{11}v_1 + G_{12}v_2 + \dots + G_{1n}v_n = i_{s1}$$

$$G_{21}v_1 + G_{22}v_2 + \dots + G_{2n}v_n = i_{s2}$$

$$\dots \quad \dots \quad \dots \quad \dots$$

$$G_{n1}v_1 + G_{n2}v_2 + \dots + G_{nn}v_n = i_{sn}$$

These equations can be written in the matrix form as :

$$\begin{bmatrix} G_{11} & G_{12} & \dots & G_{1n} \\ G_{21} & G_{22} & \dots & G_{n2} \\ \dots & \dots & \dots & \dots \\ G_{n1} & G_{n2} & \dots & G_{nn} \end{bmatrix} \begin{bmatrix} v_1 \\ v_2 \\ \dots \\ v_n \end{bmatrix} = \begin{bmatrix} i_{s1} \\ i_{s2} \\ \dots \\ i_{sn} \end{bmatrix}$$

Step 8. Solve the these equations for the desired node voltage. Find other voltages and currents in the circuit by the application of Ohm's law and KCL.

■ **EXAMPLE 11.10** Using nodal analysis, detemine all node voltages and all element currents for the network in Fig. 11.14 if $R_1 = 1\ \Omega$, $R_2 = 2\ \Omega$, $R_3 = 3\ \Omega$, $R_4 = 4\ \Omega$.

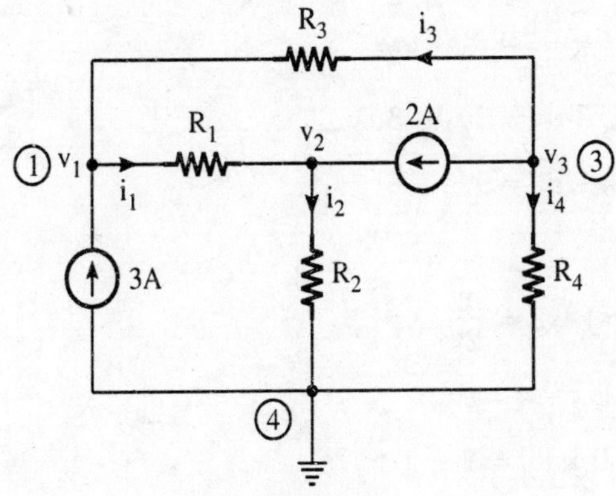

Fig 11.14

SOLUTION :

Step 1 We select 4 as the reference node. Therefore $v_4 = 0$.

Step 2 The voltages at the nonereference nodes 1, 2 and 3 are labelled as v_1, v_2, v_3 respectively.

Step 3 Assign arbitrary currents i_1, i_2, i_3 and i_4 through resistors R_1, R_2, R_3 and R_4 respectively.

Step 4 Now apply KCL at each nonreference node.

At node1 :

$$3 + i_3 = i_1 \qquad \text{(E11.10.1)}$$

A node 2 :

$$i_1 + 2 = i_2 \qquad \text{(E11.10.2)}$$

At node 3 :

$$0 = 2 + i_3 + i_4 \qquad \text{(E11.10.3)}$$

Step 5 Applying Ohm's law to express the currents i_1, i_2 and i_3 in terms of node voltages v_1, v_2 and v_3.

$$i_1 = \frac{v_1 - v_2}{R_1} = \frac{v_1 - v_2}{1} = v_1 - v_2 \tag{E11.10.4}$$

$$i_2 = \frac{v_2 - v_4}{R_2} = \frac{v_2 - 0}{2} = \frac{v_2}{2} \tag{E11.10.5}$$

$$i_3 = \frac{v_3 - v_1}{R_3} = \frac{v_3 - v_1}{3} \tag{E11.10.6}$$

$$i_4 = \frac{v_3 - v_4}{R_4} = \frac{v_3 - 0}{4} = \frac{v_3}{4} \tag{E11.10.7}$$

Step 6 Substituting the values of currents i_1, i_2, i_3 and i_4 obtained in step 5 into the KCL equations of step 4, we get the following relationships :

At node 1 :

$$3 + \frac{v_3 - v_1}{3} = v_1 - v_2$$

$$9 + v_3 - v_1 = 3v_1 - 3v_2$$

$$4v_1 - 3v_2 - v_3 = 9 \tag{E11.10.8}$$

At node 2 :

$$(v_1 - v_2) + 2 = \frac{v_2}{2}$$

$$2v_1 - 2v_2 + 4 = v_2$$

$$-2v_1 + 3v_2 - 0 \times v_3 = 4 \tag{E11.10.9}$$

At node 3 :

$$0 = 2 + \frac{v_3 - v_1}{3} + \frac{v_3}{4}$$

$$0 = 24 + 4v_3 - 4v_1 + 3v_3$$

$$-4v_1 - 0 \times v_2 + 7v_3 = -24 \tag{E11.10.10}$$

Alternative method

Node Eqs. (E11.10.8), (E11.10.9) and (E11.10.10) can be written directly by inspection as follows : The generalized node equations at node 1, 2 and 3 can be written in matrix form as

$$\begin{bmatrix} G_{11} & G_{12} & G_{13} \\ G_{21} & G_{22} & G_{23} \\ G_{31} & G_{32} & G_{33} \end{bmatrix} \begin{bmatrix} v_1 \\ v_2 \\ v_3 \end{bmatrix} = \begin{bmatrix} i_{s1} \\ i_{s2} \\ i_{s3} \end{bmatrix}$$

For the given network,

G_{11} = sum of the conductances of all branches directly connected to node 1

$$= G_1 + G_3 = \frac{1}{R_1} + \frac{1}{R_3} = 1 + \frac{1}{3} = \frac{4}{3} \text{ S}$$

G_{22} = sum of the conductances of all branches directly connected to node 2

$$= G_1 + G_2 = \frac{1}{R_1} + \frac{1}{R_2} = 1 + \frac{1}{2} = \frac{3}{2} \text{ S}$$

G_{33} = sum of the conductances of all branches directly connected to node 3

$$= G_3 + G_4 = \frac{1}{R_3} + \frac{1}{R_4} = \frac{1}{3} + \frac{1}{4} = \frac{7}{12} \text{ S}$$

$G_{12} = G_{21}$ = – (sum of the conductances of all branches directly connecting node 1 and node 2)

$$= -G_1 = -\frac{1}{R_1} = -1 \text{ S}$$

$G_{23} = G_{32}$ = – (sum of the conductances of all branches directly connecting node 2 and node 3) = 0

$G_{13} = G_{31}$ = – (sum of the conductances of all branches directly connecting node 1 and node 3)

$$= -G_3 = -\frac{1}{R_3} = -\frac{1}{3} \text{ S}$$

We shall use the convention that a source current entering a node is taken as positive.

i_{s1} = algebraic sum of all the source currents at node 1

Since there is one current source connected to node 1 and that source is forcing a current of 3 A into node 1, we have i_{s1} = 3 A.

i_{s2} = algebraic sum of all the source currents at node 2.

At node 2, the current source is forcing a current of 2 A into node 2, therefore i_{s2} = 2 A.

i_{s3} = algebraic sum of all the source currents at node 3

At node 3, the 2 A current source connected to it has a current 2 A flowing *away* from node 3, therefore, i_{s3} = – 2 A.

Substituting the values of conductances and source currents we get the node equations in matrix form as

$$\begin{bmatrix} \frac{4}{3} & -1 & -\frac{1}{3} \\ -1 & \frac{3}{2} & 0 \\ -\frac{1}{3} & 0 & \frac{7}{12} \end{bmatrix} \begin{bmatrix} v_1 \\ v_2 \\ v_3 \end{bmatrix} = \begin{bmatrix} 3 \\ 2 \\ -2 \end{bmatrix}$$

In the expanded form these equations can be written as

$$\frac{4}{3}v_1 - v_2 - \frac{1}{3}v = 3 \qquad \text{(E11.10.11)}$$

$$-v_1 + \frac{3}{2}v_2 - 0 \times v_3 = 2 \qquad \text{(E11.10.12)}$$

$$-\frac{1}{3}v_1 - 0 \times v_2 + \frac{7}{12}v_3 = -2 \qquad \text{(E11.10.13)}$$

Multiplication on both sides of Eqs. (E11.10.11), (E11.10.12) and (E11.10.13) by 3, 2 and 12 respectively gives

$$4v_1 - 3v_2 - v_3 = 9$$

$$-2v_1 + 3v_2 - 0 \times v_3 = 4$$

$$-4v_1 - 0 \times v_2 + 7v_3 = -24$$

These equations are identical to the equations we obtained earlier.

EXAMPLE 11.11 Write the node equations in matrix form for the circuit shown in Fig. 11.15.

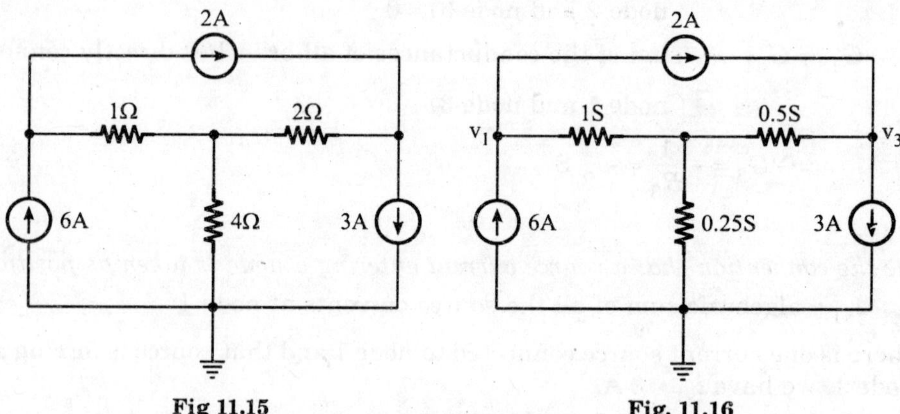

Fig 11.15 **Fig. 11.16**

SOLUTION : The circuit in Fig. 11.15 is redrawn with node voltages assigned and resistances changed to conductances. This circuit it shown in Fig. 11.16.

The node equations in matrix form are given by

$$\begin{bmatrix} G_{11} & G_{12} & G_{13} \\ G_{21} & G_{22} & G_{23} \\ G_{31} & G_{32} & G_{33} \end{bmatrix} \begin{bmatrix} v_1 \\ v_2 \\ v_3 \end{bmatrix} = \begin{bmatrix} i_{s1} \\ i_{s2} \\ i_{s3} \end{bmatrix}$$

$$\begin{bmatrix} 1 & -1 & 0 \\ -1 & (1+0.5+0.25) & -0.5 \\ 0 & -0.5 & 0.5 \end{bmatrix} \begin{bmatrix} v_1 \\ v_2 \\ v_3 \end{bmatrix} = \begin{bmatrix} 6-2 \\ 0 \\ 2-3 \end{bmatrix}$$

$$\begin{bmatrix} 1 & -1 & 0 \\ -1 & 1.75 & -0.5 \\ 0 & -0.5 & 0.5 \end{bmatrix} \begin{bmatrix} v_1 \\ v_2 \\ v_3 \end{bmatrix} = \begin{bmatrix} 4 \\ 0 \\ -1 \end{bmatrix}$$

11.10 NODAL ANALYSIS WITH A PRACTICAL VOLTAGE SOURCE

If a network contains a practical voltage source (that is, an ideal voltage source in series with a resistor) we convert it to an equivalent practical current source by source transformation. Then we can write the node equations in the usual manner.

■ **EXAMPLE 11.12** Determine the voltage across the 10 Ω resistor in Fig. 11.17 using nodal analysis.

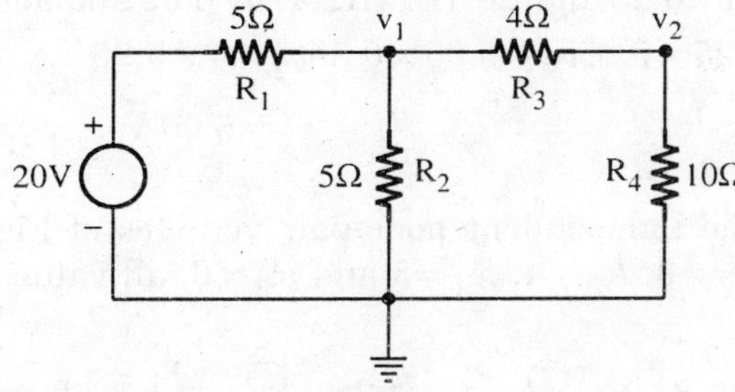

Fig 11.17

SOLUTION : The given circuit contains a voltage source of 20 V in series with a resistor of 5 Ω. We shall, therefore, convert the voltage source to the equivalent current source by source transformation. The result is shown in Fig. 11.18. The matrix equations for v_1 and v_2 can now be written as

$$\begin{bmatrix} G_1 + G_2 + G_3 & -G_3 \\ -G_3 & G_3 + G_4 \end{bmatrix} \begin{bmatrix} v_1 \\ v_2 \end{bmatrix} = \begin{bmatrix} 4 \\ 0 \end{bmatrix} \tag{E11.12.1}$$

$$G_1 = \frac{1}{R_1} = \frac{1}{5} = 0.2 \text{ S}$$

$$G_2 = \frac{1}{R_2} = \frac{1}{5} = 0.2 \text{ S}$$

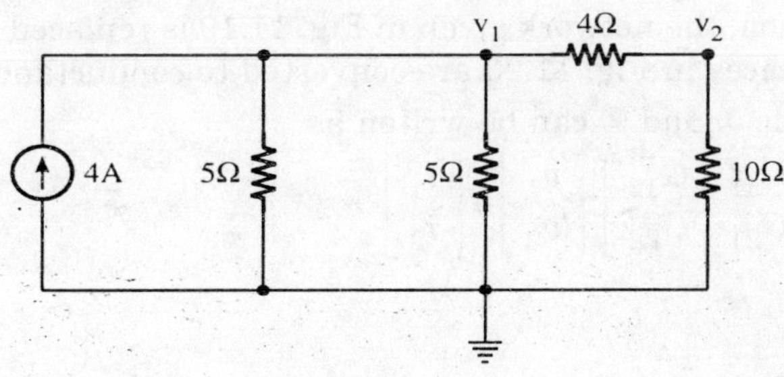

Fig 11.18

$$G_3 = \frac{1}{R_3} = \frac{1}{4} = 0.25 \text{ S}$$

$$G_4 = \frac{1}{R_4} = \frac{1}{10} = 0.1 \text{ S}$$

Substituting the values of conductances in Eq. (E11.12.1) we get

$$\begin{bmatrix} 0.65 & -0.25 \\ -0.25 & 0.35 \end{bmatrix} \begin{bmatrix} v_1 \\ v_2 \end{bmatrix} = \begin{bmatrix} 4 \\ 0 \end{bmatrix} \quad \text{(E11.12.2)}$$

In expanded form these node equations can be written as

$$0.65v_1 - 0.25v_2 = 4 \quad \text{(E11.12.3)}$$

$$-0.25v_1 + 0.35v_2 = 0 \quad \text{(E11.12.4)}$$

Multiplying Eq. (E11.12.3) by 0.25 and Eq. (E11.12.4) by 0.65 and adding the results we obtain

$$-0.25 \times 0.25v_2 + 0.65 \times 0.35v_2 = 4 \times 0.25$$

$$v_2 = 6.06 \text{ V}$$

■ **EXAMPLE 11.13** Find the independent node-pair voltages of Fig. 11.19 by nodal voltage method when $R_1 = 2$, $R_2 = 3$, $R_3 = 4$, $R_4 = 5$ and $R_5 = 6$, all values in ohms.

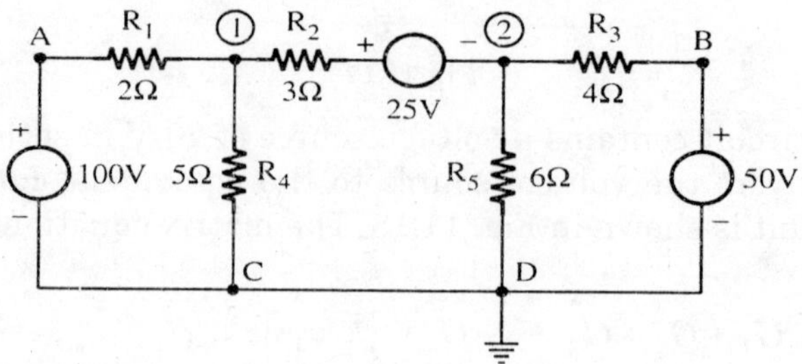

Fig 11.19

SOLUTION : The given network contains three nonideal voltage sources. We shall transform them to equivalent current sources by source transformation. The 100-V voltage source is in series with a 2 Ω resistor. It is equivalent to a current source of 50 A in parallel with 2 Ω resistor. The 25 V voltage source in series with 3 Ω resistor is equivalent to (25/3) A current source in parallel with 3 Ω resistor. Similarly, the 50 V voltage source in series with 4 Ω resistor is transformed to an equivalent (50/4) A current source in parallel with 4 Ω resistor. After source transformation, the network given in Fig. 11.19 is replaced by the network shown in Fig. 11.20. The resistances in Fig. 11.20 are converted to conductances in Fig. 11.21.

The node equation at node ① and ② can be writen as

$$\begin{bmatrix} G_{11} & G_{12} \\ G_{21} & G_{22} \end{bmatrix} \begin{bmatrix} v_1 \\ v_2 \end{bmatrix} = \begin{bmatrix} i_1 \\ i_2 \end{bmatrix}$$

$$\begin{bmatrix} \frac{1}{2}+\frac{1}{5}+\frac{1}{3} & -\frac{1}{3} \\ -\frac{1}{3} & \frac{1}{3}+\frac{1}{6}+\frac{1}{4} \end{bmatrix} \begin{bmatrix} v_1 \\ v_2 \end{bmatrix} = \begin{bmatrix} 50+\frac{25}{3} \\ \frac{50}{4}-\frac{25}{3} \end{bmatrix}$$

$$\begin{bmatrix} \frac{31}{30} & -\frac{1}{3} \\ -\frac{1}{3} & \frac{3}{4} \end{bmatrix} \begin{bmatrix} v_1 \\ v_2 \end{bmatrix} = \begin{bmatrix} \frac{175}{3} \\ \frac{25}{6} \end{bmatrix}$$

In the expanded form these equations may be written as

$$\frac{31}{30} v_1 - \frac{1}{3} v_2 = \frac{175}{3}$$

$$-\frac{1}{3} v_1 + \frac{3}{4} v_2 = \frac{25}{6}$$

Solution of these equations can be obtained by Cramer's rule as follows :

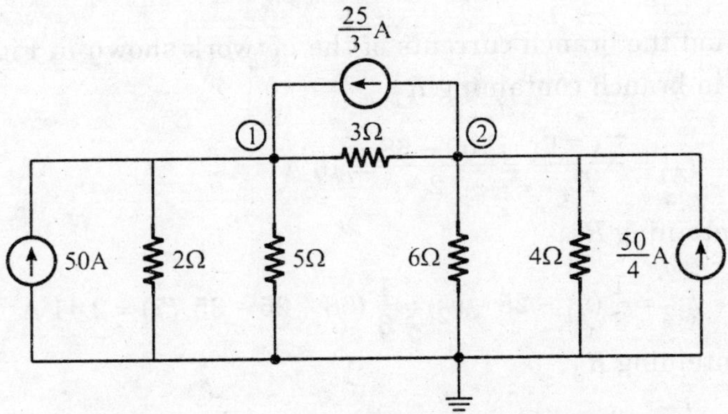

Fig 11.20 Transformed network of Fig 11.19

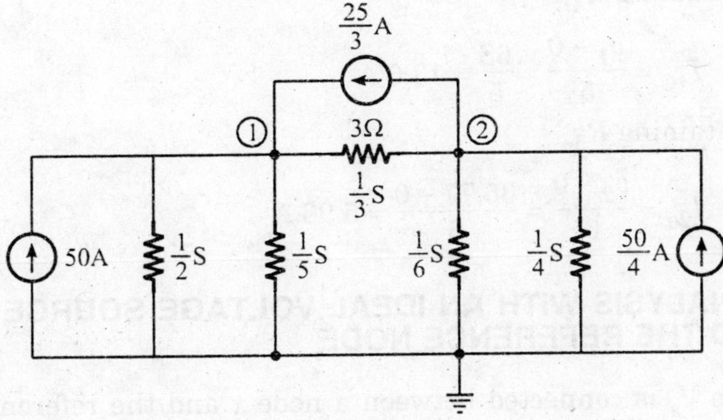

Fig 11.21

$$v_1 = \frac{\begin{vmatrix} \frac{175}{3} & -\frac{1}{3} \\ \frac{25}{6} & \frac{3}{4} \end{vmatrix}}{\begin{vmatrix} \frac{31}{30} & -\frac{1}{3} \\ -\frac{1}{3} & \frac{3}{4} \end{vmatrix}} = 68 \text{ V}$$

$$v_2 = \frac{\begin{vmatrix} \frac{31}{30} & \frac{175}{3} \\ -\frac{1}{3} & \frac{25}{6} \end{vmatrix}}{\begin{vmatrix} \frac{31}{30} & -\frac{1}{3} \\ -\frac{1}{3} & \frac{25}{6} \end{vmatrix}} = 35.77 \text{ V}$$

EXAMPLE 11.14 Find the branch currents of the network shown in Fig. 11.19.

SOLUTION : Current in branch containing R_1,

$$I_{A1} = \frac{v_A - v_1}{R_1} = \frac{100 - 68}{2} = 16 \text{ A}$$

Current in branch containing R_2,

$$I_{12} = \frac{1}{3}(v_1 - 25 - v_2) = \frac{1}{3}(68 - 25 - 35.77) = 2.41 \text{ A}$$

Current in branch containing R_3,

$$I_{B2} = \frac{v_B - v_2}{4} = \frac{1}{4}(50 - 35.77) = 3.56 \text{ A}$$

Current in branch containing R_4,

$$I_{1C} = \frac{v_1 - 0}{5} = \frac{68}{5} = 13.6 \text{ A}$$

Current in branch containing R_5

$$I_{2d} = \frac{v_2 - 0}{6} = \frac{35.77 - 0}{6} = 5.96 \text{ A}$$

11.11 NODAL ANALYSIS WITH AN IDEAL VOLTAGE SOURCE BETWEEN ONE NODE AND THE REFERENCE NODE

When an ideal voltage V_s is connected between a node x and the reference node we cannot apply KCL to node x because the current through the voltage source V_s is not known and we cannot express it in terms of v_x. However, our aim is *not* to simply write KCL equations but to obtain independent equations in terms of the node voltages. Since the ideal voltage source

V_s is connected between node x and the reference node, the node voltage v_x is equal to V_s. We now do not need to develop a node voltage equation at node x since its voltage is known . The simple node voltage equation at node x is given by $v_x = V_s$. We then write node equations at the remaining nonreference nodes in the usual manner. Thus it is found that the presence of a voltage source simplifies the node analysis by reducing the number of independent unknown node voltage equations by one.

Thus, it is found that each voltage source in a given network reduces by one the number of independent unknown node voltages. Consequently, the number of independent simultaneous node equations to be solved is reduced.

■ **EXAMPLE 11.15** Write the node equations in the networks shown in Fig. 11.22.

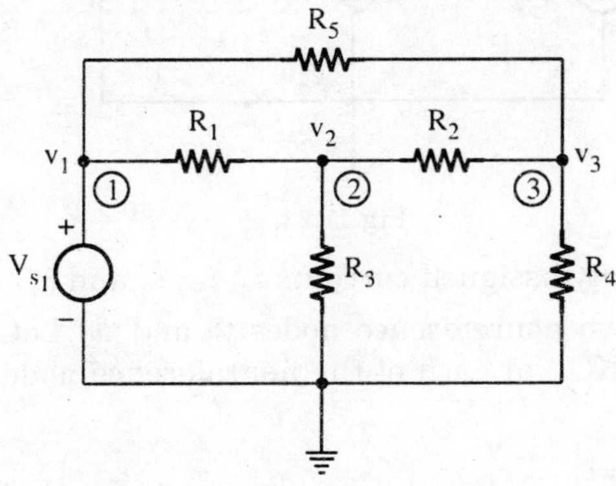

Fig. 11.22 Network of Example 11.5

SOLUTION : The voltage at node ① in given in the problem as $v_1 = v_{s1}$. Since there are two unknown node voltages v_2 and v_3 we shall write the node equations only at nodes ② and ③.

At node ②,

$$-G_1 v_{s1} + (G_1 + G_2 + G_3)\, v_2 - G_2 v_3 = 0$$

At node ③,

$$-G_5 v_{s1} - G_2 v_2 + (G_2 + G_4 + G_5)\, v_3 = 0$$

11.12 NODAL ANALYSIS WITH AN IDEAL VOLTAGE SOURCE BETWEEN TWO NONREFERENCE NODES

When an ideal voltage source V_s is connected between two non-reference nodes A and B we identify an unknown current I_x with the voltage source. Then we apply KCL at nodes A and B. These two KCL equations contain I_x , and I_x cannot be expressed in terms of V_A and V_B because Ohm's law does not apply to voltage sources. It should be noted that an ideal voltage source can pass any amount of current. The current through the voltage source depends on what is connected to the source. For this reason, the KCL equations at nodes A and B are combined to eliminate I_x. Then we apply Ohm's law to this resulting equation. In this process, we have seen that the number of node equations is reduced by one and therefore an additional equation is needed. This additional nodal equation is obtained by applying KVL to the voltage source connected between nodes A and B. We shall illustrate this method by examples.

■ **EXAMPLE 11.16** Determine the node voltages for the network shown in Fig. 11.23.

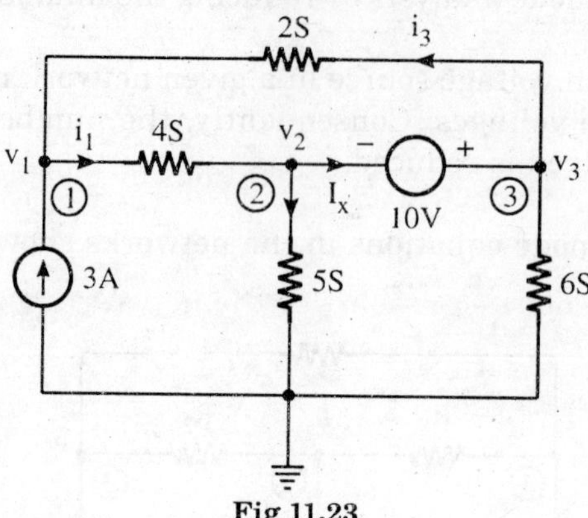

Fig 11.23

SOLUTION: We have arbitrarily assigned currents i_1, i_2, i_3 and i_4. An ideal voltage source of 10 V is connected between two nonreference nodes ② and ③. Let the current through this source be I_x. We shall apply KCL at each of the nonreference nodes ①, ② and ③. By KCL at node ①,

$$3 + i_3 = i_1 \tag{E11.16.1}$$

By Ohm's law

$$i_1 = 4\,(v_1 - v_2), \quad i_3 = 2\,(v_3 - v_1)$$

Substituting these values in Eq. (E11.16.1), we get

$$4\,(v_1 - v_2) - 2\,(v_3 - v_1) = 3$$

$$6v_1 - 4v_2 - 2v_3 = 3 \tag{E11.16.2}$$

By KCL at node 2,

$$i_1 = i_2 + I_x \tag{E11.16.3}$$

By KCL at node 3

$$I_x = i_3 + i_4 \tag{E11.16.4}$$

Since I_x is unknown and it can be expressed in terms of v_2 and v_3, we shall eliminate it by combining Eqs. (E11.16.3) and (E11.16.4). Addition of Eq. (E11.16.3) and (E11.16.4) gives

$$i_1 + I_x = i_2 + I_x + i_3 + i_4$$

$$i_1 = i_2 + i_3 + i_4 \tag{E11.16.5}$$

11.13 NODE EQUATIONS FOR NETWORKS CONTAINING DEPENDENT SOURCES

The node equations for networks containing dependent sources are written treating the dependent sources as if they were independent sources. A constraint equation (volt-ampere relationship in terms of node voltages) is then written for each of the dependent sources and substituted into the main node equations.

11.14 NODE EQUATIONS CONTAINING ONLY RESISTORS AND DEPENDENT CURRENT SOURCES

We have developed the procedure to write node equations for circuits containing only resistors and independent current sources. The method can be extended to networks containing only resistors and dependent current sources. We will treat the dependent source temporarily as if it were an independent source, and write node equations by inspection. The following example illustrates this extension.

EXAMPLE 11.17 Write the node equations for the network shown in Fig. 11.24.

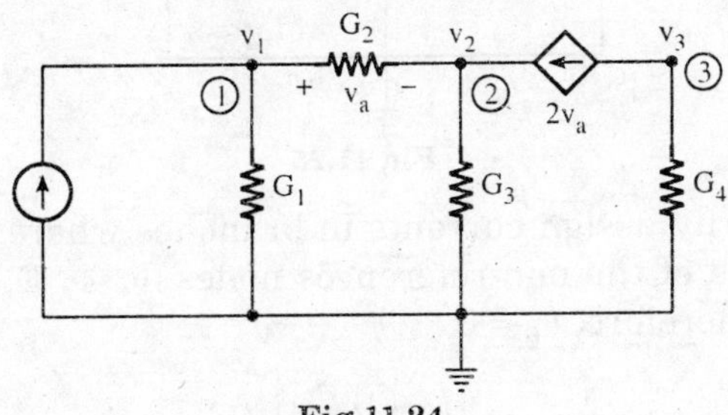

Fig 11.24

SOLUTION : Let us treat the dependent source temporarily as an independent source. We can write the node equations by inspection as

$$\begin{bmatrix} G_1+G_2 & -G_2 & 0 \\ -G_2 & G_2+G_3 & 0 \\ 0 & 0 & G_4 \end{bmatrix}\begin{bmatrix} v_1 \\ v_2 \\ v_3 \end{bmatrix} = \begin{bmatrix} i_{s1} \\ 2v_a \\ -2v_a \end{bmatrix}$$

In expanded form we can write the node equations as

$$(G_1+G_2)\,v_1 - G_2 v_2 - 0\times v_3 = i_{s1} \tag{E11.17.1}$$

$$-G_2 v_1 + (G_2+G_3)\,v_2 - 0\times v_3 = 2v_a \tag{E11.17.2}$$

$$-0\times v_1 - 0\times v_2 + G_4 v_3 = -2v_a \tag{E11.17.3}$$

By KVL $\qquad v_a = v_1 - v_2 \qquad$ (E11.17.4)

Substituting the value of v_a in the above node equations and transfering all unknows to the left hand side, we obtain the following results :

Node ① : $\qquad (G_1+G_2)\,v_1 - G_2 v_2 = i_{s1} \qquad$ (E11.17.5)

Node ② : $\qquad -(G_2+2)\,v_1 + (G_2+G_3+2)\,v_2 = 0 \qquad$ (E11.17.6)

Node ③ : $\qquad 2v_1 - 2v_2 + G_4 v_3 = 0 \qquad$ (E11.17.7)

Equations (E11.17.5), (E11.17.6) and (E11.17.7) are the required node equations. It is to be noted that these equations are not symmetrical.

EXAMPLE 11.18 Solve the circuit in Fig. 11.25 for all node voltages and all element currents. Use nodal analysis. Verify the results obtained for element currents by mesh analysis.

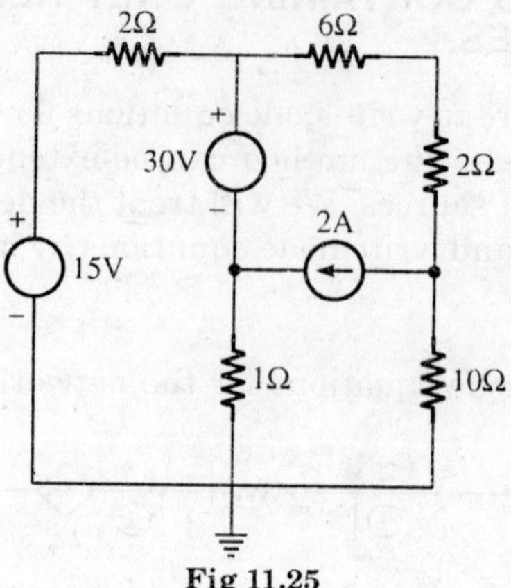

Fig 11.25

SOLUTION : Let us arbitrarily assign currents in branches where the currents are not given. We shall apply KCL at each of the non-references nodes ①, ②, ③, ④ and ⑤ in Fig. 11.26. Let ⑥ be the reference node. Therefore, $v_6 = 0$

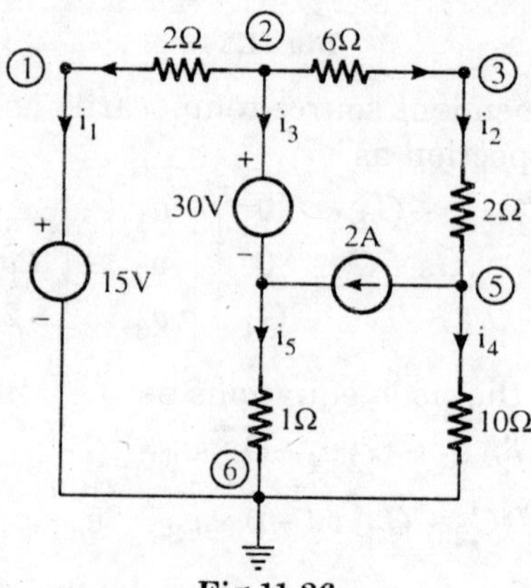

Fig 11.26

Node ① : $$i_1 = i_1 \tag{E11.18.1}$$

Node ② : $$i_1 + i_2 + i_3 = 0 \tag{E11.18.2}$$

Node ③ : $$i_2 = i_2 \tag{E11.18.3}$$

Node ④ : $$i_3 + 2 = i_5 \tag{E11.18.4}$$

Node ⑤ : $$i_4 + 2 = i_2 \tag{E11.18.5}$$

From Eqs. (E11.18.2) and (E11.18.4)

$$i_1 + i_2 + i_5 = 2 \tag{E11.18.6}$$

By Ohm's law

$$i_1 = \frac{v_2 - v_1}{2}, \quad i_2 = \frac{v_2 - v_3}{6} = \frac{v_2 - v_5}{6+2}$$

$$i_4 = \frac{v_5 - v_6}{10} = \frac{v_5 - 0}{10}, \; i_5 = \frac{v_4 - v_6}{1} = \frac{v_4 - 0}{1}$$

Here $v_1 = 15$ V, and $v_2 - v_4 = 30$ V

Substituting the values of i_1, i_2 and i_5 in Eq. (E11.18.6) we get

$$\frac{v_2 - v_1}{2} + \frac{v_2 - v_3}{6} + \frac{v_4}{1} = 2$$

or $$-3v_1 + 4v_2 - v_3 + 6v_4 = 12$$

$$-3 \times 15 + 4v_2 - v_3 + 6v_4 = 12$$

$$4v_2 - v_3 + 6v_4 = 57 \qquad \text{(E11.18.7)}$$

Also, $$v_2 - v_4 = 30$$

therefore, $$v_2 = 30 + v_4$$

Substituting this value of v_2 in Eq. (E11.18.7) we get

$$4(30 + v_4) - v_3 + 6v_4 = 57$$

$$-v_3 + 10v_4 = -63 \qquad \text{(E11.18.8)}$$

Again from Eq. (11.18.6)

$$i_1 + i_2 + i_5 = 2$$

$$\frac{v_2 - v_1}{2} + \frac{v_2 - v_5}{8} + \frac{v_4}{1} = 2$$

$$-4v_1 + 5v_2 + 8v_4 - v_5 = 16$$

Putting $v_1 = 15$ and $v_2 = 30 + v_4$

$$-4 \times 15 + 5(30 + v_4) + 8v_4 - v_5 = 16$$

$$13v_4 - v_5 = -74 \qquad \text{(E11.18.9)}$$

Again, substituting the vlaues of i_4 and i_2 in Eq. (E11.18.5) we have

$$i_4 + 2 = i_2$$

$$i_4 - i_2 = -2$$

$$\frac{v_5}{10} - \frac{v_2 - v_5}{8} = -2$$

$$9v_5 - 5v_2 = -80$$

Since $$v_2 = 30 + v_4$$

therefore, $$9v_5 - 5(30 + v_4) = -80$$

$$9v_5 - 5v_4 = 70 \qquad \text{(E11.18.10)}$$

Solving Eqs. (E11.18.9) and (E11.18.10), we get

$$v_4 = -\ 5.32\ \text{V},\ v_5 = 4.82\ \text{V}$$

$$v_2 = 30 + v_4 = 24.68\ \text{V}$$

Putting the value of v_4 in Eq. (E11.18.8)

$$-\ v_3 + 10\ (-\ 5.32) = -\ 63$$

$$v_3 = 9.8\ \text{V}$$

Thus, we have the following results :

$$v_1 = 15\ \text{V},\ \ v_2 = 24.68\ \text{V},\ \ v_3 = 9.8\ \text{V}$$

$$v_4 = -\ 5.32\ \text{V},\ \ v_5 = 4.82\ \text{V}$$

Calculation of Currents

$$i_1 = \frac{v_2 - v_1}{2} = \frac{24.68 - 15}{2} = 4.84\ \text{A}$$

$$i_2 = \frac{v_2 - v_3}{6} = \frac{24.68 - 9.8}{6} = 2.48\ \text{A}$$

$$i_4 = \frac{v_5 - 0}{10} = \frac{4.82 - 0}{10} = 0.482\ \text{A}$$

$$i_5 = \frac{v_4 - 0}{1} = \frac{-\ 5.32 - 0}{1} = -\ 5.32\ \text{A}$$

$$i_3 = i_5 - 2 = -\ 5.32 - 2 = -\ 7.32\ \text{A}$$

Solution by mesh analysis

The mesh currents are shown in Fig. 11.27.

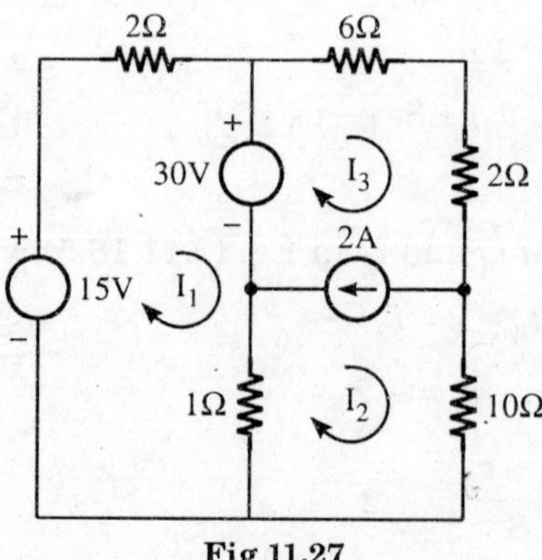

Fig 11.27

By KVL in mesh 1,

$$15 - 2I_1 - 30 - 1\ (I_1 - I_2) = 0$$

or $$-\ 3I_1 + I_2 = 15 \qquad \text{(E11.18.11)}$$

By KVL in mesh 2,

$$-1(I_2 - I_1) - v_x - 10(I_2) = 0$$

or $$I_1 - 11I_2 = v_x \quad \text{(E11.18.12)}$$

By KVL in mesh 3,

$$30 - 6I_3 - 2I_3 + v_x = 0$$

or $$8I_3 - 30 = v_x \quad \text{(E11.18.13)}$$

Also, $$I_3 - I_2 = 2 \quad \text{(E11.18.14)}$$

We have assumed the voltage across the 2 A current source as v_x. This voltage is unknown and therefore we shall eliminate it by combining the Eqs. (E11.18.12) and (E11.18.13).

$$I_1 - 11I_2 = 8I_3 - 30 \quad \text{(E11.18.15)}$$

Substituting the value of I_3 from Eq. (E11.18.14) in the above equation we get

$$I_1 - 11I_2 = 8(2 + I_2) - 30$$

or $$I_1 - 19I_2 = -14 \quad \text{(E11.18.16)}$$

It is to be noted that Eq. (E11.18.15) can also be wrtten by applying KVL in the *supermesh* formed by combining meshes 2 and 3. In this case the voltage v_x does not appear in the KVL equation. By KVL in the supermesh

$$30 - 6I_3 - 2I_3 - 10I_2 - 1 \times I_2 + 1 \times I_1 = 0$$

or $$I_1 - 11I_2 = 8I_3 - 30 \quad \text{(E11.18.15)}$$

Solution of Eqs. (E11.18.11) and (E11.18.16) gives

$$I_1 = -4.84 \text{ A}, I_2 = 0.48 \text{ A}$$

From Eq. (E11.18.14),

$$I_3 = 2 + I_2 = 2 + 0.48 = 2.48 \text{ A}$$

Comparison of results of mesh and node methods

$I_1 = -4.84$ A which is equal to i_1 (arbitrary current through 2 Ω resistor). The negative sign indicates the reverse direction of current.

$I_2 = 0.48$ A which is equal to i_4 (arbitrary current assumed through 10 Ω resistor)

$I_3 = 2.48$ A which is equal to i_2 (arbitrary current assumed through 6 Ω and 2 Ω resistors)

■ **EXAMPLE 11.19** Use nodal analysis to solve for all element voltages and currents for the network shown in Fig. 11.28.

SOLUION : Let us take node ④ as reference node. Therefore $v_4 = 0$. Here $v_1 = 20$ V. We have to determine only v_2 and v_3 for analysing the circuit by nodal anslysis.

By KCL at node ②

$$i_1 = 15 + i_4 + 15$$

or $$i_1 - i_4 = 30 \tag{E11.19.1}$$

By KCL at node ③

$$i_4 + i_2 + 15 = i_6$$

or $$i_4 + i_2 - i_6 = -15 \tag{E11.19.2}$$

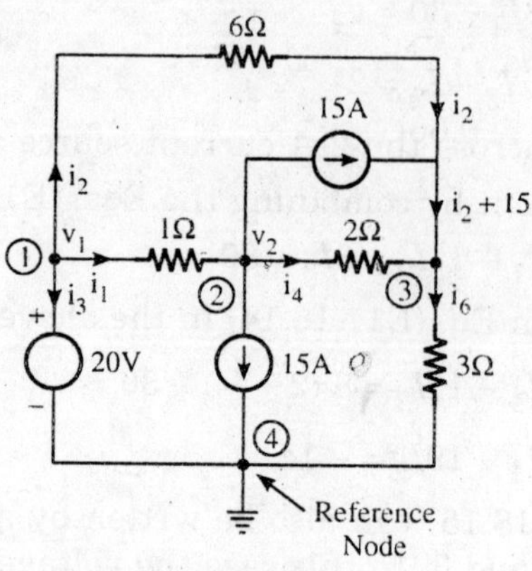

Fig 11.28

Applying Ohm's law we get

$$i_1 = \frac{v_1 - v_2}{1}$$

$$i_2 = \frac{v_1 - v_3}{6}$$

$$i_4 = \frac{v_2 - v_3}{2}$$

$$i_6 = \frac{v_3 - 0}{3}$$

Substuting these values of i_1, i_2, i_4 and i_6 in Eqs. (E11.19.1) and (E11.19.2) we get

$$(v_1 - v_2) - \frac{v_2 - v_3}{2} = 30$$

or $$2v_1 - 3v_2 + v_3 = 60 \tag{E11.19.3}$$

$$\frac{v_2 - v_3}{2} + \frac{v_1 - v_3}{6} - \frac{v_3}{3} = -15$$

$$v_1 + 3v_2 - 6v_3 = -90 \tag{E11.19.4}$$

Addition of Eqs. (E11.19.3) and (11.19.4) gives

$$3v_1 - 5v_3 = -30$$

$$3 \times 20 - 5v_3 = -30, \; v_3 = 18 \text{ V}$$

From Eq. (E11.19.3),

$$2 \times 20 - 3v_2 + 18 = 60, \ v_2 = -\frac{2}{3} \text{ V}$$

Voltage across 6 Ω resistor

$$= v_1 - v_3 = 20 - 18 = 2 \text{ V}$$

Voltage across 1 Ω resistor

$$= v_1 - v_2 = 20 + \frac{2}{3} = 20.66 \text{ V}$$

Voltage across 2 Ω resistor

$$= v_2 - v_3 = -\frac{2}{3} - 18 = -18.66 \text{ V}$$

Voltage across 3 Ω resistor

$$= v_3 - 0 = 18 - 0 = 18 \text{ V}$$

$$i_1 = \frac{v_1 - v_2}{1} = 20.66 \text{ A}$$

$$i_2 = \frac{v_1 - v_3}{6} = \frac{2}{6} = 0.33 \text{ A}$$

$$i_4 = \frac{v_2 - v_3}{2} = \frac{-18.66}{2} = -9.33 \text{ A}$$

$$i_6 = \frac{v_3}{3} = \frac{18}{3} = 6 \text{ A}$$

The remaining voltages and currents can also be found in a similar manner.

11.15 MESH ANALYSIS FOR A.C. CIRCUITS

Mesh analysis for dc circuits may also be extended to ac circuits if we replace the resistances by complex impedances. In general, mesh equations for a network having m meshs can be written as

$$\mathbf{Z}_{11}\,\mathbf{I}_1 + \mathbf{Z}_{12}\,\mathbf{I}_2 + \ldots + \mathbf{Z}_{1m}\,\mathbf{I}_m = \mathbf{V}_1$$

$$\mathbf{Z}_{21}\,\mathbf{I}_1 + \mathbf{Z}_{22}\,\mathbf{I}_2 + \ldots + \mathbf{Z}_{2m}\,\mathbf{I}_m = \mathbf{V}_2$$

$$\ldots$$

$$\ldots$$

$$\mathbf{Z}_{m1}\,\mathbf{I}_1 + \mathbf{Z}_{m2}\,\mathbf{I}_2 + \ldots + \mathbf{Z}_{mm}\,\mathbf{I}_m = \mathbf{V}_m$$

In matrix form

$$\begin{bmatrix} \mathbf{Z}_{11} & \mathbf{Z}_{12} & \cdots & \mathbf{Z}_{1m} \\ \mathbf{Z}_{21} & \mathbf{Z}_{22} & \cdots & \mathbf{Z}_{2m} \\ \cdot & \cdot & \cdots & \cdot \\ \cdot & \cdot & \cdots & \cdot \\ \cdot & \cdot & \cdots & \cdot \\ \mathbf{Z}_{m1} & \mathbf{Z}_{m2} & \cdots & \mathbf{Z}_{mm} \end{bmatrix} \begin{bmatrix} \mathbf{I}_1 \\ \mathbf{I}_2 \\ \cdot \\ \cdot \\ \cdot \\ \mathbf{I}_m \end{bmatrix} = \begin{bmatrix} \mathbf{V}_1 \\ \mathbf{V}_2 \\ \cdot \\ \cdot \\ \cdot \\ \mathbf{V}_m \end{bmatrix}$$

In a more compact form

$$[\mathbf{Z}_m]\,[\mathbf{I}_m] = [\mathbf{V}_m]$$

The matrix $[\mathbf{Z}_m]$ is called the **mesh impedance matrix.** $\mathbf{V}_j$ is the sum of the phasor voltage rises from the voltage sources in the jth mesh (the mesh being traversed in the direction of $\mathbf{I}_j$). $\mathbf{I}_k$ is the current in kth mesh. $\mathbf{Z}_{11}$, $\mathbf{Z}_{22}$, ... $\mathbf{Z}_{mm}$ are the **self impedances** of the meshes. The impedance $\mathbf{Z}_{jk}$ is the **mutual impedance** between meshes j and k. Thus,

$\mathbf{I}_1$ = mesh current in mesh 1

$\mathbf{Z}_{11}$ = sum of all the impedances through which mesh current I_1 flows

$\mathbf{Z}_{12}$ = – (sum of all the impedances common to meshes 1 and 2)

$\mathbf{V}_1$ = net source voltage tending to produce clockwise current in mesh 1

and so on.

EXAMPLE 11.20 In the network shown in Fig. 11.29 determine the branch currents and the power supplied by the source. Use mesh current method.

SOLUTION : The mesh equations are given by

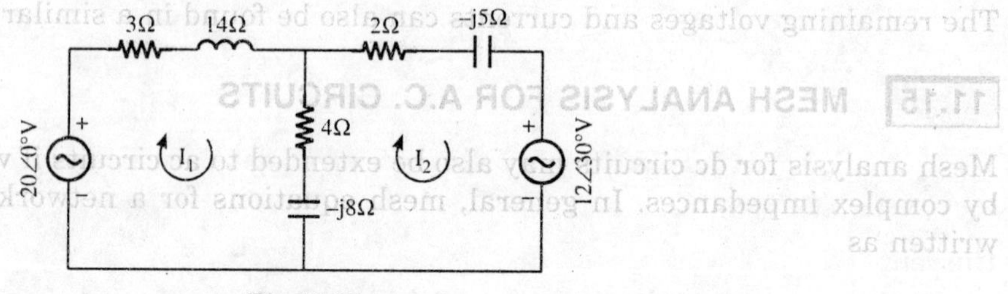

Fig 11.29

$$\begin{bmatrix} \mathbf{Z}_{11} & \mathbf{Z}_{12} \\ \mathbf{Z}_{21} & \mathbf{Z}_{22} \end{bmatrix} \begin{bmatrix} \mathbf{I}_1 \\ \mathbf{I}_2 \end{bmatrix} = \begin{bmatrix} \mathbf{V}_1 \\ \mathbf{V}_2 \end{bmatrix}$$

Here $\mathbf{Z}_{11} = 3 + j4 + 4 - j8 = 7 - j4\ \Omega$

$$\mathbf{Z}_{12} = \mathbf{Z}_{21} = -(4 - j8)\ \Omega$$

$$\mathbf{Z}_{22} = 2 - j5 + 4 - j8 = 6 - j13\ \Omega$$

$$\mathbf{V}_1 = 20\ \underline{/0^\circ}\ \text{V},\ \mathbf{V}_2 = -12\ \underline{/\ 30^\circ}\ \text{V}$$

Therefore the mesh equations can be written as

$$\begin{bmatrix} 7-j4 & -(4-j8) \\ -(4-j8) & 6-j13 \end{bmatrix}\begin{bmatrix} \mathbf{I}_1 \\ \mathbf{I}_2 \end{bmatrix} = \begin{bmatrix} 20 \,\underline{/\,0^\circ} \\ -12\,\underline{/\,30^\circ} \end{bmatrix}$$

$$\Delta_z = \begin{vmatrix} 7-j4 & -4+j8 \\ -4+j8 & 6-j13 \end{vmatrix}$$

$$= (7-j4)(6-j13) - (-4+j8)\ (-4+j8)$$

$$= 38 - j51 = 63.6\,\underline{/\,-53.31^\circ}$$

$$\Delta_1 = \begin{vmatrix} 20\,\underline{/\,0^\circ} & -4+j8 \\ -12\,\underline{/\,30^\circ} & 6-j13 \end{vmatrix}$$

$$= (20\,\underline{/\,0^\circ})\,(6-j13) + (12\,\underline{/\,30^\circ})\,(-4+j8)$$

$$= 120 - j260 + (12\,\underline{/\,30^\circ})\,(8.94\,\underline{/\,116.56^\circ})$$

$$= 120 - j260 + 107.28\,\underline{/\,146.56^\circ}$$

$$= 120 - j260 - 89.52 + j59.1$$

$$= 30.48 - j200.2 = 203.2\,\underline{/\,-81.37^\circ}$$

$$\Delta_2 = \begin{vmatrix} 7-j4 & 20\,\underline{/\,0^\circ} \\ -4+j8 & -12\,\underline{/\,30^\circ} \end{vmatrix}$$

$$= (7-j4)\,(-12\,\underline{/\,30^\circ}) + (20\,\underline{/\,0^\circ})\,(4-j8)$$

$$= (8.06\,\underline{/\,-29.74^\circ}) - (-12\,\underline{/\,30^\circ}) + 80 - j160$$

$$= -96.72\,\underline{/\,0.26^\circ} + 80 - j160$$

$$= -96.71 - j0.44 + 80 - j160$$

$$= -16.71 - j160.44$$

$$= 161.3\,\underline{/\,-95.95^\circ}$$

therefore, $$\mathbf{I}_1 = \frac{\Delta_1}{\Delta_z} = \frac{203.2\,\underline{/\,-81.37^\circ}}{63.6\,\underline{/\,-53.31^\circ}} = 3.195\,\underline{/\,-28.06^\circ}\text{ A}$$

$$\mathbf{I}_2 = \frac{\Delta_2}{\Delta_z} = \frac{161.3\,\underline{/\,-95.95^\circ}}{63.3\,\underline{/\,-53.31^\circ}} = 2.536\,\underline{/\,-42.64^\circ}\text{ A}$$

Current in branch containing the impedance $(3 + j4)\ \Omega$,

$$\mathbf{I}_1 = 3.195\,\underline{/\,-28.06^\circ}\text{ A}$$

Current in branch containing the impedance $(2 - j5)\ \Omega$,

$$\mathbf{I}_2 = 2.536\,\underline{/\,-42.64^\circ}\text{ A}$$

Current in branch containing the impedance $(4 - j8)\ \Omega$ is $I_1 - I_2$.

$$\mathbf{I}_1 - \mathbf{I}_2 = 3.195\,\underline{/\,-28.06^\circ} - 2.536\,\underline{/\,-42.64^\circ}$$

$$= 2.819 - j1.503 - 1.865 + j1.718$$

$$= 0.954 + j0.215 = 0.978\,\underline{/\,12.7^\circ}\text{ A}$$

Power supplied by the sources

$$P = V_1 I_1 \cos\phi_1 + V_2 I_2 \cos\phi_2$$

$$= 20 \times 3.195 \cos\,[0 - (28.06^\circ)] + 12\,(-2.536) \cos\,[30^\circ - (-42.64^\circ)]$$

$$= 56.38 + 9.08 = 47.4 \text{ W}$$

Check

The power supplied by the sources should be equal to the power consumed in the network.

Power consumed in the network

$$= \text{sum of the powers consumed in } 3\ \Omega,\ 2\ \Omega \text{ and } 4\ \Omega \text{ resistors}$$

$$= I_1^2 \times 3 + I_2^2 \times 2 + |\ I_1 - I_2\ |^2 \times 4$$

$$= (3.195)^2 \times 3 + (2.536)^2 \times 2 + (0.978)^2 \times 4$$

$$= 30.62 + 12.86 + 3.82 = 47.3 \text{ W}$$

EXAMPLE 11.21 Find the value of the voltage source in Fig. 11.30 if it supplies a power of 250 W to the circuit.

SOLUTION : The mesh equations of the given network are

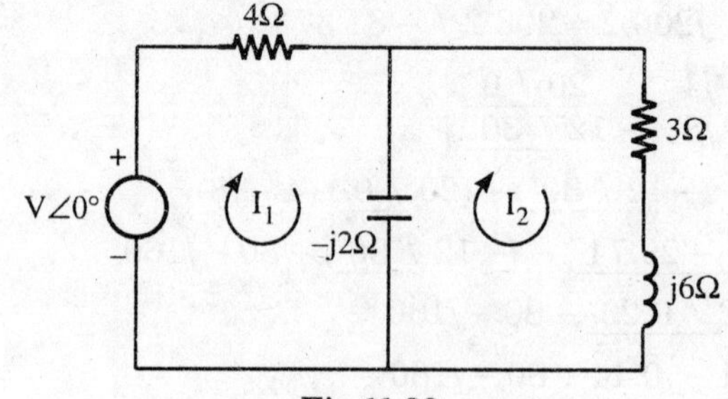

Fig 11.30

$$\begin{bmatrix} 4 - j2 & j2 \\ j2 & 3 + j4 \end{bmatrix} \begin{bmatrix} I_1 \\ I_2 \end{bmatrix} = \begin{bmatrix} V \angle 0^\circ \\ 0 \end{bmatrix}$$

$$\Delta_z = \begin{vmatrix} 4 - j2 & j2 \\ j2 & 3 + j4 \end{vmatrix} = 12 - j6 + j16 + 8 + 4$$

$$= 24 + j10$$

$$\Delta_1 = \begin{vmatrix} V & j2 \\ 0 & 3 + j4 \end{vmatrix} = V(3 + j4)$$

$$\mathbf{I}_1 = \frac{\Delta_1}{\Delta_z} = \frac{V(3 + j4)}{24 + j10} = \frac{5V \angle 53.13^\circ}{26 \angle 22.62^\circ}$$

$$= 0.1923\, V \angle 30.51^\circ \text{ A} = V(0.1657 + j0.0976) \text{ A}$$

Power supplied by the source

$$P = VI_1 \cos \phi_1$$

$$250 = V \times 0.1923\, V \cos 30.51^\circ$$

$$V^2 = \frac{250}{0.1923 \cos 30.51^\circ} = 1509$$

therefore, $V = 38.84$ volts

Alternatively $P = \text{Re}\,[\mathbf{VI}^*]$

$$250 = V \times 0.1657\,V,\; V^2 = 1508.7$$

$$V = 38.84 \text{ volts}$$

■ **EXAMPLE 11.22** In the network shown in Fig. 11.31, R and L are adjusted so that the current through Z_L is zero. Determine the values of R and L.

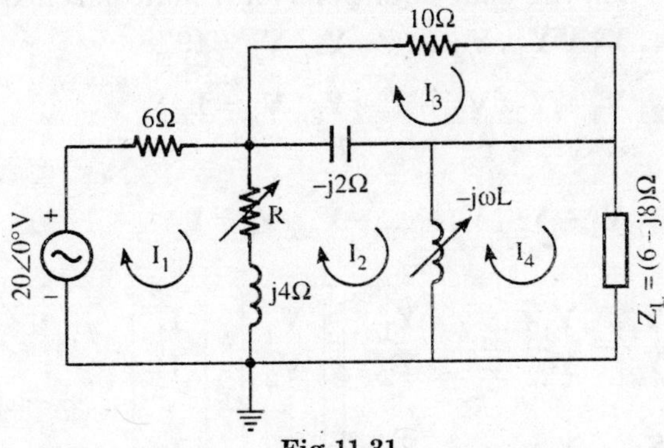

Fig 11.31

SOLUTION : The mesh equations of the given network can be written as

$$\begin{bmatrix} R+6+j4 & -(R+j4) & 0 & 0 \\ -(R+j4) & R+j(2+X_L) & j2 & -jX_L \\ 0 & j2 & 10-j2 & 0 \\ 0 & -jX_L & 0 & 6-j8+jX_L \end{bmatrix} \begin{bmatrix} \mathbf{I}_1 \\ \mathbf{I}_2 \\ \mathbf{I}_3 \\ \mathbf{I}_4 \end{bmatrix} = \begin{bmatrix} 20\,\underline{/\,0^\circ} \\ 0 \\ 0 \\ 0 \end{bmatrix}$$

$$I_4 = \frac{\Delta_4}{\Delta_z}$$

Since $\mathbf{I}_4 = 0,\ \Delta_4 = 0$. That is,

$$\begin{bmatrix} R+6+j4 & -(R+j4) & 0 & 20\,\underline{/\,0^\circ} \\ -(R+j4) & R+j(2+X_L) & j2 & 0 \\ 0 & j2 & 10-j2 & 0 \\ 0 & -jX_L & 0 & 0 \end{bmatrix} = 0$$

$$-20 \begin{vmatrix} -(R+j4) & R+j(2+X_L) & j2 \\ 0 & j2 & 10-j2 \\ 0 & -jX_L & 0 \end{vmatrix} = 0$$

$$-(R+j4)\,[jX_L\,(10-j2)] = 0$$

$$X_L\,[(2R-40)+j\,(10R+8)] = 0 \qquad \text{(E11.22.1)}$$

Either $X_L = 0$

or $(2R-40)+j\,(10R+8) = 0$

$X_L = 0 \Rightarrow \omega L = 0$, that is $L = 0$.

Equating real and imaginary parts of Eq. (E11.22.1) separately equal to zero we get $2R - 40 = 0$. That is, $R = 20\ \Omega$. Therefore, for the current through Z_L to be zero $L = 0$ and $R = 20\ \Omega$.

11.16 NODAL ANALYSIS FOR AC CIRCUITS

Nodal analysis for dc circuits may also be extended to ac circuits if we replace the conductances by complex admittances. Then the node equations for n nonreference nodes may be written as

$$\mathbf{Y}_{11}\mathbf{V}_1 + \mathbf{Y}_{12}\mathbf{V}_2 + \ldots + \mathbf{Y}_{1n}\mathbf{V}_n = \mathbf{I}_1$$

$$\mathbf{Y}_{21}\mathbf{V}_1 + \mathbf{Y}_{22}\mathbf{V}_2 + \ldots + \mathbf{Y}_{2n}\mathbf{V}_n = \mathbf{I}_2$$

$$\vdots \qquad \vdots \qquad \vdots \qquad \vdots$$

$$\mathbf{Y}_{n1}\mathbf{V}_1 + \mathbf{Y}_{n2}\mathbf{V}_2 + \ldots + \mathbf{Y}_{nn}\mathbf{V}_n = \mathbf{I}_n$$

In matrix form

$$\begin{bmatrix} \mathbf{Y}_{11} & \mathbf{Y}_{12} & \ldots & \mathbf{Y}_{1n} \\ \mathbf{Y}_{21} & \mathbf{Y}_{22} & \ldots & \mathbf{Y}_{2n} \\ \vdots & \vdots & & \vdots \\ \mathbf{Y}_{n1} & \mathbf{Y}_{n2} & \ldots & \mathbf{Y}_{nn} \end{bmatrix} \begin{bmatrix} \mathbf{V}_1 \\ \mathbf{V}_2 \\ \vdots \\ \mathbf{V}_n \end{bmatrix} = \begin{bmatrix} \mathbf{I}_1 \\ \mathbf{I}_2 \\ \vdots \\ \mathbf{I}_n \end{bmatrix}$$

In a more compact form

$$[\mathbf{Y}]\,[\mathbf{V}] = [\mathbf{I}]$$

In the above set of equations

$\mathbf{I}_i$ = phasor sum of the currents flowing from the sources *into* the ith node

$\mathbf{V}_j$ = voltage of the jth node with respect to the reference node

$\mathbf{Y}_{ii}$ = sum of all complex admittances terminating on the ith node

$\mathbf{Y}_{ii}$ is the self admittance of the ith node. $\mathbf{Y}_{ij}$ $(i \neq j)$ is the mutual admittance between the node i and node j.

$\mathbf{Y}_{ij} = \mathbf{Y}_{ji}$ = – (sum of the complex admittances of all branches *directly* connecting node i and node j)

EXAMPLE 11.23 Determine the node voltages for the network shown in Fig. 11.32a.

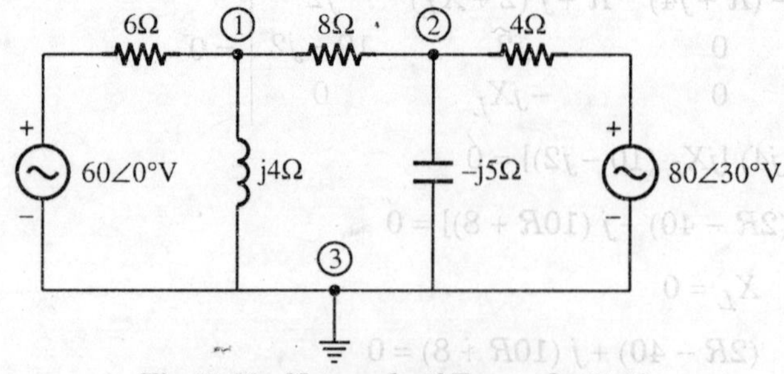

Fig 11.32a Network of Example 11.23

SOLUTION: Let us take node ③ as reference node. Since the given network has voltage sources, we transform the voltage sources to current sources by source transformation. Fig. 11.32b shows the transformed network in which the admittances are given. The nodal equations in matrix form may be written as

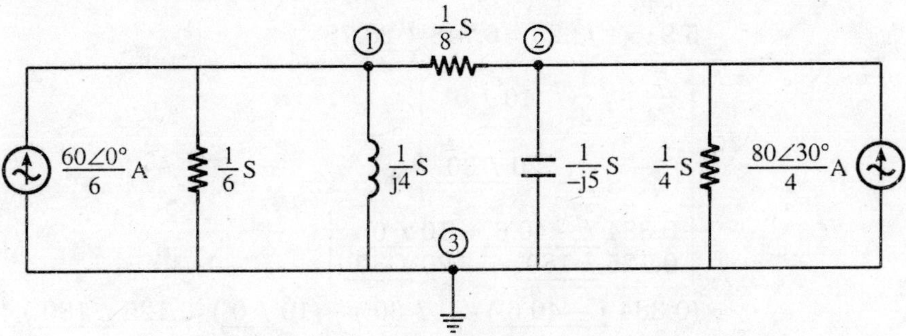

Fig 11.32b Transformed network of Fig 11.32a

$$\begin{bmatrix} \frac{1}{6}+\frac{1}{8}+\frac{1}{j4} & -\frac{1}{8} \\ -\frac{1}{8} & \frac{1}{8}+\frac{1}{4}+\frac{1}{-j5} \end{bmatrix}\begin{bmatrix} \mathbf{V}_1 \\ \mathbf{V}_2 \end{bmatrix} = \begin{bmatrix} \frac{60 \angle 0°}{6} \\ \frac{80 \angle 30°}{4} \end{bmatrix}$$

$$\left(\frac{7}{24}-j\frac{1}{4}\right)\mathbf{V}_1 - \frac{1}{8}\mathbf{V}_2 = 10 \angle 0°$$

$$-\frac{1}{8}\mathbf{V}_1 + \left(\frac{3}{8}+j\frac{1}{5}\right)\mathbf{V}_2 = 20 \angle 30°$$

$$\mathbf{V}_1 = \frac{\begin{vmatrix} 10 \angle 0° & -\frac{1}{8} \\ 20 \angle 30° & \frac{3}{8}+j\frac{1}{5} \end{vmatrix}}{\begin{vmatrix} \frac{7}{24}-j\frac{1}{4} & -\frac{1}{8} \\ -\frac{1}{8} & \frac{3}{8}+j\frac{1}{5} \end{vmatrix}}$$

$$\Delta = \begin{vmatrix} \frac{7}{24}-j\frac{1}{4} & -\frac{1}{8} \\ -\frac{1}{8} & \frac{3}{8}+j\frac{1}{5} \end{vmatrix}$$

$$= \left(\frac{7}{4}-j\frac{1}{4}\right)\left(\frac{3}{8}+j\frac{1}{5}\right) - \left(-\frac{1}{8}\right)\left(-\frac{1}{8}\right)$$

$$= 0.1437 - j0.0354 = 0.148 \angle -13.8°$$

$$\Delta_1 = \begin{vmatrix} 10 \angle 0° & -\frac{1}{8} \\ 20 \angle 30° & \frac{3}{8}+j\frac{1}{5} \end{vmatrix}$$

$$= (10 \angle 0^\circ)\left(\frac{3}{8} + j\frac{1}{5}\right) - \left(-\frac{1}{8}\right)(20 \angle 30^\circ)$$

$$= \frac{30}{8} + j2 + \frac{20}{8}\left(0.866 + j\frac{1}{2}\right)$$

$$= 5.915 + j3.25 = 6.748 \angle 28.78^\circ$$

$$\Delta_2 = \begin{vmatrix} \frac{7}{24} - j\frac{1}{4} & 10 \angle 0^\circ \\ -\frac{1}{8} & 20 \angle 30^\circ \end{vmatrix}$$

$$= \begin{vmatrix} 0.384 \angle -40.6^\circ & 10 \angle 0^\circ \\ 0.125 \angle 180^\circ & 20 \angle 30^\circ \end{vmatrix}$$

$$= (0.384 \angle -40.6^\circ)(20 \angle 30^\circ) - (10 \angle 0^\circ)(0.125 \angle 180^\circ)$$

$$= 7.68 \angle -10.6^\circ - 1.25 \angle 180^\circ$$

$$= 7.549 - j1.412 + 1.25$$

$$= 8.799 - j1.412 = 8.911 \angle -9.1^\circ$$

therefore,

$$\mathbf{V}_1 = \frac{\Delta_1}{\Delta} = \frac{6.748 \angle 28.78^\circ}{0.148 \angle -13.8^\circ} = 45.59 \angle 42.58^\circ \text{ V}$$

$$\mathbf{V}_2 = \frac{\Delta_2}{\Delta} = \frac{8.911 \angle -9.1^\circ}{0.148 \angle -13.8^\circ} = 60.2 \angle 4.7^\circ \text{ V}$$

■ **EXAMPLE 11.24** Calculate the power supplied by the source in the network shown in Fig. 11.33.

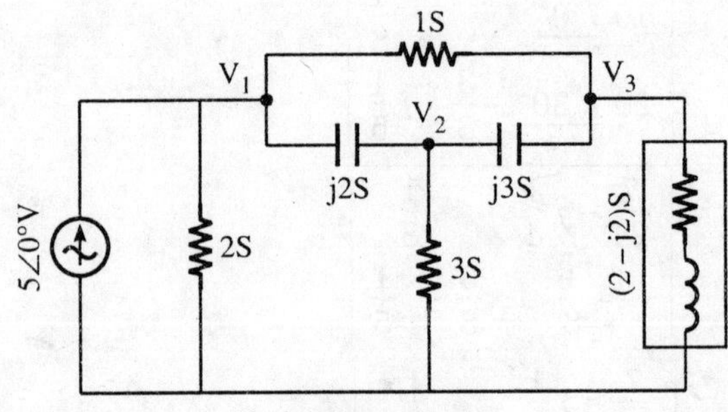

Fig 11.33

SOLUTION : The node equations can be written as

$$\begin{bmatrix} 3+j2 & -j2 & -1 \\ -j2 & 3+j5 & -j3 \\ -1 & -j3 & 3+j1 \end{bmatrix} \begin{bmatrix} \mathbf{V}_1 \\ \mathbf{V}_2 \\ \mathbf{V}_3 \end{bmatrix} = \begin{bmatrix} 5 \angle 0^\circ \\ 0 \\ 0 \end{bmatrix}$$

Let

$$\Delta_Y = \begin{vmatrix} 3+j2 & -j2 & -1 \\ -j2 & 3+j5 & -j3 \\ -1 & -j3 & 3+j1 \end{vmatrix}$$

$$= (3+j2)\begin{vmatrix} 3+j5 & -j3 \\ -j3 & 3+j1 \end{vmatrix} + j2\begin{vmatrix} -j2 & -j3 \\ -1 & 3+j1 \end{vmatrix} - 1\begin{vmatrix} -j2 & 3+j5 \\ -1 & -j3 \end{vmatrix}$$

$$= 24 + j79 = 82.56 \underline{/\ 73.1^\circ}$$

$$\Delta_{V1} = \begin{vmatrix} 5 \underline{/\ 0^\circ} & -j2 & -1 \\ 0 & 3+j5 & -j3 \\ 0 & -j3 & 3+j1 \end{vmatrix} = 5 \underline{/\ 0^\circ} \begin{vmatrix} 3+j5 & -j3 \\ -j3 & 3+j1 \end{vmatrix}$$

$$= 65 + j90 = 111 \underline{/\ 54.16^\circ}$$

$$\mathbf{V}_1 = \frac{\Delta_{V1}}{\Delta_Y} = \frac{111 \underline{/\ 54.16^\circ}}{82.56 \underline{/\ 72.1^\circ}} = 1.3445 \underline{/\ -18.94^\circ} \text{ V}$$

Power supplied by the source

$$P = V_1 I \cos\phi = 1.3445 \times 5 \times \cos 18.94^\circ$$

$$= 6.36 \text{ W}$$

■ **EXAMPLE 11.25** Determine the values of R and C so that the current through Z_L is zero in Fig. 11.34. Assume $\omega = 100$ rad/s.

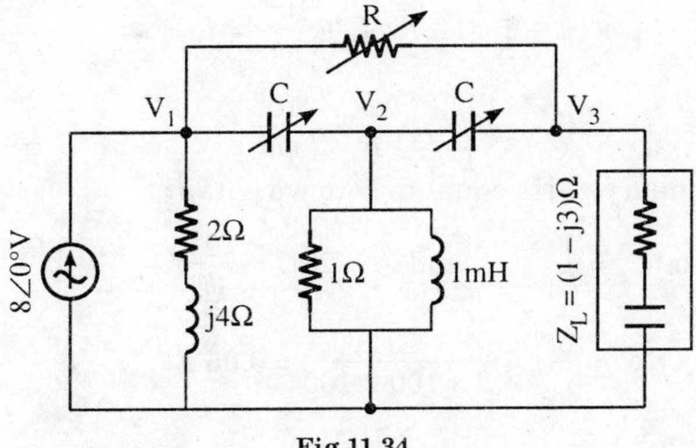

Fig 11.34

SOLUTION: The node equations in matrix form can be written as

$$\begin{bmatrix} \mathbf{Y}_{11} & \mathbf{Y}_{12} & \mathbf{Y}_{13} \\ \mathbf{Y}_{21} & \mathbf{Y}_{22} & \mathbf{Y}_{23} \\ \mathbf{Y}_{31} & \mathbf{Y}_{32} & \mathbf{Y}_{33} \end{bmatrix} \begin{bmatrix} \mathbf{V}_1 \\ \mathbf{V}_2 \\ \mathbf{V}_3 \end{bmatrix} = \begin{bmatrix} 8 \underline{/\ 0^\circ} \\ 0 \\ 0 \end{bmatrix}$$

Since the current through $\mathbf{Z}_L$ is zero, $\mathbf{V}_3$ should be equal to zero

$$\mathbf{V}_3 = \frac{\Delta_3}{\Delta}$$

$$\therefore \quad \Delta_3 = V_3 \times \Delta = 0$$

$$\Delta_3 = \begin{vmatrix} Y_{11} & Y_{12} & 8 \underline{/\ 0^\circ} \\ Y_{21} & Y_{22} & 0 \\ Y_{32} & Y_{32} & 0 \end{vmatrix} = 8 \begin{vmatrix} Y_{21} & Y_{22} \\ Y_{31} & Y_{32} \end{vmatrix}$$

$$\mathbf{Y}_{21} = -\left(\frac{1}{-jX_c}\right) = \frac{1}{jX_c} = -j\omega C$$

$$\mathbf{Y}_{22} = \frac{1}{jX_c} + \frac{10^3}{j\omega} + \frac{1}{1} - \frac{1}{jX_c}$$

$$= j\,(2\omega C) + 1 + \frac{10^3}{j\omega}$$

$$\mathbf{Y}_{31} = -\frac{1}{R},$$

$$\mathbf{Y}_{32} = -\left(-\frac{1}{jX_c}\right) = \frac{1}{jX_c} = -j\omega C$$

therefore,
$$\Delta_3 = 8\begin{vmatrix} -j\omega C & 1 + j\,(2\omega C) + \dfrac{10^3}{j\omega} \\ -\dfrac{1}{R} & -j\omega C \end{vmatrix} = 0$$

$$-\omega^2 C^2 + \frac{1}{R}\left(1 + j2\omega C + \frac{10^3}{j\omega}\right) = 0$$

$$\left(-\omega^2 C^2 + \frac{1}{R}\right) + j\left(\frac{2\omega C}{R} - \frac{10^3}{\omega R}\right) = 0$$

Equating real and imaginary parts equal to zero we obtain

$$\omega^2 C^2 = \frac{1}{R} \qquad \text{and} \qquad 2\omega C = \frac{10^3}{\omega}$$

therefore,
$$C = \frac{10^3}{2\omega^2} = \frac{10^3}{2 \times 100 \times 100} = 0.05\ \text{F}$$

$$R = \frac{1}{\omega^2 C^2} = \frac{1}{100 \times 100 \times (0.05)^2} = 0.04\ \Omega$$

EXAMPLE 11.26 Write the node equations for the network shown in Fig. 11.35.

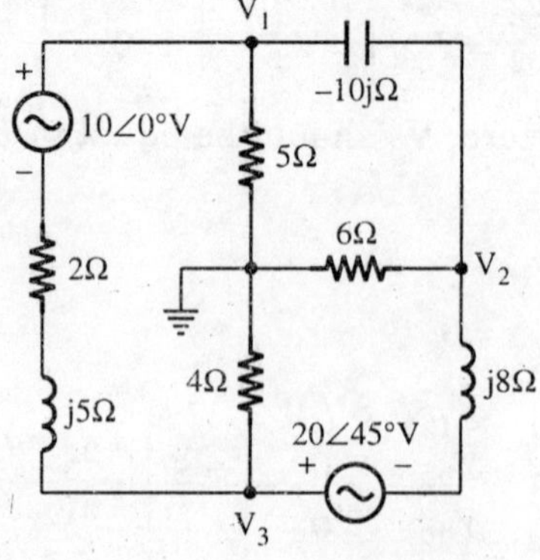

Fig 11.35 Network of Example 11.26

SOLUTION : The given network has nonideal voltage sources. These voltage sources are transformed to current sources by source transformation. The transformed network is shown in Fig. 11.36. The impedances in Fig. 11.35 are converted to admittances in Fig. 11.36.

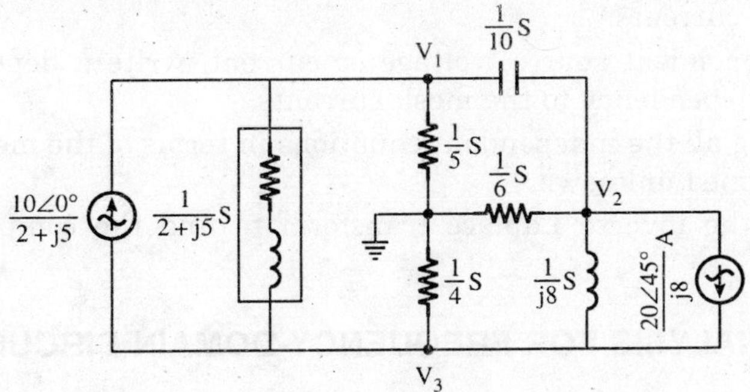

Fig 11.36 Transformed network of Fig 11.35

The nodal equations are

$$\begin{bmatrix} \frac{1}{5}+\frac{1}{2+j5}+\frac{1}{-j10} & \frac{1}{j10} & -\frac{1}{(2+j5)} \\ \frac{1}{j10} & \frac{1}{6}+\frac{1}{j8}+\frac{1}{-j10} & -\frac{1}{j8} \\ -\frac{1}{2+j5} & -\frac{1}{j8} & \frac{1}{4}+\frac{1}{j8}+\frac{1}{j5} \end{bmatrix} \begin{bmatrix} \mathbf{V}_1 \\ \mathbf{V}_2 \\ \mathbf{V}_3 \end{bmatrix} = \begin{bmatrix} \frac{10\,/\,0^\circ}{2+j5} \\ -\frac{20\,/\,45^\circ}{j8} \\ \frac{20\,/\,45^\circ}{j8}-\frac{10\,/\,0^\circ}{2+j5} \end{bmatrix}$$

11.17 MESH ANALYSIS FOR FREQUENCY-DOMAIN CIRCUITS

Mesh analysis is also applicable to s-domain circuit. The following procedure is used :

1. Replace every excitation (source) by its transform.
2. Each resistor is represented by its impedance R ohms.
3. Replace every inductor (without mutual coupling) with the series combination of its impedance sL ohms and a constant voltage source $L\,i_L\,(0_-)$ volts; the polarity of this initial-condition voltage source is in the same direction as the initial current $i_L\,(0_-)$ through the inductor.
4. Replace each capacitor by the series combination of its impedance $\frac{1}{sC}$ ohms, and a constant voltage source $\frac{v_c\,(0_-)}{s}$ volts; the polarity of this initial voltage source is in the same direction as the initial voltage $v_c\,(0_-)$ across the capacitor.
5. Write independent KVL equations for each mesh by inspection.
6. An ideal current source cannot be converted into a voltage source. An ideal current source (independent or dependent) which has two mesh currents through it, gives rise to a supermesh. A KVL equation is written for this supermesh.

7. For an ideal current source (independent or dependent) which has only one mesh through it, no KVL equation is written for that mesh.
8. Write additional independent KCL equations for each ideal current source in terms of the mesh currents.
9. For each dependent source, voltage or current, write a dependent equation that relates the dependency to the mesh currents.
10. After writing all the independent equations in terms of the mesh currents, solve for the transformed unknown.
11. Determine the inverse Laplace transform to find the complete time-domain response.

11.18 NODAL ANALYSIS FOR FREQUENCY-DOMAIN CIRCUITS

Nodal analysis is also applicable to s-domain circuits. The following steps are used :

1. Replace every resistor with an admittance $G = \frac{1}{R}$.
2. Replace every inductor (without mutual coupling) with an admittance $\frac{1}{Ls}$ in parallel with an "initial condition" current source. This source is of value $i_L(0_-)/s$ and its direction is the given direction of $i_L(0_-)$.
3. Replace every capacitor with an admittance Cs in parallel with an "initial condition" current source.
   This source is of value $q_c(0_-) = Cv_c(0_-)$, and its direction is such that the current leaves the positive terminal of the given $v_c(0_-)$.
4. Replace every given excitation (source) by its transform.
5. Write KCL equations at each numbered node by inspection.
6. An ideal voltage source cannot be converted into a current source. In this case a supernode is produced. If both ends of an ideal voltage source are at numbered nodes, a supernode KCL equation is written for that supernode by the direct application of KCL.
7. If one end of the ideal voltage is at the reference node, no KCL equation is written for that supernode.
8. Write additional independent KVL equations for each ideal voltage source in terms of node voltages.
9. For each dependent source, voltage or current, a dependent equation that relates the dependency to the node voltages, is written.
10. After writing all the independent equations in terms of node voltages, solve for the transformed unknown.
11. Determine the inverse Laplace transform to find the complete time-domain response.

EXAMPLE 11.27 Draw the circuit of Fig. 11.37a in the *s*-domain.

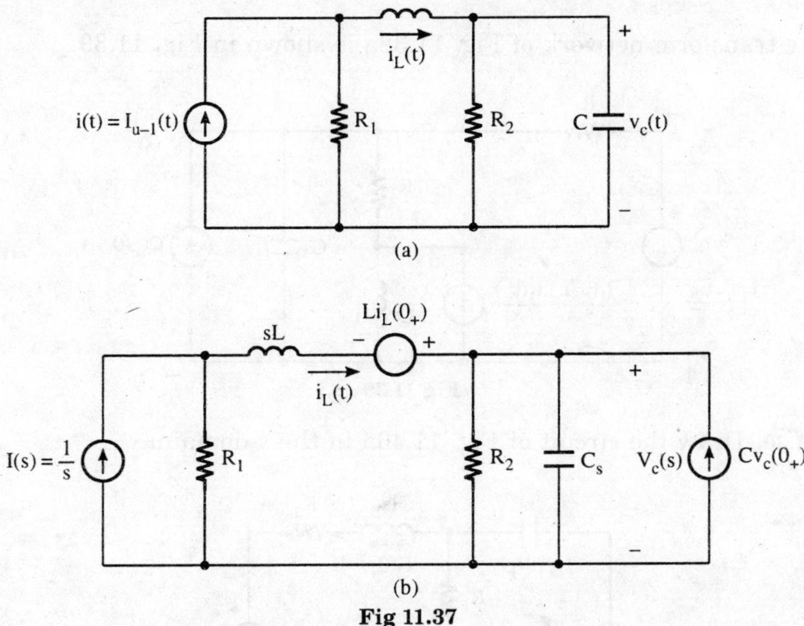

Fig 11.37

SOLUTION : The *s*-domain circuit of Fig. 11.37a is shown in Fig. 11.37b.

EXAMPLE 11.28 Find the transform network in loop form for the network shown in Fig. 11.38a.

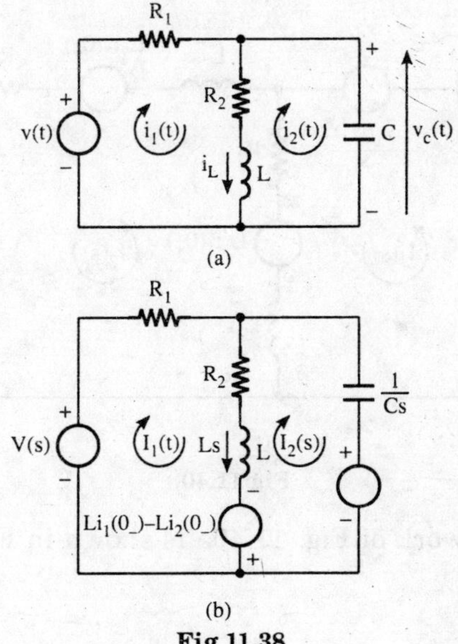

Fig 11.38

SOLUTION : The transform network of Fig. 11.38a is shown in Fig. 11.38b.

EXAMPLE 11.29 Find the transform network in node form for the network shown in Fig. 11.38a.

SOLUTION : The transform network of Fig. 11.38a is shown in Fig. 11.39.

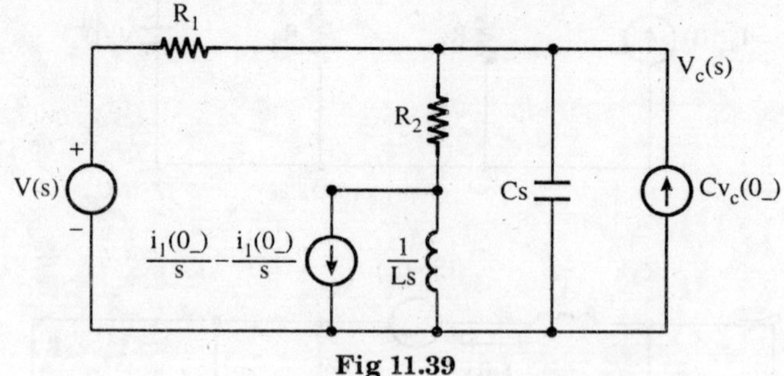

Fig 11.39

EXAMPLE 11.30 Draw the circuit of Fig. 11.40a in the s-domain.

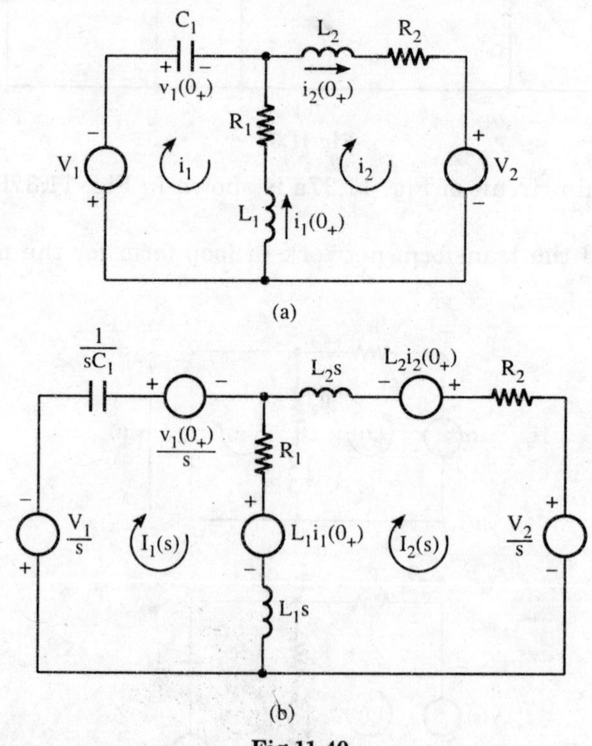

Fig 11.40

SOLUTION : The s-domain network of Fig. 11.40a is shown in Fig. 11.40b.

EXAMPLE 11.31 Draw the circuit of Fig. 11.41a in the s-domain.

SOLUTION : The s-domain circuit of Fig. 11.41a is shown in Fig. 11.41b.

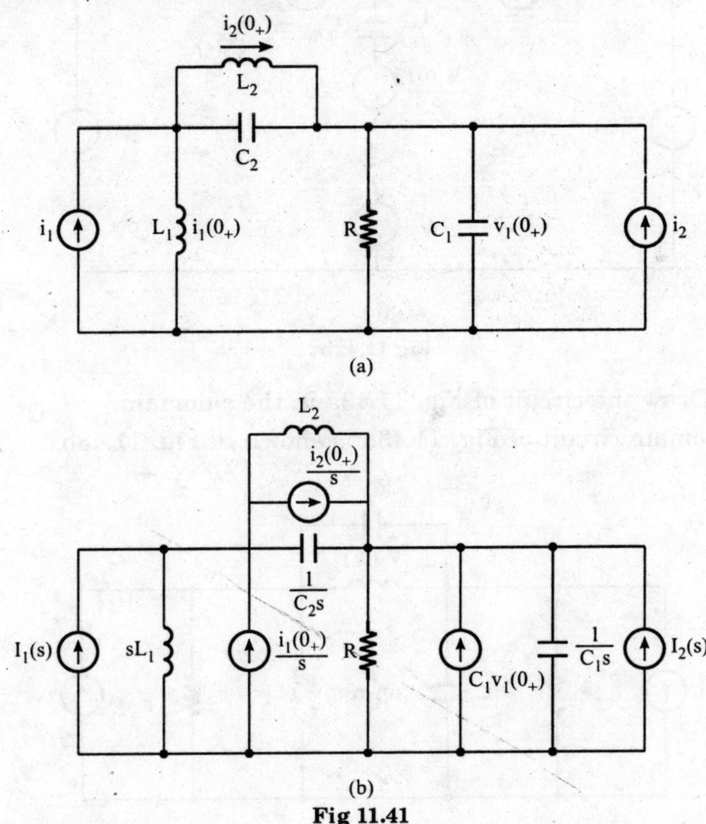

Fig 11.41

EXAMPLE 11.32 Draw the circuit of Fig. 11.42a in the s-domain.

SOLUTION : The s-domain circuit of Fig. 11.42a is shown in Fig. 11.42b.

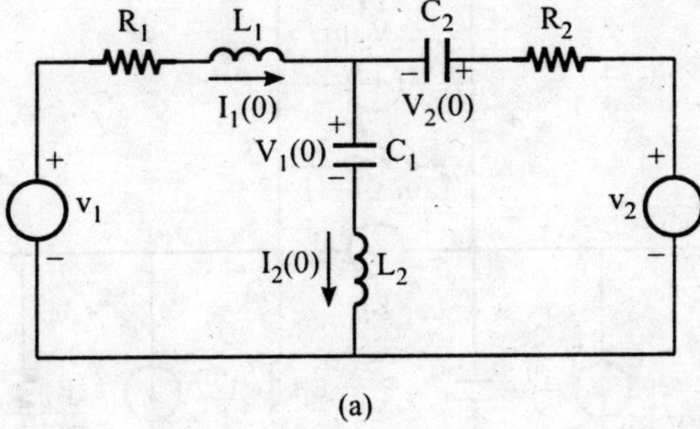

Fig 11.42a

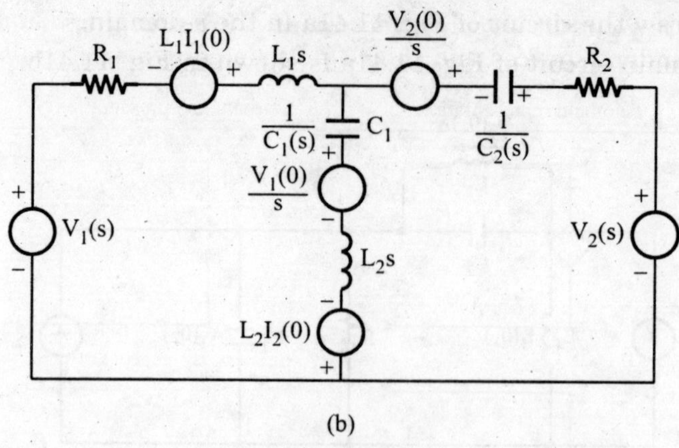

(b)

Fig 11.42b

■ **EXAMPLE 11.33** Draw the circuit of Fig. 11.43a in the s-domain.

SOLUTION : The s-domain circuit of Fig. 11.43a is shown in Fig. 11.43b.

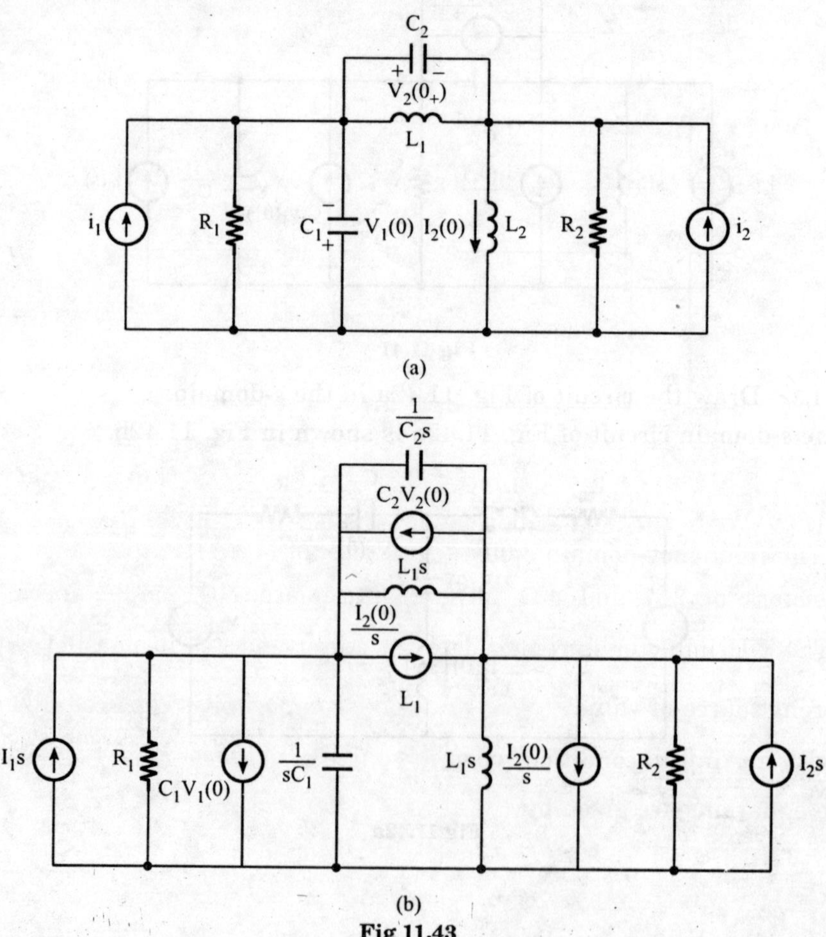

(b)

Fig 11.43

■ **EXAMPLE 11.34** Draw circuit showing dependent sources in the s-domain.

SOLUTION : The dependent sources and their s-domain equivalents are shown in Fig. 11.44.

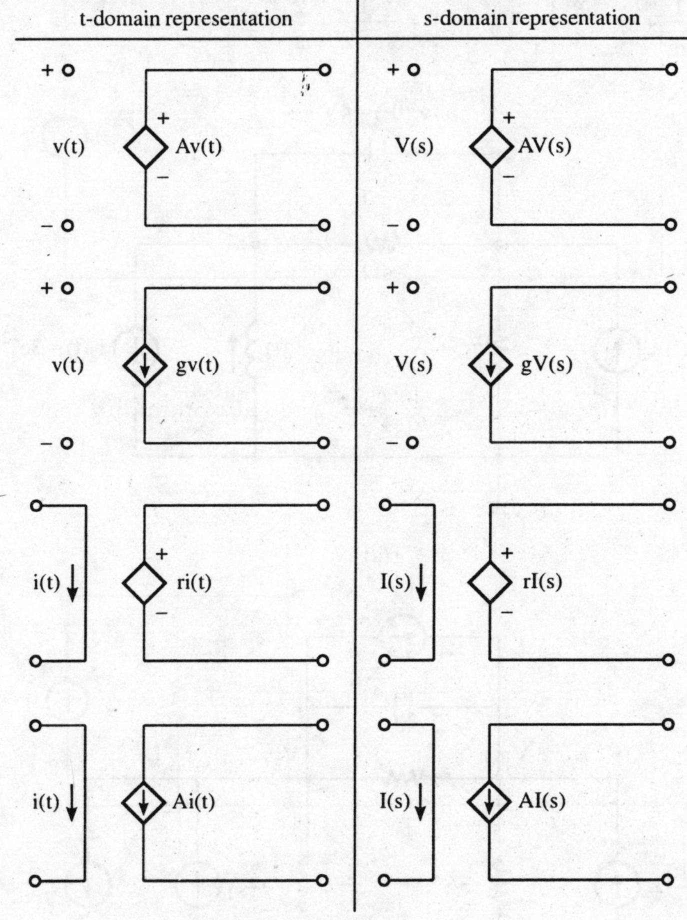

Fig 11.44

■ **EXAMPLE 11.35** Write the node equations for the network shown in Fig. 11.45a.

SOLUTION : The frequency-domain equivalent circuit of Fig. 11.45a is shown in Fig. 11.45b.

Here resistors of 2 Ω and 4 Ω have been replaced by conductances $\frac{1}{2}$ S and $\frac{1}{4}$ S respectively. The 3 H inductor is replaced by an admittane $\frac{1}{3s}$ S in parallel with an "initial condition" current source of value $\frac{i_L(0\_)}{s} = \frac{2}{s}$ A. The $\frac{1}{4}$ F capacitor is replaced by an admittance $\frac{s}{4}$ in parallel with an "initial condition" current source of value $Cv_c(0\_) = \frac{1}{4} \times 4 = 1$ A

The node equations are given by

$$\begin{bmatrix} \frac{1}{2}+\frac{1}{4}+\frac{s}{4} & -\left(\frac{1}{4}+\frac{s}{4}\right) \\ -\left(\frac{1}{4}+\frac{s}{4}\right) & \frac{1}{3s}+\frac{1}{4}+\frac{s}{4} \end{bmatrix} \begin{bmatrix} V_1 \\ V_2 \end{bmatrix} = \begin{bmatrix} \frac{10}{s}+1 \\ -\frac{3}{s+2}+\frac{2}{s}-1 \end{bmatrix}$$

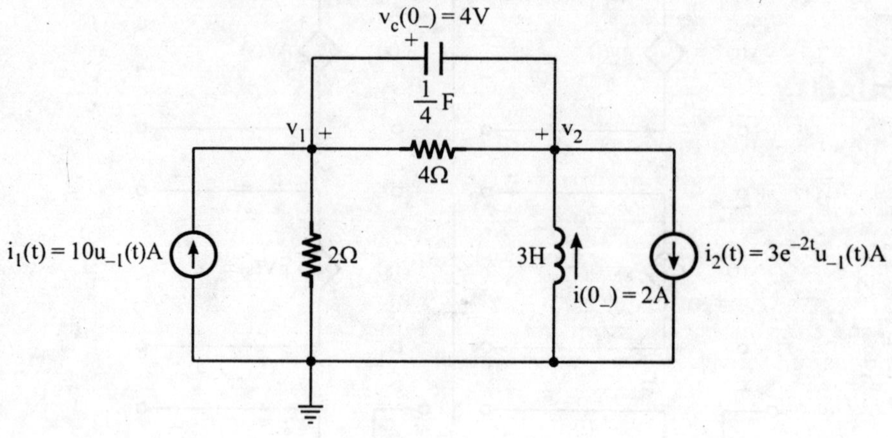

(a)

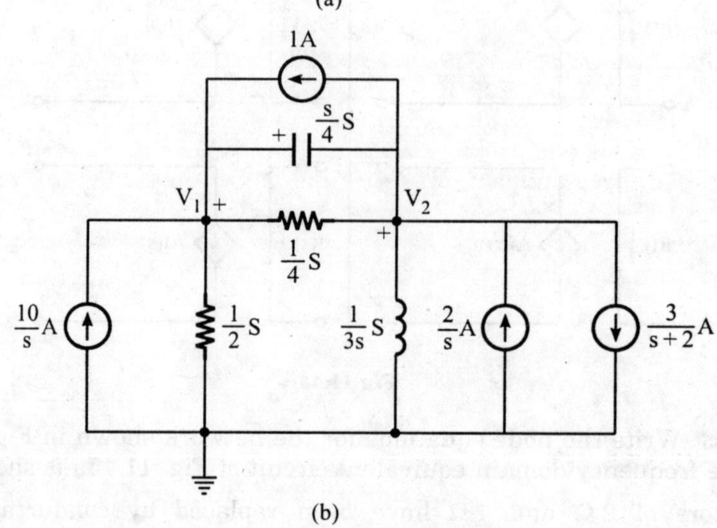

(b)

Fig 11.45

11.19 CHOICE BETWEEN MESH AND NODE ANALYSIS

Nodal analysis is applicable to planar as well as nonplanar networks, while mesh analysis is applicable to planar networks. The choice between mesh and node analysis usually depends on the number of equations to be solved for a given network. The method which results in fewer equations is preferred. Network topology gives the rules for finding the number of equations.

If b = total number of branches of the network

n_i = number of independent nodes

l = number of independent meshes

then, number of node equations = $n_i - 1$

and number of independent mesh equations

$$l = b - n_i + 1$$

11.20 DUALITY

Ohm's law may be stated in V form as

$$V = RI \tag{11.20.1}$$

or in I form as

$$I = GV \tag{11.20.2}$$

Eq. (11.20.2) may be derived from Eq. (11.20.1) by solving for I and replacing $(1/R)$ by G. But it is to be noted that either equation may be formed from the other by *replacing each symbol by its pair opposite* : I by V, V by I, R by G, G by R.

When $R_1, R_2, R_3, \ldots R_n$ are connnected in series, the equivalent resistance R_s is given by

$$R_s = R_1 + R_2 + R_3 + \ldots + R_n \tag{11.20.3}$$

That is, *series resistances add.* If we replace R by G and *series* by *parallel,*

$$G_p = G_1 + G_2 + G_3 + \ldots + G_n \tag{11.20.4}$$

It is seen that one of these equations (11.20.3) and (11.20.4) may be obtained from the other by interchanging resitances and conductances and the subscripts s and p (that is, series and parallel).

We know that the voltage across the short circuit is always zero regardless of the current through it. Mathematically, it is written as

$$V = 0, \quad I \neq 0$$

Similarly, the current in an open circuit is always zero, regardless of the voltage across it. This statement is written mathematically as,

$$I = 0, \quad V \neq 0$$

There is a systematic symmetry of opposites, or *duality*, between resistance and conductance, current and voltage, series and parallel, and short circuit and open circuit. We say that these quantities are *duals* of each other. That is R is the dual of G, G is the dual of R, voltage is the dual of current and current is the dual of voltage, series is the dual of parallel and parallel is the dual of series, short circuit is the dual of open circuit and open circuit is the dual of short circuit. Thus, the *property of duality is a mutual property*. Some of the dual relations are given in Tabel 11.1.

Table 11.1

SOME DUAL QUANTITIES		
Voltage	↔	Current
Resistance	↔	Conductance
Series connection	↔	Parallel connection
Short circuit	↔	Open circuit
Inductance	↔	Capacitance
Reactance	↔	Susceptance
Impedance	↔	Admittance
KCL	↔	KVL
Voltage source	↔	Current source
Mesh	↔	Node
Mesh Current	↔	Node voltage
Mesh equations	↔	Nodal equations
Voltage divider	↔	Current divider
Thevenin's theorem	↔	Norton's theorem
VCVS	↔	CCCS
CCVS	↔	VCCS
Tree	↔	Cotree
Twig	↔	Link
Cutset	↔	Loop

Two networks are called **duals** *of one another, if mesh equations of one have the same mathematical form as the nodal equations of the other*. If network A is dual of network B, then network B is also dual of network A. For two networks to be duals it is necessary that the variables and the elements of one network should also be the duals of the variables and elements of the other network. Since mesh equations are applicable to planar networks only, these networks have duals. For mutual inductance there is no dual relationship. Any initial charge on a capacitor has, its dual in the same value of initial current in the inductor in the dual circuit and vice versa.

The principle of duality is very useful in the sense that the solution of any network automatically gives the solution to its dual. For the three circuit elements R, L and C the duality is shown below :

$$v = Ri \qquad i = Gv$$

$$v_L = L\frac{di_L}{dt} \qquad i_c = C\frac{dv_c}{dt}$$

$$v_c = \frac{1}{C}\int i_c\, dt \quad i_L = \frac{1}{L}\int v_L\, dt$$

11.21 GRAPHICAL METHOD OF DETERMINING THE DUAL OF A NETWORK

The following steps are to be followed for finding the dual of a given network :

1. In the centre of each mesh of the given network place a dot. These dots correspond to independent nodes. Label each node with a number.
2. Place a node (node 0) *outside* of all the network meshes. This is called the *reference node* (or *datum node*).
3. Draw dashed lines between the node symbols (including references node symbol) in such a way that each line crosses only one network element, and a dashed line has been drawn through every element in the original network.
4. Draw the dual of the original network by inserting the dual of each element in the original network on the dotted line.

In dual networks, care must be taken in marking the polarity of the voltage sources (including initial voltage across each capacitor), and the direction of the current sources (including the initial current through each inductor).

The conventional positive direction for mesh currents in clockwise, while the conventional polarity for the node voltages is positive with respect to ground. Thus, a voltage or a current source, which tends to produce a clockwise current around a mesh in the original circuit should be replaced in the original circuit by the dual source element with a polarity which tends to make the node, at the centre of the mesh positive.

The method described above may be applied to any network that has a planar graph. Since it is based on a transformation of the graph of the network, it is valid for linear-time invariant as well as non-linear and time varying networks. Only planar networks without mutual inductances have duals. For mutual inductance, there is no dual relationship. It is to be noted that the dual of the dual circuit is the original circuit.

■ **EXAMPLE 11.36** Find the dual of the network shown in Fig. 11.46a.

SOLUTION : Place nodes ① to ⑤ in the centre of each mesh and a reference node ⓪ outside the network. Connect them with dashed lines each of which passes through only one element as shown in Fig. 11.46b. Replace each dashed line with the dual of the element the dashed line intersects in the original network. For example, since the line from ① to ② intersects a 4 Ω resistor, in the dual network, nodes ① and ② are connected by a 4 S conductance. The dashed line from ② to ③ passes through a 5-F capacitor in the original network, and so ② and ③ are connected by a 5-H inductor in the dual network.

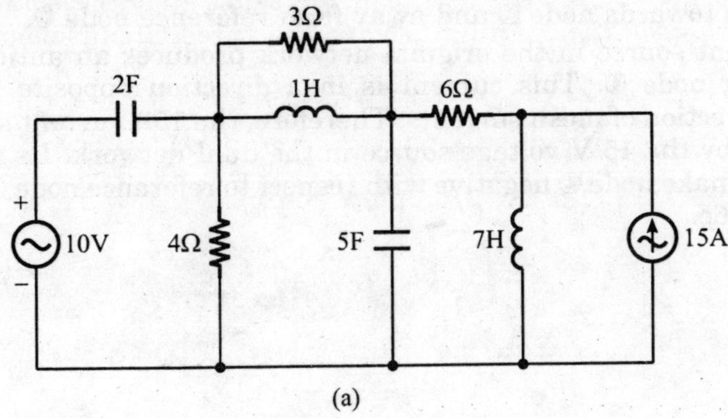

Fig. 11.46a

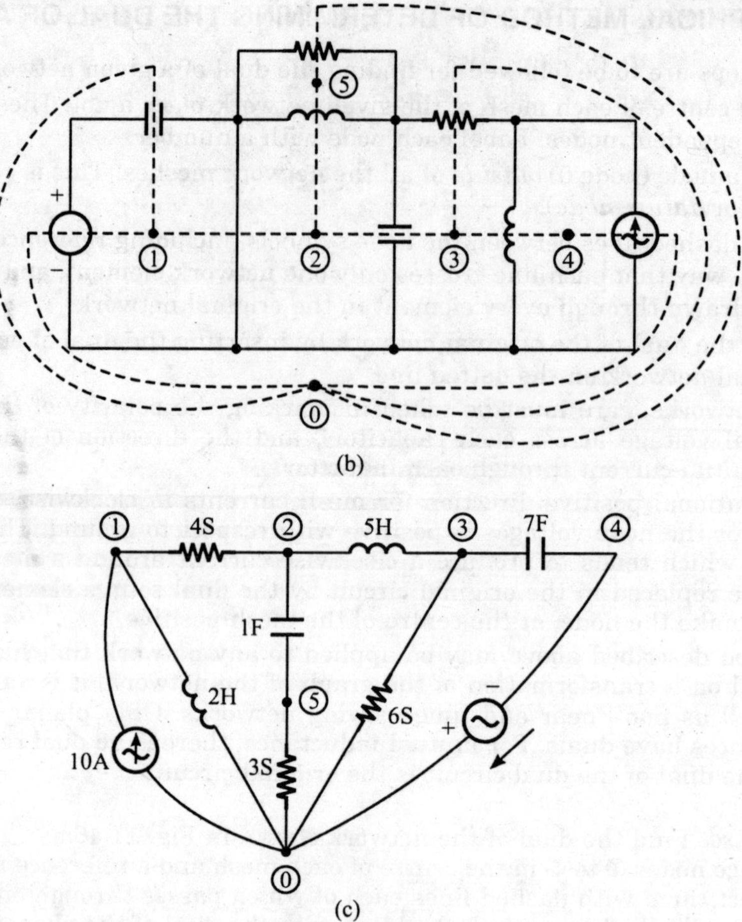

Fig. 11.46b, c

The 10 V voltage source in the original network produces a *clockwise* current in the mesh containing node ① in Fig. 11.46b. In the dual, it is replaced by a 10 A current source whose reference direction is towards node ① and away from reference node ⓪.

The 15 A current source in the original network produces an anticlockwise current in the mesh containing node ④. This current is in a direction opposite to the conventional clockwise positive direction of mesh currents. Therefore, the 15A current source of the original network is replaced by the 15 V voltage source in the dual network. Its reference polarity is such that it tends to make node ④ negative with respect to reference node ①. The dual network is shown in Fig. 11.46c.

■ **EXAMPLE 11.37** Draw the dual of the network shown in Fig. 11.47a.

SOLUTION : The procedure is shown in Fig. 11.47b, and the dual in Fig. 11.47c.

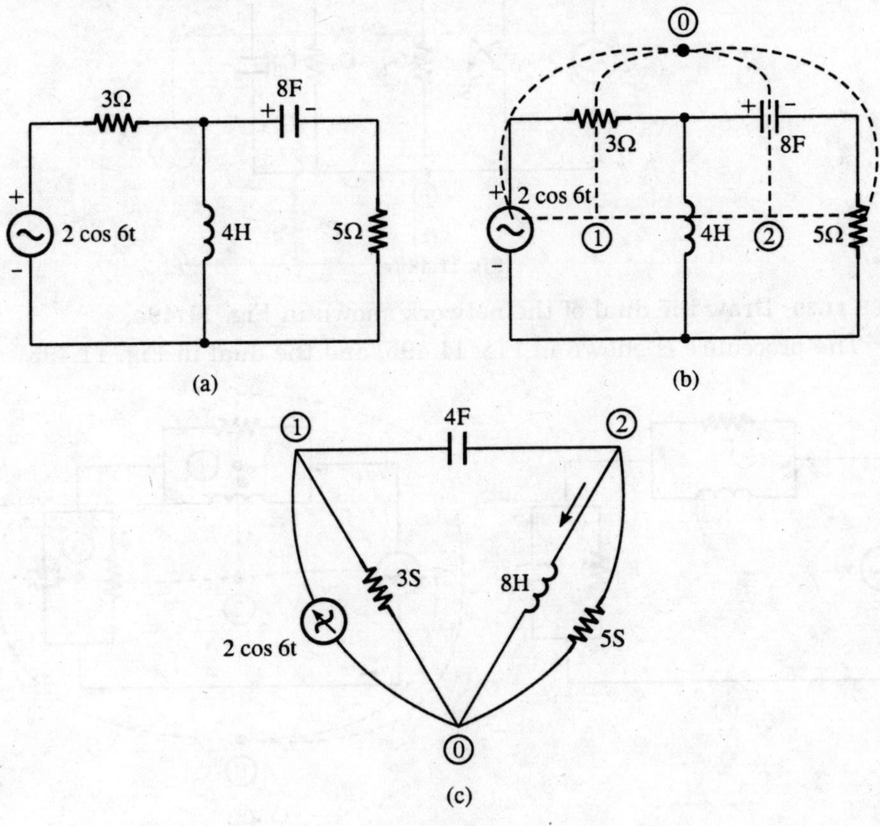

Fig 11.47

■ **EXAMPLE 11.38** Draw the dual of the network shown in Fig. 11.48a.

SOLUTION : The procedure is shown in Fig. 11.48b and the dual in Fig. 11.48c.

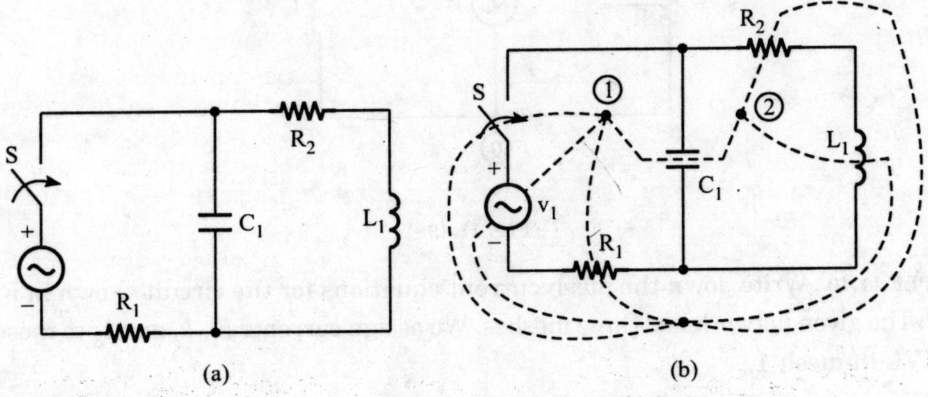

Fig 11.48 (a), (b)

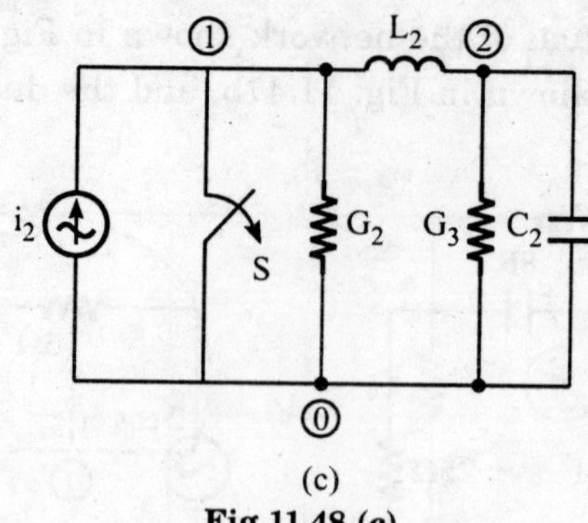

(c)

Fig 11.48 (c)

■ **EXAMPLE 11.39** Draw the dual of the network shown in Fig. 11.49a.

SOLUTION : The procedure is shown in Fig. 11.49b, and the dual in Fig. 11.49c.

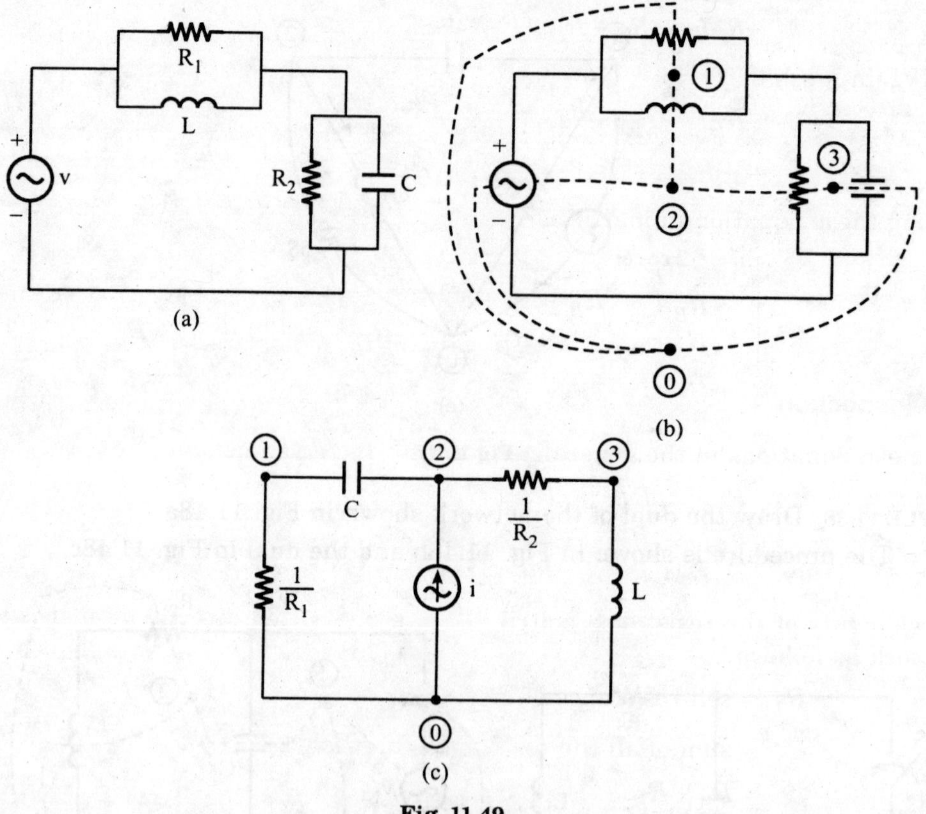

Fig. 11.49

■ **EXAMPLE 11.40** Write down the mesh current equations for the circuit shown in Fig. 11.50.

SOLUTION : The given network has three meshes. We assign currents I_1, I_2 and I_3 to these meshes.

By KVL in mesh 1,

$$E_A - R_A I_1 - R_B I_1 + R_B I_2 - E_B = 0$$

or $$(R_A + R_B)I_1 - R_B I_2 = E_A - E_B \tag{E11.40.1}$$

By KVL in mesh 2

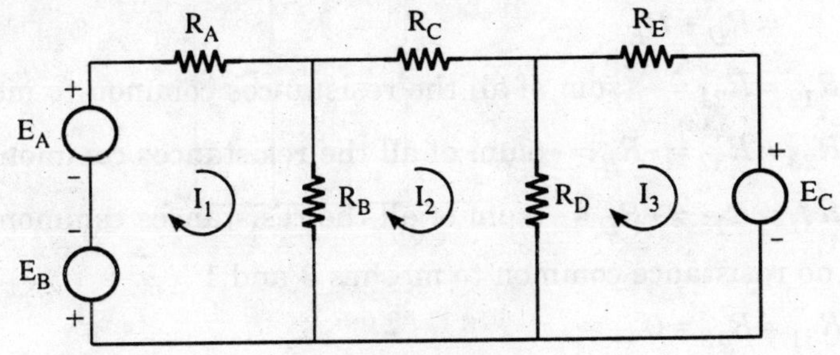

Fig. 11.50

$$R_B I_1 - R_B I_2 - R_C I_2 - R_D I_2 + R_D I_3 = 0$$

or $$-R_B I_1 + (R_B + R_C + R_D)I_2 - R_D I_3 = 0 \tag{E11.40.2}$$

By KVL in mesh 3

$$-R_D I_3 + R_D I_2 - R_3 I_3 - E_C = 0$$

or $$0 - R_D I_2 + (R_D + R_E)I_3 = -E_C \tag{E11.40.3}$$

Writing these equation in matrix form gives

$$\begin{bmatrix} R_A + R_B & -R_B & 0 \\ -R_B & R_B + R_C + R_D & -R_D \\ 0 & -R_D & R_D + R_E \end{bmatrix} \begin{bmatrix} I_1 \\ I_2 \\ I_3 \end{bmatrix} = \begin{bmatrix} E_A - E_B \\ 0 \\ -E_C \end{bmatrix} \tag{E11.40.4}$$

Method of Inspection

The mesh equations in the generalized form for the given network are

$$\begin{bmatrix} R_{11} & R_{12} & R_{13} \\ R_{21} & R_{22} & R_{23} \\ R_{31} & R_{32} & R_{33} \end{bmatrix} \begin{bmatrix} I_1 \\ I_2 \\ I_3 \end{bmatrix} = \begin{bmatrix} E_1 \\ E_2 \\ E_3 \end{bmatrix} \tag{E11.40.5}$$

The elements of the resistance matrix **[R]** can be written directly by inspection of the given network as follows :

R_{11} = self resistance of mesh 1

= sum of all the resistances in mesh 1

$= R_A + R_B$

R_{22} = self resistance of mesh 2

= sum of all the resistances in mesh 2

$= R_B + R_C + R_D$

R_{33} = self resistance of mesh 3

= sum of all the resistances in mesh 3

$= R_D + R_E$

$R_{12} = R_{21}$ = – (sum of all the resistances common to meshes 1 and 2)

$R_{23} = R_{32} = -R_B$ = –(sum of all the resistances common to meshes 2 and 3)

$R_{31} = R_{13} = -R_D$ = –(sum of all the resistances common to meshes 3 and 1)

Since there is no resistance common to meshes 3 and 1

$R_{31} = R_{13} = 0$

E_1 = algebraic sum of all the voltages of the voltage sources which produce clockwise currents in mesh 1

$= E_A - E_B$

It is to be noted that the voltage source E_A produces clockwise current and the voltage source E_B produces current in the anticlockwise direction in mesh 1.

E_2 = algebraic sum of all the voltages which produce clockwise currents in mesh 2

Since there is no voltage source in mesh 2, $E_2 = 0$

E_3 = algebraic sum of all the voltages which produce clockwise currents in mesh 3

Since the voltage source E_C in mesh current I_3 in it

$E_3 = -E_C$

Substitution of the values of resistances and voltages in Eq. (E11.40.5) gives

$$\begin{bmatrix} R_A + R_B & -R_B & 0 \\ -R_B & R_B + R_C + R_D & -R_D \\ 0 & -R_D & R_D + R_E \end{bmatrix} \begin{bmatrix} I_1 \\ I_2 \\ I_3 \end{bmatrix} = \begin{bmatrix} E_A - E_B \\ 0 \\ -E_C \end{bmatrix} \quad \text{(E11.40.6)}$$

It is seen that the matrices given by Eqs. (E11.40.4) and (E11.40.6) are the same.

EXERCISES

1. Write the mesh equations for the network shown in Fig. P11.1.

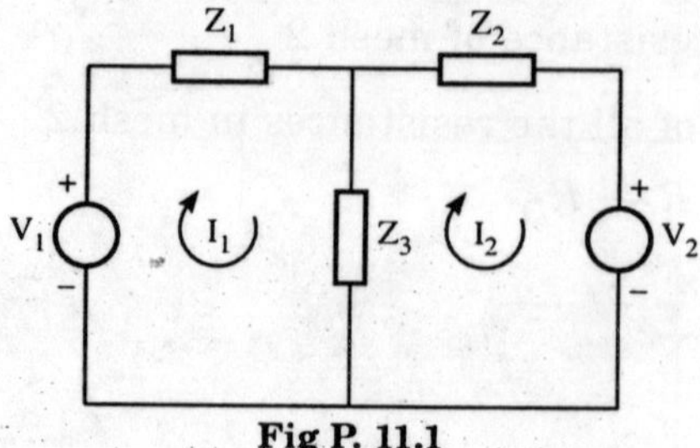

Fig P. 11.1

2. Write the mesh equations for the network shown in Fig. P11.2.

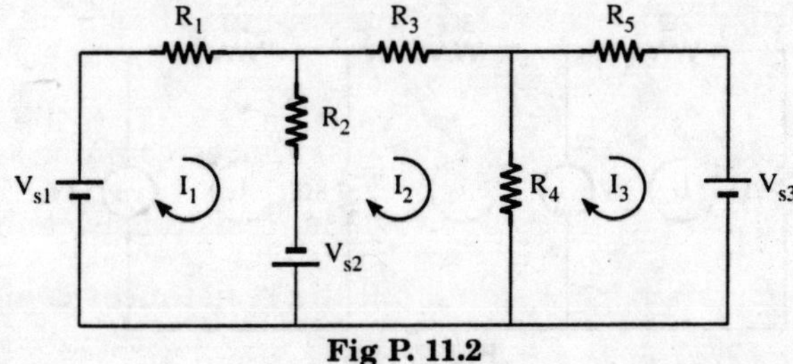

Fig P. 11.2

3. Determine the mesh currents in the network shown in Fig. P11.3.

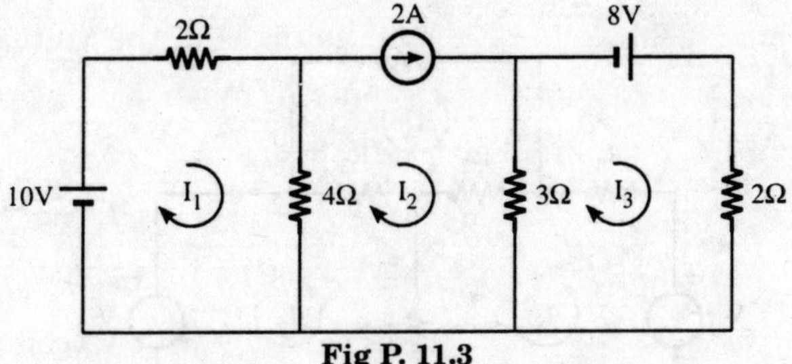

Fig P. 11.3

4. Determine the mesh currents in the network shown in Fig. P11.4.

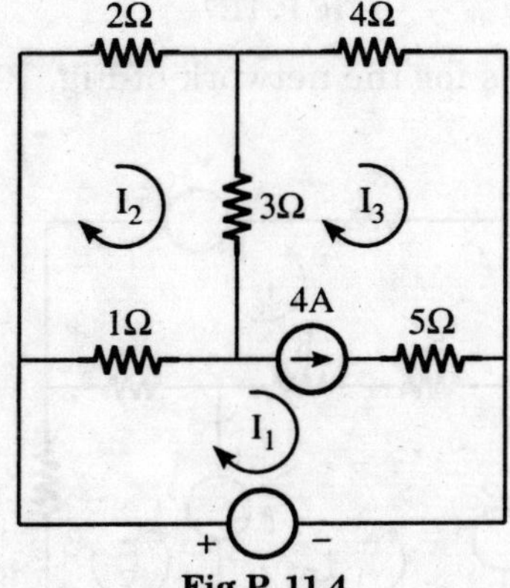

Fig P. 11.4

5. Use mesh analysis to find v_x in the circuit of Fig. P11.5.

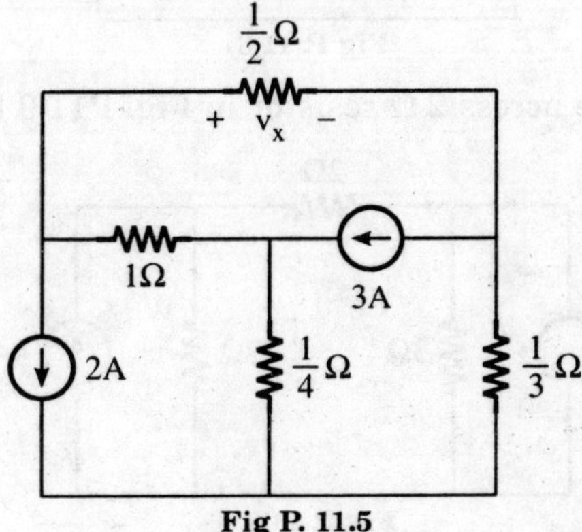

Fig P. 11.5

6. Write the mesh equations for the network shown in Fig. P11.6 in matrix form.

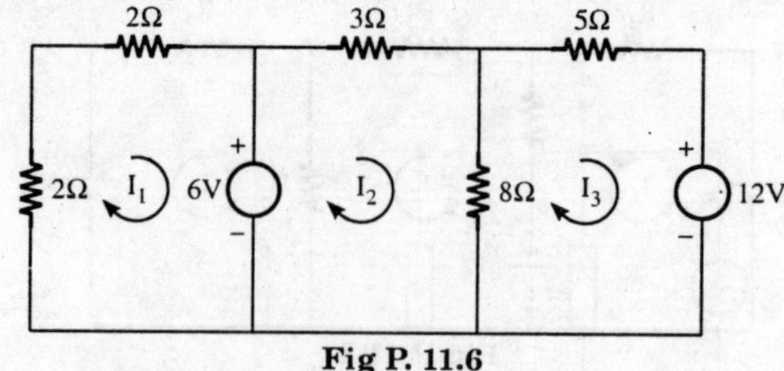

Fig P. 11.6

7. Write the mesh equations for the network of Fig. P11.7.

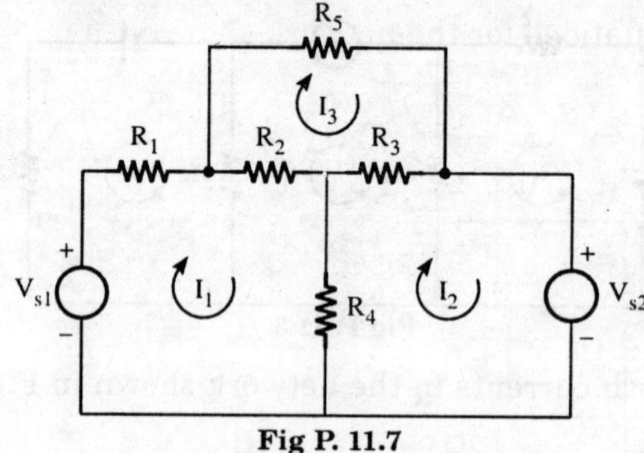

Fig P. 11.7

8. Write the mesh equations for the network of Fig. P11.8.

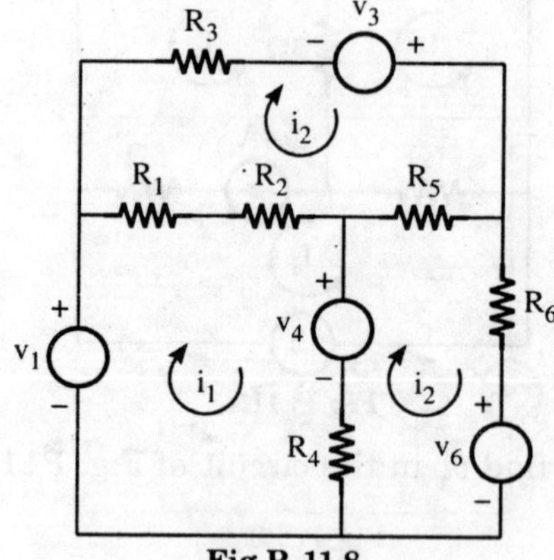

Fig P. 11.8

9. Calculate the voltage across 2 Ω resistor in Fig. P11.9 by node-voltage method.

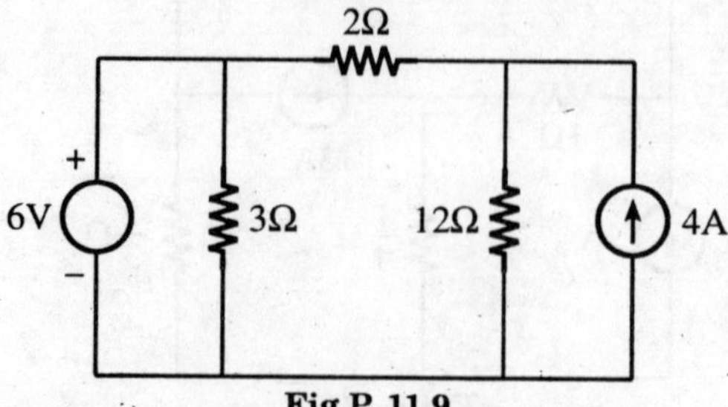

Fig P. 11.9

10. Write the node equations for the nodes 1, 2 and 3 of the network of Fig. P11.10.

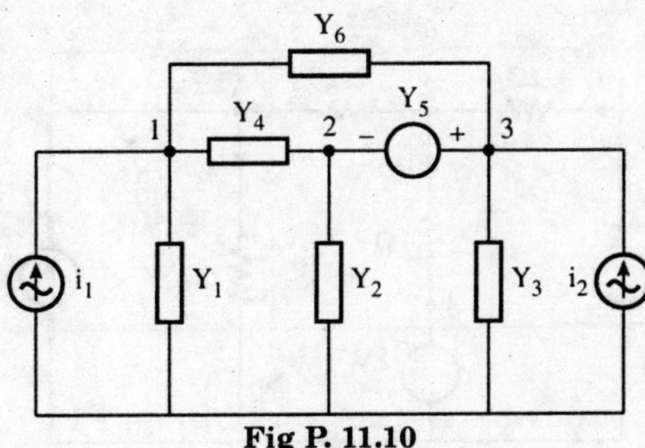

Fig P. 11.10

11. Write the node equations for the network shown in Fig. P11.11.

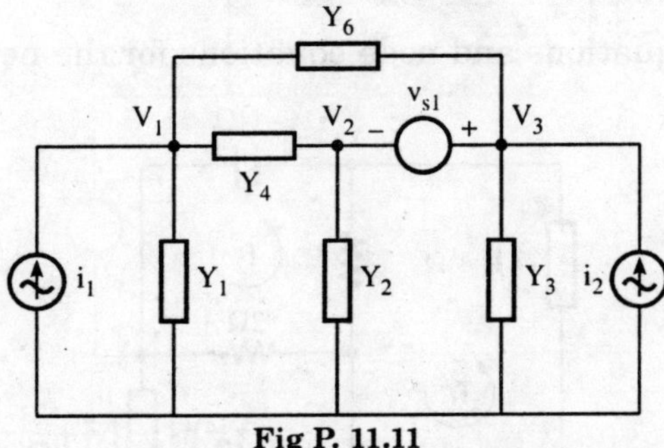

Fig P. 11.11

12. Write the mesh equations for the network of Fig. P11.12.

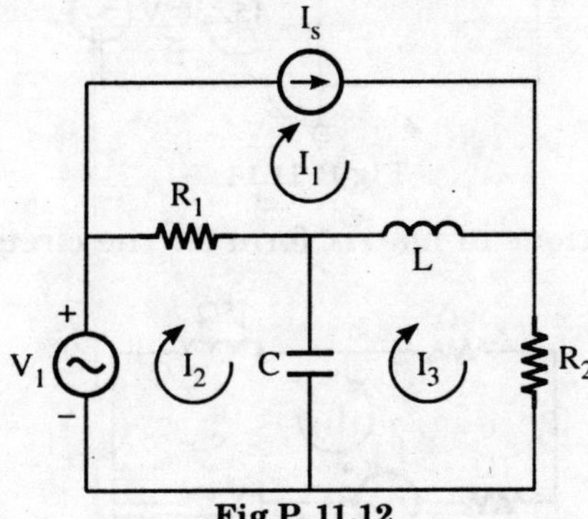

Fig P. 11.12

13. Determine the node voltages and currents in the network of Fig. P 11.13. Use nodal analysis.

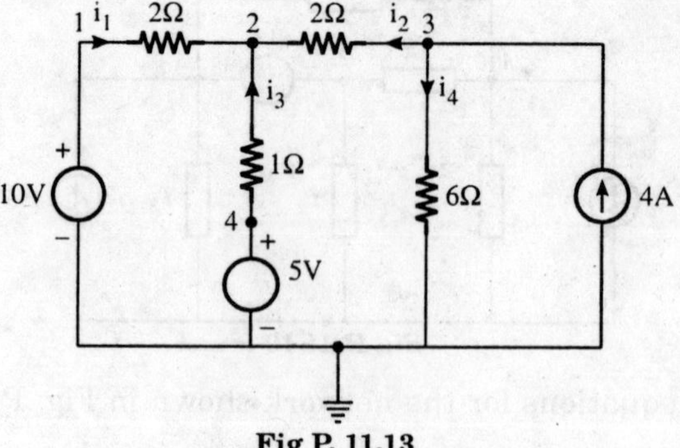

Fig P. 11.13

14. Write the mesh equations and node equations for the network of Fig. P11.14.

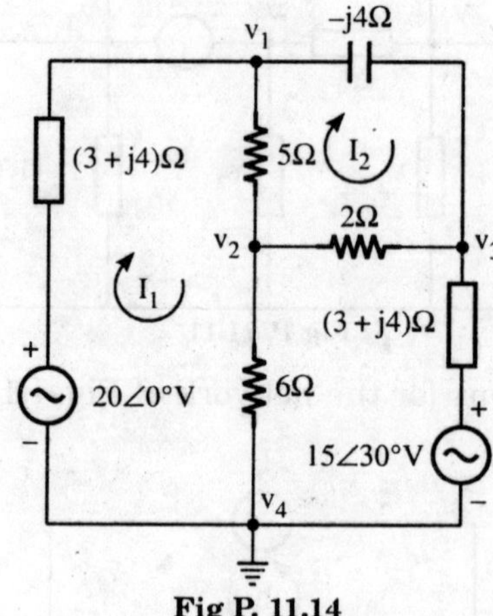

Fig P. 11.14

15. Write the mesh equations in matrix form for the circuit of Fig. P11.15.

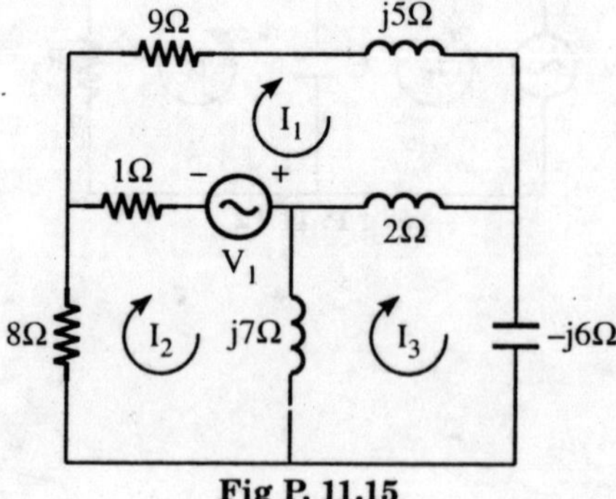

Fig P. 11.15

16. Write the mesh equations for the circuit of Fig. P11.16.

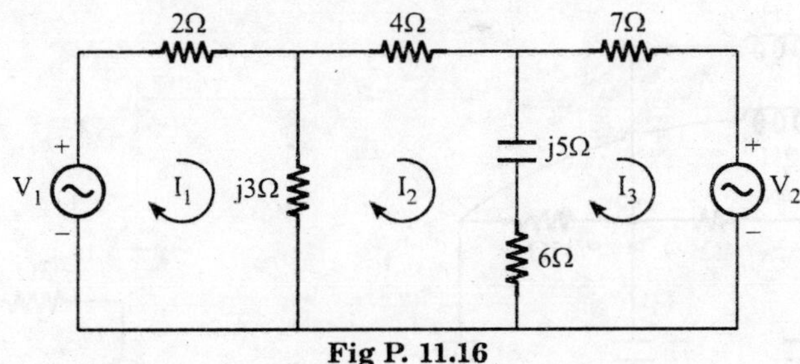

Fig P. 11.16

17. In the network shown in Fig. P11.17, the two equal capacitances C and the resistance R are adjusted so that the current I_L is zero. Assuming a source angular frequency ω, determine the values of R_1 and L_1.

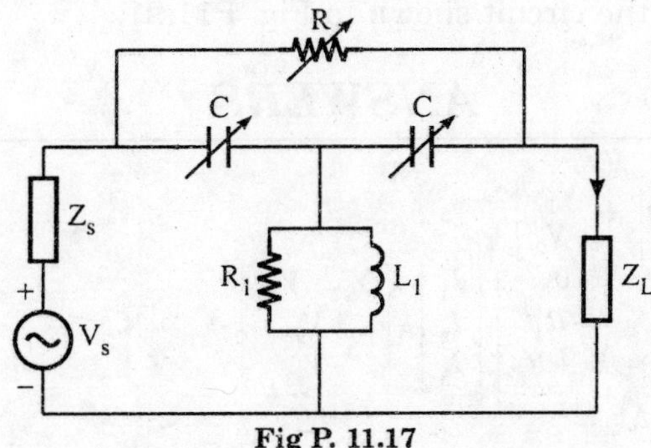

Fig P. 11.17

18. Write the mesh equations for the network shown in Fig. P11.18.

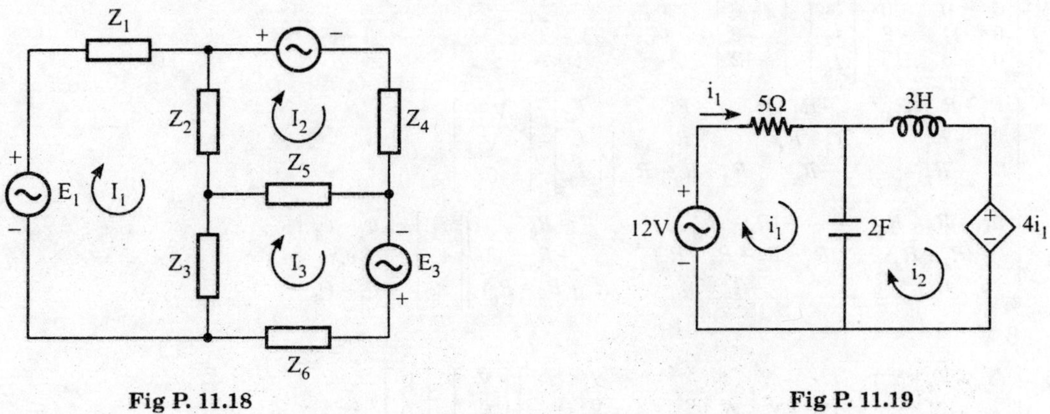

Fig P. 11.18 **Fig P. 11.19**

19. Draw the dual of the circuit shown in Fig. P11.19.

20. Draw the dual of the circuit shown in Fig. P11.20.

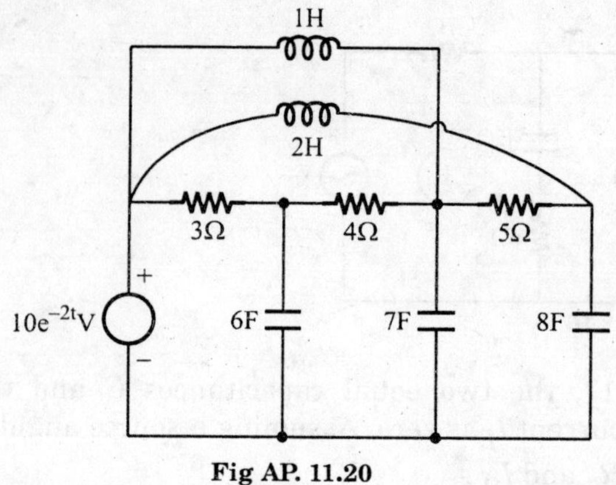

Fig AP. 11.20

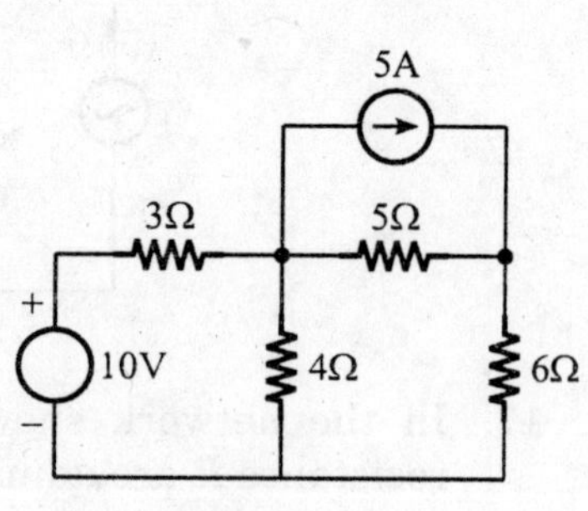

Fig P. 11.21

21. Draw the dual of the circuit shown in Fig. P11.21.

ANSWERS

1. $\begin{bmatrix} \mathbf{Z}_1+\mathbf{Z}_3 & -\mathbf{Z}_3 \\ -\mathbf{Z}_3 & \mathbf{Z}_2+\mathbf{Z}_3 \end{bmatrix}\begin{bmatrix} \mathbf{I}_1 \\ \mathbf{I}_2 \end{bmatrix}=\begin{bmatrix} \mathbf{V}_1 \\ -\mathbf{V}_2 \end{bmatrix}$

2. $\begin{bmatrix} R_1+R_2 & -R_2 & 0 \\ -R_2 & R_2+R_3+R_4 & -R_4 \\ 0 & -R_4 & R_4+R_5 \end{bmatrix}\begin{bmatrix} I_1 \\ I_2 \\ I_3 \end{bmatrix}=\begin{bmatrix} V_{s1}+V_{s2} \\ -V_{s2} \\ -V_{s3} \end{bmatrix}$

3. $I_1 = 3\text{ A}, I_2 = 2\text{ A}, I_3 = 2.8\text{ A}$

4. $I_1 = 6\text{ A}, I_2 = 2\text{ A}, I_3 = 2\text{ A}$

5. -0.18 V

6. $\begin{bmatrix} 6 & 0 & 0 \\ 0 & 11 & -8 \\ 0 & -8 & 13 \end{bmatrix}\begin{bmatrix} I_1 \\ I_2 \\ I_3 \end{bmatrix}=\begin{bmatrix} -6 \\ 6 \\ -12 \end{bmatrix}$

7. $\begin{bmatrix} R_1+R_2+R_4 & -R_4 & -R_2 \\ -R_4 & R_3+R_4 & -R_3 \\ -R_2 & -R_3 & R_2+R_3+R_5 \end{bmatrix}\begin{bmatrix} I_1 \\ I_2 \\ I_3 \end{bmatrix}=\begin{bmatrix} V_{s1} \\ -V_{s2} \\ 0 \end{bmatrix}$

8. $\begin{bmatrix} (R_1+R_2+R_4) & -(R_1+R_2) & -R_4 \\ -(R_1+R_2) & (R_1+R_2+R_3+R_5) & -R_5 \\ -R_4 & -R_5 & (R_4+R_5+R_6) \end{bmatrix}\begin{bmatrix} i_1 \\ i_2 \\ i_3 \end{bmatrix}=\begin{bmatrix} v_1-v_4 \\ v_3 \\ v_4-v_6 \end{bmatrix}$

9. 6 V

10. $\begin{bmatrix} (\mathbf{Y}_1+\mathbf{Y}_4+\mathbf{Y}_6) & -\mathbf{Y}_4 & -\mathbf{Y}_6 \\ -\mathbf{Y}_4 & (\mathbf{Y}_2+\mathbf{Y}_4+\mathbf{Y}_5) & -\mathbf{Y}_5 \\ -\mathbf{Y}_6 & -\mathbf{Y}_5 & (\mathbf{Y}_3+\mathbf{Y}_5+\mathbf{Y}_6) \end{bmatrix}\begin{bmatrix} \mathbf{V}_1 \\ \mathbf{V}_2 \\ \mathbf{V}_3 \end{bmatrix}=\begin{bmatrix} \mathbf{i}_1 \\ 0 \\ \mathbf{i}_2 \end{bmatrix}$

11. $(\mathbf{Y}_1+\mathbf{Y}_4+\mathbf{Y}_6)\,\mathbf{V}_1-\mathbf{Y}_4\,\mathbf{V}_2-\mathbf{Y}_6\,\mathbf{V}_3=\mathbf{i}_1$

$-(\mathbf{Y}_4+\mathbf{Y}_6)\,\mathbf{V}_1+(\mathbf{Y}_2+\mathbf{Y}_4)\,\mathbf{V}_2-(\mathbf{Y}_3+\mathbf{Y}_6)\,\mathbf{V}_3=\mathbf{i}_2$

$\mathbf{V}_3-\mathbf{V}_2=\boldsymbol{v}_{s1}$

12. $I_1 = I_s$

$$-R_1 I_1 + \left(R_1 + \frac{1}{j\omega C}\right) I_2 - \frac{1}{j\omega C} I_3 = V_1$$

$$-j\omega L_1 I_1 - \frac{1}{j\omega C} I_2 \left(j\omega L + R_2 + \frac{1}{j\omega C}\right) I_3 = 0$$

13. $v_1 = 10$ V, $v_2 = 8$ V, $v_3 = 12$ V, $v_4 = 5$ V

$i_1 = 1$ A, $i_2 = 2$ A, $i_3 = -3$ A, $i_4 = 2$ A

14.
$$\begin{bmatrix} 14+j4 & -5 & -6 \\ -5 & 7-j4 & -2 \\ -6 & -2 & 10+j5 \end{bmatrix} \begin{bmatrix} \mathbf{I}_1 \\ \mathbf{I}_2 \\ \mathbf{I}_3 \end{bmatrix} = \begin{bmatrix} 20 \angle 0^\circ \\ 0 \\ -15 \angle 30^\circ \end{bmatrix}$$

$$\begin{bmatrix} \left(\frac{1}{3+j4} + \frac{1}{5} + \frac{1}{-j4}\right) & -\frac{1}{5} & \frac{1}{j4} \\ -\frac{1}{5} & \left(\frac{1}{5} + \frac{1}{2} + \frac{1}{6}\right) & -\frac{1}{2} \\ \frac{1}{j4} & -\frac{1}{2} & \left(\frac{1}{2} + \frac{1}{2+j5} + \frac{1}{-j4}\right) \end{bmatrix} \begin{bmatrix} \mathbf{V}_1 \\ \mathbf{V}_2 \\ \mathbf{V}_3 \end{bmatrix} = \begin{bmatrix} \frac{20}{3+j4} \\ 0 \\ \frac{15 \angle 30^\circ}{2+j5} \end{bmatrix}$$

15.
$$\begin{bmatrix} (12+j5) & -1 & -2 \\ -1 & (9+j7) & -j7 \\ -2 & -j7 & (2+j1) \end{bmatrix} \begin{bmatrix} \mathbf{I}_1 \\ \mathbf{I}_2 \\ \mathbf{I}_3 \end{bmatrix} = \begin{bmatrix} \mathbf{V}_1 \\ \mathbf{V}_2 \\ 0 \end{bmatrix}$$

16.
$$\begin{bmatrix} (12+j3) & -j3 & 0 \\ -j3 & (10-j2) & (-6+j5) \\ 0 & (-6+j5) & 13-j5 \end{bmatrix} \begin{bmatrix} \mathbf{I}_1 \\ \mathbf{I}_2 \\ \mathbf{I}_3 \end{bmatrix} = \begin{bmatrix} \mathbf{V}_1 \\ 0 \\ -\mathbf{V}_2 \end{bmatrix}$$

17. $R_1 = \dfrac{1}{\omega^2 C^2 R}, \quad L_1 = \dfrac{1}{2\omega C}$

18.
$$\begin{bmatrix} \mathbf{E}_1 \\ -\mathbf{E}_2 \\ \mathbf{E}_3 \end{bmatrix} = \begin{bmatrix} \mathbf{Z}_1+\mathbf{Z}_2+\mathbf{Z}_3 & -\mathbf{Z}_2 & -\mathbf{Z}_3 \\ -\mathbf{Z}_2 & \mathbf{Z}_2+\mathbf{Z}_4+\mathbf{Z}_5 & -\mathbf{Z}_5 \\ -\mathbf{Z}_3 & -\mathbf{Z}_5 & \mathbf{Z}_3+\mathbf{Z}_5+\mathbf{Z}_6 \end{bmatrix} \begin{bmatrix} \mathbf{I}_1 \\ \mathbf{I}_2 \\ \mathbf{I}_3 \end{bmatrix}$$

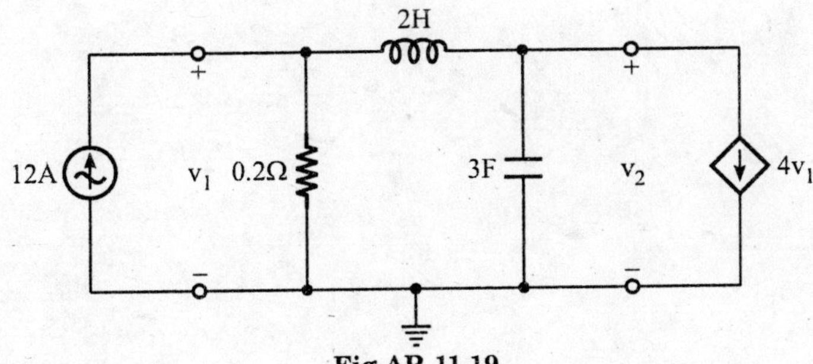

Fig AP. 11.19

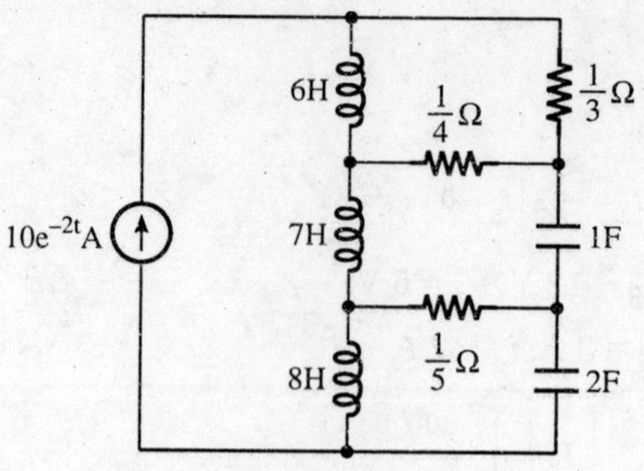

Fig AP. 11.20

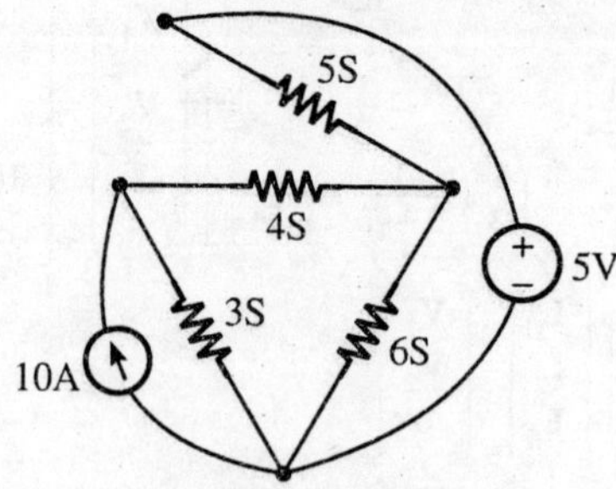

Fig P. 11.21

12

NETWORK THEOREMS

12.1 INTRODUCTION

A network may be solved by the application of the basic network laws (Ohm's law, KCL and KVL). However, in large complex networks, these methods are laborius and time consuming. In this chapter we shall study a set of linear circuit principles which provide quick and simple solutions of circuit problems. These principles are called *circuit theorems* or *network theorems* because they can be proved under very general conditions and their conclusions are relatively simple. Thus, network theorems provide short-cut methods to solve complex network. These theorems are derived from three basic laws (Ohm's law, KCL and KVL). In this chapter we shall study some important theorems, namely, superposition theorem, Thevenin's theorem, Norton's theorem, maximum power transfer theorem, Millman's theorem, substitution theorem, compensation theorem, Tellegen's theorem, and reciprocity theorem.

12.2 SUPERPOSITION THEOREM

Superposition is a general principle that allows us to determine the effect of several energy sources (current sources and voltage sources) acting simultaneously on a system by considering the effect of each source acting alone and then combining (superposing) these effects. The concept of superposition is very powerful in linear networks and linear systems. Superposition theorem can be stated as follows :

> *In a linear network, containing more than one independent energy sources (which include the equivalent sources due to initial conditions), the overall response (that is, the current through or voltage across) in any branch is equal to the sum of the response due to each independent source acting one at a time with all other ideal independent sources set equal to zero.*
>
> Setting an ideal current source to zero means that the source is replaced by an open circuit. Similarly, setting an ideal voltage source equal to zero means that the source is replaced by a short circuit. It is to be noted that an open circuit may be considered to be a current source of value zero ampere, and a short circuit may be considered to be a voltage source of value zero volt.

When the circuit contains **dependent** sources, the superposition theorem can only be used when the control variable is external to the network containing the sources.

The superposition theorem is extremely general. It is applicable to all **linear** networks time-varying or time invariant. It holds for all possible locations, types and waveforms of the independent sources. The theorem applies both in the time domain and s domain. Since the power is proportional to the square (which is nonlinear) of voltage (or current), superposition theorem cannot be applied for calculation of power.

The superposition theorem follows the assumption of linearity of circuit elements, and it is necessary for the formal proof of other theorems. It is specially very useful when the networks are exicited by different types of sources, that is, ac and dc, or ac with different frequencies.

12.3 APPLICATION OF SUPERPOSITION THEOREM

The procedure for analysing a circuit by superposition theorem is as follows :

1. Select any one source in the circuit.
2. Set all other independent sources equal to zero. That is, replace all other ideal voltage sources by short circuits, and all other ideal current sources by open circuits.
3. Keep the dependent sources in the circuit undisturbed.
4. Determine the magnitude and direction of current through the desired branch as a result of single source selected in step 1.
5. Repeat steps 1 to 4 for each source in turn until the component current through the desired branch has been calculated due to each and every source acting alone.
6. Add all the component currents to obtain the desired branch current. Algebraic addition is to be done for dc networks. For ac networks phasor addition is performed.

The voltage across any branch in the circuit can also be found in a similar manner.

The superposition theorem applies only to **independent** sources. A dependent source in a circuit is treated as just another circuit element for calculation purpose.

12.4 PROOF OF SUPERPOSITION THEOREM

Consider a linear active network consisting of m meshes with voltage sources in each mesh. The mesh equations can be written as

$$\left.\begin{aligned} z_{11} I_1 + z_{12} I_2 + \dots + z_{1m} I_m &= E_1 \\ z_{21} I_1 + z_{22} I_2 + \dots + z_{2m} I_m &= E_2 \\ \dots\dots\dots\dots\dots\dots\dots\dots & \\ z_{m1} I_1 + z_{m2} I_2 + \dots + z_{mm} I_m &= E_m \end{aligned}\right\} \qquad (12.4.1)$$

The solution of the above equations give

$$\left.\begin{aligned} I_1 &= y_{11} E_1 + y_{12} E_2 + \dots + y_{1m} E_m \\ I_2 &= y_{21} E_1 + y_{22} E_2 + \dots + y_{2m} E_m \\ \dots\dots\dots\dots\dots\dots\dots\dots & \\ I_m &= y_{m1} E_1 + y_{m2} E_2 + \dots + y_{mm} E_m \end{aligned}\right\} \qquad (12.4.2)$$

Suppose that all the E's except E_1 are set equal to zero. This gives the current in mesh 1 as $y_{11} E_1$, in mesh 2 as $y_{21} E_2$, etc. Similarly, suppose that all the E's except E_2 are set equal to zero. This gives the current in mesh 1 as $y_{12} E_2$, in mesh 2 as $y_{22} E_2$, and so on. This procedure is repeated for the remaining E's and the mesh currents are determined. Finally, suppose that all the E's are acting simultaneously, then all the solutions found separately can be added together to get the complete solution. For example, we have for the first mesh

$$I_1 = y_{11} E_1 + y_{12} E_2 + \dots y_{1m} E_m$$

But this result is the same as that obtained by solving the simultaneous equations. Thus, the form of Eq. (12.4.2) is the proof of the superpositon theorem.

A similar proof can be given using nodal equations.

The proof given above is completely general in the sense that some of the voltage terms in Eq. (12.4.2) can come from initial conditions, and the actual source, can have *any* waveshape.

The superposition principle is illustrated by the following examples :

EXAMPLE 12.1 Calculate the current in each branch of the network shown in Fig. 12.1a by the superposition principle.

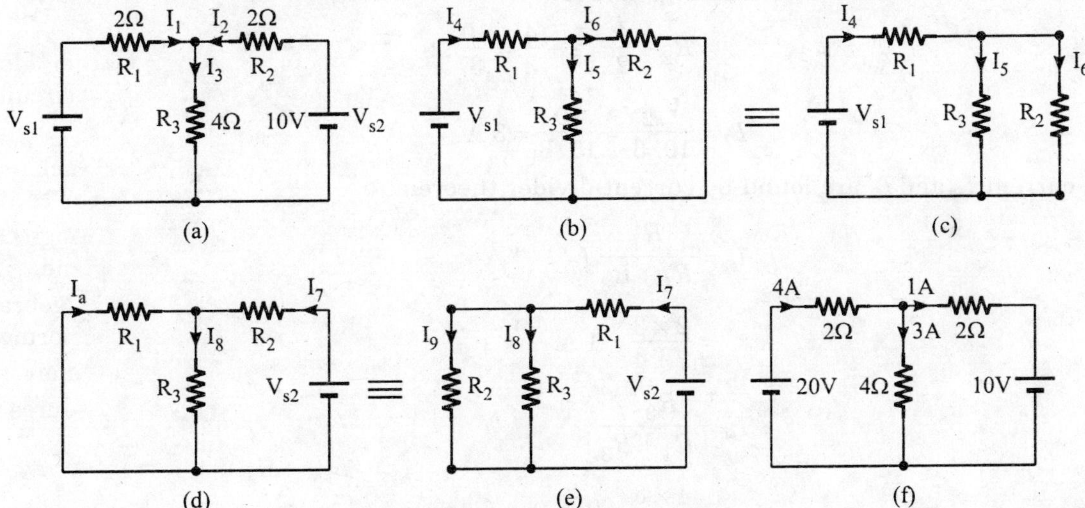

FIG. 12.1 Illustrating Example 12.1

SOLUTION Consider the 20 V source alone and replace the 10 V source by a short circuit (a zero-resistance conductor). The given network then becomes as shown in Fig. 12.1b, which is redrawn as Fig. 12.1c. Let the currents be as shown in Fig. 12.1c. Equivalent resistance of R_2 and R_3 in parallel

$$= \frac{R_2 R_3}{R_2 + R_3} = \frac{2 \times 4}{2 + 4} = \frac{4}{3}\ \Omega$$

Total resistance of the network in Fig. 12.1c

$$= R_1 + (R_2 \parallel R_3) = 2 + \frac{4}{3} = \frac{10}{3}\ \Omega$$

$$I_4 = \frac{V_{s1}}{(10/3)} = \frac{20}{10/3} = 6\ \text{A}$$

The currents I_5 and I_6 are found by current-divider theorem.

$$I_5 = \frac{R_2 I_4}{R_2 + R_3} = \frac{2 \times 6}{2 + 4} = 2\ \text{A}$$

$$I_6 = \frac{R_3 I_4}{R_2 + R_3} = \frac{4 \times 6}{2 + 4} = 4 \text{ A}$$

Alternatively, $I_6 = I_4 - I_5 = 6 - 2 = 4 \text{ A}$

Now consider the 10 V source alone and replace the 20 V source by a short circuit. The given network of Fig. 12.1a then becomes as shown in Fig. 12.1d, which is redrawn as Fig. 12.1e. Let the currents be as shown in Fig. 12.1e.

Equivalent resistance of R_1 and R_3 in parallel

$$= \frac{R_1 R_3}{R_1 + R_3} = \frac{2 \times 4}{2 + 4} = \frac{4}{3}\ \Omega$$

Total resistance of the network in Fig. 12.1 (e)

$$= R_2 + \frac{4}{2} = 2 + \frac{4}{3} = \frac{10}{3}\ \Omega$$

$$I_7 = \frac{V_{s2}}{10/3} = \frac{10}{10/3} = 3 \text{ A}$$

The current I_8 and I_9 are found by current-divider theorem.

$$I_8 = \frac{R_1}{R_1 + R_3} I_7$$

$$= \frac{2 \times 3}{2 + 4} = 1 \text{ A}$$

$$I_9 = \frac{R_3 I_7}{R_1 + R_3}$$

$$= \frac{4 \times 3}{2 + 4} = 2 \text{ A}$$

Alternatively,

$$I_9 = I_7 - I_8 = 3 - 1 = 2 \text{ A}$$

According to the superposition theorem, the current through any branch is found by taking the algebraic sum of the currents through it. Thus,

I_1 = current through R_1 in Fig. 12.1a

= algebraic sum of the currents through R_1 in Fig. 12.1c and Fig. 12.1e

= algebraic sum of the currents through R_1 in Fig. 12.1b and Fig. 12.1d

$= I_4 + (-I_9) = I_4 - I_9 = 6 - 2 = 4 \text{ A}$

I_2 = current through R_2 in Fig. 12.1a

= algebraic sum of the currents through R_2 in Fig. 12.1c and Fig. 12.1e

= algebraic sum of the currents through R_2 in Fig. 12.1b and Fig. 12.1d

$= I_7 - I_6 = 3 - 4 = -1 \text{ A}$

I_3 = current through R_3 in Fig. 12.1a

= algebraic sum of the currents through R_3 in Fig. 12.1c and Fig. 12.1e

= algebraic sum of the currents through R_3 in Fig. 12.1b and Fig. 12.1d

$= I_5 + I_8 = 2 + 1 = 3$ A

The current in R_2 in Fig. 12.1a is – 1 A. The negative sign indicates that the direction of I_2 in Fig. 12.1a is opposite to the assumed direction. Hence the corrected currents are shown in Fig. 12.1f.

The results obtained for I_1, I_2 and I_3 can be verified by application of KCL and KVL.

■ **EXAMPLE 12.2** Determine the current I in the network shown in Fig. 12.2 by the principle of superposition.

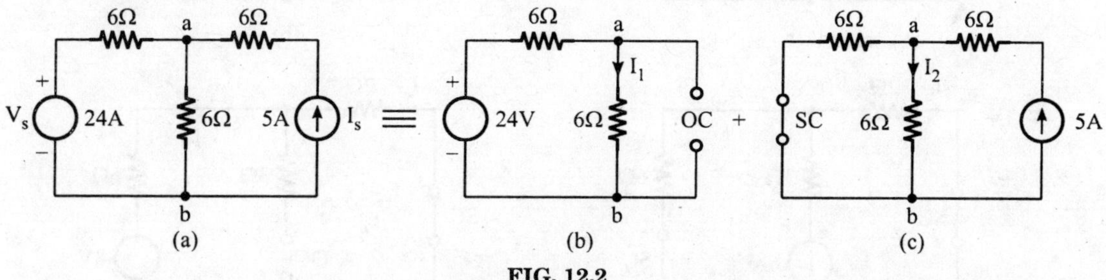

FIG. 12.2

SOLUTION The given circuits consist of two sources, namely, 24 V voltage source and 5 A current source. We shall determine current I due to each source acting alone.

(a) *Only 24 V voltage source in the circuit*

The 5 A current source is open-circuited as shown in Fig. 12.2b. The current I_1 through $R_{ab} = 6$, is determined by the 24 V voltage source alone. By Ohm's law

$$I_1 = \frac{24}{6+6} = 2 \text{ A}$$

(b) *Only 5 A current source in the circuit*

In Fig. 12.2c, the voltage source of 24 V is short-circuited. The current I_2 through R_{ab} is determined due to 5 A current source alone.

$$I_2 = \frac{5}{2} = 2.5 \text{ A}$$

By superposition theorem, the total current through R_{ab} is

$$I = I_1 + I_2 = 2 + 2.5 = 4.5 \text{ A}$$

■ **EXAMPLE 12.3** Determine the current I_1 in the circuit of Fig. 12.3a by the superposition theorem.

SOLUTION There are three sources in the given circuit. We shall determine the current through the 2 Ω resistor due to each source acting alone.

(a) *Only 12 V source in the circuit*

The 6 A current source is open-circuited and the 8 V voltage source is short-circuited. The resulting circuit is shown in Fig. 12.3b. There is no current through the open circuit. Hence the current through the 3 Ω resistor is zero. By Ohm's law

$$I_a = -\frac{12}{2+6} = -1.5 \text{ A}$$

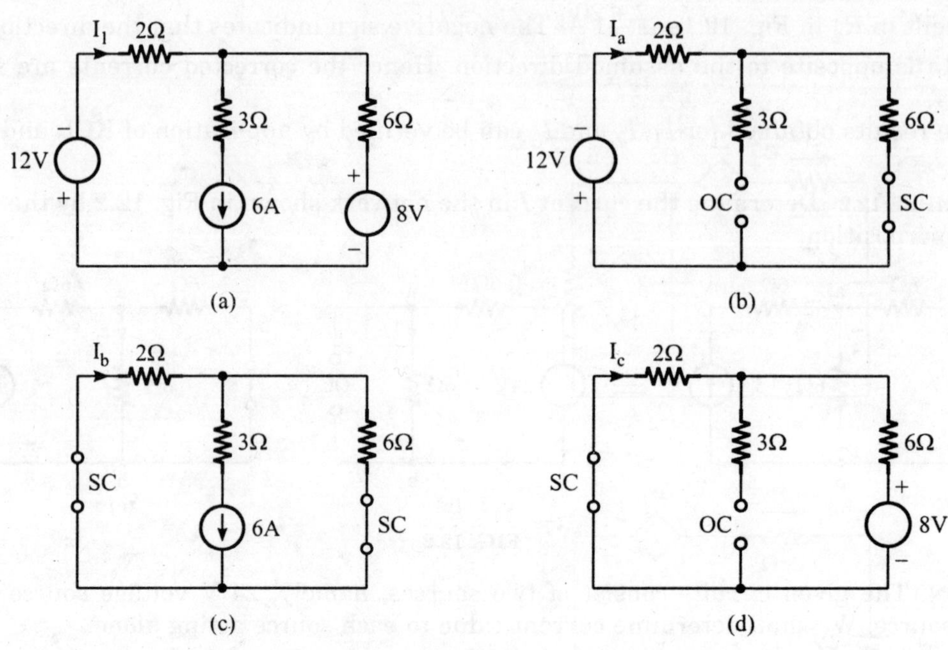

FIG. 12.3

(b) *Only 6 A current source in the circuit*

The 12 V voltage source and the 8 V voltage source are short circuited as shown in Fig. 12.3c. The 6 A current source is supplying current to 2 Ω and 6 Ω resistors in parallel. By current division rule in Fig. 12.3c.

$$I_b = \frac{6}{2+6} \times 6 = 4.5 \text{ A}$$

(c) *Only 8 V source in the circuit*

The 6 A current source is open-circuited and the 12 V voltage source is short-circuited as shown in Fig. 12.3d. By Ohm's law

$$I_c = -\frac{8}{6+2} = -1 \text{ A}$$

By superposition principle, the total current through the 2 Ω resistor is

$$I = I_a + I_b + I_c = -1.5 + 4.5 - 1 = 2 \text{ A}$$

■ **EXAMPLE 12.4** Use superposition theorem to find the current i for the circuit in Fig. 12.4a.

SOLUTION The given circuit contains two independent voltage sources of 2 V and 10 V and one dependent voltage source of $2i$ volts. Since the superposition theorem applies only to independent sources, we shall treat the dependent source as just another circuit element for calculation purposes. First we shall short circuit the 10 V voltage source. The resulting circuit is shown in Fig. 12.4b. Since the 5 Ω resistor is in parallel with a short circuit (zero resistance), this parallel combination results in a zero resistance, so the circuit of Fig. 12.4b is reduced to that of Fig. 12.4c. The current i is shown as i_2 to indicate that it is the component of i due to 2 V source alone. By KVL in the closed path of Fig. 12.4c, we have

$$2 + 20i_2 - 2i_2 = 0$$

or

$$i_2 = -\frac{1}{9}\text{ A}$$

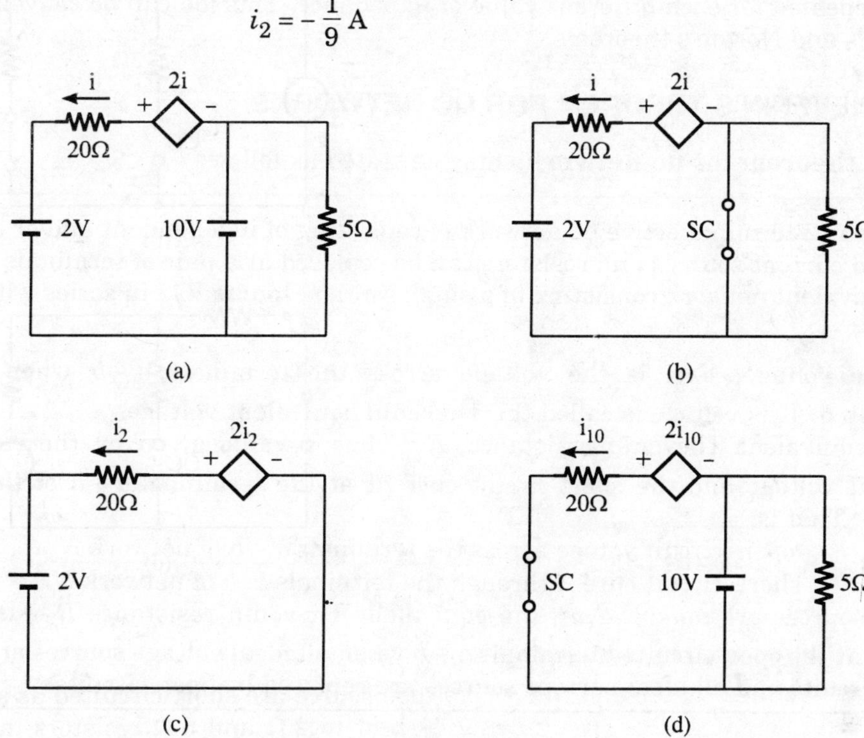

FIG. 12.4

The next step is to short circuit the 2 V source. This gives the circuit shown in Fig. 12.4d in which the current i is shown as i_{10} to indicate that it is the component of i due to 10 V source alone. By KVL aound the path containing the dependent source we get

$$20i_{10} - 2i_{10} - 10 = 0, \quad i_{10} = \frac{5}{9}\text{ A}$$

By superposition theorem,

$$i = i_2 + i_{10} = -\frac{1}{9} + \frac{5}{9} = \frac{4}{9}\text{ A}$$

12.5 THEVENIN'S AND NORTON'S THEOREM

Frequently, in analysis of a network, we are not interested in the solution for all of the unknown variables, but only in the solution for one of them. Thus, we may wish to determine the current through or voltage across any one element in a network. In such cases it is not necessary to write complete set of equations for the whole network and then solving these equations for this single curent or voltage in particular branch. In other words, it is always not necessary to analyse the complete circuit. Thevenin's and Norton's theorems may conveniently be used for such problems. In addition we may require the calculation of the variables in a given branch for several values of the branch impedance, while all other elements in the network are kept constant. In such a case it is desirable to be able to separate a branch from the rest of the network and apply some kind of solution technique that will not have to be repeated for each different value of impedance. This job can be conveniently done by Thevenin's and Norton's theorems.

12.6 THEVENIN'S THEOREM FOR DC NETWORKS

Thevenin's theorem for dc networks may be stated as follows :

A linear 2-terminal active dc network N consisting of independent and/or dependent voltage and current sources and resistors can be replaced at a pair of terminals $a - b$ by a simple equivalent network consisting of a single voltage source V_{Th} in series with a single resistance R_{Th}.

The source voltage, V_{Th}, is the voltage across the terminals $a - b$ when they are open-circuited. This voltage is called the Thevenin equivalent voltage.

The equivalent Thevenin resistance, R_{Th}, has a value given by the ratio of the open-circuit voltage and the short-circuit current at the terminals $a - b$ of the original network N. That is,

$$R_{Th} = \frac{\text{open–circuit votage across the terminals } a - b \text{ of network } N}{\text{short–circuit curent through the terminals } a\text{–}b \text{ of network } N}$$

If all the sources are *independent*, the equivalent Thevenin resistance R_{Th} is the total resistance at the open-circuited terminals $a - b$ when all ideal voltage sources are replaced by short circuits and all ideal current sources are replaced by open circuits.

Consider a linear active network N connected to an arbitrary load resistance R_L as shown in Fig. 12.5. For the purpose of determining the current through or voltage across the resistor R_L connected between the terminals $a - b$, the rest of the network may be replaced by a simple equivalent network containing V_{Th} in series with R_{Th}.

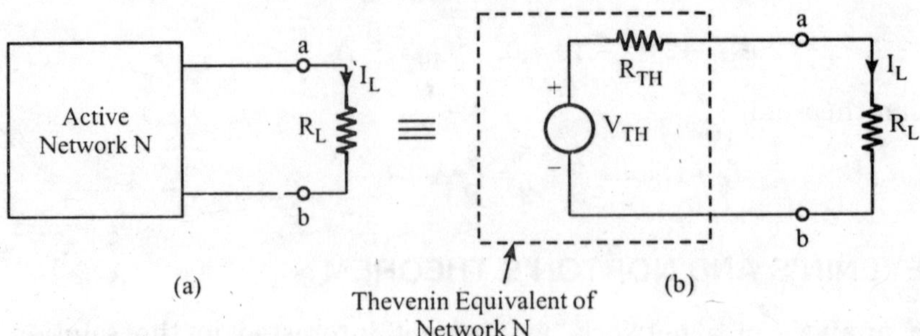

FIG. 12.5 (a) Original network (b) Thevenin equivalent network

12.7 APPLICATION OF THEVENIN'S THEOREM

The procedure for the application of Thevenin's theorem is as follows :

1. Open-circuit the terminals a–b with respect to which the Thevenin equivalent circuit is desired. In other words, remove all the circuitry that will not be replaced by a Thevenin equivalent. Thus, an open circuit at terminals a–b is created.
2. Find the voltage across the terminals a–b when they are open-circuited. This voltage is V_{Th} or V_{OC}. It may be calculated by using KCL, KVL, superposition principle, or node and mesh analysis depending on the complexity of the circuit.
3. Find the current I_{SC} through the terminals a–b when they are short-circuited.
4. Determine the value of the Thevenin equivalent resistance R_{Th} with the help of the relation

$$R_{Th} = \frac{V_{Th}}{I_{SC}} = \frac{V_{OC}}{I_{SC}}$$

   If all the sources in the original network are independent, the equivalent Thevenin resistance is equal to the total resistance at the open-circuited terminals a–b looking back into the network when all ideal voltage sources are replaced by short circuits and all ideal current sources are replaced by open circuits.
5. Draw the Thevenin equivalent network between the Thevenin terminals a–b by connecting the voltage source V_{Th} in series with R_{Th}.
6. The external circuitry that was removed in step 1 is now reconnected across the Thevenin terminals a–b. It is to be noted that the polarity of V_{Th} is such that it produces the current in the external circuitry in the same direction as the original circuit produced it.

12.8 NORTON'S THEOREM FOR DC NETWORKS

Norton's theoem is also an equivalent circuit theorem which is applicable for a portion of the circuit. With the help of this theorem we can replace a complex network by a *current source* in *parallel* with a single resistor. Norton's theorem is the *dual* of Thevenin's theorem and can be written simply by paraphrasing Thevenin's theorem.

Norton's theorem for dc networks may be stated as follows :

> A linear 2-terminal active dc network N consisting of independent and or/dependent voltage and current sources and resistors can be replaced at a pair of terminals a–b by a simple equivalent network consisting of a single current source I_N in parallel with a single resistor R_N.
>
> The source current, I_N , is the current through the terminals a–b when they are short-circuited. This current is called the *Norton equivalent current.* Since the current I_N is the current through the terminals a–b when they are short-circuited, it may be also be denoted by I_{SC}. That is, $I_N = I_{SC}$.
>
> The equivalent Norton resistance, R_N, is the resistance between the load terminals a–b of the original network N, with the load removed and all sources replaced by their internal resistances.

The linear active network N and its Norton eqivalent circuit are shown in Fig. 12.6.

The Norton equivalent resistance is calculated in exactly the same way as the Thevenin equivalent resistance.

$$\therefore \qquad R_N = R_{Th}$$

In the Norton equivalent circuit, R_N is connected in parallel with the current source I_N. The load current I_L is calculated by the curent division rule,

$$I_L = \frac{R_N}{R_N + R_L} I_N$$

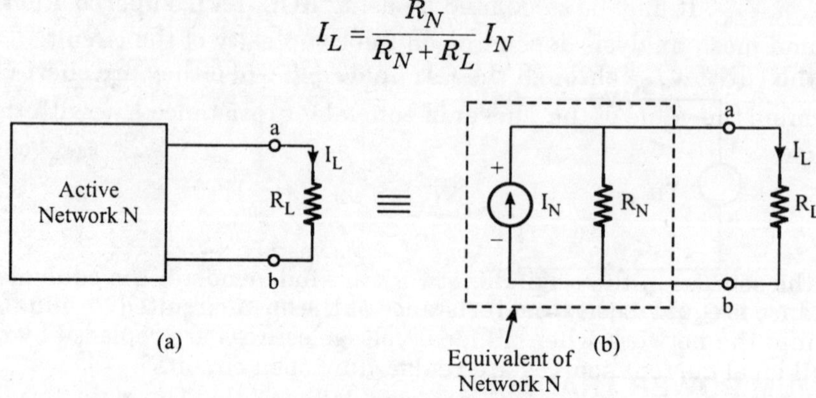

FIG. 12.6 (a) Original network (b) Norton equivalent network

12.9 APPLICATION OF NORTON'S THEOREM

The procedure for the application of Norton's theorem is as follows :

1. Remove the exernal load R_L and then short-circuit the load terminals a–b with respect to which the Norton equivalent circuit is desired.
2. Determine the current through the terminals when they are short-circuited. This current is I_N or I_{SC}.

   There is *alternative method* of determining I_N.

$$I_N = \frac{V_{Th}}{R_{Th}}$$

3. Redraw the network with each ideal voltage source replaced by a short circuit and each ideal current source is replaced by an open circuit.
4. Calculate the resistance R_N of the redrawn network at the open-circuited terminals a–b. It is to be noted that $R_N = R_{Th}$.
5. Draw the Norton equivalent circuit between the terminals a–b by connecting the current source I_N in parallel with R_N.
6. The external load R_L that was short-circuited in step 1 is now reconnected across the Norton terminals a–b. it is to be noted that the direction of I_N is such that it circulates the current in the load R_L in the same direction as the original circuit circulated it.
7. The load current I_L through R_L is calculated from the formula

$$I_L = \frac{R_N}{R_N + R_L} I_N$$

It is to be noted that the Norton equivalent network can be obtained from the Thevenin equivalent network and *vice versa* simply by source transformation as shown in Fig. 12.7.

Here $$I_N = \frac{V_{Th}}{R_{Th}}, \quad R_N = R_{Th}$$

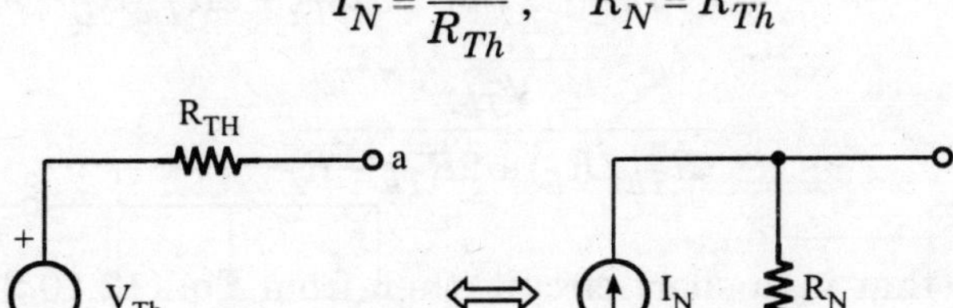

FIG. 12.7 Equivalent of Thevenin and Norton networks

12.10 MAXIMUM POWER TRANSFER THEOREM FOR DC NETWORKS

The **maximum power transfer theorem for dc networks** is stated as follows :

A resistive load connected to a dc network receives maximum power when the load resistance is equal to the Thevenin equivalent resistance of the network as seen from the load terminals.

Proof : Consider a dc network N Fig. 12.8 (*a*) connected to a resistor load R_L at terminals *a–b*. The Thevenin equivalent network is shown in Fig. 12.8 (*b*).

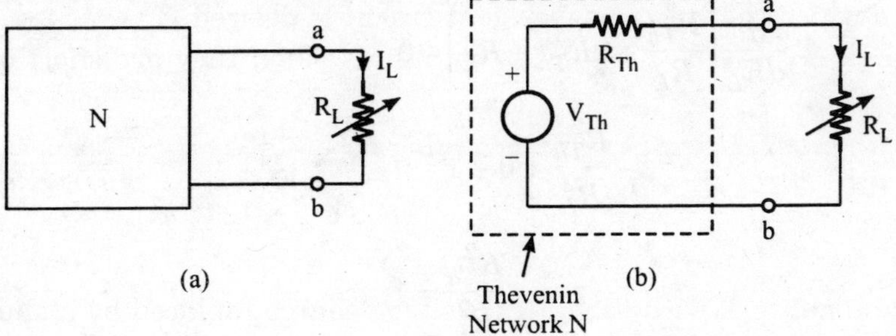

FIG. 12.8

Here V_{Th} represents the Thevenin voltage and R_{Th} the Thevenin resistance. We are interested in determining the value of R_L when it reveives maximum power from the network N.

The load current is given by

$$I_L = \frac{V_{Th}}{R_{Th} + R_L} \tag{12.10.1}$$

The power delivered to the load is

$$P_L = I_L^2 R_L \tag{12.10.2}$$

$$= \left(\frac{V_{Th}}{R_{Th} + R_L}\right)^2 R_L = \frac{V_{Th}^2 R_L}{R_{Th}^2 + 2R_{Th} R_L + R_L^2}$$

$$= \frac{V_{Th}^2}{(R_{Th}^2/R_L) + 2R_{Th} + R_L} \tag{12.10.3}$$

It is seen that if $R_L = 0$, that is, a short circuit, then from Eq. (12.10.2), $P_L = 0$. When R_L is infinite, that is, an open circuit, $I_L = 0$ and hence $P_L = 0$. Between these extreme conditions P_L is always finite and positive. Consequently, we expect that for a particular value of R_L, the power P_L will be a maximum. This occurs when the right-hand side of Eq. (12.10.3) is a maximum. Since V_{Th} is a constant, the right-hand side of Eq. (12.10.3) will be a maximum when its denominator is a minimum. Let

$$\frac{R_{Th}^2}{R_L} + 2R_{Th} + R_L = x \tag{12.10.4}$$

For a minimum value of x

$$\frac{dx}{dR_L} = 0 \text{ and } \frac{d^2x}{dR_L^2} \text{ is positive}$$

therefore,

$$\frac{d}{dR_L}\left(\frac{R_{Th}^2}{R_L} + 2R_{Th} + R_L\right) = 0$$

$$-\frac{R_{Th}^2}{R_L^2} + 0 + 1 = 0$$

$$\frac{R_{Th}^2}{R_L^2} = 1$$

or

$$R_L = R_{Th} \tag{12.10.5}$$

Now

$$\frac{d^2x}{dR_L^2} = \frac{d}{dx}\left(-\frac{R_{Th}^2}{R_L^2} + 1\right) = \frac{2R_{Th}^2}{R_L^3} \tag{12.10.6}$$

Substituting $R_L = R_{Th}$ in Eq. (12.10.6)

$$\frac{d^2x}{dR_L^2} = \frac{2R_{Th}^2}{R_{Th}^3} = \frac{2}{R_{Th}}$$

We find that $\frac{d^2x}{dR_L^2}$ is positive, showing that $R_L = R_{Th}$ is a condition for a minimum value of x. In other words, it is a condition for a maximum value of R_L. Thus, the theorem is proved.

When $R_L = R_{Th}$, the load is said to be matched to the source network. The maximum value of power consumed by the load is obtained by putting $R_L = R_{Th}$ is Eq. (12.10.3).

$$P_{L\,(max)} = \frac{V_{Th}^2 R_{Th}}{(R_{Th} + R_{Th})^2} = \frac{V_{Th}^2}{4R_{Th}} \tag{12.10.7}$$

Also, the power consumed by R_{Th}

$$P_{Th} = I_L^2 R_{Th} = I_L^2 R_L = \frac{V_{Th}^2}{4R_{Th}} \tag{12.10.8}$$

therefore, $$P_{Th} = P_{L\,(max)}$$

Total power supplied by the network

$$P_T = P_{Th} + P_{L\,(max)} = \frac{V_{Th}^2}{4R_{Th}} + \frac{V_{Th}^2}{4R_{Th}} = \frac{V_{Th}^2}{2R_{Th}} \tag{12.10.9}$$

We define the *efficiency* of power transfer as the ratio of the average power consumed by the load and the total power supplied to the load.

$$\eta = \frac{P_{L\,(max)}}{P_T} = 0.5 \text{ p.u. or } 50\%$$

In electronic and communication circuits, where the power involved is sufficiently small, this low efficiency is tolerable for the sake of maximum power to be transferred to the load. In power systems, the power losses and voltage drops in the generators and the transmission systems are kept as small as possible. Also, the load resistance is many times greater than the internal resistance of the source. Hence, power circuits are never operated under optimum matching conditions. It should be noted that in power systems, efficiency is the prime consideration, while in electronic and communication systems, maximum power transfer is more important.

■ **EXAMPLE 12.5** Use Thevenin's theorem to determine the current through and the voltage across the 25 Ω resistor given in Fig. 12.9a.

SOLUTION :

Step 1 : Since we want to determine the current and the voltage across the 25 Ω resistor, our first step is to disconnect the 25 Ω resistor from the original network. The resulting network is shown in Fig. 12.9b. Then we determine V_{TH} and R_{TH}.

Step 2 : Determination of V_{TH}

Since the circuit is open between a–b, there is no current in R_2. Hence there is no voltage drop in R_2.

Thus, V_{ab} = voltage drop across $R_3 = R_3 I_1 = R_3 \cdot \dfrac{V_s}{R_1 + R_3} = 10 \times \dfrac{20}{10 + 10} = 10\text{ V}$

V_{TH} = open-circuit voltage at a–b = V_{ab} = 10 V

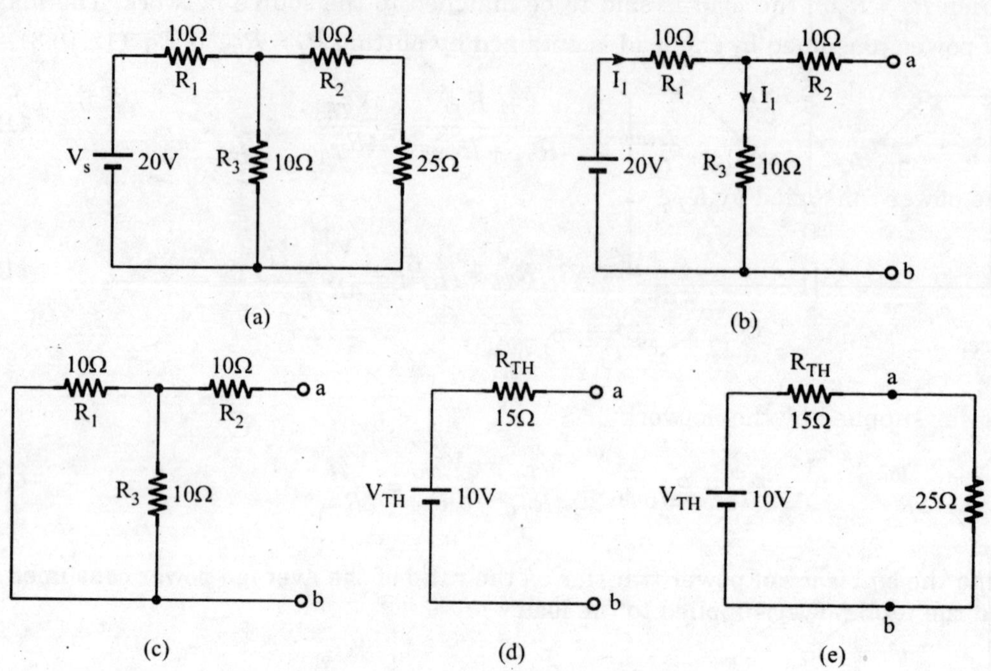

FIG. 12.9 Illustrating Example 12.5

Step 3 : Determination of R_{TH}

The 20 V voltage source is shorted. The network of Fig. 12.9b, then reduces to that shown in Fig. 12.9c.

R_{TH} = resistance of the network of Fig. 12.9c measured at the terminals a–b

$$= R_2 + (R_1 \,||\, R_3) = R_2 + \frac{R_1 R_3}{R_1 + R_3} = 10 + \frac{10 \times 10}{10 + 10} = 15\ \Omega$$

The Thevenin's equivalent is shown in Fig. 12.9 (d).

Step 4 : Determination of current through 25 Ω resistor

The 25 Ω resistor is now connected between a–b of the Thevenin's equivalent as shown in Fig. 12.9e. Let I be the current through 25 Ω resistor.

$$I = \frac{V_{TH}}{R_{TH} + 25} = \frac{10}{15 + 25} = 0.25\text{ A}$$

Determination of voltage across 25 Ω resistor

Voltage across 25 Ω resistor

$$= 25I = 25 \times 0.25 = 6.25\text{ A}$$

EXAMPLE 12.6 Determine the current in 20 Ω resistor of the network shown in Fig. 12.10.a by Thevenin's theorem.

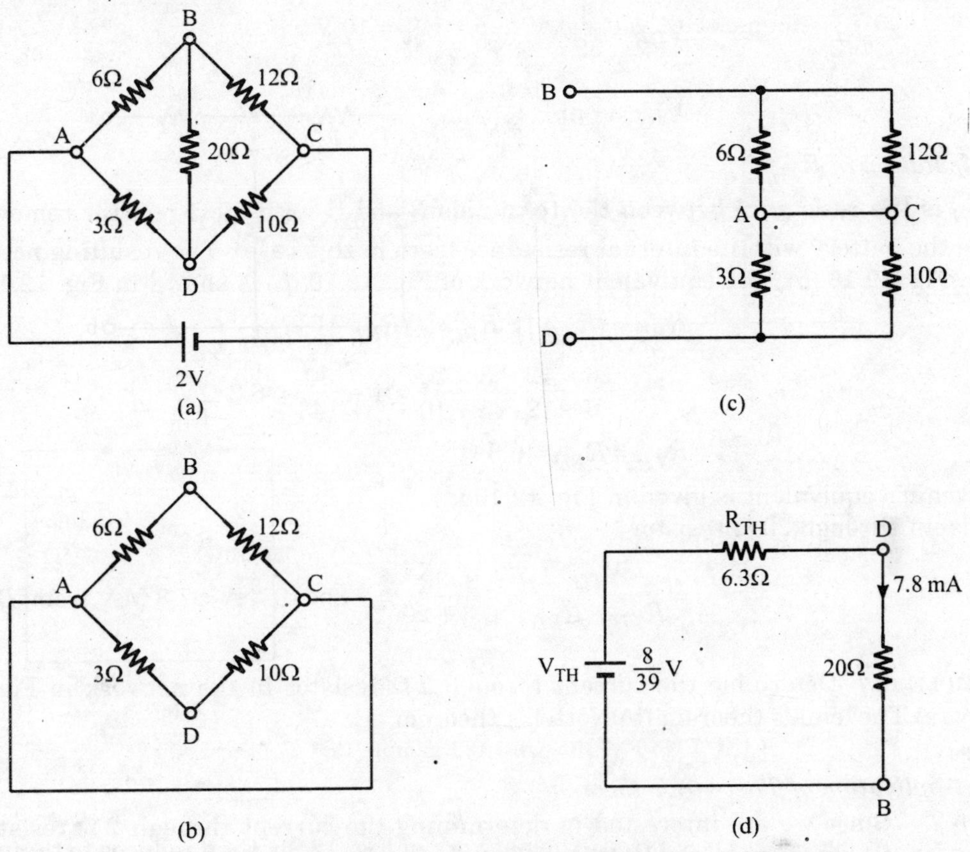

FIG. 12.10 Illustrating Example 12.6

SOLUTION : *Calculation of V_{TH}*

Imagine the resistor between B and D removed.

$$\text{Current via } ABC = \frac{2}{6+12} = \frac{1}{9}\text{ A}$$

Voltage drop across R_{AB}

$$V_{AB} = R_{AB}\, I_{AB} = 6 \times \frac{1}{9} = \frac{2}{3}\text{ V}$$

$$\text{Current via } ADC = \frac{2}{3+10} = \frac{2}{13}\text{ A}$$

Voltage drop in AD, $\quad V_{AD} = R_{AD}\, I_{AD} = 3 \times \frac{2}{13} = \frac{6}{13}\text{ V}$

In mesh $ABDA$, by KVL

$$V_{AB} + V_{BD} + V_{DA} = 0$$

$$V_{BD} = -V_{DA} - V_{AB} = V_{AD} - V_{AB} = \frac{6}{13} - \frac{2}{3} = -\frac{8}{39}\text{ V}$$

$$V_{DB} = \frac{8}{39}\text{ V}$$

$$V_{TH} = V_{DB} = \frac{8}{39}\text{ V}$$

Calculation of R_{TH}

R_{TH} is the resistance between the terminals B and D with 20 Ω resistor removed and replacing the battery with its internal resistance (zero in this case). The resulting network is shown in Fig. 12.10 (*b*). The equivalent network of Fig. 12.10 (*b*) is shown in Fig. 12.10 (*c*).

$$R_{BD} = (R_{AB} \mid\mid R_{BC}) + (R_{CD} \mid\mid R_{DA})$$

$$= \frac{6 \times 12}{6 + 12} + \frac{3 \times 10}{3 + 10} = 4 + \frac{30}{143} = 6.3\ \Omega$$

$$R_{TH} = R_{BD} = 6.3\ \Omega$$

The Thevenin's equivalent is given in Fig. 12.10d.

Current through 20 Ω resistor

$$= \frac{V_{TH}}{R_{TH} + R_{BD}} = \frac{8/39}{6.3 + 20} = 7.8 \times 10^{-3}\text{ A} = 7.8\text{ MA from } D \text{ to } B$$

■ **EXAMPLE 12.7** Determine the current through 2 Ω resistor in the network in Fig. 12.11 (a) by (a) Thevenin's theorem (b) Norton's theorem.

SOLUTION :

(A) *Application of Thevenin's theorem*

Step 1 : Since we are interested in determining the current through 2 Ω resistor, our first step is to disconnect the 2 Ω resistor from the original network as shown in Fig. 12.11 (b).

Step 2 : Determination of Thevenin voltage V_{Th}

V_{Th} = open-circuit voltage between A and B

= voltage between A and B when R_3 is removed from the circuit

By KVL in the outer path MANPQM of Fig. 12.11 (b),

$$12 - R_1 I - R_2 I - 6 = 0$$

$$12 - 1 \times I - 1 \times I - 6 = 0, I = 3\text{ A}$$

By KVL in path MABQM, $12 - R_1 I - V_{Th} = 0$

$$12 - 1 \times 3 - V_{Th} = 0,\ V_{Th} = 9\text{ V}$$

Step 3 : Determination of Thevenin resistance R_{Th}

The 12 V and 6 V voltage sources are short-circuited. The network of Fig. 12.11b the reduces to that shown in Fig. 12.11c, which can be redrawn as Fig. 12.11d. The terminals A and B are taken out.

Thevenin resistance

R_{Th} = resistance of the network of Fig. 12.11d looking back into the network from the terminals AB

$$= R_1 \,||\, R_2 = \frac{R_1 R_2}{R_1 + R_2} = \frac{1 \times 1}{1+1} = 0.5\ \Omega$$

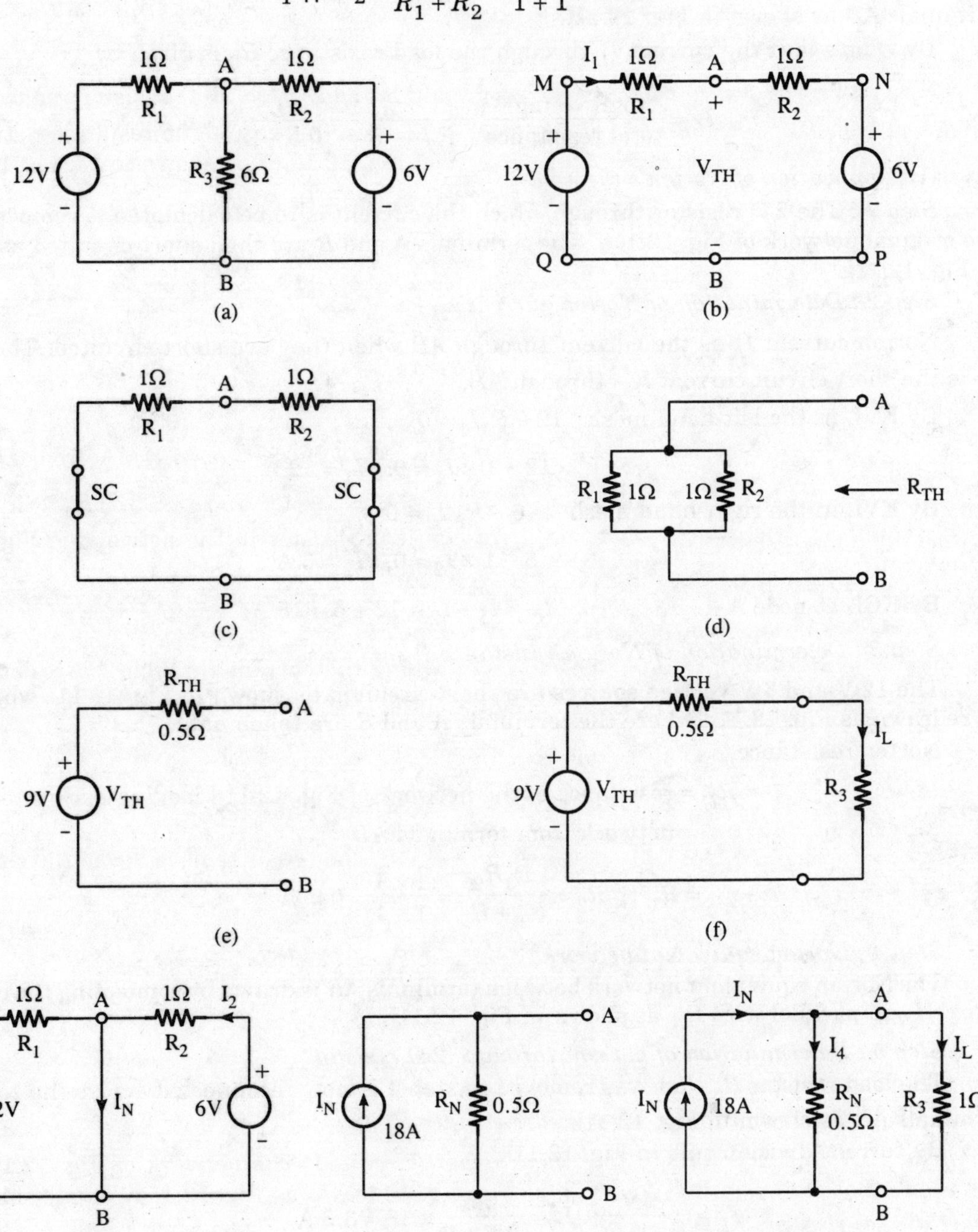

FIG. 12.11

Step 4 : Thevenin equivalent network

The Thevenin equivalent network between terminals AB is drawn by connecting the voltage source V_{Th} in series with R_{Th} as shown in Fig. 12.11e.

Step 5 : Determination of current through 2 Ω *resistor*

The load resistor R_3 that was removed in step 1 is now reconnected across the Thevenin terminals AB as shown in Fig. 12.11f.

By Ohm's law, the current I_L through the load resistance R_3 is given by

$$I_L = \frac{\text{total voltage}}{\text{total resistance}} = \frac{V_{Th}}{R_{Th} + R_3} = \frac{9}{0.5 + 2} = 3.6 \text{ A}$$

(B) *Application of Norton's theorem*

Step 1 : The 2 Ω resistor through which the current is to be calculated is removed from the original network of Fig. 12.11a. The terminals A and B are then short-circuited as shown in Fig. 12.11g.

Step 2 : Determination of Norton current I_N

Norton current I_N is the current through AB when they are short-circuited. Therefore. I_N is the short circuit current I_{SC} through AB.

By KVL in the left-hand mesh $\quad 12 - R_1 I_1 = 0$

$$12 - 1 \times I_1 = 0, \quad I_1 = 12 \text{ A}$$

By KVL in the right-hand mesh $\quad 6 - R_2 I_2 = 0$

$$6 - 1 \times I_2 = 0, \quad I_2 = 6 \text{ A}$$

By KCL at node A $\quad I_N = I_1 + I_2 = 12 + 6 = 18 \text{ A}$

Step 3 : Determination of Norton resistance R_N

The 12 V and 6 V voltage sources are short-circuited as shown in Fig. 12.11c which can be redrawn as Fig. 12.11d where the terminals A and B are taken out.

Norton resistance

R_N = resistance of the network of Fig. 12.11d looking back into the network from terminals AB

$$= R_1 \,||\, R_2 = \frac{R_1 R_2}{R_1 + R_2} = \frac{1 \times 1}{1 + 1} = 0.5 \text{ Ω}$$

Step 4 : Norton-equivalent network

The Norton equivalent network between terminals AB is drawn by connecting the current source I_N in parallel with R_N as shown in Fig. 12.11h.

Step 5 : Determination of current through 2 Ω *resistor*

The load resistor R_3 that was removed in step 1 is now reconnected across the Norton's terminals AB as shown in Fig. 12.11k,

By current-division rule in Fig. 12.11k,

$$I_L = \frac{R_N}{R_N + R_3} \cdot I_N = \frac{0.5}{0.5 + 2} \times 18 = 3.6 \text{ A}$$

■ **EXAMPLE 12.8** Determine the current I in the network of Fig. 12.12a by Thevenin's theorem.

SOLUTION :

Step 1 : The 3 Ω resistor through which I is flowing is removed as shown in Fig. 12.12b.

Step 2 : Determination of Thevenin voltage V_{Th}

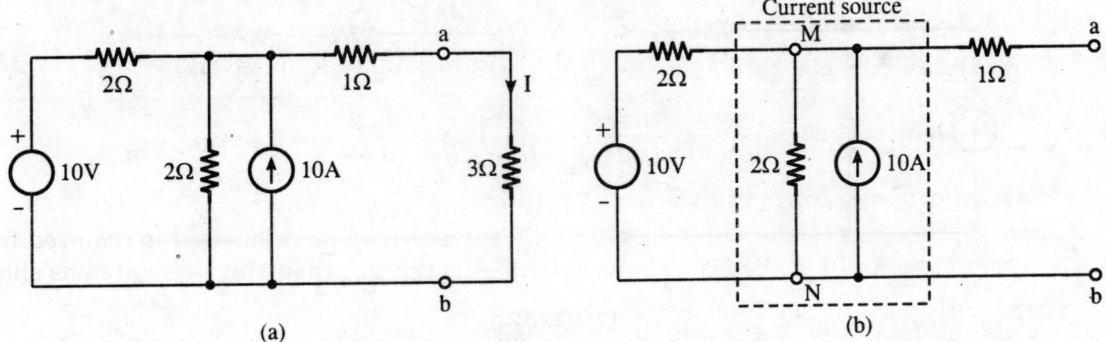

FIG. 12.12

Method 1

The 10 A current source in parallel with 2 Ω resistor is transformed to 10 × 2 (= 20) V source in series with a 2 Ω resistor as shown in Fig. 12.12c. The current through 1 Ω resistor is zero because of the open circuit between terminals a–b. The current distribution is shown in Fig. 12.12c. By KVL in the left-hand mesh

$$10 - 2I - 2I - 20 = 0, \quad I = -\frac{10}{4} = -2.5 \text{ A}$$

V_{Th} = open-circuit voltage between terminals a–b

By KVL in the right-hand mesh, $20 + 2I - V_{Th} = 0$

$$V_{Th} = 20 + 2I = 20 + 2(-2.5) = 15 \text{ V}$$

Method 2

The 10 V voltage source in series with 2 Ω resistor is transformed to a $\frac{10}{2} = 5$ A current source in parallel with a 2 Ω resistor as shown in Fig. 12.12d. The two current sources of 10 A and 5 A supplying current in the same direction are equivalent to a single current source of 10 + 5 = 15 A.

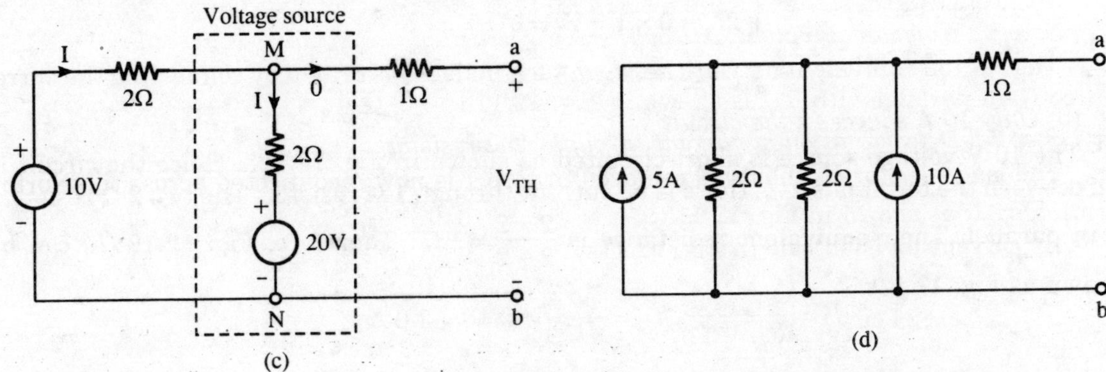

FIG. 12.12

The equivalent resistance of two 2 Ω resistors in parallel is $\frac{2 \times 2}{2+1} = 1\ \Omega$.

Therefore the circuit in Fig. 12.12d can be redrawn as in Fig. 12.12e. Since the terminals *a–b* are open, the current through 1 Ω resistor in parallel with the 15 A current source is 15 A as shown in Fig. 12.12e.

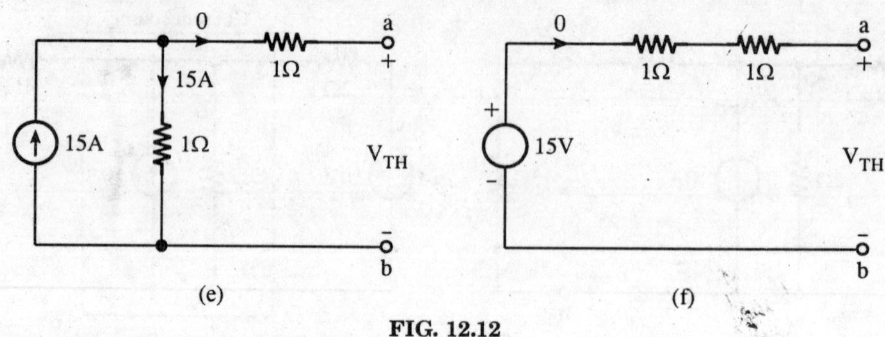

FIG. 12.12

V_{Th} = open-circuit voltage between *a–b*

By KVL in the right-hand mesh

$$+15 \times 1 - 0 \times 1 - V_{Th} = 0$$

therefore, $$V_{Th} = 15\ \text{V}$$

By source transformation, the 15 A current source in parallel with 1 Ω resistor of Fig. 12.12e can be transformed to a 15 × 1 = 15 A voltage source in series with an 1 Ω resistor as shown in Fig. 12.12f. Since the terminals *a* and *b* are open, there is no current in the circuit of 12.12f. By KVL,

therefore, $$15 - 0 \times 1 - 0 \times 1 - V_{Th} = 0, \quad V_{Th} = 15\ \text{V}$$

Method 3

The Thevenin voltage may also be determined by using superposition principle.

(a) *Only 10 V source in the circuit*

The 10 A current source is open-circuited as shown in Fig. 12.12g. By KVL in the left-hand mesh,

$$10 - 2I_1 - 2I_1 = 0, \quad I_1 = 2.5\ \text{A}$$

By KVL in the right-hand mesh

$$+2I_1 - 0 \times 1 - V_1 = 0$$

$$V_1 = 2I_1 = 2 \times 2.5 = 5\ \text{V}$$

(b) *Only 10 A source in the circuit*

The 10 V voltage source is short-circuited as shown in Fig. 12.12h. Since the circuit is open between the terminals *a–b*, there is no current through 1 Ω resistor. The two 2 Ω resistors are in parallel. Their equivalent resistance is $\frac{2 \times 2}{2+2} = 1\ \Omega$. Therefore, Fig. 12.12 (*h*) can be redrawn as Fig. 12.12k.

By KVL in the right-hand mesh

$$+10 \times 1 - 0 \times 1 - V_2 = 0, \quad V_2 = 10 \text{ V}$$

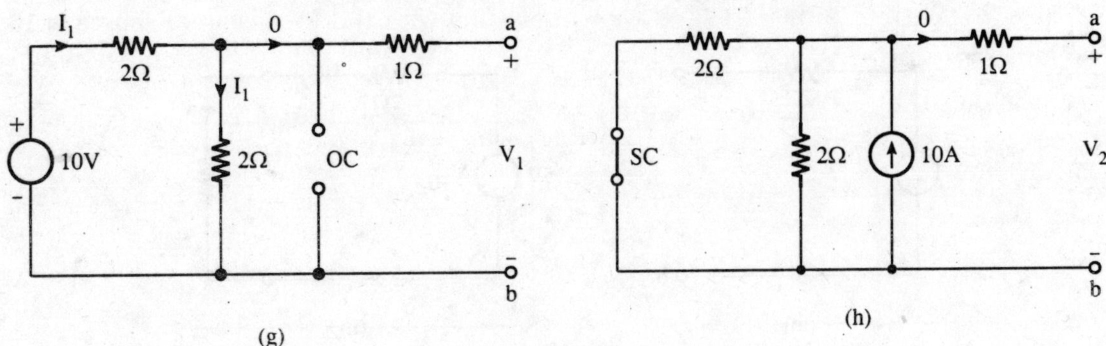

FIG. 12.12

By superposition principle, the open-circuit voltage between a–b is

$$V_{ab} = V_1 + V_2 = 5 + 10 = 15 \text{ V}$$

Also, $V_{Th} = V_{ab} = 15$ V

Step 3 : Determination of Thevenin resistance R_{Th}

The 10 V voltage source is short-circuited and the 10 A current source is open-circuited as shown in Fig. 12.12l.

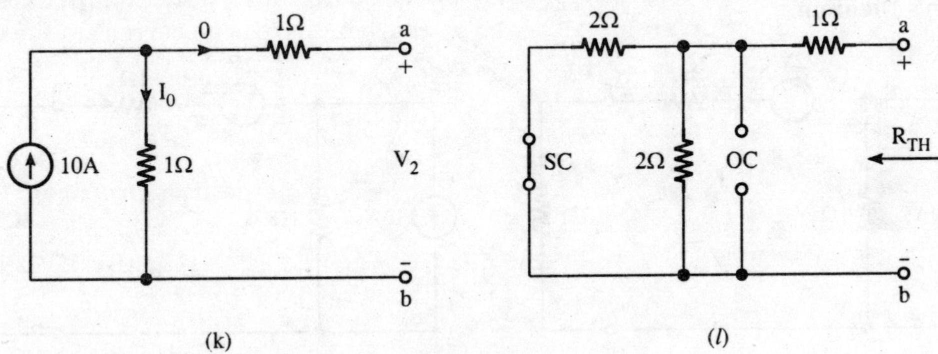

FIG. 12.12

Thevenin resistance

R_{Th} = resistance between the terminals a–b looking back into the network

$$= 1 + 2 \,||\, 2$$

$$= 1 + \frac{2 \times 2}{2 + 2} = 1 + 1 = 2\ \Omega$$

Step 4 : Thevenin equivalent network

The Thevenin equivalent network between terminals *a–b* is drawn by connecting the voltage source V_{Th} in series with R_{Th} as shown in Fig. 12.12m.

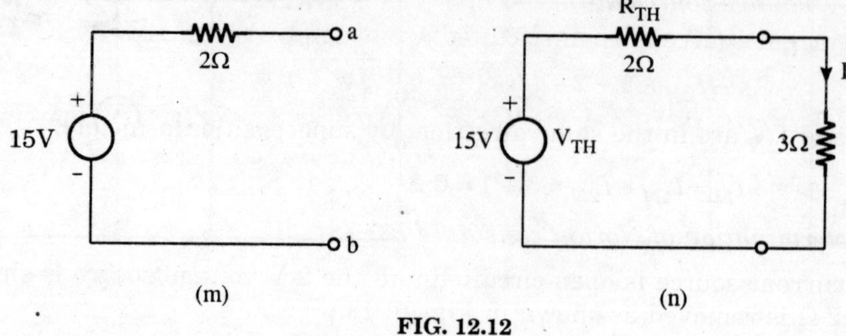

FIG. 12.12

Step 5 : Determination of current through 3 Ω resistor

The load resistance of 3 Ω which was removed in step 1 is now connected across the terminals *a–b* of the Thevenin equivalent network of Fig. 12.12m as shown in Fig. 12.12n. By Ohm's law

$$I = \frac{V_{Th}}{R_{Th} + R_L} = \frac{15}{2+3} = 3 \text{ A}$$

■ **EXAMPLE 12.9** Determine the voltage across the 2 Ω resistor of Fig. 12.13a by using Norton's theorem.

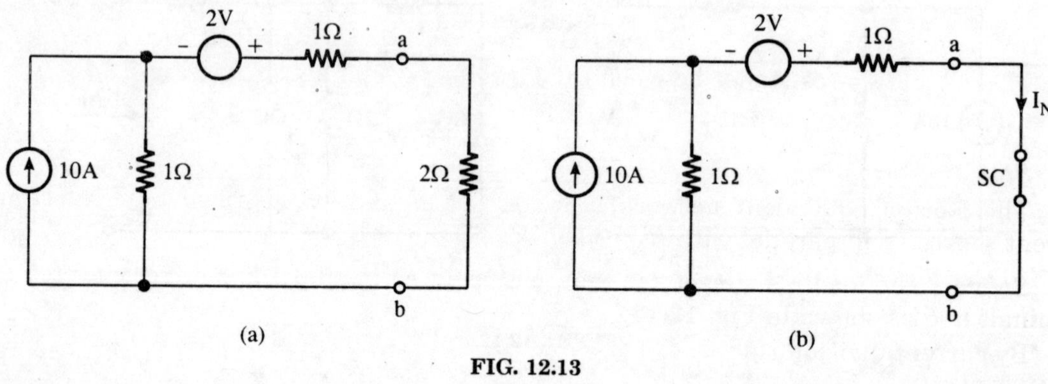

FIG. 12.13

SOLUTION

Step 1 : The 2 Ω resistor across which the voltage is to be calculated is disconnected from the original network of Fig. 12.13a. The terminals *a* and *b* are then short-circuited as shown in Fig. 12.13b.

Step 2 : Determination of Norton current I_N

The current I_N through the short circuit may be determined by superposition.

(a) *Only 10 A current source in the circuit*

The 2 V voltage source is short-circuited as shown in Fig. 12.13c.

By current division rule

$$I_{1N} = \frac{1}{1+1} \times 10 = 5 \text{ A}$$

(b) *Only 2 V voltage source in the circuit*

The 10 A current source is open-circuited as shown in Fig. 12.13d. By KVL

$$-I_{2N} \times 1 + 2 - I_{2N} \times 1 = 0, \quad I_{2N} = 1 \text{ A}$$

Since I_{1N} and I_{2N} are in the same direction, by superposition principle,

$$I_N = I_{1N} + I_{2N} = 5 + 1 = 6 \text{ A}$$

Step 3 : Determination of Norton resistance R_N

The 10 A current source is open-circuited and the 2 V voltage source is short-circuited and the load of 2 Ω is removed as shown in Fig. 12.13e.

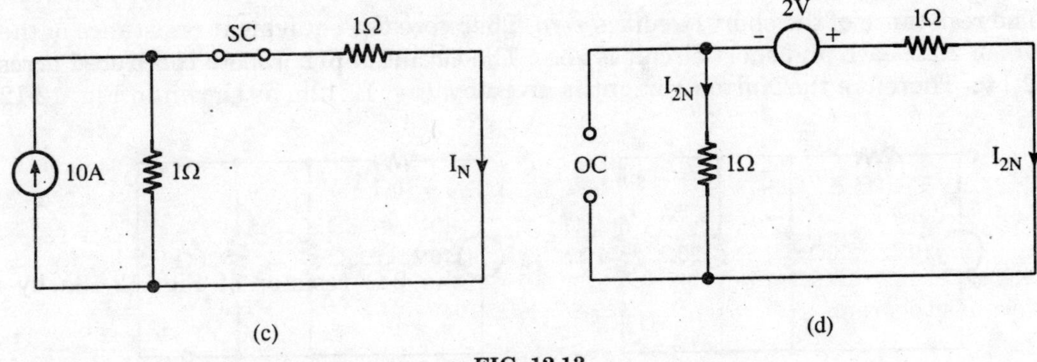

FIG. 12.13

$$R_N = \text{resistance between the terminals } a\text{–}b \text{ of network of Fig. 12.13e.}$$

$$= 1 + 1 = 2\ \Omega$$

Step 4 : Norton equivalent network

The Norton equivalent network between terminals a–b is drawn by connecting the current source I_N in parallel with R_N as shown in Fig. 12.13f.

Step 5 : The load resistor of 2 Ω that was removed in step 1 is now reconnected across terminals a–b as shown in Fig. 12.13g.

By current division rule

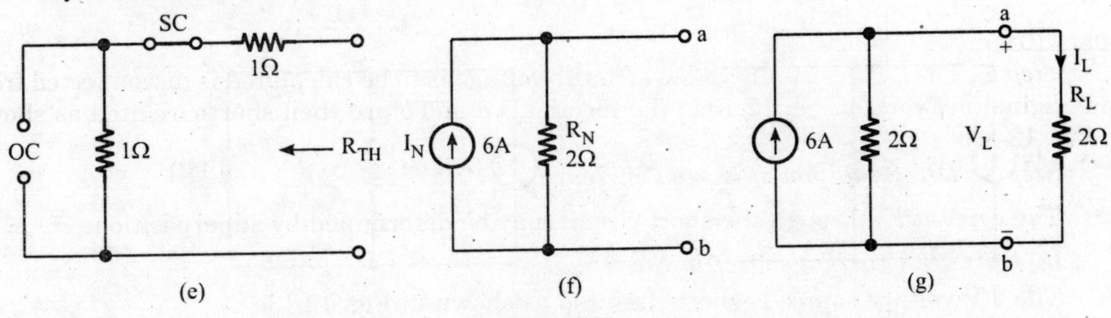

FIG. 12.13

$$I_L = \frac{2}{2+2} \times 6 = 3 \text{ A}$$

∴ Voltage across the 2 Ω resistor of the original network

$$V_L = I_L R_L = 3 \times 2 = 6 \text{ V}$$

EXAMPLE 12.10 Determine the current I_L through the 15 Ω resistor in the network given in Fig. 12.14a by Norton's theorem.

SOLUTION

Step 1 : Since we are interested in determining the current through the 15 Ω resistor, our first step is to remove the 15 Ω resistor from the original network of Fig. 12.14a and to replace it by a short circuit. When the terminals *a–b* are short-circuited, the original circuit is reduced to that of Fig. 12.14b.

Step 2 : Determination of Norton Current I_N

The resistance of the short circuit is zero. Therefore the equivalent resistance of the two resistors of 20 Ω each and short-circuit is zero. The circuit of Fig. 12.14b is reduced to that of Fig. 12.14c. Therefore the Norton current is given by

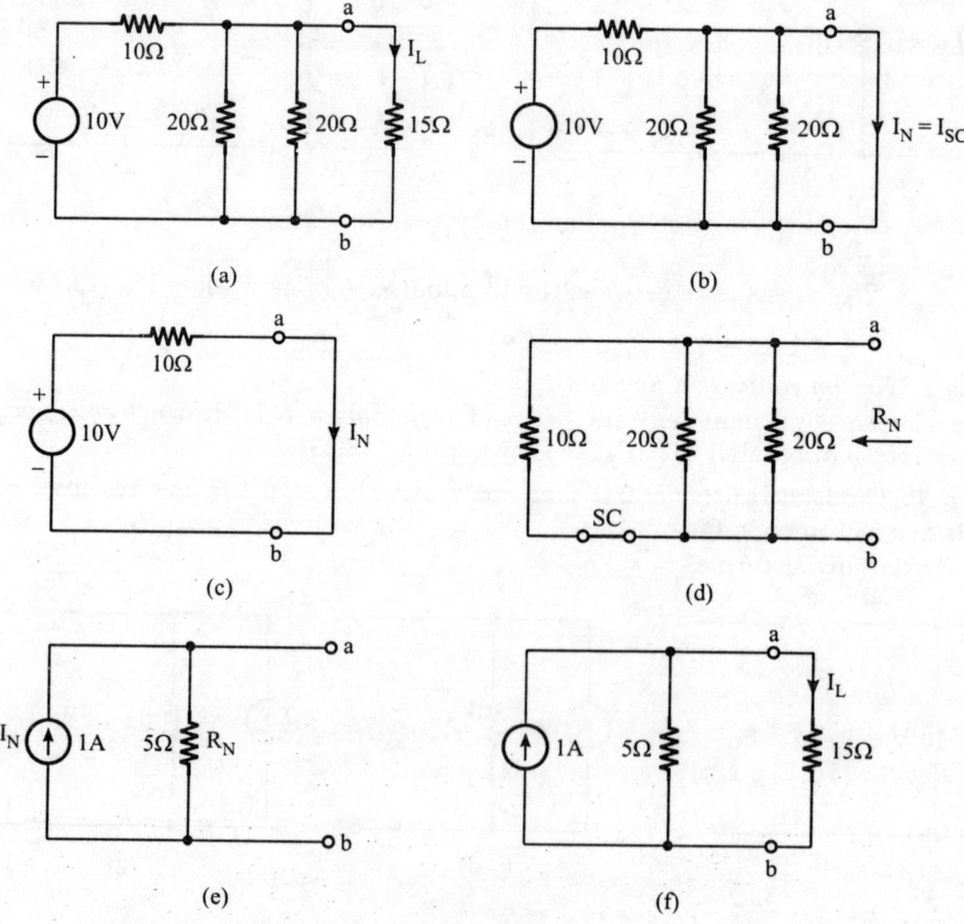

FIG. 12.14

$$I_N = I_{SC} = \frac{\text{total voltage}}{\text{total resistance}} = \frac{10}{10} = 1 \text{ A}$$

Step 3 : Determination of Norton-resistance R_N

The 10 V voltage source is short-circuited. The network to the left of the terminals a–b of Fig. 12.14a is then reduced to that of Fig. 12.14d. Thus, R_N is equal to the equivalent resistance of 10 Ω, 20 Ω and 20 Ω resistors in parallel.

$$\frac{1}{R_N} = \frac{1}{10} + \frac{1}{20} + \frac{1}{20}, \quad R_N = 5\ \Omega$$

Step 4 : Norton equivalent network

The Norton equivalent network between terminals a–b is drawn by connecting a current source I_N in parallel with R_N as shown in Fig. 12.14e.

Step 5 : Determination of current through 15 Ω *resistor*

The 15 Ω resistor that was removed in step 1 is now reconnected between the terminals a–b as shown in Fig. 12.14f.

By current-division rule $I_L = \dfrac{5}{5+15} \times 1 = 0.25$ A

■ **EXAMPLE 12.11** In the network shown in Fig. 12.15a determine (a) the value of the load resistance to give maximum power transfer and (b) the power delivered to the load.

SOLUTION : In order to determine maximum power transfer, we have to determine the Thevenin equivalent network.

Determination of V_{Th}

Disconnect the load resistor R_L from the original network as shown in Fig. 12.15b.

Since the circuit is open between the terminals a–b, there is no current through R_3 and the current through R_1 and R_2 is the same. By KVL in the left-hand mesh,

$$+10 - I_1R_1 - I_1R_2 = 0$$

$$10 - I_1 \times 6 - I_1 \times 6 = 0$$

therefore, $$I_1 = \frac{10}{12} = \frac{5}{6} \text{ A}$$

By KVL in the right-hand mesh

$$+I_1 \times 6 - 0 \times 2 - V_{Th} = 0$$

therefore, $$V_{Th} = 6I_1 = 6 \times \frac{5}{6} = 5 \text{ V}$$

Determination of R_{Th}

The 10 V voltage source is short-circuited. The network of Fig. 12.15b is reduced to that shown in Fig. 12.15c. Fig. 12.15c can be redrawn as Fig. 12.15d.

R_{Th} = resistance of the network of Fig. 12.15d looking back at terminals a–b

$$= R_3 + R_1 \,||\, R_2 = 2 + 6 \,||\, 6 = 2 + \frac{6 \times 6}{6+6} = 5\ \Omega$$

For maximum power transfer $R_L = R_{Th} = 5\ \Omega$

$$P_{L\max} = I_L^2 R_{Th} = \frac{V_{Th}^2}{4R_{Th}} = \frac{5^2}{4 \times 5} = 1.25 \text{ W}$$

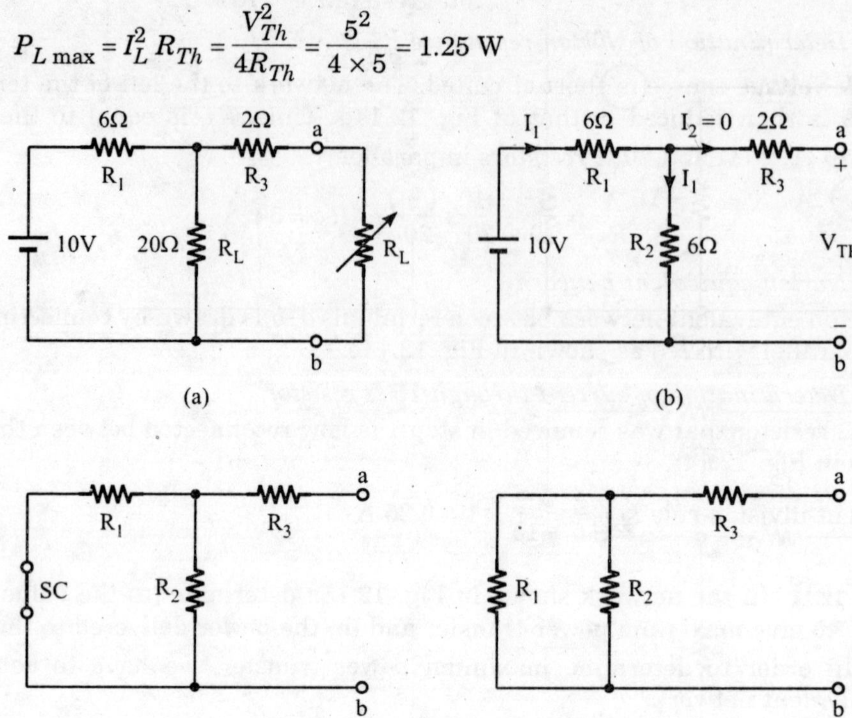

FIG. 12.15

■ **EXAMPLE 12.12** Determine the voltage across the $\frac{1}{3}\ \Omega$ resistor in Fig. 12.16a using Thevenin's theorem.

SOLUTION :

Determination of V_{Th}

First we remove the $\frac{1}{3}\ \Omega$ resistor as shown in Fig. 12.16b. By KCL at node *c*, $i = 2$ A. By KVL in the closed path *cdabc*,

$$-\frac{1}{4}i + 3 - V_{Th}$$

$$V_{Th} = 3 - \frac{1}{4} \times 2 = 2.5 \text{ V}$$

Determination of R_{Th}

In order to find R_{Th} we remove the 1/3 Ω resistor, open circuit the 2 A current source, and short circuit the 3 V voltage source as shown in Fig. 12.16c. The two 1/2 Ω resistors are connected in series are equivalent to 1 Ω resistor. This 1 Ω resistor is in parallel with a short circuit (Fig. 12.16d). The equivalent resistance of this parallel combination is zero ohm. From Fig. 12.16e,

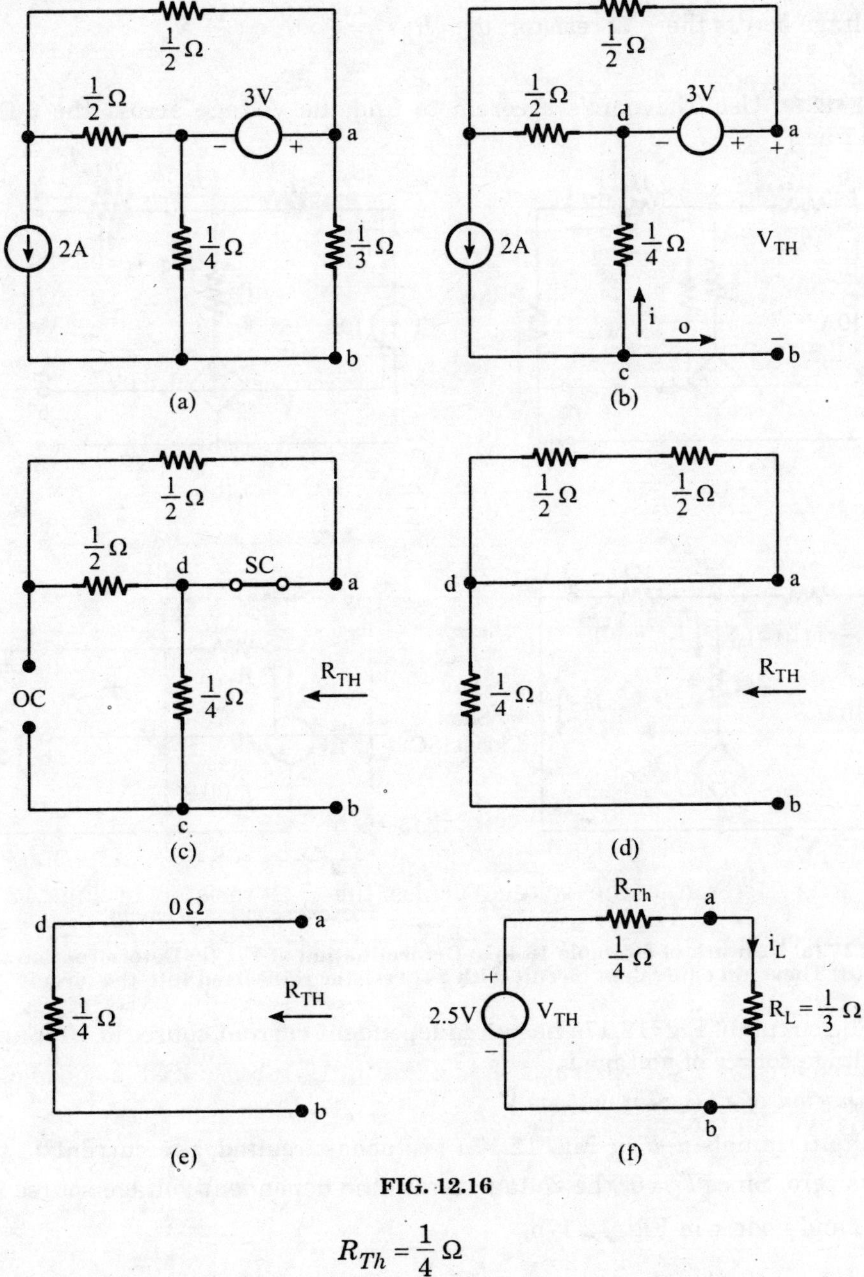

FIG. 12.16

$$R_{Th} = \frac{1}{4}\,\Omega$$

Determination of the voltage across the $\frac{1}{3}\,\Omega$ *resistor*

The Thevenin equivalent circuit is shown in Fig. 12.16f in which the $\frac{1}{3}\,\Omega$ resistor has been reconnected. The current through the $\frac{1}{3}\,\Omega$ resistor, $i_L = \dfrac{V_{Th}}{R_{Th} + R_L} = \dfrac{2.5}{\frac{1}{4} + \frac{1}{3}} = \dfrac{30}{7}$ A

The voltage across the $\frac{1}{3}$ Ω resistor, $v_L = R_L i_L = \frac{1}{3} \times \frac{30}{7} = \frac{10}{7}$ V

EXAMPLE 12.13 Use Thevenin's theorem to find the voltage across the 3 Ω resistor in Fig. 12.17 a.

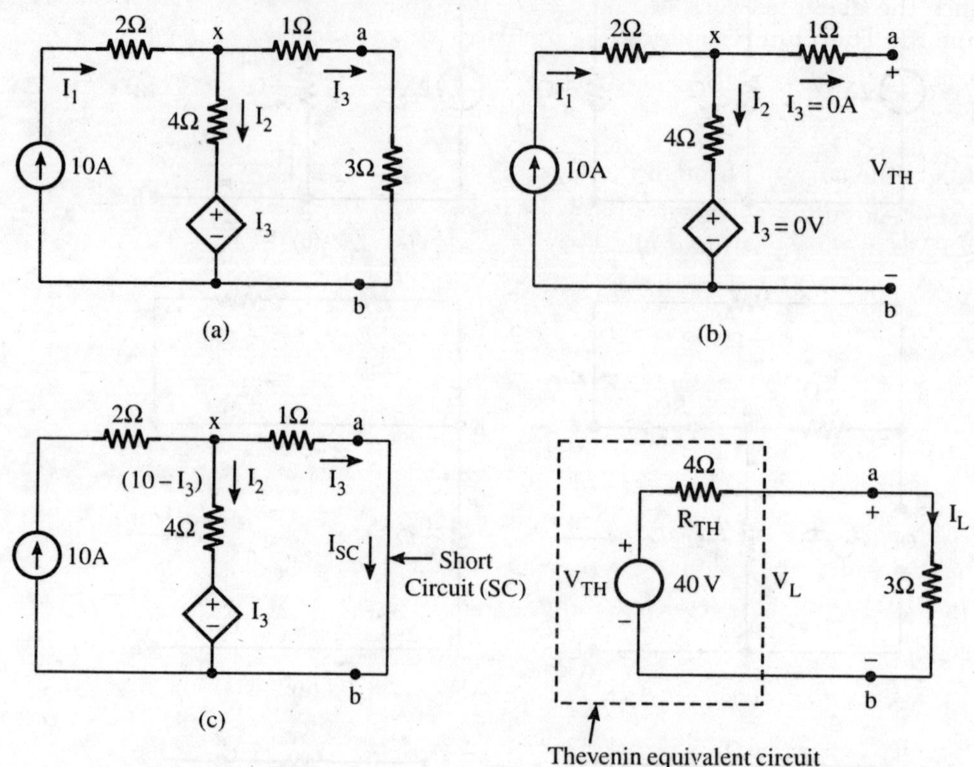

FIG. 12.17 (a) Network of Example 12.4 (b) Determination of V_{Th} (c) Determination of R_{Th} (d) Thevenin equivalent circuit with 3-Ω resistor reinserted into the circuit

SOLUTION: The circuit of Fig. 12.17a has an independent current source of 10 A and a current controlled voltage source of voltage i_3.

Determination of Thevenin voltage V_{Th}

When the terminals *a–b* in Fig. 12.17a are open-circuited, the current I_3 through the 3 Ω resistor is zero. Since $I_3 = 0$, the voltage across the dependent voltage source is also zero.

By KCL and node *x* in Fig. 12.17b,

$$I_1 = I_2 + I_3$$

Since the $I_1 = 10$ A and $I_3 = 0$ (current through the open circuit)

therefore,

$$I_1 = I_2 = 10 \text{ A}$$

The Thevenin voltage V_{Th} is the voltage across the open-circuit terminals *a–b*.

By KVL in the right-hand mesh of Fig. 12.17b,

$$I_3 + 4I_2 - I_3 \times 1 - V_{Th} = 0$$

$$0 + 4 \times 10 - 0 \times 1 - V_{Th} = 0$$

therefore, $$V_{Th} = 40 \text{ V}$$

Determination of R_{Th}

Since the given network of Fig. 12.17a contains a dependent voltage source, we shall determine the Thevenin resistance R_{Th} by using the relationship

$$R_{Th} = \frac{V_{Th}}{I_{SC}}$$

where I_{SC} is the current through the short circuit between the terminals *a–b*. The terminals *a–b* are short circuited as shown in Fig. 12.17c.

By KVL in the right-hand mesh of Fig. 12.17c.

$$I_3 + 4\,(10 - I_3) - I_3 \times 1 = 0$$

therefore, $$I_3 = 10 \text{ A}$$

Also, $$I_{SC} = I_3 = 10 \text{ A}$$

therefore, $$R_{Th} = \frac{V_{Th}}{I_{SC}} = \frac{40}{10} = 4\ \Omega$$

Determination of the voltage across the 3 Ω resistor

The Thevenin equivalent network is drawn by connecting V_{Th} in series with R_{Th} as shown in Fig. 12.17d. The load resistance $R_L = 3\ \Omega$ is reconnected. The current through the load resistor,

$$I_L = \frac{V_{Th}}{R_{Th} + R_L} = \frac{40}{4+3} = \frac{40}{7} \text{ A}$$

The voltage across the 3 Ω resistor, $V_L = I_L R_L = \frac{40}{7} \times 3 = 17.14 \text{ V}$

■ **EXAMPLE 12.14** For the circuit shown in Fig. 12.18a, find the Thevenin equivalent circuit at terminals *a–b* wiht R_L disconnected.

SOLUTION : The circuit of Fig. 12.18a has an independent voltage source and a current-controlled current source αI.

Determination of Thevenin voltage V_{Th}

The Thevenin voltage V_{Th} is the voltage across the terminals *a–b* when R_L is removed. With the removal of R_L, the current I becomes zero, and therefore, the current source αI becomes an open circuit. The circuit of Fig. 12.18a now becomes the circuit as shown in Fig. 12.18b. By Ohm's law, the current through the resistor R_3 is given by

$$I_1 = \frac{V_1}{R_1 + R_2 + R_3} \qquad \text{(E12.14.1)}$$

The Thevenin voltage is given by

$$V_{Th} = V_{OC} = \text{voltage across } R_3$$

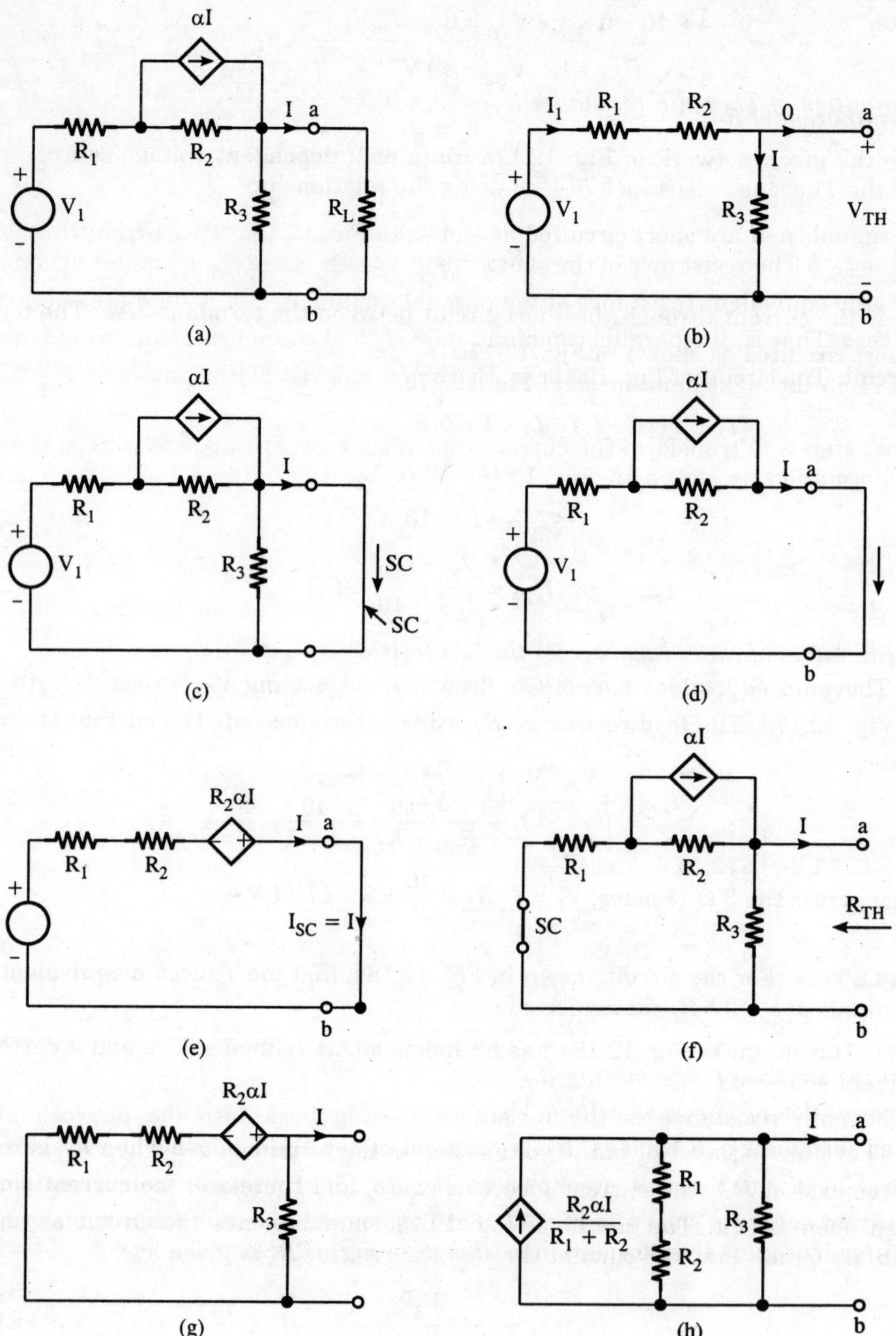
αI
I a
R1
R2
+
V1
-
R3
RL
b
(a)
I1
R1
R2
0
a
+
I
V1
R3
VTH
-
b
(b)
αI
I
R1
R2
+
V1
R3
SC
SC
-
(c)
αI
I a
R1
R2
+
V1
-
b
(d)
R2αI
I a
R1
R2
+
ISC = I
-
b
(e)
αI
I a
R1
R2
SC
R3
RTH
b
(f)
R2αI
I
R1
R2
R3
(g)
I a
R1
R2αI
R1 + R2
R3
R2
b
(h)

FIG. 12.18

$$= I_1 R_3 = \frac{R_3 V_1}{R_1 + R_2 + R_3} \tag{E12.14.2}$$

Determination of Thevenin resistance R_{Th}

$$R_{Th} = \frac{V_{Th}}{I_{SC}} \tag{E12.14.3}$$

The terminals a–b are short circuited as shown in Fig. 12.18c. The current through the short circuit is I_{SC}. The resistance of the short circuit is zero. Since R_3 is connected across the short circuit, the equivalent resistance of the parallel combination of R_3 and the short circuit ($R_{SC} = 0$) is zero. That is, the parallel combination of R_3 and the short circuit is equivalent to the short circuit. The circuit of Fig. 12.18c is, therefore, reduced to that shown in Fig. 12.18d. It is seen that $I_{SC} = I$.

The next step is to transform the current-controlled current source to voltage source by source transformation as shown in Fig. 12.18e. Now by KVL around the closed circuit of Fig. 12.18e,

$$V_1 - R_1 I - R_2 I + R_2 \alpha i = 0$$

$$V_1 = (R_1 + R_2 - R_2 \alpha) I$$

$$I = \frac{V_1}{R_1 + R_2 - R_2 \alpha} = \frac{V_1}{R_1 + (1 - \alpha) R_2}$$

But $$I_{SC} = I$$

therefore, $$I_{SC} = \frac{V_1}{R_1 + (1 - \alpha) R_2} \tag{E 12.14.4}$$

From Eqs. (E12.14.2), (E12.14.3) and (E12.14.4)

$$R_{Th} = \frac{V_{Th}}{I_{SC}} = \left(\frac{R_3 V_1}{R_1 + R_2 + R_3}\right) \div \frac{V_1}{R_1 + (1 - \alpha) R_2}$$

or $$R_{Th} = \frac{R_3 [R_1 + (1 - \alpha) R_2]}{R_1 + R_2 + R_3}$$

Alternative method of determining R_{Th}

The Thevenin resistance is the resistance looking back into the network at the open-circuited terminals a–b (that is, R_L is removed) when the independent voltage source V_1 is replaced by a short circuit as shown in Fig. 12.18f. By source transformation, the controlled-current source is shown in Fig. 12.18g. The voltage source is again transformed to a current source by source transformation as shown in Fig. 12.18h.

Let $$R_4 = R_3 \,||\, (R_1 + R_2)$$

therefore, $$R_4 = \frac{R_3 (R_1 + R_2)}{R_3 + R_1 + R_2}$$

Therefore, the circuit in Fig. 12.18 (*h*) is reduced to that shown in Fig. 12.18k. Again by source transformation, the current source in Fig. 12.18k is transformed to the voltage source as shown in Fig. 12.18l. Let V be the voltage across the terminals a–b. By KVL in Fig. 12.18l.

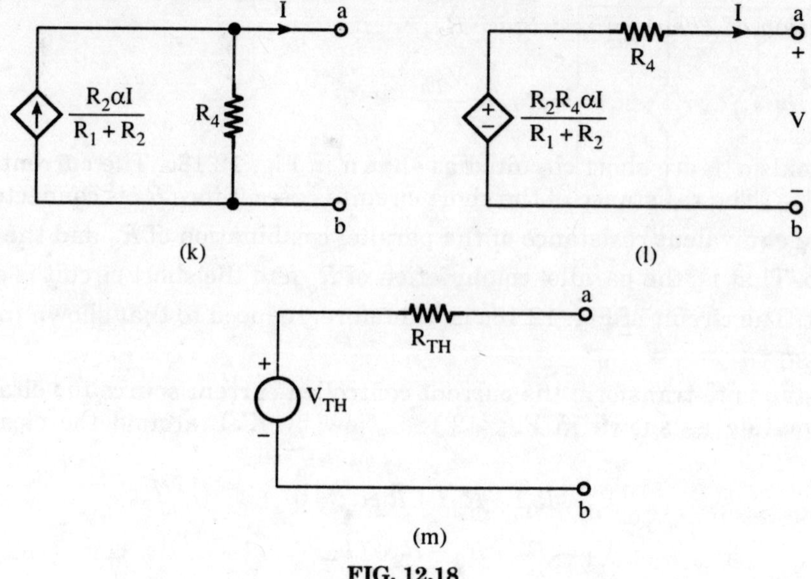

FIG. 12.18

$$\frac{R_2 R_4 \alpha I}{R_1 + R_2} - R_4 I - V = 0$$

$$V = -IR_4 \left(1 - \frac{\alpha R_2}{R_1 + R_2}\right)$$

$$= \frac{V}{-I} = \frac{R_3 (R_1 + R_2)}{R_1 + R_2 + R_3} \left[\frac{R_1 + (1-\alpha) R_2}{R_1 + R_2}\right]$$

But $$\frac{V}{-I} = R_{in} = R_{Th}$$

therefore, $$R_{Th} = \frac{R_3 [R_1 + (1-\alpha) R_2]}{R_1 + R_2 + R_3}$$

which is the same as obtained before.

The Thevenin equivalent circuit for the network of Fig. 12.18a is shown in Fig. 12.18m in which

$$V_{Th} = \frac{R_3 V_1}{R_1 + R_2 + R_3}$$

and $$R_{Th} = \frac{R_3 [R_1 + (1-\alpha) R_2]}{R_1 + R_2 + R_3},$$

■ **EXAMPLE 12.15** For the circuit shown in Fig. 12.19a, use the node-voltage method to find V_x and find the Thevenin and Norton equivalent circuits at the port a–b.

SOLUTION :

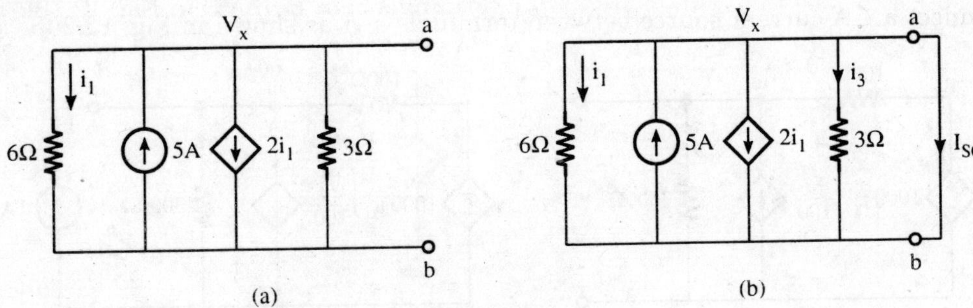

FIG. 12.19

We have $$i1 = \frac{V_x}{6}$$

The current through the 3 Ω resistor $i_3 = \frac{V_x}{3}$

Applying KCL at node A,

sum of the incoming currents = sum of the outgoing currents

$$5 = i_1 + 2i_1 + i_3$$

$$5 = \frac{V_x}{6} + 2\frac{V_x}{6} + \frac{V_x}{3}$$

therefore, $V_x = 6$ V

In order to determine Thevenin and Norton equivalent circuits, we will find V_{OC} and I_{SC}.

But $$V_{OC} = V_x = 6 \text{ V}$$

When the port a–b is short-circuited, the point A is at zero volt. Therefore

$$i_1 = 0, \ 2i_1 = 0 \ \text{ and } \ i_3 = 0$$

By KCL at node a when the port a–b is short circuited (Fig. 12.19b)

$$5 = i_1 + 2i_1 + i_3 + I_{SC}$$

$$5 = 0 + 0 + 0 + I_{SC}$$

therefore, $$I_{SC} = 5 \text{ A}$$

$$R_{Th} = R_N = \frac{V_{OC}}{I_{SC}} = \frac{6}{5} = 1.2 \ \Omega$$

therefore, $$V_{Th} = 6 \text{ V}, \ R_{Th} = 1.2 \ \Omega$$

and $$I_N = I_{sc} = 5 \text{ A}, \ R_N = 1.2 \ \Omega$$

■ **EXAMPLE 12.16** Find the Thevenin and Norton equivalent circuits at terminal a–b of the network of Fig. 12.20a.

SOLUTION : Since there are no independent sources in the given network,

$$V_{Th} = 0 \text{ and } I_N = 0$$

Determination of R_{Th}

Connect a 1 A current source between terminals *a–b* as shown in Fig. 12.20b.

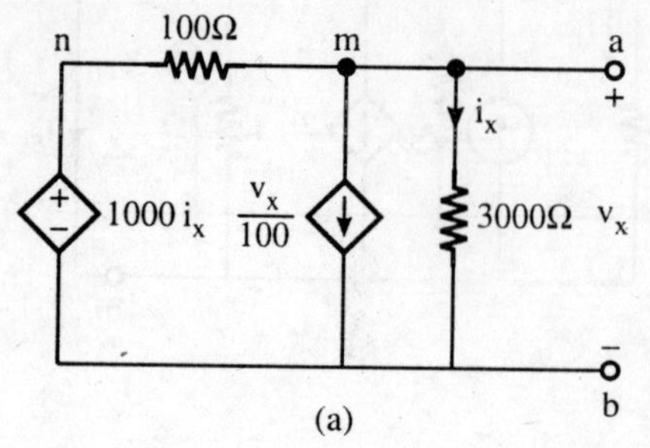

(a)

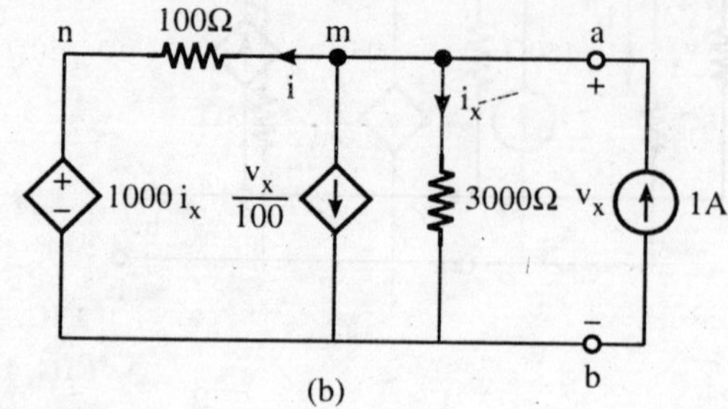

(b)

FIG. 12.20

By KCL at node *m* Fig. 12.20b,

$$i_x + \frac{v_x}{100} + i = 1 \tag{E12.16.1}$$

$$\frac{v_x}{3000} + \frac{v_x}{100} + \frac{v_m - v_n}{100} = 1 \tag{E12.16.2}$$

But
$$v_m = v_x,\ v_n = 1000 i_x = 1000\frac{v_x}{3000} = \frac{v_x}{3}$$

Substituting the values of v_m and v_n in Eq. (E12.16.2), we get

$$\frac{v_x}{3000} + \frac{v_x}{100} + \frac{v_x - \frac{1}{3}v_x}{100} = 1$$

$$v_x = \frac{3000}{51} = 58.82 \text{ V}$$

therefore,
$$R_{Th} = \frac{v_x}{I_s} = \frac{v_x}{1} = \frac{58.82}{1} = 58.82\ \Omega$$

Also,
$$R_N = R_{Th} = 58.82\ \Omega$$

■ **EXAMPLE 12.17** Find the Thevenin equivalent for the circuit shown in Fig. 12.21a.

SOLUTION : Since there is no independent source in the given circuit, $V_{Th} = 0$

In order to determine R_{Th}, we connect a current source $I_S = 1$ A between terminals *a–b* as shown in Fig. 12.21b.

By applying KCL at node *n* in Fig. 12.21b.

$$I_s + 0.01v_x = i_2$$

$$1 + 0.1v_x = i_2$$

By current-division rule,

$$i_1 = \frac{150}{(200 + 50) + 150} i_2 = \frac{150}{400}(1 + 0.01v_x)$$

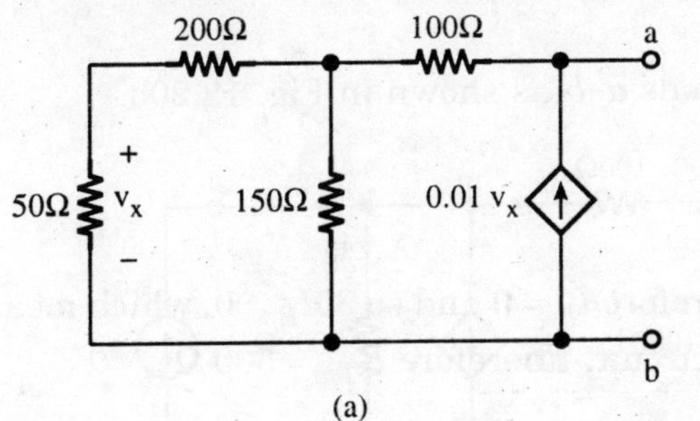

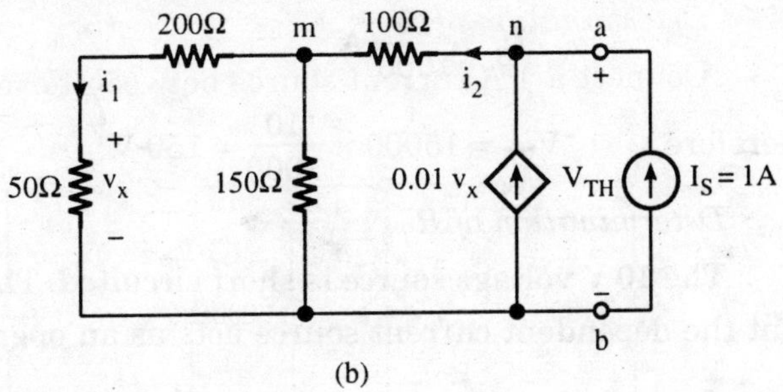

FIG. 12.21

$$i_1 = \frac{3}{8}(1 + 0.01v_x)$$

By Ohm's law, $i_1 = \dfrac{v_x}{50}$

therefore, $$\frac{3}{8}(1 + 0{\cdot}01\, v_x) = \frac{v_x}{50}$$

$$150 + 1.5v_x = 8v_x,$$

$$\therefore \quad v_x = \frac{150}{6.5}\text{ V} = 23.07\text{ V}$$

By KVL in the outer loop

$$v_x + 200i_1 + 100i_2 - V_{Th} = 0$$

$$V_{Th} = v_x + 200 \times \frac{v_x}{50} + 100\,(1 + 0.01v_x)$$

$$= v_x + 4v_x + 100 + v_x$$

$$= 100 + 6v_x = 100 + 6 \times 23.07 = 238.42\text{ V}$$

therefore, $$R_{Th} = \frac{V_{Th}}{I_S} = \frac{238.42}{1} = 238.42\ \Omega$$

■ **EXAMPLE 12.18** Find the Thevenin equivalent circuit at terminals a and b of the circuit shown in Fig. 12.22a.

SOLUTION :

Determination of V_{Th}

Since V_{Th} is the open-circuit voltage across the terminals a–b

$$V_{Th} = \text{voltage across } 500\ \Omega \text{ resistor.}$$

By KCL, current through 500 Ω resistor is $30\, I_B$.

therefore, $$V_{Th} = 500 \times (30I_B) = 15000\, I_B$$

with positive terminal at b.

From the base circuit,

$$I_B = \frac{10}{1000} \text{ A}$$

therefore, $$V_{Th} = 15000 \times \frac{10}{1000} = 150 \text{ V}$$

Determination of R_{Th}

The 10 V voltage source is short circuited. Therefore $I_B = 0$ and so $30I_B = 0$, which means that the dependent current source acts as an open circuit. Therefore $R_{Th} = 500\ \Omega$.

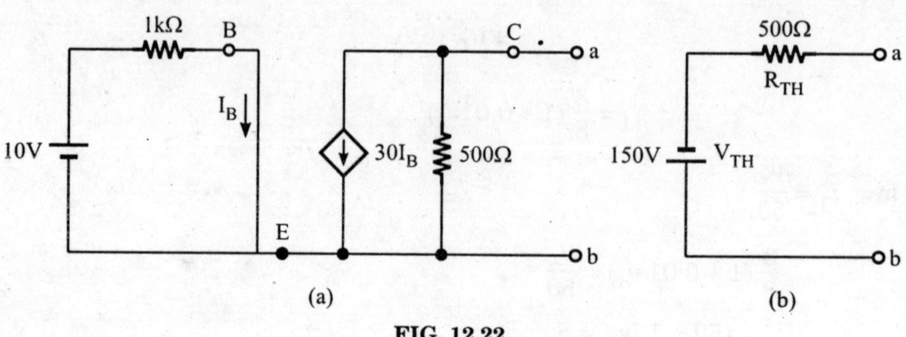

FIG. 12.22

The Thevenin equivalent circuit is shown in Fig. 12.22b.

The 150 V source has its positive terminal towards the point *b*.

■ **EXAMPLE 12.19** Find the Thevenin equivalent of the circuit of Fig. 12.23a, across *a–b*.

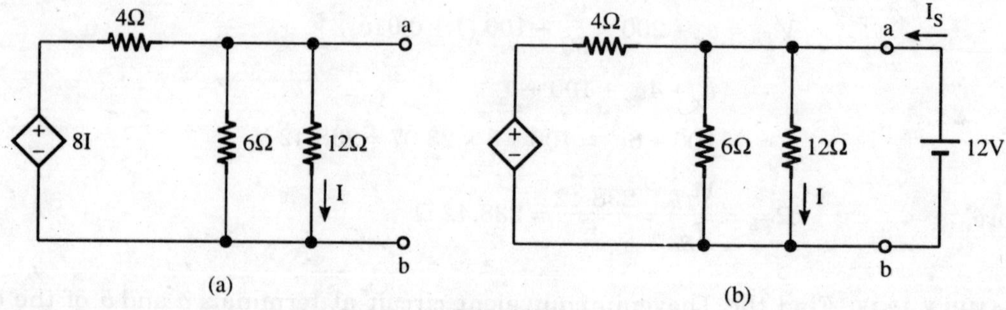

FIG. 12.23

SOLUTION Since the circuit does not contain any independent source, $V_{Th} = 0$. In order to determine R_{Th}, we shall apply an independent voltage source V_s and determine the source current I_s. The Thevenin resistance R_{Th} will be given by

$$R_{Th} = \frac{V_s}{I_s}$$

For the sake of convenience let us take $V_s = 12$ V as shown in Fig. 12.23b.

therefore, $$I = \frac{12}{12} = 1 \text{ A}$$

The voltage provided by the dependent voltage source is $8I = 8 \times 1 = 8$ V By KCL

$$I_s = \frac{12}{12} + \frac{12}{6} + \frac{12-8}{4} = 4 \text{ A}$$

and
$$R_{Th} = \frac{V_s}{I_s} = \frac{12}{4} = 3\ \Omega$$

■ **EXAMPLE 12.20** For the circuit of Fig. 12.24a, obtain the Thevenin equivalent at the terminals a–b.

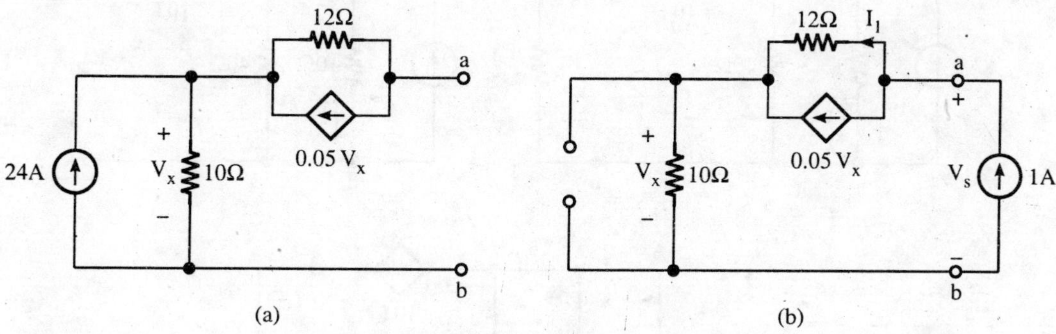

FIG. 12.24

SOLUTION :

Determination of V_{Th}

Since the circuit is open at the terminals, there is no current through a–b. Therefore, the 24 A current supplied by the independent current source flows through the 10 Ω resistor and $V_x = 24 \times 10 = 240$ V. The current supplied by the dependent source $= 0.1\ V_x = 0.05 \times 240 = 12$ A. This current flows through 12 Ω resistor. The voltage across 12 Ω resistor = 12 × 12 = 144 V.

By KVL, the Thevenin's voltage is given by

$$V_{Th} = V_x - 144 = 240 - 144 = 96 \text{ V}$$

Determination of R_{Th}

The independent current source of 24 A is open circuited. The resulting circuit is shown in Fig. 12.24b.

Since a dependent source is present in the circuit, R_{Th} is found by applying an independent source across a–b and determining the ratio of the source voltage to the source current. For convenience, let us apply a current source of 1 A as shown in Fig. 12.24b. The current through 10 Ω resistor will be 1 A and $V_x = 1 \times 10 = 10$ V.

The current supplied by the dependent current source

$$= 0.05\ V_x = 0.05 \times 10 = 0.5 \text{ A}$$

The current through 12 Ω resistor,

$$I_1 = I_s - 0.05V_x = 1 - 0.5 = 0.5 \text{ A}$$

By KVL,
$$V_s - 12I_1 - V_x = 0$$

$$V_s - 12 \times 0.5 - 10 = 0,\ V_s = 16 \text{ V}$$

therefore, $$R_{Th} = \frac{V_{s'}}{I_s} = \frac{16}{1} = 16\ \Omega$$

EXAMPLE 12.21 In the circuit of Fig. 12.25a, what resistor R_L connected across terminals a and b will absorb maximum power and what is this power ?

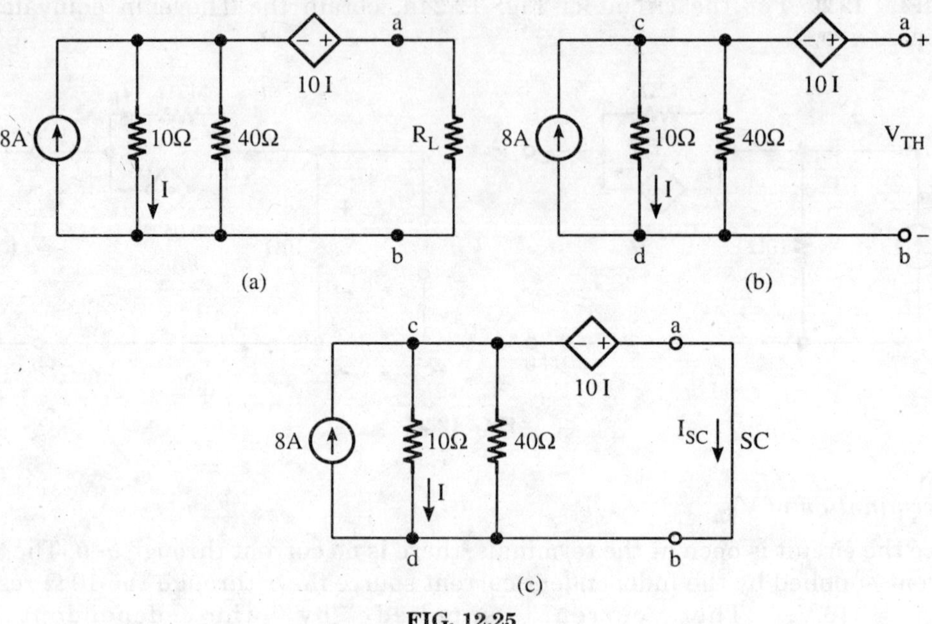

FIG. 12.25

SOLUTION : For maximum power transfer,

$$R_L = R_{Th} \text{ and } P_{\max} = \frac{V_{Th}^2}{4R_{Th}}$$

Therefore, we shall find the Thevenin equivalent of the portion of the circuit to the left of the terminals a–b. In the first step, the load resistor R_L is removed, and there is open circuit across the terminals a–b. Therefore, the 8 A current source supplies current to 10 Ω and 40 Ω resistors only. By current division rule, the currrent through the 10 Ω resistor

$$I = \frac{40}{40 + 10} \times 8 = 6.4 \text{ A}$$

The voltage supplied by the dependent source

$$= 10I = 10 \times 6.4 = 64 \text{ V}$$

By KVL in the loop *acdba* of Fig. 12.25b.

$$-10I - 10 \times I + V_{Th} = 0$$

$$V_{Th} = 20I = 20 \times 6.4 = 128 \text{ V}$$

Determination of R_{Th}

Let us assume that the terminals a–b are short circuited (Fig. 12.25c) so that the current through the terminals a–b is I_{SC}. By KVL in the loop *dcabd*,

$$+ 10I + 10I = 0, \therefore I = 0.$$

The voltage drop across both the resistors becomes zero, so that the total current of 8 A supplied by the current source flows through the short circuit. That is, $I_{Sc} = 8$ A.

therefore, $$R_{Th} = \frac{V_{Th}}{I_{SC}} = \frac{128}{8} = 16\ \Omega$$

For maximum power transfer, $R_L = R_{Th}$, and maximum power abosorbed by the load resistor R_L is

$$P_{max} = \frac{V_{Th}^2}{4R_{Th}} = \frac{(128)^2}{4 \times 16} = 256\text{ W}$$

■ **EXAMPLE 12.22** Find the Thevenin equivalent of the circuit shown in Fig. 12.26a.

SOLUTION It is to be noted that the current i_x shown in the circuit must be zero. This is due to the fact that there is no return path for i_x to enter the left-hand portion of the circuit.

Determination of V_{Th}

$$V_{Th} = v_{ab} = \text{voltage across the } 25\ \Omega \text{ resistor} = v$$

$$= (-20i)(25) = -500i$$

By KVL in the left-hand closed path

$$5 - 2000i - 3v = 0$$

therefore, $$i = \frac{5 - 3v}{2000} = \frac{5 - 3V_{Th}}{2000}$$

and $$V_{Th} = -500i = -500 \times \frac{5 - 3V_{Th}}{2000}$$

$$4V_{Th} = -5 + 3V_{Th},\ V_{Th} = -5\text{ V}$$

Determination of R_{Th}

$$R_{Th} = \frac{V_{Th}}{I_{SC}}$$

In order to calculate the short-circuit current I_{SC}, we place a short circuit across the terminals ab as shown in Fig. 12.26b.

When the terminals ab are short circuited, the control voltage v becomes zero and the circuit of Fig. 12.26a becomes the circuit shown in Fig. 12.26b. The 25 Ω resistor in parallel with the short circuit is equivalent to a resistance of zero ohm as shown in Fig. 12.26c. From this figure,

$$I_{SC} = -20i$$

Since the voltage controlling the dependent voltage source has become zero, the current controlling the dependent current source is found from the left-hand closed path of Fig. 12.26c as

$$5 - 2000i = 0,\ i = \frac{5}{2000}\text{ A}$$

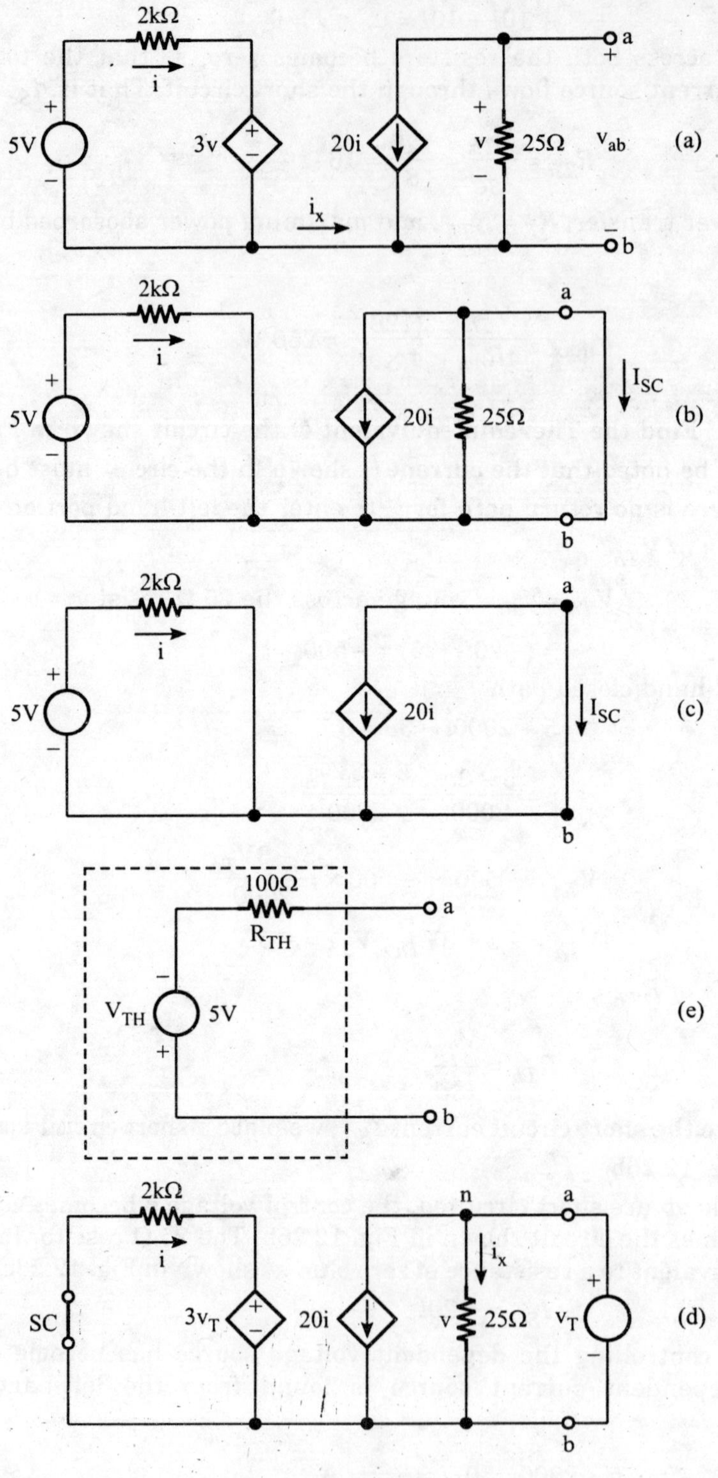

2kΩ
i
5V
3v
20i
v
25Ω
v_{ab}
a
b
i_x
(a)
2kΩ
i
5V
20i
25Ω
I_{SC}
a
b
(b)
2kΩ
i
5V
20i
I_{SC}
a
b
(c)
100Ω
R_{TH}
V_{TH}
5V
a
b
(e)
2kΩ
i
SC
$3v_T$
20i
n
i_x
v
25Ω
v_T
a
b
(d)

FIG. 12.26

therefore, $$I_{SC} = -20i = -20 \times \frac{5}{2000} = -50 \times 10^{-3} \text{ A}$$

and $$R_{Th} = \frac{V_{Th}}{I_{SC}} = \frac{-5}{-50 \times 10^{-3}} = 100\ \Omega$$

Alternative method of determining R_{Th}

We first set the 5 V independent voltage source to zero by replacing it with a short circuit. Then we apply either a test voltage source or a test current source between the terminals a–b. If we apply a test voltage source, we will know the voltage of the dependent voltage source and hence the controlling current i. Therefore, we apply a test voltage source between the terminals a–b as shown in Fig. 12.26d.

Let v_T = externally applied test voltage source

i_T = current delivered by the test voltage source to the circuit

therefore, $R_{Th} = \dfrac{v_T}{i_T}$

From Fig. 12.26d,

$$i_x = \frac{v_T}{25}$$

By KCL at node n,

$$i_T = i_x + 20i$$

$$= \frac{v_T}{25} + 20i$$

By KVL in the left-hand closed path

$$-2000i - 3v_T = 0$$

$$i = -\frac{3v_T}{2000}$$

therefore, $$i_T = \frac{v_T}{25} + 20\left(-\frac{3v_T}{2000}\right)$$

$$i_T = \frac{v_T}{100}$$

and $$R_{Th} = \frac{v_T}{i_T} = 100\ \Omega$$

The Thevenin equivalent circuit is shown in Fig. 12.26e. The reference polarity sign (+ and –) on the Thevenin voltage source in Fig. 12.26e are marked according to the relation $V_{Th} = -5$V.

12.11 THEVENIN'S THEOREM FOR AC NETWORKS

Thevenin's theorem for dc networks allows us to replace a linear dc circuit by a single equivalent dc voltage source in series with a single equivalent resistance. The ac version of Thevenin's theorem is similar and can be stated as follows :

> A linear 2 terminal active network N may be replaced at any pair of terminls a–b and at a given frequency by a simple equivalent network consisting of ac voltage source $\mathbf{V}_{Th}$ in series with an impedenace $\mathbf{Z}_{Th}$.
>
> The source voltage $\mathbf{V}_{Th}$ is the voltage across the terminals a–b when they are open circuited.
>
> The series impedance $\mathbf{Z}_{Th}$ is the impedance looking back into the network at the open-circuited terminals a–b when all voltage sources are replaced by short circuits and all current sources are replaced by open circuits. (Any internal impedance associated with a source remains in the circuit). Dependent sources are left unaltered and they are not set to zero. The quantity $\mathbf{V}_{Th}$ is called the *Thevenin equivalent voltage* and $\mathbf{Z}_{Th}$ is called the *Thevenin equivalent impedance.*

Consider a linear active network N connected to an arbitrary load $\mathbf{Z}_L$ as shown in Fig. 12.27a. The network contains any configuration of impedances with constant values of R, L and C together with independent and/or dependent voltage and current sources. The initial conditions of inductors and capacitors are included in the sources. According to Thevenin's theorem, the *complex* network to the left of terminals a–b can be replaced by a *simple* network consisting of an ac voltage source $\mathbf{V}_{Th}$ in *series* with an impedenace $\mathbf{Z}_{Th}$ (Fig. 12.27b). The voltage, $\mathbf{V}_{Th}$ is the open-circuit voltage that is the voltage across the terminals a–b when $\mathbf{Z}_L$ is removed (Fig. 12.28).

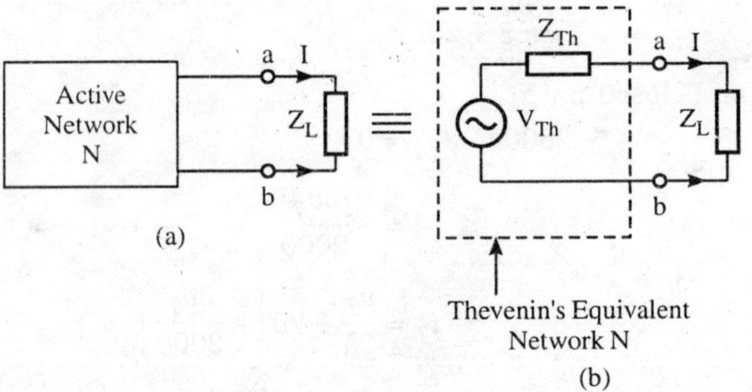

FIG. 12.27

The series impedance $\mathbf{Z}_{Th}$ is the input impedance measured (or calculated) when all ideal *independent* sources and all initial conditions of the inductors and capacitors are reduced to zero (Fig. 12.29). The source voltage $\mathbf{V}_{Th}$ is called the *Thevenin voltage,* and the series impedance $\mathbf{Z}_{Th}$ is called the *Thevenin impedance.*

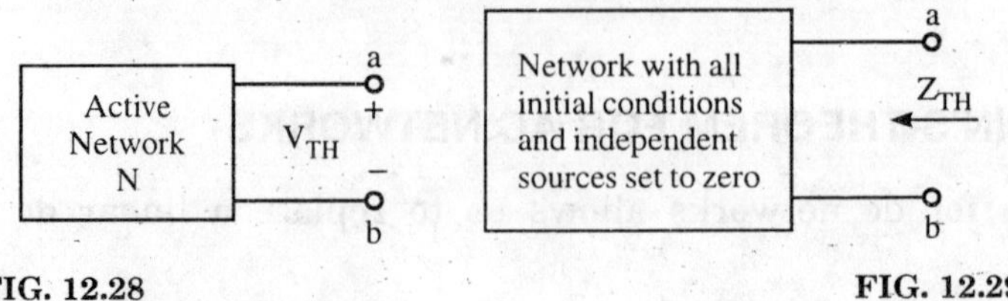

FIG. 12.28 FIG. 12.29

12.12 PROOF OF THEVENIN'S THEOREM

Consider a network consisting of two subnetworks (parts) M and N connected at two nodes a and b as shown in Fig. 12.30a. Let **V** and **I** be the voltage and current at the terminals a–b. Part M of the circuit is called the *load*. Part M may be a complex circuit containing voltage and current sources, constant values of R, L and C connected in any configuration. M may also be a simple circuit such as resistor, an impedance or an open circuit. N is a linear active network containing any configuration of impedances with constant values of R, L and C together with independent and/or dependent voltage and crrent sources. We are interested in finding an equivalent circuit of N. A Thevenin equivalent circuit of N consists of an equivalent voltage source $\mathbf{V}_{Th}$ in series with an equivalent impedance $\boldsymbol{Z}_{Th}$ as shown in Fig. 12.30b. The voltages and currents in all branches of M are not affected.

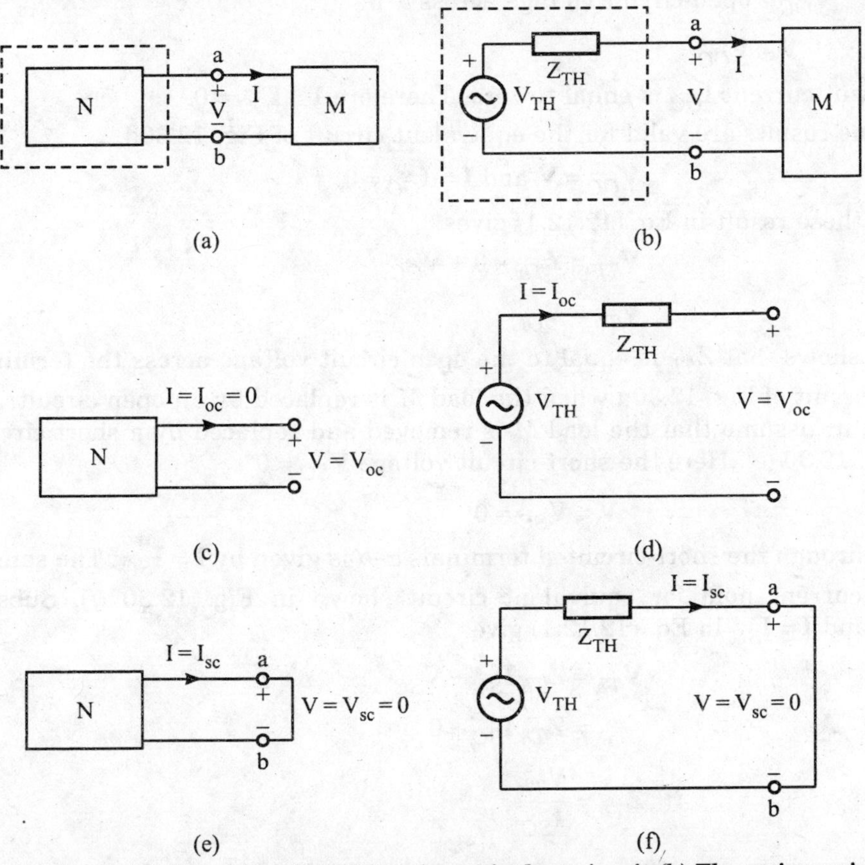

FIG. 12.30 (a) Original circuit for deriving Thevenin equivalent circuit (b) Thevenin equivalent circuit (c) Original circuit with an open circuit at terminals $a-b$ (d) Equivalent circuit with an open circuit at terminals $a-b$ (e) Original circuit with a short circuit at $a-b$ (f) Equivalent circuit with a short circuit at $a-b$

Also, the network M must not have controlled source coupling to network N. Since it has been assumed that the interaction between the networks N and M is only through terminals, the dependent sources (if any) in N and/or M should be controlled by the voltages or currents with the corresponding subnetworks itself, and the values of the dependent sources should not depend on the voltages or currents in the other subnetwork.

The voltage and current at the terminals *a* and *b* in the circuits of Fig. 12.30a and Fig. 12.30b must be the same irrespective of the nature of load *M*. The only restriction is that any element in *M* must not be magnetically coupled to any element in *N*. Also the load *M*, must not have any controlled source coupling to network *N*. The equivalence must be valid for all possible loads. By KVL for the circuit shown in Fig. 12.30b for any arbitrary load

$$\mathbf{V}_{Th} = \mathbf{Z}_{Th}\,\mathbf{I} + \mathbf{V} \tag{12.12.1}$$

The values of $\mathbf{V}_{Th}$ and $\mathbf{Z}_{Th}$ are found from Eq. (12.12.1) by considering two simple loads, an open circuit and a short circuit.

Let the load *M* be removed and replaced by an open circuit (*OC*). For the original circuit with an open circuit at terminals *a–b* (Fig. 12.30c), let

$$\mathbf{V}_{OC} = \text{open-circuit voltage across } a\text{–}b.$$

therefore, $\mathbf{V} = \mathbf{V}_{OC}$

The open-circuit current $\mathbf{I}_{OC}$ is equal to zero. Therefore $\mathbf{I} = \mathbf{I}_{OC} = 0$

The same results are valid for the equivalent circuit of Fig. 12.30d.

therefore, $\mathbf{V}_{OC} = \mathbf{V}$ and $\mathbf{I} = \mathbf{I}_{OC} = 0$

Substituting these result in Eq. (12.12.1) gives

$$\mathbf{V}_{Th} = \mathbf{Z}_{Th} \times 0 + \mathbf{V}_{OC}$$

therefore,
$$\mathbf{V}_{Th} = \mathbf{V}_{OC} \tag{12.12.2}$$

Eq. (12.12.2) shows that $\mathbf{Z}_{Th}$ is equal to the open-circuit voltage across the terminals *a–b* in the original circuit of Fig. 12.30a when the load *M* is replaced by an open circuit.

Now let us assume that the load *M* is removed and replaced by a short circuit (*SC*) as shown in Fig. 12.30 (*e*). Here the short-circuit voltage $\mathbf{V}_{SC} = 0$

therefore, $\mathbf{V} = \mathbf{V}_{SC} = 0$

The current through the short-circuited terminals *a–b* is given by $\mathbf{I} = \mathbf{I}_{SC}$. The same values of voltage and current hold for equivalent circuit shown in Fig. 12.30 (*f*). Substitution of $\mathbf{V} = \mathbf{V}_{SC} = 0$, and $\mathbf{I} = \mathbf{I}_{SC}$ in Eq. (12.12.1) gives

$$\mathbf{V}_{Th} = \mathbf{Z}_{Th}\,\mathbf{I}_{SC} + \mathbf{V}_{SC}$$

$$= \mathbf{Z}_{Th}\,\mathbf{I}_{SC} + 0$$

therefore,
$$\mathbf{Z}_{Th} = \frac{\mathbf{V}_{Th}}{\mathbf{I}_{SC}} \tag{12.12.3.}$$

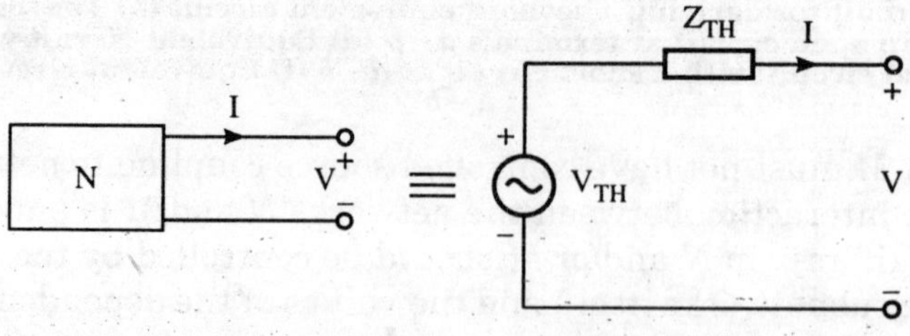

FIG. 12.31 Network *N* and its Thevenin equivalent

Substitution of Eq. (12.12.2) into Eq. (12.12.3) gives

$$\boxed{\mathbf{Z}_{Th} = \frac{\mathbf{V}_{OC}}{\mathbf{I}_{SC}}} \qquad (12.12.4)$$

Eq. (12.12.4) shows that the Thevenin impedance $\mathbf{Z}_{Th}$ can be found by the ratio of the open-circuit voltage and the short-circuit current at the terminals *a–b* of the original circuit *N*. Fig. 12.31 shows the network *N* and its Thevenin equivalent.

12.13 ALTERNATIVE METHOD OF FINDING Z_{Th}

Let the load *M* be disconnected temporarily. The two networks *N* and ts equivalent still remain equivalent if all the independent sources are set to zero. Setting of nonideal sources to zero is equivalent to replacing them by their internal impedances. Thus, every ideal independent voltage source is short circuited and every ideal current source is open circuited. It should be noted that the dependent (controlled) sources are left unaltered and they should not be set to zero. Let us rewrite Eq. (12.12.1)

$$\mathbf{V}_{Th} = \mathbf{Z}_{Th}\boldsymbol{I} + \mathbf{V}$$

When $\quad \mathbf{V}_{Th} = 0,\ \mathbf{Z}_{Th} = \dfrac{\mathbf{V}}{-\mathbf{I}} \qquad (12.13.1)$

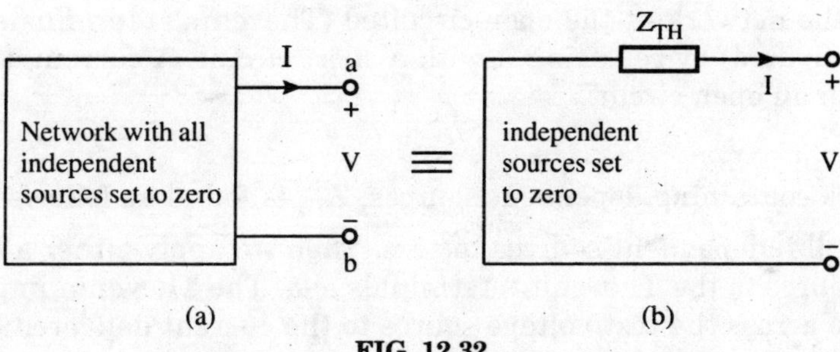

FIG. 12.32

The same result can be obtained directly from Fig. 12.32b by applying Ohm's law. The voltage-current relationship must hold for Fig. 12.32a, so that

$$\mathbf{Z}_{Th} = \frac{\mathbf{V}}{-\mathbf{I}} \qquad (12.13.2)$$

when all the independent sources in the network *N* are set to zero (that is, deactivated). Eq. (12.13.2) can be interpreted easily if we assume that $\mathbf{I}_1 = -\mathbf{I}$ as shown in Fig. 12.33. With this assumption Eq. (12.13.2) becomes

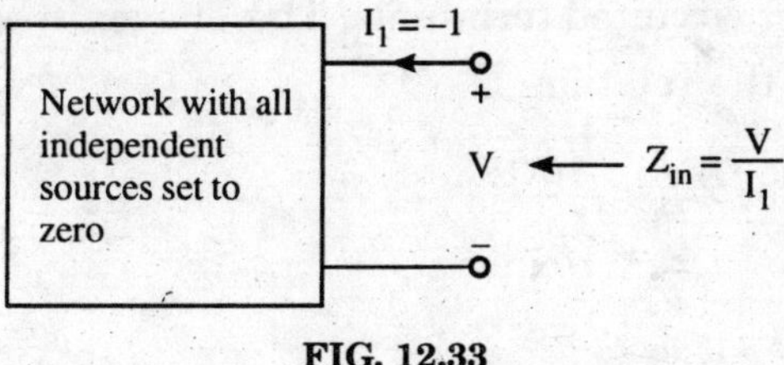

FIG. 12.33

$$\mathbf{Z}_{Th} = \frac{\mathbf{V}}{\mathbf{I}_1} \tag{12.13.3}$$

From Fig. 12.33, the ratio V/I_1 is the input inpedance $\mathbf{Z}_{in}$ so that

$$\mathbf{Z}_{in} = \frac{\mathbf{V}}{\mathbf{I}_1} \tag{12.13.4}$$

Combination of Eqs. (12.13.3) and (12.13.4) gives

$$\mathbf{Z}_{Th} = \mathbf{Z}_{in} = \frac{\mathbf{V}}{\mathbf{I}_1} \tag{12.13.5}$$

Thus, the equivalent impedance Z_{Th} of the Thevenin equivalent network is the impedance as seen from the open-circuited terminals a–b when all the independent sources of the network N are replaced by their internal impedances and dependent (controlled) sources are not altered.

12.14 METHODS OF DETERMINING Z_{Th}

The following methods may be used for determining the Thevenin impedance Z_{Th} :

Method 1

For a network containing only *independent sources*, the following method is used to determine $\mathbf{Z}_{Th}$. First we set all *independent sources* to zero and then determine the impedance seen looking into the network at the open-circuited (Thevenin's) terminals. A voltage source is set to zero (deactivated) by replacing it with a short circuit. A current source is set to zero by replacing it with an open circuit.

Method 2

For a network containing dependent sources, $\mathbf{Z}_{Th}$ is found as follows :

First we set all independent sources to zero. Then we apply either a test voltage source or a test current source to the Thevenin's terminals a–b. The Thevenin impedance equals the ratio of the voltage across the test voltage source to the current delivered by the test voltage source.

$$\mathbf{Z}_{Th} = \frac{\mathbf{V}_{Test}}{\mathbf{I}_{Test}}$$

Method 3

In this method, the load M is removed leaving the circuit N with terminals a–b. Thus, an open circuit at terminals a–b is created. Find the voltage $\mathbf{V}_{OC}$ across the terminals a–b when they are open-circuited. This voltage is $\mathbf{V}_{Th}$. Short circuit the terminals a–b and find the current through the short circuited terminals. The Thevenin equivalent impedance $\mathbf{Z}_{Th}$ is determined with the help of the relation

$$\mathbf{Z}_{Th} = \frac{\mathbf{V}_{OC}}{\mathbf{I}_{SC}}$$

12.15 NORTON'S THEOREM FOR AC NETWORKS

Let us perform the source transformation for the Thevenin equivalent circuit of Fig. 12.34a. The result is shown in Fig. 12.34b. Since in the application of Thevenin's theorem, network *M* can be any load, consider the case when the load is a short circuit as shown in Fig. 12.35a. Let the current through the short circuit be I_{SC}.

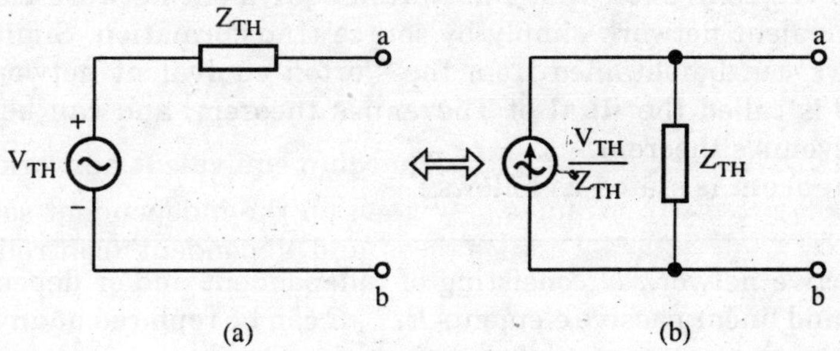

FIG. 12.34

From Fig. 12.35b,

$$\mathbf{I}_{SC} = \frac{\mathbf{V}_{Th}}{\mathbf{Z}_{Th}} \tag{12.15.1}$$

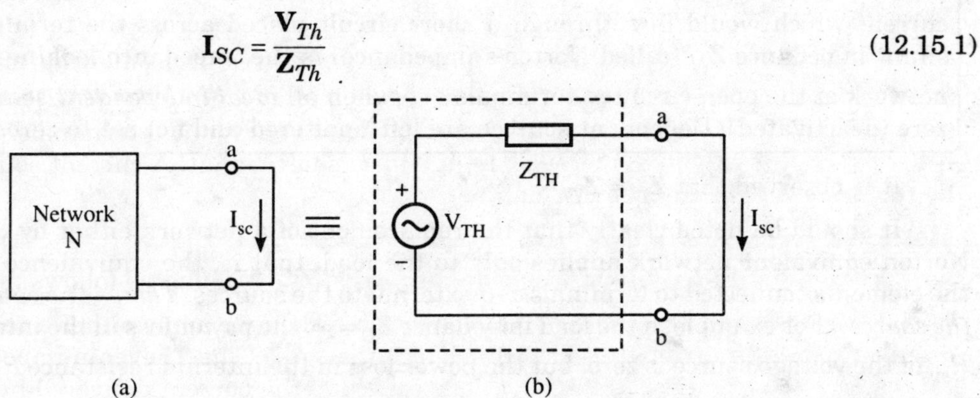

FIG. 12.35

Thus, a circuit equivalent to Thevenin equivalent is shown in Fig. 12.36, where

$\mathbf{I}_{SC}$ is the current through the short circuit between the terminals *a–b*. This circuit is called Norton equivalent of network *N*. The Thevenin and Norton equivalent circuits can be drawn as shown in Fig. 12.37.

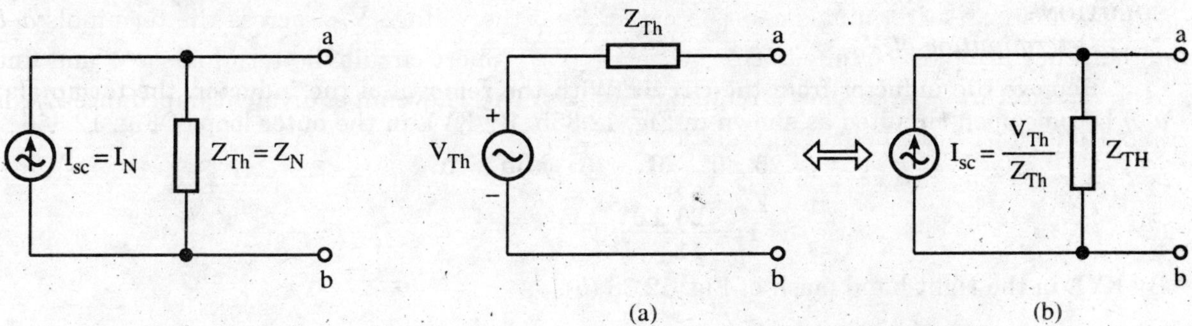

FIG. 12.36 Norton equivalent network

FIG. 12.37 Equivalence of Thevenin and Norton networks

The equivalence relationships can be written as

$$\mathbf{V}_{Th} = \mathbf{V}_{OC} = \mathbf{Z}_{Th}\,\mathbf{I}_{SC} \tag{12.15.2}$$

$$\mathbf{I}_{SC} = \frac{\mathbf{V}_{OC}}{\mathbf{Z}_{Th}} = \frac{\mathbf{V}_{Th}}{\mathbf{Z}_{Th}} \tag{12.15.3}$$

Eqs. (12.15.2) and (12.15.3) show that the Norton equivalent network can be obtained from the Thevenin equivalent network simply by source transformation. Similarly, the Thevenin equivalent network can be obtained from the Norton equivalent network. For this reason Norton's theorem is called the **dual** of Thevenin's theorem, and can be written simply by paraphrasing Thevenin's theorem.

Norton's theorem is stated as follows :

> A linear active network N consisting of independent and/or dependent voltage and current sources and linear passive elements R, L, C can be replaced at any pair of terminals a–b by a simple equivalent network consisting of a current source $\mathbf{I}_N$ in parallel with an impedance $\mathbf{Z}_N$.
>
> The magnitude of the source current $\mathbf{I}_N$ (called Norton's current) is equal to the current which would flow through a short circuit placed across the terminals a–b. The shunt impedance $\mathbf{Z}_N$ (called Norton's impedance) is the impedance looking back into the network at the open-circuited terminals a–b when all *ideal independent* sources are set to zero (deactivated). Dependent sources are left unaltered and not set to zero.

It is observed that $\mathbf{Z}_N = \mathbf{Z}_{Th}$

It should be noted clearly that the replacement of a network either by the Thevenin or Norton equivalent network applies only to the load, that is, the equivalence applies only to the elements connected to terminals a–b external to the sources. *There is no equivalence within the sources.* For example, if the load impedance $Z_L = \infty$, the power loss in the internal resistance R_{sv} of the voltage source is zero, but the power loss in the internal resistance R_{si} of the current source is $I_{sc}^2 R_{si}$.

The procedure for the application of Norton's theorem for ac networks is similar to that for dc networks.

■ **EXAMPLE 12.23** Determine the voltage across and current through the inductor in Fig. 12.38a using Thevenin's theorem.

SOLUTION :

Determination of $\mathbf{V}_{Th}$

Remove the inductor from the circuit. With the removal of the inductor, the terminals a–b become open circuited as shown in Fig. 12.38b. By KVL in the outer loop of Fig. 12.38b.

$$20\angle 0^\circ - 5\mathbf{I}_1 - (6 - j3)\,\mathbf{I}_1 = 0$$

$$\mathbf{I}_1 = \frac{20\angle 0^\circ}{11 - j3}$$

By KVL in the right-hand mesh of Fig. 12.38 (*b*).

$$\mathbf{V}_{Th} = \mathbf{V}_{ab} = (6 - j3)\,\mathbf{I}_1$$

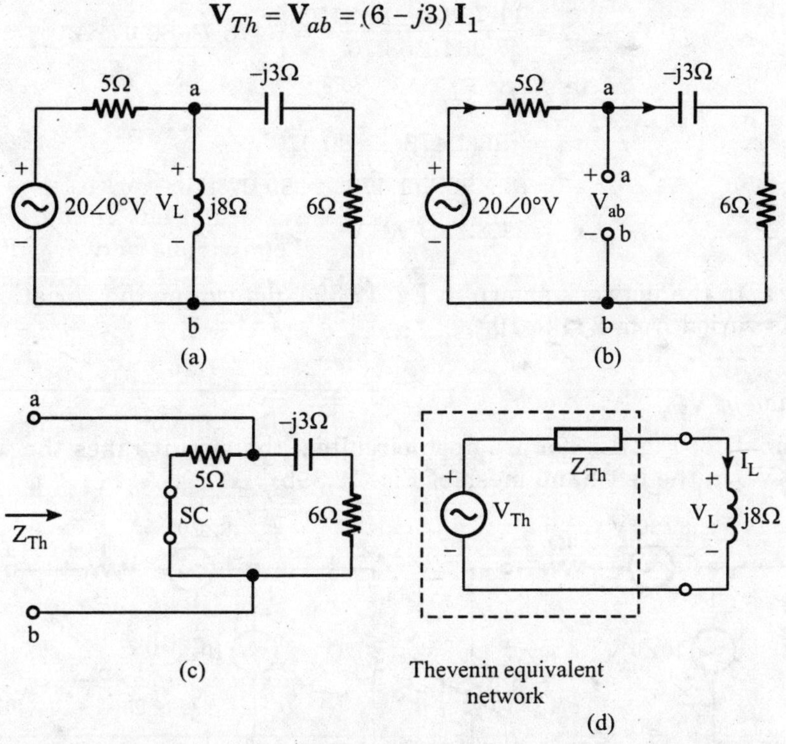

FIG. 12.38 (a) Network of Example 12.23 (b) Determination of V_{Th} (c) Determination of Z_{Th} (d) Thevenin equivalent circuit with $j8\ \Omega$ reconnected

$$= (6 - j3)\,\frac{20\ \angle 0^\circ}{11 - j3}$$

$$= \frac{6.708\ \angle -26.56^\circ}{11.40} \times 20\ \angle 0^\circ$$

$$= 11.768\ \underline{/-11.31^\circ}\ \text{V}$$

Determination of $\mathbf{Z}_{Th}$

In order to determine $\mathbf{Z}_{Th}$, we short circuit the voltage source of $20\ \angle 0^\circ$ V, as shown in Fig. 12.38c. From this figure,

$$\mathbf{Z}_{Th} = 5\ ||\ (6 - j3)$$

$$= \frac{5\,(6 - j3)}{5 + 6 - j3} = 2.942\ \underline{/-11.31^\circ}\ \Omega$$

$$= (2.885 - j0.577)\ \Omega$$

Determination of $\mathbf{I}_L$ *and* $\mathbf{V}_L$

Fig. 12.38d shows the Thevenin equivalent circuit in which the load inductor of $j8\ \Omega$ is reconnected. By Ohm's in this figure,

$$\mathbf{I}_L = \frac{\mathbf{V}_{Th}}{\mathbf{Z}_{Th} + j8} = \frac{11.768\ \angle -11.31^\circ}{2.885 - j0.577 + j8}$$

$$= \frac{11.768 \angle -11.31^\circ}{7.964 \angle 68.76^\circ} = 1.478 \underline{/-80.07^\circ} \text{ A}$$

$$\mathbf{V}_L = (jX_L)\, I_L$$

$$= (j8)\,(1.478 \angle -80.07^\circ)$$

$$= (8 \angle 90^\circ)\,(1.478 \angle -80.07^\circ)$$

$$= 11.82 \angle 9.93^\circ \text{ V}$$

■ **EXAMPLE 12.24** In the network shown in Fg. 12.39a, determine the variation of curent in R_{ab} when it is varied from 2 Ω to 10 Ω.

SOLUTION :

Determination of $\mathbf{V}_{Th}$

When terminals of Fig. 12.39a are open circuited, the circuit takes the form shown in Fig. 12.39b. By KVL in the left-hand mesh of Fig. 12.39b.

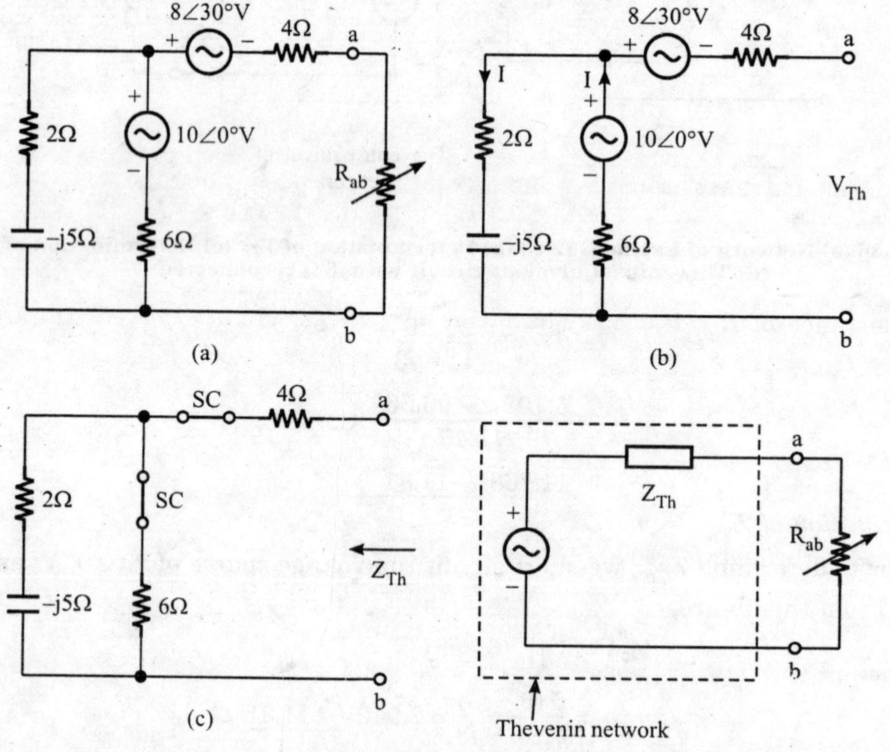

FIG 12.39 (a) Network of Example 12.24 (b) Determination of V_{Th} (c) Determination of Z_{Th} (d) Thevenin equivalent network with R_{ab} reconnected

$$(2 - j5)\,\mathbf{I} - 10 \angle 0^\circ + 6\mathbf{I} = 0$$

$$I = \frac{10 \angle 0^\circ}{8 - j5} = \frac{10 \angle 0^\circ}{9.434 \angle -32^\circ} = 1.06 \angle 32^\circ \text{ A}$$

By KVL in the right-hand mesh of Fig. 12.39b

$$-6\mathbf{I} + 10\angle 0^\circ - 8\angle 30^\circ - 0 \times 4 - \mathbf{V}_{Th} = 0$$

$$\mathbf{V}_{Th} = -6 \times 1.06\angle 32^\circ + 10\angle 0^\circ - 8\angle 30^\circ$$

$$= -5.39 - j3.37 + 10 - 6.928 - j4$$

$$= -2.318 - j7.37 = 7.726\angle -107.46^\circ \text{ V}$$

Determination of $\mathbf{Z}_{Th}$

The voltage sources are short circuited as shown in Fig. 12.39c. The Thevenin impedance $\mathbf{Z}_{Th}$ is the impedance between the open-circuited terminals a–b.

$$\mathbf{Z}_{Th} = 4 + 6 \,||\, (2 - j5)$$

$$= 4 + \frac{6\,(2 - j5)}{6 + 2 - j5}$$

$$= 4 + \frac{32.31\angle -68.2^\circ}{9.434\angle -32^\circ} = 4 + 3.4248\angle -36.2^\circ$$

$$= 4 + 2.7637 - j2.02$$

$$= (6.7637 - j2.02)\ \Omega$$

Determination of current through R_{ab}

Fig. 12.39d shows the Thevenin equivalent network in which R_{ab} is reconnected at terminals a–b. The current through R_{ab} is given by

$$\mathbf{I}_{ab} = \frac{\mathbf{V}_{Th}}{\mathbf{Z}_{Th} + R_{ab}}$$

Since the resistance of R_{ab} is varied from 2 Ω to 10 Ω maximum current through R_{ab} is obtained when $R_{ab} = 2\ \Omega$.

$$(\mathbf{I}_{ab})_{\max} = \frac{7.726\angle -107.46^\circ}{6.7637 - j2.02 + 2}$$

$$= \frac{7.726\angle -107.46^\circ}{8.993\angle -12.98^\circ} = 0.859\angle -94.48^\circ \text{ A}$$

Minimum current through R_{ab} is obtained when $R_{ab} = 10\ \Omega$.

$$(\mathbf{I}_{ab})_{\min} = \frac{7.726\angle -107.46^\circ}{6.7637 - j2.02 + 10} = 0.457\angle -100.59^\circ \text{ A}$$

Hence the current through R_{ab} varies from $0.457\angle -100.59^\circ$ A to $0.859\angle -94.48^\circ$ A.

■ **EXAMPLE 12.25** Use Thevenin's theorem to determine the current through the impedance $(6 + j8)\ \Omega$ in the network of Fig. 12.40a.

SOLUTION :

Determination of $\mathbf{V}_{Th}$

Remove the load impedance of $(6 + j8)\ \Omega$ as shown in Fig. 12.40b. Since the circuit is open between the terminals a–b, the current through the branch containing $-j5\ \Omega$ is zero.

Therefore $\qquad \mathbf{I}_1 = \mathbf{I}_2 = \mathbf{I}$ (say),

By KVL in the left-hand mesh Fig. 12.40b

$$20 \angle 0^\circ - 2\mathbf{I} - (4 + j3)\,\mathbf{I} = 0$$

$$\mathbf{I} = \frac{20}{6 + j3} = 2.98 \angle -25.56^\circ \text{ A}$$

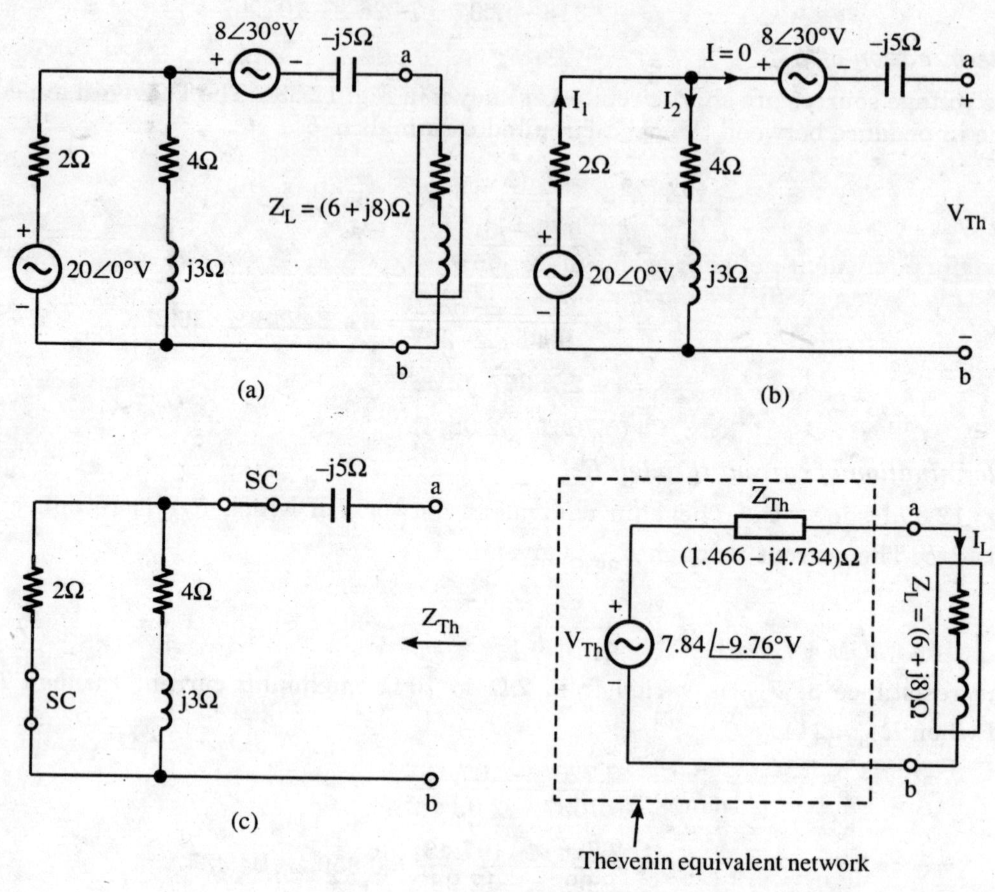

FIG. 12.40

By KVL in the right-hand mesh of Fig. 12.40b

$$(4 + j3)\,\mathbf{I} - 8 \angle 30^\circ - (-j5) \times 0 - \mathbf{V}_{Th} = 0$$

$$(5 \angle 36.87^\circ)\,(2.98 \angle -26.56^\circ) - 8 \angle 30^\circ = \mathbf{V}_{Th}$$

$$14.90 \angle 10.31^\circ - 8 \angle 30^\circ = \mathbf{V}_{Th}$$

$$14.66 + j2.67 - (6.93 + j4) = \mathbf{V}_{Th}$$

$$\mathbf{V}_{Th} = 7.73 - j1.33 = 7.84 \angle -9.76^\circ \text{ V}$$

Determination of $\mathbf{Z}_{Th}$

In order to determine Z_{Th} we short circuit the voltage sources $20 \angle 0^\circ$ V and $8 \angle 30^\circ$ V as shown in Fig. 12.40c. The Thevenin impedance is equal to the impedance between the terminals a–b. From Fig. 12.40c.

$$\mathbf{Z}_{Th} = -j5 + 2 \;||\; (4 + j3)$$

$$= -j5 + \frac{2\,(4 + j3)}{2 + 4 + j3}$$

$$= -j5 + \frac{10\,\angle 36.87^\circ}{6.71\,\angle 26.57^\circ}$$

$$= -j5 + 1.49\,\angle 10.3^\circ$$

$$= -j5 + 1.466 + j0.266$$

$$= 1.466 - j4.734 = 4.955\,\angle -72.79^\circ\ \Omega$$

The Thevenin equivalent network is shown in Fig. 12.40d where the removed load impedance of $(6 + j8)\ \Omega$ is reconnected. The current through the load impedance is

$$\mathbf{I}_L = \frac{\mathbf{V}_{Th}}{\mathbf{Z}_{Th} + \mathbf{Z}_L}$$

$$= \frac{7.84\,\angle -9.76^\circ}{1.466 - j4.734 + 6 + j8}$$

$$= \frac{7.84\,\angle -9.76^\circ}{7.466 + j3.266} = \frac{7.84\,\angle -9.76^\circ}{8.149\,\angle 23.62^\circ}$$

$$= 0.962\,\angle -33.38^\circ\ \text{A}$$

12.16 MAXIMUM POWER TRANSFER THEOREM FOR AC NETWORKS

The power transferred from an active network to a load depends on a number of factors. The important factors are as follows :

1. The load is a pure resistance.
2. The load resistance and the load reactance are independently variable.
3. The load has a variable impedance but a constant power factor.

Consider an ac network N (Fig. 12.41a) connected to a load impedance Z_L at terminals a–b. The Thevenin equivalent network is shown in Fig. 12.41b. Here $\mathbf{V}_{Th}$ represents the Thevenin voltage and $\mathbf{Z}_{Th}$ the Thevenin equivalent impedance of the network at terminals a–b.

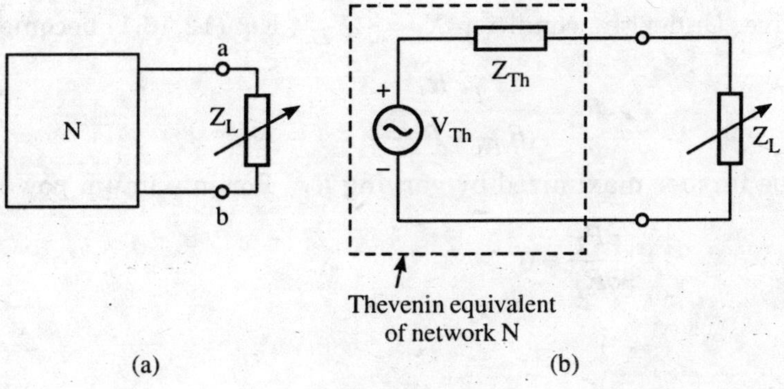

FIG. 12.41

We are interested in determining the value of $\mathbf{Z}_L$ when it receives maximum power from the network N.

Let
$$\mathbf{Z}_{Th} = R_{Th} + jX_{Th}$$
$$\mathbf{Z}_L = R_L + jX_L$$

The current through the load is

$$\mathbf{I}_L = \frac{\mathbf{V}_{Th}}{\mathbf{Z}_{Th} + \mathbf{Z}_L} = \frac{\mathbf{V}_{Th}}{(R_{Th} + R_L) + j(X_{Th} + X_L)}$$

$$I_L = |\mathbf{I}_L| = \frac{V_{Th}}{\sqrt{[(R_{Th} + R_L)^2 + (X_{Th} + X_L)^2]}}$$

The average power transferred to the load is

$$P_L = I_L^2 R_L = \frac{V_{Th}^2 R_L}{(R_{Th} + R_L)^2 + (X_{Th} + X_L)^2} \tag{12.16.1}$$

For a given network N the values of V_{Th}, R_{Th} and X_{Th} are fixed. For maximum power transfer, R_L and X_L can be varied independently. The general procedure of finding the maximum value of P_L is to differentiate P_L with respect to the variable and set the derivative equal to zero. That is,

$$\frac{\partial P_L}{\partial R_L} = 0, \quad \text{and} \quad \frac{\partial P_L}{\partial X_L} = 0$$

Let us consider the case of varying X_L and keeping R_L constant.

From Eq. (12.16.1)

$$\frac{\partial P_L}{\partial X_L} = \frac{0 - 2V_{Th}^2 R_L (X_{Th} + X_L)}{[(R_{Th} + R_L)^2 + (X_{Th} + X_L)^2]^2} = 0$$

$$-2V_{Th}^2 R_L (X_{Th} + X_L) = 0$$

$$\therefore \quad X_L = -X_{Th} \tag{12.16.2}$$

Hence the *first condition* for maximum power transfer is that the reactance of the load must have exactly the same magnitude, but must be of opposite phase difference to the reactance component of the Thevenin impedance. This condition implies that if X_{Th} is inductive, then X_L must be capacitive. Under the condition $X_L = -X_{Th}$, Eq. (12.16.1) becomes

$$P_L = \frac{V_{Th}^2 R_L}{(R_{Th} + R_L)^2} \tag{12.16.3}$$

The power P_L can be further maximized by varying R_L. For maximum power transfer,

$$\frac{\partial P_L}{\partial R_L} = 0$$

From Eq. (12.16.3)

$$\frac{\partial P_L}{\partial R_L} = \frac{\partial}{\partial R_L} [V_{Th}^2 R_L (R_{Th} + R_L)^{-2}]$$

$$0 = V_{Th}^2 [(R_{Th} + R_L)^{-2} - 2R_L (R_{Th} + R_L)^{-3}]$$

or $$2R_L = R_{Th} + R_L$$

$$R_L = R_{Th} \tag{12.16.4}$$

Hence the *second condition* for maximum power transfer is that the resistance of the load is made equal to the resistance component of the Thevenin impedance.

From Eqs. (12.16.3) and (12.16.4) we get

$$P_{L\,max} = \frac{V_{Th}^2 R_{Th}}{(R_{Th} + R_L)^2} = \frac{V_{Th}^2}{4R_{Th}} \tag{12.16.5}$$

For the two conditions given in Eqs. (12.16.2) and (12.16.4) to be satisfied simultaneously, the impedance of the load should be adjusted so that

$$\mathbf{Z}_L = R_L + jX_L = R_{Th} - jX_{Th} = \mathbf{Z}_{Th}^* \tag{12.16.6}$$

Hence the **maximum power transfer theorem of ac networks** may be stated as follows :

> *Maximum power is transferred from an ac network to a load if the load impedance is equal to the complex conjugate of the Thevenin equivalent impedance of the network.*

When this condition is satisfied, we say that the load impedance is *conjugately matched* with the Thevenin equivalent impedance.

Let us now consider the case when R_L and X_L are not independently varied but only the *magnitude* of the load is varied. The phase angle ϕ of the load is constant and cannot be varied. The load impedance can be expressed as

$$\mathbf{Z}_L = Z_L \angle\phi = Z_L \cos\phi + jZ_L \sin\phi = R_L + jX_L \tag{12.16.7}$$

Substituting $R_L = Z_L \cos\phi_L$ and $X_L = Z_L \sin\phi_L$ in Eq. (12.16.1) we obtain

$$P_L = \frac{V_{Th}^2 Z_L \cos\phi}{(R_{Th} + Z_L \cos\phi)^2 + (X_{Th} + Z_L \sin\phi_L)^2} \tag{12.16.8}$$

For maximum power transfer

$$\frac{dP_L}{dZ_L} = 0$$

$$\frac{V_{Th}^2 \cos\phi}{D^r}\left[\left(\frac{d}{dZ_L} Z_L\right) D^r - Z_L \frac{d}{dZ_L}(D^r)\right] = 0 \tag{12.16.9}$$

where $$D^r = (R_{Th} + Z_L \cos\phi)^2 + (X_{Th} + Z_L \sin\phi)^2$$

Therefore Eq. (12.16.9) becomes

$$(R_{Th} + Z_L \cos \phi)^2 + (X_{Th} + Z_L \sin \phi)^2$$
$$- Z_L [2 (R_{Th} + Z_L \cos \phi) \cos \phi + 2 (X_{Th} + Z_L \sin \phi) \sin \phi] = 0$$

$$R_{Th}^2 + 2R_{Th} Z_L \cos \phi + Z_L^2 \cos^2 \phi + X_{Th}^2 + 2X_{Th} Z_L \sin \phi$$
$$+ Z_L^2 \sin^2 \phi - 2R_{Th} Z_L \cos \phi - 2Z_L^2 \cos^2 \phi - 2X_{Th} Z_L \sin \phi - 2Z_L^2 \sin^2 \phi = 0$$

$$Z_L^2 (\cos^2 \phi + \sin^2 \phi) = R_{Th}^2 + X_{Th}^2$$

$$Z_L^2 = R_{Th}^2 + X_{Th}^2 \quad (12.16.10)$$

$$| \mathbf{Z}_L | = | \mathbf{Z}_{Th} | \quad (12.16.11)$$

It follows from Eq. (12.16.11) that when the magnitude of the load impedance is varied while its phase angle remains constant, maximum power is transferred to the load when the magnitude of the load impedance is equal to the magnitude of the Thevenin equivalent impedance of the network.

This is an important concept in using a transformer to achieve maximum power transfer. In this case Z_L can be adjusted but ϕ cannot be adjusted.

In a special case when the load is a pure resistance R_L and it is variable. Then $| \mathbf{Z}_L | = R_L$. Therefore from Eq. (12.16.11) the condition for maximum power transfer is

$$R_L = | \mathbf{Z}_{Th} | \quad (12.16.12)$$

Thus, for maximum power transfer, the load resistance is made equal to the magnitude of the Thevenin equivalent impedance of the network.

■ **EXAMPLE 12.26** A source of 100 V feeds a load impedance $\mathbf{Z}_L$ through a series impedance $\mathbf{Z}_s = (25 + j40)\ \Omega$.

(*a*) Determine the load impedance for maximum power transfer and the value of the maximum power.

(*b*) If the load consists of a pure resistance R_L, find the value of R_L for which maximum power is transferred to the load. Determine the value of the maximum power.

SOLUTION

(*a*) For maximum power trnsfer

$$\mathbf{Z}_L = \mathbf{Z}_{Th}^* = (25 + j40)^* = (25 - j40)\ \Omega$$

Load current

$$\mathbf{I}_L = \frac{\mathbf{V}_{Th}}{\mathbf{Z}_{Th} + \mathbf{Z}_L}$$

$$= \frac{100}{25 + j40 + 25 - j40} = 2 \text{ A}$$

$$P_{L\max} = I_L^2 R_L = (2)^2 \times 25 = 100 \text{ W}$$

Check

$$P_{L\max} = \frac{V_{Th}^2}{4R_{Th}} = \frac{100^2}{4 \times 25} = 100 \text{ W}$$

(*b*) For maximum power transfer to a purely resistive load R_L

$$= \sqrt{25^2 + 40^2} = 47\ \Omega$$

Load current

$$I_L = \frac{V_{Th}}{|\ \mathbf{Z}_{Th} + R_L\ |} = \frac{100}{|\ (25 + j40 + 47)\ |}$$

$$= \frac{100}{\sqrt{(72^2 + 40^2)}} = 1.214 \text{ A}$$

$$P_{L\,\max} = I_L^2 R_L = (1.214)^2 \times 47 = 69.3 \text{ W}$$

■ **EXAMPLE 12.27** For the circuit shown in Fig. 12.42a, what load impedance $\mathbf{Z}_L$ absorbs maximum average power and what is this power?

SOLUTION First we remove the load $\mathbf{Z}_L$ Fig. 12.42b and determine the Thevenin equivalent to the left of the terminals a–b. From Fig. 12.42b,

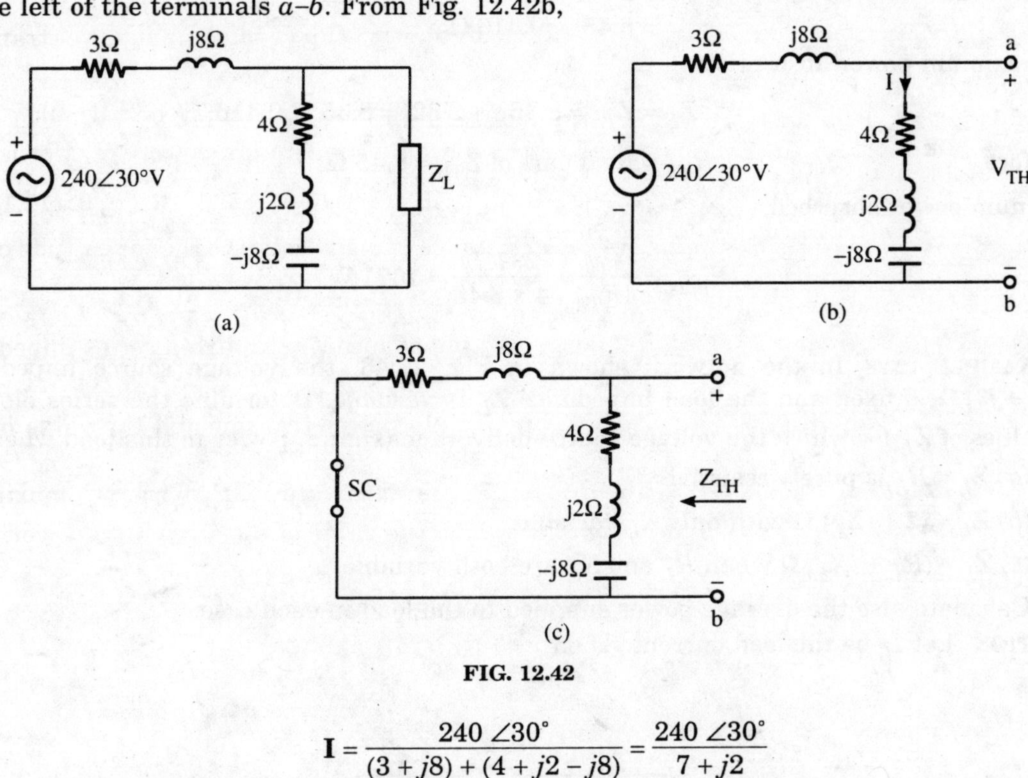

FIG. 12.42

$$\mathbf{I} = \frac{240\ \angle 30^\circ}{(3 + j8) + (4 + j2 - j8)} = \frac{240\ \angle 30^\circ}{7 + j2}$$

$$= \frac{240\ \angle 30^\circ}{7.28\ \angle 15.95^\circ} = 32.97\ \angle 14.05^\circ \text{ A}$$

By KVL in the right-hand mesh of Fig. 12.42b,

$$\mathbf{V}_{Th} = \mathbf{I}\,(4 + j2 - j8)$$

$$= (32.97\ \angle 14.05^\circ)\,(4 - j6)$$

$$= (32.97\ \angle 14.05^\circ)\,(7.21\ \angle -56.31^\circ)$$

$$= 237.7 \angle -42.26° \text{ V}$$

Determination of $\mathbf{Z}_{Th}$

In order to determine $\mathbf{Z}_{Th}$, we remove $\mathbf{Z}_L$ and the voltage source is short circuited as shown in Fig. 12.42c. From this figure

$$\mathbf{Z}_{Th} = (3 + j8) \,||\, (4 + j2 - j8)$$

$$= \frac{(3 + j8)(4 + j2 - j8)}{(3 + j8) + (4 + j2 - j8)}$$

$$= \frac{(3 + j8)(4 - j6)}{7 + j2}$$

$$= \frac{(8.54 \angle 69.44°)(7.21 \angle -56.31°)}{7.28 \angle 15.95°}$$

$$= 8.46 \angle -2.82° \ \Omega$$

$$= 8.45 - j0.416 \ \Omega$$

For maximum power absorption

$$\mathbf{Z}_L = \mathbf{Z}^*_{Th} = 8.46 \underline{/+2{\cdot}82°} = 8.45 + j0.416 \Omega$$

therefore, $R_{Th} = \text{Real part of } \mathbf{Z}_{\text{Th}} = 8.45 \ \Omega$

Maximum power absorbed

$$P_{\max} = \frac{V^2_{Th}}{4R_{Th}} = \frac{237.7^2}{4 \times 8.45} = 1671 \text{ W}$$

■ **EXAMPLE 12.28** In the network shown in Fig. 12.43, the voltage source impedance, $(4 + j3)\ \Omega$ is fixed and the load impedance $\mathbf{Z}_L$ is variable. Determine the series element values of $\mathbf{Z}_L$ for which the voltage source delivers maximum power to this load when

(*a*) $\mathbf{Z}_L = R_L$ is purely resistive

(*b*) $\mathbf{Z}_L = (2 + jX_L)\ \Omega$ with only X_L variable

(*c*) $\mathbf{Z}_L = (R_L + jX_L)\ \Omega$ when R_L and X_L are both variable.

Calculate also the average power supplied to the load in each case.

SOLUTION Let I_L be the load current. Then

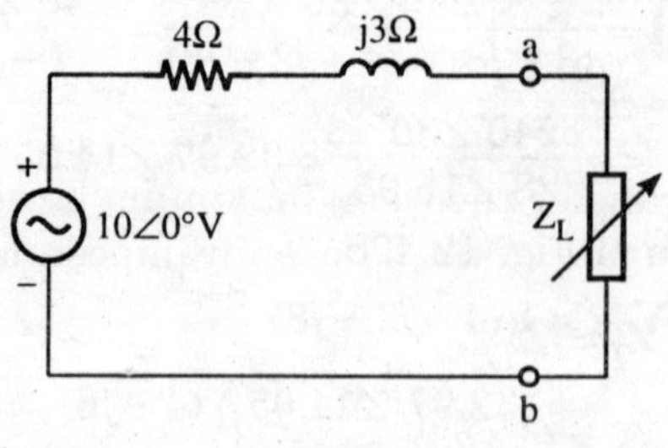

FIG. 12.43

$$\mathbf{I}_L = \frac{10\angle 0^\circ}{4 + j3 + R_L + jX_L} = \frac{10}{(4 + R_L) + j(3 + X_L)} \qquad \text{(E12.28.1)}$$

The average power supplied to the load is

$$P_L = I_L^2 R_L = \frac{(10)^2 R_L}{(4 + R_L)^2 + (3 + X_L)^2} \qquad \text{(E12.28.2)}$$

(*a*) Here $X_L = 0$

$$P_L = \frac{100R_L}{(4 + R_L)^2 + 9} \qquad \text{(E12.28.3)}$$

For P_L to be a maximum, $\frac{dP_L}{dR_L} = 0$

$$\frac{[(4 + R_L)^2 + 9]\,100 - 100R_L \times 2\,(4 + R_L)}{[(4 + R_L)^2 + 9]^2} = 0$$

or

$$100\,(16 + 8R_L + R_L^2 + 9 - 8R_L - 2R_L^2) = 0$$

or

$$R_L^2 = 25,\ R_L = 5\ \Omega$$

Under the condition, the maximum power transfer is found by putting $R_L = 5\ \Omega$ in Eq. (E12.28.3)

$$P_{L\,\max} = \frac{100 \times 5}{(4 + 5)^2 + 9} = 5.56\ \text{W}$$

(*b*) Here $R_L = 2\ \Omega$and X_L is variable.

Putting $R_L = 2\ \Omega$ in Eq. (E12.28.2) gives

$$P_L = \frac{100 \times 2}{(4 + 2)^2 + (3 + X_L)^2} \qquad \text{(E12.28.4)}$$

From Eq. (E12.28.4), P_L is a maximum when the denominator on the right-hand side of Eq. (E12.28.4) is a minimum. The denominator becomes minimum when $(3 + X_L)^2 = 0$. That is, $X_L = -3\ \Omega$

The negative sign indicates that X_L is a capacitive reactance.

The maximum power in this case is found from Eq. (E12.28.4) by putting $X_L = -3$.

$$P_{L\,\max} = \frac{100 \times 2}{36} = 5.56\ \text{W}$$

(*c*) By maximum power transfer theorem, maximum power is delivered to the load when the load impedance is the conjugate of the Thevenin impedance.

$$\mathbf{Z}_L = \mathbf{Z}_{Th}^*$$

$$R_L + jX_L = (4 + j3)^* = 4 - j3$$

therefore, $R_L = 4\ \Omega,\ X_L = -3\ \Omega$

$$P_{L\max} = \frac{V_{Th}^2}{4R_{Th}} = \frac{(10)^2}{4 \times 4} = 6.25\text{ W}$$

■ **EXAMPLE 12.29** **In the network of Fig. 12.44a, determine components of the load impedance that will develop maximum output power.**

SOLUTION : Inductive reactance of the inductor

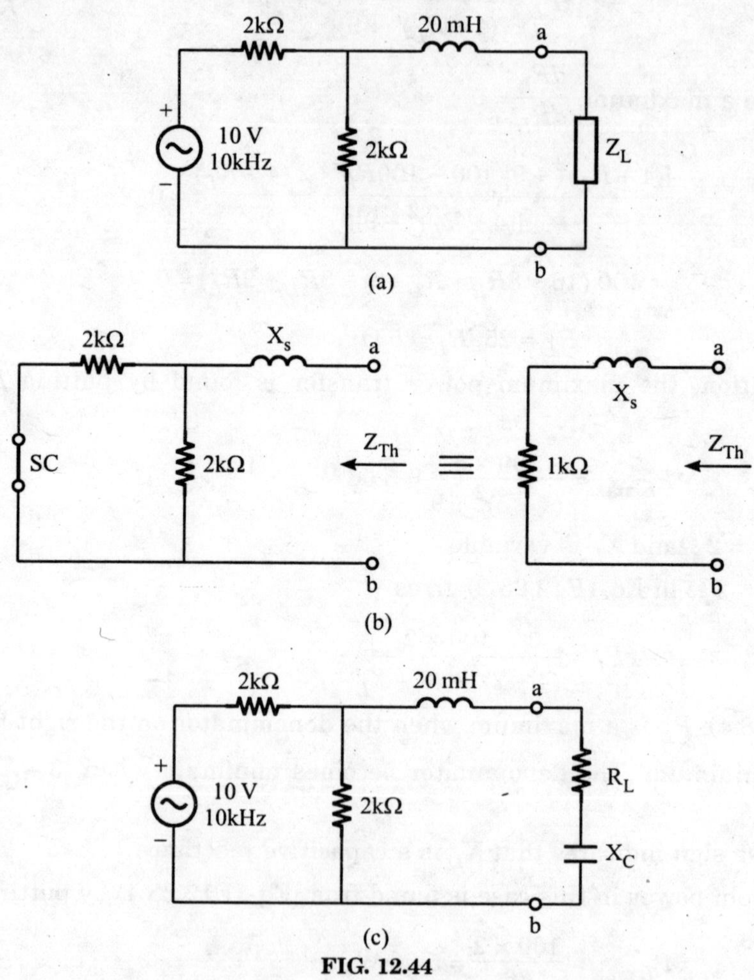

FIG. 12.44

$$X_s = 2\pi fL = 2\pi\,(10 \times 10^3)\,(20 \times 10^{-3}) = 1257\ \Omega$$

The Thevenin equivalent $\mathbf{Z}_{Th}$ of the network to the left of the terminals a–b is found from Fig. 12.44b, where $\mathbf{Z}_L$ has been removed and the voltage source of 10 V has been short circuited in Fig. 12.44a. From Fig. 12.44b

$$\mathbf{Z}_{Th} = R_s + jX_s = (1000 + j1257)\ \Omega$$

For maximum power transfer, $\mathbf{Z}_L$ is the conjugate of $\mathbf{Z}_{Th}$

therefore, $$\mathbf{Z}_L = \mathbf{Z}_{Th}^* = (1000 - j1257)\ \Omega$$

Therefore, the load consists of a resistance R_L of 1000 Ω and capacitive reactance X_C of 1257 Ω at 10 kHz.

$$\mathbf{Z}_L = R_L - jX_C,\ X_C = \frac{1}{2\pi fC}$$

$$C = \frac{1}{2\pi f X_C} = \frac{1}{2\pi \times 10 \times 10^3 \times 1257} = 1.266 \times 10^{-8}\ \text{F}$$

The complete circuit for maximum power transfer is shown in Fig. 12.44c.

■ **EXAMPLE 12.30** Determine the load $\mathbf{Z}_L$ in Fig. 12.45a for maximum power transfer. Also calculate the value of this power.

SOLUTION :

Determination of Thevenin impedance $\mathbf{Z}_{Th}$

First we remove the load $\mathbf{Z}_L$ and determine the Thevenin equivalent network. The Thevenin impedance is found by looking back from the load terminals a–b when the voltage source has been short circuited as shown in Fig. 12.45b.

From Fig. 12.45b,

$$\mathbf{Z}_{Th} = \mathbf{Z}_3 + \mathbf{Z}_1 \ ||\ \mathbf{Z}_2 = \mathbf{Z}_3 + \frac{\mathbf{Z}_1\,\mathbf{Z}_2}{\mathbf{Z}_1 + \mathbf{Z}_2}$$

$$= j5 + \frac{(j3)\,(4)}{j3 + 4} = \frac{15 - j8}{4 + j3}$$

$$= \frac{17\ \angle -28^\circ}{5\ \angle 36.87^\circ} = 3.4\ \angle -64.87^\circ$$

$$= (1.44 - j3.08)\ \Omega$$

When the load $\mathbf{Z}_L$ is removed, the current through the impedance $\mathbf{Z}_3$ is zero because the circuit is open.

The current $\mathbf{I}_1$ in Fig. 12.45c is given by

$$\mathbf{I}_1 = \frac{\mathbf{V}_s}{\mathbf{Z}_1 + \mathbf{Z}_2} = \frac{30\ \angle 20^\circ}{j3 + 4} = \frac{30\ \angle 20^\circ}{5\ \angle 36.87^\circ}$$

$$= 6\ \angle -16.87^\circ\ \text{A}$$

Determination of Thevenin voltage $\mathbf{V}_{Th}$

The Thevenin voltage $\mathbf{V}_{Th}$ is the open-circuit voltage across the terminals a–b, that is, the voltage across a–b when $\mathbf{Z}_L$ is removed. By KVL in the right-hand mesh of Fig. 12.45c,

$$\mathbf{Z}_2\mathbf{I}_1 - \mathbf{Z}_3 \times 0 - \mathbf{V}_{Th} = 0$$

therefore, $$\mathbf{V}_{Th} = \mathbf{Z}_2\,\mathbf{I}_1 = 4 \times 6\ \angle -16.87^\circ$$

$$= 24\ \angle -16.87^\circ\ \text{V}$$

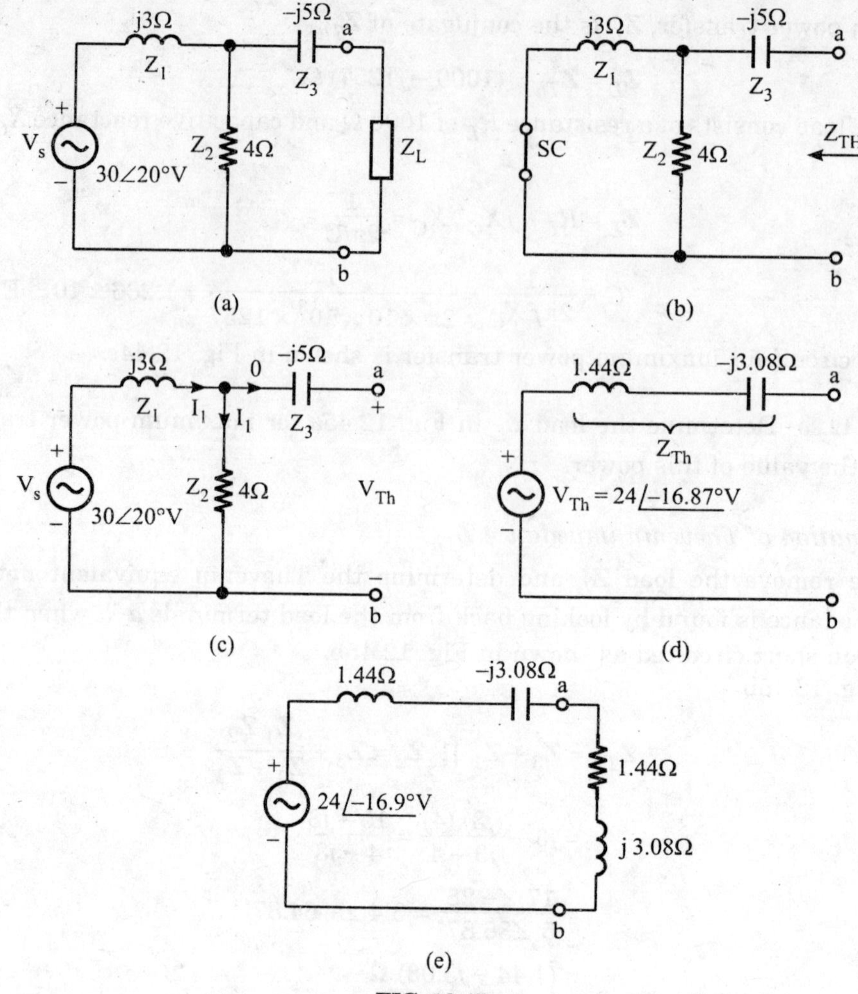

FIG. 12.45

The Thevenin equivalent network is shown in Fig. 12.45d.

According to maximum power transfer theorem, the load impedance $\mathbf{Z}_L$ is given by

$$\mathbf{Z}_L = \mathbf{Z}_{Th}^* = (1.44 - j3.08)^* = (1.44 + j3.08)\ \Omega$$

The positive reactive term in $\mathbf{Z}_L$ indicates that the reactance is inductive. Figure 12.45e shows the Thevenin equivalent network along with the load.

The maximum power

$$P_{L\,\max} = \frac{|\,V_{Th}\,|^2}{4R_L} = \frac{(24)^2}{4 \times 1.44} = 100\ \text{W}$$

■ **EXAMPLE 12.31** Determine the maximum power transferred to a pure resistive load connected across the terminals a–b of the network shown in Fig. 12.46a.

SOLUTION : In order to use nodal analysis, we transform the voltage source into a current source. The given network is redrawn in Fig. 12.46b where the admittance values of the elements are shown. Let the node voltages be $\mathbf{V}_1$, $\mathbf{V}_2$ and $\mathbf{V}_3$.

Determination of Thevenin voltage $\mathbf{V}_{Th}$

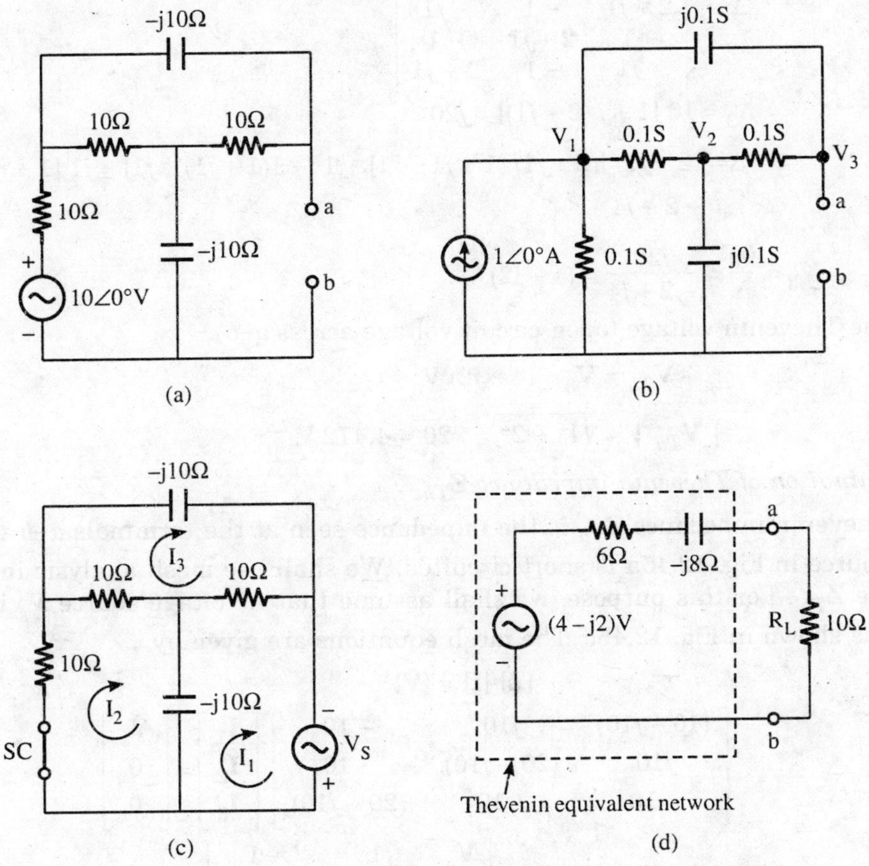

FIG. 12.46

In Fig. 12.46b, the node-voltage equations are given by

$$[\mathbf{Y}]\,[\mathbf{V}] = [\mathbf{I}]$$

$$\begin{bmatrix} (0.1+0.1+j0.1) & -0.1 & -j0.1 \\ -0.1 & (0.1+0.1+j0.1) & -j0.1 \\ -j0.1 & -0.1 & (0.1+j0.1) \end{bmatrix} \begin{bmatrix} \mathbf{V}_1 \\ \mathbf{V}_2 \\ \mathbf{V}_3 \end{bmatrix} = \begin{bmatrix} 1 \\ 0 \\ 0 \end{bmatrix}$$

or

$$\begin{bmatrix} (2+j1) & -1 & -j1 \\ -1 & (2+j1) & -1 \\ -j1 & -1 & (1+j1) \end{bmatrix} \begin{bmatrix} \mathbf{V}_1 \\ \mathbf{V}_2 \\ \mathbf{V}_3 \end{bmatrix} = \begin{bmatrix} 10 \\ 0 \\ 0 \end{bmatrix}$$

$$\mathbf{V}_3 = \frac{\Delta_3}{\Delta} = \frac{\begin{vmatrix} 2+j1 & -1 & 10 \\ -1 & 2+j1 & 0 \\ -j1 & -1 & 0 \end{vmatrix}}{\begin{vmatrix} 2+j1 & -1 & -j1 \\ -1 & 2+j1 & -1 \\ -j1 & -1 & 1+j1 \end{vmatrix}}$$

$$\Delta_3 = 10\,[1 + j\,(2 + j1)] = j20$$

$$\Delta = (2+j1)\,[(2+j1)\,(1+j1) - 1] + 1\,[-1(1+j1) - j1] - j1\,[1 + j1\,(2+j1)]$$

$$= -2 + j4$$

therefore, $\mathbf{V}_3 = \dfrac{\Delta_3}{\Delta} = \dfrac{j20}{-2+j4} = (4 - j2)\text{ V}$

But $\mathbf{V}_3$ is the Thevenin voltage (open-circuit voltage across a–b).

therefore, $\mathbf{V}_{Th} = \mathbf{V}_3 = (4 - j2)\text{ V}$

$$|\,\mathbf{V}_{Th}\,| = \sqrt{4^2 + 2^2} = \sqrt{20} = 4.472\text{ V}$$

Determination of Thevenin impedance $\mathbf{Z}_{Th}$

The Thevenin impedance $\mathbf{Z}_{Th}$ is the impedance seen at the terminals a–b when the 10 V voltage source in Fig. 12.46a is short circuited. We shall use mesh analysis in Fig. 12.46c to determine $\mathbf{Z}_{Th}$. For this purpose, we shall assume that a voltage source $\mathbf{V}_s$ is connected across a–b as shown in Fig. 12.46c. The mesh equations are given by

$$[\mathbf{Z}]\,[\mathbf{I}] = [\mathbf{V}]$$

$$\begin{bmatrix} (10-j10) & j10 & -10 \\ j10 & (20-j10) & -10 \\ -10 & -10 & (20-j10) \end{bmatrix} \begin{bmatrix} \mathbf{I}_1 \\ \mathbf{I}_2 \\ \mathbf{I}_3 \end{bmatrix} = \begin{bmatrix} \mathbf{V}_s \\ 0 \\ 0 \end{bmatrix}$$

$$\mathbf{I}_1 = \frac{\Delta_1}{\Delta} = \frac{(10)^2 \begin{vmatrix} \mathbf{V}_s & j1 & -1 \\ 0 & (2-j1) & -1 \\ 0 & -1 & (2-j1) \end{vmatrix}}{(10)^3 \begin{vmatrix} (1-j1) & j1 & -1 \\ j1 & (2-j1) & -1 \\ -1 & -1 & 2-j1 \end{vmatrix}}$$

$$\mathbf{I}_1 = \frac{\mathbf{V}_s\,(2-j4)}{-10\,(2+j4)}$$

$$\mathbf{Z}_{Th} = \mathbf{Z}_{ab} = \frac{\mathbf{V}_s}{\mathbf{I}_1} = \frac{-10\,(2+j4)}{2-j4} = (6 - j8)\ \Omega$$

$$|\,\mathbf{Z}_{Th}\,| = \sqrt{6^2 + 8^2} = 10\ \Omega$$

Also, $\mathbf{Z}_{Th} = R_{Th} - jX_{Th} = (6 - j8)\ \Omega$

therefore, $R_{Th} = 6\ \Omega,\ X_{Th} = 8\ \Omega$ and $R_L = |\,Z_{Th}\,| = 10\ \Omega$

The Thevenin equivalent network along with the load R_L is shown in Fig. 12.46d.

$$P_{L\,max} = \frac{V_{Th}^2 R_L}{(R_{Th} + R_L)^2 + X_{Th}^2} = \frac{20 \times 10}{(6 + 10)^2 + 8^2} = 0.625 \text{ W}$$

■ **EXAMPLE 12.32** Determine the load $\mathbf{Z}_L$ for maximum power transfer in the network of Fig. 12.47a.

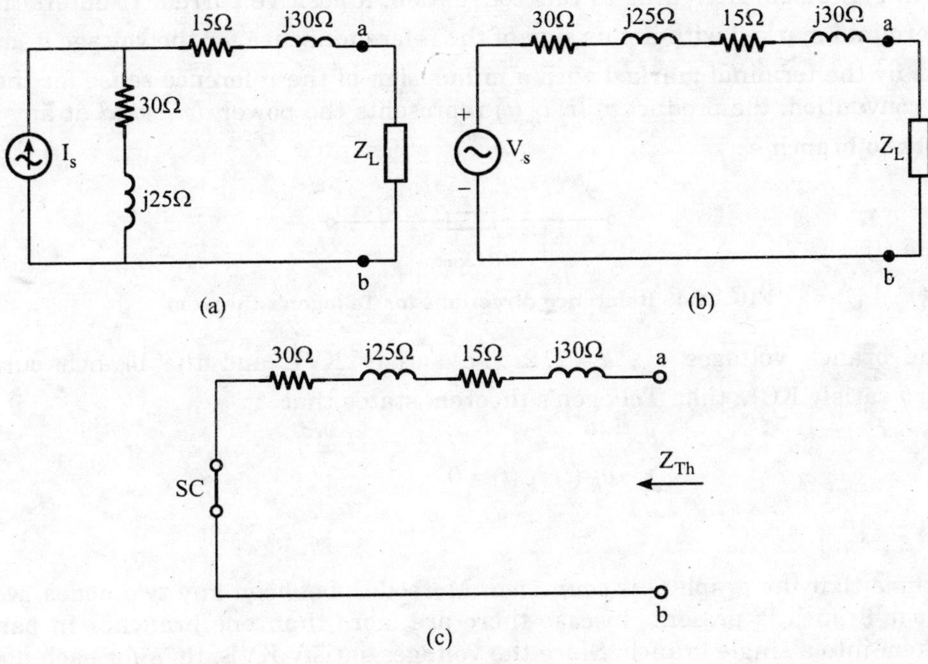

FIG. 12.47

SOLUTION : The current source in Fig. 12.47a is transformed to its equivalent voltage source as shown in Fig. 12.47b, where $\mathbf{V}_s = (30 + j25)\,\mathbf{I}_s$. In order to determine Thevenin equivalent impedance $\mathbf{Z}_{Th}$, the voltage source in Fig. 12.47b is short circuited.

The Thevenin equivalent impedance $\mathbf{Z}_{Th}$ is the impedance between the terminals a–b when they are open circuited and the voltage source is short circuited as shown in Fig. 12.47 c. From this figure,

$$\mathbf{Z}_{Th} = (30 + j25) + (15 + j30) = (45 + j55)\ \Omega$$

For maximum power transfer,

$$\mathbf{Z}_L = \mathbf{Z}_{Th}^* = (45 + j55)^*$$

$$= 45 - j55 = 71.06 \angle -50.71^\circ\ \Omega$$

12.17 TELLEGEN'S THEOREM

Tellegen's theorem is one of most general theorems in circuit theory. It applies to *any* network made up of *lumped* two-terminal elements, regardless of their nature. The elements may be linear or nonlinear, passive or active, time invariant or time varying. The circuit may contain independent or dependent sources. The theorem is valid both in the time domain and s domain.

Consider an arbitrary network N made up of lumped 2-terminal elements. The network N has n_i nodes and b branches. Let G be the oriented graph of N. Let v_k and i_k deonte the voltage across and current through branch k for $k = 1, 2, \ldots, b$. For the sake of convenience we shall choose *associated reference directions* for the branch voltages v_k and branch currents i_k as shown in Fig. 12.29. According to this convention, a positive current i_k enters the branch k by the terminal marked with a plus sign of the reference sense for the voltage v_k and leaves the branch by the terminal marked with a minus sign of the reference sense for the voltage. With this convention, the product $v_k(t)\, i_k(t)$ represents the power *delivered* at any time t by the network to branch k.

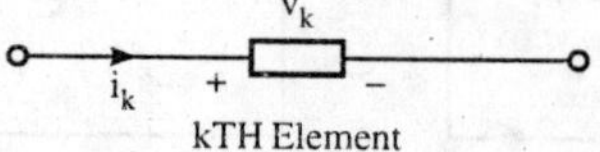

FIG. 12.48 Reference directions for Tellegen's theorem

If the branch voltages v_k, $k = 1, 2, \ldots b$ satisfy KVL and the branch currents i_k, $k = 1, 2, \ldots b$ satisfy KCL, then Tellegen's theorem states that

$$\sum_{k=1}^{b} v_k(t)\, i_k(t) = 0$$

Proof

Let us assume that the graph G is *connected.* Moreover, between any two nodes, we assume that only *one* branch is present. In case there are more than one branches in parallel, we combine them into a single branch. Since the voltages satisfy KVL, then for each node in the network we can specify a voltage with respect to some fixed reference or datum node. Let the nodes be numbered $1, 2, \ldots, n_t$. Suppose that node 1 is taken as the reference node, then its voltage is zero. That is, $e_1 = 0$. It should be noted that once the branch voltages $v_1, v_2, \ldots, v_b$ are selected, then by KVL the node voltages $e_1, e_2, \ldots e_x, \ldots, e_y, \ldots$ are uniquely specified. Let e_x and e_y be the voltages of nodes x and y with respect to node 1. The branch k is connected between the nodes x and y with respect to node 1. The branch k is connected between the nodes x and y (Fig. 12.49)

Therefore,

$$v_k = e_x - e_y$$

Node x — Branch k — Node y
e_x i_{xy} e_y

FIG. 12.49

The branch current is i_{xy} flowing from x to y. Then we have

$$v_k\, i_k = (e_x - e_y)\, i_{xy} \tag{12.17.1}$$

This relation can also be written as

$$v_k\, i_k = (e_y - e_x)\, i_{yx} \tag{12.17.2}$$

where i_{yx} is the current from node y to node x. Obviously, $i_{xy} = -i_{yx}$. Addition of Eqs. (12.17 .1) and (12.17.2) gives

$$v_k\, i_k = \tfrac{1}{2}\,[(e_x - e_y)\, i_{xy} + (e_y - e_x)\, i_{yx}] \tag{12.17.3}$$

Such an equation can be written for every branch of the network. Summing over all the b branches of graph G we get

$$\sum_{k=1}^{b} v_k\, i_k = \frac{1}{2} \sum_{x=1}^{nt} \sum_{y=1}^{nt} (e_x - e_y)\, i_{xy} \tag{12.17.4}$$

which covers all the nodes of the graph. In case a particular branch is not connected, then its current $i_{xy} = 0$ and $i_{yx} = 0$. Equation (12.17.4) can be written as

$$\sum_{k=1}^{b} v_k\, i_k = \frac{1}{2} \sum_{x=1}^{nt} e_x \left(\sum_{y=1}^{nt} i_{xy} \right) - \frac{1}{2} \sum_{y=1}^{nt} e_y \left(\sum_{x=1}^{nt} i_{xy} \right) \tag{12.17.5}$$

According to KCL, the algebraic sum of currents at each node is equal to zero. Hence

$$\sum_{x=1}^{nt} i_{xy} = 0$$

and

$$\sum_{y=1}^{nt} i_{xy} = 0$$

Therefore Eq. (12.17.5) becomes

$$\sum_{k=1}^{b} v_k\, i_k = 0 \tag{12.17.6}$$

If the graph G is not connected, Eq. (12.17.6) still holds good for each subgraph and hence it is valid for the complete graph. This proves Tellegen's theorem.

Alternative Proof

Let **v** and **i** denote column vectors whose elements are branch voltages and branch currents of network N respectively.

$$\mathbf{v} = \begin{bmatrix} v_1 \\ v_2 \\ . \\ . \\ v_b \end{bmatrix} \text{ and } \mathbf{i} = \begin{bmatrix} i_1 \\ i_2 \\ . \\ . \\ i_b \end{bmatrix}$$

Therefore,

$$\mathbf{v} = [v_1, v_2, \ldots, v_b]^{\tau} \tag{12.17.7}$$

$$\mathbf{i} = [i_1, i_2, \ldots, i_b]^{\tau} \tag{12.17.8}$$

where τ denotes the transpose.

From Eq. (13.33.1) we have

$$\mathbf{v} = \mathbf{A}^{\tau}\, \mathbf{e} \tag{12.17.9}$$

Taking the transpose of both the sides of Eq. (12.17.9) we get

$$\mathbf{v}^{\tau} = (\mathbf{A}^{\tau}\,\mathbf{e}) = \mathbf{e}^{\tau}\,(\mathbf{A}^{\tau})^{\tau} = \mathbf{e}^{\tau}\,\mathbf{A} \tag{12.17.10}$$

Post multiplying Eq. (12.17.10) with **i** we get

$$\mathbf{v}^{\tau}\,\mathbf{i} = \mathbf{e}^{\tau}\,(\mathbf{A}\,\mathbf{i}) \tag{12.17.11}$$

The matrix representation of KCL is given by

$$\mathbf{A}\,\mathbf{i} = 0 \tag{12.17.12}$$

Combination of Eqs. (12.17.11) and (12.17.12) gives

$$v^{\tau}\,\mathbf{i} = 0 \tag{12.17.13}$$

$$\sum_{k=1}^{b} v_k\,\mathbf{i}_k = 0 \tag{12.17.14}$$

12.18 CONSERVATION OF POWER

Consider an arbitrary network. By applying Tellegen's theorem to the branch voltages and currents at the same instant we have

$$\sum_{k=1}^{b} v_k\,(t)\,i_k\,(t) = 0 \text{ for all } t$$

Since $v_k\,(t)\,i_k\,(t)$ is the power delivered at time by the network to branch k, Tellegen's theorem can be interpreted as follows :

At any time t the sum of the power delivered to each branch of the network is zero. This is the **law of conservation of power**. For a network containing independent sources and other passive elements we conclude that the *sum of power delivered by the independent sources to the network is equal to the sum of the power absorbed by all the passive elements of the network.* Hence for lumped circuits KCL and KVL imply law of conservation of power.

■ **EXAMPLE 12.33** Verify Tellegen's theorem for the network shown in Fig. 12.50 in which the following sets of branch voltages and branch currents exist.

Branch voltages in volts :

$$v_1 = 10,\ v_2 = 6,\ v_3 = 4,\ v_3 = 3,$$

$$v_5 = 1,\ v_6 = 9$$

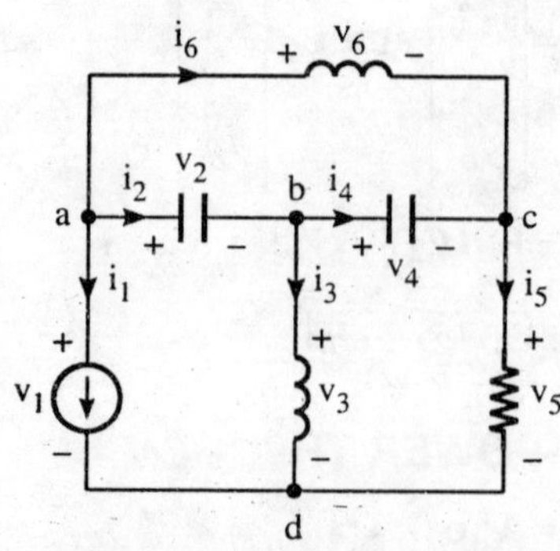

FIG. 12.50

Branch curents in amperes :

$$i_1 = 1,\ i_2 = 2,\ i_3 = 4$$

$$i_4 = -2,\ i_5 = -5,\ i_6 = -3$$

SOLUTION : **Before applying Tellegen's theorem we shall verify that the branch voltages satisfy KVL and the branch currents satisfy KCL.**

Verification of KVL for branch voltages

For the mesh *abda*

$$-v_2 - v_3 + v_1 = 0$$

$$-6 - 4 + 10 = 0$$

For the mesh *bcdb*

$$-v_4 - v_5 + v_3 = 0$$

$$-3 - 1 + 4 = 0$$

For the mesh *abca*

$$-v_2 - v_4 + v_6 = -6 - 3 + 9 = 0$$

Hence the branch voltages satisfy KVL.

Verification of KCL for branch currents

Take the incoming currents positive and outgoing currents negative.

For the node *a*

$$-(i_1 + i_2 + i_6) = -(1 + 2 - 3) = 0$$

For the node *b*

$$i_2 - i_3 - i_4 = 2 - 4 + 2 = 0$$

For the node *c*

$$i_4 - i_5 + i_6 = -2 + 5 - 3 = 0$$

For the node *d*

$$i_1 + i_3 + i_5 = 1 + 4 - 5 = 0$$

Hence the branch currents satisfy KCL.

Verification of Tellegen's theorem

$$\sum_{k=1}^{b} v_k\, i_k = \sum_{k=1}^{6} v_k\, i_k$$

$$= v_1 i_1 + v_2 i_2 + v_3 i_3 + v_4 i_4 + v_5 i_5 + v_6 i_6$$

$$= 10 \times 1 + 6 \times 2 + 4 \times 4 + 3\,(-2) + 1\,(-5) + 9\,(-3) = 0$$

Since $\sum_{k=1}^{b} v_k\, i_k = 0,$

the Tellegen's therorem is verified.

12.19 TELLEGEN'S QUASI-POWER THEOREM

Consider two networks *N* and *N'* made up of lumped two-terminal elements whose graphs are

identical. Hence both the graphs have the same number of node and branches; corresponding branches are incident on corresponding nodes. The node and branch numbering and the reference directions are also the same for the two networks. Let v_k, i_k ($k = 1, 2, ..., b$) be the branch voltages and branch currents of the network N and v_k', i'_k be the corresponding branch voltages and branch currents of the network N'.

Since the branch voltages v_k' s and v_k's satisfy the *same* set of KVL equations and since the branch currents i_k's and i_k' s satisfy the *same* set of KCL equations,then

(*a*)
$$\sum_{k=1}^{b} v_k \, i_k' = 0 \tag{12.19.1}$$

or equivalently,

$$\mathbf{v}^{\tau} \, \mathbf{i}' = 0 \tag{12.19.2}$$

and (*b*)
$$\sum_{k=1}^{b} v_k' \, i_k = 0 \tag{12.19.3}$$

or equivalently,

$$\mathbf{v}'^{\tau} \, \mathbf{i} = 0 \tag{12.19.4}$$

where $\mathbf{v}$ = branch voltage vector of N

$\mathbf{i}$ = branch current vector of N

$\mathbf{v}'$ = branch voltage vector of N'

$\mathbf{i}'$ = branch current vector of N'

The statements given by Eqs. (12.19.1) to (12.19.4) are called **Tellegen's quasi-power theorem**.

Proof

For the network N

$$\mathbf{v} = \mathbf{A}^{\tau} \, \mathbf{e} \tag{12.19.5}$$

Taking the transpose of both the sides of Eq. (12.19.5) we get

$$\mathbf{v}^{\tau} = \mathbf{e}^{\tau} \, \mathbf{A} \tag{12.19.6}$$

where $\mathbf{A}$ is the incidence matrix of the network N. Post multiplying both the sides of Eq. (12.19.6) by $\mathbf{i}'$ we get

$$\mathbf{v}^{\tau} \, \mathbf{i} = \mathbf{e}^{\tau} \, \mathbf{A} \, \mathbf{i}' \tag{12.19.7}$$

If $\mathbf{A}'$ denotes the incidence matrix of the network N', then

$$\mathbf{A}' = \mathbf{A} \tag{12.19.8}$$

since the graphs of N and N' are identical. Hence from Eqs. (12.19.7) and (12.19.8) we get

$$\mathbf{v}^{\tau} \, \mathbf{i}' = \mathbf{e}^{\tau} \, (\mathbf{A}' \, \mathbf{i}') \tag{12.19.9}$$

But
$$\mathbf{A}' \, \mathbf{i}' = 0 \tag{12.19.10}$$

Therefore Eq. (12.19.9) becomes

$$\mathbf{v}^{\tau} \, \mathbf{i}' = 0 \tag{12.19.11}$$

or equivalently,

$$\sum_{k=1}^{b} v_k i_k' = 0 \tag{12.19.12}$$

The second part of the theorem can be proved in a similar manner. For the network N'

$$\mathbf{v}' = (\mathbf{A}')^{\tau}\, \mathbf{e}'$$

therefore, $$(\mathbf{v}')^{\tau} = (\mathbf{e}')^{\tau}\, \mathbf{A}$$

$$(\mathbf{v}')^{\tau}\, \mathbf{i} = (\mathbf{e}')^{\tau}\, \mathbf{A}'\, \mathbf{i} = (\mathbf{e}')^{\tau}\, (\mathbf{A}\, \mathbf{i}) = 0$$

therefore, $$(\mathbf{v}')^{\tau}\, \mathbf{i} = 0 \tag{12.19.13}$$

or equivalently

$$\sum_{k=1}^{b} v_k' i_k = 0 \tag{12.19.14}$$

The left-hand side of Eq. (12.19.12) has the dimensions of watts but it does not represent physical power since v_k and i_k' exist in two different circuits. This equation shows a mathematical relationship between the branch voltages of one circuit and the branch currents of another circuit with the same *topology*. The two circuits may have different elements. For the two networks, no constraints are put on the element types and their values. For example, consider the two circuits shown in Fig. 12.51 with the *same topology* but having different elements. Let the branch voltages of the network of Fig. 12.51a denoted by $v_1, v_2, \ldots, v_6$ and the branch currents of the network of Fig. 12.51b be denoted by $i'_1, i'_2 \ldots, i'_6$. Then by Eq. (12.19.12) we can write

$$v_1 i'_1 + v_2 i'_2 + v_3 i'_3 + v_4 i'_4 + v_5 i'_5 + v_6 i'_6 = 0$$

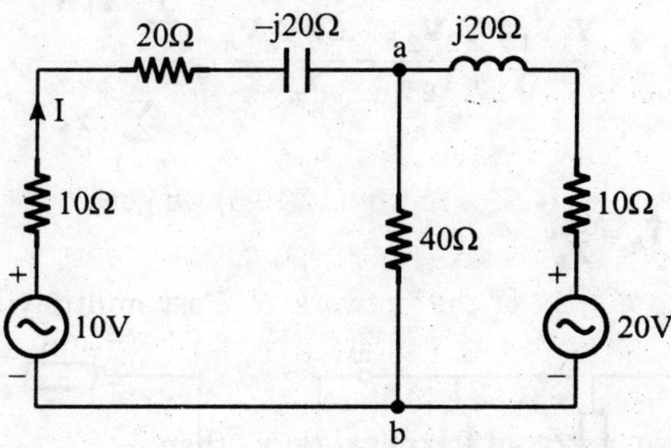

FIG. 12.51 Two networks with the same topology

12.20 APPLICATION OF TELLEGEN'S THEOREM TO PHASORS AND LAPLACE TRANSFORMS

We have stated Tellegen's theorem in terms of instantaneous voltages and currents. This theorem is based only on the requirements that the branch voltages satisfy KVL and the branch currents satisfy KCL. Since the phasors of branch voltages and branch currents also satisfy KVL and KCL, this theorem also holds for the phasors of branch voltages and currents. Hence we can write

$$\sum_{k=1}^{b} \mathbf{V}_k \mathbf{I}_k = 0 \tag{12.20.1}$$

where $\mathbf{V}_k$ = phasor branch r.m.s. voltages

$\mathbf{I}_k$ = phasor branch r.m.s. currrents

Similarly, the Laplace transforms of the branch voltages and branch currents also satisfy Kirchhoff's laws. Hence Tellegen's theorem also holds for the Laplace transforms of the branch voltages and branch currents. We can write

$$\sum_{k=1}^{b} \mathbf{V}_k(s)\, \mathbf{I}_k(s) = 0 \tag{12.20.2}$$

where $\mathbf{V}_k(s) = £\, v_k(t)$

$\mathbf{I}_k(s) = £\, i_k(t)$

12.21 MILLMAN'S THEOREM

Consider the circuit in Fig. 12.52a where several voltage sources $\mathbf{V}_1, \mathbf{V}_2 \ldots, \mathbf{V}_n$ having internal impedances $\mathbf{Z}_1, \mathbf{Z}_2, \ldots \mathbf{Z}_n$ are connected in **parallel** supplying a common load $\mathbf{Z}_L$. Millman theorem states that this arrangement can be replaced by a single voltage source $\mathbf{V}_{eq}$ in series with an impedance $\mathbf{Z}_{eq}$ as shown in Fig. 12.52b where

$$\mathbf{V}_{eq} = \frac{\mathbf{Y}_1\mathbf{V}_1 + \mathbf{Y}_2\mathbf{V}_2 + \ldots + \mathbf{Y}_n\mathbf{V}_n}{\mathbf{Y}_1 + \mathbf{Y}_2 + \ldots + \mathbf{Y}_n} = \frac{\sum_{k=1}^{n} \mathbf{Y}_k \mathbf{V}_k}{\sum_{k=1}^{n} \mathbf{Y}_k} \tag{12.21.1}$$

and

$$\mathbf{Y}_k = \frac{1}{\mathbf{Z}_k} \tag{12.21.2}$$

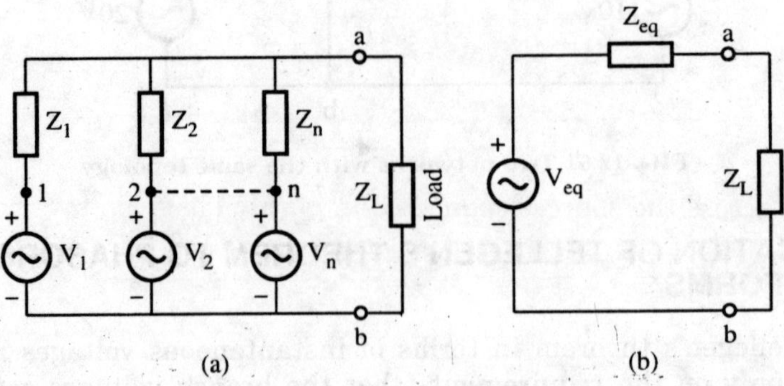

FIG. 12.52

Proof

By source transformation a voltage source **V** with series impedance **Z** may be replaced by an equivalent current source **I** in parallel with an admittance **Y** where **Y** = 1/**Z** and **I** = **YV**. Accordingly, the network of Fig. 12.52a may be replaced by an equivalent network consisting of n current sources in parallel as shown in Fig. 12.53. The source currents are given by

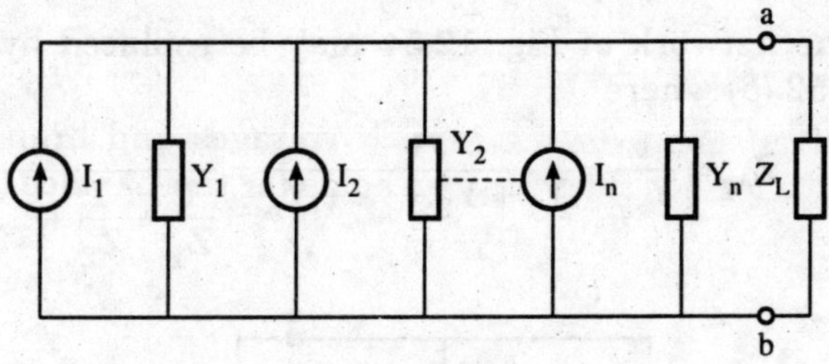

FIG. 12.53

$$\mathbf{I}_1 = \mathbf{Y}_1\mathbf{V}_1,\ \mathbf{I}_2 = \mathbf{Y}_2\mathbf{V}_2, \ldots, \mathbf{I}_n = \mathbf{Y}_n\mathbf{V}_n$$

By KCL the total current towards $\mathbf{Z}_L$ is

$$\mathbf{I}_{eq} = \mathbf{I}_1 + \mathbf{I}_2 + \ldots + \mathbf{I}_n \tag{12.21.3}$$

Since all the source admittances in Fig. 12.53 are in parallel, the total equivalent admittance is

$$\mathbf{Y}_{eq} = \mathbf{Y}_1 + \mathbf{Y}_2 + \ldots + \mathbf{Y}_n = \sum_{k=1}^{n} \mathbf{Y}_k \tag{12.21.4}$$

Thus, the current sources in Fig. 12.53 may be replaced by a single current source with a current $\mathbf{I}_{eq}$ in parallel with a single admittance $\mathbf{Y}_{eq}$, where $\mathbf{I}_{eq}$ and $\mathbf{Y}_{eq}$ are given by Eqs. (12.21.3) and (12.21.4) respectively.

Eq. (12.21.3) can be written as

$$\mathbf{I}_{eq} = \frac{\mathbf{V}_1}{\mathbf{Z}_1} + \frac{\mathbf{V}_2}{\mathbf{Z}_2} + \ldots + \frac{\mathbf{V}_n}{\mathbf{Z}_n} \tag{12.21.5}$$

$$= \mathbf{Y}_1\mathbf{V}_1 + \mathbf{Y}_2\mathbf{V}_2 + \ldots + \mathbf{Y}_n\mathbf{V}_n = \sum_{k=1}^{n} \mathbf{Y}_k\mathbf{V}_k \tag{12.21.6}$$

Hence the voltage across the sources connected in parallel is given by

$$\mathbf{V}_{ab} = \mathbf{V}_{eq} = \frac{\mathbf{I}_{eq}}{\mathbf{Y}_{eq}} = \frac{\mathbf{Y}_1\mathbf{V}_1 + \mathbf{Y}_2\mathbf{V}_2 + \ldots + \mathbf{Y}_n\mathbf{V}_n}{\mathbf{Y}_1 + \mathbf{Y}_2 + \ldots + \mathbf{Y}_n}$$

$$= \frac{\sum_{k=1}^{n} \mathbf{Y}_k \mathbf{V}_k}{\sum_{k=1}^{n} \mathbf{Y}_k} \tag{12.21.7}$$

The current source of the network of Fig. 12.54 may be replaced by an equivalent voltage source shown in Fig. 12.52 (*b*) where

$$\mathbf{Z}_{eq} = \frac{1}{\mathbf{Y}_{eq}} = \frac{1}{\mathbf{Y}_1 + \mathbf{Y}_2 + \ldots + \mathbf{Y}_n} = \frac{1}{\frac{1}{\mathbf{Z}_1} + \frac{1}{\mathbf{Z}_2} + \ldots + \frac{1}{\mathbf{Z}_n}} \tag{12.21.8}$$

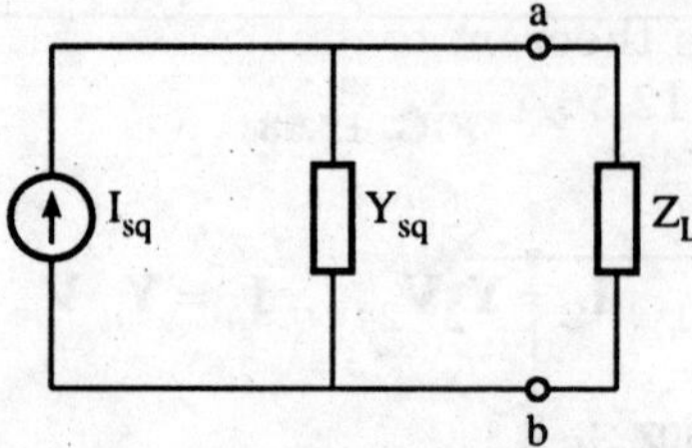

FIG. 12.54 Equivalent current source derived from Millman's theorem

Thus, Millman's theorem enables a number of voltage (or current) sources to be combined into a single equivalent voltage (or current) source. This theorem can also be applied to a combination of voltage and current sources. It is to be noted that *this theorem applies only to the sources connected directly in parallel.* It does not apply where there are impedances between the sources.

From Fig. 12.52a, it is seen that the impedances $\mathbf{Z}_1, \mathbf{Z}_2, \ldots \mathbf{Z}_n$ terminate at a common point *a* and the voltages from another point *b* to the free ends 1, 2, ..., *n* of these impedances are $\mathbf{V}_1, \mathbf{V}_2, \ldots, \mathbf{V}_n$ respectively. The voltage $\mathbf{V}_{ab}$ is given by Eq. (12.21.7).

EXAMPLE 12.34 Use Millman's theorem to determine the current in $\mathbf{Z}_2$ in Fig. 12.55.

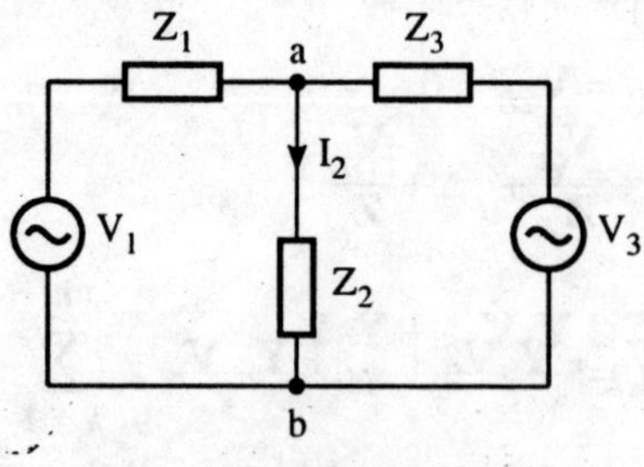

FIG. 12.55

SOLUTION : The given circuit may be considered to consist of three sources in parallel. The first is a voltage source $\mathbf{V}_1$ with an internal impedance $\mathbf{Z}_1$, the second a voltage source of zero volt with an internal impedance $\mathbf{Z}_2$, and the third a voltage source $\mathbf{V}_3$ having an internal impedance $\mathbf{Z}_3$, By Millman's theorem

$$\mathbf{V}_{ab} = \frac{\mathbf{Y}_1\mathbf{V}_1 + \mathbf{Y}_2\mathbf{V}_2 + \mathbf{Y}_3\mathbf{V}_3}{\mathbf{Y}_1 + \mathbf{Y}_2 + \mathbf{Y}_3}$$

$$= \frac{(\mathbf{V}_1/\mathbf{Z}_1) + (\mathbf{V}_2/\mathbf{Z}_2) + (\mathbf{V}_3/\mathbf{Z}_3)}{\frac{1}{\mathbf{Z}_1} + \frac{1}{\mathbf{Z}_2} + \frac{1}{\mathbf{Z}_3}}$$

Here $\mathbf{V}_2 = 0$, therefore

$$\mathbf{V}_{ab} = \frac{(\mathbf{V}_1/\mathbf{Z}_1) + (\mathbf{V}_3/\mathbf{Z}_3)}{\frac{1}{\mathbf{Z}_1} + \frac{1}{\mathbf{Z}_2} + \frac{1}{\mathbf{Z}_3}} \cdot \mathbf{I}_2 = \frac{\mathbf{V}_{ab}}{\mathbf{Z}_2}$$

■ **EXAMPLE 12.35** Use Millman's theorem to determine the voltage across the 50 Ω resistor in the network shown in Fig. 12.56.

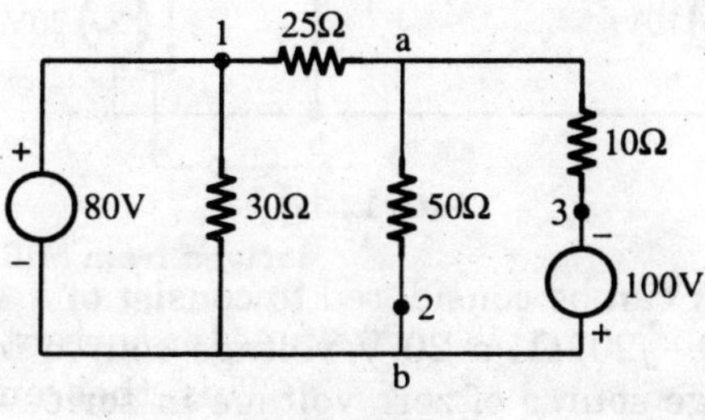

FIG. 12.56

SOLUTION : Let the two ends of the 50 Ω resistor be marked a and b. In order to determine the voltage $\mathbf{V}_{ab}$, we may consider three sources in parallel. The first is a 80 V voltage source having an internal resistance of 25 Ω, The second is zero-volt voltage source with an internal resistance of 50 Ω. The third source is a voltage source of 100 V with an internal resistance of 10 Ω. Hence we have

$$\mathbf{Z}_1 = 25\ \Omega,\ \mathbf{Z}_2 = 50\ \Omega,\ \mathbf{Z}_3 = 10\ \Omega$$

$$\mathbf{V}_1 = \mathbf{V}_{1b} = 80\text{ V}$$

$$\mathbf{V}_2 = \mathbf{V}_{2b} = 0$$

$$\mathbf{V}_3 = \mathbf{V}_{3b} = -100\text{ V}$$

The sign of $\mathbf{V}_3$ is to be noted.

By Millman's theorem,

$$\mathbf{V}_{ab} = \frac{\mathbf{Y}_1\mathbf{V}_{1b} + \mathbf{Y}_2\mathbf{V}_{2b} + \mathbf{Y}_3\mathbf{V}_{3b}}{\mathbf{Y}_1 + \mathbf{Y}_2 + \mathbf{Y}_3}$$

$$= \frac{(\mathbf{V}_1/\mathbf{Z}_1) + (\mathbf{V}_2/\mathbf{Z}_2) + (\mathbf{V}_3/\mathbf{Z}_3)}{\frac{1}{\mathbf{Z}_1} + \frac{1}{\mathbf{Z}_2} + \frac{1}{\mathbf{Z}_3}}$$

$$= \frac{(80/25) + (0/50) + (-100/10)}{(1/25) + (1/50) + (1/10)} = -42.5\text{ V}$$

Hence point a is at a lower potential than b.

Alternatively, we may consider that the three impedances $\mathbf{Z}_1$, $\mathbf{Z}_2$ and $\mathbf{Z}_3$ meet at common point a. Their other ends are marked 1, 2 and 3. Therefore, as before, $\mathbf{Z}_1 = 25\ \Omega$, $\mathbf{Z}_2 = 50\ \Omega$, $\mathbf{Z}_3 = 10\ \Omega$ and $\mathbf{V}_{1b} = 80$ V, $\mathbf{V}_{2b} = 0$, $\mathbf{V}_{3b} = -100$ V. The other steps are the same as before.

■ **EXAMPLE 12.36** Use Millman's theorem to calculate I in the network shown in Fig. 12.57.

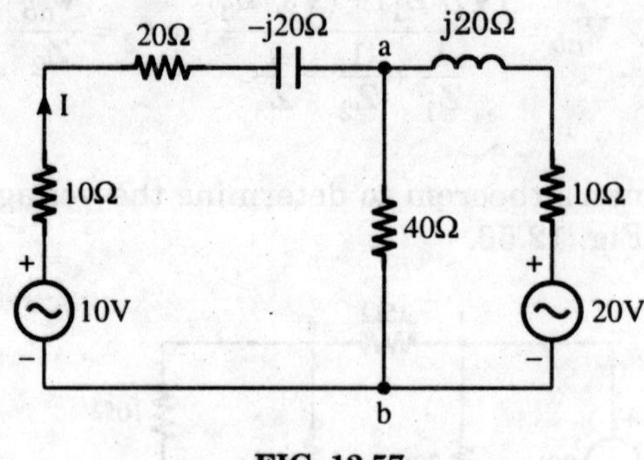

FIG. 12.57

SOLUTION: The given network can be considered to consist of a 10 V voltage source in series with an impedance of $(10 + 20 - j20)\ \Omega$, a 20 V voltage source in series with an impedance $(10 + j20)\ \Omega$, and a third voltage source of zero voltage in series with an impedance of $40\ \Omega$. By Millman's theorem

$$\mathbf{V}_{ab} = \frac{\mathbf{Y}_1\mathbf{V}_1 + \mathbf{Y}_2\mathbf{V}_2 + \mathbf{Y}_3\mathbf{V}_3}{\mathbf{Y}_1 + \mathbf{Y}_2 + \mathbf{Y}_3}$$

Here $\mathbf{V}_1 = 10,\ \mathbf{V}_2 = 20,\ \mathbf{V}_3 = 0$

$$\mathbf{Y}_1 = \frac{1}{\mathbf{Z}_1} = \frac{1}{30 - j20},\ \mathbf{Y}_2 = \frac{1}{\mathbf{Z}_2} = \frac{1}{10 + j20}$$

$$\mathbf{Y}_3 = \frac{1}{\mathbf{Z}_3} = \frac{1}{40}$$

$$\mathbf{V}_{ab} = \frac{\dfrac{10}{30 - j20} + \dfrac{20}{10 + j20} + \dfrac{0}{40}}{\dfrac{1}{30 - j20} + \dfrac{1}{10 + j20} + \dfrac{1}{40}} = \frac{280 - j80}{23 + j4}$$

By KVL in the left-hand mesh

$$10 - \mathbf{I}\,(20 - j20) - \mathbf{V}_{ab} = 0$$

$$\mathbf{I} = \frac{10 - \mathbf{V}_{ab}}{30 - j20}$$

$$= \left(10 - \frac{280 - j80}{23 + j4}\right) \div (30 - j20)$$

$$= \frac{-5 + j12}{(23 + j4)(3 - j2)} = \frac{(13 \angle 112.6^\circ)}{(23.34 \angle 9.865^\circ)(3.6 \angle -33.69^\circ)}$$

$$= 0.1547 \angle 136.424^\circ = (-0.112 + j0.106) \text{ A}$$

12.22 SUBSTITUTION THEOREM

Sometimes it is convenient to replace an impedance branch of network by another branch having different circuit components, without changing any voltage or any current in the entire network. Consider a branch k of an arbitrary network. Suppose that this branch is not coupled to any other branch of the network. That is, this branch is neither the branch of a coupled inductor nor a branch of a dependent source. Let V_k be the branch voltage and I_k the branch current.

Subsitutition theorem states that *the branch k may be substituted (replaced) by (1) an independent voltage source of value V_k without affecting the branch current I_k, or (2) a current source of value I_k without affecting the branch voltage V_k.*

The polarity of the voltage source should be the same as the branch voltage V_k. The direction of the current in the current source should be same as I_k.

Figure 12.58 gives the equivalent representations. It should be noted that unless V_k is known we cannot replace it by a voltage source. Similarly, unless I_k is known we cannot replace it by a current source. In other words, a *known* voltage in a branch can be replaced by an ideal voltage source, and a *known* current can be replaced by an ideal current source. Thus, substitution theorem cannot be applied unless solution to the problem is already known. For the application of this theorem, it is important that both the original and the substituted networks have unique solutions. This indicates that the application of substitution theorem is limited. However, this theorem applies to ***any*** network, linear or nonlinear, time-varying or time invariant. It is applicable both in the time domain and in the s domain. It is generally used to replace an impedance branch by an independent source without disturbing the branch voltages and currents in the entire network. The basic requirement is that the new substituted branch must have the *same* branch voltage and current as the original branch.

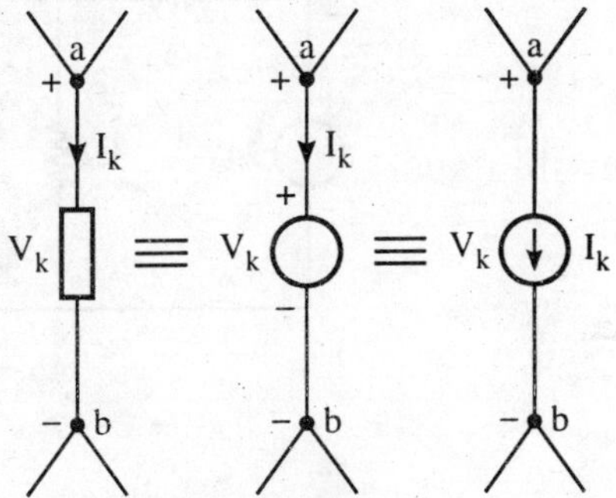

FIG. 12.58.
Illustrating substitution theorem

12.23 PROOF OF THE SUBSTITUTION THEOREM

Let the number of branches in a given network be b. It is assumed that the branch voltages, $V_1, V_2, \ldots, V_b$ satisfy KVL and the branch currents satisfy KCL. These quantities also satisfy the initial conditions and the branch equations of the network. Let us assume that the branch k is substituted (replaced) by a voltage source of value V_k. Since the branch voltage V_k and the branch current I_k of the replaced branch k remain unaltered, therefore, all the branch voltages and all the branch currents of the modified network satisfy all the conditions assumed for the original network. This proves the substitution theorem. With similar reasoning, we can prove the substitution theorem when the branch k is replaced by a current source of value I_k.

■ **EXAMPLE 12.37** For the network shown in Fig. 12.59a, substitute the branch *ab* by (*a*) a voltage source, (*b*) a current source.

SOLUTION : Total resistance of the circuit

$$= 2 + [4 \; || \; (2+2)] = 2 + 2 = 4\ \Omega$$

Total current $= \dfrac{24}{4} = 6$ A

By current-division rule, $I_{ab} = \dfrac{6}{2} = 3$ A

$\therefore$ $V_{ab} = 3 \times 2 = 6$ V

By substitution theorem, the branch *ab* may be substituted either by an independent voltage source of value 6 V with the branch curent $I_{ab} = 3$ A or by an independent current source of value 3 A with the branch voltage $V_{ab} = 6$ V. The two substitutions are shown in Fig. 12.59b and Fig. 12.59c respectively.

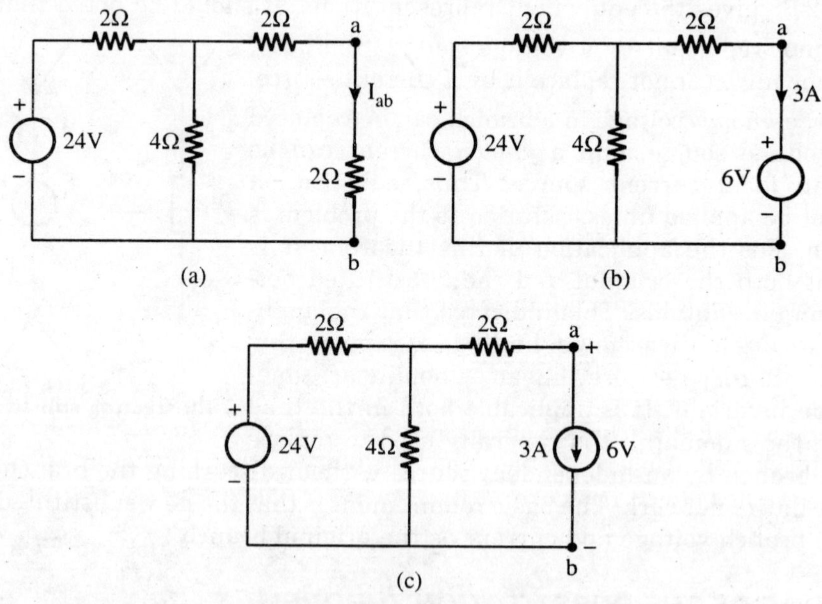

FIG. 12.59 (a) Original network (b) Modified network with the branch *ab* replaced by a voltage source (c) Modified network with the branch *ab* replaced by a current source

12.24 COMPENSATION THEOREM

Consider a linear time-invariant network as shown in Fig. 12.60a. Suppose that we are interested in determining the change of the current in the kth branch due to a small change $\Delta \mathbf{Z}$ in the impedance $\mathbf{Z}$ of the uncoupled kth branch. The given network can be replaced by its Thevenin equivalent network with the kth branch as the load as shown in Fig. 12.60b.

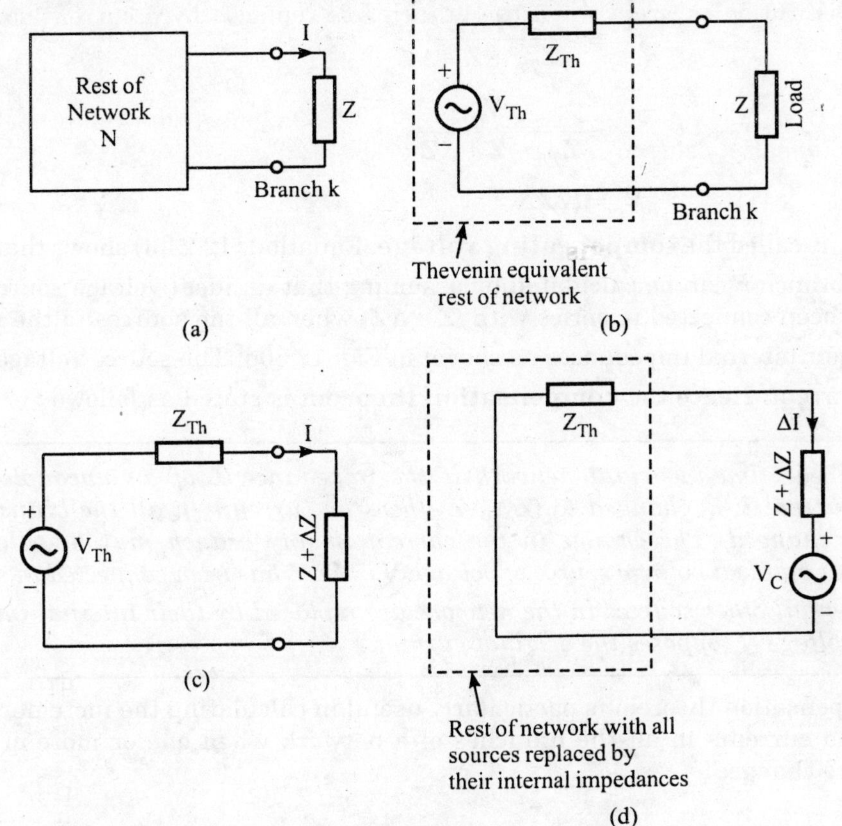

FIG. 12.60

The current through branch k is given by

$$\mathbf{I} = \frac{\mathbf{V}_{Th}}{\mathbf{Z}_{Th} + \mathbf{Z}} \tag{12.24.1}$$

Let the impedance **Z** of branch k be changed to $(\mathbf{Z} + \Delta\mathbf{Z})$ as shown in Fig. 12.60c

Since the rest of the circuit remains unchanged, the Thevenin equivalent network remains the same as before. The current in branch k is given by

$$\mathbf{I}' = \frac{\mathbf{V}_{Th}}{\mathbf{Z}_{Th} + (\mathbf{Z} + \Delta\mathbf{Z})} \tag{12.24.2}$$

Hence the change in current in branch k is

$$\Delta\mathbf{I} = \mathbf{I}' - \mathbf{I} = \frac{\mathbf{V}_{Th}}{\mathbf{Z}_{Th} + \mathbf{Z} + \Delta\mathbf{Z}} - \frac{\mathbf{V}_{Th}}{\mathbf{Z}_{Th} + \mathbf{Z}}$$

$$= \frac{\mathbf{V}_{Th}\,[\mathbf{Z}_{Th} + \mathbf{Z} - (\mathbf{Z}_{Th} + \mathbf{Z} + \Delta\mathbf{Z})]}{(\mathbf{Z}_{Th} + \mathbf{Z} + \Delta\mathbf{Z})\,(\mathbf{Z}_{Th} + \mathbf{Z})}$$

$$= -\left(\frac{\mathbf{V}_{Th}}{\mathbf{Z}_{Th} + \mathbf{Z}}\right)\frac{\Delta\mathbf{Z}}{\mathbf{Z}_{Th} + \mathbf{Z} + \Delta\mathbf{Z}}$$

$$= -\frac{\mathbf{I}\,\Delta\mathbf{Z}}{\mathbf{Z}_{Th} + \mathbf{Z} + \Delta\mathbf{Z}}$$

$$= \frac{-\mathbf{V}_c}{\mathbf{Z}_{Th} + \mathbf{Z} + \Delta\mathbf{Z}} \tag{12.24.3}$$

where $\mathbf{V}_c = \mathbf{I}\,\Delta\mathbf{Z}$ (12.24.4)

The voltage $\mathbf{V}_c$ is called the **compensating voltage**. Equation (12.24.3) shows that the change in current in branch k can be calculated by assuming that an ideal voltage source of voltage $\mathbf{V}_c = \mathbf{I}\,\Delta\mathbf{Z}$ has been connected in series with $(\mathbf{Z} + \Delta\mathbf{Z})$ when all the sources in the network are replaced by their internal impedances as shown in Fig. 12.60d. The source voltage $\mathbf{V}_c$ opposes the original current. Hence the **compensation theorem** is stated as follows :

> *In a linear time-invariant network if the impedance* **Z** *of an uncoupled branch, carrying a current* **I**, *is changed to* $(\mathbf{Z} + \Delta\mathbf{Z})$ *then the currents in all the branches of the network are changed. The change in the current in any branch may be calculated by assuming that an ideal voltage source of voltage* $\mathbf{V}_c = \mathbf{I}\,\Delta\mathbf{Z}$ *has been connected in series with* $(\mathbf{Z} + \Delta\mathbf{Z})$ *when all other sources in the network are replaced by their internal impedances. The source voltage* $\mathbf{V}_c$ *opposes the original current.*

The compensation theorem is particularly useful in calculating the incremental changes in voltages and currents in all the branches of a network when one or more of the branch impedances are changed.

■ **EXAMPLE 12.38** By means of compensation theorem determine the new values of the branch currents in the network shown in Fig. 12.61a when the value of 2 Ω resistor is increased by 10 per cent.

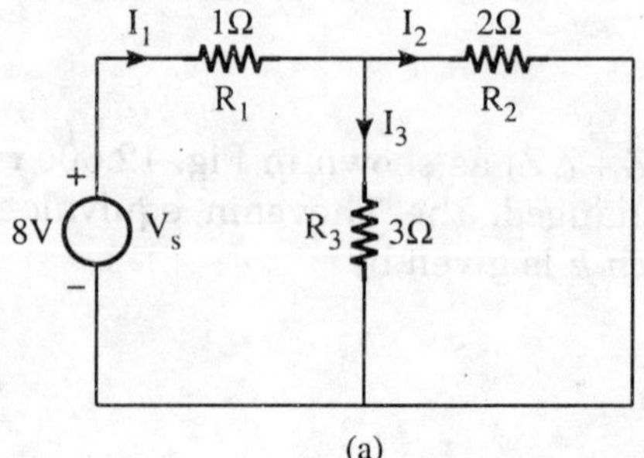

(a)

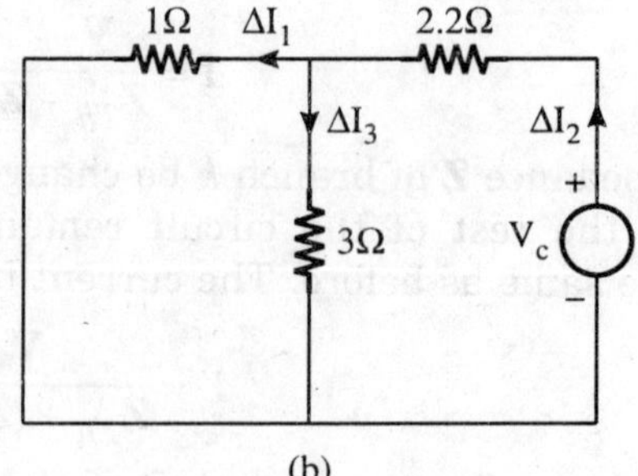

(b)

FIG. 12.61

SOLUTION :

Original current distribution

Total resistance of the network = 1+ 2 || 3

$$= 1 + \frac{2 \times 3}{2 + 3} = 2.2\ \Omega$$

$$I_1 = \frac{8}{2.2} = \frac{40}{11}\ \text{A}$$

By current-division rule

$$I_2 = \frac{3}{3+2} I_2 = \frac{3}{5} \times \frac{40}{11} = \frac{24}{11} \text{ A}$$

$$I_3 = \frac{2}{3+2} I_1 = \frac{2}{5} \times \frac{40}{11} = \frac{16}{11} \text{ A}$$

When the value of 2 Ω resistor is increased by 10 per cent its new value becomes 2.2 Ω. Therefore

$$R_2 + \Delta R_2 = 2.2 \ \Omega \text{ and } \Delta R_2 = 0.2 \ \Omega$$

When the resistor R_2 is changed from 2 Ω to 2.2 Ω the current in any branch may be calculated by assuming that a compensation voltage $V_c = I_2 \Delta R_2$ is connected in series with 2.2 Ω and the voltage source V_s is replaced by a short circuit. Such an arrangement is shown in Fig. 12.61 b. The polarity of the compensation voltage is such that it opposes the original current.

$$V_c = I_2 \Delta R_2 = \frac{24}{11} \times 0.2 = 0.436 \text{ V}$$

From Fig. 12.61b,

$$\Delta I_2 = \frac{V_c}{R'_T}$$

where
$$R'_T = 2.2 + 1 \ || \ 3 = 2.2 + \frac{1 \times 3}{1+3} = 2.95 \ \Omega$$

Therefore, by current-division rule

$$\Delta I_2 = \frac{0.436}{2.95} = 0.148 \text{ A}$$

$$\Delta I_3 = \frac{1}{1+3} \Delta I_2 = 0.037 \text{ A}$$

$$\Delta I_1 = \frac{3}{1+3} \Delta I_2 = 0.111 \text{ A}$$

New current distribution

With the increase of R_2 by 10 per cent, the new current distribution may be found by superposing the currents in Fig.s 12.61a and 12.61b. New current in R_1 is

$$I'_1 = I_1 - \Delta I_1 = \frac{40}{11} - 0.111 = 3.525 \text{ A}$$

New current in R_2 is $I'_2 = I_2 - \Delta I_2 = \frac{24}{11} - 0.148 = 2..033 \text{ A}$

New current in R_3 is $I'_3 = I_3 + \Delta I_3 = \frac{16}{11} + 0.037 = 1.491 \text{ A}$

■ **EXAMPLE 12.39** By use of compensation theorem find the voltage across BD in the bridge circuit shown in Fig. 12.62a when the resistor in the branch BC is changed from R to $(R + \Delta R)$.

SOLUTION : The circuit shown in Fig. 12.62a is balanced. Hence $V_{BD} = 0$. Current in each resistor $I = (V_s/2R)$. When the resistor in branch BC is changed from R to $(R + \Delta R)$, the change in current in any branch may be calculated by assuming that an ideal voltage source of voltage $(I\Delta R)$ is connected in series with $(R + \Delta R)$ and the voltage source V_s is replaced by a short circuit. Such an arrangement is shown in Fig. 12.62b, which is simplified to Fig. 12.62c. Figs. 12.62d and 12.62e show further simplification. From these figures,

$$R_1 = R \;||\; \left(r + \frac{R}{2}\right)$$

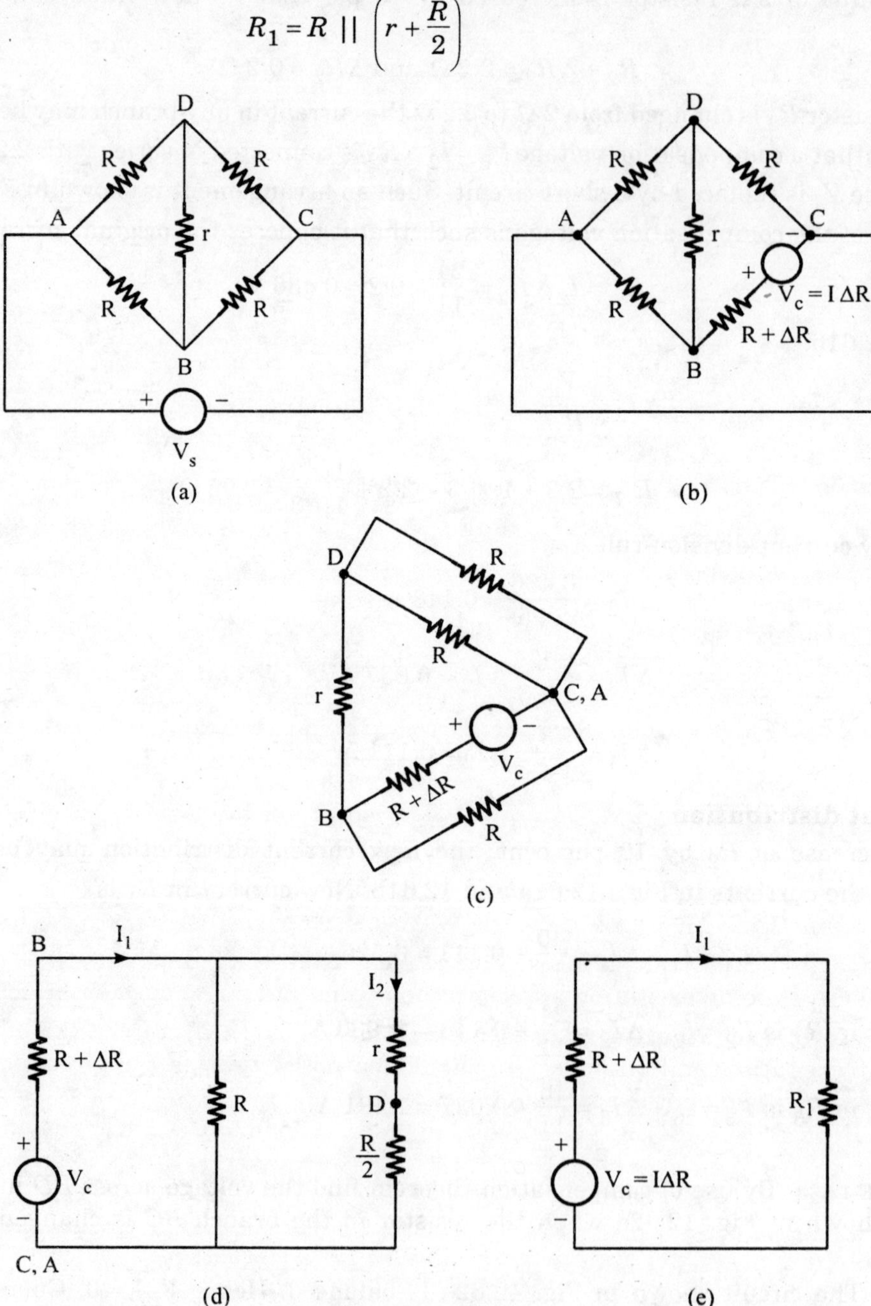

FIG. 12.62

$$= \frac{R\,(r + R/2)}{R + r + R/2} = \frac{R\,(R + 2r)}{3R + 2r}$$

From Fig. 12.62c, $$I_1 = \frac{I\,\Delta R}{R + \Delta R + R_1}$$

From Fig. 12.62d, $$I_2 = \frac{R}{R + r + R/2} I_1 = \frac{2R}{3R + 2r} I_1$$

Voltage drop across BD is

$$V_{BD} = r I_2 = \frac{2RrI\,\Delta R}{(3R + 2r)\,(R + \Delta R + R_1)}$$

$$= \frac{2Rr\,(V_s/2R)\,\Delta R}{(3R + 2r)\left[R + \Delta R + \dfrac{R\,(R + 2r)}{3R + 2r}\right]}$$

$$= \frac{V_s\, r\,\Delta R}{(R + \Delta R)\,(3R + 2r) + R\,(R + 2r)}$$

$$= \frac{V_s\, r\,\Delta R}{4R\,(R + r) + \Delta R\,(3R + 2r)}$$

12.25 RECIPROCITY

Suppose that a system is excited at point A and its response is observed at point B Next consider that the positions of the source and the observation points are interchanged. That is, the system is excited at B and its response is observed at point A. Reciprocity gives the relation between the responses in the two cases. In circuit theory reciprocity applies to any linear time-invariant network consisting of resistor, inductors, coupled inductors, capacitors and transformers. The principle of reciprocity applies to many physical systems. It is very important in the analysis and design of systems. It is also useful in the measurement techniques.

12.26 RECIPROCITY THEOREM

Consider a linear time-invariant network N_R consisting or resistors, inductors, coupled inductors, capacitors and transformers. The general reciprocity theorem states that in the network N_R with a single excitation applied at one point and a response observed at another point, the ratio of excitation to response is not altered by an interchange of the points (positions) of excitation and response. It should be noted that this interchange produces changes in the response in other parts of the network.

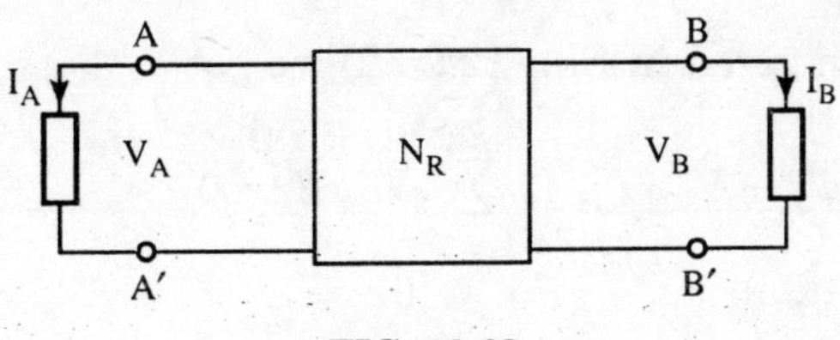

FIG. 12.63

Consider a linear, time-invariant RLCM network N_R with two terminal pairs A, A' and B, B' at which voltage or current sources can be connected, or voltage or current responses can be measured. In other words, these pair of terminals are accessible for applying input and for observing the response. Let us denote these terminal voltages and currents that appear due to excitations or responses as (V_A, I_A) and (V_B, I_B) corresponding to the branches A and B respectively as shown in Fig. 12.63. Let b denote the number of branches of N_R. Voltages and currents, denoted by V_k, I_k $(k = 1, 2, ..., b)$ are set up in each branch of N_R. Thus, there will be $(b + 2)$ branches in all since there are two branches A and B connected between the terminal pairs A, A' and B, B'.

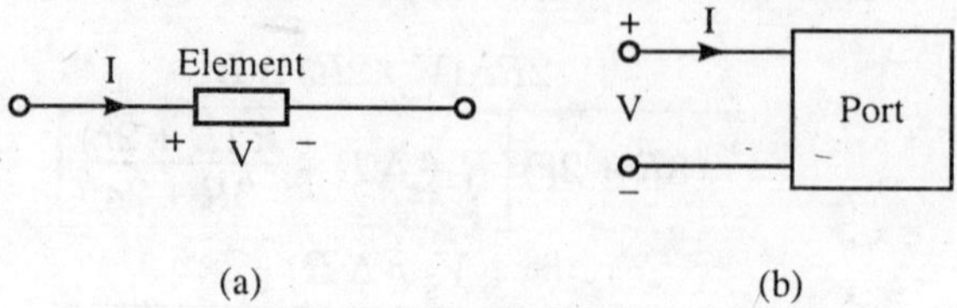

FIG. 12.64 $V - I$ conventions for element and port

Consider the network N_R connected to two different excitation conditions at the terminal pairs A, A' and B, B'.

The resulting voltages and currents in branch k are denoted by (V'_k, I'_k) and (V''_k, I''_k) respectively. If two networks N_1 and N_2 have the *same* graphs, then by Tellegen's theorem

$$\sum_k V'_k I_k = 0 \tag{12.26.1}$$

and

$$\sum_k V_k I'_k = 0 \tag{12.26.2}$$

where the summation is taken over all elements. In Eq. (12.26.1) I_k s are the branch currents in N_1 satisfying KCL and V'_k s are branch voltages in N_2 satisfying KVL. Similarly, in Eq. (12.26.2) I'_k s are the branch currents in N_2 satisfying KCL and V_k s are the branch voltages in N_1 satisfying KVL.

It should be noted that for all branches the associated references directions are used. The *V-I* convention for the element is that current I is entering into positive end of V as shown in Fig. 12. 64a. For the port the current I leaves at the positive end of V as shown in Fig. 12.64b. Thus, the port current direction with respect to the port voltage is opposite to the convention for the element. Applying Tellegen's theorem given in Eq. (12.26.1) we get

$$V'_A I''_A + V'_B I''_B + \sum_{k=1}^{b} V'_k I''_k = 0 \tag{12.26.3}$$

Applying Tellegen's theorem given in Eq. (12.26.2) we get

$$V''_A I'_A + V''_B I'_B + \sum_{k=1}^{b} V''_k I'_k = 0 \tag{12.26.4}$$

It is assumed that the network N_R does not contain any independent source and that each branch is described by $V_k = Z_k I_k$, $(k = 1, 2, ..., b)$

Thus, $$V'_k = Z_k I'_k$$

and $$V''_k = Z_k I''_k$$

where Z_k is the impedance of branch k.

Now $$V'_k I''_k = (Z_k I'_k) I''_k = (Z_k I''_k) I'_k = V''_k I'_k$$

and $$\sum_{k=1}^{b} V' I''_k = \sum_{k=1}^{b} V''_k I'_k \tag{12.26.5}$$

Combination of Eqs. (12.26.3), (12.26.4) and (12.26.5) gives the general volt-ampere relationship for a 2-port reciprocal network as

$$V'_A I''_A + V'_B I''_B = V''_A I'_A + V''_B I'_B \tag{12.26.6}$$

Any 2-port network for which relation given in Eq. (12.26.6) holds is said to be *reciprocal* two-port network.

Consider the general volt-ampere relationship given in Eq. (12.26.6) for a 2-port network, under different specific terminal conditions. The four voltages and four currents in Eq. (12.26.6) correspond to either the excitations or the resulting responses at the appropriate terminal pairs. For a given condition of the network, only two of the four (V_A, I_A, V_B, I_B) quantities will be excitations and the remaining two will become the responses. Thus, particular cases of Eq. (12.26.6) can be considered corresponding to different terminal conditions of excitations and the responses. The reciprocity theorem can be stated in three ways.

Statement 1

Consider a reciprocal network N_R shown in Fig. 12.65. Let the input be a voltage source V_1 connected across the terminals A, A' and let the zero-state current response I_2 be observed in a short circuit connected to the terminals B, B' as shown in Fig. 12.65.

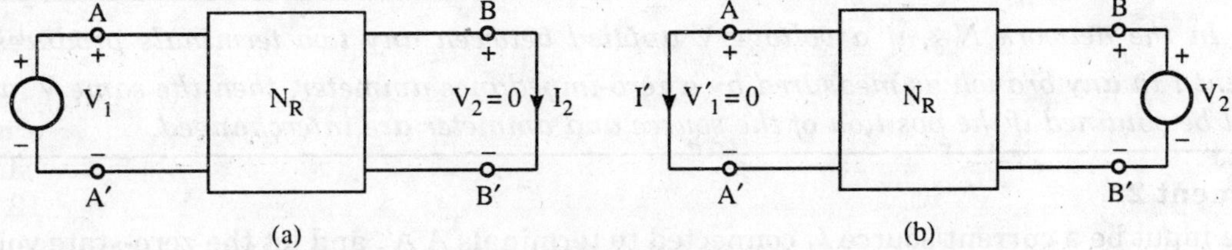

FIG. 12.65

Hence the ratio of response to excitation is $\left.\dfrac{I_2}{V_1}\right|_{V_2 = 0}$

Next, connect a voltage source V'_2 across the terminals B, B' and observe the zero-state current response I'_1 in a short circuit connected to the terminals A, A' as shown in Fig. 12.66. Hence the ratio of response to encitation is

$$\left.\frac{I'_1}{V'_2}\right|_{V'_1=0}$$

Then by reciprocity theorem

$$\left.\frac{I_2}{V_1}\right|_{V_2=0} = \left.\frac{I'_1}{V'_2}\right|_{V'_1=0} \tag{12.26.7}$$

We define the transfer admittance from A, A' to B, B' as

$$y_{21} \stackrel{\Delta}{=} \left.\frac{I_2}{V_1}\right|_{V_2=0} \tag{12.26.8}$$

Similarly, we define the transfer admittance from B, B' to A, A' as

$$y_{12} \stackrel{\Delta}{=} \left.\frac{I'_1}{V'_2}\right|_{V'_1=0} \tag{12.26.9}$$

The reciprocity theorem implies that

$$y_{21} = y_{12} \tag{12.26.10}$$

In particular, if $V_1 = V_2 = V$ (say), we can write,

$$I_2 = I_1 = I \text{ (say)}$$

That is, if a voltage source V across the terminals AA' produces a current $I_2 = I$ through terminals BB', then the same voltage source V across BB' acting in the same reference sense will produce the same current I through the terminals AA'

An ideal voltage source has zero internal impedance. A short circuit has also zero impedance. Hence for the transfer admittance measurement we may interchange the positions of the voltage source and the zero-impedance ammeter to measure the short-circuit current. This result leads to the **classical statement of reciprocity theorem :**

> *In the network N_R, if a voltage V applied between any two terminals produces a current I in any branch as measured by a zero-impedance ammeter, then the same V and I will be obtained if the position of the source and ammeter are interchanged.*

Statement 2

Let the input be a current source I_1 connected to terminals AA', and let the zero-state voltage

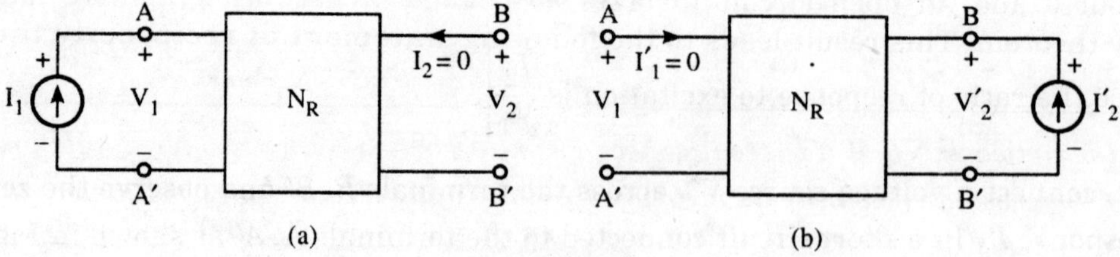

FIG. 12.66

response V_2 be observed across the open-circuited terminals as shown in Fig. 12.66a. Hence the ratio of response to excitation is $\left.\frac{V_2}{I_1}\right|_{I_2=0}$

Next, connect a current source I'_2 to the terminals B, B' and observe the zero-state voltage response V'_1 across the open-circuited $A A'$ as shown in Fig. 12.66b. Hence the ratio of response to excitation is

$$\left.\frac{V'_1}{I'_2}\right|_{I'_1=0}$$

Then by reciprocity theorem

$$\left.\frac{V_2}{I_1}\right|_{I_2=0} = \left.\frac{V'_1}{I'_2}\right|_{I'_1=0} \tag{12.26.10}$$

We define the transfer impedance from port 1 to port 2

$$z_{21} \triangleq \left.\frac{V_2}{I_1}\right|_{I_2=0} \tag{12.26.11}$$

Similarly, we define the transfer impedance from port 2 to port 1

$$z_{12} \triangleq \left.\frac{V'_1}{I'_2}\right|_{I'_1=0} \tag{12.26.12}$$

The reciprocity implies that

$$z_{21} = z_{12} \tag{12.26.13}$$

Eq. (12.26.10) shows that the ratio of response to excitation is invariant to an interchange of the positions in the network of the excitation and the response.

If $\quad I_1 = I'_2 = I \;$ (say)

then from Eq. (12.26.10), we can write $V_2 = V'_1 = V$ (say)

That is, if a current source I connected to terminals $A A'$ produces a voltage V across the open-circuited terminals BB' will produce the same voltage V across the open-circuited terminals $A A'$.

An ideal current source has infinite internal impedance. An open circuit has also infinite impedance. Thus, an ideal current source is a generalized open circuit. Since interchanging a current source and an open circuit involves no change in network graph we may apply reciprocity theorem. This result leads to the following **statement of reciprocity theorem :**

> *In the network N_R, if a current source I applied between any two terminals produces a voltage V across any branch as measured by an infinite impedance voltmeter, then the same I and V will be obtained if the positions of the current source and voltmeter are interchanged.*

Statement 3

Let the input be a voltage source V_1 connected across the terminals $A A'$, and let the zero-state voltage response V_2 be observed across the open-circuited terminals BB' (Fig. 12.67a). Hence the ratio of response to excitation is $\left.\frac{V_2}{V_1}\right|_{V_2=0}$. Next, connect a current source I'_2 to the terminals BB' and observe the zero-state current response I'_1 in a short circuit connected to terminals $A A'$ (Fig. 12.67b). Hence the ratio of response to excitation is $\left.\frac{I'_1}{I'_2}\right|_{V'_1=0}$. Then by reciprocity theorem

$$\left.\frac{V_2}{V_1}\right|_{I_2=0} = \left.\frac{I'_1}{I'_2}\right|_{V'_1=0}$$

Hence for both sets of measurement, a zero impedance is connected to terminals $A A'$ and, an infinite impedance is connected to terminals BB'.

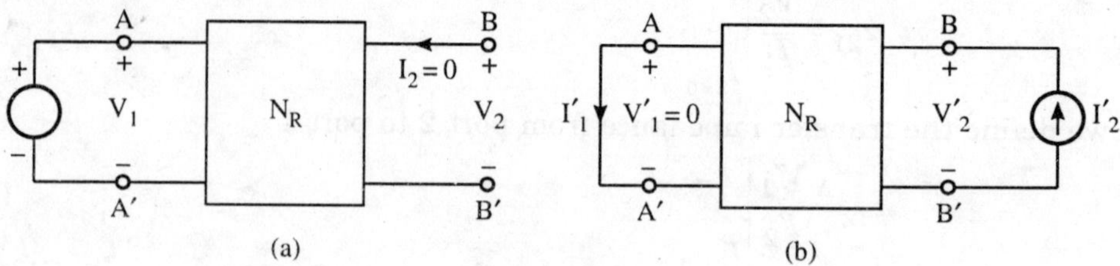

FIG. 12.67

It should be noted that if interchanging excitation and response involves a change in network graph, reciprocity theorem cannot be applied. Therefore reciprocity does *not* apply to a voltage response from a voltage excitation or a current response from a current excitation. If we apply the reciprocity theorem to a voltage ratio, we would be interchanging a voltmeter of infinite impedance with a voltage source of zero impedance. This would result in a change in the network and therefore a different transfer function would be obtained. Similarly, if we apply the reciprocity theorem to a current ratio, we would be interchanging an ammeter of zero impedance with a current source of infinite impedance. Thus the network would change to give a different transfer function. Hence the reciprocity theorem applies to transfer impedances and transfer admittances but *not* to transfer ratios.

Remarks

The reciprocity theorem is applicable to a linear network N_R that satisfies the following conditions :

1. The network N_R contains only resistors, inductors, coupled inductors, capacitors, and transformers.
2. The network N_R does not contain dependent and independent sources.
3. There are no time-varying elements in the network N_R.

4. While applying the reciprocity theorem the network topology should not change when the voltage (or current) source is moved. For example, it is not permissible to move the voltage source from its position in *series* with a branch to a new position in *parallel* with another branch, since the network graph would change.
5. When the voltage source is moved from one position to another, the currents in other parts of the network will change. Similarly, when the current source is moved, the voltages in other parts of the network will change. In general, the interchange of an excitation and its response produces changes in the response in other parts of the network.
6. Only single excitation is applied at a time. If additional sources are present, the superposition theorem is used in conjunction with the reciprocity theorem to find the total response. Thus, all initial conditions in the network must be zero unless superposition theorem is used to take their account as separate sources.

EXAMPLE 12.40 Find the voltage **V** in the network shown in Fig. 12.68 (*a*). Verify reciprocity theorem by interchanging the current source and the resulting voltage **V**.

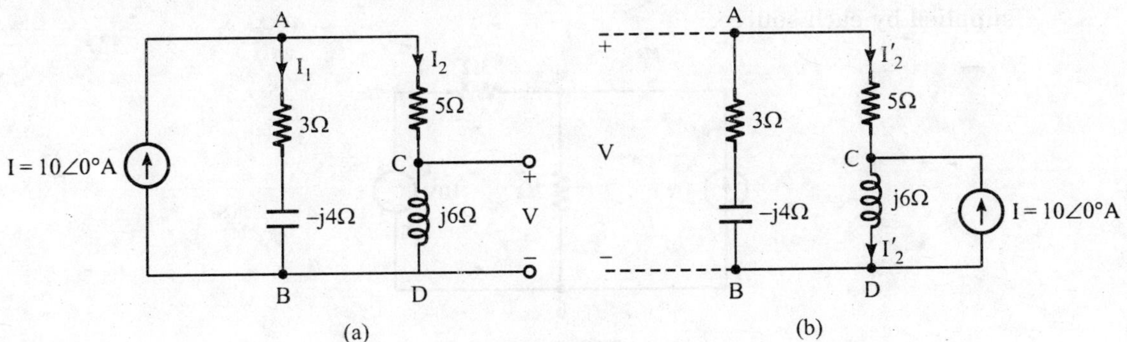

FIG. 12.68

SOLUTION : From Fig. 12.68a, by current-division rule

$$\mathbf{I}_2 = \frac{3 - j4}{(3 - j4) + (5 + j6)} 10 \angle 0^\circ$$

$$= \frac{3 - j4}{4 + j1} 5 \angle 0^\circ = \frac{5 \,/\!- 53.1^\circ}{4.12 \angle 14^\circ} \times 5 \angle 0^\circ$$

$$= 6.068 \angle -67.1^\circ \text{ A}$$

$$\mathbf{V} = (j6)\, \mathbf{I}_2 = (6 \angle 90^\circ)(6.0.68 \angle -67.1^\circ)$$

$$= 36.4 \angle 22.9^\circ \text{ V}$$

The current source and the terminals at which the voltage **V** is measured are interchanged as shown in Fig. 12.68b. From Fig 12.68b

$$\mathbf{I'}_1 = \frac{j6}{8 + j2} 10 \angle 0^\circ = \frac{6 \angle 90^\circ \times 10 \angle 0^\circ}{8.246 \angle 14^\circ} = 7.276 \angle 76^\circ \text{ A}$$

$$\mathbf{V} = \mathbf{Z}_{AB}\, \mathbf{I'}_1$$

$$= (3 - j4)(7.276 \angle 76^\circ) = 36.4 \angle 22.9^\circ \text{ V}$$

Since the value of **V** remains the same in both the cases, the reciprocity theorem is verified

EXERCISES

1. Use the principle of superposition to determine the current through the 3 Ω resistor in the network of Fig. P12.1.

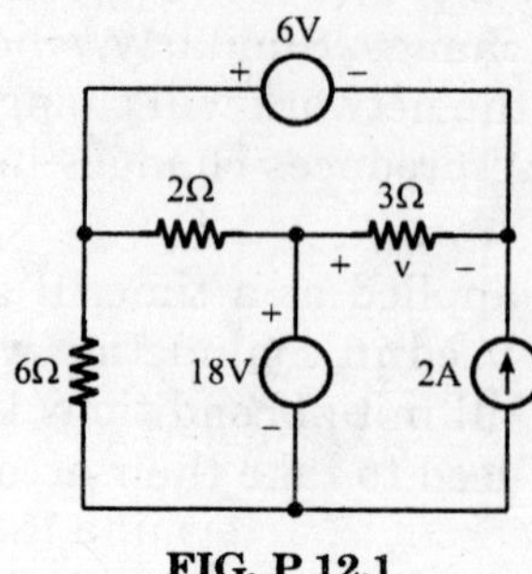

FIG. P 12.1

2. Use the superposition theorem to calculate V_{ab} in Fig. P12.2. Calculate the power supplied by each source.

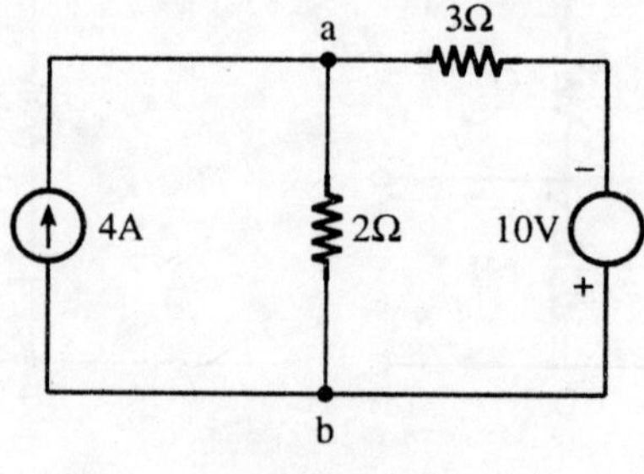

FIG. P 12.2

3. Use the superposition theorem to calculate 1 in the network of Fig. P12.3.

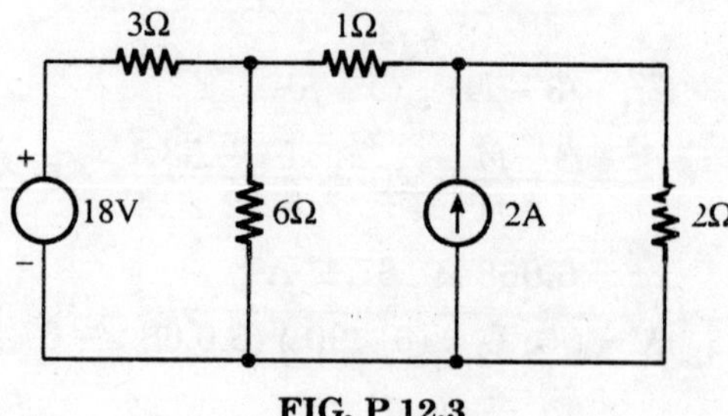

FIG. P 12.3

4. Use the superposition theorem to determine the voltage v in the circuit of Fig. P12.4.

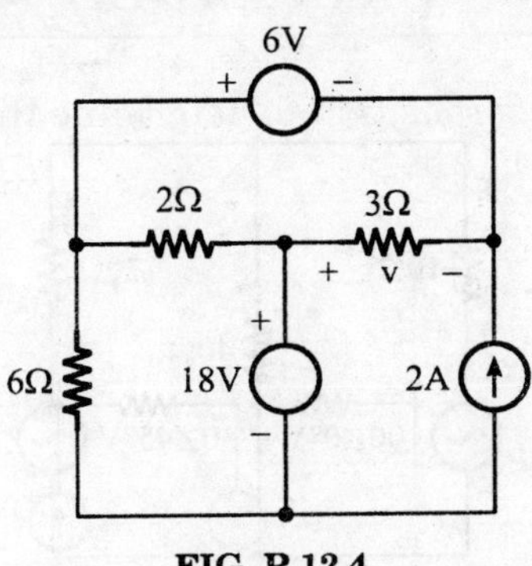

FIG. P 12.4

5. Use the superposition theorem to determine the voltage v and power dissipated by the 3 Ω resistor of the network shown in Fig. P12.5.

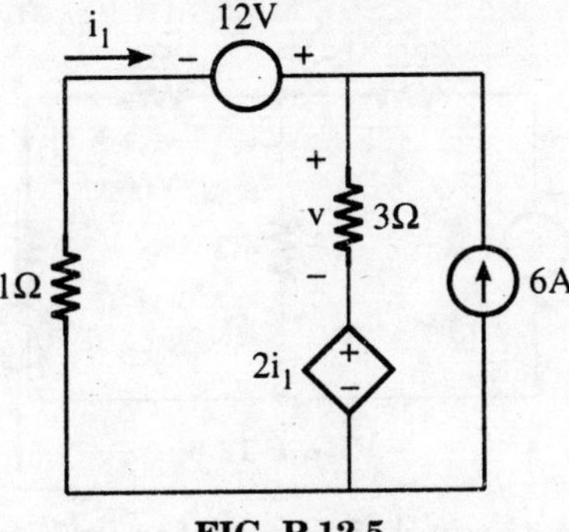

FIG. P 12.5

6. Using the principle of superposition find i_1 and v_2 in the circuit of Fig. P12.6.

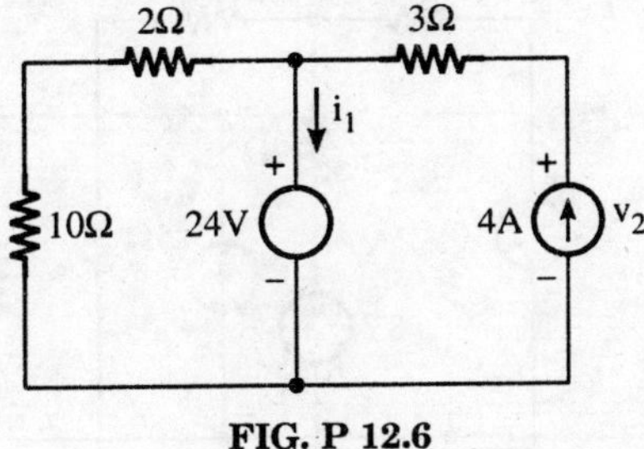

FIG. P 12.6

7. Using the superposition theorem, determine the current in each branch of the network in Fig. P12.7.

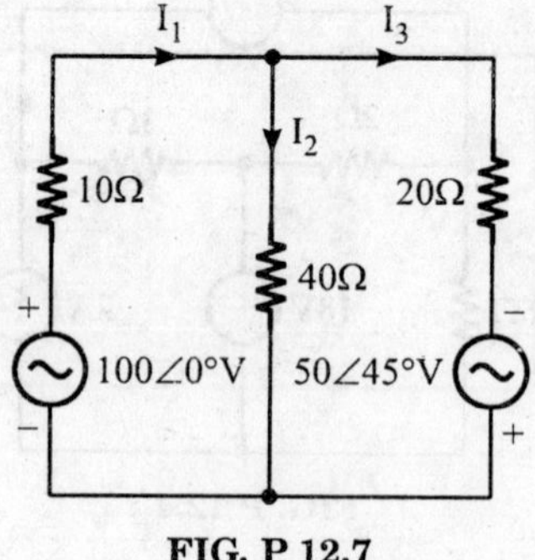

FIG. P 12.7

8. Determine the current I in the circuit of Fig. P12.8 by using (*a*) source transformation, (*b*) Thevenin's theorem, (*c*) Norton's theorem.

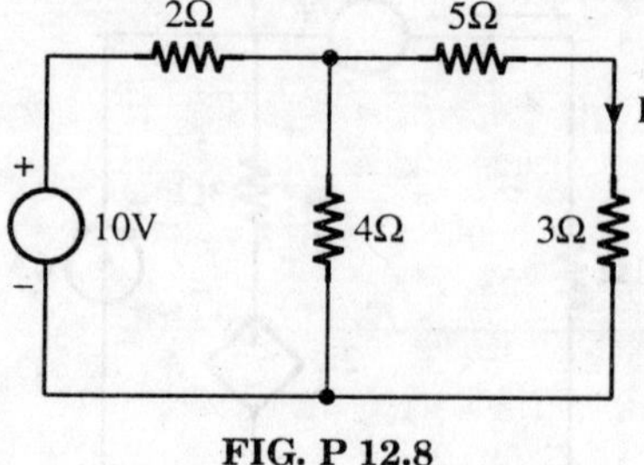

FIG. P 12.8

9. Determine the current through 5 Ω resistor in Fig. P.12.9 by using (*a*) Thevenin's theorem, (*b*) Norton's theorem.

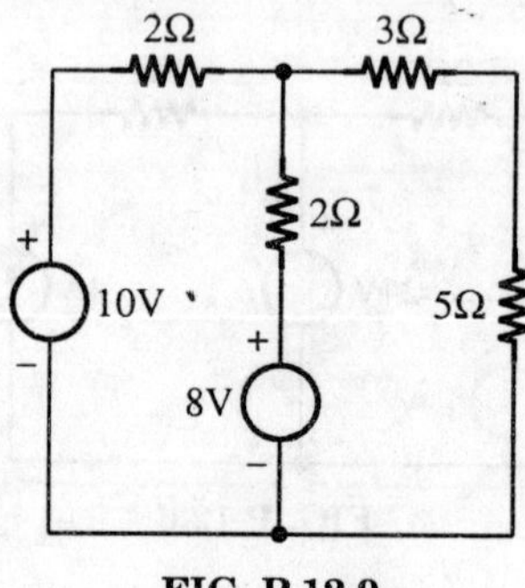

FIG. P 12.9

10. For the circuit shown in Fig. 12.10, draw the Norton's equivalent circuit at terminals *a–b*.

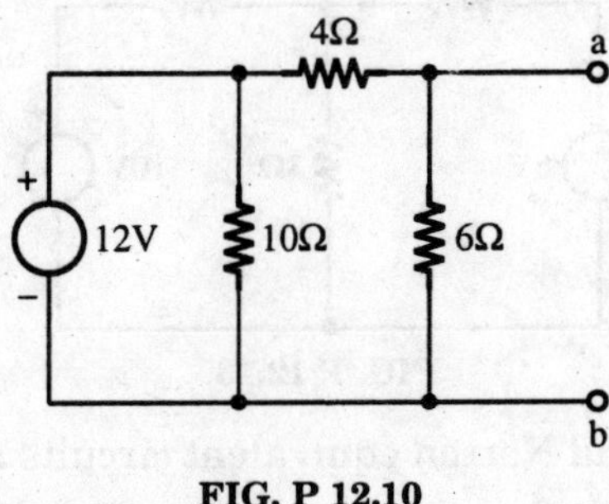

FIG. P 12.10

11. Find the Thevenin equivalent network at terminals *a–b* of the network of Fig. P12.11.

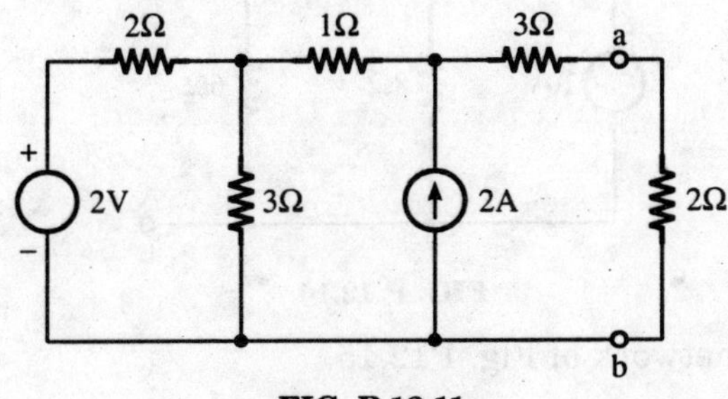

FIG. P 12.11

12. State Thevenin's and Norton's theorems. Deduce Thevenin's and Norton's equivalent circuits for terminals *a–b* in the network shown in Fig. P12.12.
    A 100 Ω load resistor is connected across *a–b*. Use the Thevenin equivalent circuit to calculate the load current and the Norton's equivalent circuit to calculate the load voltage, and compute the load power.

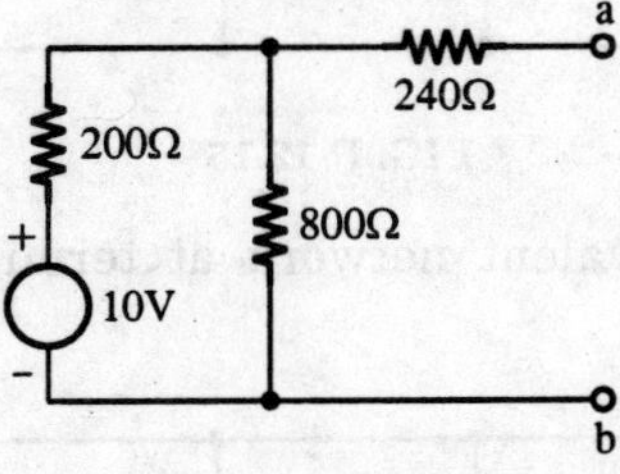

FIG. P 12.12

13. Determine the current through 3 Ω resistor in Fig. P12.13 using Norton's theorem.

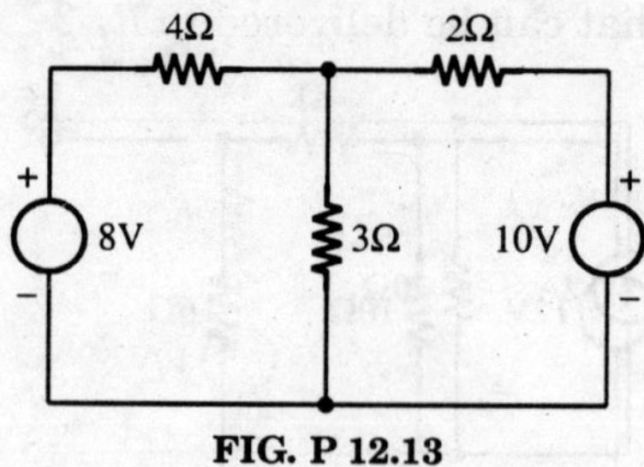

FIG. P 12.13

14. Obtain the Thevenin and Norton equivalent circuits at terminals *ab* for the network of Fig. P12.14.

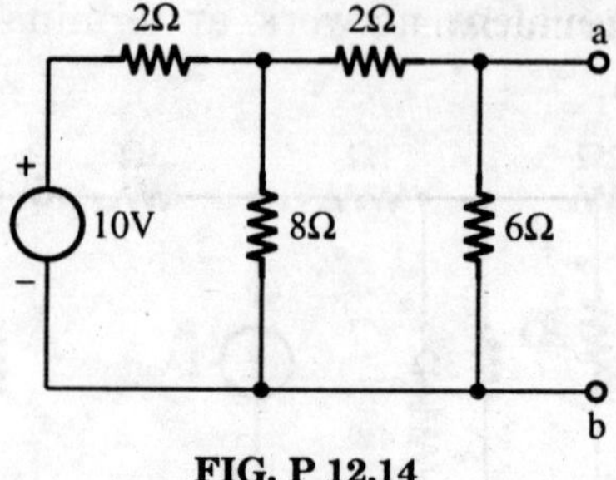

FIG. P 12.14

15. Find V_{ab} for the network of Fig. P12.15.

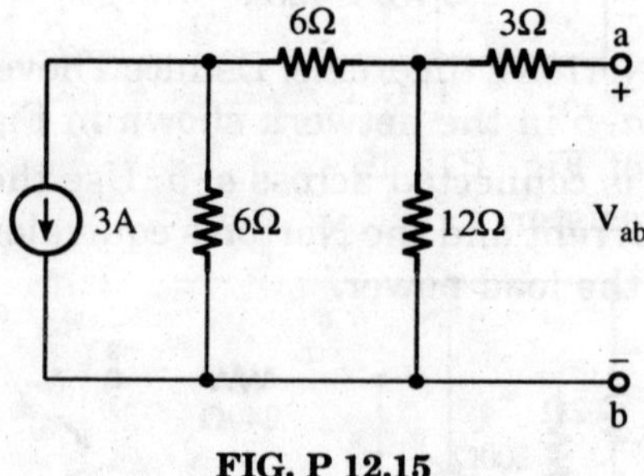

FIG. P 12.15

16. Find the Thevenin equivalent network at terminals *ab* for the network of Fig. P12.16.

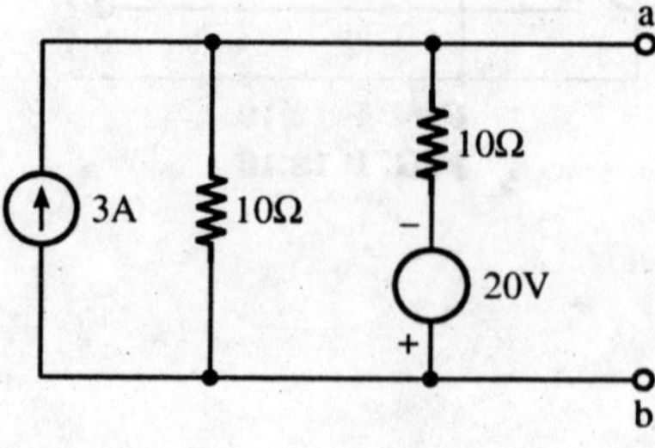

FIG. P 12.15

17. In the circuit of Fig. P12.17, a resistance R_L is connected between a and b. What is the maximum power that can be delivered to R_L ?

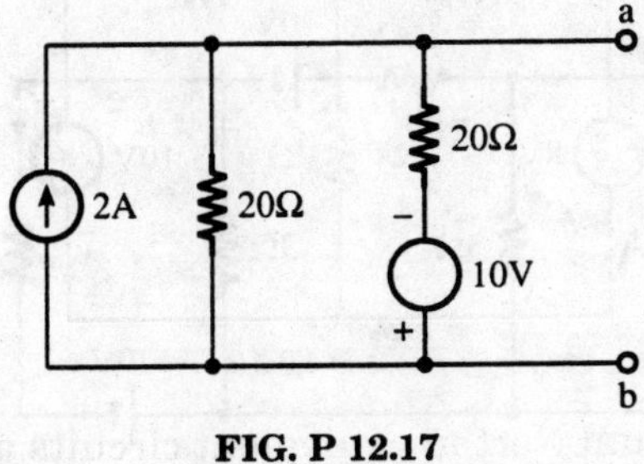

FIG. P 12.17

18. Determine the current through 1 Ω resistor in the network of Fig. P12.18 by using (*a*) source transformations,
(*b*) Thevenin's theorem,
(*c*) Norton's theorem

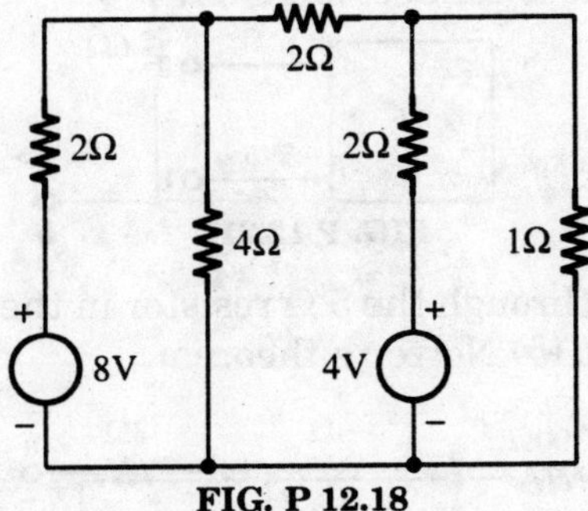

FIG. P 12.18

19. Reduce the network of Fig. P12.19 to a single current source and calculate the current through 5 Ω resistor.

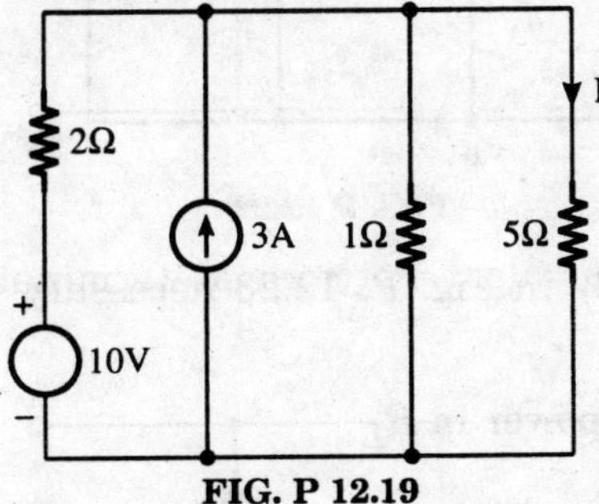

FIG. P 12.19

20. Use Thevenin's theorem to determine the current through the 2 Ω resistor of the network shown in Fig. P12.20.

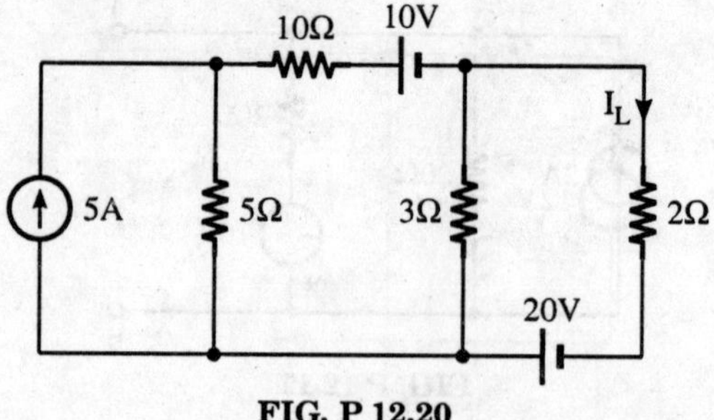

FIG. P 12.20

21. The network N of Fig. P12.21 contains energy sources and resistances. The open-circuit voltage across terminals a–b is 5 V. When the terminals a–b are short circuited, the short-circuit is 1 A. Determine the current that flows through a resitance of 5 Ω when it is connected across a–b.

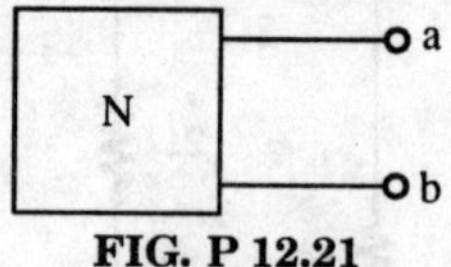

FIG. P 12.21

22. Determine the current through the 5 Ω resistor in the circuit of Fig. P12.22 by using (*a*) Thevenin's theorem, (*b*) Norton's theorem.

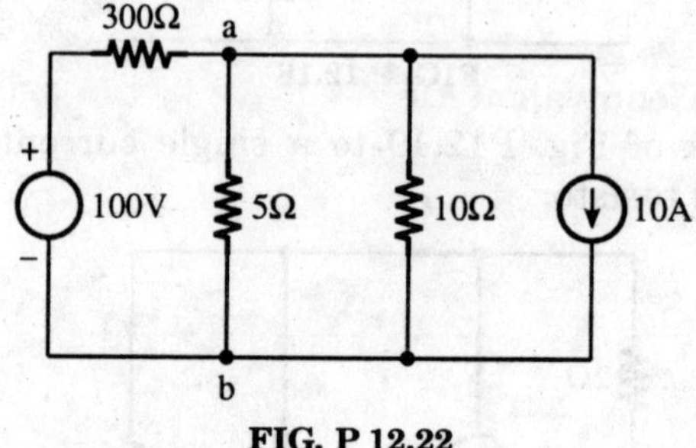

FIG. P 12.22

23. (*a*) Find the value of R_L in Fig. P. 12.23 necessary to obtain maximum power in R_L.

(*b*) Find the maximum power in R_L.

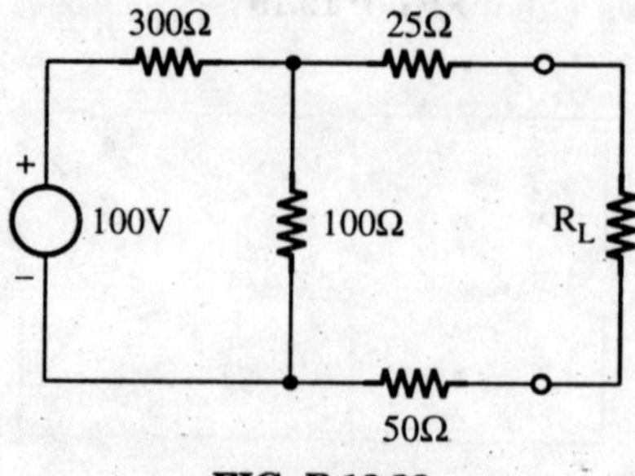

FIG. P 12.23

24. In the network of Fig. P12.24, calculate the resistance R which allows maximum power dissipated in it. Also calculate the maximum power.

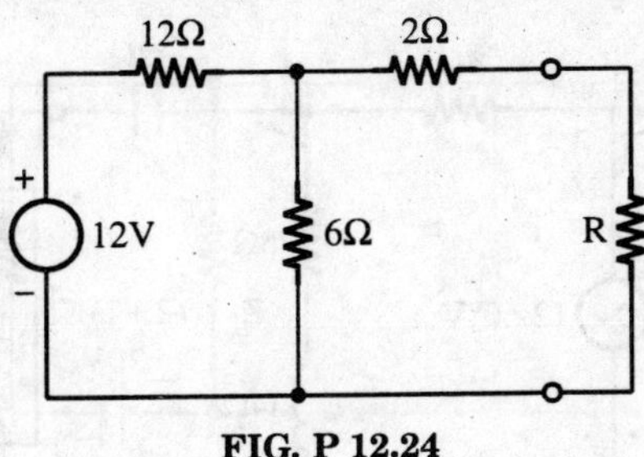

FIG. P 12.24

25. Obtain the Thevenin equivalent across a–b for the network shown in Fig. P12.25.

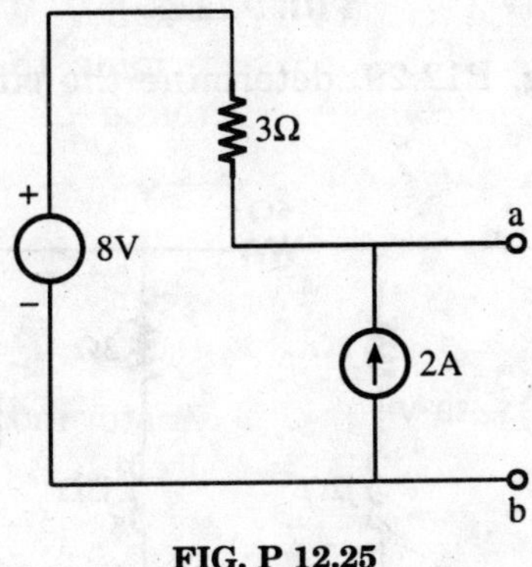

FIG. P 12.25

26. Find the Thevenin equivalent circuit lying to the right of terminals a–b in Fig. P12.26.

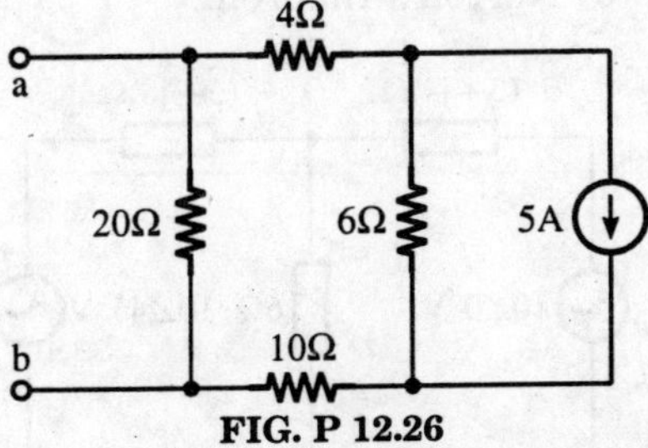

FIG. P 12.26

27. Determine the current through and voltage across 30 Ω resistor in Fig. P12.27 by applying (a) Thevenin's theorem, (b) Norton's theorem.

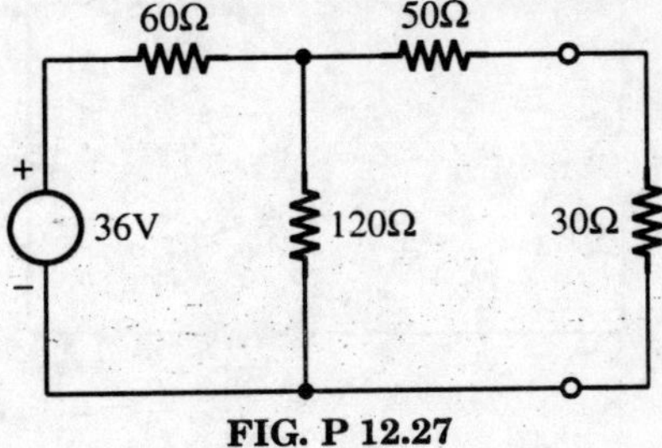

FIG. P 12.27

28. In the circuit of Fig. P12.28, determine the current through Z_L using (*a*) Thevenin's theorem, (*b*) Norton's theorem.

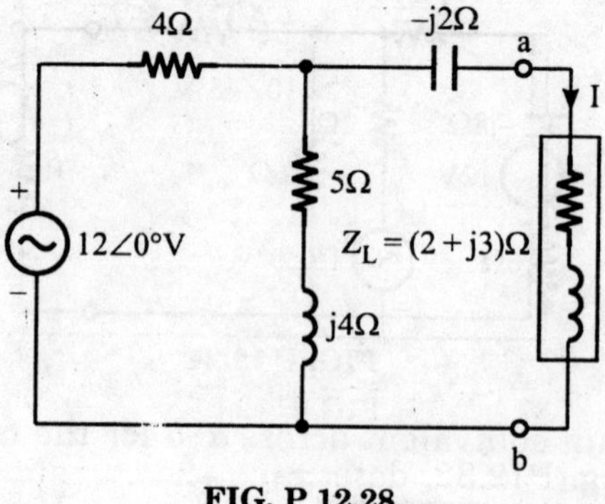

FIG. P 12.28

29. In the network of Fig. P12.29, determine the current through 5 Ω resistor using Thevenin's theorem.

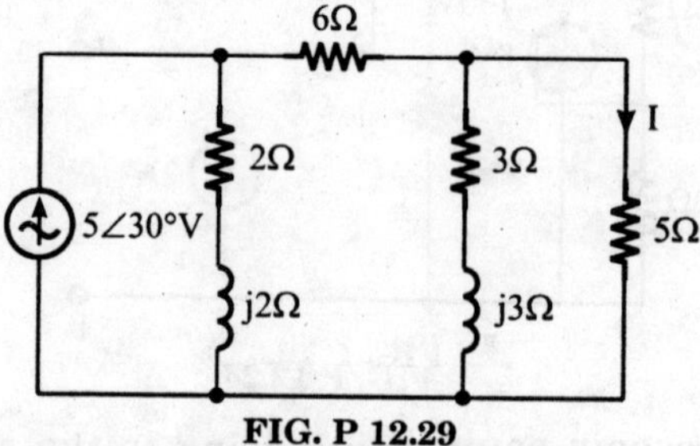

FIG. P 12.29

30. In the circuit of Fig. P12.30 determine the current through 6 Ω resistor using (*a*) Thevenin's theorem, (*b*) Norton's theorem.

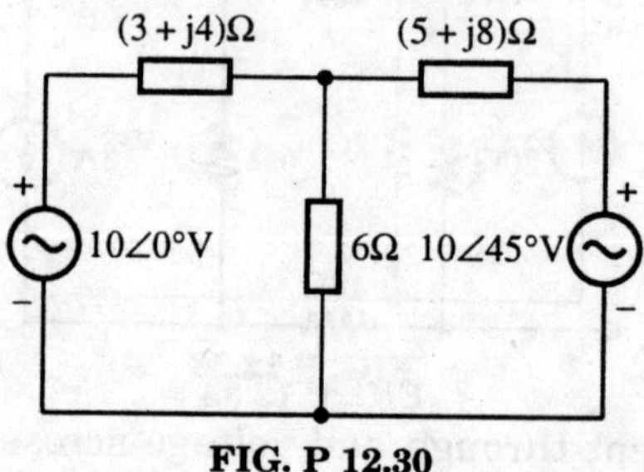

FIG. P 12.30

31. Obtain the Thevenin equivalent circuit at terminals *a–b* of the network shown in Fig. P12.31.

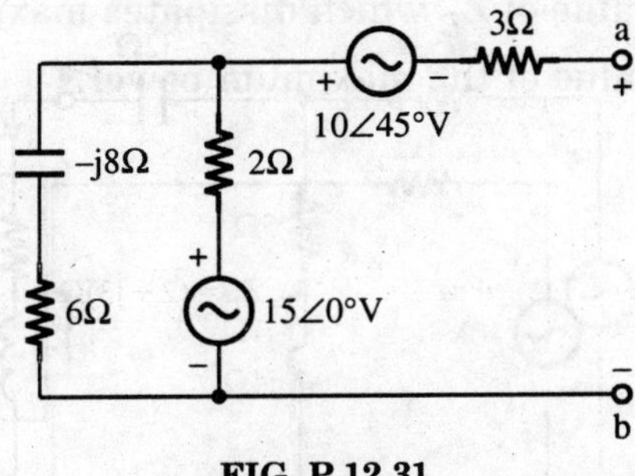

FIG. P 12.31

32. In the circuit of Fig. P12.32, determine the current through 3 Ω resistor using (*a*) Thevenin's theorem, (*b*) Norton's theorem.

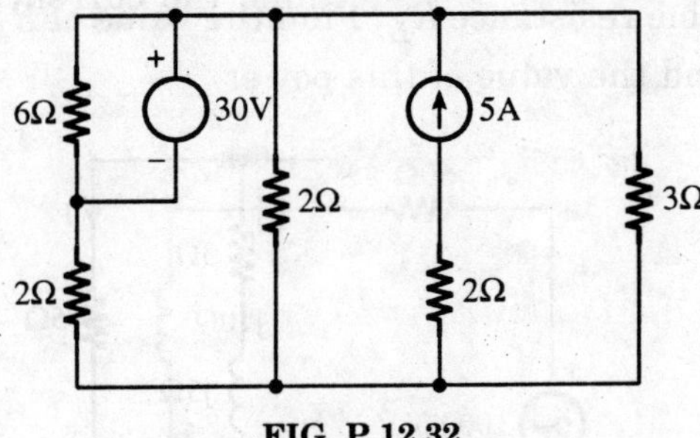

FIG. P 12.32

33. Determine the Thevenin equivalent network in Fig. P12.33 with reference to the terminals *a–b*.

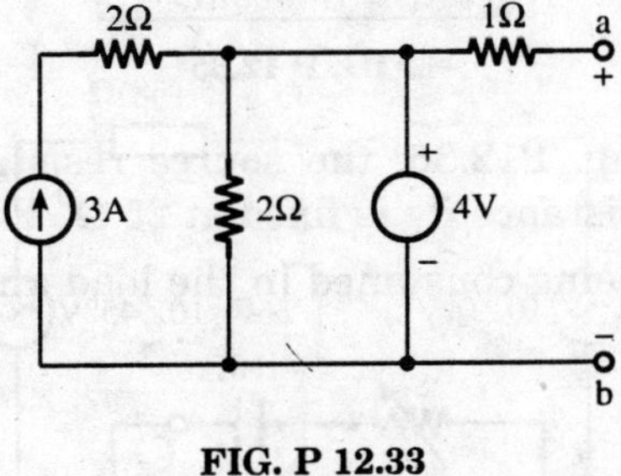

FIG. P 12.33

34. A current source $\mathbf{I}_s$ with its internal admittance $\mathbf{Y}_s$ supplies power to a load admittance $\mathbf{Y}_L$. Show that the maximum power transferred to the load is given by

$$P_{L\,\max} = \frac{I^2 G_L}{(G_S + G_L)^2 + (B_S + B_L)^2}$$

where

$$\mathbf{Y}_S = G_S + jB_S$$

$$\mathbf{Y}_L = G_L + jB_L$$

35. In the network of Fig. P12.34, both the resistive and reactive parts of the impedance $\mathbf{Z}_L$ can be varied.

(*a*) Determine the value of $\mathbf{Z}_L$ which dissipates maximum power.

(*b*) Determine the value of the maximum power.

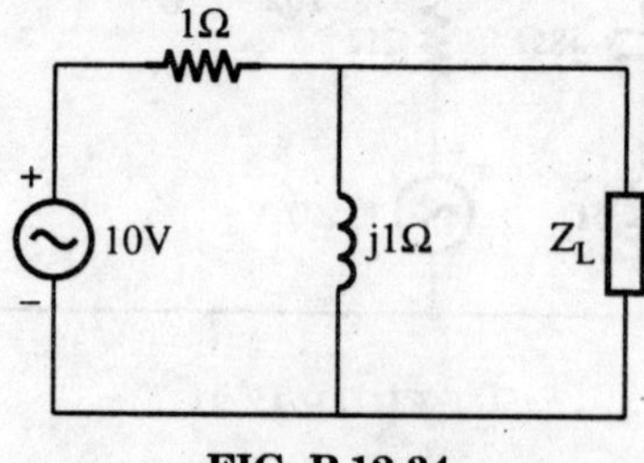

FIG. P 12.34

36. In the network of Fig. P12.35, the load consists of a fixed inductance of reactance $j10\ \Omega$ and a variable resistance R_L. Find the value of R_L for which maximum power transfer occurs and the value of this power.

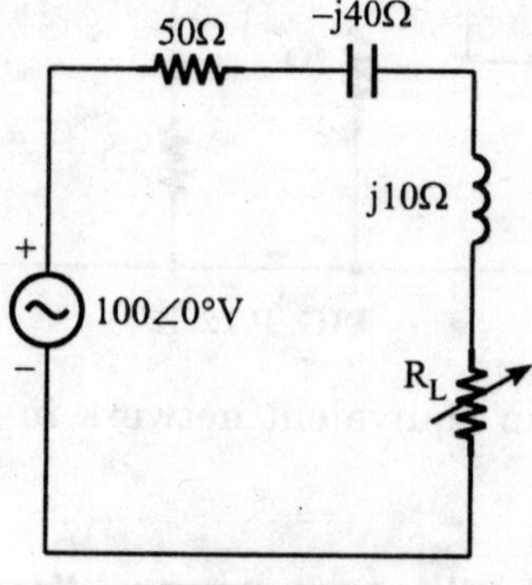

FIG. P 12.35

37. In the network of Fig. P12.36, the source resistance is variable between 5 and 50 Ω and the load resistance R_L is fixed at 25 Ω. Find the value of R_L which results in maximum power being consumed in the load and the value of this power.

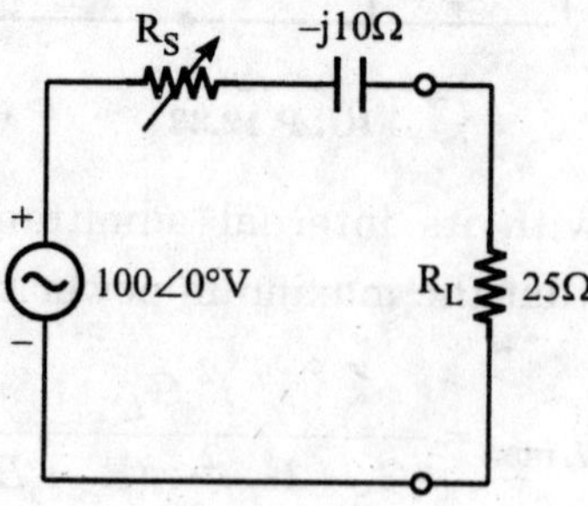

FIG. P 12.36

38. What should be the value of $\mathbf{Z}_L$ in Fig. P12.37 connected across the terminals a–b so that it will draw maximum power ? What is the amount of maximum power ?

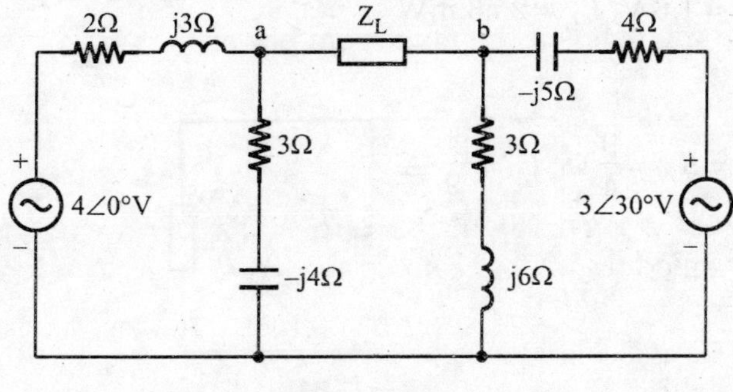

FIG. P 12.37

39. In the network of Fig. P12.38 what load $\mathbf{Z}_L$ will receive maximum power ?

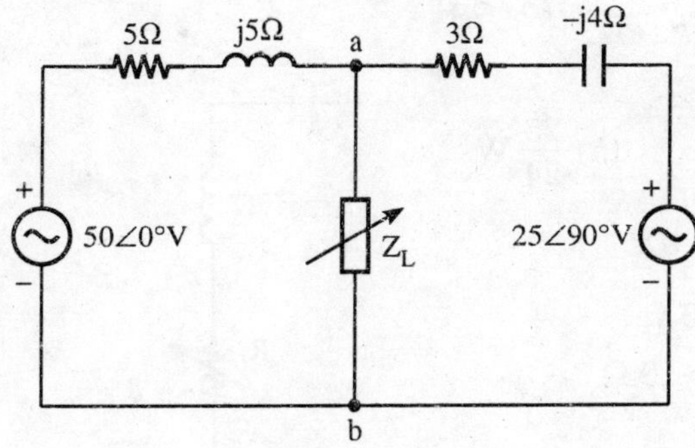

FIG. P 12.38

ANSWERS

1. 4A
2. 8.8 V, current source supplies 35.2 W, voltage source supplies 4 W
3. 1.6 A
4. 5 V
5. v = 15 V, P = 75 W
6. 2 A, 36 V
7. $I_1 = 5.4 \angle 10.79°$ A

   $I_2 = 1.7 \angle 8.55°$ A

   $I_3 = 4.32 \angle 17.09°$ A
8. $\frac{5}{7}$ A
9. 1 A

10. $I_N = 3$ A, $R_N = 2.4\ \Omega$
11. $V_{Th} = 5.6$ V, $R_{Th} = 5.2\ \Omega$
12. $I_L = 16$ mA, $V_L = 1.6$ V, $P_L = 25.6$ mW
13. $\frac{28}{13}$ A
14. $V_{Th} = 5$ V, $R_{Th} = R_N = \frac{9}{4}\ \Omega$, $I_N = \frac{20}{9}$ A
15. – 9 A
16. $V_{Th} = 5$ V, $R_{Th} = 5\ \Omega$
17. 5.625 W
18. 2 A
19. $I = \frac{16}{17}$ A
20. $I_L = 5$ A
21. 0.5 V
22. $\frac{20}{11}$ A from b to a
23. (*a*) 150 Ω ; (*b*) $\frac{25}{24}$ W
24. 6 Ω, $\frac{2}{3}$ W
25. $V_{Th} = 14$ V, $R_{Th} = 3\ \Omega$
26. $V_{Th} = 25.5$ V, $R_{Th} = 10\ \Omega$
27. 0.2 A, 6 V
28. $1.62 \angle -5.5^\circ$ A
29. $0.65 \angle 83^\circ$ A
30. $1.145 \angle -2.6^\circ$ A
31. $V_{Th} = 10.79 \angle 55.8^\circ$ V, $Z_{Th} = 4.756 \angle -3^\circ\ \Omega$
32. 5 A
33. $V_{Th} = 4$ V, $Z_{Th} = 1\ \Omega$
35. (*a*) $\mathbf{Z}_L = (0.5 - j0.5)\ \Omega$, (*b*) $P_{L\,\max} = 25$ W
36. $R_L = 58.31\ \Omega$, $P_{L\,\max} = 46.16$ W
37. $R_L = 5\ \Omega$, $P_{L\,\max} = 250$ W
38. $\mathbf{Z}_L = 9.56 \angle -7.8^\circ\ \Omega$, $P_{L\,\max} = 0.962$ W
39. $\mathbf{Z}_L = (4.22 + j1.15)\ \Omega$, $P_{L\,\max} = 5.63$ W

13

NETWORK GRAPH THEORY

13.1 INTRODUCTION

Kirchhoff's current and voltage laws (KCL and KVL) form the basis of network analysis. Since these laws do not depend on the nature of the network elements, it seems natural to overlook the types of the elements that are used in it. Kirchhoff's laws depend only on the geometry or *topology* of the network. **Topology** is the inner structure of the network. Network topology is the branch of the network theory which deals with the study of the properties of the network that arise out of the geometrical structure of the network. The topology of a network is important in the systematic formulation of network equations.

The topological properties of a network are developed by replacing every two-terminal network element (regardless of its nature) by a line segment (—) called a **branch**, and each end point of the element by a dot (•) called a **node**. *A **network graph** is defined as the set of nodes and branches with the condition that each branch terminates at each end into a node. In graph theory each branch must have two end points given by nodes.*

13.2 TERMINOLOGY AND DEFINITIONS

In this section some of the terms used in graph theory are introduced.

Branch. A branch is a line segment which represents a network element or a combination of elements connected between two points. A line segment together with its two distinct end points is called a branch. A branch is sometimes also called an **edge.** The line segments do not say anything about the type of elements.

Node. A node is defined to be an end point of a line segment or an isolated point. A node is sometimes also called a **vertex** or a **junction.** Thus a node exists at the junction between two branches or at the ends of an isolated branch.

In general, a branch may represent a single network element such as a resistor R, an inductor L, or a capacitor C or any other network element (excluding short circuits and open circuits). A branch may also represent a combination of these elements. The choice of branches and nodes with respect to network elements is flexible. Fig. 13.1 shows the flexibility of the choice. For example, the parallel combination of RC elements can be considered as either a single branch connected between two nodes in Fig. 13.1f, or as two parallel branches in

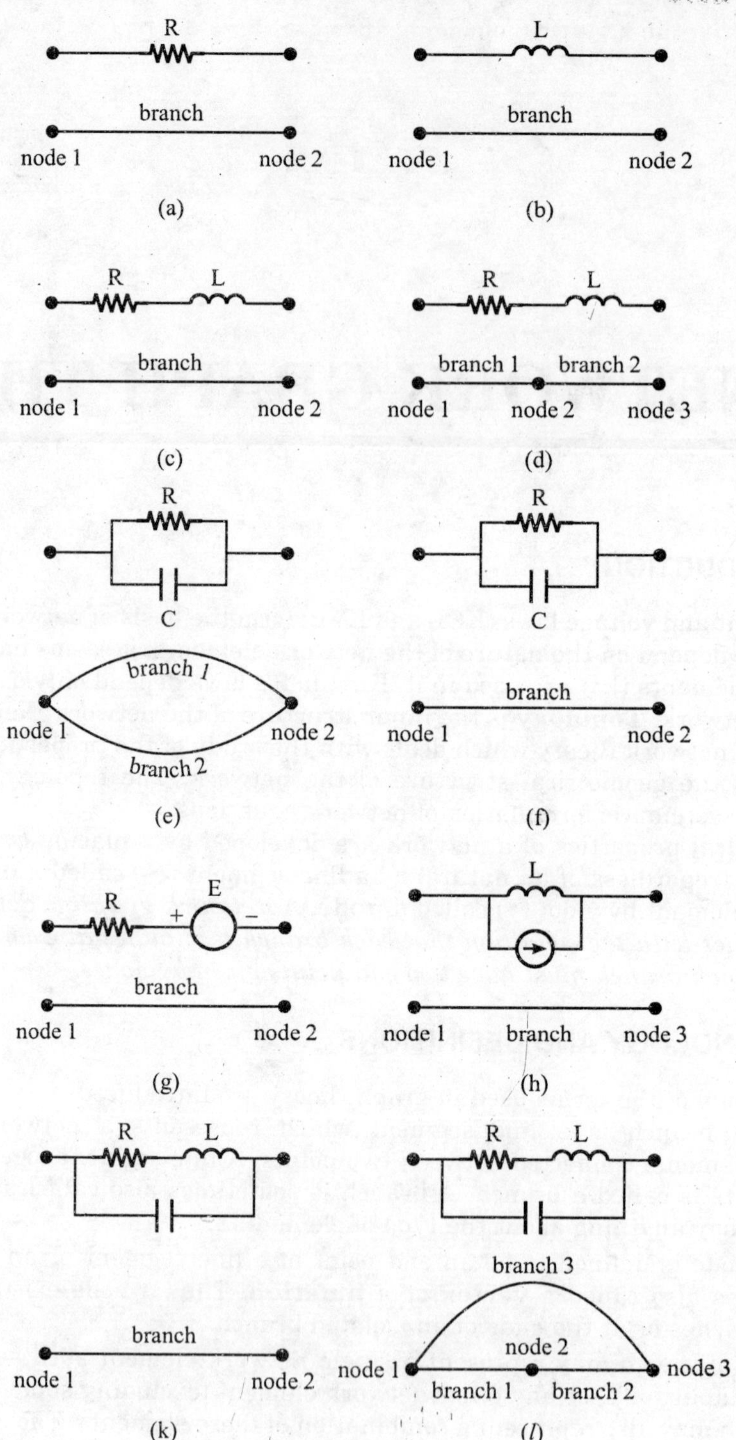

FIG. 13.1 Examples of branches and nodes.

Fig. 13.1e. Similarly, the series RL elements shown in Figs. 13.1c and 13.1d can be regarded as either a single branch in Fig. 13.1c or two branches in Fig. 13.1d. Fig. 13.1k and 13.1l also show the flexibility of choosing branches and nodes with respect to network elements.

Networks without sources are called ***basic networks***. Two basic networks having the same number of nodes and branches connected in the same manner have the same graph. Fig. 13.2a and 13.2b show two different basic networks. Fig. 13.2c shows their graph which is the same for the two networks.

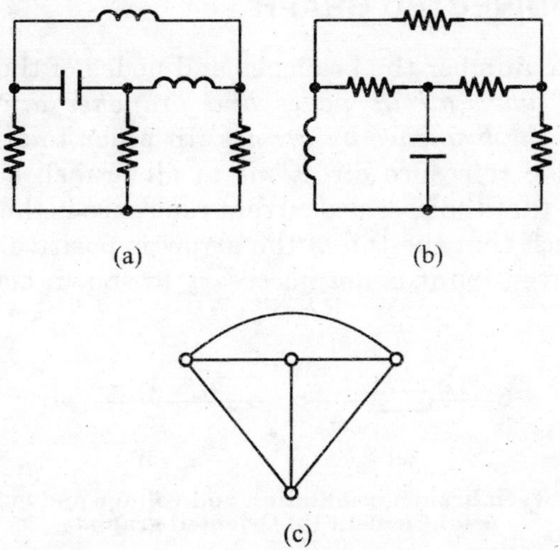

FIG. 13.2 Common graph *c* of network (*a*) and (*b*)

13.3 GRAPHS OF NETWORKS WITH VOLTAGE AND CURRENT SOURCES

In constructing the graph of a linear network an ideal voltage source is replaced by a short circuit (shorted branch). This is because of the fact that an ideal voltage source has zero

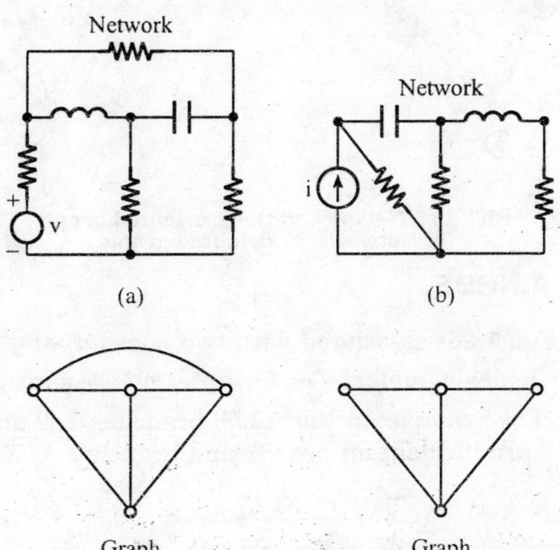

FIG. 13.3 Examples of graphs of network with voltage and current sources.

internal impedance and only the geometry of the source is being represented. Similarly, *an ideal current source is replaced by an open circuit* (*open branch*) since it has infinite impedance. In other words, *ideal sources do not appear in the graph of a linear network and they are set to zero.* Consequently, a passive element in series with a current source is not a branch. Similarly, a passive element in parallel with a voltage source is not a branch. Fig. 13.3 shows examples of graphs of networks with ideal sources.

13.4 ORIENTED OR DIRECTED GRAPH

In network applications we number the branches and nodes of the graph. *A graph is said to be* ***oriented*** *or* ***directed*** *when all its nodes and branches are numbered and arbitrary directions are assigned to the branches by arrows.* In order to form a directed graph, it is necessary to assign positive reference directions to all branch currents and voltages. The convention will be adopted that the positive current reference is the direction of the arrow and the voltage reference is such that the tail of the arrow is positive. This convention is shown in Fig. 13.4. With this convention it is not necessary to shown both references, since one of them implies the other.

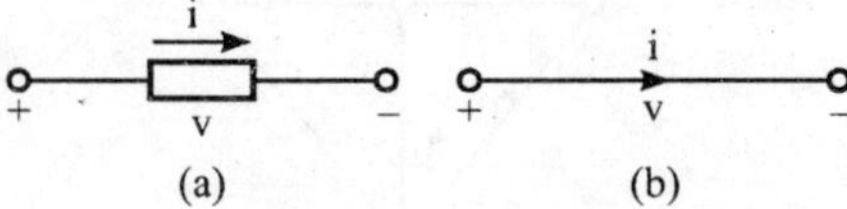

FIG. 13.4 Relationship between branch orientation and voltage and current reference directions (a) Element (b) Oriented graph.

Fig. 13.5a shows a network with reference directions. Fig. 13.5b shows its oriented graph in which ①✍⑤ are nodes and 1-7 are branches.

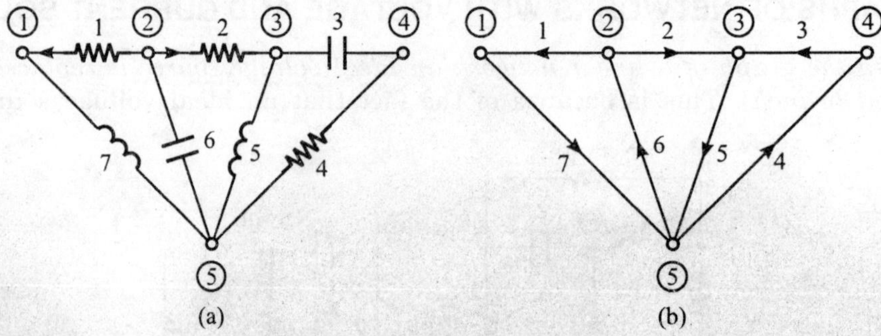

FIG. 13.5 Network with its oriented graph (a) Network (b) Oriented graph.

13.5 DEGREE OF A NODE

If there is a branch b_k (Fig. 13.6) associated with two nodes n_i and n_j, then the branch b_k is said to be *incident* to the nodes n_i and n_j. *The branches whose ends fall on a node are said to be incident on that node.* For example, in Fig. 13.5b branches 1, 2 and 6 are incident on node ②. Branches 4, 5, 6 and 7 are incident on node ⑤ and branches 1, 7 are incident on node ①.

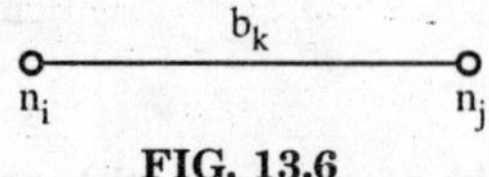

FIG. 13.6

The **degree** *of a node is the number of branches incident to it.* For example, in Fig. 13.5b, nodes ① and ④ and are of degree two, nodes ② and ③ are of degree three and node ⑤ is of degree four. *A node of degree two is called a simple node and a node of degree zero is called an isolated node.*

13.6 SUBGRAPH

A portion of a graph is called a **subgraph**. A graph G_1 is said to be a **subgraph** of a graph G if every node of G_1 is a node of G and every branch of G_1 is a branch of G. In order words, a subgraph G_1 of a graph G is a collection branches and nodes, of G such that every branch and node in G_1 is also contained in G. Thus, if a graph G is given, we can find the subgraph G_1 by deleting from G some branches and/or some nodes. Fig. 13.7 shows five subgraphs G_1, G_2, G_3, G_4 and G_5 of the graph G.

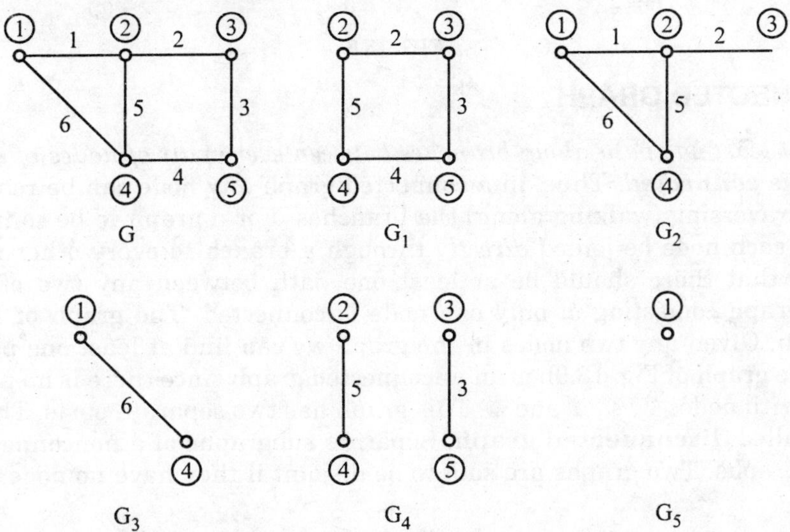

FIG. 13.7 Subgraphs G_1, G_2, G_3, G_4 and G_5 of graph G.

There is no lower limit on the number of nodes or branches in a subgraph. For example, the subgraph G_5 consists of only one node, and is called a *degenerate subgraph.* A subgraph may be connected or unconnected. *A subgraph is said to be a* **proper subgraph** if *it does not contain all the branches and nodes of the graph.*

13.7 PATH

For a graph G an ordered sequence of branches traversed in going from one node to another is called a **path**. In Fig. 13.8b, branches 1, 4, 3 together with nodes ①, ②, ④ and ③ form a path of graph G of Fig. 13.8a from node ① to node ③. Thus, the terminal nodes for this path are ① and ③. The nodes in a path other than terminal nodes are called internal nodes. In Fig. 13.8b, ② and ④ are internal nodes. It is seen that at terminal node ① only branch 1 is incident. Similarly, at the other terminal node ③, only branch 3 is incident. Thus, at each of the terminal nodes, there is incident exactly one branch of the subgraph. At internal node ②, branches 1 and 4 are incident. Similarly, at internal node ④, branches 3 and 4 are incident. Thus, at each internal node, there are incident exactly two branches of the subgraph. In Fig. 13.8c branches 4 and 2 together with nodes ④, ②and ③ form a path from node ④ to node ③. Thus the terminal nodes are ④ and ③ and internal node is ②. It is again seen that the degree of the terminal nodes is 1 and the degree of the internal node is 2.

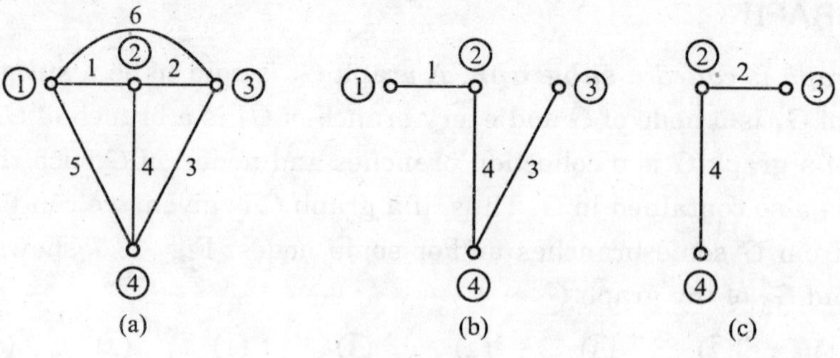

FIG. 13.8

13.8 CONNECTED GRAPH

If there exists at least one path along branches between every pair of nodes of a graph we say that the graph is ***connected***. Thus, in a connected graph any node can be reached from any other node by traversing (walking along) the branches. For a graph to be connected it is *not* necessary that each node be joined *directly* through a branch to every other node. The only requirement is that there should be at least one path between any two of its nodes. By convention, a graph consisting of only one node is connected. The graph of Fig. 13.9a is a connected graph. Given any two nodes in the graph we can find at least one path connecting these nodes. The graph of Fig. 13.9b is an unconnected graph since there is no path connecting nodes ①and ② with nodes ③, ④, ⑤ and ⑥. This graph has two separate parts. The unconnected graph is also called **disconnected graph**. Separate subgraphs of a nonconnected graph are called disjoint graphs. Two graphs are said to be disjoint if they have no node in common.

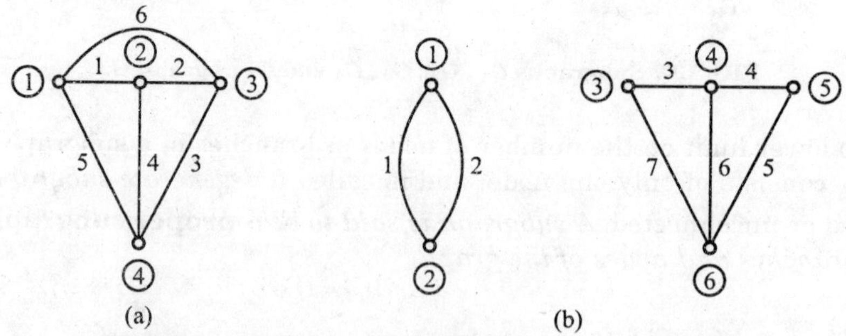

FIG. 13.9 (a) Connected graph (b) Unconnected graph.

13.9 REMOVAL OF A BRANCH

To remove a branch from a graph means to delete the branch but not the nodes to which the branch is connected. Consider the graph shown in Fig. 13.10a. In Fig. 13.10b, branch 3 has been removed leaving the nodes ③ and ④ (at the two ends of the branch) in their proper positions in the graph. Similarly, branch 5 has been removed in Fig. 13.10b leaving the nodes ① and ④ in their proper positions in the graph.

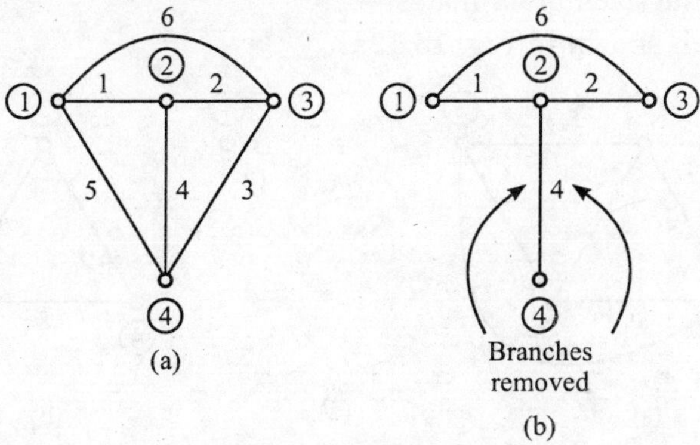

FIG. 13.10 Removal of branches.

13.10 LOOP

In a connected graph, there may be more than one path from one node to another. *A **closed path** is formed if we start at a node, traverse a set of branches and return to the same node without traversing any branch more than once.* In a closed path the terminal nodes coincide.

Let us consider some node n_1 in a graph. Suppose that we start at n_1 and traverse a set of branches such that we return at n_1. If we do not traverse any branch more than once and if we do not encounter any node except n_1 more than once and n_1 exactly twice, then the resulting set of branches traversed is called a *loop*. *A* **loop** *is, therefore, a collection of branches in a graph (oriented or unoriented) which form a closed path with the restriction that exactly two branches of the path are connected to each of the nodes. More precisely, a loop is defined as a connected subgraph of a graph, which has exactly two branches of the subgraph connected to each of its nodes.* This definition requires (*a*) connectedness and (*b*) the existence of exactly two branches of a subgraph at each node.

A loop is specified by either of the following notations :

(1) the set of all branches in the loop

(2) the set of all nodes in the loop when no two branches are in parallel. Each of these sets uniquely specify the loop. For example, nodes {②, ③, ④} of branches {2, 4, 5} specify a loop in Fig. 13.11.

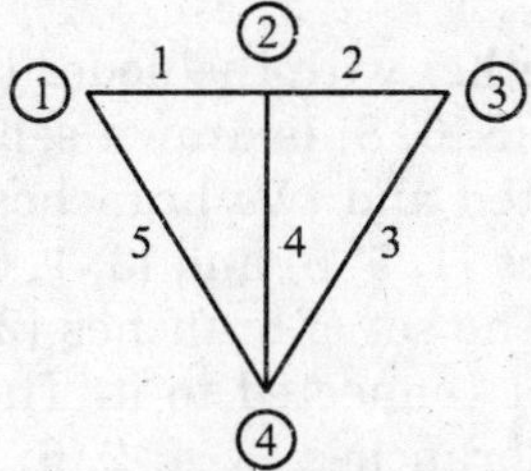

FIG. 13.11 Specification of loops.

To check whether a subgraph is a loop, the following procedure may be adopted :

(*a*) Draw the subgraph separately.

(*b*) Check the conditions of connectedness and the existence of exactly two branches of the subgraph at each of its nodes.

Consider the graph shown in Fig. 13.12a.

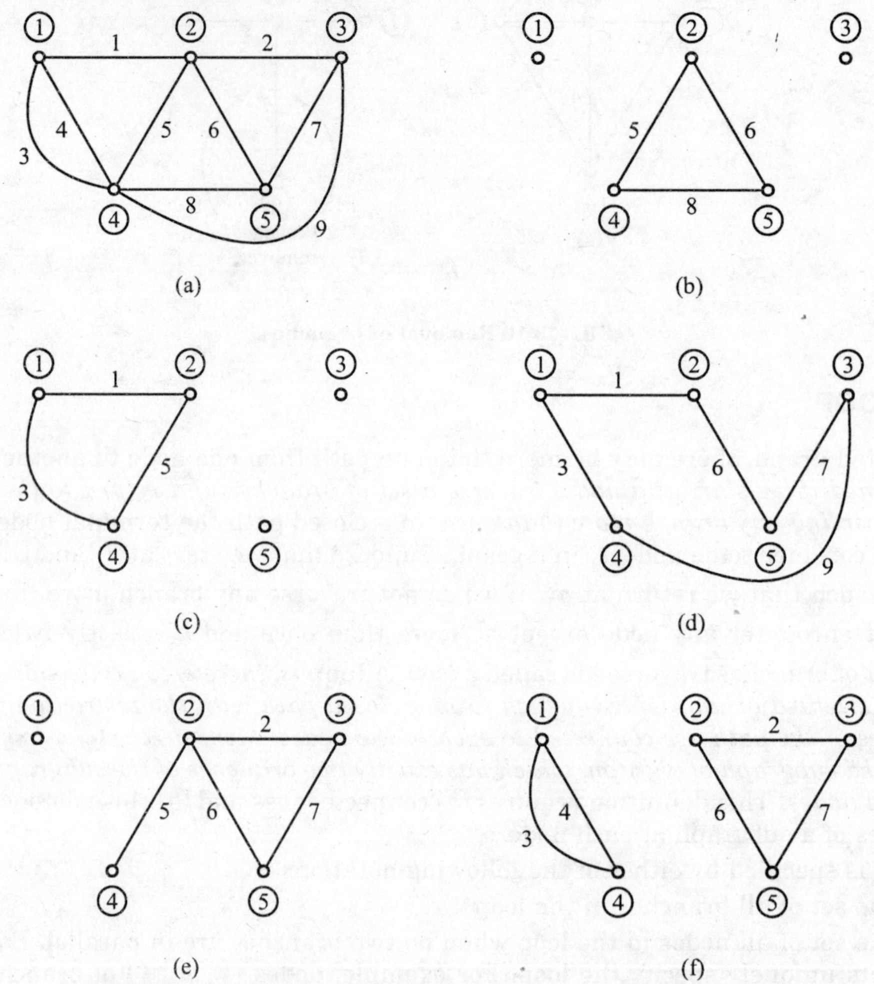

FIG. 13.12 Identification of loops (a) Graph (b)-(f) Subgraphs.

The set of branches {5, 6, 8} together with the nodes {②, ④, ⑤} form a loop. To verify this, the subgraph consisting of branches {5, 6, 8} is drawn separately as shown in Fig. 13.12b. It is seen that this subgraph is connected and two branches of the subgraph are connected to each of its nodes. The sets of branches {1, 3, 5} and {3, 1, 6, 7, 9} also form loops as shown in Fig. 13.12c and 13.12d respectively. The set of branches {2, 5, 6, 7} does not form a loop since node ② has three branches of the set connected to it. The subgraph of this set is shown in Fig. 13.12e. Now consider the set of branches {3, 4, 2, 6, 7}. The subgraph corresponding to this set is given in Fig. 13.12f. Since this subgraph is not connected, the set of branches (3, 4, 2, 6, 7) do not form a loop.

13.11 TREES

*A **tree** of a connected graph is defined as a set of branches which together connect **all** the nodes of the graph without forming any loops.* Thus, a tree T of a connected graph G is a connected subgraph of G with the following properties :

(*a*) T contains *all* the nodes of G

(*b*) All the nodes must be connected by branches called the tree branches. That is, T is a connected subgraph.

(*c*) T does not contain any loop.

*The branches of a tree are called **tree branches** or **twigs.***

The set of all remaining branches in the graph G. (together with their corresponding nodes) which are not in the T, form a subgraph of G. This subgraph of G is called the complement of the tree or the cotree corresponding to T. The branches of a cotree are called links or chords.

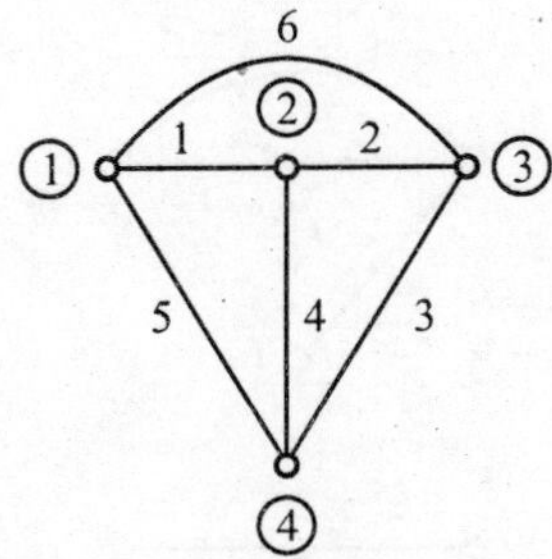

FIG. 13.13

It should be noted that even a simple graph has many different trees. Consider the graph shown in Fig. 13.13. It contains 16 trees. They are shown in Fig 13.14

Since a graph may have several trees, the link branches depend upon the selection of a particular tree. Once the tree branches of a graph are specified, the remaining branches become link branches or chords.

Consider three trees T_{10}, T_{15} and T_{16} of graph of Fig. 13.13. For these trees the twigs and links are shown below :

Tree	Twigs	Links
T_{10}	1, 2, 3	4, 5, 6
T_{15}	2, 4, 6	1, 3, 5
T_{16}	3, 4, 5	1, 2, 6

For the same graph a particular branch may be a twig for one tree of the graph whereas the same branch may be link for another tree of the same graph. For example, branch 4 is a twig for T_{15} whereas the same branch 4 is a link for T_{10}. Similarly, branch 3 is a twig for T_{16} and it is a link for T_{15}.

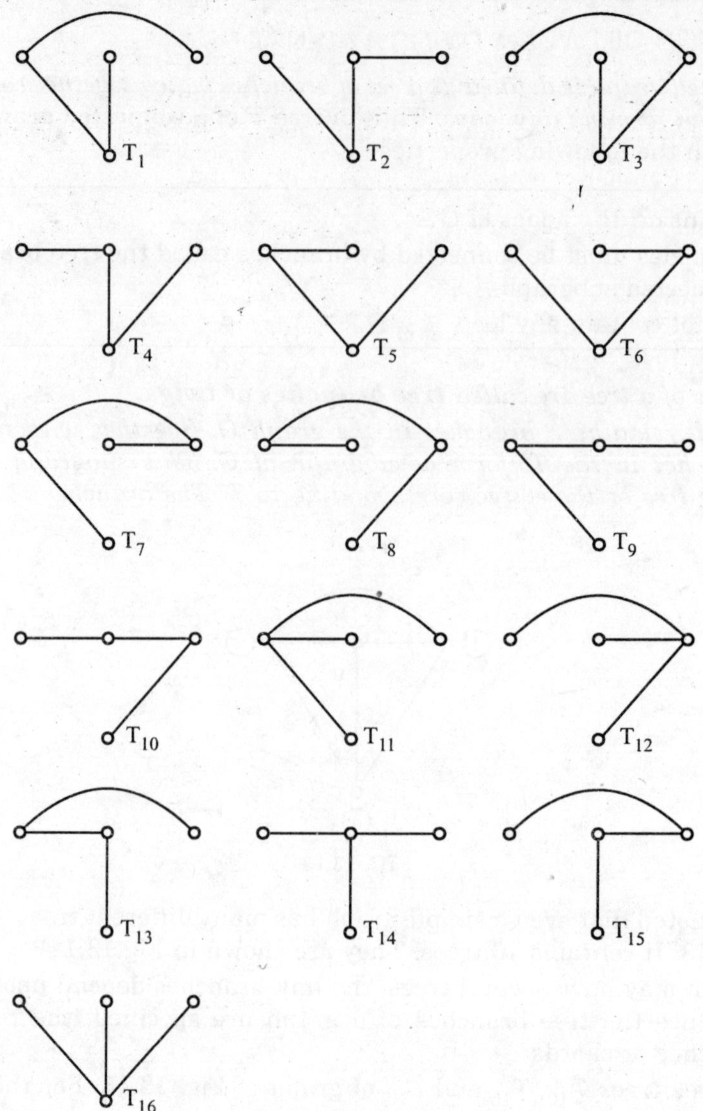

FIG. 13.14

The construction of the tree follows from the definition :

First, draw all the nodes, then fill in enough branches to connect all the nodes without creating any loops.

The branches of a cotree are called links or chords. In other words, the branches not belonging to a tree are called links or chords, and the set of all links form a cotree.

The branches of a cotree may or may not be connected, whereas the branches of a tree are always connected. Also, cotrees may have loops. By definition, a tree contains no loops.

13.12 RELATION BETWEEN TWIGS AND NODES

The number of twigs on a tree is one less than the number of nodes.

That is, number of twigs = number of nodes – 1

Let n_t = number of nodes of the graphs G

b = total number of branches of the graph G

Therefore, *the number of tree branches or twigs on a tree is equal to* $(n_t - 1)$.

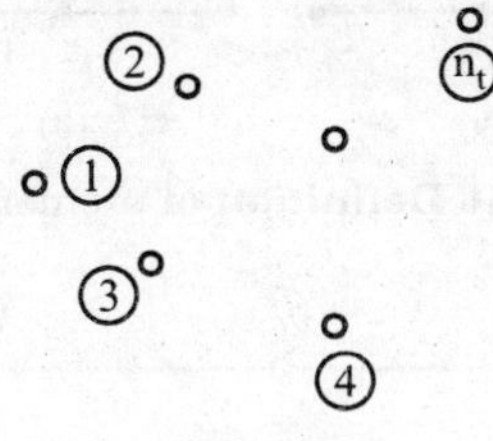

FIG. 13.15

This result can be proved as follows :

Suppose we mark the n_t nodes as ①, ②, ③, ..., n_t as shown in Fig. 13.15. Let a tree be constructed branch by branch starting from node ①. The first branch 1 connects node ① and any other node say node ②. Next, we take a new node ③. We can find a branch 2 connecting ① or ② to the new node ③. Thus, we find that one branch is added and a new node is connected into the growing tree. This process continues till a tree is completed. In this process each added branch brings a new node.

Since the first pair of nodes, when connected, resulted in only one branch, we find that the total number of tree branches of a graph containing n_t nodes is equal to $(n_t - 1)$. Let $n_t - 1 = n$. We also know that the total number of branches in a graph is equal to the sum of the number of tree branches (twigs) and the links. That is,

total number of branches in the graph = twigs + links

$b = (n_t - 1)$ + links

therefore, number of links $= b - (n_t - 1) = b - n$

13.13 COMPLETE INCIDENCE MATRIX A_a

An oriented graph may be described completely in a compact and convenient matrix. In this matrix, called the complete incidence matrix, we specify the orientation of each branch in the graph and the nodes at which this branch is incident. Branches whose ends fall on a node are said to be incident at the node. This matrix $\mathbf{A}_a$ shows which branch is incident to which node. Each row of this matrix corresponds to a node of the graph and each column corresponds to a branch. For graph having n_t nodes and b branches, the complete incidence matrix $\mathbf{A}_a = [a_{ij}]$ is an $n_t \times b$ rectangular matrix whose elements a_{ij} are defined as follows :

$a_{ij} = 1$	if branch j is incident at node i and is oriented *away* from the node, that is, branch j *leaves* node i. (Fig. 13.16a)
$a_{ij} = -1$	if branch j is incident at node i and is oriented *towards* node i, that is, branch j *enters* node i. (Fig. 13.16b)
$a_{ij} = 0$	if branch j is *not incident* at node i.

The complete incidence matrix is also called the **augmented incidence matrix**.

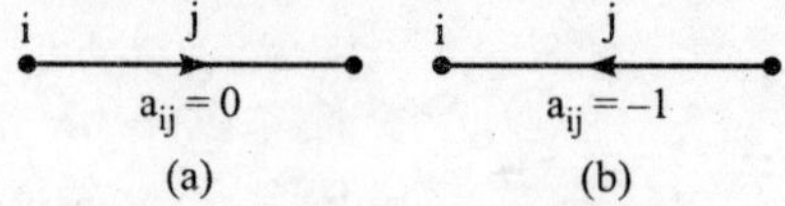

FIG. 13.16 Definition of elements a_{ij}.

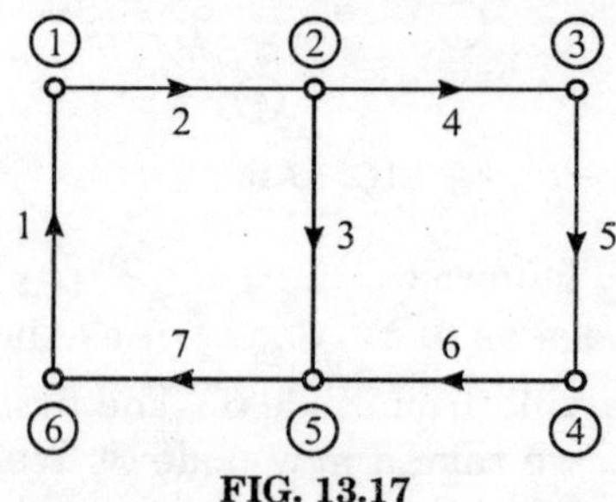

FIG. 13.17

Consider the connected oriented graph shown in Fig. 13.17. There are 6 nodes and 7 branches in this graph. The complete incidence matrix $\mathbf{A}_a$ of this graph will have 6 rows and 7 columns.

$$\begin{array}{c} \text{nodes} \\ \downarrow \end{array} \quad \text{branches} \rightarrow$$

$$\mathbf{A}_a = \begin{array}{c} ① \\ ② \\ ③ \\ ④ \\ ⑤ \\ ⑥ \end{array} \begin{bmatrix} -1 & 1 & 0 & 0 & 0 & 0 & 0 \\ 0 & -1 & 1 & 1 & 0 & 0 & 0 \\ 0 & 0 & 0 & -1 & 1 & 0 & 0 \\ 0 & 0 & 0 & 0 & -1 & 1 & 0 \\ 0 & 0 & -1 & 0 & 0 & -1 & 1 \\ 1 & 0 & 0 & 0 & 0 & 0 & -1 \end{bmatrix} \tag{13.13.1}$$

(columns correspond to branches 1 2 3 4 5 6 7)

The rows and columns are numbered. The elements in the first row correspond to node ①. It is seen from the graph that on node ① only two branches 1 and 2 are incident. Branch 1 is oriented towards node ① hence the entry a_{11} is – 1. Branch 2 is oriented away from node ① and therefore the entry $a_{12} = 1$. Since branches 3, 4, 5, 6 and 7 are not incident on node ①, the entries a_{13}, a_{14}, a_{15}, a_{16} and a_{17} are all equal to zero. Next we focus our attention on node ②, Branches 2, 3 and 4 are incident at node ②. Branch 2 is towards node ② therefore the element $a_{22} = -1$. Branches 3 and 4 are oriented away from node ② therefore $a_{23} = 1$ and $a_{24} = 1$. Since the remaining branches 1, 5, 6 and 7 are not incident on node ②, the entires a_{21}, a_{25}, a_{26} and a_{27} are all equal to zero. Similarly, we focus our attention on nodes ③, ④, ⑤ and ⑥ one at a time and fill their elements. The complete incidence matrix for the graph in Fig. 13.17 is given in Eq. (13.13.1)

Since the arrow in each branch *leaves* one node and *enters* another, each column of $\mathbf{A}_a$ has exactly two nonzero elements, one of which is + 1 and the other is – 1. Thus, it is also possible to focus our attention on the *branches* of the graph one at a time and fill the elements of the complete incidence matrix. For example, branch 1 enters node ① and leaves node ⑥. Therefore $a_{11} = -1$, and $a_{16} = 1$. All the remaining elements of column 1 are zero. Similarly, branch 2 leaves node ① and enters node ②, therefore $a_{12} = 1$ and $a_{22} = -1$. All the remaining elements of column 2 are zero. Let each row of the complete incidence matrix be represented by R with a subscript. For example, the first row is R_1 and the second row is R_2. Suppose that all the rows of $\mathbf{A}_a$ are added, the sum will be a row of zeros.

Thus,

$$\Sigma R = R_1 + R_2 + R_3 + R_4 + R_5 + R_6$$
$$= [0 \quad 0 \quad 0 \quad 0 \quad 0 \quad 0]$$

Also,

$$R_2 = -(R_1 + R_3 + R_4 + R_5 + R_6)$$
$$R_3 = -(R_1 + R_2 + R_4 + R_5 + R_6)$$
$$R_6 = -(R_1 + R_2 + R_3 + R_4 + R_5)$$

Thus, any one row of $\mathbf{A}_a$ is the negative of the sum of all the other rows. Although this result is true for $\mathbf{A}_a$ of graph in Fig. 13.17, it is true for all connected oriented graphs. Hence, we say that the rows of $\mathbf{A}_a$ form a *linearly dependent* set. Since the rows of $\mathbf{A}_a$ are linearly dependent, *any* one row can be obtained from the remaining rows as shown for the rows R_2, R_3 and R_6.

13.14 INCIDENCE MATRIX *A*

We have seen that *any* one of the rows of a complete incidence matrix can be obtained from the remaining rows. Thus, it is possible to delete any one row from $\mathbf{A}_a$ without loosing any information contained in $\mathbf{A}_a$. When one row is deleted from the complete incidence matrix $\mathbf{A}_a$, the remaining matrix is called a **reduced incidence matrix**, or simply an **incidence matrix**.

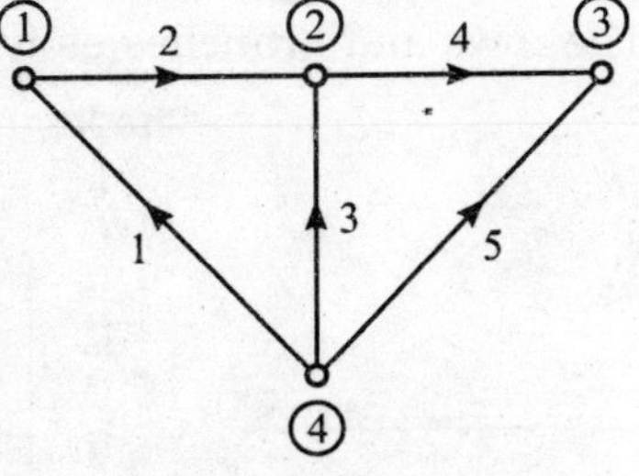

FIG. 13.18

The incidence matrix is represented by the symbol **A**, without a subscript. It should be noted that the subscript a on $\mathbf{A}_a$ indicates *all nodes*, since all rows are included.

For a graph with n_t nodes and b branches, the order of the incidence matrix is $(n_t - 1) \times b$. That is, A has $(n_t - 1)$ rows and b columns.

The node corresponding to the deleted row is called the **reference node** or **datum node**.

> The number of possible trees of a graph = det (**AA′**) where **A** is the reduced incidence matrix and **A′** is the transpose of the matrix **A**.

Consider the oriented graph shown in Fig. 13.18. The complete incidence matrix $\mathbf{A}_a$ for this graph is given by

$$\begin{array}{c} \text{nodes} \\ \downarrow \end{array} \quad \begin{array}{c} \text{branches} \rightarrow \\ \begin{array}{ccccc} 1 & 2 & 3 & 4 & 5 \end{array} \end{array}$$

$$\mathbf{A}_a = \begin{array}{c} ① \\ ② \\ ③ \\ ④ \end{array} \begin{bmatrix} -1 & 1 & 0 & 0 & 0 \\ 0 & -1 & -1 & 1 & 0 \\ 0 & 0 & 0 & -1 & -1 \\ 1 & 0 & 1 & 0 & 1 \end{bmatrix}$$

If we assume that node ④ is the datum node, the corresponding incidence matrix can be obtained from $\mathbf{A}_a$ by deleting the last row :

$$\mathbf{A} = \begin{bmatrix} -1 & 0 & 0 & 0 & 0 \\ 0 & -1 & -1 & 1 & 0 \\ 0 & 0 & 0 & -1 & -1 \end{bmatrix}$$

It is to be noted that *the algebraic sum of each column entries of an augmented incidence matrix is zero.*

13.15 DRAWING A GRAPH FROM AN INCIDENCE MATRIX

Given any incidence matrix we can draw its corresponding graph. We know that each column of $\mathbf{A}_a$ has exactly two nonzero elements, one of which is + 1 and the other is – 1. Indicate n_t nodes on a paper and number them in a random manner. Connect these nodes according to the above property. Consider the following complete incidence matrix :

$$\mathbf{A}_a = \begin{bmatrix} -1 & 0 & 1 & 1 & 0 \\ 0 & 1 & 0 & -1 & 0 \\ 1 & -1 & 0 & 0 & -1 \\ 0 & 0 & -1 & 0 & 1 \end{bmatrix}$$

The given matrix is of the order 4 × 5. Thus, it has 4 nodes and 5 branches. Let us number the rows and branches as follows :

$$\begin{array}{c} \text{nodes} \\ \downarrow \end{array} \quad \begin{array}{c} \text{branches} \rightarrow \\ \begin{array}{ccccc} 1 & 2 & 3 & 4 & 5 \end{array} \end{array}$$

$$\begin{array}{c} ① \\ ② \\ ③ \\ ④ \end{array} \begin{bmatrix} -1 & 0 & 1 & 1 & 0 \\ 0 & 1 & 0 & -1 & 0 \\ 1 & -1 & 0 & 0 & -1 \\ 0 & 0 & -1 & 0 & 1 \end{bmatrix}$$

Mark four nodes ①, ②, ③ and ④ arbitrarily as shown in Fig. 13.19.

The first column of $\mathbf{A}_a$ corresponds to branch 1. It is seen from this column that branch 1 is incident at nodes ① and ③. The entries of – 1 at node ① and + 1 at node ③ show that branch 1 enters node ① and leaves node ③. Therefore, connect 1 and 3 by branch 1 with the orientation as shown in Fig. 13.19. The second column of $\mathbf{A}_a$ corresponds to branch 2. The elements of this column indicate that branch 2 is incident at nodes ② and ③. At ② it leaves and at ③ it enters. Therefore, connect nodes ② and ③ by branch 2 with the orientation as shown in Fig. 13.19. It is seen from third column of $\mathbf{A}_a$ that branch 3 is connected between nodes ① and ④. The fourth column of $\mathbf{A}_a$ shows that branch 4 is connected between nodes ① and ②. Similarly, the fifth column of $\mathbf{A}_a$ corresponds to branch 5. It shows that branch 5 is connected between nodes ③ and ④. The oriented graph of $\mathbf{A}_a$ is shown in Fig. 13.19.

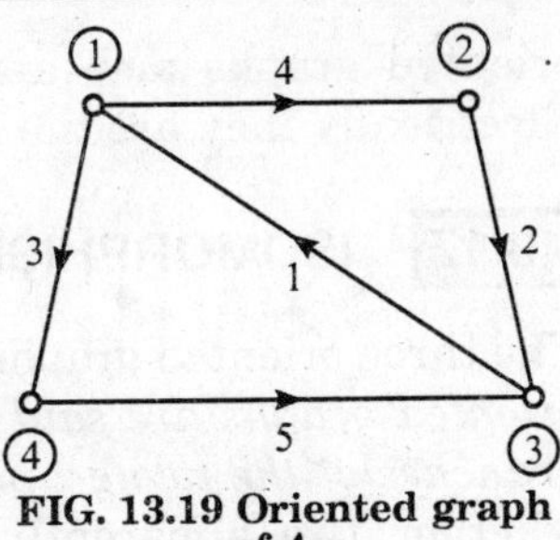

FIG. 13.19 Oriented graph of A_a.

13.16 DRAWING A GRAPH FROM A REDUCED INCIDENCE MATRIX

The following rule is used for obtaining an oriented graph from a reduced incidence matrix :

If the reduced incidence matrix **A** has got n nodes then write down $(n + 1)$ nodes on the paper and number the nodes according to rows. Then consider the columns one at a time. There are *almost* two nonzero elements in each column. Draw a branch between the two nodes corresponding to the two rows having nonzero elements in that column. If there is only one nonzero element in a column, the branch goes between the node corresponding to this row and the extra node (or reference node). The orientation is determined by the signs of the elements of **A**.

■ **EXAMPLE 13.1** Draw the oriented graph corresponding to the following incidence matrix :

$$[\mathbf{A}] = \begin{bmatrix} -1 & 1 & 0 & 0 & 0 & 0 & 0 \\ 0 & -1 & 1 & 1 & 0 & 0 & 0 \\ 0 & 0 & 0 & -1 & 1 & 0 & 0 \\ 0 & 0 & 0 & 0 & -1 & 1 & 0 \\ 0 & 0 & -1 & 0 & 0 & -1 & 1 \end{bmatrix}$$

SOLUTION : It is seen that the first column of [A] has a single nonzero element 1. There are five rows in [A]. Therefore, the number of nodes to be marked on the paper for drawing the graph will be (5 + 1), that is 6. We shall take node ⑥ as the reference node. The complete incidence matrix can, therefore, be written as follows :

nodes ↓ \ branches →	1	2	3	4	5	6	7
①	−1	1	0	0	0	0	0
②	0	−1	1	1	0	0	0
③	0	0	0	−1	1	0	0
④	0	0	0	0	−1	1	0
⑤	0	0	−1	0	0	−1	1
⑥	1	0	0	0	0	0	−1

$$[\mathbf{A}_a] = \begin{bmatrix} -1 & 1 & 0 & 0 & 0 & 0 & 0 \\ 0 & -1 & 1 & 1 & 0 & 0 & 0 \\ 0 & 0 & 0 & -1 & 1 & 0 & 0 \\ 0 & 0 & 0 & 0 & -1 & 1 & 0 \\ 0 & 0 & -1 & 0 & 0 & -1 & 1 \\ 1 & 0 & 0 & 0 & 0 & 0 & -1 \end{bmatrix}$$

Now draw the graph corresponding to this complete incidence matrix showing the branches and their orientations.

The nodes may be placed arbitrarily on the paper according to convenience. We shall place them in three ways as shown in Fig. 13.20. There is no resemblance among the three oriented graphs a, b and c although they are obtained from the same incidence matrix. Electrically they are the same but physically they appear to be different.

13.17 ISOMORPHISM

The three oriented graphs shown in Figs. 13.20a, b and c are called **isomorphic graphs**. *Two oriented graphs are said to be* ***isomorphic*** *if there is one-to-one correspondence between the branches of the two graphs so that both have got the same incidence matrix.* In order to test whether two apparently dissimilar graphs (but having the same number of nodes) are isomorphic or not, we have to prepare a table showing one-to-one correspondence between nodes and branches.

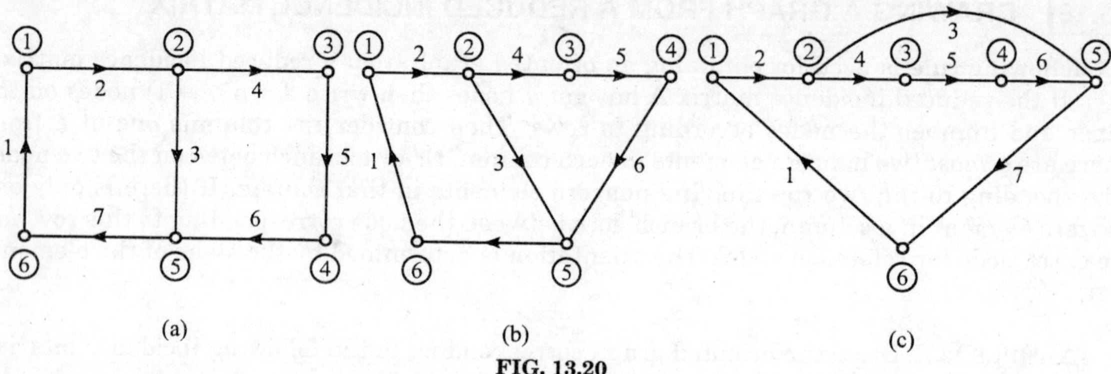

(a) (b) (c)

FIG. 13.20

13.18 KIRCHHOFF'S CURRENT LAW (KCL)

Consider the graph of Fig. 13.21. It has four nodes. Let node 4 be taken as the reference node. The positive reference direction of the branch currents corresponds to the orientation of the graph branches. Let the branch currents be denoted by $i_1, i_2, \ldots, i_6$.

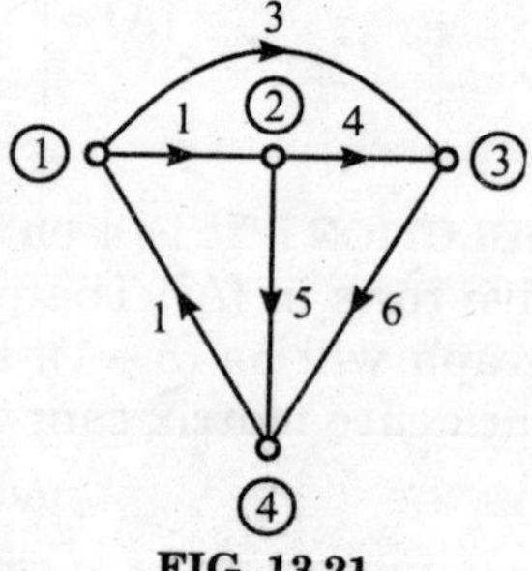

FIG. 13.21

Applying KCL at nodes ①, ② and ③, we get

$$-i_1 + i_2 + i_3 = 0$$

$$-i_2 - i_4 + i_5 = 0$$

$$-i_3 + i_4 + i_6 = 0$$

In matrix form, these equations can be written as

$$\begin{bmatrix} -1 & 1 & 1 & 0 & 0 & 0 \\ 0 & -1 & 0 & -1 & 1 & 0 \\ 0 & 0 & -1 & 1 & 0 & 1 \end{bmatrix} \begin{bmatrix} i_1 \\ i_2 \\ i_3 \\ i_4 \\ i_5 \\ i_6 \end{bmatrix} = \begin{bmatrix} 0 \\ 0 \\ 0 \end{bmatrix}$$

or
$$\mathbf{A}\,\mathbf{i}_b = 0 \tag{13.18.1}$$

Eq. (13.18.1) gives the matrix representation of KCL. In this equation $\mathbf{i}_b$ represents the vector of branch currents. The matrix **A** is called the **reduced incidence matrix**. The **branch current vector** or the **branch matrix** $\mathbf{i}_b$ is given by

$$\mathbf{i}_b = \begin{bmatrix} i_1 \\ i_2 \\ i_3 \\ \cdot \\ \cdot \\ \cdot \\ \cdot \\ i_b \end{bmatrix}$$

The reduced incidence matrix **A** is written in a compact form as

$$\mathbf{A} = (\mathbf{a}_{ij})$$

The element a_{ij} of **A** are defined as follows :

$a_{ij} = 1$ if branch j is incident at node i and *leaves* the node i

$a_{ij} = -1$ if branch j is incident at node i and *enters* node i

$a_{ij} = 0$ if branch j is *not incident* on node i.

The incident matrix **A** of a graph G with n_t nodes and b branches is a $(n_t - 1) \times b$ matrix obtained from the augmented incidence matrix of G deleting one of its rows. The node corresponding to the deleted row is called the **reference node** or **datum node**.

It is to be noted that Eq. (13.18.1) gives as set of $(n_t - 1)$ linearly independent equations in the variables $i_1, i_2, \ldots, i_b$.

13.19 FUNDAMENTAL LOOPS OF A GRAPH

The addition of a link between any two nodes of a tree forms a loop called the fundamental loop. Consider a connected graph G of Fig. 13.22a. Let T be a tree of G as shown in Fig. 13.22b. The twigs of this tree are branches 2, 4 and 5. The links of this tree are branches 1, 3 and 6. Out of these three links let us replace link 1 in its proper place as shown in Fig. 13.22c. It is seen that a loop {1, 2, 5} designated by l_1 is formed. This loop has one *link* 1 and a unique set of branches 2 and 5. This loop is called the *fundamental loop* or *f-loop* of G with respect to a tree T. It is to be noted that no other combination of the twigs with link 1 forms a fundamental loop.

Here link 1 is called the **defining link**. *By convention, a fundamental loop is given the same orientation as its defining link.* Fig. 13.22c shows the orientation of loop l_1. It is seen

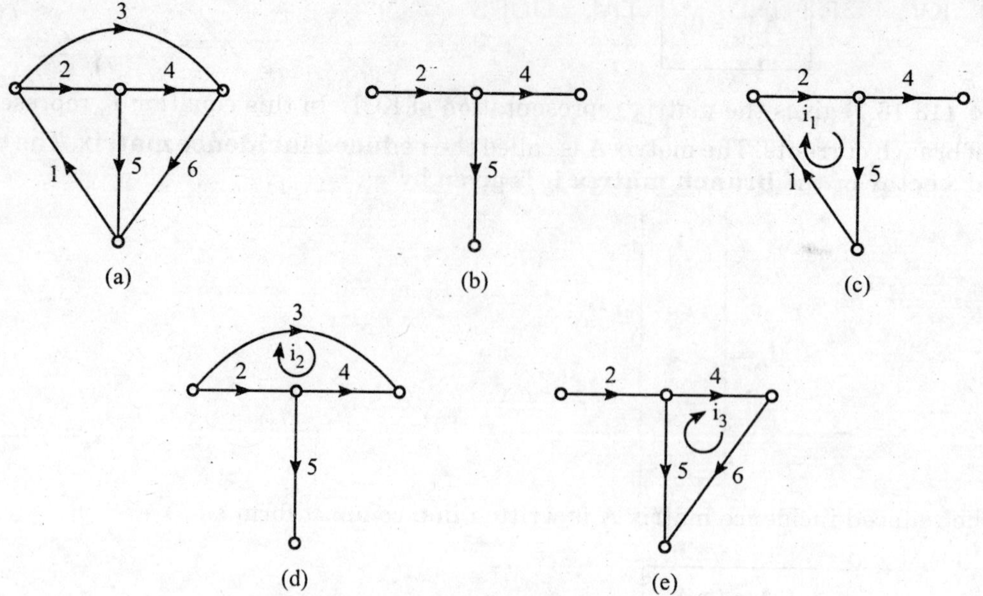

FIG. 13.22 Formation of f-loops.

that the current in link 1 will be the same as the current in branches 2 and 5 of the f-loop (1, 2, 5).

Next, if we replace link 3 in its proper position we find that a loop {3, 4, 2} is formed. This f-loop is indicated as loop l_2 in Fig. 13.22d. The orientation of l_2 coincides with the orientation of its defining link 3. In a similar manner replacing link 6 in its proper position results in the f-loop {6, 5, 4}. This loop is shown as l_3 in Fig. 13.22e. Now we can define a fundamental loop as follows :

*A **fundamental loop (or f-loop)** of a graph G with respect to a tree T is a loop that is formed by one link and the unique path in the tree joining the two nodes of the link. A fundamental loop is also called an **f-circuit** or **tie-set.***

Since for each link of T there corresponds a fundamental loop, the number of f-loops is equal to the number of links of a chosen tree. Thus for a tree T of a connected graph G with n_t nodes and b branches the number of f-loops is equal to $b - (n_t - 1)$.

13.20 PROCEDURE OF OBTAINING f-LOOPS OF A GRAPH

1. Select a tree for the given graph.
2. Replace one link in its proper place to form a loop. This is the first f-loop.
3. Replace the second link in its proper place to form the second f-loop.
4. Repeat the above procedure till each of the remaining links of the chosen tree is replaced in its proper place *one at a time.*

In ths way as many f-loops are formed as there are links in the chosen tree.

By convention, a fundamental loop is assigned the same orientation as its defining link.

13.21 KVL FOR FUNDAMENTAL LOOPS

Consider a connected graph G with b branches and n_t nodes. Select any arbitrary tree. The tree will contain $n = n_t - 1$ tree branches (twigs) and $l = (b - n)$ link branches. Every link defines a fundamental loop of the network. Let us take the example of the graph shown in Fig. 13.23a.

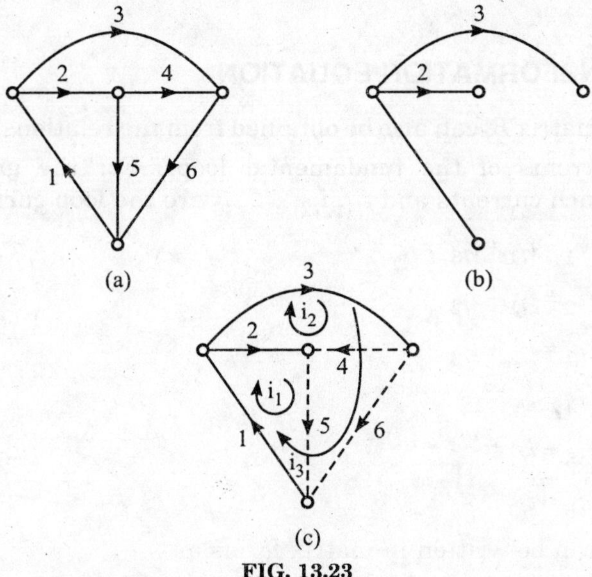

FIG. 13.23

Let T be a tree of G as shown in Fig. 13.23b. The number of fundamental loops of this graph will be $(b - n) = 3$. The three f-loop l_1, l_2 and l_3 are shown in Fig. 13.23c. In order to apply KVL to each fundamental loop, we take reference direction of the loop which coincides with the reference direction of the link defining the loop.

$$l_1 : v_1 + v_2 + v_5 = 0$$
$$l_2 : -v_2 + v_3 + v_4 = 0$$
$$l_3 : v_1 + v_3 + v_6 = 0$$

In matrix form, we can write

$$\begin{array}{c} \text{loop} \\ \downarrow \\ 1 \\ \\ 2 \\ \\ 3 \end{array} \begin{array}{c} \text{branch} \rightarrow \\ \begin{array}{cccccc} 1 & 2 & 3 & 4 & 5 & 6 \end{array} \\ \begin{bmatrix} 1 & 1 & 0 & 0 & 1 & 0 \\ 0 & -1 & 1 & 1 & 0 & 0 \\ 1 & 0 & 1 & 0 & 0 & 1 \end{bmatrix} \end{array} \begin{bmatrix} v_1 \\ v_2 \\ v_3 \\ v_4 \\ v_5 \\ v_6 \end{bmatrix} = \begin{bmatrix} 0 \\ 0 \\ 0 \\ 0 \\ 0 \\ 0 \\ 0 \\ 0 \end{bmatrix}$$

or $\boxed{\mathbf{B}_f v_b = \mathbf{0} \quad \text{KVL}}$ (13.21.1)

where $\mathbf{B}_f$ is an $l \times b$ matrix called the **fundamental loop matrix** or **tie-set matrix**.

$$\mathbf{B}_f = [b_{kj}]$$

where b_{kj} are the elements of $\mathbf{B}_f$. The (k, j) element of the matrix is defined as follows :

$b_{kj} = 1$ when branch b_j is in the f-loop l_k and their reference directions (orientations) coincide

$b_{kj} = -1$ when branch b_j is in the f-loop l_k and has opposite orientation

$b_{kj} = 0$ when branch b_j is not in the f-loop l_k

13.22 LOOP-TRANSFORMATION EQUATION

The fundamental loop matrix B_f can also be obtained from the relationship between the branch currents and loop currents of the fundamental loops. For the graph in Fig. 13.24, if $i_1, i_2, \ldots, i_6$ are the branch currents and $i_{l1}, i_{l2}, \ldots, i_{l6}$ are the loop currents, we can write

$$i_1 = i_{l1} + i_{l3}$$

$$i_2 = i_{l1} - i_{l2}$$

$$i_3 = i_{l2} + i_{l3}$$

$$i_4 = i_{l2}$$

$$i_5 = i_{l1}$$

$$i_6 = i_{l3}$$

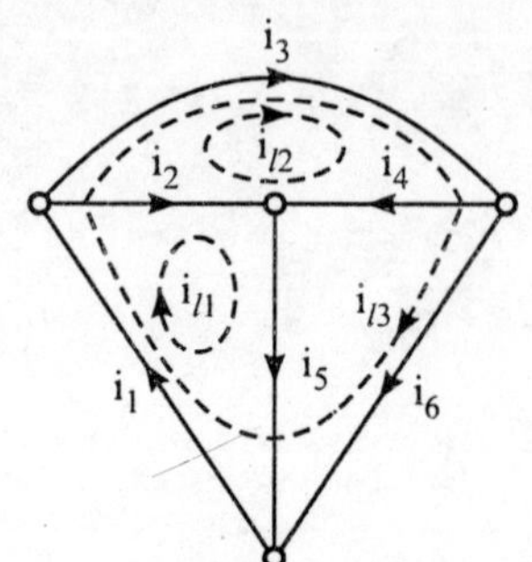

FIG. P 13.24

These equations can be written in matrix forms as

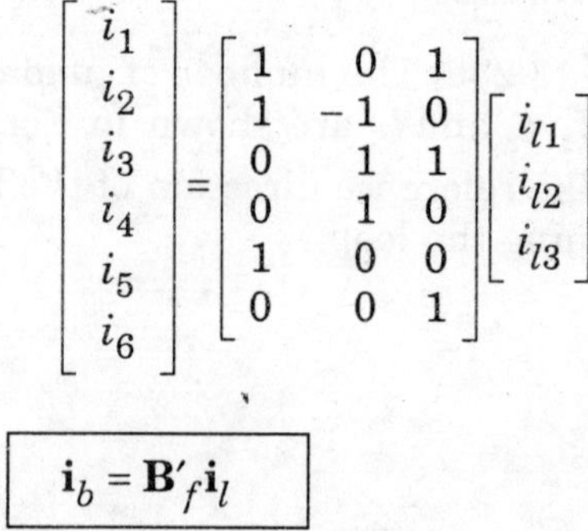

$$\begin{bmatrix} i_1 \\ i_2 \\ i_3 \\ i_4 \\ i_5 \\ i_6 \end{bmatrix} = \begin{bmatrix} 1 & 0 & 1 \\ 1 & -1 & 0 \\ 0 & 1 & 1 \\ 0 & 1 & 0 \\ 1 & 0 & 0 \\ 0 & 0 & 1 \end{bmatrix} \begin{bmatrix} i_{l1} \\ i_{l2} \\ i_{l3} \end{bmatrix}$$

or

$$\boxed{\mathbf{i}_b = \mathbf{B}'_f \mathbf{i}_l} \qquad (13.22.1)$$

Eq. (13.22.1) expresses the relationship between the loop currents and the branch currents. This equation is known as the *loop-transformation equation*.

■ **EXAMPLE 13.2** In the graph shown in Fig. 13.25a, a tree consisting of branches 4, 5, 6, 7, 8 is chosen, as shown by heavy lines. Write the fundamental loop matrix of the graph.

SOLUTION The fundamental loops defined by links {1, 2, 3} and their orientations are shown in Figs. 13.25b; c and d. Consider loop l_1. It contains branches {1, 6, 8, 5}. The orientation of loop l_1 is given the same orientation as its defining link 1. Therefore the element b_{11} is written 1. The directions of branches 6, 8 and 5 in l_1 are the same as l_1. Therefore the entries b_{16}, b_{18} and b_{15} are each equal to 1. Since branches 2, 3, 4 and 7 are not in loop l_1, $b_{12} = 0$, $b_{13} = 0$, $b_{14} = 0$, $b_{17} = 0$.

Loop l_2 contains branches {2, 6, 7}. The orientation of l_2 is given the same orientation as its defining link. 2. Therefore the element $b_{22} = 1$. The orientations of branches 6 and 7 are opposite to the orientation of l_2. Consequently, $b_{26} = -1$ and $b_{27} = -1$. Since branches {1, 3, 4, 5, 8} are not contained in l_2, $b_{21} = 0$, $b_{23} = 0$, $b_{24} = 0$, $b_{25} = 0$, $b_{28} = 0$.

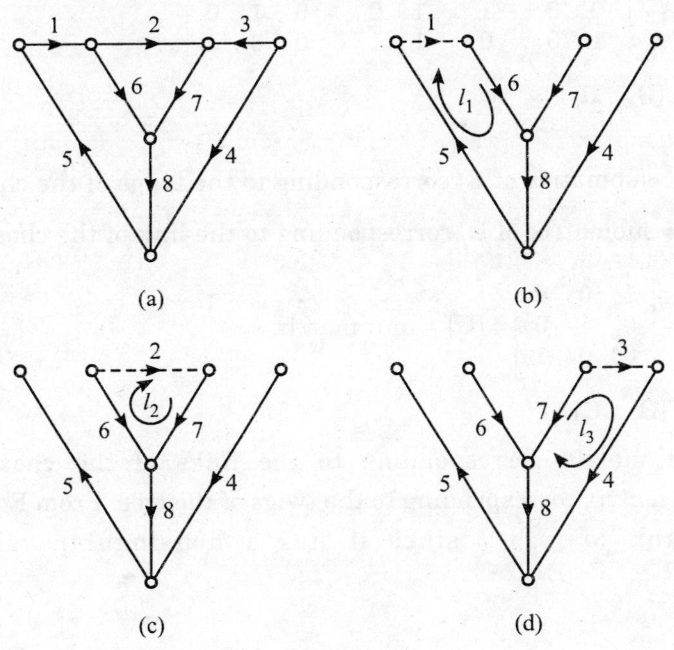

FIG. 13.25

Loop l_3 has branches (3, 4, 8, 7). The orientation of l_3 is the same as that of its defining link 3.Therefore the element $b_{33} = 1$. The directions of branches 4 and 7 coincide with the orientation of l_3. Hence $b_{34} = 1$, $b_{37} = 1$. The orientation of branch 8 does not coincide with the orientation of l_3. Hence $b_{38} = -1$. Since branches {1, 2, 5, 6} are not contained in l_3, $b_{31} = 0$, $b_{32} = 0$, $b_{35} = 0$, $b_{36} = 0$. Thus, we obtain the following matrix :

$$\begin{array}{c} \text{loops} \\ \downarrow \end{array} \quad \begin{array}{c} \text{branches} \rightarrow \\ \begin{array}{cccccccc} 1 & 2 & 3 & 4 & 5 & 6 & 7 & 8 \end{array} \end{array}$$

$$\mathbf{B}_f = \begin{array}{c} l_1 \\ l_2 \\ l_3 \end{array} \begin{bmatrix} 1 & 0 & 0 & 0 & 1 & 1 & 0 & 1 \\ 0 & 1 & 0 & 0 & 0 & -1 & -1 & 0 \\ 0 & 0 & 1 & 1 & 0 & 0 & 1 & -1 \end{bmatrix}$$

13.23 RANK OF FUNDAMENTAL LOOP MATRIX

Let the fundamental loop matrix in Example 13.2 be partitioned in the form

$$\begin{array}{c} \text{branches} \rightarrow \\ \begin{array}{cc} \text{loops} & \text{twigs} \qquad\qquad \text{links} \\ \downarrow & 4 \quad 5 \quad 6 \quad 7 \quad 8 \quad 1 \quad 2 \quad 3 \end{array} \end{array}$$

$$\mathbf{B}_f = \begin{array}{c} l_1 \\ l_2 \\ l_3 \end{array} \begin{bmatrix} 0 & 1 & 1 & 0 & 1 & 1 & 0 & 0 \\ 0 & 0 & -1 & -1 & 0 & 0 & 1 & 0 \\ 1 & 0 & 0 & 1 & 1 & 0 & 0 & 1 \end{bmatrix}$$

$$\mathbf{B}_f = [\mathbf{B}_{ft}, \mathbf{B}_{fl}]$$

where

$\mathbf{B}_{ft}$ = submatrix of $\mathbf{B}_f$ corresponding to the twigs of the chosen tree

$\mathbf{B}_{fl}$ = submatrix of $\mathbf{B}_f$ corresponding to the link of the chosen tree

$$\mathbf{B}_{fl} = \begin{bmatrix} 1 & 0 & 0 \\ 0 & 1 & 0 \\ 0 & 0 & 1 \end{bmatrix} = [\mathbf{U}] = \text{unit matrix}$$

$$\therefore \qquad \mathbf{B}_f = [\mathbf{B}_{ft}, \mathbf{U}] \tag{13.23.1}$$

$\mathbf{B}_{fl}$ is a unit matrix corresponding to the links of the chosen tree. $\mathbf{B}_{ft}$ is a $(n_t - 1) \times (b - n_t + 1)$ matrix corresponding to the twigs of the tree. From Eq. (13.23.1) it is seen that B_f is of rank $(b - n_t + 1)$ since it has a nonsingular submatrix of order $(b - n_t + 1) \times (b - n_t + 1)$.

13.24 CUT-SET

By removing certain branches of a graph, a connected graph can be separated into two parts. This operation is equivalent to cutting a graph into two parts.The term cut-set is derived from this property.

Consider the graph shown in Fig. 13.26a. Let us remove branches 1, 3, 4 and 5 from this graph. After removal of these branches we see that the connected graph of Fig. 13.26a is separated into two distinct parts each of which is connected as shown in Fig. 13.26b. Now suppose each removed branch is replaced one at a time. Fig. 13.26c shows the result if branch 1 is replaced in Fig. 13.26b. The graph is now connected. Similarly, if we replace the removed branches 3, 4 and 5 of the set {1, 3, 4, 5} one at a time, all other ones remaining removed, we get the resulting graphs shown in Figs. 13.26d, e and f. The set formed by the branches 1, 3, 4 and 5 is called the cut-set of the connected graph of Fig. 13.26a. It is represented by {1, 3, 4, 5} where the numerals 1, 3, 4, 5 enclosed by the curl bracket, stand for the respective branches of the cut-set. We can define a cut-set as follows :

A **cut-set** *is a set of minimum number of branches of a connected graph whose removal causes the graph to be cut (separated) into exactly two parts (subgraphs), each of which is a connected graph, with the condition that replacing any one branch from the cut-set renders the graph connected.*

This implies that if any branch of the cut-set is not removed (deleted), the graph remains connected. In other words, restoring any *one* of the branches of the cut-set will destroy the separation property of the two parts.

For testing whether a set of branches of a connected graph constitute a cut-set we have replaced each removed branch one at a time, while we have kept all the other branches of the set removed. Then we have seen that the resulting graph is connected or not. This procedure is very lengthy for testing a set of branches to determine if it is a cut-set. A more convenient

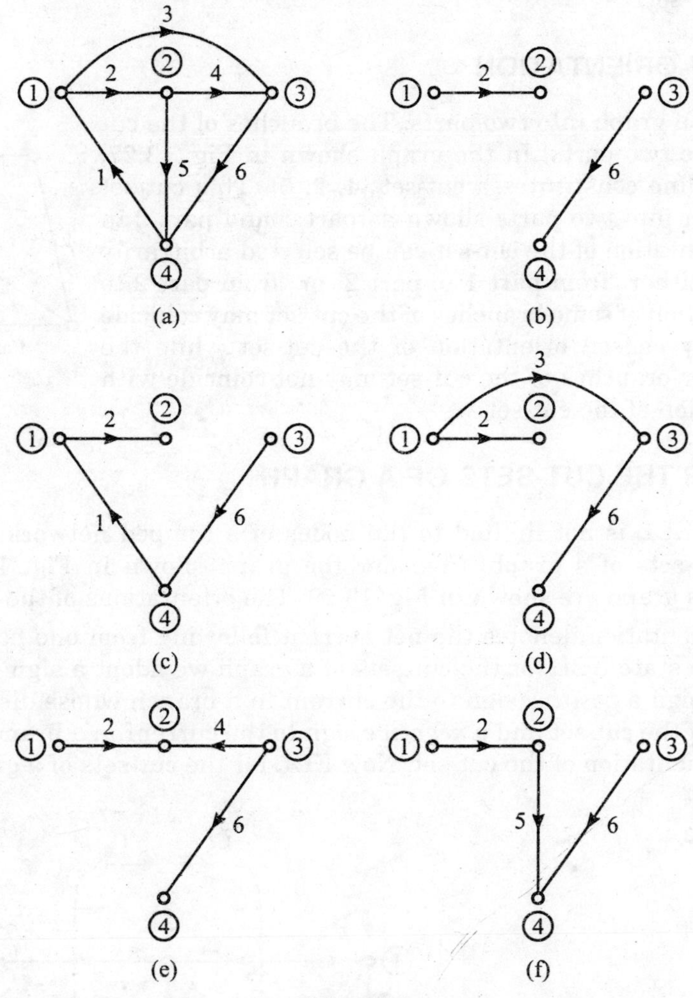

FIG. 13.26

method is to check the condition that removal of any of the proper *subset* of this set (called cut-set) leaves the graph connected. Let us consider the set {1, 3, 4, 5}. Some of its subsets are (1, 3, 4), (3, 4, 5), (1, 4, 5). If the subset (1, 3, 4) is removed from the graph of Fig. 13.26a, the resulting graph is shown in Fig. 13.26f. Thus, removal of the subset (1, 3, 4) does not cut the graph into two parts. Similarly, if the subsets (3, 4, 5) and (1, 4, 5) are removed from the graph, the resulting graphs are shown in Fig. 13.26c and d respectively. In both these cases also the graphs are not cut into two distinct parts. Thus, we can also define a cut-set as follows :

A **cut-set** *is a set of minimum number of branches of a connected graph whose removal causes the graph to be cut into exactly two connected subgraphs with the restriction that the removal of any of the proper subset of this set (called cut-set) reunites the graph into a single* *connected graph. Each cut-set contains one and only one twig, the remaining elements being tree links.*

A cut-set is shown on a graph by a dashed line, where the dashed line passes through the branches defining the cut-set. A graph can have more than one cut-set. Also, every graph has at least one cut-set.

13.25 CUT-SET ORIENTATION

A cut-set separates a graph into two parts. The branches of the cut set run between the two parts. In the graph shown in Fig. 13.27, cut by the dashed line constitutes a cut-set {4, 2, 5}. This cut-set separates the graph into two parts shown as part 1 and part 2 in Fig. 13.27. The orientation of the cut-set can be selected arbitrarily as the orientation either "from part 1 to part 2" or "from part 2 to part 1". The orientation of some branches of the cut-set may coincide with the arbitrarily chosen orientation of the cut-set while the corientation of other branches of the cut-set may not coincide with the chosen orientation of the cut-set.

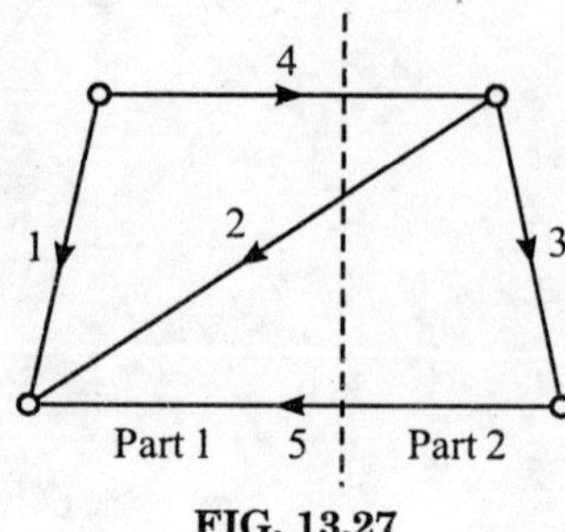

FIG. 13.27

13.26 KCL FOR THE CUT-SETS OF A GRAPH

The application of KCL is not limited to the nodes of a lumped network. KCL can also be applied for the cut-sets of a graph. Consider the graph shown in Fig. 13.28. Six cut-sets $C_1, C_2, ..., C_6$ of this graph are shown in Fig. 13.29. The orientations of the cut-set are chosen arbitrarily. This orientation denotes the net current following from one part of the graph to another. In order to state KCL for the cut-sets of a graph we adopt a sign convention for the currents. Let us assign a *positive* sign to the current in a branch whose direction is the same as the orientation of the cut-set and a *negative* sign to the current in a branch whose direction is opposite to the orientation of the cut-set. Now KCL for the cut-sets of a graph can be stated as follows :

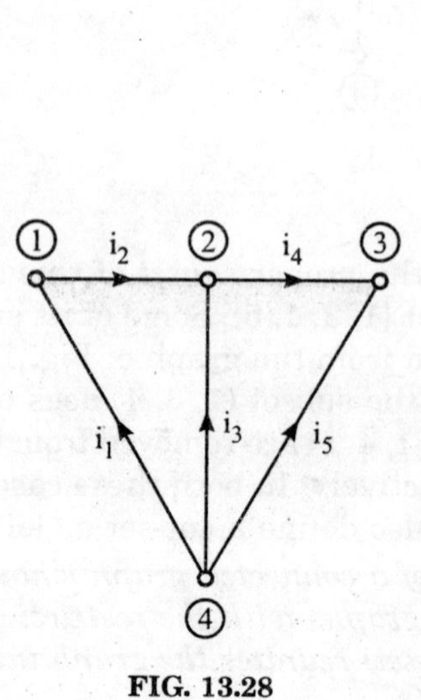

FIG. 13.28

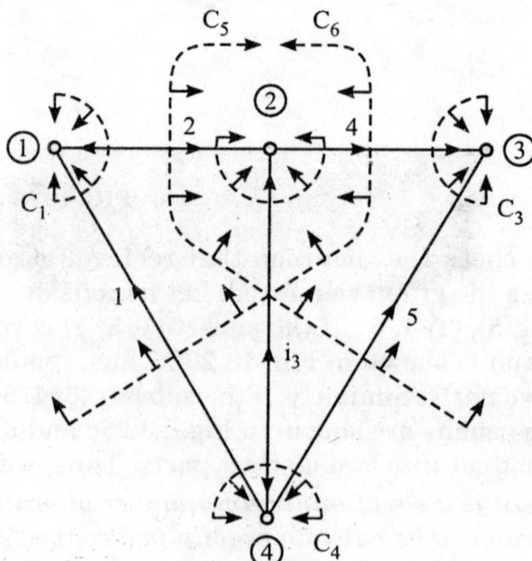

FIG. 13.29

> *For any lumped electrical network the algebraic sum of all the currents entering or leaving a cut-set is equal to zero.*

Applying KCL to each of the six cut-sets of Fig.13.29, we get

$$\left.\begin{array}{ll} \text{Cut–set 1 :} & i_1 - i_2 = 0 \\ \text{Cut–set 2 :} & i_2 + i_3 - i_4 = 0 \\ \text{Cut–set 3 :} & i_4 + i_5 = 0 \\ \text{Cut–set 4 :} & -i_1 - i_3 - i_5 = 0 \\ \text{Cut–set 5 :} & i_2 + i_3 + i_5 = 0 \\ \text{Cut–set 6 :} & i_1 + i_3 - i_4 = 0 \end{array}\right\} \tag{13.26.1}$$

These equations can be put in matrix form as

$$\begin{bmatrix} 1 & -1 & 0 & 0 & 0 \\ 0 & 1 & 1 & -1 & 0 \\ 0 & 0 & 0 & 1 & 1 \\ -1 & 0 & -1 & 0 & -1 \\ 0 & 1 & 1 & 0 & 1 \\ 1 & 0 & 1 & -1 & 0 \end{bmatrix} \begin{bmatrix} i_1 \\ i_2 \\ i_3 \\ i_4 \\ i_5 \\ i_6 \end{bmatrix} = \begin{bmatrix} 0 \\ 0 \\ 0 \\ 0 \\ 0 \\ 0 \end{bmatrix} \tag{13.26.2}$$

Eq. (13.6.2) can be put in a more compact form as

$$\mathbf{Q}_a \, \mathbf{i}_b = \mathbf{0} \tag{13.26.3}$$

where $\mathbf{Q}_a$ is called the **augmented cut-set matrix** of the graph under consideration. Here subscript a means that *all* possible cut-sets of the graph are included. Eqs. (13.26.3) are called the **cut-set equations** of the network. $\boldsymbol{i}_b$ is the column matrix of all cut-set currents.

The augmented cut-set matrix $\mathbf{Q}_a$ can be written in a compact form as

$$\mathbf{Q}_a = [q_{ij}]$$

where q_{ij} is an element in the augmented cut-set matrix. The order of $\mathbf{Q}_a$ is (number of cut-sets) × (number of branches). The subscript i stands for cut-set number and subscript j branch number. The elements have the following values :

> $q_{ij} = +1$ if jth branch is in cut-set i and the orientations of the cut-set and branch coincide
>
> $q_{ij} = -1$ if the jth branch is in cut-set i and their orientations do not coincide
>
> $q_{ij} = 0$ if the jth branch is not in ith cut-set.

The augmented cut-set matrix $\mathbf{Q}_a$ tells us the presence or absence of a branch in a cut-set and the orientation of the branches present in the cut-set.

■ **EXAMPLE 13.3** Write the augmented cut-set matrix $\mathbf{Q}_a$ for the network graph shown in Fig. 13.30a.

SOLUTION There are six branches and seven cut-sets in the graph of Fig. 13.30. The cut-set are

$C_1 : \{1, 2, 3\}; C_2 : \{3, 4, 6\}$

$C_3 : \{1, 5, 6\}; C_4 : \{2, 4, 5\}$

$C_5 : \{1, 3, 4, 5\}; C_6 : \{2, 3, 5, 6\}$

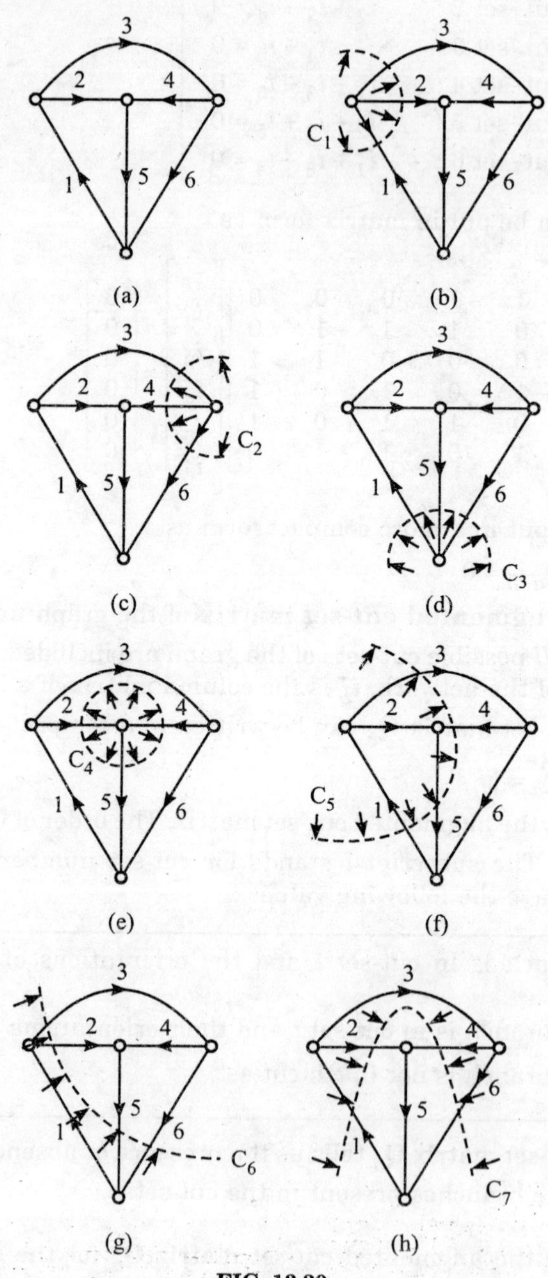

FIG. 13.30

$C_7 : \{1, 2, 4, 6\}$

These cut-sets are shown in Fig. 13.30b-h.

Each row of the $\mathbf{Q}_a$ matrix corresponds to a cut set of the graph. The columns of the matrix correspond to the branches of the graph. The elements of $\mathbf{Q}_a$ are written as follow :

Consider cut-set C_1. It contains branches 1, 2 and 3. The orientation of branch 1 does not coincide with the orientation of cut-set 1. Therefore the element q_{11} is written – 1. The direction of branch 2 is the same as the direction of cut-set 1. Hence the element q_{12} of matrix $\mathbf{Q}_a$ is written + 1. That is, the entry in the first row and second column is + 1. Similarly, branch 3 is in cut-set 1 and the orientation of branch 3 is the same as the orientation of cut-set 1. Hence the element q_{13} of matrix $\mathbf{Q}_a$ is + 1. Since branches 4, 5 and 6 are not contained in cut-set 1, the elements q_{14}, q_{15} and q_{16} are zero.

Next, we consider cut-set 2. It contains branches 3, 4 and 6. Here $q_{23} = -1$, $q_{24} = 1$, $q_{26} = 1$. Since branches 1, 2 and 5 are not contained in cut-set 2, $q_{21} = 0$, $q_{22} = 0$, $q_{25} = 0$. Similarly, we consider cut-sets 3, 4, 5, 6 and 7 and fill the entries of matrix $\mathbf{Q}_a$ with the help of the procedure used for cut-set 1 and 2. Thus we obtain the following matrix :

Cut–sets Branches →

$$\begin{array}{c} \downarrow \\ \\ \\ \\ \mathbf{Q}_a = \end{array} \begin{array}{c} \\ 1\\2\\3\\4\\5\\6\\7 \end{array} \begin{array}{c} \begin{matrix} 1 & 2 & 3 & 4 & 5 & 6 \end{matrix} \\ \begin{bmatrix} -1 & 1 & 1 & 0 & 0 & 0 \\ 0 & 0 & -1 & 1 & 0 & 1 \\ 1 & 0 & 0 & 0 & -1 & -1 \\ 0 & -1 & 0 & -1 & 1 & 0 \\ -1 & 0 & 1 & -1 & 1 & 0 \\ 0 & 1 & 1 & 0 & -1 & -1 \\ -1 & 1 & 0 & 1 & 0 & 1 \end{bmatrix} \end{array}$$

Rank of $\mathbf{Q}_a = n_t - 1 = 4 - 1 = 3$

13.27 FUNDAMENTAL CUT-SETS OR f-CUT-SETS

It is seen from the set of Eq. (13.26.1) that only three equations are linearly independent. The last three equations can be obtained by linear combinations of the first three. In order to obtain a set of linearly independent cut-set equations we use the concept of fundamental cut-sets. A fundamental cut-set is a special type of cut-set which is formed according to the following rule:

For a given connected graph choose any arbitrary tree and then fix attention on any twig (tree branch) b_k of the chosen tree. If this twig is removed from the tree, then the tree will be cut into two parts (one of the parts may contain only one isolated node). Some (or all) links of the chosen tree together with the twig b_k are found to form a cut-set of the graph of which the chosen tree is a part. This cut-set formed by some (or all) the links of the chosen tree of the graph together with the twig b_k of the chosen tree is called the fundamental cut-set or f-cut set. Thus we can define the fundamental cut-set as follows :

*A **fundamental cut-set** of a graph G with respect to a tree T is a cut-set that is formed by one twig and a unique set of links (chords).*

It is apparent that for each twig (of a chosen tree), there will be a fundamental cut-set. Hence for a graph having n_t nodes there will be $(n_t - 1)$ number of f-cut-sets, because any tree of the graph having n_t nodes always consists of $(n_t - 1)$ twigs.

13.28 ORIENTATION OF FUNDAMENTAL CUT-SETS

As a convention the orientation of a cut-set is chosen to coincide with that of its ***defining twig*** *b_k. In other words, a fundamental cut-set is assigned the same orientation as that of the twig it contains.*

13.29 FUNDAMENTAL CUT-SET MATRIX Q_f

A fundamental cut-set matrix $\mathbf{Q}_f$ describes the branches contained in a fundamental cut-set and their orientations. Each row of this matrix corresponds to a fundamental cut-set of the graph, the columns of the matrix correspond to the branches of the graph.

The fundamental cut-set matrix $\mathbf{Q}_f$ of a graph G with n_t nodes and b branches corresponding to a tree T is an $(n_t - 1) \times b$ matrix. It is given by

$$\mathbf{Q_f} = [\mathbf{q_{kj}}]$$

where

$q_{kj} = 1$ when b_j is in c_k and has the same orientation

$q_{kj} = -1$ when b_j is in c_k and has the opposite orientation

$q_{kj} = 0$ when b_j is *not* in c_k

EXAMPLE 13.4 Write the fundamental cut-set matrix for the network graph shown in Fig. 13.30a.

SOLUTION. Let us choose an arbitrary tree T of this graph. Suppose that the tree T is shown by solid lines and the corresponding chords (links) by broken lines as shown in Fig. 13.31.

The twigs of this tree are {1, 2, 4}. The links of T are {3, 5, 6}. Let us focus our attention on twig 1, that is, $b_k = 1$. It is seen that corresponding to twig 1, the f-cut-set is {1, 5, 6} which is cut-set c_1. Its orientation coincides with the defining twig 1. Corresponding to twig 2, the f-cut-set is {2, 3, 5, 6} which is cut-set c_2 having the same orientation as 2. Corresponding to twig 4 the f-cut-set is {3, 4, 6} which is cut-set c_3. The orientation of c_3 is coincident with the direction of twig 4. The f-cut-set matrix is written as follows :

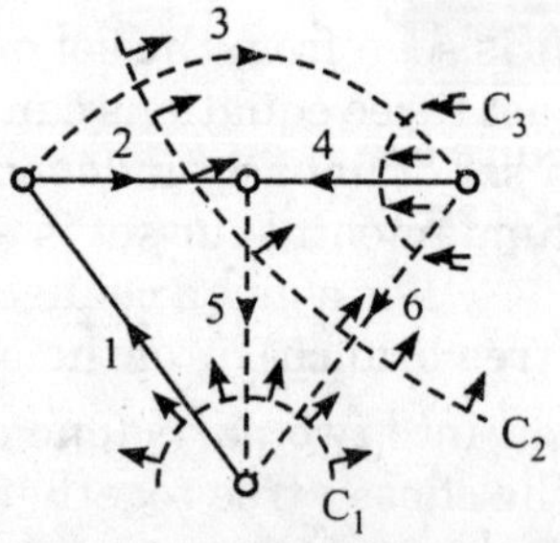

FIG. 13..31 Fundamental cut-sets corresponding to a tree T of the graph of Fig. 13.30.a

$$\begin{array}{c} \text{f–cut–sets} \\ \downarrow \end{array} \quad \begin{array}{c} \text{branches} \rightarrow \end{array}$$

$$\mathbf{Q}_f = \begin{array}{c} c_1 \\ c_2 \\ c_3 \end{array} \begin{bmatrix} 1 & 0 & 0 & 0 & -1 & -1 \\ 0 & 1 & 1 & 0 & -1 & -1 \\ 0 & 0 & -1 & 1 & 0 & 1 \end{bmatrix} \quad (\text{columns: branches } 1\ 2\ 3\ 4\ 5\ 6)$$

13.30 KCL FOR THE FUNDAMENTAL CUT-SETS

The basic property of the fundamental cut-sets is that they give linearly independent KCL equations. Since there are $(n_t - 1)$ fundamental cut-sets in a graph of n_t nodes, to obtain *all* the linearly independent cut-set equations, it is sufficient to apply KCL to the f-cut-sets of the graph of Fig. 13.31.

$$c_1 : i_1 - i_5 - i_6 = 0$$

$$c_2 : i_2 + i_3 - i_5 - i_6 = 0$$

$$c_3 : - i_3 + i_4 + i_6 = 0$$

These equations can be written in matrix form as

f–cut–sets ↓ branches → 1 2 3 4 5 6

$$\begin{array}{c} c_1 \\ c_2 \\ c_3 \end{array} \begin{bmatrix} 1 & 0 & 0 & 0 & -1 & -1 \\ 0 & 1 & 1 & 0 & -1 & -1 \\ 0 & 0 & -1 & 1 & 0 & 1 \end{bmatrix} \begin{bmatrix} i_1 \\ i_2 \\ i_3 \\ i_4 \\ i_5 \\ i_6 \end{bmatrix} = \begin{bmatrix} 0 \\ 0 \\ 0 \end{bmatrix}$$

or $\boxed{\mathbf{Q}_f \mathbf{i}_b = 0 \qquad \text{KCL}}$ (13.30.1)

Eq. (13.30.1) is the KCL for the f-cut-sets.

13.31 RANK OF THE f-CUT-SET MATRIX

Since the orientation of the f-cut-sets is the same as the orientation of the tree branches (twigs) $\mathbf{Q}_f$ can be partitioned in the form

f–cut–sets ↓ twigs 1 2 4 links 3 5 6

$$\mathbf{Q}_f = \begin{array}{c} 1 \\ 2 \\ 3 \end{array} \begin{bmatrix} 1 & 0 & 0 & 0 & -1 & -1 \\ 0 & 1 & 0 & 1 & -1 & -1 \\ 0 & 0 & 1 & -1 & 0 & 1 \end{bmatrix}$$

$$\mathbf{Q}_f = [\mathbf{Q}_t,\ \mathbf{Q}_l] = [\mathbf{U},\ \mathbf{Q}_l] \qquad (13.31.1)$$

where $\mathbf{Q}_t$ = submatrix of $\mathbf{Q}_f$ corresponding to twigs of a chosen a tree

$\mathbf{Q}_l$ = submatrix of $\mathbf{Q}_f$ corresponding to links of the chosen tree

U = unit matrix

In our case

$$\mathbf{Q}_t = \begin{bmatrix} 1 & 0 & 0 \\ 0 & 1 & 0 \\ 0 & 0 & 1 \end{bmatrix} = \mathbf{U}, \quad \mathbf{Q}_l = \begin{bmatrix} 0 & -1 & -1 \\ 1 & -1 & -1 \\ -1 & 0 & 1 \end{bmatrix}$$

$\mathbf{Q}_t$ is an $(n_t - 1) \times (n_t - 1)$ unit matrix corresponding to the twigs of the tree and $\mathbf{Q}_l$ is a $(n_t - 1) \times (b - n_t + 1)$ matrix corresponding to the links of the chosen tree. From Eq. (13.31.1) it is seen that $\mathbf{Q}_f$ is of rank $(n_t - 1)$, since it has a nonsingular submatrix of order $(n_t - 1) \times (n_t - 1)$.

13.32 RELATION BETWEEN BRANCH VOLTAGES AND TREE-BRANCH VOITAGES

Let the tree-branch voltages be denoted by $v_{t1}, v_{t2}, ..., v_{tn}$. By applying KVL to each fundamental loop of corresponding to the chosen tree T of graph shown in Fig. 13.32.

$$\text{Loop } l_1 : -v_2 + v_3 + v_4 = 0$$

$$\text{Loop } l_2 : v_1 + v_2 + v_5 = 0$$

$$\text{Loop } l_3 : v_1 + v_2 - v_4 + v_6 = 0$$

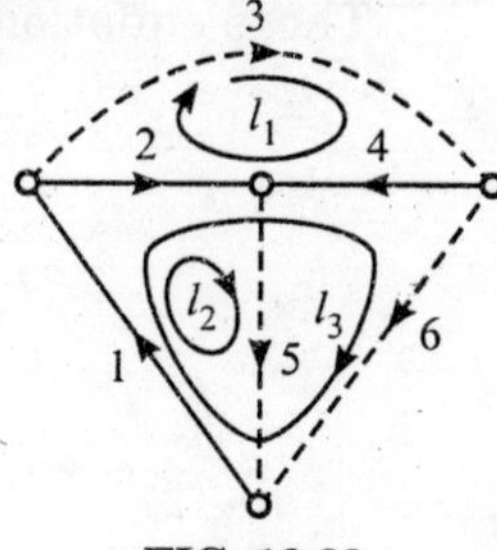

FIG. 13.32

Let the tree-branch voltages for Fig. 13.32 be

$$v_1 = v_{t1}, v_2 = v_{t2}, v_4 = v_{t3}$$

We can express each branch voltage in terms of tree-branch voltages as follows :

$$v_1 = v_{t1}$$

$$v_2 = v_{t2}$$

$$v_3 = v_2 - v_4 = v_{t2} - v_{t3}$$

$$v_4 = v_{t3}$$

$$v_5 = -v_1 - v_2 = -v_{t1} - v_{t2}$$

$$v_6 = -v_1 - v_2 + v_4 = -v_{t1} - v_{t2} + v_{t3}$$

Putting these relationships in matrix form, we get

$$\begin{bmatrix} v_1 \\ v_2 \\ v_3 \\ v_4 \\ v_5 \\ v_6 \end{bmatrix} = \begin{bmatrix} 1 & 0 & 0 \\ 0 & 1 & 0 \\ 0 & 1 & -1 \\ 0 & 0 & 1 \\ -1 & -1 & 0 \\ -1 & -1 & 1 \end{bmatrix} \begin{bmatrix} v_{t1} \\ v_{t2} \\ v_{t3} \end{bmatrix}$$

or $$\mathbf{v}_b = \mathbf{Q}'_f \mathbf{v}_t \tag{13.32.1}$$

Eq. (13.32.1) shows that each branch voltage can be expressed as a linear combination of the tree-branch voltages.

13.33 NODE-TANSFORMATION EQUATION

Let us determine the relationship between the branch voltages and the independent node voltages. The branch voltage is defined as the voltage across the branch of a graph. The polarity of the branch voltage is positive at the tail end of the arrow on the branch. Since the branch direction also represents the positive direction of curent in the branch, the branch voltage is interpreted as a voltage drop.

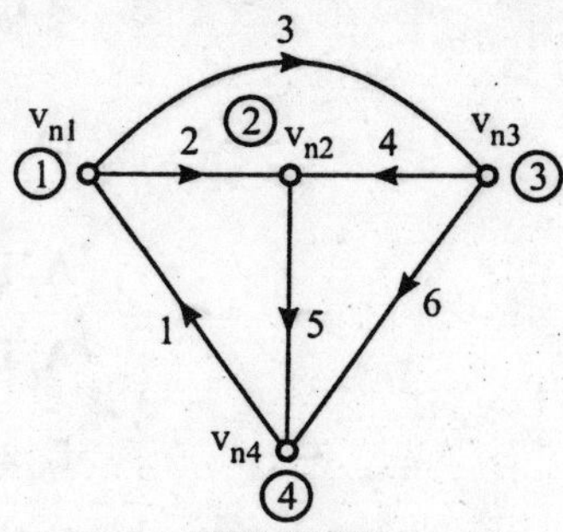

FIG. 13.33

Consider the graph of Fig. 13.33. It has four nodes. Let node 4 be taken as the reference node and the node voltages be v_{n1}, v_{n2}, v_{n3} and v_{n4}. Here $v_{n4} = 0$. The branch voltages are assumed to be v_1, v_2, v_3, v_4, v_5 and v_6. For the given graph the relationships between branch voltages and node voltages are

$$v_1 = v_{n4} - v_{n1} = -v_{n1}$$

$$v_2 = v_{n1} - v_{n2}$$

$$v_3 = v_{n1} - v_{n3}$$

$$v_4 = v_{n3} - v_{n2}$$

$$v_5 = v_{n2} - v_{n4} = v_{n2}$$

$$v_6 = v_{n3} - v_{n4} = v_{n3}$$

These relationships can be expressed in matrix form as follows :

$$\begin{bmatrix} v_1 \\ v_2 \\ v_3 \\ v_4 \\ v_5 \\ v_6 \end{bmatrix} = \begin{bmatrix} -1 & 0 & 0 \\ 1 & -1 & 0 \\ 1 & 0 & -1 \\ 0 & -1 & 1 \\ 0 & 1 & 0 \\ 0 & 0 & 1 \end{bmatrix} \begin{bmatrix} v_{n1} \\ v_{n2} \\ v_{n3} \end{bmatrix}$$

or $$\mathbf{v}_b = \mathbf{A}'\mathbf{v}_n \qquad \text{KVL} \tag{13.33.1}$$

Eq. (13.33.1) gives the relationship between node-to-datum voltages $\mathbf{v}_n$ and branch voltages $\mathbf{v}_b$. It is known as **node-tansformation equation**.

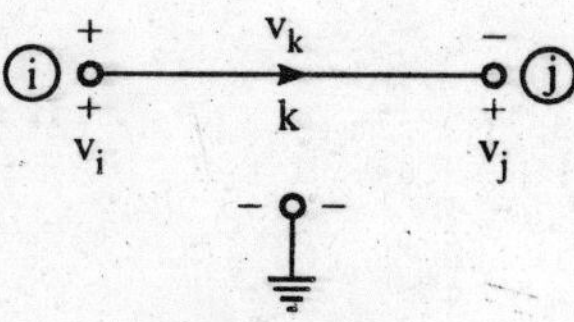

FIG. 13.34

Let v_k be the kth branch voltage, $k = 1, 2, ..., b$. Suppose that v_i is the node-to-datum voltage of node i, $i = 1, 2, ..., (n_t - 1)$. Thus if branch k connects the ith node to datum node we have

$v_k = v_n$ if branch k leaves node i

$v_k = -v_n$ if branch k enters node i

If branch k leaves node i and enters node j, then from Fig. 13.34 we have

$$v_k = v_i - v_j$$

13.34 RELATIONSHIP BETWEEN LINK AND TWIG CURRENTS

KCL can be written as

$$\mathbf{A}\,\mathbf{i}_b = \mathbf{0}$$

$$[\mathbf{A}_t \vdots \mathbf{A}_l]\begin{bmatrix}\mathbf{i}_{bt} \\ \mathbf{i}_{bl}\end{bmatrix} = \mathbf{0}$$

$$\mathbf{A}_t\,\mathbf{i}_{bt} + \mathbf{A}_l\,\mathbf{i}_{bl} = \mathbf{0}$$

$$\mathbf{A}_t\,\mathbf{i}_{bt} = -\mathbf{A}_l\,\mathbf{i}_{bl}$$

$$\mathbf{i}_{bt} = -\mathbf{A}_t^{-1}\,\mathbf{A}_l\,\mathbf{i}_{bl} \tag{13.34.1}$$

This relation shows that if the link currents can be determined by some method the twig curents can be calculated.

13.35 RELATIONSHIPS BETWEEN SUBMATRICES OF *Q*, *A* AND *B*

We have

$$\mathbf{AB}' = \mathbf{0} \tag{13.35.1}$$

$$[\mathbf{A}_t \vdots \mathbf{A}_l]\begin{bmatrix}\mathbf{B}'_t \\ \mathbf{B}'_l\end{bmatrix} = \mathbf{0}$$

$$\mathbf{A}_t\,\mathbf{B}'_t + \mathbf{A}_l\,\mathbf{B}'_l = \mathbf{0} \tag{13.35.2}$$

Since $\mathbf{A}_t$ is nonsingular,

$$\mathbf{B}'_t = -(\mathbf{A}_t^{-1}\,\mathbf{A}_l)\,\mathbf{B}'_l$$

$$\mathbf{B}_t = -\mathbf{B}_l\,(\mathbf{A}_t^{-1}\,\mathbf{A}_l)'$$

$$\mathbf{B} = [\mathbf{B}_t \vdots \mathbf{B}_l]$$

$$= [-\mathbf{B}_l\,(\mathbf{A}_t^{-1}\,\mathbf{A}_l)' \vdots \mathbf{B}_l]$$

$$\mathbf{B} = \mathbf{B}_l\,[-(\mathbf{A}_t^{-1}\,\mathbf{A}_l)' \vdots \mathbf{U}] \tag{13.35.3}$$

From Eq. (13.5.2) we have

$$\mathbf{A}_l\,\mathbf{B}'_l = -\,\mathbf{A}_t\,\mathbf{B}'_t$$

Since B'_1 is nonsingular

$$\mathbf{A}_l = -\,\mathbf{A}_t\,\mathbf{B}'_t\,(\mathbf{B}_l^{-1})'$$

$$= -\,\mathbf{A}_t\,(\mathbf{B}_l^{-1}\,\mathbf{B}_t)'$$

$$\mathbf{A} = [\mathbf{A}_t \,\vdots\, \mathbf{A}_l]$$

$$= [\mathbf{A}_t \,\vdots\, -\,\mathbf{A}_t\,(\mathbf{B}_l^{-1}\,\mathbf{B}_t)']$$

or

$$\mathbf{A} = \mathbf{A}_t\,[\mathbf{U} \,\vdots\, -\,(\mathbf{B}_l^{-1}\,\mathbf{B}_t)'] \tag{13.35.4}$$

$$\mathbf{Q}_f\,\mathbf{B}'_f = \mathbf{0} \tag{13.35.5}$$

$$[\mathbf{Q}_t \,\vdots\, \mathbf{Q}_l]\begin{bmatrix}\mathbf{B}'_{ft}\\ \mathbf{B}'_{fl}\end{bmatrix} = \mathbf{0}$$

$$[\mathbf{U}_t \,\vdots\, \mathbf{Q}_l]\begin{bmatrix}\mathbf{B}'_{ft}\\ \mathbf{B}'_{fl}\end{bmatrix} = \mathbf{0}$$

where $\mathbf{U}_t$ is a unit matrix

$$\mathbf{U}\,\mathbf{B}'_{ft} + \mathbf{Q}_l\,\mathbf{B}'_{fl} = \mathbf{0}$$

$$\mathbf{B}'_{ft} + \mathbf{Q}_l\,\mathbf{B}'_{fl} = \mathbf{0}$$

$$\mathbf{Q}_l\,\mathbf{B}'_{fl} = -\,\mathbf{B}'_{ft}$$

$$\mathbf{Q}_l = -\,\mathbf{B}'_{ft}\,(\mathbf{B}'_{fl})^{-1}$$

$$= -\,(\mathbf{B}_{fl}^{-1}\;\mathbf{B}_t)' \tag{13.35.6}$$

$$\mathbf{Q}_f = [\mathbf{Q}_t \,\vdots\, \mathbf{Q}_l]$$

$$= [\mathbf{U} \,\vdots\, -\,(\mathbf{B}_{fl}^{-1}\,\mathbf{B}_{ft})'] \tag{13.35.7}$$

From Eq. (13.35.4), we have

$$\mathbf{A} = \mathbf{A}_t\,[\mathbf{U} \,\vdots\, -\,(\mathbf{B}_{fl}^{-1}\;\mathbf{B}_{ft})'] \tag{13.35.8}$$

Combination of Eqs. (13.35.7) and (13.35.8) gives

$$\mathbf{A} = \mathbf{A}_t \mathbf{Q}_f \tag{13.35.9}$$

Premultiplying both the sides of Eq. (13.35.9) by $\mathbf{A}_t^{-1}$ gives

$$\mathbf{A}_t^{-1} \mathbf{A} = \mathbf{A}_t^{-1} \mathbf{A}_t \mathbf{Q}_f$$

$$\mathbf{Q}_f = \mathbf{A}_t^{-1} \mathbf{A}$$

$$= \mathbf{A}_t^{-1} [\mathbf{A}_t \ \ \mathbf{A}_l]$$

$$= [\mathbf{A}_t^{-1} \mathbf{A}_t \ \vdots \ \mathbf{A}_t^{-1} \mathbf{A}_l]$$

or

$$\mathbf{Q}_f = [\mathbf{U} \ \vdots \ \mathbf{A}_t^{-1} \mathbf{A}_l] \tag{13.35.10}$$

13.36 RELATION BETWEEN LINK-BRANCH VOLTAGES AND TREE-BRANCH VOLTAGES

By KVL,

$$\mathbf{B}_f \mathbf{v}_b = \mathbf{0}$$

$$\mathbf{B}_f = [\mathbf{B}_t \ \vdots \ \mathbf{B}_l] \tag{13.36.1}$$

where $\mathbf{B}_t$ corresponds to tree branches and $\mathbf{B}_l$ to the link branches. Since each loop has a different link, $\mathbf{B}_l$ is a unit matrix

$$\mathbf{B}_l = \mathbf{U}$$

We can partition $\mathbf{v}_b$ into tree-branch voltages and link-branch voltages so that

$$\mathbf{v}_b = \begin{bmatrix} \mathbf{v}_t \\ \mathbf{v}_l \end{bmatrix}$$

Substituting the values of $\mathbf{B}_f$ and $\mathbf{v}_b$ in Eq. (13.36.1), we get

$$[\mathbf{B}_t \ \vdots \ \mathbf{U}] \begin{bmatrix} \mathbf{v}_t \\ \mathbf{v}_l \end{bmatrix} = \mathbf{0}$$

$$\mathbf{B}_t \mathbf{v}_t + \mathbf{v}_l = \mathbf{0}$$

$$\mathbf{v}_l = -\mathbf{B}_t \mathbf{v}_t \tag{13.36.2}$$

Eq. (13.36.2) gives the relation between link-branch voltages and tree-branch voltage.

13.37 NODE EQUATIONS

Consider the general branch of a network shown in Fig. 13.35. As before, the branch voltage is denoted by v_b and the branch current by i_b.

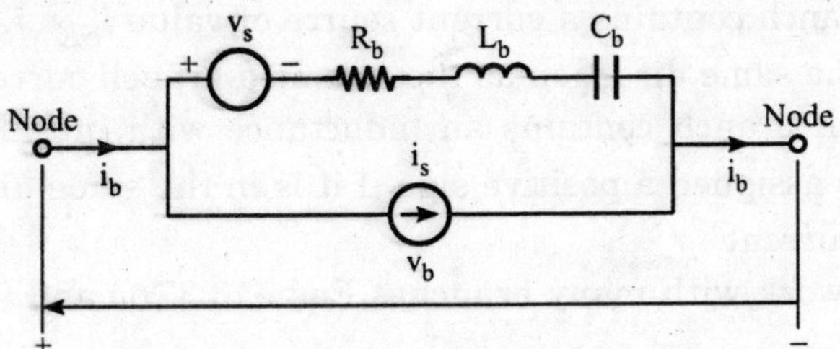

FIG. 13.35 A general branch of a network

Let

R_b = total resistance in the branch

L_b = total inductance in the branch

C_b = total capacitance in the branch

v_s = effective voltage source in the branch (including initial capacitor voltages in loop analysis)

i_s = total current source across the branch (including initial inductor currents in nodal analysis)

The voltage-current relations for the branch can be written as

$$v_b = v_s + \left(R_b + L_b \frac{d}{dt} + \frac{1}{C_b}\int_{t_0^+}^{t} d\tau\right)(i_b - i_s)$$

$$i_b = i_s + \left(G_b + C_b \frac{d}{dt} + \frac{1}{L_b}\int_{t_0^+}^{t} d\tau\right)(v_b - v_s)$$

$$v_b = v_s + \left(R_b + L_b D + \frac{1}{C_b} D^{-1}\right)(i_b - i_s) \qquad (13.37.1)$$

$$i_b = i_s + \left(G_b + C_b D + \frac{1}{L_b} D^{-1}\right)(v_b - v_s) \qquad (13.37.2)$$

where D is the differential operator and D^{-1} is the integral operator.

$$D = \frac{d}{dt} = s,\ D^{-1} = \frac{1}{D} = \int_{0^+}^{t} d\tau = \frac{1}{s}$$

The values of voltage source and current source are specified as follows :

$v_s = 0$ if the kth branch does not contain any voltage source or capacitance

$v_s = v_{SK}$ if the kth branch contains a voltage source of value v_{SK} and its plus side is at the tail end of the branch arrow

$v_s = -v_{SK}$ if the kth branch contains a voltage source of value v_{SK} and its negative side is at the tail end of the branch arrow

$v_s = \pm v_{CK}(0)$ if the kth branch contains a capacitance with initial voltage $v_{CK}(0)$. Here also, + sign is taken if v_{CK} decreases in the direction of the branch.

$i_s = 0$ if the kth branch does not contain any current source and inductance.

$i_s = \pm i_{SK}$

if kth branch contains a current source of value i_{SK}. i_{SK} has a positive sign if it is in the same direction as the assumed branch current.

$i_s = i_{LK}(0)$ if the kth branch contains an inductance with initial current $i_{LK}(0)$. Again $i_{LK}(0)$ is assigned a positive sign if it is in the same direction as the assumed branch current.

For a general network with many branches Eqs. (13.37.1) and (13.37.2) can be written in matrix form as :

$$\mathbf{v}_b = \mathbf{v}_s + \mathbf{Z}_b(\mathbf{i}_b - \mathbf{i}_s) \tag{13.37.3}$$

$$\mathbf{i}_b = \mathbf{i}_s + \mathbf{Y}_b(\mathbf{v}_b - \mathbf{v}_s) \tag{13.37.4}$$

where $\mathbf{v}_b$, $\mathbf{v}_s$, $\mathbf{i}_b$ and $\mathbf{i}_s$ are $b \times 1$ matrices and $\mathbf{Z}_b$ and $\mathbf{Y}_b$ are the branch impedance and branch admittance matrices respectively. They are $b \times b$ diagonal matrices, if the network does not contain mutual inductances. The (k, k) element of the main diagonal of $\mathbf{Z}_b$ is the total impedance of the kth branch. Similarly, the (k, k) element on the main diagonal of $\mathbf{Y}_b$ is the total admittance of the kth branch.

KCL can be written as

$$\mathbf{A}\,\mathbf{i}_b = \mathbf{0} \tag{13.37.5}$$

The node-transformation equation is given by

$$\mathbf{v}_b = \mathbf{A}'\,\mathbf{v}_n \tag{13.37.6}$$

Substituting Eq. (13.37.4) in Eq. (13.37.5), we get

$$\mathbf{A}\,\mathbf{i}_s + \mathbf{A}\mathbf{Y}_b(\mathbf{v}_b - \mathbf{v}_s) = \mathbf{0} \tag{13.37.7}$$

Now substituting Eq. (13.37.6) in Eq. (13.37.7) gives

$$\mathbf{A}\,\mathbf{i}_s + \mathbf{A}\,\mathbf{Y}_b(\mathbf{A}'\,\mathbf{v}_n - \mathbf{v}_s) = \mathbf{0}$$

or

$$\mathbf{A}\,\mathbf{Y}_b\,\mathbf{v}_s - \mathbf{A}\,\mathbf{i}_s = \mathbf{A}\,\mathbf{Y}_b\,\mathbf{A}'\,\mathbf{v}_n \tag{13.37.8}$$

Eq. (13.37.8) is of the same form as the node equation

$$\mathbf{i}_n = \mathbf{Y}_n\,\mathbf{v}_n \tag{13.37.9}$$

If we introduce the notations

$$\mathbf{i}_n \triangleq \mathbf{A}\,\mathbf{Y}_b\,\mathbf{v}_s - \mathbf{A}\,\mathbf{i}_s \tag{13.37.10}$$

and

$$\mathbf{Y}_n \triangleq \mathbf{A}\,\mathbf{Y}_b\,\mathbf{A}' \tag{13.37.11}$$

then Eq. (13.37.8) also becomes

$$\mathbf{i}_n = \mathbf{Y}_n\,\mathbf{v}_n \tag{13.37.12}$$

The set of Eq. (13.37.12) is called the **node equation**; $\mathbf{Y}_n$ is called the **node-admittance matrix**. Eq. (13.37.12) represents a set of $(n_t - 1)$ independent node equations of a network with n_t nodes. The node-pair voltages can be found by solving Eq. (13.37.12).

Therefore $$\mathbf{v}_n = \mathbf{Y}_n^{-1}\,\mathbf{i}_n \tag{13.37.13}$$

if $\mathbf{Y}_n^{-1}$ is nonsingular.

The KCL equation can be written as

$$\mathbf{Q}_f\,\mathbf{i}_b = \mathbf{0} \tag{13.30.1}$$

Also, $$\mathbf{i}_b = \mathbf{i}_s + \mathbf{Y}_b\,(\mathbf{v}_b - \mathbf{v}_s) \tag{13.37.4}$$

Substituting the value of $\mathbf{v}_b$ from Eq. (13.32.1) in Eq. (13.37.4) we get

$$\mathbf{i}_b = \mathbf{i}_s + \mathbf{Y}_b\,(\mathbf{Q}'_f\,\mathbf{v}_t - \mathbf{v}_s) \tag{13.32.2}$$

Combination of Eqs. (13.30.1) and (13.32.2) gives

$$\mathbf{Q}_f\,\mathbf{i}_s + \mathbf{Q}_f\,\mathbf{Y}_b\,\mathbf{Q}'_f\,\mathbf{v}_t - \mathbf{Q}_f\,\mathbf{Y}_b\,\mathbf{v}_s = \mathbf{0} \tag{13.32.3}$$

Let $$\mathbf{Y}_t \triangleq \mathbf{Q}_f\,\mathbf{Y}_b\,\mathbf{Q}'_f \tag{13.32.4}$$

$\mathbf{Y}_t$ is called the **node-pair admittance matrix.**

Thus, we have the generalised KCL equations for the twig voltages

$$\mathbf{Y}_t\,\mathbf{v}_t = \mathbf{Q}_f\,\mathbf{Y}_b\,\mathbf{v}_s - \mathbf{Q}_f\,\mathbf{i}_s \tag{13.32.5}$$

13.37.1 Nodal Analysis Based on Graph Theory

The following steps are required for nodal analysis of a given network based on graph theory:

1. Choose an arbitrary node as the datum node and write down the incidence matrix $\mathbf{A}$ from the network graph.
2. Determine the branch admittance matrix $\mathbf{Y}_b$.
3. Find the node-admittance matrix from $\mathbf{Y}_n = \mathbf{A}\,\mathbf{Y}_b\,\mathbf{A}'$
4. Obtain the voltage and current source vectors $\mathbf{v}_s$ and $\mathbf{i}_s$ from the network.
5. Determine $\mathbf{i}_n$ from the relation
$$\mathbf{i}_n = \mathbf{A}\,\mathbf{Y}_b\,\mathbf{v}_s - \mathbf{A}\,\mathbf{i}_s$$
6. Write the node equations of the network from the relation
$$\mathbf{i}_n = \mathbf{Y}_n\,\mathbf{v}_n$$
7. Use steps 5 and 6 to determine $\mathbf{v}_n$ as follows :
$$\mathbf{v}_n = \mathbf{Y}_n^{-1}\,\mathbf{A}\,\mathbf{Y}_b\,\mathbf{v}_s - \mathbf{Y}^{-1}\,\mathbf{A}\mathbf{i}_s$$
8. Use the node-transformation equation
$$\mathbf{v}_b = \mathbf{A}'\,\mathbf{v}_n \text{ and}$$
$$\mathbf{i}_b = \mathbf{i}_s + \mathbf{Y}_b\,(\mathbf{v}_b - \mathbf{v}_s)$$
to determine branch voltage and current vectors $\mathbf{v}_b$ and $\mathbf{i}_b$.

■ **EXAMPLE 13.5** Write the node equations for the network shown in Fig. 13.36a.

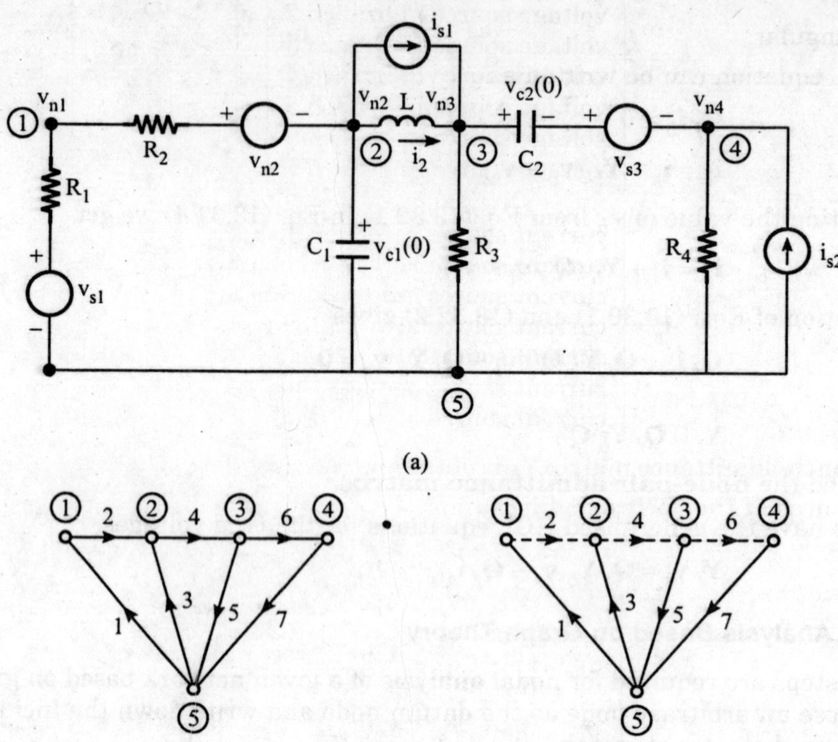

(a)

(b) (c)

FIG. 13.36

SOLUTION The network graph is shown in Fig. 13.36b. It is to be noted that ideal voltage sources have been replaced by short circuits and ideal current sources are replaced by open circuits in the graph. The direction of branches is chosen arbitrarily. From inspection.

$$\begin{array}{c} \text{nodes} \\ \downarrow \end{array} \quad \begin{array}{c} \text{branches} \rightarrow \\ \begin{array}{ccccccc} 1 & 2 & 3 & 4 & 5 & 6 & 7 \end{array} \end{array}$$

$$\mathbf{A} = \begin{array}{c} ① \\ ② \\ ③ \\ ④ \end{array} \begin{bmatrix} -1 & 1 & 0 & 0 & 0 & 0 & 0 \\ 0 & -1 & -1 & 1 & 0 & 0 & 0 \\ 0 & 0 & 0 & -1 & 1 & 1 & 0 \\ 0 & 0 & 0 & 0 & 0 & -1 & 1 \end{bmatrix}$$

Therefore

$$\mathbf{A}' = \begin{bmatrix} -1 & 0 & 0 & 0 \\ 1 & -1 & 0 & 0 \\ 0 & -1 & 0 & 0 \\ 0 & 1 & -1 & 0 \\ 0 & 0 & 1 & 0 \\ 0 & 0 & 1 & -1 \\ 0 & 0 & 0 & 1 \end{bmatrix}$$

$$\mathbf{v}_s = \begin{bmatrix} \text{voltage source in branch 1} \\ \text{voltage source in branch 2} \\ \text{voltage source in branch 3} \\ \text{voltage source in branch 4} \\ \text{voltage source in branch 5} \\ \text{voltage source in branch 6} \\ \text{voltage source in branch 7} \end{bmatrix} = \begin{bmatrix} -v_{s1} \\ v_{s2} \\ v_{c1\,(0)} \\ 0 \\ 0 \\ v_{c2\,(0)} + v_{s3} \\ 0 \end{bmatrix}$$

$$\mathbf{i}_s = \begin{bmatrix} \text{current source across branch 1} \\ \text{current source across branch 2} \\ \text{current source across branch 3} \\ \text{current source across branch 4} \\ \text{current source across branch 5} \\ \text{current source across branch 6} \\ \text{current source across branch 7} \end{bmatrix} = \begin{bmatrix} 0 \\ 0 \\ 0 \\ i_{2\,(0)} + i_{s1} \\ 0 \\ 0 \\ -i_{s2} \end{bmatrix}$$

The branch-admittance matrix $\mathbf{Y}_b$ is obtained by filling the positions of its main diagonal with admittances of the corresponding branch. The off-diagonal elements of this matrix are zero. Thus

$$\mathbf{Y}_b = \begin{bmatrix} \frac{1}{R_1} & 0 & 0 & 0 & 0 & 0 & 0 \\ 0 & \frac{1}{R_2} & 0 & 0 & 0 & 0 & 0 \\ 0 & 0 & C_1 s & 0 & 0 & 0 & 0 \\ 0 & 0 & 0 & \frac{1}{sL} & 0 & 0 & 0 \\ 0 & 0 & 0 & 0 & \frac{1}{R_3} & 0 & 0 \\ 0 & 0 & 0 & 0 & 0 & C_2 s & 0 \\ 0 & 0 & 0 & 0 & 0 & 0 & \frac{1}{R_4} \end{bmatrix}$$

$$\mathbf{AY}_b = \begin{bmatrix} -\frac{1}{R_1} & \frac{1}{R_2} & 0 & 0 & 0 & 0 & 0 \\ 0 & -R_2 & -C_1 s & \frac{1}{Ls} & 0 & 0 & 0 \\ 0 & 0 & 0 & \frac{-1}{sL} & \frac{1}{R_3} & C_2 s & 0 \\ 0 & 0 & 0 & 0 & 0 & -C_2 s & \frac{1}{R_4} \end{bmatrix}$$

The node-admittance matrix is given by

$$\mathbf{Y}_n = \mathbf{A}\mathbf{Y}_b\mathbf{A}' = \begin{bmatrix} \frac{1}{R_1}+\frac{1}{R_2} & -\frac{1}{R_2} & 0 & 0 \\ -\frac{1}{R_2} & \frac{1}{R_2}+C_1 s+\frac{1}{sL} & -\frac{1}{sL} & 0 \\ 0 & -\frac{1}{sL} & \frac{1}{sL}+\frac{1}{R_3}+C_2 s & -C_2 s \\ 0 & 0 & -C_2 s & C_2 s+\frac{1}{R_4} \end{bmatrix}$$

From Eq. (13.37.10), we have

$$\mathbf{i}_n = \mathbf{A}\mathbf{Y}_b\,\mathbf{v}_s - \mathbf{A}\mathbf{i}_s = \begin{bmatrix} -\frac{1}{R_1} & \frac{1}{R_2} & 0 & 0 & 0 & 0 & 0 \\ 0 & -\frac{1}{R_2} & -C_1 s & \frac{1}{Ls} & 0 & 0 & 0 \\ 0 & 0 & 0 & \frac{-1}{sL} & \frac{1}{R_3} & C_2 s & 0 \\ 0 & 0 & 0 & 0 & 0 & -C_2 s & \frac{1}{R_4} \end{bmatrix} \begin{bmatrix} -v_{s1} \\ v_{s2} \\ v_{c1(0)} \\ 0 \\ 0 \\ v_{c2(0)}+v_{s3} \\ 0 \end{bmatrix}$$

$$- \begin{bmatrix} -1 & 1 & 0 & 0 & 0 & 0 & 0 \\ 0 & -1 & -1 & 1 & 0 & 0 & 0 \\ 0 & 0 & 0 & -1 & 1 & 1 & 0 \\ 0 & 0 & 0 & 0 & 0 & -1 & 1 \end{bmatrix} \begin{bmatrix} 0 \\ 0 \\ 0 \\ i_{2(0)}+i_{s1} \\ 0 \\ 0 \\ -i_{s2} \end{bmatrix}$$

$$= \begin{bmatrix} \frac{v_{s1}}{R_1}+\frac{v_{s2}}{R_2} \\ -\frac{v_{s2}}{R_2}-v_{C1(0)}\,C_1 s \\ v_{C2(0)}\,C_2 s+v_{s3}\,C_2 s \\ -v_{C2(0)}\,C_2 s-v_{s3}\,C_2 s \end{bmatrix} - \begin{bmatrix} 0 \\ i_{L(0)}+i_{S1} \\ -i_{L(0)}-i_{S1} \\ -i_{S2} \end{bmatrix}$$

$$= \begin{bmatrix} \frac{v_{s1}}{R_1}+\frac{v_{s2}}{R_2} \\ -\frac{v_{s2}}{R_2}-v_{C1(0)}\,C_1 s-i_{L(0)}-i_{s1} \\ v_{C2(0)}C_2 s+v_{s3}\,C_2 s+i_{L(0)}+i_{s1} \\ -v_{C2(0)}C_2 s-v_{s3}C_2 s+i_{s2} \end{bmatrix}$$

If v_{n1}, v_{n2}, v_{n3} and v_{n4} are the node voltages of nodes ①, ②, ③ and ④ with respect to node ⑤, then the node-voltage vector is given by

$$\mathbf{v}_n = \begin{bmatrix} v_{n1} \\ v_{n2} \\ v_{n3} \\ v_{n4} \end{bmatrix}$$

The node equations of the network can be written as

$$\mathbf{i}_n = \mathbf{Y}_n \mathbf{v}_n = \begin{bmatrix} \frac{v_{s1}}{R_1} + \frac{v_{s2}}{R_2} \\ -\frac{v_{s2}}{R_2} - v_{C1(0)} C_1 s - i_{L(0)} - i_{s1} \\ v_{C2(0)} C_2 s + v_{s3} C_2 s + i_{L(0)} + i_{s1} \\ - V_{C2(0)} C_2 s - v_{s3} C_2 s + i_{s2} \end{bmatrix}$$

$$= \begin{bmatrix} \frac{1}{R_1} + \frac{1}{R_2} & -\frac{1}{R_2} & 0 & 0 \\ -\frac{1}{R_2} & \frac{1}{R_2} + C_1 s + \frac{1}{sL} & -\frac{1}{sL} & 0 \\ 0 & -\frac{1}{sL} & \frac{1}{sL} + \frac{1}{R_3} + C_2 s & 0 \\ 0 & 0 & -C_2 s & C_2 s + \frac{1}{R_4} \end{bmatrix} \begin{bmatrix} v_{n1} \\ v_{n2} \\ v_{n3} \\ v_{n4} \end{bmatrix}$$

13.38 LOOP EQUATIONS

KVL can be written as

$$\mathbf{B}_f \mathbf{v}_b = \mathbf{0} \tag{13.38.1}$$

The loop-transformation equation is given by

$$\mathbf{i}_b = \mathbf{B}'_f \mathbf{i}_L \tag{13.38.2}$$

The branch voltage-current relation is given by

$$\mathbf{v}_b = \mathbf{v}_s + \mathbf{Z}_b (\mathbf{i}_b - \mathbf{i}_s) \tag{13.38.3}$$

Premultiplication of Eq. (13.38.3) by B_f gives

$$\mathbf{B}_f \mathbf{v}_b = \mathbf{B}_f \mathbf{v}_s + \mathbf{B}_f \mathbf{Z}_b (\mathbf{i}_b - \mathbf{i}_s) \tag{13.38.4}$$

Substituting Eq. (13.38.1) in Eq. (13.38.4) gives

$$\mathbf{B}_f \mathbf{v}_s + \mathbf{B}_f \mathbf{Z}_b (\mathbf{i}_b - \mathbf{i}_s) = \mathbf{0} \tag{13.38.5}$$

Substituting Eq. (13.38.2) in Eq. (13.38.5) gives

$$\mathbf{B}_f \mathbf{v}_s + \mathbf{B}_f \mathbf{Z}_b (\mathbf{B}'_f \mathbf{i}_L - \mathbf{i}_s) = \mathbf{0}$$

$$\mathbf{B}_f \mathbf{Z}_b \mathbf{i}_s - \mathbf{B}_f \mathbf{v}_s = \mathbf{B}_f \mathbf{Z}_b \mathbf{B}'_f \mathbf{i}_L \tag{13.38.6}$$

Let

$$\mathbf{v}_l \triangleq \mathbf{B}_f \mathbf{Z}_b \mathbf{i}_s - \mathbf{B}_f \mathbf{v}_s \tag{13.38.7}$$

and $$\mathbf{Z}_L \triangleq \mathbf{B}_f \mathbf{Z}_b \mathbf{B}'_f \tag{13.38.8}$$

Eq. 13.38.6 can be written as

$$\mathbf{v}_L = \mathbf{Z}_L \mathbf{i}_L \tag{13.38.9}$$

Eq. (13.38.9) is known as **matrix loop eqution** and $\mathbf{Z}_L$ is called the **loop impedance matrix**. The loop currents are obtained by solving for $\mathbf{i}_L$ in Eq. (13.38.9). Thus

$$\mathbf{i}_L = \mathbf{Z}_L^{-1} \mathbf{v}_L \tag{13.38.10}$$

13.38.1 Loop Analysis Based on Graph Theory

The following steps are required for loop analysis of a given network based on graph theory :

1. Choose a tree T of the graph of the network. Determine the fundamental loop matrix $\mathbf{B}_f$ corresponding to this tree.
2. Determine the branch admittance matrix $\mathbf{Z}_b$.
3. Determine the loop-impedance matrix from the relationship
$$\mathbf{Z}_L = \mathbf{B}_f \mathbf{Z}_b \mathbf{B}'_f$$
4. Write down the voltage-source vector $\mathbf{v}_s$ and the current-source vector $\mathbf{i}_s$ from the network.
5. Determine $\mathbf{v}_L$ from the relationship
$$\mathbf{v}_L = \mathbf{B}_f \mathbf{Z}_b \mathbf{i}_s - \mathbf{B}_f \mathbf{v}_s$$
6. Write down the matrix loop equation from
$$\mathbf{v}_L = \mathbf{Z}_L \mathbf{i}_L$$

■ **EXAMPLE 13.6** Write the matrix loop equation for the network shown in Fig. 13.37a using the above method.

SOLUTION The oriented graph is shown in Fig. 13.37b. It is to be noted that the ideal voltage source is replaced by a short circuit. The chosen tree of the graph is shown in Fig. 13.37c. Fig. 13.37d shows the fundamental loops l_1 and l_2. The fundamental loop matrix $\mathbf{B}_f$ is written by inspection of the oriented graphs as follows :

f–loops ↓ branches →

$$\mathbf{B}_f = \begin{matrix} & 1 & 2 & 3 & 4 \\ 1 & 1 & 1 & 0 & 0 \\ 2 & 0 & -1 & 1 & 1 \end{matrix}$$

The branch impedance matrix $\mathbf{Z}_b$ is obtained by setting the elements in the main diagonal with the branch impedances and the remaining elements with zeros. Thus,

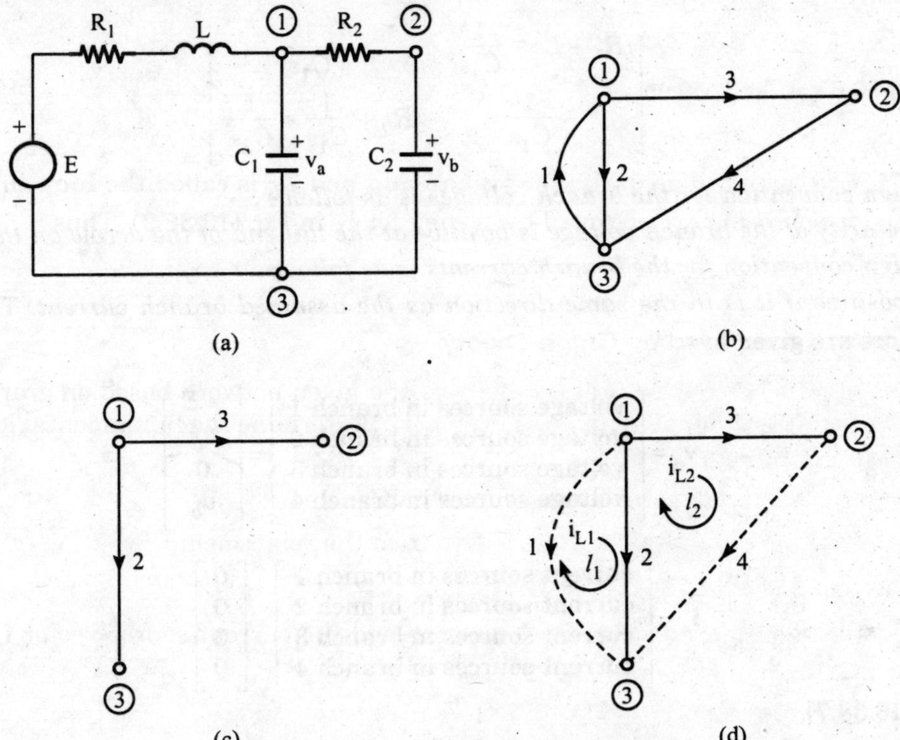

FIG. 13.37 (a) Network for Example 13.4 (b) Oriented graph (c) A tree of the graph (d) Fundamental loops

$$\mathbf{Z}_b = \begin{bmatrix} R_1 + Ls & 0 & 0 & 0 \\ 0 & \dfrac{1}{C_1 s} & 0 & 0 \\ 0 & 0 & R_2 & 0 \\ 0 & 0 & 0 & \dfrac{1}{C_2 s} \end{bmatrix}$$

From Eq. (13.38.8), the loop-impedance matrix is given by

$$\mathbf{Z}_L = \mathbf{B}_f \mathbf{Z}_b \mathbf{B}'_f$$

$$= \begin{bmatrix} 1 & 1 & 0 & 0 \\ 0 & -1 & 1 & 1 \end{bmatrix} \begin{bmatrix} R_1 + Ls & 0 & 0 & 0 \\ 0 & \dfrac{1}{C_1 s} & 0 & 0 \\ 0 & 0 & R_2 & 0 \\ 0 & 0 & 0 & \dfrac{1}{C_2 s} \end{bmatrix} \begin{bmatrix} 1 & 0 \\ 1 & -1 \\ 0 & 1 \\ 0 & 1 \end{bmatrix}$$

$$= \begin{bmatrix} R_1 + Ls + \dfrac{1}{C_1 s} & -\dfrac{1}{C_1 s} \\ -\dfrac{1}{C_1 s} & R_2 + \dfrac{1}{C_1 s} + \dfrac{1}{C_2 s} \end{bmatrix}$$

The sign convention for the branch voltages is as follows :

The polarity of the branch voltage is positive at the tail end of the arrow on the branch.

The sign convention for the branch currents is as follows :

i_s is positive if it is in the same direction as the assumed branch current. The branch source vectors are given by

$$\mathbf{v}_s = \begin{bmatrix} \text{voltage sources in branch 1} \\ \text{voltage sources in branch 2} \\ \text{voltage sources in branch 3} \\ \text{voltage sources in branch 4} \end{bmatrix} = \begin{bmatrix} -E \\ v_a \\ 0 \\ v_b \end{bmatrix}$$

$$\mathbf{i}_s = \begin{bmatrix} \text{current sources in branch 1} \\ \text{current sources in branch 2} \\ \text{current sources in branch 3} \\ \text{current sources in branch 4} \end{bmatrix} = \begin{bmatrix} 0 \\ 0 \\ 0 \\ 0 \end{bmatrix}$$

From Eq. (13.38.7),

$$\mathbf{v}_L = \mathbf{B}_f \mathbf{Z}_b \, \mathbf{i}_s - \mathbf{B}_f \mathbf{v}_s$$

$$= \begin{bmatrix} R_1 + Ls & \dfrac{1}{C_1 s} & 0 & 0 \\ 0 & -\dfrac{1}{C_1 s} & R_2 & \dfrac{1}{C_2 s} \end{bmatrix} \begin{bmatrix} 0 \\ 0 \\ 0 \\ 0 \end{bmatrix} - \begin{bmatrix} 1 & 1 & 0 & 0 \\ 0 & -1 & 1 & 1 \end{bmatrix} \begin{bmatrix} -E \\ v_a \\ 0 \\ v_b \end{bmatrix}$$

$$= \begin{bmatrix} E - v_a \\ v_a - v_a \end{bmatrix}$$

Hence the matrix loop equation can be written as

$$\mathbf{v}_L = \mathbf{Z}_L \, \mathbf{i}_L$$

$$\begin{bmatrix} E - v_a \\ v_a - v_b \end{bmatrix} = \begin{bmatrix} R_1 + Ls + \dfrac{1}{C_1 s} & -\dfrac{1}{C_1 s} \\ -\dfrac{1}{C_1 s} & R_2 + \dfrac{1}{C_1 s} + \dfrac{1}{C_2 s} \end{bmatrix} \begin{bmatrix} i_{L1} \\ i_{L2} \end{bmatrix}$$

■ **EXAMPLE 13.7** Write the loop equations for the network shown in Fig. 13.36a.

SOLUTION Choose the tree as shown in Fig. 13.36c. The fundamental loop matrix B_f is written by inspection as follows :

$$\mathbf{B}_f = \begin{array}{c} \text{loops} \downarrow \\ 1 \\ 2 \\ 3 \end{array} \overset{\begin{array}{c}\text{branches} \rightarrow \\ \overbrace{1 \quad 3 \quad 5 \quad 7}^{\text{twigs}} \quad \overbrace{2 \quad 4 \quad 6}^{\text{links}}\end{array}}{\begin{bmatrix} 1 & -1 & 0 & 0 & 1 & 0 & 0 \\ 0 & 1 & 1 & 0 & 0 & 1 & 0 \\ 0 & 0 & -1 & 1 & 0 & 0 & 1 \end{bmatrix}}$$

$$\mathbf{B}'_f = \begin{bmatrix} 1 & 0 & 0 \\ -1 & 1 & 0 \\ 0 & 1 & -1 \\ 0 & 0 & 1 \\ 1 & 0 & 0 \\ 0 & 1 & 0 \\ 0 & 0 & 1 \end{bmatrix}$$

branches ↓ branches ↓

$$\mathbf{v}_s = \begin{array}{c} 1 \\ 3 \\ 5 \\ 7 \\ 2 \\ 4 \\ 6 \end{array} \begin{bmatrix} -v_{s1} \\ v_{C1}(0) \\ 0 \\ 0 \\ v_{s2} \\ 0 \\ v_{s2}(0) + v_{s3} \end{bmatrix};$$

$$\mathbf{i}_s = \begin{array}{c} 1 \\ 3 \\ 5 \\ 7 \\ 2 \\ 4 \\ 6 \end{array} \begin{bmatrix} 0 \\ 0 \\ 0 \\ -i_{s2} \\ 0 \\ i_{s1} + i_L(0) \\ 0 \end{bmatrix}$$

The branch impedance matrix $\mathbf{Z}_b$ is obtained by filling the main diagonal elements with impedances of the corresponding branch. The off-diagonal elements of this matrix are zero. Thus

$$\mathbf{Z}_b = \begin{array}{c|ccccccc|} & 1 & 3 & 5 & 7 & 2 & 4 & 6 \\ \hline 1 & R_1 & 0 & 0 & 0 & 0 & 0 & 0 \\ 3 & 0 & \frac{1}{C_1 s} & 0 & 0 & 0 & 0 & 0 \\ 5 & 0 & 0 & R_3 & 0 & 0 & 0 & 0 \\ 7 & 0 & 0 & 0 & R_4 & 0 & 0 & 0 \\ 2 & 0 & 0 & 0 & 0 & R_2 & 0 & 0 \\ 4 & 0 & 0 & 0 & 0 & 0 & Ls & 0 \\ 6 & 0 & 0 & 0 & 0 & 0 & 0 & \frac{1}{C_2 s} \end{array}$$

$$\mathbf{B}_f \mathbf{Z}_b = \begin{bmatrix} R_1 & -\frac{1}{C_1 s} & 0 & 0 & R_2 & 0 & 0 \\ 0 & \frac{1}{C_1 s} & R_3 & 0 & 0 & Ls & 0 \\ 0 & 0 & -R_3 & R_4 & 0 & 0 & \frac{1}{C_2 s} \end{bmatrix}$$

$$\mathbf{Z}_L = \mathbf{B}_f \mathbf{Z}_b \mathbf{B}'_f = \begin{bmatrix} R_1 & -\frac{1}{C_1 s} & 0 & 0 & R_2 & 0 & 0 \\ 0 & \frac{1}{C_1 s} & R_3 & 0 & 0 & Ls & 0 \\ 0 & 0 & -R_3 & R_4 & 0 & 0 & \frac{1}{C_2 s} \end{bmatrix} \begin{bmatrix} 1 & 0 & 0 \\ -1 & 1 & 0 \\ 0 & 1 & -1 \\ 0 & 0 & 1 \\ 1 & 0 & 0 \\ 0 & 1 & 0 \\ 0 & 0 & 1 \end{bmatrix}$$

$$= \begin{bmatrix} R_1 + \frac{1}{C_1 s} + R_2 & -\frac{1}{C_1 s} & 0 \\ -\frac{1}{C_1 s} & \frac{1}{C_1 s} + R_3 + Ls & -R_3 \\ 0 & -R_3 & R_3 + R_4 + \frac{1}{C_2 s} \end{bmatrix}$$

From Eq. (13.38.7)

$$\mathbf{v}_L = \mathbf{B}_f \mathbf{Z}_b \mathbf{i}_s - \mathbf{B}_f \mathbf{v}_s = \begin{bmatrix} R_1 & -\frac{1}{C_1 s} & 0 & 0 & R_2 & 0 & 0 \\ 0 & \frac{1}{C_1 s} & R_3 & 0 & 0 & Ls & 0 \\ 0 & 0 & -R_3 & R_4 & 0 & 0 & \frac{1}{C_2 s} \end{bmatrix} \begin{bmatrix} 0 \\ 0 \\ 0 \\ -i_{s2} \\ 0 \\ i_{s1} + i_L(0) \\ 0 \end{bmatrix}$$

$$- \begin{bmatrix} 1 & -1 & 0 & 0 & 1 & 0 & 0 \\ 0 & 1 & 1 & 0 & 0 & 1 & 0 \\ 0 & 0 & -1 & 1 & 0 & 0 & 1 \end{bmatrix} \begin{bmatrix} -v_{s1} \\ v_{C1}(0) \\ 0 \\ 0 \\ v_{s2} \\ 0 \\ v_{C2}(0) + v_{s3} \end{bmatrix}$$

$$= \begin{bmatrix} 0 \\ i_L(0)\,Ls + i_{s1}\,Ls \\ -R_4\, i_{s2} \end{bmatrix} - \begin{bmatrix} -v_{s1} - v_{C1}(0) + v_{s2} \\ v_{C1}(0) \\ v_{C2}(0) + v_{s3} \end{bmatrix}$$

$$= \begin{bmatrix} v_{s1} + v_{C1}(0) - v_{s2} \\ i_L(0) + i_{S1}\,Ls - v_{C1}(0) \\ -R_4\, i_{S2} - v_{C2}(0) - v_{s3} \end{bmatrix}$$

The loop equations of the network can be written as

$$\mathbf{v}_L = \mathbf{Z}_L \mathbf{i}_L \quad \begin{bmatrix} v_{s1} + v_{C1}(0) - v_{s2} \\ i_L(0) + i_{S1}\,Ls - v_{C1}(0) \\ -R_4\, i_{S2} - v_{C2}(0) - v_{s3} \end{bmatrix}$$

$$= \begin{bmatrix} R_1 + \frac{1}{C_1 s} + R_2 & -\frac{1}{C_1 s} & 0 \\ -\frac{1}{C_1 s} & \frac{1}{C_1 s} + R_3 + Ls & -R_3 \\ 0 & -R_3 & R_3 + R_4 + \frac{1}{C_2 s} \end{bmatrix} \begin{bmatrix} i_{l1} \\ i_{l2} \\ i_{l3} \end{bmatrix}$$

■ **EXAMPLE 13.8** Write the matrix loop equation for the network shown in Fig. 13.38a.

SOLUTION The network graph is shown in Fig. 13.38b. Choose the tree as shown in Fig. 13.38c. The fundamental loop matrix $\mathbf{B}_f$ is written by inspection of the oriented graph as follows :

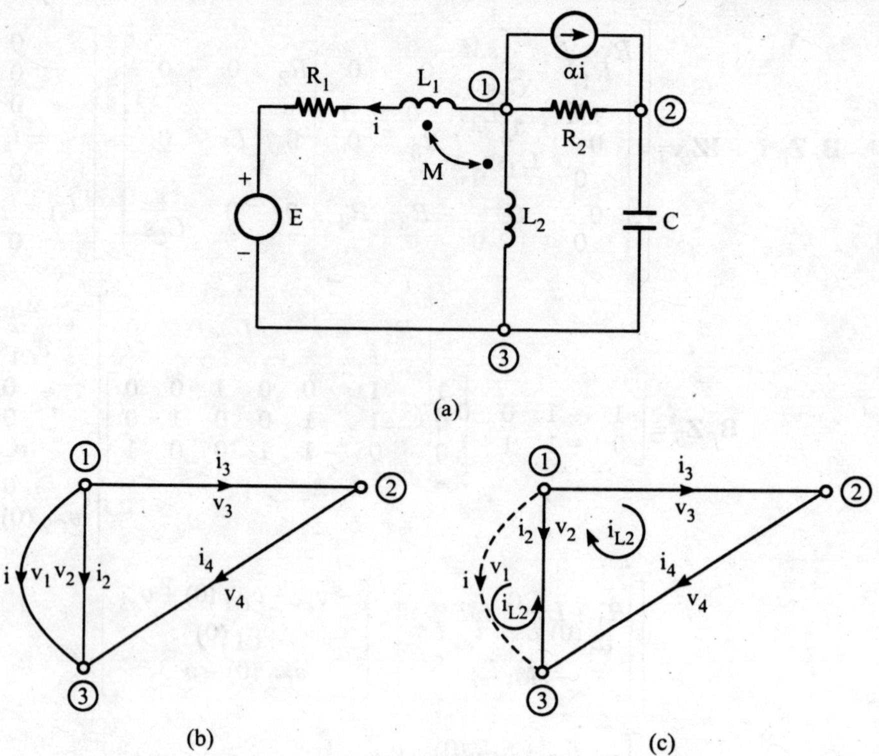

FIG. 13.38

$$
\begin{array}{c} \text{f-loops} \\ \downarrow \end{array} \quad \begin{array}{c} \text{branches} \rightarrow \\ 1 \quad 2 \quad 3 \quad 4 \end{array}
$$

$$
\mathbf{B}_f = \begin{array}{c} 1 \\ 2 \end{array} \begin{bmatrix} 1 & -1 & 0 & 0 \\ 0 & -1 & 1 & 1 \end{bmatrix}
$$

$$
\mathbf{B}'_f = \begin{bmatrix} 1 & 0 \\ -1 & -1 \\ 0 & 1 \\ 0 & 1 \end{bmatrix}
$$

$$
\mathbf{v}_s = \begin{bmatrix} E \\ 0 \\ 0 \\ 0 \end{bmatrix}, \quad \mathbf{i}_s = \begin{bmatrix} 0 \\ 0 \\ \alpha\, i \\ 0 \end{bmatrix}
$$

Since the network contains mututal inductance, the branch impedance $\mathbf{Z}_b$ will no longer be a diagonal matrix. The diagonal elements will be the impedances of the corresponding branches. The off-diagonal elements (1, 2) and (2, 1) of the matrix will be the impedance sM of the mutual inductance. We adopt the following convention for the polarities of the voltages :

When the currents in two coils enter or leave the dotted terminals, the voltage due to mutual inductance M is positive with respect to the voltage due to self inductance L; otherwise, the voltage is negative. With this convention for mutual inductance, the branch impedance matrix is given by

$$\mathbf{Z}_b = \begin{bmatrix} R_1 + L_1 s & sM & 0 & 0 \\ sM & L_2 s & 0 & 0 \\ 0 & 0 & R_2 & 0 \\ 0 & 0 & 0 & \frac{1}{Cs} \end{bmatrix}$$

$$\mathbf{B}_f \mathbf{Z}_b = \begin{bmatrix} 1 & -1 & 0 & 0 \\ 0 & -1 & 1 & 1 \end{bmatrix} \begin{bmatrix} R_1 + L_1 s & sM & 0 & 0 \\ sM & L_2 s & 0 & 0 \\ 0 & 0 & R_2 & 0 \\ 0 & 0 & 0 & \frac{1}{Cs} \end{bmatrix}$$

$$= \begin{bmatrix} R_1 + L_1 s - sM & sM - L_2 s & 0 & 0 \\ -sM & -L_2 s & R_2 & \frac{1}{Cs} \end{bmatrix}$$

$$\mathbf{Z}_L = \mathbf{B}_f \mathbf{Z}_b \mathbf{B}'_f = \begin{bmatrix} R_1 + L_1 s - sM & sM - L_2 s & 0 & 0 \\ -sM & -L_2 s & R_2 & \frac{1}{Cs} \end{bmatrix} \begin{bmatrix} 1 & 0 \\ -1 & -1 \\ 0 & 1 \\ 0 & 1 \end{bmatrix}$$

$$= \begin{bmatrix} R_1 + (L_1 + L_2 - 2M)\, s & (L_2 - M)\, s \\ (L_2 - M)\, s & R_2 + L_2 s + \frac{1}{Cs} \end{bmatrix}$$

$$\mathbf{v}_L = \mathbf{B}_f \mathbf{Z}_b\, \mathbf{i}_s - \mathbf{B}_f \mathbf{v}_s = \begin{bmatrix} R_1 + (L_1 - M)s & sM - L_2 s & 0 & 0 \\ -sM & -L_2 s & R_2 & \frac{1}{Cs} \end{bmatrix} \begin{bmatrix} 0 \\ 0 \\ \alpha i \\ 0 \end{bmatrix}$$

$$- \begin{bmatrix} 1 & -1 & 0 & 0 \\ 0 & -1 & 1 & 1 \end{bmatrix} \begin{bmatrix} E \\ 0 \\ 0 \\ 0 \end{bmatrix}$$

$$= \begin{bmatrix} 0 \\ \alpha R_2 i \end{bmatrix} + \begin{bmatrix} E \\ 0 \end{bmatrix} = \begin{bmatrix} E \\ \alpha R_2 i \end{bmatrix}$$

$$\mathbf{v}_L = \mathbf{Z}_L\, \mathbf{i}_L$$

$$\begin{bmatrix} E \\ \alpha R_2 i \end{bmatrix} = \begin{bmatrix} R_1 + (L_1 + L_2 - 2M)s & -sM + L_2 s \\ -sM + L_2 s & R_2 + L_2 s + \dfrac{1}{Cs} \end{bmatrix} \begin{bmatrix} i_{L1} \\ i_{L2} \end{bmatrix}$$

But $i = i_{L1}$ therefore $\alpha R_2 i = R_2 i_{L1}$

Since i_{L1} is a dependent variable it is to be transferred to the right hand side of the above equation. Therefore the matrix loop equation becomes

$$\begin{bmatrix} E \\ 0 \end{bmatrix} = \begin{bmatrix} R_1 + (L_1 + L_2 - 2M)s & (L_2 - M)s \\ (L_2 - M)s - \alpha R_2 & R_2 + L_2 s + \dfrac{1}{Cs} \end{bmatrix} \begin{bmatrix} i_{L1} \\ i_{L2} \end{bmatrix}$$

It is to be noted that for a network with dependent sources, the loop impedance matrix $\mathbf{Z}_L$ is not symmetrical.

EXAMPLE 13.9 Draw the graph of the network shown in Fig. 13.39a. How many trees are possible for this graph? Draw the trees.

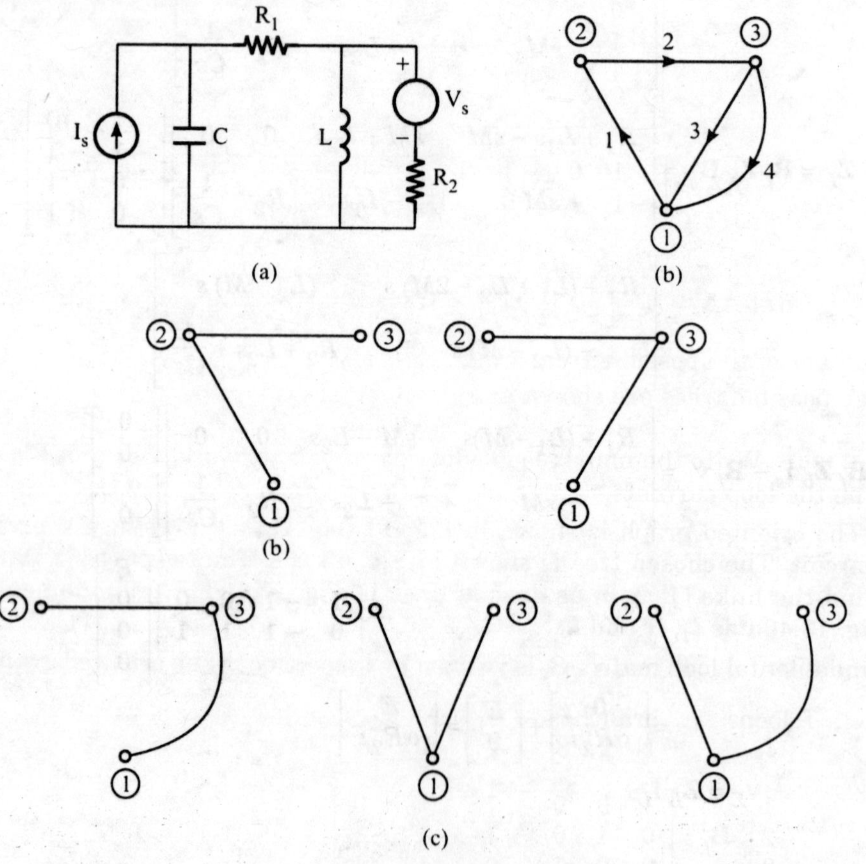

FIG. 13.39

SOLUTION : In constructing the graph of the network, the current source I_s is replaced by an open circuit, the voltage source V_s is replaced by a short circuit, and all other branches containing linear elements are shown by line segments. The network graph is shown Fig. 13.39b. This graph has 3 nodes and 4 branches. The complete incidence matrix is given by

nodes ↓ branches → 1 2 3 4

$$\mathbf{A}_a = \begin{matrix} ① \\ ② \\ ③ \end{matrix} \begin{bmatrix} 1 & 0 & -1 & 1 \\ -1 & 1 & 0 & 0 \\ 0 & -1 & 1 & -1 \end{bmatrix}$$

The reduced incidence matrix is given by

$$\mathbf{A} = \begin{bmatrix} 1 & 0 & -1 & 1 \\ -1 & 1 & 0 & 0 \end{bmatrix}$$

The transpose of **A** is given by

$$\mathbf{A}' = \begin{bmatrix} 1 & -1 \\ 0 & 1 \\ -1 & 0 \\ 1 & 0 \end{bmatrix}$$

$$\mathbf{AA}' = \begin{bmatrix} 1 & 0 & -1 & 1 \\ -1 & 1 & 0 & 0 \end{bmatrix} \begin{bmatrix} 1 & -1 \\ 0 & 1 \\ -1 & 0 \\ 1 & 0 \end{bmatrix} = \begin{bmatrix} 3 & -1 \\ -1 & 2 \end{bmatrix}$$

$$\textbf{det AA}' = \begin{vmatrix} 3 & -1 \\ -1 & 2 \end{vmatrix} = 3 \times 2 - (-1)(-1) = 5$$

The number of the possible trees = det **AA′** = 5

The five possible trees are shown in Fig. 13.39c.

■ **EXAMPLE 13.10** Write the matrix loop equation for the network shown in Fig. 13.40a, and determine the loop currents.

SOLUTION The oriented graph is shown in Fig. 13.40b. The ideal voltage source is replaced by a short circuit. The chosen tree is shown in Fig. 13.40c. The twigs (4, 5, 6) are shown as solid lines and the links (1, 2, 3) as dashed lines in Fig. 13.40d. The fundamental loops are shown in Fig. 13.40d as l_1, l_2 and l_3.

The fundamental loop matrix $\mathbf{B}_f$ is written by inspection of the oriented graph as follows:

f–loops ↓ branches → 1 2 3 4 5 6

$$\begin{matrix} 1 \\ 2 \\ 3 \end{matrix} \quad \mathbf{B}_f = \begin{bmatrix} 1 & 0 & 0 & 1 & 1 & 0 \\ 0 & 1 & 0 & -1 & 0 & -1 \\ 0 & 0 & 1 & 0 & -1 & 1 \end{bmatrix}$$

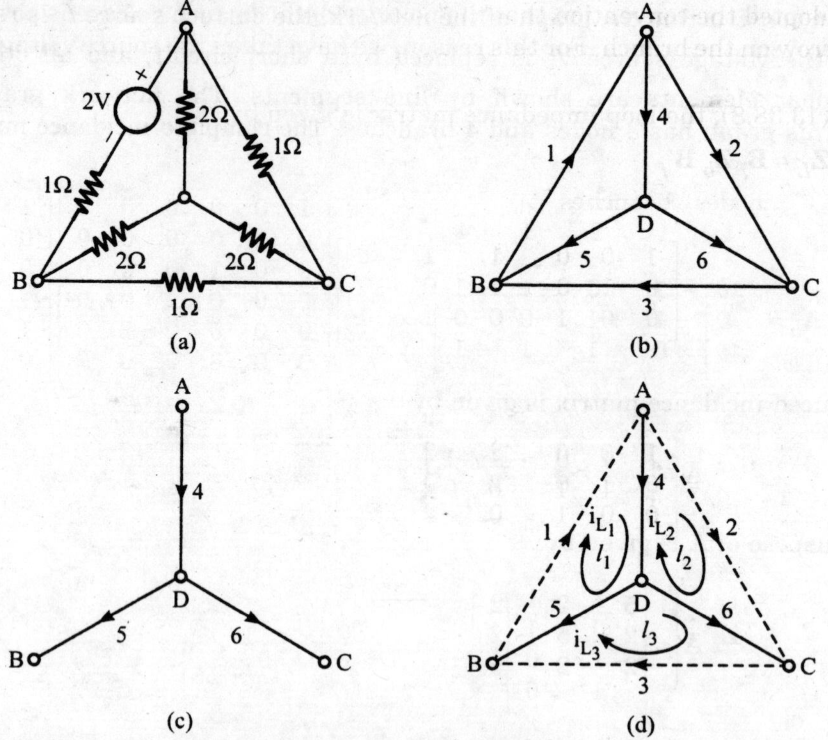

FIG. 13.40 (a) Network for Example 13.9 (b) Oriented graph (c) A tree of the graph (d) Fundamental loops.

The branch impendance matrix $\mathbf{Z}_b$ is obtained by setting the main diagonal with the branch impedances and the remaining elements with zeros. Thus,

$$\mathbf{Z}_b = \begin{bmatrix} 1 & 0 & 0 & 0 & 0 & 0 \\ 0 & 1 & 0 & 0 & 0 & 0 \\ 0 & 0 & 1 & 0 & 0 & 0 \\ 0 & 0 & 0 & 2 & 0 & 0 \\ 0 & 0 & 0 & 0 & 2 & 0 \\ 0 & 0 & 0 & 0 & 0 & 2 \end{bmatrix}$$

$$\mathbf{B}'_f = \begin{bmatrix} 1 & 0 & 0 \\ 0 & 1 & 0 \\ 0 & 0 & 1 \\ 1 & -1 & 0 \\ 1 & 0 & -1 \\ 0 & -1 & 1 \end{bmatrix}$$

$$\mathbf{v}_s = \begin{bmatrix} -2 \\ 0 \\ 0 \\ 0 \\ 0 \\ 0 \end{bmatrix} \qquad \mathbf{i}_s = \begin{bmatrix} 0 \\ 0 \\ 0 \\ 0 \\ 0 \\ 0 \end{bmatrix}$$

We have adopted the convention that the polarity of the branch voltage is positive at the tail end of the arrow on the branch. For this reason, we have taken the source voltage in branch 2 as – 2 V.

From Eq. (13.38.8), the loop impedance matrix is given by

$$\mathbf{Z}_L = \mathbf{B}_f \mathbf{Z}_b \mathbf{B}'_f$$

$$= \begin{bmatrix} 1 & 0 & 0 & 1 & 1 & 0 \\ 0 & 1 & 0 & -1 & 0 & -1 \\ 0 & 0 & 1 & 0 & -1 & 1 \end{bmatrix} \begin{bmatrix} 1 & 0 & 0 & 0 & 0 & 0 \\ 0 & 1 & 0 & 0 & 0 & 0 \\ 0 & 0 & 1 & 0 & 0 & 0 \\ 0 & 0 & 0 & 2 & 0 & 0 \\ 0 & 0 & 0 & 0 & 2 & 0 \\ 0 & 0 & 0 & 0 & 0 & 2 \end{bmatrix} \begin{bmatrix} 1 & 0 & 0 \\ 0 & 1 & 0 \\ 0 & 0 & 1 \\ 1 & -1 & 0 \\ 1 & 0 & -1 \\ 0 & -1 & 1 \end{bmatrix}$$

$$= \begin{bmatrix} 1 & 0 & 0 & 2 & 2 & 0 \\ 0 & 1 & 0 & -2 & 0 & -2 \\ 0 & 0 & 1 & 0 & -2 & 2 \end{bmatrix} \begin{bmatrix} 1 & 0 & 0 \\ 0 & 1 & 0 \\ 0 & 0 & 1 \\ 1 & -1 & 0 \\ 1 & 0 & -1 \\ 0 & -1 & 1 \end{bmatrix}$$

$$\mathbf{Z}_L = \begin{bmatrix} 5 & -2 & -2 \\ -2 & 5 & -2 \\ -2 & -2 & 5 \end{bmatrix}$$

$$\mathbf{B}_f \mathbf{v}_s = \begin{bmatrix} 1 & 0 & 0 & 1 & 1 & 0 \\ 0 & 1 & 0 & -1 & 0 & -1 \\ 0 & 0 & 1 & 0 & -1 & 1 \end{bmatrix} \begin{bmatrix} -2 \\ 0 \\ 0 \\ 0 \\ 0 \\ 0 \end{bmatrix} = \begin{bmatrix} -2 \\ 0 \\ 0 \end{bmatrix}$$

From Eq. (13.38.7)

$$\mathbf{v}_L = \mathbf{B}_f \mathbf{Z}_b \, \mathbf{i}_s - \mathbf{B}_f \mathbf{v}_s$$

$$\mathbf{v}_L = -\mathbf{B}_f \mathbf{v}_s = -\begin{bmatrix} -2 \\ 0 \\ 0 \end{bmatrix} = \begin{bmatrix} 2 \\ 0 \\ 0 \end{bmatrix}$$

Hence the loop equations in matrix form can be written as

$$\mathbf{v}_L = \mathbf{Z}_L \, \mathbf{i}_L$$

$$\begin{bmatrix} 2 \\ 0 \\ 0 \end{bmatrix} = \begin{bmatrix} 5 & -2 & -2 \\ -2 & 5 & -2 \\ -2 & -2 & 5 \end{bmatrix} \begin{bmatrix} i_{L1} \\ i_{L2} \\ i_{L3} \end{bmatrix}$$

In expanded form, these equations can be written as

$$5i_{L1} - 2i_{L2} - 2i_{L3} = 2$$

$$-2i_{L1} + 5i_{L2} - 2i_{L3} = 0$$

$$-2i_{L1} - 2i_{L2} + 5i_{L3} = 0$$

These equations can be solved to give

$$i_{L1} = \frac{6}{7} A, i_{L2} = \frac{4}{7} A, i_{L3} = \frac{4}{7} A$$

■ **EXAMPLE 13.11** For the network shown in Fig. 13.41a, write the fundamental cut-set matrix and hence obtain the network equilibrium equations on node basis.

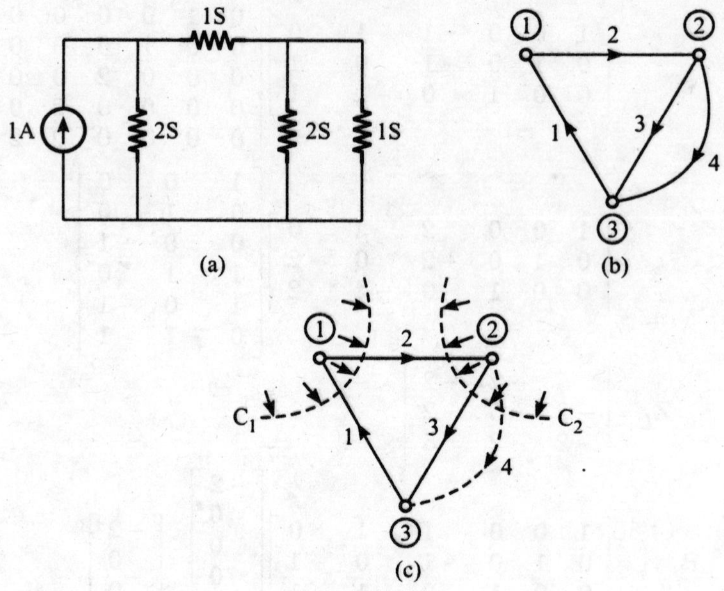

FIG. 13.41

SOLUTION The oriented graph of the given network is shown in Fig. 13.41b. The ideal current source is replaced by an open circut. Fig. 13.41c shows the chosen tree. The twigs (1, 3) are shown as solid lines and the links (2, 4) as dashed lines. The cut-sets C_1 and C_2 are shown as curved lines in Fig. 13.41c.

The fundamental cut-sets are written as :

$$C_1 : (1, 2)$$

$$C_2 : (3, 2, 4)$$

The fundamental cut-set matrix is given by

f–cut–sets ↓ branches →

$$\begin{matrix} & & 1 & 2 & 3 & 4 \\ 1 & \mathbf{Q}_f = & \Bigg[1 & 1 & 0 & 0 \\ 2 & & 0 & -1 & 1 & 1 \Bigg] \end{matrix}$$

therefore,

$$\mathbf{Q'}_f = \begin{bmatrix} 1 & 0 \\ 1 & -1 \\ 0 & 1 \\ 0 & 1 \end{bmatrix}, \quad \mathbf{i}_s = \begin{bmatrix} -1 \\ 0 \\ 0 \\ 0 \end{bmatrix}, \quad \mathbf{v}_s = \begin{bmatrix} 0 \\ 0 \\ 0 \\ 0 \end{bmatrix}$$

It is to be noted that only branch 1 contains a current source of value 1*A*. We have adopted the convention that the current i_{SK} in the *k*th branch has a positive sign if it is in the same direction as the assumed branch curent. Therefore $i_{s1} = -1$. There is no voltage source in any branch. Therefore all the elements of the matrix $\mathbf{v}_s$ are zero.

The generalised KCL equation for the twig voltages are

$$\mathbf{Y}_t\,\mathbf{v}_t = \mathbf{Q}_f\,\mathbf{Y}_b\,\mathbf{i}_s - \mathbf{Q}_f\,\mathbf{i}_s \tag{13.32.5}$$

where

$$\mathbf{Y}_t = \mathbf{Q}_f\,\mathbf{Y}_b\,\mathbf{Q}'_f \tag{13.32.4}$$

$$\mathbf{Q}_f\,\mathbf{i}_s = \begin{bmatrix} 1 & 1 & 0 & 0 \\ 0 & -1 & 1 & 1 \end{bmatrix}\begin{bmatrix} -1 \\ 0 \\ 0 \\ 0 \end{bmatrix} = \begin{bmatrix} -1 \\ 0 \end{bmatrix}$$

$$\mathbf{Y}_t = \mathbf{Q}_f\,\mathbf{Y}_b\,\mathbf{Q}_f = \begin{bmatrix} 1 & 1 & 0 & 0 \\ 0 & -1 & 1 & 1 \end{bmatrix}\begin{bmatrix} 2 & 0 & 0 & 0 \\ 0 & 1 & 0 & 0 \\ 0 & 0 & 2 & 0 \\ 0 & 0 & 0 & 1 \end{bmatrix}\begin{bmatrix} 1 & 0 \\ 1 & -1 \\ 0 & 1 \\ 0 & 1 \end{bmatrix}$$

$$= \begin{bmatrix} 2 & 1 & 0 & 0 \\ 0 & -1 & 2 & 1 \end{bmatrix}\begin{bmatrix} 1 & 0 \\ 1 & -1 \\ 0 & 1 \\ 0 & 1 \end{bmatrix} = \begin{bmatrix} 3 & -1 \\ -1 & 4 \end{bmatrix}$$

Therefore Eq. (13.32.5) becomes

$$\begin{bmatrix} 3 & -1 \\ -1 & 4 \end{bmatrix}\begin{bmatrix} v_{t1} \\ v_{t2} \end{bmatrix} = -\begin{bmatrix} -1 \\ 0 \end{bmatrix}$$

or

$$\begin{bmatrix} 3 & -1 \\ -1 & 4 \end{bmatrix}\begin{bmatrix} v_{t1} \\ v_{t2} \end{bmatrix} = \begin{bmatrix} 1 \\ 0 \end{bmatrix}$$

EXAMPLE 13.12 Draw the oriented graph of the network shown in Fig. 13.42a. Obtain the tie-set matrix and write the network equations on KVL basis.

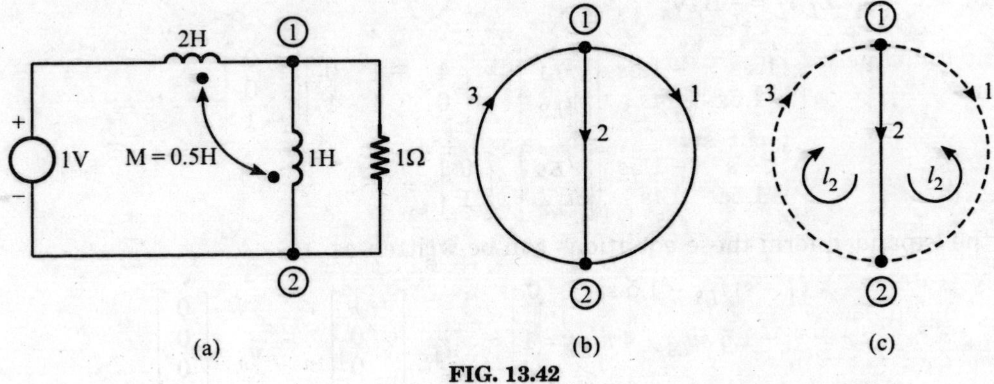

FIG. 13.42

SOLUTION Choose the directions of currents as shown in Fig. 13.42a. The currents are assumed to leave the dotted terminals of the inductances. With this choice, the mutual inductance is positive. The oriented graph is shown in Fig. 13.42b.

A tree of this graph is shown in Fig. 13.42c. The twig 2 is shown by a solid line and the links 1 and 3 are shown by dotted lines. The tie-set matrix $\boldsymbol{B}_f$ can be written as follows :

$$\begin{array}{c} \text{f-loops} \\ \downarrow \\ \\ 1 \\ 3 \end{array} \quad \begin{array}{c} \text{branches} \rightarrow \\ \begin{array}{ccc} 1 & 2 & 3 \end{array} \\ \mathbf{B}_f = \begin{bmatrix} 1 & -1 & 0 \\ 0 & 1 & 1 \end{bmatrix} \end{array}$$

The branch impedance matrix is given by

$$\mathbf{Z}_b = \begin{bmatrix} 1 & 0 & 0 \\ 0 & s & 0.5s \\ 0 & 0.5s & 2s \end{bmatrix}$$

The loop impedance matrix is given by

$$\mathbf{Z}_L = \mathbf{B}_f \mathbf{Z}_b \mathbf{B}'_f$$

$$= \begin{bmatrix} 1 & -1 & 0 \\ 0 & 1 & 1 \end{bmatrix} \begin{bmatrix} 1 & 0 & 0 \\ 0 & s & 0.5s \\ 0 & 0.5s & 2s \end{bmatrix} \begin{bmatrix} 1 & 0 \\ -1 & 1 \\ 0 & 1 \end{bmatrix} = \begin{bmatrix} 1+s & -1.5s \\ -1.5s & 4s \end{bmatrix}$$

$$\mathbf{i}_s = \begin{bmatrix} 0 \\ 0 \\ 0 \end{bmatrix}, \quad \mathbf{v}_s = \begin{bmatrix} 0 \\ 0 \\ -1 \end{bmatrix}$$

It is to be noted that branch 3 contains a voltage source of value 1V and its negative side is at the tail end of the branch arrow. Therefore, we have taken $v_{s3} = -1$ V. The network loop equation is

$$\mathbf{v}_L = \mathbf{Z}_L \, \mathbf{i}_L \tag{13.38.9}$$

where
$$\mathbf{v}_L = \mathbf{B}_f \mathbf{Z}_b \, \mathbf{i}_s - \mathbf{B}_f \mathbf{v}_s \tag{13.38.7}$$

As
$$\mathbf{i}_s = \mathbf{0}$$

$$\mathbf{Z}_L \, \mathbf{i}_L = -\mathbf{B}_f \mathbf{v}_s$$

$$\begin{bmatrix} 1+s & -1.5s \\ -1.5s & 4s \end{bmatrix} \begin{bmatrix} i_{L1} \\ i_{L2} \end{bmatrix} = -\begin{bmatrix} 1 & -1 & 0 \\ 0 & 1 & 1 \end{bmatrix} \begin{bmatrix} 0 \\ 0 \\ -1 \end{bmatrix}$$

or

$$\begin{bmatrix} 1+s & -1.5s \\ -1.5s & 4s \end{bmatrix} \begin{bmatrix} i_{L1} \\ i_{L2} \end{bmatrix} = \begin{bmatrix} 0 \\ 1 \end{bmatrix}$$

In the expanded form these equations can be written as

$$(1+s)\, i_{L1} - 1.5\, s i_{L2} = 0$$

$$-1.5\, s i_{L1} + 4\, s i_{L2} = 1$$

■ **EXAMPLE 13.13** Draw the fundamental loops and the fundamental cut-sets coresponding to a tree T of the network graph which is shown by solid lines in Fig. 13.43a. Write the fundamental loop matrix $\boldsymbol{B_f}$ and the cut-set matrix $\boldsymbol{Q_f}$ of the graph. Write all linearly independent KVL and KCL equations of the network.

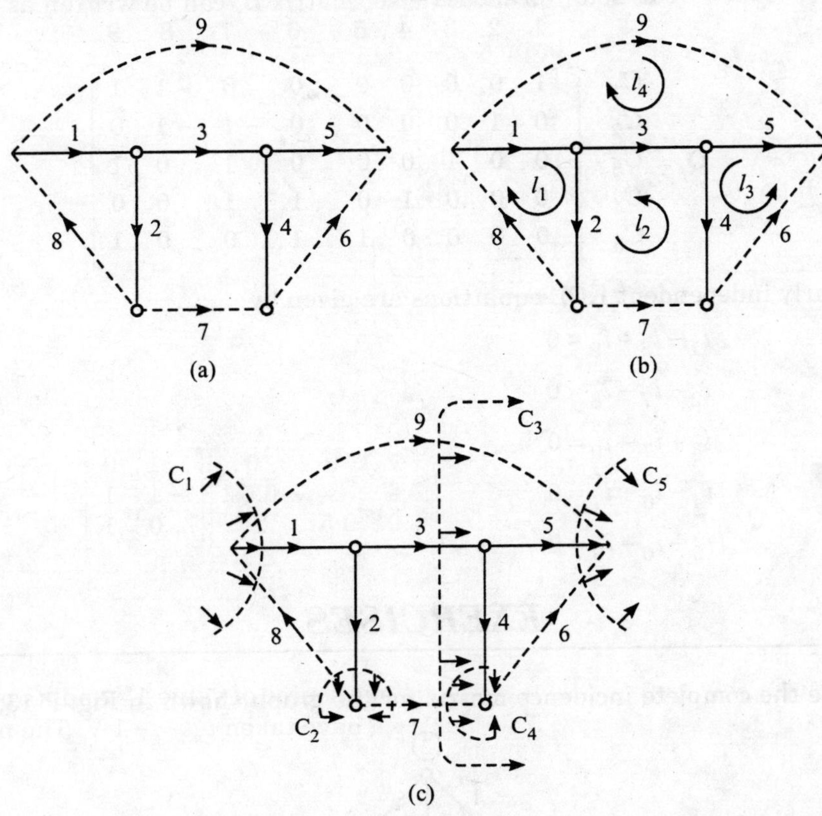

FIG. 13.43

SOLUTION The fundamental loops are shown in Fig. 14.43b. The fundamental loop matrix can be written as follows :

$$\begin{array}{c} \text{f-loops} \\ \downarrow \end{array} \quad \begin{array}{c} \text{branches} \rightarrow \\ 1 \;\; 2 \;\; 3 \;\; 4 \;\; 5 \;\; 6 \;\; 7 \;\; 8 \;\; 9 \end{array}$$

$$\mathbf{B}_f = \begin{array}{c} l_1 \\ l_2 \\ l_3 \\ l_4 \end{array} \begin{bmatrix} 1 & 1 & 0 & 0 & 0 & 0 & 0 & 1 & 0 \\ 0 & 1 & -1 & -1 & 0 & 0 & 1 & 0 & 0 \\ 0 & 0 & 0 & 1 & -1 & 1 & 0 & 0 & 0 \\ -1 & 0 & -1 & 0 & -1 & 0 & 0 & 0 & 1 \end{bmatrix}$$

The linearly independent KVL equations can be written as

$$v_1 + v_2 + v_8 = 0$$

$$v_2 - v_3 - v_4 + v_7 = 0$$

$$v_4 - v_5 + v_6 = 0$$

$$-v_1 - v_3 - v_5 + v_9 = 0$$

The fundamental cut-sets are shown in Fig. 13.43c. The fundamental cut-set matrix is given by

$$\begin{array}{c} \text{cut-sets} \\ \downarrow \end{array} \quad \begin{array}{l} \text{branches} \rightarrow \\ \;1\;\;2\;\;3\;\;4\;\;5\;\;\;6\;\;\;7\;\;\;8\;\;9 \end{array}$$

$$\mathbf{Q}_f = \begin{array}{c} C_1 \\ C_2 \\ C_3 \\ C_4 \\ C_5 \end{array} \begin{bmatrix} 1 & 0 & 0 & 0 & 0 & 0 & 0 & -1 & 1 \\ 0 & 1 & 0 & 0 & 0 & 0 & -1 & -1 & 0 \\ 0 & 0 & 1 & 0 & 0 & 0 & 1 & 0 & 1 \\ 0 & 0 & 0 & 1 & 0 & -1 & 1 & 0 & 0 \\ 0 & 0 & 0 & 0 & 1 & 1 & 0 & 0 & 1 \end{bmatrix}$$

The linearly independent KCL equattions are given by

$$i_1 - i_8 + i_9 = 0$$

$$i_2 - i_7 - i_8 = 0$$

$$i_3 + i_7 + i_9 = 0$$

$$i_4 - i_6 + i_7 = 0$$

$$i_5 + i_6 + i_9 = 0$$

EXERCISES

1. Write the complete incidence matrix for the graph shown in Fig. P 13.1.

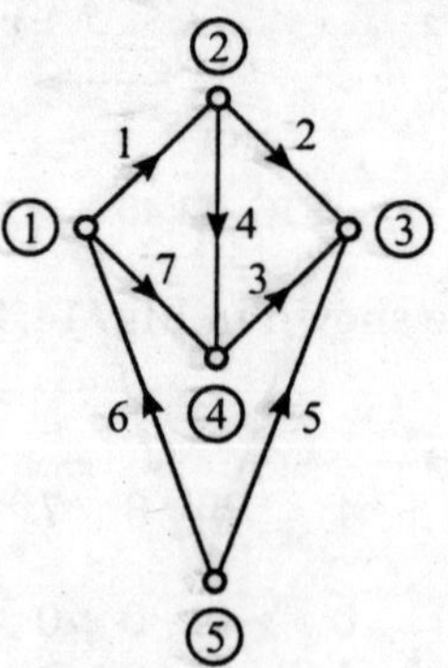

FIG. P .13.1

2. Draw the oriented graph corresponding to the reduced incidence matrix given by

$$\mathbf{A} = \begin{bmatrix} 1 & 1 & 0 & 0 & 0 & 0 & 0 \\ 0 & -1 & 1 & 1 & 0 & 0 & 0 \\ 0 & 0 & 0 & -1 & 1 & 0 & 0 \\ 0 & 0 & 0 & 0 & 0 & -1 & 1 \end{bmatrix}$$

3. **Draw the oriented graph of the network shown in Fig. P 13.3a and write the *A* matrix by inspection.**

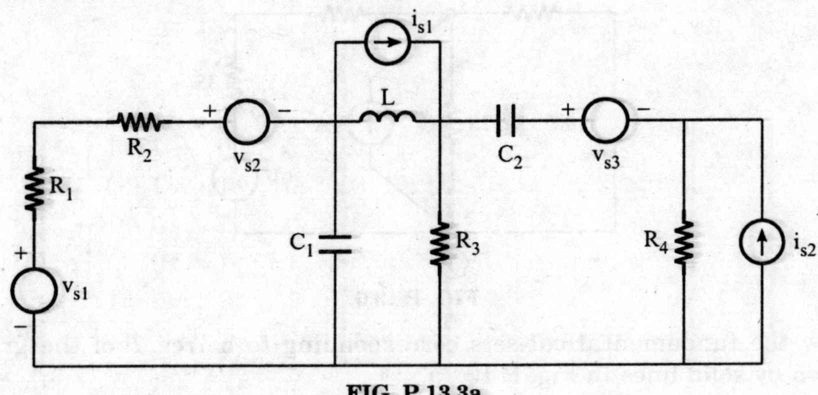

FIG. P 13.3a

4. **Show that cut-sets of the network graph shown in Fig. P 13.4 and write its cut-sets matrix for all the cut-sets.**

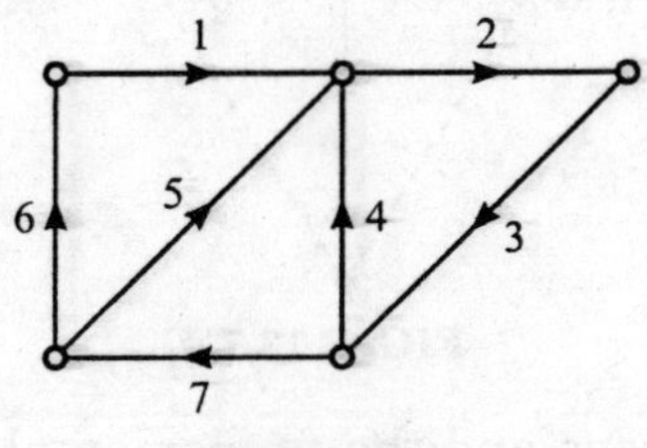

FIG. P 13.4

5. **Write the node equations for the network shown in Fig. P 13.5.**

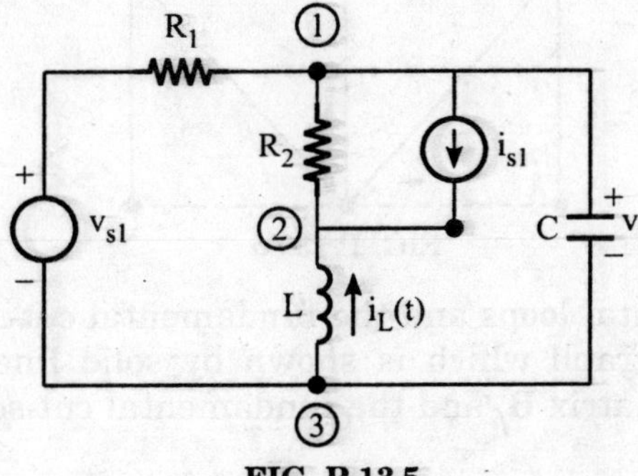

FIG. P 13.5

6. Write the node equation for the network shown in Fig. P 13.6.

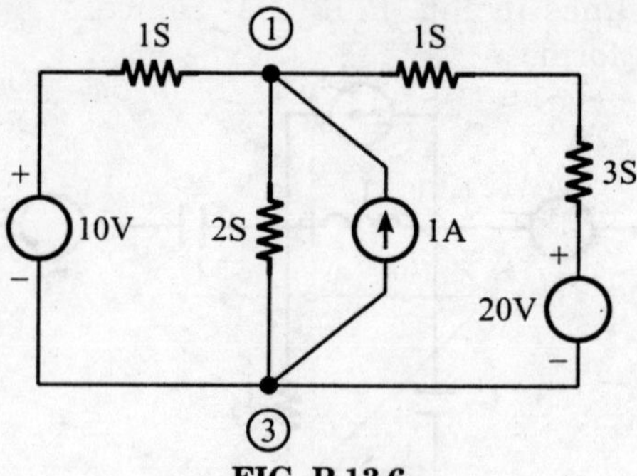

FIG. P 13.6

7. Draw the fundamental cut-sets corresponding to a tree T of the graph which is shown by solid lines in Fig. P 13.7a.

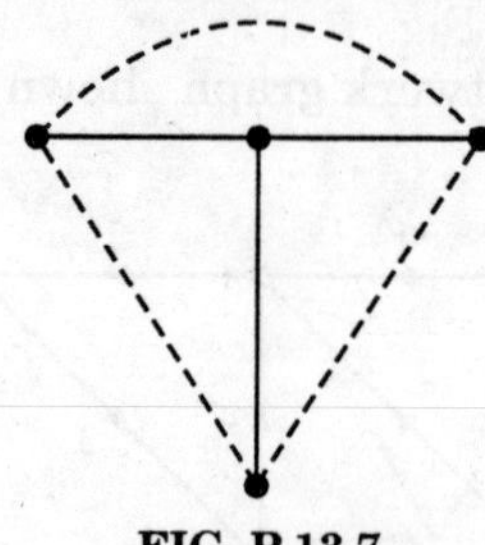

FIG. P 13.7

8. Draw the fundametal loops and the fundamental cut-sets corresponding to a tree T of the graph which is shown by solid lines in Fig. P 13.8a.

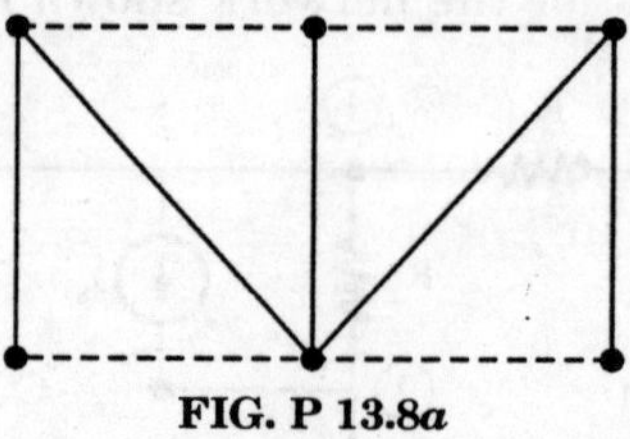

FIG. P 13.8*a*

9. Draw the fundamental loops and the fundamental cut-sets corresponding to a tree T of the network graph which is shown by solid lines in Fig. 13.9a. Write the fundamental loop matrix $\mathbf{B}_f$ and the fundamental cut-set matrix $\mathbf{Q}_f$ of the graph.

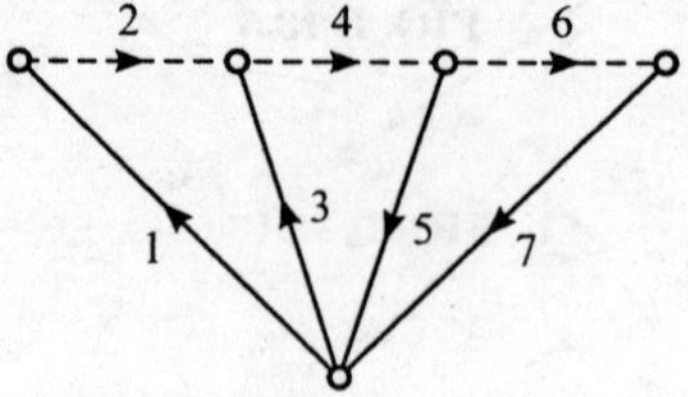

FIG. P 13.9*a*

10. Draw the fundamental cut-sets corresponding to a tree *T* of the network graph which is shown by solid lines in Fig. P 13.10a. Hence, write the KCL equations for the network in matrix form.

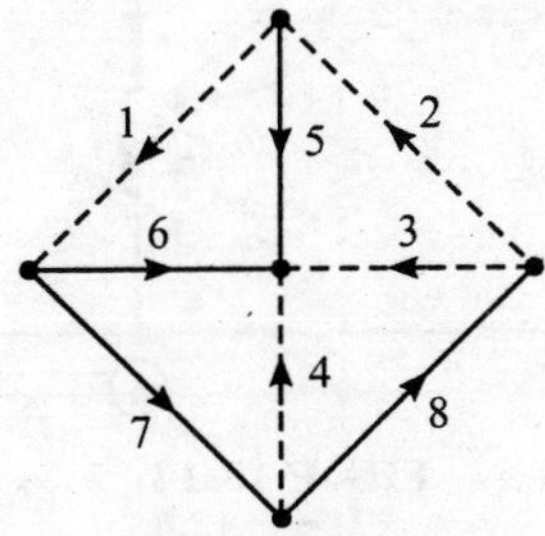

FIG. P 13.10a

11. Draw the fundamental loops corresponding to a tree *T* of the network graph which is shown by solid lines in Fig. P 13.11a. Hence write the KVL equations for the network in matrix form.

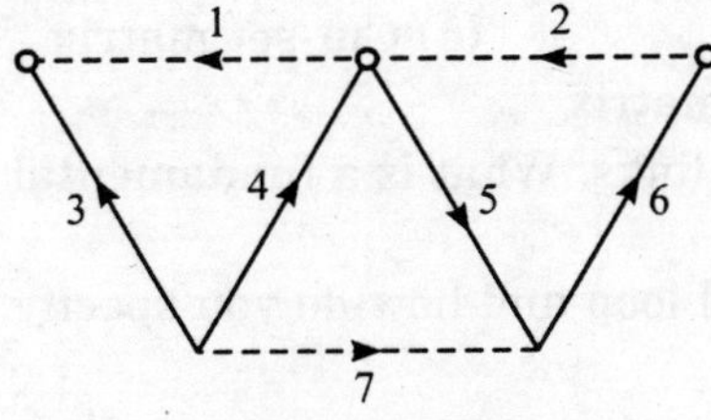

FIG. P 13.11a

12. Draw the fundamental cut-sets corresponding to a tree *T* of the network graph which is shown by solid lines in Fig. P 13.11a. Hence write the KCL equations for the network in matrix form.
13. Draw the fundamental loops corresponding to a tree *T* of the network graph which is shown by solid lines in Fig. P 13.13a. Hence write the KVL equations for the network in matrix form.

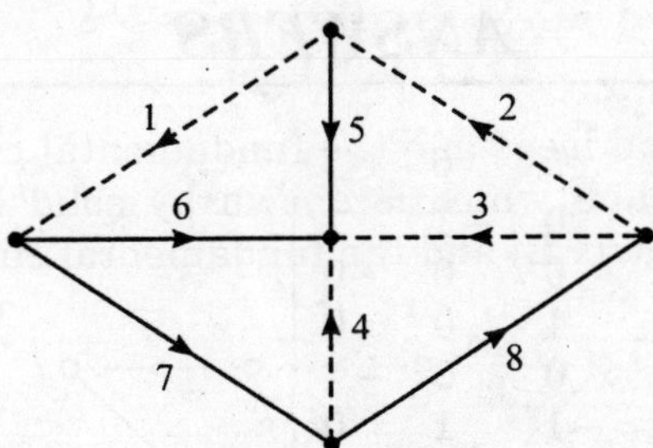

FIG. P 13.13*a*

14. Use graph theory to write the loop equations for the network shown in Fig. P 13.14.

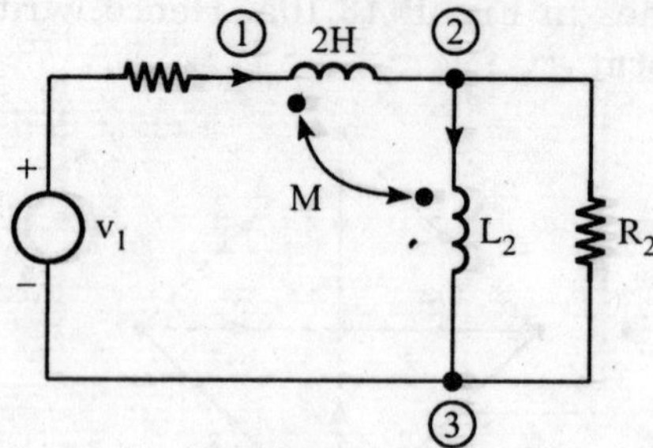

FIG. P 13..14

15. Define precisely and explain with the help of a simple example in each case the folowing topological terms used in network analysis :
    (*a*) Tree (*b*) Graph
    (*c*) Cut-set (*d*) Tie-set
16. Explain with illustrative examples the meaning of the following terms :
    (*a*) Incidence matrix, (*b*) Cut-set matrix
    (*c*) Fundamental loop matrix
17. Define tree, twigs and links. What is a fundamental cut-set and how do you specify its orientation?
18. What is a fundamental loop and how do you specify its orientation.
19. Prove that

$$\mathbf{A} = \mathbf{A}_t\,[\mathbf{U} - (\mathbf{B}_l^{-1}\ \mathbf{B}_t)']$$

20. Define fundamental cut-set matrix and fundamental loop matrix. Prove that

    (*a*) $\mathbf{Q}_f = [\mathbf{U} - (\mathbf{B}_{fl}^{-1}\ \mathbf{B}_{ft})']$

    (*b*) $\mathbf{Q}_f\,\mathbf{i}_b = \mathbf{0}$

ANSWERS

1.

	1	2	3	4	5	6	7
①	1	0	0	0	0	−1	1
②	−1	1	0	1	0	0	0
③	0	−1	−1	0	1	0	0
④	0	0	1	−1	0	0	−1
⑤	0	0	0	0	−1	1	0
	1	2	3	4	5	6	7

2. $$\mathbf{A}_a = \begin{matrix} ① \\ ② \\ ③ \\ ④ \\ ⑤ \end{matrix}\begin{bmatrix} 1 & 1 & 0 & 0 & 0 & 0 & 0 \\ 0 & -1 & 1 & 1 & 0 & 0 & 0 \\ 0 & 0 & 0 & -1 & 1 & 1 & 0 \\ 0 & 0 & 0 & 0 & 0 & -1 & 1 \\ -1 & 0 & -1 & 0 & -1 & 0 & -1 \end{bmatrix}$$
$$\begin{matrix} 1 & 2 & 3 & 4 & 5 & 6 & 7 \end{matrix}$$

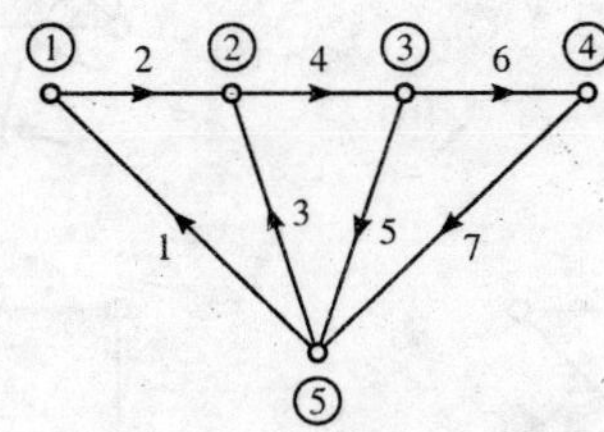

Fig. P 13.2b

FIG. P 13.2b

3. $$[\mathbf{A}] = \begin{matrix} ① \\ ② \\ ③ \\ ④ \end{matrix}\begin{bmatrix} -1 & 1 & 0 & 0 & 0 & 0 & 0 \\ 0 & -1 & -1 & 1 & 0 & 0 & 0 \\ 0 & 0 & 0 & -1 & 1 & 1 & 0 \\ 0 & 0 & 0 & 0 & 0 & -1 & 1 \end{bmatrix}$$

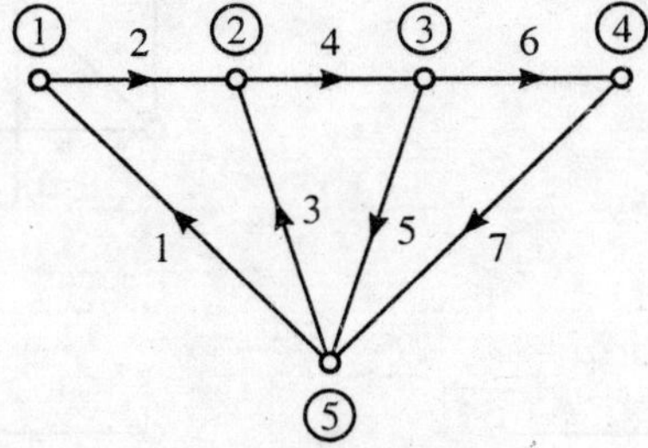

FIG. P 13.3b

4. cut–sets ↓ branches →

$$\begin{matrix} & 1 & 2 & 3 & 4 & 5 & 6 & 7 \\ C_1 & 1 & 0 & 0 & 0 & 0 & -1 & 0 \\ C_2 & 0 & 0 & 0 & 0 & 1 & 1 & -1 \\ C_3 & 0 & 0 & -1 & 1 & 1 & 1 & 0 \\ C_4 & 0 & -1 & 1 & 0 & 0 & 0 & 0 \\ C_5 & 0 & 0 & -1 & 1 & 0 & 0 & 1 \\ C_6 & 1 & 0 & 0 & 0 & 1 & 0 & -1 \\ C_7 & -1 & 0 & 1 & -1 & -1 & 0 & 0 \\ C_8 & 0 & 1 & 0 & -1 & 0 & 0 & -1 \\ C_9 & 0 & 1 & 0 & -1 & -1 & -1 & 0 \\ C_{10} & -1 & 1 & 0 & -1 & -1 & 0 & 0 \end{matrix}$$

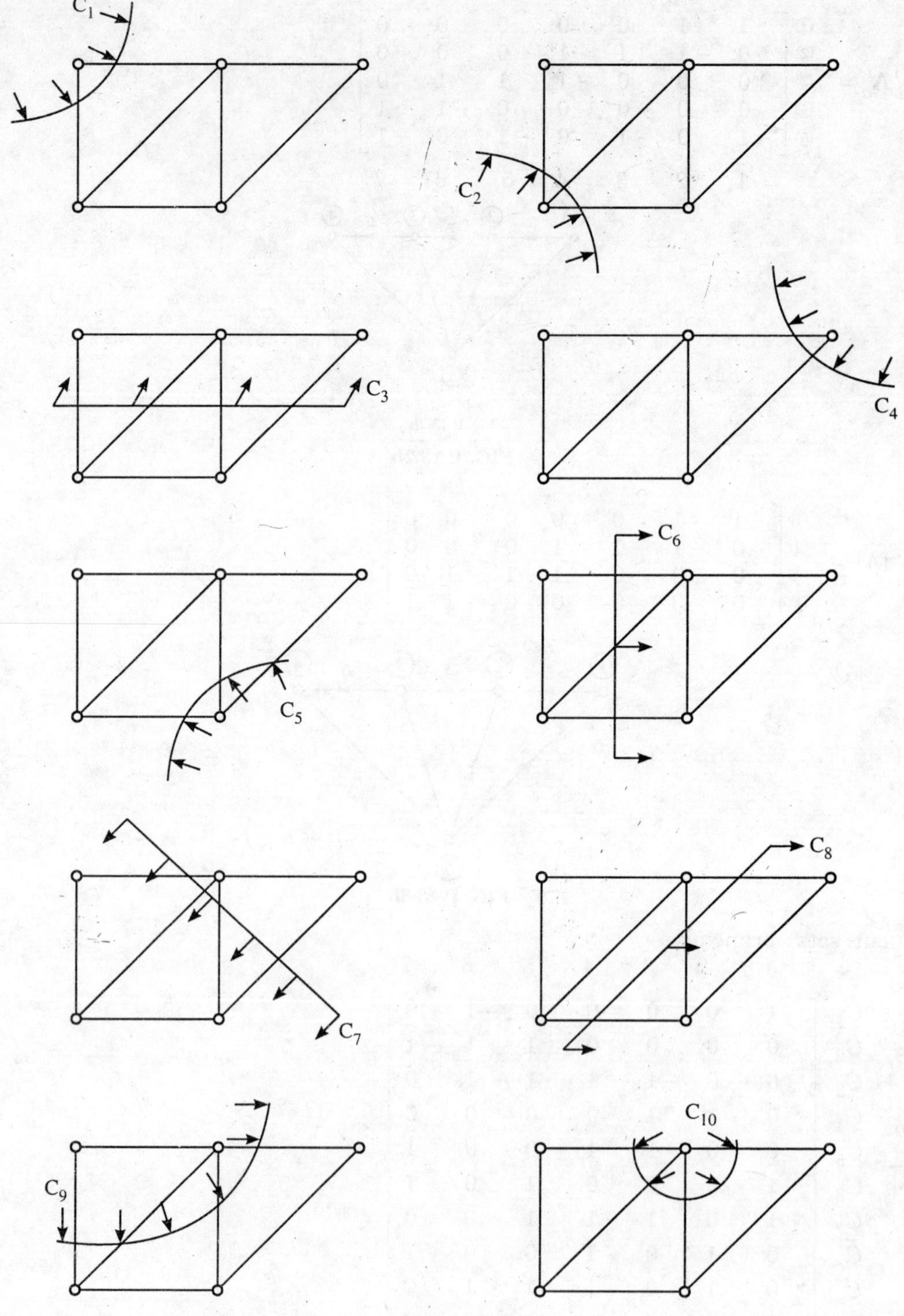

FIG. P 13.4b

5. $$\begin{bmatrix} \dfrac{v_{s1}}{R_1} - i_{s1} \\ -i_L(t_0^+) + i_{s1} \end{bmatrix} = \begin{bmatrix} \dfrac{1}{R_1} + \dfrac{1}{R_2} + Cs & -\dfrac{1}{R_2} \\ -\dfrac{1}{R_2} & \dfrac{1}{R_2} + \dfrac{1}{sL} \end{bmatrix} \begin{bmatrix} v_{n1} \\ v_{n2} \end{bmatrix}$$

6. $$\begin{bmatrix} 11 \\ 60 \end{bmatrix} = \begin{bmatrix} 4 & -1 \\ -1 & 4 \end{bmatrix} \begin{bmatrix} v_{n1} \\ v_{n2} \end{bmatrix}$$

7.

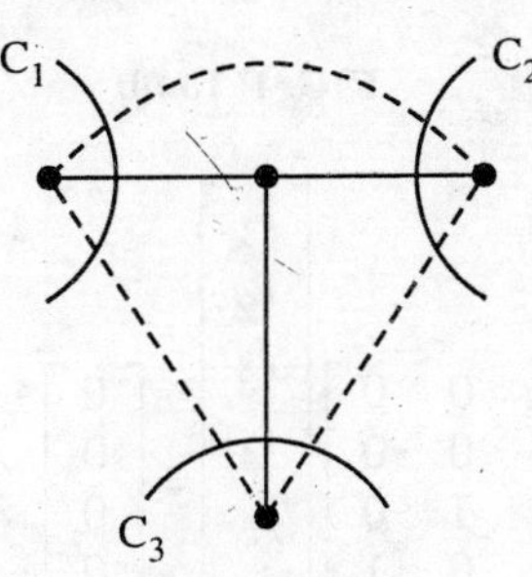

FIG. P 13.7b

8.

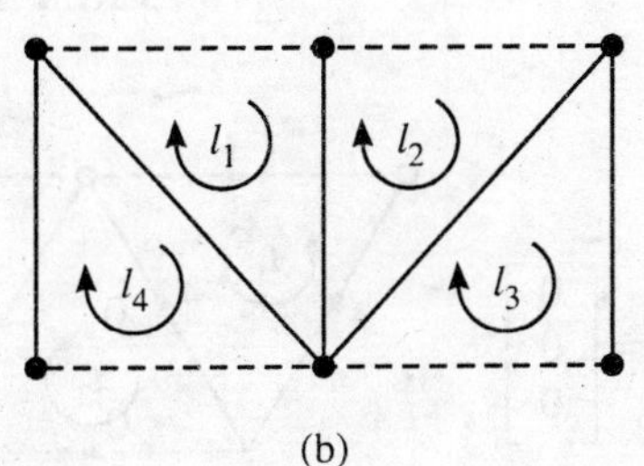

(b)

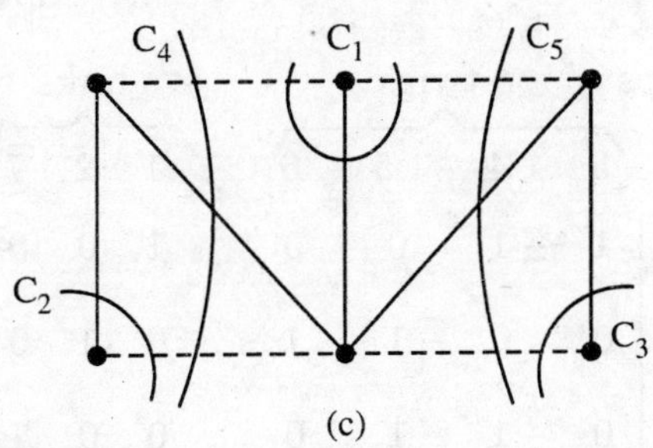

(c)

FIG. P 13.8

9. f-loops ↓, twigs (1 3 5 7), links (2 4 6)

$$\mathbf{B}_f = \begin{array}{c} 1 \\ 2 \\ 3 \end{array} \begin{bmatrix} 1 & -1 & 0 & 0 & 1 & 0 & 0 \\ 0 & 1 & 1 & 0 & 0 & 1 & 0 \\ 0 & 0 & -1 & 1 & 0 & 0 & 1 \end{bmatrix}$$

cut-sets ↓, twigs (1 3 5 7), links (2 4 6)

$$\mathbf{Q}_f = \begin{array}{c} C_1 \\ C_2 \\ C_3 \\ C_4 \end{array} \begin{bmatrix} 1 & 0 & 0 & 0 & -1 & 0 & 0 \\ 0 & 1 & 0 & 0 & 1 & -1 & 0 \\ 0 & 0 & 1 & 0 & 0 & -1 & 1 \\ 0 & 0 & 0 & 1 & 0 & 0 & -1 \end{bmatrix}$$

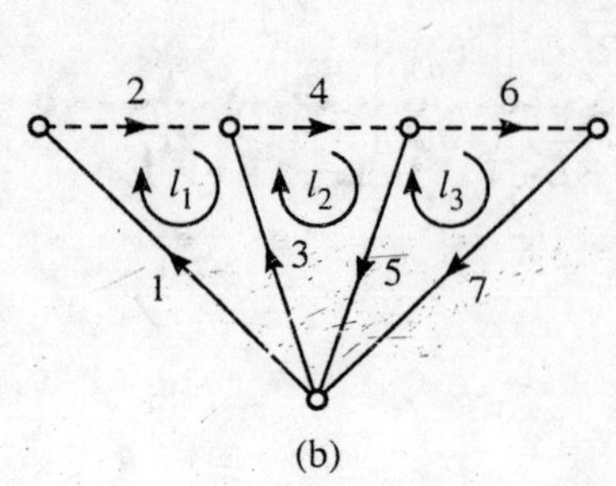

(b)

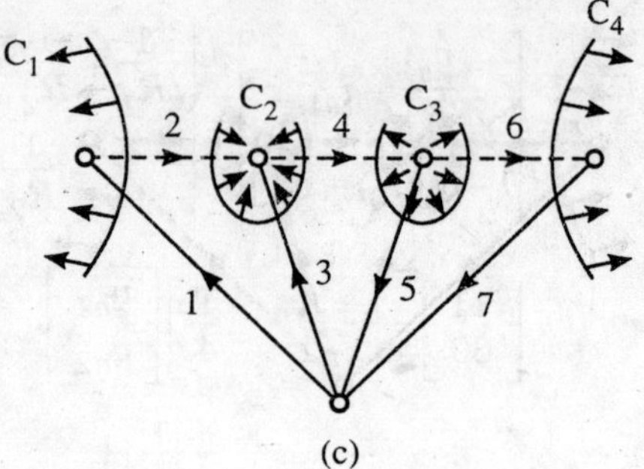

(c)

FIG. P 13.9b

10.

$$\begin{bmatrix} 1 & -1 & 0 & 0 & 1 & 0 & 0 & 0 \\ -1 & 1 & 1 & 1 & 0 & 1 & 0 & 0 \\ 0 & -1 & -1 & -1 & 0 & 0 & 1 & 0 \\ 0 & -1 & -1 & 0 & 0 & 0 & 0 & 1 \end{bmatrix} \begin{bmatrix} i_1 \\ i_2 \\ i_3 \\ i_4 \\ i_5 \\ i_6 \\ i_7 \\ i_8 \end{bmatrix} = \begin{bmatrix} 0 \\ 0 \\ 0 \\ 0 \end{bmatrix}$$

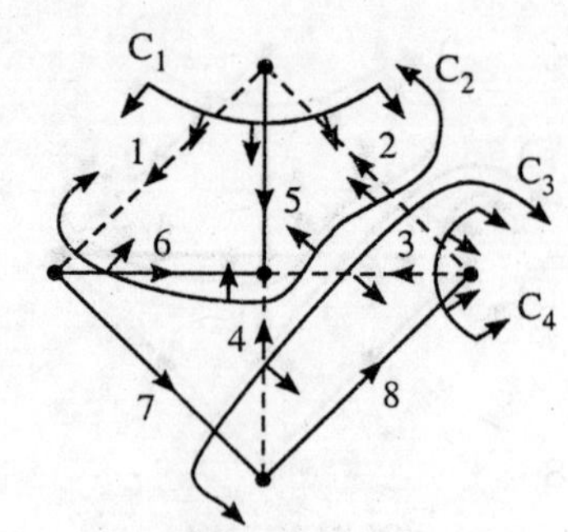

FIG. P 13.10b

11.

f-loops ↓ — twigs: 3 4 5 6 — links: 1 2 7

$$\begin{matrix} l_1 \\ l_2 \\ l_3 \end{matrix} \begin{bmatrix} 1 & -1 & 0 & 0 & 1 & 0 & 0 \\ 0 & 0 & -1 & -1 & 0 & 1 & 0 \\ 0 & -1 & -1 & 0 & 0 & 0 & 1 \end{bmatrix} \begin{bmatrix} v_3 \\ v_4 \\ v_5 \\ v_6 \\ v_1 \\ v_2 \\ v_7 \end{bmatrix} = \begin{bmatrix} 0 \\ 0 \\ 0 \end{bmatrix}$$

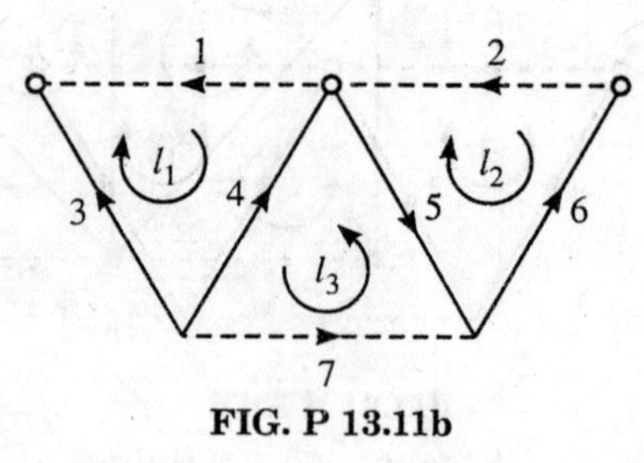

FIG. P 13.11b

12.

f-cut-sets ↓ — twigs: 3 4 5 6 — links: 1 2 7

$$\begin{matrix} C_1 \\ C_2 \\ C_3 \\ C_4 \end{matrix} \begin{bmatrix} 1 & 0 & 0 & 0 & -1 & 0 & 0 \\ 0 & 1 & 0 & 0 & 1 & 0 & 1 \\ 0 & 0 & 1 & 0 & 0 & 1 & 1 \\ 0 & 0 & 0 & 1 & 0 & 1 & 0 \end{bmatrix} \begin{bmatrix} i_3 \\ i_4 \\ i_5 \\ i_6 \\ i_1 \\ i_2 \\ i_7 \end{bmatrix} = \begin{bmatrix} 0 \\ 0 \\ 0 \\ 0 \end{bmatrix}$$

C1 C2 C3 C4 1 2 3 4 5 6 7

FIG. P 13.12b

13.

$$\begin{array}{c} \text{f-loops} \quad \text{branches}\rightarrow \\ \downarrow \quad 1\ 2\ 3\ 4\ \ 5\ \ 6\ 7\ 8 \end{array}$$

$$\begin{matrix} l_1 \\ l_2 \\ l_3 \\ l_4 \end{matrix}\begin{bmatrix} 1 & 0 & 0 & 0 & -1 & 1 & 0 & 0 \\ 0 & 1 & 0 & 0 & 1 & -1 & 1 & 1 \\ 0 & 0 & 1 & 0 & 0 & -1 & 1 & 1 \\ 0 & 0 & 0 & 1 & 0 & -1 & 1 & 0 \end{bmatrix}\begin{bmatrix} v_1 \\ v_2 \\ v_3 \\ v_4 \\ v_5 \\ v_6 \\ v_7 \\ v_8 \end{bmatrix} = \begin{bmatrix} 0 \\ 0 \\ 0 \\ 0 \end{bmatrix}$$

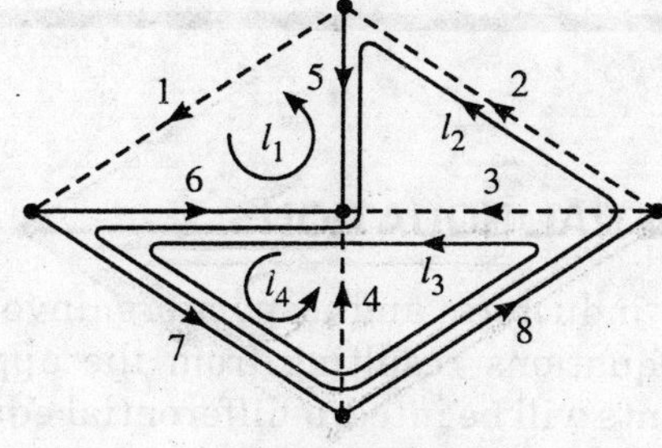

FIG. P 13.13b

14. $$\begin{bmatrix} v_1 \\ 0 \end{bmatrix} = \begin{bmatrix} R_1 + s\,(L_1 + L_2 + 2M) & -s\,(L_2 + M) \\ -s\,(L_2 + M) & R_2 + sL_2 \end{bmatrix}\begin{bmatrix} i_{L1} \\ i_{L2} \end{bmatrix}$$

14

CIRCUIT TRANSIENTS

14.1 CIRCUIT DIFFERENTIAL EQUATIONS

The *v-i* relationships across inductors and capacitors involve integral and differential relationships. Therefore the equations resulting from the application of KCL and KVL to circuits containing these elements will be integro-differential equations. Differential equations contain functions and their derivatives. Integral equations contain functions and their integrals.

Consider the differential equation

$$a^2 \frac{d^2}{dt^2} x(t) + a_1 \frac{d}{dt} x(t) + a_0 x(t) = f(t) \tag{14.1.1}$$

This equation involves a relationship between, two variables. One is the dependent variable $x(t)$ and the other is the independent variable t. The function $f(t)$ is called the **forcing function.**

Let us define the terms related to differential equations. The **order** of a differential equation represents the highest derivative present in the equation. For example, Eq. (14.1.1) is the second-order differential equation, since the second derivative, $\frac{d^2}{dt^2} x(t)$, is the highest derivative involved. Since the coefficients of the terms (a_0, a_1, a_2) on the left-hand side of Eq. (14.1.1) are constants (not functions of time), the equation is considered as time invariant or one having *constant coefficients*. An **ordinary differential equation** is one which contains ordinary derivatives instead of partial derivatives. A **linear differential equation** is one in which the dependent variable and its derivatives appear in first power only. For example, Eq. (14.1.1) is a linear differential equation. In Eq. (14.1.1), if $f(t)$ representing the input (or forcing function), is zero, the differential equation is said to be **homogeneous**. If $f(t)$ is not equal to zero, the equation is non-homogeneous.

A number of methods are available for solving differential equations. We will use the simplest of these methods in our applications.

14.2 TRANSIENT

DC circuit are in a "steady state" since the current and voltages do not change in time. We also use the term steady state to describe currents and voltages that display repetitive temporal waveform. Thus, systems in which the currents and voltages can be described by constant-amplitude, constant-frequency sinusoidal functions are also considered to be in a steady state. Neither the amplitude nor the frequency of the sinusoid changes in such a steady-state circuit. The term "transient" refers to the behaviour of the voltage or current when it is in transition between one steady state and another.

If a circuit containing one or more energy-storage elements such as inductance and capacitance is excited by a source which abruptly changes its value, the energy state of the circuit is disturbed. After a certain time the circuit settles down to a new steady-state operation and the node voltages and element currents reach their final steady-state values. The time required for the network to change from one steady state to another steady state is called **transient period.** The values of voltage and current during the transient period are known as the **transient responses** or simply **transients**. Transient means short lived.

14.3 NATURAL AND FORCED RESPONSES

On applying KCL and KVL to a network containing resistance, inductance, and capacitance, we obtain a linear differential equation with constant coefficients. The solution of the differential equation represents the response of the network. The response or behaviour of the network is governed by its internal stored energy and the external energy supplied by a source. The behaviour determined by the internal energy stored in the network is called the **natural response**. It depends upon the network elements alone and independent of the source. The behaviour determined by the application of external energy from the source is known as **forced response.** In a network energy may be stored internally in the electric field of a capacitor, or the magnetic field of an inductor. A natural response dies out gradually, that is, it approaches zero as the time becomes infinite. This is due to the fact that the stored energy is dissipated in the resistance associated with practical capacitors and inductors. For this reason, the natural response is also called the **transient response,** that is, a response with a very short life.

The external energy source or a forcing function may be a direct voltage or current source, a sinusoidal source, an exponential function, a ramp function, or any other function. A forced response may be maintained indefinitely if energy is supplied continuously to make up losses. The **complete** or **total response** of a network, is the sum of the natural response and the forced response.

The complete solution of a linear differential equation with constant coefficients describing the network, consists of two parts, namely, the particular integral and the complementary function. The particular integral represents the forced response due to the particular driving source. It satisfies the differential equation but not the initial conditions. The forced response depends on both the network elements and the source. The **complementary function** is the solution of the differential equation with forcing function set to zero. Therefore, the complementary function represents the **source-free response.** This response is also called the **natural response** or **transient response.** It contains arbitrary constants of integration. The number of constants of integration is equal to the order of the differential equation. Thus, the first-order differential equation contains one constant and the second-order differential equation contains two constants of integration. These constants in the complete solution are determined from initial conditions of the network.

14.4 INITIAL CONDITIONS

Initial conditions of a network are the values of currents and voltages just before switching. The instant just before switching is denoted by $t = 0_-$. The instant exactly at switching is denoted by $t = 0$ and the instant just after switching is denoted by $t = 0_+$. It should be noted that $t = 0_-$ is as close as we please but just before $t = 0$. The instant $t = 0$ is *exactly* at $t = 0$ and $t = 0_+$ is as close as we please but just *after* $t = 0$. Thus, 0_- and 0_+ represent infinitesimally small times, and as a limit their difference approaches zero. That is, as a limit the time that elapses between 0_- and 0_+ is zero.

The response of a memory circuit. [a circuit containing memory elements(s)] also depends upon the initial conditions. That is, the response is related to what has happened in the past. In other words, the circuit can have nonzero response even if there is no input signal.

14.5 ZERO-INPUT RESPONSE

The zero-input response of a network is the response that results from initial conditions when the excitation (input) is zero.

14.6 ZERO-STATE RESPONSE

The **state** of a network is the minimal amount of information needed at $t = t_0$, so that, given an input for $t > t_0$, the output can be determined for all $t > t_0$.

The **zero-state response** for a network is the response (output) for an excitation (input) which is applied when all the initial conditions are zero. Such a network is also said to be **at rest** or **initially relaxed.**

14.7 INITIAL (OR BOUNDARY) CONDITIONS FOR INDUCTANCE

For an inductor the $v - i$ relationship is

$$v_L = L\frac{di_L}{dt}$$

$$i_L(t) = \frac{1}{L}\int_{-\infty}^{t} v_L\,dt \tag{14.7.1}$$

Let us define the following notation:

$i_L(0_-)$ = the value of i_L at $t = 0_-$

$i_L(0_+)$ = the value of i_L at $t = 0_+$

Therefore, at $t = 0_+$

$$i_L(0_+) = \frac{1}{L}\int_{-\infty}^{0_+} v_L\,dt$$

or $$i_L(0_+) = \frac{1}{L}\int_{-\infty}^{0_-} v_L\, dt + \frac{1}{L}\int_{0_-}^{0_+} v_L\, dt$$

$$= i_L(0_-) + \frac{1}{L}\int_{0_-}^{0_+} v_L\, dt \tag{14.7.2}$$

$i_L(0_-)$ is the initial state of the inductor. It gives the *past history* (before $t = 0$) of the inductor.

The interval from 0_- to 0_+ is almost zero and if v_L is finite and does not contain impulses

$$\frac{1}{L}\int_{0_-}^{0_+} v_L\, dt = 0 \tag{14.7.3}$$

Therefore, from Eqs. (14.7.2) and (14.7.3)

$$i_L(0_+) = i_L(0_-) \tag{14.7.4}$$

That is, *the current through the inductor just after the switch is closed must be the same as that just before closing the switch.*

Let the current i change to a finite value in the time from 0_- to 0_+. Then

di = the change in current in the time from 0_- to 0_+

dt = the change in the time from 0_- to 0_+

The voltage across the inductor

$$v_L = L\frac{di}{dt}$$

$$= L \times \frac{\text{change in current}}{\text{change in time}}$$

$$= L \times \frac{\text{finite value of current}}{0} = \infty$$

This infinite voltage should be supplied by the voltage source. According to our assumption a practical voltage source will not supply infinite voltage. Hence, we conclude that the current in an inductance cannot change instantaneously. Instantaneously means in zero time, from $t = 0_-$ to $t = 0_+$. This fact is called the **continuity relationship for inductance.** It is written mathematically as

$$\boxed{i_L(0_+) = i_L(0_-)} \tag{14.7.5}$$

Inductor with no initial current

If an inductor has no initial current, then $i_L(0_-) = 0$. From Eq. (14.7.4) $i_L(0_+) = 0$. Hence an inductor with no initial current behaves like an *open circuit at* $t = 0_+$.

An inductor with initial current $i_L(0_-) = I_{Lo}$ is considered as a current source I_{Lo} at $t = 0_+$ as shown in Fig. 14.1

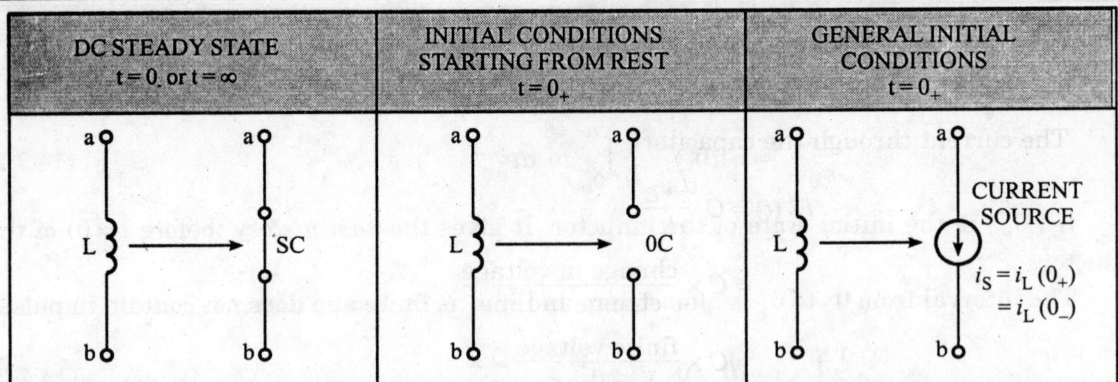

FIG. 14.1 Conditions at switching an inductor

14.8 INITIAL (OR BOUNDARY) CONDITIONS FOR CAPACITANCE

A capacitor is a memory element. The v-i relationship or element constraint for a capacitor is given by

$$i_C(t) = C\frac{d\,v_C(t)}{dt} \tag{14.8.1}$$

In order to determine the capacitor voltage in terms of capacitor current, we can write

$$v_C(t) = \frac{1}{C}\int_{-\infty}^{t} i_C\,dt \tag{14.8.2}$$

This expression states that the voltage at any time t is proportional to the integral of the current through the capacitor from time immemorial to time t. At $t = 0_+$,

$$v_C(0_+) = \frac{1}{C}\int_{-\infty}^{0_-} i_C\,dt + \frac{1}{C}\int_{0_-}^{0_+} i_C\,dt$$

$$= v_C(0_-) + \frac{1}{C}\int_{0_-}^{0_+} i_C\,dt \tag{14.8.3}$$

The interval from 0_- to 0_+ is almost zero. If i_C is finite and does not contain impulses or derivatives of impulses,

$$\frac{1}{C}\int_{0_-}^{0_+} i_C\,dt = 0 \tag{14.8.4}$$

From Eq. (14.8.3) and (14.8.4)

$$v_C(0_+) = v_C(0_-) \tag{14.8.5}$$

Since $\quad v_C = \dfrac{q}{C}$

$$q(0_-) = q(0_+) \tag{14.8.6}$$

These results can also be deduced as follows :

Let the voltage v_C change to a finite value in the time from 0_- to 0_+, then

$d\,v_C$ = the change in voltage in the time from 0_- to 0_+

dt = the change in the time from 0_- to 0_+

The current through the capacitor.

$$i_C(t) = C\frac{d\,v_C}{dt}$$

$$= C \times \frac{\text{change in voltage}}{\text{change in time}}$$

$$= C \times \frac{\text{finite voltage}}{0} = \infty$$

This infinite current should be supplied by the current source. Since it is assumed that our current sources will not supply infinite currents, we conclude that *the voltage across a capacitor cannot change instantaneously* since it would require an infinite current. This fact is called the **continuity relationship for capacitance.** It is written mathematically as

$$\boxed{v_C(0_+) = v_C(0_-)} \qquad (14.8.7)$$

Eq. (14.8.6) shows that the charge q on a capacitor is the same immediately before and after switching.

It should be noted that although the voltage and charge on a capacitor cannot change instantaneously during switching operation, the capacitor current has no such restriction. In general,

$$i_C(0_+) \neq i_C(0_-)$$

Capacitor with no initial charge or voltage

Here $q(0_-) = 0$ and $v_C(0_-) = 0$. Therefore from Eqs. (14.8.5) and (14.8.6)

$$v_C(0_+) = 0 \text{ and } q(0_+) = 0$$

Thus a capacitor with no initial voltage or charge behaves like a short circuit at $t = 0_+$.

A capacitor with an initial charge $q_C(0_-) = Q_0$ is equivalent to a voltage source of value $V_{co} = Q_0/C$ at $t = 0_+$ as shown in Fig. 14.2.

DC STEADY STATE $t = 0_-$ or $t = \infty$	INITIAL CONDITIONS STARTING FROM REST $t = 0_+$	GENERAL INITIAL CONDITIONS $t = 0_+$
a, C, b → a, OC, b	a, C, b → a, SC, b	a, C, b → a, VOLTAGE SOURCE, $v_S = v_C(0_+) = v_C(0_-)$, b

FIG. 14.2 Conditions at switching a capacitor

14.9 EFFECT OF SWITCHING ON A RESISTOR

The $v-i$ relationship of a linear resistor is given by $V_R = R\,i$. Theoretically, the current through or voltage across a resistor can change instantaneously, that is, in zero time. Hence there will be no transients in purely resistive networks. The network changes from one steady state to another without any transient. A resistor does not store energy, and, therefore, for the resistor in general

$$i_R(0_+) \neq i_R(0_-)$$

and $$v_R(0_+) \neq v_R(0_-)$$

14.10 DETERMINATION OF INITIAL CONDITIONS

All initial conditions are determined simply by the application of Ohm's law, KCL and KVL at the instant $t = 0_+$ and the following facts :

(1) In dc steady state, a pure inductor can be represented by a short circuit.

(2) In dc steady state, a pure capacitor is represented by an open circuit.

(3) The voltage across a capacitor cannot change instantaneously. That is,

$$v_C(0_+) = v_C(0_-) \tag{14.10.1}$$

Since charge, q, on a capacitor is related linearly to the voltage on the capacitor $q = C\,v$, the boundary condition of Eq. (14.10.1) may also be written

$$q(0_+) = q(0_-) \tag{14.10.2}$$

(4) The current through an inductor cannot change instantaneously. That is,

$$i_L(0_+) = i_L(0_-) \tag{14.10.3}$$

14.11 NATURAL RESPONSE OF AN RL NETWORK

Consider a circuit containing a single inductor L as shown in Fig. 14.3a. The resistors and sources can be replaced by a Thevenin equivalent circuit (Fig. 14.3b). Here v_T is the Thevenin voltage and R_T is the Thevenin resistance as seen from the terminals ab between which the energy storage element L is connected. By KVL

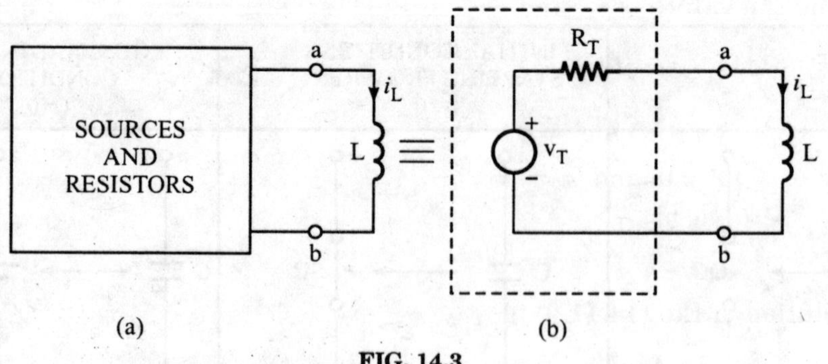

FIG. 14.3

$$v_{RT} + v_L = v_T \text{ for } t > 0$$

$$R_T i_L + L\frac{di_L}{dt} = v_T \tag{14.11.1}$$

If the source is switched out of the network, $v_T = 0$ and Eq. (14.11.1) becomes

$$L\frac{di_L}{dt} + R_T i_L = 0$$

$$\frac{di_L}{dt} + \frac{R_T}{L} i_L = 0$$

or

$$\frac{di_L}{dt} + k\, i_L = 0 \text{ for } t > 0 \tag{14.11.2}$$

where $k = R_T/L$

Eq. (14.11.2) is a first order, linear, homogeneous, ordinary differential equation. Since the behaviour of the network is described by a first order differential equation, this network is called the **first-order network.** Since the source has been switched out of the network, there is no source or forcing function in this differential equation. The solution of such a differential equation is called the *source-free response* or *natural response* of the network. *The solution is called the natural response because this is what the system does naturally—without being forced to.*

Eq. (14.11.2) suggests that we should find a function $i\,(t)$ such that its derivative when added to itself (multiplied by a constant k) totals zero for all t. This means that the function $i\,(t)$ and its derivative $\frac{di}{dt}$ must have the same wave shape. This property is possessed by an exponential function. The solution of Eq. (14.11.1) may, therefore, be assumed as

$$i_L\,(t) = A\,e^{st} \tag{14.11.3}$$

where A and s are to be determined.

Differentiation of Eq. (14.11.3) gives

$$\frac{di_L}{dt} = A\,s\,e^{st} \tag{14.11.4}$$

Substituting the values of i_L and $\frac{di_L}{dt}$ in Eq. (14.11.2), we get

$$A\,s\,e^{st} + k\,A\,e^{st} = 0$$

$$(s + k)\,A\,e^{st} = 0$$

Since $A \neq 0$ and $s \neq -\infty$

$$s + k = 0 \tag{14.11.5}$$

Thus

$$s = -k \tag{14.11.6}$$

and the solution of Eq. (14.11.2) is

$$i_L = A\,e^{-kt} \tag{14.11.7}$$

or

$$i_L = A\,e^{-\frac{R_T}{L}t} \tag{14.11.8}$$

We still have not evaluated the unknown constant A. This will be determined from initial conditions.

Eq. (14.11.5) is called the **characteristic equation** or **eigen equation.** The solution of this characteristic equation is given by Eq. (14.11.6). It is called the **characteristic value** or **eigen value**. It is also called the **characteristic root** of the circuit, since its value determines the nature or character of the source-free response.

Time Constant

If we put $t = 1/k$ in Eq. (14.11.7), we get

$$i = A\,e^{-k\,(1/k)} = A\,e^{-1} = 0{\cdot}3679A \tag{14.11.9}$$

It is seen that when $t = 1/k$, the value of i has decayed to e^{-1} times its initial value at $t = 0_+$. The quantity i/k is called the *time constant* of the system. The time constant is usually denoted by the symbol τ, Therefore, $\tau = 1/k$. Similarly, at $t = 2\tau$,

$$i = A\,e^{-t/\tau} = A\,e^{-2\tau/\tau} = A\,e^{-2} \tag{14.11.10}$$

That is, at two time constants after $t = 0$, the response is e^{-2} times what it was at $t = 0_+$. It is to be noted that the time constant is the negative reciprocal of the characteristic value.

The shape of the curve given by Eq. (14.11.7) can easily be plotted with the help of Table 14.1.

Table 14.1

t	0	τ	2τ	3τ	4τ	5τ
i	A	$0.3679\,A$	$0.135\,A$	$0.0498\,A$	$0.0183\,A$	$0.0067\,A$

The shape of i versus t curve is a decaying exponential. It is seen from Table 14.1 that each value is e^{-1} times (that is, 0.3679 times) the previous value. At $t = 5\,\tau$, the current is $A\,e^{-5}$ which is approximately $0.00674\,A$. Thus, the current becomes practically zero after five time constants. The time constant is of special significance since it determines the decay rate.

The time constant is also related to the shape of the decay curve. Differentiation of Eq. (14.11.7) gives

$$\frac{di}{dt} = \frac{d}{dt}(A\,e^{-t/\tau}) = -\frac{1}{\tau}A\,e^{-t/\tau} \tag{14.11.11}$$

The initial slope of the curve is found by putting $t = 0_+$ in Eq. (14.11.11).

$$\left.\frac{di}{dt}\right|_{t=0_+} = \left.-\frac{1}{\tau}A\,e^{-t/T}\right|_{t=0_+} = -\frac{1}{\tau}\;A$$

The expression $\left.\dfrac{di}{dt}\right|_{t=0.}$ is read "the value of the derivative of i evaluated at time zero plus." Thus, if the current continued to fall at its initial rate it would reach zero in a time equal to one time constant τ. In Fig. 14.4, a line is drawn tangential to the initial portion of the curve. It meets the time axis at $t = \tau$. The time constant of the RL network is L/R_T. Here R_T is the Thevenin equivalent resistance of the rest of the network as seen from the terminals between which the energy-storage element L is connected. The time constant of the RL network is denoted by τ_L. Therefore, $\tau_L = L/R_T$.

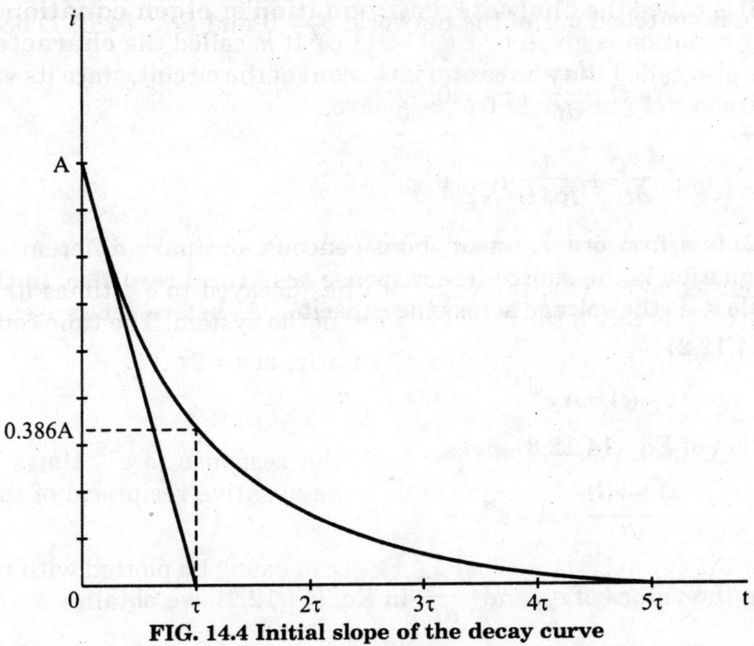

FIG. 14.4 Initial slope of the decay curve

14.12 NATURAL RESPONSE OF RC NETWORK

Consider a circuit containing a single capacitor C as shown in Fig. 14.5a. The resistors and sources can be replaced by a Thevenin equivalent circuit as shown in Fig. 14.5b.

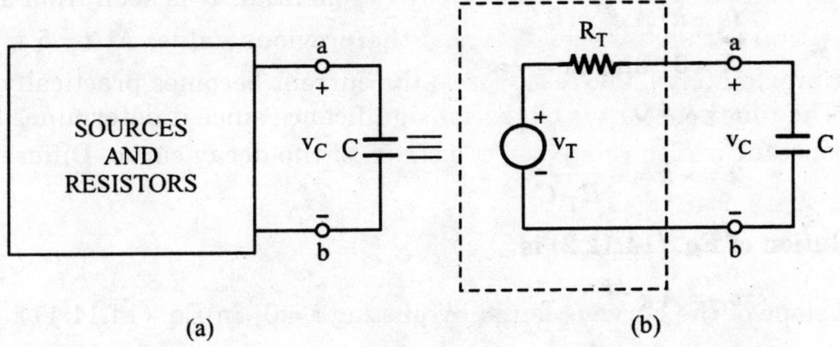

FIG. 14.5

Here v_T is the Thevenin voltage and R_T is the Thevenin resistance as seen from the terminals $a - b$ between which the energy-storage element C is connected.

By KVL $$v_{R_T} + v_C = v_T \quad \text{for } t > 0$$

$$R_T\, i + v_c = v_T$$

$$R_T\left(C\,\frac{d\,v_c}{dt}\right) + v_C = v_T$$

$$R_T C \frac{d v_C}{dt} + v_C = v_T \text{ for } t > 0 \quad (14.12.1)$$

If the source is switched out of the network, $v_T = 0$ and Eq. (14.12.1) becomes

$$R_T C \frac{d v_C}{dt} + v_C = 0$$

or

$$\frac{d v_C}{dt} + \frac{1}{R_T C} v_C = 0 \quad (14.12.2)$$

Eq. (14.12.2) is a first order, linear, homogeneous, ordinary differential equation. The solution of this equation is the source-free response or natural response. In this equation the dependent variable v_C is the voltage across the capacitor. As before, let us assume the following solution of Eq. (14.12.2) :

$$v_C(t) = A\, e^{st} \quad (14.12.3)$$

Differentiation of Eq. (14.12.3) gives

$$\frac{d v_C(t)}{dt} = A\, s\, e^{st} \quad (14.12.4)$$

Substituting the values of v_C and $\frac{d v_C}{dt}$ in Eq. (14.12.2), we obtain

$$A\, s\, e^{st} + \frac{1}{R_T C} A\, e^{st} = 0$$

If we put $\frac{1}{R_T C} = k$, the above equation becomes

$$(s + k) A\, e^{st} = 0$$

Since $\quad A \neq 0 \text{ and } s \neq -\infty$

$$s + k = 0 \quad (14.12.5)$$

Thus

$$s = -k = -\frac{1}{R_T C} \quad (14.12.6)$$

and the solution of Eq. (14.12.2) is

$$v_C = A\, e^{-kt} \quad (14.12.7)$$

or

$$v_C = A\, e^{-\frac{1}{R_T C} t} \quad (14.12.8)$$

Eq. (14.12.5) is called the **characteristic equation** or **eigen equation.** The solution of this characteristic equation is given by Eq. (14.12.6). It is called the **characteristic value** or **eigen value.** It is also called the **characteristic root** of the circuit, since its value determines the nature or character of the source-free response.

The quantity $1/k$ is called the time constant. It is denoted by τ_C. Therefore, $\tau_C = 1/k = R_T C$. It is to be noted that the time constant is the negative reciprocal of the characteristic value.

The zero-input response of the RC network is given by Eq. (14.12.8). It is the familiar exponential waveform studied in Chapter 10 and shown in Fig. 14.6. The time constant of the waveform is $\tau_C = R_T C$ and the initial amplitude is A. From our previous study we know that the waveform will decay to about 37 percent of its initial amplitude in one time constant. It will become practically zero in about five time constants ($5\ \tau_C$). It is to be noted that in an RC network, $\tau_C = R_T C$. Here R_T is the Thevenin equivalent resistance of the rest of the network as seen from the terminals between which the energy-storage element C is connected.

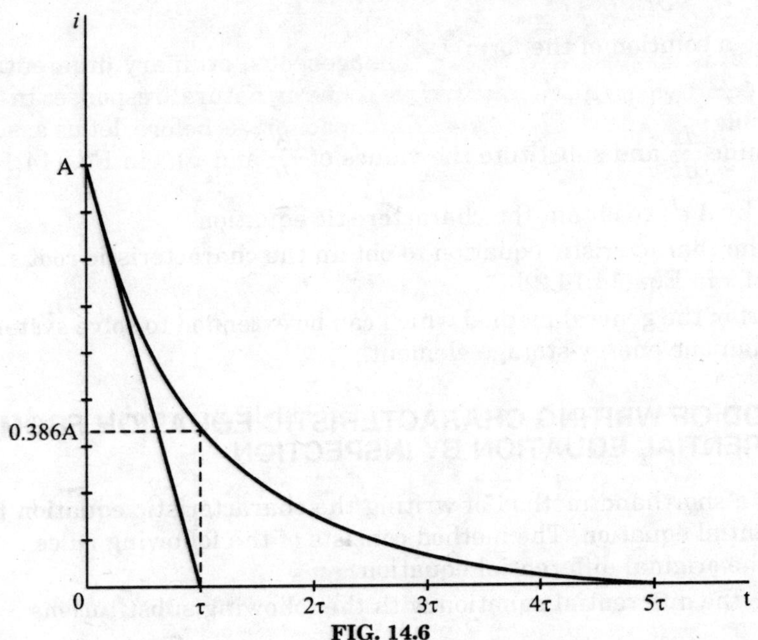

FIG. 14.6

14.13 GENERAL DIFFERENTIAL EQUATION OF A SINGLE-MEMORY CIRCUIT

We have studied the simplest memory circuits containing a single inductor or a single capacitor. The resistors and sources were replaced by a Thevenin equivalent circuit. In general, a single-memory circuit is one which contains the equivalent of a single inductor or a single capacitor. The general form of the differential equation describing a single-memory circuit can be written as

$$\tau \frac{dx}{dt} + x = f(t)$$

where $x(t)$ is a current or a voltage, τ is the circuit time constant, and $f(t)$ is the equivalent source term. Sometimes it may not be convenient to formulate this differential equation in terms of the state variables (inductor, current or capacitor voltage). That is, it may be inconvenient to determine the Thevenin equivalent seen by the energy-storage element. In such cases we apply KCL, KVL and v–i relationships of the elements of the network to derive the differential equation in terms of any single variable that is convenient.

14.14 GENERAL METHOD OF DETERMINING NATURAL RESPONSE OF FIRST ORDER SYSTEMS

A source-free system containing only one energy-storage element can be solved for its natural response as follows :

1. Write the differential equation describing the system during the time interval of interest (usually for $t > 0$). This will be a homogeneous equation of the form

$$\frac{dx}{dt} + \alpha x = 0 \tag{14.14.1}$$

2. Assume a solution of the form

$$x(t) = A e^{st} \tag{14.14.2}$$

3. Determine $\frac{dx}{dt}$ and substitute the values of $\frac{dx}{dt}$ and $x(t)$ in Eq. (14.14.1).

   Divide by $A e^{st}$ to obtain the characteristic equation.
4. Solve the characteristic equation to obtain the characteristic root s. Substitute this value of s in Eq. (14.14.2).

This is a part of the general method which can be extended to solve systems with sources and with more than one energy-storage element.

14.15 METHOD OF WRITING CHARACTERISTIC EQUATION FROM DIFFERENTIAL EQUATION BY INSPECTION

We shall develop a shorthand method of writing the characteristic equation by inspection of the circuit differential equation. The method consists of the following rules :

1. Write the original differential equation.
2. Rewrite the differential equation with the following substitutions :

   (*a*) Replace $\frac{d}{dt}$ by s

   Replace $\frac{d^2}{dt^2}$ by s^2

   Replace $\frac{d^n}{dt^n}$ by s^n

   Replace $\int \{\ \}\, dt$ by $\frac{1}{s}$

   Replace $\int\int \{\ \cdot\ \}\, dt^2$ by $\frac{1}{s^2}$

   Replace $\int\int \cdots \int \{\ \cdot\ \}\, dt^n$ by $\frac{1}{s^n}$

   (*b*) Write the constant coefficients just as they appear in the original differential equation.

   (*c*) Write 1 for the dependent variable.

   (*d*) Write 0 for the forcing function.

14.16 GENERAL METHOD OF DETERMINING FORCED RESPONSE OF FIRST-ORDER SYSTEMS

The following steps are used to determine the forced response of first order systems :

1. Write the differential equation describing the system for $t > 0$.
2. Solve the homogeneous (source-free) equation for the natural response by :
   (*a*) Assuming $x_n(t) = A\,e^{st}$
   (*b*) Determining $\frac{dx_n}{dt}$ and substituting $\frac{dx_n}{dt}$ and $x_n(t)$ into the homogeneous differential equation and solving for the value of s.
   It should be noted that the unknown constant A remains to be evaluated.
3. Solve the forced differential equation for the forced response $x_f(t)$ by assuming
   $x_f(t)$ = a linear sum of the forcing function and all its nonzero derivatives.
4. Determine the complete response by using the relationship
   $$x(t) = x_n(t) + x_f(t)$$
5. Apply initial conditions to determine the constant A.

It should be clearly noted that the initial conditions should *not* be applied immediately after step 2 to determine A.

14.17 LAPLACE TRANSFORM METHOD

The classical method of solving circuit differential equations is rather involved. In the classical method the complementary function and particular integral have to be determined and finally the arbitrary constants have to be obtained from known initial conditions. The Laplace transform method of solving integro-differential equations is better than the classical method due to the following advantages :

(*a*) It transforms exponential and trigonometric functions into simple algebraic functions.

(*b*) It transforms integro-differential equations into algebraic equations which are easier to solve.

(*c*) The arbitrary constant need not be determined separately. In solving a differential equation, by the Laplace transform method, the solution obtained is the complete solution.

In this chapter, various methods have been used to solve circuit differential equations keeping in mind that mathematics should be used as a tool and not as a technique.

■ **EXAMPLE 14.1** Write the characteristic equation by the shorthand method for the differential equation

$$20\frac{dv_C}{dt} + 3\,v_C = 6$$

SOLUTION :

$$20 \frac{d}{dt} v_C + 3 v_C = 6$$

$$\downarrow \quad \downarrow \quad \downarrow \quad \downarrow \quad \downarrow \quad \downarrow \quad \downarrow$$

$$20 \cdot s \cdot 1 + 3 \cdot 1 = 0$$

or $$20s + 3 = 0$$

which is the required characteristic equation.

■ **EXAMPLE 14.2** Write the characteristic equations for the following differential equations :

(*a*) $$5 \frac{d^2 x}{dt^2} + 6 \frac{dx}{dt} + 4x = 30$$

(*b*) $$10 \frac{d^2 v}{dt^2} + 7 \int v \, dt = 15 t^2 + e^{-12t}$$

SOLUTION :

(*a*) $$5 \frac{d^2}{dt^2} x + 6 \frac{d}{dt} x + 4 x = 30$$

$$\downarrow \quad \downarrow \quad \downarrow \quad \downarrow \downarrow \quad \downarrow \quad \downarrow \downarrow \downarrow \quad \downarrow \quad \downarrow \quad \downarrow$$

$$5 \cdot s^2 \cdot 1 + 6 \cdot s \cdot 1 + 4 \cdot 1 = 0$$

or $$5 s^2 + 6s + 4 = 0$$

(*b*) $$10 \frac{d^2}{dt^2} v + 7 \int v \, dt = 15 t^2 + e^{-12t}$$

$$\downarrow \quad \downarrow \quad \downarrow \downarrow \downarrow \quad \downarrow \quad \downarrow \quad \downarrow$$

$$10 \cdot s^2 \cdot 1 + 7 \cdot \frac{1}{s} = 0$$

$$10 s^2 + \frac{7}{s} = 0$$

or $$10 s^3 + 7 = 0$$

14.18 COMPLETE RESPONSE OF FIRST-ORDER SYSTEMS

We shall develop a general method of determining the complete response of first-order systems. This method is based on the principle of superposition and will be illustrated by examples.

■ **EXAMPLE 14.3** Find the complete response $i_L(t)$ for $t > 0$ for the network shown in Fig. 14.7 if $i_L(0_-) = 0$ and $v_S(t) = 20$ V.

SOLUTION : By KVL

$$v_L + v_R = v_S \quad \text{for} \quad t > 0$$

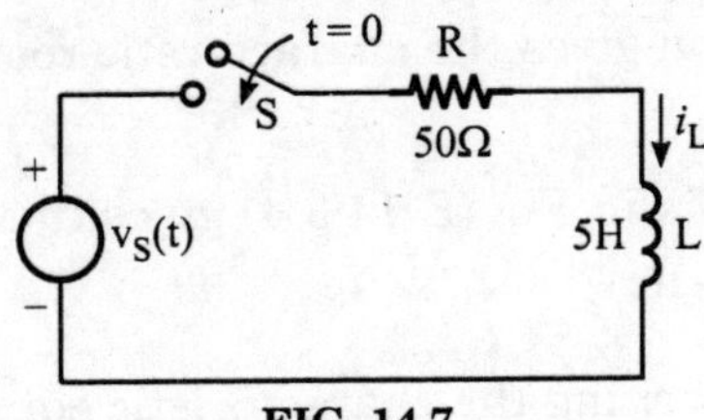

FIG. 14.7

$$L\frac{di_L}{dt} + R\,i_L = v_S$$

$$5\frac{di_L}{dt} + 50\,i_L = 20$$

or $$\frac{di_L}{dt} + 10\,i_L = 4 \text{ for } t > 0 \tag{E 14.3.1}$$

This is not a homogeneous differential equation because it has a nonzero forcing function (4) on the right-hand side.

In order to solve this differential equation let us add zero to its right-hand side. That is,

$$\frac{di_L}{dt} + 10\,i_L = 4 + 0 \tag{E.14.3.2}$$

Actually, this equation would be obtained if a second but zero-valued voltage source $v_0(t) = 0$ V were added to the circuit in series with $v_S(t)$. Addition of such a source has no electrical effect on the circuit.

Now we shall solve for $i_L(t)$ due to source $v_0(t)$ alone and then for $i_L(t)$ due to source $v_S(t)$ alone. By superposition principle, the sum of these component responses gives the complete response $i_L(t)$ for $t > 0$. Therefore, to solve Eq. (E 14.3.2) we proceed as follows : Set the first term on the right-hand side of Eq. (E 14.3.2) to zero and determine the response due to this zero-valued forcing function. This is actually the natural response. Let this natural response be denoted by i_{Ln}. Therefore,

$$\frac{di_{Ln}}{dt} + 10\,i_{Ln} = 0 \tag{E 14.3.3}$$

Assume a solution of the form

$$i_{Ln} = A\,e^{st} \tag{E 14.3.4}$$

$$\therefore \quad \frac{di_{Ln}}{dt} = \frac{d}{dt}A\;e^{st} = A\;se^{st} \tag{E 14.3.5}$$

Substitution of Eqs. (E 14.3.4) and (E 14.3.5) in Eq. (E 14.3.3) gives

$$A\,s\,e^{st} + 10\,A\,e^{st} = 0$$

$$A\,e^{s\,t}\,(s + 10\,) = 0$$

Since A and e^{st} are not zero, we get the characterstic equation

$$s + 10 = 0 \tag{E 14.3.6}$$

The solution of this equation gives the characterstic root

$$s = -10 \tag{E 14.3.7}$$

Substitution of the value of s in Eq. (E 14.3.4) gives the natural response

$$i_{Ln} = A\ e^{-10t} \tag{E 14.3.8}$$

The short-hand method of writing the characteristic equation may be applied as follows:

$$\begin{array}{ccccccc} \frac{d}{dt} & i_{Ln} & + & 10 & i_{Ln} & = & 0 \\ \downarrow & \downarrow & \downarrow & \downarrow & \downarrow & \downarrow & \downarrow \\ s & \cdot\ 1 & + & 10 & \cdot\ 1 & = & 0 \end{array}$$

or $$s + 10 = 0$$

Determination of forced response

The forced response $i_{Lf}(t)$ depends on the forcing function (4) on the right-hand side of Eq. (14.3.1).

Thus, the function $i_{Lf}(t)$ will satisfy

$$\frac{di_{Lf}}{dt} + 10\, i_{Lf} = 4 \tag{E 14.3.9}$$

The function $i_{Lf}(t)$ is also called the *particular response* because it depends on the particular source used. We have to solve Eq. (E 14.3.9) for i_{Lf}. Eq. (E 14.3.9) suggests that we should find a function i_{Lf} (t) such that its derivative when added to itself (multiplied by 10) equals a constant (4) for all $t > 0$. For a forced solution, that is, a function that makes the left-hand side equal to the right-hand side, let us try a constant value:

$$i_{Lf}(t) = k \tag{E 14.3.10}$$

Then $$\frac{d}{dt} i_{Lf} = 0$$

Substitution of these values in Eq. (E 14.3.9) gives

$$0 + 10\, k = 4$$

therefore, $$k = \frac{4}{10} = 0{\cdot}4$$

Thus the forced solution is

$$i_{Lf}(t) = 0{\cdot}4\ A \tag{E 14.3.11}$$

This example is an application of the following rule:

When the forcing function is a constant, the forced component of the total solution is also a constant whose value is found directly from the system differential equation by dividing the value of the forcing function by the coefficient of the dependent variable.

That is,

$$\text{Forced component} = \frac{\text{constant forcing function}}{(\text{ coefficient of the dependent variable })}$$

In our case,

$$i_{Lf} = \frac{20}{50} = 0{\cdot}4$$

Complete or total response

By superposition principle the complete response is equal to the sum of the natural and the forced responses.

Thus,

$$\text{complete response} = \text{natural response} + \text{forced response} \tag{E 14.3.12}$$

In our case

$$i_L(t) = i_{Ln}(t) + i_{Lf}(t)$$

Substitution of the values from Eqs. (E 14.3.8) and (E 14.3.11) gives

$$i_L(t) = A\,e^{-10\,t} + 0{\cdot}4 \tag{E 14.3.13}$$

The complete response $i_L(t)$ is determined from Eq. (E 14.3.13) if the constant A is known. This constant is determined by applying initial conditions. We are given that $i_L(0_-) = 0$.

For an inductor

$$i_L(0_+) = i_L(0_-)$$

therefore, $\quad i_L(0_+) = 0$

Applying initial conditions in Eq. (E 14.3.13) we get

$$i_L(0_+) = A_e^{-10 \times 0} + 0{\cdot}4$$

$$0 = A + 0{\cdot}4 \qquad (\because e^0 = 1)$$

$$A = -0{\cdot}4$$

By substitution this value of A in Eq. (E 14.3.13) gives the complete response.

$$i_L(t) = -\,0{\cdot}4e^{-10\,t} + 0{\cdot}4 \ \text{ for } t > 0$$

or $\quad i_L(t) = 0{\cdot}4\,(1 - e^{-10\,t})$

Alternative method

Laplace transform of Eq. (E 14.3.1) gives

$$[s\,I_L(s) - i_L(0_-)] + 10\,I_L(s) = \frac{4}{s}$$

Since $\quad i_L(0_-) = 0$

$$(s + 10)\,I_L(s) = \frac{4}{s}$$

$$I_L(s) = \frac{4}{s\,(s+10)} = \frac{4}{10}\left(\frac{1}{s} - \frac{1}{s+10}\right)$$

Inverse laplace transform of this relation gives

$$i_L(t) = 0{\cdot}4\,(1 - e^{-10t}),$$

as obtained before.

■ **EXAMPLE 14.4** In the circuit of Fig. 14.8 a, the switch S has been in position 1 for a long time. At $t = 0$, the switch is moved to position 2. Find $i_L(t)$ for $t > 0$.

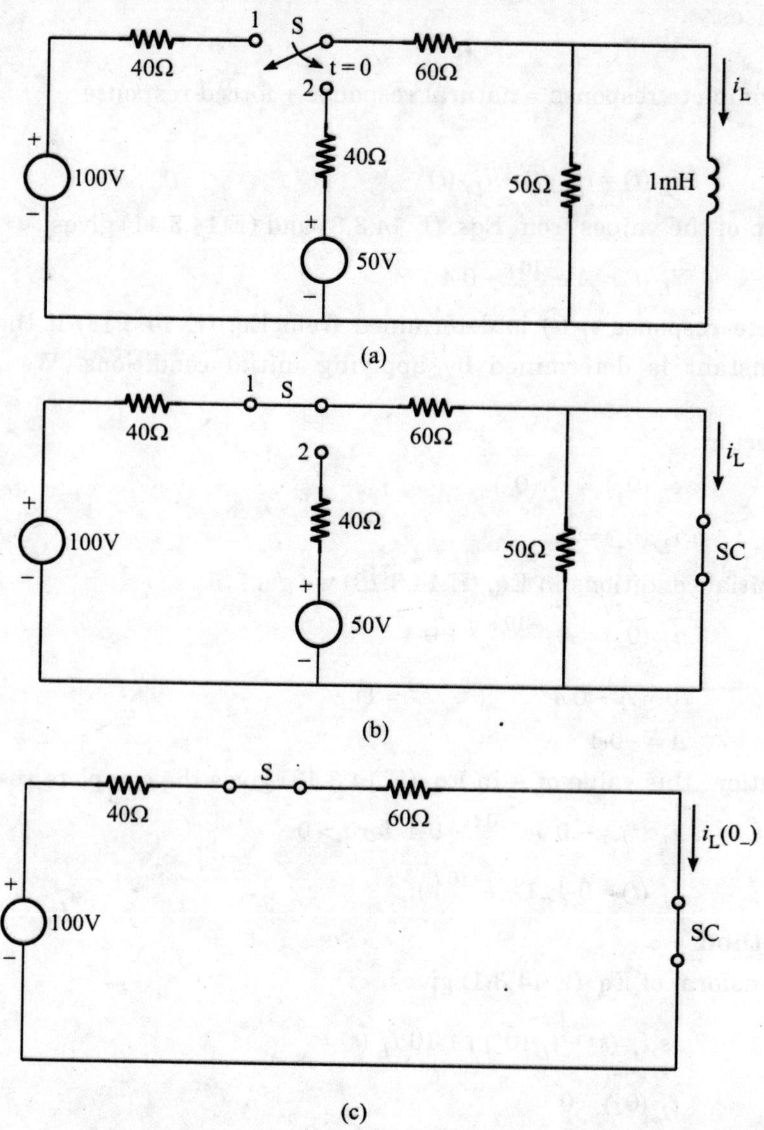

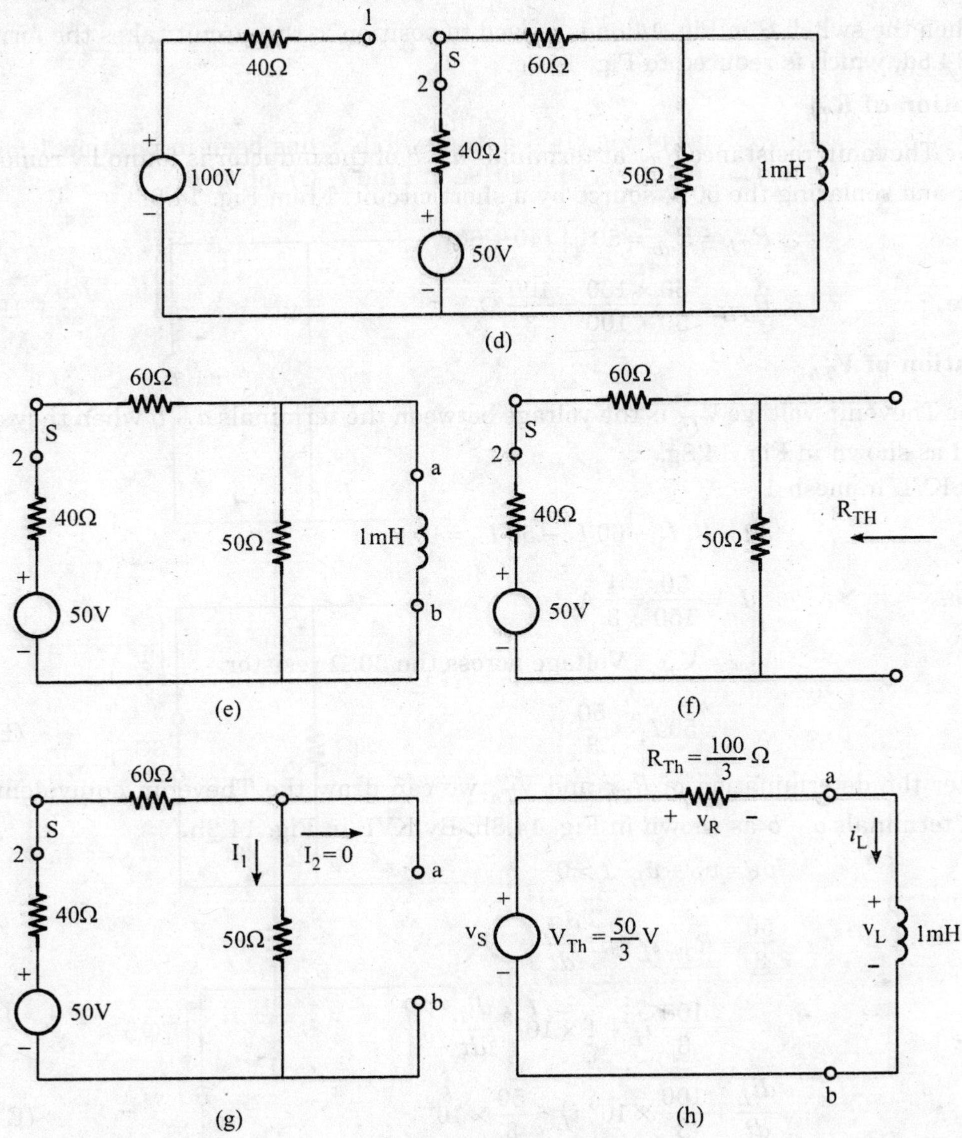

FIG. 14.8

SOLUTION : **Determination of initial conditions $i_L(0\_)$**

Since the switch S has been in position 1 for a long time, the circuit is in the dc steady state. Under this condition the inductor can be replaced by a short circuit as shown in Fig. 14.8 b. Since a short circuit is placed across the 50 Ω resistor, the total resistance of the short circuit ($R_{sc} = 0$) and the 50 Ω resistor is $\frac{50 \times 0}{50 + 0}$, that is, zero. The circuit in Fig. 14.8b is therefore reduced to that shown in Fig. 14.8c. This circuit is used to determine the initial condition, that is, $i_L(0\_)$. Fig. 14.8c.

$$i_L(0\_) = \frac{100}{40 + 60} = 1\,A \tag{E 14.4.1}$$

When the switch S in Fig. 14.8a is moved to position 2, the circuit takes the form shown in Fig. 14.8d, which is reduced to Fig. 14.8e.

Calculation of R_{Th}

The Thevenin resistance R_{Th} at terminals $a - b$ of the inductor is found by removing the inductor and replacing the 50 V source by a short circuit. From Fig. 14.8f

$$R_{Th} = R_{ab} = 50 \; || \; (40 + 60)$$

therefore,
$$R_{TH} = \frac{50 \times 100}{50 + 100} = \frac{100}{3} \, \Omega \qquad \text{(E 14.4.2)}$$

Calculation of V_{Th}

The Thevenin voltage V_{Th} is the voltage between the terminals $a - b$ when they are open circuited as shown in Fig. 14.8g.

By KVL in mesh 1,

$$50 - 40\, I_1 - 60\, I_1 - 50\, I_1 = 0$$

therefore,
$$I_1 = \frac{50}{150} = \frac{1}{3} A$$

$$V_{Th} = V_{ab} = \text{Voltage across the 50 } \Omega \text{ resistor}$$

$$= 50\, I_1 = \frac{50}{3} V \qquad \text{(E 14.4.3)}$$

After the determination of R_{Th} and V_{Th} we can draw the Thevenin equivalent circuit between terminals $a - b$ as shown in Fig. 14.8h. By KVL in Fig. 14.8h,

$$v_S = v_R + v_L \quad t > 0$$

$$\frac{50}{3} = R_{Th}\, i_L + L \frac{d_{iL}}{dt}$$

$$= \frac{100}{3} i_L + 1 \times 10^{-3} \frac{di_L}{dt}$$

$$\frac{di_L}{dt} + \frac{100}{3} \times 10^3\, i_L = \frac{50}{3} \times 10^3 \qquad \text{(E 14.4. 4)}$$

Calculation of natural response i_{Ln}

$$\frac{di_{Ln}}{dt} + \frac{10^5}{3} i_{Ln} = 0$$

Let
$$i_{Ln} = A\, e^{st}$$

$$\frac{di_{Ln}}{d_n} = A\, s\, e^{st}$$

therefore,
$$A\, s\, e^{st} + \frac{10^5}{3} A\, e^{st} = 0$$

$$A\,e^{st}\left(s+\frac{10^5}{3}\right)=0$$

therefore, $s=\dfrac{-10^5}{3}$

and $$i_{Ln}=A\,e^{\frac{-10^5}{3}t} \qquad \text{(E 14.4.5)}$$

Calculation of forced response i_{L_f}

$$\frac{di_{Lf}}{dt}+\frac{10^5}{3}\,i_{Lf}=\frac{50}{3}\times 10^3$$

Let $i_{Lf}=k$

$$\frac{di_{Lf}}{dt}=0$$

therefore, $0+\dfrac{10^5}{3}k=\dfrac{50}{3}\times 10^3$

therefore, $k=0{\cdot}5$ and $i_{Lf}=k=0{\cdot}5$ (E 14.4.6)

Total response $i_L(t)$

$$i_L(t)=i_{Ln}(t)+i_{Lf}(t)$$

$$=A\,e^{\frac{-10^5}{3}t}+0.5 \qquad \text{(E 14.4.7)}$$

We shall determine the value of A by applying the initial condition $i_L(0_-)=1$

At $t=0_-$, $i_L(t)=i_L(0_-)=1$

By continuity equation, $i_L(0_+)=i_L(0_-)=1$

Substituting these values in Eq. (E 14.4.7), we get

$$1=Ae^0+0{\cdot}5$$

therefore, $A=0.5$

and $$i_L(t)=\left(0.5\,e^{\frac{-10^5}{3}t}+0.5\right)A$$

■ **EXAMPLE 14.5** For the circuit shown in Fig. 14.9a, the switch S **has been in position 1 for a long time and is moved to position 2 at** $t=0$**.** Find $i(t)$ for $t\geq 0$.

SOLUTION : Determination of initial conditions.

Since the switch S is in position 1 for a long time, the circuit is in the steady state and the capacitor is represented by an open circuit as shown in Fig, 14.9b. Under this condition, voltage across the capacitor = voltage across the 10 Ω resistor

$$v_c(0_-)=\frac{20}{5+5+10}\times 10=10\text{ V}$$

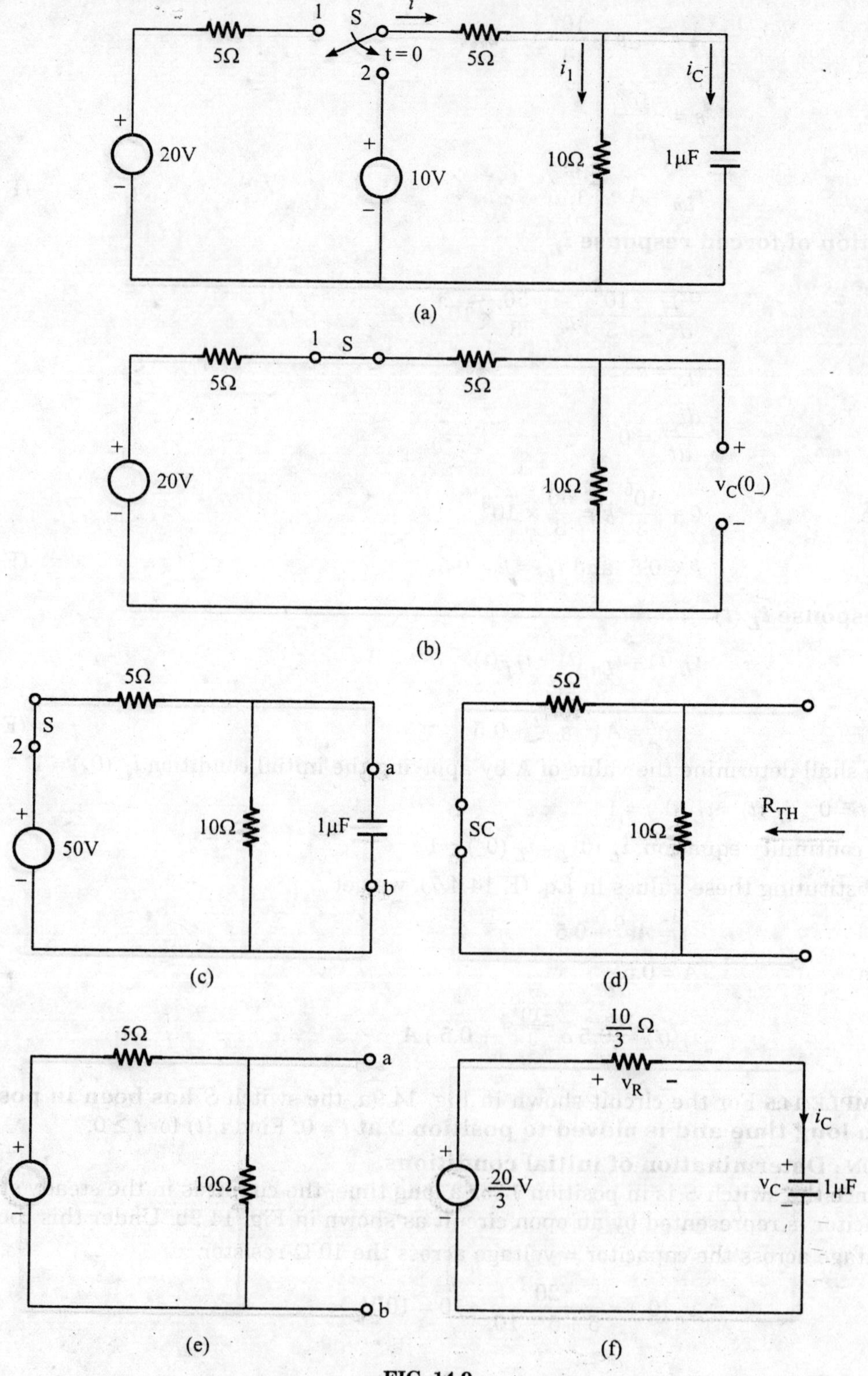
5Ω
1
S
i
t = 0
5Ω
2
i_1
i_C
20V
10V
10Ω
1μF
(a)
1
S
5Ω
5Ω
20V
10Ω
$v_C(0_-)$
(b)
5Ω
S
2
a
50V
10Ω
1μF
b
(c)
5Ω
SC
10Ω
R_{TH}
(d)
5Ω
a
10Ω
b
(e)
$\frac{10}{3}$ Ω
v_R
i_C
$\frac{20}{3}$ V
v_C
1μF
(f)

FIG. 14.9

When the switch S is in position 2, the circuit of Fig. 14.9a is reduced to Fig. 14.9c. We shall find the Thevenin equivalent at terminals $a - b$ of the capacitor.

Calculation of R_{Th}

The Thevenin resistance R_{Th} at terminals $a - b$ of the capacitor is found by removing the capacitor and replacing the 10 V source by a short circuit as shown in Fig. 14.9d. From this figure,

$$R_{Th} = R_{ab} = 10 \mid\mid 5 = \frac{10 \times 5}{10 + 5} = \frac{10}{3} \ \Omega$$

Calculation of v_{Th}

The Thevenin voltage v_{Th} is the voltage between the terminals $a - b$ when they are open circuited as shown in Fig. 14.9e.

$$v_{Th} = \text{voltage across the 10 } \Omega \text{ resistor}$$

$$= \frac{10}{5 + 10} \times 10 = \frac{20}{3} \text{ V}$$

The Thevenin equivalent network is shown in Fig. 14.9f.

By KVL,

$$v_{Th} = v_R + v_C$$

$$\frac{20}{3} = \frac{10}{3} i_C + v_C$$

$$\frac{20}{3} = \frac{10}{3} C \frac{dv_C}{dt} + v_C$$

$$\frac{10}{3} \times 1 \times 10^{-6} \frac{dv_C}{dt} + v_C = \frac{20}{3}$$

or

$$\frac{dv_C}{dt} + 3\times 10^5 \, v_C = 2 \times 10^6 \quad \text{(E 14. 5.1)}$$

Calculation of natural response v_{cn}

$$\frac{dv_{cn}}{dt} + 3 \times 10^5 \, v_{cn} = 0 \quad \text{(E 14.5.2)}$$

Let

$$v_{cn} = Ae^{st}$$

$$\frac{dv_{cn}}{dt} = A \, s \, e^{st}$$

Substitution of these values in Eq. (14.5.2) gives

$$A \, s \, e^{st} + 3 \times 10^5 \, A \, e^{st} = 0$$

$$A \, e^{st} \, (s + 3 \times 10^5) = 0$$

therefore,

$$s = -3 \times 10^5$$

and $$v_{cn} = A\,e^{-3\times 10^5 t} \qquad \text{(E 14.5.3)}$$

Calculation of forced response v_{cf}

$$\frac{dv_{cf}}{dt} + 3\times 10^5\, v_{cf} = 2\times 10^6 \qquad \text{(E 14.5.4)}$$

Let $$v_{cf} = k$$

therefore, $$\frac{dv_{cf}}{dt} = 0$$

Substitution of these values in Eq. (E 14.5.4) gives

$$0 + 3\times 10^5\, k = 2\times 10^6$$

therefore, $$k = \frac{20}{3}$$

Total response $v_C(t)$

$$v_C(t) = v_{cn} + v_{cf}$$

or $$v_C(t) = A\,e^{-3\times 10^5 t} + \frac{20}{3} \qquad \text{(E 14.5.5)}$$

The value of unknown A is determined by applying the initial condition

$$v_C(0_-) = 10.$$

By continuity equation

$$v_C(0_+) = v_C(0_-) = 10$$

Putting $t = 0_+$ in Eq. (14.5.5), we get

$$v_C(0_+) = A\,e^0 + \frac{20}{3}$$

$$10 = A + \frac{20}{3}$$

therefore, $$A = 10 - \frac{20}{3} = \frac{10}{3}$$

and $$v_C(t) = \frac{10}{3}\,e^{-3\times 10^5 t} + \frac{20}{3} \qquad \text{(E 14.5.6)}$$

$$i_C(t) = C\,\frac{dv_C}{dt}$$

$$= 10^{-6}\left[-\frac{10}{3}\times 3\times 10^5\, e^{-3\times 10^5 t} + 0\right]$$

or $$i_C(t) = -e^{-3\times 10^5 t}$$

$$i_1 = \frac{v_C}{10} = \frac{1}{3}\,e^{-3\times 10^5 t} + \frac{2}{3}$$

$$i = i_C + i_1$$

$$= -e^{-3\times 10^5 t} + \frac{1}{3}e^{-3\times 10^5 t} + \frac{2}{3}$$

$$i = \frac{2}{3} - \frac{2}{3}e^{-3\times 10^5 t} \quad \text{for } t \ge 0$$

EXAMPLE 14.6 Determine the current through the inductor L for $t \ge 0$ of a parallel RL circuit shown in Fig. 14.10. The switch S has been in position 1 for a long time and is then moved to position 2 at $t = 0$.

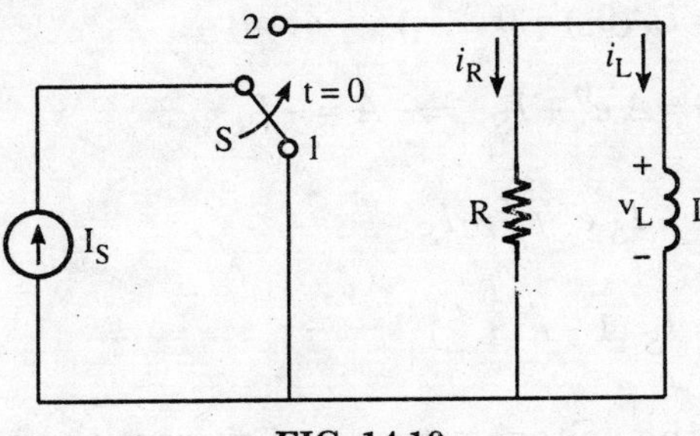

FIG. 14.10

SOLUTION : Initially the switch S is in position 1 and the current through the inductor is zero. That is, $i_L(0\_) = 0$.

At $t > 0$, the KCL equation of the network is

$$i_R + i_L = I_S$$

$$\frac{v_L}{R} + i_L = I_S$$

$$\frac{1}{R} \times L\frac{di_L}{dt} + i_L = I_S$$

$$\frac{di_L}{dt} + \frac{R}{L}i_L = \frac{R}{L}I_S \tag{E 14.6.1}$$

If I_{Ln} is the natural response, then

$$\frac{di_{Ln}}{dt} + \frac{R}{L}i_{Ln} = 0 \tag{E 14.6.2}$$

As usual, the solution of this equation is

$$i_{Ln} = A\,e^{st} = A\,e^{-\frac{R}{L}t} \tag{E 14.6.3}$$

If i_{Lf} is the forced response, then

$$\frac{di_{Lf}}{dt} + \frac{R}{L}i_{Lf} = \frac{R}{L}I_S \tag{E 14.6.4}$$

The solution of this equation is obtained by assuming $i_{Lf} = k$.

therefore, $$\frac{di_{Lf}}{dt} = 0$$

and $$0 + \frac{R}{L} k = \frac{R}{L} I_s \Rightarrow k = I_S$$

The complete response is given by

$$i_L(t) = i_{Ln} + i_{Lf}$$

or $$i_L(t) = A\, e^{-\frac{R}{L}t} + I_S \tag{E 14.6.5}$$

At $t = 0_+$, $i_L(t) = i_L(0_+) = i_L(0_-) = 0$

and $$i_L(0_+) = A\, e^0 + I_S \Rightarrow A = -I_S$$

therefore, $$i_L(t) = -I_S\, e^{-\frac{R}{L}t} + I_S$$

or $$i_L(t) = I_S(1 - e^{-\frac{R}{L}t})$$

■ **EXAMPLE 14.7** Determine the voltage across and current through the capacitor C for $t \geq 0$ of a parallel RC network shown in Fig. 14.11. The switch S has been in position 1 for a long time and is then moved to position 2.

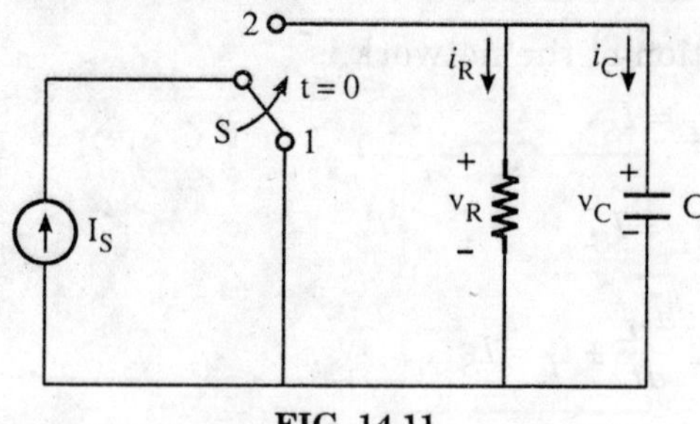

FIG. 14.11

SOLUTION : Initially the switch is in position 1 and the voltage across the capacitor is zero. That is, $v_C(0_-) = 0$.

At $t > 0$, the KCL equation of the network is given by

$$i_C + i_R = I_S$$

$$C\frac{dv_C}{dt} + \frac{v_C}{R} = I_S$$

$$\frac{dv_C}{dt} + \frac{1}{RC} v_C = \frac{1}{C} I_S \tag{E 14.7.1}$$

As before, $v_{cn} = A\, e^{-\frac{1}{RC}t}$, $v_{Cf} = R\, I_S$

$$v_C(t) = v_{cn} + v_{cf}$$

$$v_C(t) = A\,e^{-\frac{t}{RC}} + R\,I_S \qquad \text{(E 14.7.2)}$$

$$v_C(0_+) = v_C(0_-) = 0$$

Putting $t = 0_+$ in Eq. (E 14.7.2), we get

$$v_C(0_+) = A\,e^0 + R\,I_S$$

therefore, $A = -R\,I_S$

and $v_C(t) = -R\,I_S\,e^{-\frac{t}{RC}} + R\,I_S$

or $v_C(t) = R\,I_S\,(1 - e^{-\frac{t}{RC}})$

$$i_C(t) = C\frac{dv_C}{dt}$$

$$= C\,R\,I_S\,(0 + \frac{1}{RC}e^{-\frac{t}{RC}})$$

or $i_C(t) = I_S\,e^{-\frac{t}{RC}}$

■ **EXAMPLE 14.8.** The switch S in the circuit of Fig. 14.12 has been open for a long time before being closed at $t = 0$. Find $i(t)$ for $t \geq 0$.

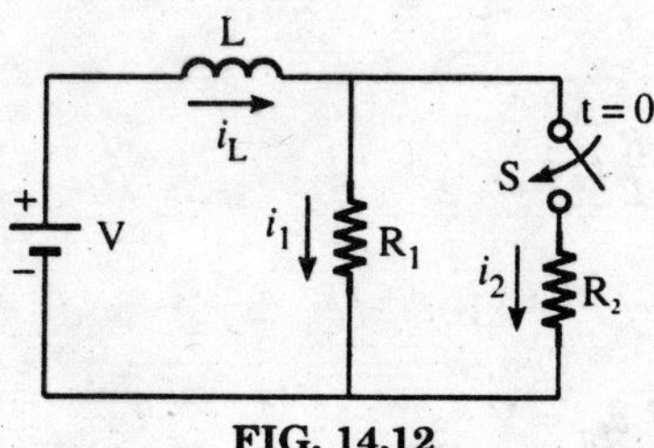

FIG. 14.12

SOLUTION : Since the switch has been open for a long time before being **closed,** the circuit is in its steady state with L acting as a short circuit. Thus, at $t = 0_-$

$$i_L(0_-) = \frac{V}{R_1}$$

At $t \geq 0$ when S is closed, the circuit equations can be written as

$$L\frac{di_L}{dt} + R_1\,i_1 = V \qquad \text{(E 14.8.1)}$$

$$i_L = i_1 + i_2 \qquad \text{(E 14.8.2)}$$

$$R_1\,i_1 = R_2\,i_2 \qquad \text{(E 14.8.3)}$$

Eliminating i_1 from Eq. (E 14.8.1) using Eqs. (E 14.8.2) and (E 14.8.3), we get

$$L\frac{di_L}{dt}+\frac{R_1R_2}{R_1+R_2}i_L=V \quad \text{(E 14.8.4)}$$

For the sake of convenience let

$$\frac{R_1R_2}{R_1+R_2}=R \quad \text{(E 14.8.5)}$$

Therefore, Eq. (E 14.8.4) becomes

$$L\frac{di_L}{dt}+R\,i_L=V \quad \text{(E 14.8.6)}$$

Natural response i_{Ln}

$$L\frac{di_{Ln}}{dt}+R\,i_{Ln}=0 \quad \text{(E 14.8.7)}$$

Let the solution be

$$i_{Ln}=A\,e^{st}$$

therefore, $$\frac{di_{Ln}}{dt}=s\,A\,e^{st}$$

therefore, $$L\,s\,A\,e^{st}+R\,A\,e^{st}=0$$

$$A\,e^{st}\,(s\,L+R)=0$$

$$s=-\frac{R}{L}$$

therefore, $$i_{Ln}=A\,e^{-\frac{R}{L}t} \quad \text{(E 14.8.8)}$$

Forced response i_{Lf}

$$L\frac{di_{Lf}}{dt}+R\,i_{Lf}=V \quad \text{(E 14.8.9)}$$

Let the solution be

$$i_{Lf}=k$$

$$\frac{di_{Lf}}{dt}=0$$

therefore, $$0+R\,k=V,\ k=\frac{V}{R}$$

and $$i_{Lf}=\frac{V}{R} \quad \text{(E 14.8.10)}$$

The complete solution is given by

$$i_L\,(t)=i_{Ln}+i_{Lf}$$

$$i_L(t) = A\,e^{-\frac{R}{L}t} + \frac{V}{R} \qquad \text{(E 14.8.11)}$$

The value of unknown A is found by the application of initial condition

$$i_L(0_-) = \frac{V}{R_1}$$

By continuity equation

$$i_L(0_+) = i_L(0_-) = \frac{V}{R_1}$$

From Eq. (E 14.8.11), at $t = 0_+$,

$$i_L(0_+) = A\,e^{-\frac{R}{L}\times 0} + \frac{V}{R}$$

$$\frac{V}{R_1} = A + \frac{V}{R}$$

therefore, $$A = \frac{V}{R_1} - \frac{V}{R}$$

$$= \frac{V}{R_1} - \frac{V(R_1+R_2)}{R_1 R_2} = -\frac{V}{R_2}$$

Substitution of the value of A in Eq. (E 14.8.11) gives

$$i_L(t) = -\frac{V}{R_2}e^{-\frac{R}{L}t} + \frac{V}{R}$$

$$= -\frac{V}{R_2}e^{-\frac{R}{L}} + \frac{V(R_1+R_2)}{R_1 R_2}$$

$$= V\left[\frac{1}{R_1} + \frac{1}{R_2}(1 - e^{-(R/L)t}\right]$$

where $$R = \frac{R_1 R_2}{R_1 + R_2}$$

■ **EXAMPLE 14.9.** Determine the inductor current for $t > 0$ in the network shown in Fig. 14.13a. The switch S has been open for a long time and is then closed at $t = 0$.

SOLUTION : *Determination of initial condition* $[\,i_L(0_-)\,]$

Since the switch S has been open for a long time, the circuit is in the dc steady state. In dc steady state, an inductor can be represented by a short circuit. Therefore, the equivalent circuit valid for $t = 0_-$ is shown in Fig. 14.13b. From Fig. 14.13b.

$$i_L(0_-) = \frac{V_s}{R_1 + R_2}$$

By KVL

$$v_L + v_{R_1} = V_S \text{ for } t > 0 \tag{E 14.9.1}$$

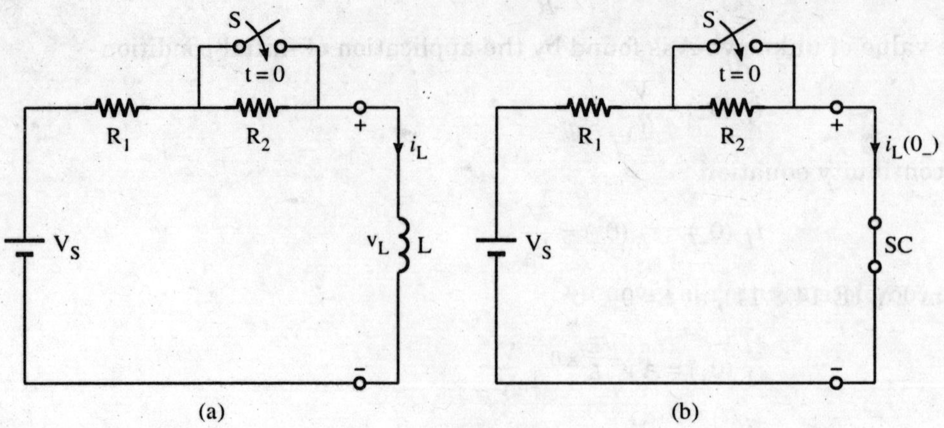

FIG. 14.13

It is to be noted that when the switch S is closed at $t = 0$, the resistor R_2 is short circuited and in Eq. (E 14.9.1) no term for v_{R_2} appears. Equation (E 14.9.1) can be written as

$$L\frac{di_L}{dt} + R_1\, i_L = V_S$$

therefore, for $t > 0$

$$\frac{di_L}{dt} + \frac{R_1}{L}\, i_L = \frac{V_S}{L} \tag{E 14.9.2}$$

Natural response i_{Ln}

$$\frac{di_{Ln}}{dt} + \frac{R_1}{L}\, i_{Ln} = 0 \tag{E 14.9.3}$$

Let $i_{Ln} = A\, e^{st}$ be the solution.

$$\frac{di_{Ln}}{dt} = A\, e^{st}$$

therefore, $$s\, A\, e^{st} + \frac{R_1}{L} A\, e^{st} = 0$$

$$A\, e^{st}\left(s + \frac{R_1}{L}\right) = 0\ ,\ s = -\frac{R_1}{L}$$

therefore, $$i_{Ln} = A\, e^{-\frac{R_1}{L}t} \tag{E 14.9.4}$$

Forced response i_{Lf}

$$\frac{di_{Lf}}{dt} + \frac{R_1}{L}\, i_{Lf} = \frac{V_s}{L}$$

Let $i_{Lf} = k$ be the solution

$$\frac{di_{Lf}}{dt} = 0$$

therefore, $$0 + \frac{R_1}{L} k = \frac{V_s}{L}, \; k = \frac{V_s}{R_1}$$

and $$i_{Lf} = \frac{V_s}{R_1} \tag{E 14.9.5}$$

The complete solution is given by

$$i_L(t) = i_{Ln} + i_{Lf}$$

or $$i_L(t) = A\, e^{-\frac{R_1}{L}t} + \frac{V_s}{R_1} \tag{E 14.9.6}$$

We shall determine the value of A by applying the initial condition

$$i_L(0_-) = \frac{Vs}{R_1 + R_2}$$

At $t = 0_-$.

$$i_L(t) - i_L(0_-) = \frac{V_S}{R_1 + R_2}$$

By continuity equation,

$$i_L(0_+) = i_L(0_-) = \frac{V_s}{R_1 R_2}$$

Substituting these values in Eq. (E 14.9.6), we get

$$\frac{V_s}{R_1 + R_2} = A\, e^0 + \frac{V_s}{R}$$

therefore, $$A = \frac{V_s}{R_1 + R_2} - \frac{V_s}{R_1}$$

Substitution of the value of A in Eq. (E 14.9.6) gives the current through the inductor as

$$i_L(t) = \frac{V_s}{R_1} + \left(\frac{V_s}{R_1 + R_2} - \frac{V_s}{R_1} \right) e^{-\frac{R_1}{L}t} \tag{E 14.9.7}$$

Alternative method

Equation (E 14.9.2) can be written as

$$\frac{di_L}{dt} + \frac{R_1}{L} i_L = \frac{V_s}{L} u_{-1}(t) \tag{E 14.9.8}$$

Taking the Laplace transform of both the sides of Eq. (E 14.9.8), we obtain

$$[s\,I_L(s) - i_L(0_-)] + \frac{R_1}{L} I_L(s) = \frac{V_s}{L} \times \frac{1}{s}$$

$$\left[s \frac{1}{L}(s) - \frac{V_s}{R_1 + R_2}\right] + \frac{R_1}{L} I_L(s) = \frac{V_s}{sL}$$

$$\left(s + \frac{R_1}{L}\right) I_L(s) = \frac{V_s}{R_1 + R_2} + \frac{V_s}{sL}$$

$$I_L(s) = \frac{V_s}{R_1 + R_2} \times \frac{1}{\left(s + \frac{R_1}{L}\right)} + \frac{V_s/L}{s\left(s + \frac{R_1}{L}\right)}$$

$$= \frac{V_s}{R_1 + R_2} \times \frac{1}{\left(s + \frac{R_1}{L}\right)} + \frac{V_s}{L} \times \frac{L}{R_1}\left[\frac{1}{s} - \frac{1}{\left(s + \frac{R_1}{L}\right)}\right] \quad \text{(E 14.9.9)}$$

Taking the inverse Laplace transform of Eq. (E 14.9.9), we get

$$i_L(t) = \frac{V_s}{R_1 + R_2} e^{-\frac{R_1}{L}t} + \frac{V_s}{R_1}\left(1 - e^{-\frac{R_1}{L}t}\right)$$

or

$$i_L(t) = \frac{V_s}{R_1} + \left(\frac{V_s}{R_1 + R_2} - \frac{V_s}{R_1}\right) e^{-\frac{R_1}{L}t} \quad \text{(E 14.9.10)}$$

This result is the same as obtained in Eq. (E 14.9.7).

■ **EXAMPLE 14.10** Find the unit-step response for i_L in the network shown in Fig. 14.14a. It is given that $i_L(0_-) = 0$.

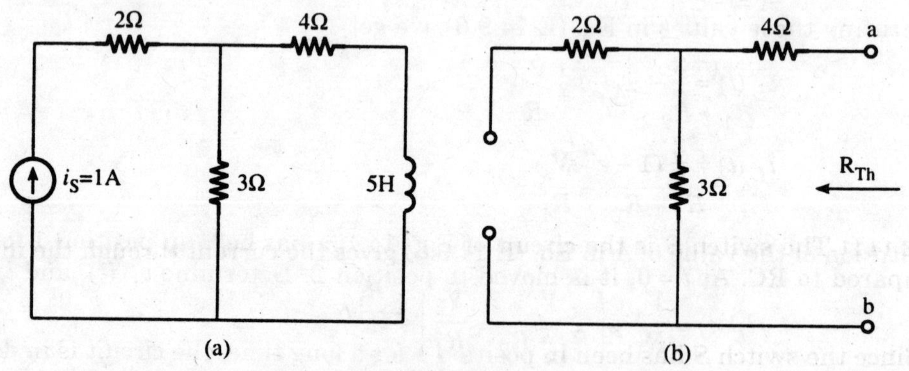

FIG. 14.14

SOLUTION : Calculation of R_{Th}.

The Thevenin resistance R_{Th} at terminals $a - b$ of the inductor is found by removing the inductor and replacing the 1 A current source by an open circuit as shown in Fig. 14.14b. From this figure,

$$R_{Th} = 4 + 3 = 7\ \Omega$$

It is implied that $i_s(t) = u_{-1}(t)$

The natural response can be written as

$$i_{Ln} = A\,e^{-\frac{R_{Th}}{L}t}$$

$$i_{Ln} = A\,e^{\frac{-7}{5}t} \qquad \text{(E 14.10.1)}$$

For a constant 1 A source i_s, the inductor is an effective short circuit under steady-state conditions. Thus, the current of 1 A is divided into the 3 Ω and 4 Ω resistors. The forced response can, therefore, be written as

$$i_{Lf} = \frac{3}{3+4} \times 1 = \frac{3}{7}\text{ A} \qquad \text{(E 14.10.2)}$$

Total response

$$i_L(t) = i_{Ln} + i_{Lf}$$

$$i_L(t) = \frac{3}{7} + A\,e^{-\frac{7}{5}t} \qquad \text{(E 14.10.3)}$$

At $t = 0_-$, $i_L(t) = i_L(0_-) = 0$

By continuity equation

$$i_L(0_+) = i_L(0_-) = 0$$

Applying the initial conditions in Eq. (E 14.10.3), we get

$$0 = \frac{3}{7} + A\,e^{-\frac{7}{5}\times 0}$$

therefore, $$A = -\frac{3}{7}$$

and $$i_L(t) = \frac{3}{7} - \frac{3}{7}e^{-\frac{7}{5}t}$$

$$i_L(t) = \frac{3}{7}\left(1 - e^{-\frac{7}{5}t}\right)$$

■ **EXAMPLE 14.11** The switch S is the circuit of Fig. 14.15a has been in position 1 for a long time compared to RC. At $t = 0$, it is moved to position 2. Determine $v_C(t)$ and $i_C(t)$ for $t \geq 0$.

SOLUTION : Since the switch S has been in position 1 for a long time, the circuit is in dc steady state. In dc steady state a capacitor can be represented by an open circuit as shown in Fig. 14.15b. The initial condition can be written from Fig. 14.15b as

$$v_C(0_-) = V_A \qquad \text{(E 14.11.1)}$$

When the switch S is moved to position 2, the circuit takes the form shown in Fig. 14.15c. By KVL

$$V_B = v_R + v_C \quad t > 0$$

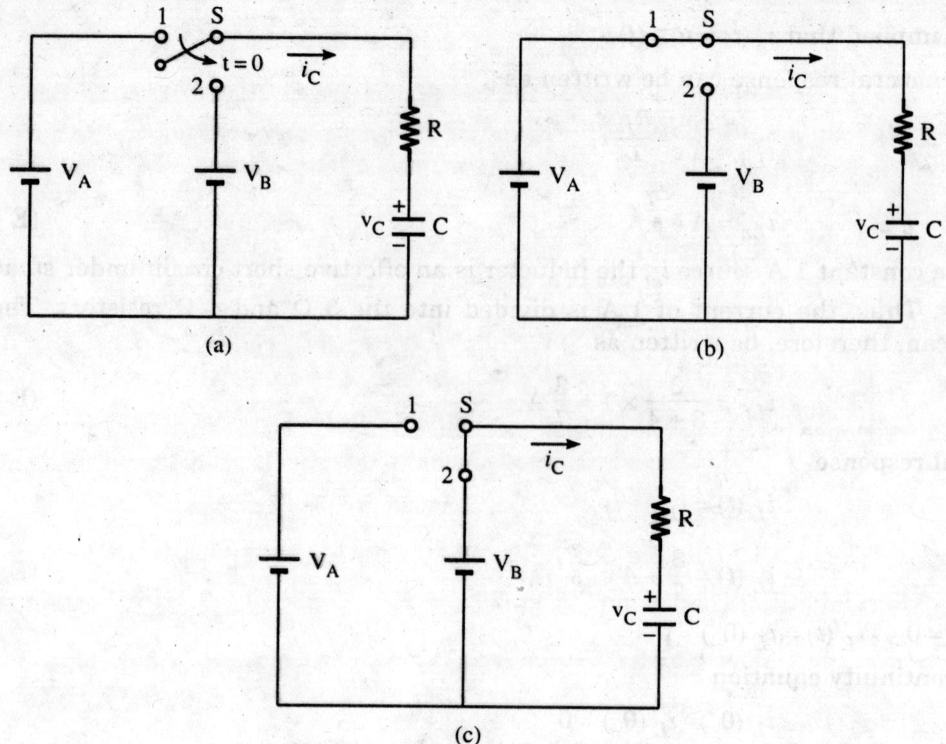

FIG. 14.15

$$= R\, i_C + v_C$$

$$V_B = RC \frac{dv_C}{dt} + v_C$$

therefore, $$\frac{dv_C}{dt} + \frac{1}{RC} v_C = \frac{1}{RC} V_B \qquad t > 0 \qquad \text{(E 14.11.2)}$$

Calculation of natural response v_{cn}

$$\frac{dv_{cn}}{dt} + \frac{1}{RC} v_{cn} = 0$$

Let $$v_{cn} = A\, e^{st}$$

therefore, $$\frac{dv_C}{dt} = A\, s\, e^{st}$$

$$A\, s\, e^{st} + \frac{1}{RC} A\, e^{st} = 0$$

This equation gives

$$v_{cn} = A\, e^{-\frac{1}{RC}t} \qquad \text{(E 14.11.3)}$$

Calculation of forced response v_{cf}

$$\frac{dv_{cf}}{dt} + \frac{1}{RC} v_{cf} = \frac{1}{RC} V_B$$

Let $v_{cf} = k$

therefore, $\frac{dv_{cf}}{dt} = 0$

$$0 + \frac{1}{RC} k = \frac{1}{RC} V_B$$

therefore, $k = V_B = v_{cf}$

Total response $v_C(t)$ is given by

$$v_C(t) = v_{cn} + v_{cf}$$

$$v_C(t) = A\, e^{-\frac{1}{RC} t} + V_B \quad \text{(E 14.11.4)}$$

Applying the initial condition

$$v_C(0_-) = A\, e^{0} + V_B$$

$$V_A = A + V_B$$

therefore, $A = V_A - V_B$

and $$v_C(t) = V_B + (V_A - V_B)\, e^{-\frac{t}{RC}} \quad \text{(E 14.11.5)}$$

$$i_C(t) = C \frac{dv_C}{dt} = C\left[0 + \left(-\frac{1}{RC}\right)(V_A - V_B)\, e^{-\frac{t}{RC}}\right]$$

or $$i_C(t) = \frac{V_B - V_A}{R} e^{-\frac{t}{RC}} \quad \text{(E 14.11.6)}$$

Alternative method

Eq. (E 14.11.2) can be written as

$$\frac{dv_C}{dt} + \frac{1}{RC} v_C = \frac{1}{RC} V_B\, u_{-1}(t) \quad \text{(E 14.11.7)}$$

Taking the Laplace transform of both the sides of Eq. (E 14.11.7), we get

$$[\, s\, V_C(s) - v_C(0_-)\,] + \frac{1}{RC} V_C(s) = \frac{V_B}{RC} \times \frac{1}{s}$$

$$[\, s\, V_C(s) - V_A\,] + \frac{1}{RC} V_C(s) = \frac{V_B}{s\, RC}$$

$$\left(s + \frac{1}{Rc}\right) V_C(s) = \frac{V_B}{s\; RC} + V_A$$

$$V_C(s) = \frac{V_B}{s\,RC\left(s + \frac{1}{RC}\right)} + \frac{V_A}{\left(s + \frac{1}{RC}\right)}$$

$$= V_B\left(\frac{1}{s} - \frac{1}{s + \frac{1}{RC}}\right) + \frac{V_A}{\left(s + \frac{1}{RC}\right)} \quad \text{(E 14.11.8)}$$

Taking the inverse Laplace transform of Eq. (14.11.8), we get

$$v_C(t) = V_B\left(1 - e^{-\frac{t}{RC}}\right) + V_A\, e^{-\frac{t}{RC}}$$

$$= V_B + (V_A - V_B)\, e^{-\frac{t}{RC}}$$

■ **EXAMPLE 14.12** In the circuit shown in Fig. 14.16a, the switch S has been open for a long time and is then closed at $t = 0$. Determine the current $i(t)$.

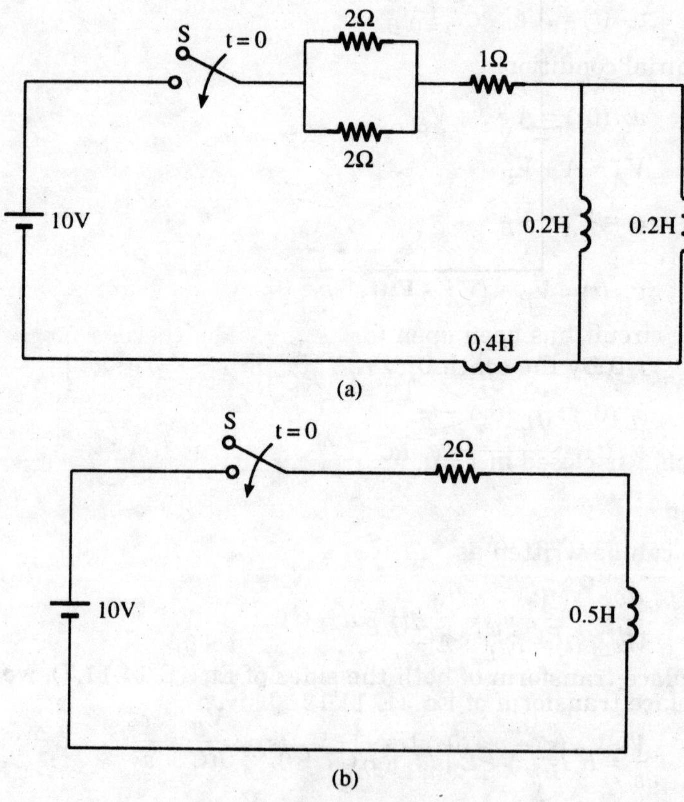

FIG. 14.16

SOLUTION: The given network can be reduced to a simple RL network. Here all resistance can be combined into a single resistance. The equivalent resistance is given by

$$R = (2\,||\,2) + 1 = \frac{2 \times 2}{2 + 2} + 1 = 2\ \Omega$$

The equivalent inductance in given by

$$L = (0.2 \,||\, 0.2) + 0.4 = 0.5 \text{ H}$$

Hence the network of Fig. 14.16a is equivalent to the network shown in Fig. 14.16b.

therefore, $$i_L(t) = \frac{V_s}{R}\left(1 - e^{-\frac{R}{L}t}\right)$$

$$= \frac{10}{2}\left(1 - e^{-\frac{2}{0.5}t}\right)$$

or $$i_L(t) = 5\left(1 - e^{-4t}\right)$$

■ **EXAMPLE 14.13** In the circuit shown in Fig. 14.17, the switch S has been open for a long time and is closed at $t = 0$. Use Laplace transform method to determine $i_L(t)$ for $t \geq 0$.

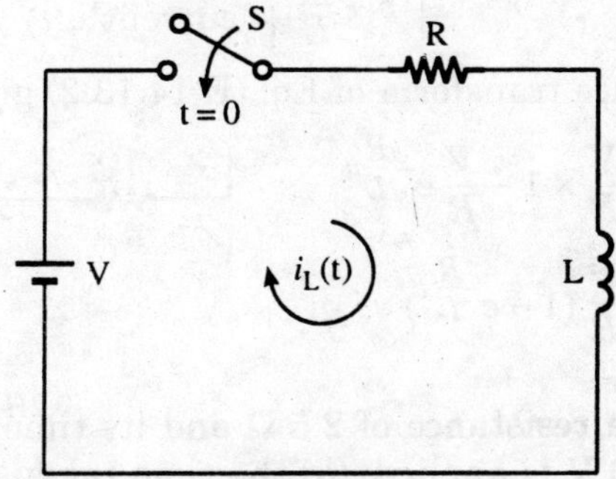

FIG. 14.17

SOLUTION : Since the circuit has been open for a long time, there is no current in the circuit at $t = 0$. That is, $i_L(0_-) = 0$ By the continuity relationship

$$i_L(0_+) = i_L(0_-) = 0$$

When the switch S is closed at $t = 0$, a current $i_L(t)$ flows in the circuit.

By KVL

$$V u_{-1}(t) = R i_L + L\frac{di_L}{dt} \qquad t > 0 \qquad \text{(E 14.13.1)}$$

Taking the Laplace transform of Eq. (E 14.13.1) gives

$$\frac{V}{s} = R\, I_L(s) + L\,[s\, I_L(s) - i_L(0_-)]$$

$$\frac{V}{s} = (R + s\,L)\, I_L(s)$$

$$I_L(s) = \frac{V}{s\,(R + s\,L)}$$

$$= \frac{V/L}{s\left(s + \frac{R}{L}\right)}$$

$$= \frac{k_1}{s} + \frac{k_2}{s + \frac{R}{L}}$$

$$k_1 = sI_L(s)\,|_{s=0} = \frac{V}{R}$$

$$k_2 = I_L(s)\left(s + \frac{R}{L}\right)\Bigg|_{s=-\frac{R}{L}} = -\frac{V}{R}$$

therefore, $$I_L(s) = \frac{V}{R}\left(\frac{1}{s}\right) - \frac{V}{R}\left(\frac{1}{s + \frac{R}{L}}\right) \quad \text{(E 14. 13.2)}$$

Taking the inverse Laplace transform of Eq. (E 14.13.2) gives

$$i_L(t) = \frac{V}{R} \times 1 - \frac{V}{R} e^{-\frac{R}{L}t}$$

$$= \frac{V}{R}(1 - e^{-\frac{R}{L}t}) \quad \text{(E 14.13.3)}$$

■ **EXAMPLE 14.14** **A coil has a resistance of 2.5 Ω and its time constant is 1.6 s. Determine (*a*) the current 1 s after 10 V is applied; (*b*) the time taken for the current to attain half its final value; (*c*) the initial rate of increase of current.**

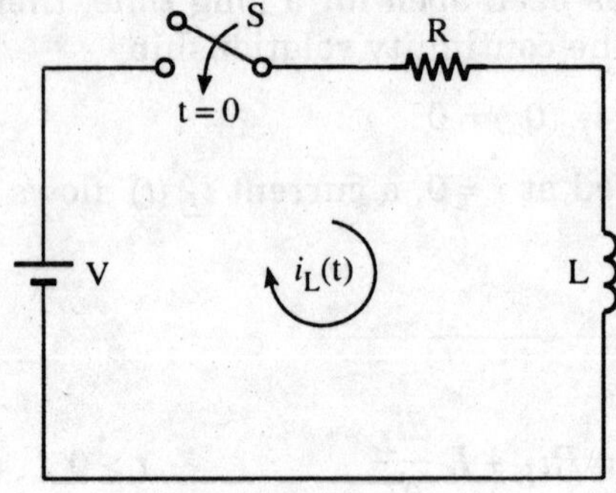

FIG. 14.17

SOLUTION :

(*a*) $$i_L = \frac{V}{R} = (1 - e^{-\frac{R}{L}t})$$

$V = 10$ volts, $R = 2.5\ \Omega$, $\tau_L = \dfrac{L}{R} = 1.6$ s

At $t = 1$ s

$$i_L = \frac{10}{2.5}\left(1 - e^{-\frac{1}{1.6}}\right)$$

$$= 4\,(1 - 0.535) = 1.86\text{ A}$$

(*b*) Final value of current

$$= \frac{V}{R} = \frac{10}{2.5} = 4\text{ A}$$

Half of the final value of current = 2 A

$$i_L = \frac{V}{R}\left(1 - e^{-\frac{t}{1.6}}\right)$$

$$2 = 4\left(1 - e^{-\frac{t}{1.6}}\right)$$

$$e^{-\frac{t}{1.6}} = 1/2,\ t = 1.1\text{ s}$$

(*c*) The initial rate of increase of current

$$= \frac{V}{L} = \frac{V}{R} \times \frac{R}{L} = \frac{10}{2.5} \times \frac{1}{1.6} = 2.5\text{ A/s}$$

■ **EXAMPLE 14.15** In the circuit shown in Fig. 14.18, there is an initial voltage v_{co} on the capacitor. The switch is closed at $t = 0$. Use Laplace transform method to determine $v_C(t)$, v_R and $i_C(t)$ for $t \geq 0$.

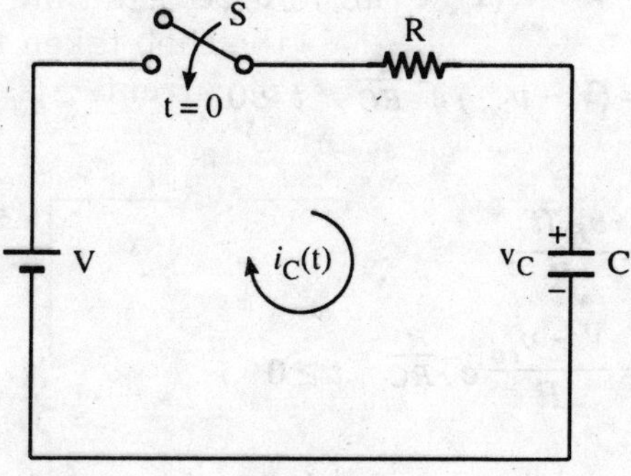

FIG. 14.18

SOLUTION : By KVL

$$Vu_{-1}(t) = v_R + v_C \quad t > 0$$

$$= R\ i_C + v_C$$

$$= RC\frac{dv_c}{dt} + v_C \qquad \text{(E 14.15.1)}$$

Taking the Laplace transform of both the sides of Eq. (E 14.15.1), we get

$$\frac{V}{s} = RC\,[\,s\,V_C(s) - v_C(0_-)\,] + V_C(s)$$

Since $v_C(0_-) = v_{co}$

$$\frac{V}{s} = RC\,[s\,V_C(s) - v_{co}] + V_C(s)$$

$$(RCs + 1)\,V_C(s) = \frac{V}{s} + RC\;v_{co}$$

$$V_C(s) = \frac{V}{s\,(RC\;\;s+1)} + \frac{RC\;v_{co}}{(RC\;s+1)}$$

$$= \frac{\frac{V}{RC}}{s\left(s+\frac{1}{RC}\right)} + \frac{v_{co}}{s+\frac{1}{RC}}$$

$$= V\left(\frac{1}{s} - \frac{1}{s+\frac{1}{RC}}\right) + v_{co} \times \frac{1}{s+\frac{1}{RC}} \qquad \text{(E 14.15.2)}$$

Taking the inverse Laplace transform of Eq. (E 14.15.2), we get

$$v_C(t) = V(1 - e^{-\frac{t}{RC}}) + v_{co}\,e^{-\frac{t}{RC}} \qquad \text{(E 14.15.3)}$$

Voltage drop across the resistor

$$v_R(t) = V - v_C(t)$$

$$= V - V(1 - e^{-\frac{t}{RC}}) - v_{co}\,e^{-\frac{t}{RC}}$$

$$= (V - v_{co})\,e^{-\frac{t}{RC}} \quad t \geq 0 \qquad \text{(E 14.15.4)}$$

Current in the circuit

$$i_C(t) = \frac{v_R(t)}{R}$$

$$= \frac{V - v_{co}}{R}\,e^{-\frac{t}{RC}} \quad t \geq 0 \qquad \text{(E 14.15.5)}$$

■ **EXAMPLE 14.16. Find the complete response $i_L(t)$ for $t > 0$ for the network shown in Fig. 14.7 if $i_L(0_-) = 0$ and $v_S(t) = 20 \sin(4t + 60^\circ)$.**

SOLUTION : By KVL

$$v_L + v_R = v_s \quad \text{for } t > 0$$

$$L\frac{di_L}{dt} + R\;i_L = v_s$$

$$5\frac{di_L}{dt} + 50\,i_L = 20 \sin(4\,t + 60^\circ)$$

$$\frac{di_L}{dt} + 10\;i_L = 4 \sin(4\,t + 60^\circ) \quad \text{for } t > 0 \qquad \text{(E 14.16.1)}$$

Determination of natural response

As in Example 14.3, the natural response is

$$i_{Ln} = A\,e^{-10\,t} \qquad \text{(E 14.16.2)}$$

Determination of forced response

Here the forcing function is a *sinusoid* 4 sin (4 *t* + 60°). Let $i_{Lf}(t)$ be the forced response. It will satisfy the equation

$$\frac{di_{Lf}}{dt} + 10\,i_{Lf} = 4\sin(4\,t + 60^\circ) \qquad \text{(E 14.16.3)}$$

We shall assume the forced solution $i_{Lf}(t)$ as a linear sum of the forcing function and all its nonzero derivatives. That is,

$$i_{Lf}(t) = k_1 \sin 4\,t + k_2 \cos 4\,t \qquad \text{(E 14.16.4)}$$

therefore,

$$\frac{di_{Lf}}{dt} = 4\,k_1 \cos 4\,t - 4\,k_2 \sin 4\,t \qquad \text{(E 14.16.5)}$$

Substituting the values from Eqs. (E 14.16.4) and (E 14.16.5) in Eq. (E 14.16.3) gives

$$4\,k_1 \cos 4\,t - 4\,k_2 \sin 4\,t + 10\,(k_1 \sin 4\,t + k_2 \cos 4\,t) = 4\sin(4\,t + 60^\circ)$$

$$(4\,k_1 + 10\,k_2)\cos 4\,t + (10\,k_1 - 4\,k_2)\sin 4t$$

$$= (4\cos 60^\circ)\sin 4\,t + (4\sin 60^\circ)\cos 4\,t$$

Comparing the coefficients of cos 4 *t* and sin 4 *t* on both the sides gives

$$4\,k_1 + 10\,k_2 = 4\sin 60^\circ$$

$$10\,k_1 - 4\,k_2 = 4\cos 60^\circ$$

These equations are solved for k_1 and k_2.

$$k_1 = 0.2918,\; k_2 = 0.2296$$

Substitution of the values of k_1 and k_2 in Eq. (E 14.16.4) gives

$$i_{Lf}(t) = 0.2918\sin 4\,t + 0.2296\cos 4\,t = 0.3713\sin(4\,t + 38.2^\circ) \qquad \text{(E 14.16.6)}$$

The complete response is given by

$$i_L(t) = i_{Ln}(t) + i_{Lf}(t)$$

$$= A\,e^{-10\,t} + 0.3713\sin(4\,t + 38.2^\circ) \qquad \text{(E 14.16.7)}$$

Applying initial conditions $i_L(0_+) = i_L(0_-) = 0$ in Eq. (E 14.16.7) we get

$$0 = A\,e^{-10\times 0} + 0.3713\sin(4\times 0 + 38.2^\circ)$$

$$0 = A + 0.3713\;\sin\;38.2^\circ$$

$$A = -\,0.2296$$

Substituting the value of *A* in Eq. (E 14.16.7) we get the complete response as

$$i_L(t) = -0.2296\,e^{-10\,t} + 0.3713\sin(4\,t + 38.2^\circ)\quad \text{for } t > 0$$

or $$i_L(t) = [\,0.3713 \sin(4t + 38.2°) - 0.2296\, e^{-10t}\,]\, u_{-1}(t)$$

Alternative method of determining the forced response by phasors

The given circuit is the simple series RL circuit connected across a sinusoidal source of rms voltage equal to $20/\sqrt{2}$ volts and radian frequency $\omega = 4$ rad / s. In phasor form

$$\mathbf{V} = \frac{20}{\sqrt{2}} \underline{/\ 60°}$$

$$\mathbf{Z} = R + jX_L = R + j\,wL = 50 + j\,(4) \times 5$$

$$= 53.85 \underline{/\ 21.8°}\ \Omega$$

$$\mathbf{I} = \frac{\mathbf{V}}{\mathbf{Z}} = \frac{20 \underline{/\ 60°}}{\sqrt{2} \times 53.85 \underline{/ 21.80°}} = \frac{0.3714}{\sqrt{2}} \underline{/\ 38.2°}\ \text{A}$$

The instantaneous value of current in steady state is given by

$$i_{Lf} = (\sqrt{2}\, I) \sin(4\,t + 38.2°)$$

or $$i_{Lf} = 0.3714 \sin(4\,t + 38.2°)$$

as obtained before

14.19 COMPLETE RESPONSE OF A GENERAL SERIES RL NETWORK TO SINUSOIDAL INPUT

Figure 14.19 shows a series RL network where a sinusoidal voltage is applied at $t = 0$ through a switch S. The voltage function could be at any particular point in the cycle at which the switch is closed. Consider the sinusoidal voltage

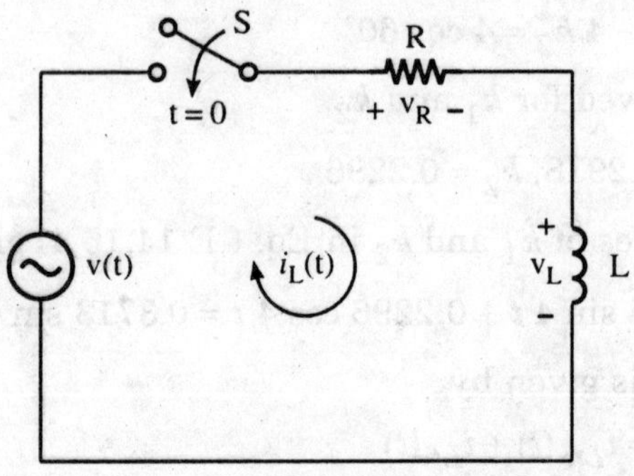

FIG. 14.19

$$v = V_m \sin(\omega\,t + \alpha_0) \tag{14.19.1}$$

where the phase angle α_0 can take any value from 0 to 2π radians. When the switch is closed at $t = 0$, we have by KVL

$$v_L + v_R = v \quad \text{for} \quad t > 0$$

or $$L\frac{di_L}{dt} + R\,i_L = V_m \sin(\omega\,t + \alpha_0) \quad \text{for } t > 0 \tag{14.19.2}$$

Natural response or transient response

$$i_{Ln} = A\,e^{st}$$

or
$$i_{Ln} = A\,e^{-\frac{R}{L}t} \tag{14.19.3}$$

Forced response or steady-state response i_{Lf}

Here the forcing function is a sinusoid $V_m \sin(\omega t + \alpha_0)$. The forced response i_{Lf} will satisfy the equation

$$L\frac{di_{Lf}}{dt} + R\,i_{Lf} = V_m \sin(\omega t + \alpha_0) \tag{14.19.4}$$

The following methods are generally used to determine the forced response:

(*a*) Method of undetermined coefficients.

(*b*) Method of phasor transforms.

Method of undetermined coefficients

Since the input is a sinusoid, we shall try a solution in the form of a sinusoid. We shall assume the forced solution $i_{Lf}(t)$ as a linear sum of the forcing function and all its nonzero derivatives. That is,

$$i_{Lf}(t) = k_1 \sin\omega t + k_2 \cos\omega t \tag{14.19.5}$$

therefore,
$$\frac{di_{Lf}}{dt} = k_1\omega\cos\omega t - k_2\omega\sin\omega t \tag{14.19.6}$$

Substituting the values from Eqs. (14.19.5) and (14.19.6) in Eq. (14.19.4) gives

$$L(k_1\omega\cos\omega t - k_2\omega\sin\omega t) + R(k_1\sin\omega t + k_2\cos\omega t)$$

$$= V_m\sin\omega t\cos\alpha_0 + V_m\cos\omega t\sin\alpha_0$$

$$(R k_1 - \omega L k_2)\sin\omega t + (R k_2 + \omega L k_1)\cos\omega t$$

$$= (V_m\cos\alpha_0)\sin\omega t + (V_m\sin\alpha_0)\cos\omega t \tag{14.19.7}$$

Comparing the coefficients of $\sin\omega t$ and $\cos\omega t$ on both the sides of Eq. (14.19.7) gives

$$Rk_1 - \omega L k_2 = V_m\cos\alpha_0 \tag{14.19.8}$$

$$\omega L k_1 + R k_2 = V_m\sin\alpha_0 \tag{14.19.9}$$

These equations are solved for k_1 and k_2.

$$k_1 = \frac{V_m}{R^2+\omega^2L^2}(R\cos\alpha_0 + \omega L\sin\alpha_0)$$

$$k_2 = \frac{V_m}{R^2+\omega^2L^2}(R\sin\alpha_0 - \omega L\cos\alpha_0)$$

Substitution of the values of k_1 and k_2 in Eq. (14.19.5) gives

$$i_{Lf}(t) = \frac{V_m}{R^2+\omega^2L^2}[R\cos\alpha_0 + \omega L\sin\alpha_0]\sin\omega t$$

$$+ (R \sin \alpha_0 - \omega L \cos \alpha_0) \cos \omega t] \qquad (14.19.10)$$

Let $\quad R \cos \alpha_0 + \omega L \sin \alpha_0 = A_1 \cos \phi \cos \alpha_0 + A_1 \sin \phi \sin \alpha_0$

where $\quad A_1 \cos \phi = R$

$A_1 \sin \phi = \omega L$

therefore, $\quad A_1 = \sqrt{R^2 + \omega^2 L^2}\,,\ \phi = \tan^{-1} \dfrac{\omega L}{R}$

and $\quad R \cos \alpha_0 + \omega L \sin \alpha_0 = A_1 \cos (\alpha_0 - \phi)$

$$= \sqrt{R^2 + \omega^2 L^2} \cos (\alpha_0 - \phi)$$

Let $\quad R \sin \alpha_0 - \omega L \cos \alpha_0 = A_2 \cos \Psi \sin \alpha_0 - A_2 \sin \Psi \cos \alpha_0$

where $\quad A_2 \cos \Psi = R$

$A_2 \sin \Psi = \omega L$

therefore, $\quad A_2 = \sqrt{R^2 + \omega^2 L^2}\,,\ \Psi = \tan^{-1} \dfrac{\omega L}{R}$

and $\quad R \sin \alpha_0 - \omega L \cos \alpha_0 = A_2 \sin (\alpha_0 - \Psi)$

$$= \sqrt{R^2 + \omega^2 L^2} \sin (\alpha_0 - \Psi)$$

It is seen that

$$A_1 = A_2 = \sqrt{R^2 + \omega^2 L^2}$$

and $\quad \Psi = \Phi = \tan^{-1} \dfrac{\omega L}{R}$

Therefore Eq. (14.19.10) can be written as

$$i_{Lf}(t) = \frac{V_m}{R^2 + \omega^2 L^2} \times \sqrt{R^2 + \omega^2 L^2}\, [\sin \omega t \cos (\alpha_0 - \Phi) + \cos \omega t \sin (\alpha_0 - \Phi)]$$

$$= \frac{V_m}{\sqrt{R^2 + \omega^2 L^2}} \sin (\omega t + \alpha_0 - \Phi) \qquad (14.19.11)$$

The complete response in given by

$$i_L(t) = i_{Ln}(t) + i_{Lf}(t)$$

$$= A e^{-\frac{R}{L} t} + \frac{V_m}{\sqrt{R^2 + \omega^2 L^2}} \sin (\omega t + \alpha_0 - \Phi) \qquad (14.19.12)$$

The value of unknown A is found by the application of initial condition $i_L(0_-) = 0$

By continuity equation

$$i_L(0_+) = i_L(0_-) = 0$$

Putting $t = 0_+$ in Eq. (14.19.12) we get

$$i_L(0_+) = A\,e^0 + \frac{V_m}{\sqrt{R^2\,\omega^2 L^2}} \sin(0 + \alpha_0 - \Phi)$$

$$0 = A + \frac{V_m}{\sqrt{R^2 + \omega^2 L^2}} \sin(\alpha_0 - \Phi)$$

$$A = -\frac{V_m}{\sqrt{R^2 + \omega^2 L^2}} \sin(\alpha_0 - \Phi)$$

Substituting this value of A in Eq. (14.19.12) we get

$$i_L(t) = \frac{V_m}{\sqrt{R^2 + \omega^2 L^2}} \sin(\omega t + \alpha_0 - \Phi) - \left[\frac{V_m}{\sqrt{R^2 + \omega^2 L^2}} \sin(\alpha_0 - \Phi)\right] e^{-\frac{R}{L}t]} \qquad (14.19.13)$$

From Eq. (14.19.13) it is seen that the natural response depends on the value of α_0, that is on the instant in the cycle at which the switch is closed.

Let $$\frac{V_m}{\sqrt{R^2 + \omega^2 L^2}} = I_m$$

Method of phasor transforms

The given circuit is the simple series RL circuit connected across a sinusoidal source of r.m.s. voltage equal to $V_m / \sqrt{2}$ volts.

Here $\mathbf{Z} = R + j X_L = R + j\omega L$

Circuit current

$$\mathbf{I_L} = \frac{\mathbf{V}}{\mathbf{Z}} = \frac{V \underline{/0^\circ}}{R + j\omega L}$$

$$= \frac{V \underline{/0^\circ}}{\sqrt{R^2 + \omega L^2}\, \underline{/\Phi}}$$

where $$\Phi = \tan^{-1}\frac{\omega L}{R}$$

therefore, $$\mathbf{I_L} = \frac{V \underline{/-\Phi}}{\sqrt{R^2 + \omega^2 L^2}}$$

The instantaneous value of current in the steady state is given by

$$i_{Lf}(t) = (\sqrt{2}\, I_L) \sin(\omega t + \alpha_0 - \Phi)$$

$$= \frac{V\sqrt{2}}{\sqrt{R^2 + \omega^2 L^2}} \sin(\omega t + \alpha_0 - \Phi)$$

$$= \frac{V_m}{\sqrt{R^2 + \omega^2 L^2}} \sin(\omega t + \alpha_0 - \Phi)$$

as obtained before in Eq. (14.19.11).

Special cases

Let us consider three cases:

Case 1 $t = 0$, and $\alpha_0 = 0$

The voltage is passing through zero and is positive going

$$i_L(t) = I_m \sin(\omega t - \Phi) + (I_m \sin \Phi)\, e^{-\frac{R}{L}t}$$

The peak value of the total current will be greater than its normal steady-state peak current I_m.

Case 2 $t = 0$, and $\alpha_0 = \Phi$

The voltage is passing through $V_m \sin \Phi$. Here $A = 0$ and $i_{Ln} = 0$. Hence under this condition, the natural response current (that is, the transient current) is zero. At this instant of closing the switch, the forced response (that is, the steady state current) will also be zero.

Case 3 $t = 0$, and $\alpha_0 = \Phi = \pm\frac{\pi}{2}$

$$i_{Ln} = -I_m \sin\left(\pm\frac{\pi}{2}\right) e^{-\frac{R}{L}t} = \pm I_m\, e^{-\frac{R}{L}t}$$

Thus, the natural response has the maximum possible initial value.

14.20 COMPLETE RESPONSE OF A SERIES RC NETWORK TO SINUSOIDAL INPUT

Fig. 14.20 shows a series RC network where a sinusoidal voltage is applied at $t = 0$ through a switch S. Consider the sinusoidal voltage

$$v = V_m \sin(\omega t + \alpha_0) \tag{14.20.1}$$

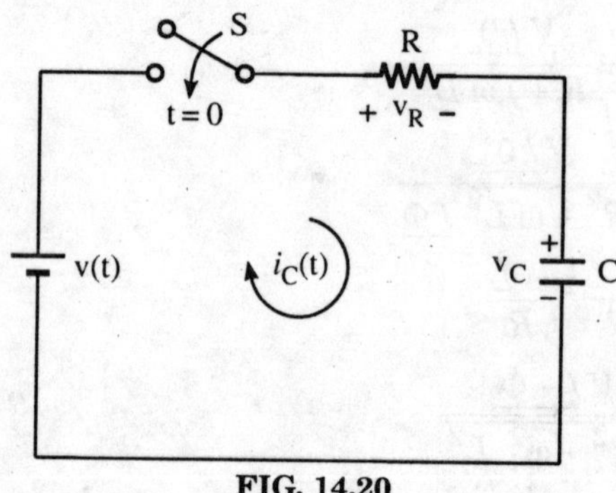

FIG. 14.20

By KVL

$$v_R + v_C = v \quad \text{for } t > 0 \tag{14.20.2}$$

or

$$RC\frac{dv_C}{dt} + v_C = V_m \sin(\omega t + \alpha_0)$$

Eq. (14.20.2) may also be written as follows:

$$R\, i_C + \frac{1}{C}\int i_C\, dt = V_m \sin(\omega t + \alpha_0) \tag{14.20.4}$$

Natural response or transient response

$$i_{cn} = A\, e^{st}$$

$$i_{cn} = A\, e^{-\frac{1}{RC}} \tag{14.20.5}$$

Forced response or steady-state response i_{cf}

The following methods are generally used to determine the forced response due to sinusoidal forcing function:

(*a*) Method of undetermined coefficients.

(*b*) Method of phasor transforms

Here we shall use the method of phasor transforms. The given circuit is the simple series RC circuit connected across a sinusoidal source of r.m.s voltage equal to $V_m / \sqrt{2}$ volts. Here

$$\mathbf{Z}_C = R - j\,X_C = R - j\,\frac{1}{\omega C}$$

$$= \left(\sqrt{R^2 + \frac{1}{\omega^2\, C^2}}\right) \underline{/ -\Psi}$$

where $\Psi = \tan^{-1} \dfrac{1}{\omega\, CR}$

The steady-state current in given by

$$\mathbf{I}_{cf} = \frac{\mathbf{V}}{\mathbf{Z}_C} = \frac{V\,\underline{/\,0^\circ}}{\left(\sqrt{R^2 + \frac{1}{\omega^2\, C^2}}\right) \underline{/ -\Psi}}$$

$$= \frac{V\,\underline{/\,\Psi}}{\sqrt{R^2 + \frac{1}{\omega^2\, C^2}}}$$

The instantaneous value of this current is given by

$$i_{cf} = (\sqrt{2}\, I_{cf})\, \sin\,(\omega\, t + \alpha_0 + \Psi) \tag{14.20.6}$$

Complete response $i_C\,(t)$

$$i_C\,(t) = i_{cn} + i_{cf}$$

$$= A\, e^{-\frac{t}{RC}} + (\sqrt{2}\, I_{cf})\, \sin\,(\omega\, t + \alpha_0 + \Psi) \tag{14.20.7}$$

Evaluation of constant A

The voltage on a capacitor cannot change instantaneously. That is,

$$v_C\,(0_+) = v_C\,(0_-)$$

Since $v_C\,(0_-) = 0$

$$v_C\,(0_+) = v_C\,(0_-) = 0$$

Applying the initial condition in Eq. (14.20.2) we get

$$R\, i_C\,(0_+) + v_C\,(0_+) = V_m\, \sin\,(0 + \alpha_0)$$

$$i_C(0_+) = \frac{V_m}{R}\sin\alpha_0 \tag{14.20.8}$$

Putting $t = 0_+$ in Eq. (14.20.7) we obtain

$$i_C(0_+) = A\,e^0 + \sqrt{2}\,I_{cf}\sin(0 + \alpha_0 + \Psi) \tag{14.20.9}$$

Combination of Eqs. (14.20.8) and (14.20.9) gives

$$\frac{V_m}{R}\sin\alpha_0 = A + \sqrt{2}\,I_{cf}\sin(\alpha_0 + \Psi)$$

$$A = \frac{V_m}{R}\sin\alpha_0 - \sqrt{2}\,I_{cf}\sin(\alpha_0 + \Psi) \tag{14.20.10}$$

Substitution of the value of A from Eq. (14.20.10) in Eq. (14.20.7) gives the total response as

$$i_C(t) = \left[\frac{V_m}{R}\sin\alpha_0 - I_m\sin(\alpha_0 + \Psi)\right]e^{-\frac{t}{RC}} + I_m\sin(\omega t + \alpha_0 + \Psi) \tag{14.20.11}$$

where $I_m = \sqrt{2}\,I_{cf}$

■ **EXAMPLE 14.17** In the series RL circuit shown in Fig. 14.19, a sinusoidal voltage $v = V_m \sin\omega t$ is applied at $t = 0$ through a switch S. The switch has been open for a long time. Use Laplace transform method to determine $i_L(t)$ for $t > 0$.

SOLUTION : When the switch is closed at $t = 0$, we have by KVL

$$v_R + v_L = v$$

$$Ri_L + L\frac{di_L}{dt} = V_m\sin\omega t \quad \text{for } t > 0 \tag{E 14.17.1}$$

Taking Laplace transform on both the sides of Eq. (E 14.17.1) we get

$$R\,I_L(s) + L\,[\,s\,I_L(s) - i_L(0_-)\,] = V_m \times \frac{\omega}{s^2 + \omega^2}$$

Since $i_L(0_-) = 0$,

$$(sL + R)\,I_L(s) = \frac{\omega V_m}{s^2 + \omega^2}$$

$$I_L(s) = \frac{\omega V_m}{(s - j\omega)(s + j\omega)(R + sL)} \tag{E 14.17.2}$$

$$= \frac{V_m}{2j\,(s - j\omega)(R + j\omega L)} - \frac{V_m}{2j\,(s + j\omega)(R - j\omega L)} + \frac{(V_m)\,\omega L}{(R^2 + \omega^2 L^2)\left(s + \frac{R}{L}\right)} \tag{E 14.17.3}$$

Taking inverse Laplace transform we get

$$i_L(t) = \frac{-jV_m}{2(R + j\omega L)}e^{j\omega t} + \frac{j\,V_m}{2(R - j\omega L)}e^{-j\omega t} + \frac{\omega L V_m}{R^2 + \omega^2 L^2}e^{-\frac{Rt}{L}}$$

$$= \frac{V_m}{2(R^2+\omega^2 L^2)} [-j(R-j\omega L)e^{j\omega t} + j(R+j\omega L)e^{-j\omega t}] + \frac{\omega L V_m}{R^2+\omega^2 L^2} e^{-\frac{Rt}{L}}$$

$$= \frac{V_m}{R^2+\omega^2 L^2}\left[R\left(\frac{e^{j\omega t}-e^{-j\omega t}}{2j}\right) - \omega L\left(\frac{e^{j\omega t}+e^{-j\omega t}}{2}\right)\right] + \frac{\omega L V_m}{R^2+\omega^2 L^2} e^{-\frac{Rt}{L}}$$

$$= \frac{V_m}{(R^2+\omega^2 L^2)} (R \sin\omega t - \omega L \cos\omega t)) + \frac{\omega L V_m}{R^2+\omega^2 L^2} e^{-\frac{Rt}{L}} \qquad \text{(E 14.17.4)}$$

Let $\quad R = A_1 \cos\phi$

$\omega L = A_1 \sin\phi$

therefore, $\quad A_1 = \sqrt{R^2+\omega^2 L^2}$

$$\tan\phi = \frac{\omega L}{R}, \ \phi = \tan^{-1}\frac{\omega L}{R}$$

$$R\sin\omega t - \omega L\cos\omega t = A_1(\sin\omega t\cos\phi - \cos\omega t\sin\phi)$$

$$= A_1 \sin(\omega t - \phi) = \sqrt{(R^2+\omega^2 L^2)} \sin(\omega t - \phi)$$

therefore,
$$i_L(t) = \frac{V_m}{R^2+\omega^2 L^2} \times \sqrt{(R^2+\omega^2 L^2)} \sin(\omega t-\phi) + \frac{\sqrt{(R^2 \ \omega^2 L^2)}}{R^2+\omega^2 L^2}(V_m \sin\phi)(e^{-\frac{R}{L}t})$$

$$= \frac{V_m}{\sqrt{R^2+\omega^2 L^2}} \sin(\omega t-\phi) + \frac{V_m \sin\phi}{\sqrt{R^2+\omega^2 L^2}} e^{-\frac{R}{L}t} \qquad \text{(E 14.17.5)}$$

■ **EXAMPLE 14.18** Determine the capacitor current $i_C(t)$ for $t > 0$ in the network shown in Fig. 14.20 when a sinusoidal voltage $v = V_m \sin\omega t$ is applied at $t = 0$ through a switch S. Use the Laplace transform method.

SOLUTION : At $t > 0$, when the switch is closed, by KVL

$$v_R + v_C = v$$

$$R\, i_C + \frac{1}{C}\int_0^t i_C\, dt = V_m \sin\omega t \quad \text{for } t > 0 \qquad \text{(E 14.18.1)}$$

Laplace transformation of Eq. (E 14.18.1) gives

$$R\, I_C(s) + \frac{1}{C}\left[\frac{1}{s} I_C(s) + \frac{q_C(0_-)}{s}\right] = \frac{V_m \omega}{s^2+\omega^2}$$

We shall assume that the circuit is initially relaxed, that is,

$$q_C(0_-) = 0.$$

therefore,
$$\left(R + \frac{1}{sC}\right) I_C(s) = \frac{\omega V_m}{s^2+\omega^2}$$

$$I_C(s) = \frac{\omega V_m s}{R\left(s + \frac{1}{RC}\right)(s^2 + \omega^2)} \tag{E 14.18.2}$$

$$= \frac{A_1}{s + \frac{1}{RC}} + \frac{A_2}{s + j\omega} + \frac{A_3}{s - j\omega} \tag{E 14.18.3}$$

$$A_1 = \frac{-V_m \frac{\omega}{RC}}{R\left(\omega^2 + \frac{1}{R^2 C^2}\right)} = \frac{-\omega C V_m}{1 + \omega^2 C^2 R^2}$$

$$A_2 = \frac{\omega C V_m (1 + j\omega C R)}{2(1 + \omega^2 C^2 R^2)}$$

$$A_3 = A_2^* = \frac{\omega C V_m (1 - j\omega C R)}{2(1 + \omega^2 C^2 R^2)}$$

$$I_C(s) = -\frac{\omega C V_m}{1 + \omega^2 C^2 R^2} \times \frac{1}{\left(s + \frac{1}{RC}\right)} + \frac{\omega C V_m (1 + j\omega C R)}{2(1 + \omega^2 C^2 R^2)(s + j\omega)}$$

$$+ \frac{\omega C V_m (1 - j\omega C R)}{2(1 + \omega^2 C^2 R^2)(s - j\omega)} \tag{E 14.18.4}$$

On taking the inverse Laplace transform of Eq. (E 14.18.4) we get

$$i_C(t) = \frac{\omega C V_m}{1 + \omega^2 C^2 R^2}\left[-e^{-\frac{t}{RC}} + \frac{1}{2}(1 + j\omega C R)\, e^{-j\omega t} + \frac{1}{2}(1 - j\omega C R)\, e^{j\omega t}\right]$$

$$= V_m\left(R^2 + \frac{1}{\omega^2 C^2}\right)\left[-\frac{1}{\omega C} e^{-\frac{t}{RC}} + \frac{1}{\omega C}\left(\frac{e^{j\omega t} + e^{-j\omega t}}{2}\right) + R\frac{1}{2j}\left(e^{j\omega t} - e^{-j\omega t}\right)\right]$$

$$= \frac{V_m}{\left(R^2 + \frac{1}{\omega^2 C^2}\right)}\left(\frac{1}{\omega C}\cos\omega t + R\sin\omega t\right) - \frac{\frac{V_m}{\omega C}}{R^2 + \frac{1}{\omega^2 C^2}}\, e^{-\frac{t}{RC}} \tag{E 14.18.5}$$

Let

$$R = A\cos\phi$$

$$\frac{1}{\omega C} = A\sin\phi$$

$$A = \sqrt{R^2 + \frac{1}{\omega^2 C^2}}, \quad \phi = \tan^{-1}\frac{1}{\omega C R}$$

$$\sin\phi = \frac{1}{\omega C\sqrt{R^2 + \frac{1}{\omega^2 C^2}}}$$

Therefore, Eq. (E 14.18.5) can be written as

$$i_C(t) = \frac{V_m}{Z}\sin(\omega t + \phi) - \left(\frac{V_m}{Z}\sin\phi\right)e^{-\frac{t}{RC}} \quad \text{(E 14.18.6)}$$

where $Z = \sqrt{R^2 + \dfrac{1}{\omega^2 C^2}}$

■ **EXAMPLE 14.19** A series circuit of resistance 10 Ω and inductance 0.1 H is connected across a 50 Hz sinusoidal voltage of maximum value 200 V. (*a*) Find an expression for the value of current at any instant after the voltage is applied, assuming that the voltage is zero at the instant of application. (*b*) Calculate the value of current 0.02 s after switching on.

SOLUTION : Since the voltage is zero at switching

$$v = V_m \sin\omega t$$

$$\mathbf{Z} = R + jX_L$$

$$= 10 + j\,2\pi \times 50 \times 0.1$$

$$= 10 + j\,31.4 = 33\,\underline{/72.3^\circ}\ \Omega$$

$$I_m = \frac{V_m}{Z} = \frac{200}{33} = 6.07\text{ A}$$

(*a*) Natural response

$$i_{Ln} = A\,e^{-\frac{R}{L}t} = A\,e^{-100t}$$

Forced response

$$i_{Lf} = I_m \sin(\omega t - \phi)$$

$$= 6.07\sin(314t - 72.3^\circ)$$

Complete response

$$i(t) = i_{Ln} + i_{Lf}$$

$$= A\,e^{-100t} + 6.07\sin(314t - 72.3^\circ)$$

At $t = 0$, $i = 0$

Therefore

$$0 = A\,e^0 + 6.07\sin(-72.3^\circ)$$

$$A = 6.07\sin 72.3^\circ = 5.78$$

therefore, $i(t) = 5.78\,e^{-100t} + 6.07\sin(314t - 72.3^\circ)$

(*b*) At $t = 0.02$ s

$$\omega t = 2 \times 50 \times 0.02 \text{ radians}$$

$$= 2\pi \text{ radians} = 360^\circ$$

therefore, $\sin(\omega t - 72.3^\circ) = \sin(360^\circ - 72.3^\circ)$

$$= -\sin 72.3^\circ = -0.9527$$

The value of current after 0.02 s is

$$i = 5.78\, e^{-100 \times 0.02} + 6.07\,(-0.9527)$$

$$= 5.78 \times 0.1353 - 5.78 = -5 \text{ A}$$

14.21 RESPONSE OF A SERIES RL CIRCUIT TO EXPONENTIAL EXCITATION

Fig. 14.21 shows a series RL circuit with exponential excitation given by

$$v_S(t) = V e^{-\alpha t} \quad \text{for } t > 0$$

where V and α are constants. The initial current through the inductor is zero, that is, $i_L(0_-) = 0$

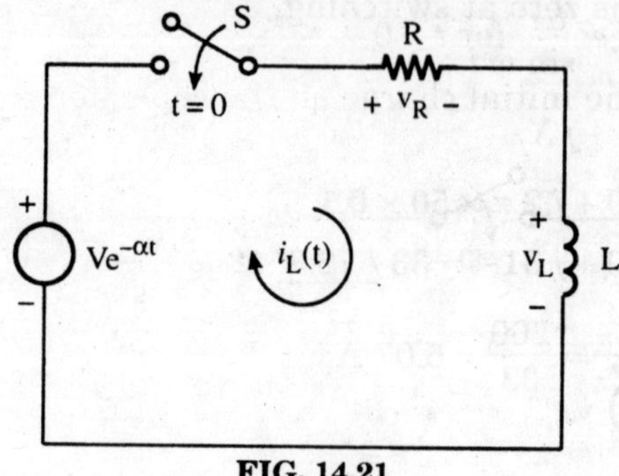

FIG. 14.21

By KVL

$$v_L + v_R = v_s \quad \text{for } t > 0$$

$$L \frac{di_L}{dt} + R\, i_L = V e^{-\alpha t} \tag{14.21.1}$$

The Laplace transform of both the sides gives

$$L\,[\,s\, I_L(s) - i_L(0_-)\,] + R\, I_L(s) = \frac{V}{s+\alpha}$$

Since $i_L(0_-) = 0$,

$$I_L(s) = \frac{v}{(s+\alpha)\,(s\,L + R)} = \frac{V}{L}\left[\frac{1}{(s+\alpha)\left(s + \frac{R}{L}\right)}\right] \tag{14.21.2}$$

(*a*) If $\alpha \neq \dfrac{R}{L}$

$$I_L(s) = \frac{V}{L}\left[\frac{1}{(s+\alpha)\left(s+\frac{R}{L}\right)}\right] = \frac{V}{(R - \alpha\, L)}\left[\frac{1}{s+\alpha} - \frac{1}{s + \frac{R}{L}}\right] \tag{14.21.3}$$

By taking the inverse Laplace transform of Eq. (14.21.3) we get

$$i_L(t) = \frac{V}{(R - \alpha\, L)}\,[\,e^{-\alpha t} - e^{-(R/L)\,t}\,] \tag{14.21.4}$$

(*b*) If $\alpha = \frac{R}{L}$, Eq. (14.21.2) becomes

$$I_L(s) = \frac{V}{L(s+\alpha)^2} \tag{14.21.5}$$

This inverse Laplace transform of Eq. (14.21.5) gives

$$i_L(t) = \frac{V}{L} t\, e^{-\alpha t} \tag{14.21.6}$$

14.22 RESPONSE OF A SERIES RC CIRCUIT TO EXPONENTIAL EXCITATION

Fig. 14.22 shows a series RC circuit excited by an exponential function given by

$$v_S(t) = V e^{-\alpha t} \text{ for } t > 0$$

where V and α are constants. The initial charge on the capacitor is zero, that is., $q(0_-) = 0$.

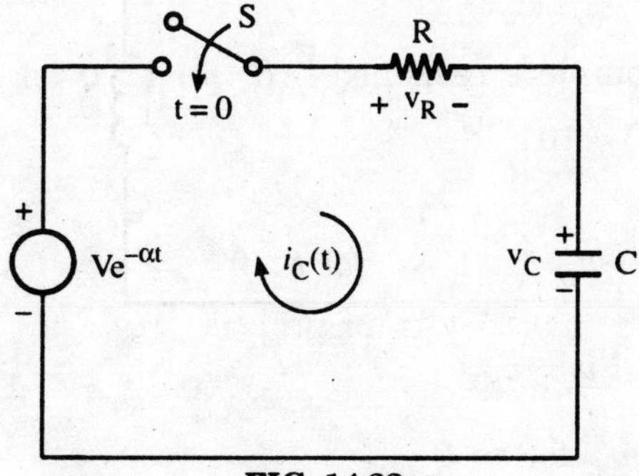

FIG. 14.22

By KVL

$$v_R + v_C = v_s \text{ for } t > 0$$

$$R\, i_C + \frac{1}{C}\int_0^t i_C(t)\, dt = V e^{-\alpha t} \tag{14.22.1}$$

By taking the Laplace transform of both the sides of Eq. (14.22.1) we get

$$R\, I_C(s) + \frac{1}{C}\left[\frac{I_C(s)}{s} + \frac{q(0_-)}{s}\right] = \frac{V}{s+\alpha}$$

since $q(0_-) = 0$

$$I_C(s) = \frac{V}{R}\left[\frac{s}{(s+\alpha)\left(s+\frac{1}{RC}\right)}\right] \tag{14.22.2}$$

(*a*) If $\alpha \neq \frac{1}{RC}$ Eq. (14.22.2) becomes

$$I_C(s) = \frac{V}{R\left(\alpha - \frac{1}{RC}\right)}\left[\frac{\alpha}{s+\alpha} - \frac{1}{RC\left(s+\frac{1}{RC}\right)}\right] \tag{14.22.3}$$

By taking the inverse Laplace transform of Eq. (14.22.3) we obtain

$$i_C(t) \frac{V}{R\left(\alpha - \frac{1}{RC}\right)}\left[\alpha e^{-\alpha t} - \frac{1}{RC} e^{-\frac{t}{RC}}\right] \tag{14.22.4}$$

(*b*) If $\alpha = \frac{1}{RC}$, Eq. (14.22.2) becomes

$$I_C(s) = \frac{V}{R}\frac{s}{(s+\alpha)^2}$$

$$= \frac{V}{R}\left[\frac{1}{s+\alpha} - \frac{\alpha}{(s+\alpha)^2}\right] \tag{14.22.5}$$

The inverse Laplace transform of Eq. (14.22.5) gives

$$i_C(t) = \frac{V}{R}[e^{-\alpha t} - \alpha t e^{-\alpha t}] \tag{14.22.6}$$

■ **EXAMPLE 14.20** Find the complete response $i_L(t)$ for $t > 0$ for the network shown in Fig. 14.7 if $i_L(0\_) = 0$ and $v_S(t) = 10 e^{-6t}$.

SOLUTION : By KVL

$$v_L + v_R = v_S \qquad \text{for } t > 0$$

$$L\frac{di_L}{dt} + R i_L = v_S$$

$$5\frac{di_L}{dt} + 50 i_L = 10 e^{-6t}$$

$$\frac{di_L}{dt} + 10 i_L = 2 e^{-6t} \quad \text{for } t > 0 \tag{E 14.20.1}$$

Determination of natural response

As in Example 14.3, the natural response is

$$i_{Ln} = A e^{-10t} \tag{E 14.20.2}$$

Determination of forced response

Here the driving source is an exponential function. Assume that $i_{Lf}(t)$ is a linear sum of the forcing function $10 e^{-6t}$ and all its derivatives. Thus

$$i_{Lf}(t) = k_1 e^{-6t} + k_2(-6) e^{-6t} + k_3(36) e^{-6t} + \ldots.$$

$$= (k_1 - 6k_2 + 36k_3 + \ldots\ldots) e^{-6t} = k_0 e^{-6t} \tag{E 14.20.3}$$

We observe that the forced response is simply a single term because all derivatives of the exponential function has the same time dependence as the original forcing function.

$$\frac{d}{dt} i_{Lf} = -6 k_0 e^{-6t}$$

We have

$$\frac{di_{Lf}}{dt} + 10\, i_{Lf} = 2\, e^{-6t} \text{ for } t > 0 \tag{E 14.20.4}$$

$$-6\, k_0\, e^{-6t} + 10\, k_0\, e^{-6t} = 2\, e^{-6t}$$

$$4\, k_0 = 2,\; k_0 = 0{\cdot}5$$

therefore,

$$i_{Lf}(t) = k_0\, e^{-6t}$$

$$= 0{\cdot}5\, e^{-6t}$$

The complete solution is

$$i_L(t) = i_{Ln}(t) + i_{Lf}(t)$$

or

$$i_L(t) = A\, e^{-10t} + 0{\cdot}5\, e^{-6t}$$

Applying the initial conditions

$$i_L(0_+) = i_L(0_-) = 0$$

$$0 = A\, e^{-10 \times 0} + 0{\cdot}5\, e^{-6 \times 0}$$

$$A + 0{\cdot}5 = 0,\quad A = -\,0{\cdot}5$$

Therefore the complete response is given by

$$i_L(t) = -\,0{\cdot}5\, e^{-10t} + 0{\cdot}5\, e^{-6t}$$

$$= 0{\cdot}5\,(e^{-6t} - e^{-10t}) \text{ for } t > 0$$

or

$$i_L(t) = 0{\cdot}5\,(e^{-6t} - e^{-10t})\, u_{-1}(t)$$

Alternative method

By taking the Laplace transform of Eq.(E 14.20.1) we get

$$[\,s\, I_L(s) - i_L(0_-)\,] + 10\, I_L(s) = \frac{2}{s+6}$$

Since $i_L(0_-) = 0$,

$$I_L(s) = \frac{2}{(s+6)(s+10)} = \frac{1}{2}\left[\frac{1}{s+6} - \frac{1}{s+10}\right]$$

$$i_L(t) = \frac{1}{2}(e^{-6t} - e^{-10t}) \text{ for } t > 0$$

14.23 SPECIAL CASE OF EXPONENTIAL INPUT

If the exponential forcing function is identical to the exponential term in the natural response, the forced response has a modified form, which can be derived as follows :

$$\frac{dx_f}{dt} + \alpha\, x_f = k_1\, e^{-\alpha t}$$

$$e^{\alpha t}\frac{dx_f}{dt}+\alpha\, e^{\alpha t}\, x_f = k_1$$

$$\frac{d}{dt}(x_f e^{\alpha t}) = k_1$$

Integration on both the sides gives

$$x_f e^{\alpha t} = k_1 t + k_2$$

$$x_f(t) = e^{-\alpha t}(k_1 t + k_2)$$

EXAMPLE 14.21. Find the complete response $i_L(t)$ for $t>0$ for the network shown in Fig. 14.7 if $i_L(0_-)=0$ and $v_s(t) = 20\, e^{-10t}$

SOLUTION : By KVL

$$v_L + v_R = v_S \text{ for } t > 0$$

$$L\frac{di_L}{dt} + R\, i_L = v_S$$

$$5\frac{di_L}{dt} + 50\, i_L = 20\, e^{-10t}$$

$$\frac{di_L}{dt} + 10\, i_L = 4\, e^{-10t} \text{ for } t>0 \tag{E 14.21.1}$$

As in Example 14.3, the natural response is

$$i_{Ln} = A\, e^{-10t} \tag{E 14.21.2}$$

In this example the exponential forcing function is identical to the exponential term in the natural response. Therefore we shall assume that the forced response is of the form

$$i_{Lf}(t) = e^{-10t}(k_1 t + k_2) \tag{E 14.21.3}$$

$$\frac{di_{Lf}}{dt} = e^{-10t}(k_1 + 0) + (k_1 t + k_2)(-10\, e^{-10t})$$

$$= (k_1 - 10\, k_2)\, e^{-10t} - 10\, k_1\, t\, e^{-10\, t} \tag{E 14.21.4}$$

Substituting the values from Eqs. (E 14.21.3) and (E 14.21.4) in Eq. (E 14.21.1) we get

$$(k_1 - 10\, k_{2)}\, e^{-10\, t} - 10\, k_1\, t\, e^{-10t} + 10\, e^{-10t}\, (k_1\, t + k_{2)} = 4\, e^{-10t}$$

$$k_1\, e^{-10t} = 4\, e^{-10t},\; k_1 = 4$$

therefore, $$i_{Lf}(t) = (4\, t + k_2)\, e^{-10t} \tag{E 14.21.5}$$

which is obtained by substituting the value of k_1 in Eq. (E 14.21.3)

The complete response

$$i_L(t) = i_{Ln}(t) + i_{Lf}(t)$$

$$i_L(t) = A\,e^{-10t} + (4t + k_2)\,e^{-10t} \qquad \text{(E 14.21.6)}$$

Applying the initial condition,

$$i_L(0_+) = i_L(0_-) = 0$$

$$0 = A\,e^0 + (4 \times 0 + k_2)\,e^0$$

$$0 = A + k_2 \qquad \text{(E 14.21.7)}$$

Combination of Eqs. (E 14.21.6) and (E 14.21.7) gives

$$i_L(t) = (A + k_2)\,e^{-10t} + 4\,t\,e^{-10t}$$

$$= 0 \times e^{-10t} + 4\,t\,e^{-10t}$$

or

$$i_L(t) = 4\,t\,e^{-10t} \text{ for } t > 0$$

Alternative method

Laplace transform of Eq. (E 14.21.1) gives

$$[\,s\,I_L(s) - i_L(0_-)\,] + 10\,I_L(s) = \frac{4}{s+10}$$

Since $i_L(0_-) = 0$

$$I_L(s) = \frac{4}{(s+10)\,(s+10)} = \frac{4}{(s+10)^2}$$

The inverse Laplace of this relation gives

$$i_L(t) = 4\,t\,e^{-10t} \quad \text{for } t > 0$$

14.24 RESPONSE OF A SERIES RL CIRCUIT TO AN IMPULSE INPUT

Consider a series RL circuit with an impulse input given by

$$v_S(t) = A\,u_0(t)$$

This circuit is shown in Fig. 14.23.

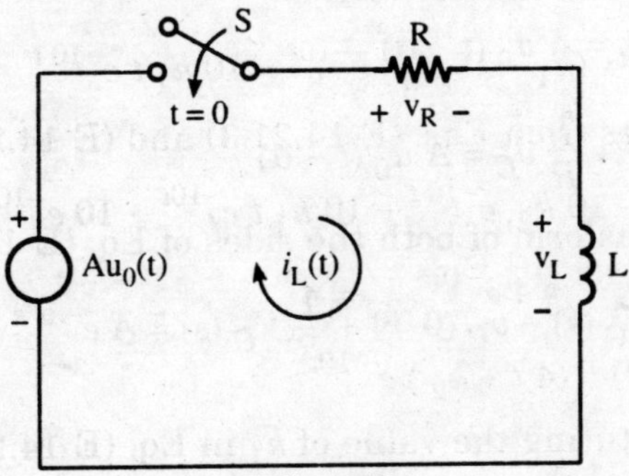

FIG. 14.23

By KVL,

$$v_L + v_R = v_S \qquad \text{for } t > 0$$

$$L\frac{di_L}{dt} + R\,i_L = A\,u_0\,(t) \qquad (14.24.1)$$

Taking the Laplace transform we get

$$L\,[s\,I_L\,(s) - i_L\,(0\_)] + R\,I_L\,(s) = A \times 1$$

Since $i_L\,(0\_) = 0$

$$s\,L\,I_L\,(s) + RI_L\,(s) = A$$

$$I_L\,(s) = \frac{A}{R + s\,L} = \frac{A}{L\left(s + \frac{R}{L}\right)} \qquad (14.24.2)$$

Taking the inverse Laplace transform we get

$$i_L\,(t) = \frac{A}{L}\,e^{-\frac{R}{L}t} \text{ for } t > 0$$

or

$$i_L\,(t) = \left(\frac{A}{L}\,e^{-\frac{R}{L}t}\right) u_{-1}\,(t) \qquad (14.24.3)$$

■ **EXAMPLE 14.22** Determine the voltage across the capacitor of a parallel RC circuit of Fig. 14.24 where an impulse current $i_S\,(t) = A\,u_0\,(t - \alpha)$ excites the circuit if $v_C\,(0\_) = 0$.

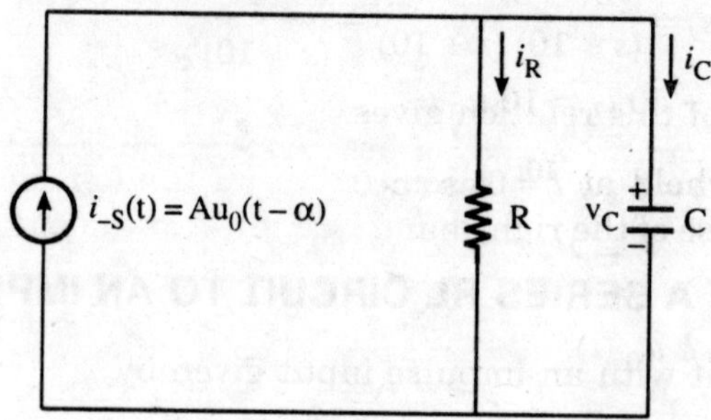

FIG. 14.24

SOLUTION : The KCL equation of the network is

$$i_C + i_R = A\,u_0\,(t - \alpha)$$

$$C\frac{dv_C}{dt} + \frac{1}{R}v_C = A\,u_0\,(t - \alpha) \qquad \text{(E 14.22.1)}$$

Taking the Laplace transform of both the sides of Eq. (E 14.22.1) we get

$$C\,[s\,V_C\,(s) - v_C\,(0\_)] + \frac{1}{R}V_C\,(s) = A\,e^{-\alpha s}$$

Since $v_C\,(0\_) = 0$

$$\left(s\,C + \frac{1}{R}\right)V_C\,(s) = A\,e^{-\alpha s}$$

$$V_C(s) = \frac{A\,e^{-\alpha s}}{s\,C + \frac{1}{R}}$$

or
$$V_C(s) = \frac{A\,e^{-\alpha s}}{C\left(s + \frac{1}{RC}\right)} \quad \text{(E 14.22.2)}$$

Taking the inverse Laplace transform of both the sides of Eq. (E 14.22.2) gives

$$v_C(t) = \frac{A}{C}\,e^{-\frac{t}{RC}} \quad \text{for } t > 0$$

or
$$v_C(t) = \left(\frac{A}{C}\,e^{-\frac{t}{RC}}\right) u_{-1}(t - \alpha)$$

■ **EXAMPLE 14.23.** Find the complete response $i_L(t)$ for the network shown in Fig. 14.7 if the circuit is initially at rest and $v_S(t) = 10\,u_0(t)$.

SOLUTION : If a KVL equation is written for $t > 0$, the driving function will not appear, since $10\,u_0(t)$ occurs at $t = 0$. Therefore, an equation that is valid for all time must be written.

$$L\frac{di_L}{dt} + R \cdot i_L = v_S \qquad \text{for all } t$$

$$5\frac{di_L}{dt} + 50\,i_L = 10\,u_0(t) \qquad \text{for all } t \quad \text{(E 14.23.1)}$$

For Eq. (E 14.23.1) to hold at $t = 0$; some term on the left-hand side must contain an impulse to balance the impulse of the right-hand side. If we assume that i_L contains an impulse at $t = 0$

$$i_L = k\,u_0(t) \quad \text{(E 14.23.2)}$$

therefore,
$$\frac{di_L}{dt} = k\,u_{+1}(t) \quad \text{(E 14.23.3)}$$

Since there is no $u_{+1}(t)$ term on the right-hand side of Eq. (E 14.23.1), our assumption that i_L contains an impulse at $t = 0$ is not true. Therefore, i_L does not contain an impulse, but it does contain a discontinuity at $t = 0$.

Since Eq. (E 14.23.1) is written to be valid for all values of time, the assumed solution must also be valid for all t. Therefore the assumed solution may be written as

$$i_L(t) = k\,e^{-10t} \quad \text{(E 14.23.4)}$$

To find the derivative of i_L, we must be careful. At $t = 0_-$, $i_L(0_-) = 0$, but at $t = 0_+$, $i_L(0_+) = k$. Therefore at $t = 0$, i_L has a discontinuity equal to k, which means the derivative of i_L contains an impulse at $t = 0$ of magnitude k in addition to the derivative in the usual sense. That is,

$$\frac{di_L}{dt} = k\,u_0(t) - 10\,k\,e^{-10\,t} \quad \text{(E 14.23.5)}$$

Substituting the values from Eqs. (E 14.23.4) and (E 14.23.5) in Eq. (E 14.23.1) gives

$$5\,[\,k\,u_0\,(t) - 10\,k\,e^{-10t}\,] + 50\,k\,e^{-10t} = 10\,u_0\,(t) \qquad \text{(E 14.23.6)}$$

The exponential terms cancel, leaving

$$5\,k\,u_0\,(t) = 10\,u_0\,(t)$$

therefore, $k = 2$

Substituting the values of k in Eq. (E 14.23.4) gives

$$i_L\,(t) = 2\,e^{-10t} \text{ for } t > 0$$

To show that this equation has meaning only for positive time, we write

$$i_L\,(t) = 2\,e^{-10t}\,u_{-1}\,(t) \qquad \text{for all } t$$

Alternative method

Laplace transform of Eq. (E 14.22.1) gives

$$5\,[s\,I_L\,(s) - i_L\,(0\_)\,] + 50\,I_L\,(s) = 10 \times 1$$

Since $i_L\,(0\_) = 0$

$$s\,I_L\,(s) + 10\,I_L\,(s) = 2$$

$$I_L\,(s) = \frac{2}{s + 10}$$

Inverse Laplace transform of this relation gives

$$i_L\,(t) = 2\,e^{-10t} \text{ for } t > 0$$

or

$$i_L\,(t) = 2\,e^{-10t}\,u_{-1}\,(t) \qquad \text{for all } t$$

■ **EXAMPLE 14.24.** Find the complete response $i_L\,(t)$ for $t > 0$ for the network shown in Fig. 14.7 if $i_L\,(0\_) = 0$ and $v_S\,(t) = 25\,t$.

SOLUTION : By KVL

$$v_L + v_R = v_S \text{ for } t > 0$$

$$L\frac{di_L}{dt} + R\,i_L = v_S$$

$$5\frac{di_L}{dt} + 50\,i_L = 25\,t$$

$$\frac{di_L}{dt} + 10\,i_L = 5\,t \text{ for } t > 0 \qquad \text{(E 14.24.1)}$$

Determination of natural response

Here we have to solve the homogeneous (source-free) equation

$$\frac{di_{Ln}}{dt} + 10\,i_{Ln} = 0 \qquad \text{(E 14.24.2)}$$

We have already found the solution of this equation as

$$i_{Ln} = A\,e^{-10t} \qquad \text{(E 14.24.3)}$$

Determination of forced response

Let $i_{Lf}(t)$ be the forced response. Here the forcing function is a ramp input $v_S(t) = 25\,t$. The time function $i_{Lf}(t)$ will satisfy the equation

$$\frac{di_{Lf}}{dt} + 10\,i_{Lf} = 5\,t \qquad \text{(E 14.24.4)}$$

We shall assume the forced solution $i_{Lf}(t)$ as linear sum of the forcing function and all its nonzero derivatives. That is,

$$i_{Lf}(t) = k_1\,t + k_2 \qquad \text{(E 14.24.5)}$$

Therefore
$$\frac{di_{Lf}}{dt} = k_1 \qquad \text{(E 14.24.6)}$$

Substitution of Eqs. (E 14.24.5) and (E 14.24.6) in Eq. (E 14.24.4) gives

$$k_1 + 10(k_1\,t + k_2) = 5\,t$$

$$10\,k_1\,t + (k_1 + 10\,k_2) = 5\,t$$

Equating coefficients of t^1 and t^0 on both sides and solving gives

$$10\,k_1 = 5,\ k_1 = 0{\cdot}5$$

and
$$k_1 + 10\,k_2 = 0,\ k_2 = -\,0{\cdot}05$$

Substituting the values of k_1 and k_2 in Eq. (E 14.24.5) gives

$$i_{Lf} = 0{\cdot}5\,t - 0{\cdot}05$$

Determination of complete response

$$i_L(t) = i_{Ln}(t) + i_{Lf}(t)$$

$$= A\,e^{-10t} + 0{\cdot}5\,t - 0{\cdot}05 \qquad \text{(E 14.24.7)}$$

Applying initial conditions,

$$i_L(0_+) = i_L(0_-) = 0$$

$$i_L(0) = A\,e^{-10\times 0} + 0{\cdot}5 \times 0 - 0{\cdot}05$$

$$0 = A - 0{\cdot}05,\ A = 0{\cdot}05$$

therefore,
$$i_L(t) = 0{\cdot}05\,e^{-10t} + 0{\cdot}5\,t - 0{\cdot}05 \text{ for } t > 0$$

or
$$i_L(t) = (0{\cdot}05\,e^{-10t} + 0{\cdot}5\,t - 0{\cdot}05)\,u_{-1}(t)$$

Alternative method

Laplace transform of Eq. (E 14.24.1) gives

$$[s\,I_L(s) - i_L(0_-)] + 10\,I_L(s) = \frac{5}{s^2}$$

Since $i_L(0\_) = 0$

$$(s+10)\, I_L(s) = \frac{5}{s^2}$$

$$I_L(s) = \frac{5}{(s+10)\, s^2}$$

$$= \frac{5}{100}\left(\frac{1}{s+10}\right) + \frac{5}{10\, s^2} - \frac{5}{100\, s}$$

Inverse Laplace transform of this relation gives

$$i_L(t) = (0{\cdot}05\, e^{-10t} + 0{\cdot}5\, t - 0{\cdot}05) \text{ for } t > 0$$

or $$i_L(t) = (0{\cdot}05\, e^{-10t} + 0{\cdot}5\, t - 0{\cdot}05)\, u_{-1}(t) \text{ for all } t$$

14.25 RESPONSE OF RL CIRCUIT TO PULSE EXCITATION

Fig. 14.25 shows a series RL circuit excited by a pulse of width a and magnitude A given by

$$v_S(t) = A\,[\,u_{-1}(t) - u_{-1}(t-a)\,]$$

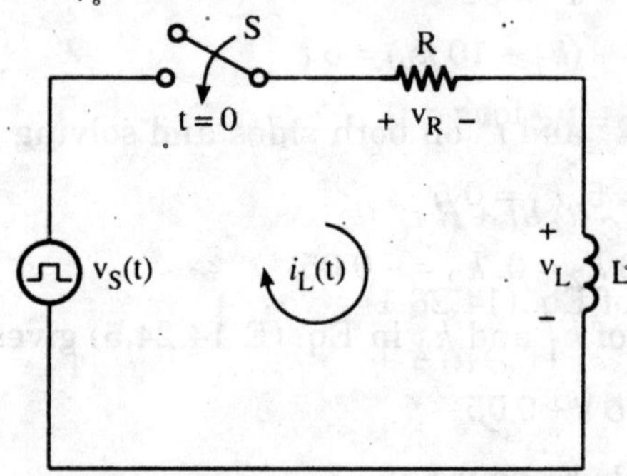

FIG. 14.25 Series RL circuit with pulse excitation

By KVL

$$v_L + v_R = v_S \text{ for } t > 0$$

$$L\frac{di_L}{dt} + R\, i_L = A\,[u_{-1}(t) - u_{-1}(t-a)] \tag{14.25.1}$$

Laplace transformation of this equation gives

$$L\,[s\, I_L(s) - i_L(0\_)\,] + R\, I_L(s) = A\left[\frac{1}{s} - \frac{1}{s}e^{-as}\right]$$

if $i_L(0\_) = 0$,

$$I_L(s) = A\,\frac{1-e^{-as}}{s\,(R+s\,L)} = \frac{A}{L}\left[\frac{1-e^{-as}}{s\left(s+\dfrac{R}{L}\right)}\right] = \frac{A}{R}\left[\frac{1}{s} - \frac{1}{s+\dfrac{R}{L}} - \frac{e^{-as}}{s} + \frac{e^{-as}}{s+\dfrac{R}{L}}\right]$$

Inverse Laplace transform of this equation gives

$$i_L(t) = \frac{A}{R}(1 - e^{-\frac{Rt}{L}})\, u_{-1}(t) - \frac{A}{R}\left[1 - e^{-\frac{R}{L}(t-a)}\right] u_{-1}(t-a)$$

14.26 RESPONSE OF RC CIRCUIT TO PULSE EXCITATION

Fig. 14.26 shows a series RL circuit excited by a pulse is given by

$$v_S(t) = A\,[u_{-1}(t) - u_{-1}(t-a)]$$

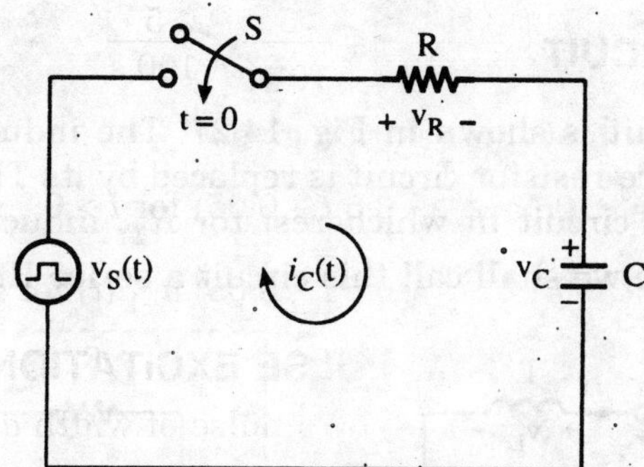

FIG. 14.26 Series RC circuit with pulse excitation.

BY KVL,

$$v_C + v_R = v_S \text{ for } t > 0$$

$$\frac{1}{C}\int_0^{\infty} i_C(t)\,dt + R\,i_C(t) = A\,[u_{-1}(t) - u_{-1}(t-a)] \tag{14.26.1}$$

Laplace transformation of Eq. (14.26.1) gives

$$\frac{1}{C}\left[\frac{1}{s} I_C(s) - \frac{q(0_-)}{s}\right] + R\,I_C(s) = A\left[\frac{1}{s} - \frac{1}{s}e^{-as}\right]$$

Assuming that the initial charge on the capacitor is zero, $q(0_-) = 0$.

$$I_C(s)\left[\frac{1}{sC} + R\right] = A\left(\frac{1 - e^{-as}}{s}\right)$$

$$I_C(s) = \frac{A(1 - e^{-as})}{s\left(\dfrac{1}{sC} + R\right)} = \frac{A(1 - e^{-as})}{R\left(s + \dfrac{1}{RC}\right)} A(1 - e^{-as}) = \frac{A}{R}\left[\frac{1}{\left(s + \dfrac{1}{RC}\right)} - \frac{e^{-as}}{s + \dfrac{1}{RC}}\right]$$

therefore, $$i_C(t) = \mathcal{L}^{-1} I_C(s) = \frac{A}{R}\left[e^{-\frac{t}{RC}} u_{-1}(t) - e^{-\frac{t-a}{RC}} u_{-1}(t-a)\right]$$

14.27 SECOND-ORDER CIRCUITS

We have investigated circuits containing single energy-storage elements and found that the analysis of these circuits required the solution of a first-order differential equation. We shall now consider the solution of circuits which contain at least two independent energy-storage elements (inductors and/or capacitors) and which cannot be combined to produce a single equivalent energy-storage element. Analysis of this circuit will require the solution of a second-order differential equation(or two simultaneous first order equations). For this reason, *circuits containing two independent energy storage elements are called* ***second order circuits.***

Although the analysis of second order circuits is more complex than the analysis of first-order circuits, we will use the same approach taken for solving first-order circuits. Second order circuits find wide use in engineering applications. Any physical system that can be described by a second-order differential equation can be modelled by a series RLC circuit. This circuit illustrates almost all the basic concepts of double-memory circuits.

14.28 SERIES RLC CIRCUIT

The general series RLC circuit is shown in Fig. 14.27. The inductor L and capacitor C are connected in series. The source-resistor circuit is replaced by its Thevenin's equivalent circuit (v_T and R_T). The result is a circuit in which resistor R_T, inductor L, and capacitor C are connected in series. Therefore we shall call this circuit a series RLC circuit.

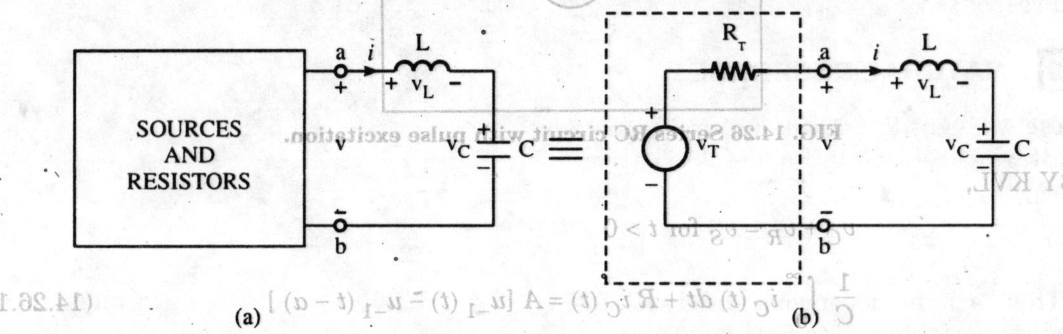

FIG. 14.27 The series RLC circuit (a) Given circuit (b) Thevenin's equivalent circuit

We can write the following equilibrium equations for the given circuit.

$$v_T = v + R_T i \tag{14.28.1}$$

$$v = v_L + v_C \tag{14.28.2}$$

$$v_L = L\frac{di}{dt} \tag{14.28.3}$$

$$i = C\frac{dv_C}{dt} \tag{14.28.4}$$

Thus we have four equations in four unknowns (i, v, v_L, v_C). We eliminate three unknown variables and retain one variable for solution. For the sake of convenience let us select the capacitor voltage v_C as the solution variable. Substituting Eqs. (14.28.2) and (14.28.4) in Eq. (14.28.1) gives

$$v_L + v_C + R_T C\frac{dv_C}{dt} = v_T \tag{14.28.5}$$

Substitutiton of Eq. (14.28.4) in Eq. (14.28.3) gives

$$v_L = LC\frac{d^2 v_C}{dt^2} \tag{14.28.6}$$

In order to eliminate v_L, we substitute Eq. (14.28.6) in Eq. (14.28.5)

$$L\,C\frac{d^2 v_C}{dt^2} + R_T\,C\frac{dv_C}{dt} + v_C = v_T \tag{14.28.7}$$

This is a second order, linear differential equation in which the dependent variable is the capacitor voltage v_C. After solving for v_C we can determine every other desired voltage or current in the circuit.

The solution of Eq. (14.28.7) depends on the following :

(*a*) The nature of input signal (v_T).

(*b*) The values of the circuit parameters.

(*c*) The initial values of the capacitor voltage and the inductor current.

As with first order circuit, we start by considering the natural response and then the forced response.

14.29 NATURAL RESPONSE

Suppose we want to find the natural (source-free) response of a second-order circuit for $t > 0$. Thus, if we put $v_T = 0$ in Eq. (14.28.7), it becomes

$$L\,C\frac{d^2 v_C}{dt^2} + R_T\,C\frac{dv_C}{dt} + v_C = 0 \tag{14.29.1}$$

This is a homogeneous differential equation. It shows that the linear sum of the dependent variable v_C, its first and second derivatives is equal to zero for all $t > 0$. As we did in the case of first-order circuits, let us assume the solution of the form

$$v_C = A\,e^{st} \tag{14.29.2}$$

therefore, $\dfrac{dv_C}{dt} = A\,s\,e^{st}$, $\dfrac{d^2 v_C}{dt^2} = A\,s^2\,e^{st}$

Substituting these values in Eq. (14.29.1) we get

$$L\,C\,A\,s^2\,e^{st} + R_T\,C\,A\,s\,e^{st} + A\,e^{st} = 0$$

$$\left(s^2 + \frac{R_T}{L}s + \frac{1}{LC}\right)A\,e^{st} = 0 \tag{14.29.3}$$

For Eq. (14.29.3) to be satisfied for all $t > 0$, it is necessary that either $A = 0$ or $s = -\infty$, or $s^2 + \dfrac{R_T}{L}s + \dfrac{1}{LC} = 0$

But if $A = 0$ or $s = -\infty$, then every response is zero. Therefore, for Eq. (14.29.3) to be satisfied for all $t > 0$, it is necessary that

$$s^2 + \frac{R_T}{L}s + \frac{1}{LC} = 0 \tag{14.29.4}$$

This is the *characteristic equation* of the circuit. It is quadratic in s because the circuit contains two noncombinable memory elements. The characteristic roots s_1 and s_2 of this equation are given by

$$s_1 = -\frac{R_T}{2L} + \sqrt{\left(\frac{R_T}{2L}\right)^2 - \frac{1}{LC}} \tag{14.29.5}$$

$$s_2 = -\frac{R_T}{2L} - \sqrt{\left(\frac{R_T}{2L}\right)^2 - \frac{1}{LC}} \tag{14.29.6}$$

Therefore, there must be two solutions, one using s_1 in Eq. (14.29.2), and the other using s_2 in Eq. (14.29.2). Since either solution will give zero for the sum of three terms in Eq. (14.29.1), their sum must also be a solution. Therefore, the homogeneous solution is given by

$$v_C(t) = A_1 e^{s_1 t} + A_2 e^{s_2 t} \tag{14.29.7}$$

where A_1 and A_2 are the constants to be determined from initial conditions. The roots s_1 and s_2 are determined from the circuit parameters R_T, L and C.

These roots are in general complex numbers, known as the **complex natural frequencies** of the system.

Since there are two constants A_1 and A_2 to be determined, we need *two independent* linear equations. The two equations are usually found from initial conditions. Therefore we will need two initial conditions.

To obtain the initial conditions the following continuity relationships are applied in the absence of impulses :

$$v_C(0_+) = v_c(0_-)$$

$$i_L(0_+) = i_L(0_-)$$

Since the dependent variable of the differential equation is $v_C(t)$, it is advisable to convert the initial conditions to that variable. In our case the statement about $i_L(t)$ is converted to a corresponding statement about $v_C(t)$.

Since $i_L(t) = i_C(t)$ for $t > 0$, at $t = 0_+$ (which is included in $t > 0$), this becomes

$$i_L(0_+) = i_C(0_+)$$

Also, in general,

$$i_C(t) = C\frac{dv_C}{dt}$$

or

$$\frac{dv_C}{dt} = \frac{i_C(t)}{C}$$

therefore,

$$\left.\frac{dv_C}{dt}\right|_{t=0_+} = \frac{i_C(0_+)}{C} \tag{14.29.8}$$

This is called the **derived initial condition**.

14.30 CLASSIFICATION OF RESPONSES

We found that the natural response of series RLC circuit is given by

$$L\,C\frac{d^2 v_C}{dt^2}+R_T\,C\frac{dv_C}{dt}+v_C=0 \tag{14.30.1}$$

or

$$\frac{d^2 v_C}{dt^2}+\frac{R_T}{L}\frac{dv_C}{dt}+\frac{1}{LC}v_C=0 \tag{14.30.2}$$

This is a second-order differential equation. Actually, any circuit variable of a second-order circuit can be described by a second-order differential equation. Let us choose the general form of this second-order differential equation to be

$$\frac{d^2 x}{dt^2}+2\,\alpha\frac{dx}{dt}+\omega_n^2\,x=0 \tag{14.30.3}$$

where x is our circuit variable.

If we compare Eqs. (14.30.2) and (14.30.3) we see that

$$x=v_c,\ \alpha=\frac{R_T}{2L},\ \omega_n^2=\frac{1}{LC}$$

Then the roots of Eq. (14.30.3) can be expressed as

$$s_1=-\alpha+\sqrt{\alpha^2-\omega_n^2}=-\alpha+\beta \tag{14.30.4}$$

$$s_2=-\alpha-\sqrt{\alpha^2-\omega_n^2}=-\alpha-\beta \tag{14.30.5}$$

where $$\beta=\sqrt{\alpha^2-\omega_n^2}=\sqrt{\frac{R^2}{4L^2}-\frac{1}{LC}}$$

The relative values of α and ω_n determine the nature of the roots s_1 and s_2.

There are three types of solution of Eq. (14.30.2) depending upon whether $\alpha>\omega_n$, $\alpha=\omega_n$, or $\alpha<\omega_n$. These three solutions are called the overdamped, the critically damped, and the underdamped solutions respectively.

14.31 NATURAL RESPONSE : OVERDAMPED CASE

If $\alpha>\omega_n$, then $\alpha^2>\omega_n^2$ and $(\alpha^2-\omega_n^2)$ is a **positive** number s_1 and s_2 are real numbers. This condition is called the *overdamped case.* Since α and ω_n must be positive numbers for all physically realizable values of R, L and C, we have that if $\alpha>\omega_n$

then $$\alpha>\sqrt{(\alpha^2-\omega_n^2)}$$

and $$[-\alpha-\sqrt{\alpha^2-\omega_n^2}\,]<[-\alpha+\sqrt{\alpha^2-\omega_n^2}\,]<0$$

That is, s_1 and s_2 are both *negative* real numbers. The natural response in the overdamped condition is, therefore, given by the algebraic sum of two exponentials as

$$v_{cn}(t)=A_1\,e^{s_1 t}+A_2\,e^{s_2 t} \tag{14.31.1}$$

Since s_1 and s_2 are negative real numbers, both the exponentials in Eq. (14.31.1) decay with time. Also, since $| s_2 | > | s_1 |$, the term containing s_2 has more rapid rate of decrease. The damping takes very long time. Hence the response is called overdamped.

14.32 NATURAL RESPONSE : UNDERDAMPED CASE

If $\alpha < \omega_n$, the response is called *underdamped*. Since $(\alpha^2 - \omega_n^2)$ is negative let us put

$$\alpha^2 - \omega_n^2 = -\omega_d^2 \tag{14.32.1}$$

where ω_d is called the **damped natural angular frequency** or simply **damped frequency** in underdamped condition.

If $\alpha < \omega_n$, we have

$$\sqrt{\alpha^2 - \omega_n^2} = \sqrt{-(\omega_n^2 - \alpha^2)} = \sqrt{-\omega_d^2} = j\,\omega_d$$

Therefore, for the undamped case

$$s_1 = -\alpha + j\,\omega_d \tag{14.32.2}$$

$$s_2 = -\alpha - j\,\omega_d \tag{14.32.3}$$

The natural response for $v_c(t)$ is

$$v_{cn}(t) = A_1 e^{s_1 t} + A_2\, e^{s_2 t} \tag{14.32.4}$$

Substituting the values of s_1 and s_2 from Eqs. (14.32.2) and (14.32.3) in Eq. (14.32.4) gives

$$v_{cn}(t) = A_1\, e^{(-\alpha + j\omega_d)t} + A_2\, e^{(-\alpha - j\omega_d)t}$$

$$= e^{-\alpha t}(A_1\, e^{j\omega_d t} + A_2\, e^{-j\omega_d t})$$

$$= e^{-\alpha t}\,[A_1(\cos \omega_d t + j \sin \omega_d t) + A_2(\cos \omega_d t - j \sin \omega_d t)]$$

or

$$v_{cn}(t) = e^{-\alpha t}\,[(A_1 + A_2)\cos \omega_d t + j(A_1 - A_2)\sin \omega_d t] \tag{14.32.5}$$

where A_1 and A_2 are constants that may be complex numbers. Since $v_c(t)$ is real and α and ω_d are real numbers, $(A_1 + A_2)$ and $j(A_1 - A_2)$ must be real numbers.

Let $\quad A_1 + A_2 = B_1$

and $\quad j(A_1 - A_2) = B_2$

so that Eq. (14.32.5) becomes

$$v_{cn}(t) = e^{-\alpha t}(B_1 \cos \omega_d t + B_2 \sin \omega_d t) \tag{14.32.6}$$

where B_1 and B_2 are real numbers to be determined from initial conditions.

Eq. (14.32.6) may also be put in an alternative form :

$$v_{cn}(t) = ke^{-\alpha t} \sin(\omega_d t + \varphi) \tag{14.32.7}$$

where $k = \sqrt{B_1^2 + B_2^2}$, $\tan \varphi = B_1/B_2$.

The constants k and φ are to be determined from initial conditions. Equation (14.32.7) shows that $v_{cn}(t)$ has a sinusoidal component and an exponential component. The exponential component causes the response to decay gradually and the rate of decay is dependent on the value of α. For this reason, α is called the **damping coefficient or damping factor.** The value of α determines how fast the sinusoid's amplitude diminishes. If α is large the response dies out rapidly. To understand ω_d, first consider the case when $R_T = 0$. Under this condition, $\alpha = 0$, $\omega_d = \omega_n$ and $v_{cn}(t)$ is a sinusoid at radian frequency ω_n. Therefore, ω_n is the radian frequency at which the circuit response would oscillate if the resistance of the circuit were zero. Consequently, ω_n is often called the **undamped natural frequency** of the circuit. The radian frequency ω_d is always less than ω_n if the resistance is greater than zero so ω_d is called the **damped natural frequency**. Since Eq. (14.32.7) is a sinusoidal function with an exponentially decaying amplitude, it is called a **damped sinusoid.**

14.33 NATURAL RESPONSE : CRITICALLY DAMPED CASE

If $\alpha = \omega_n$, the response is called *critically damped.* It represents the boundary point between the underdamped (oscillatory) case and the overdamped case (two different exponentials). Any infinitesimal change in the value of any element in a critically damped circuit will result in one of the underdamped and overdamped cases.

When $\alpha = \omega_n$, $s_1 = s_2 = -\alpha$. That is, both characteristic roots are real, negative and equal. For this condition,

$$v_{cn}(t) = A_1 e^{s_1 t} + A_2 e^{s_2 t} \tag{14.33.1}$$

$$= A_1 e^{-\alpha t} + A_2 e^{-\alpha t}$$

$$= (A_1 + A_2)\, e^{-\alpha t} = B\, e^{-\alpha t}$$

This cannot be the solution to Eq. (14.30.2) since in general it is impossible to satisfy two initial conditions $v_c(0)$ and $\left.\dfrac{dv_c}{dt}\right|_{t=0}$ with the single constant B. The solution of Eq. (14.30.2) for this special case is done as follows :

Rewriting Eq. (14.30.3) in the form

$$\frac{d^2 x}{dt^2} + 2\alpha \frac{dx}{dt} + \alpha^2 x = 0 \tag{14.33.2}$$

$$\frac{d^2 x}{dt^2} + \frac{\alpha\, dx}{dt} + \frac{\alpha\, dx}{dt} + \alpha^2 x = 0$$

$$\frac{d}{dt}\left(\frac{dx}{dt} + \alpha x\right) + \alpha\left(\frac{dx}{dt} + \alpha x\right) = 0$$

Let
$$\frac{dx}{dt} + \alpha x = u \tag{14.33.3}$$

therefore,
$$\frac{du}{dt} + \alpha u = 0 \tag{14.33.4}$$

The solution of Eq. (14.33.4) is

$$u = A_1 e^{-\alpha t} \tag{14.33.5}$$

Putting the value of u in Eq. (14.33.3)

$$\frac{dx}{dt} + \alpha x = A_1 e^{-\alpha t}$$

$$e^{\alpha t} \frac{dx}{dt} + \alpha e^{\alpha t} x = A_1$$

$$\frac{d}{dt}(xe^{\alpha t}) = A_1 \tag{14.33.6}$$

The integration of Eq. (14.33.6) gives

$$xe^{\alpha t} = A_1 t + A_2$$

$$x = e^{-\alpha t}(A_1 t + A_2) \tag{14.33.7}$$

where A_1 and A_2 are constants to be determined from initial conditions.

The circuit resistance that leads to the critically damped condition is called the **critical resistance R_c**. It is obtained by putting $R_T = R_C$ in

$$\left(\frac{R_T}{2L}\right)^2 - \frac{1}{LC} = 0$$

That is,

$$\left(\frac{R_C}{2L}\right)^2 - \frac{1}{LC} = 0$$

therefore,

$$R_C = 2\sqrt{\frac{L}{C}} \tag{14.33.8}$$

A convenient way of comparing the circuit resistance, R_T to the critical resistance R_C is by calculating their ratio. This ratio is called the **damping ratio** ζ (zeta). Thus

$$\boxed{\zeta \triangleq \frac{\text{circuit resistance}}{\text{critical resistance}} = \frac{R_T}{R_C} = \frac{R_T}{2}\sqrt{\frac{C}{L}}} \tag{14.33.9}$$

$$\frac{\alpha}{\omega_n} = \frac{R}{2L} \div \frac{1}{\sqrt{LC}} = \frac{R}{2}\sqrt{\frac{C}{L}}$$

therefore,

$$\boxed{\zeta = \frac{\alpha}{\omega_n}} \tag{14.33.10}$$

14.34 GENERAL FORM OF CHARACTERISTIC EQUATION

The general form of the characteristic equation is

$$s^2 + 2\zeta\omega_n s + \omega_n^2 = 0 \tag{14.34.1}$$

Solving this equation we get

$$s_1 = -\zeta\,\omega_n + \omega_n\sqrt{\zeta^2 - 1} \tag{14.34.2}$$

$$s_2 = -\zeta\,\omega_n - \omega_n\sqrt{\zeta^2 - 1} \tag{14.34.3}$$

For the underdamped case ($R < R_C$), $\zeta < 1$ and the roots become complex numbers given by

$$s_1 = -\zeta\,\omega_n + j\omega_n\sqrt{1 - \zeta^2} \tag{14.34.4}$$

$$s_1 = -\zeta\,\omega_n - j\omega_n\sqrt{1 - \zeta^2} \tag{14.34.5}$$

For the underdamped case, the natural response is given by

$$x_n(t) = ke^{-\alpha t}\sin(\omega_n\sqrt{1 - \zeta^2}\,t + \phi) \tag{14.34.6}$$

We can see the effect of the damping factor ζ. If $\zeta = 0$, $s_1 = j\omega$, $s_2 = -j\omega$, the exponential decaying factor $e^{-\alpha t}$ becomes unity and the natural response is a pure sinusoid with radian frequency ω_n given by

$$x_n(t) = k\sin(\omega_n t + \phi) \tag{14.34.7}$$

Thus the response does not decay with time and it is *undamped*. For this reason ω_n is called the undamped natural frequency. When $0 < \zeta < 1$, the exponential damping term is present in the natural response. In such cases the *damped* natural frequency is

$$\omega_d = \omega_n\sqrt{1 - \zeta^2} \tag{14.34.8}$$

When $\zeta > 1$ ($R > R_C$) there are two unequal real roots given by Eqs. (14.34.1) and (14.34.2). The natural response will be of the form

$$x_n(t) = A_1 e^{s_1 t} + A_2 e^{s_2 t}$$

This is the overdamped case.

When $\zeta = 1$ ($R = R_C$) there are two real and equal roots

$$s_1 = s_2 = -\zeta\,\omega_n$$

The natural response will be of the form

$$x_n(t) = e^{-\alpha t}(A_1 t + A_2)$$

$$x_n(t) = e^{-\zeta\omega_n t}(A_1 t + A_2) \tag{14.34.9}$$

This is the critically damped case.

For the undamped case $\zeta = 0$ ($R = 0$).

The roots are conjugate pure imaginary

$$s_1 = j\omega_n\,,\; s_2 = -j\omega_n$$

The natural response is of the form

$$x_n(t) = k\sin(\omega_n t + \phi)$$

For underdamped case $\zeta < 1$ ($R < R_C$) and the response is given by Eq. (14.34.6).

14.35 PARALLEL RLC CIRCUIT

The general parallel RLC circuit is shown in Fig. 14.28. The inductor and capacitor are connected in parallel, the source-resistor circuit is replaced by its Norton's equivalent circuit (i_T and G_T). The result is a circuit in which the resistance, inductance, and capacitance are connected in parallel. Therefore we call this circuit a parallel RLC circuit.

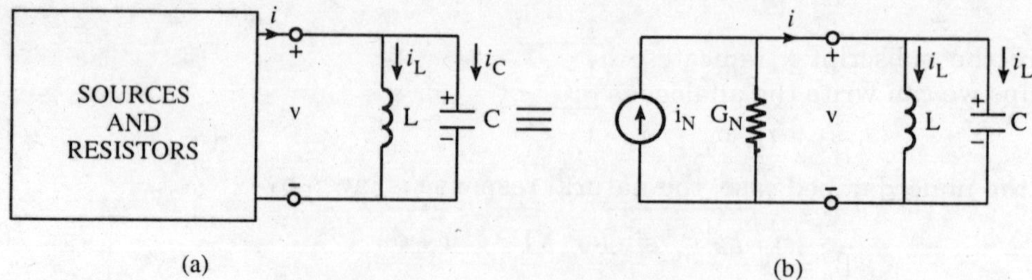

FIG. 14.28 The parallel RLC circuit (a) Given circuit (b) Norton's equivalent circuit

We can write the following equilibrium equations for the given circuit :

$$i_N = i + G_N v \tag{14.35.1}$$

$$i = i_L + i_C \tag{14.35.2}$$

$$i_C = C\frac{dv}{dt} \tag{14.35.3}$$

$$v = L\frac{di_L}{dt} \tag{14.35.4}$$

Thus we have four equations in four unknowns (v, i, i_C, i_L). We eliminate three unknown variables and retain one variable for solution. For the sake of convenience let us select the inductor current i_L as the solution variable. Substituting Eqs. (14.35.2) and (14.35.4) in Eq. (14.35.1) gives

$$i_C + i_L + G_N L\frac{di_L}{dt} = i_N \tag{14.35.5}$$

Substitution of Eq. (14.35.4) in Eq. (14.35.3) gives

$$i_C = LC\frac{d^2 i_L}{dt^2} \tag{14.35.6}$$

Now substitute Eq. (14.35.6) in Eq. (14.35.5) i_C

$$LC\frac{d^2 i_L}{dt^2} + G_N L\frac{di_L}{dt} + i_L = i_N \tag{14.35.7}$$

This is a second-order, linear differential equation of the same form we obtained for the series RLC circuit (Eq. 14.28.7).

This is an example of the concept of duality. If in Eq. (14.28.7) we replace $v_T \to i_N$, $L \to C$, $R_T \to G_N$, $C \to L$, $i \to v$ we obtain Eq. (14.35.7). Hence we need not study the parallel

RLC circuit in detail since all of the results obtained for the series RLC circuit apply for the parallel RLC circuit.

For the series RLC circuit, α_s and β_s are defined as

$$\alpha_s = \frac{R}{2L}, \beta_s = \sqrt{\frac{R^2}{4L^2} - \frac{1}{LC}}$$

where the subscript s indicates the RLC elements in series. If we use the duality relationships we can write the analogous quantities for the parallel RLC circuits as

$$\alpha_p = \frac{G}{2C} = \frac{1}{2RC}$$

$$\beta_p = \sqrt{\frac{G^2}{4C^2} - \frac{1}{CL}} = \sqrt{\frac{1}{4R^2C^2} - \frac{1}{LC}}$$

where the subscript p indicates the RLC elements are in parallel.

Initial Conditions

The initial conditions in the circuit are initial inductor current and initial capacitor voltage. The continuity relationships are

$$i_L(0_+) = i_L(0_-)$$

$$v_C(0_+) = v_C(0_-)$$

In our case the dependent variable is $i_L(t)$. Hence the first initial condition can be applied directly. The second initial condition should be converted to that variable. From the circuit of Fig. 14.28

$$v_C(t) = L\frac{di_L(t)}{dt}$$

$$\frac{di_L(t)}{dt} = \frac{v_C(t)}{L}$$

$$\left.\frac{di_L(t)}{dt}\right|_{t=0_+} = \frac{v_C(0_+)}{L}$$

These results are the dual of those obtained for the series RLC circuit.

14.36 ZERO-INPUT RESPONSE OF PARALLEL RLC CIRCUIT

In order to determine the zero-input response, we put $i_N = 0$ in Eq. (14.35.7) to get

$$LC\frac{d^2 i_L}{dt^2} + G_N L\frac{di_L}{dt} + i_L = 0 \tag{14.36.1}$$

The solution of this homogeneous equation gives the zero-input response. The characteristic equation is

$$LCs^2 + G_N Ls + 1 = 0 \tag{14.36.2}$$

This characteristic equation has two roots :

$$s_1, s_2 = \frac{1}{2LC}\,[-G_N L \pm \sqrt{(G_N L)^2 - 4LC}\,] \qquad (14.36.3)$$

As before, we have three distinct types of roots.

Overdamped Case

If $(G_N L)^2 > 4LC$, the quantity under the radical sign is positive and there will be two, unequal, real roots. This is the overdamped case. The zero-input response will be of the form $i_L(t) = A_1 e^{s_1 t} + A_2 e^{s_2 t}$

Critically Damped Case

If $(G_N L)^2 = 4\,LC$, the radicand is zero, and there will be two real, equal roots $s_1 = s_2 = -\alpha$. This is the critically damped case. The zero-input response will be of the form

$$i_L(t) = e^{-\alpha t}(A_1 t + A_2)$$

Underdamped Case

If $(G_N L)^2 < 4\,LC$, the radicand is negative and there will be a pair of complex, conjugate roots

$$s_1 = -\alpha + j\omega_d$$

$$s_2 = -\alpha + j\omega_d$$

This is the underdamped case. The natural response will be of the form

$$i_L(t) = e^{-\alpha t}(A_1 \cos \omega_d t + A_2 \sin \omega_d t)$$

Thus we see that all the results derived in the case of the series RLC apply to the parallel RLC circuit.

■ **EXAMPLE 14.25** For the circuit of Fig. 14.29, determine $v_C(t)$ for $t > 0$. The input is a voltage step function $v_S(t) = 100\,u_{-1}(t)$.

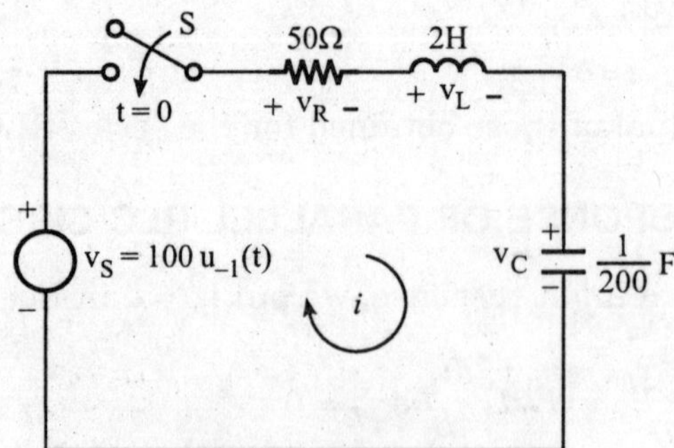

FIG. 14.29 Circuit for Example 14.25

SOLUTION : From KCL,

$$i_R = i_L = i_C = i \qquad \text{for } t > 0 \qquad (E\ 14.25.1)$$

From KVL,

$$v_L + v_R + v_C = v_S \quad \text{for } t > 0 \tag{E 14.25.2}$$

$$L\frac{di_L}{dt} + Ri_R + v_C = 100 \quad \text{for } t > 0 \tag{E 14.25.3}$$

$$L\frac{d}{dt}\left[C\frac{dv_C}{dt}\right] + R\left[C\frac{dv_C}{dt}\right] + v_C = 100$$

$$LC\frac{d^2 v_C}{dt^2} + RC\frac{dv_C}{dt} + v_C = 100 \text{ for } t > 0 \tag{E 14.25.4}$$

Determination of initial conditions

Since the capacitor is initially uncharged, $v_C(0_-) = 0$. By the continuity relationship for the capacitor

$$v_C(0_+) = v_C(0_-) = 0 \tag{E 14.25.5}$$

Since the switch is initially open, $i_L(0_-) = 0$ and by the continuity relationship for the inductor.

$$i_L(0_+) = i_L(0_-) = 0 \tag{E 14.25.6}$$

In this problem the dependent variable is $v_C(t)$. Therefore, the initial conditions should be converted to $v_C(t)$. We have

$$i_C(t) = C\frac{dv_C}{dt}$$

$$\frac{dv_C}{dt} = \frac{i_C(t)}{C}$$

$$\left.\frac{dv_C}{dt}\right|_{t=0.} = \frac{i_C(0_+)}{C} \tag{E 14.25.7}$$

But $i_C(0_+) = i_L(0_+)$, so that

$$\left.\frac{dv_C}{dt}\right|_{t=0.} = \frac{0}{C} = 0 \tag{E 14.25.8}$$

In this problem,

$$R = 50\ \Omega,\ L = 2\text{ H},\ C = \frac{1}{200}\text{ F}$$

Substitution of these values in Eq. (E 14.25.4) gives

$$2 \times \frac{1}{200}\frac{d^2 v_C}{dt^2} + 50 \times \frac{1}{200}\frac{dv_C}{dt} + v_C = 100$$

$$0{\cdot}01\frac{d^2 v_C}{dt^2} + 0{\cdot}25\frac{dv_C}{dt} + v_C = 100 \quad \text{for } t > 0 \tag{E 14.25.9}$$

As with first-order systems, the complete solution (response) of the second-order system given by Eq. (E 14.25.9) is equal to the sum of the natural response and the forced response.

therefore, $$v_C(t) = v_{cn}(t) + v_{cf}(t) \qquad \text{(E 14.25.10)}$$

Determination of the natural response

We have to solve the equation

$$0{\cdot}01\,\frac{d^2 v_{cn}}{dt^2} + 0{\cdot}25\,\frac{dv_{cn}}{dt} + v_{cn} = 0 \qquad \text{(E 14.25.11)}$$

The characterstic equation is

$$0{\cdot}01\,s^2 + 0{\cdot}25\,s + 1 = 0$$

or $$s^2 + 25\,s + 100 = 0 \qquad \text{(E 14.25.12)}$$

The general form of the characteristic equation is

$$s^2 + 2\,\zeta\,\omega_n\,s + \omega_n^2 = 0 \qquad \text{(E 14.25.13)}$$

Comparison of Eqs. (E 14.25.12) and (E 14.25.13) shows that

$$\omega_n^2 = 100$$

and $$2\,\zeta\,\omega_n = 25$$

$\therefore$ $$\omega_n = 10,\ \zeta = 1{\cdot}25$$

Since $\zeta > 1$, our circuit is *overdamped.* The natural response of the overdamped case is given by

$$v_{cn}(t) = A_1\,e^{s_1 t} + A_2\,e^{s_2 t} \qquad \text{(E 14.25.14)}$$

The roots of Eq. (14.25.12) are

$$s_{1,2} = \frac{1}{2}\left[-25 \pm \sqrt{(25)^2 - 4 \times 100}\,\right]$$

therefore, $$s_1 = -5,\ s_2 = -20$$

Substituting the values of s_1 and s_2 in Eq. (E 14.25.14) gives

$$v_{cn}(t) = A_1\,e^{-5t} + A_2\,e^{-20t} \qquad \text{(E 14.25.15)}$$

Determination of the forced response

In this case, the forcing function is a constant. Let $v_{cf}(t) = k$, be the forced response. This must satisfy Eq. (E 14.25.9). That is.

$$0{\cdot}01\,\frac{d^2 v_{cf}}{dt^2} + 0{\cdot}25\,\frac{dv_{cf}}{dt} + v_{cf} = 100$$

Since $v_{cf}(t) = k$, $\dfrac{dv_{cf}}{dt} = 0,\ \dfrac{d^2 v_{cf}}{dt^2} = 0$

$$0 + 0 + k = 100 \text{ or } k = 100$$

and $$v_{cf}(t) = 100 \qquad \text{(E 14.25.16)}$$

Determination of the complete response

The complete response is

$$v_C(t) = v_{cn}(t) + v_{cf}(t)$$

$$v_C(t) = A_1 e^{-5t} + A_2 e^{-20t} + 100 \qquad \text{(E 14.25.17)}$$

We shall evaluate the constants A_1 and A_2 by using initial conditions.

The first initial condition is

$$v_C(0_+) = 0$$

When we put $t = 0_+$ in Eq. (E 14.25.17), we get

$$v_C(0_+) = A_1 e^{-5 \times 0} + A_2 e^{-20 \times 0} + 100$$

$$0 = A_1 + A_2 + 100$$

or $$A_1 + A_2 = -100 \qquad \text{(E 14.25.18)}$$

The second initial condition to be applied is given by Eq. (E 14.25.8). Differentiating both sides of Eq. (E 14.25.17) gives

$$\frac{dv_C}{dt} = -5 A_1 e^{-5t} - 20 A_2 e^{-20t} + 0$$

$$\left.\frac{dv_C}{dt}\right|_{t=0_+} = -5 A_1 e^{-5 \times 0} - 20 A_2 e^{-20 \times 0}$$

$$0 = -5A_1 - 20A_2$$

or $$A_1 + 4A_2 = 0 \qquad \text{(E 14.25.19)}$$

Solving Eqs. (14.25.18) and (E 14.25.19) simultaneously gives

$$A_2 = \frac{100}{3} = 33{\cdot}33, A_1 = -133{\cdot}33$$

Substitution of the values of A_1 and A_2 in Eq. (E 14.25.17) gives the complete response as

$$v_C(t) = 100 - 133{\cdot}33\, e^{-5t} + 33{\cdot}33\, e^{-20t} \text{ V} \quad \text{for } t > 0$$

■ **EXAMPLE 14.26** For the circuit of Fig. 14.30, determine $v_C(t)$ for $t > 0$. The input is a voltage step function $v_S(t) = 0{\cdot}4\, u_{-1}(t)$

SOLUTION : From KCL,

$$i_R = i_L = i_C = i \quad \text{for } t > 0 \qquad \text{(E 14.26.1)}$$

From KVL,

$$v_L + v_R + v_C = v_S \quad \text{for } t > 0$$

$$L\frac{di_L}{dt} + Ri_R + v_C = v_S \quad \text{for } t > 0$$

$$L\frac{d}{dt}\left[C\frac{dv_C}{dt}\right] + R\left[C\frac{dv_C}{dt}\right] + v_C = 0{\cdot}4 \quad \text{for } t > 0$$

$$LC\frac{d^2 v_C}{dt^2} + RC\frac{dv_C}{dt} + v_C = 0{\cdot}4 \quad \text{for } t > 0 \qquad \text{(E 14.26.2)}$$

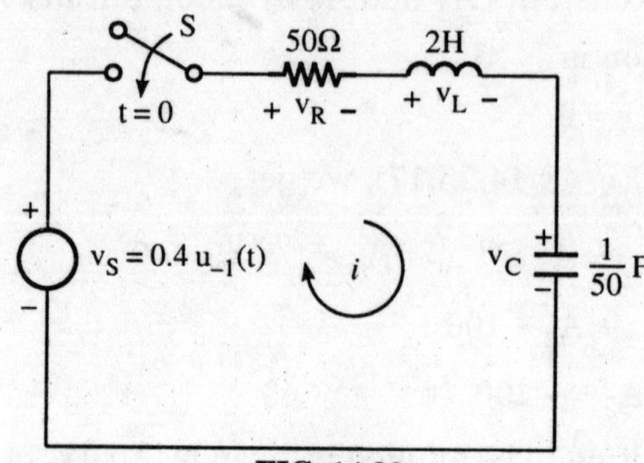

FIG. 14.30

In this problem,

$$R = 20\ \Omega,\ L = 2\ \text{H},\ C = \frac{1}{50}\ \text{F}$$

Substitution of these values in Eq. (E 14.26.2) gives

$$2 \times \frac{1}{50}\frac{d^2 v_C}{dt^2} + 20 \times \frac{1}{50}\frac{dv_C}{dt} + v_C = 0{\cdot}4$$

or

$$\frac{d^2 v_C}{dt^2} + 10\frac{dv_C}{dt} + 25\, v_C = 10 \qquad \text{(E 14.26.3)}$$

As in Example 14.25, the initial conditions are

$$v_C(0_+) = 0 \text{ and } \left.\frac{dv_C}{dt}\right|_{t=0_+} = 0$$

Determination of the natural response

In order to find the natural response $v_{cn}(t)$, we have to solve the equation

$$\frac{d^2 v_{cn}}{dt^2} + 10\frac{dv_{cn}}{dt} + 25\, v_{cn} = 0 \qquad \text{(E 14.26.4)}$$

The characteristic equation is given by

$$s^2 + 10\,s + 25 = 0 \qquad \text{(E 14.26.5)}$$

$$(s+5)^2 = 0$$

therefore,

$$s_1 = s_2 = -5$$

Since the characteristic roots are real, negative, and equal our circuit is critically damped. The natural response is given by

$$v_{cn}(t) = e^{-5t}(A_1 t + A_2) \qquad \text{(E 14.26.6)}$$

Determination of the forced response

In this problem, the forcing function is a constant. Let the forced response be given by $v_{cf}(t) = k$. This must satisfy the equation

$$\frac{d^2 v_{cf}}{dt^2} + 10\frac{dv_{cf}}{dt} + 25\, v_{cf} = 10 \qquad \text{(E 14.26.7)}$$

Since $v_{cf}(t) = k$, $\dfrac{dv_{cf}}{dt} = 0$, $\dfrac{d^2 v_{cf}}{dt^2} = 0$

Substituting these values in Eq. (E 14.26.7) gives

$$0 + 0 + 25k = 10,\ k = \frac{10}{25} = 0{\cdot}4$$

and $\quad v_{cf}(t) = 0{\cdot}4$

Determination of the complete response

The complete response is

$$v_C(t) = v_{cn}(t) + v_{cf}(t)$$

$$v_C(t) = e^{-5t}(A_1 t + A_2) + 0{\cdot}4 \qquad \text{(E 14.26.8)}$$

The constants A_1 and A_2 are determined by using initial conditions. The first initial condition is $v_c(0_+) = 0$. When we put $t = 0_+$ in Eq. (E 14.26.8) we get

$$v_C(0_{+}) = e^{-5\times 0}(A_1 \times 0 + A_2) + 0{\cdot}4$$

$$0 = A_2 + 0{\cdot}4,\ A_2 = -\,0{\cdot}4$$

In order to apply the second initial condition $\left.\dfrac{dv_C}{dt}\right|_{t=0_+} = 0$, we differentiate Eq. (E 14.26.8).

$$\frac{dv_C}{dt} = e^{-5t}(A_1 + 0) + (A_1 t + A_2)(-5e^{-5t}) + 0$$

At $t = 0_+$,

$$\left.\frac{dv_C}{dt}\right|_{t=0_+} = e^{-5\times 0} A_1 + (A_1 \times 0 + A_2)(-5e^{-5\times 0})$$

$$0 = A_1 - 5A_2,\ A_1 = 5A_2$$

or $\quad A_1 = -\,5 \times 0{\cdot}4 = -\,2$

Substitution of the values of A_1 and A_2 in Eq. (E 14.26.8) gives the complete response as

$$v_C(t) = e^{-5t}(-2t - 0{\cdot}4) + 0{\cdot}4$$

or
$$v_C(t) = 0{\cdot}4 - (2t + 0{\cdot}4)\,e^{-5t}$$

■ **EXAMPLE 14.27** **For the circuit of Fig. 14.30, determine $v_C(t)$ and $i(t)$ for $t > 0$ if the 20 Ω resistor is replaced by a 12 Ω resistor.**

SOLUTION : We have

$$LC\frac{d^2 v_C}{dt^2} + RC\frac{dv_C}{dt} + v_C = v_S \qquad \text{(E 14.27.1)}$$

Here $R = 12\ \Omega$, $L = 2$ H, $C = \frac{1}{50}$ F

Putting these values in Eq. (E 14.27.1) we get

$$2 \times \frac{1}{50}\frac{d^2 v_C}{dt^2} + 12 \times \frac{1}{50}\frac{dv_C}{dt} + v_C = 0{\cdot}4$$

$$\frac{d^2 v_C}{dt^2} + 6\frac{dv_C}{dt} + 25\,v_C = 10 \qquad \text{(E 14.27.2)}$$

The initial conditions are

$$v_C(0_+) = 0, \quad \left.\frac{dv_C}{dt}\right|_{t=0_+} = 0$$

The characteristic equation is

$$s^2 + 6s + 25 = 0$$

$$s = \frac{-6 \pm \sqrt{36 - 100}}{2} = -3 \pm j\,4$$

$$s_1 = -3 + j\,4,\ s_2 = -3 - j\,4$$

Also
$$s_1 = -\alpha + j\,\omega_d,\ s_2 = -\alpha - j\omega_d$$

$$\alpha = 3,\ \omega_d = 4$$

$$\omega_n = \sqrt{\alpha^2 + \omega_d^2} = \sqrt{3^2 + 4^2} = 5$$

Since $\alpha < \omega_n$, the response is *underdamped.*

The natural response is

$$v_{cn}(t) = e^{-\alpha t}(A_1 \cos \omega_d t + A_2 \sin \omega_d t)$$

$$= e^{-3t}(A_1 \cos 4t + A_2 \sin 4t)$$

The forced response is

$$v_{cf}(t) = 0{\cdot}4$$

The complete response is

$$v_C(t) = e^{-3t}(A_1 \cos 4t + A_2 \sin 4t) + 0{\cdot}4$$

Applying the first initial condition $v_C(0_+) = 0$ we get

$$v_C(0_+) = e^{-3\times 0}(A_1 \cos 4\times 0 + A_2 \sin 4\times 0) + 0{\cdot}4$$

$$0 = A_1 + 0{\cdot}4, A_1 = -0{\cdot}4$$

Differentiating $v_C(t)$, we get

$$\frac{dv_C}{dt} = -3A_1 e^{-3t} \cos 4t - 4A_1 e^{-3t} \sin 4t - 3A_2 e^{-3t} \sin 4t + 4A_2 e^{-3t} \cos 4t + 0$$

therefore, $$\left.\frac{dv_C}{dt}\right|_{t=0_+} = -3A_1 + 4A_2$$

$$0 = -3A_1 + 4A_2$$

$$0 = -3(-0{\cdot}4) + 4A_2, A_2 = -0{\cdot}3$$

Therefore the complete response is

$$v_C(t) = 0{\cdot}4 - 0{\cdot}4e^{-3t} \cos 4t - 0{\cdot}3e^{-3t} \sin 4t$$

$$= 0{\cdot}4 - 0{\cdot}1\, e^{-3t}(4\cos 4t + 3 \sin 4t)$$

Since $v_C(t) = 0$ for $t < 0$, the zero-state voltage response can be written as

$$v_C(t) = [\,0{\cdot}4 - 0{\cdot}1\, e^{-3t}(4 \cos 4t + 3 \sin 4t)\,]\, u_{-1}(t)$$

The zero-state current step response is

$$i(t) = C\frac{dv_C}{dt}$$

$$= \frac{1}{50}\frac{d}{dt}[(0{\cdot}4 - 0{\cdot}4\, e^{-3t} \cos 4t - 0{\cdot}3\, e^{-3t} \sin 4t)\, u_{-1}(t)]$$

$$= \frac{1}{50}(0{\cdot}4 - 0{\cdot}4\, e^{-3t} \cos 4t - 0{\cdot}3\, e^{-3t} \sin 4t)\, u_0(t)$$

$$+ \frac{1}{50}(0 + 1{\cdot}2\, e^{-3t} \cos 4t + 1{\cdot}6\, e^{-3t} \sin 4t$$

$$\text{sss} + 0{\cdot}9\, e^{-3t} \sin 4t - 1{\cdot}2\, e^{-3t} \cos 4t)\, u_{-1}(t)$$

$$= \frac{1}{50}(0{\cdot}4 - 0{\cdot}4)\, u_0\, t + \frac{1}{50}(2{\cdot}5\, e^{-3t} \sin 4t)\, u_{-1}(t)$$

$$= (0{\cdot}05\, e^{-3t} \sin 4t)\, u_{-1}(t)$$

■ **EXAMPLE 14.28** In Example 14.27 if the input is changed from $0{\cdot}4u_{-1}(t)$ volts to the impulse function $0{\cdot}4\, u_0(t)$ determine

(*a*) the zero-state voltage impulse response across the capacitor.

(*b*) The zero-state current impulse response.

SOLUTION : The impulse response of a system is the complete response of the desired (output) variable of that system to the impulse (input) when all initial conditions are zero.

We know that the $u_0(t)$ is the derivative of $u_{-1}(t)$, that is, the (zero-state) impulse response is the derivative of the (zero-state) step response. Hence the resulting voltage across the capacitor is

$$v_{u0}(t) = \frac{dv(t)}{dt} = (2{\cdot}5e^{-3t} \sin 4t)\, u_{-1}(t)$$

The zero-state current impulse response is

$$i_{u0}(t) = \frac{di(t)}{dt} = \frac{d}{dt}[(0{\cdot}05\, e^{-3t} \sin 4t)\, u_{-1}(t)]$$

$$= (0{\cdot}05\, e^{-3t} \sin 4t)\, u_0(t)$$

$$+ 0{\cdot}05\,(-3\, e^{-3t} \sin 4t + 4e^{-3t} \cos 4t)\, u_{-1}(t)$$

$$= 0 + 0.05\,(4e^{-3t} \cos 4t - 3e^{-3t} \sin 4t)\, u_{-1}(t)$$

$$= 0.05\,(4e^{-3t} \cos 4t - 3e^{-3t} \sin 4t)\, u_{-1}(t)$$

Put $r \cos \alpha = 4$, $r \sin \alpha = 3$

therefore, $r = \sqrt{4^2 + 3^2} = 5$, $\tan \alpha = \dfrac{3}{4}$

$$\alpha = 36{\cdot}86° = 0{\cdot}6435 \text{ radian}$$

therefore, $i_{u0}(t) = 0{\cdot}05\, e^{-3t}(r \cos \alpha \cos 4t - r \sin \alpha \sin 4t)\, u_{-1}(t)$

$$= 0{\cdot}05\, e^{-3t}\, r \cos(4t + \alpha)\, u_{-1}(t)$$

or $i_{u0}(t) = 0{\cdot}25\, e^{-3t} \cos(4t + 0{\cdot}6435)\, u_{-1}(t)$

■ **EXAMPLE 14.29** A series RLC network with $R = 5\ \Omega$, $L = 0{\cdot}1$ H, and $C = 500\ \mu$F has a constant voltage $V = 10$ volts applied at $t = 0$. Determine the resulting current.

SOLUTION : $\alpha = \dfrac{R}{2L} = \dfrac{5}{2 \times 0{\cdot}1} = 25$

$$\omega_n = \frac{1}{\sqrt{LC}} = \frac{1}{\sqrt{(0{\cdot}1 \times 500 \times 10^{-6})}} = 141{\cdot}4$$

Since $\alpha < \omega_n$, an underdamped response is indicated

$$\omega_d = \sqrt{\omega_n^2 - \alpha^2} = \sqrt{20\,000 - 625} = 139$$

therefore, and $i = e^{-\alpha t}(B_1 \cos \omega_d t + B_2 \sin \omega_d t)$

At $t = 0_+$, $i = 0$ $\quad 0 = e^0 (B_1 + 0) \Rightarrow B_1 = 0$

therefore, $i = e^{-\alpha t}(B_2 \sin \omega_d t)$ (E 14.29.1)

For the given network

$$Ri + L\frac{di}{dt} + \frac{1}{C}\int i\,dt = 10 \qquad \text{(E 14.29.2)}$$

At $t = 0, i = 0, \frac{1}{C}\int i\,dt = 0$

Substituting these values in Eq. (E 14.29.2) we get

$$\left.\frac{di}{dt}\right|_{t=0} = \frac{10}{L} = \frac{10}{0.1} = 100$$

Differentiation of Eq. (E 14.29.1) gives

$$\frac{di}{dt} = B_2\,[e^{-\alpha t}\,\omega_d \cos \omega_d t - \alpha e^{-\alpha t} \sin \omega_d t]$$

$$= B_2\,[e^{-25t} \times 139 \cos 139\,t - 25e^{-25t} \sin 139\,t]$$

At $t = 0$,

$$100 = B_2\,[e^0 \times 139 \cos 0° - 0]$$

$$B_2 = \frac{100}{139} = 0{\cdot}72$$

Substituting the values of α, ω_d, and B_2 in Eq. (E 14.29.1) gives

$$i = 0{\cdot}72e^{-25t} \sin 139\,t.$$

EXAMPLE 14.30 A series RLC network with $R = 5\ \Omega$, $L = 0{\cdot}1$ H, and $C = 500\ \mu$F has a sinusoidal voltage $v = 100 \sin (250\,t + \phi)$ volts applied at a time when $\phi = 0°$. Determine the resulting current.

SOLUTION : The natural response is given by

$$i_n = e^{-\alpha t}(B_1 \cos \omega_d t + B_2 \sin \omega_d t)$$

From Example 14.29,

$$i_n = e^{-25t}(B_1 \cos 139\,t + B_2 \sin 139\,t) \qquad \text{(E 14.30.1)}$$

Determination of forced response

$$X_L = 250 \times 0{\cdot}1 = 25\ \Omega$$

$$X_C = \frac{1}{2\pi fC} = \frac{1}{250 \times 500 \times 10^{-6}} = 8\ \Omega$$

$$\mathbf{Z} = R + j(X_L - X_C)$$

$$= 5 + j(25 - 8) = 5 + j\,17$$

$$= 17{\cdot}72 \,\underline{/\ 73{\cdot}6°}$$

$$\mathbf{I}_m = \frac{\mathbf{V}_m}{\mathbf{Z}} = \frac{100\,\underline{/\ 0°}}{17{\cdot}72\,\underline{/\ 73{\cdot}6°}} = 5{\cdot}64\,\underline{/\ -73{\cdot}6°}\ \text{A}$$

therefore, $i_f = I_m \sin(\omega t - \phi)$

$$= 5{\cdot}64 \sin (250\, t - 73{\cdot}6°) \tag{E 14.30.2}$$

The complete response is given by

$$i = i_n + i_f$$

$$= e^{-25t} (B_1 \cos 139\, t + B_2 \sin 139\, t) + 5{\cdot}64 \sin (250\, t - 73{\cdot}6°) \tag{E 14.30.3}$$

To determine the constants B_1 and B_2 we have to determine i and $\frac{di}{dt}$ at $t = 0$.

By KVL

$$v_R + v_L + v_C = v_S$$

$$Ri + L\frac{di}{dt} + \frac{1}{C}\int idt = 100 \sin (250\, t + 0°) \tag{E 14.30.4}$$

At $t = 0$, $i = 0$, $\frac{1}{C}\int idt = 0$

therefore, $L\left.\frac{di}{dt}\right|_{t=0} = 0$

or $\left.\frac{di}{dt}\right|_{t=0} = 0$

Differentiation of Eq. (E 14.30.3) gives

$$\frac{di}{dt} = e^{-25t} (-\,139\, B_1 \sin 139t + 139\, B_2 \cos 139t)$$

$$+ (B_1 \cos 139t + B_2 \sin 139t) (-25e^{-25t})$$

$$+ 5{\cdot}64 \times 250 \cos (250\, t - 73{\cdot}6°) \tag{E 14.30.5}$$

At $t = 0$, $\frac{di}{dt} = 0$

Putting these values in Eq. (E 14.30.5) gives

$$0 = e^0 (0 + 139B_2) + B_1(-\,25\, e^0) + 5{\cdot}64 \times 250 \cos (-\,73{\cdot}6°)$$

or $$139B_2 - 25B_1 + 398 = 0 \tag{E 14.30.6}$$

Substitution of $i = 0$ at $t = 0$ in Eq. (E 14.30.3) gives

$$0 = e^0 (B_1 \cos 0° + B_2 \sin 0°) + 5{\cdot}64 \times \sin (-\,73.6°)$$

or $$B_1 = 5{\cdot}41$$

Substitution of the value of B_1 in Eq. (E 14.30.6) gives

$$139\, B_2 - 25 \times 5{\cdot}41 + 398 = 0$$

$$B_2 = -\,1{\cdot}89$$

Substituting the values of B_1 and B_2 in Eq. (E 14.30.3) gives the complete response as

$$i = e^{-25t} (5{\cdot}41 \cos 139\, t - 1{\cdot}89 \sin 139\, t) + 5{\cdot}64 \sin (250\, t - 73{\cdot}6°)\, A$$

14.37 TRANSFORMED EQUIVALENT CIRCUITS

The complete time response of a circuit to any dc or time-varying input can be obtained by the use of **Laplace transformed equivalent circuit.** Such a circuit is also known as **s-domain circuit.** The following steps are used :

1. Determine all initial inductor currents $i_L(0_-)$.
2. Determine all initial capacitor voltages $v_C(0_-)$.
3. Draw the s-domain circuit, for $t > 0$ with :
   (*a*) All time-domain independent sources represented by their transformed values;
   (*b*) All time-domain dependent sources represented by their transformed values ;
   (*c*) Each resistor represented by its impedance, $R\ \Omega$;
   (*d*) Each inductor represented by the series combination of its impedance, $sL\ \Omega$, and a constant voltage source $L\, i_L(0_-)$ volts; the polarity of this initial-condition voltage source is the same as the initial current, $i_L(0_-)$, through the inductor
   (e) Each capacitor represented by the series combination of its impedance, $\frac{1}{sC}\ \Omega$ and a constant voltage source $\frac{v_C(0_-)}{s}$ volts; the polarity of this initial-condition voltage source is in the same direction as the initial voltage, $v_C(0_-)$, across the capacitor.
4. Solve the s-domain circuit for the unknown transformed voltage or current by applying circuit techniques (KCL, KVL, Ohm's law, mesh analysis, node analysis, etc.)
5. Take the inverse Laplace transform of the unknown transformed voltage or current. This gives the response in the time domain.

■ **EXAMPLE 14.31** For the circuit shown in Fig. 14.31a, the switch S has been open for an infinite time and is then closed at $t = 0$. Determine the nodal voltage $v(t)$ for $t > 0$ by using the Laplace transformed equivalent circuit method. Check the result by the method using Thevenin equivalent circuit.

SOLUTION : The equivalent circuit for $t < 0$ is shown in Fig. 14.31b. By current-division rule in this circuit.

$$i_L(0_-) = \frac{10}{3+10} \times 10 = \frac{100}{13}\text{ A}$$

The s-domain network of the circuit in Fig. 14.31a for $t > 0$ is shown in Fig. 14.31c. By nodal analysis,

$$Y(s)\, V(s) = I(s)$$

$$\left(\frac{1}{10} + \frac{1}{10} + \frac{1}{3+4s}\right) V(s) = \frac{10}{s} - \frac{100}{13s}\left[\frac{4s}{3+4s}\right]$$

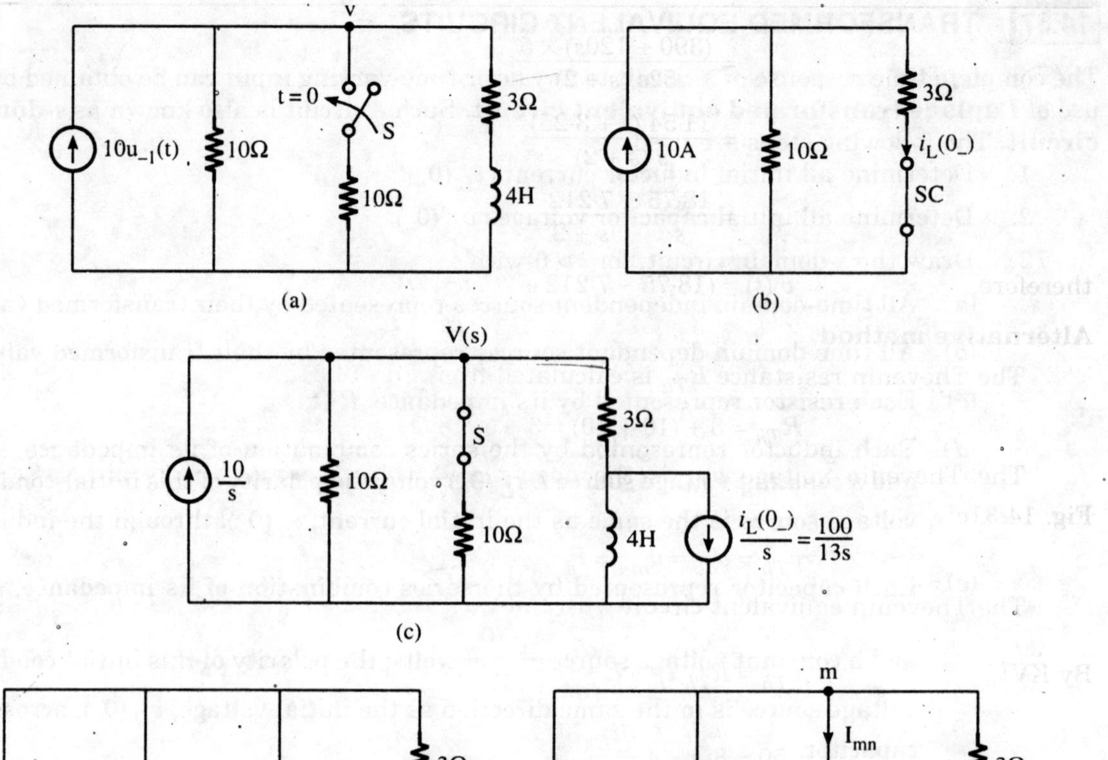

FIG. 14.31

$$V(s) = \frac{\dfrac{10}{s} - \dfrac{400}{13\,(3+4s)}}{\left(\dfrac{1}{10} + \dfrac{1}{10} + \dfrac{1}{3+4s}\right)} = \frac{\dfrac{390 + 520s - 400s}{13s\,(3+4s)}}{\dfrac{3+4s+5}{5\,(3+4s)}}$$

$$= \frac{(390 + 120s) \times 5}{52s.(s + 2)}$$

$$= \frac{11{\cdot}54\,(s + 3{\cdot}25)}{s\,(s + 2)}$$

$$= \frac{18{\cdot}75}{s} - \frac{7{\cdot}212}{s + 2}$$

therefore, $v\,(t) = (18{\cdot}75 - 7{\cdot}212\,e^{-2t})\,u_{-1}\,(t)$

Alternative method

The Thevenin resistance R_{Th} is calculated from Fig. 14.31d.

$$R_{Th} = 3 + (10\,||10) = 3 + 5 = 8\ \Omega$$

The Thevenin voltage v_{Th} is the open-circuit voltage across the terminals $a - b$ in Fig. 14.31e.

$$v_{Th} = v_{ab} = v_{mn} = I_{mn} \times 10 = 5 \times 10 = 50 \text{ V}$$

The Thevenin equivalent circuit with the load inductor of 4 H is shown in Fig. 14·31f.

By KVL, $v_{Th} = R_{Th}\,i_L + L\dfrac{di_L}{dt}$

$$50 = 8i_L + 4\frac{di_L}{dt}$$

This equation can be solved by the method given earlier in this chapter. The solution is given by

$$i_L\,(t) = i_{Ln} + i_{Lf}$$

$$= Ae^{-\frac{Rt}{L}} + \frac{50}{8}$$

$$= Ae^{-\frac{8}{4}t} + \frac{50}{8}$$

$$= Ae^{-2t} + \frac{25}{4} \qquad \text{(E 14.31.1)}$$

By the continuity relationship of an inductor

$$i_L\,(0_+) = i_L\,(0_-) = \frac{100}{13}$$

Substituting $t = 0_+$ in Eq. (E 14.31.1), we get

$$i_L\,(0_+) = A\,e^0 + \frac{25}{4}$$

$$\frac{100}{13} = A + \frac{25}{4}$$

therefore, $A = \dfrac{100}{13} - \dfrac{25}{3} = \dfrac{75}{52}$

and $$i_L(t) = \frac{75}{52}e^{-2t} + \frac{25}{4}$$

From Fig. 14.31g,

$$10 = i_1 + i_1 + i_L$$

therefore, $$i_t = \frac{1}{2}(10 - i_L)$$

$$= 5 - \frac{1}{2}\left(\frac{75}{52}e^{-2t} + \frac{25}{4}\right)$$

$$v = 10\, i_1$$

$$= 50 - \frac{5 \times 75}{52}e^{-2t} - \frac{250}{8}$$

therefore, $$v(t) = 18{\cdot}75 - 7{\cdot}211\, e^{-2t}$$

■ **EXAMPLE 14.32** In the circuit shown in Fig. 14.32a, the switch S has been in position 1 for a long time and is moved to position 2 at $t = 0$. Determine $i_L(t)$ for $t > 0$. The source voltage is $v = 200 \sin(314\, t + 30°)$.

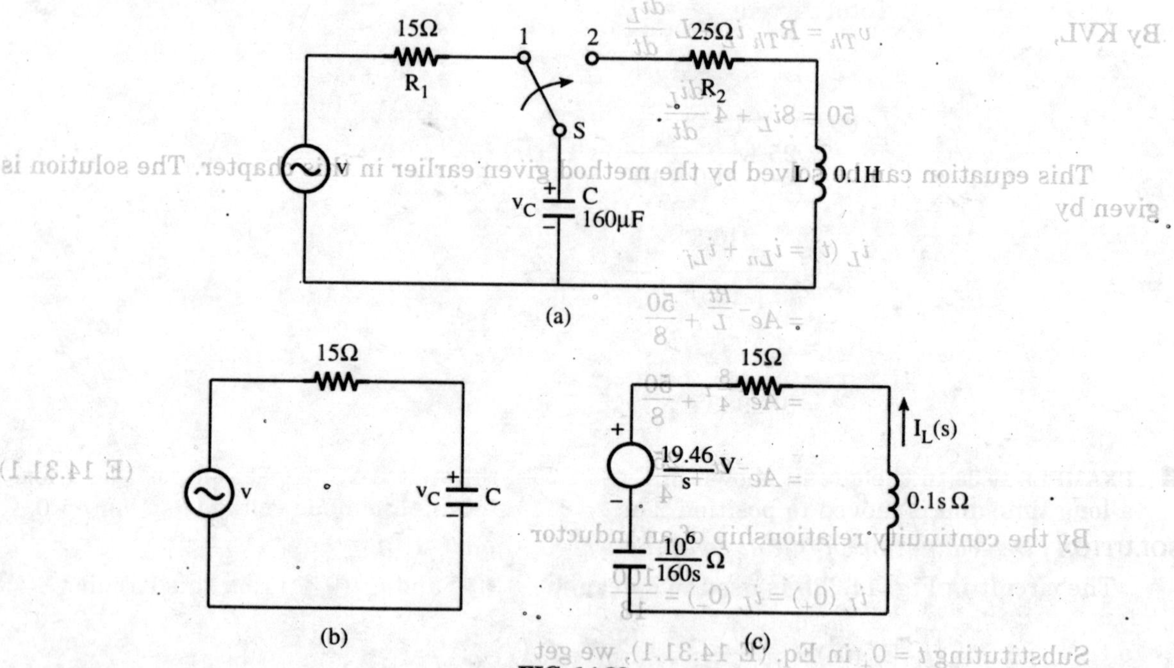

FIG. 14.32

SOLUTION : *Determination of initial conditions* $i_L(0\_)$ and $v_C(0\_)$.

The inductor current $i_L(0\_) = 0$, because the switch is in position 1 at $t = 0\_$. It is to be noted carefully that $v_C(0\_)$ is the voltage across C when the switch is in position 1 and the source voltage is the sinusoidal voltage given by

$$v = 200 \sin(31\,4t + 30°)$$

Therefore, the capacitor should *not* be treated as an open circuit, which is permissible only with a dc source. Since the switch S has been in position 1 for a long time, steady-state condition has reached and there is no transient. Therefore, we should determine the steady-state expression for $v_C(t)$ at $t = 0_-$. We shall use the phasor method to determine $v_C(0_-)$. The circuit for $t < 0$ s shown in Fig. 14.32b.

$$X_C = \frac{1}{2\pi f C} = \frac{1}{314 \times 160 \times 10^{-6}} = 19{\cdot}90\ \Omega$$

$$\mathbf{Z} = R - jX_c = 15 - j\,19{\cdot}9 = 24{\cdot}92 \angle -52{\cdot}99°\ \Omega$$

$$\mathbf{I_m} = \frac{\mathbf{V_m}}{\mathbf{Z}} = \frac{200 \angle 30°}{24.92 \angle -52.99°} = 8.026 \angle 82.99°\ \text{A}$$

$$\mathbf{V_{cm}} = -j I_m X_c = 8.026 \times 19.9 \angle -90° + 82.99° = 159.71 \angle -7.01°\ \text{V}$$

$$v_C(t) = 159{\cdot}71 \sin(\omega t - 7{\cdot}01°)$$

$$v_C(0_-) = 159{\cdot}71 \sin(0° - 7{\cdot}01°) = -19{\cdot}46\ \text{V}$$

To determine $i_L(t)$ for $t > 0$, the s-domain equivalent circuit shown in Fig. 14.32c is used.

$$\text{Total current} = \frac{\text{total voltage}}{\text{total impedance}}$$

$$I_L(s) = \frac{\dfrac{19{\cdot}46}{s}}{25 + \dfrac{1}{160 \times 10^{-6} s} + 0{\cdot}1\,s} = \frac{194{\cdot}6}{s^2 + 250s + 62\,500}$$

$$= \frac{216{\cdot}5 \times \dfrac{194{\cdot}6}{216{\cdot}5}}{(s+125)^2 + (216{\cdot}5)^2} = \frac{0{\cdot}8988 \times 216{\cdot}5}{(s+125)^2 + (216{\cdot}5)^2}$$

$$i_L(t) = \pounds^{-1} I_2(s)$$

$$= (0{\cdot}8988\, e^{-125t} \sin 216{\cdot}5t)\, u_{-1}(t)$$

■ **EXAMPLE 14.33** In the circuit shown in Fig. 14.33a, the switch S has been in position 1 for a long time and is moved to position 2 at $t = 0$. Determine the node voltage $v(t)$ for $t > 0$.

SOLUTION : *Determination of initial conditions* $i_L(0_-)$ and $v_C(0_-)$.

The circuit in Fig. 14.33b is used to determine $i_L(0_-)$ and $v_C(0_-)$. From this circuit

$$i_L(0_-) = \frac{V}{R_1 + R_2} = \frac{30}{2+4} = 5\ \text{A}$$

$$v_C(0_-) = R_2\, i_L(0_-) = 4 \times 5 = 20\ \text{V}$$

The determine $v(t)$ for $t > 0$, the s-domain equivalent curcuit shown in Fig. 14.33c is obtained. By source transformation, we get the circuit shown in Fig. 14.33d.

The nodal equations at nodes (1) and (2) can be written as

$$[\mathbf{Y}]\,[\mathbf{V}] = [\mathbf{I}]$$

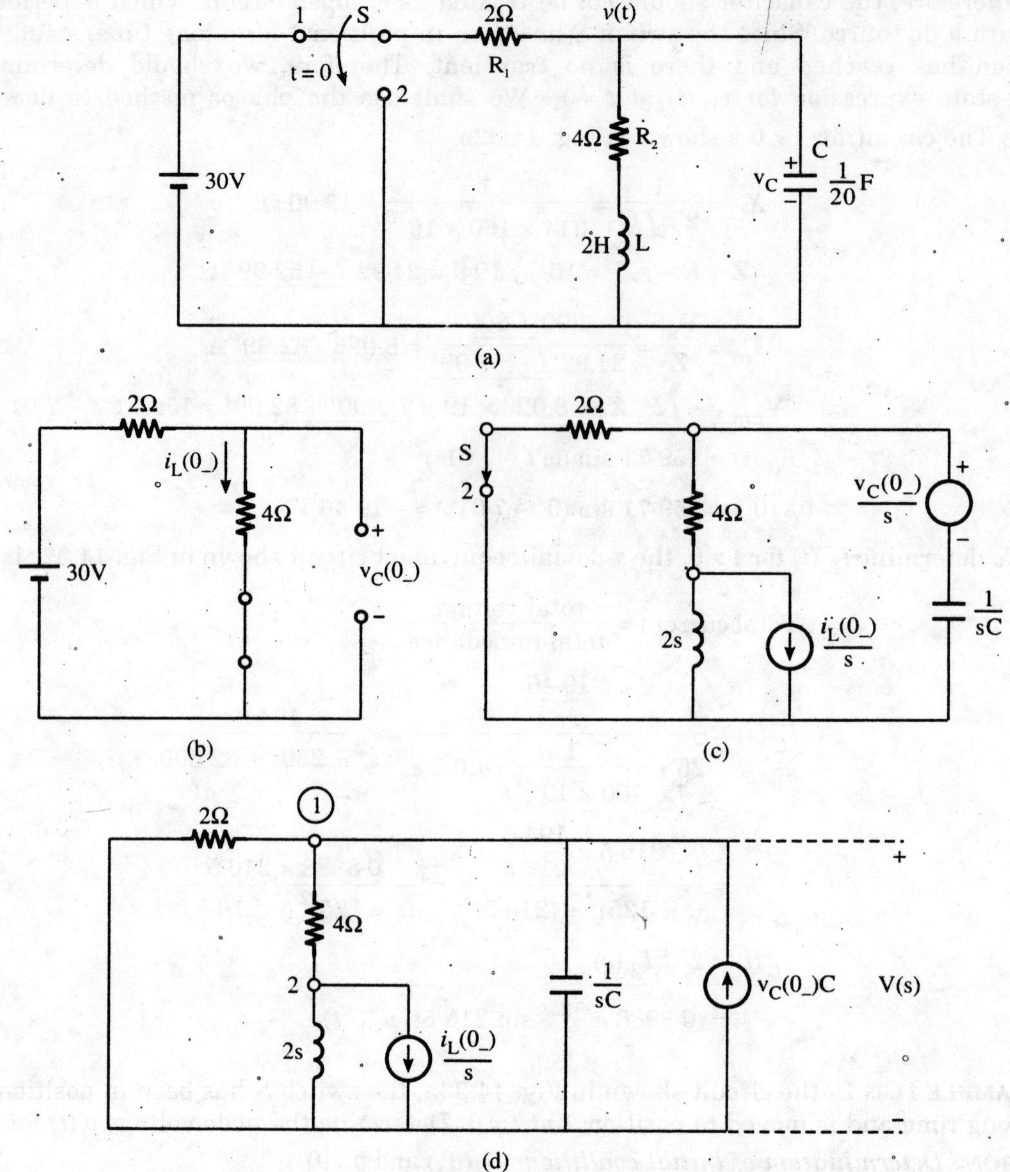

FIG. 14.33

$$s\begin{bmatrix} \frac{1}{2}+\frac{1}{4}+\frac{1}{20}s & -\frac{1}{4} \\ -\frac{1}{4} & \frac{1}{4}+\frac{1}{2s} \end{bmatrix}\begin{bmatrix} V_1(s) \\ V_2(s) \end{bmatrix} = \begin{bmatrix} v_C(0_-)\,C \\ -\frac{i_L(0_-)}{s} \end{bmatrix} = \begin{bmatrix} 1 \\ -\frac{5}{s} \end{bmatrix}$$

$$|Y| = \left(\frac{1}{2}+\frac{1}{4}+\frac{1}{20}s\right)\left(\frac{1}{4}+\frac{1}{2s}\right) - \left(\frac{1}{4}\right)^2 = \frac{s^2+12s+30}{80\,s}$$

By Cramer's rule,

$$V(s) = V_1(s) = \frac{\begin{vmatrix} 1 & -\frac{1}{4} \\ -\frac{5}{s} & \frac{1}{4} + \frac{1}{2s} \end{vmatrix}}{|Y|} = \frac{\frac{1}{4} + \frac{1}{2s} - \frac{5}{4s}}{\frac{s^2 + 12s + 30}{80s}}$$

$$= \frac{20(s-3)}{s^2 + 12s + 30} = \frac{20(s-3)}{(s + 8{\cdot}45)(s + 3{\cdot}55)}$$

therefore,

$$V(s) = \frac{46{\cdot}73}{s + 8{\cdot}45} - \frac{26{\cdot}73}{s + 3{\cdot}55}$$

$$v(t) = 46{\cdot}73\, e^{-8{\cdot}45t} - 26{\cdot}73\, e^{-3{\cdot}55\, t}$$

EXAMPLE 14.34 In the circuit shown in Fig. 14.34a, the switch S has been open for a long time and is closed at $t = 0$. Determine $i_2(t)$ for $t > 0$.

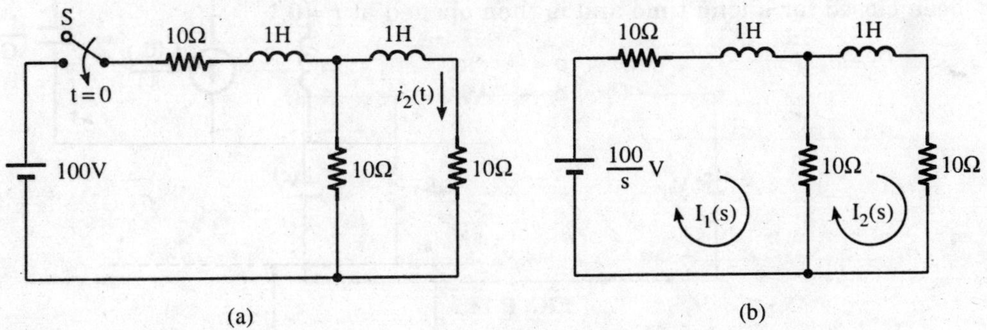

FIG. 14.34

SOLUTION : To determine $i_2(t)$ for $t > 0$, the s-domain circuit shown in Fig. 14.34b is used. The mesh equations can be written as

$$(s + 20)\, I_1(s) - 10\, I_2(s) = \frac{100}{s}$$

$$-10\, I_1(s) + (s + 20)\, I_2 = (s) = 0$$

$$I_2(s) = \frac{\begin{vmatrix} s + 20 & \frac{100}{s} \\ -10 & 0 \end{vmatrix}}{\begin{vmatrix} s + 20 & -10 \\ -10 & s + 20 \end{vmatrix}} = \frac{1000}{s(s^2 + 40s + 300)} = \frac{1000}{s(s + 10)(s + 30)}$$

$$= \frac{3{\cdot}33}{s} - \frac{5}{s + 10} + \frac{1{\cdot}67}{s + 30}$$

therefore,

$$i_2(t) = (3{\cdot}33 - 5e^{-10t} + 1{\cdot}67\, e^{-30t})\, u_{-1}(t)$$

EXERCISES

1. Determine the inductor current for $t > 0$ in the network shown in Fig. P 14.1 The switch S has been open for a long time and is then closed at $t = 0$.

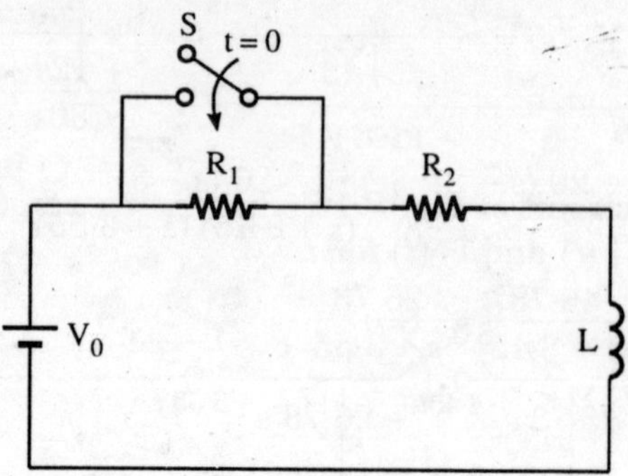

FIG. P 14.1

2. Determine the voltage across the capacitor for $t > 0$ in Fig. P 14.2. The switch S has been closed for a long time and is then opened at $t = 0$.

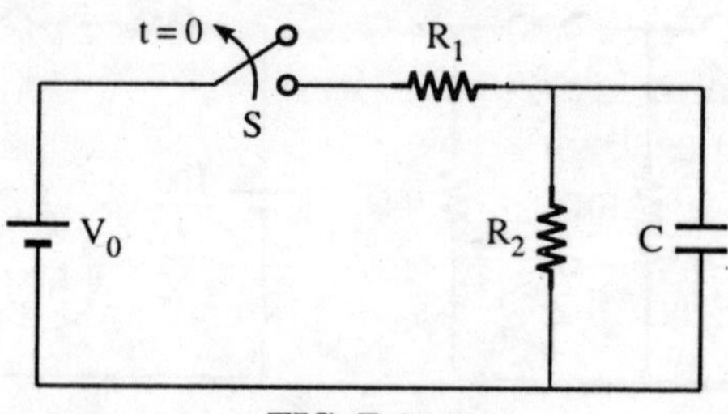

FIG. P 14.2

3. In the circuit shown in Fig. P 14.3, the switch S is moved from position 1 to position 2 at $t = 0$, the steady-state current having previously established in the circuit. Find $i_L(t)$ after switching.

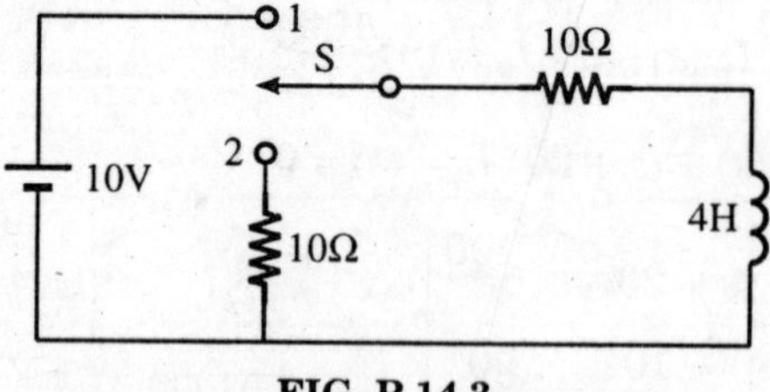

FIG. P 14.3

4. A ramp voltage $v = kt$ is applied across a RL circuit at $t = 0$. Determine the current $i_L(t)$ in the circuit at $t \geq 0$ assuming $i_L(0_-) = 0$.
5. In the circuit shown in Fig. P 14.4, the switch S has been open for a long time and is closed at $t = 0$. Determine i_L and v_L for $t > 0$.

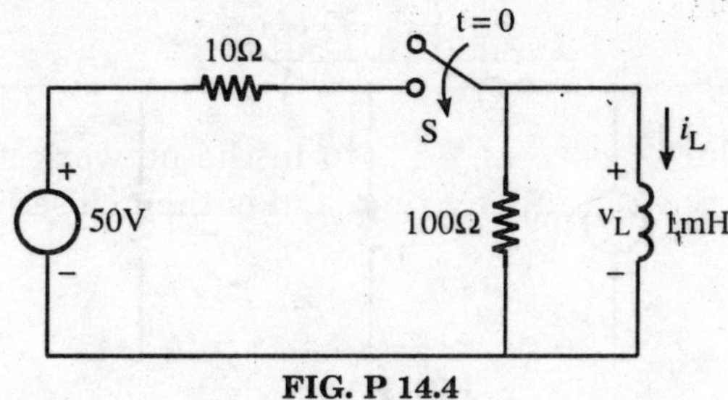

FIG. P 14.4

6. The switch S in the circuit of Fig. P 14.5 has been open for a long time and is closed at $t = 0$. Determine $i_1(t)$ and $i_2(t)$ for $t > 0$.

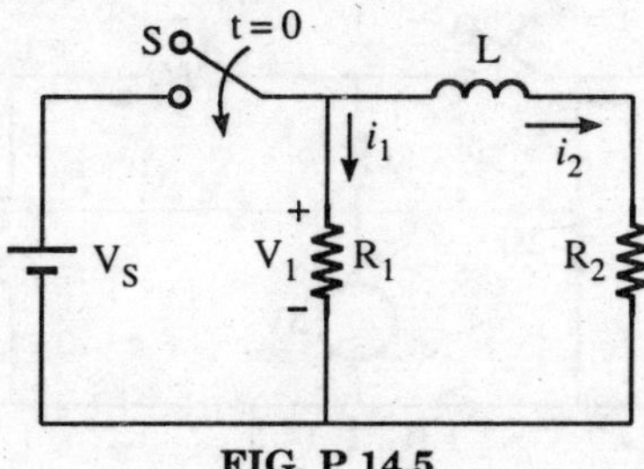

FIG. P 14.5

7. For the circuit shown in Fig. P 14.6, the switch S has been in position 1 for a long time and is moved to position 2 at $t = 0$. Find $i(t)$ for $t \geq 0$.

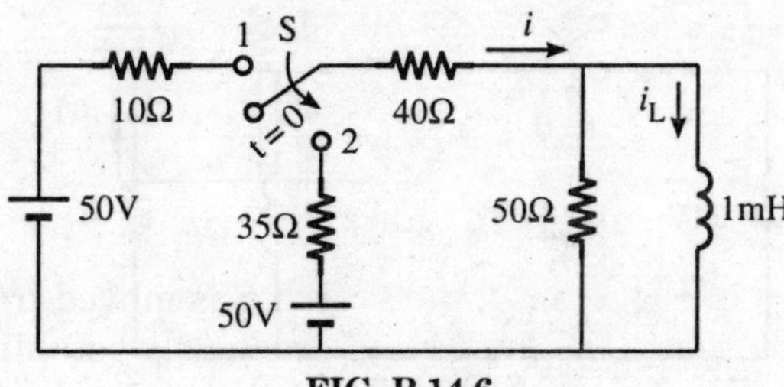

FIG. P 14.6

8. For the circuit shown in Fig. P 14.7, the switch S has been in position 1 for a long time and is moved to position 2 at $t = 0$. Find $i_L(t)$ for $t \geq 0$.

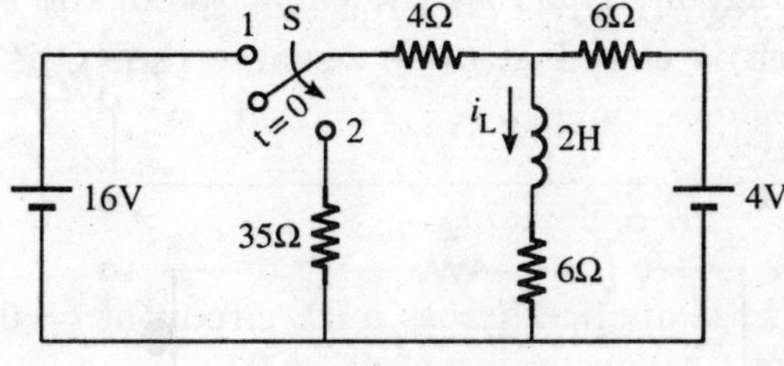

FIG. P 14.7

9. For the circuit shown in Fig. P 14.8, determine the current i_1 for $t > 0$. The switch S has been closed for a long time and is opened at $t = 0$.

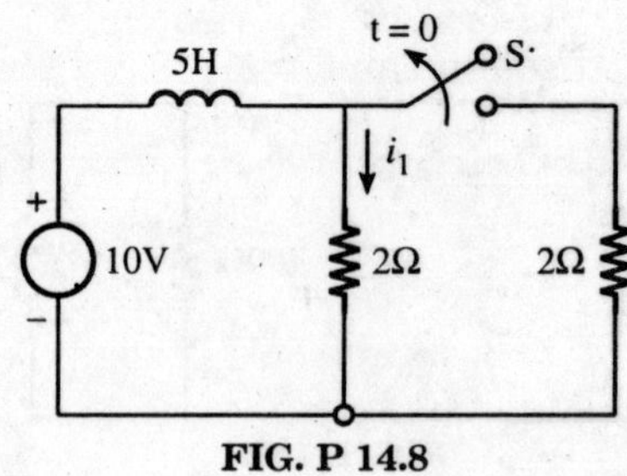

FIG. P 14.8

10. For the circuit shown in Fig. P 14.9, the switch S has been closed for a long time and is then opened at $t = 0$. Determine the current through the inductor at $t > 0$.

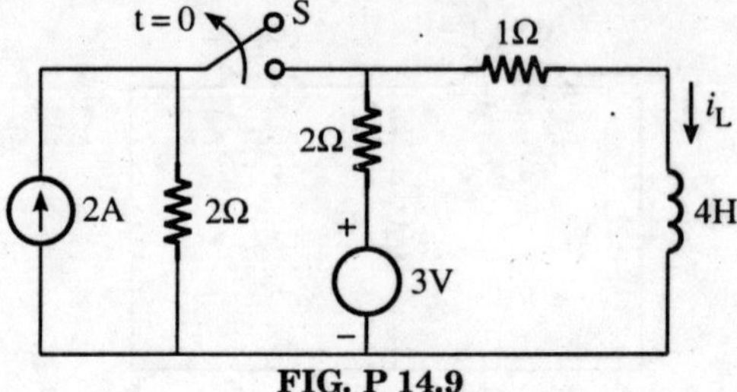

FIG. P 14.9

11. In the circuit shown in Fig. P 14.10, the switch S has been open for a long time and is closed at $t = 0$. Determine the current through the 4Ω resistor for $t > 0$.

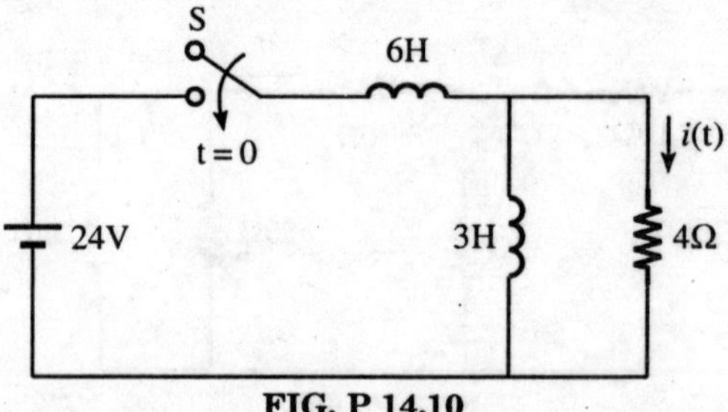

FIG. P 14.10

12. Determine the variation of voltage v across a parallel G-C circuit if a current source $i = I_m \sin(\omega t + \alpha)$ is applied across its terminals at $t = 0$. Assume $v_C(0\_) = 0$.
13. Determine the voltage $v_C(t)$ across the capacitor in the series RLC circuit of Fig. P 14.11. The switch is closed at $t = 0$. Assume that C is initially uncharged and $R < 2\sqrt{L/C}$.

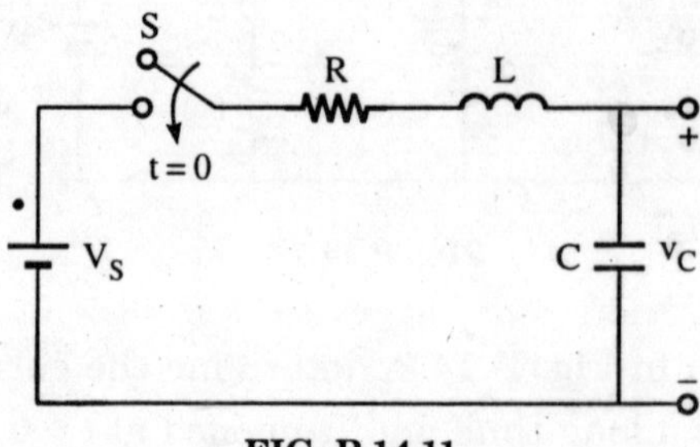

FIG. P 14.11

14. Determine the current through the circuit at $t > 0$ when a voltage $v_S(t) = 10\,e^{-2t}$ is applied across a series RLC circuit at $t = 0$ if $R = 8\ \Omega$, $L = 2\ H$, and $C = \frac{1}{6}$ F. Assume $i_L(0_-) = 0$.

ANSWERS

1. $i(t) = \dfrac{V_0}{R_2}\left(1 - \dfrac{R_1}{R_1 + R_2}\, e^{-\frac{R_2}{L}t}\right)$

2. $v_C(t) = \dfrac{R_2 V_0}{R_1 + R_2}\, e^{-\frac{t}{R_2 C}}$

3. $i_L(t) = e^{-5t}$

4. $i_L(t) = \dfrac{k}{R}\left(1 - \dfrac{L}{R}\right) + \dfrac{kL}{R^2}\, e^{-\frac{Rt}{L}}$

5. $i_L(t) = 5 - 5\,e^{-9.1 \times 10^3\, t}$ A

$v_L(t) = 45.5\,e^{-9.1 \times 10^3\, t}$ V

6. $i_1(t) = \dfrac{V_S}{R_1}$

$i_2(t) = \dfrac{V_S}{R_2} - \dfrac{V_S}{R_2}\, e^{-\frac{R_2}{L}t}$

7. $i(t) = 0.67 + 0.13\,e^{-3 \times 10^4\, t}$ A

8. $i_L(t) = \dfrac{2}{9} + \dfrac{10}{9}\, e^{-4.5\,t}$ A

9. $i_1(t) = 5 + 5\,e^{-\frac{t}{2.5}}$ A

10. $i_L(t) = \dfrac{5}{4}\, e^{-3\frac{t}{4}} - 1$ A

11. $i(t) = 2 - 2\,e^{-2t}$ A

12. $v(t) = \dfrac{I_m}{\sqrt{(G^2 + \omega^2 C^2)}}\,[\sin(\omega t + \alpha - \Psi) - \sin(\alpha - \Psi)\, e^{-\frac{G}{C}t}]$

where $\Psi = \tan^{-1}\left(\dfrac{\omega C}{G}\right)$

13. $v_C(t) = V_S\,[1 - e^{-\alpha t}(\cos \omega_n t + \dfrac{\alpha}{\omega_n} \sin \omega_n t)]$ V

where $\alpha = \dfrac{R}{2L}$, $\omega_n = \sqrt{\dfrac{1}{LC} - \left(\dfrac{R}{2L}\right)^2}$

14. $i(t) = -\,2.5\,e^{-t} + 10\,e^{-2t} - 7.5\,e^{-3t}$ A

15

STATE-VARIABLE ANALYSIS

15.1 INTRODUCTION

The technique for determining the state of a system is called **state-variable analysis**. The ***state*** *of a dynamic system is the smallest set of variables such that the knowledge of these variables at* $t = t_0$ *together with the knowledge of the input for* $t \geq t_0$ *completely determines the behaviour of the system for any time* $t \geq t_0$. This set of variables is called **state variables** or **phase variables.**

If we need n variables to completely describe the behaviour of a given system, then these n state variables may be considered as n components of a vector $\boldsymbol{x}$. Such a vector is called a **state vector**. *A* ***state vector*** *is thus a vector which determines uniquely the system state* $\boldsymbol{x}(t)$ *for any time* $t \geq t_0$*, once the state at* $t = t_0$ *is given and the input* $\mathbf{u}(t)$ *for* $t \geq t_0$ *is specified.*

The n-dimensional space whose coordinate axes consist of the x_1 axis, x_2 axis, ..., x_n axis is called the **state space**. Any state can be represented by a point in the state space.

15.2 STATE EQUATIONS

Consider an RLC network shown in Fig. 15.1. Let the current at time $t = 0$ be $i_L(0)$ and the capacitor voltage at time $t = 0$ be $v_C(0)$. Thus, we can say that the **condition** or the **state** of the network at time $t = 0$ is specified by the inductor current and the capacitor voltage. Hence the pair $[i_L(0), v_C(0)]$ is called the **initial state** of the network. Similarly, the pair $[i_L(t), v_C(t)]$ may be called the state of the network at time t. The variables i_L and v_C are called *state variables* of the network.

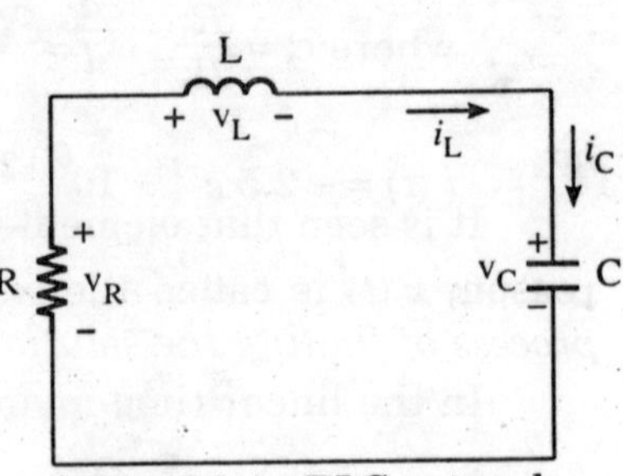

FIG. 15.1 An RLC network

Applying KVL to the circuit we get

$$v_R + v_L + v_C = 0$$

$$R\, i_L + L\frac{di_L}{dt} + v_C = 0 \qquad (15.2.1)$$

Also, $$i_C = i_L = C\frac{dv_C}{dt} \tag{15.2.2}$$

In Eqs. (15.2.1) and (15.2.2) we find that only variables present are state variables i_L and v_C. Let us write these equations in such a way that the coefficients of the derivatives are unity, and the derivatives alone are on one side of the equation. Thus, we can write Eqs. (15.2.1) and (15.2.2) in the form

$$\frac{di_L}{dt} = -\frac{R}{L}i_L - \frac{1}{L}v_C \tag{15.2.3}$$

$$\frac{dv_C}{dt} = \frac{1}{C}i_L \tag{15.2.4}$$

Equations of this form are called **state equations**. In such equations all the variables present are state variables.

The above two state equations may be written as a single matrix state equation as

$$\begin{bmatrix} \dfrac{di_L}{dt} \\ \dfrac{dv_C}{dt} \end{bmatrix} = \begin{bmatrix} -\dfrac{R}{L} & -\dfrac{1}{L} \\ \dfrac{1}{C} & 0 \end{bmatrix} \begin{bmatrix} i_L \\ v_C \end{bmatrix} \tag{15.2.5}$$

Eq. (15.2.5) may be written as

$$\frac{d}{dt}\begin{bmatrix} i_L \\ v_C \end{bmatrix} = \begin{bmatrix} -\dfrac{R}{L} & -\dfrac{1}{L} \\ \dfrac{1}{C} & 0 \end{bmatrix} \begin{bmatrix} i_L \\ v_C \end{bmatrix} \tag{15.2.6}$$

Let us define the matrices $\boldsymbol{x}(t)$ and $\boldsymbol{A}$ as follows :

$$\boldsymbol{x}(t) \overset{\Delta}{=} \begin{bmatrix} i_L \\ v_C \end{bmatrix}$$

$$\boldsymbol{A} = \begin{bmatrix} -\dfrac{R}{L} & -\dfrac{1}{L} \\ \dfrac{1}{C} & 0 \end{bmatrix}$$

Eq. (15.2.6) then takes the form

$$\frac{d}{dt}\boldsymbol{x}(t) = \boldsymbol{A}\,\boldsymbol{x}(t) \tag{15.2.7}$$

or $$\dot{\boldsymbol{x}}(t) = \boldsymbol{A}\,\boldsymbol{x}(t) \tag{15.2.8}$$

It is seen that elements of the column matrix $\boldsymbol{x}(t)$ are state variables i_L and v_C. For this reason, $\boldsymbol{x}(t)$ is called the *state-vector*. The vector $\boldsymbol{x}(0)$ is called the *initial-state vector. The process of finding the solution to the state equations is called* ***state-variable analysis***.

In the linear time-invariant systems, the general form of state equations are

$$\dot{\boldsymbol{x}} = \boldsymbol{A}\,\boldsymbol{x}(t) + \boldsymbol{B}\,\boldsymbol{u}(t) \tag{15.2.9}$$

$$\boldsymbol{y} = \boldsymbol{C}\, x\,(t) + \boldsymbol{D}\,\boldsymbol{u}\,(t) \tag{15.2.10}$$

There equations are vector differential equations where $\boldsymbol{x}$ is the state vector, $\boldsymbol{u}$ is the input vector, and $\boldsymbol{y}$ is the output vector.

Eq. (15.2.9) is usually called the state equation and Eq. (15.2.10) is called the **output equation**. Together they are sometimes known as the **state equations**. The matrices **A** , **B** , **C** and **D** should not be confused with the transmissions parameters (scalars) A, B, C and D of 2-port networks.

15.3 CHOICE OF STATE VARIABLES

Since the state equations represent the dynamical behaviour of a given network, it is logical to choose the state variables as the charges or voltages on the capacitors and the fluxes or currents through the inductors. Such a choice ensures that the resulting dynamical equations are a set of first-order differential equations.

State equations follow a specific format. On the left-hand side each state equation has the derivative of one (and only one) of the state variables. On the right-hand side is a mathematical function involving any or all the state variables and the sources. No derivatives appear on the right-hand side.

15.4 CAPACITOR-ONLY LOOP

*A loop is called a **capacitor-only loop** if it is made up of capacitors only (two or more) and possibly some independent voltage sources.*

15.5 INDUCTOR-ONLY CUTSET

*A cutset is called an **inductor-only cutset** if it is made up of inductors only (two or more) and possibly some current sources.*

15.6 PROCEDURE OF WRITING THE STATE EQUATIONS OF SIMPLE NETWORKS BY INSPECTION

The following steps are involved in writing the state equations of simple networks. This method is based on KCL, KVL and *v-i* relationships of the network elements.

Step 1. If there are no capacitor-only loops and no inductor-only cut-sets, choose either
(*a*) the capacitor voltages and inductor currents *or*
(*b*) the capacitor charges and inductor fluxes as the state variables.

Step 2. Write down the linearly independent KVL and KCL equations.

Step 3. Use the branch *v-i* relationship to express all the network variables that are not the state variables in terms of the state variables chosen in step 1.

Step 4. Use the equations obtained in step 3 to eliminate all the variables that are not the chosen state variables in the equations obtained in step 2.

The resulting equations contain only the state variables and their derivatives and sources. Then transferring all the derivatives to the left-hand side, the resulting equations will be the desired state equations in the **normal form**, or **standard form**.

$$\dot{\boldsymbol{x}}\,(t) = \boldsymbol{A}\,\boldsymbol{x}\,(t) + \boldsymbol{B}\,\boldsymbol{u}\,(t)$$

■ **EXAMPLE 15.1** Write the matrix state equations for the network of Fig. 15.2.

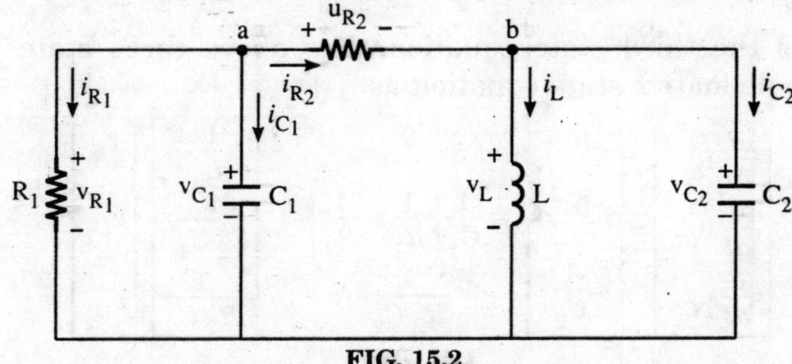

FIG. 15.2

SOLUTION : Capacitor voltages and inductor currents provide a suitable set of state variables for electrical networks. In the cirucit of Fig. 15.2, the state variables are i_L , v_{C_1} and v_{C_2}.

We have, $$L\frac{di_L}{dt} = v_L = v_{C_2}$$

Therefore, the first state equation can be written as

$$\frac{di_L}{dt} = \frac{1}{L}v_{C_2} \qquad \text{(E 15.1.1)}$$

By KCL at node a ,

$$i_{C_1} + i_{R_1} + i_{R_2} = 0$$

$$i_{C_1} = -i_{R_1} - i_{R_2}$$

$$C_1\frac{dv_{C_1}}{dt} = \frac{-v_{R_1}}{R_1} - \frac{v_{R_2}}{R_2}$$

$$= -\frac{v_{C_1}}{R_1} - \frac{(v_{C_1} - v_{C_2})}{R_2}$$

$$\frac{dv_{C_1}}{dt} = -\frac{1}{C_1}\left(\frac{1}{R_1} + \frac{1}{R_2}\right)v_{C_1} + \frac{1}{R_2 C_1}v_{C_2} \qquad \text{(E 15.1.2)}$$

Eq. (E 15.1.2) is the second state equation.

By KCL at node b,

$$i_{C_2} + i_L = i_{R_2}$$

$$i_{C_2} = -i_L + i_{R_2}$$

$$C_2\frac{dv_{C_2}}{dt} = -i_L + \frac{v_{R_2}}{R_2}$$

$$= -i_L + \frac{v_{C_1} - v_{C_2}}{R_2}$$

therefore,
$$\frac{dv_{C_2}}{dt} = -\frac{1}{C_2} i_L + \frac{1}{R_2 C_2} v_{C_1} - \frac{1}{R_2 C_2} v_{C_2} \quad \text{(E 15.1.3)}$$

Eq. (E 15.1.3) is the third state equation. The above three state equations can be combined to give a single matrix state equation as

$$\begin{bmatrix} \frac{d}{dt} i_L \\ \frac{d}{dt} v_{C_1} \\ \frac{d}{dt} v_{C_2} \end{bmatrix} = \begin{bmatrix} 0 & 0 & \frac{1}{L} \\ 0 & -\frac{1}{C_1}\left(\frac{1}{R_1} + \frac{1}{R_2}\right) & \frac{1}{R_2 C_1} \\ -\frac{1}{C_2} & \frac{1}{R_2 C_2} & -\frac{1}{R_2 C_2} \end{bmatrix} \begin{bmatrix} i_L \\ v_{C_1} \\ v_{C_2} \end{bmatrix}$$

$$\frac{d}{dt} \begin{bmatrix} i_L \\ v_{C_1} \\ v_{C_2} \end{bmatrix} = \begin{bmatrix} 0 & 0 & \frac{1}{L} \\ 0 & -\frac{R_1 + R_2}{R_1 R_2 C_1} & \frac{1}{R_2 C_1} \\ -\frac{1}{C_2} & \frac{1}{R_2 C_2} & \frac{-1}{R_2 C_2} \end{bmatrix} \begin{bmatrix} i_L \\ v_{C_1} \\ v_{C_2} \end{bmatrix} \quad \text{(E 15.1.4)}$$

■ **EXAMPLE 15.2** Write the matrix state equations for the circuit of Fig. 15.3.

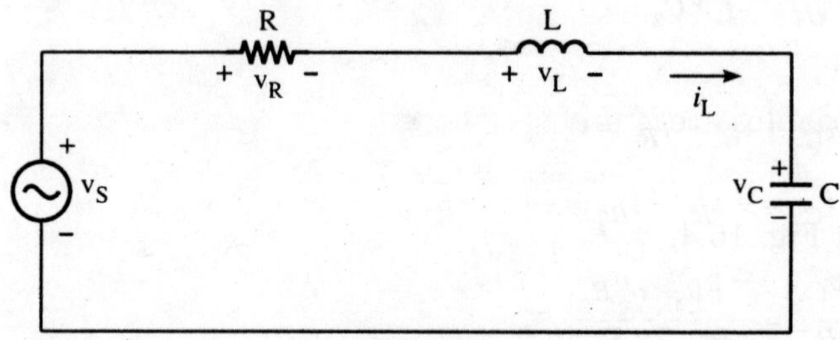

FIG. 15.3

SOLUTION : Let us choose the inductor current i_L and the capacitor voltage v_C as the state variables. By KVL in the circuit of Fig. 15.3,

$$v_S = v_R + v_L + v_C$$

$$v_L = -v_R - v_C + v_S$$

$$L \frac{di_L}{dt} = -Ri_L - v_C + v_S$$

$$\frac{di_L}{dt} = -\frac{R}{L} i_L - \frac{1}{L} v_C + \frac{1}{L} v_S \quad \text{(E 15.2.1.)}$$

Also,
$$C \frac{dv_C}{dt} = i_L$$

$$\frac{dv_C}{dt} = \frac{1}{C} i_L \quad \text{(E 15.2.2)}$$

Combining Eqs. (E 15.2.1) and (E 15.2.2) we obtain

$$\frac{d}{dt}\begin{bmatrix} i_L \\ v_C \end{bmatrix} = \begin{bmatrix} -\frac{R}{L} & -\frac{1}{L} \\ \frac{1}{C} & 0 \end{bmatrix}\begin{bmatrix} i_L \\ v_C \end{bmatrix} + \begin{bmatrix} \frac{1}{L} \\ 0 \end{bmatrix} v_S \qquad \text{(E 15.2.3)}$$

Eq. (E 15.2.3) is of the form

$$\frac{d}{dt}\boldsymbol{x}(t) = \boldsymbol{A}\,\boldsymbol{x}(t) + \boldsymbol{B}\,\boldsymbol{u}(t)$$

or

$$\dot{\boldsymbol{x}}(t) = \boldsymbol{A}\,\boldsymbol{x}(t) + \boldsymbol{B}\,\boldsymbol{u}(t)$$

■ **EXAMPLE 15.3** Write the matrix state equation for the zero-input circuit shown in Fig. 15.4.

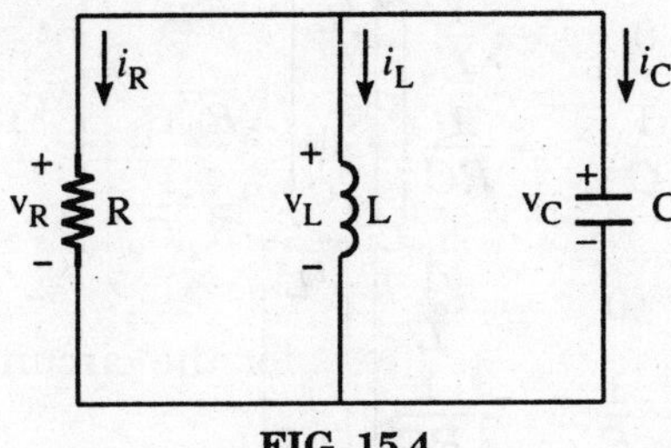

FIG. 15.4

SOLUTION : The state variables are i_L and v_C. Therefore we shall express $\frac{di_L}{dt}$ and $\frac{dv_C}{dt}$ in terms of i_L and v_C.

From the circuit of Fig. 15.4,

$$v_L = v_C$$

$$L\frac{di_L}{dt} = v_C$$

$$\frac{di_L}{dt} = \frac{1}{L}v_C \qquad \text{(E 15.3.1)}$$

Eq. (E 15.3.1) is in terms of the state variables i_L and v_C. It is therefore the first state equation. Now

$$C\frac{dv_C}{dt} = i_C \qquad \text{(E 15.3.2)}$$

Since i_C is not a state variable, we shall eliminate it. By KCL

$$i_R + i_L + i_C = 0$$

Eq. (E 15.3.2) now becomes

$$C\frac{dv_C}{dt} = -i_R - i_L \qquad \text{(E 15.3.3)}$$

In Eq. (15.3.3), v_C and i_L are state variables but i_R is not a state variable. Therefore i_R is to be expressed in terms of state variables. By Ohm's law

$$i_R = \frac{v_R}{R} = \frac{v_C}{R}$$

Substituting the value of i_R in Eq. (E 15.3.3) we obtain

$$C\frac{dv_C}{dt} = -i_L - \frac{v_C}{R}$$

$$\frac{dv_C}{dt} = -\frac{1}{C}i_L - \frac{1}{RC}v_C \quad \text{(E 15.3.4)}$$

Eq. (E 15.3.4) is the second state equation. The two state Eqs. (E 15.3.1) and (E 15.3.4) can be combined into a single matrix state equation as

$$\begin{bmatrix} \frac{di_L}{dt} \\ \frac{dv_C}{dt} \end{bmatrix} = \begin{bmatrix} 0 & \frac{1}{L} \\ -\frac{1}{C} & -\frac{1}{RC} \end{bmatrix} \begin{bmatrix} i_L \\ v_C \end{bmatrix}$$

or

$$\frac{d}{dt}\begin{bmatrix} i_L \\ v_C \end{bmatrix} = \begin{bmatrix} 0 & \frac{1}{L} \\ -\frac{1}{C} & -\frac{1}{RC} \end{bmatrix} \begin{bmatrix} i_L \\ v_C \end{bmatrix} \quad \text{(E 15.3.5)}$$

Eq. (E 15.3.5) is of the form

$$\frac{d}{dt}\boldsymbol{x}(t) = \boldsymbol{A}\,\boldsymbol{x}(t)$$

■ **EXAMPLE 15.4** Write the state equation for the circuit of Fig. 15.5.

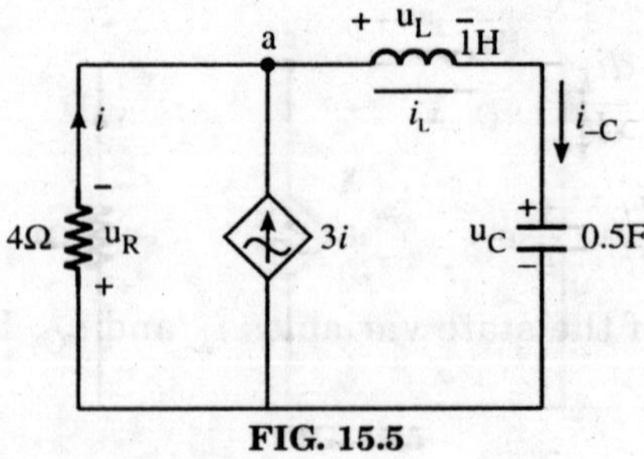

FIG. 15.5

SOLUTION : By KCl at node a.

$$i_L = i + 3i = 4i$$

By KVL in the outer loop

$$-v_R - v_L - v_C = 0$$

$$v_L = -v_R - v_C$$

$$= -4i - v_C$$

$$L\frac{di_L}{dt} = -i_L - v_C$$

since $\qquad L = 1H,$

$$\frac{di_L}{dt} = -i_L - v_C \tag{E 15.4.1}$$

Also, $\qquad i_C = i_L$

$$C\frac{dv_C}{dt} = i_L$$

$$0.5\frac{dv_C}{dt} = i_L$$

$$\frac{dv_C}{dt} = 2i_L \tag{E 15.4.2}$$

Eq. (E 15.4.1) and (E 15.4.2) are the state equations. In matrix form these equations can be written as

$$\frac{d}{dt}\begin{bmatrix} i_L \\ v_C \end{bmatrix} = \begin{bmatrix} -1 & -1 \\ 2 & 0 \end{bmatrix}\begin{bmatrix} i_L \\ v_C \end{bmatrix}$$

■ **EXAMPLE 15.5** Write the state equation of the cirucit shown in Fig. 15.6.

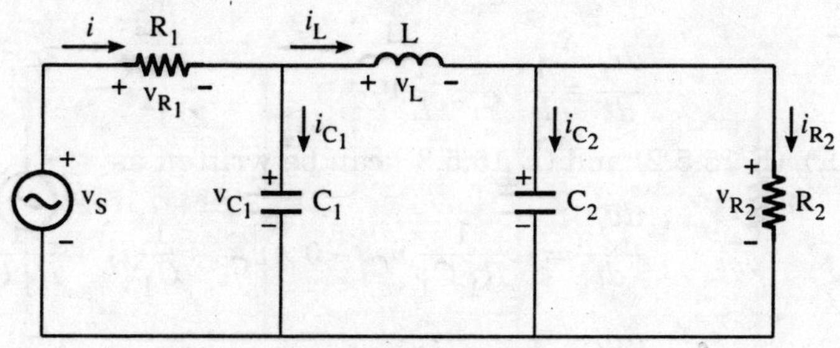

FIG. 15.6

SOLUTION : The state variables will be i_L, v_{C_1}, and v_{C_2}.

By KVL in the left-hand mesh $v_S - v_{R_1} - v_{C_1} = 0$

But $v_{R_1} = R_1 i$ and $i = i_L + i_{C_1}$

therefore, $\qquad v_S - (i_L + i_{C_1})R_1 - v_{C_1} = 0$

$$i_{C_1} = \frac{1}{R_1}(v_S - v_{C_1}) - i_L$$

$$C_1\frac{dv_{C_1}}{dt} = \frac{1}{R_1}(v_S - v_{C_1}) - i_L$$

$$\frac{dv_{C_1}}{dt} = -\frac{1}{R_1 C_1}v_{C_1} - \frac{1}{C_1}i_L + \frac{1}{R_1 C_1}v_S \tag{E 15.5.1}$$

By KVL in the right-hand mesh

$$-v_{R_2} + v_{C_2} = 0$$

But $v_{R_2} = R_2\, i_{R_2}$ and $i_L = i_{C_2} + i_{R_2}$

therefore,
$$-(i_L - i_{C_2})\, R_2 + v_{C_2} = 0$$

$$i_{C_2} = -\frac{1}{R_2} v_{C_2} + i_L$$

$$C_2 \frac{dv_{C_2}}{dt} = -\frac{1}{R_2} v_{C_2} + i_L$$

$$\frac{dv_{C_2}}{dt} = -\frac{1}{R_2\, C_2} v_{C_2} + \frac{1}{C_2} i_L \qquad \text{(E 15.5.2)}$$

By KVL in the middle mesh

$$-v_L - v_{C_2} + v_{C_1} = 0$$

$$v_L = v_{C_1} - v_{C_2}$$

$$L \frac{di_L}{dt} = v_{C_1} - v_{C_2}$$

$$\frac{di_L}{dt} = \frac{1}{L} v_{C_1} - \frac{1}{L} v_{C_2} \qquad \text{(E 15.5.3)}$$

Eqs. (E 15.5.1.), (E 15.5.2) and (E 15.5.3) can be written as

$$\frac{dv_{C_1}}{dt} = -\frac{1}{R_1\, C_1} v_{C_1} - 0 \times v_{C_2} - \frac{1}{C_1} i_L + \frac{1}{R_1\, C_1} v_S$$

$$\frac{dv_{C_2}}{dt} = 0 \times v_{C_1} - \frac{1}{R_2\, C_1} v_{C_2} + \frac{1}{C_2} i_L + 0 \times v_S$$

$$\frac{di_L}{dt} = \frac{1}{L} v_{C_1} - \frac{1}{L} v_{C_2} + 0 \times i_L + 0 \times v_S$$

These equations can be combined into a single matrix equation as

$$\frac{d}{dt}\begin{bmatrix} v_{C_1} \\ v_{C_2} \\ i_L \end{bmatrix} = \begin{bmatrix} -\frac{1}{R_1\, C_1} & 0 & -\frac{1}{C_1} \\ 0 & -\frac{1}{R_2\, C_2} & \frac{1}{C_2} \\ \frac{1}{L} & -\frac{1}{L} & 0 \end{bmatrix} \begin{bmatrix} v_{C_1} \\ v_{C_2} \\ i_L \end{bmatrix} + \begin{bmatrix} \frac{1}{R_1\, C_1} \\ 0 \\ 0 \end{bmatrix} [v_S] \qquad \text{(E 15.5.4)}$$

This equation is of the form

$$\dot{\boldsymbol{x}} = \boldsymbol{A}\, \boldsymbol{x} + \boldsymbol{B}\, \boldsymbol{u}$$

■ **EXAMPLE 15.6** Write the state equations of the circuit shown in Fig. 15.7.

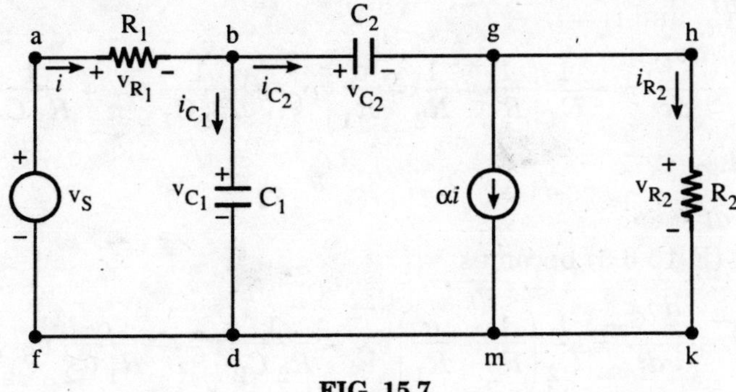

FIG. 15.7

SOLUTION : By KCL at node b.

$$i_{C_1} = i - i_{C_2} \tag{E 15.6.1}$$

By KVL in loop *bhkdb*

$$-v_{C_2} - v_{R_2} + v_{C_1} = 0$$

$$-v_{C_2} - (i_{C_2} - \alpha i)\,R_2 + v_{C_1} = 0$$

$$(i_{C_2} - \alpha i)\,R_2 = v_{C_1} - v_{C_2}$$

$$i_{C_2} = \frac{1}{R_2}(v_{C_1} - v_{C_2}) + \alpha i \tag{E 15.6.2}$$

By KVL in mesh *abdfa*

$$-i\,R_1 - v_{C_1} + v_S = 0$$

$$i = \frac{1}{R_1}(v_S - v_{C_1}) \tag{E 15.6.3}$$

Substituting the value of i from Eq. (E 15.6.3) in Eqs. (E 15.6.1) and (E 15.6.2) we get

$$i_{C_1} = \frac{1}{R_1}(v_S - v_{C_1}) - i_{C_2} \tag{E 15.6.4}$$

$$i_{C_2} = \frac{1}{R_2}(v_{C_1} - v_{C_2}) + \frac{\alpha}{R_1}(v_S - v_{C_1})$$

or

$$i_{C_2} = \left(\frac{1}{R_2} - \frac{\alpha}{R_1}\right)v_{C_1} - \frac{1}{R_2}v_{C_2} + \frac{\alpha}{R_1}v_S \tag{E 15.6.5}$$

Substituting the value of i_{C_2} from Eq. (E 15.6.5) in Eq. (E 15.6.4) we get

$$i_{C_1} = \frac{1}{R_1}(v_S - v_{C_1}) - \left(\frac{1}{R_2} - \frac{\alpha}{R_1}\right)v_{C_1} + \frac{1}{R_2}v_{C_2} - \frac{\alpha}{R_1}v_S$$

$$i_{C_1} = -\left(\frac{1}{R_1} + \frac{1}{R_2} - \frac{\alpha}{R_1}\right)v_{C_1} + \frac{1}{R_2}v_{C_2} + \left(\frac{1}{R_1} - \frac{\alpha}{R_1}\right)v_S \tag{E 15.6.6}$$

But $i_{C_1} = C_1 \dfrac{dv_{C_1}}{dt}$

$$\frac{dv_{C_1}}{dt} = -\frac{1}{C_1}\left(\frac{1}{R_1} + \frac{1}{R_2} - \frac{\alpha}{R_1}\right) v_{C_1} + \frac{1}{R_2 C_1} v_{C_2} + \frac{1}{R_1 C_1}(1-\alpha)\, v_S \quad \text{(E 15.6.7)}$$

Also $i_{C_2} = C_2 \dfrac{dv_{C_2}}{dt}$

Therefore, Eq. (E 15.6.5) becomes

$$\frac{dv_{C_2}}{dt} = \frac{1}{C_2}\left(\frac{1}{R_2} - \frac{\alpha}{R_1}\right) v_{C_1} - \frac{1}{R_2 C_2} v_{C_2} + \frac{\alpha}{R_1 C_2} v_S \quad \text{(E 15.6.8)}$$

Combining Eqs. (E 15.5.7) and (E 15.5.8) we obtain

$$\frac{d}{dt}\begin{bmatrix} v_{C_1} \\ v_{C_2} \end{bmatrix} = \begin{bmatrix} \frac{1}{C_1}\left(\frac{1}{R_1} + \frac{1}{R_2} - \frac{\alpha}{R_1}\right) & \frac{1}{R_2 C_1} \\ \frac{1}{C_2}\left(\frac{1}{R_2} - \frac{\alpha}{R_1}\right) & -\frac{1}{R_2 C_2} \end{bmatrix} \begin{bmatrix} v_{C_1} \\ v_{C_2} \end{bmatrix} + \begin{bmatrix} \frac{1-\alpha}{R_1 C_1} \\ \frac{\alpha}{R_1 C_2} \end{bmatrix} [v_S] \quad \text{(E 15.6.9)}$$

EXAMPLE 15.7 Write the state equation of the circuit shown in Fig. 15.8.

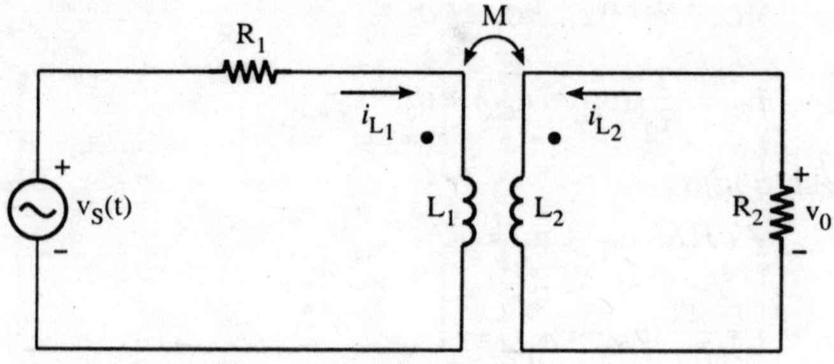

FIG. 15.8

SOLUTION : The inductor currents i_{L_1} and i_{L_2} are selected as the state variables.

The two loop equations can be written as

$$R_1 i_{L_1} + L_1 \frac{di_{L_1}}{dt} + M \frac{di_{L_2}}{dt} = v_S(t)$$

$$R_2 i_{L_2} + L_2 \frac{di_{L_2}}{dt} + M \frac{di_{L_1}}{dt} = 0$$

These two equations can be solved for $\dfrac{di_{L_1}}{dt}$ and $\dfrac{di_{L_2}}{dt}$. The result in matrix form is given by

$$\frac{d}{dt}\begin{bmatrix} i_{L_1} \\ i_{L_2} \end{bmatrix} = \begin{bmatrix} -\dfrac{R_1 L_2}{\Delta} & \dfrac{R_2 M}{\Delta} \\ \dfrac{R_2 M}{\Delta} & -\dfrac{R_2 L_1}{\Delta} \end{bmatrix}\begin{bmatrix} i_{L_1} \\ i_{L_2} \end{bmatrix} + \begin{bmatrix} \dfrac{L_2}{\Delta} \\ -\dfrac{M}{\Delta} \end{bmatrix}[v_S]$$

Where $\Delta = L_1 L_2 - M^2 \neq 0$

The output relation is given by

$$v_0 = -R_2 i_{L_2}$$

15.7 EQUIVALENT SOURCE METHOD

In the equivalent-source method of writing the state equations, each capacitor is replaced by a voltage source and each inductor by a current source. Thus, the circuit will contain only resistors, current sources, and voltage sources. Superposition principle is then applied to determine the current i_C through C and the voltage v_L across L.

15.8 ORDER OF COMPLEXITY OF A NETWORK

*The **order of complexity** of a network is equal to the number of independent initial conditions that can be specified in the network, or equivalently,*

*The **order of complexity** of a network is the number of the independent state variables that completely specify the dynamical behaviour of the network; or equivalently,*

*The **order of complexity** of a network is the number of independent first-order differential equations that completely specify the behavior of the network.*

The order of complexity of a network is equal to the total number of its energy-storing elements minus the total number of independent capacitor-only loops and inductor-only cutsets. Thus,

$$n_0 = n_{ic} - (n_c + n_i)$$

where

n_0 = order of complexity of the network

n_{ic} = total number of inductors and capacitors

n_c = number of independent capacitor-only loops

n_i = number of independent inductor-only cutsets

In counting n_c, the capacitor-only loops must be independent of each other. For this, it is sufficient that all capacitor-only loops contain at least one new capacitor. Similarly, the inductor-only cutsets must be independent, and each cutset should contain at least one new inductor.

To obtain n_c by inspection, open circuit all the resistors, inductors and current sources. The remaining network will show clearly the capacitor-only loops. Similarly, short circuit all the resistors, capacitors and voltage sources to obtain by inspection the inductor-only cutsets.

Initial conditions are used to evaluate the arbitrary constants in the solution of a set of differential equations. That is, n_0 is also the number of these constants, and therefore also the degree of the characteristic equation of the network. But the degree of the characteristic equation is also equal to the number of natural frequencies (modes) of the network. Thus, the order of complexity of a network is also the number of independent initial conditions and is also the number of its natural frequencies.

■ **EXAMPLE 15.8** Determine the order of complexity of the network shown in Fig. 15.9.a

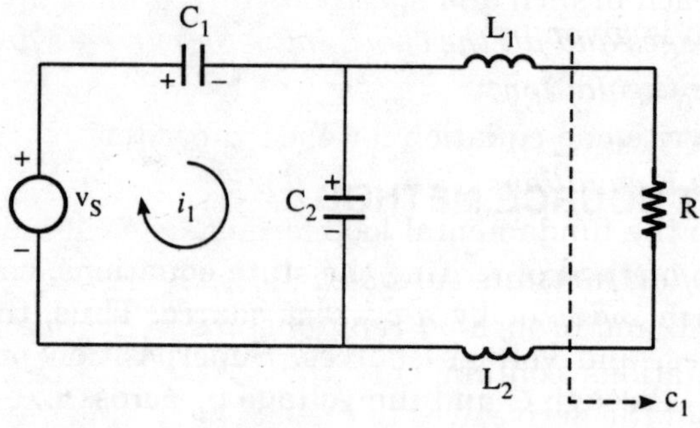

FIG. 15.9

SOLUTION : There are four energy-storing elements (2 capacitors and 2 inductors) in the network. This network has one capacitor-only loop, and one inductor-only cutset. Therefore the order of complexity of the network.

$$= 4 - 1 - 1 = 2$$

Number of indequendent state variables = order of complexity of the network = 2

15.9 WRITING STATE EQUATIONS USING NETWORK GRAPHS

The method of writing state equations by inspection is applicable for networks which have no capacitor-only loops and no inductor-only cutsets. Now we shall consider the general formulation of state equations based on network graph theory. A systematic procedure is given for obtaining the state equations of any network containing two-terminal resistors, inductors, capacitors and voltage and current sources. In order to do this we first define a normal tree as follows :

A ***normal tree*** *of a lumped network comprised of two-terminal resistors, inductors, capacitors, and voltage and current sources is a tree that contains all the voltage sources, no current sources, maximum possible number of capacitors, minimum possible inductors and some resistors.*

It is to be noted that in this definition we excluded the possibility of having any loops of voltage sources only or any cutsets of current sources only.

With this definition, it is now possible to give a formal procedure for obtaining state equations :

Step 1. Draw the graph of the network. Since identification of individual elements is important, a parallel of series combination must not be drawn as a single edge.

Step 2. *Choice of a normal tree*

Choose a normal tree that contains all the independent voltage sources, the maximum possible number of capacitors, some resistors and the minimum number of inductors. All the current sources and a maximum number of inductors are assigned to the cotree. As far as possible, an element whose voltage controls a dependent source should be placed in the tree, and the element whose current controls a dependent source should be placed in the cotree. More than one normal tree may be possible. Certain networks do not permit any normal tree to be drawn.

Step 3. *Assignment of voltage and current variables*

Arbitrarily, assign a voltage to each branch and mark its polarity. Similarly, assign a current to each branch and specify its direction by an arrow. *All the three branch capacitor voltages and all the chord inductor currents are selected as state variables.*

Step 4. *Writing of the C equation*

Use KCL to write one equation for each capacitor.

Step 5. *Writing of the L equations*

Apply KVL to the fundamental loop formed by each inductor.

Step 6. *Elimination of all nonstate variables*

Use KCL, KVL and branch *v-i* relationships to eliminate all nonstate variables.

The resulting equations contain only the state variables, their derivative, and the sources. Transforming all the derivatives to the left-hand side, the resulting equations will be the desired equations in the normal form

$$\dot{\boldsymbol{x}}(t) = \boldsymbol{A}\,\boldsymbol{x}(t) + \boldsymbol{B}\,\boldsymbol{u}(t)$$

It is to be noted that the utility of graph theory in the formulation of the state equations is limited to the selection of a normal tree. Then for a particular normal tree, *all* the tree branch capacitor voltages and *all* the chord inductor currents are selected as the state variables. Once this selection is done, the step-by-step procedure given above is applicable.

The procedure is illustrated by the following examples :

■ **EXAMPLE 15.9** Write the state equations for the circuit shown in Fig. 15.10a.

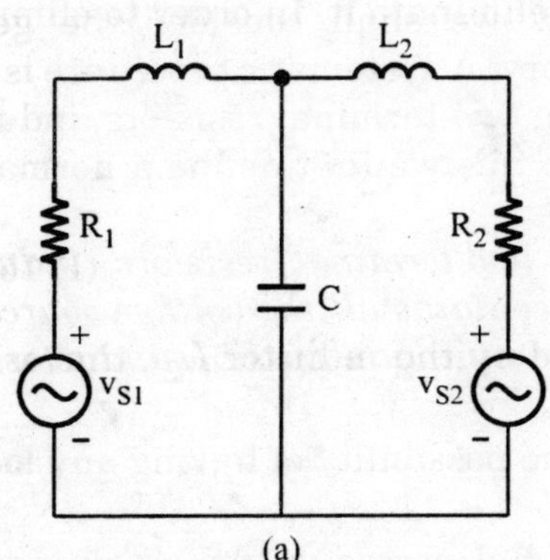

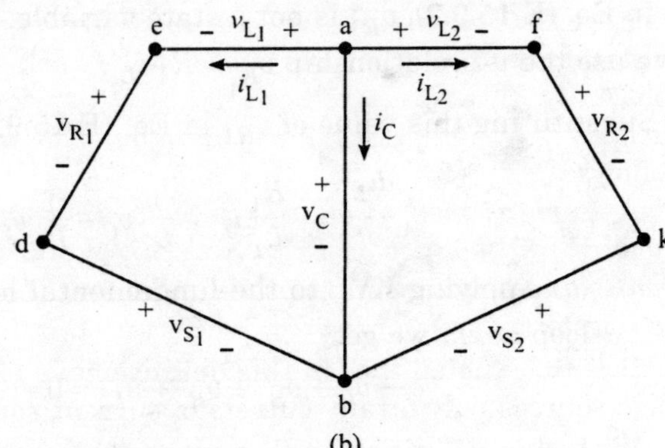

FIG. 15.10

SOLUTION :

Step 1. *Choice of the normal tree*

Choose the normal tree containing the voltage sources v_{S_1}, v_{S_2}, capacitor C and the resitors R_1 and R_2 as shown in Fig. 15.10b by full lines. The two inductors are in the cotree shown by dashed lines in Fig 15.10b.

Step 2. *Assignment of voltage and current variables*

Let v_C be the voltage across the capacitor C and i_{L_1}, i_{L_2} be the currents through the inductors L_1 and L_2. The currents and voltages are marked *arbitrarily* as shown in Fig. 15.10b. The tree branch capacitor voltage v_C and the link inductor current i_{L_1} and i_{L_2} are selected as state variables.

Step 3. We have to write one KCL equation for the capacitor C. By KCL at node a

$$i_C + i_{L_1} + i_{L_2} = 0$$

$$C\frac{dv_C}{dt} = -i_{L_1} - i_{L_2}$$

$$\frac{dv_C}{dt} = -\frac{1}{C}i_{L_1} - \frac{1}{C}i_{L_2} \quad \text{(E 15.9.1)}$$

Step 4. (*a*) Applying KVL to the fundamental loop formed by the inductor L_1, that is, the loop *abdea*, we have

$$-v_C + v_{S_1} + v_{R_1} + v_{L_1} = 0$$

$$v_{L_1} = v_C - v_{S_1} - v_{R_1}$$

$$L_1\frac{di_{L_1}}{dt} = v_C - v_{S_1} - v_{R_1}$$

$$\frac{di_{L_1}}{dt} = \frac{1}{L_1}v_C - \frac{1}{L_1}v_{S_1} - \frac{1}{L_1}v_{R_1} \quad \text{(E 15.9.2)}$$

In Eq. (E 15.9.2), v_{R1} is not a state variable. We shall eliminate it. In order to eliminate v_{R1} we use the *v-i* relationship $v_{R_1} = R_1 i_{L_1}$.

Substituting this value of v_{R1} in Eq. (E 15.9.2) we get

$$\frac{di_{L_1}}{dt} = -\frac{R_1}{L_1}i_{L_1} + \frac{1}{L_1}v_C - \frac{1}{L_1}v_{S_1} \quad \text{(E 15.9.3)}$$

(*b*) Applying KVL to the fundamental loop formed by the inductor L_2, that is, the loop *abkfa* we get

$$-v_C + v_{S_2} + v_{R_2} + v_{L_2} = 0$$

$$v_{L_2} = -v_{R_2} + v_C - v_{S_2}$$

$$L_2\frac{di_{L_2}}{dt} = -v_{R_2} + v_C - v_{S_2}$$

$$\frac{di_{L_2}}{dt} = -\frac{1}{L_2}v_{R_2} + \frac{1}{L_2}v_C - \frac{1}{L_2}v_{S_2} \quad \text{(E 15.9.4)}$$

In Eq. (E 15.9.4), v_{R_2} is not a state variable. We shall eliminate it. In order to eliminate this nonstate variable we use the *v-i* relationship $v_{R_2} = R_2\, i_{L_2}$ and substitute in Eq. (E 15.9.4) to obtain

$$\frac{di_{L_2}}{dt} = -\frac{R_2}{L_2} i_{L_2} + \frac{1}{L_2} v_C - \frac{1}{L_2} v_{S_2} \tag{E 15.9.5}$$

We can write Eqs. (E 15.9.3), (E 15.9.5) and (E 15.9.1) as follows :

$$\frac{di_{L_1}}{dt} = -\frac{R_1}{L_1} i_{L_1} + 0 \times i_{L_2} + \frac{1}{L_1} v_C - \frac{1}{L_1} v_{S_1} \tag{E 15.9.6}$$

$$\frac{di_{L_2}}{dt} = 0 \times i_{L_1} - \frac{R_2}{L_2} i_{L_2} + \frac{1}{L_2} v_C - \frac{1}{L_2} v_{S_2} \tag{E 15.9.7}$$

$$\frac{dv_C}{dt} = -\frac{1}{C} i_{L_1} - \frac{1}{C} i_{L_2} + 0 \times v_C + 0 \tag{E 15.9.8}$$

Eqs. (E 15.9.6), (E 15.9.7) and (E 15.9.8) are the state equations. In matrix form these equations can be written as

$$\frac{d}{dt}\begin{bmatrix} i_{L_1} \\ i_{L_2} \\ v_C \end{bmatrix} = \begin{bmatrix} -\frac{R_1}{L_1} & 0 & \frac{1}{L_1} \\ 0 & -\frac{R_2}{i_{L_2}} & \frac{1}{L_2} \\ -\frac{1}{C} & -\frac{1}{C} & 0 \end{bmatrix} \begin{bmatrix} i_{L_1} \\ i_{L_2} \\ v_C \end{bmatrix} + \begin{bmatrix} -\frac{1}{L_1} v_{S_1} \\ -\frac{1}{L_2} v_{S_2} \\ 0 \end{bmatrix}$$

■ **EXAMPLE 15.10** Write the state equation for the circuit shown in Fig. 15.11a.

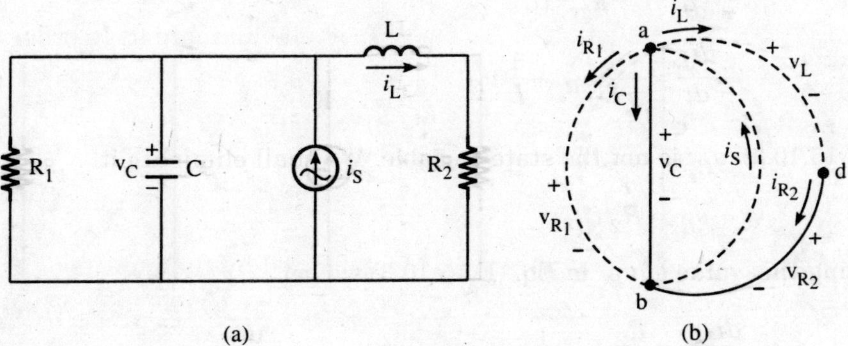

FIG. 15.11

SOLUTION :

Step 1. *Choice of normal tree*

Choose the normal tree containing the capacitor C and the resistor R_2. This tree is shown by full lines in Fig. 15.11b. The inductor L, current source i_S and the resitor R_1 are left for the cotree shown by dashed lines in Fig. 15.11b.

Step 2. *Assignment of voltage and current variables*

The currents and voltages are *arbitrarily* marked as shown in Fig. 15.11b. The tree branch capacitor voltage v_C and the link inductor current i_L are chosen as the state variables.

Step 3. We have to write one KCL equation for the capacitor C. By KCL at node a,

$$i_C + i_L + i_{R_1} - i_S = 0$$

$$C\frac{dv_C}{dt} + i_L + i_{R_1} - i_S = 0$$

$$\frac{dv_C}{dt} = -\frac{1}{C}i_L - \frac{1}{C}i_{R_1} + \frac{1}{C}i_S \quad \text{(E 15.10.1)}$$

In Eq. (E 15.10.1) i_{R_1} is not the state variable. We shall eliminate it.

$$i_{R_1} = \frac{v_{R_1}}{R_1} = \frac{v_C}{R_1}$$

Substituting the value of i_{R_1} in Eq. (E 15.10.1) we get

$$\frac{dv_C}{dt} = -\frac{1}{C}i_L - \frac{1}{R_1 C}v_C + \frac{1}{C}v_S \quad \text{(E 15.10.2)}$$

Step 4. Applying KVL to the fundamental loop *badb* formed by the inductor L we get

$$v_C - v_L - v_{R_2} = 0$$

$$v_L = -v_{R_2} + v_C$$

$$L\frac{di_L}{dt} = -v_{R_2} + v_C$$

$$\frac{di_L}{dt} = -\frac{1}{L}v_{R_2} + \frac{1}{L}v_C \quad \text{(E 15.10.3)}$$

In Eq. (E 15.10.3), v_{R_2} is not the state variable. We shall eliminate it.

$$v_{R_2} = R_2 i_L$$

Substituting this value of v_{R_2} in Eq. (E 15.10.3) we get

$$\frac{di_L}{dt} = -\frac{R_2}{L}i_L + \frac{1}{L}v_C + 0 \quad \text{(E 15.10.4)}$$

Eqs. (E 15.10.2) and (E 15.10.4) are state equations. In matrix form these equations can be written as

$$\frac{d}{dt}\begin{bmatrix} i_L \\ v_C \end{bmatrix} = \begin{bmatrix} -\frac{R_2}{L} & \frac{1}{L} \\ -\frac{1}{C} & -\frac{1}{R_1 C} \end{bmatrix}\begin{bmatrix} i_L \\ v_C \end{bmatrix} + \begin{bmatrix} 0 \\ \frac{1}{C}i_S \end{bmatrix}$$

■ **EXAMPLE 15.11** Write the state equations for the circuit shown in Fig 15.12a.

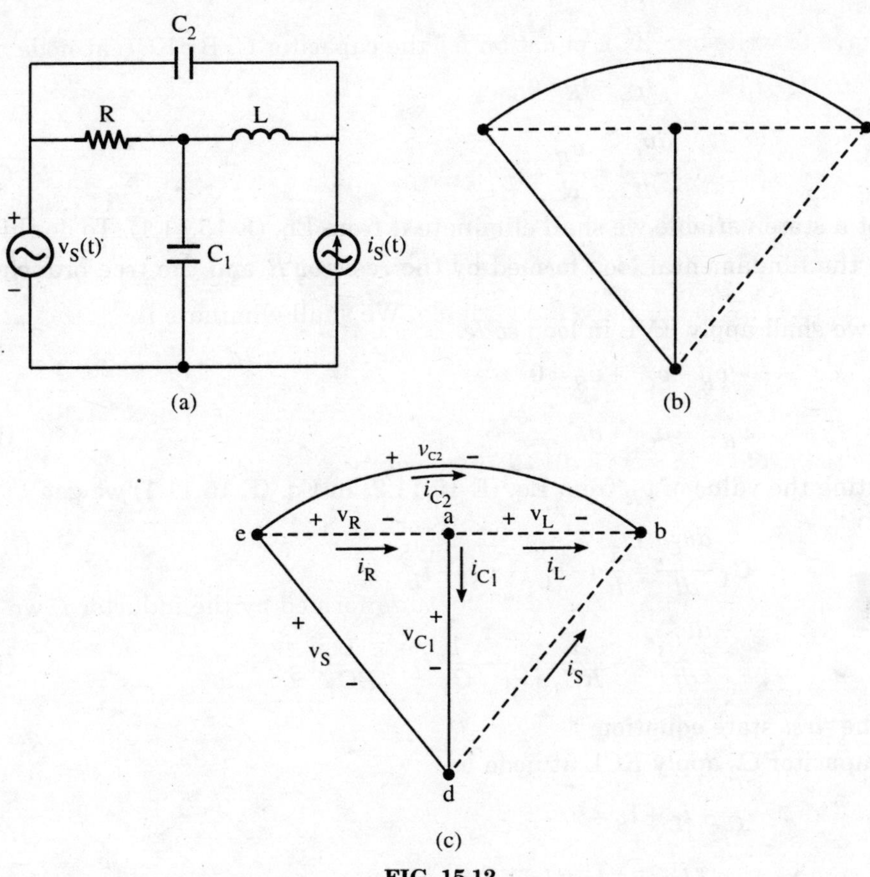

FIG. 15.12

SOLUTION :

Step 1. *Choice of the normal tree*

Choose the normal tree containing the voltage source v_S and the two capacitors C_1 and C_2. This tree is shown by full lines in the graph of the network in Fig. 15.12b. The current source i_S, the inductor L and the resistor R are in the cotree shown by dashed lines in Fig. 15.12b.

Step 2. *Assignment of voltage and current variables*

Let v_{C_1}, v_{C_2} be the voltages across the capacitors C_1 and C_2, and the current through the inductor L be i_L. The voltages v_{C_1} and v_{C_2} and the source voltage v_S are shown across the tree branches. The source current i_S and the inductor current i_L are marked on the links of the cotree. The capacitor voltages v_{C_1} and v_{C_2} and inductor current i_L are chosen as the state variables. The polarity of all voltages and directions of all currents have been chosen *arbitrarily* as shown in Fig. 15.12c.

Step 3. Using KCL we have to write one equation for each capacitor.

(*a*) For capacitor C_1 apply KCL at node a

$$i_R = i_{C_1} + i_L$$

$$i_{C_1} = i_R - i_L$$

$$C\frac{dv_{C_1}}{dt} = \frac{v_R}{R} - i_L \tag{E 15.11.1}$$

Since v_R is not a state variable we shall eliminate it from Eq. (E 15.11.1). To do this we shall apply KVL to the fundamental loop formed by the resistor R and the tree branches *ad* and *de*.

That is, we shall apply KVL in loop *eade*.

$$-v_R - v_{C_1} + v_S = 0$$

therefore,

$$v_R = -v_{C_1} + v_S \tag{E 15.11.2}$$

Substituting the value of v_R from Eq. (E 15.11.2) in Eq. (E 15.11.1) we get

$$C_1\frac{dv_{C_1}}{dt} = \frac{1}{R}(-v_{C_1} + v_S) - i_L$$

$$\frac{dv_{C_1}}{dt} = -\frac{1}{RC_1}v_{C_1} - \frac{1}{C_1}i_L + \frac{1}{RC_1}v_S \tag{E 15.11.3}$$

This is the *first* state equation.

(*b*) For capacitor C_2 apply KCL at node b.

$$i_{C_2} + i_L + i_S = 0$$

$$i_{C_2} = -i_L - i_S$$

$$C_2\frac{dv_{C_2}}{dt} = -i_L - i_S$$

$$\frac{dv_{C_2}}{dt} = -\frac{1}{C_2}i_L - \frac{1}{C_2}i_S \tag{E 15.11.4}$$

This is the *second* state equation.

Step 4. Applying KVL to the fundamental loop formed by the inductor link and the entire tree branches, that is, the loop *abeda*.

$$-v_L + v_{C_2} - v_S + v_{C_1} = 0$$

$$-L\frac{di_L}{dt} = -v_{C_1} - v_{C_2} + v_S$$

$$\frac{di_L}{dt} = \frac{1}{L}v_{C_1} + \frac{1}{L}v_{C_2} - \frac{1}{L}v_S \tag{E 15.11.5}$$

This is the *third* state equation.

EXAMPLE 15.12 Write the state equations for the network shown in Fig. 15.13a.

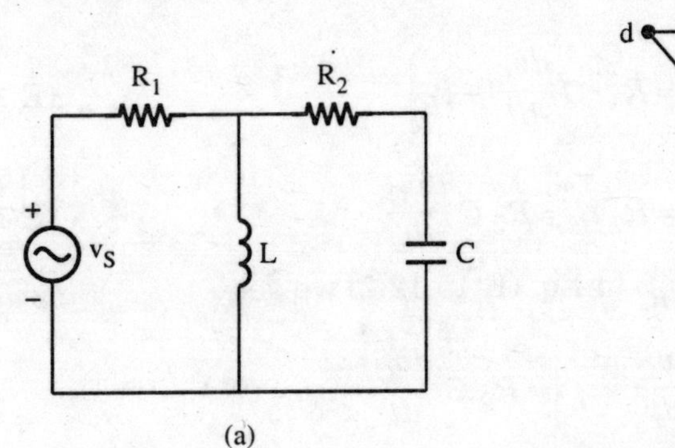

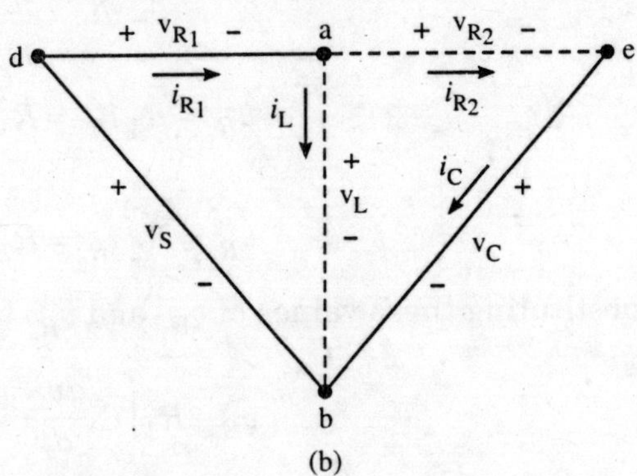

FIG. 15.13

SOLUTION : Choose the normal tree containing the voltage source v_S, the capacitor C and the resistor R_1. We have selected R_1 as a tree element arbitrarily. We can also choose R_2 as a tree element instead of R_1. The normal tree is shown in Fig. 15.13b by full lines. The inductor L and the resistor R_2 are left for the cotree shown by dashed lines in Fig. 15.13b. Let us *arbitrarily* choose the polarity of the voltage across the capacitor C and the direction of current i_L through the inductor L as shown in Fig. 15.13b. The tree branch capacitor voltage v_C and the link inductor current i_L are chosen as the state variables. Let us assign various voltages and currents as indicated in Fig. 15.13b.

Applying KVL to the fundamental loop determined by the link inductor (that is, the left-hand mesh *bdab*) in Fig. 15.13b we get

$$v_S - v_{R_1} - v_L = 0$$

$$v_L = - v_{R_1} + v_S$$

$$L\frac{di_L}{dt} = - v_{R_1} + v_S$$

$$\frac{di_L}{dt} = -\frac{1}{L} v_{R_1} + \frac{1}{L} v_S \qquad \text{(E 15.12.1)}$$

Applying KVL to the fundamental loop determined by the resistor R_2 (that is, the outer loop *bdaeb*) in Fig. 15.13b we get

$$v_S - v_{R_1} - v_{R_2} - v_C = 0 \qquad \text{(E 15.12.2)}$$

In Eqs. (E 15.12.1) and (E 15.12.2), v_{R_1} and v_{R_2} are not the state variables. These nonstate variables are to be eliminated from Eqs. (E 15.12.1) and (E 15.12.2).

By KCL at node a,

$$i_{R_1} = i_{R_2} + i_L$$

$$i_{R_1} = i_C + i_L$$

$$= C\frac{dv_C}{dt} + i_L$$

$$v_{R_1} = R_1 i_{R_1} = R_1\left(C\frac{dv_C}{dt} + i_L\right) \quad \text{(E 15.12.3)}$$

$$v_{R_2} = R_2 i_{R_2} = R_2 i_C = R_2 C\frac{dv_C}{dt} \quad \text{(E 15.12.4)}$$

Substituting these values of v_{R_1} and v_{R_2} in Eq. (E 15.12.2) we get

$$v_S - R_1\left(C\frac{dv_C}{dt} + i_L\right) - R_2 C\frac{dv_C}{dt} - v_C = 0$$

$$C(R_1 + R_2)\frac{dv_C}{dt} = -R_1 i_L - v_C + v_S$$

$$\frac{dv_C}{dt} = -\frac{R_1}{C(R_1+R_2)} i_L - \frac{1}{C(R_1+R_2)} v_C + \frac{1}{C(R_1+R_2)} v_S \quad \text{(E 15.12.5)}$$

In order to eliminate the nonstate variable v_{R_1} from Eq. (E 15.10.1) we substitute the value of v_{R_1} from Eq. (E 15.12.3) in Eq. (E 15.12.1)

$$\frac{di_L}{dt} = -\frac{1}{L}R_1\left(C\frac{dv_C}{dt} + i_L\right) + \frac{1}{L}v_S$$

or

$$\frac{di_L}{dt} = -\frac{R_1}{L}i_L - \frac{R_1 C}{L}\frac{dv_C}{dt} + \frac{1}{L}v_S \quad \text{(E 15.12.6)}$$

Substituting the value of $\frac{dv_C}{dt}$ from Eq. (E 15.12.5) in Eq. (15.12.6) we get

$$\frac{di_L}{dt} = -\frac{R_1}{L}i_L - \frac{R_1 C}{L}\left[\frac{-R_1}{C(R_1+R_2)}i_L - \frac{1}{C(R_1+R_2)}v_C + \frac{1}{C(R_1+R_2)}v_S\right] + \frac{1}{L}v_S$$

therefore,

$$\frac{di_L}{dt} = -\frac{R_1}{L}i_L\left(1 - \frac{R_1}{R_1+R_2}\right) + \frac{R_1}{L(R_1+R_2)}v_C + \frac{R_2}{L(R_1+R_2)}v_S$$

$$= \frac{R_1 R_2}{L(R_1+R_2)}i_L + \frac{R_1}{L(R_1+R_2)}v_C + \frac{R_2}{L(R_1+R_2)}v_S \quad \text{(E 15.12.7)}$$

Eqs. (E 15.12.5) and (E 15.12.7) are the state equations. In matrix form these equations can be written as

$$\begin{bmatrix} \frac{di_L}{dt} \\ \frac{dv_C}{dt} \end{bmatrix} = \begin{bmatrix} -\frac{R_1 R_2}{L(R_1+R_2)} & \frac{R_1}{L(R_1+R_2)} \\ -\frac{R_1}{C(R_1+R_2)} & -\frac{1}{C(R_1+R_2)} \end{bmatrix} \begin{bmatrix} i_L \\ v_C \end{bmatrix} + \begin{bmatrix} \frac{R_2}{L(R_1+R_2)} \\ \frac{1}{C(R_1+R_2)} \end{bmatrix} v_S$$

EXAMPLE 15.13 Write the state equations for the circuit in Fig 15.14a.

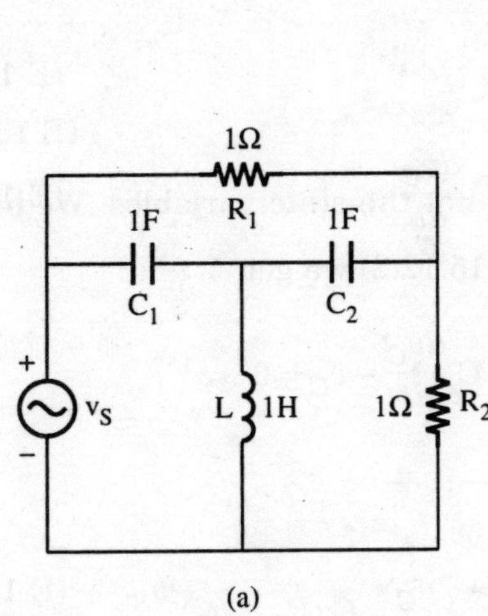

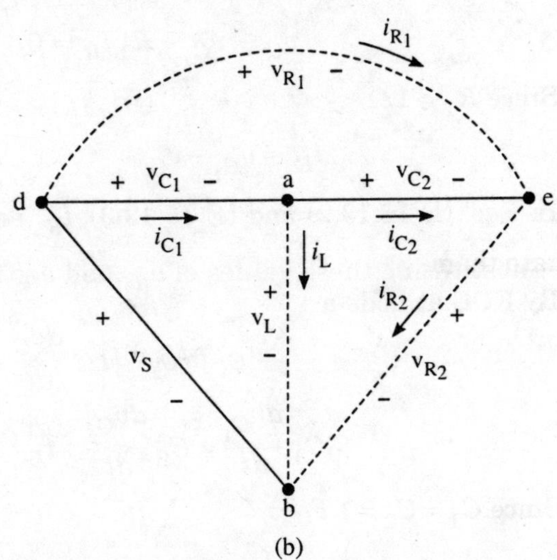

FIG. 15.14

SOLUTION : Choose the normal tree containing the voltage source v_S and both the capacitors C_1 and C_2. This tree is shown by full lines in Fig. 15.14b. The inductor L, and resistors R_1 and R_2 are left for the cotree shown by dashed lines. The currents and voltages are marked as shown in Fig. 145.14b. The tree-branch capacitor voltages v_{C_1} and v_{C_2} and the link inductor current i_L are selected as the state variables.

Applying KVL to the fundamental loop *bdab* determined by the link inductor L we get,

$$v_S - v_{C_1} - v_L = 0$$

$$v_L = - v_{C_1} + v_S$$

$$L\frac{di_L}{dt} = - v_{C_1} + v_S$$

Since $L = 1$ H

$$\frac{di_L}{dt} = - v_{C_1} + v_S \tag{E 15.13.1}$$

Eqs. (E 15.13.1) is the state equation.

By KVL in the lower outer loop *bdaeb*

$$v_S - v_{C_1} - v_{C_2} - v_{R_2} = 0$$

$$v_S - v_{C_1} - v_{C_2} - R_2\, i_{R_2} = 0$$

Since $R_2 = 1\ \Omega$

$$v_S - v_{C_1} - v_{C_2} - i_{R_2} = 0 \tag{E 15.13.2}$$

By KVL in the upper mesh *daed*

$$-v_{C_1} - v_{C_2} + v_{R_1} = 0$$

$$-v_{C_1} - v_{C_2} + R_1 i_{R_1} = 0$$

Since $R_1 = 1\ \Omega$

$$i_{R_1} = v_{C_1} + v_{C_2} \quad \text{(E 15.13.3)}$$

In Eqs. (E 15.13.2) and (E 15.13.3), i_{R_2} and i_{R_1} are not the state variables. We have to eliminate them.

By KCL at node a

$$i_{C_1} = i_{C_2} + i_L$$

$$C_1 \frac{dv_{C_1}}{dt} = C_2 \frac{dv_{C_2}}{dt} + i_L$$

Since $C_1 = C_2 = 1$ F

$$\frac{dv_{C_1}}{dt} = \frac{dv_{C_2}}{dt} + i_L \quad \text{(E 15.13.4)}$$

By KCL at node e

$$i_{R_1} + i_{C_2} = i_{R_2} \quad \text{(E 15.13.5)}$$

Substituting the values of i_{R_1} and i_{R_2} from Eqs. (E 15.13.3) and (E 15.13.2) in Eq. (E 15.13.5) we get

$$v_{C_1} + v_{C_2} + C_2 \frac{dv_{C_2}}{dt} = v_S - v_{C_1} - v_{C_2}$$

Since $C_2 = 1$ F

$$\frac{dv_{C_2}}{dt} = -2\,v_{C_1} - 2\,v_{C_2} + v_S \quad \text{(E 15.13.6)}$$

Eq. (E 15.13.6) is the state equation

Substituting the value of $\frac{dv_{C_2}}{dt}$ from Eq. (E 15.13.6) in Eq. (E 15.13.4) we get

$$\frac{dv_{C_1}}{dt} = -2\,v_{C_1} - 2\,v_{C_2} + v_S + i_L \quad \text{(E 15.13.7)}$$

We can write the state equations (E 15.13.7), (E 15.13.6), and (E 15.13.1) as

$$\frac{dv_{C_1}}{dt} = -2\,v_{C_1} - 2\,v_{C_2} + i_L + v_S$$

$$\frac{dv_{C_2}}{dt} = -2\,v_{C_1} - 2\,v_{C_2} + 0 \times i_L + v_S$$

$$\frac{di_L}{dt} = -v_{C_1} + 0 \times v_{C_2} + 0 \times i_L + v_S$$

In matrix form, these equations can be written as

$$\frac{d}{dt}\begin{bmatrix} v_{C_1} \\ v_{C_2} \\ i_L \end{bmatrix} = \begin{bmatrix} -2 & -2 & 1 \\ -2 & -2 & 0 \\ -1 & 0 & 0 \end{bmatrix}\begin{bmatrix} v_{C_1} \\ v_{C_2} \\ i_L \end{bmatrix} + \begin{bmatrix} 1 \\ 1 \\ 1 \end{bmatrix} v_S$$

■ **EXAMPLE 15.14** Write the state equations for the network shown in Fig. 15.15a if $v_S(t) = 5e^{-2t}$.

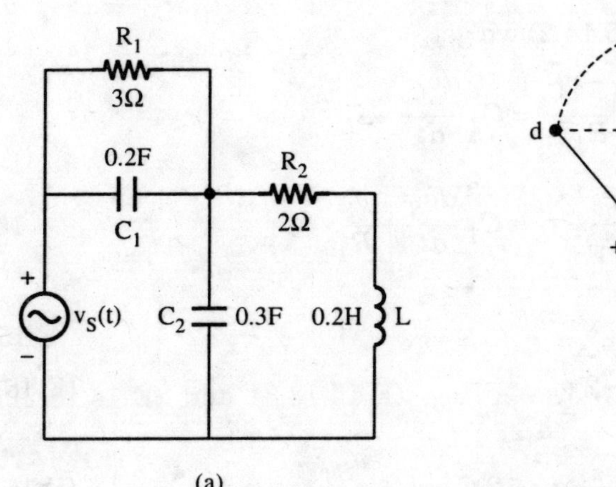

(a)

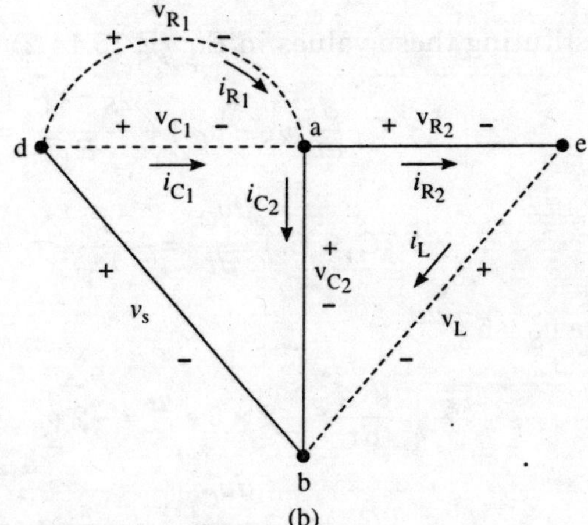

(b)

FIG. 15.15

SOLUTION : Choose the normal tree containing the voltage source v_S, the capacitor C_2 and the resistor R_2. This tree is shown in Fig. 15.15b by full lines. It is to be noted that the two capacitors and the voltage source cannot all be placed in a normal tree because C_1, C_2 and v_S form a mesh. The inductor L, capacitor C_1 and resistor R_1 are put in the cotree as shown in Fig. 15.15b by dashed lines. The tree branch capacitor voltage v_{C_2} and the link inductor current i_L are selected as the state variables. The currents and voltages are marked arbitrarily as shown in Fig. 15.15b.

By KVL in the right-hand mesh *baeb*

$$v_{C_2} - v_{R_2} - v_L = 0$$

$$v_L = -v_{R_2} + v_{C_2}$$

$$L\frac{di_L}{dt} = -R_2 i_L + v_{C_2}$$

$$\frac{di_L}{dt} = -\frac{R_2}{L} i_L + \frac{1}{L} v_{C_2} \qquad \text{(E 15.14.1)}$$

By KCL at node a

$$i_{C_1} + i_{R_1} = i_{C_2} + i_L$$

$$C_1 \frac{dv_{C_1}}{dt} + \frac{v_{R_1}}{R_1} = C_2 \frac{dv_C}{dt} + i_L \quad \text{(E 15.14.2)}$$

By KVL in the left-hand mesh

$$v_S - v_{C_1} - v_{C_2} = 0$$

$$v_{C_1} = v_S - v_{C_2}$$

Also, $$v_{R_1} = v_{C_1}$$

Substituting these values in Eq. (E 15.14.2) we get

$$C_1 \frac{d}{dt}(v_S - v_{C_2}) + \frac{v_S - v_{C_2}}{R_1} = C_2 \frac{dv_{C_2}}{dt} + i_L$$

$$(C_1 + C_2)\frac{dv_{C_2}}{dt} = -\frac{v_{C_2}}{R_1} - i_L + C_1 \frac{dv_S}{dt} + \frac{v_S}{R_1}$$

Since $v_S = 5\,e^{-2t}$

$$\frac{dv_S}{dt} = -2 \times 5\,e^{-2t} = -2\,v_S \quad \text{(E 15.14.3)}$$

therefore, $$(C_1 + C_2)\frac{dv_{C_2}}{dt} = -\frac{v_{C_2}}{R_1} - i_L - 2\,C_1\,v_S + \frac{v_S}{R_1} \quad \text{(E 15.14.4)}$$

We have

$$C_1 = 0.2\text{ F},\ C_2 = 0.3\text{ F},\ R_1 = 3\ \Omega,\ R_2 = 2\ \Omega,\ L = 0.2\text{ H}$$

Substituting these values in Eqs. (E 15.14.1)'and (E 15.14.4) we get

$$\frac{di_L}{dt} = -\frac{2}{0.2} i_L + \frac{1}{0.2} v_{C_2}$$

$$0.5 \frac{dv_{C_2}}{dt} = -\frac{1}{3} v_{C_2} - i_L - 0.4\,v_S + \frac{v_S}{3}$$

On simplification, these equations become

$$\frac{di_L}{dt} = -10\,i_L + 5\,v_{C_2} + 0 \times v_S$$

$$\frac{dv_{C_2}}{dt} = -2\,i_L - \frac{2}{3} v_C - \frac{2}{15} v_S$$

In matrix form these equations can be written as

$$\frac{d}{dt}\begin{bmatrix} i_L \\ v_{C_2} \end{bmatrix} = \begin{bmatrix} -10 & 5 \\ -2 & -\frac{2}{3} \end{bmatrix}\begin{bmatrix} i_L \\ v_{C_2} \end{bmatrix} + \begin{bmatrix} 0 \\ -\frac{2}{15} \end{bmatrix} v_S$$

15.10 ADVANTAGES OF STATE-VARIABLE ANALYSIS

State variable approach (or state space technique) enables us to describe the nth-order system by a set of n simultaneous first-order differential equations.

This set may be obtained either directly or from the nth-order equation of the system.

Since the state equations are first-order differential equations, the state-variable approach has a number of advantages over conventional methods. The following are its important advantages :

(1) The solution of first-order differential equation is much easier and a number of methods, both analytical and numerical, are available.

(2) First-order differential equations are much easier to programme on digital computers, therefore, computer aided design is possible by state-variable approach.

(3) If the describing equations of a network are given in the form of first-order differential equations, the techniques available in systems theory can be applied to determine stability, controllability and observability of the network.

(4) The state-space technique is equally applicable to the time-varying systems and nonlinear systems as it is applicable to time invariant and linear systems.

(5) The solutions of the state equations are obtained in the time domain. Therefore, the state-variable analysis is sometimes called the *time-domain analysis*.

EXERCISES

Write the state equations in matrix form for the network shown in the following figures :

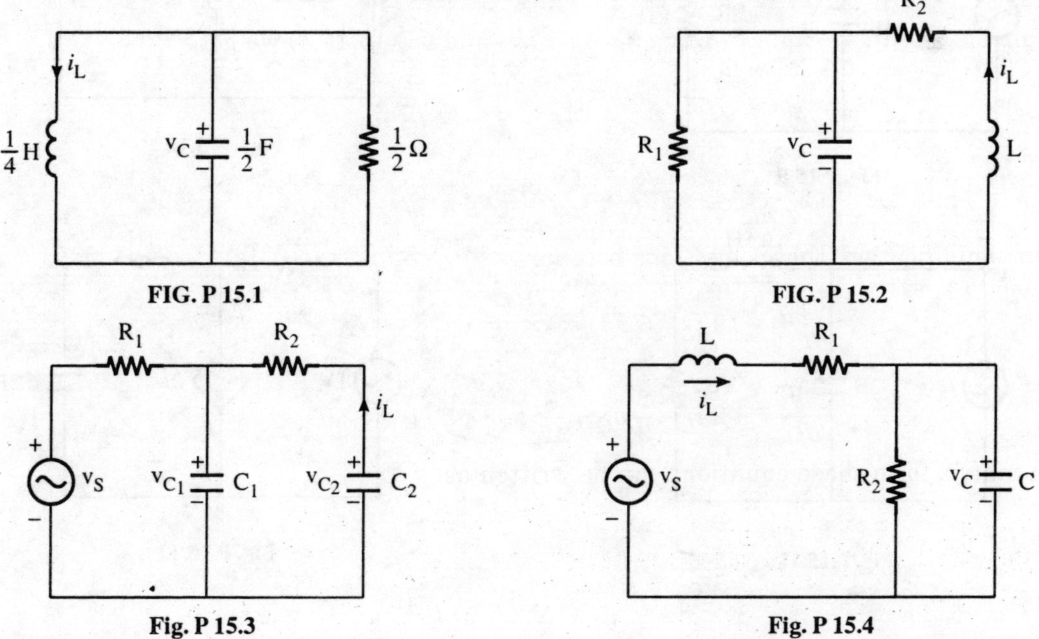

FIG. P 15.1

FIG. P 15.2

Fig. P 15.3

Fig. P 15.4

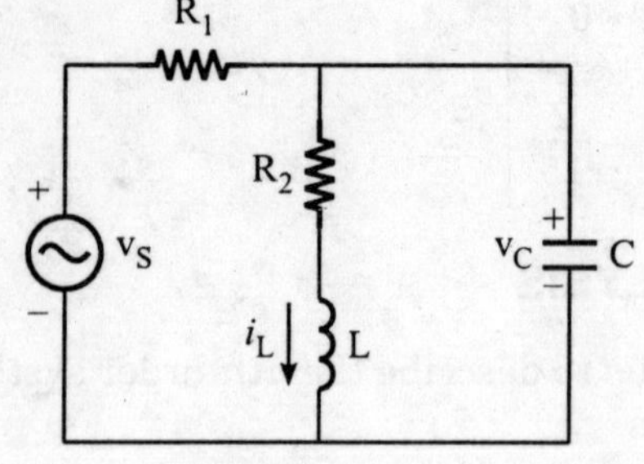

Fig. P 15.5

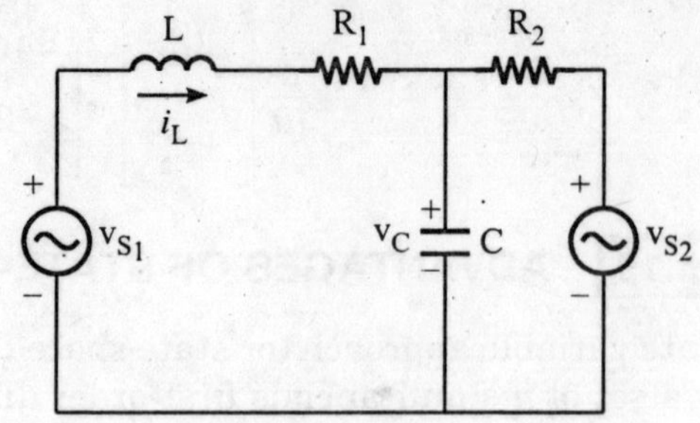

Fig. P 15.6

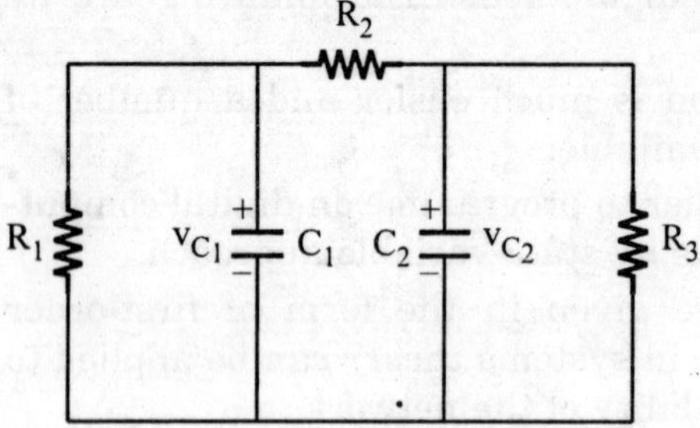

Fig. P 15.7

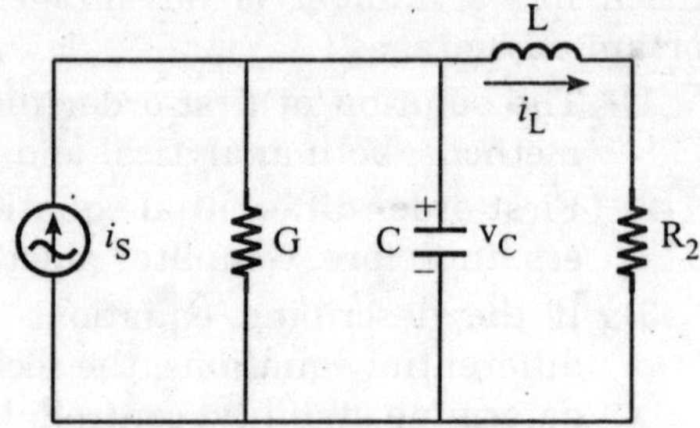

Fig. P 15.8

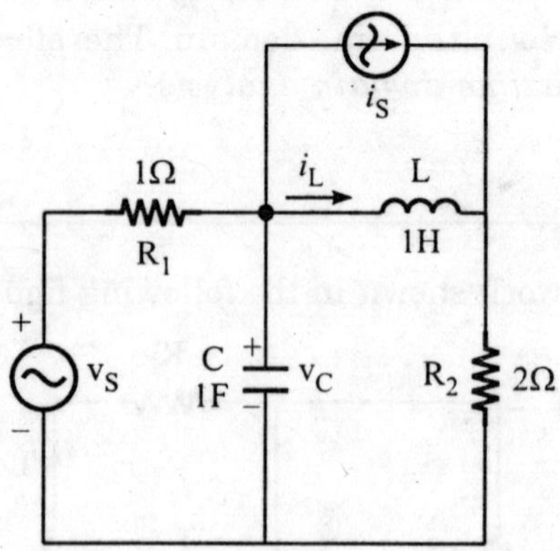

Fig. P 15.9

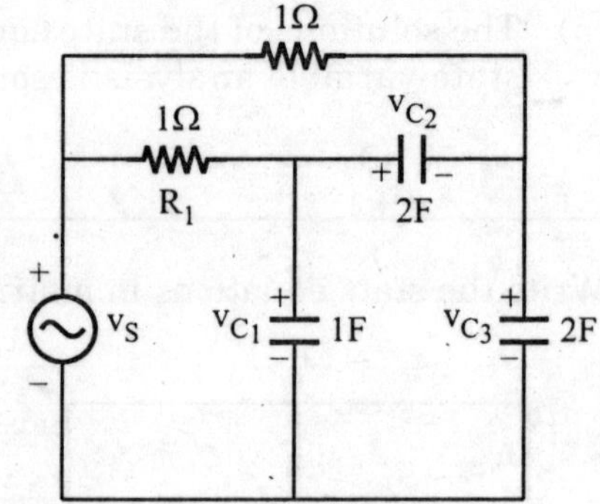

Fig. P 15.10

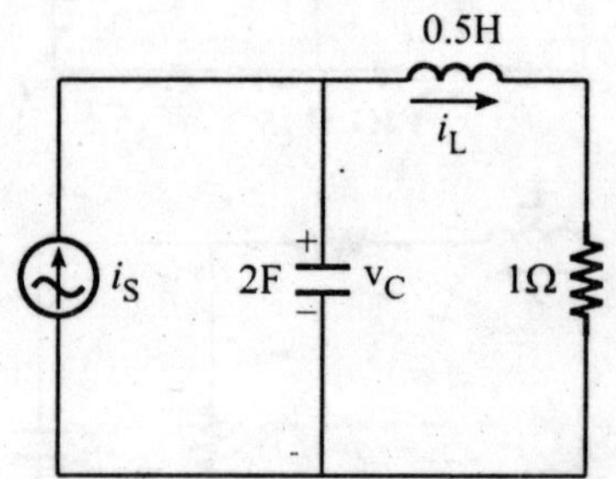

Fig. P 15.11

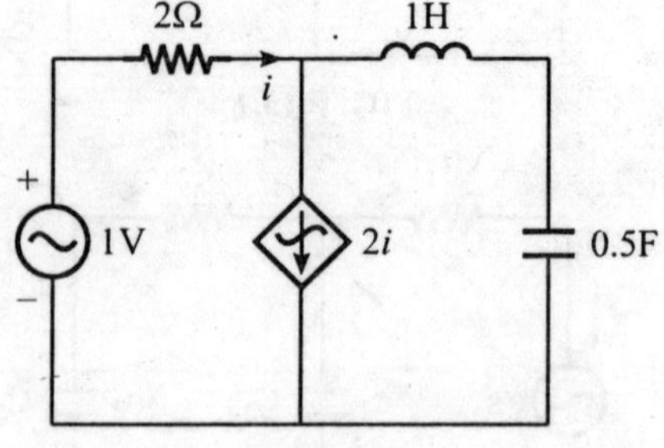

Fig. P 15.12

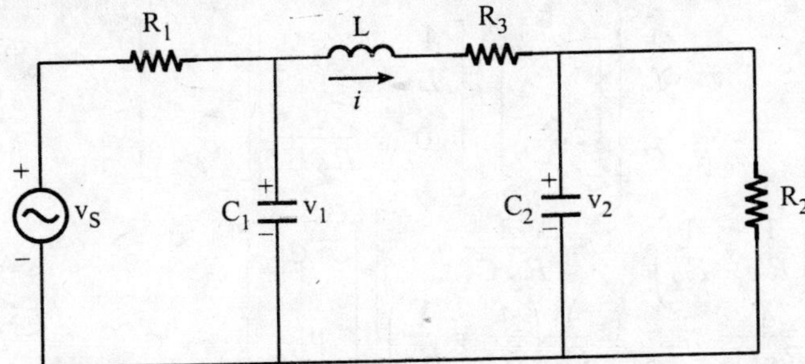

Fig. P 15.13

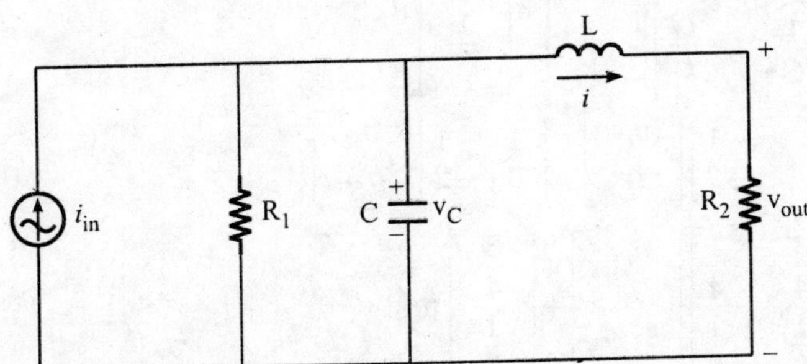

Fig. P 15.14

ANSWERS

1. $\frac{d}{dt}\begin{bmatrix} i_L \\ v_C \end{bmatrix} = \begin{bmatrix} 0 & -4 \\ 2 & -4 \end{bmatrix}\begin{bmatrix} i_L \\ v_C \end{bmatrix}$

2. $\frac{d}{dt}\begin{bmatrix} i_L \\ v_C \end{bmatrix} = \begin{bmatrix} -\frac{R_2}{L} & -\frac{1}{L} \\ \frac{1}{C} & -\frac{1}{R_1 C} \end{bmatrix}\begin{bmatrix} i_L \\ v_C \end{bmatrix}$

3. $\frac{d}{dt}\begin{bmatrix} v_{C_1} \\ v_{C_2} \end{bmatrix} = \begin{bmatrix} -\frac{1}{R_1 C_1} - \frac{1}{R_2 C_1} & \frac{1}{R_2 C_1} \\ \frac{1}{R_2 C_2} & -\frac{1}{R_2 C_2} \end{bmatrix}\begin{bmatrix} v_{C_1} \\ v_{C_2} \end{bmatrix} + \begin{bmatrix} \frac{v_S}{R_1 C_1} \\ 0 \end{bmatrix}$

4. $\frac{d}{dt}\begin{bmatrix} i_L \\ v_C \end{bmatrix} = \begin{bmatrix} -\frac{R_1}{L} & -\frac{1}{L} \\ \frac{1}{C} & -\frac{1}{R_2 C} \end{bmatrix}\begin{bmatrix} i_L \\ v_C \end{bmatrix} + \begin{bmatrix} \frac{1}{L} \\ 0 \end{bmatrix}[v_S]$

5. $\frac{d}{dt}\begin{bmatrix} v_C \\ i_L \end{bmatrix} = \begin{bmatrix} -\frac{1}{R_1 C} & -\frac{1}{C} \\ \frac{1}{L} & -\frac{R_2}{L} \end{bmatrix}\begin{bmatrix} v_C \\ i_L \end{bmatrix} + \begin{bmatrix} \frac{1}{R_1 C} \\ 0 \end{bmatrix}[v_S]$

6. $$\frac{d}{dt}\begin{bmatrix} i_L \\ v_C \end{bmatrix} = \begin{bmatrix} -\dfrac{R_1}{L} & -\dfrac{1}{L} \\ \dfrac{1}{C} & -\dfrac{1}{R_2 C} \end{bmatrix}\begin{bmatrix} i_L \\ v_C \end{bmatrix} + \begin{bmatrix} \dfrac{1}{L} & 0 \\ 0 & \dfrac{1}{R_2 C} \end{bmatrix}\begin{bmatrix} v_{S_1} \\ v_{S_2} \end{bmatrix}$$

7. $$\frac{d}{dt}\begin{bmatrix} v_{C_1} \\ v_{C_2} \end{bmatrix} = \begin{bmatrix} -\dfrac{1}{C}\left(\dfrac{1}{R_1}+\dfrac{1}{R_2}\right) & \dfrac{1}{R_2 C_1} \\ \dfrac{1}{R_2 C_2} & -\dfrac{1}{C_2}\left(\dfrac{1}{R_3}+\dfrac{1}{R_2}\right) \end{bmatrix}\begin{bmatrix} v_{C_1} \\ v_{C_2} \end{bmatrix}$$

8. $$\frac{d}{dt}\begin{bmatrix} i_L \\ v_C \end{bmatrix} = \begin{bmatrix} -\dfrac{R_2}{L} & \dfrac{1}{L} \\ -\dfrac{1}{C} & -\dfrac{G}{C} \end{bmatrix}\begin{bmatrix} i_L \\ v_C \end{bmatrix} + \begin{bmatrix} 0 \\ \dfrac{1}{C} \end{bmatrix}[i_S]$$

9. $$\frac{d}{dt}\begin{bmatrix} v_C \\ i_L \end{bmatrix} = \begin{bmatrix} -1 & -1 \\ 1 & -2 \end{bmatrix}\begin{bmatrix} v_C \\ i_L \end{bmatrix} + \begin{bmatrix} 1 & -1 \\ 0 & -2 \end{bmatrix}\begin{bmatrix} v_S \\ i_S \end{bmatrix}$$

10. $$\frac{d}{dt}\begin{bmatrix} v_{C_1} \\ v_{C_2} \end{bmatrix} = \begin{bmatrix} -\dfrac{3}{4} & \dfrac{1}{4} \\ -\dfrac{1}{8} & -\dfrac{1}{8} \end{bmatrix}\begin{bmatrix} v_{C_1} \\ v_{C_2} \end{bmatrix} + \begin{bmatrix} \dfrac{3}{4} \\ \dfrac{1}{8} \end{bmatrix}[v_S]$$

11. $$\frac{d}{dt}\begin{bmatrix} v_C \\ i_L \end{bmatrix} = \begin{bmatrix} 0 & -\dfrac{1}{2} \\ 2 & -2 \end{bmatrix}\begin{bmatrix} i_L \\ v_C \end{bmatrix} + \begin{bmatrix} 0 \\ \dfrac{1}{2} \end{bmatrix}[i_S]$$

12. $$\frac{d}{dt}\begin{bmatrix} i_L \\ v_C \end{bmatrix} = \begin{bmatrix} 2 & -1 \\ 1 & 0 \end{bmatrix}\begin{bmatrix} i_L \\ v_C \end{bmatrix} + \begin{bmatrix} 1 \\ 0 \end{bmatrix}1$$

13. $$\frac{d}{dt}\begin{bmatrix} i \\ v_1 \\ v_2 \end{bmatrix} = \begin{bmatrix} -\dfrac{R_3}{L} & \dfrac{1}{L} & -\dfrac{1}{L} \\ -\dfrac{1}{C_1} & -\dfrac{1}{R_1 C_1} & 0 \\ \dfrac{1}{C_2} & 0 & -\dfrac{1}{R_2 C_2} \end{bmatrix}\begin{bmatrix} i \\ v_1 \\ v_2 \end{bmatrix} + \begin{bmatrix} 0 \\ \dfrac{1}{R_1 C_1} \\ 0 \end{bmatrix}v_S$$

14. $$\begin{bmatrix} \dfrac{dv_C}{dt} \\ \dfrac{di_L}{dt} \end{bmatrix} = \begin{bmatrix} -\dfrac{1}{R_1 C} & -\dfrac{1}{C} \\ \dfrac{1}{L} & -\dfrac{R_2}{L} \end{bmatrix}\begin{bmatrix} v_C \\ i_L \end{bmatrix} + \begin{bmatrix} \dfrac{1}{C} \\ 0 \end{bmatrix}[i_{in}]$$

$$v_{\text{out}} = \begin{bmatrix} 0 & R_2 \end{bmatrix}\begin{bmatrix} v_C \\ i_L \end{bmatrix} + \begin{bmatrix} 0 \end{bmatrix} i_{in}$$

16

TWO-PORT NETWORKS

16.1 INTRODUCTION

A pair of terminals at which an electrical signal or electrical energy enters or leaves a network is called a **port.** A network that has only one such pair of terminals is known as **one-port network.** It is shown in Fig. 16·1 a. For the one-port network, the current entering into one terminal is equal to the current leaving the other terminal. The requirement that the current into the circuit at one terminal the current into the circuit at one terminal of the port is equal to the current flowing out of the other terminal of the port is called the **port condition** or **port-current requirement**. It is shown in Fig. 16.1 b.

FIG. 16.1 One-port network

A network that has two pairs of terminals is called a **two-port network**. One pair forms the input terminals and the other the output terminals. A general 2-port network is shown in Fig. 16.2

The terminals 1 and 1′ from the input port, and the terminals 2 and 2′ form the output port.

The box shown in Fig. 16.2 is often called a **black box**. The word **black** was presumably chosen to indicate something unknown or mysterious was inside the box. We are not interested to calculate the voltages and currents inside the box.

Transformers, transmission lines, amplifiers, filters, and attenuators are some of the examples of 2-port networks. Of all possible networks, 2-port networks are encountered most frequently in practice.

16.2 PARAMETERS OF TWO-PORT NETWORKS

The standard convention adopted in a 2-port analysis for the polarities of the terminal voltages V_1 and V_2 and for the directions of the terminal currents I_1 and I_2 is shown in Fig. 16.2a. As per convention, the currents entering the upper terminals of the ports are taken positive. It is to be noted that the terminals at the input port carry equal and opposite currents. Similarly, the currents at the output terminals are equal and opposite. According to the port-current, requirement, $I_1 = I_1'$ and $I_2 = I_2'$. This feature distinguishes a 2-port network from a general 4-terminal network. Usually, the current flowing out of the lower terminal at a port is not shown in the figure, but it is understood to be equal and opposite to the current at the upper terminal of that port. Thus, a general 2-port network is shown in Fig. 16.2b.

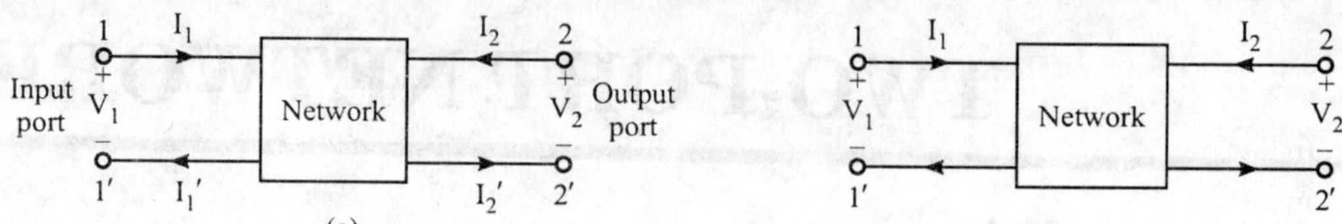

FIG. 16.2 A general 2-port network

In the 2-port network analysis, it is assumed that the network consists of linear elements including the dependent sources, but there are no independent sources inside the network. Moreover, the network is in the zero state, that is, the initial conditions on inductors and capacitors are set to zero. Also, no external connections exist between the input and output ports.

As mentioned earlier, we are not interested to calculate the voltage, and currents inside the network. We are interested in the terminal behavior. The **terminal behavior** of a 2-port network may be described by two independent equations relating the four variables V_1, I_1, V_2 and I_2 at the terminals of the ports. Any two of the four variables may be taken as independent (known) variables, and the remaining two as dependent (unknown) variables in writing these equation. There are six possible ways of selecting two independent variables from four variables. Thus, there are six different pairs of equations that might be written to describe a linear 2-port network. The constants of the variables are called parameters. Each of there six pairs of equations gives its own set of **parameters** or **constants**. Thus, each pair of equations is called the **defining equation pair** for a particular parameter. The two-port parameters are comprised of immittance parameters (z and y), the hybrid parameters (h and g), and the transmission parameters (a and h).

Any one of the six parameter sets may be found directly, either by calculation or by measurement.

16.2.1 Three-terminal Networks

A **three-terminal network** is a subclass of a 2-port network. Here the minus reference terminals of the input and output ports are at the same potential, that is, a short circuit exists in the network between these terminals. Such a network can be drawn with the short circuit external to the box representing the network as shown in Fig. 16·3. Thus, a **three-terminal network** is a 2-port network. Since the only modification required to change a 3-terminal network to a 2-port network is to connect an extra wire to the third terminal, all the treatments regarding 2-port networks apply equally well to 3-terminal networks.

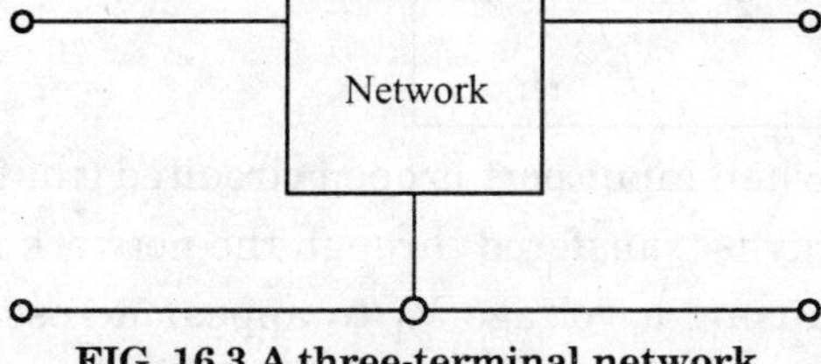

FIG. 16.3 A three-terminal network

16.3 IMPEDANCE (or z) PARAMETERS

Let us take I_1 and I_2 as independent variables. The voltages V_1 and V_2 may be expressed as

$$V_1 = z_{11} I_1 + z_{12} I_2 \quad (16.3.1)$$

$$V_2 = z_{21} I_1 + z_{22} I_2 \quad (16.3.2)$$

In matrix form these equations may be written as

$$\begin{bmatrix} V_1 \\ V_2 \end{bmatrix} = \begin{bmatrix} z_{11} & z_{12} \\ z_{21} & z_{22} \end{bmatrix} \begin{bmatrix} I_1 \\ I_2 \end{bmatrix}$$

The constants z_{11}, z_{12}, z_{21} and z_{22} are called the **impedance (or z) parameters of the 2-port network.**

Eqs. (16·3·1) and (16·3·2) are called the **defining equations of z parameters.** The matrix

$$[z] = \begin{bmatrix} z_{11} & z_{12} \\ z_{21} & z_{22} \end{bmatrix}$$

is called the **impedance (or z) matrix.**

The impedance parameters may be defined as follows :

To define z_{11}, put $I_2 = 0$ in Eq. (16·3·1).

Thus, $V_1 = z_{11} I_1$ when $I_2 = 0$.

$$z_{11} \triangleq \left. \frac{V_1}{I_1} \right|_{I_2 = 0} \quad (16.3.3)$$

Since V_1/I_1 is the input impedance and $I_2 = 0$ corresponding to the output port being open circuited, z_{11} is called the **open-circuit input impedance** or the **driving-point impedance at port 1.**

To define z_{12}, put $I_1 = 0$ in Eq. (16.3.1).

Thus, $V_1 = z_{12} I_2$ when $I_1 = 0$.

That is,

$$z_{12} \overset{\Delta}{=} \left.\frac{V_1}{I_2}\right|_{I_1=0} \tag{16.3.4}$$

This expression shows that when input port is open circuited (that is $I_1 = 0$), there is a current I_2 at the output and the energy is transfered through the network in the *reverse* direction (that is, from port 2 to port 1) causing a voltage V_1 to appear across the input terminals. Since z_{12} is the ratio of the input voltage V_1 to the output current I_2 under the conditon $I_1 = 0$, it is called the **open-circuit reverse transfer impedance.**

To define z_{21}, put $I_2 = 0$ in Eq. (16.3.2). Thus, $V_2 = z_{21}I_1$ when $I_2 = 0$.

That is,

$$z_{21} \overset{\Delta}{=} \left.\frac{V_2}{I_1}\right|_{I_2=0} \tag{16.3.5}$$

This expression shows that when the output port is open circuit (that is, $I_2 = 0$), there is a current I_1 at the input port and the energy is transferred through the network in the forward direction (that is, from port 1 to port 2) causing a voltage V_2 to appear across the output terminals, Since z_{21} is the ratio of the output voltage V_2 to the input current I_1 under the condition $I_2 = 0$, it is called the **open-circuit forward transfer impedance.**

To define z_{22}, put $I_1 = 0$ in Eq. (16.3.2). Thus $V_2 = z_{22}I_2$ when $I_1 = 0$.

That is,

$$z_{22} = \left.\frac{V_2}{I_2}\right|_{I_1=0} \tag{16.3.6}$$

Since V_2/I_2 is the output impedance and $I_1 = 0$ corresponds to input port being open circuited, z_{22} is called the **open-circuit output impedance**.

It is seen that dimensionally each parameter is an impedance. Also, each parameter is calculated or measured by open-circuiting either the input or the output port. Therefore, these parameters are called the **open-circuit impedance parameters.** It is to be noted that the **open circuit** is at the port at which excitation is **not** being applied. The parameters z_{12} and z_{21} are called the **open-circuit transfer impedances** because each of them relates the current at one port and the voltage at the other port. It is precisely the **transfer** impedance that makes the two ports more powerful circuits than one port.

The biggest advantage of the z parameters is due to the fact that they can be determined experimentally by open-circuiting one port at a time. This can rarely do any damage to the network inside the box.

Eq. (16.3.3) to (16.3.6) show that the four z parameters can be described as follows:

1. z_{11} is the impedance seen looking into port 1 when port 2 is open.
2. z_{12} is the transfer impedance. It is the ratio of the port 1 voltage to the port 2 current when port 1 is open.
3. z_{21} is the transfer impedance. It is the ratio of port 2 voltage to port 1 current when port 1 is open.
4. z_{22} is the impedance seen looking into port 2 when port 1 is open.

Therefore, the impedance parameters may be either calculated or measured by first opening port 2 and then determining the ratios (V_1/I_1) and (V_2/I_1), and then opening port 1 and determining the ratios (V_1/I_2) and (V_2/I_2).

In general, it is easier to use Eqs. (16.3.3), (16.3.4), (16.3.5) or (16.3.6) when only one parameter is desired. If we are interested to calculate all of them, it is usually easier to write loop equations and carry the general solution.

16.4 EQUIVALENT NETWORKS (MODELS) OF z PARAMETERS

We have seen how to find the z parameters for a given network. This process is called **two-port analysis**. Now let us consider the inverse problem : Given a set of z parameters, how do we find a two-port network which has such a set of parameters. This process is called **two-port synthesis** or **modelling**. Since this is a synthesis problem, there is no unique solution.

Two-port models are useful in several ways :

(1) They permit us to visualize a circuit by a set of two-port parameters, to consider the effects of interconnecting them with other elements or subcircuits.

(2) In the case of coupled coils, they permit us to model a circuit with mutual inductance by one that does not contain mutual inductance.

In order to derive the 2-port models of z parameters we use their defining equations :

$$V_1 = z_{11}I_1 + z_{12}I_2$$

$$V_2 = z_{21}I_1 + z_{22}I_2$$

Any circuit that satisfies these two equations is called the **ciruit model** for this two-port description.

We may use any circuit configuration that agrees with these equations. Usually we select a simple circuit configuration. One of the most common network representation for a given set of z parameters is the T-network representation. This is applicable to all reciprocal networks, that is, the networks in which $z_{12} = z_{21}$. A reciprocal network usually contains only resistors, inductors and capacitors. Nonreciprocal networks contain dependent (controlled) sources and/or nonlinear devices also. A linear, passive, reciprocal 2-port network may be replaced by an equivalent T network (model). A nonreciprocal network may be replaced by either a two-controlled source model or a one controlled-source model.

16.5 EQUIVALENT T NETWORK OF A TWO PORT RECIPROCAL NETWORK

A linear, passive, 2-port reciprocal network may be replaced by an equivalent T network (model) as shown in Fig. 16·4 a. The elements of the equivalent network may conveniently be expressed in terms of z parameters of the 2-port network. For a reciprocal network $z_{12} = z_{21}$.

The z parameters of the equivalent network of Fig. 16·4 a may be found from their definitions. The current in Z_c is $I_1 + I_2$.

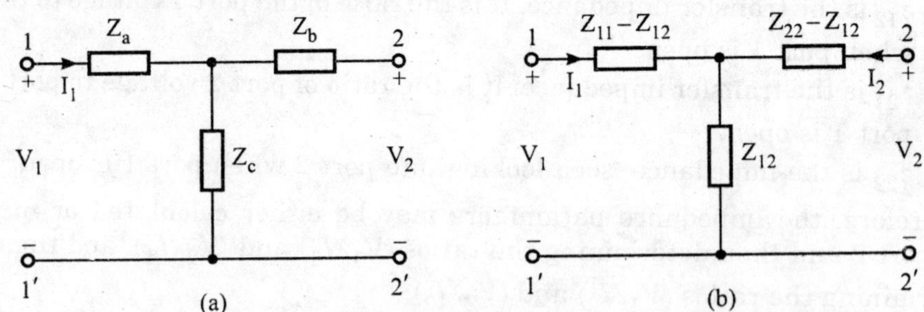

FIG. 16.4 (a) Equivalent *T* network of a 2-port reciprocal network (b) Equivalent *T* network in terms of *z* parameters.

When port 2 is open circuited

$I_2 = 0$ and there is no voltage drop in Z_b. The current in Z_c becomes I_1.

$$V_1 = Z_a I_1 + Z_c I_1 = (Z_a + Z_c)\, I_1$$

$$\frac{V_1}{I_1} = Z_a + Z_c$$

$$z_{11} = \left.\frac{V_1}{I_1}\right|_{I_2=0} = Z_a + Z_c \tag{16.5.1}$$

Also $V_2 = Z_c I_1$, $\quad \dfrac{V_2}{I_1} = Z_c$

$$z_{21} = \left.\frac{V_2}{I_1}\right|_{I_2=0} = Z_c \tag{16.5.2}$$

When port 1 is open circuited

$I_1 = 0$, and the voltage drop in Z_a is equal to zero. The current in Z_c becomes I_2.

$$V_1 = Z_c I_2, \quad \frac{V_1}{I_2} = Z_c$$

$$z_{12} = \left.\frac{V_1}{I_2}\right|_{I_1=0} = Z_c \tag{16.5.3}$$

Also $V_2 = Z_b I_2 + Z_c I_2 = (Z_b + Z_c)\, I_2$

$$z_{22} = \left.\frac{V_2}{I_2}\right|_{I_1=0} = Z_b + Z_c \tag{16.5.4}$$

From Eqs. (16.5.1) to (16.5.4) the values of Z_a, Z_b, and Z_c can be written as

$$Z_a = z_{11} - z_{12} \tag{16.5.5}$$

$$Z_b = z_{22} - z_{12} \tag{16.5.6}$$

$$Z_c = z_{12} = z_{21} \tag{16.5.7}$$

The equivalent T network of a 2-port network having its elements in terms of the z parameters is shown in Fig. 16.4b.

Alternative Method

The z parameters of the network shown in Fig. 16.3 may also be found by writing mesh equations and comparing them with the defining equations of the z parameters.

By KCL, the current in Z_c is $I_1 + I_2$.

By KVL in the left mesh of Fig. 16.3,

$$V_1 = Z_a I_1 + Z_c (I_1 + I_2)$$

or

$$V_1 = (Z_a + Z_c) I_1 + Z_c I_2 \tag{16.5.8}$$

By KVL in the right mesh

$$V_2 = Z_b I_2 + Z_c (I_1 + I_2)$$

or

$$V_2 = Z_c I_1 + (Z_b + Z_c) I_2 \tag{16.5.9}$$

The defining equations of the z parameters are

$$V_1 = z_{11} I_1 + z_{12} I_2 \tag{16.5.10}$$

$$V_2 = z_{21} I_1 + z_{22} I_2 \tag{16.5.11}$$

Comparison of Eq. (16.5.8) with Eq. (16.5.10) and of Eq. (16.5.9) with Eq. (16.5.11) gives the z parameters of the T network.

$$\boxed{\begin{aligned} z_{11} &= Z_a + Z_c \\ z_{12} &= Z_c \\ z_{21} &= Z_c \\ z_{22} &= Z_b + Z_c \end{aligned}} \quad \text{or} \quad [z] = \begin{bmatrix} Z_a + Z_c & Z_c \\ Z_c & Z_b + Z_c \end{bmatrix}$$

Therefore,

$$\boxed{\begin{aligned} Z_a &= z_{11} - z_{12} \\ Z_b &= z_{22} - z_{12} \\ Z_c &= z_{12} = z_{21} \end{aligned}}$$

as obtained before.

16.6 EQUIVALENT NETWORKS (MODELS) OF A TWO-PORT NON-RECIPROCAL NETWORK USING z PARAMETERS

The defining equations of z parameters are

$$V_1 = z_{11} I_1 + z_{12} I_2 \tag{16.3.1}$$

$$V_2 = z_{21}I_1 + z_{22}I_2 \qquad (16.3.2)$$

The right-hand side of Eq. (16.3.1) is the sum of two terms $z_{11}I_1$ and $z_{12}I_2$. The first term $z_{11}I_1$ is obtained by multiplying the impedance z_{11} with the input current I_1 The second term $z_{12}I_2$ is dependent upon I_2 which does not flow in the input circuit. Hence this term can be replaced by a **current-controlled voltage source** (CCVS).

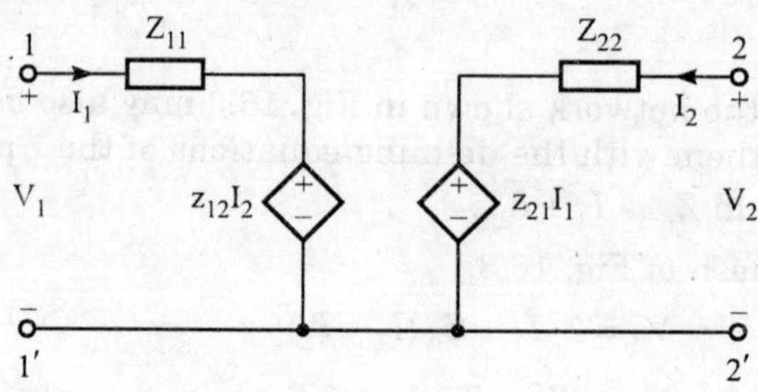

FIG. 16.5 A two-current controlled voltage source model of a nonreciprocal set of *z* parameters.

Similarly, Eq. (16.3.2) shows that port 2 can be represented by an impedance z_{22} and a current-controlled voltage source $z_{21}I_1$. Fig. 16.5 shows the equivalent circuit (model) of a two-port nonreciprocal network using the z parameters. The circuit has Thevenin equivalent representations at both ports. It is a two current controlled voltage source model.

We have

$$V_1 = z_{11}I_1 + z_{12}I_2 \qquad (16.3.1)$$

By adding and subtracting $z_{12}I_1$ in Eq. (16.3.1) we have

$$V_1 = z_{11}I_1 + z_{12}I_2 + z_{12}I_1 - z_{12}I_1$$

or

$$V_1 = (z_{11} - z_{12})\, I_1 + z_{12}\, (I_1 + I_2) \qquad (16.6.1)$$

Also,

$$V_2 = z_{21}I_1 + z_{22}I_2 \qquad (16.3.2)$$

Equation (16.3.2) can be written as

$$V_2 = (z_{21} - z_{12})\, I_1 + (z_{22} - z_{12})\, I_2 + z_{12}\, (I_1 + I_2) \qquad (16.6.2)$$

From Eqs. (16.6.1) and (16.6.2) we can draw an alternate network with one current controlled voltage source (CCVS) as shown in Fig. 16.6. This network is sometimes useful in determining the z parameters directly. It is to be noted that when this representation is applied to a reciprocal network in which $z_{12} = z_{21}$, the controlled source disappears, and this representation becomes a T network shown in Fig. 16.4 b.

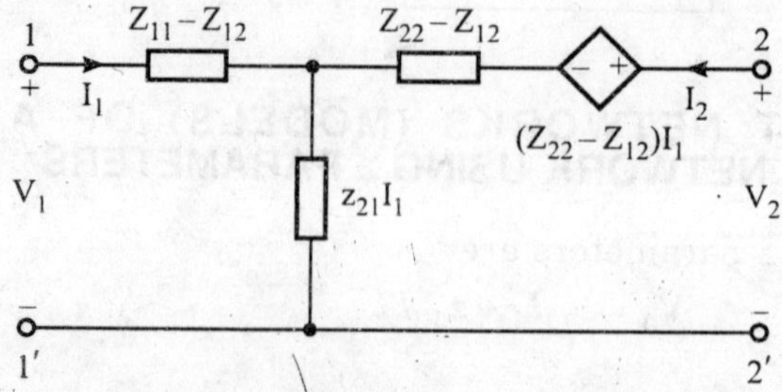

FIG. 16.6 A one-controlled source model of a nonreciprocal set of *z* parameters.

16.7 ADMITTANCE (or y) PARAMETERS

If we take V_1 and V_2 as independent variables and I_1 and I_2 as dependent variables, the 2-port equations may be written as

$$I_1 = y_{11} V_1 + y_{12} V_2 \tag{16.7.1}$$

$$I_2 = y_{21} V_1 + y_{22} V_2 \tag{16.7.2}$$

In matrix form these equations are written as

$$\begin{bmatrix} I_1 \\ I_2 \end{bmatrix} = \begin{bmatrix} y_{11} & y_{12} \\ y_{21} & y_{22} \end{bmatrix} \begin{bmatrix} V_1 \\ V_2 \end{bmatrix}$$

The constants y_{11}, y_{12}, y_{21}, and y_{22} are called the **admittance (or y) parameters** of the 2-port network. Eqs. (16.7.1) and (16.7.2) are called the **defining equations of the y parameters.**

The matrix

$$[y] = \begin{bmatrix} y_{11} & y_{12} \\ y_{21} & y_{22} \end{bmatrix}$$

is called the admittance (or y) matrix.

The y parameters are defined as follows :

$$y_{11} \stackrel{\Delta}{=} \left. \frac{I_1}{V_1} \right|_{V_2 = 0} = \text{short-circuit input admittance} \tag{16.7.3}$$

$$y_{12} \stackrel{\Delta}{=} \left. \frac{I_1}{V_2} \right|_{V_1 = 0} = \text{short-circuit reverse transfer admittance} \tag{16.7.4}$$

$$y_{21} \stackrel{\Delta}{=} \left. \frac{I_2}{V_1} \right|_{V_2 = 0} = \text{short-circuit forward transfer admittance} \tag{16.7.5}$$

$$y_{22} \stackrel{\Delta}{=} \left. \frac{I_2}{V_2} \right|_{V_1 = 0} = \text{short-circuit output admittance} \tag{16.7.6}$$

It is seen that each parameter is an admittance. Also, each parameter is obtained by short circuiting either the input or output port. Therefore, these parameters are called the **short-circuit admittance parameters.** It is to be noted that the *short circuit* is applied to the port at which the excitation is not being applied. The parameters y_{12} and y_{21} are called the **short-circuit transfer admittances** because each of them relates the current at one port and the voltage at the other port.

16.8 EQUIVALENT NETWORKS (MODELS) USING y PARAMETERS

We shall find a network representation for a given set of y parameters. For a reciprocal network (for which $y_{12} = y_{21}$) the π representation is a convenient model. When the network is not reciprocal (when $y_{12} \neq y_{21}$), then a controlled source realization is used. Again a nonreciprocal network may be replaced by either a two controlled-source model or a one controlled-source model.

16.9 EQUIVALENT Π NETWORK OF A RECIPROCAL 2-PORT NETWORK

A linear, passive, reciprocal 2-port network may also be replaced by an equivalent Π network as shown in Fig. 16.7a. The elements of this equivalent network may conveniently be expressed in terms of the y parameters of the 2-port network. The y parameters of the equivalent network may be found from their definitions as follows :

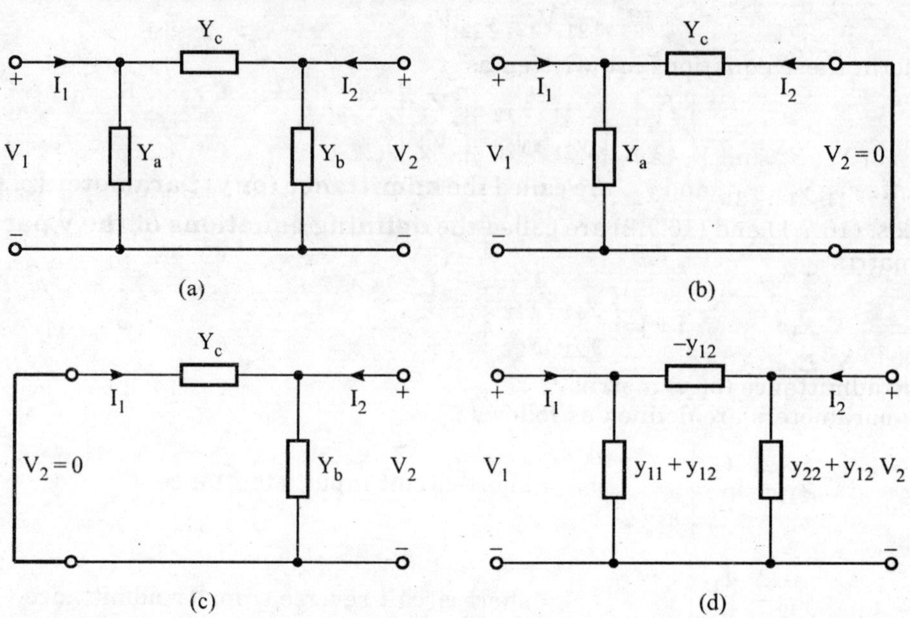

FIG. 16.7 (a) Equivalent π network of reciprocal 2-port network (b) Port 2 short circuited (c) Port 1 short circuited (d) Equivalent π network of a reciprocal 2-port network in terms of y parameters

Port 2 short circuited ($V_2 = 0$)

A π network with output port short circuited is shown in Fig. 16.7b. The admittances Y_a and Y_c are now in parallel. From Fig. 16.7b,

$$I_1 = (Y_a + Y_c)\, V_1$$

therefore,
$$y_{11} = \left.\frac{I_1}{V_1}\right|_{V_2=0} = Y_a + Y_c \tag{16.9.1}$$

Again from Fig. 16.7b.

$$-I_2 = V_1 Y_c$$

therefore,
$$y_{21} = \left.\frac{I_2}{V_1}\right|_{V_2=0} = -Y_c \tag{16.9.2}$$

Port 1 short circuited ($V_1 = 0$)

A π network with port 1 short circuited is shown in Fig. 16.7c. The admittances Y_b and Y_c are now in parallel.

From Fig. 16.7c.

$$I_2 = (Y_b + Y_c)\, V_2$$

therefore,
$$y_{22} = \left.\frac{I_2}{V_2}\right|_{V_1=0} = Y_b + Y_c \tag{16.9.3}$$

Again from Fig. 16.7c

$$-I_1 = Y_c V_2$$

therefore,
$$y_{12} = \left.\frac{I_1}{V_2}\right|_{V_1=0} = -Y_c \tag{16.9.4}$$

The values of Y_a, Y_b and Y_c can be found from Eqs (16.9.1) to (16.9.4) as

$$Y_a = y_{11} + y_{12}$$

$$Y_b = y_{22} + y_{12}$$

$$Y_c = -y_{12} = -y_{21}$$

The equivalent π network of a 2 port network having its elements in terms of the y parameters is shown in Fig. 16.7d.

Alternative method

The y parameters of the π network shown in Fig. 16.7a may also be found by writing node equations and comparing them with the defining equations of the y parameters. The node equations are given by

$$I_1 = (Y_a + Y_c)\, V_1 - Y_c V_2$$

$$I_2 = -Y_c V_1 + (Y_b + Y_c)\, V_2$$

Comparison of these equation with the defining equations of the y parameters gives

$$y_{11} = Y_a + Y_c$$

$$y_{12} = y_{21} = -Y_c$$

$$y_{22} = Y_b + Y_c$$

or
$$[\,y\,] = \begin{bmatrix} Y_a + Y_c & -Y_c \\ -Y_c & Y_b + Y_c \end{bmatrix}$$

16.10 EQUIVALENT NETWORKS OF A TWO-PORT NONRECIPROCAL NETWORK USING y PARAMETERS

A circuit model using y parameters can be developed from Eqs. (16.7.1) and (16.7.2). It is shown in Fig. 16.8 a. It has two admittances and two voltage controlled voltage sources (VCVS).

We have

$$I_1 = y_{11}V_1 + y_{12}V_2 \tag{16.7.1}$$

By adding and subtracting $y_{12}V_1$ in Eq. (16.7.1) we get

$$I_1 = y_{11}V_1 + y_{12}V_2 + y_{12}V_1 - y_{12}V_1$$

or
$$I_1 = (y_{11} + y_{12})\, V_1 - y_{12}\,(V_1 - V_2) \tag{16.10.1}$$

Also,
$$I_2 = y_{21}V_1 + y_{22}V_2 \tag{16.7.2}$$

Eq. (16.7.2) can be written as

$$I_2 = (y_{21} - y_{12}) V_1 + (y_{22} + y_{12}) V_2 - y_{12} (V_2 - V_1) \tag{16.10.2}$$

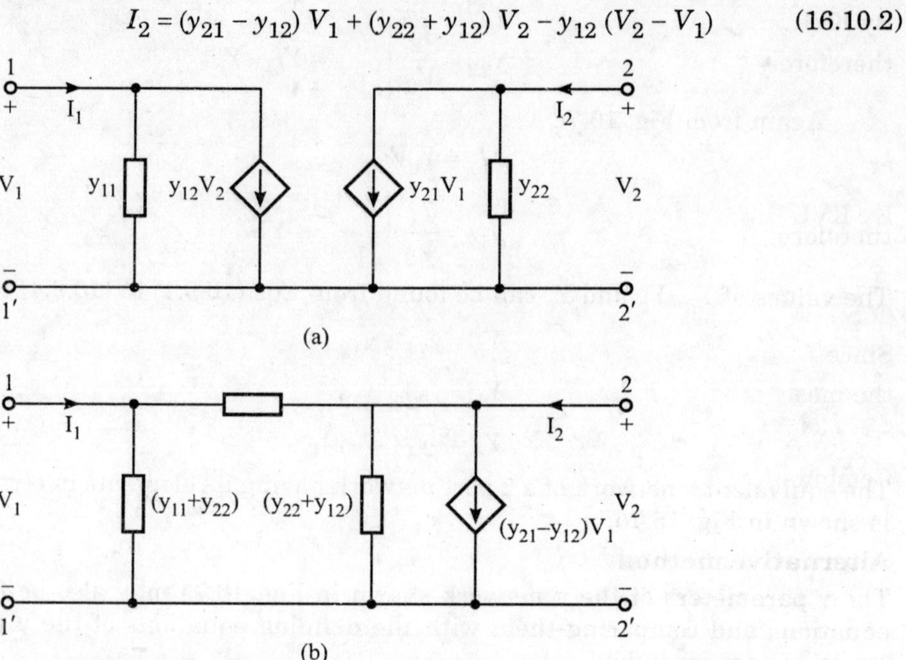

FIG. 16.8 (a) A two-controlled source model of a nonreciprocal set of *y* parameters. (b) A one-controlled source *y* parameter model

From Eqs. (16.10.1) and (16.10.2) we draw an alternate network with one voltage controlled current source (VCCS) as shown in Fig. 16.8b. This network is sometimes useful in determining the *y* parameters directly. For a reciprocal network [that is for the case when $y_{12} = y_{21}$], this network becomes the π-network model shown in Fig. 16.10.

■ **EXAMPLE 16.1** Find the *z* parameters for the symmetrical Π network of Fig. 16.9a

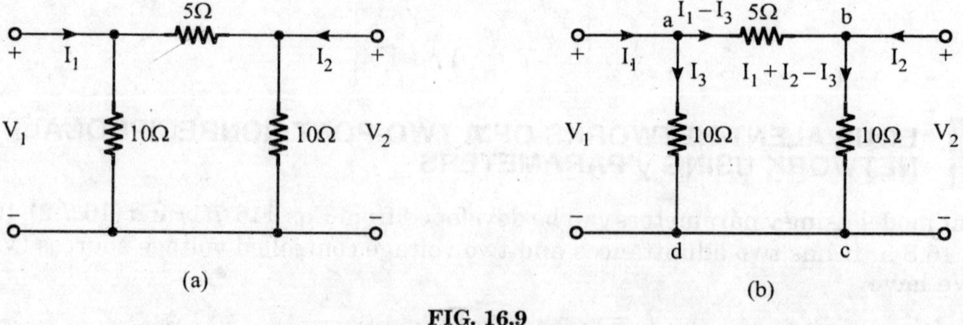

FIG. 16.9

SOLUTION Let I_3 be the current through the 10 Ω resistor on the left-hand side. Fig. 16.9b. shows the currents in different branches of the network after application of KCL at nodes a and b.

By KVL in the left-hand mesh

$$V_1 = 10I_3 = 0$$

or $$V_1 = 10I_3 \tag{E 16.1.1}$$

By KVL in mesh *a b c a*

$$-5\,(I_1 - I_3) - 10\,(I_1 + I_2 - I_3) + 10I_3 = 0$$

$$-15I_1 - 10I_2 + 25I_3 = 0$$

or $$3I_1 + 2I_2 - 5I_3 = 0 \tag{E 16.1.2}$$

By KVL in the right-hand mesh

$$V_2 - 10\,(I_1 + I_2 - I_3) = 0$$

or $$V_2 = 10I_1 + 10I_2 - 10I_3 \tag{E 16.1.3}$$

Since I_3 does not appear in the defining equations of the z parameters, we eliminate it from the mesh equations. Multiplying Eq. (E 16.1.2) by 2 on both sides

$$6I_1 + 4I_2 = 10I_3 \tag{E 16.1.4}$$

Combination of Eqs. (E 16.1.1) and (E 16.1.4) gives

$$V_1 = 6I_1 + 4I_2 \tag{E 16.1.5}$$

Substituting the value of $10I_3$ from Eq. (E 16.1.4) in Eq. (E 16.1.3) we get

$$V_2 = 10I_1 + 10I_2 - (6I_1 + 4I_2)$$

or $$V_2 = 4I_1 + 6I_2 \tag{E 16.1.6}$$

The defining equations of z parameters are

$$V_1 = z_{11}I_1 + z_{12}I_2 \tag{E 16.1.7}$$

$$V_2 = z_{12}I_1 + z_{22}I_2 \tag{E 16.1.8}$$

Comparison of Eq. (E 16.1.5) with Eq. (E 16.1.7) and of Eq. (E 16.1.6) with Eq. (E 16.1.8) gives

$$z_{11} = 6\ \Omega,\ z_{12} = 4\ \Omega,\ z_{21} = 4\ \Omega$$

$$z_{22} = 6\ \Omega$$

EXAMPLE 16.2 Find the z parameters for the circuit in Fig. 16.10a.

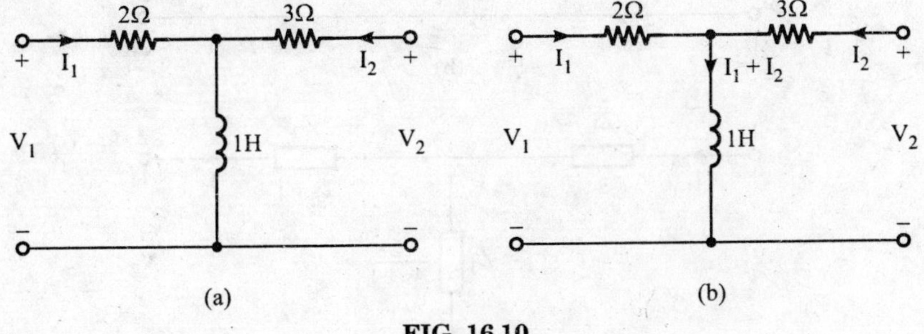

FIG. 16.10

SOLUTION In the s domain the given circuit is shown in Fig. 16.10b

By KVL in the left-hand mesh

$$V_1 - 2I_1 - s\,(I_1 + I_2) = 0$$

or $$V_1 = (2 + s)\,I_1 + sI_2 \tag{E 16.2.1}$$

By KVL in the right-hand mesh

$$s\,(I_1 + I_2) + 3I_2 - V_2 = 0$$

$$V_2 = sI_1 + (3 + s)\,I_2 \tag{E 16.2.2}$$

The defining equations of the z parameters are

$$V_1 + z_{11}I_I + z_{12}I_2 \tag{E 16.2.3}$$

$$V_2 = z_{21}I_1 + z_{22}I_2 \tag{E 16.2.4}$$

Comparison of Eq. (E 16.2.1) with Eq. (E 16.2.3) and of Eq. (E 16.2.2) with Eq. (E 16.2.4) gives

$$z_{11} = s + 2$$

$$z_{12} = z_{21} = s,\ z_{22} = s + 3$$

■ **EXAMPLE 16.3** Determine the z parameters of the network shown in Fig. 16.11a.

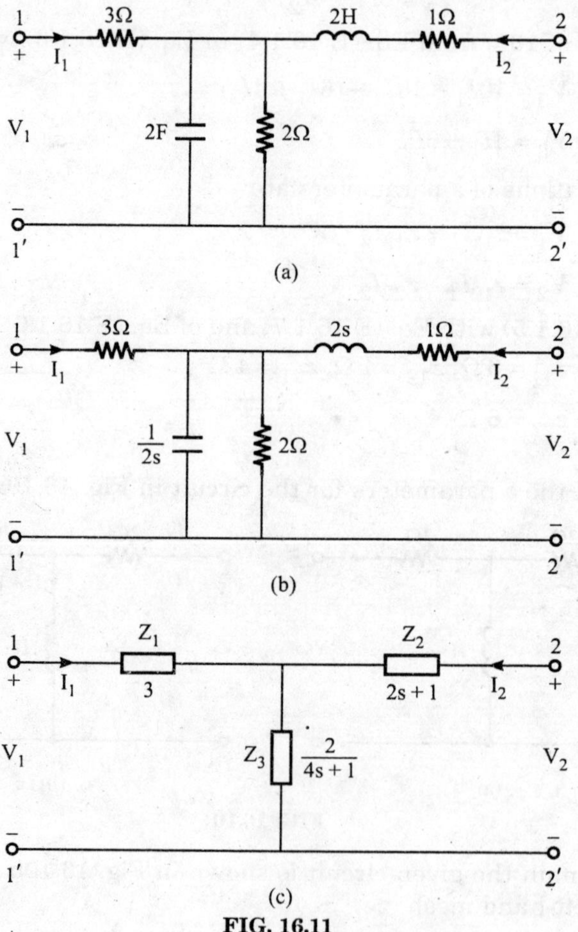

FIG. 16.11

SOLUTION In the s domain the given circuit is shown in Fig. 16.11b. This is a T network having three impedances Z_1, Z_2 and Z_3 as shown in Fig. 16.11c.

Here $$Z_1 = 3,\ Z_2 = 2s + 1$$

$$Z_3 = 2 \,\Big|\Big|\, \left(\frac{1}{2s}\right) = \frac{2 \times \frac{1}{2s}}{2 + \frac{1}{2s}} = \frac{2}{4s+1}$$

$$z_{11} = \left.\frac{V_1}{I_1}\right|_{I_2=0} = Z_1 + Z_3 = 3 + \frac{2}{4s+1} = \frac{12s+5}{4s+1}$$

$$z_{12} = \left.\frac{V_1}{I_2}\right|_{I_1=0} = Z_3 = \frac{2}{4s+1}$$

$$z_{21} = \left.\frac{V_2}{I_1}\right|_{I_2=0} = Z_3 = \frac{2}{4s+1}$$

$$z_{22} = \left.\frac{V_2}{I_2}\right|_{I_1=0} = Z_2 + Z_3 = 2s + 1 + \frac{2}{4s+1}$$

$$= \frac{8s^2 + 6s + 3}{4s+1}$$

■ **EXAMPLE 16.4** Find the open circuit parameters of the network shown in Fig. 16.12a.

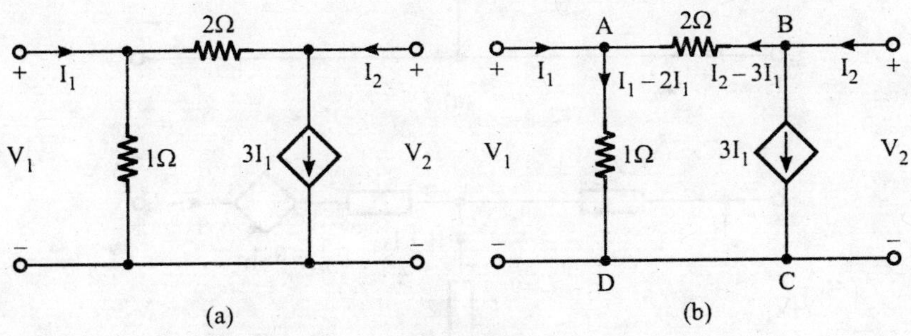

FIG. 16.12

SOLUTION By KCL at node B, the current through the 2 Ω resistor is $(I_2 - 3I_1)$. By KCL at node A, the current through 1 Ω resistor is $I_2 - 3I_1 + I_1 = I_2 - 2I_1$. These currents are shown in Fig. 16.12b. From this figure

$$V_1 = V_{AD} = 1 \times (I_2 - 2I_1)$$

therefore, $$V_1 = -2I_1 + I_2 \tag{E 16.4.1}$$

By Ohm's law

$$V_{BA} = 2 \times (I_2 - 3I_1)$$

$$V_B - V_A = -6I_1 + 2I_2$$

$$V_2 - V_1 = -6I_1 + 2I_2$$

$$V_2 = V_1 - 6I_1 + 2I_2 \tag{E 16.4.2}$$

Substituting the value of V_1 from Eq. (E 16.4.1) in Eq. (E 16.4.2) we get

$$V_2 = -2I_1 + I_2 - 6I_1 + 2I_2$$

or

$$V_2 = -8I_1 + 3I_2 \tag{E 16.4.3}$$

The defining equations of z parameters are

$$V_1 = z_{11}I_1 + z_{12}I_2 \tag{E 16.4.4}$$

$$V_2 = z_{21}I_1 + z_{22}I_2 \tag{E 16.4.5}$$

By comparing Eq. (E 16.4.1) with Eq. (E 16.4.4) and Eq. (E 16.4.3) with Eq. (E 16.4.5) we get

$$z_{11} = -2,\ z_{12} = 1$$

$$z_{21} = -8,\ z_{22} = 3$$

■ **EXAMPLE 16.5** Determine the open-circuit impedance parameters of the network of Fig. 16.13a.

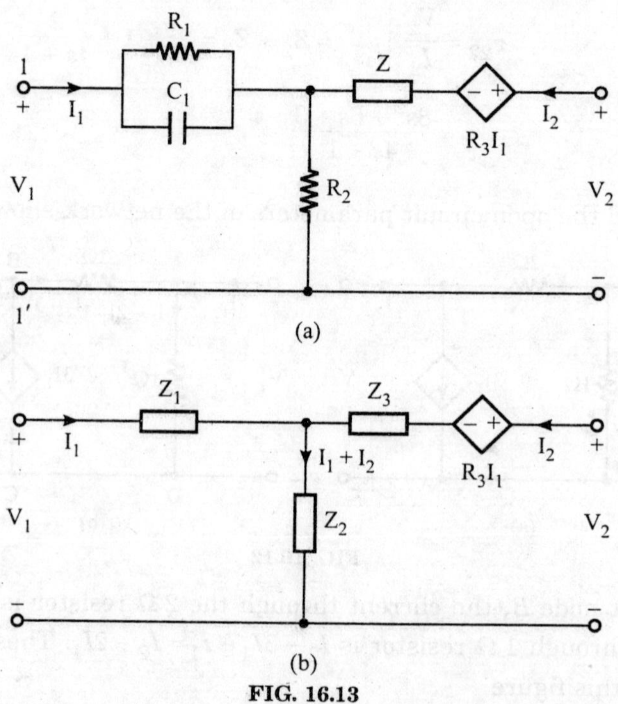

FIG. 16.13

SOLUTION For convenience, we draw the network of Fig. 16.13a as shown in Fig. 16.13b.

Here

$$Z_1 = R_1 \,\Big|\Big|\, \frac{1}{sC_1} = \frac{R_1 \times \dfrac{1}{sC_1}}{R_1 + \dfrac{1}{sC_1}} = \frac{R_1}{1 + sR_1C_1}$$

$$Z_2 = R_2, \qquad Z_3 = Z$$

By KVL in the left-hand mesh

$$V_1 - Z_1 I_1 - Z_2 (I_1 + I_2) = 0$$

$$V_1 = (Z_1 + Z_2) I_1 + Z_2 I_2$$

Also $$V_1 = z_{11} I_1 + z_{12} I_2$$

By comparing these equations we get

$$z_{11} = Z_1 + Z_2 = \frac{R_1}{1 + sR_1C_1} + R_2$$

and $$z_{12} = Z_2 = R_2$$

By KVL in the right-hand mesh

$$V_2 - R_3 I_1 - Z_3 I_2 - Z_2 (I_1 + I_2) = 0$$

$$V_2 = (R_3 + Z_2) I_1 + (Z_2 + Z_3) I_2$$

Also, $$V_2 = z_{21} I_1 + z_{22} I_2$$

By comparing these equations we get

$$z_{21} = R_3 + Z_2 = R_3 + R_2$$

$$z_{22} = Z_2 + Z_3 = R_2 + Z$$

EXAMPLE 16.6 Specify the elements of a T network so that

$$[z] = \begin{bmatrix} 20 & 5 \\ 5 & 30 \end{bmatrix} \Omega$$

SOLUTION For a T network we have

$$[z] = \begin{bmatrix} Z_a + Z_c & Z_c \\ Z_c & Z_b + Z_c \end{bmatrix}$$

therefore, $$Z_a + Z_c = 20$$

$$Z_c = 5$$

$$Z_b + Z_c = 30$$

Solving, we get $$Z_a = 15\ \Omega,\ Z_b = 25\ \Omega,\ Z_c = 5\ \Omega$$

EXAMPLE 16.7 The z parameters of a 2-port network N are given by

$$z_{11} = 2s + \frac{1}{s},\ z_{12} = z_{21} = 2s,\ z_{22} = 2s + 4$$

Find the T equivalent of N.

SOLUTION Since $z_{12} = z_{21}$, the given network is reciprocal. The three branches of the T equivalent network of Fig. 16.4a are

$$Z_a = z_{11} - z_{12} = 2s + \frac{1}{s} - 2s$$

$$= \frac{1}{s} \text{ (a capacitor)}$$

$$Z_b = z_{22} - z_{12} = 2s + 4 - 2s = 4\ \Omega \text{ (a resistor)}$$

$$Z_c = z_{12} = z_{21} = 2s \text{ (an inductor)}$$

The circuit is shown in Fig. 16.14

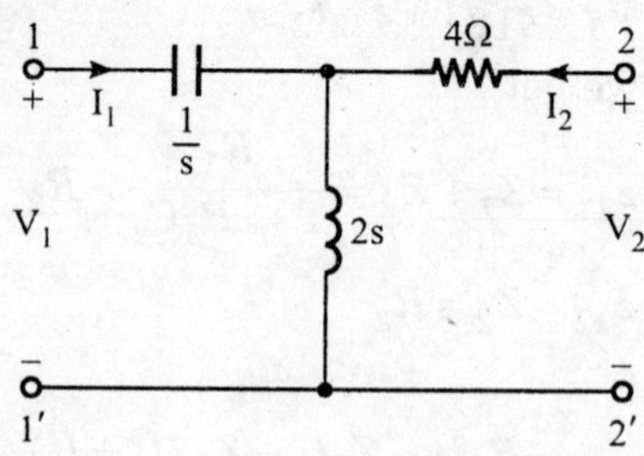

FIG. 16.14

EXAMPLE 16.8 Find a two port with

$$[z] = \begin{bmatrix} 6 + \dfrac{1}{2s} & 1 + \dfrac{1}{2s} \\ 4 + \dfrac{1}{2s} & 4s + 1 + \dfrac{1}{2s} \end{bmatrix} \Omega$$

SOLUTION Since $z_{12} \neq z_{21}$, this is a nonreciprocal network. The element values of Fig. 16.6 are

$$z_{11} - z_{12} = 6 + \frac{1}{2s} - \left(1 + \frac{1}{2s}\right)$$

$$= 5\ \Omega \text{ (a resistance)}$$

$$z_{22} - z_{12} = 4s + 1 + \frac{1}{2s} - \left(1 + \frac{1}{2s}\right)$$

$$= 4s \text{ (an inductance)}$$

$$z_{12} = 1 + \frac{1}{2s} \text{ (a resistance in series with a capacitance)}$$

$$(z_{21} - z_{12})\, I_1 = \left[4 + \frac{1}{2s} - \left(1 + \frac{1}{2s}\right)\right] I_1$$

$$= 3I_1 \text{ (a CCVS).}$$

The circuit is shown in Fig. 16.15.

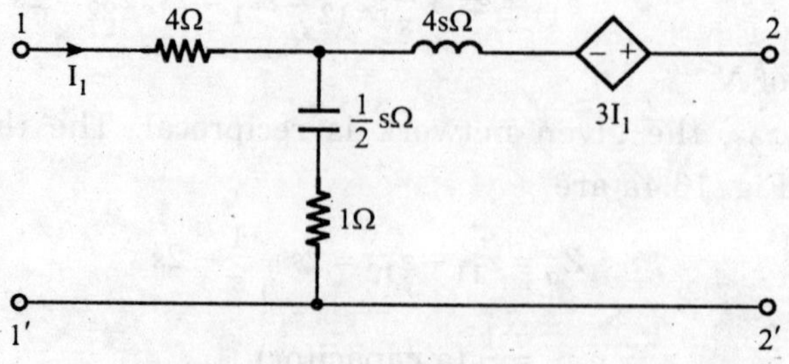

FIG. 16.15

EXAMPLE 16.9 Determine the y parameters of the T network shown in Fig. 16.16a.

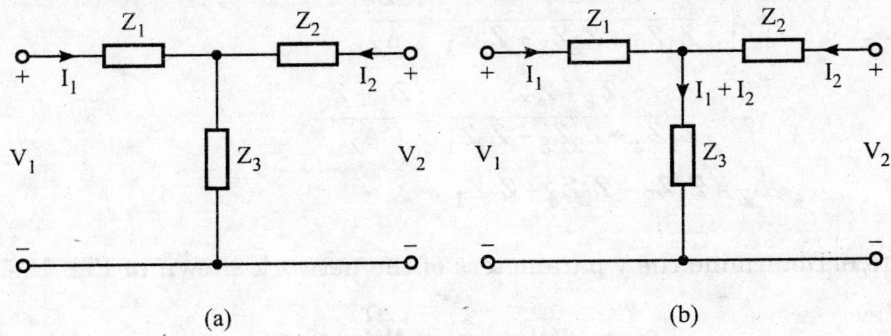

FIG. 16.16

SOLUTION By KCL the current through Z_3 is $(I_1 + I_2)$ as shown in Fig. 16.16b.

By KVL in the left-hand mesh of Fig. 16.16*b*,

$$V_1 - Z_1 I_1 - Z_3 (I_1 + I_2) = 0$$

$$V_1 = (Z_1 + Z_3) I_1 + Z_3 I_2 \qquad \text{(E 16.9.1)}$$

By KVL in the right-hand mesh of Fig. 16.16*b*,

$$V_2 - Z_2 I_2 - Z_3 (I_1 + I_2) = 0$$

$$V_2 = Z_3 I_1 + (Z_2 + Z_3) I_2 \qquad \text{(E 16.9.2)}$$

Multiplying Eq. (E 16.9.1) by $(Z_2 + Z_3)$ and Eq. (E 16.9.2) by Z_3 and subtracting

$$V_1 (Z_2 + Z_3) - V_2 Z_3 = (Z_1 + Z_3)(Z_2 + Z_3) I_1 - Z_3^2 I_1$$

$$(Z_1 Z_2 + Z_2 Z_3 + Z_3 Z_1) I_1 = (Z_2 + Z_3) V_1 - Z_3 V_2$$

$$I_1 = \frac{Z_2 + Z_3}{Z_1 Z_2 + Z_2 Z_3 + Z_3 Z_1} V_1 - \frac{Z_3}{Z_1 Z_2 + Z_2 Z_3 + Z_3 Z_1} V_2$$

Also $\qquad I_1 = y_{11} V_1 + y_{12} V_2$

therefore, $\qquad y_{11} = \dfrac{Z_2 + Z_3}{Z_1 Z_2 + Z_2 Z_3 + Z_3 Z_1} = \dfrac{Z_2 + Z_3}{\Delta_Z}$

$$y_{12} = \frac{-Z_3}{Z_1 Z_2 + Z_2 Z_3 + Z_3 Z_1} = \frac{-Z_3}{\Delta_Z}$$

In order to determine y_{21} and y_{22} we eliminate I_1 from Eqs. (E 16.9.1) and (E 16.9.2). To eliminate I_1 we multiply Eq. (E 16.9.1) by Z_3 and Eq. (E 16.9.2) by $(Z_1 + Z_3)$ and subtract.

$$Z_3 V_1 - (Z_1 + Z_3) V_2 = Z_3^2 I_2 - (Z_1 + Z_3)(Z_2 + Z_3) I_2$$

$$(Z_1 Z_2 + Z_2 Z_3 + Z_3 Z_1) I_2 = -Z_3 V_1 + (Z_1 + Z_3) V_2$$

$$I_2 = \frac{-Z_3 v_1}{Z_1 Z_2 + Z_2 Z_3 + Z_3 Z_1} + \frac{Z_1 + Z_3}{Z_1 Z_2 + Z_2 Z_3 + Z_3 Z_1} V_2$$

Also $\qquad I_2 = y_{21} V_1 + y_{22} V_2$

therefore, $$y_{21} = \frac{-Z_3}{Z_1Z_2 + Z_2Z_3 + Z_3Z_1} = \frac{-Z_3}{\Delta_Z}$$

$$y_{22} = \frac{Z_1 + Z_3}{Z_1Z_2 + Z_2Z_3 + Z_3Z_1} = \frac{Z_1 + Z_3}{\Delta_Z}$$

where $$\Delta_Z = Z_1Z_2 + Z_2Z_3 + Z_3Z_1$$

■ **EXAMPLE 16.10** Determine the y parameters of the network shown in Fig. 16.17.

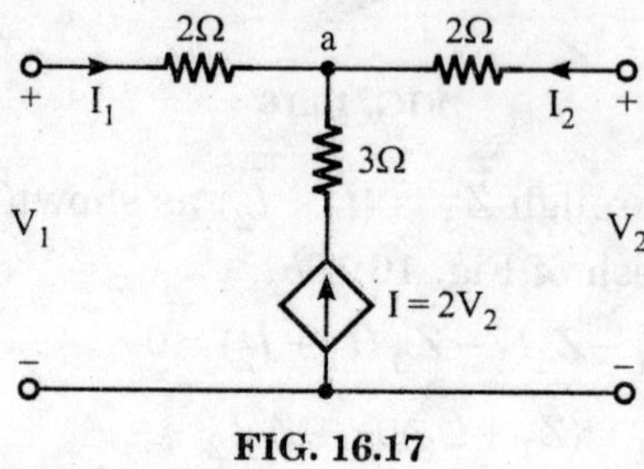

FIG. 16.17

SOLUTION By KVL in the outer loop,

$$V_1 - V_2 - 2I_1 + 2I_2 = 0 \quad \text{(E 16.10.1)}$$

By KCL at node a

$$I_1 + I_2 + I = 0 \quad \text{(E 16.10.2)}$$

For the current source,

$$I_1 = 2V_2 \quad \text{(E 16.10.3)}$$

therefore, $$I_1 + I_2 + 2V_2 = 0$$

$$I_1 = -I_2 - 2V_2 \quad \text{(E 16.10.4)}$$

and $$I_2 = -I_1 - 2V_2 \quad \text{(E 16.10.5)}$$

Substituting the value of I_2 from Eq. (E 16.10.5) in Eq. (E 16.10.1) we obtain

$$V_1 - V_2 - 2I_1 + 2(-I_1 - 2V_2) = 0$$

$$V_1 - 4I_1 - 5V_2 = 0$$

$$I_1 = \frac{1}{4}V_1 - \frac{5}{4}V_2$$

Also $$I_1 = y_{11}V_1 + y_{12}V_2$$

By comparing these equations we get

$$y_{11} = \frac{1}{4}S, \; y_{12} = \frac{-5}{4}S$$

Now substituting the value of I, from Eq. (E 16.10.4) in Eq. (E 16.10.1) we get

$$V_1 - V_2 - 2(I_2 - 2V_2) + 2I_2 = 0$$

$$4I_2 = -V_1 - 3V_2$$

$$I_2 = -\frac{1}{4}V_1 - \frac{3}{4}V_2$$

Also, $$I_2 = y_{21}V_1 + y_{22}V_2$$

therefore, $$y_{21} = -\frac{1}{4}\,S,\; y_{22} = -\frac{3}{4}\,S$$

EXAMPLE 16.11 Determine the z and y parameters of the network shown in Fig. 16.18a.

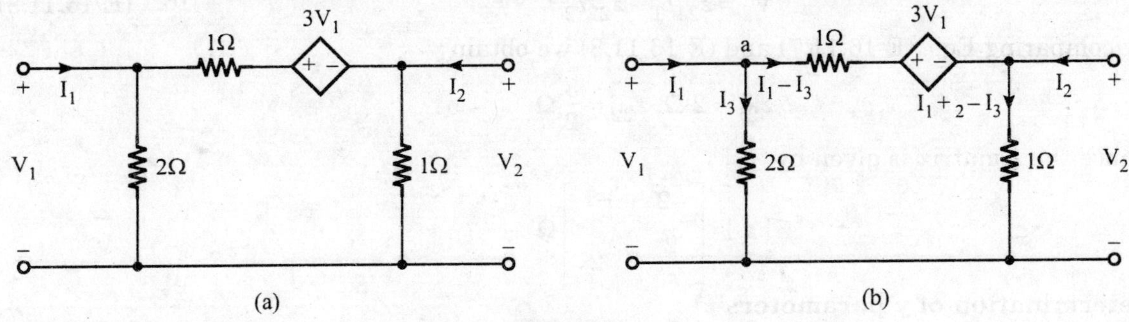

FIG. 16.18

SOLUTION The currents in different branches are found by the application of KCL at nodes a and b as shown in Fig. 16.18b.

By KVL in the left-hand mesh of Fig. 16.18b.

$$V_1 = 2I_3 \tag{E 16.11.1}$$

By KVL in the right-hand mesh

$$V_2 = 1 \times (I_1 + I_2 - I_3) \tag{E 16.11.2}$$

By KVL in the middle mesh

$$2I_3 - 1 \times (I_1 - I_3) - 3V_1 - 1 \times (I_1 + I_2 - I_3) = 0$$

or $$3V_1 = -2I_1 - I_2 + 4I_3 \tag{E 16.11.3}$$

From Eq. (E 16.11.1), $I_3 = \dfrac{V_1}{2}$

Substituting the value of I_3 in Eqs. (E 16.11.2) and (E 16.11.3) we get

$$V_2 = I_1 + I_2 - \frac{V_1}{2} \tag{E 16.11.4}$$

$$3V_1 = -2I_1 - I_2 + 2V_1$$

$$V_1 = -2I_1 - I_2 \tag{E 16.11.5}$$

Equations. (E 16.11.4) and (E 16.11.5) are sufficient to determine all the parameters by simply rearranging them in the form of standard defining equations of the parameters.

Determination of z parameters

Equation (E 16.11.5) is of the form

$$V_1 = z_{11}I_1 + z_{12}I_2 \tag{E 16.11.6}$$

Comparison of Eqs. (E 16.11.5) and (E 16.11.6) gives

$$z_{11} = -2\ \Omega,\ z_{12} = -1\ \Omega$$

Substituting the value of V_1 from Eq. (E 16.11.5) in Eq. (E 16.11.4) we get

$$V_2 = I_1 + I_2 - \frac{1}{2}(-2I_1 - I_2)$$

or

$$V_2 = 2I_1 + \frac{3}{2}I_2 \qquad \text{(E 16.11.7)}$$

Eq. (E 16.11.7) is of the form

$$V_2 = z_{21}I_1 + z_{22}I_2 \qquad \text{(E 16.11.8)}$$

By comparing Eqs. (E 16.11.7) and (E 16.11.8) we obtain

$$z_{21} = 2\ \Omega,\ z_{22} = \frac{3}{2}\ \Omega$$

Hence the z matrix is given by

$$[\boldsymbol{z}] = \begin{bmatrix} -2 & -1 \\ 2 & \frac{3}{2} \end{bmatrix} \Omega$$

Determination of y parameters

We shall use Eqs. (E 16.11.4) and (E 16.11.5) again to determine the y parameters. We shall add Eqs. (E 16.11.4) and (E 16.11.5) to eliminate I_2.

$$V_2 + V_1 = -I_1 - \frac{V_1}{2}$$

or

$$I_1 = -\frac{3}{2}V_1 - V_2 \qquad \text{(E 16.11.9)}$$

Eq. (E 16.11.9) is of the form

$$I_1 = y_{11}V_1 + y_{12}V_2 \qquad \text{(E 16.11.10)}$$

By comparing Eqs. (E 16.11.9) and (E 16.11.10) we get

$$y_{11} = -\frac{3}{2}\ S,\ y_{12} = -1\ S$$

Now we shall eliminate I_1 from Eqs. (E 16.11.4) and (E 16.11.5).

Multiplying Eq. (E 16.11.4) by 2 and adding the result to Eq. (E 16.11.5)

$$2V_2 + V_1 = 2I_2 - V_1 - I_2$$

$$I_2 = 2V_1 + 2V_2 \qquad \text{(E 16.11.10)}$$

Also

$$I_2 = y_{21}V_1 + y_{22}V_2$$

therefore,

$$y_{21} = 2\ S,\ y_{22} = 2\ S$$

Hence the y matrix is given by

$$[y] = \begin{bmatrix} y_{11} & y_{12} \\ y_{21} & y_{22} \end{bmatrix} = \begin{bmatrix} -\frac{3}{2} & -1 \\ 2 & 2 \end{bmatrix} S$$

EXAMPLE 16.12 Determine the short-circuit admittance parameters of the network shown in Fig. 16.19a. Show that the network is not reciprocal.

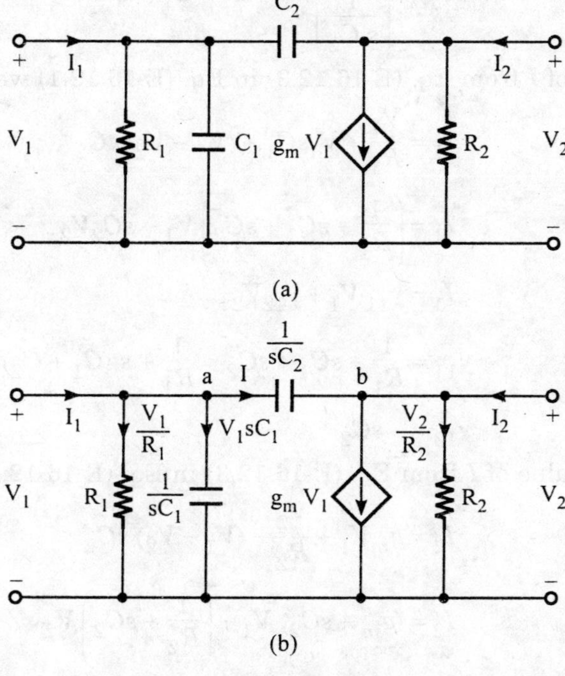

FIG. 16.19

SOLUTION In the s domain, the network is shown in Fig. 16.19b. The voltage across R_1 and $\frac{1}{s\,C_1}$ is V_1. Therefore, by Ohm's law

$$\text{current through } R_1 = \frac{V_1}{R_1}$$

$$\text{current through } \frac{1}{sC_1} = \frac{V_1}{\frac{1}{sC_1}} = V_1 sC_1$$

The voltage across R_2 is V_2. Therefore by Ohm's law, the current through R_2 is $\frac{V_2}{R_2}$.

Let I be the current through $\frac{1}{sC_2}$

By KCL in Fig. 16.19b.

$$I_1 = \frac{V_1}{R_1} + V_1 s\,C_1 + I \qquad \text{(E 16.12.1)}$$

$$I_2 + I = \frac{V_2}{R_2} + g_m V_1 \qquad \text{(E 16.12.2)}$$

Since node a is at potential V_1 and node b at potential V_2, by Ohm's law

$$I = \frac{V_a - V_b}{\left(\frac{1}{s\,C_2}\right)} = (V_1 - V_2)s\,C_2 \qquad \text{(E 16.12.3)}$$

Substituting the value of I from Eq. (E 16.12.3) in Eq. (E 16.12.1) we get

$$I_1 = \frac{V_1}{R_1} + V_1 sC_1 + (V_1 - V_2)\,sC_2$$

or

$$I_1 = \left(\frac{1}{R_1} + sC_1 + sC_2\right) V_1 - sC_2V_2 \qquad \text{(E 16.12.4)}$$

Also,

$$I_1 = y_{11}\,V_1 + y_{12}V_2 \qquad \text{(E 16.12.5)}$$

therefore,

$$y_{11} = \frac{1}{R_1} + sC_1 + sC_2 = \frac{1}{R_1} + s\,(C_1 + C_2)$$

$$y_{12} = -\,sC_2$$

Now substituting the value of I from Eq. (E 16.12.3) in Eq. (E 16.12.2) we get

$$I_2 = g_m V_1 + \frac{V_2}{R_2} - (V_1 - V_2)\,sC_2$$

or

$$I_2 = (g_m - sC_2)\,V_1 + \left(\frac{1}{R_2} + sC_2\right) V_2 \qquad \text{(E 16.12.6)}$$

Also,

$$I_2 = y_{21}V_1 + y_{22}V_2 \qquad \text{(E 16.12.7)}$$

therefore,

$$y_{21} = g_m - sC_2,\; y_{22} = \frac{1}{R_2} + sC_2$$

Since $y_{12} \neq y_{21}$, the network is not reciprocal.

16.11 HYBRID (or *h*) PARAMETERS

If I_1 and V_2 are taken as the independent variables, the 2-port equations may be written as

$$V_1 = h_{11}I_1 + h_{12}V_2 \qquad \text{(16.11.1)}$$

$$I_2 = h_{21}I_1 + h_{22}V_2 \qquad \text{(16.11.2)}$$

In matrix form these equations may be written as

$$\begin{bmatrix} V_1 \\ I_2 \end{bmatrix} = \begin{bmatrix} h_{11} & h_{12} \\ h_{21} & h_{22} \end{bmatrix} \begin{bmatrix} I_1 \\ V_2 \end{bmatrix} \qquad \text{(16.11.3)}$$

The constants h_{11}, h_{12}, h_{21}, and h_{22} are called the **hybrid (or *h*) parameters of the 2-port network.**

Eqs. (16.11.1) and (16.11.2) are called the **defining equations of *h* parameters**. The matrix

$$[\mathbf{h}] = \begin{bmatrix} h_{11} & h_{12} \\ h_{21} & h_{22} \end{bmatrix}$$

is called the **hybrid (or *h*) matrix**. The h parameters are defined as follows :

$$h_{11} \stackrel{\Delta}{=} \left.\frac{V_1}{I_1}\right|_{V_2=0} = \text{short-circuit input impedance}$$

$$h_{12} \stackrel{\Delta}{=} \left.\frac{V_1}{V_2}\right|_{I_1=0} = \text{open-circuit reverse voltage transfer ratio}$$

$$h_{21} \stackrel{\Delta}{=} \left.\frac{I_2}{I_1}\right|_{V_2=0} = \text{short-circuit forward current ratio}$$

$$h_{22} \stackrel{\Delta}{=} \left.\frac{I_2}{V_2}\right|_{I_1=0} = \text{open-circuit output admittance}$$

From the definitions of h parameters it is seen that h_{11} has the dimensions of impedance, h_{12} and h_{21} are dimensionless, and h_{22} has the dimensions of admittance. Thus, these parameters are dimensionally inhomogeneous. It is also seen that h_{11} and h_{21} are short-circuit parameters while h_{12} and h_{22} are open-circuit parameters. That is why, these parameters are called hybrid parameters. These parameters are most suitable for transistor testing because of the following reasons.

(*a*) The output impedance of a transistor is very high and short circuiting of port 2–2′ is easier to measure h_{11} and h_{21}.

(*b*) The input impedance is low and the open-circuit test can be performed conveniently to measure h_{12} and h_{22}.

In transistor applications, the symbols h_{11}, h_{21}, h_{12}, h_{22} are replaced by h_i, h_f, h_r, h_o respectively to denote input, forward, reverse and output.

There are three modes in which a transistor may be operated : the common emitter CE, the common base CB and the common collector CC.

Treating the transistor as a 2-port network, the h parameters will change with each configuration. To distinguish which parameter has been used or which is available, a second subscript has been added to the h-parameter notation. For common-base configuration the lower case letter b is added, while for the common-emitter and common-collector configurations the letters e and c are added, respectively. Table 16.1. gives the typical values of the h parameters for all three configurations.

TABLE 16.1

h parameters	CE configuration	CB configuration	CC configuration
h_{11}	h_{ie}	h_{ib}	h_{ic}
h_{12}	h_{re}	h_{rb}	h_{rc}
h_{21}	h_{fe}	h_{fb}	h_{fc}
h_{22}	h_{oe}	h_{ob}	h_{oc}

16.12 EQUIVALENT NETWORK USING *h* PARAMETERS

The defining equations may be used to determine the equivalent circuit of a 2-port network in terms of the *h* parameters. This circuit is shown in Fig. 16.20. It is to be noted that the input is a Thevenin equivalent network and the output is Norton equivalent network.

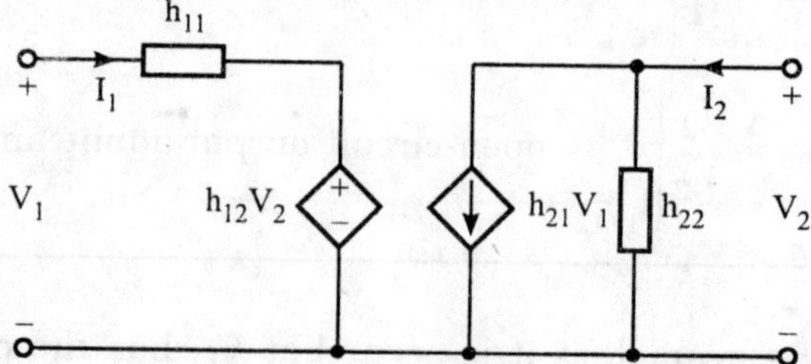

FIG. 16.20 Equivalent circuit of a 2-port network in terms of *h* parameters

16.13 INVERSE HYBRID (or *g*) PARAMETERS

If V_1 and I_2 are chosen as independent variables the 2-port network equations may be written as

$$I_1 = g_{11}V_1 + g_{12}I_2 \tag{16.13.1}$$

$$V_2 = g_{21}V_1 + g_{22}I_2 \tag{16.13.2}$$

In matrix form, these equations are written as

$$\begin{bmatrix} I_1 \\ V_2 \end{bmatrix} = \begin{bmatrix} g_{11} & g_{12} \\ g_{21} & g_{22} \end{bmatrix} \begin{bmatrix} V_1 \\ I_2 \end{bmatrix}$$

The constants g_{11}, g_{12}, g_{21}, and g_{22} are called the **inverse hybrid (or *g*) parameters.** Eqs. (16.13.1) and (16.13.2) are called **the defining equations of the *g* parameters**. The matrix

$$[\mathbf{g}] = \begin{bmatrix} g_{11} & g_{12} \\ g_{21} & g_{22} \end{bmatrix}$$

is the ***g*** matrix

The ***g*** parameters are defined as follows :

$$g_{11} \stackrel{\Delta}{=} \left.\frac{I_1}{V_1}\right|_{I_2=0} = \text{open-circuit input admittance}$$

$$g_{12} \stackrel{\Delta}{=} \left.\frac{I_1}{I_2}\right|_{V_1=0} = \text{short-circuit reverse current gain}$$

$$g_{21} \stackrel{\Delta}{=} \left.\frac{V_2}{V_1}\right|_{I_2=0} = \text{open-circuit forward voltage gain}$$

$$g_{22} \stackrel{\Delta}{=} \left.\frac{V_2}{I_2}\right|_{V_1=0} = \text{short-circuit output impedance}$$

Eqs. (16.13.1) and (16.13.2) may be used to develop the equivalent circuit of a 2-port network as shown in Fig. 16.21.

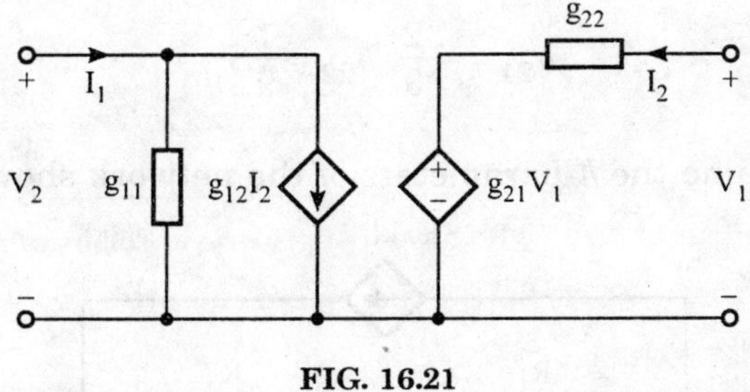

FIG. 16.21

■ **EXAMPLE 16.13** **Determine the h parameters for the T network shown in Fig. 16.22a.**

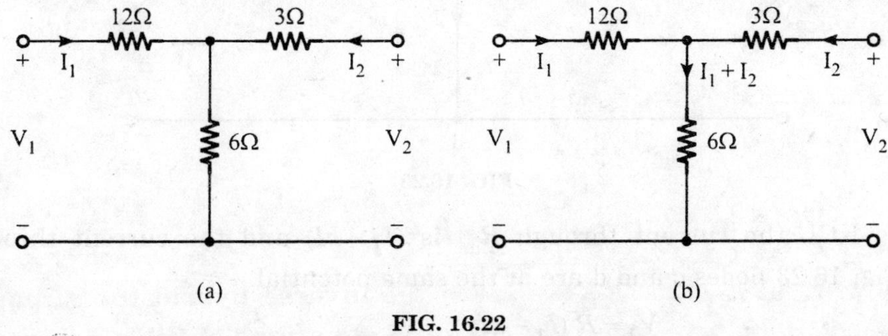

FIG. 16.22

SOLUTION **By KCL the current through the 6–Ω resistor is $(I_1 + I_2)$ as shown in Fig. 16.21b.**

By KVL in the left-hand mesh

$$V_1 - 12I_1 - 6\,(I_1 + I_2) = 0$$

or

$$18\,I_1 + 6\,I_2 = V_1 \qquad \text{(E 16.13.1)}$$

By KVL in the right-hand mesh

$$V_2 - 3I_2 - 6(I_1 + I_2) = 0$$

$$6I_1 + 9I_2 = V_2 \quad \text{(E 16.13.2)}$$

Multiplying Eq. (E 16.13.1) by 3 and Eq. (E 16.13.2) by 2 and subtracting we obtain

$$42I_1 = 3V_1 - 2V_2$$

$$V_1 = 14I_1 + \frac{2}{3}V_2 \quad \text{(E 16.13.3)}$$

Also, $$V_1 = h_{11}I_1 + h_{12}V_2 \quad \text{(E 16.13.4)}$$

therefore, $$h_{11} = 14\ \Omega,\ h_{12} = \frac{2}{3}$$

From Eq. (E 16.13.2)

$$9I_2 = -6I_1 + V_2$$

$$I_2 = -\frac{2}{3}I_1 + \frac{1}{9}V_2$$

Also, $$I_2 = h_{21}I_1 + h_{22}V_2$$

therefore, $$h_{21} = -\frac{2}{3},\ h_{22} = \frac{1}{9}S$$

■ **EXAMPLE 16.14.** Determine the h-parameters of the network shown in Fig. 16.23.

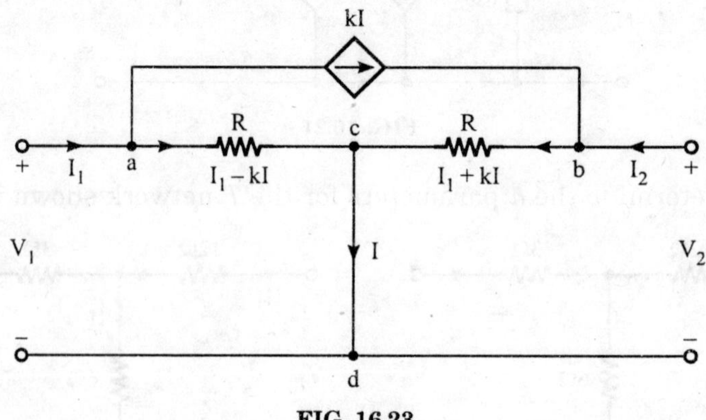

FIG. 16.23

SOLUTION By KCL the current through R_{ac} is $(I_1 - kI)$ and the current through R_{bc} is $(I_2 + kI)$. In Fig. 16.23 nodes c and d are at the same potential.

$$V_1 = R(I_1 - kI) \quad \text{(E16.14.1)}$$

$$V_2 = R(I_2 + kI) \quad \text{(E16.14.2)}$$

By KCL at node c $$I_1 - kI + I_2 + kI = I$$

therefore, $$I = I_1 + I_2 \quad \text{(E16.14.3)}$$

Substitution of the value of I from Eq. (E16.14.3) in Eqs (E16.14.1) and (E16.14.2) gives

$$V_1 = R(1-k)I_1 - RkI_2 \tag{E16.14.4}$$

$$V_2 = R(1+k)I_2 + RkI_1 \tag{E16.14.5}$$

From Eq. (E16.14.5) we have

$$I_2 = -\frac{k}{1+k}I_1 + \frac{1}{R(1+k)}V_2 \tag{E16.14.6}$$

This equation is of the form

$$I_2 = h_{21}I_1 + h_{22}V_2 \tag{E16.14.7}$$

Comparison of Eqs. (E16.14.6) and (E16.14.7) gives

$$h_{21} = -\frac{k}{1+k}, \qquad h_{22} = \frac{1}{R(1+k)}$$

Substituting the value of I_2 from Eq. (E16.14.6) in Eq. (E 16.14.4) we get

$$V_1 = R(1-k)I_1 - Rk\left[-\frac{k}{1+k}I_1 + \frac{1}{R(1+k)}V_2\right]$$

$$= R\left[1-k+\frac{k^2}{1+k}\right]I_1 - \frac{k}{1+k}V_2$$

$$V_1 = \frac{R}{1+k}I_1 - \frac{k}{1+k}V_2 \tag{E16.14.8}$$

Eq. (E16.14.8) is of the form

$$V_1 = h_{11}I_1 + h_{12}V_2 \tag{E16.14.9}$$

By comparing Eqs. (E16.14.8) and (E16.14.9) we get

$$h_{11} = \frac{R}{1+k}, \qquad h_{12} = \frac{-k}{1+k}$$

16.14 TRANSMISSION (or ABCD) PARAMETERS

If we select V_2 and I_2 as independent variables, the 2-port equations may be written as

$$V_1 = AV_2 + B(-I_2) \tag{16.14.1}$$

$$I_1 = CV_2 + D(-I_2) \tag{16.14.2}$$

In matrix form

$$\begin{bmatrix} V_1 \\ I_1 \end{bmatrix} = \begin{bmatrix} A & B \\ C & D \end{bmatrix}\begin{bmatrix} V_2 \\ -I_2 \end{bmatrix} = \begin{bmatrix} a_{11} & a_{12} \\ a_{21} & a_{22} \end{bmatrix}\begin{bmatrix} V_2 \\ -I_2 \end{bmatrix}$$

The constants A, B, C, D are called the **transmission parameters** or *ABCD* parameters. They are also sometimes known as **chain parameters.** It is to be noted that

$$A = a_{11},\ B = a_{12},\ C = a_{21}, \text{ and } D = a_{22}$$

The matrix

$$[\mathbf{a}] = \begin{bmatrix} a_{11} & a_{12} \\ a_{21} & a_{22} \end{bmatrix} = \begin{bmatrix} A & B \\ C & D \end{bmatrix}$$

is called the **transfer matrix** of the network.

As per convention, the currents entering the upper terminals of the ports are taken positive. In a power transmission network the output current I_2 is coming *out* of the upper terminal of the output port. Hence the reference direction of the output current I_2 is opposite to that shown in Fig. 16.2. This accounts for the current I_2 to carry a negative sign in Eqs (16.14.1) and (16.14.2). In other words, the output current is $-I_2$.

The ABCD parameters are defined as follows :

$$A \stackrel{\Delta}{=} \left.\frac{V_1}{V_2}\right|_{I_2=0} = \text{open-circuit reverse voltage transfer ratio}$$

$$B \stackrel{\Delta}{=} \left.\frac{V_1}{-I_2}\right|_{V_2=0} = \text{short-circuit reverse transfer impedance}$$

$$C \stackrel{\Delta}{=} \left.\frac{I_1}{V_2}\right|_{I_2=0} = \text{open-circuit reverse transfer admittance}$$

$$D \stackrel{\Delta}{=} \left.\frac{I_1}{-I_2}\right|_{V_2=0} = \text{short-circuit reverse current transfer ratio}$$

It is seen that all the four parameters are transfer functions. Hence these parameters are appropriately called **transmission parameters**. The ABCD parameters are widely used in transmission-line calculations. In transmission-line theory, the input port is called the *sending end* and the output port is called the *receiving* end.

16.15 ABCD PARAMETERS OF CASCADED NETWORKS

Two networks are said to be connected in **cascade** if the output port of one is the input port of the second as shown in Fig. 16.24.

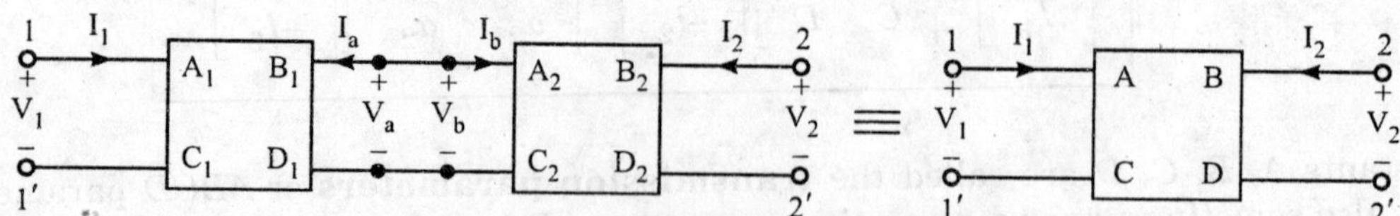

FIG. 16.24 Two cascaded networks and their equivalent network

The overall parameters for several two-port networks connected in cascade may conveniently be found using ABCD parameters. Figure 16.24 shows two cascaded networks and their equivalent network.

For the first network,

$$\begin{bmatrix} V_1 \\ I_1 \end{bmatrix} = \begin{bmatrix} A_1 & B_1 \\ C_1 & D_1 \end{bmatrix} \begin{bmatrix} V_a \\ -I_a \end{bmatrix} \quad (16.15.1)$$

For the second network,

$$\begin{bmatrix} V_b \\ I_b \end{bmatrix} = \begin{bmatrix} A_2 & B_2 \\ C_2 & D_2 \end{bmatrix} \begin{bmatrix} V_2 \\ -I_2 \end{bmatrix} \quad (16.15.2)$$

For the equivalent network,

$$\begin{bmatrix} V_1 \\ I_1 \end{bmatrix} = \begin{bmatrix} A & B \\ C & D \end{bmatrix} \begin{bmatrix} V_2 \\ -I_2 \end{bmatrix} \quad (16.15.3)$$

From Fig. 16.23,

$$\begin{bmatrix} V_a \\ -I_a \end{bmatrix} = \begin{bmatrix} V_b \\ I_b \end{bmatrix} \quad (16.15.4)$$

Combination of Eqs (16.15.1), (16.15.2), and (16.15.4) gives

$$\begin{bmatrix} V_1 \\ I_1 \end{bmatrix} = \begin{bmatrix} A_1 & B_1 \\ C_1 & D_1 \end{bmatrix} \begin{bmatrix} A_2 & B_2 \\ C_2 & D_2 \end{bmatrix} \begin{bmatrix} V_2 \\ -I_2 \end{bmatrix} \quad (16.15.5)$$

Comparison of Eqs (16.15.3) and (16.15.5) gives

$$\begin{bmatrix} A & B \\ C & D \end{bmatrix} = \begin{bmatrix} A_1 & B_1 \\ C_1 & D_1 \end{bmatrix} \begin{bmatrix} A_2 & B_2 \\ C_2 & D_2 \end{bmatrix} \quad (16.15.6)$$

Thus the *overall ABCD matrix for two cascaded networks is equal to the product of ABCD matrices of the individual networks* ***taken in order.*** The result may be extended to any number of cascaded networks.

16.16 OUTPUT IN TERMS OF INPUT

Eqs. (16.14.1) and (16.14.2) give the voltage and current at the sending end in terms of the voltage and current at the receiving end. The relations for the receiving end voltage and current can be found in terms of voltage and current at the sending end as follows :

For convenience we rewrite Eqs. (16.14.1) and (16.14.2).

$$V_1 = AV_2 - BI_2 \quad (16.14.1)$$

$$I_1 = CV_2 - DI_2 \quad (16.14.2)$$

Multiplying Eq. (16.14.1) by D and Eq. (16.14.2) by B and subtracting the results we obtain

$$DV_1 - BI_1 = (AD - BC)\, V_2$$

For a reciprocal network

$$AD - BC = 1$$

Therefore, $$V_2 = DV_1 - BI_1 \quad (16.16.1)$$

Again multiplying Eq. (16.14.1) by C and Eq. (16.14.2) by A and subtracting the results we get

$$CV_1 - AI_1 = (AD - BC)\, I_2$$

Again, $AD - BC = 1$

therefore, $$I_2 = CV_1 - AI_1 \tag{16.16.2}$$

In matrix form Eqs. (16.16.1) and (16.16.2) may be written as

$$\begin{bmatrix} V_2 \\ I_2 \end{bmatrix} = \begin{bmatrix} D & B \\ C & A \end{bmatrix} \begin{bmatrix} V_1 \\ -I_1 \end{bmatrix} = \begin{bmatrix} b_{11} & b_{12} \\ b_{21} & b_{22} \end{bmatrix} \begin{bmatrix} V_1 \\ -I_1 \end{bmatrix} \tag{16.16.3}$$

■ **EXAMPLE 16.15** Find the transmission matrices of simple series and shunt elements shown in Figs. 16.25a and 16.25b.

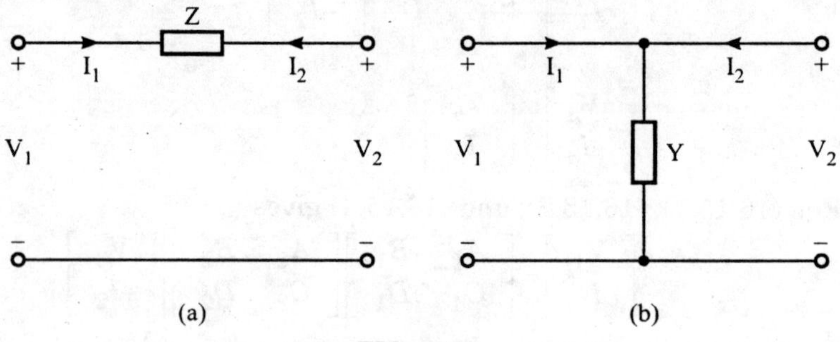

FIG. 16.25

SOLUTION : For the network in Fig. 16.25a,

$$V_1 = ZI_1 + V_2$$

$$I_1 = -I_2$$

$$\begin{bmatrix} V_1 \\ I_1 \end{bmatrix} = \begin{bmatrix} 1 & Z \\ 0 & 1 \end{bmatrix} \begin{bmatrix} V_2 \\ -I_2 \end{bmatrix}$$

Therefore the transmission matrix of a simple 2-port network consisting of a single series impedance Z is given by

$$[\, a \,] = \begin{bmatrix} 1 & Z \\ 0 & 1 \end{bmatrix}$$

For the network shown in Fig. 16.25b,

$$V_1 = V_2$$

$$I_1 = YV_2 - I_2$$

$$\begin{bmatrix} V_1 \\ I_1 \end{bmatrix} = \begin{bmatrix} 1 & 0 \\ Y & 1 \end{bmatrix} \begin{bmatrix} V_2 \\ -I_2 \end{bmatrix}$$

Therefore the transmission matrix of a 2-port network consisting of a single shunt admittance Y is given by

$$[\, a \,] = \begin{bmatrix} 1 & 0 \\ Y & 1 \end{bmatrix}$$

■ **EXAMPLE 16.16** Find the *ABCD* prameters of the *T* network shown in Fig. 16.26 by considering the network as a cascade of three 2-port networks containing only one element each.

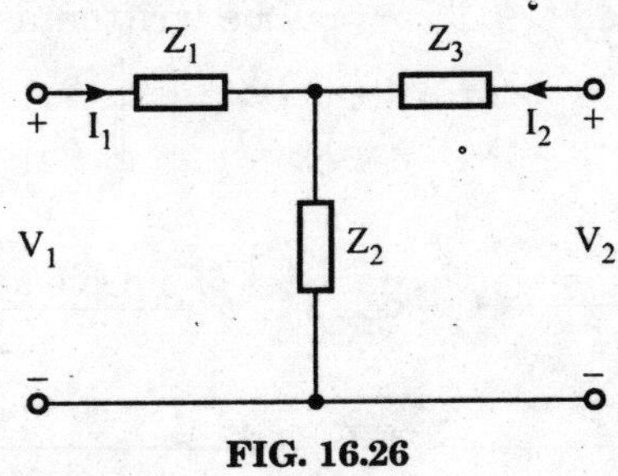

FIG. 16.26

SOLUTION : The transmission matrix of a simple 2-port network consisting of a single series element Z_1 is

$$[a_1] = \begin{bmatrix} 1 & Z_1 \\ 0 & 1 \end{bmatrix}$$

The transmission matrix of a simple 2-port network consisting of a single shunt element Z_2 is

$$[a_2] = \begin{bmatrix} 1 & 0 \\ \dfrac{1}{Z_2} & 1 \end{bmatrix}$$

The transmission matrix of a simple 2-port network consisting of a single series element Z_3 is

$$[a_3] = \begin{bmatrix} 1 & Z_3 \\ 0 & 1 \end{bmatrix}$$

The overall transmission matrix $[a]$ of the network of Fig. 16.26 is equal to the product of the transmission matrices of the individual networks taken in order.

therefore, $[a] = [a_1][a_2][a_3]$

$$= \begin{bmatrix} 1 & Z_1 \\ 0 & 1 \end{bmatrix} \begin{bmatrix} 1 & 0 \\ \dfrac{1}{Z_2} & 1 \end{bmatrix} \begin{bmatrix} 1 & Z_3 \\ 0 & 1 \end{bmatrix}$$

$$= \begin{bmatrix} 1 & Z_1 \\ 0 & 1 \end{bmatrix} \begin{bmatrix} 1 & Z_3 \\ \dfrac{1}{Z_2} & \dfrac{Z_3}{Z_2} + 1 \end{bmatrix} = \begin{bmatrix} 1 + \dfrac{Z_1}{Z_2} & Z_1 + Z_3 + \dfrac{Z_1}{Z_2} Z_3 \\ \dfrac{1}{Z_2} & 1 + \dfrac{Z_3}{Z_2} \end{bmatrix}$$

EXAMPLE 16.17 Find the transmission matrix of the 2-port shown in Fig. 16.27*a*.

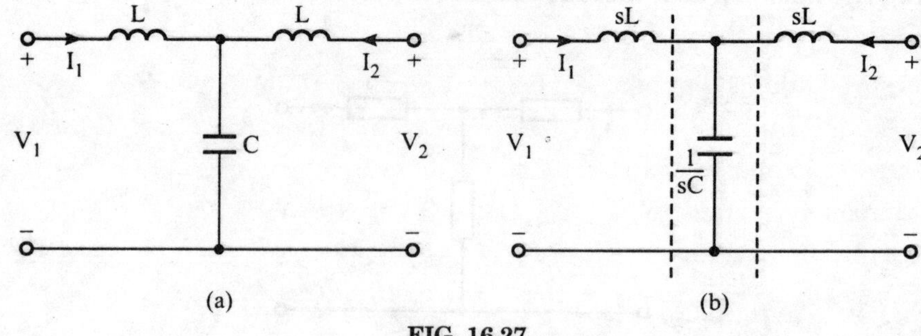

FIG. 16.27

SOLUTION: In the *s* domain the circuit is shown in Fig. 16.27b. The overall transmission matrix of the network can be obtained by writing the product of three **a** matrices. Using the results of Example 16.15, we have

$$[\mathbf{a}] = [\mathbf{a_1}][\mathbf{a_2}][\mathbf{a_3}]$$

$$= \begin{bmatrix} 1 & sL \\ 0 & 1 \end{bmatrix}\begin{bmatrix} 1 & 0 \\ sC & 1 \end{bmatrix}\begin{bmatrix} 1 & sL \\ 0 & 1 \end{bmatrix}$$

$$= \begin{bmatrix} 1 & sL \\ 0 & 1 \end{bmatrix}\begin{bmatrix} 1 & sL \\ sC & s^2LC+1 \end{bmatrix}$$

$$= \begin{bmatrix} 1+s^2LC & sL\,(2+s^2LC) \\ sC & 1+s^2LC \end{bmatrix}$$

16.17 INTER-RELATIONSHIP OF PARAMETERS

Since all sets of parameters describe the same 2-port network, they are inter-related. One set of parameters may be expressed in terms of the other set. Suppose that the parameters of the first set are to be expressed in terms of the parameters of the second set.

General Procedure

The general procedure for the transformation is as follows :

1. Write the defining equations of both the sets.
2. Solve the equations of the second set for the dependent (unknown) variables of the first set of equations.
3. Express the resulting equations in the forms of the equations of the first set.
4. Compare the equations obtained in step 3 with those of the first set to obtain the required parameters.

The procedure is illustrated to express the y parameters in terms of the z parameters. The defining equations of the *y* parameters are

$$I_1 = y_{11}V_1 + y_{12}V_2 \tag{16.17.1}$$

$$I_2 = y_{21}V_1 + y_{22}V_2 \tag{16.17.2}$$

The defining equations of the *z* parameters are

$$V_1 = z_{11}I_1 + z_{12}I_2 \tag{16.17.3}$$

$$V_2 = z_{21}I_1 + z_{22}I_2 \qquad (16.17.4)$$

Determination of y_{11} and y_{12}

The parameters y_{11} and y_{12} appear in Eq. (16.17.1). We have to solve Eqs (16.17.3) and (16.17.4) for I_1 to determine y_{11} and y_{12}. If we examine Eq. (16.17.1), it is found that out of the four variables I_2 is missing in it.. Hence we have to eliminate I_2 from Eqs. (16.17.3) and (16.17.4). For this purpose, multiplying Eq. (16.17.3) by the coefficient z_{22} of I_2 in Eq. (16.17.4) and Eq. (16.17.4) by the coefficient z_{12} of I_2 in Eq. (16.17.3) and then subtracting we get

$$z_{22}V_1 - z_{12}V_2 = (z_{11}z_{22} - z_{12}z_{21})\, I_1$$

$$= \Delta_z I_1$$

where

$$\Delta_z = z_{11}z_{22} - z_{12}z_{21} = \begin{vmatrix} z_{11} & z_{12} \\ z_{21} & z_{22} \end{vmatrix} \qquad (16.17.5)$$

$$I_1 = \frac{z_{22}}{\Delta_z} V_1 - \frac{z_{12}}{\Delta_z} V_2 \qquad (16.17.6)$$

Also,

$$I_1 = y_{11}V_1 + y_{12}V_2 \qquad (16.17.1)$$

Comparison of Eqs. (16.17.6) and (16.17.1) gives

$$y_{11} = \frac{z_{22}}{\Delta_z}, \qquad y_{12} = \frac{-z_{12}}{\Delta_z}$$

Determination of y_{21} and y_{22}

The parameters y_{21} and y_{22} appear in Eq. (16.17.2). Examination of this equation shows that out of four variables I_1 is missing in it. Hence I_1 is to be eliminated from Eqs. (16.17.3) and (16.17.4).

Multiplying Eq. (16.17.3) by the coefficient z_{21} of I_1 in Eq. (16.17.4) and Eq. (16.17.4) by the coefficient z_{11} of I_1 in Eq. (16.17.3) and subtracting

$$z_{21}V_1 - z_{11}V_2 = (z_{12}z_{21} - z_{11}z_{22})\, I_2 = -\Delta_z I_2$$

therefore,

$$I_2 = \frac{-z_{21}}{\Delta_z} V_1 + \frac{z_{11}}{\Delta_z} V_2 \qquad (16.17.7)$$

Also,

$$I_2 = y_{21}V_1 + y_{22}V_2 \qquad (16.17.2)$$

Comparison of Eqs. (16.17.7) and (16.17.2) gives

$$y_{21} = -\frac{z_{21}}{\Delta_z}, \qquad y_{22} = \frac{z_{11}}{\Delta_z}$$

Relationship, between other sets of parameters may be derived in a similar mananer. Matrix method may also be used to derive the interrelationships of the parameters. Table 16.2 gives the relationships between different sets of parameters. For conversion to be possible, the determinant for the source parameter must be nonzero.

Table 16.2

	z		y		h		g		**a**	
z	z_{11}	z_{12}	$\frac{y_{22}}{\Delta_y}$	$-\frac{y_{12}}{\Delta_y}$	$\frac{\Delta_h}{h_{22}}$	$\frac{h_{12}}{h_{22}}$	$\frac{1}{g_{11}}$	$-\frac{g_{12}}{g_{11}}$	$\frac{a_{11}}{a_{21}}$	$\frac{\Delta_a}{a_{21}}$
	z_{21}	z_{22}	$-\frac{y_{21}}{\Delta_y}$	$\frac{y_{11}}{\Delta_y}$	$-\frac{h_{21}}{h_{22}}$	$\frac{1}{h_{22}}$	$\frac{g_{21}}{g_{11}}$	$\frac{\Delta_g}{g_{11}}$	$\frac{1}{a_{21}}$	$\frac{a_{22}}{a_{21}}$
y	$\frac{z_{22}}{\Delta_z}$	$-\frac{z_{12}}{\Delta_z}$	y_{11}	y_{12}	$\frac{1}{h_{11}}$	$-\frac{h_{12}}{h_{11}}$	$\frac{\Delta_g}{g_{22}}$	$\frac{g_{12}}{g_{22}}$	$\frac{a_{22}}{a_{12}}$	$-\frac{\Delta_a}{a_{12}}$
	$-\frac{z_{21}}{\Delta_z}$	$\frac{z_{11}}{\Delta_z}$	y_{21}	y_{22}	$\frac{h_{21}}{h_{11}}$	$\frac{1}{h_{11}}$	$\frac{g_{21}}{g_{22}}$	$\frac{1}{g_{22}}$	$-\frac{1}{a_{12}}$	$\frac{a_{11}}{a_{12}}$
h	$\frac{\Delta_z}{z_{22}}$	$\frac{z_{12}}{z_{22}}$	$\frac{1}{y_{11}}$	$-\frac{y_{12}}{y_{11}}$	h_{11}	h_{12}	$\frac{g_{22}}{\Delta_g}$	$\frac{-g_{12}}{\Delta_g}$	$\frac{a_{12}}{a_{22}}$	$\frac{\Delta_a}{a_{22}}$
	$-\frac{z_{21}}{z_{22}}$	$\frac{1}{z_{22}}$	$\frac{y_{21}}{y_{11}}$	$\frac{\Delta_y}{y_{11}}$	h_{21}	h_{22}	$-\frac{g_{21}}{\Delta_g}$	$\frac{g_{11}}{\Delta_g}$	$-\frac{1}{a_{22}}$	$\frac{a_{21}}{a_{22}}$
g	$\frac{1}{z_{11}}$	$-\frac{z_{12}}{z_{11}}$	$\frac{\Delta_y}{y_{22}}$	$\frac{y_{12}}{y_{22}}$	$\frac{h_{22}}{\Delta_h}$	$-\frac{h_{12}}{\Delta_h}$	g_{11}	g_{12}	$\frac{a_{21}}{a_{11}}$	$-\frac{\Delta_a}{a_{11}}$
	$\frac{z_{21}}{z_{11}}$	$\frac{\Delta_z}{z_{11}}$	$-\frac{y_{21}}{y_{22}}$	$\frac{1}{y_{22}}$	$-\frac{h_{21}}{\Delta_h}$	$\frac{h_{11}}{\Delta_h}$	g_{21}	g_{22}	$\frac{1}{a_{11}}$	$\frac{a_{12}}{a_{11}}$
a	$\frac{z_{11}}{z_{21}}$	$\frac{\Delta_z}{z_{21}}$	$-\frac{y_{22}}{y_{21}}$	$-\frac{1}{y_{21}}$	$-\frac{\Delta_h}{h_{21}}$	$-\frac{h_{11}}{h_{21}}$	$\frac{1}{g_{21}}$	$\frac{g_{22}}{g_{21}}$	a_{11}	a_{12}
	$\frac{1}{z_{21}}$	$\frac{z_{22}}{z_{21}}$	$-\frac{\Delta_y}{y_{21}}$	$-\frac{y_{11}}{y_{21}}$	$-\frac{h_{22}}{h_{21}}$	$-\frac{1}{h_{21}}$	$\frac{g_{11}}{g_{21}}$	$\frac{\Delta_g}{g_{21}}$	a_{21}	a_{22}
a'	$\frac{z_{22}}{z_{12}}$	$\frac{\Delta_z}{z_{12}}$	$-\frac{y_{11}}{y_{21}}$	$-\frac{1}{y_{12}}$	$\frac{1}{h_{12}}$	$\frac{h_{11}}{h_{12}}$	$-\frac{\Delta_g}{g_{12}}$	$-\frac{g_{22}}{g_{12}}$	$\frac{a_{22}}{\Delta_a}$	$\frac{a_{12}}{\Delta_a}$
	$\frac{1}{z_{12}}$	$\frac{z_{11}}{z_{12}}$	$-\frac{\Delta_y}{y_{12}}$	$-\frac{y_{22}}{y_{12}}$	$\frac{h_{22}}{h_{12}}$	$\frac{\Delta_h}{h_{12}}$	$-\frac{g_{11}}{g_{12}}$	$-\frac{1}{g_{12}}$	$\frac{a_{21}}{\Delta_a}$	$\frac{a_{11}}{\Delta_a}$

■ **EXAMPLE 16.18** Determine the z parameters for the network in Fig 16.28a.

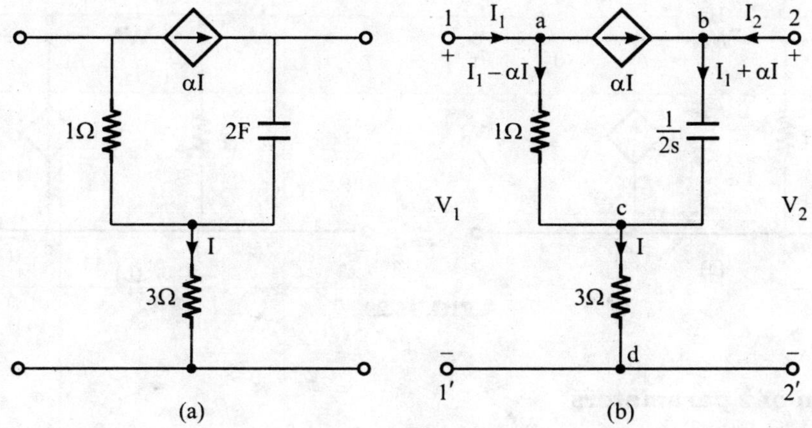

FIG. 16.28 (a) Network for Example 16.18 (b) Network for Example 16.18 in the *s* domain

SOLUTION In the s domain the circuit is shown in Fig. 16.28b. The currents in different branches are found by the application of KCL at the nodes of the network. By KCL at node c,

$$I = I_1 - \alpha I + I_2 + \alpha I = I_1 + I_2$$

By KVL in mesh $1\,a\,c\,d\,1'1$

$$-1 \times (I_1 - \alpha I) - 3I + V_1 = 0$$

$$V_1 - I_1 + \alpha\,(I_1 + I_2) - 3\,(I_1 + I_2) = 0$$

$$V_1 = (4 - \alpha)\,I_1 + (3 - \alpha)\,I_2$$

Also, $$V_1 = z_{11}I_1 + z_{12}I_2$$

Comparison of these two equations gives

$$z_{11} = 4 - \alpha, \quad z_{12} = 3 - \alpha$$

By KVL in mesh $2\,b\,c\,d\,2'2$,

$$-\frac{1}{2s}\,(I_2 + \alpha I) - 3I + V_2 = 0$$

$$V_2 - \frac{1}{2s}\,I_2 - \left(\frac{\alpha}{2s} + 3\right)I = 0$$

$$V_2 - \frac{1}{2s}\,I_2 - \left(\frac{\alpha}{2s} + 3\right)(I_1 + I_2) = 0$$

$$V_2 = \left(\frac{\alpha}{2s} + 3\right)I_1 + \left(\frac{1 + \alpha}{2s} + 3\right)I_2$$

Also, $$V_2 = z_{21}I_1 + z_{22}I_2$$

therefore, $$z_{21} = \left(\frac{\alpha}{2s} + 3\right), \qquad z_{22} = \left(\frac{1 + \alpha}{2s} + 3\right)$$

EXAMPLE 16.19 Determine the z and $ABCD$ parameters of the network shown in Fig. 16.29a.

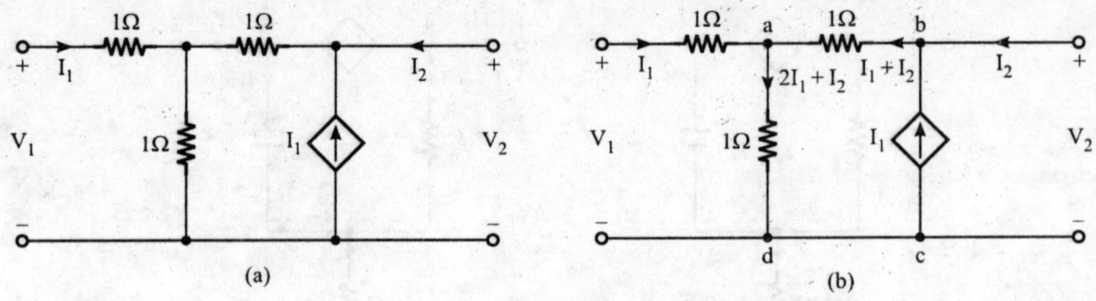

FIG. 16.29

SOLUTION :

Determination of z parameters

Figure 16.29b shows the currents in different branches of the network after the application of KCL at nodes a and b.

Application of KVL in the left-hand mesh gives

$$V_1 - 1 \times I_1 - 1 \times (2I_1 + I_2) = 0$$

or $$V_1 = 3I_1 + I_2 \qquad \text{(E16.19.1)}$$

Also, $$V_1 = z_{11}I_1 + z_{12}I_2 \qquad \text{(E16.19.2)}$$

Comparison of Eqs. (E16.19.1) and (E16.19.2) gives

$$z_{11} = 3\ \Omega, \qquad z_{12} = 1\ \Omega$$

Application of KVL in mesh $a\ b\ c\ a$ gives

$$1 \times (I_1 + I_2) - V_{bc} + 1 \times (2I_1 + I_2) = 0$$

But $$V_{bc} = V_2$$

therefore, $$I_1 + I_2 - V_2 + 2I_1 + I_2 = 0$$

or $$V_2 = 3I_1 + 2I_2 \qquad \text{(E16.19.3)}$$

Also, $$V_2 = z_{21}I_1 + z_{22}I_2 \qquad \text{(E16.19.4)}$$

Comparison of Eqs. (E16.19.3) and (E16.19.4) gives

$$z_{21} = 3\ \Omega, \qquad z_{22} = 2\ \Omega$$

Determination of ABCD parameters

In order to determine ABCD parameters we shall rearrange Eqs. (E16.19.1) and (E16.19.3) in the form of the defining equations of the ABCD parameters. Elimination of I_1 from Eqs. (E16.19.1) and (E16.19.3) gives

$$V_1 = V_2 - I_2 \qquad \text{(E16.19.5)}$$

Also, $$V_1 = AV_2 - BI_2 \qquad \text{(E16.19.6)}$$

Comparison of Eqs. (E16.19.5) and (E16.19.6) gives

$$A = 1,\ B = 1\ \Omega$$

From Eq. (E 16.19.3) we have

$$3I_1 = V_2 - 2I_2$$

$$I_1 = \frac{1}{3}V_2 - \frac{2}{3}I_2 \qquad \text{(E16.19.7)}$$

Also $$I_1 = CV_2 - DI_2 \qquad \text{(E16.19.8)}$$

Comparison of Eqs. (E16.19.7) and (E16.19.8) gives

$$C = \frac{1}{3}\,S, \qquad D = \frac{2}{3}$$

■ **EXAMPLE 16.20** The z parameters of a 2-port network are as follows :

$$z_{11} = 5\ \Omega,\ z_{12} = z_{21} = 2\ \Omega,\ z_{22} = 1\ \Omega$$

Determine the y, h and transmission parameters.

SOLUTION The defining equations of the z parameters are

$$V_1 = z_{11}I_1 + z_{12}I_2$$

$$V_2 = z_{21}I_1 + z_{22}I_2$$

therefore, $$V_1 = 5I_1 + 2I_2 \qquad \text{(E16.20.1)}$$

$$V_2 = 2I_1 + I_2 \qquad \text{(E16.20.2)}$$

We shall rearrange these equations in the form of the defining equations of the parameters to be determined.

Determination of y parameters

To eliminate I_2, let us multiply Eq. (E16.20.2) by 2 and subtract it from Eq. (E16.20.1)

$$V_1 - 2V_2 = I_1$$

or $$I_1 = V_1 - 2V_2 \qquad \text{(E16.20.3)}$$

Also $$I_1 = y_{11}V_1 + y_{12}V_2 \qquad \text{(E16.20.4)}$$

Comparison of Eq. (E16.20.3) with Eq. (E16.20.4) gives

$$y_{11} = 1S, \qquad y_{12} = -2S$$

To eliminate I_1, we multiply Eq. (E16.20.1) by 2 and Eq. (E16.20.2) by 5 and subtract

$$2V_1 - 5V_2 = -I_2$$

or $$I_2 = -2V_1 + 5V_2 \qquad \text{(E16.20.5)}$$

Also, $$I_2 = y_{21}V_1 + y_{22}V_2 \qquad \text{(E16.20.6)}$$

Comparison of Eq. (E16.20.5) with Eq. (E16.20.6) gives

$$y_{21} = -2S, \qquad y_{22} = 5S$$

Determination of *h* parameters

Eq. (E16.20.3) may be written as

$$V_1 = I_1 + 2V_2 \qquad \text{(E16.20.7)}$$

Also, $$V_1 = h_{11}I_1 + h_{12}V_2 \qquad \text{(E16.20.8)}$$

Comparison of Eq. (E16.20.7) with Eq. (16.20.8) gives

$$h_{11} = 1, h_{12} = 2$$

Eq. (E16.20.2) can be rearranged as

$$I_2 = -2I_1 + V_2 \quad \text{(E16.20.9)}$$

Also

$$I_2 = h_{21}I_1 + h_{22}V_2 \quad \text{(E16.20.10)}$$

Comparison of Eqs. (E16.20.9) and (E16.20.10) gives

$$h_{21} = -2, \qquad h_{22} = 1$$

Determination of transmission parameters

Eq. (E16.20.5) can be rearranged as

$$2V_1 = 5V_2 - I_2$$

$$V_1 = \frac{5}{2}V_2 - \frac{1}{2}I_2 \quad \text{(E16.20.11)}$$

Also

$$V_1 = AV_2 - BI_2 \quad \text{(E16.20.12)}$$

Comparison of these equations gives

$$A = \frac{5}{2}, \qquad B = \frac{1}{2}$$

Rearranging Eq. (E16.20.2) we get

$$2I_1 = V_2 - I_2$$

$$I_1 = \frac{1}{2}V_2 - \frac{1}{2}I_2$$

Also

$$I_1 = CV_2 - DI_2$$

therefore,

$$C = \frac{1}{2}, \qquad D = \frac{1}{2}$$

Alternative Method

Two-port parameter conversion Table 16.2 may also be used to determine the parameters.

We have

$$\begin{bmatrix} z_{11} & z_{12} \\ z_{21} & z_{22} \end{bmatrix} = \begin{bmatrix} 5 & 2 \\ 2 & 1 \end{bmatrix}$$

$$\Delta_z = \begin{vmatrix} z_{11} & z_{12} \\ z_{21} & z_{22} \end{vmatrix} = z_{11}z_{22} - z_{12}z_{21} = (5 \times 1) - 2 \times 2 = 1$$

$$\begin{bmatrix} y_{11} & y_{12} \\ y_{21} & y_{22} \end{bmatrix} = \begin{bmatrix} \frac{z_{22}}{\Delta_z} & -\frac{z_{12}}{\Delta_z} \\ -\frac{z_{21}}{\Delta_z} & \frac{z_{11}}{\Delta_z} \end{bmatrix} = \begin{bmatrix} 1 & -2 \\ -2 & 5 \end{bmatrix}$$

$$\begin{bmatrix} h_{11} & h_{12} \\ h_{21} & h_{22} \end{bmatrix} = \begin{bmatrix} \dfrac{\Delta_z}{z_{22}} & \dfrac{z_{12}}{z_{22}} \\ -\dfrac{z_{21}}{z_{22}} & \dfrac{1}{z_{22}} \end{bmatrix} = \begin{bmatrix} 1 & 2 \\ -2 & 1 \end{bmatrix}$$

$$\begin{bmatrix} A & B \\ C & D \end{bmatrix} = \begin{bmatrix} \dfrac{z_{11}}{z_{21}} & \dfrac{\Delta_z}{z_{21}} \\ \dfrac{1}{z_{21}} & \dfrac{z_{22}}{z_{21}} \end{bmatrix} = \begin{bmatrix} \dfrac{5}{2} & \dfrac{1}{2} \\ \dfrac{1}{2} & \dfrac{1}{2} \end{bmatrix}$$

16.18.1 Choice Of Parameter Type

The choice of parameter type depends upon the nature of the problem. Several factors affect the choice of parameters. (1) In certain cases, some parameters may not exist. For example. in case of an ideal transformer z and y parameters do not exist, and therefore some other parameters are used. (2) Sometimes a particular choice of parameters may provide a very simple representation. (3) Some parameters are more convenient to work with when the network is connected to other networks, as shown in section 16.21.

16.18.2 Nonexistence of Parameters

There are networks in which z and y parameters do not both exist. For example, consider a 2-port network consisting of a single series element as shown in Fig. 16.30a. The y-parameter matrix for this network is

$$[y] = \begin{bmatrix} Y & -Y \\ -Y & Y \end{bmatrix}$$

$$\Delta_y = \begin{vmatrix} Y & -Y \\ -Y & Y \end{vmatrix} = Y^2 - Y^2 = 0$$

Since Δ_y is zero, the z parameters as determined from the y parameters do not exist.

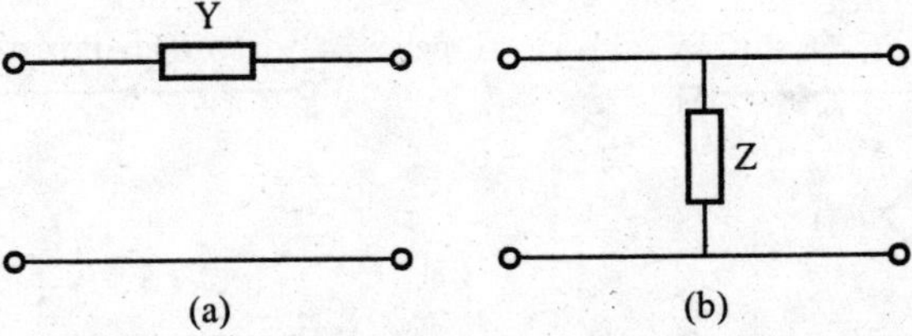

FIG. 16.30 Single-element networks (a) Single series element (b) Single shunt element

As another example of a network which does not have both z and y parameters, consider a 2-port consisting of a single shunt element as shown in Fig. 16.30b. The z parameters matrix for this network is

$$[z] = \begin{bmatrix} Z & Z \\ Z & Z \end{bmatrix}$$

$$\Delta_z = Z \times Z - Z \times Z = 0$$

Since the determinant Δ_z is zero, the y parameters as determined from the z parameters do not exist.

For some networks, neither the z nor the y parameters exist. As an example of this, consider the ideal transformer shown in Fig. 16.31.

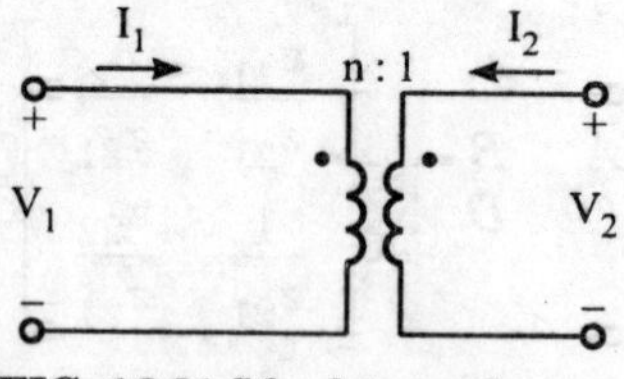

FIG. 16.31 Ideal transformer

The defining equations are

$$V_1 = nV_2$$

$$I_1 = -\frac{1}{n} I_2$$

Since these equations require that the voltage at one port be specified as a function of voltage at the other port, and similarly that the current at one port be expressed in terms of the current at the other port, neither the z parameters nor the y parameters exist for an ideal transformer.

In general, except for some very unusual degenerate networks, there will be at least one parameter set that will exist for any given network configuration.

The simplest description of the voltage controlled voltage source (VCVS) shown in Fig. 16.32a. is in terms of the g parameters. It is given as follows :

$$\begin{bmatrix} i_1 \\ v_2 \end{bmatrix} = \begin{bmatrix} 0 & 0 \\ \mu & 0 \end{bmatrix} \begin{bmatrix} v_1 \\ i_2 \end{bmatrix}$$

Here $$g_{11} = g_{12} = g_{22} = 0 \text{ and } g_{21} = \mu$$

(a) VCVs

(c) CCCs

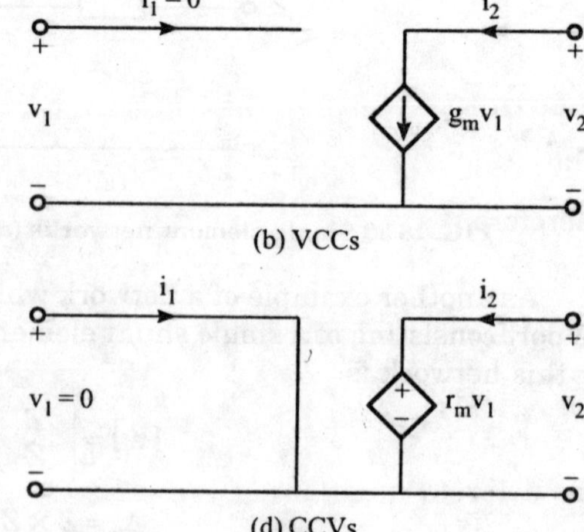

FIG. 16.32 Controlled sources (a) VCVS (b) VCCS (c) CCCS (d) CCVS

The voltage-controlled current source (VCCS) shown in Fig. 16.32b has the simplest description in terms of the y parameters.

$$\begin{bmatrix} i_1 \\ i_2 \end{bmatrix} = \begin{bmatrix} 0 & 0 \\ g_m & 0 \end{bmatrix} \begin{bmatrix} v_1 \\ v_2 \end{bmatrix}$$

Here $\qquad y_{11} = y_{22} = y_{12} = 0, \text{ and } y_{21} = g_m.$

For the current-controlled current source (CCCS) shown in Fig. 16.32c the simplest description is given in terms of the h parameters.

$$\begin{bmatrix} v_1 \\ i_2 \end{bmatrix} = \begin{bmatrix} 0 & 0 \\ \alpha & 0 \end{bmatrix} \begin{bmatrix} i_1 \\ v_2 \end{bmatrix}$$

Here $\qquad h_{11} = h_{22} = h_{12} = 0 \text{ and } h_{21} = \alpha$

For the current-controlled voltage source (CCVS) shown in Fig. 16.32d, the simplest description is given in terms of the z parameters.

$$\begin{bmatrix} v_1 \\ v_2 \end{bmatrix} = \begin{bmatrix} 0 & 0 \\ r_m & 0 \end{bmatrix} \begin{bmatrix} i_1 \\ i_2 \end{bmatrix}$$

Here, $\qquad z_{11} = z_{22} = z_{12} = 0 \text{ and } z_{21} = r_m$

EXAMPLE 16.21 Find the z parameters of the 2-port network shown in Fig. 16.33. Hence prove that the y parameters of this network do not exist.

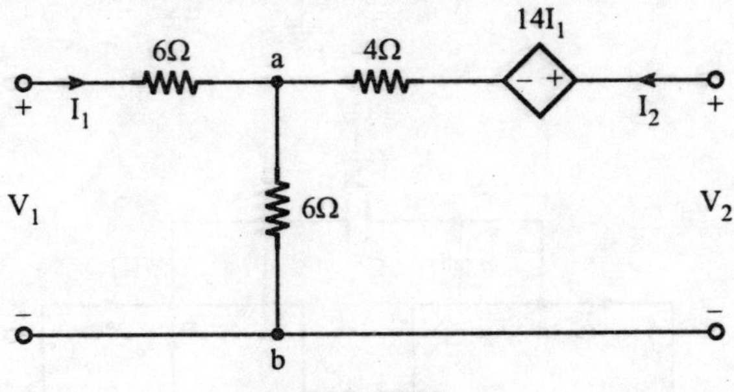

FIG. 16.33

SOLUTION : By KVL in the left mesh, $V_1 = 6I_1 + 6I_{ab}$

But $\qquad I_{ab} = I_1 + I_2$

therefore, $\qquad V_1 = 6I_1 + 6(I_1 + I_2)$

or $\qquad V_1 = 12I_1 + 6I_2 \qquad$ (E16.21.1)

By KVL in the right mesh

$$V_2 = 14I_1 + 4I_2 + 6(I_1 + I_2)$$

or $\qquad V_2 = 20I_1 + 10I_2 \qquad$ (E16.21.2)

The defining equations of z parameters are

$$V_1 = z_{11}I_1 + z_{12}I_2 \qquad \text{(E16.21.3)}$$

$$V_2 = z_{21}I_2 + z_{22}I_2 \qquad \text{(E16.21.4)}$$

By comparing Eq. (E16.21.1) with Eq. (E16.21.3) and Eq. (E16.21.2) with Eq. (E16.21.4) we get

$$z_{11} = 12\ \Omega,\ z_{12} = 6\ \Omega$$

$$z_{21} = 20\ \Omega,\ z_{22} = 10\ \Omega$$

$$\Delta_z = \begin{vmatrix} z_{11} & z_{12} \\ z_{21} & z_{22} \end{vmatrix} = z_{11}z_{12} - z_{12}z_{21} = (12 \times 10) - (6 \times 20) = 0$$

Since the determinant Δ_z is zero, the y parameters are not defined.

■ **EXAMPLE 16.22** Determine the h parameters of the two port network shown in Fig. 16.34a.

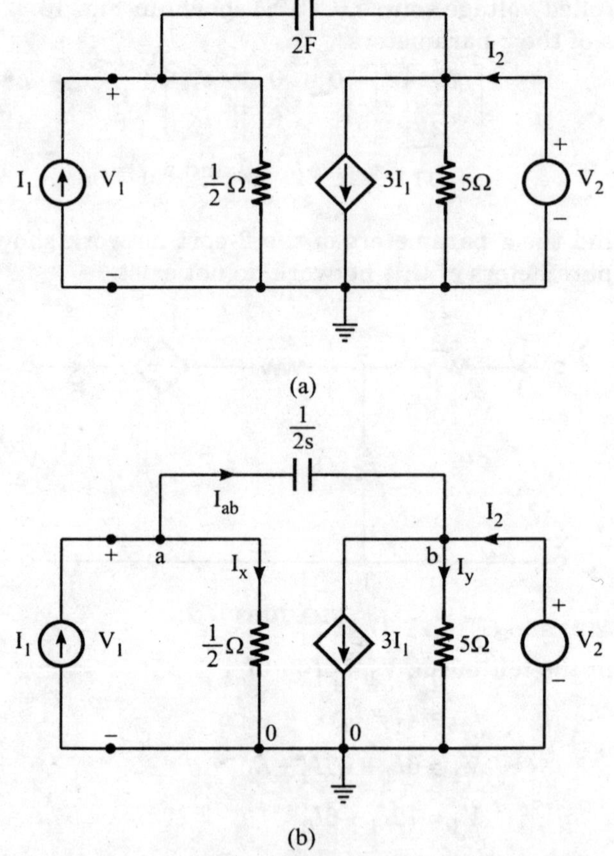

FIG. 16.34

SOLUTION : Let us take the ground node as the datum node. In the s domain, the circuit is shown in Fig. 16.34b. The currents in different branches are found by the application of KCL at nonreference nodes a and b in Fig. 16.34b.

By KCL at node a

$$I_1 = I_{ab} + I_x$$

By Ohm's law, $$I_{ab} = \frac{V_{ab}}{Z_{ab}} = \frac{V_a - V_b}{\frac{1}{2s}} = (V_1 - V_2)\, 2s$$

$$I_x = \frac{V_a - V}{1/2} = 2V_1$$

therefore, $$I_1 = (V_1 - V_2)\, 2s + 2V_1$$

or $$I_1 = (2 + 2s)\, V_1 - 2sV_2 \qquad \text{(E16.22.1)}$$

By KCL at node b

$$I_2 + I_{ab} = I_y + 3I_1$$

By Ohm's law, $$I_y = \frac{V_b - 0}{5} = \frac{V_2}{5}$$

therefore, $$I_2 = I_y + 3I_1 - I_{ab}$$

$$I_2 = \frac{V_2}{5} + 3I_1 - (V_1 - V_2)\, 2s$$

or $$I_2 = 3I_1 + (\frac{1}{5} + 2s)\, V_2 - 2sV_1 \qquad \text{(E16.22.2)}$$

Eqs. (E16.22.1) and (E16.22.2) will be rearranged to express V_1 and I_2 in terms of I_1 and V_2.

From Eq. (E16.22.1)

$$V_1 = \frac{1}{2\,(s+1)} I_1 + \frac{s}{s+1} V_2 \qquad \text{(E16.22.3)}$$

Also $$V_1 = h_{11}I_1 + h_{12}V_2 \qquad \text{(E16.22.4)}$$

By comparing Eqs. (E16.22.3) and (E16.22.4) we get

$$h_{11} = \frac{1}{2\,(s+1)}, \qquad h_{12} = \frac{s}{s+1}$$

From Eqs. (E16.22.2) and (E16.22.3)

$$I_2 = 3I_1 + \left(\frac{1}{5} + 2s\right) V_2 - 2s\left(\frac{1}{2+2s} I_1 + \frac{s}{1+s} V_2\right)$$

$$= \left(3 - \frac{s}{1+s}\right) I_1 + \left(\frac{1}{5} + 2s - \frac{2s^2}{1+s}\right) V_2$$

$$I_2 = \frac{3+2s}{1+s} I_1 + \frac{1+11s}{5\,(1+s)} V_2 \qquad \text{(E16.22.5)}$$

Also, $$I_2 = h_{21}I_1 + h_{22}I_2 \qquad \text{(E16.22.6)}$$

By comparing Eqs. (E16.22.5) and (E16.22.6) we get

$$h_{21} = \frac{2s+3}{s+1}, \qquad h_{22} = \frac{11s+1}{5\,(s+1)}$$

16.19 PARAMETERS OF TERMINATED TWO-PORT NETWORK

External networks that may be connected at the input and output ports are called **terminations**. Suppose that a generator (source) of voltage V_G and impedance Z_G is connected to the input terminals and a load of Z_L is connected at the output terminals as shown in Fig. 16.35.

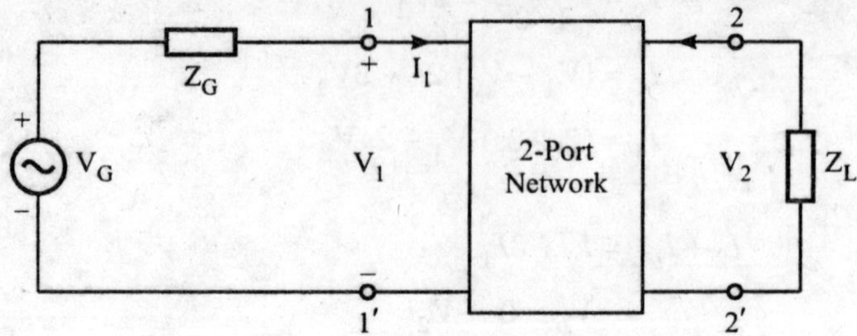

FIG. 16.35 A terminated 2-port network

The equations at the two ports are

$$V_1 = V_G - Z_G I_1 \tag{16.19.1}$$

$$V_2 = -Z_L I_2 \tag{16.19.2}$$

Eqs. (16.9.1) and (16.19.2) describe constraints imposed by terminations. These two equations can be combined with a given pair of parametric equations to determine all the four terminal variables V_1, I_1, V_2, I_2. The following important parameters define the terminal behavior of a terminated 2-port network.

1. Input Impedance

$$Z_{in} \triangleq \frac{V_1}{I_1} \tag{16.19.3}$$

2. Output Impedance (Z_{out}) or Thevenin Impedance (Z_{Th})

Output impedance is the impedance seen at the output terminals, when the load impedance Z_L is removed and Z_G is replaced by a short circuit (that is, $Z_G = 0$).

$$\therefore \quad Z_{out} = Z_{Th} \triangleq \frac{V_2}{I_2}\bigg|_{V_G = 0,\ Z_L = 0} \tag{16.9.4}$$

3. Current Gain or Current Transfer Ratio A_i

$$A_i \triangleq \frac{I_2}{I_1} \tag{16.19.5}$$

4. Voltage Gain or Voltage Transfer Ratio A_v

$$A_v \triangleq \frac{V_2}{V_1} \tag{16.19.6}$$

These parameters may be found in terms of Z_G, Z_L and any of the six sets of 2-port parameters. The procedure of finding out these parameters in illustrated in Example 16.23. Table 16.3 shows the values of these parameters in terms of y, z, h, g and a parameters.

TABLE 16.3

	z	y	h	g	a
Z_{in}	$\dfrac{\Delta_z + z_{11}Z_L}{z_{22}+Z_L}$	$\dfrac{1+y_{22}Z_L}{y_{11}+\Delta_y Z_L}$	$\dfrac{h_{11}+\Delta_h Z_L}{1+h_{22}Z_L}$	$\dfrac{g_{22}+Z_L}{\Delta_g + g_{11}Z_L}$	$\dfrac{B+AZ_L}{D+CZ_L}$
Z_{out}	$\dfrac{\Delta_z + z_{22}Z_G}{z_{11}+Z_G}$	$\dfrac{1+y_{11}Z_G}{y_{22}+\Delta_y Z_G}$	$\dfrac{h_{11}+Z_G}{\Delta_h + h_{22}Z_G}$	$\dfrac{g_{22}+\Delta_g Z_G}{1+g_{11}Z_G}$	$\dfrac{B+DZ_G}{A+CZ_G}$
$\dfrac{I_2}{I_1}$	$\dfrac{-z_{21}}{z_{22}+Z_L}$	$\dfrac{y_{21}}{y_{11}+\Delta_y Z_G}$	$\dfrac{h_{21}}{1+h_{22}Z_L}$	$\dfrac{-g_{21}}{\Delta_g + g_{11}Z_L}$	$\dfrac{-1}{D+CZ_L}$
$\dfrac{V_2}{V_1}$	$\dfrac{z_{21}Z_L}{\Delta_z + z_{11}Z_L}$	$\dfrac{-y_{21}Z_L}{1+y_{22}Z_L}$	$\dfrac{-h_{21}Z_L}{h_{11}+\Delta_h Z_L}$	$\dfrac{g_{21}Z_L}{g_{22}+Z_L}$	$\dfrac{Z_L}{B+AZ_L}$

■ **EXAMPLE 16.23** For the network shown in Fig. 16.35, determine in terms of the z parameters.

(*a*) the current gain,
(*b*) the voltage gain,
(*c*) the input impedance,
(*d*) the output impedance.

SOLUTION : The defining equations of the network in terms of the z parameters are

$$V_1 = z_{11}I_1 + z_{12}I_2 \tag{E16.23.1}$$

$$V_2 = z_{21}I_1 + z_{22}I_2 \tag{E16.23.2}$$

At the input terminals,

$$V_1 = V_G - Z_G I_1 \tag{E16.23.3}$$

At the output terminals

$$V_2 = -Z_L I_2 \tag{E16.23.4}$$

(*a*) **Current gain A_i**

$$A_i = \frac{I_2}{I_1}$$

Eliminating V_2 from Eqs (E16.23.2) and (E16.23.4) we get

$$-Z_L I_2 = z_{21}I_1 + z_{22}I_2$$

$$-(Z_L + z_{22})\, I_2 = z_{21}I_1$$

$$A_i = \frac{I_2}{I_1} = \frac{-z_{21}}{Z_L + z_{22}} \tag{E16.23.5}$$

(*b*) **Voltage gain A_v**

$$A_v = \frac{V_2}{V_1} = \frac{z_{21}I_1 + z_{22}I_2}{z_{11}I_1 + z_{12}I_2}$$

$$= \frac{z_{21} + z_{22}\,(I_2/I_1)}{z_{11} + z_{12}\,(I_2/I_1)}$$

$$= \frac{z_{21} + z_{22}\left(\dfrac{-z_{21}}{Z_L + z_{22}}\right)}{z_{11} + z_{12}\left(\dfrac{-z_{21}}{Z_L + z_{22}}\right)}$$

$$= \frac{z_{21}Z_L + z_{21}z_{22} - z_{21}z_{22}}{z_{11}Z_L + z_{11}z_{22} - z_{12}z_{21}}$$

$$= \frac{z_{21}Z_L}{z_{11}Z_L + \Delta_Z} \tag{E16.23.6}$$

(*c*) **Input impedance Z_{in}**

$$Z_{in} = \frac{V_1}{I_1}$$

From Eq. (E16.23.1)

$$\frac{V_1}{I_1} = z_{11} + z_{12}\frac{I_2}{I_1}$$

Substituting the value of I_2/I_1 from Eq. (E16.23.5) in the above equation

$$\frac{V_1}{I_1} = z_{11} - \frac{z_{12}z_{21}}{Z_L + z_{22}}$$

therefore,
$$Z_{in} = \frac{V_1}{I_1} = \frac{\Delta_z + z_{11}Z_L}{z_{22} + Z_L}$$

(*d*) **Output impedance Z_{out}**

Output impedance is the impedance seen at the output terminals with V_G and Z_L removed (see Fig. 16.36).

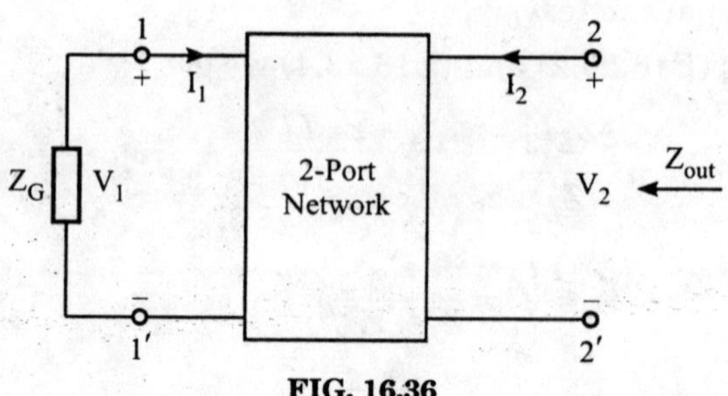

FIG. 16.36

$$Z_{out} = \left.\frac{V_2}{I_2}\right|_{V_G=0,\, Z_{L=0}} = Z_{Th}$$

With $V_G = 0$. Eq. (E16.23.3) becomes

$$V_1 = -Z_G I_1 \qquad \text{(E16.23.7)}$$

Putting the value of V_1 in Eq. (E16.23.1) gives

$$-Z_G I_1 = z_{11} I_1 + z_{12} I_2$$

$$-(z_{11} + Z_G) I_1 = z_{12} I_2$$

$$\frac{I_1}{I_2} = \frac{-z_{12}}{z_{11} + Z_G} \qquad \text{(E16.23.8)}$$

From Eq. (E16.23.2)

$$\frac{V_2}{I_2} = z_{21}\frac{I_1}{I_2} + z_{22}$$

Putting the value of I_1/I_2 from Eq. (E16.23.8) in the above equation we get

$$\frac{V_2}{I_2} = z_{21}\left(\frac{-z_{12}}{z_{11} + Z_G}\right) + z_{22}$$

$$= \frac{-z_{12}z_{21} + z_{11}z_{22} + z_{22}Z_G}{z_{11} + Z_G}$$

therefore,

$$Z_{out} = \frac{V_2}{I_2} = \frac{\Delta_z + z_{22}Z_G}{z_{11} + Z_G} \qquad \text{(E16.23.9)}$$

16.20 SYMMETRY AND RECIPROCITY OF TWO-PORT NETWORKS

A passive, linear 2-port network is said to be **symmetrical** if the two ports can be interchanged (that is, the input port is made the output port and the output port is made the input port) without changing the port voltages and currents. The condition of symmetry is conveniently written in terms of z parameters as

$$\boxed{z_{11} = z_{22}} \qquad \text{(16.20.1)}$$

or

$$\left.\frac{V_1}{I_1}\right|_{I_2=0} = \left.\frac{V_2}{I_2}\right|_{I_1=0} \qquad \text{(16.20.2)}$$

The conditions of symmetry may also be derived in terms of other parameters. These conditions are given in Table 16.4

Table 16.4

Parameter relationships for symmetrical and reciprocal networks

	Parameter	Condition of symmetry	Condition of reciprocity
	z	$z_{11} = z_{22}$	$z_{12} = z_{21}$
	y	$y_{11} = y_{12}$	$y_{12} = y_{21}$
	h	$\Delta_h = 1$	$h_{12} = -h_{21}$
	g	$\Delta_g = 1$	$g_{12} = -g_{21}$
	a	$A = D$	$AD - BC = 1$
or	a	$a_{11} = a_{22}$	$\Delta_a = 1$
	b	$b_{11} = b_{22}$	$\Delta_b = 1$

A passive linear, 2-port network is said to be **reciprocal** if reciprocity principle holds. That is,

$$\left.\frac{I_1}{V_1}\right|_{V_1=0} = \left.\frac{I_2}{V_2}\right|_{V_2=0} \tag{16.20.3}$$

or

$$\boxed{y_{12} = y_{21}} \tag{16.20.4}$$

The conditions of reciprocity may also be derived in terms of other parameters. These conditions are given in Table 16.4.

Thus, it is seen that a passive, linear reciprocal, 2-port network may be characterized by three z parameters, or three-y parameters. Similarly, three $ABCD$ parameters are sufficient to describe a passive, linear, reciprocal 2-port network. Further if the network is symmetrical (for which $A = D$), only two parameters are needed to describe the network.

■ **EXAMPLE 16.24** Establish the conditions of (*a*) symmetry, (*b*) reciprocity in terms of $ABCD$ parameters of a linear, passive, 2-port network.

SOLUTION : In terms of z parameter

$$A = \frac{z_{11}}{z_{21}}, B = \frac{\Delta_z}{z_{21}} = \frac{z_{11}z_{22} - z_{21}z_{12}}{z_{21}}$$

$$C = \frac{1}{z_{21}}, D = \frac{z_{22}}{z_{21}}$$

(*a*) For a symmetrical network, $z_{11} = z_{22}$

therefore, $A = D$

(*b*) $$AD - BC = \frac{z_{11}}{z_{21}} \times \frac{z_{22}}{z_{21}} - \frac{z_{11}z_{22} - z_{21}z_{12}}{z_{12}^2} = \frac{z_{12}}{z_{21}}$$

Since for a reciprocal network, $z_{12} = z_{21}$

$$AD - BC = 1$$

16.21 INTERCONNECTION OF TWO-PORT NETWORKS

Simple two-port networks may be interconnected to form larger complex networks. The two ports can be interconnected in five basic ways. These are series-series, parallel-parallel, series-parallel, parallel-series, and cascade. We are interested in determining the parameters of the overall network in terms of those of the component networks. Before interconnecting two ports, it is important to verify that for the overall two port, the current entering one terminal of a port is still the same as the current leaving the other terminal of the same port. It is to be noted that this is the basis for the definition of a two port.

The tests to establish whether each terminal pair still satisfies the definition of a port after interconnection is made are called **validity tests** or **Brune's tests** after their inventor O. Brune. The cascade interconnection does not require the satisfaction of any validity tests, but these tests are necessary for the remaining four interconnections.

16.22 SERIES-SERIES INTERCONNECTION

Two 2-port networks are said to be connected in series if the corresponding ports are connected in series, as shown in Fig. 16.37.

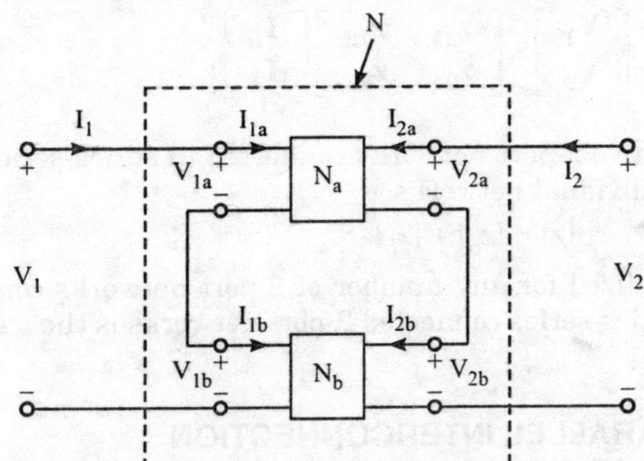

FIG. 16.37 Series-series interconnection of two 2-ports

Let N_a be described by z_a parameters, N_b by z_b and the overall two port network N by z parameters. Before the networks are interconnected, the relationship between voltages and currents for N_a are given by

$$\begin{bmatrix} \mathbf{V}_{1a} \\ \mathbf{V}_{2a} \end{bmatrix} = \begin{bmatrix} \mathbf{z}_{11a} & \mathbf{z}_{12a} \\ \mathbf{z}_{21a} & \mathbf{z}_{22a} \end{bmatrix} \begin{bmatrix} \mathbf{I}_{1a} \\ \mathbf{I}_{2a} \end{bmatrix} \tag{16.22.1}$$

Similarly for the network N_b,

$$\begin{bmatrix} \mathbf{V}_{1b} \\ \mathbf{V}_{2b} \end{bmatrix} = \begin{bmatrix} \mathbf{z}_{11b} & \mathbf{z}_{12b} \\ \mathbf{z}_{21b} & \mathbf{z}_{22b} \end{bmatrix} \begin{bmatrix} \mathbf{I}_{1b} \\ \mathbf{I}_{2b} \end{bmatrix} \tag{16.22.2}$$

Let us assume that the external properties of the networks N_a and N_b are not altered by their interconnection and at a port the current entering is the same as the current leaving. From Fig. 16.37, it is seen that after interconnection

$$\mathbf{I}_1 = \mathbf{I}_{1a} = \mathbf{I}_{1b}, \qquad \mathbf{I}_2 = \mathbf{I}_{2a} = \mathbf{I}_{2b}$$

$$\mathbf{V}_1 = \mathbf{V}_{1a} + \mathbf{V}_{1b}, \qquad \mathbf{V}_2 = \mathbf{V}_{2a} + \mathbf{V}_{2b}$$

Now,

$$\begin{aligned}\mathbf{V}_1 &= \mathbf{V}_{1a} + \mathbf{V}_{1b} \\ &= (\mathbf{z}_{11a}\mathbf{I}_{1a} + \mathbf{z}_{12a}\mathbf{I}_{2a}) + (\mathbf{z}_{11b}\mathbf{I}_{1b} + \mathbf{z}_{12b}\mathbf{I}_{2b}) \\ &= (\mathbf{z}_{11a} + \mathbf{z}_{11b})\,\mathbf{I}_1 + (\mathbf{z}_{12a} + \mathbf{z}_{12b})\,\mathbf{I}_2\end{aligned}$$

and

$$\begin{aligned}\mathbf{V}_2 &= \mathbf{V}_{2a} + \mathbf{V}_{2b} \\ &= (\mathbf{z}_{21a}\mathbf{I}_{1a} + \mathbf{z}_{22a}\mathbf{I}_{2a}) + (\mathbf{z}_{21b}\mathbf{I}_b + \mathbf{z}_{22b}\mathbf{I}_b) \\ &= (\mathbf{z}_{21a} + \mathbf{z}_{21b})\,\mathbf{I}_1 + (\mathbf{z}_{22a} + \mathbf{z}_{22b})\,\mathbf{I}_2\end{aligned}$$

In matrix form we can write

$$\begin{bmatrix}\mathbf{V}_1 \\ \mathbf{V}_2\end{bmatrix} = \begin{bmatrix}\mathbf{z}_{11a} + \mathbf{z}_{11b} & \mathbf{z}_{12a} + \mathbf{z}_{12b} \\ \mathbf{z}_{21a} + \mathbf{z}_{21b} & \mathbf{z}_{22a} + \mathbf{z}_{22b}\end{bmatrix}\begin{bmatrix}\mathbf{I}_1 \\ \mathbf{I}_2\end{bmatrix}$$

For the overall network N,

$$\begin{bmatrix}\mathbf{V}_1 \\ \mathbf{V}_2\end{bmatrix} = \begin{bmatrix}\mathbf{z}_{11} & \mathbf{z}_{12} \\ \mathbf{z}_{21} & \mathbf{z}_{22}\end{bmatrix}\begin{bmatrix}\mathbf{I}_1 \\ \mathbf{I}_2\end{bmatrix}$$

Thus, the **z** matrix of the two 2-port networks connected in series-series is equal to the sum of the **z** matrices of the individual networks,

$$[\mathbf{z}] = [\mathbf{z}_a] + [\mathbf{z}_b]$$

This result may be generalized for any number of 2-port networks connected in series. The overall z matrix for the series-series connected 2-port networks is the sum of z matrices of the individual networks.

16.23 PARALLEL-PARALLEL INTERCONNECTION

The method of interconnecting two 2-port networks shown in Fig. 16.38 is called a parallel-parallel connection of two ports because the corresponding ports of each network are connected in parallel. Let us find the y parameters of the overall 2-port network N. Before the networks are interconnected, the network N_a satisfies

$$\begin{bmatrix}\mathbf{I}_{1a} \\ \mathbf{I}_{2a}\end{bmatrix} = \begin{bmatrix}y_{11a} & y_{12a} \\ y_{21a} & y_{22a}\end{bmatrix}\begin{bmatrix}\mathbf{V}_{1a} \\ \mathbf{V}_{2a}\end{bmatrix}$$

Similarly, for the network N_b

$$\begin{bmatrix}\mathbf{I}_{1b} \\ \mathbf{I}_{2b}\end{bmatrix} = \begin{bmatrix}y_{11b} & y_{12b} \\ y_{21b} & y_{22b}\end{bmatrix}\begin{bmatrix}\mathbf{V}_{1b} \\ \mathbf{V}_{2b}\end{bmatrix}$$

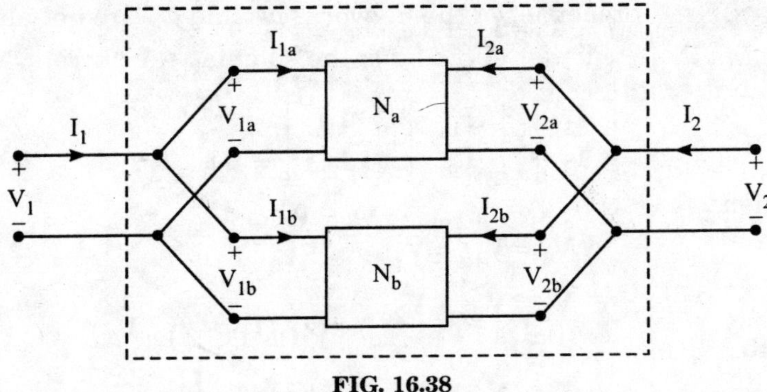

FIG. 16.38

In order to connect two networks N_a and N_b in parallel they must satisfy certain conditions. One test is performed at the output end and the other at the input end. For the Brune's test at the output end, the input ports 1 and 1′ are connected in parallel and the output ports are short circuited individually as shown in Fig. 16.39a. The short circuiting is done because the y parameters characterizing the individual two ports and the overall two port are the short-circuited admittance parameters. Then the condition under which the two networks N_a and N_b can be connected in parallel at the input end is that $V_x = 0$. The network in Fig. 16.39a can be redrawn as in Fig. 16.39b.

By KCL at the ports

$$\mathbf{I}_1 = \mathbf{I}_{1a} + \mathbf{I}_{1b}$$

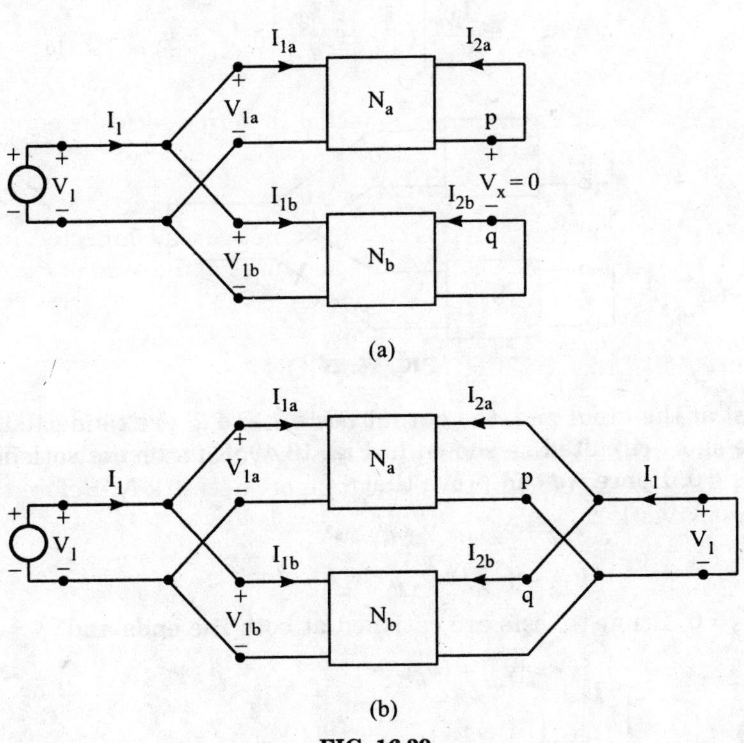

FIG. 16.39

$$\mathbf{I}_2 = \mathbf{I}_{2a} + \mathbf{I}_{2b}$$

In matrix form

$$\begin{bmatrix} \mathbf{I}_1 \\ \mathbf{I}_2 \end{bmatrix} = \begin{bmatrix} \mathbf{I}_{1a} \\ \mathbf{I}_{2a} \end{bmatrix} + \begin{bmatrix} \mathbf{I}_{1b} \\ \mathbf{I}_{2b} \end{bmatrix}$$

For parallel connection

$$\mathbf{V}_1 = \mathbf{V}_{1a} = \mathbf{V}_{1b}$$

$$\mathbf{V}_2 = \mathbf{V}_{2a} = \mathbf{V}_{2b}$$

Now from Fig. 16.34b,

$$\mathbf{V}_2 = \mathbf{V}_{2a} = \mathbf{V}_{2b} = 0; \text{ hence}$$

$$\mathbf{y}_{11} = \frac{\mathbf{I}_1}{\mathbf{V}_1}\bigg|_{\mathbf{V}_2 = 0} = \frac{\mathbf{I}_{1a} + \mathbf{I}_{1b}}{\mathbf{V}_1}\bigg|_{\mathbf{V}_2 = 0}$$

$$= \frac{\mathbf{I}_{1a}}{\mathbf{V}_{1a}}\bigg|_{\mathbf{V}_{2a} = 0} + \frac{\mathbf{I}_{1b}}{\mathbf{V}_{1b}}\bigg|_{\mathbf{V}_{2b} = 0} = \mathbf{y}_{11a} + \mathbf{y}_{11b}$$

and

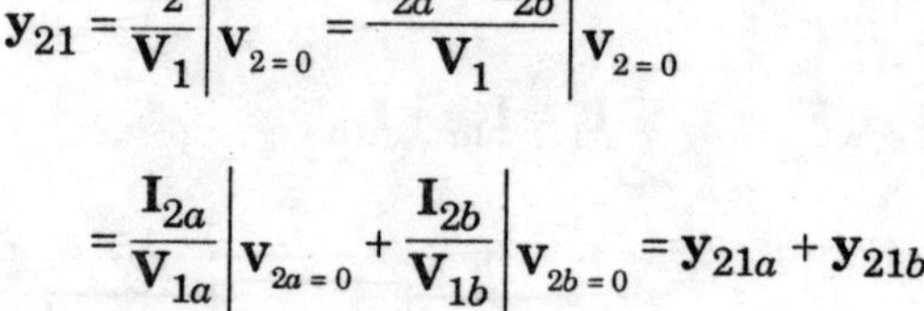

$$\mathbf{y}_{21} = \frac{\mathbf{I}_2}{\mathbf{V}_1}\bigg|_{\mathbf{V}_2 = 0} = \frac{\mathbf{I}_{2a} + \mathbf{I}_{2b}}{\mathbf{V}_1}\bigg|_{\mathbf{V}_2 = 0}$$

$$= \frac{\mathbf{I}_{2a}}{\mathbf{V}_{1a}}\bigg|_{\mathbf{V}_{2a} = 0} + \frac{\mathbf{I}_{2b}}{\mathbf{V}_{1b}}\bigg|_{\mathbf{V}_{2b} = 0} = \mathbf{y}_{21a} + \mathbf{y}_{21b}$$

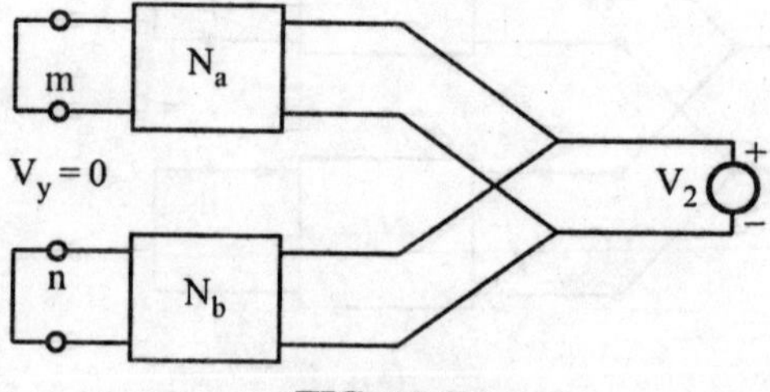

FIG. 16.40

For the Brune's test at the input end, the output ports 2 and 2′ are connected in parallel and the input ports are short circuited as shown in Fig. 16.40. This test is satisfied at the input end when $\mathbf{V}_{mn} = \mathbf{V}_y = 0$. Hence we can prove that

$$\mathbf{y}_{22} = \mathbf{y}_{22a} + \mathbf{y}_{22b}$$

$$\mathbf{y}_{12} = \mathbf{y}_{12a} + \mathbf{y}_{12b}$$

Therefore if $V_x = V_y = 0$, Brune'ts tests are satisfied at both the ends and

$$[\mathbf{y}] = [\mathbf{y}_a] + [\mathbf{y}_b]$$

$$\begin{bmatrix} \mathbf{y}_{11} & \mathbf{y}_{12} \\ \mathbf{y}_{21} & \mathbf{y}_{22} \end{bmatrix} = \begin{bmatrix} \mathbf{y}_{11a} + \mathbf{y}_{11b} & \mathbf{y}_{12a} + \mathbf{y}_{12b} \\ \mathbf{y}_{21a} + \mathbf{y}_{21b} & \mathbf{y}_{22a} + \mathbf{y}_{22b} \end{bmatrix}$$

Thus, the y parameters of two 2-port networks connected in parallel-parallel are found by adding the y parameters of the component networks. This result may be generalized for any number of networks connected in parallel.

The following conclusions are drawn from the above discussion :

1. When two ports are connected in parallel it is easier to find the y parameters. From the y parameters, other 2-port parameters can be derived.
2. When two ports are connected in series, it is easier to find the z parameters.
3. When two ports are connected in cascade, the transmission matrix is generally easier to obtain. The overall chain matrix is obtained by multiplying the individual chain matrices together **in the order of cascading**.

Table 16.5 shows overall parameters for various interconnected 2 port.

Table 16.5

Interconnection	Individual parameters	Overall parameters
Series-series	$\mathbf{z}_a, \mathbf{z}_b$	$\mathbf{z}_a + \mathbf{z}_b$
Parallel-parallel	$\mathbf{y}_a, \mathbf{y}_b$	$\mathbf{y}_a + \mathbf{y}_b$
Series-parallel	$\mathbf{h}_a, \mathbf{h}_b$	$\mathbf{h}_a + \mathbf{h}_b$
Parallel-series	$\mathbf{g}_a, \mathbf{g}_b$	$\mathbf{g}_a + \mathbf{g}_b$
Cascade	$[\mathbf{a}_a], [\mathbf{a}_b]$	$[\mathbf{a}_a]\,[\mathbf{a}_b]$

■ **EXAMPLE 16.25** Determine the z parameters for the bridged T network shown in Fig. 16.41a.

SOLUTION : The given network N may be redrawn as a series interconnection of two simpler networks N_a and N_b. In the s domain these networks are shown in Fig. 16.41b. It is seen that Brune's tests are satisfied at both ends. The network N_a is a Π network. The network N_b consists of a single shunt element Z_D of impedance $\frac{1}{s}$ Ω.

The y parameters of network N_a are given by

$$y_{11a} = Y_A + Y_C = \frac{1}{Z_A} + \frac{1}{Z_C} = \frac{1}{1} + \frac{1}{s} = \frac{s+1}{s}$$

$$y_{12a} = y_{21a} = -Y_C = -\frac{1}{Z_C} = -\frac{1}{s}$$

$$y_{22a} = Y_B + Y_C = \frac{1}{Z_B} + \frac{1}{Z_C} = \frac{1}{1} + \frac{1}{s} = \frac{s+1}{s}$$

$$\Delta_{ya} = \begin{vmatrix} y_{11a} & y_{12a} \\ y_{21a} & y_{22a} \end{vmatrix} = y_{11a}\,y_{22a} - y_{12a}\,y_{21a}$$

$$= \frac{(s+1)}{s} \times \frac{s+1}{s} - \frac{1}{s} \times \frac{1}{s} = \frac{1}{s^2} + \frac{2}{s} + 1 - \frac{1}{s^2} = \frac{s+2}{s}$$

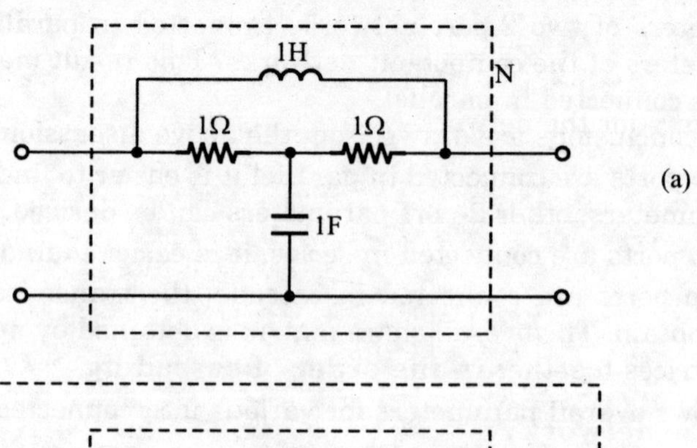

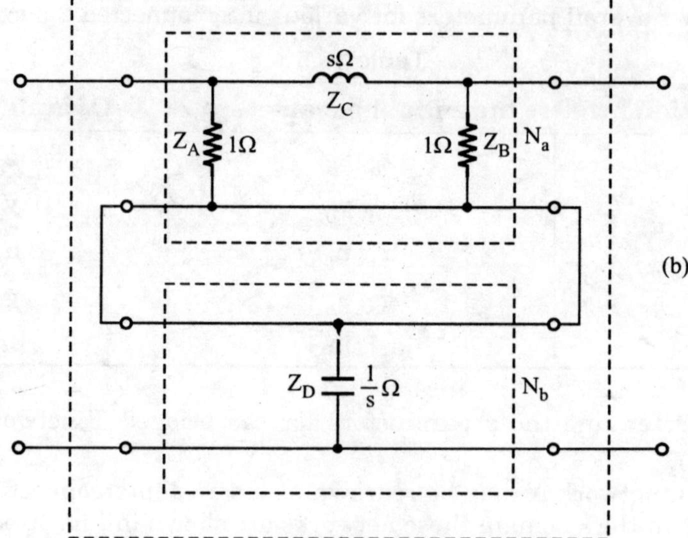

FIG. 16.41

z parameters of the network N_a

From the parameter conversion Table 16.2

$$z_{11a} = \frac{y_{22a}}{\Delta_{ya}} = \left(\frac{s+1}{s}\right) \div \frac{s+2}{s} = \frac{s+1}{s+2}$$

$$z_{12a} = z_{21a} = \frac{-y_{12a}}{\Delta_{ya}} = \left(\frac{1}{s}\right) \div \frac{s+2}{s} = \frac{1}{s+2}$$

$$z_{22a} = \frac{y_{11a}}{\Delta_{ya}} = \left(\frac{s+1}{s}\right) \div \left(\frac{s+2}{s}\right) = \frac{s+1}{s+2}$$

therefore,
$$[\mathbf{z}_a] = \begin{bmatrix} z_{11a} & z_{12a} \\ z_{21a} & z_{22a} \end{bmatrix} = \begin{bmatrix} \frac{s+1}{s+2} & \frac{1}{s+2} \\ \frac{1}{s+2} & \frac{s+1}{s+2} \end{bmatrix}$$

z parameters of the network N_b

The network N_b consists of a single shunt element of impedance

$$Z_D = \frac{1}{s}\ \Omega.$$

The z parameters for the network N_b are given by

$$z_{11} = z_{12} = z_{21} = z_{22} = Z_D = \frac{1}{s}$$

$$[\,\mathbf{z}_b\,] = \begin{bmatrix} z_{11b} & z_{12b} \\ z_{21b} & z_{22b} \end{bmatrix} = \begin{bmatrix} \frac{1}{s} & \frac{1}{s} \\ \frac{1}{s} & \frac{1}{s} \end{bmatrix}$$

z parameters of the network N

Since the networks N_a and N_b are connected in series, the overall **z** matrix is the sum of **z** matrices of N_a and N_b. Therefore the **z** parameters for the network N are

$$[\,\mathbf{z}\,] = [\,\mathbf{z}_a\,] + [\,\mathbf{z}_b\,]$$

$$= \begin{bmatrix} \frac{s+1}{s+2} & \frac{1}{s+2} \\ \frac{1}{s+2} & \frac{s+1}{s+2} \end{bmatrix} + \begin{bmatrix} \frac{1}{s} & \frac{1}{s} \\ \frac{1}{s} & \frac{1}{s} \end{bmatrix}$$

$$= \begin{bmatrix} \frac{s+1}{s+2} + \frac{1}{s} & \frac{1}{s+2} + \frac{1}{s} \\ \frac{1}{s+2} + \frac{1}{s} & \frac{s+1}{s+2} + \frac{1}{s} \end{bmatrix}$$

16.24 INTERCONNECTION THROUGH TRANSFORMERS

If the 2-port networks to be interconnected do not satisfy the Brune's tests, an ideal transformer of 1 : 1 turns ratio is connected at one of the ports as shown in Fig. 16.42. This transformer provides the necessary isolation and ensures that 2-port network does not alter to any extent the character of the other after their interconnection.

(*a*) Series interconnection employing an ideal transformer

(*b*) Parallel interconnection employing an ideal transformer

(*c*) Series-parallel interconnection employing an ideal transformer

(*d*) Parallel-series interconnection using an ideal transformer.

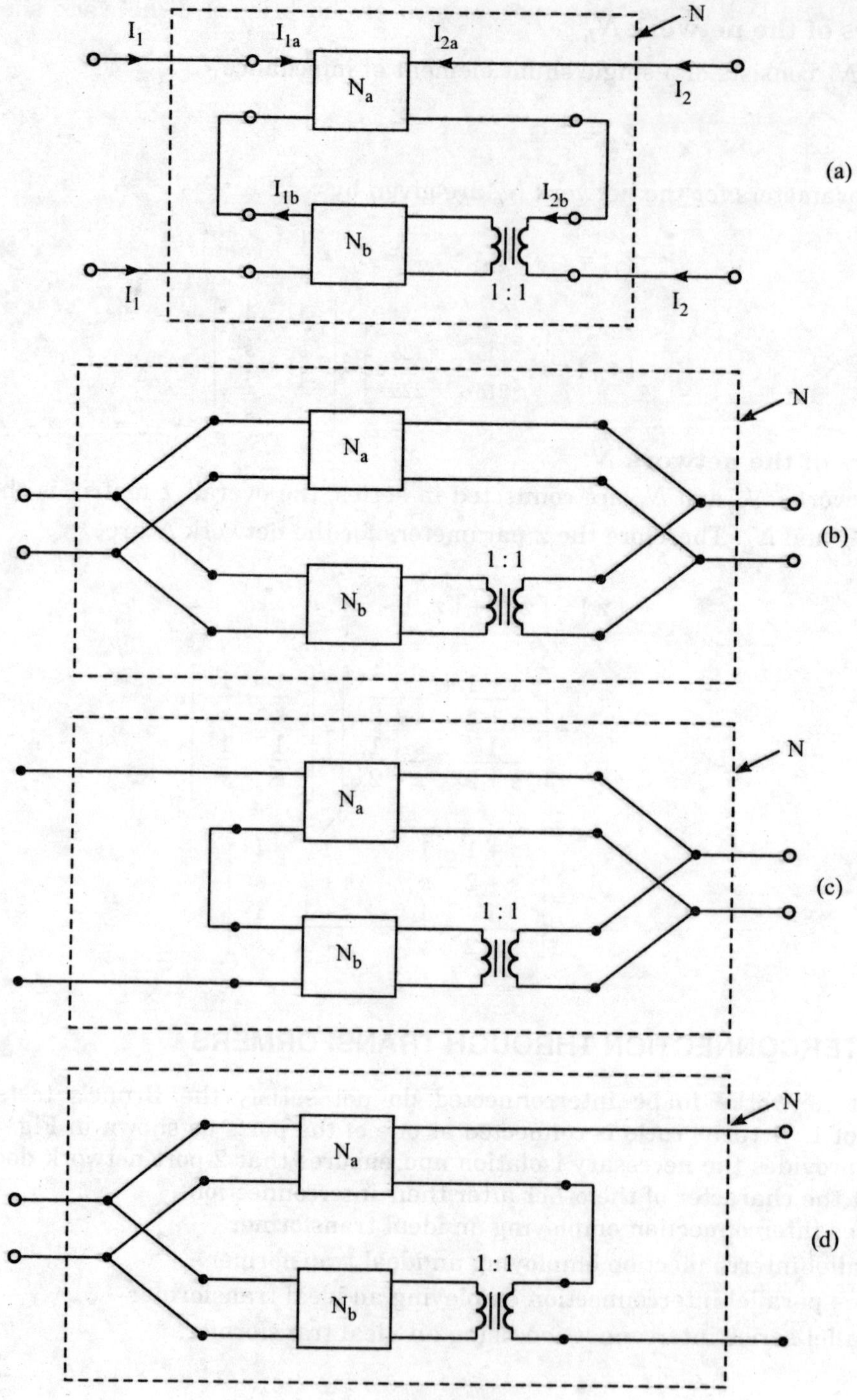
N
I_1
I_{1a}
I_{2a}
N_a
I_2
(a)
I_{1b}
I_{2b}
N_b
1 : 1
I_1
I_2
N
N_a
(b)
1 : 1
N_b
N
N_a
(c)
1 : 1
N_b
N
N_a
(d)
1 : 1
N_b

FIG. 16.42

■ **EXAMPLE 16.26** Determine the y parameters of the bridged T network shown in Fig. 16.43a

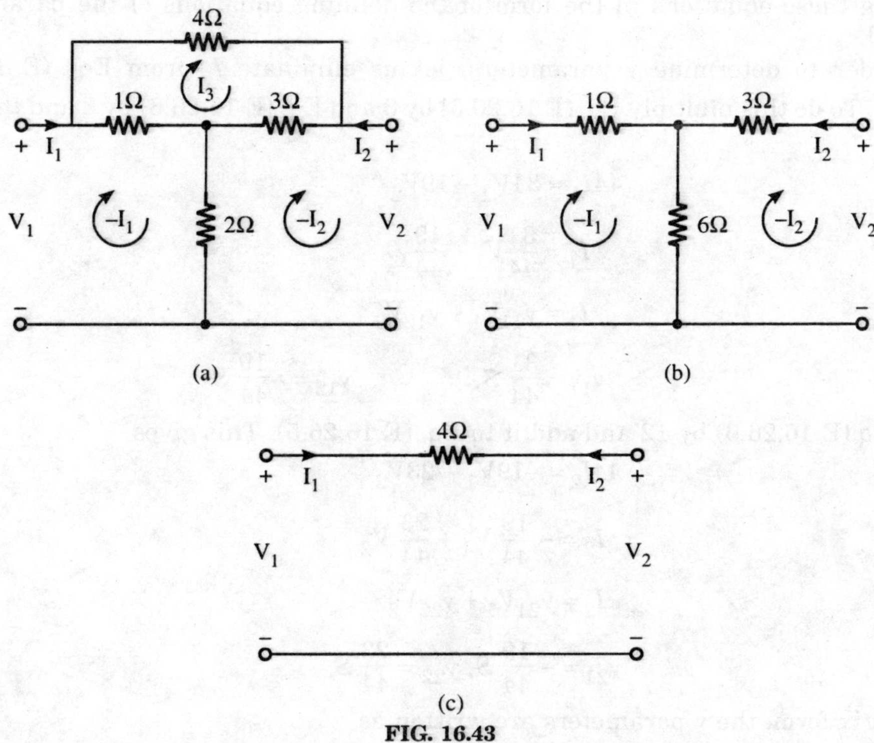

FIG. 16.43

SOLUTION : The mesh equations for the given network may be written by inspection as

$$\begin{bmatrix} 3 & -2 & -1 \\ -2 & 5 & -3 \\ -1 & -3 & 8 \end{bmatrix} \begin{bmatrix} I_1 \\ -I_2 \\ I_3 \end{bmatrix} = \begin{bmatrix} V_1 \\ -V_2 \\ 0 \end{bmatrix}$$

In the expanded form these mesh equations are

$$3I_1 + 2I_2 - I_3 = V_1 \tag{E 16.26.1}$$

$$-2I_1 - 5I_2 - 3I_3 = -V_2 \tag{E 16.26.2}$$

$$-I_1 + 3I_2 + 8I_3 = 0 \tag{E 16.26.3}$$

Addition of these three equations gives

$$4I_3 = V_1 - V_2 \tag{E 16.26.4}$$

Multiplying Eq. (E 16.26.1) by 4 and substituting the value of $4I_3$ from Eq. (E 16.26.4) we obtain

$$12I_1 + 8I_2 - (V_1 - V_2) = 4V_1$$

$$12I_1 + 8I_2 = 5V_1 - V_2 \tag{E 16.26.5}$$

Combination of Eqs. (E 16.26.3) and (E 16.26.4) gives

$$-I_1 + 3I_2 + 2\,(V_1 - V_2) = 0$$

$$-I_1 + 3I_2 = -2V_1 + 2V_2 \quad \text{(E 16.26.6)}$$

Eqs. (E 16.26.5) and (E 16.26.6) may be used to determine all the 2-port parameters by rearranging these equations in the form of the defining equations of the parameters to be determined.

In order to determine y parameters, let us eliminate I_2 from Eqs (E 16.26.5) and (E 16.26.6). To do this multiply Eq. (E 16.26.5) by 3 and Eq. (E 16.26.6) by 8 and then subtract. This gives

$$44I_1 = 31V_1 - 19V_2$$

$$I_1 = \frac{31}{44}V_1 - \frac{19}{44}V_2$$

Also $$I_1 = y_{11}V_1 + y_{12}V_2$$

therefore, $$y_{11} = \frac{31}{44}S, \qquad y_{12} = -\frac{19}{44}S$$

Multiply Eq.(E 16.26.6) by 12 and add it to Eq. (E 16.26.5). This gives

$$44I_2 = -19V_1 + 23V_2$$

$$I_2 = -\frac{19}{44}V_1 + \frac{23}{44}V_2$$

Also $$I_2 = y_{21}V_1 + y_{22}V_2$$

therefore, $$y_{21} = -\frac{19}{44}S, \; y_{22} = \frac{23}{44}S$$

In the matrix form, the y parameters are written as

$$[y] = \begin{bmatrix} \frac{31}{44} & -\frac{19}{44} \\ -\frac{19}{44} & \frac{23}{44} \end{bmatrix} S$$

Alternative Method

The y parameters can also be determined by considering the given network as a parallel combination of two networks shown in Figs. 16.43b and 16.43c.

The mesh equations for the network in Fig. 16.43b are written by inspection as

$$\begin{bmatrix} 3 & -2 \\ -2 & 5 \end{bmatrix} \begin{bmatrix} I_1 \\ -I_2 \end{bmatrix} = \begin{bmatrix} V_1 \\ -V_2 \end{bmatrix}$$

In the expanded form

$$3I_1 + 2I_2 = V_1 \quad \text{(E 16.26.7)}$$

$$-2I_1 - 5I_2 = -V_2 \quad \text{(E 16.26.8)}$$

Multiplying Eq. (E 16.26.7) by 5 and Eq. (E 16.26.8) by 2 and adding the results we get

$$11I_1 = 5V_1 - 2V_2$$

$$I_1 = \frac{5}{11}V_1 - \frac{2}{11}V_2$$

Also $$I_1 = y_{11}V_1 + y_{12}V_2$$

therefore, $$y_{11} = \frac{5}{11} S, \; y_{12} = -\frac{2}{11} S$$

Now multiplying Eq. (E 16.26.7) by 2 and Eq. (E 16.26.8) by 3 and adding the results we get

$$-11I_2 = 2V_1 - 3V_2$$

$$I_2 = -\frac{2}{11} V_1 + \frac{3}{11} V_2$$

Also, $$I_2 = y_{21}V_1 + y_{22}V_2$$

therefore, $$y_{21} = -\frac{2}{11} S, \; y_{22} = \frac{3}{11} S$$

For the network in Fig. 16.43c, we have

$$V_1 - 4I_1 = V_2$$

and $$V_2 - 4I_2 = V_1$$

From these equations

$$I_1 = \frac{1}{4} V_1 - \frac{1}{4} V_2$$

$$I_2 = -\frac{1}{4} V_1 + \frac{1}{4} V_2$$

Therefore the y matrix of Fig. 16.43b

$$[\mathbf{y}_b] = \begin{bmatrix} \frac{1}{4} & -\frac{1}{4} \\ -\frac{1}{4} & \frac{1}{4} \end{bmatrix}$$

The y matrix of Fig. 16.43a

$$[\mathbf{y}_a] = \begin{bmatrix} \frac{5}{11} & -\frac{2}{11} \\ -\frac{2}{11} & \frac{3}{11} \end{bmatrix}$$

Since the network in Fig. 16.43a is a parallel combination of the networks in Figs. 16.43*b* and 16.43c

$$[\mathbf{y}] = [\mathbf{y_a}] + [\mathbf{y_b}]$$

$$= \begin{bmatrix} \frac{5}{11} & -\frac{2}{11} \\ -\frac{2}{11} & \frac{3}{11} \end{bmatrix} + \begin{bmatrix} \frac{1}{4} & -\frac{1}{4} \\ -\frac{1}{4} & \frac{1}{4} \end{bmatrix}$$

$$= \begin{bmatrix} \frac{5}{11} + \frac{1}{4} & -\frac{2}{11} - \frac{1}{4} \\ -\frac{2}{11} - \frac{1}{4} & \frac{3}{11} + \frac{1}{4} \end{bmatrix} = \begin{bmatrix} \frac{31}{44} & -\frac{19}{44} \\ -\frac{19}{44} & \frac{23}{44} \end{bmatrix}$$

This result agrees with the one found earlier.

EXAMPLE 16.27 Determine the y parameters for the twin T network shown in Fig. 16.44

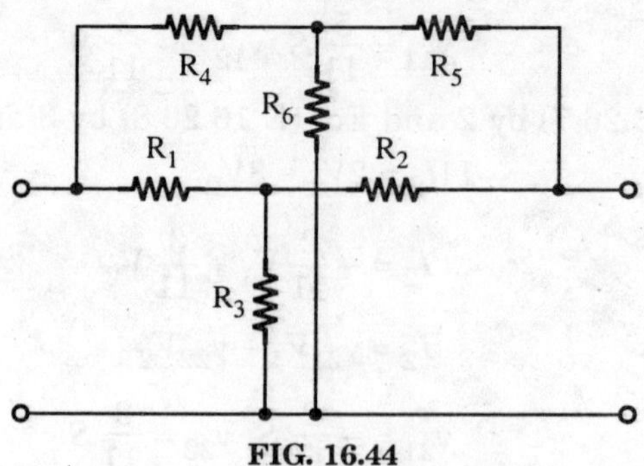

FIG. 16.44

SOLUTION : The given network can be treated as a parallel interconnection of two T networks N_a and N_b. Network N_a is composed of R_1, R_2 and R_3. Similarly, network N_b is composed of R_4, R_5 and R_6.

From Example 16.9 the y parameters of the T networks are given by

$$y_{11a} = \frac{R_2 + R_3}{\Delta R_a}, \; y_{12a} = y_{21a} = \frac{-R_3}{\Delta R_a}, \; y_{22a} = \frac{R_1 + R_3}{\Delta R_a}$$

$$\Delta R_a = R_1R_2 + R_2R_3 + R_3R_1$$

$$y_{11b} = \frac{R_4 + R_5}{\Delta R_b}, \; y_{12b} = y_{21b} = \frac{-R_6}{\Delta R_b}, \; y_{22b} = \frac{R_4 + R_6}{\Delta R_b}$$

$$\Delta R_b = R_4R_5 + R_5R_6 + R_6R_4$$

Hence the overall y parameters of the parallel combination of N_a and N_b are given by

$$[\mathbf{y}] = [\mathbf{y}_a] + [\mathbf{y}_b] = \begin{bmatrix} y_{11a} + y_{11b} & y_{12a} + y_{12b} \\ y_{21a} + y_{21b} & y_{22a} + y_{22b} \end{bmatrix}$$

$$= \begin{bmatrix} \dfrac{R_2 + R_3}{\Delta R_a} + \dfrac{R_5 + R_6}{\Delta R_b} & -\dfrac{R_3}{\Delta R_a} - \dfrac{R_6}{\Delta R_b} \\ -\dfrac{R_3}{\Delta R_a} - \dfrac{R_6}{\Delta R_b} & \dfrac{R_1 + R_3}{\Delta R_a} + \dfrac{R_4 + R_6}{\Delta R_b} \end{bmatrix}$$

EXAMPLE 16.28 Test results for a 2-port network are follows :

(*i*) Port 2 open circuited and a voltage of $10/\underline{/\ 0°}$ V applied to port 1 resulted in $\mathbf{I}_1 = 2\underline{/0°}\ A$ and $\mathbf{V}_2 = 4\ \underline{/\ 0°}$ V

(*ii*) Port 1 open circuited and a voltage of $6\ \underline{/\ 0°}$ V applied to port 2 resulted in $\mathbf{I}_2 = 2\ \underline{/\ 0°}$ A and $\mathbf{V}_1 = 3\ \underline{/\ 0°}$ V.

(*a*) Determine the z parameters of network and write the 2-port equations.

(*b*) What will be the voltage across a 2 Ω resistor connected at the output terminals if a $10\ \underline{/\ 0°}$ V source is connected across port 1 ?

SOLUTION :

(*a*)
$$\mathbf{z}_{11} \triangleq \left.\frac{\mathbf{V}_1}{\mathbf{I}_1}\right|_{I_2=0} = \frac{10}{2} = 5\ \Omega$$

$$\mathbf{z}_{21} \triangleq \left.\frac{\mathbf{V}_2}{\mathbf{I}_1}\right|_{I_2=0} = \frac{4}{2} = 2\ \Omega$$

$$\mathbf{z}_{12} \triangleq \left.\frac{\mathbf{V}_1}{\mathbf{I}_2}\right|_{I_1=0} = \frac{3}{2} = 1.5\ \Omega$$

$$\mathbf{z}_{22} \triangleq \left.\frac{\mathbf{V}_2}{\mathbf{I}_2}\right|_{I_1=0} = \frac{6}{2} = 3\ \Omega$$

The port equations in terms of z parameters are

$$V_1 = z_{11}I_1 + z_{12}I_2$$

$$V_2 = z_{21}I_1 + z_{22}I_2$$

Substitution of the values of the z parameters gives

$$V_1 = 5I_1 + 1.5I_2$$

$$V_2 = 2I_1 + 3I_2$$

(*b*) Let the source impedance be zero.

therefore, $\mathbf{V}_1 = 10\ \underline{/0°}\,\mathbf{V}$, $R_L = 2\ \Omega$

$$\mathbf{V}_2 = -I_2R_L = -2I_2$$

With these values, the port equations become

$$10 = 5I_1 + 1.5I_2$$

$$-2I_2 = 2I_1 + 3I_2$$

Solving these equations we get

$$I_2 = -\frac{10}{11}\,\mathbf{A}$$

Therefore the voltage across the 2 Ω resistor

$$= -I_2R_L = \frac{10}{11} \times 2 = \frac{20}{11}\,\mathbf{V}$$

16.25 LADDER NETWORK

Figure 16.45 shows a ladder network. Let $Z_1, Z_3, \ldots, Z_{n-1}$ be the series impedances and $Y_2, Y_4, \ldots, Y_n$ be the shunt admittances. We shall calculate the input impedance Z_{in} of the ladder.

From Fig. 16.45, we have

$$Z_{\text{In}} = Z_1 + \frac{1}{Y_1'}$$

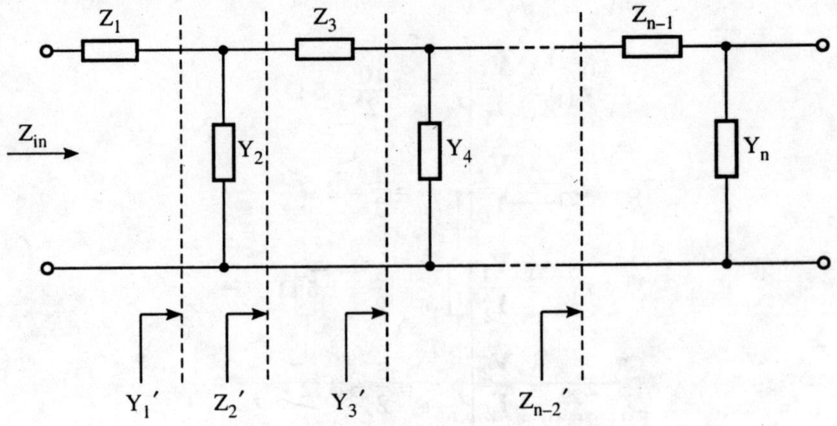

FIG. 16.45

$$Y_1' = Y_2 + \frac{1}{Z_2'}$$

$$Z_2' = Z_3 + \frac{1}{Y_3'}$$

$$\cdots\cdots\cdots\cdots\cdots\cdots$$

$$Z'_{n-2} = Z_{n-1} + \frac{1}{Y_n}$$

Combination of these equation gives

$$Z_{in} = Z_1 + \cfrac{1}{Y_2 + \cfrac{1}{Z_3 + \cfrac{1}{Y_4 + \cfrac{1}{\vdots \; + \cfrac{1}{Z_{n-1} + \cfrac{1}{Y_n}}}}}}$$

This relation shows that the input impedance of a ladder can be written in the form of a continued fraction.

To calculate the currents and voltages for the ladder, we start at the output end work back to the input end. Let us illustrate the procedure for a four branch ladder shown in Fig. 16.4b. We have

$$I_3 = Y_4 V_4$$

$$V_2 = Z_3 I_3 + V_4$$

$$I_1 = Y_2 V_2 + I_3$$

$$V_1 = Z_1 I_1 + V_2$$

FIG. 16.46 A four-branch ladder network

16.26 LATTICE NETWORKS

A lattice network is shown in Fig. 16.47a. This network is actually a bridge network as shown in Fig. 16.47b. The arms Z_a and Z_d are called **series arms,** and Z_b and Z_c are called **diagonal arms**.

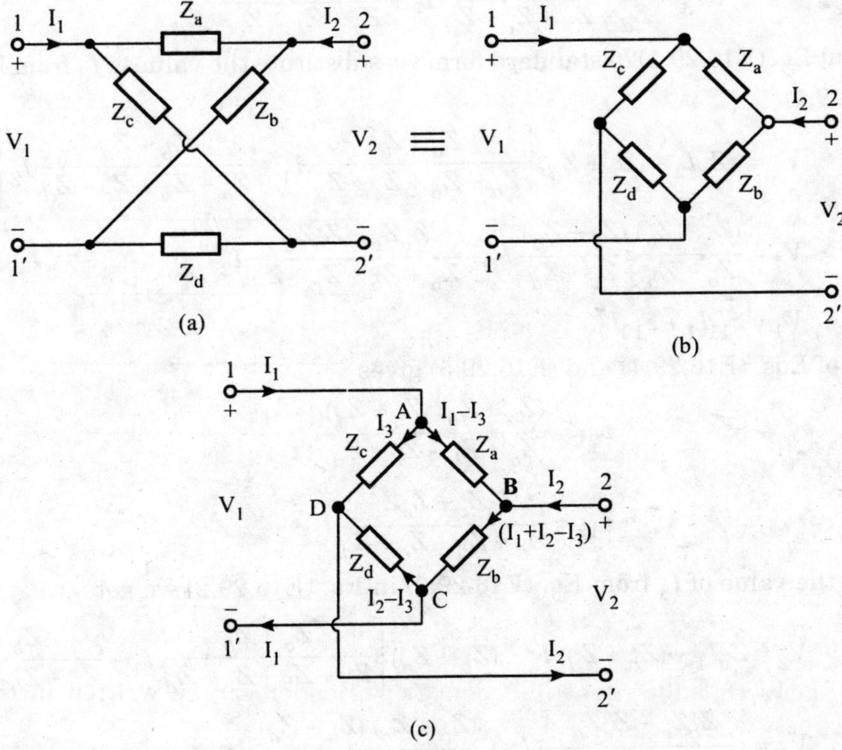

FIG. 16.47 (A) A lattice network (b) Its bridge form

For a **symmetrical lattice**

$$Z_a = Z_d,\ Z_b = Z_c$$

■ **EXAMPLE 16.29** Determine the z parameters for the lattice network shown in Fig. 16.47a.

SOLUTION : The lattice network given in Fig. 16.47a is redrawn as a bridge network as shown in Fig. 16.47c.

Let the current through Z_c be I_3. By the application of KCL at nodes A, B, C and D, the currents through Z_a, Z_b and Z_d are found. They are shown in Fig. 16.47c.

By KVL in mesh ADC1′1A

$$-Z_c I_3 + Z_d (I_2 - I_3) + V_1 = 0$$

or
$$V_1 = -Z_d I_2 + (Z_c + Z_d) I_3 \tag{E16.29.1}$$

By KVL in mesh BCD2′2B

$$-Z_b (I_1 + I_2 - I_3) - Z_d (I_2 - I_3) + V_2 = 0$$

or
$$V_2 = Z_b I_1 + (Z_b + Z_d) I_2 - (Z_b + Z_d) I_3 \tag{E16.29.2}$$

By KVL in mesh ABCDA

$$-Z_a (I_1 - I_3) - Z_b (I_1 + I_2 - I_3) - Z_d (I_2 - I_3) + Z_c I_3 = 0$$

$$(Z_a + Z_b + Z_c + Z_d) I_3 = (Z_a + Z_b) I_1 + (Z_b + Z_d) I_2$$

or
$$I_3 = \frac{Z_a + Z_b}{Z_a + Z_b + Z_c + Z_d} I_1 + \frac{Z_b + Z_d}{Z_a + Z_b + Z_c + Z_d} \tag{E16.29.3}$$

In order to put Eq. (E16.29.1) in standard form we substitute the value of I_3 from Eq. (E16.29.3) in Eq. (E16.29.1)

$$V_1 = -Z_d I_2 + (Z_c + Z_d) \left[\frac{Z_a + Z_b}{Z_a + Z_b + Z_c + Z_d} I_1 + \frac{Z_b + Z_d}{Z_a + Z_b + Z_c + Z_d} I_2 \right]$$

or
$$V_1 = \frac{(Z_a + Z_b)(Z_c + Z_d)}{Z_a + Z_b + Z_c + Z_d} I_1 + \frac{Z_b Z_c - Z_a Z_d}{Z_a + Z_b + Z_c + Z_d} I_2 \tag{E16.29.4}$$

Also,
$$V_1 = z_{11} I_1 + z_{12} I_2 \tag{E16.29.5}$$

Comparison of Eqs. (E16.29.4) and (E16.29.5) gives

$$z_{21} = \frac{(Z_a + Z_b)(Z_c + Z_d)}{Z_a + Z_b + Z_c + Z_d} \tag{E16.29.6}$$

and
$$z_{12} = \frac{Z_b Z_c - Z_a Z_d}{Z_a + Z_b + Z_c + Z_d} \tag{E16.29.7}$$

Substituting the value of I_3 from Eq. (E16.29.3) in Eq. (E16.29.2) we get

$$V_2 = Z_b I_1 + (Z_b + Z_d) I_2 - (Z_b + Z_d) \left[\frac{Z_a + Z_b}{Z_a + Z_b + Z_c + Z_d} I_1 + \frac{Z_b + Z_d}{Z_a + Z_b + Z_c + Z_d} I_2 \right]$$

or
$$V_2 = \frac{Z_b Z_c - Z_a Z_d}{Z_a + Z_b + Z_c + Z_d} I_1 + \frac{(Z_a + Z_c)(Z_b + Z_d)}{Z_a + Z_b + Z_c + Z_d} I_2 \tag{E16.29.8}$$

Also,
$$V_2 = z_{21} I_1 + z_{22} I_2 \tag{E16.29.9}$$

Comparison of Eqs. (E16.29.8) and (E16.29.9) gives

$$z_{21} = \frac{Z_b Z_c - Z_a Z_d}{Z_a + Z_b + Z_c + Z_d} \quad \text{(E16.29.10)}$$

$$z_{22} = \frac{(Z_a + Z_c)(Z_b + Z_d)}{Z_a + Z_b + Z_c + Z_d} \quad \text{(E16.29.11)}$$

For a **symmetrical lattice network,**

$$Z_a = Z_d \text{ and } Z_b = Z_c$$

therefore,

$$z_{11} = z_{22} = \frac{Z_a + Z_b}{2} \quad \text{(E16.29.12)}$$

$$z_{12} = z_{21} = \frac{Z_b - Z_a}{2} \quad \text{(E16.29.13)}$$

and

$$Z_a = z_{11} - z_{12} \quad \text{(E16.29.14)}$$

$$Z_b = z_{11} + z_{12} \quad \text{(E16.29.15)}$$

■ **EXAMPLE 16.30** Find the z, y, h and ABCD parameters of the lattice network shown in Fig. 16.48a

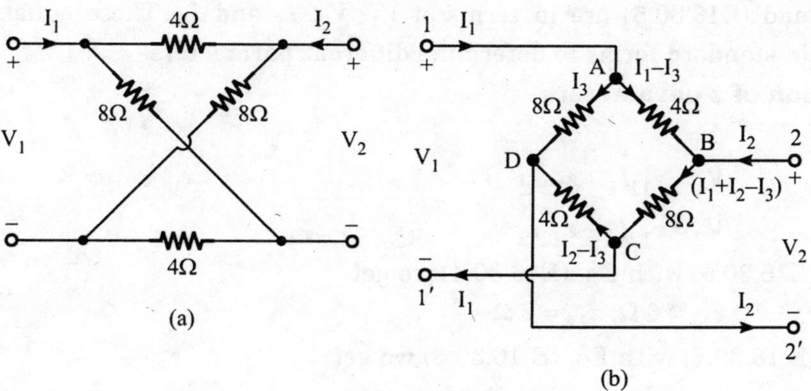

FIG. 16.48

SOLUTION The lattice network given in Fig. 16.48a is redrawn as a bridge network shown in Fig. 16.48b.

Let the current through 8 Ω resistor between A and D in Fig. 16.48b be I_3. The currents through other resistors are marked by the application of KCL at nodes A, B, C and D. We shall write KVL equations in closed paths in terms of V_1, V_2, I_1, I_2 and I_3. Since we have introduced an additional current I_3, it will be eliminated from these equations to get two KVL equations in terms of V_1, V_2, I_1 and I_2. These two equations will be arranged to put in standard forms to determine different parameters.

By KVL in mesh ADC1′1A,

$$-8I_3 + 4(I_2 - I_3) + V_1 = 0$$

or

$$V_1 = -4I_2 + 12I_3 \quad \text{(E16.30.1)}$$

By KVL in mesh BCD2′2B,

$$-8\,(I_1 + I_2 - I_3) - 4\,(I_2 - I_3) + V_2 = 0$$

or $$V_2 = 8I_1 + 12I_2 - 12I_3 \qquad \text{(E16.30.2)}$$

By KVL in mesh ABCDA

$$-4\,(I_1 - I_3) - 8\,(I_1 + I_2 - I_3) - 4\,(I_2 - I_3) + 8I_3 = 0$$

$$24I_3 = 12I_1 + 12I_2$$

or $$I_3 = \frac{1}{2}\,(I_1 + I_2) \qquad \text{(E16.30.3)}$$

Substituting the value of I_3 from Eq. (E16.30.3) in Eq. (E16.30.1) we get

$$V_1 = -4I_2 + 6\,(I_1 + I_2)$$

or $$V_1 = 6I_1 + 2I_2 \qquad \text{(E16.30.4)}$$

Substituting the value of I_3 from Eq. (E16.30.3) in Eq. (E16.30.2) we get

$$V_2 = 8I_1 + 12I_2 - 6\,(I_1 + I_2)$$

$$V_2 = 2I_1 + 6I_2 \qquad \text{(E16.30.5)}$$

Eqs. (E16.30.4) and (E16.30.5) are in terms of V_1, V_2, I_1 and I_2. These equations will be arranged to put in standard forms to determine different parameters.

(*a*) **Determination of z parameters**

We have

$$V_1 = z_{11}I_1 + z_{12}I_2 \qquad \text{(E16.30.6)}$$

$$V_2 = z_{12}I_1 + z_{22}I_2 \qquad \text{(E16.30.7)}$$

Comparing Eq. (E16.30.6) with Eq. (E16.30.4) we get

$$z_{11} = 6\ \Omega,\ z_{12} = 2\ \Omega$$

Comparing Eq. (E 16.30.7) with Eq. (E 16.30.5) we get

$$z_{12} = 2\ \Omega\ ,\ z_{22} = 6\ \Omega$$

(*b*) **Determination of y parameters**

We have

$$I_1 = y_{11}V_1 + y_{12}V_2 \qquad \text{(E16.30.8)}$$

$$I_2 = y_{21}V_1 + y_{22}V_2 \qquad \text{(E16.30.9)}$$

In order to find *y* parameters Eqs. (E16.30.4) and (E16.30.5) will be arranged in the form of Eqs. (E16.30.8) and (E16.30.9). For eliminating I_2 from Eqs. (E16.30.4) and (E16.30.5), substitute the value of I_2 from Eq. (E16.30.4) in Eq. (E16.30.5)

$$V_2 = 2I_1 + 3\,(V_1 - 6I_1)$$

$$I_1 = \frac{3}{16}\,V_1 - \frac{1}{16}\,V_2 \qquad \text{(E16.30.10)}$$

Comparing Eq. (E16.30.10) with Eq. (E16.30.8) we get

$$y_{11} = \frac{3}{16}, y_{12} = -\frac{1}{16}$$

Again from Eqs. (E16.30.4) and (E16.30.5)

$$V_1 - 3V_2 = 2I_2 - 18I_2$$

$$16I_2 = -V_1 + 3V_2$$

$$I_2 = -\frac{1}{16} V_1 + \frac{3}{16} V_2 \quad \text{(E16.30.11)}$$

Comparing Eq. (E16.30.11) with Eq. (E16.30.9) we get

$$y_{21} = -\frac{1}{16}, y_{22} = \frac{3}{16}$$

(*c*) **Determination of *h* parameters**

We have

$$V_1 = h_{11}I_1 + h_{12}V_2 \quad \text{(E16.30.12)}$$

$$I_2 = h_{21}I_1 + h_{22}V_2 \quad \text{(E16.30.13)}$$

From Eqs. (E16.30.4) and (E16.30.5)

$$V_1 = 6I_1 + \frac{1}{3}(V_2 - 2I_1)$$

or

$$V_1 = \frac{16}{3} I_1 + \frac{1}{3} V_2$$

therefore,

$$h_{11} = \frac{16}{3}, h_{12} = \frac{1}{3}$$

From Eq. (E16.30.5) we have

$$6I_2 = -2I_1 + V_2$$

$$I_2 = -\frac{1}{3} I_1 + \frac{1}{6} V_2$$

therefore,

$$h_{21} = -\frac{1}{3}, h_{22} = \frac{1}{6}$$

(*d*) **Determination of A, B, C. D parameters**

We have

$$V_1 = AV_2 - BI_2 \quad \text{(E16.30.14)}$$

$$I_1 = CV_2 - DI_2 \quad \text{(E16.30.15)}$$

From Eqs. (E16.30.4) and (E16.30.5)

$$V_1 = 3(V_2 - 6I_2) + 2I_2$$

or

$$V_1 = 3V_2 - 16I_2 \quad \text{(E16.30.16)}$$

Comparison of Eq. (E16.30.16) with Eq. (E16.30.4) gives

$$A = 3, B = 16$$

From Eq. (E16.30.5)

$$I_1 = \frac{1}{2}V_2 - 3I_2 \tag{E16.30.17}$$

Comparison of Eq. (E16.30.17) with Eq. (16.30.15) gives $C = \frac{1}{2}$, $D = 3$.

EXAMPLE 16.31 Obtain the lattice equivalent of a symmetrical T network shown in Fig. 16.49a.

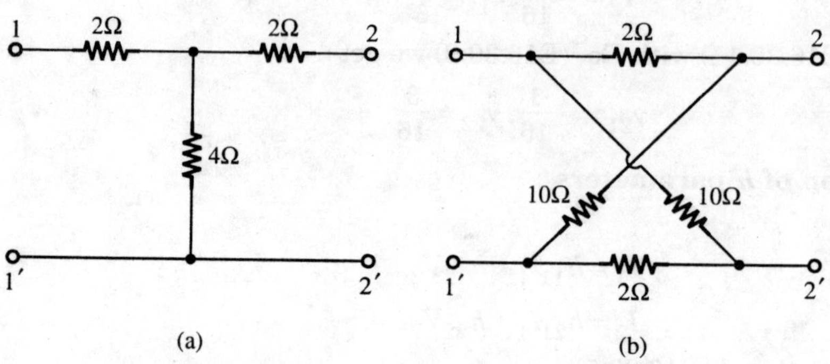

FIG. 16.49 (a) *T* network (b) Lattice network

SOLUTION : A given 2-port network can be realised as a symmetric lattice if it is symmetric and reciprocal. The z parameters of the T network are given by

$$z_{11} = 2 + 4 = 6\ \Omega,\ z_{12} = z_{21} = 4\ \Omega$$

$$z_{22} = 2 + 4 = 6\ \Omega$$

Since $z_{11} = z_{22}$ and $z_{12} = z_{21}$, the given network is symmetrical and reciprocal. The parameters of the lattice network are given by

$$Z_a = z_{11} - z_{12} = 6 - 4 = 2\ \Omega$$

$$Z_b = z_{11} + z_{12} = 6 + 4 = 10\ \Omega$$

The lattice equivalent network is shown in Fig. 16.49b.

EXAMPLE 16.32 Obtain the lattice equivalent of a symmetrical Π network shown in Fig. 16.50a. The z parameters of the given network are found in Example 16.1 as

$$z_{11} = z_{22} = 6\ \Omega$$

$$z_{12} = z_{21} = 4\ \Omega$$

(a) (b)

FIG. 16.50

Hence the parameters of the lattice network are

$$Z_a = z_{11} - z_{12} = 6 - 4 = 2\ \Omega$$

$$Z_b = z_{11} + z_{12} = 6 + 10 = 10\ \Omega$$

The lattice equivalent network is shown in Fig. 16.50b

16.27 INDEFINITE ADMITTANCE MATRIX

Consider a network having n terminals 1, 2, ..., n. Let us choose the reference (datum) point of zero potential arbitrarily **outside** the n-terminal network. The voltage sources connected between n terminals and this datum are $V_1, V_2, \ldots, V_n$ as shown in Fig. 16.51.

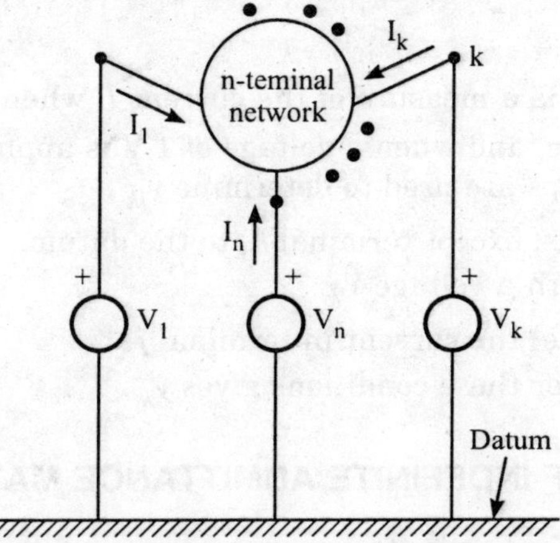

FIG. 16.51 An outside datum for n-terminal network

The currents entering the terminals 1, 2, ..., n are $I_1, I_2, \ldots, I_n$ respectively. Superposition principle is applicable to linear networks and each current is equal to the sum of the partial contributions due to the n sources. That is,

$$\left.\begin{aligned} I_1 &= y_{11}V_1 + y_{12}V_2 + \ldots + y_{1n}V_n \\ I_2 &= y_{21}V_1 + y_{22}V_2 + \ldots + y_{2n}V_n \\ &\vdots \\ I_n &= y_{n1}V_1 + y_{n2}V_2 + \ldots + y_{nn}V_n \end{aligned}\right\} \tag{16.27.1}$$

Although these equations resemble the short-circuit parameters, there is a distinct basic difference. In this case, the datum (reference) node is **arbitrary** and **outside** the network. Eq. (16.27.1) can be written in matrix form as

$$\begin{bmatrix} I_1 \\ I_2 \\ \vdots \\ I_n \end{bmatrix} = \begin{bmatrix} y_{11} & y_{12} \cdots y_{1n} \\ y_{21} & y_{22} \cdots y_{2n} \\ \vdots & \\ y_{n1} & y_{n2} \cdots y_{nn} \end{bmatrix} \begin{bmatrix} V_1 \\ V_2 \\ \vdots \\ V_n \end{bmatrix}$$

In compact form,

$$\mathbf{I} = \mathbf{Y}_i\ \mathbf{V} \tag{16.27.2}$$

where
$$\mathbf{Y}_i = = \begin{bmatrix} y_{11} & y_{12} \cdots y_{1n} \\ y_{21} & y_{22} \cdots y_{2n} \\ \vdots & \\ y_{n1} & y_{n2} \cdots y_{nn} \end{bmatrix} \tag{16.27.3}$$

The matrix $\mathbf{Y}_i$ is called the **indefinite admittance matrix** (IAM). The word "indefinite" indicates that the reference node for voltages is **arbitrary** and **outside** the network.

A typical element y_{jk} of $\mathbf{Y}_i$ is defined as follows :

$$y_{jk} = \frac{I_j}{V_k}\bigg|_{V_1 = V_2 \ldots = V_{k-1} = V_{k+1} = \ldots = V_n = 0} \tag{16.27.4}$$

for
$$j = k \text{ or } j \neq k.$$

In words, the parameter y_{jk} is a measure of the current I_j when all but the kth terminal are shorted to the reference node, and when a voltage of 1 V is applied at terminal k.

Thus, the following steps are used to determine y_{jk} :

1. Connect all terminals, except terminal k, to the datum.
2. Excite terminal k with a voltage V_k.
3. Measure (or calculate) the current in terminal j
4. The ratio I_j / V_k under these conditions gives y_{jk}.

16.28 PROPERTIES OF INDEFINITE ADMITTANCE MATRIX Y_i

The main properties of indefinite admittance matrix are as follows :

Property 1.

The sum of the elements of every column of an indefinite admittance matrix is zero.

Proof

If we add all the equations in Eq. (16.27.1), we get

$$I_1 + I_2 + \ldots + I_n = 0 = (y_{11} + y_{21} + \ldots + y_{n1})\, V_1 + (y_{12} + y_{22} + \ldots + y_{n2})\, V_2 + \ldots + (y_{1n} + y_{2n} + \ldots + y_{nn})\, V_n \tag{16.28.1}$$

This must be valid for any arbitrary $V_1, V_2, \ldots, V_n$. In particular, for $V_1 \neq 0$,

$V_2 = V_3 = \ldots = V_n = 0$. Next, for $V_2 \neq 0$,

$V_1 = V_3 = \ldots = 0$, and so on. Therefore,

we conclude that

$$\sum_{j=1}^{n} y_{j1} = \sum_{j=1}^{n} y_{j2} = \ldots = \sum_{j=1}^{n} y_{jn} = 0 \tag{16.28.2}$$

That is, the sum of the admittances of every column in $\mathbf{Y}_i$ is zero.

Property 2.

The sum of the elements of every row of an indefinite admittance matrix is zero

Proof

Suppose that

$$V_1 = V_2 = \dots = V_n = V_0 \neq .$$

Since all the nodes are at the same potential, no currents will flow in the terminals, that is

$$\begin{aligned} 0 &= y_{11}V_0 + y_{12}V_0 + \dots + y_{1n}V_0 \\ 0 &= y_{21}V_0 + y_{22}V_0 + \dots + y_{2n}V_0 \\ &\vdots \\ 0 &= y_{n1}V_0 + y_{n2}V_0 + \dots + y_{nn}V_0 \end{aligned}$$

Since $V_0 \neq 0$, we get

$$\left.\begin{aligned} y_{11} + y_{12} + \dots + y_{1n} &= 0 \\ y_{21} + y_{22} + \dots + y_{2n} &= 0 \\ &\vdots \\ y_{n1} + y_{n2} + \dots + y_{nn} &= 0 \end{aligned}\right\} \quad (16.28.3)$$

That is, the sum of the admittances of every row in $\mathbf{Y}_i$ is zero.

The significance of properties 1 and 2 is that any one row and its corresponding column of an indefinite admittance matrix are redundant. Thus, we need at most one parameter to characterize a 2-terminal circuit, for parameters for a 3-terminal circuit, and $(n - 1) \times (n - 1)$ parameters for an n-terminal circuit.

Property 3

If an isolated node is added to an n-terminal network, the resultant indefinite admittance matrix is $(n + 1) \times (n + 1)$ and is obtained by adding a row and column of zeros to the $n \times n$ matrix of the original circuit. The position of the new row and column corresponds to the number assigned to the new node.

Suppose that we take a 3-terminal network and add to it an isolated node 4 as shown in Fig. 16.52.

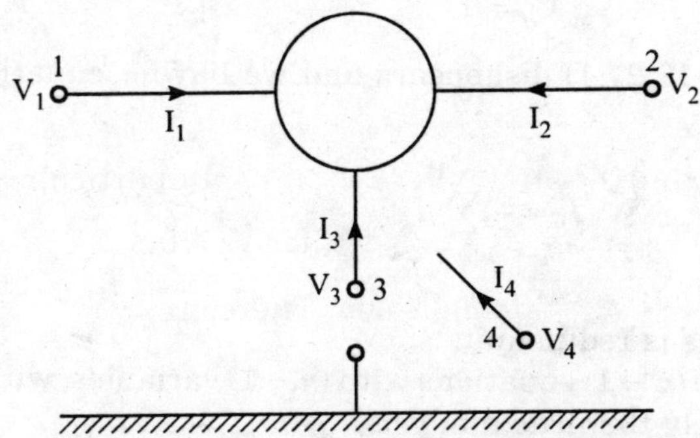

FIG. 16.52 Addition of an isolated node to a 3-terminal network

Since this node is not connected to any other, we can write the following equations :

$$I_1 = y_{11}V_1 + y_{12}V_2 + y_{13}V_3 + 0V_4$$

$$I_2 = y_{21}V_1 + y_{22}V_2 + y_{23}V_3 + 0V_4$$

$$I_3 = y_{31}V_1 + y_{32}V_2 + y_{33}V_3 + 0V_4$$
$$I_4 = 0V_1 + 0V_2 + 0V_3 + 0V_4$$

Thus, the indefinite admittance matrix of the 4-terminal network is simply that of the 3-terminal network appended by a row and column of zeros.

The property 3 of the indefinite admittance matrix seems trivial, but it is the basis on which we may regard all interconnections of multiterminal circuits as parallel connections.

If two n-terminal circuits are connected in parallel, the indefinite admittance matrix of the overall circuit is the sum of the admittance matrices of the two circuits.

Property 4.

As a result of properties 1 and 2. $\mathbf{Y}_i$ is a singular matrix. That is, its determinant is zero. Its rank is $(n - 1)$, which means that only $(n - 1)$ of its rows-columns are independent. Its nth row is dependent upon the remaining rows. From KCL for n-terminal network, $\sum_{k=1}^{n} I_k = 0$, that is,

$$I_1 + I_2 + \ldots + I_n = 0 \tag{16.28.4}$$
$$I_n = -(I_1 + I_2 + \ldots + I_{n-1})$$
$$I_n = -\sum_{k=1}^{n-1} I_k \tag{16.28.5}$$

Property 5.

For a reciprocal n-terminal network, $\mathbf{Y}_i$ is symmetric

$$y_{jk} = y_{kj} \tag{16.28.6}$$

Property 6.

If we take the arbitrary reference (datum) as one of the terminals, say n^{th}, then

$$V_n = 0 \tag{16.28.7}$$

The nth column in Eq. (16.27.1) disappears and we have n equations with $(n - 1)$ variables. Since

$$I_n = -\sum_{k=1}^{n-1} I_k$$

and so the nth equations is redundant.

Therefore, we have $(n - 1)$ equations with $(n - 1)$ variables, with nth node as the reference node. But this is precisely the node analysis.

Hence we conclude that if one of the nodes of an n-terminal network is made the reference node, the row and the column corresponding to the reference node are removed from the indefinite admittance matrix $\mathbf{Y}_i$. In this way, the short-circuit admittance is obtained.

■ **EXAMPLE 16.32** Find $\mathbf{Y}_i$ for the 3-terminal network shown in Fig. 16.53a.

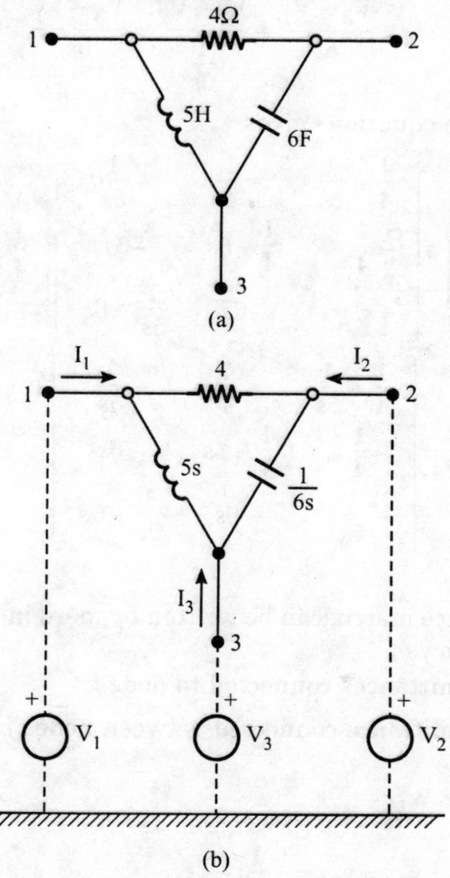

FIG. 16.53

SOLUTION No independent sources and no initial conditions exist and, therefore, the external voltage sources are shown in dotted lines in Fig. 16.53b for the sake of convenience. To calculate $\mathbf{Y}_i$, we write KCL at each node.

By KCL, sum of incoming currents = sum of outgoing currents

Node 1.

$$I_1 = \frac{V_1 - V_2}{4} + \frac{V_1 - V_3}{5s}$$

Node 2

$$I_2 = \frac{V_2 - V_1}{4} + \frac{V_2 - V_3}{\frac{1}{6s}} = \frac{V_2 - V_1}{4} + 6s\,(V_2 - V_3)$$

Node 3

$$I_3 = \frac{V_3 - V_1}{5s} + \frac{V_3 - V_2}{\frac{1}{6s}} = \frac{V_3 - V_1}{5s} + 6s\,(V_3 - V_2)$$

Rearrangement of these equations gives

$$\begin{bmatrix} I_1 \\ I_2 \\ I_3 \end{bmatrix} = \begin{bmatrix} \frac{1}{4} + \frac{1}{5s} & -\frac{1}{4} & -\frac{1}{5s} \\ -\frac{1}{4} & \frac{1}{4} + 6s & -6s \\ -\frac{1}{5s} & -6s & \frac{1}{5s} + 6s \end{bmatrix} \begin{bmatrix} V_1 \\ V_2 \\ V_3 \end{bmatrix}$$

therefore,

$$\mathbf{Y}_i = \begin{bmatrix} \frac{1}{4} + \frac{1}{5s} & -\frac{1}{4} & -\frac{1}{5s} \\ -\frac{1}{4} & \frac{1}{4} + 6s & -6s \\ -\frac{1}{5s} & -6s & \frac{1}{5s} + 6s \end{bmatrix}$$

Alternative method

The indefinite admittance matrix can be written by mere inspection. The elements of the matrix can be written as follows :

y_{kk} = sum of all the admittances connected to node k

y_{jk} = negative of the admittance connected between nodes j and k.

therefore,

$$y_{11} = \frac{1}{4} + \frac{1}{5s}$$

$$y_{12} = y_{21} = -\frac{1}{4}$$

$$y_{22} = \frac{1}{4} + 6s$$

$$y_{33} = \frac{1}{5s} + 6s$$

$$y_{23} = y_{32} = -6s$$

$$y_{13} = y_{31} = -\frac{1}{5s}$$

■ **EXAMPLE 16.33** Find the indefinite admittance matrix of the circuit shown in Fig. 16.54a.

SOLUTION : By KCL in Fig. 16.54b

sum of incoming currents = sum of outgoing currents

Node 1

$$I_1 = I + \alpha I$$

Node 2

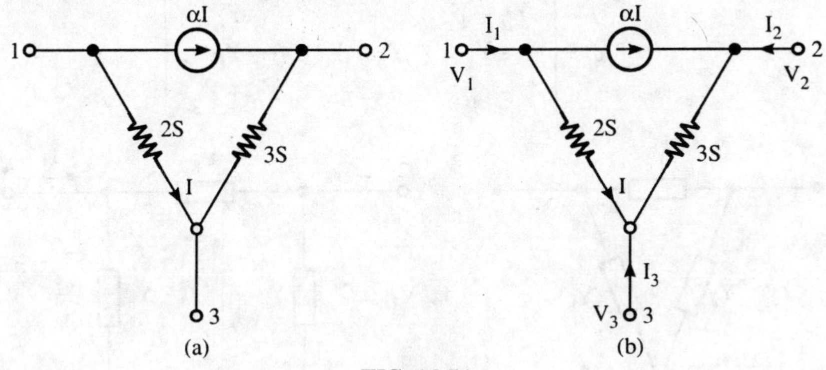

FIG. 16.54

$\because$ $$I_2 + \alpha I = 3\,(V_2 - V_3)$$

Node 3,

$$I_3 + I_1 = 3\,(V_3 - V_2)$$

Also, $$I = 2\,(V_1 - V_3)$$

Substituting the value of I in the above equation

$$I_1 = (1+\alpha)\,I = 2\,(1+\alpha)\,(V_1 - V_3)$$

or $$I_1 = 2\,(1+\alpha)\,V_1 - 0V_2 - 2\,(1+\alpha)\,V_3$$

$$I_2 = 3\,(V_2 - V_3) - \alpha I$$

$$= 3\,(V_2 - V_3) - 2\alpha\,(V_1 - V_3)$$

or $$I_2 = -2\alpha V_1 + 3V_2 + (-3 + 2\alpha)\,V_3$$

$$I_3 = 3\,(V_3 - V_2) - I$$

$$= 3\,(V_3 - V_2) - 2\,(V_1 - V_3)$$

or $$I_3 = -2V_1 - 3V_2 + 5V_3$$

In matrix form,

$$\begin{bmatrix} I_1 \\ I_2 \\ I_3 \end{bmatrix} = \begin{bmatrix} 2\,(1+\alpha) & 0 & -2\,(1+\alpha) \\ -2\alpha & 3 & -3+2\alpha \\ -2 & -3 & 5 \end{bmatrix} \begin{bmatrix} V_1 \\ V_2 \\ V_3 \end{bmatrix}$$

therefore, $$\mathbf{Y}_i = \begin{bmatrix} 2\,(1+\alpha) & 0 & -2\,(1+\alpha) \\ -2\alpha & 3 & -3+2\alpha \\ -2 & -3 & 5 \end{bmatrix}$$

■ **EXAMPLE 16.34** For the network shown in Fig. 16.55a, write the node equations when :

(*a*) Node 1 is grounded,

(*b*) Node 3 is grounded.

The indefinite admittance matrix $\mathbf{Y}_i$ of the network is

$$\mathbf{Y}_i = \begin{bmatrix} Y_a + Y_c & -Y_c & -Y_a \\ -Y_c & Y_b + Y_c & -Y_b \\ -Y_a & -Y_b & Y_a + Y_b \end{bmatrix}$$

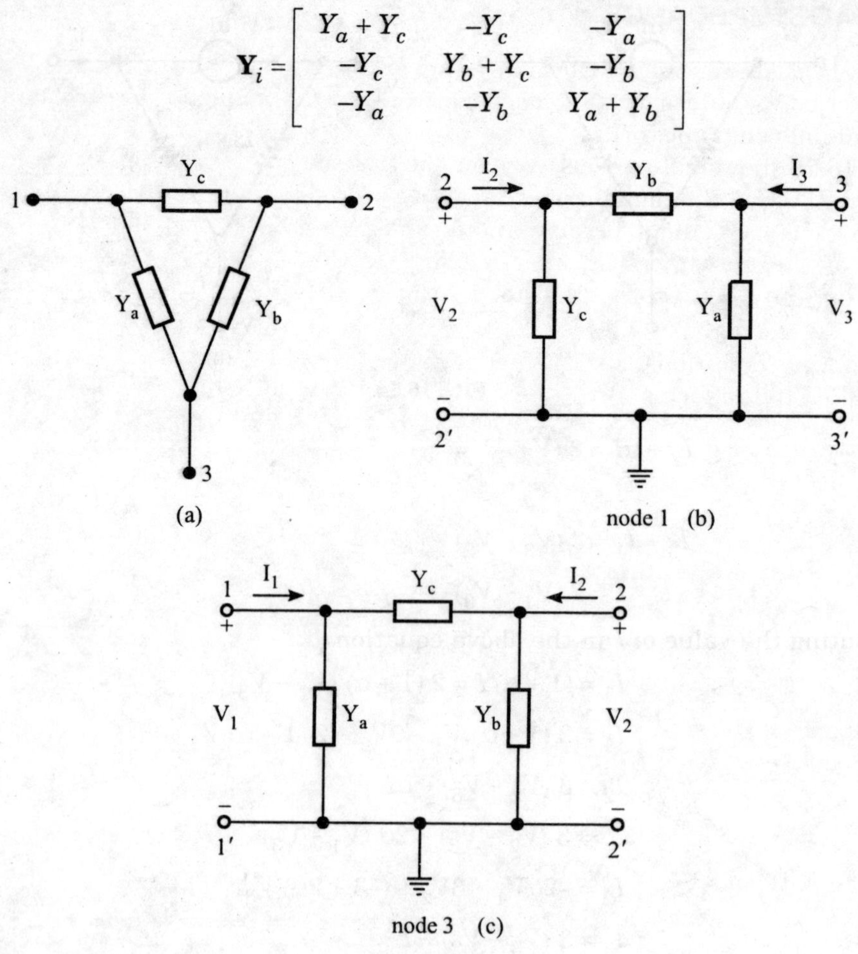

FIG. 16.55

SOLUTION :

(*a*) When node 1 is grounded we delete row 1 and column 1, and the node equations become

$$\begin{bmatrix} I_2 \\ I_3 \end{bmatrix} = \begin{bmatrix} Y_b + Y_c & -Y_b \\ -Y_b & Y_a + Y_b \end{bmatrix} \begin{bmatrix} V_2 \\ V_3 \end{bmatrix}$$

The network for these equations is shown in Fig. 16.55b. It is a 2-port network.

(*b*) When node 3 is grounded, we delete row 3 and column 3, and the node equations become

$$\begin{bmatrix} I_1 \\ I_2 \end{bmatrix} = \begin{bmatrix} Y_a + Y_c & -Y_c \\ -Y_c & Y_b + Y_c \end{bmatrix} \begin{bmatrix} V_1 \\ V_2 \end{bmatrix}$$

The configuration for these equations is shown in Fig. 16.55c. It is also a 2-port network.

16.29 IMAGE IMPEDANCE

Two port networks are cascaded to form a system for the transmission of signals. For maximum power transfer, it is necessary that each network in the chain is properly matched to its preceding and following networks.

Figure 16.2b shows a 2-port network for the transmission of a signal. Since the signals pass along a system in one direction, it is convenient to show the output current coming **out** of the network. The modified diagram of a network passing information from left to right is shown in Fig. 16.56

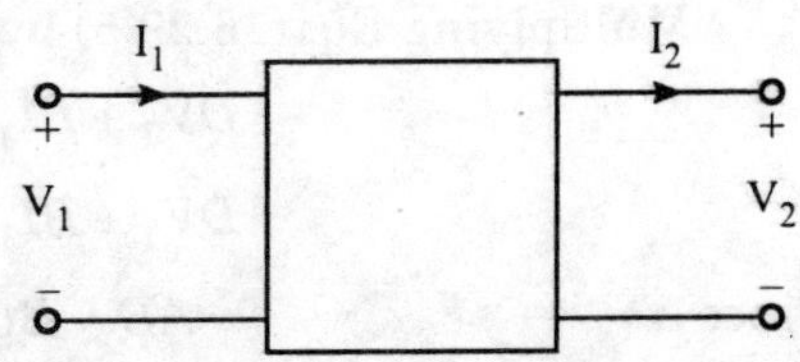

FIG. 16.56 A passive, linear 2-port network

The defining equations for the network are consequently modified to the forms

$$V_1 = AV_2 + BI_2$$

$$I_1 = CV_2 + DI_2$$

In a 2-port network there are two image impedances, namely, the **input image impedance Z_{i1}** and the **output image impedance Z_{i2}**.

These impedances are the functions of the network such that when the network is terminated at the output terminals by Z_{i2}, the impedance seen at the input is Z_{i1}. When the input is terminated by an impedance Z_{i1}, the impedance seen at the output terminals is Z_{i2}. This is shown in Fig. 16.57.

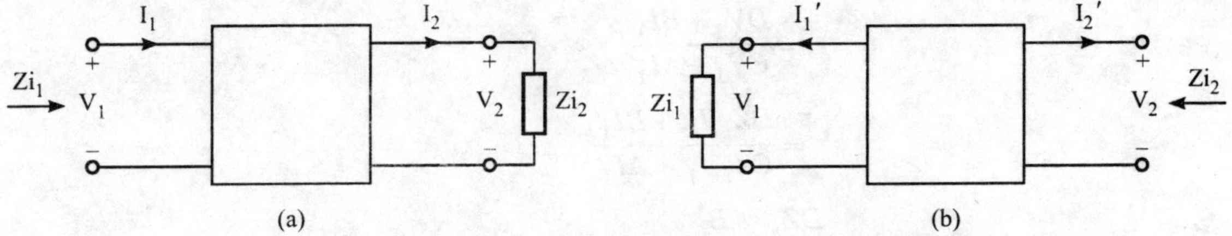

FIG. 16.57 Image impedances of a 2-port network

Image impedances may be expressed in terms of ABCD parameters.

Fig. 16.57a shows a 2-port network with impedance Z_{i2} connected across port 2. From this figure,

$$V_1 = AV_2 + BI_2 \tag{16.29.1}$$

$$I_1 = CV_2 + DI_2 \tag{16.29.2}$$

$$V_2 = Z_{i2} I_2 \tag{16.29.3}$$

$$Z_{i1} = \frac{V_1}{I_1} = \frac{AV_2 + BI_2}{CV_2 + DI_2} = \frac{AZ_{i2}I_2 + BI_2}{CZ_{i2}I_2 + DI_2}$$

or

$$Z_{i1} = \frac{AZ_{i2} + B}{CZ_{i2} + D} \tag{16.29.4}$$

Fig. 16.57b shows the 2-port network with impedance Z_{i1} connected across port 1.

From Fig. 16.57b

$$V_1 = AV_2 - BI_2' \tag{16.29.5}$$

$$-I_1' = CV_2 - DI_2' \tag{16.29.6}$$

Multiplying Eq. (16.29.5) by D and Eq. (16.29.6) by B and subtracting them

$$DV_1 + BI_1' = ADV_2 - BDI_2' - BCV_2 + BDI_2'$$

$$DV_1 + BI_1' = (AD - BC)\,V_2$$

Since $AD - BC = 1$

therefore,

$$V_2 = DV_1 + BI_1' \tag{16.29.7}$$

Now multiplying Eq. (16.29.5) by C and Eq. (16.29.6) by A and subtracting

$$CV_1 + AI_1' = ACV_2 - BCI_2' - ACV_2 + ADI_2'$$

$$= (AD - BC)\,I_2'$$

Since $AD - BC = 1$

$$I_2' = CV_1 + AI_1' \tag{16.29.8}$$

Also,

$$V_1 = Z_{i1}\,I_1' \tag{16.29.9}$$

$$Z_{i2} = \frac{V_2}{I_2'}$$

$$= \frac{DV_1 + BI_1'}{CV_1 + AI_1'}$$

$$= \frac{DZ_{i1}I_1' + BI_1'}{CZ_{i1}I_1' + AI_1'}$$

$$= \frac{DZ_{i1} + B}{CZ_{i1} + A} \tag{16.29.10}$$

From Eq. (16.29.4)

$$CZ_{i1}Z_{i2} + DZ_{i1} - AZ_{i2} - B = 0 \tag{16.29.11}$$

From Eq. (16.29.10)

$$CZ_{i1}Z_{i2} + AZ_{i2} - DZ_{i1} - B = 0 \tag{16.29.12}$$

Subtraction of Eq. (16.29.12) from Eq. (16.29.11) gives

$$2DZ_{i1} - 2AZ_{i2} = 0$$

therefore,

$$\frac{Z_{i1}}{Z_{i2}} = \frac{A}{D} \tag{16.29.13}$$

Addition of Eqs. (16.29.11) and (16.29.12) gives

$$2CZ_{i1}Z_{i2} - 2B = 0$$

therefore,

$$Z_{i1}Z_{i2} = \frac{B}{C} \tag{16.29.14}$$

Multiplying Eqs. (16.29.13) and (16.29.14) we get

$$\frac{Z_{i1}}{Z_{i2}} \times Z_{i1} Z_{i2} = \frac{AB}{CD}$$

therefore,
$$Z_{i1} = \sqrt{\frac{AB}{CD}} \tag{16.29.15}$$

Dividing Eq. (16.26.14) by Eq. (16.29.13) we obtain

$$Z_{i2}^2 = \frac{BD}{AC}$$

therefore,
$$Z_{i2} = \sqrt{\frac{BD}{AC}} \tag{16.29.16}$$

Eqs. (16.29.15) and (16.29.16) give the values of image impedances Z_{i1} and Z_{i2} in terms of *ABCD* parameters.

For a symmetrical network $A = D$

therefore,
$$Z_{i1} = Z_{i2} = \sqrt{\frac{B}{C}}$$

16.30 IMAGE IMPEDANCES IN TERMS OF OPEN-CIRCUIT AND SHORT-CIRCUIT IMPEDANCES

When the output terminals of a 2-port network are open circuited, $I_2 = 0$ and the input impedance is given by

$$Z_{OC1} = \left.\frac{V_1}{I_1}\right|_{I_2=0} = \left.\frac{AV_2 + BI_2}{CV_2 + DI_2}\right|_{I_2=0} = \frac{A}{C} \tag{16.30.1}$$

When the output terminals are short circuited, $V_2 = 0$ and the impedance measured at the input terminals is

$$Z_{SC1} = \left.\frac{V_1}{I_1}\right|_{V_2=0} = \left.\frac{AV_2 + BI_2}{CV_2 + DI_2}\right|_{V_2=0} = \frac{B}{D} \tag{16.30.2}$$

From Eqs. (16.30.1) and (16.30.2)

$$Z_{OC1} Z_{SC1} = \frac{AB}{CD}$$

But from Eq. (16.29.15)

$$\frac{AB}{CD} = Z_{i1}^2$$

therefore,
$$Z_{i1} = \sqrt{Z_{OC1} Z_{SC1}} \tag{16.30.4}$$

When the input terminals are open circuited $I_1 = 0$ and the impedance measured at the output terminals is

$$Z_{OC2} = \frac{V_1}{I_2'} = \left.\frac{DV_1 + BI_1'}{CV_1 + AI_1'}\right|_{I'_1=0} = \frac{D}{C} \tag{16.30.5}$$

When the input terminals are short circuited $V_1 = 0$ and the impedance measured at the output terminals is

$$Z_{SC2} = \left.\frac{DV_1 + BI_1'}{CV_1 + AI_1'}\right|_{V_1=0} = \frac{B}{A} \tag{16.30.6}$$

therefore, $$Z_{OC2}\, Z_{SC2} = \frac{BD}{AC} \tag{16.30.7}$$

But from Eq. (16.29.16)

$$Z_{i2} = \sqrt{\frac{BD}{AC}} \tag{16.30.8}$$

Therefore from Eqs. (16.30.7) and (16.30.8)

$$Z_{i2} = \sqrt{Z_{OC2}\, Z_{SC2}} \tag{16.30.9}$$

Eqs. (16.30.4) and (16.30.9) may be used to determine the image parameters Z_{i1} and Z_{i2} from physical measurements of open-circuit and short-circuit impedances.

16.31 *T* NETWORK

Consider a symmetrical *T* network as shown in Fig. 16.58. We know that, if the image, impedances at ports 1–1′ and port 2–2′ are equal to each other, the image impedances are known as **characteristic impedances**. Thus, if the network in Fig. 16.58 is terminated in its characteristic impedance Z_0, the input impedance will also be Z_0. The input impedance of the *T* network when it is terminated in Z_0 is given by

$$Z_{in} = \frac{Z_1}{2} + Z_2 \,\Big\|\, \left(\frac{Z_1}{2} + Z_0\right)$$

$$= \frac{Z_1}{2} + \frac{Z_2\left(\frac{Z_1}{2} + Z_0\right)}{Z_2 + \frac{Z_1}{2} + Z_0}$$

Also, $$Z_{in} = Z_0$$

therefore, $$Z_0 = \frac{Z_1}{2} + \frac{Z_2\left(\frac{Z_1}{2} + Z_0\right)}{Z_2 + \frac{Z_1}{2} + Z_0}$$

$$Z_0 Z_2 + Z_0\frac{Z_1}{2} + Z_0^2 = \frac{Z_1 Z_2}{2} + \frac{Z_1^2}{4} + \frac{Z_1 Z_0}{2} + \frac{Z_1 Z_2}{2} + Z_2 Z_0$$

$$Z_0^2 = Z_1 Z_2 + \frac{Z_1^2}{4}$$

For a *T* network, let $Z_0 = Z_{OT}$

therefore, $$Z_{OT}^2 = Z_1 Z_2 + \frac{Z_1^2}{4}$$

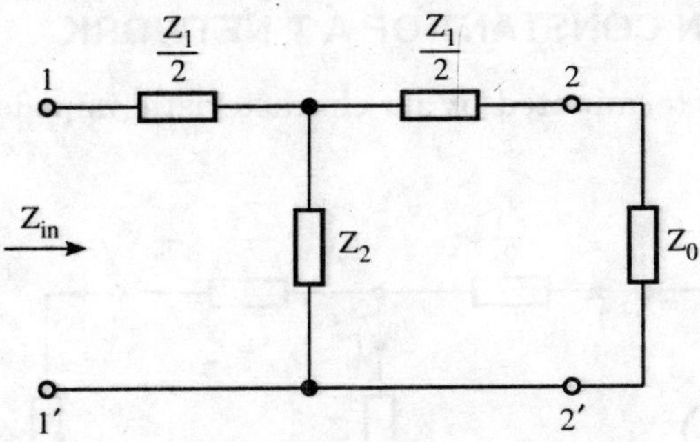

FIG. 16.58 A symmetrical T network

$$Z_{OT} = \sqrt{Z_1 Z_2 \left(1 + \frac{Z_1}{4Z_2}\right)} \tag{16.31.1}$$

The characteristic impedance Z_{OT} for a symmetrical T network can also be found from the relation

$$Z_{OT} = \sqrt{Z_{OC} Z_{SC}} \tag{16.31.2}$$

When the terminals 2–2′ of Fig. 16.58 are open circuited, the input impedance is given by

$$Z_{OC} = \frac{Z_1}{2} + Z_2 \tag{16.31.3}$$

When the terminals 2–2′ of Fig. 16.58 are short circuited, the input impedance is given by

$$Z_{SC} = \frac{Z_1}{2} + \left(Z_2 \;||\; \frac{Z_1}{2}\right)$$

$$= \frac{Z_1}{2} + \frac{Z_2 \times \frac{Z_1}{2}}{Z_2 + \frac{Z_1}{2}}$$

or

$$Z_{SC} = \frac{Z_1^2 + 4Z_1 Z_2}{2Z_1 + 4Z_2} \tag{16.31.4}$$

therefore,

$$Z_{OT} = \sqrt{Z_{OC} Z_{SC}}$$

$$= \sqrt{\left(\frac{Z_1}{2} + Z_2\right)\left(\frac{Z_1^2 + 4Z_1 Z_2}{2Z_1 + 4Z_2}\right)}$$

$$= \sqrt{Z_1 Z_2 + \frac{Z_1^2}{4}}$$

or

$$Z_{OT} = \sqrt{Z_1 Z_2 \left(1 + \frac{Z_1}{4Z_2}\right)} \tag{16.31.5}$$

16.32 PROPAGATION CONSTANT OF A T NETWORK

A symmetrical T network terminated by its characteristic impedance Z_{OT} is shown in Fig. 16.59a.

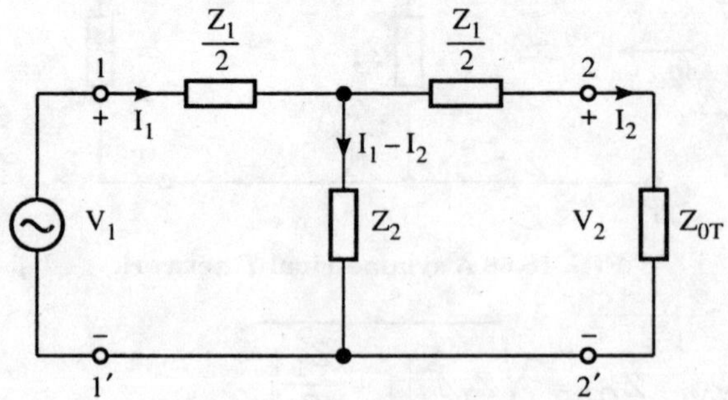

FIG. 16.59 Symmetrical network terminated in Z_{OT}

By KVL in the right-hand mesh

$$Z_2(I_1 - I_2) - \frac{Z_1}{2} I_2 - Z_{OT} I_2 = 0$$

$$Z_2 I_1 = \left(\frac{Z_1}{2} + Z_{OT} + Z_2\right) I_2$$

therefore,
$$\frac{I_1}{I_2} = \frac{1}{Z_2}\left(\frac{Z_1}{2} + Z_{OT} + Z_2\right) \tag{16.32.1}$$

Also,
$$\frac{I_1}{I_2} = e^{\gamma} \tag{16.32.2}$$

therefore,
$$e^{\gamma} = \frac{1}{Z_2}\left(\frac{Z_1}{2} + Z_{OT} + Z_2\right)$$

or
$$Z_{OT} = Z_2 e^{\gamma} - Z_2 - \frac{Z_1}{2}$$

$$Z_{OT} = Z_2(e^{\gamma} - 1) - \frac{Z_1}{2} \tag{16.32.3}$$

The characteristic impedance of a symmetrical T network is given by

$$Z_{OT} = \sqrt{Z_1 Z_2 \left(1 + \frac{Z_1}{4Z_2}\right)} \tag{16.32.4}$$

therefore,
$$Z_2(e^{\gamma} - 1) - \frac{Z_1}{2} = \sqrt{Z_1 Z_2 \left(1 + \frac{Z_1}{4Z_2}\right)}$$

$$\left[Z_2(e^{\gamma} - 1) - \frac{Z_1}{2}\right]^2 = Z_1 Z_2 + \frac{Z_1^2}{4}$$

$$Z_2^2 (e^{\gamma} - 1)^2 + \frac{Z_1^2}{4} - Z_1 Z_2 (e^{\gamma} - 1) = Z_1 Z_2 + \frac{Z_1^2}{4}$$

$$Z_2^2 (e^{\gamma} - 1)^2 = Z_1 Z_2 e^{\gamma}$$

$$e^{2\gamma} - 2e^{\gamma} + 1 = \frac{Z_1}{Z_2} e^{\gamma}$$

$$e^{\gamma} - 2 + e^{-\gamma} = \frac{Z_1}{Z_2}$$

$$\frac{e^{\gamma} + e^{-\gamma}}{2} = 1 + \frac{Z_1}{2Z_2}$$

$$\cos h\gamma = 1 + \frac{Z_1}{2Z_2} \qquad (16.32.5)$$

Also,

$$\cos h\gamma = 1 + 2 \sin h^2 \frac{\gamma}{2}$$

therefore,

$$1 + 2 \sin h^2 \frac{\gamma}{2} = 1 + \frac{Z_1}{2Z_2}$$

$$\sin h \frac{\gamma}{2} = \sqrt{\frac{Z_1}{4Z_2}} \qquad (16.32.6)$$

16.33 Π NETWORK

Consider a symmetrical Π section shown in Fig. 16.60. When the network is terminated in its characteristic impedance Z_0, its input impedance is given by

$$Z_{in} = 2Z_2 \,\Big|\Big|\, \left[Z_1 + (2Z_2 \,||\, Z_0)\right]$$

$$= 2Z_2 \,\Big|\Big|\, \left[Z_1 + \frac{2Z_2 Z_0}{2Z_2 + Z_0}\right]$$

$$= 2Z_2 \,\Big|\Big|\, \left[\frac{2Z_1 Z_2 + Z_0 Z_1 + 2Z_2 Z_0}{2Z_2 + Z_0}\right]$$

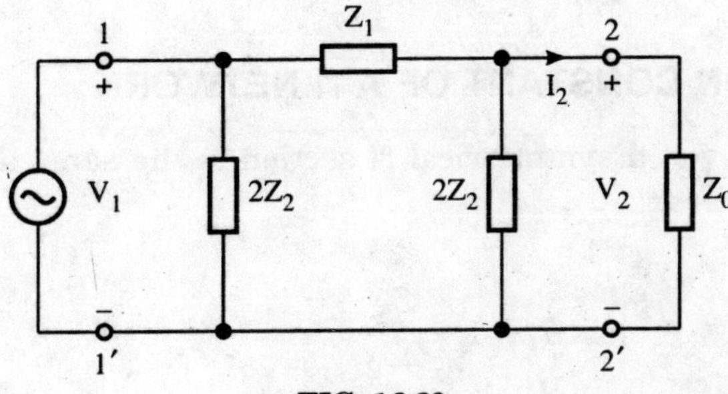

FIG. 16.60

$$Z_0 = \frac{2Z_2\left[\dfrac{2Z_1Z_2 + Z_0Z_1 + 2Z_2Z_0}{2Z_2 + Z_0}\right]}{2Z_2 + \dfrac{2Z_1Z_2 + Z_0Z_1 + 2Z_2Z_0}{2Z_2 + Z_0}}$$

(by definition of characteristic impedance, $Z_{in} = Z_0$)

$$Z_0 = \frac{2Z_2\,(2Z_1Z_2 + Z_0Z_1 + 2Z_2Z_0)}{4Z_2^2 + 2Z_2Z_0 + 2Z_1Z_2 + Z_0Z_1 + 2Z_2Z_0}$$

$$4Z_2^2Z_0 + 2Z_2Z_0^2 + 2Z_1Z_2Z_0 + Z_0^2Z_1 + 2Z_2Z_0^2$$
$$= 4Z_1Z_2^2 + 2Z_0Z_1Z_2 + 4Z_2^2Z_0$$

$$Z_0^2\,(Z_1 + 4Z_2) = 4Z_1Z_2^2$$

$$Z_0^2 = \frac{4Z_1Z_2^2}{Z_1 + 4Z_2}$$

$$Z_0 = \sqrt{\frac{Z_1Z_2}{1 + \dfrac{Z_1}{4Z_2}}} \tag{16.33.1}$$

Equation (16.33.1) gives the characteristic impedance of a symmetrical Π network denoted by $Z_{O\Pi}$.

therefore,
$$Z_{O\Pi} = \sqrt{\frac{Z_1Z_2}{1 + \dfrac{Z_1}{4Z_2}}} \tag{16.33.2}$$

For a symmetrical T network $Z_{OT} = \sqrt{Z_1Z_2\left(1 + \dfrac{Z_1}{4Z_2}\right)}$

therefore,
$$Z_{O\Pi} = \frac{Z_1Z_2}{Z_{OT}} \tag{16.33.3}$$

The characteristic impedance of a symmetrical Π network can also be found from the relation

$$Z_{O\Pi} = \sqrt{Z_{OC}\,Z_{SC}} \tag{16.33.4}$$

16.34 PROPAGATION CONSTANT OF A Π NETWORK

The propagation constant γ of a symmetrical Π section is the same as that of a symmetrical T section.

That is, γ is given by

$$\cos h\gamma = 1 + \frac{Z_1}{2Z_2}$$

or
$$\sin h\frac{\gamma}{2} = \sqrt{\frac{Z_1}{4Z_2}}$$

■ **EXAMPLE 16.35** Find the characteristic impedances of symmetrical *T* and Π networks shown in Fig. 16.61.

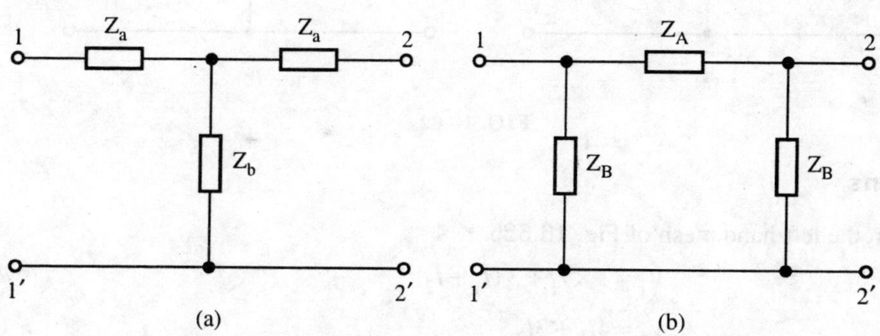

FIG. 16.61 (a) Symmetrical *T* network (b) Symmetrical Π network

SOLUTION :

(*a*) Symmetrical *T* network

When the terminals 22′ of Fig. 16.61a are open circuited, the input impedance is

$$Z_{OC} = Z_a + Z_b$$

When the terminals 22′ of Fig. 16.61 are short circuited, the input impedance is given by

$$Z_{SC} = Z_a + (Z_a \,||\, Z_b)$$

$$= Z_a + \frac{Z_a Z_b}{Z_a + Z_b} = \frac{Z_a(Z_a + 2Z_b)}{Z_a + Z_b}$$

therefore,
$$Z_0 = \sqrt{Z_{OC} Z_{SC}} = \sqrt{Z_a(Z_a + 2Z_b)}$$

(*b*) Symmetrical Π network

When the terminals 22′ of Fig. 16.61b are open circuited, the input impedance is

$$Z_{OC} = Z_B \,||\, (Z_A + Z_B) = \frac{Z_B(Z_A + Z_B)}{Z_A + 2Z_B}$$

When the terminals 22′ of Fig. 16.61b are short circuited, the input impedance is

$$Z_{SC} = Z_B \,||\, Z_A = \frac{Z_A Z_B}{Z_A + Z_B}$$

therefore,
$$Z_0 = \sqrt{Z_{OC} Z_{SC}} = Z_B \sqrt{\frac{Z_A}{Z_A + 2Z_B}}$$

■ **EXAMPLE16.36** For the network shown in Fig. 16.62a calculate *z*, *y*, *h*, *g* and transmission parameters.

SOLUTION :

By *KCL* the current through the 3 – Ω resistor is $(I_1 + I_2)$ as shown in Fig. 16.62b.

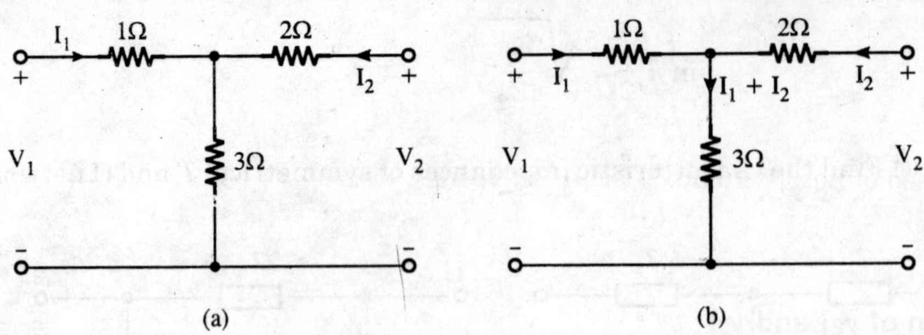

FIG. 16.62

Mesh Equations

By *KVL* in the left-hand mesh of Fig. 16.62b.

$$V_1 = 1 \times I_1 + 3\,(I_1 + I_2)$$

or

$$V_1 = 4I_1 + 3I_2 \quad \text{(E16.36.1)}$$

By KVL in the right-hand mesh of Fig. 16.62b.

$$3\,(I_1 + I_2) + 2I_2 - V_2 = 0$$

or

$$V_2 = 3I_1 + 5I_2 \quad \text{(E16.36.2)}$$

Determinationof *z* Parameters

The defining equations of *z* parameters are

$$V_1 = z_{11}I_1 + z_{12}I_2 \quad \text{(E 16.36.3)}$$

$$V_2 = z_{21}I_1 + z_{22}I_2 \quad \text{(E 16.36.4)}$$

Comparison of Eqs. (E 16.36.1) and (E 16.36.3) gives

$$z_{11} = 4\Omega,\; z_{12} = 3\Omega$$

Comparison of Eqs. (E 16.36.2) and (E 16.36.6) gives

$$z_{21} = 3\Omega,\; z_{22} = 5\Omega$$

Determination of *y* parameters

The defining equations of *y* parameters are

$$I_1 = y_{11}V_1 + y_{12}V_2 \quad \text{(E 16.36.5)}$$

$$I_2 = y_{21}V_1 + y_{22}V_2 \quad \text{(E 16.36.6)}$$

In order to find out the *y* parameters, the mesh equations (E 16.36.1) and (E 16.36.2) are arranged in the form of Eqs. (E 16.36.5) and (E 16.36.6) respectively as follows :

Determination of y_{11} and y_{12}

Equation (E 16.36.5) does not contain I_2. Therefore we climinate I_2 from Eqs. (E 16.36.1) and (E 16.36.2). Multiplying Eq. (E 16.36.1) by 5 and Eq. (E. 16.36.2) by 3 and subtracting we obtain

$$11I_1 = 5V_1 - 3V_2 \quad \text{(E 16.36.7)}$$

or

$$I_1 = \frac{5}{11} V_1 - \frac{3}{11} V_2 \quad \text{(E 16.36.7a)}$$

Comparison of Eqs. (E 16.36.5) and (E 16.36.7a) gives

$$y_{11} = \frac{5}{11} S, \quad y_{12} = -\frac{3}{11} S$$

Determination of y_{21} and y_{22}

Equation (E 16.36.6) does not contain I_1. Therefore, we eliminate I_1 from Eqs. (E 16.36.1) and (E 16.36.2). Multiplying Eq. (E 16.36.1) by 3 and Eq. (E 16.36.2) by 4 and subtracting we get

$$3V_1 - 4V_2 = -11I_2 \quad \text{(E 16.36.8)}$$

or

$$I_2 = -\frac{3}{11} V_1 + \frac{4}{11} V_2 \quad \text{(E 16.36.8a)}$$

Comparison of Eqs. (E 16.36.8a) and (E 16.36.6) gives

$$y_{21} = -\frac{3}{11} S, \quad y_{22} = \frac{4}{11} S$$

Determination of h parameters

The defining equations of h parameters are

$$V_1 = h_{11} I_1 + h_{12} V_2 \quad \text{(E 16.36.9)}$$

$$I_2 = h_{21} I_1 + h_{22} V_2 \quad \text{(E 16.36.10)}$$

Determination of h_{11} and h_{12}

Equation (E 16.36.9) does not contain I_2. By arranging Eq. (E 16.36.7) we get

$$5V_1 = 11I_1 + 3V_2$$

or

$$V_1 = \frac{11}{5} I_1 + \frac{3}{5} V_2 \quad \text{(E 16.36.11)}$$

Comparison of (E 16.36.11) and (E 16.36.9) gives

$$h_{11} = \frac{11}{5} \Omega, \quad h_{12} = \frac{3}{5}$$

Determination of h_{21} and h_{22}

The defining equation (E 16.36.10) for h_{21} and h_{22} does not contain V_1. Therefore, we arrange Eq. (E 16.36.2) in the form of Eq. (E 16.36.10) as follows :

$$5I_2 = -3I_1 + V_2 \quad \text{(E 16.36.12)}$$

or

$$I_2 = -\frac{3}{5} I_1 + \frac{1}{5} V_2 \quad \text{(E 16.35.12a)}$$

Comparing Eq. (E 16.36.12a) with Eq. (E 16.36.10) we obtain

$$h_{21} = -\frac{3}{5}, \quad h_{22} = \frac{1}{5}\,S$$

Determination of *g* Parameters

The defining equations of g parameters are

$$I_1 = g_{11}V_1 + g_{12}I_2 \qquad \text{(E 16.36.13)}$$

$$V_2 = g_{21}V_1 + g_{22}I_2 \qquad \text{(E 16.36.14)}$$

Determination of g_{11} and g_{12}

The defining equation (E 16.36.13) for g_{11} and g_{12} does not contain V_2. Therefore we arrange Eq. (E 16.36.1) in the form of Eq. (16.36.13).

$$4I_1 = V_1 - 3I_2$$

or

$$I_1 = \frac{1}{4}V_1 - \frac{3}{4}I_2 \qquad \text{(E 16.36.15)}$$

Comparison of Eqs. (E 16.36.15) and (E 16.36.13) gives

$$g_{11} = \frac{1}{4}\,S, \quad g_{12} = -\frac{3}{4}$$

Determination of g_{21} and g_{22}

The defining equation (E 16.36.14) for g_{21} and g_{22} does not contain I_1. Therefore we arrange Eq. (E 16.36.8) in the form of Eq. (E 16.36.14).

$$4V_2 = 3V_1 + 11I_2 \qquad \text{(E 16.36.16)}$$

or

$$V_2 = \frac{3}{4}V_1 + \frac{11}{4}I_2 \qquad \text{(E 16.36.16a)}$$

Comparison of Eqs. (E 16.36.16a) and (E 16.36.14) gives

$$g_{21} = \frac{3}{4}, \quad g_{22} = \frac{11}{4}\,\Omega$$

Determination of Transmission Parameters

The defining equations of transmission *ABCD* parameters

$$V_1 = AV_2 + B(-I_2) \qquad \text{(E 16.36.17)}$$

$$I_1 = CV_2 + D(-I_2) \qquad \text{(E 16.36.18)}$$

Determination of *A* and *B* parameters

The defining equation (E 16.36.17) for A and B parameters does not contain I_1. Therefore, we arrange Eq. (E 16.36.16) in the form of Eq. (E 16.36.17).

$$3V_1 = 4V_2 - 11I_2$$

$$V_1 = \frac{4}{3} V_2 - \frac{11}{3} I_2 \tag{E 16.36.19}$$

Comparison of Eqs. (E 16.36.19) and (E 16.36.17) gives

$$A = \frac{4}{3}, B = \frac{11}{3} \Omega$$

Determination of *C* and *D* parameters

The defining equation (E 16.36.18) for *C* and *D* parameters does not contain V_1. Therefore, we arrange Eq. (E 16.36.12) in the form of Eq. (E 16.36.18)

$$3I_1 = V_2 - 5I_2$$

or

$$I_1 = \frac{1}{3} V_2 + \frac{5}{3} (-I_2) \tag{E 16.36.20}$$

Comparison of Eqs. (E 16.36.20) and (E 16.36.18) gives

$$C = \frac{1}{3} S, D = \frac{5}{3}$$

■ **EXAMPLE 16.37** For the network shown in Fig. 16.63a, find the *z* and *y* parameters. Assume ideal transformer.

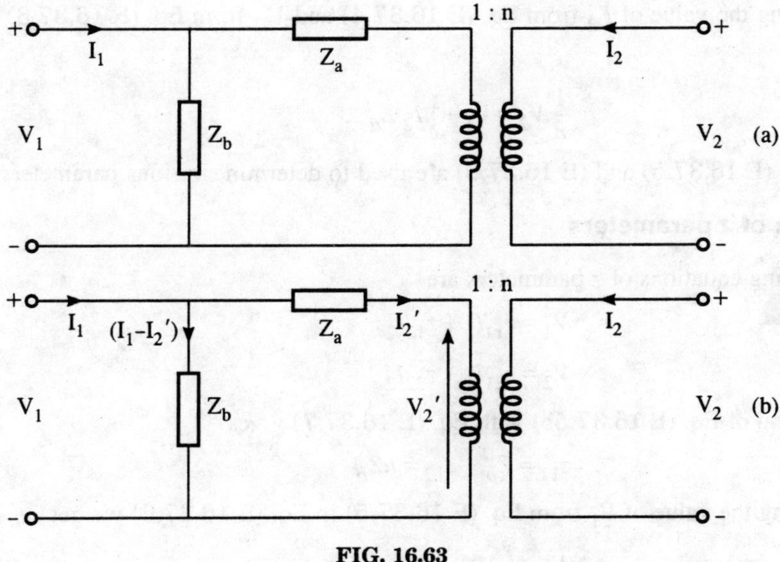

FIG. 16.63

SOLUTION :

Let V'_2 and I'_2 be the input voltage and input current respectively, as shown in Fig. 16.63b. The KVL equation in the left-hand mesh is

$$V_1 = (I_1 - I_2') Z_b \tag{E 16.37.1}$$

The KVL equation in the left-hand loop is

$$V_1 - I'_2 Z_a - V_2' = 0$$

or
$$V'_2 = V_1 - I'_2 Z_a \quad \text{(E 16.37.2)}$$

By transformer action

$$\frac{V'_2}{V_2} = \frac{1}{n}$$

therefore,
$$V'_2 = \frac{1}{n} V_2 \quad \text{(E 16.37.3)}$$

$$\frac{I'_2}{-I_2} = \frac{n}{1}$$

or
$$I'_2 = -nI_2 \quad \text{(E 16.37.4)}$$

In order to find the two-port parameters, the KVL equations (E 16.37.1) and (E 16.37.2) are to be expressed in terms of V_1, I_1, V_2 and I_2. The assumed values V'_2 and I'_2 are to be eliminated.

Substituting the value of I'_2 from Eq. (E 16.37.4) in Eq. (E 16.37.1) we get

$$V_1 = (I_1 + nI_2)\, Z_b \quad \text{(E 16.37.5a)}$$

or
$$V_1 = Z_b I_1 + nZ_b I_2 \quad \text{(E 16.37.5b)}$$

Substituting the value of I'_2 from Eq. (E 16.37.4) and V'_2 from Eq. (E 16.37.3) in Eq. (E 16.37.2) we get

$$\frac{1}{n} V_2 = V_1 + nI_2 Z_a \quad \text{(E 16.37.6)}$$

Equations (E 16.37.5) and (E 16.37.6) are used to determine various parameters of the network.

Determination of *z* parameters

The defining equations of *z* parameters are

$$V_1 = z_{11} I_1 + z_{12} I_2 \quad \text{(E 16.37.7)}$$

$$V_2 = z_{21} I_1 + z_{22} I_2 \quad \text{(E 16.37.8)}$$

Comparison of Eq. (E 16.37.5b) with Eq. (E 16.37.7) gives

$$z_{11} = Z_b, \quad z_{12} = nZ_b$$

Substituting the value of V_1 from Eq. (E 16.37.5) in Eq. (E 16.37.6) we get

$$\frac{1}{n} V_2 = (Z_b I_1 + nZ_b I_2) + nZ_a I_2$$

or
$$V_2 = nZ_b I_1 + n^2 (Z_a + Z_b) I_2 \quad \text{(E. 16.37.a)}$$

Comparison of Eqs. (E. 16.37.9) and (E. 16.37.8) gives

$$z_{21} = nZ_b, \quad z_{22} = n^2 (Z_a + Z_b)$$

Determination of *y* parameters

By putting $V_1 = 0$ in Eq. (E. 16.37.5a) and (E. 16.37.6) we get

$$nI_2 = -I_1 \tag{E. 16.37.10}$$

$$\frac{V_2}{n} = nI_2 Z_a \tag{E. 16.37.11}$$

$$= -I_1 Z_a$$

$$\frac{I_1}{V_2} = -\frac{1}{nZ_a}$$

therefore,

$$y_{12} = \frac{I_1}{V_2} V_1 \bigg|_{V_1 = 0} = -\frac{1}{nZ_a}$$

From Eq. (E. 16.37.11)

$$\frac{I_2}{V_2} = \frac{1}{n^2 Z_a}$$

therefore,

$$y_{22} = \frac{I_2}{V_2}\bigg|_{V_1 = 0} = \frac{1}{n^2 Z_a}$$

By putting $V_2 = 0$ in Eq. (E. 16.37.6), we get

$$0 = V_1 + nZ_a I_2 \tag{E. 16.37.12}$$

or

$$\frac{I_2}{V_1} = -\frac{1}{nZ_a}$$

therefore,

$$y_{21} = \frac{I_2}{V_1}\bigg|_{V_2 = 0} = -\frac{1}{nZ_a}$$

From Eq. (E. 16.37.5)

$$\frac{V_1}{Z_b} - I_1 = nI_2 \tag{E. 16.37.13}$$

From Eq. (E. 16.37.12)

$$nI_2 = -\frac{V_1}{Z_a} \tag{E. 16.37.14}$$

From Eqs. (E. 16.37.13) and (E. 16.37.14)

$$\frac{V_1}{Z_b} - I_1 = -\frac{V_1}{Z_a}$$

$$\left(\frac{1}{Z_a} + \frac{1}{Z_b}\right) V_1 = I_1$$

therefore,
$$y_{11} = \left.\frac{I_1}{V_1}\right|_{V_2=0} = \frac{1}{Z_a} + \frac{1}{Z_b}$$

■ **EXAMPLE 16.38** Two identical sections of the network shown in Fig. 16.64a are cascaded. Calculate the transmission parameters of the resulting network.

SOLUTION :

By KCL the current through the resistor R_C is $(I_1 + I_2)$ as shown in Fig. 16.64b. By KVL in the left-hand mesh

$$V_1 - I_1R_a - (I_1 + I_2)\,R_C = 0$$

or,
$$V_1 = 2I_1 + I_2 \tag{E. 16.38.1}$$

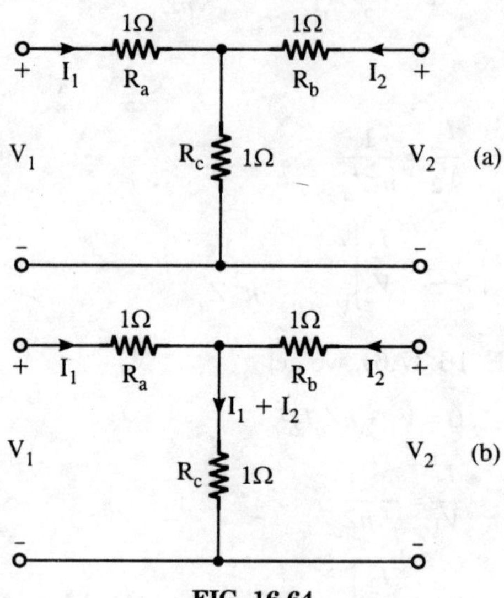

FIG. 16.64

By KVL in the right-hand mesh

$$+ (I_1 + I_2)\,R_C + I_2R_b - V_2 = 0$$

or
$$V_2 = I_1 + 2I_2 \tag{E. 16.38.2}$$

In order to find out the transmission parameters, we arrange Eqs. (E. 16.38.1) and (E. 16.38.2) in the form of the defining equations of the transmission parameters. The defining equation of the transmission parameters are

$$V_1 = AV_2 + B(-I_2) \tag{E. 16.38.3}$$

$$I_1 = CV_2 + D(-I_2) \tag{E. 16.38.4}$$

From Eqs. (E. 16.38.1) and (E. 16.38.2)

$$V_1 - 2V_2 = -3I_2$$

$$V_1 = 2V_2 - 3I_2 \tag{E. 16.37.5}$$

Comparison of Eqs. (E. 16.37.1) and (E. 16.37.5) gives

$$A = 2, B = 3$$

From Eq. (E. 16.37.2)

$$I_1 = V_2 - 2I_2 \tag{E. 16.37.6}$$

Comparison of Eqs. (E. 16.37.6) and (E. 16.37.4) gives

$$C = 1, D = 2$$

The transmission matrix of the section of the given network is

$$[a] = \begin{bmatrix} A & B \\ C & D \end{bmatrix} = \begin{bmatrix} 2 & 3 \\ 1 & 2 \end{bmatrix}$$

The overall matrix of the cascaded network is

$$\begin{bmatrix} A' & B' \\ C' & D' \end{bmatrix} = \begin{bmatrix} A & B \\ C & D \end{bmatrix}\begin{bmatrix} A & B \\ C & D \end{bmatrix}$$

$$= \begin{bmatrix} 2 & 3 \\ 1 & 2 \end{bmatrix}\begin{bmatrix} 2 & 3 \\ 1 & 2 \end{bmatrix}$$

$$= \begin{bmatrix} 2\times 2+3\times 1 & 2\times 3+3\times 2 \\ 1\times 2+2\times 1 & 1\times 3+2\times 2 \end{bmatrix}$$

$$= \begin{bmatrix} 7 & 12 \\ 4 & 7 \end{bmatrix}$$

■ **EXAMPLE 16.39** Determine the z and transmission parameters of the gyrator shown in Fig. 16.65.

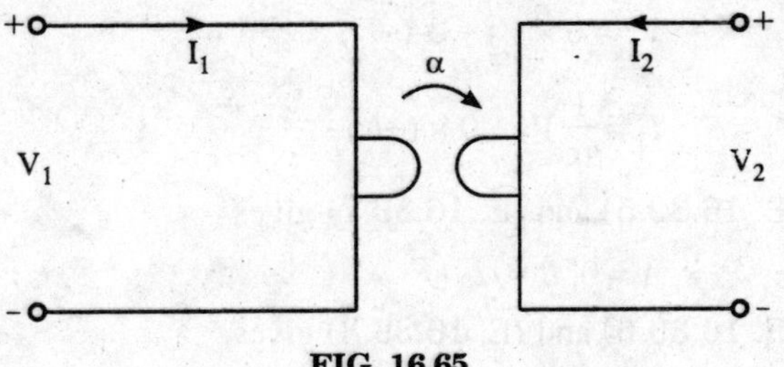

FIG. 16.65

SOLUTION :

The defining equations of the gyrator are

$$V_1 = \alpha(-I_2) \tag{E. 16.39.1}$$

$$V_2 = \alpha I_1 \tag{E. 16.39.2}$$

Determination of *z* parameters

These equations may be written as

$$V_1 = 0 \times I_1 + \alpha(-I_2)$$

$$V_2 = \alpha I_1 + 0 \times I_2$$

In matrix form

$$\begin{bmatrix} V_1 \\ V_2 \end{bmatrix} = \begin{bmatrix} 0 & -\alpha \\ \alpha & 0 \end{bmatrix} \begin{bmatrix} I_1 \\ I_2 \end{bmatrix} \quad \text{(E. 16.39.3)}$$

The defining equations of the z parameters are

$$\begin{bmatrix} V_1 \\ V_2 \end{bmatrix} \begin{bmatrix} z_{11} & z_{12} \\ z_{21} & z_{22} \end{bmatrix} \begin{bmatrix} I_1 \\ I_2 \end{bmatrix} \quad \text{(E. 16.39.4)}$$

Comparison of Eqs. (E. 16.39.3) and (E. 16.39.4) gives

$$z_{11} = 0,\ z_{12} = -\alpha$$

$$z_{21} = \alpha,\ z_{22} = 0$$

Determination of transmission parameters

The defining equation of the transmission parameters are written as

$$V_1 = AV_2 - BI_2 \quad \text{(E. 16.39.5)}$$

$$I_1 = CV_2 - DI_2 \quad \text{(E. 16.39.6)}$$

The defining equations (E. 16.39.1) and (E. 16.39.2) can be written as

$$V_1 = 0 \times V_2 + \alpha(-I_2) \quad \text{(E. 16.39.7)}$$

$$I_1 = \frac{1}{\alpha} V_2 + 0 \times (-I_2) \quad \text{(E. 16.39.8)}$$

Comparison of Eqs. (E. 16.39.5) and (E. 16.39.7) gives

$$A = 0,\ B = \alpha$$

Comparison of Eqs. (E. 16.39.6) and (E. 16.39.8) gives

$$C = \frac{1}{\alpha},\ D = 0$$

The transmission matrix of the gyrator may be written as

$$[a_G] = \begin{bmatrix} 0 & \alpha \\ \frac{1}{\alpha} & 0 \end{bmatrix}$$

EXAMPLE 16.40 Detrmine the transmission parameters of the ideal transformer shown in Fig. 16.66.

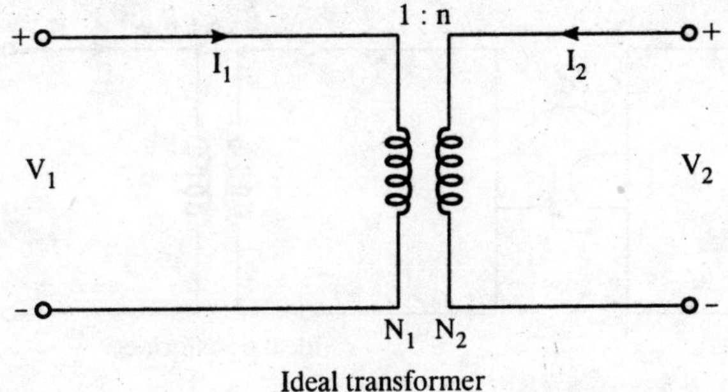

FIG. 16.66

SOLUTION :

The defining equations of the ideal transformer are

$$\frac{V_1}{V_2} = \frac{N_1}{N_2}$$

or
$$V_1 = \frac{N_1}{N_2} V_2 = nV_2 \quad \text{(E. 16.40.1)}$$

where $n = \frac{V_1}{V_2}$

$$\frac{I_1}{(-I_2)} = \frac{N_2}{N_1} = \frac{1}{n}$$

therefore,
$$I_1 = -\frac{1}{n} I_2 \quad \text{(E. 16.40.2)}$$

The defining equations of the transmission parameters are

$$V_1 = AV_2 - BI_2 \quad \text{(E. 16.40.3)}$$

$$I_1 = CV_2 - DI_2 \quad \text{(E. 16.40.4)}$$

Equations (E. 16.40.1) and (E. 16.40.2) may be written as

$$V_1 = nV_2 - 0 \times I_2 \quad \text{(E. 16.40.5)}$$

$$I_1 = 0 \times V_2 - \frac{1}{n} I_2 \quad \text{(E. 16.40.6)}$$

Comparison of Eqs. (E. 16.40.3) and (E. 16.40.5) gives $A = n, B = 0$

Comparison of Eqs. (E. 16.40.4) and (E. 16.40.6) gives $C = 0, D = \frac{1}{n}$

therefore,
$$[a] = \begin{bmatrix} 0 & 0 \\ 0 & 1/n \end{bmatrix}$$

EXAMPLE 16.41 Find the *ABCD* parameters for the network shown in Fig. 16.67 a.

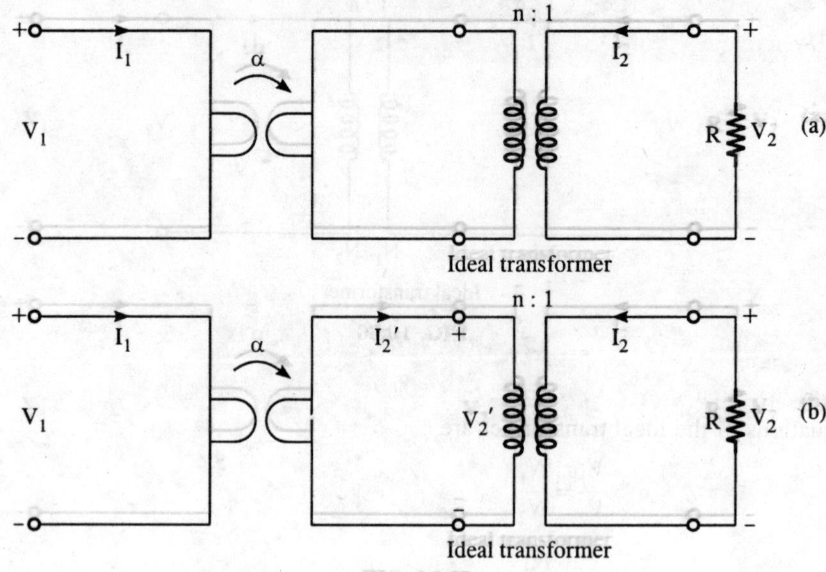

FIG. 16.67

SOLUTION :

Let V'_2 and I'_2 be the primary voltage and primary current of the transformer as shown in Fig. 16.67b.

$$V_2 = -RI_2 \tag{E. 16.41.1}$$

$$I_2 = -nI'_2 \tag{E. 16.41.2}$$

$$V'_2 = nV_2 \tag{E. 16.41.3}$$

Combining these equations we get

$$V_2 = -R\,(-nI'_2)$$

therefore,

$$I'_2 = \frac{1}{nR}\,V_2 \tag{E. 16.41.4}$$

$$V'_2 = nV_2$$

or

$$V'_2 = -nRI_2 \tag{E. 16.41.5}$$

The defining equations of the *ABCD* parameters of the given transformer are

$$V'_2 = AV_2 - BI_2 \tag{E. 16.41.6}$$

$$I'_2 = CV_2 - DI_2 \tag{E. 16.41.7}$$

Comparison of Eqs. (E. 16.41.5) and (E. 16.41.6) gives $A = 0$, $B = nR$

Comparison of Eqs. (E. 16.41.4) and (E. 16.41.7) gives

$$C = \frac{1}{nR},\ D = 0$$

Therefore, the transmission matrix of the given transformer can be written as

$$[a_T] = \begin{bmatrix} A & B \\ C & D \end{bmatrix} = \begin{bmatrix} 0 & nR \\ \dfrac{1}{nR} & 0 \end{bmatrix}$$

The *ABCD* parameters of the gyrator are

$$[a_G] = \begin{bmatrix} 0 & \alpha \\ \dfrac{1}{\alpha} & 0 \end{bmatrix}$$

The overall *ABCD* parameters of the given network

$$[a_0] = [a_g]\,[a_T]$$

$$= \begin{bmatrix} 0 & \alpha \\ \dfrac{1}{\alpha} & 0 \end{bmatrix} \begin{bmatrix} 0 & nR \\ \dfrac{1}{nR} & 0 \end{bmatrix}$$

$$= \begin{bmatrix} \dfrac{\alpha}{nR} & 0 \\ 0 & \dfrac{nR}{\alpha} \end{bmatrix}$$

EXERCISES

1. Determine the z, y and transmission parameters of the network shown in Fig. P 16.1.

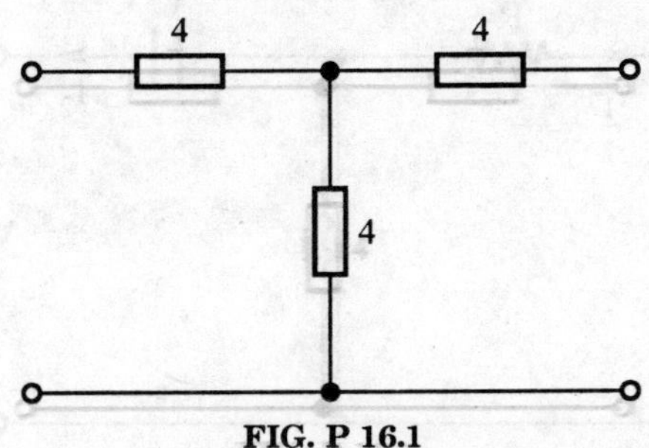

FIG. P 16.1

2. Determine the z parameters of the network shown in Fig. P 16.2.

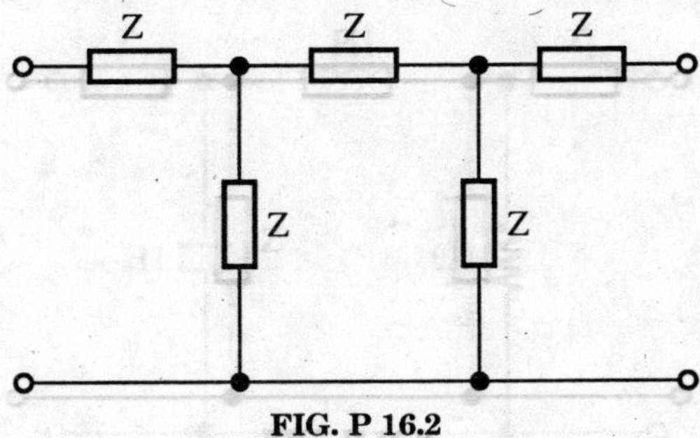

FIG. P 16.2

3. Determine the y, z, h and a parameters of the π network of Fig P 16.3.

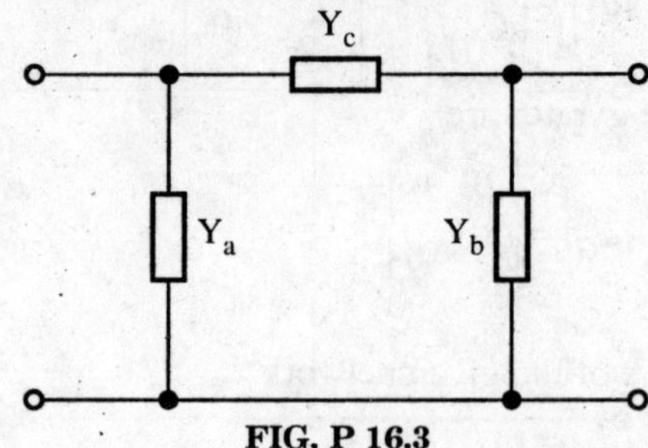

FIG. P 16.3

4. Determine the z parameters of the network of Fig. P 16.4.

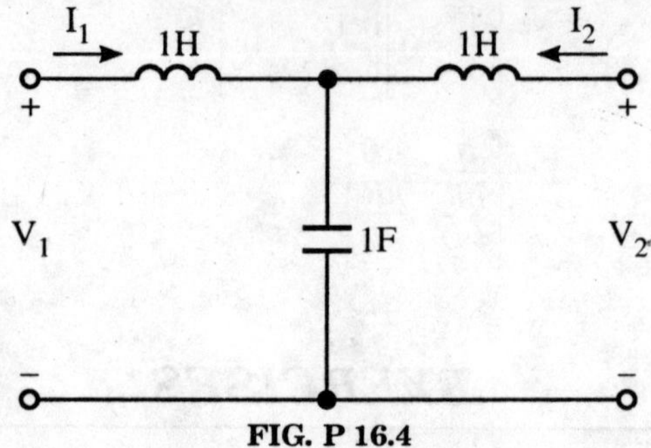

FIG. P 16.4

5. Find the z matrix for the 2-port in Fig. P 16.5.

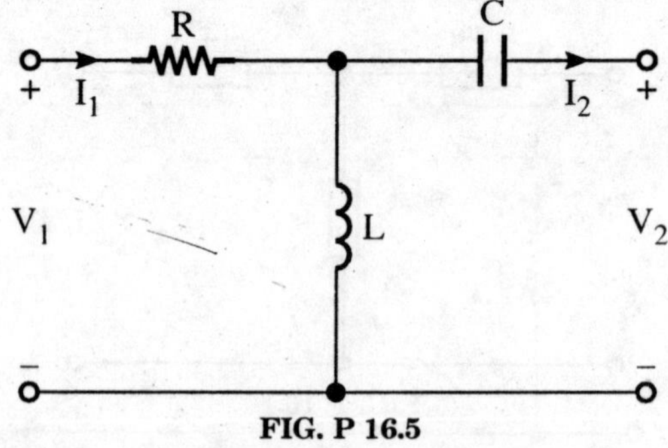

FIG. P 16.5

6. Find the z parameters for the network in Fig. P 16.6.

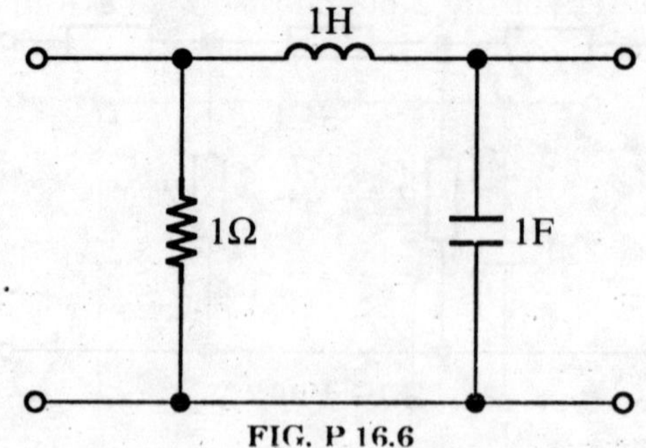

FIG. P 16.6

7. Determine the y parameters for the network in Fig. P 16.7.

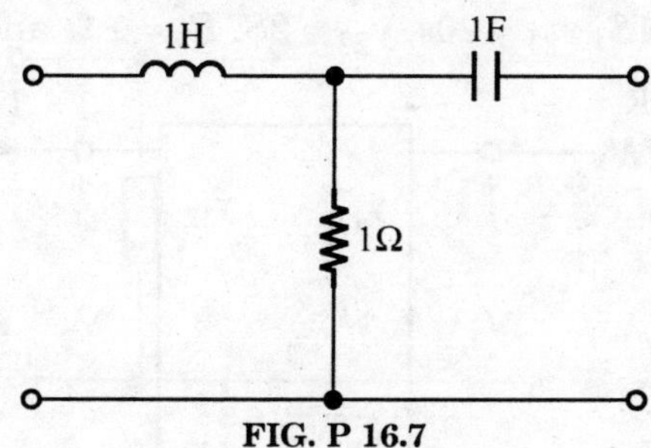

FIG. P 16.7

8. Determine the y parameters of the network shown in Fig. P 16.8.

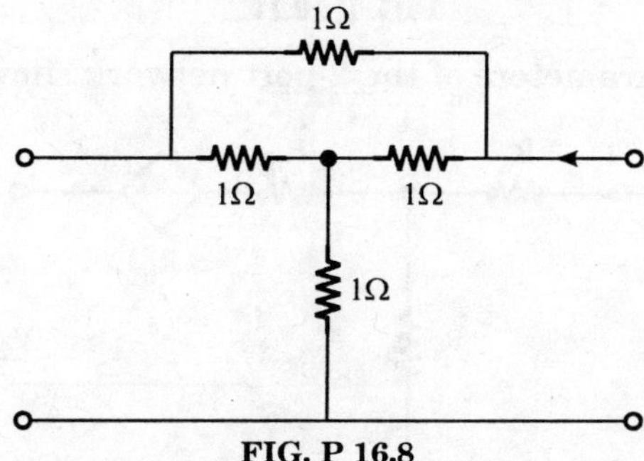

FIG. P 16.8

9. Determine the y and z parameters for the 2-port shown in Fig. P 16.9.

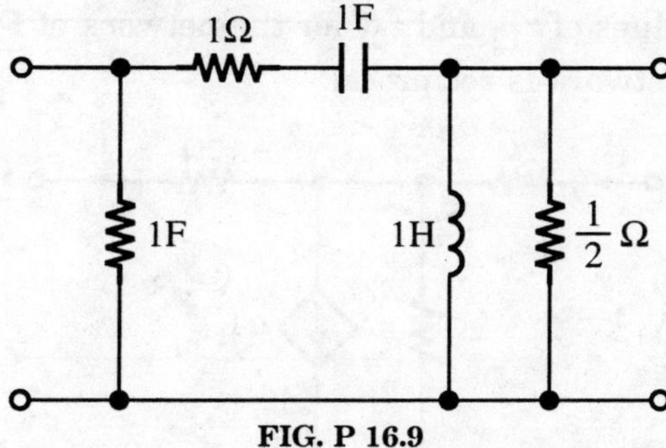

FIG. P 16.9

10. Find the y parameters of the 2-port network shown in Fig. P 16.10.

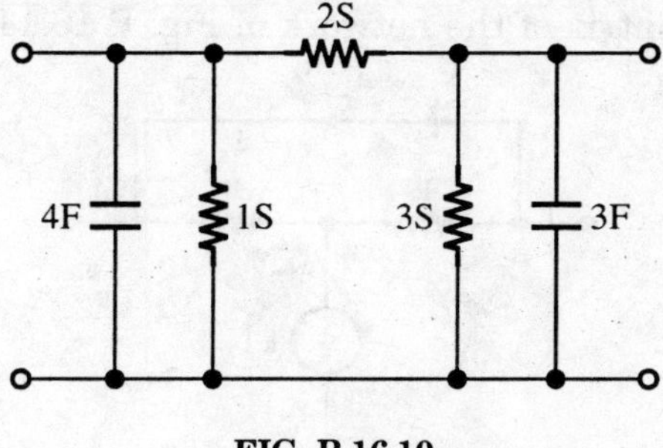

FIG. P 16.10

11. For the network shown in Fig. P 16.11, find V_2/V_s if $y_{11} = 3S$,

$$y_{12} = -1S,\ y_{21} = 20s,\ y_{22} = 2S,\ R_s = 1\ \Omega \text{ and } R_L = 0.1\ \Omega$$

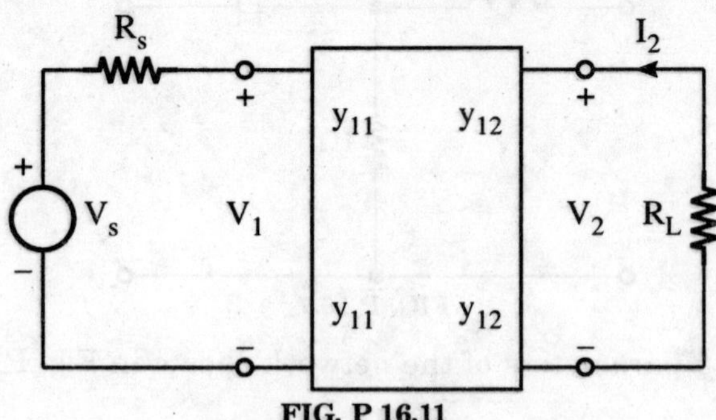

FIG. P 16.11

12. Determine the *h* parameters of the 2-port network shown in Fig. P 16.12.

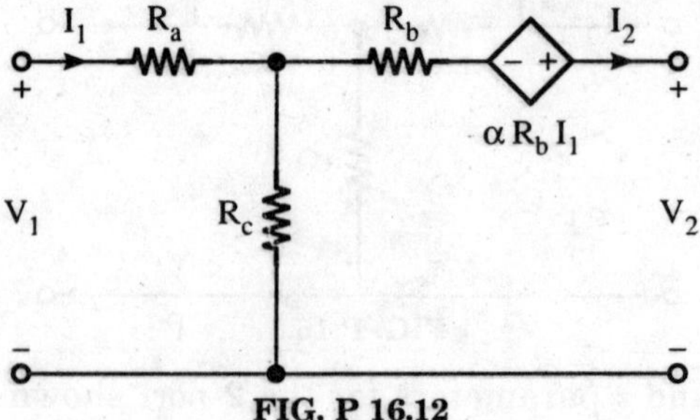

FIG. P 16.12

13. Determine the values of z_{12} and z_{21} for the network of Fig. P 16.13. Find the value of α so that the network is reciprocal.

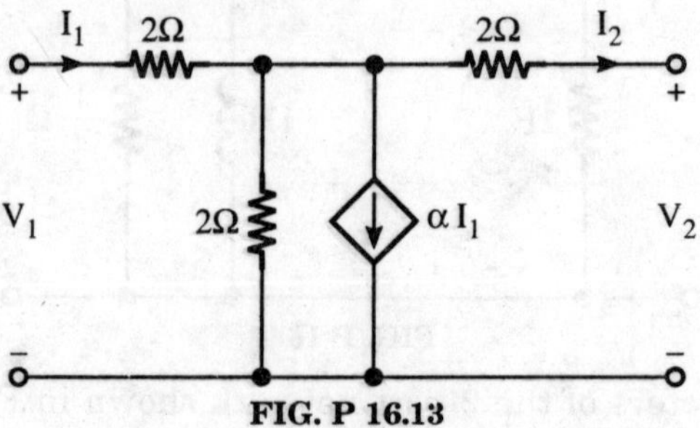

FIG. P 16.13

14. Determine the *h* matrix of the network of Fig. P 16.14.

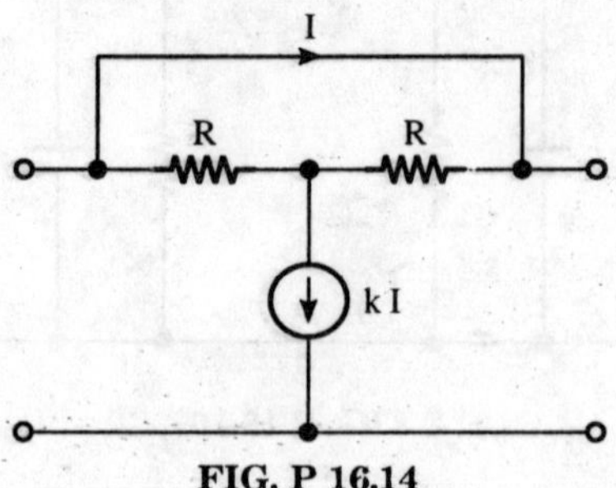

FIG. P 16.14

15. Figure P 16.15 shows the equivalent circuit of a common base transistor. Determine the z and h parameters.

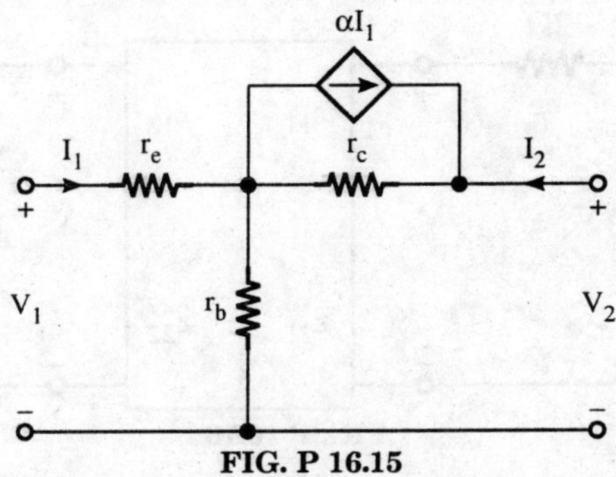

FIG. P 16.15

16. Determine the y parameters of the bridged T network shown in Fig. P 16.16. Show that the network is symmetrical and reciprocal.

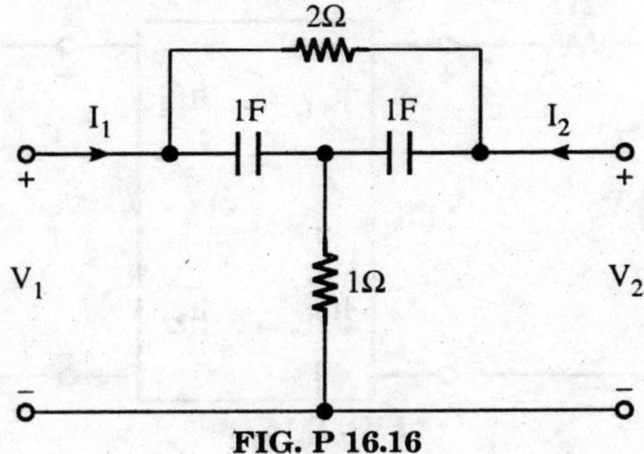

FIG. P 16.16

17. Determine the output admittance at the terminals 2–2′ for the network shown in Fig. P 16.17.

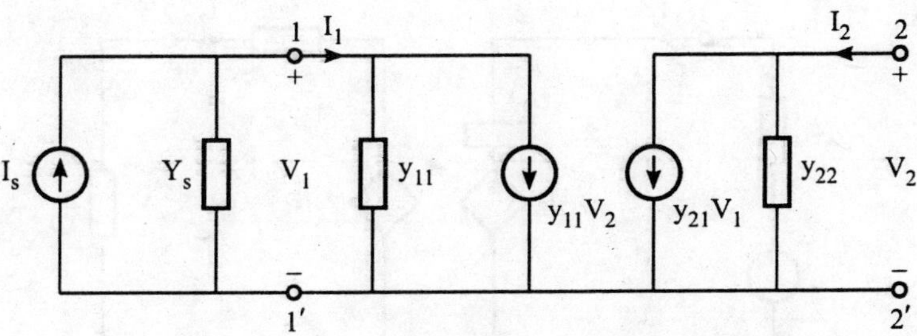

FIG. P 16.17

18. For the network shown in Fig. P 16.18, find V_2/V_S given that $z_{11} = 2\ \Omega$, $z_{12} = z_{21}\ 3\Omega$, and $z_{22} = 3\ \Omega$,

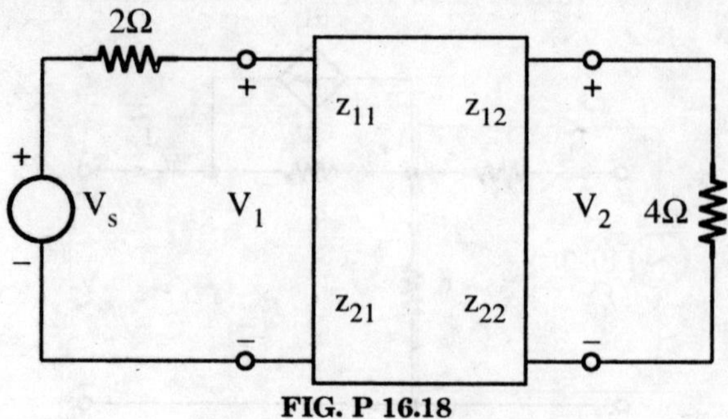

FIG. P 16.18

19. For the network shown in Fig. P 16.19, find V_2/V_S if $h_{11} = 2\ \Omega$, $h_{12} = 0.6$, $h_{21} = 4$, and $h_{22} = 0.5\ S$.

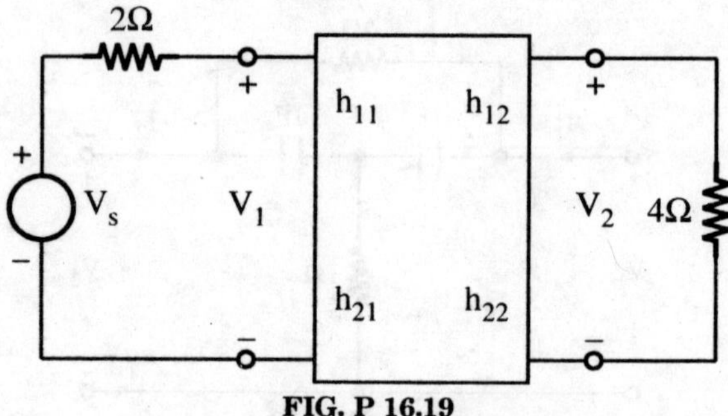

FIG. P 16.19

20. Determine the current gain I_2/I_1, the voltage gain V_2/V_1 and the input impedance at the terminals 1–1′ for the network shown in Fig. P 16.20.

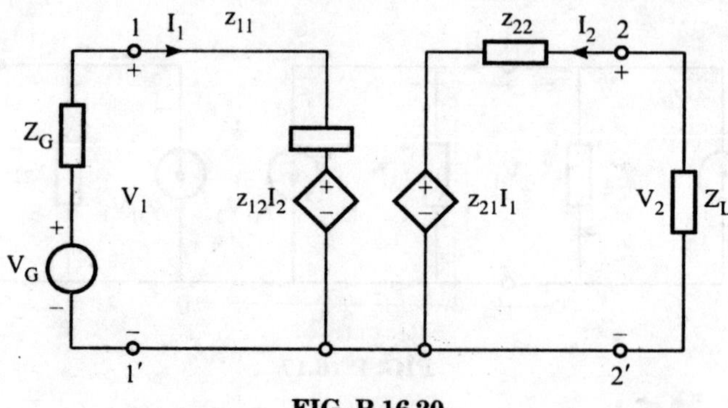

FIG. P 16.20

21. The 2-port network N shown in Fig. P 16.21 is linear and bilateral. With the switch S open $I_1 = \sqrt{2} \angle 45^\circ$ A and $V_2 = 10\sqrt{2} \angle -45^\circ$ V. With S closed $I_1 = 1 \angle 0^\circ$ A. Determine the z parameters of the network N and draw an equivalent T for the network N.

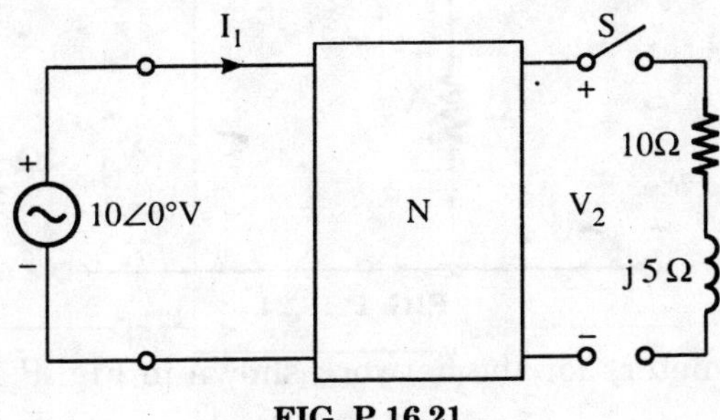

FIG. P 16.21

22. Define the h parameters of a 2-port network. Indicate the dimensions of each parameter.
Find the z and y parameters for the circuit shown in Fig. P 16.22.

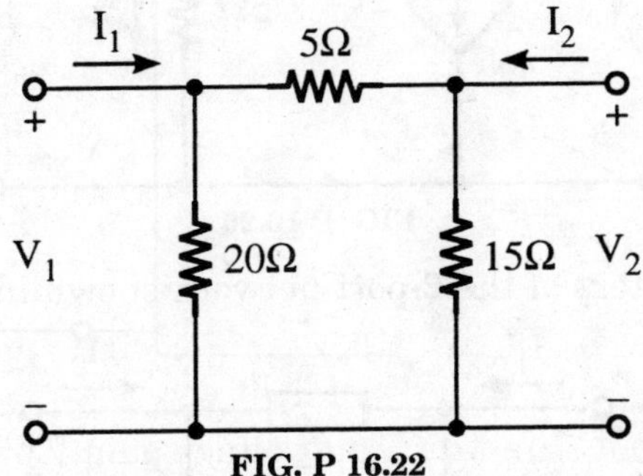

FIG. P 16.22

23. Find the transmission parameters for the network shown in Fig. P 16.23.

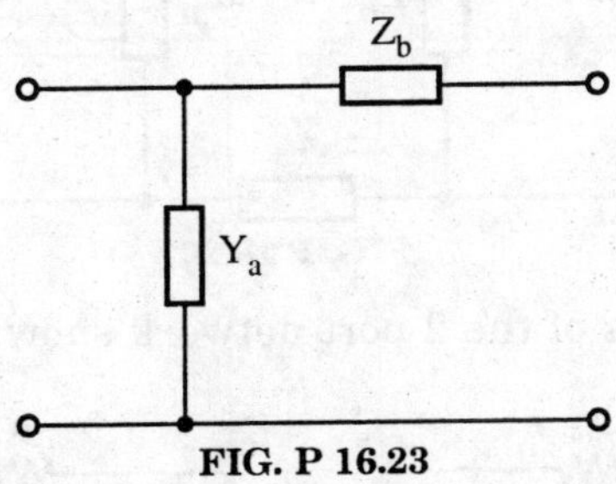

FIG. P 16.23

24. Find the y parameters for the circuit shown in Fig. P 16.24.

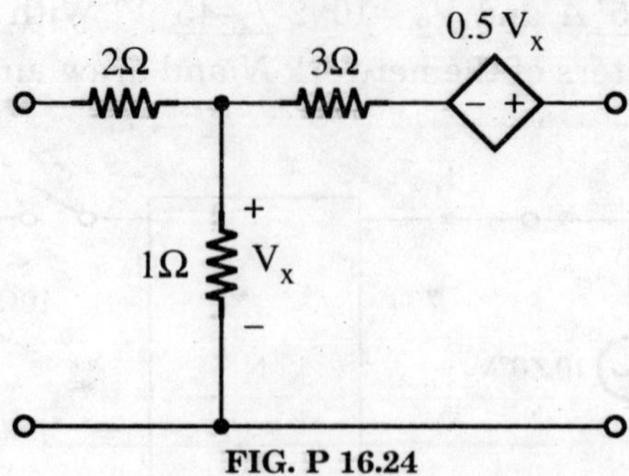

FIG. P 16.24

25. Find the z parameters for the network shown in Fig. P 16.25.

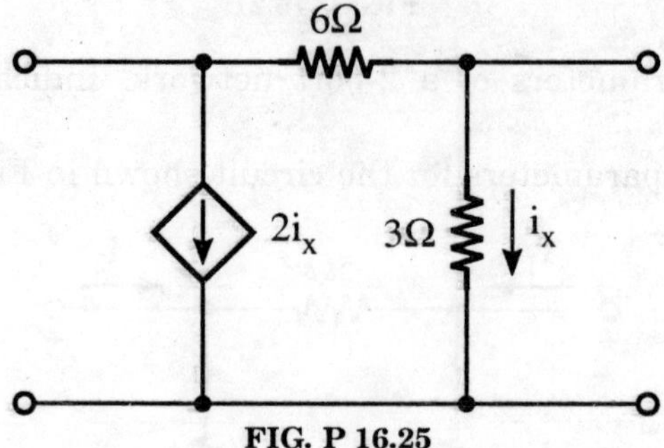

FIG. P 16.25

26. Find the z parameters of the 2-port network shown in Fig. P 16.26.

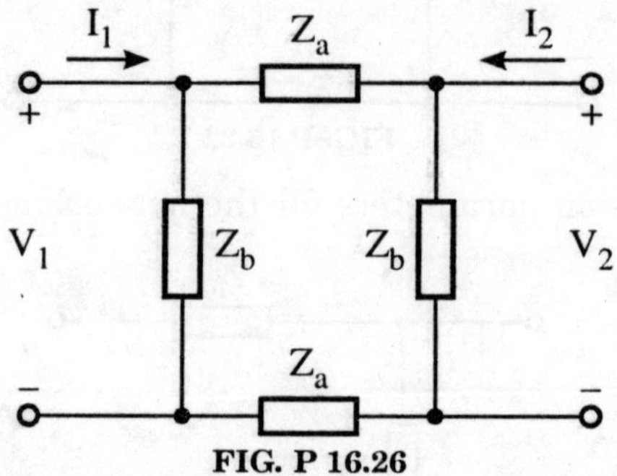

FIG. P 16.26

27. Find the z parameters of the 2-port network shown in Fig. P 16.27.

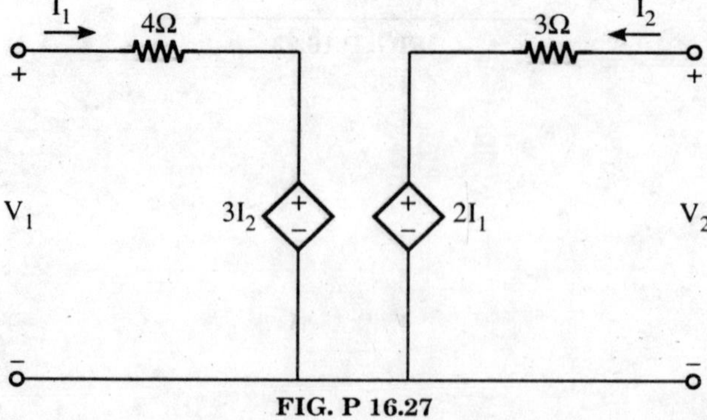

FIG. P 16.27

28. Find the z parameters of the circuit shown in Fig. P 16.28.

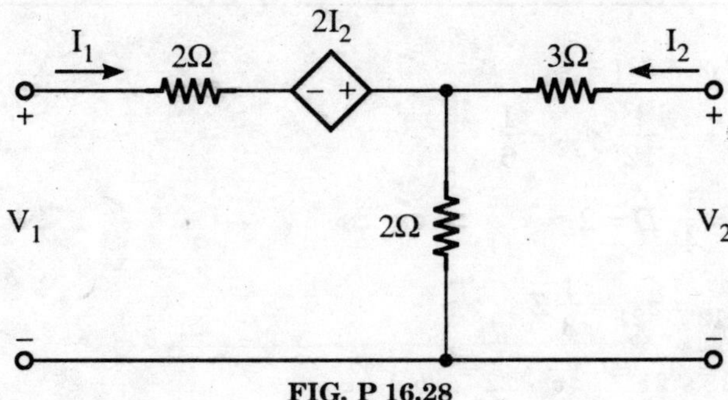

FIG. P 16.28

29. Find the z parameters of the circuit shown in Fig. P 16.29.

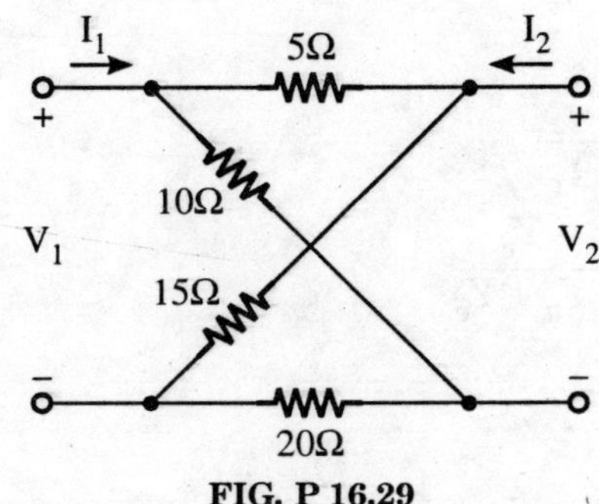

FIG. P 16.29

30. Find the z parameters of the network shown in Fig. P 16.30.

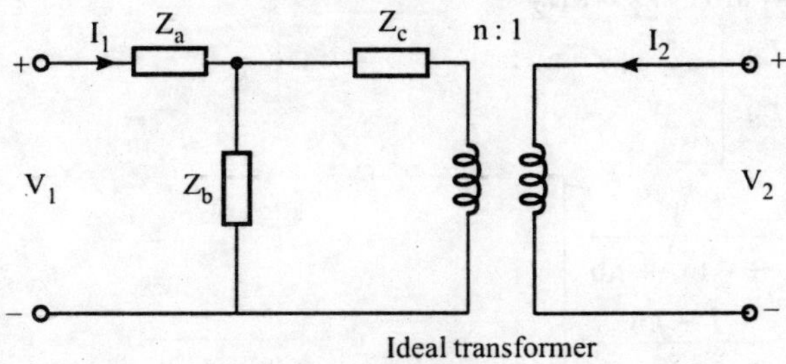

Fig. 16.30

31. Find the z parameters of the network shown in Fig. P 16.31.

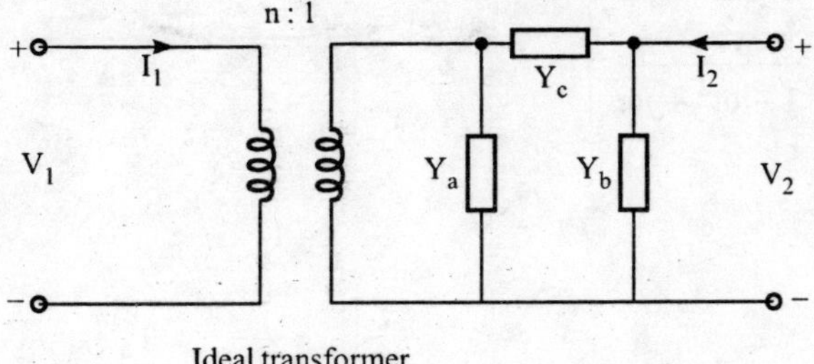

Fig. 16.31

ANSWERS

1. $z_{11} = 8, z_{22} = 8, z_{12} = z_{21} = 4$

$y_{11} = \frac{1}{6}, y_{12} = y_{21} = -\frac{1}{12}, y_{22} = \frac{1}{6}$

$A = 2, B = 12, C = \frac{1}{4}, D = 2$

2. $z_{11} = z_{22} = \frac{5}{3} Z, z_{12} = z_{21} = \frac{1}{3} Z$

3. $y_{11} = Y_a + Y_c, y_{12} = y_{21} = -Y_c, y_{22} = Y_b + Y_c$

$z_{11} = \frac{Y_b + Y_c}{\Delta_y}, z_{12} = z_{21} = \frac{Y_c}{\Delta_y}, z_{22} = \frac{Y_a + Y_c}{\Delta_y}$

$\Delta_y = Y_a Y_b + Y_b Y_c + Y_c Y_a$

$h_{11} = \frac{1}{Y_a + Y_c}, h_{12} = \frac{Y_c}{Y_a + Y_c}$

$h_{21} = \frac{-Y_c}{Y_a + Y_c}, h_{22} = Y_b + \frac{Y_c}{Y_a + Y_c}$

$A = \frac{Y_b + Y_c}{Y_c}, B = \frac{1}{Y_c}$

$C = \frac{1}{Y_c}(Y_a Y_b + Y_b Y_c + Y_c Y_a), D = \frac{1}{Y_c}(Y_a + Y_c)$

4. $z_{11} = sL_1, z_{12} = z_{21}\ sM, z_{22} = sL_2$

5. $\begin{bmatrix} R + Ls & Ls \\ Ls & \frac{1}{Cs} + Ls \end{bmatrix}$

6. $\begin{bmatrix} \frac{1-\omega^2}{1-\omega^2+j\omega} & \frac{1}{1-\omega^2+j\omega} \\ \frac{1}{1-\omega^2+j\omega} & \frac{1+j\omega}{1-\omega^2+j\omega} \end{bmatrix}$

7. $\begin{bmatrix} \frac{1+j\omega}{1-\omega^2+j\omega} & \frac{-j\omega}{1-\omega^2+j\omega} \\ \frac{-j\omega}{1-\omega^2+j\omega} & \frac{-\omega^2+j\omega}{1-\omega^2+j\omega} \end{bmatrix}$

8. $\begin{bmatrix} \frac{5}{3} & -\frac{4}{3} \\ -\frac{4}{3} & \frac{5}{3} \end{bmatrix}$

9. $y_{11} = \dfrac{s^2 + 2s}{s+1}$, $y_{12} = y_{21} = \dfrac{-s}{s+1}$

$y_{22} = \dfrac{3s^2 + 3s + 1}{s\,(s+1)}$

$z_{11} = \dfrac{(3s^2 + 3s + 1)\,(s+1)}{s\,(3s^3 + 8s^2 + 7s + 2)}$

$z_{12} = z_{21} = \dfrac{s\,(s+1)}{3s^3 + 8s^2 + 7s + 2}$

$z_{22} = \dfrac{(s^2 + 2s)\,(s+1)}{3s^3 + 8s^2 + 7s + 2}$

10. $\begin{bmatrix} 3+4s & -2 \\ -2 & 5+3s \end{bmatrix}$

11. $-\dfrac{5}{17}$

12. $h_{11} = \dfrac{1}{R_b + R_c}\,[R_aR_b + R_aR_c + R_bR_c\,(1-\alpha)]$

$h_{12} = \dfrac{R_c}{R_b + R_c}$, $h_{21} = -\dfrac{\alpha R_b + R_c}{R_b + R_c}$

$h_{22} = \dfrac{1}{R_b + R_c}$

13. $z_{12} = 2$, $z_{21} = 2\,(1-\alpha)$

For the network to be reciprocal $\alpha = 0$

14. $\begin{bmatrix} 0 & 1 \\ \dfrac{k-2}{k+2} & 0 \end{bmatrix}$

15. $z_{11} = r_e + r_b$, $z_{12} = r_b$

$z_{21} = \alpha r_e + r_b$, $z_{22} = r_c + r_b$

$h_{11} = -\dfrac{r_br_c\,(1-\alpha)}{r_b + r_c}$, $h_{12} = \dfrac{r_b}{r_b + r_c}$

$h_{21} = -\dfrac{r_b + \alpha r_c}{r_b + r_c}$, $h_{22} = \dfrac{1}{r_b + r_c}$

16. $y_{11} = y_{22} = \dfrac{2s^2 + 4s + 1}{2s\,(1+2s)}$

$y_{12} = y_{21} = -\dfrac{2s^2 + 2s + 1}{2s\,(1+2s)}$

17. $Y_{out} = \dfrac{1}{y_{11} + Y_S}\,(y_{11}y_{22} - y_{12}y_{21} + y_{22}Y_S)$

18. 12/19

19. $-20/3$

20. $\frac{I_2}{I_1} = \frac{-z_{21}}{z_{22} + Z_L}$;

$\frac{V_2}{V_1} = \frac{z_{21} Z_L}{z_{11} z_{22} - z_{12} z_{21} + z_{11} Z_L}$;

$$Z_{in} = \frac{z_{11} z_{22} - z_{12} z_{21} + z_{11} Z_L}{z_{22} + Z_L}\ \Omega$$

21. $z_{11} = 5\,(1 - j)\ \Omega$, $z_{21} = z_{12} = -j\ 10\ \Omega$,

$z_{22} = -j\ 15\ \Omega$.

Elements of T : $5\,(1 + j)\ \Omega$, $-j\ 5\ \Omega$ (series) : $-j\ 10\ \Omega$ (shunt)

22. $z_{11} = 10\ \Omega$, $z_{12} = 7.5\ \Omega$, $z_{21} = 7.5\ \Omega$, $z_{22} = 9.375\ \Omega$

$y_{11} = 0.25\ S$, $y_{12} = y_{21} = -0.2\ S$, $y_{22} = \frac{4}{15}\ S$

23. $\begin{bmatrix} 1 & Z_b \\ Y_a & 1 + Y_a Z_b \end{bmatrix}$

24. $\begin{bmatrix} \frac{3}{8} & -\frac{1}{12} \\ -\frac{1}{8} & \frac{1}{4} \end{bmatrix} S$

25. $\begin{bmatrix} 3 & -3 \\ 1 & 1 \end{bmatrix} \Omega$

26. $z_{11} = z_{22} = \frac{1}{2} Z_b \frac{(2Z_a + Z_b)}{Z_a + Z_b}$

$z_{12} = z_{21} = \frac{1}{2} \frac{Z_b^2}{Z_a + Z_b}$

27. $z_{11} = 4$, $z_{12} = -3$

$z_{21} = 2$, $z_{22} = 3$

28. $z_{11} = 4$, $z_{12} = 0$, $z_{21} = 2$, $z_{22} = 5$

29. $z_{11} = 12\ \Omega$, $z_{12} = 1\ \Omega$, $z_{21} = 1\ \Omega$, $z_{22} = 10.5\ \Omega$

30. $[z] = \begin{bmatrix} z_a + z_b & \frac{z_b}{n} \\ \frac{z_b}{n} & \frac{z_b + z_a}{n^2} \end{bmatrix}$

31. $[z] = \begin{bmatrix} \frac{n^2(Y_b + Y_c)}{Y_a Y_b + Y_b Y_c + Y_c Y_a} & \frac{n Y_c}{Y_a Y_b + Y_b Y_c + Y_c Y_a} \\ \frac{n Y_c}{Y_a Y_b + Y_b Y_c + Y_c Y_a} & \frac{Y_a + Y_c}{Y_a Y_b + Y_b Y_c + Y_c Y_a} \end{bmatrix}$

17

COUPLED ELEMENTS AND CIRCUITS

17.1 INTRODUCTION

Two circuits are said to be coupled when all or part of the electrical energy supplied to one circuit is transferred to the other. Certain coupled elements are frequently used in network analysis and synthesis. Among these are the transformer, the negative impedance converter, the gyrator and the transistor etc. Each of these elements may be represented as a 2-port. In this chapter the properties of these elements are investigated and it is shown that each involves a coupling of one kind or the other.

17.2 MUTUAL INDUCTANCE

When two circuits are so placed that a portion of the magnetic flux produced by one links with the turns of both the circuits, they are said to be mutually coupled magnetically. ***Mutual inductance*** is the property of two circuits when a voltage is induced in one circuit by a change of current in the other circuit. It is a circuit parameter.

Consider two coils 1 and 2 with N_1 and N_2 turns respectively as shown in Fig. 17.1a. The coils

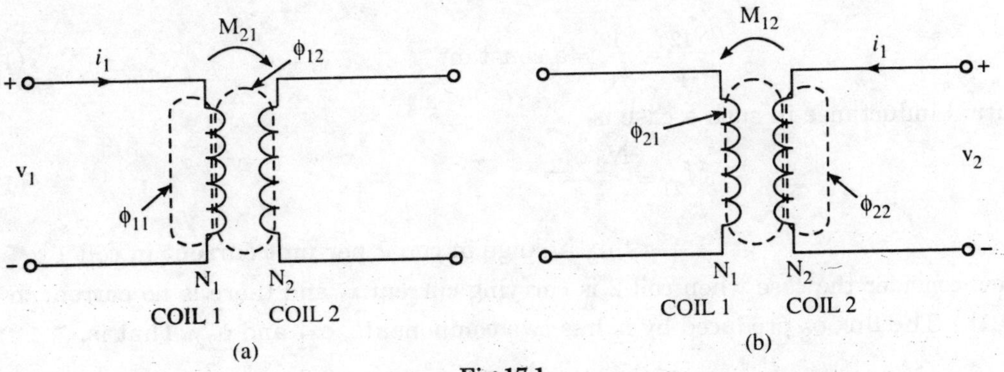

Fig 17.1

are located physically close to one another so that a part of the flux produced by current in one coil links the second coil also. Let the current in coil 1 be i_1. It produces a magnetic flux ϕ_1. This flux has two components ϕ_{11} and ϕ_{12}, that is $\phi_1 = \phi_{11} + \phi_{12}$. The component ϕ_{11} links with the turns of coil 1 only and does not link with the turns of coil 2. The flux ϕ_{11} is called the **self flux** or **leakage flux** of coil 1. The other component ϕ_{12} links with the turns of *both* the coils. It is called the **mutual flux** produced by coil 1. It is also called the **useful flux.**

Mutual flux If the current i_1 changes with time, the flux produced by it will also vary with time. Thus, the change of current in coil 1 changes the flux linkages in both the coils. The voltage induced in coil 1 is called the **self induced voltage**, and the voltage induced in coil 2 is called the **mutually induced voltage**. The voltage induced in coil 2 due to change of current in coil 1 is proportional to the rate of change of i_1. That is,

$$v_{21} \propto \frac{di_1}{dt}$$

$$v_{21} = M_{21} \frac{di_1}{dt} \tag{17.2.1}$$

where M_{21} is a constant called the **coefficient of mutual inductance** or *simply the* **mutual inductance** between the two coils. The order of subscripts on M_{21} indicates that a voltage is induced in coil 2 by changing current in coil 1. The unit of mutual inductance is henry (H). Two coils have a mutual inductance of one 1H when a current changing at the rate of 1 A/s in one coil induces a voltage of 1 V in the other. By Faraday's law the voltage induced in coil 2 is given by

$$v_{21} = N_2 \frac{d\phi_{12}}{dt} \tag{17.2.2}$$

Combination of equations (17.2.1) and (17.2.2) gives

$$M_{21} \frac{di_1}{dt} = N_2 \frac{d\phi_{12}}{dt}$$

$$M_{21} = N_2 \frac{d\phi_{12}}{di_1} \tag{17.2.3}$$

If the coils are linked with air as the medium, then the flux is directly proportional to the current. Hence

$$\frac{d\phi_{12}}{di_1} = \frac{\phi_{12}}{i_1} = \text{a constant} \tag{17.2.4}$$

The mutual inductance in such a case is

$$M_{21} = \frac{N_2 \, \phi_{12}}{i_1} \tag{17.2.5}$$

$$= \text{flux linkage in coil 2 per unit current in coil 1}$$

Now consider the case when coil 2 is carrying current i_2 and there is no current in coil 1 (Fig. 17.1b). The flux ϕ_2 produced by i_2 has two components, ϕ_{21} and ϕ_{22}. That is,

$$\phi_2 = \phi_{21} + \phi_{22} \tag{17.2.6}$$

The component ϕ_{22} links with the turns of coil 2 only and does not link with turns of coil 1. It is called the *self flux* or *leakage flux* of coil 2. The other component ϕ_{21} links with the turn of *both* the *coils*. It is called the *mutual flux* produced by coil 2.

The voltage induced in coil 1 due to change in current in coil 2

$$v_{12} = M_{12}\frac{di_2}{dt} \tag{17.2.7}$$

and the mutual inductance

$$M_{12} = N_1\frac{d\phi_{21}}{di_2} \tag{17.2.8}$$

If the coils are linked with air as the medium then the flux is directly proportional to the current. Hence

$$M_{12} = N_1\frac{\phi_{21}}{i_2} \tag{17.2.9}$$

= flux linkage in coil 1 per unit current in coil 2

It should be noted that

$$M_{12} = M_{21} = M \tag{17.2.10}$$

The proof of this result is as follows :

Consider the path taken by the flux that links both the coils. The flux ϕ_{12} passes through this reluctance and ϕ_{21} passes through the *same* reluctance. That is, the reluctance of the mutual flux path is R_{12} or R_{21} such that $R_{12} = R_{21}$

Now, reluctance $= \dfrac{\text{mmf}}{\text{flux}}$

therefore, $$\frac{N_1 i_1}{\phi_{12}} = \frac{N_2 i_2}{\phi_{21}}$$

or $$\frac{N_1\phi_{21}}{i_2} = \frac{N_2\phi_{12}}{i_1}$$

That is, $$M_{12} = M_{21} = M \text{ say}$$

This relation is true for coils wound on nonmagnetic cores.

17.3 COUPLING COEFFICIENT k

In a practically coupled circuit, not all the magnetic flux produced by one circuit links with the other circuit. If most of the flux produced by one coil links with both coils, the coils are said to be **tightly coupled** or **closely coupled** as in the case of a power transformer. If only a small portion of the flux produced by one coil links with the other, the coils are said to be **loosely coupled**. The degree of magnetic coupling is given by the coupling coefficient. The **coupling coefficient** ***k*** is defined as the ratio of mutual flux to the total flux.

$$k \overset{\Delta}{=} \frac{\text{linking flux}}{\text{total flux}} \tag{17.3.1}$$

Thus, $$k = \frac{\phi_{12}}{\phi_1} = \frac{\phi_{21}}{\phi_2} \tag{17.3.2}$$

Since $\phi_{12} \le \phi_1$ and $\phi_{21} \le \phi_2$, the value of k lies between 0 and 1. That is, $0 \le k \le 1$. The maximum value of k is 1, when the entire flux of each coil links with the other. The value of k depends on the spacing and orientation of the coils and on the permeability of the medium. The closer the two coils are to each other physically, the larger the value of k. Multiplication of Eqs. (17.2.5) and (17.2.9) gives

$$M_{21} \times M_{12} = \frac{N_2 \phi_{12}}{i_1} \times \frac{N_1 \phi_{21}}{i_2} = \frac{N_2 k \phi_1}{i_1} \times \frac{N_1 k \phi_2}{i_2}$$

or $$M^2 = k^2 \frac{N_1 \phi_1}{i_1} \times \frac{N_2 \phi_2}{i_2} = k^2 L_1 L_2$$

$$M = k\sqrt{L_1 L_2} \tag{17.3.3}$$

$$\boxed{k = \frac{M}{\sqrt{L_1 L_2}}} \tag{17.3.4}$$

The maximum value of M occurs when there is perfect coupling ($k = 1$) and there is no leakage.

$$M_{max} = \sqrt{L_1 L_2} \tag{17.3.5}$$

If $k = 0$, there is no coupling between the coils.

■ **EXAMPLE 17.1** The number of turns in two coupled coils are 600 and 1200 respectively. When a current of 4A flows in coil 1, the total flux in this coil is 0.5 mWb and the flux linking coil 2 is 0.4 mWb. Determine L_1, L_2, M, and k.

SOLUTION :

$$L_1 = \frac{\phi_1 N_1}{I_1} = \frac{0.5 \times 10^{-3} \times 600}{4} = 0.075 \text{ H}$$

Since the self inductance is directly proportional to the square of the number of turns, $L \propto N^2$ and

$$\frac{L_1}{L_2} = \frac{N_1^2}{N_2^2}$$

or $$L_2 = \left(\frac{N_2}{N_1}\right)^2 \times L_1 = \left(\frac{1200}{600}\right)^2 \times 0.075 = 0.3 \text{ H}$$

$$M = \frac{N_2 \phi_{12}}{L_1} = \frac{1200 \times 0.4 \times 10^{-3}}{4} = 0.12 \text{ H}$$

$$k = \frac{\phi_{12}}{\phi_1} = \frac{0.4}{0.5} = 0.8$$

Alternatively,

$$k = \frac{M}{\sqrt{L_1 L_2}} = \frac{0.12}{\sqrt{0.075 \times 0.3}} = 0.8$$

17.3.1 Mutual Reactance

Let $\qquad i_1 = I_m \sin \omega t$

From Eq. (17.2.1), $v_{21} = M_{21}\dfrac{di_1}{dt} = \dfrac{M\,di_1}{dt}$

$$v_{21} = M\frac{d}{dt}(I_m \sin \omega t)$$

$$= M\,\omega I_m \cos \omega t$$

or $\qquad v_{21} = M\omega I_m \sin(\omega t + 90°)$

Thus, the phasor V_{21} leads phasor I_1 by 90°. The quantity ωM is called **mutual inductive reactance.**

The complex mutual inductive reactance is $j\omega M$.

17.4 DOT NOTATION AND THE SIGN OF MUTUAL INDUCTANCE

When the voltages induced by self and mutual inductance occur simultaneously, the relative polarities of these two voltages must be determined before making any circuit calculations. The voltage induced in a coil by mutual inductance may either aid or oppose the voltage of self inductance depending upon the sense of the windings and the directions of the currents in the coils. Algebraically, this means that the voltage drop $M\frac{di_2}{dt}$ in coil 1, resulting from the mutual inductance may either be positive or negative with respect to the drop, $L_1\frac{di_2}{dt}$ in coil 1, resulting from self inductance L_1. This condition is taken into account by assigning to M either a positive or a negative sign.

Consider a pair of coupled coils wound on a rectangular core as shown in Fig. 17.2a.

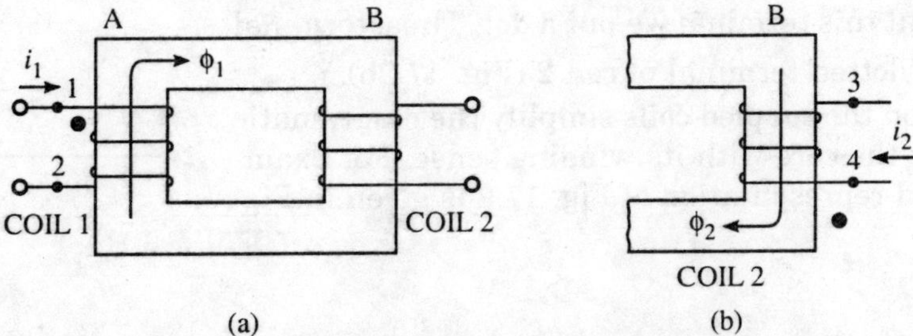

Fig 17.2 Coupled coils on a rectangular core

To determine the proper signs on the voltages of mutual inductance apply the right-hand rule to each coil. The **right-hand rule** is stated as follows :

Imagine the coil to be grasped by the right hand in such a way that the fingers point in the direction of the current in the coil, then the thumb will point in the direction of flux.

This rule is illustrated in Fig. 17.3.

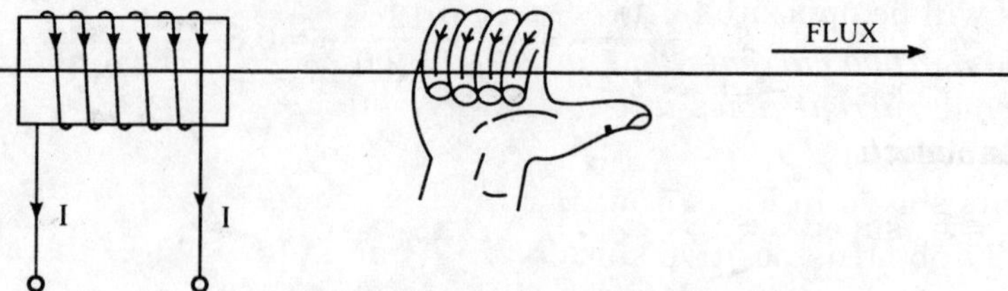

Fig 17.3 Right - hand rule

If the fluxes produced by the two coils aid one another, then the signs on the voltages of mutual inductance are the same as the signs on the voltages of self inductance.

It is inconvenient to draw the windings in electrical networks. To obtain the correct sign of M, the coils are marked with dots at the terminals which are instantaneously of the same polarity.

To assign the dots to a pair of coupled coils, we choose a current direction in one coil and put a dot at the terminal where the current *enters* the winding. Now we apply the right-hand rule to determine the flux produced by this current. Let this flux be ϕ_1. It links coil 2 also. Now apply the right-hand rule to determine the direction of current in coil 2 which produces a flux ϕ_2 such that ϕ_2 aids ϕ_1. Put a dot at the terminal of coil 2 where the current *enters* coil 2. This terminal of coil 2 is positive simultaneously with the terminal of coil 1 where the initial current entered.

Procedure for determining dot markings

The procedure of assigning the dots to a pair of coupled coils in Fig. 17.2 is as follows :

Arbitrarily drive current i at terminal 1 of coil L_1. Put a dot as this terminal. This current will produce a flux ϕ_1 (Fig. 17.2a). The direction of ϕ_1 is found by the application of the right-hand rule. It is seen that this flux is upwards in leg A and downwards in leg B. Now, in coil 2, we have to find the direction of current which should produce a flux ϕ_2. The flux ϕ_2 should aid the flux ϕ_1. By the application of right-hand rule to coil 2, it is found that a current *entering* terminal 4 will produce flux ϕ_2, which will aid flux ϕ_1. Here at this terminal we put a dot. Thus, terminal 4 is the properly dotted terminal of coil 2 (Fig. 17.2b).

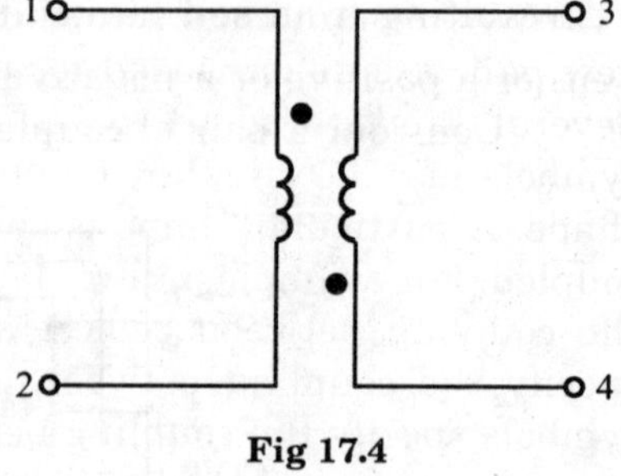

Fig 17.4

Dot marks on the coupled coils simplify the diagramatic representation of the core with its winding sense. For example, the simplified representation of Fig. 17.2 is given in Fig. 17.4.

Dot Rule

The dot rule may be stated as follows :

(1) If both currents enter dotted ends of coupled coils or if both current enter undotted ends, then the signs on the M-terms will be the same as the signs on the L-terms; but

(2) If one current enters a dotted end and the other an undotted end, the signs on the M-terms will be opposite to the signs on the L-terms

In other words, *if two currents flow into each dot, or out of each dot, M is positive.* On the other hand, *if one current flows into a dotted terminal and the other leaves by a dotted terminal, then M is negative.*

For the circuits shown in Fig. 17.5a, M is positive since both I_1 and I_2 flow into dots. In the circuit of Fig. 17.5b, M is positive since both I_1 and I_2 flow out of the dots. Figure 17.5c

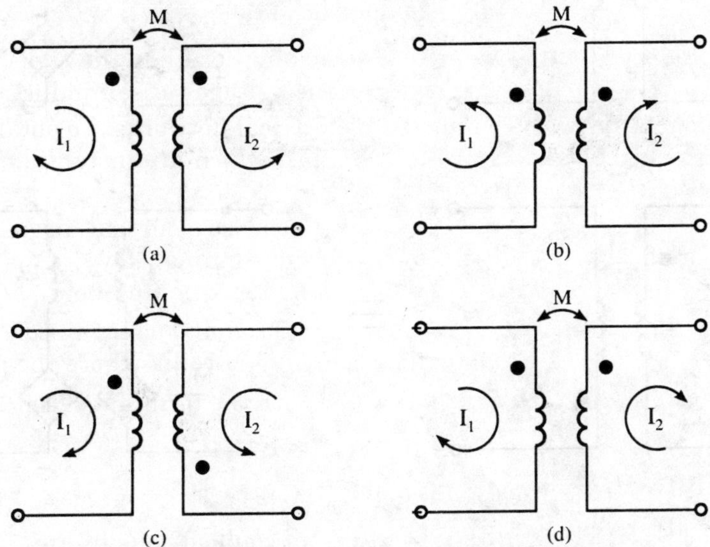

Fig 17.5 Dot convention for induced voltages

and 17.5d show the cases where M is negative since I_1 flows into the dot whereas I_2 flows out of the dot.

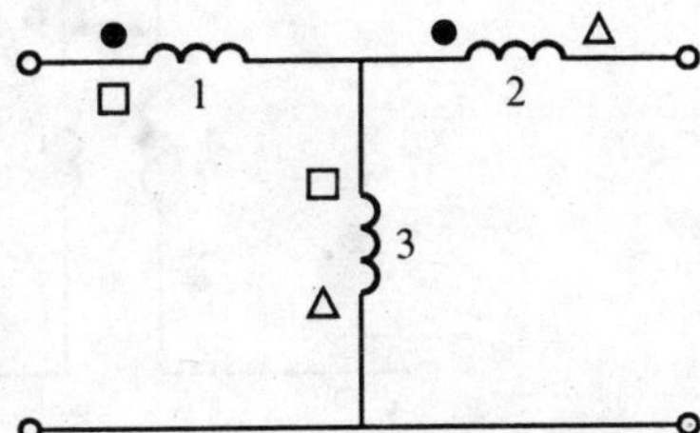

Fig 17.6 Use of several kinds of dots to specify couplings

When more than two inductors are coupled together a new pair of dots is needed for each pair of inductors coupled. Several kinds of dots (polarity marks) are used. The dot symbols may be circular, triangular, square or of some other shape. A particular shape is chosen for each pair of inductors coupled. For example, in Fig. 17.6 the circular symbols indicate the coupling between coils 1 and 2; the triangular symbols specify the coupling between coils 2 and 3, and the square symbols specify the coupling between coils 1 and 3.

17.15 CIRCUIT EQUATIONS FOR COUPLED INDUCTORS

Consider a coupled circuit consisting of two inductors L_1 and L_2 as shown in Fig. 17.7. Assume that voltages v_1 and v_2 are applied to the respective inductors. Each of the two currents i_1 and i_2 are arbitrarily assigned directions *into* the dots.

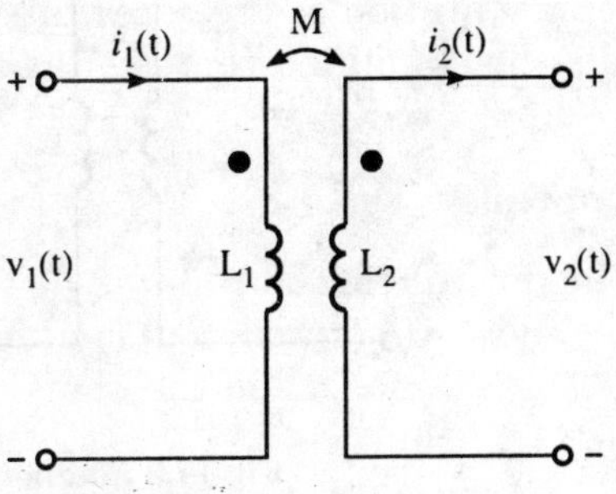

Fig 17.7 Coupled inductors

Since i_1 enters the dot of L_1 and i_2 enters the dot of L_2, the sign of mutual inductance M is positive. By applying KVL to each of the inductors we get

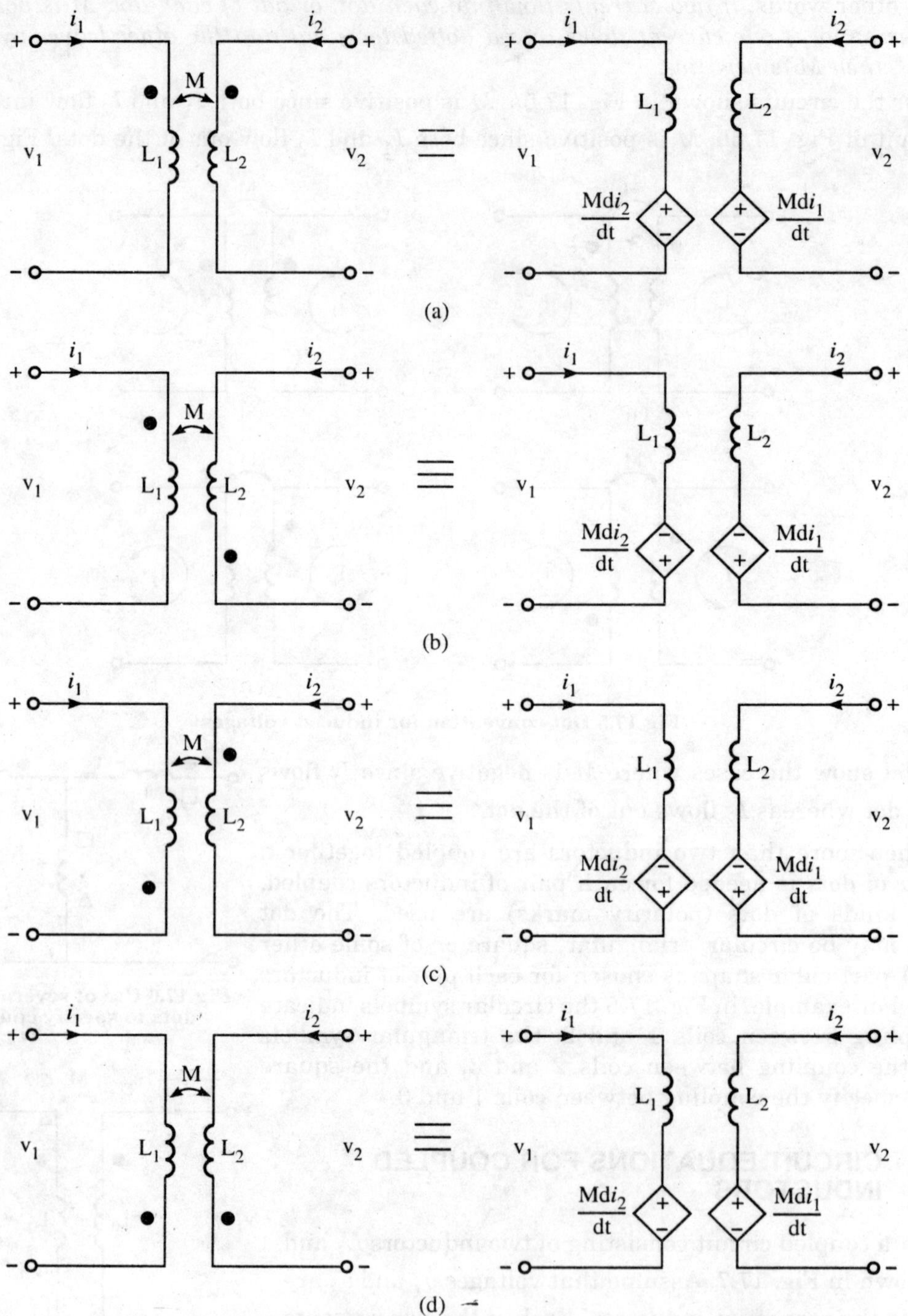

Fig 17.8 Magnetically coupled circuits and their equivalent circuits

$$v_1 = L_1 \frac{di_1}{dt} + M \frac{di_2}{dt}$$

$$v_2 = M \frac{di_1}{dt} + L_2 \frac{di_2}{dt}$$

The mutually induced voltages may be shown as controlled voltage sources.

Figure 17.8 shows all possible combinations of the dot convention between two magnetically coupled windings. Their equivalent circuits show the mutually induced voltages as controlled voltage sources with proper polarities.

For Fig. 17.8a,

$$v_1 = L_1 \frac{di_1}{dt} + M \frac{di_2}{dt}$$

$$v_2 = L_2 \frac{di_2}{dt} + M \frac{di_1}{dt}$$

For Fig. 17.8b,

$$v_1 = L_1 \frac{di_1}{dt} - M \frac{di_2}{dt}$$

$$v_2 = L_2 \frac{di_2}{dt} - M \frac{di_1}{dt}$$

For Fig. 17.8c,

$$v_1 = L_1 \frac{di_1}{dt} - M \frac{di_2}{dt}$$

$$v_2 = L_2 \frac{di_2}{dt} - M \frac{di_1}{dt}$$

For Fig. 17.8d,

$$v_1 = L_1 \frac{di_1}{dt} + M \frac{di_2}{dt}$$

$$v_2 = L_2 \frac{di_2}{dt} + M \frac{di_1}{dt}$$

17.6 FREQUENCY-DOMAIN MODEL OF A COUPLED CIRCUIT

Consider the coupled time-domain circuit shown in Fig. 17.7. The pair of equations describing this circuit in the time domain for the general case of $i_1 \neq 0$ and $i_2 \neq 0$ is given by

$$v_1 = L_1 \frac{di_1}{dt} + M \frac{di_2}{dt} \tag{17.6.1}$$

$$v_2 = M \frac{di_1}{dt} + L_2 \frac{di_2}{dt} \tag{17.6.2}$$

For an inductor in the time domain

$$v = L\frac{di}{dt} \tag{17.6.3}$$

The phasor **V** leads **I** by 90° and therefore in the frequency domain, Eq. (17.6.3) becomes

$$\mathbf{V} = j\omega L\mathbf{I} \tag{17.6.4}$$

Mutual inductance transforms to the frequency domain in a similar manner. If

$$v_2 = M\frac{di_1}{dt} \tag{17.6.5}$$

the phasor $\mathbf{V}_2$ leads phasor $\mathbf{I}_1$ by 90°. Therefore in the frequency domain, Eq. (17.6.5) becomes

$$\mathbf{V}_2 = j\omega M\mathbf{I}_1 \tag{17.6.6}$$

The quantity ωM is called the mutual inductive reactance. It is denoted by X_m. Therefore,

$$X_m = \omega M \tag{17.6.7}$$

The mutual impedance is given by

$$\mathbf{Z}_m = jX_m = j\omega M \tag{17.6.8}$$

In the frequency domain, Eqs. (17.6.1) and (17.6.2) become

$$\mathbf{V}_1 = j\omega L_1\,\mathbf{I}_1 + j\omega M\mathbf{I}_2 \tag{17.6.9}$$

$$\mathbf{V}_2 = j\omega M\mathbf{I}_1 + j\omega L_2\mathbf{I}_2 \tag{17.6.10}$$

Eqs. (17.6.9) and (17.6.10) relate the phasor voltages and phasor currents of the coupled coils. Since the current in each circuit induces a voltage in the other circuit, there is a dependent voltage source in each circuit. The frequency-domain circuit of the coupled circuit of Fig. 17.7 is shown in Fig. 17.9a.

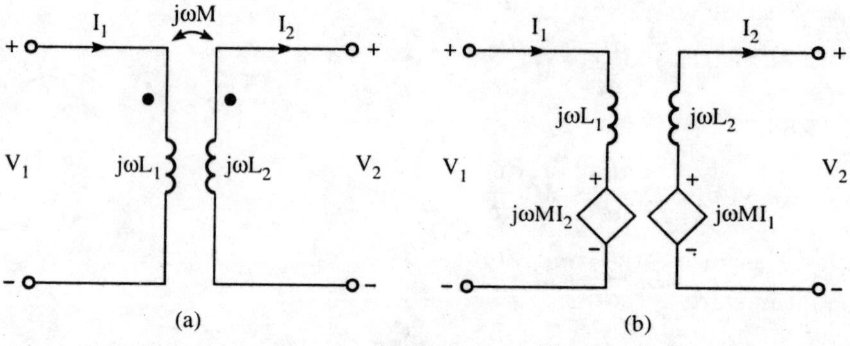

Fig 17.9

It is to be noted that *the sign convention and the meaning of the dot notation are not changed by the transformation from the time domain to the frequency domain.*

It is often convenient to show the mutually induced voltages as controlled voltage sources with proper polarities as shown in Fig. 17.9b. Here both the voltage sources are positive at the top because both of the dots are at the top of the windings. Fig. 17.9b is drawn on the basis of Eqs. (17.6.9) and (17.6.10).

■ **EXAMPLE 17.2** Write the voltage equation for the networks shown in Figs 17.10a, b and c.

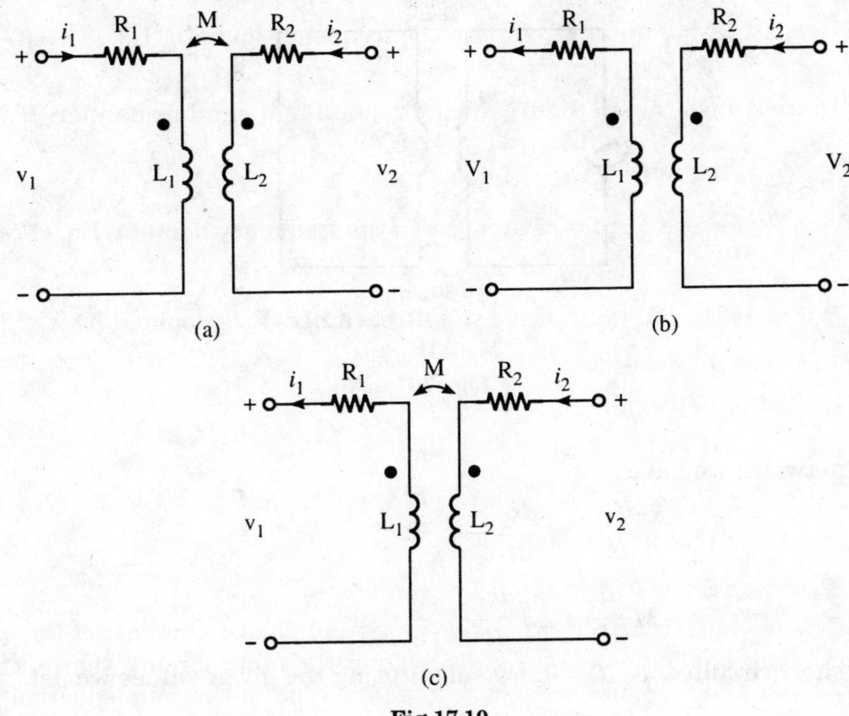

Fig 17.10

SOLUTION :

For the network of Fig. 17.10a, the voltage equations are

$$-v_1 = R_1 i_1 + L_1 \frac{di_1}{dt} + M\frac{di_2}{dt}$$

$$v_2 = R_2 i_2 + L_2 \frac{di_2}{dt} + M\frac{di_1}{dt}$$

Here M is positive because both i_1 and i_2 enter dotted terminals. For the network of Fig. 17.10b, the voltage equations are

$$v_1 = -R_1 i_1 - L_1 \frac{di_1}{dt} + M\frac{di_2}{dt}$$

$$v_2 = R_2 i_2 + L_2 \frac{di_2}{dt} - M\frac{di_1}{dt}$$

The voltage equations of the network in Fig. 17.10c are

$$v_1 = R_1 i_1 + L_1 \frac{di_1}{dt} - M\frac{di_2}{dt}$$

$$v_2 = R_2 i_2 + L_2 \frac{di_2}{dt} - M\frac{di_1}{dt}$$

EXAMPLE 17.3 Determine the current $i_2(t)$ in the coupled circuit of Fig. 17.11.

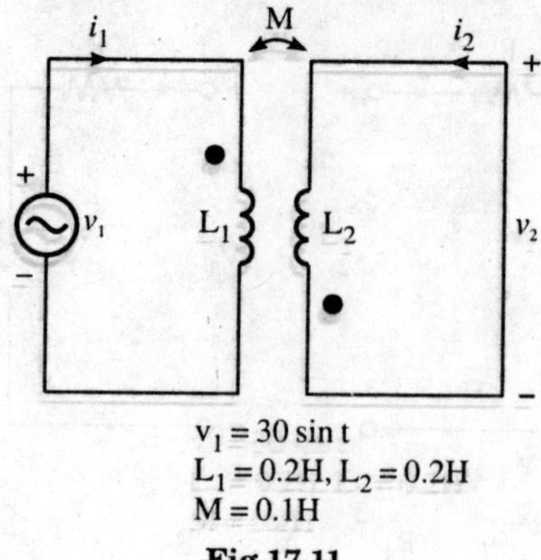

$v_1 = 30 \sin t$
$L_1 = 0.2H, L_2 = 0.2H$
$M = 0.1H$

Fig 17.11

SOLUTION :

For the given network, we have

$$v_1 = L_1 \frac{di_1}{dt} - M \frac{di_2}{dt} \tag{E 17.3.1}$$

$$v_2 = -M \frac{di_1}{dt} + L_2 \frac{di_2}{dt} \tag{E 17.3.2}$$

Since coil 2 is short circuited, $v_2(t) = 0$. By substituting the given values we get

$$30 \sin t = 0.2 \frac{di_1}{dt} - 0.1 \frac{di_2}{dt} \tag{E 17.3.3}$$

$$0 = -0.1 \frac{di_1}{dt} + 0.2 \frac{di_2}{dt} \tag{E 17.3.4}$$

Eq. (E 17.3.4) gives

$$\frac{di_1}{dt} = 2 \frac{di_2}{dt}$$

Substituting this value in Eq. (E 17.3.3) we get

$$30 \sin t = 0.4 \frac{di_2}{dt} - 0.1 \frac{di_2}{dt}$$

$$\frac{di_2}{dt} = 100 \sin t$$

Integrating both sides we get

$$i_2(t) = \int_0^t 100 \sin t \, dt$$

$$= \Big[-100 \cos t\Big]_0^t$$

or

$$i_2(t) = 100(1 - \cos t)$$

■ **EXAMPLE 17.4** Two mutually coupled coils have inductances as indicated in Fig. 17.12. Determine the voltage induced in coil 2.

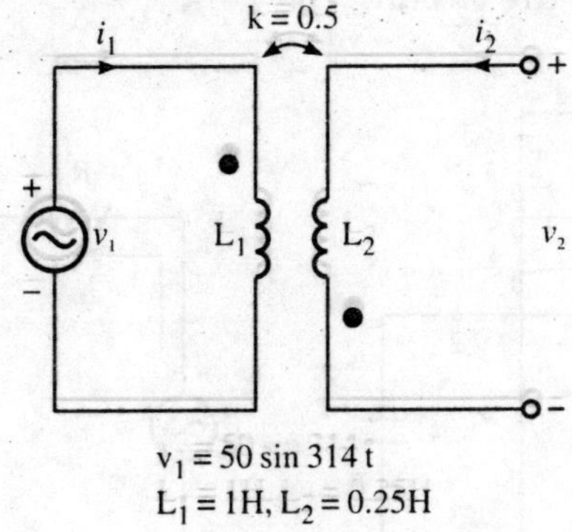

Fig 17.12

SOLUTION :

$$M = k\sqrt{L_1 L_2} = 0.5 \times \sqrt{1 \times 0.25} = 0.25$$

The voltage induced in coil 2,

$$v_2 = M\frac{di_1}{dt} + L_2\frac{di_2}{dt} \qquad \text{(E 17.4.1)}$$

Since coil 2 is open circuited,

$$i_2 = 0 \text{ and, therefore, } \frac{di_2}{dt} = 0$$

therefore,
$$v_2 = M\frac{di_1}{dt} \qquad \text{(E 17.4.2)}$$

The voltage induced on coil 1

$$v_1 = L_1\frac{di_1}{dt} + M\frac{di_2}{dt}$$

$$= L_1\frac{di_1}{dt} + M \times 0$$

therefore,
$$\frac{di_1}{dt} = \frac{v_1}{L_1} = \frac{50 \sin 314\,t}{1} \qquad \text{(E 17.4.3)}$$

Substituting the values of M and $\frac{di_1}{dt}$ in Eq. (E 17.4.2),

$$v_2 = M\frac{di_1}{dt}$$

$$= 0.25 \times 50 \sin 314\,t$$

or
$$v_2 = 12.5 \sin 314\,t$$

EXAMPLE 17.5 Draw the dotted equivalent of the circuit shown in Fig. 17.13a. Draw the equivalent circuit showing the mutually induced voltages as controlled voltage sources. Write the KVL equation of the circuit.

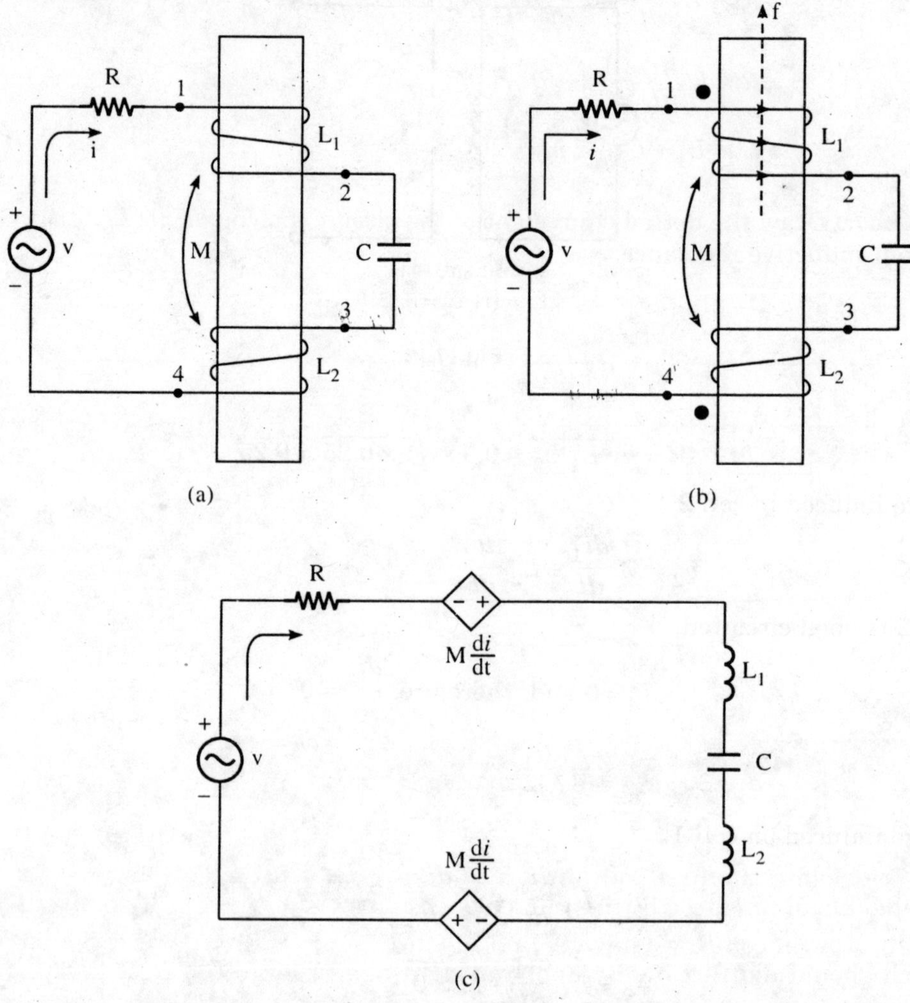

Fig 17.13 (a) Circuit of Example 17.5 (b) Dotted equivalent network of Fig. 17.13a (c) circuit showing the mutually induced voltages as controlled voltage sources

SOLUTION :

Put a dot at terminal 1 of coil L_1 where i_1 *enters*. Determine the flux produced by this current by applying the right-hand rule. This flux is in upward direction as shown in Fig. 17.13b. A current entering terminal 4 will produce a flux upwards. This flux will aid the flux produced by L_1 and will link both the coils L_1 and L_2. Therefore we put a dot at terminal 4. The dotted equivalent circuit is shown in Fig. 17.13b. The equivalent circuit showing the mutually induced voltages as voltage-controlled sources is shown in Fig. 17.13c.

Since the current i enters L_1 at the dotted the and leaves L_2 at the dotted end, the sign of M is negative. The KVL equation is

$$Ri + L_1 \frac{di}{dt} - M\frac{di}{dt} + \frac{1}{C}\int i\,dt + L_2 \frac{di}{dt} - M\frac{di}{dt} = v$$

or
$$Ri + L' \frac{di}{dt} + \frac{1}{C}\int i\,dt = v$$

where
$$L' = L_1 + L_2 - 2M$$

■ **EXAMPLE 17.6** Draw the dotted equivalent of the circuit shown in Fig. 17.14a and find the equivalent inductive reactance.

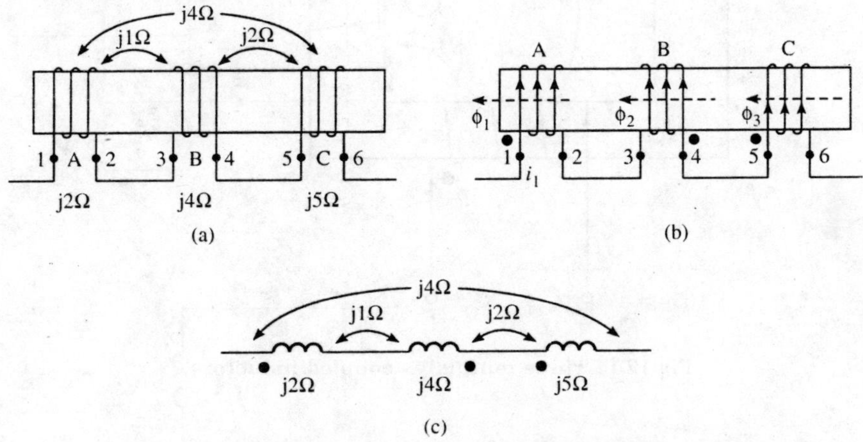

Fig 17.14 (a) Circuit of Example 17.6 (b) Application of right - hand rule (c) Dotted equivalent of Fig 17.14a

SOLUTION :

Draw a current i into the first coil and put a dot at the terminal 1 where i enters coil A. Determine the flux ϕ_1 produced by this current by applying right-hand rule. This flux is shown in Fig. 17.14b. Now in coil B, we have to find the direction of current which should produce a flux ϕ_2 which should aid flux ϕ_1. By applying right-hand rule to coil B, it is found that the current entering terminal 4 will produce flux ϕ_2 which will aid flux ϕ_1. We, therefore, put a dot at terminal 4. Similarly, we put a dot at terminal 5, because a current entering at terminal 5 will produce a flux ϕ_3 which will aid flux ϕ_1. It is to be clearly understood that series connections of the coils are ignored while determining the positions of the dots. The dotted equivalent circuit is shown in Fig. 17.14c.

The equivalent impedance of the circuit is

$$\mathbf{Z} = j2 + j4 + j5 - j1 - j2 + j4 - j1 - j2 + j4 = j13\ \Omega$$

Therefore the equivalent inductive reactance is 13 Ω.

■ **EXAMPLE 17.7** Assign the proper dot notation to the three mutually coupled inductors shown in Fig. 17.15 and write the equation for terminal voltages.

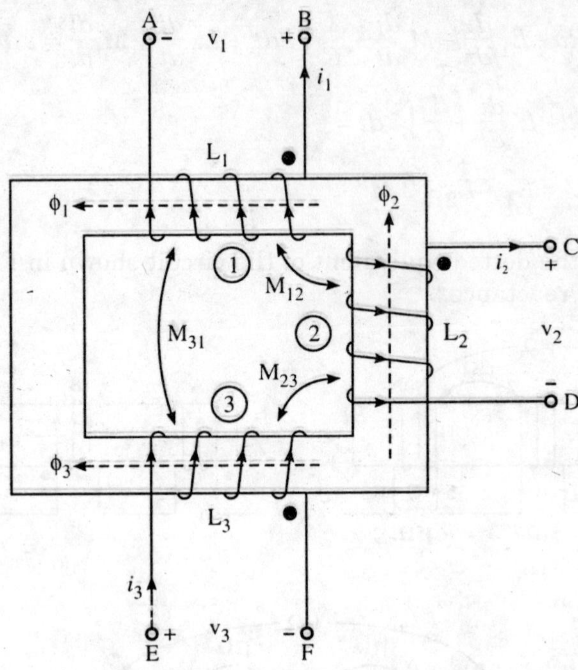

Fig 17.15 Three mutually - coupled inductors

SOLUTION :

We shall apply the right-hand rule to determine the fluxes ϕ_1, ϕ_2 and ϕ_3 due to current i_1, i_2, and i_3 respectively. These fluxes are shown in Fig. 17.15. Since current i_1 is entering coil 1 at terminal B, we put a dot at B. Fluxes ϕ_1 and ϕ_2 are in the same direction, so we place a dot at terminal C. Since flux ϕ_3 is opposing flux ϕ_1 we shall place the third dot at terminal F where current i_3 leaves the coil 3.

We can write the voltage equations as follows :

$$v_1 = L_1 \frac{di_1}{dt} + M_{12} \frac{di_2}{dt} - M_{31} \frac{di_3}{dt}$$

M_{12} is positive since both currents i_1 and i_2 enter at the dotted terminals. M_{31} is negative since i_1 enters the dotted terminal while i_3 flows out of the dotted terminal.

$$v_2 = M_{12} \frac{di_1}{dt} + L_2 \frac{di_2}{dt} - M_{23} \frac{di_3}{dt}$$

M_{21} is positive as indicated above. M_{23} is negative since i_2 enters the dotted terminal and i_3 leaves the dotted terminal.

$$v_3 = -M_{31} \frac{di_1}{dt} - M_{23} \frac{di_2}{dt} + L_3 \frac{di_3}{dt}$$

■ **EXAMPLE 17.8** Write the voltage equation for the coupled coils shown in Fig. 17.16 assuming that $i \neq 0$, $i_2 \neq 0$.

SOLUTION : $M = k\sqrt{L_1 L_2} = 0.6 \times \sqrt{2 \times 8} = 2.4$ H

The voltage equations are

$$v_1 = L_1 \frac{di_1}{dt} - M\frac{di_2}{dt}$$

$$= 2\frac{di_1}{dt} - 2.4\frac{di_2}{dt}$$

$$v_2 = +M\frac{di_1}{dt} - L_2\frac{di_2}{dt}$$

$$= 2.4\frac{di_1}{dt} - 8\frac{di_2}{dt}$$

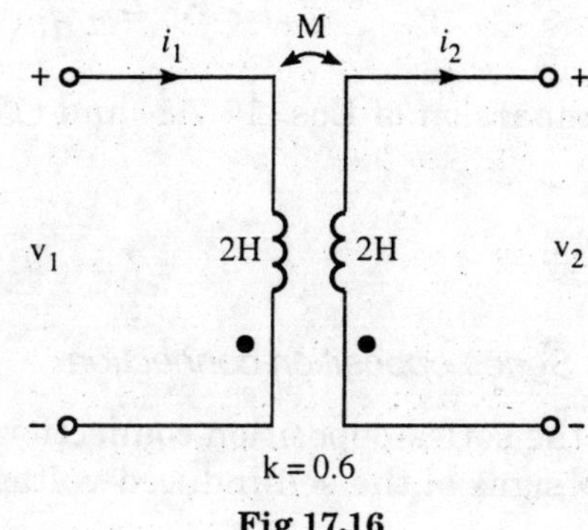

Fig 17.16

17.7 INDUCTANCE OF COUPLED COILS IN SERIES

Consider a circuit containing two mutually coupled coils connected in series. These are two possible ways to connect the coils in series. In one connection (Fig. 17.17a), the fluxes produced by the coils act in the same direction. These coils are said to be connected is *series aiding*. If the current is reversed in the other coil so that the fluxes produced by the coils oppose each other, the coils are said to be connected in *series-opposition*. Such a connection is shown in Fig. 17.17b. Let the self inductances of coils 1 and 2 be L_1 and L_2 respectively. The two coils are coupled by mutual inductance M.

Fig 17.17 (a) Series-aiding connection (b) Series-opposing connection

(a) Series-aiding connection

In the series-aiding connection, the mutually induced voltages have the same sign as the self induced voltages. By KVL

$$v = v_{S1} + v_{M1} + v_{S2} + v_{M2}$$

$$= L_1\frac{di}{dt} + M\frac{di}{dt} + L_2\frac{di}{dt} + M\frac{di}{dt}$$

or

$$v = (L_1 + L_2 + 2M)\frac{di}{dt} \tag{17.7.1}$$

If L_{eq} is the equivalent inductance of two coils series-aiding connection, then the total supplied voltage

$$v = L_{eq}\frac{di}{dt} \tag{17.7.2}$$

Comparsion of Eqs (17.7.1) and (17.7.2) gives

$$\boxed{L_{eq} = L_1 + L_2 + 2M} \tag{17.7.3}$$

(b) Series-opposition connection

In the series-opposition connection, the signs of the mutually induced voltages are opposite to the signs of the self-induced voltages. By KVL,

$$v = v_{S1} + v_{M1} + v_{S2} + v_{M2}$$

$$= L_1\frac{di}{dt} - M\frac{di}{dt} + L_2\frac{di}{dt} - M\frac{di}{dt}$$

or $$v = (L_1 + L_2 - 2M)\frac{di}{dt} \tag{17.7.4}$$

If L'_{eq} is the equivalent inductance of the two coils in series-opposition connection, then the total supplied voltage is

$$v = L'_{eq}\frac{di}{dt} \tag{17.7.5}$$

Comparison of Eqs (17.7.4) and (17.7.5) gives

$$\boxed{L'_{eq} = L_1 + L_2 - 2M}$$

■ **EXAMPLE 17.9** When two identical coupled coils are connected in series, the inductance is found to be 80 mH. When the connections to one of the coils are reversed, a similar measurement indicates 20 mH. Find the mutual inductance and the coupling coefficent between the coils.

SOLUTION :

$$L_{eq} = L_1 + L_2 + 2M$$

$$80 = L_1 + L_2 + 2M$$

$$L'_{eq} = L_1 + L_2 - 2M$$

$$20 = L_1 + L_2 - 2M$$

$$L_{eq} - L'_{eq} = 4M$$

$$80 - 20 = 4M\,,\ M = 15\text{ mH}$$

Since the two coils are identical, $L_1 = L_2$. Therefore

$$L_{eq} = 2L_1 + 2M$$

$$80 = 2L_1 + 2 \times 15\ ,\ L_1 = 25 \text{ mH}$$

$$L_2 = L_1 = 25 \text{ mH}$$

Coupling coefficient

$$k = \frac{M}{\sqrt{L_1 L_2}} = \frac{15}{\sqrt{25 \times 25}} = 0.6$$

17.8 WRITING MESH EQUATIONS OF MUTUALLY COUPLED CIRCUITS BY INSPECTION

Mesh equations of mutually coupled circuits (without supermeshes) also can be written by inspection in the same manner as before, except for the mutual terms. In mesh analysis the unknowns are mesh currents. The coefficients of these unknowns are dependent upon the position numbers. For example, in the 1-1 position, instead of adding all the impedances in mesh 1 and multiplying by $I_1(s)$ we add all the voltages in mesh 1 due only to $I_1(s)$. In order to account for mutual voltages, we consider the mutual terms in terms of their position numbers. For example, in 1-2 position, instead of adding all the impedance, common to meshes 1 and 2 multiplying by $I_2(s)$, we add all the voltages in mesh 1 due only to $I_2(s)$. In general, in the j–k position, instead of adding all the impedances common to meshes j and k and multiplying by $I_k(s)$, we add all the voltages in mesh j due only to $I_k(s)$.

In order to write mesh equations for networks containing super-meshes, KVL is applied directly.

■ **EXAMPLE 17.9** Write the mesh equations for the circuit shown in Fig. 17.18.

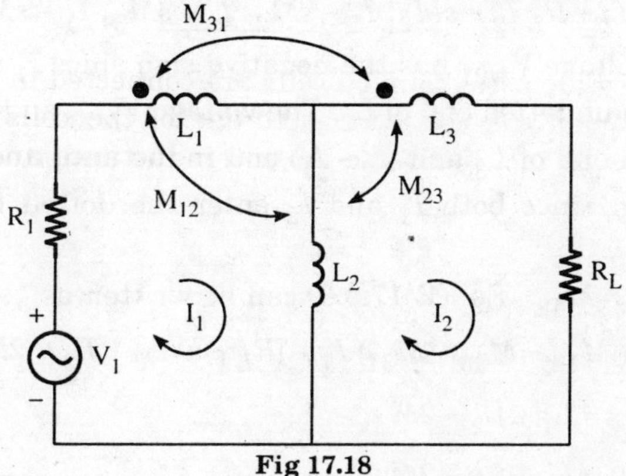

Fig 17.18

SOLUTION :

By KVL in mesh 1,

$$V_1 = V_{R1} + V_{L1} + V_{L2} + V_{M12} + V_{M21} + V_{M13} + V_{M23} \qquad \text{(E 17.9.1)}$$

where V_{R1} = voltage drop in $R_1 = R_1 I_1$

V_{L1} = voltage drop in $L_1 = sL_1 I_1$

V_{L2} = voltage drop in $L_2 = sL_2 (I_1 - I_2)$

V_{m12} = mutually induced voltage in L_1 due to current $(I_1 - I_2)$ in L_2 ,

$= s\, M_{12}(I_1 - I_2)$

The voltage V_{M12} has the positive sign since both I_1 and $(I_1 - I_2)$ flow into the dotted terminals of inductors L_1 and L_2 respectively.

V_{M21} = mutually induced voltage in L_2 due to current I_1 in $L_1 = sM_{21} I_1$

By similar reasoning as above, this voltage is also positive.

V_{M13} = mutually induced voltage in L_1 due to current I_2 in $L_3 = sM_{13} I_2$

This voltage is positive since both I_1 and I_2 flow into the dotted terminals of L_1 and L_3 respectively.

V_{M23} = mutually induced voltage in L_2 due to current I_2 in $L_3 = sM_{23} I_2$

This voltage is positive since both $(I_1 - I_2)$ and I_2 flow into the dotted terminals of L_2 and L_3 respectively.

Substituting these values in Eq. (E 17.9.1) we get

$$V_1 = R_1 I_1 + sL_1 I_1 + sL_2 (I_1 - I_2) + sM_{12} (I_1 - I_2) + sM_{21} I_1 + sM_{13} I_2 + sM_{23} I_2$$

Since $M_{21} = M_{12}$, and $M_{13} = M_{31}$

$$V_1 = [R_1 + s (L_1 + L_2 + 2M_{12})] I_1 - s (L_2 + M_{12} - M_{23} - M_{31}) I_2 \quad \text{(E 17.9.2)}$$

Similarly, by KVL in mesh 2

$$0 = V_{L2} + V_{L3} + V_{RL} + V_{M21} + V_{M23} + V_{M31} + V_{M32}$$

$$0 = sL_2 (I_2 - I_1) + sL_3 I_2 + R_L I_2 - sM_{21} I_1 - sM_{23} I_2 + sM_{31} I_1 - sM_{32} (I_2 - I_1) \quad \text{(E 17.9.3)}$$

It is to be noted that the voltage V_{M21} has the negative sign since I_1 enters the dotted end of L_1 while $(I_2 - I_1)$ enters the undotted end of L_2. The voltage V_{M23} and V_{M32} have the negative signs since I_2 enters dotted end of L_3 and $(I_2 - I_1)$ enters the undotted end of L_2. The voltage V_{M31} has the positive sign since both I_1 and I_2 enter the dotted terminals of L_1 and L_3 respectively.

Since $M_{32} = M_{23}$, and $M_{21} = M_{12}$, Eq. (E 17.9.3) can be written as

$$0 = - s (L_2 + M_{12} - M_{23} - M_{31}) I_1 + [R_L + s (L_2 + L_3 - 2M_{23})] I_2 \quad \text{(E 17.9.4)}$$

Alternative method

The two mesh equations can be written by inspection.

Mesh 1 Position 1-1 Position 1-2

$$[(R_1 + sL_1 + sL_2 + sM_{12} + sM_{12}) I_1 (s)] + [(- sL_2 - sM_{12} + sM_{23} + sM_{31}) I_2 (s)] = V_1 (s)$$

Mesh 2 Position 2-1 Position 2-2

$$[(-sL_2 - sM_{12} + sM_{23} + sM_{31})\, I_1(s)] + [(R_L + sL_2 + sL_3 - sM_{23} - sM_{23})\, I_2(s)] = 0$$

The procedure of writing the above equations is as follows :

Terms in Position 1–1

In the 1-1 position, each term represents the voltage in mesh 1 only due to $I_1(s)$. The terms $R_1 I_1(s)$, $sL_1 I_1(s)$ and $sL_2 I_1(s)$ represent the self voltages due to $I_1(s)$ in R_1, sL_1 and sL_2 in mesh 1. Their signs are positive due to 1-1 position. There are two mutual voltages in mesh 1. The first term sM_{12} is the mutual voltage across sL_1 due to $I_2(s)$ in the sL_2 coil. The second term sM_{12} is the mutual voltage across sL_2 due to $I_2(s)$ in the sL_1 coil. Since $I_1(s)$ enters the dot of sL_1 and also the dot of sL_2, the sign of each mutual term is positive.

Terms in Position 1–2

In the 1-2 position each term represents the voltage in mesh 1 only due to $I_2(s)$. The impedance common to meshes 1 and 2 is sL_2. The voltage, induced across it is $sL_2 I_2(s)$ due to current $I_2(s)$ in sL_2. This is the self voltage. Its sign is negative due to 1-2 position. The term $sM_{12} I_2(s)$ is the mutual voltage in sL_2 due to current $I_1(s)$ in sL_1. The sign of this mutual voltage is negative because $I_1(s)$ enters the dotted end of sL_1 and $I_2(s)$ enter the undotted end of sL_2. The term $sM_{23} I_2(s)$ represents the mutual voltage in sL_2 due to current $I_2(s)$ in sL_3. Since $I_1(s)$ is entering the dot of sL_2 and $I_2(s)$ is entering the dot of sL_3, the sign of $sM_{23} I_2(s)$ is positive. The term $sM_{31} I_2(s)$ is the mutually induced voltage in sL_1 due to current $I_2(s)$ in sL_3. The sign of $sM_{31} I_2(s)$ is positive because $I_1(s)$ enters the dot of sL_1 and $I_2(s)$ enters the dot of sL_3.

Terms in 2–1 Position

Here each term represents the voltage in mesh 2 due only to $I_1(s)$. The impedance common to meshes 1 and 2 is sL_2. The self voltage induced across it is $sL_2 I_1(s)$ due to current $I_1(s)$ in sL_2. Its sign is negative due to the 2-1 position. The term $sM_{12} I_1(s)$ is the mutual voltage in sL_2 due to current $I_1(s)$ in sL_1. The sign of this mutual voltage is negative because $I_1(s)$ and $I_2(s)$ enter 'unlike' ends of sL_1 and sL_2. The term $sM_{23} I_1(s)$ is the mutual voltage in sL_2 due to current $I_2(s)$ in sL_3. Its sign is positive because $I_1(s)$ and $I_2(s)$ enter like ends (dots) of sL_2 and sL_3 respectively. The term $sM_{31} I_1(s)$ is the mutual voltage in sL_3 due to current $I_1(s)$ in sL_1. Its sign is also positive because $I_1(s)$ and $I_2(s)$ enter like ends (dots) of sL_1 and sL_3 respectively.

Terms in 2–2 Position

Here each term represents the voltage in mesh 2 due only to $I_2(s)$. The terms $R_L I_2(s)$, $sL_2 I_2(s)$, and $sL_3 I_2(s)$ are self voltages due to R_L, sL_2, and sL_3. Their signs are positive due to 2-2 position. There are two mutual voltages in mesh 2 due to $I_2(s)$. The first

$sM_{23} I_2(s)$ is the mutual voltage across sL_2 due to $I_2(s)$ in the sL_3 coil, and the second $sM_{23} I_2(s)$ in the mutual voltage across sL_3 coil due to $I_2(s)$ in sL_2 coil. Since $I_2(s)$ enters the dotted end of sL_3 and the undotted end of the sL_2 coil, the sign of each mutual term is negative.

The right-hand side of mesh equation 1 is the sum of all the voltage sources in mesh 1. Similarly, the right-hand side of mesh equation 2 is the sum of all the voltage sources in mesh 2.

It is to be noted that the impedance terms in 1-2 position are the same as the impedance terms in the 2-1 position.

■ **EXAMPLE 17.10** Write the mesh equations for the circuit shown in Fig. 17.19.

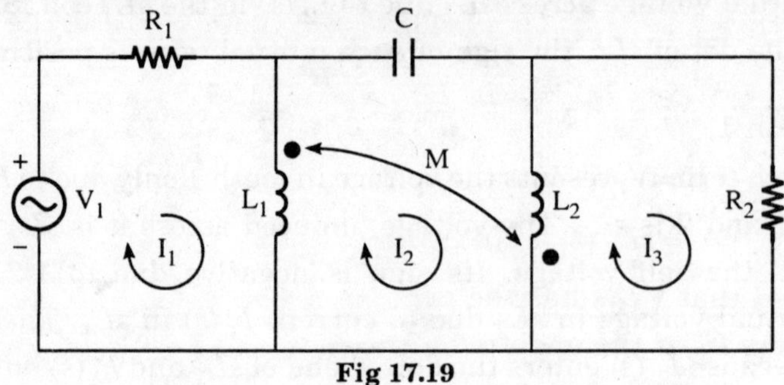

Fig 17.19

SOLUTION :

By KVL in mesh 1,

$$V_1 = V_{R1} + V_{L1} + V_{M12}$$

$$= R_1 I_1 + sL_1 (I_1 - I_2) + sM (I_3 - I_2)$$

$$V_1 = (R_1 + sL_1) I_1 - s (L_1 + M) I_2 + sMI_3 \qquad \text{(E 17.10.1)}$$

By KVL in mesh 2,

$$0 = V_{L1} + V_C + V_{L2} + V_{M12} + V_{M21}$$

$$= sL_1 (I_2 - I_1) + \frac{1}{sC} I_2 + sL_2 (I_2 - I_3) + sM (I_2 - I_3) + sM (I_2 - I_1)$$

or

$$0 = -s(L_1 + M) I_1 + \left(sL_1 + sL_2 + sM + sM + \frac{1}{sC}\right) I_2 - s (L_2 + M) I_3 \qquad \text{(E 17.10.2)}$$

By KVL in mesh 3,

$$0 = V_{L2} + V_{R2} + V_{M31}$$

$$= sL_2 (I_3 - I_2) + R_2 I_3 + sM (I_1 - I_2)$$

$$= sM I_1 - s (L_2 + M) I_2 + (sL_2 + R_2) I_3 \qquad \text{(E 17.10.3)}$$

■ **EXAMPLE 17.11** Find an expression for the input impedance of the circuit shown in Fig. 17.20.

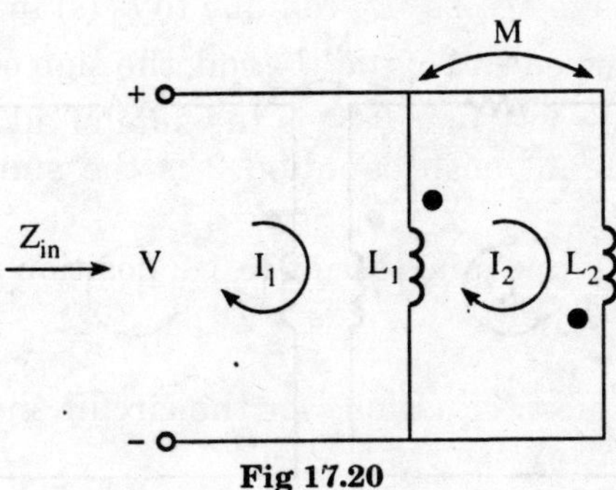

Fig 17.20

SOLUTION :

By KVL in mesh 1,

$$V = V_{L1} + V_{M12}$$
$$= sL_1 (I_1 - I_2) - sMI_2$$

It is to be noted that V_{M12} has the negative sign since $(I_1 - I_2)$ enters L_1 at the dotted terminal and I_2 enters L_2 at the undotted terminal. Therefore

$$V = sL_1 I_1 - s(L_1 + M) I_2 \tag{E 17.11.1}$$

By KVL in mesh 2,

$$0 = V_{L1} + V_{L2} + V_{M12} + V_{M21}$$
$$= sL_1 (I_2 - I_1) + sL_2 I_2 + sMI_2 + sM(I_2 - I_1)$$
$$= -s(L_1 + M) I_1 + s(L_1 + L_2 + 2M) I_2 \tag{E 17.11.2}$$

Multiplying Eq. (E 17.11.1) by $(L_1 + L_2 + 2M)$ and Eq. (E 17.11.2) by $(L_1 + M)$ and adding the results to eliminate I_2 we obtain,

$$V(L_1 + L_2 + 2M) = sL_1 (L_1 + L_2 + 2M) I_1 - s(L_1 + M)^2 I_1$$
$$= s(L_1 L_2 - M^2) I_1$$

Therefore the input impedance of the circuit is given by

$$Z_{in} = \frac{V}{I_1} = \frac{s(L_1 L_2 - M^2)}{L_1 + L_2 + 2M} \tag{E 17.11.3}$$

For sinusoidal steady state, $s = j\omega$ and

$$Z_{in} = \frac{j\omega (L_1 L_2 - M^2)}{L_1 + L_2 + 2M} \tag{E 17.11.4}$$

■ **EXAMPLE 17.12** Write the mesh equations for the circuit shown in Fig. 17.21 and determine the input impedance of the circuit.

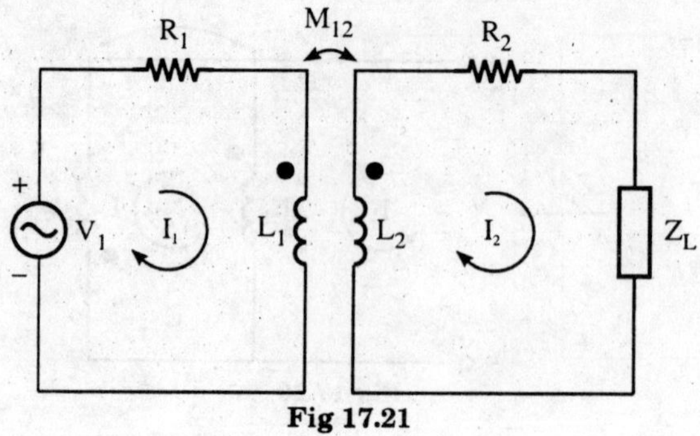

Fig 17.21

SOLUTION :

By KVL in mesh 1,

$$V_1 = V_{R1} + V_{L1} + V_{M12}$$

$$= R_1 I_1 + sL_1 I_1 - sMI_2$$

$$V_1 = (R_1 + sL_1) I_1 - sMI_2 \quad \text{(E 17.12.1)}$$

or $$V_1 = z_{11} I_1 - sMI_2 \quad \text{(E 17.12.2)}$$

where $$z_{11} = R_1 + sL_1 \quad \text{(E 17.12.3)}$$

By KVL in mesh 2,

$$0 = V_{L2} + V_{R2} + V_{ZL} + V_{M21}$$

$$= sL_2 I_2 + R_2 I_2 + Z_L I_2 - sMI_1$$

$$0 = -sMI_1 + (R_2 + sL_2 + Z_L) I_2 \quad \text{(E 17.12.4)}$$

or $$0 = -sMI_1 + z_{22} I_1 \quad \text{(E 17.12.5)}$$

where $$z_{22} = R_2 + sL_2 + Z_L \quad \text{(E 17.12.6)}$$

Multiplying Eq. (E 17.12.2) be z_{22} and Eq. (E 17.12.5) by sM and adding the results to eliminate I_2 we get

$$z_{22} V_1 = (z_{11} z_{22} - s^2 M^2) I_1$$

Therefore the input impedance of the circuit is given by

$$Z_{in} = \frac{V_1}{I_1} = z_{11} - \frac{s^2 M^2}{z_{22}} \quad \text{(E 17.12.7)}$$

For sinusoidal steady state $s = j\omega$

$$Z_{in} = z_{11} + \frac{\omega^2 M^2}{z_{22}} \quad \text{(E 17.12.8)}$$

■ **EXAMPLE 17.13** Write the mesh equations for the circuit shown in Fig. 17.22.

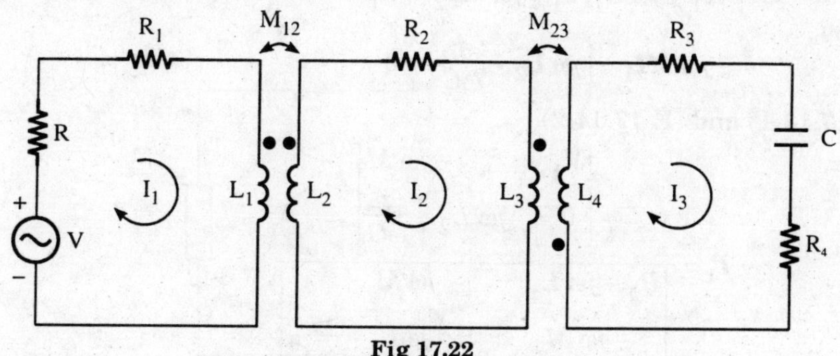

Fig 17.22

SOLUTION :

By KVL in mesh 1,

$$V = V_R + V_{R1} + V_{L1} + V_{M12}$$

$$= RI_1 + R_1 I_1 + sL_1 I_1 - sM_{12} I_2$$

$$= (R + R_1 + sL_1) I_1 - sM_{12} I_2 + 0 \times I_3 \quad \text{(E 17.13.1)}$$

By KVL in mesh 2,

$$0 = V_{L2} + V_{M21} + V_{R12} + V_{L3} + V_{M23}$$

$$= sL_2 I_2 - sM_{21} I_1 + R_2 I_2 + sL_3 I_2 + sM_{23} I_3$$

$$= - sM_{21} I_1 + (R_2 + sL_2 + sL_3) I_2 + sM_{23} I_3 \quad \text{(E 17.13.2)}$$

By KVL in mesh 3,

$$0 = V_{L4} + V_{M32} + V_{R3} + V_c + V_{R4}$$

$$= sL_4 I_3 + sM_{23} I_2 + R_3 I_3 + \frac{1}{sC} I_3 + R_4 I_3$$

$$= 0 \times I_1 + sM_{23} I_2 + \left(R_3 + R_4 + sL_4 + \frac{1}{sC} \right) I_3 \quad \text{(E 17.13.3)}$$

Alternative method

The above mesh equations can also be written directly by inspection.

■ **EXAMPLE 17.14** Find the value of C in Fig. 17.23, given that the supply voltage and current are in phase.

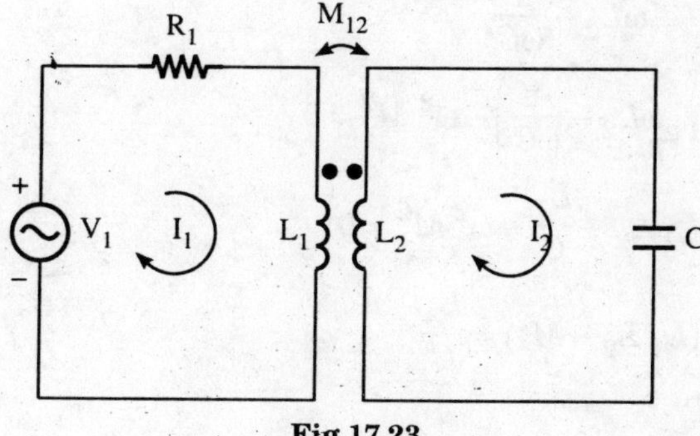

Fig 17.23

SOLUTION : The mesh equations of the network can be written as

$$(R_1 + j\omega L_1)\, I_1 - j\omega M I_2 = V_1 \tag{E 17.14.1}$$

$$-j\omega M I_1 + \left(j\omega L_2 - \frac{j}{\omega C}\right) = 0 \tag{E 17.14.2}$$

From Eqs. (E 17.14.1) and (E 17.14.2)

$$I_1 = \frac{\begin{vmatrix} V_1 & -j\omega M \\ 0 & j\omega L_2 - \dfrac{j}{\omega C} \end{vmatrix}}{\begin{vmatrix} R_1 + j\omega L_1 & -j\omega M \\ -j\omega M & j\omega L_2 - \dfrac{j}{\omega C} \end{vmatrix}}$$

$$I_1 = \frac{V_1 \left(j\omega L_2 - \dfrac{j}{\omega C}\right)}{(R_1 + j\omega L_1)\left(j\omega L_2 - \dfrac{j}{\omega C}\right) + \omega^2 M^2}$$

$$\frac{V_1}{I_1} = \frac{(R_1 + j\omega L_1)\left(j\omega L_2 - \dfrac{j}{\omega C}\right) + \omega^2 M^2}{\left(j\omega L_2 - \dfrac{j}{\omega C}\right)}$$

$$= (R_1 + j\omega L_1) + \frac{\omega^2 M^2}{j\left(\omega L_2 - \dfrac{1}{\omega C}\right)}$$

$$= R_1 + j\omega L_1 - j\,\frac{\omega^2 M^2}{\omega L_2 - \dfrac{1}{\omega C}}$$

$$= R_1 + j\left(\omega L_1 - \frac{\omega^2 M^2}{\omega L_2 - \dfrac{1}{\omega C}}\right)$$

Since V_1 and I_1 are in phase, the j term of V_1/I_1 should be zero.

therefore,

$$\omega L_1 - \frac{\omega^2 M^2}{\omega L_2 - \dfrac{1}{\omega C}} = 0$$

$$\omega L_1\left(\omega L_2 - \frac{1}{\omega C}\right) - \omega^2 M^2 = 0$$

$$\omega^2 L_1 L_2 - \frac{L_1}{C} - \omega^2 M^2 = 0$$

$$\omega^2 (L_1 L_2 - M^2) = \frac{L_1}{C}$$

$$C = \frac{L_1}{\omega^2 (L_1 L_2 - M^2)}$$

17.9 EQUIVALENT CIRCUITS FOR MAGNETICALLY COUPLED COILS

Sometimes it is convenient to convert a magnetically coupled circuit which does not involve magnetic coupling. Such a circuit is called *conductively coupled equivalent circuit*. This technique is used in many computer simulations of magnetically coupled circuits.

Consider a pair of magnetically coupled inductors shown in Fig. 17.24a.

We can replace the magnetically coupled inductors with a set of inductors that are not magnetically coupled. The important restriction is that the voltage between the points, b and d in Fig 17.24a must be zero. In other words, the terminals b and d can be shorted together without disturbing the voltages and currents in the original circuit.

For the given references and polarity dots, the KVL equations for the circuit of Fig. 17.24a are

$$v_1 = L_1 \frac{di_1}{dt} + M \frac{di_2}{dt} \tag{17.9.1}$$

$$v_2 = M \frac{di_1}{dt} + L_2 \frac{di_2}{dt} \tag{17.9.2}$$

The *T*-Equivalent Circuit

By adding and subtracting $M \frac{di_1}{dt}$ on the right side of Eq. (17.9.1) we get

$$v_1 = (L_1 - M) \frac{di_1}{dt} + M \frac{d}{dt} (i_1 + i_2) \tag{17.9.3}$$

By adding and subtracting $M \frac{di_2}{dt}$ on the right side of Eq. (17.9.2) we get

$$v_2 = (L_2 - M) \frac{di_2}{dt} + M \frac{d}{dt} (i_1 + i_2) \tag{17.9.4}$$

From Eqs. (17.9.3) and (17.9.4) we can draw a T network shown in Fig 17.24b. Thus, the magnetically mutually coupled circuit of Fig. 17.24a has been replaced by a conductively coupled circuit.

It is to be noted that *the dot rule is not needed in the conductively coupled circuit*, and the common circuit analysis methods are applicable. It is also to be noted that the voltage between points b and d is zero.

The Π-Equivalent Circuit

In the s domain, Eqs. (17.9.1) and (17.9.2) can be written as

$$V_1 = sL_1 I_1 + sMI_2 \tag{17.9.5}$$

$$V_2 = sMI_1 + sL_2 I_2 \tag{17.9.6}$$

The Π-equivalent for the magnetically coupled inductors of circuit Fig. 17.24a can be obtained

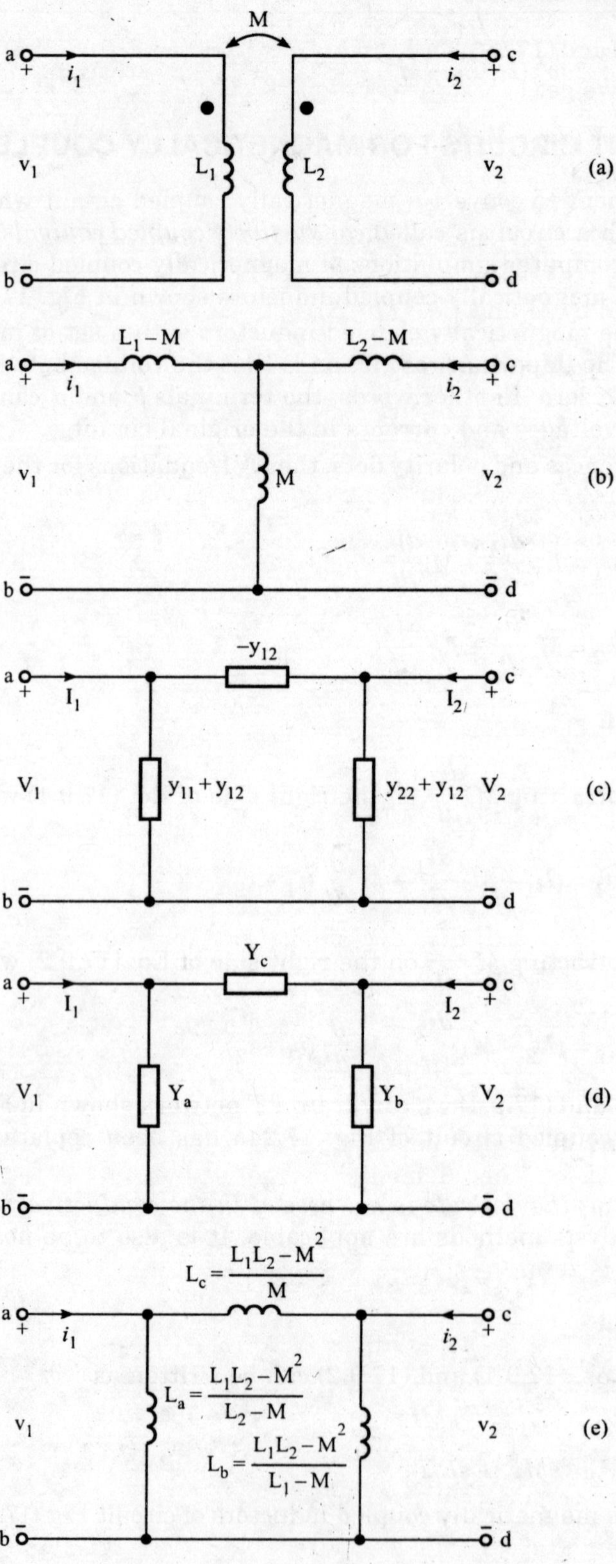
M
a
+
i_1
c
+
i_2
v_1
L_1
L_2
v_2
(a)
b
d
$L_1 - M$
$L_2 - M$
M
(b)
$-y_{12}$
I_1
I_2
V_1
$y_{11} + y_{12}$
$y_{22} + y_{12}$
V_2
(c)
Y_c
Y_a
Y_b
(d)
$L_c = \frac{L_1L_2 - M^2}{M}$
$L_a = \frac{L_1L_2 - M^2}{L_2 - M}$
$L_b = \frac{L_1L_2 - M^2}{L_1 - M}$
(e)

by solving Eqs. (17.9.5) and (17.9.6) for I_1 and I_2. Multiplying Eq. (17.9.5) by L_2 and Eq. (17.9.6) by M and subtracting we get

$$V_1 L_2 - V_2 M = s\,(L_1 L_2 - M^2)\, I_1$$

therefore,

$$I_1 = \frac{L_2}{s(L_1 L_2 - M^2)} V_1 - \frac{M}{s(L_1 L_2 - M^2)} V_2 \tag{17.9.7}$$

Multiplying Eq. (17.9.5) by M and Eq. (17.9.6) by L_1 and subtracting we get

$$V_1 M - V_2 L_1 = s\,(M^2 - L_1 L_2)\, I_2$$

$$I_2 = \frac{-M}{s\,(L_1 L_2 - M^2)} V_1 + \frac{L_1}{s\,(L_1 L_2 - M^2)} V_2 \tag{17.9.8}$$

The defining equations of the y parameters are

$$I_1 = y_{11} V_1 + y_{12} V_2 \tag{17.9.9}$$

$$I_2 = y_{21} V_1 + y_{22} V_2 \tag{17.9.10}$$

Comparison of Eq. (17.9.7) with Eq. (17.9.9) and Eq. (17.9.8) with Eq. (17.9.10) gives

$$y_{11} = \frac{L_2}{s\,(L_1 L_2 - M^2)},$$

$$y_{12} = \frac{-M}{s\,(L_1 L_2 - M^2)}$$

$$y_{21} = \frac{-M}{s\,(L_1 L_2 - M^2)},$$

$$y_{22} = \frac{L_1}{s\,(L_1 L_2 - M^2)}$$

The equivalent Π network of a 2-port network having its elements in terms of the y parameters are shown in Fig. 17.24c. The elements of the equivalent Π network are shown in Fig. 17.24d.

Here

$$Y_a = y_{11} + y_{12} = \frac{L_2 - M}{s\,(L_1 L_2 - M^2)}$$

$$Y_b = y_{22} + y_{12} = \frac{L_1 - M}{s\,(L_1 L_2 - M^2)}$$

$$Y_c = -\,y_{12} = \frac{M}{s\,(L_1 L_2 - M^2)}$$

therefore,
$$Z_a = \frac{1}{Y_a} = \frac{s\,(L_1 L_2 - M^2)}{L_2 - M} = sL_a$$

$$Z_b = \frac{1}{Y_b} = \frac{s\,(L_1 L_2 - M^2)}{L_1 - M} = sL_b$$

$$Z_c = \frac{1}{Y_c} = \frac{s\,(L_1 L_2 - M^2)}{M} = sL_c$$

where
$$L_a = \frac{(L_1 L_2 - M^2)}{L_2 - M}$$

$$L_b = \frac{(L_1 L_2 - M^2)}{L_1 - M}$$

$$L_c = \frac{(L_1 L_2 - M^2)}{M}$$

The Π equivalent circuit for the magnetically coupled inductors of Fig. 17.24a is shown in Fig. 17.24e. It has three inductors L_a , L_b and L_c. It is to be noted that there is no magnetic coupling between the inductors L_a , L_b and L_c.

In the T and Π equivalent circuit, the mutual inductance M carries its own algebraic sign. That is, if the magnetic polarity of the coupled coils is reversed from that given in Fig. 17.24a, the algebraic sign of M reverses. A reversal in magnetic polarity requires moving one polarity dot without changing the reference polarities of terminal currents and voltages. Thus, if the dot on L_2 in the circuit shown in Fig. 17.24a is moved to the lower terminal as

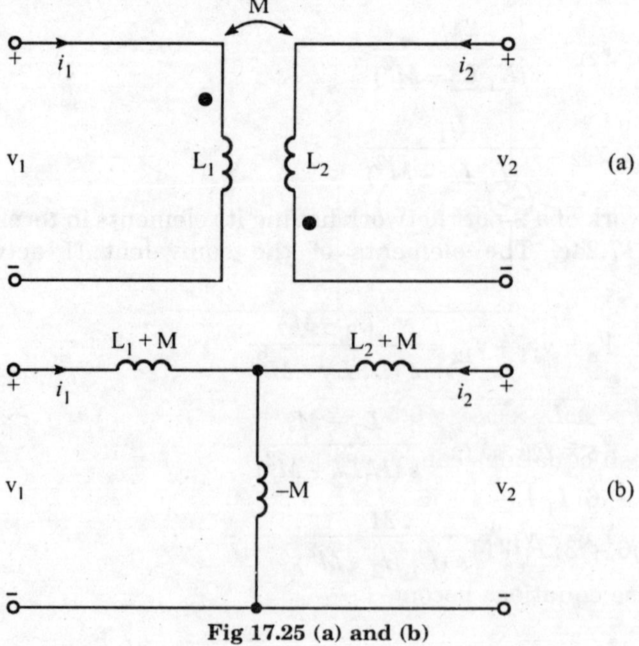

Fig 17.25 (a) and (b)

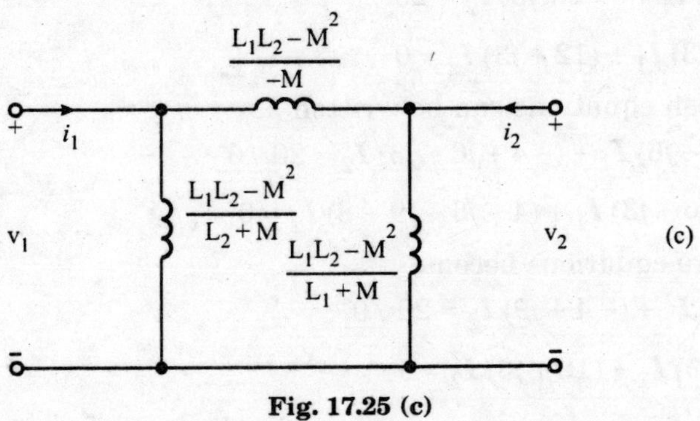

Fig. 17.25 (c)

shown in Fig. 17.25a, the T and Π equivalent circuits of Fig. 17.25a are as shown in Fig. 17.25b and Fig. 17.25c respectively.

■ **EXAMPLE 17.15** Write the mesh equations for the networks shown in Fig. 17.26 and Fig. 17.27.

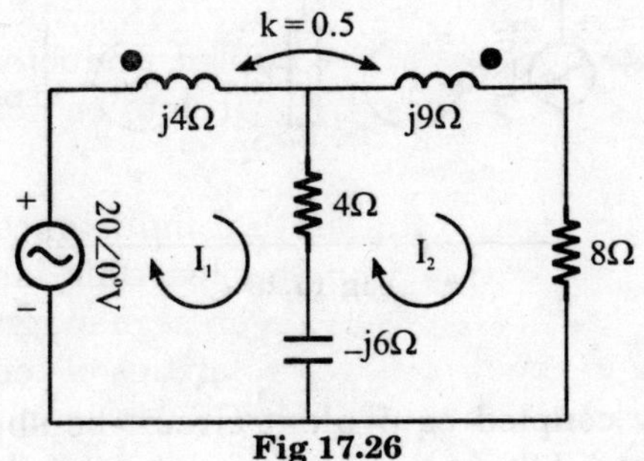

Fig 17.26

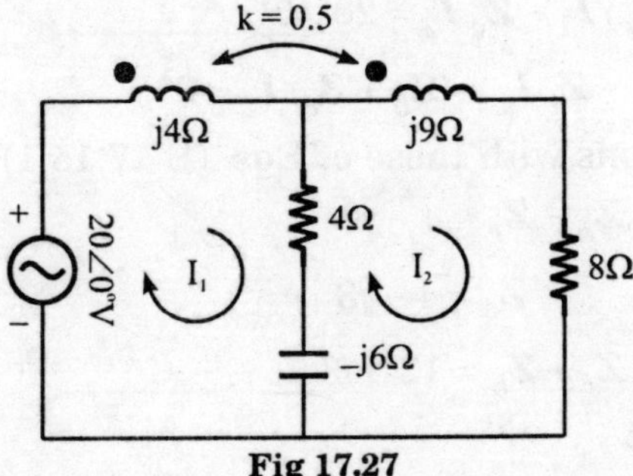

Fig 17.27

SOLUTION :

$$\omega M = k \times \sqrt{\omega L_1 \times \omega L_2} = 0.5 \times \sqrt{4 \times 9} = 3\ \Omega$$

For Fig. 17.26, the mesh equations can be written as

$$(4 + j4 - j6)\, \boldsymbol{I_1} + (-4 + j6 - j3)\, \boldsymbol{I_2} = 20\ \underline{/0^\circ}$$

$$(-4 + j6 - j3)\, \boldsymbol{I_1} + (4 - j6 + j9 + 8)\, \boldsymbol{I_2} = 0$$

On simplification, these equations become

$$(4-j2)\,\boldsymbol{I}_1+(-4+j3)\,\boldsymbol{I}_2=20 \qquad \text{(E 17.15.1)}$$

$$(-4+j3)\,\boldsymbol{I}_1+(12+j3)\,\boldsymbol{I}_2=0 \qquad \text{(E 17.15.2)}$$

For Fig. 17.27, the mesh equations can be written as

$$(4-j4-j6)\,\boldsymbol{I}_1+(-4+j6+j3)\,\boldsymbol{I}_2=20\ \underline{/0^\circ}$$

$$(-4+j6+j3)\,\boldsymbol{I}_1+(4-j6+j9+8)\,\boldsymbol{I}_2=0$$

On simplification, these equations become

$$(4-j2)\,\boldsymbol{I}_1+(-4+j9)\,\boldsymbol{I}_2=20\ \underline{/0^\circ}$$

$$(-4+j9)\,\boldsymbol{I}_1+(12+j3)\,\boldsymbol{I}_2=0$$

■ **EXAMPLE 17.16** Obtain a conductively coupled equivalent circuit for magnetically coupled circuit of Fig. 17.28.

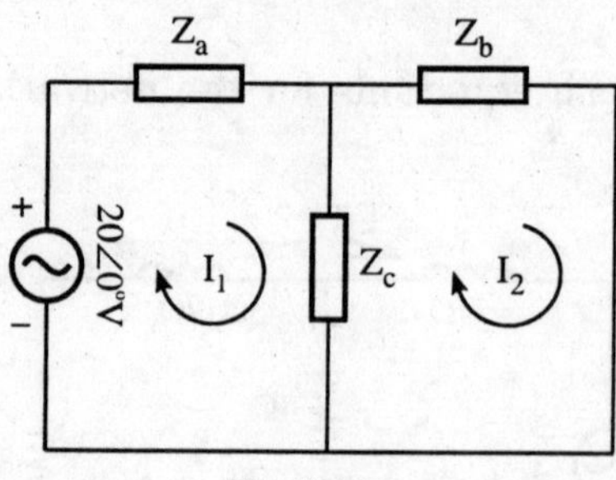

Fig 17.28

SOLUTION :

Let the required conductively coupled equivalent circuit be shown in Fig. 17.28. The mesh equations for the circuit in Fig. 17.28 are

$$(\boldsymbol{Z}_a+\boldsymbol{Z}_c)\,\boldsymbol{I}_1-\boldsymbol{Z}_c\,\boldsymbol{I}_2=20\ \underline{/0^\circ}$$

$$-\boldsymbol{Z}_c\,\boldsymbol{I}_1+(\boldsymbol{Z}_b+\boldsymbol{Z}_c)\,\boldsymbol{I}_2=0$$

Comparing these mesh equations with those of Eqs. (E 17.15.1) and (E 17.15.2), we obtain.

$$\boldsymbol{Z}_a+\boldsymbol{Z}_c=4-j2$$

$$\boldsymbol{Z}_c=4-j3$$

$$\boldsymbol{Z}_c+\boldsymbol{Z}_b=12+j3$$

Solving these equations we get

$$\boldsymbol{Z}_a=j1\ \Omega,$$

$$\boldsymbol{Z}_b=(8+j6)\ \Omega$$

$$\boldsymbol{Z}_c=(4-j3)\ \Omega$$

■ **EXAMPLE 17.17** For the coupled circuit shown in Fig. 17.29 show that the current I_1 is the same irrespective of the positions of the dots.

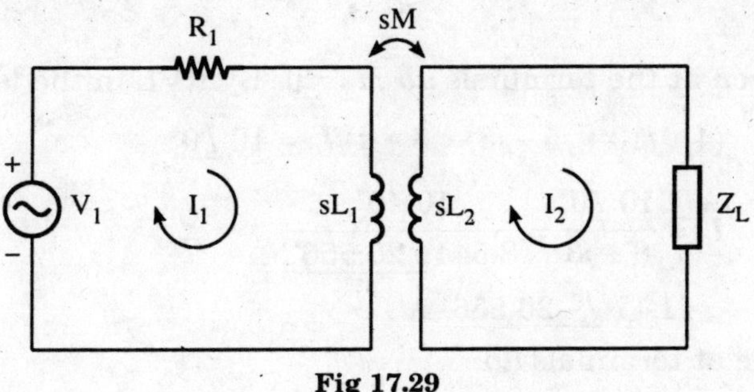

Fig 17.29

SOLUTION :

The mesh equations of the network are

$$(R_1 + sL_1)\, I_1 \pm sMI_2 = V_1$$

$$\pm sMI_1 + (sL_2 + Z_L)\, I_2 = 0$$

$$I = \frac{\begin{vmatrix} V_1 & \pm sM \\ 0 & Z_L + sL_2 \end{vmatrix}}{\begin{vmatrix} R_1 + sL_1 & \pm sM \\ \pm sM & Z_L + sL_2 \end{vmatrix}} = \frac{V_1\,(Z_L + sL_2)}{(R_1 + sL_1)\,(Z_L + sL_2) - s^2 M^2}$$

This relation shows that I_1 is the same irrespective of the dot positions because of the presence of M^2.

■ **EXAMPLE 17.18** Obtain Thevenin and Norton equivalent circuits at terminals *ab* for the coupled circuit shown in Fig. 17.30a.

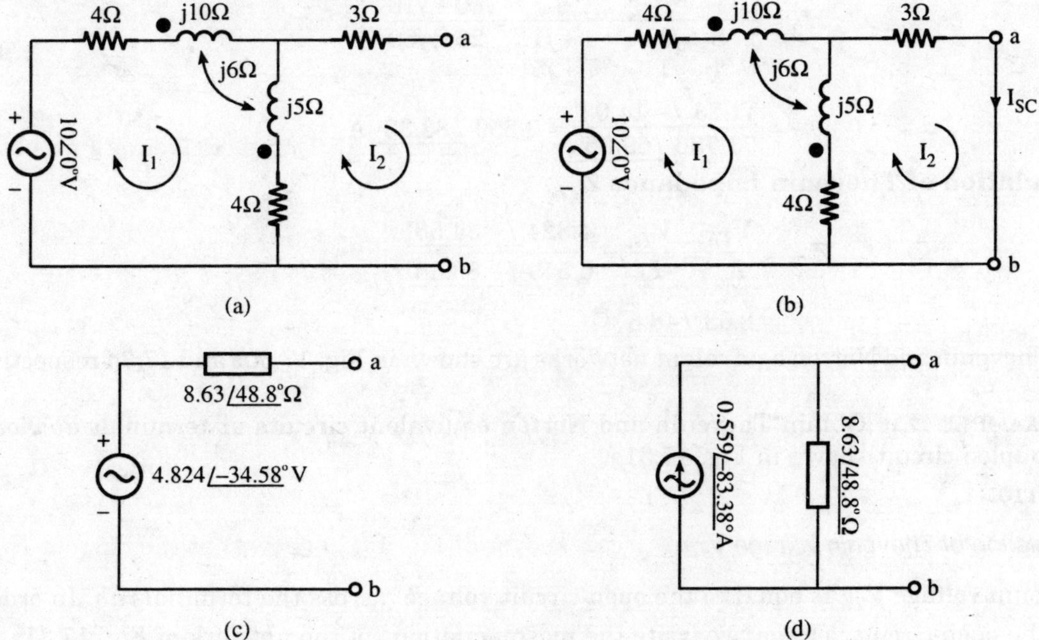

Fig 17.30

SOLUTION :

Since the circuit is open at the terminals ab , $I_2 = 0$. By KVL in the left-hand mesh

$$(4 + j10 + j5 - j6 - j6 + 4)\, \boldsymbol{I}_1 = 10 \underline{/0^\circ}$$

$$\boldsymbol{I}_1 = \frac{10 \underline{/0^\circ}}{8 + j3} = \frac{10 \underline{/0^\circ}}{8.544 \underline{/20.556^\circ}}$$

$$= 1.17 \underline{/-20.556^\circ} \text{ A}$$

Thevenin voltage at terminals ab

= open circuit voltage at terminals ab

$$= (4 + j5)\, \boldsymbol{I}_1 - (j6)\, I_1 = (4 - j1)\, \boldsymbol{I}_1$$

$$= (4.123 \underline{/-14.036^\circ})\,(1.17 \underline{/-20.556^\circ}) = 4.824 \underline{/-34.58^\circ} \text{ V}$$

Determination of short-circuit current I_{sc}

In order to find the short-circuit current, we short circuit the terminals ab. The current through ab is the short-circuit current $\boldsymbol{I}_{sc}$ as shown in Fig. 17.30b. It is to be noted that $\boldsymbol{I}_{sc} = \boldsymbol{I}_2$. The mesh equations of Fig. 17.30b are

$$(8 + j3)\, \boldsymbol{I}_1 - (4 + j5 - j6)\, \boldsymbol{I}_2 = 10 \underline{/0^\circ}$$

$$-(4 + j5 - j6)\, \boldsymbol{I}_1 + (4 + j5 + 3)\, \boldsymbol{I}_2 = 0$$

or

$$(8 + j3)\, \boldsymbol{I}_1 + (-4 + j1)\, \boldsymbol{I}_2 = 10 \underline{/0^\circ}$$

and

$$(-4 + j1)\, \boldsymbol{I}_1 + (7 + j5)\, \boldsymbol{I}_2 = 0$$

$$\boldsymbol{I}_2 = \frac{\begin{vmatrix} 8 + j3 & 10 \\ -4 + j1 & 0 \end{vmatrix}}{\begin{vmatrix} 8 + j3 & -4 + j1 \\ -4 + j1 & 7 + j5 \end{vmatrix}} = \frac{40 - j10}{26 + j69}$$

$$= \frac{41.23 \underline{/-14.03^\circ}}{73.736 \underline{/69.35^\circ}} = 0.559 \underline{/83.38^\circ} \text{ A}$$

Calculation of Thevenin impedance Z_{Th}.

$$Z_{Th} = \frac{V_{Th}}{I_{sc}} = \frac{V_{Th}}{I_2} = \frac{4.824 \underline{/-34.58^\circ}}{0.559 \underline{/-83.38^\circ}}$$

$$= 8.63 \underline{/48.8^\circ}\ \Omega$$

The Thevenin and Norton equivalent networks are shown in Fig. 17.30c and 17.3d respectively.

■ **EXAMPLE 17.19** Obtain Thevenin and Norton equivalent circuits at terminals ab for the coupled circuit shown in Fig. 17.31a.

SOLUTION :

Calculation of Thevenin voltage V_{Th}

Thevenin voltage V_{Th} is equal to the open-circuit voltage across the terminals ab. In order to find the open-circuit voltage, we write the mesh equations of the network of Fig. 17.31a.

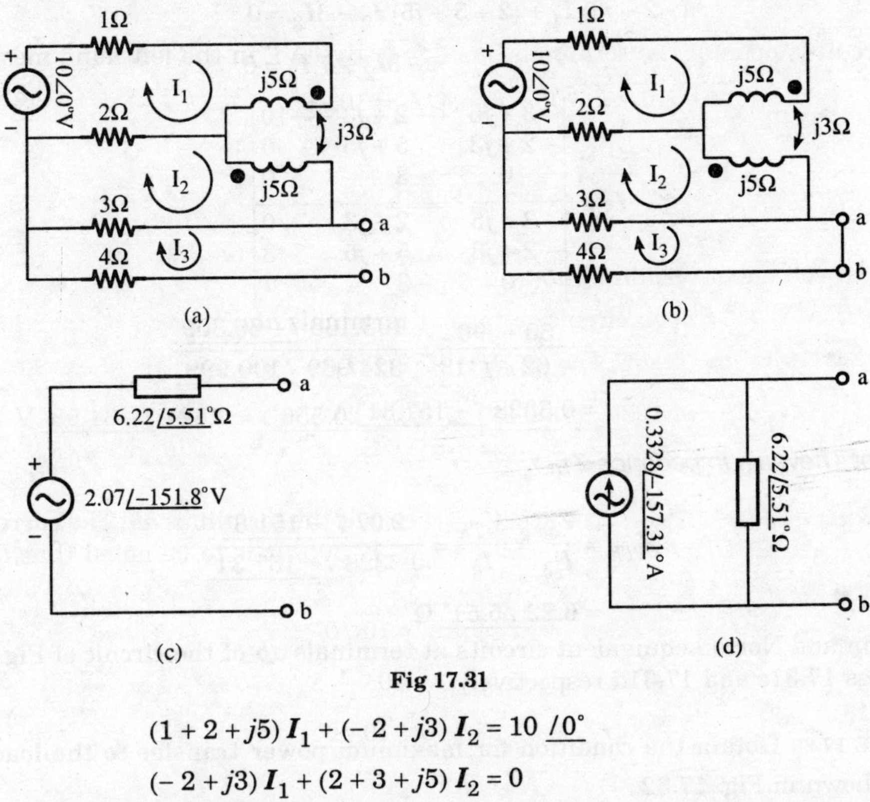

Fig 17.31

$$(1+2+j5)\,I_1 + (-2+j3)\,I_2 = 10\ \underline{/0^\circ}$$
$$(-2+j3)\,I_1 + (2+3+j5)\,I_2 = 0$$

It is to be noted that the sign of mutual inductance is positive in both mesh equations because both I_1 and I_2 enter the dotted ends. Since the circuit is open between the terminals *ab*, the mesh current $I_3 = 0$. By KVL in mesh 3,

$$3I_2 - V_{ab} = 0$$
$$3I_2 - V_{oc} = 0$$

or

$$V_{oc} = 3I_2$$

$$I_2 = \frac{\begin{vmatrix} 3+j5 & 10\ \underline{/0^\circ} \\ -2+j3 & 0 \end{vmatrix}}{\begin{vmatrix} 3+j5 & -2+j3 \\ -2+j3 & 5+j5 \end{vmatrix}} = \frac{20-j30}{-5+j52}$$

$$= \frac{36.0555\ \underline{/-56.309^\circ}}{52.2398\ \underline{/95.492^\circ}} = 0.69\ \underline{/-151.8^\circ}\ \text{A}$$

therefore,

$$V_{Th} = V_{oc} = 3I_2 = 2.07\ \underline{/-151.8^\circ}\ \text{V}$$

Calculation of short-circuit current I_{sc}

Fig. 17.31b shows the circuit when the terminals *ab* are short circuited. The mesh equations for this network are

$$(1+2+j5)\,I_1 + (-2+j3)\,I_2 = 10$$

$$(-2+j3)\,I_1+(2+3+j5)\,I_2-3I_3=0$$

$$-3I_2+7\,I_3=0$$

$$I_3=\frac{\begin{vmatrix}3+j5 & -2+j3 & 10\\ -2+j3 & 5+j5 & 0\\ 0 & -3 & 0\end{vmatrix}}{\begin{vmatrix}3+j5 & -2+j3 & 0\\ -2+j3 & 5+j5 & -3\\ 0 & -3 & 7\end{vmatrix}}$$

$$=\frac{60-j90}{-62+j319}=\frac{108.166\,\underline{/-56.309^\circ}}{324.969\,\underline{/100.999^\circ}}$$

$$=0.3328\,\underline{/-157.31^\circ}\text{ A}$$

Calculation of Thevenin Impedance Z_{Th}

$$Z_{Th}=\frac{V_{oc}}{I_{sc}}=\frac{V_{Th}}{I_3}=\frac{2.07\,\underline{/-151.8^\circ}}{0.3328\,\underline{/-157.31^\circ}}$$

$$=6.22\,\underline{/5.51^\circ}\ \Omega$$

The Thevenin and Norton equivalent circuits at terminals *ab* of the circuit of Fig. 17.31a are shown in Figs 17.31c and 17.31d respectively.

■ **EXAMPLE 17.20** Obtain the condition for maximum power transfer to the load Z_L in the circuit shown in Fig. 17.32.

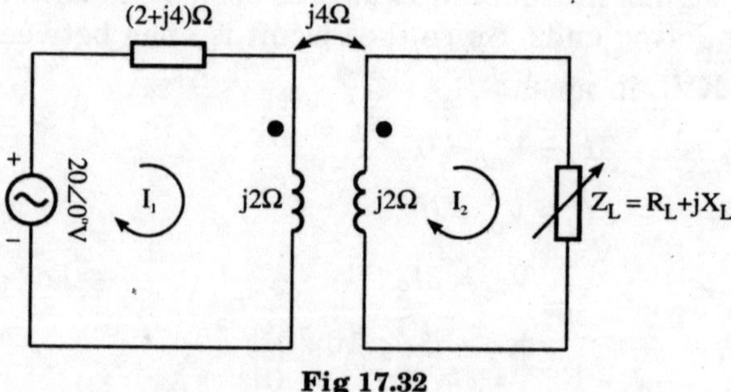

Fig 17.32

SOLUTION :

The mesh equations of the given network are

$$(2+j4+j2)\,I_1-j4\,I_2=20\,\underline{/0^\circ}$$

$$-j4\,I_1+[R_L+j\,(X_L+2)]\,I_2=0$$

$$I_2=\frac{\begin{vmatrix}2+j6 & 20\\ -j4 & 0\end{vmatrix}}{\begin{vmatrix}2+j6 & -j4\\ -j4 & R_L+j\,(X_L+2)\end{vmatrix}}$$

$$= \frac{j80}{(4 + 2R_L - 6X_L) + j\,(4 + 6R_L + 2X_L)}$$

$$= \frac{j40}{(2 + R_L - 3X_L) + j\,(2 + 3R_L + X_L)}$$

$$|I_2| = \frac{40}{\sqrt{[(2 + R_L - 3X_L)^2 + (2 + 3R_L + X_L)^2]}}$$

Power supplied to the load

$$P_L = I_2^2 R_L$$

$$= \frac{1600\,R_L}{(2 + R_L - 3X_L)^2 + (2 + 3R_L + X_L)^2}$$

For P_L to be a maximum,

$$\frac{\partial P_L}{\partial X_L} = 0 \quad \text{and} \quad \frac{\partial P_L}{\partial R_L} = 0$$

Let $\quad (2 + R_L - 3X_L)^2 + (2 + 3R_L + X_L)^2 = D^r$

therefore, $\quad \dfrac{\partial P_L}{\partial X_L} = \dfrac{\partial}{\partial X_L} \dfrac{1600\,R_L}{D^r}$

or $\quad 0 = \dfrac{D^r \dfrac{\partial}{\partial X_L}(1600\,R_L) - 1600\,R_L \dfrac{\partial}{\partial X_L} D^r}{(D^r)^2}$

or $\quad \dfrac{\partial}{\partial X_L} D^r = 0$

or $\quad \dfrac{\partial}{\partial X_L}\,[(2 + R_L - 3X_L)^2 + (2 + 3R_L + X_L)^2] = 0$

$$2\,(2 + R_L - 3X_L)\,(-3) + 2\,(2 + 3R_L + X_L) = 0$$

$$-6 - 3R_L + 9X_L + 3R_L + 2 + X_L = 0$$

$$-4 + 10X_L = 0\,,\ X_L = 0.4\ \Omega$$

Substituting the value of X_L in the expression for power we get

$$P_L = \frac{1600\,R_L}{(2 + R_L - 1.2)^2 + (2 + 3R_L + 0.4)^2} = \frac{160\,R_L}{(0.8 + R_L)^2}$$

Applying the second condition for maximum power transfer

$$\frac{\partial P_L}{\partial R_L} = 0$$

or
$$\frac{(0.8+R_L)^2 \times 160 - 160\,R_L \times 2\,(0.8+R_L)}{(0.8+R_L)^4} = 0$$

$$1 - 2R_L = 0\,,\; R_L = 0.5\,\Omega$$

Substituting the value of R_L and X_L in the expresion for power we get the maximum power.

$$P_{Lmax} = \frac{160\,R_L}{(0.8+R_L)^2} = \frac{160 \times 0.5}{(0.8+0.5)^2} = 47.34\text{ W}$$

17.10 TRANSFORMERS

Transformers are coupled coils with minimum flux leakage. A 2-winding transformer and its 2-port representation is shown in Fig. 17.33. Electrical energy is transferred from the primary circuit (port 1) to the secondary circuit (port 2) by means of magnetic flux. Thus, the primary and secondary circuits are magnetically coupled.

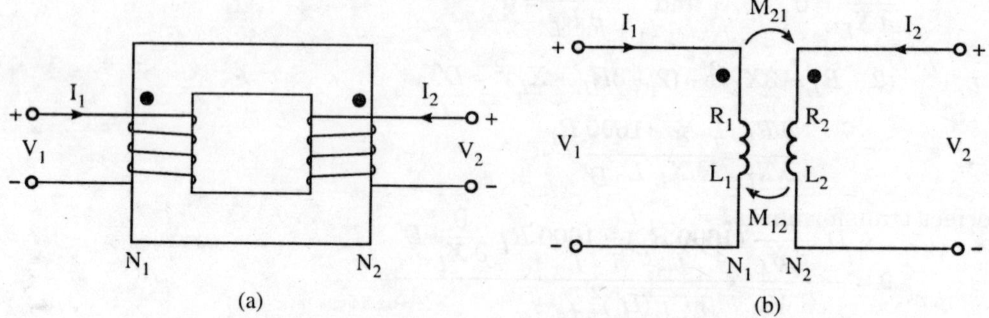

Fig 17.33 (a) Transformer (b) 2-port representation

17.11 PERFECT TRANSFORMER

A **perfect transformer** is a model of a transformer where we neglect the winding resistances and the coefficient of coupling is unity. For a perfect transformer

$$R_1 = 0\,,\; R_2 = 0,\; k = 1$$

Since
$$k = \frac{M}{\sqrt{L_1 L_2}}\,,\; M = \sqrt{L_1 L_2}$$

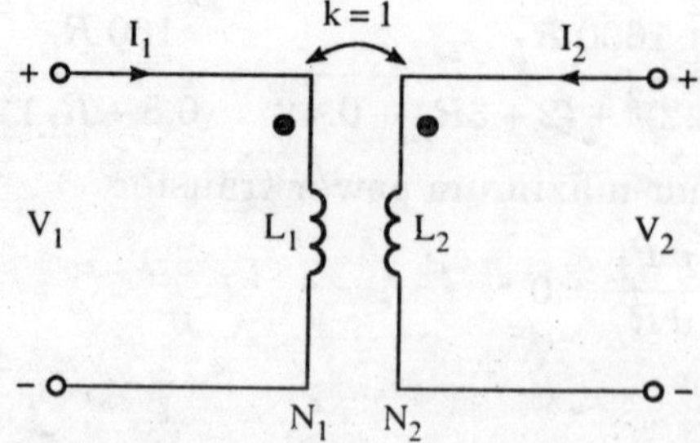

Fig 17.34 Two-port model of a perfect transformer

and $$L_1 L_2 - M^2 = 0$$

for perfect coupling. A perfect transformer is also called a **unit coupling transformer**. The 2-port model of a perfect transformer is shown in Fig. 17.34.

17.12 IDEAL TRANSFOMER

A perfect transformer can be converted into an ideal transformer by imposing an additional constraint on the self inductance of each winding approaches infinity but their ratio remains finite and constant. Thus, an **ideal transformer** possesses the following characteristics :

1. No leakage flux, that is, $k = 1$ (perfect coupling)
2. No power loss
3. L_1 and L_2 approach infinity but $\frac{L_1}{L_2} = \left(\frac{N_1}{N_2}\right)^2$

Consider the case of the perfect transformer shown in Fig. 17.34. The mesh equations in th frequency domain can be written as

$$V_1 = j\omega L_1 I_1 + j\omega M I_2 \quad (17.12.1)$$

$$V_2 = j\omega M I_1 + j\omega L_2 I_2 \quad (17.12.2)$$

$$\frac{V_1}{V_2} = \frac{L_1 I_1 + M I_2}{M I_1 + L_2 I_2}$$

For a perfect transformer, $M = \sqrt{L_1 L_2}$

$$\therefore \quad \frac{V_1}{V_2} = \frac{L_1 I_1 + \sqrt{L_1 L_2}\, I_2}{\sqrt{L_1 L_2}\, I_1 + L_2 I_2}$$

$$= \frac{\sqrt{L_1}}{\sqrt{L_2}} \frac{[\sqrt{L_1}\, I_1 + \sqrt{L_2}\, I_2]}{[\sqrt{L_1}\, I_1 + \sqrt{L_2}\, I_2]}$$

or $$\frac{V_1}{V_2} = \sqrt{\frac{L_1}{L_2}} \quad (17.12.3)$$

Since the inductance L is proportional to the square of the number of turns, $L = \beta N^2$ where β is a constant. If N_1 and N_2 are the number of turns of the primary and secondary windings,

$$L_1 = \beta N_1^2, \; L_2 = \beta N_2^2$$

$$\frac{L_1}{L_2} = \frac{N_1^2}{N_2^2} \quad (17.12.4)$$

and $$\frac{V_1}{V_2} = \frac{N_1}{N_2} \quad (17.12.5)$$

If we denote

$$\frac{N_1}{N_2} = n \quad (17.12.6)$$

then $$V_1 = nV_2 \tag{17.12.7}$$

The parameter $n = N_1/N_2$ is called the *turns ratio* of the transformer. From Eq. (17.12.1)

$$I_1 = \frac{V_1}{j\omega L_1} - \frac{M}{L_1} I_2$$

For perfect coupling, $M = \sqrt{L_1 L_2}$, therefore

$$I_1 = \frac{V_1}{j\omega L_1} - \frac{\sqrt{L_1 L_2}}{L_1} I_2$$

$$= \frac{V_1}{j\omega L_1} - \sqrt{\frac{L_2}{L_1}} I_2 \tag{17.12.8}$$

When L_1 and L_2 approach infinity such that the ratio (L_2/L_1) remains finite and constant, along with the perfect coupling, we say that the transformer is *ideal*. For an ideal transformer, Eq. (17.12.8) becomes

$$I_1 = -\sqrt{\frac{L_2}{L_1}} I_2 = -\frac{N_2}{N_1} I_2 \tag{17.12.9}$$

$$I_1 = -\frac{1}{n} I_2 \tag{17.12.10}$$

Ideal transformers are not physically realizable, but within limits many practical transformers can be treated as ideal transformers.

To obtain the effect of approximate perfect coupling *iron-cored* transformers are used. An ideal transformer is represented as shown in Fig. 17.35. The relationships between the voltages and the currents are given by

$$V_1 = n\, V_2 \Leftrightarrow V_2 = \frac{1}{n} V_1 \tag{17.12.11}$$

$$I_1 = -\frac{1}{n} I_2 \Leftrightarrow I_2 = -n\, I_1 \tag{17.12.12}$$

where $n = N_1/N_2$

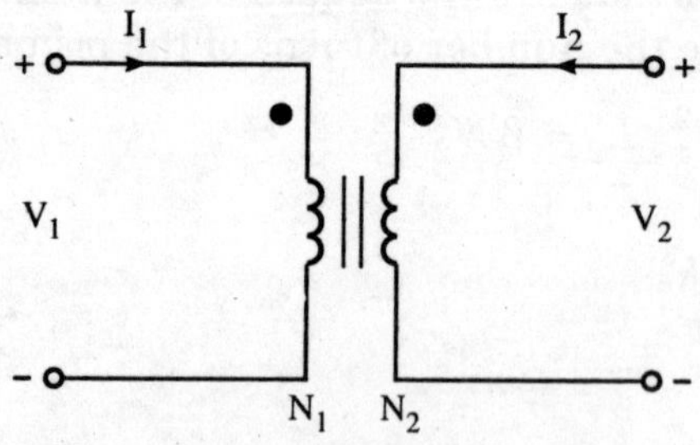

Fig 17.35 Ideal transformer

If one of the currents enters a dot and the other leaves as shown in Fig. 17.36, the defining equations for the ideal transformer can be written as

$$\frac{V_1}{V_2} = -\frac{N_1}{N_2} \tag{17.12.13}$$

$$\frac{I_1}{I_2} = \frac{N_2}{N_1} \tag{17.12.14}$$

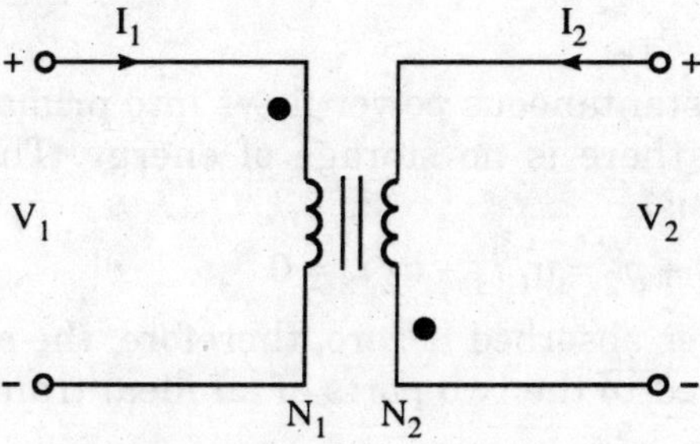

Fig 17.36 Ideal transformer

It is to be noted that in ideal transformers self and mutual inductances do not appear in the *v-i* relations.

An ideal transformer is basically different from a perfect transformer. In the symbolic representation of the ideal transformer the mutual inductance M is removed from the diagram and the coupling of the two coils is indicated by the number of turns N_1 and N_2 of the windings.

Since Eqs (17.12.11) and (17.12.12) are linear functions, and N_1 and N_2 are independent of time, the ideal transformer is a *linear time-invariant circuit element*.

An ideal transformer may be represented by either of the two controlled source models shown in Fig. 17.37. For the circuit shown in Fig. 17.37a, $V_2 = \frac{1}{n} V_1$ and $I_1 = -\frac{1}{n} I_2$; for the circuit in Fig. 17.37b. $V_1 = nV_2$ and $I_2 = -nI_1$.

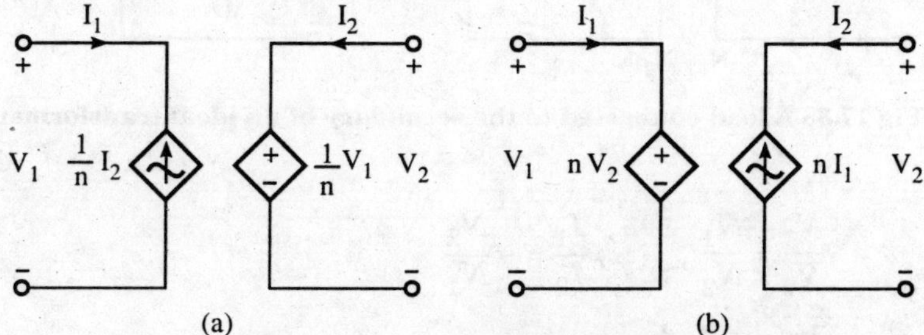

Fig 17.37 Controlled-source presentations of an ideal transformer

■ **EXAMPLE 17.21** Show that an ideal transformer is a passive device.

SOLUTION :

The instantaneous power absorbed by the primary winding of the ideal transformer, $p_1 = v_1 i_1$.

The instantaneous power absorbed by the secondary, $p_2 = v_2 i_2$.

For an ideal transformer

$$v_2 = \frac{n_2}{n_1} v_1 \, , i_2 = -\frac{n_1}{n_2} i_1$$

$\therefore$
$$p_2 = v_2 \, i_2 = \frac{n_2}{n_1} v_1 \left(-\frac{n_1}{n_2} i_1 \right) = -v_1 \, i_1$$

and
$$p_2 = -p_1$$

This shows that whatever instantaneous power flows into primary terminals flows out at the secondary terminals. Hence there is no storage of energy. The total instantaneous power absorbed by the transformer is

$$p = p_1 + p_2 = v_1 \, i_1 + v_2 \, i_2 = 0$$

Since the instantaneous power absorbed is zero, therefore, the average power is also zero.

The total energy supplied to the two ports of an ideal transformer

$$W(t) = \int_0^t p dt = \int_0^t (v_1 \, i_1 + v_2 \, i_2) \, dt = \int_0^t 0 dt = 0$$

Since $W(t) = 0$, an ideal transformer is a passive device.

17.13 IMPEDANCE TRANSFORMATION

Consider a load Z_L connected to the secondary winding of an ideal transformer as shown in Fig. 17.38.

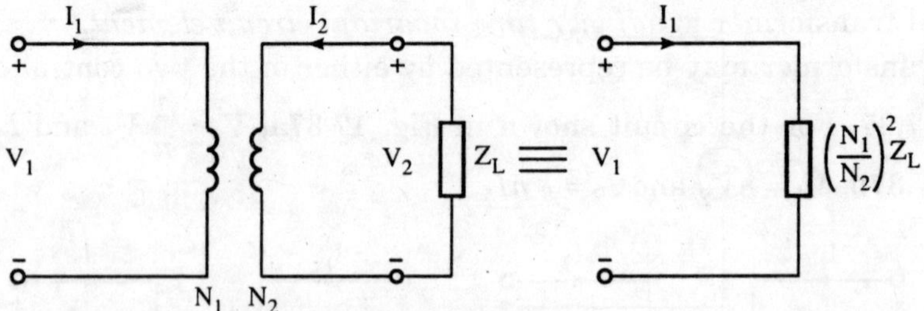

Fig 17.38 A load connected to the secondary of an ideal transformer

We have

$$\frac{V_1}{V_2} = \frac{N_1}{N_2} \, , \qquad \frac{I_1}{I_2} = -\frac{N_2}{N_1}$$

$$V_2 = -Z_L \, I_2$$

Therefore,

$$V_1 = \frac{N_1}{N_2} V_2 = \frac{N_1}{N_2} (-Z_L \, I_2)$$

$$= \frac{N_1}{N_2} \times \frac{N_1}{N_2} Z_L \, I_1$$

The input impedance seen at port 1

$$Z_{in} = \frac{V_1}{I_1} = \left(\frac{N_1}{N_2}\right)^2 Z_L = n^2 Z_L$$

Thus, the input impedance of the ideal transformer terminated in a load impedance Z_L is equal to the product of Z_L and the square of the turns ratio. In other words, the load impedance Z_L appears at the primary as an equivalent impedance of $(N_1/N_2)^2 Z_L$. Thus, an ideal transformer may be called a positive impedance converter (*PIC*).

The impedance changing property of an ideal transformer is used in situations where the *matching* of circuits with different impedances is required or where maximum power is to be transferred from a source with a given source impedance to a fixed load.

17.14 REFLECTED IMPEDANCE

Consider the circuit shown in Fig. 17.39a. The input impedance of the circuit is given by Eq. (17.12.8) as

$$Z_{in} = z_{11} + \frac{\omega^2 M^2}{z_{22}} \tag{17.14.1}$$

where $$z_{11} = R_1 + j\omega L_1 \tag{17.14.2}$$

and $$z_{22} = R_2 + j\omega L_2 + Z_L \tag{17.14.3}$$

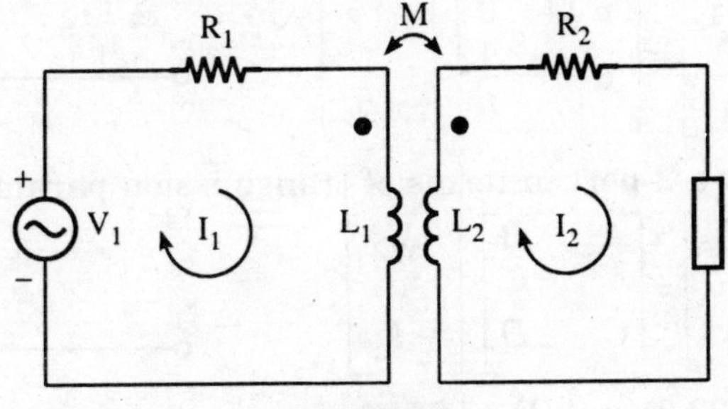

Fig 17.39

It is observed that the input impedance consists of two terms.

1. The first term is $z_{11} = R_1 j\omega L_1 = Z_p$. It is the impedance of the primary circuit.
2. The second term is $\dfrac{\omega^2 M^2}{z_{22}}$

It is due to the presence of the secondary circuit. This term is called the *reflected impedance* from the secondary circuit to the primary circuit. It is the equivalent impedance of the secondary circuit and load impedance transmitted or reflected to the primary or the input circuit. Thus, the reflected impedance from the secondary to the primary is given by

$$Z_{2ref} = \frac{\omega^2 M^2}{\text{total impedance of the secondary}} \tag{17.14.4}$$

and $$Z_{in} = Z_p + Z_{2ref}$$

It is to be noted that the reflected impedance is due solely to the existence of mutual inductance. That is, if the two coils are decoupled, M becomes zero and

$$Z_{2ref} = 0 \text{ and } Z_{in} = R_1 + j\omega L_1$$

Similarly,

Z_{out} = impedance of the output circuit + reflected impedance from the primary circuit to the secondary circuit

It is to be noted that Z_{in} is independent of the location of dot, on either winding because of the presence of M^2.

■ **EXAMPLE 17.22** Write down the defining equations for an ideal transformer in terms of transmission and hybrid 2-port parameters. Show that an ideal transformer is a reciprocal device.

SOLUTION :

The defining equations of an ideal transformer of Fig. 17.35 are given by

$$V_1 = nV_2 \tag{E 17.22.1}$$

$$I_1 = -\frac{1}{n} I_2 \tag{E 17.22.2}$$

These equations can be written in matrix form as

$$\begin{bmatrix} V_1 \\ I_1 \end{bmatrix} = \begin{bmatrix} n & 0 \\ 0 & \frac{1}{n} \end{bmatrix} \begin{bmatrix} V_2 \\ -I_2 \end{bmatrix} \tag{E 17.22.3}$$

The defining equations for a 2-port in terms of transmission parameters are

$$\begin{bmatrix} V_1 \\ I_1 \end{bmatrix} = \begin{bmatrix} A & B \\ C & D \end{bmatrix} \begin{bmatrix} V_2 \\ -I_2 \end{bmatrix} \tag{E 17.22.4}$$

Comparison of Eqs. (E 17.22.3) and (E 17.22.4) gives

$$A = n \text{ , } B = 0 \text{ , } C = 0, D = \frac{1}{n}$$

Also

$$AD - BC = n \times \frac{1}{n} - 0 = 1$$

Eqs. (E 17.22.1) and (E 17.22.2) of an ideal transformer can also be written as

$$\begin{bmatrix} V_1 \\ I_2 \end{bmatrix} = \begin{bmatrix} 0 & n \\ -n & 0 \end{bmatrix} \begin{bmatrix} I_1 \\ V_2 \end{bmatrix} \tag{E 17.22.5}$$

The defining equations for a 2-port in terms of hybrid parameters are

$$\begin{bmatrix} V_1 \\ I_2 \end{bmatrix} = \begin{bmatrix} h_{11} & h_{12} \\ h_{21} & h_{22} \end{bmatrix} \begin{bmatrix} I_1 \\ V_2 \end{bmatrix} \tag{E 17.22.6}$$

Comparison of Eqs. (E 17.22.5) and (E 17.22.6) gives

$$h_{11} = 0 \quad , h_{22} = 0$$

$$h_{12} = n \quad , h_{21} = -n$$

$$\therefore \qquad h_{21} = -h_{12}$$

Either of the two conditions should be satisfied for reciprocity.

(1) $AD - BC = 1$

(2) $h_{21} = -h_{12}$

In an ideal transformer both these conditions are satisfied. Hence the transformer is a reciprocal device.

■ **EXAMPLE 17.23** **Convert the 2-port network shown in Fig. 17.40 into a 1-port. Calculate (*a*) the impedance seen at port 1-1′, (*b*) the impedance seen between terminals *a-b*, and (*c*) the load impedance at terminals 2-2′.**

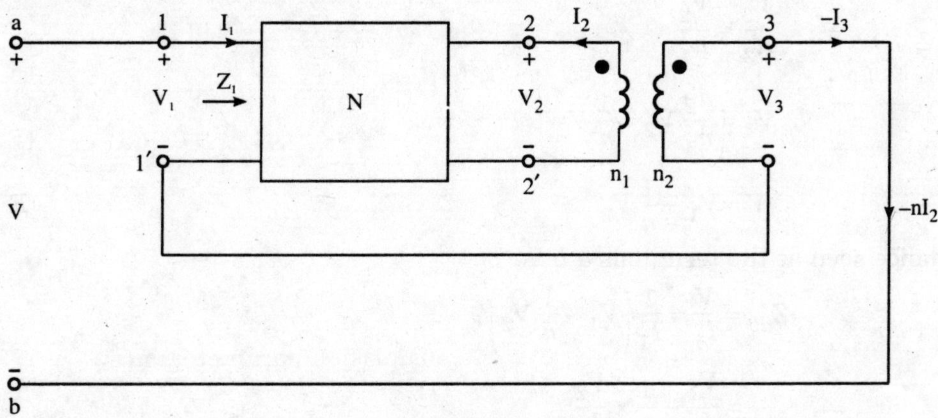

Fig 17.40

SOLUTION :

The *z*-parameter equations of the 2-port network *N* are given by

$$V_1 = z_{11} I_1 + z_{12} I_2 \qquad \text{(E 17.23.1)}$$

$$V_2 = z_{21} I_1 + z_{22} I_2 \qquad \text{(E 17.23.2)}$$

For the ideal transformer in Fig. 17.33,

$$-I_2 = -\left(-I_3 \frac{n_2}{n_1}\right)$$

$$\therefore \qquad I_3 = -\frac{n_1}{n_2} I_2 = -nI_2$$

By $\qquad I_3 = I_1$, therefore $I_1 = -nI_2$

and $\qquad I_2 = -\frac{1}{n} I_1$

Substituting the value of I_2 in Eqs. (E 17.23.1) and (E 17.23.2) we get

$$V_1 = z_{11} I_1 - \frac{1}{n} z_{12} I_1 \quad \text{(E 17.23.3)}$$

$$V_2 = z_{21} I_1 - \frac{1}{n} z_{22} I_1 \quad \text{(E 17.23.4)}$$

Impedance seen at port 1-1′

$$Z_1 = \frac{V_1}{I_1} = z_{11} - \frac{1}{n} z_{12} \quad \text{(E 17.23.5)}$$

By KVL

$$V_{ba} + V_{a1} + V_{11'} + V_{1'3'} + V_{3'3} + V_{3b} = 0$$

$$V + 0 + (-V_1) + 0 + V_3 + 0 = 0$$

or $$V = V_1 - V_3 \quad \text{(E 17.23.6)}$$

For the ideal transformer of the network

$$\frac{V_3}{n_2} = \frac{V_2}{n_1}$$

$$V_3 = \frac{n_2}{n_1} V_2 = \frac{1}{n} V_2$$

∴ $$V = V_1 - \frac{1}{n} V_2 \quad \text{(E 17.23.7)}$$

The impedance seen at the terminals *a-b* is

$$Z_{ab} = \frac{V}{I_1} = \frac{1}{I_1}\left(V_1 - \frac{1}{n} V_2\right)$$

$$= \frac{V_1}{I_1} - \frac{1}{n} \times \frac{V_2}{I_1}$$

$$= \left(z_{11} - \frac{1}{n} z_{12}\right) - \frac{1}{n}\left(z_{21} - \frac{1}{n} z_{22}\right)$$

$$= z_{11} - \frac{1}{n}(z_{12} + z_{21}) + \frac{1}{n^2} z_{22} \quad \text{(E 17.23.8)}$$

The load impedance at the terminals 2-2′ is given by

$$Z_L = -\frac{V_2}{I_2} = -\frac{1}{I_2}(-nz_{21} I_2 + z_{22} I_2)$$

or $$Z_L = nz_{21} - z_{22} \quad \text{(E 17.23.9)}$$

17.15 GYRATOR

An ideal gyrator is a 2-port element. The symbolic reprentation of an ideal gyrator is shown in Fig. 17.41. The ideal gyrator has the property that it *gyrates* the *current* of one port into the *voltage* of the other port and vice versa, that is,

$$v_1 = -\alpha\, i_2 \quad (17.15.1)$$

$$v_2 = \alpha\, i_1 \tag{17.15.2}$$

In matrix form

$$\begin{bmatrix} v_1 \\ v_2 \end{bmatrix} = \begin{bmatrix} 0 & -\alpha \\ \alpha & 0 \end{bmatrix} \begin{bmatrix} i_1 \\ i_2 \end{bmatrix} \tag{17.15.3}$$

Fig 17.41 Symbolic representation of an ideal gyrator

These equations are called the *defining equations of the gyrator.* The coefficient α , which has the dimension of a resistance is called the **gyration constant, gyration resistance**, or **gyration ratio.** The name gyrator has been given by its originator Tellegen.

Gyrators can be constructed using transistors, operational amplifiers (OA), or operational transconductance amplifiers (OTA). The equivalent circuit of a gyrator is shown in Fig. 17.42.

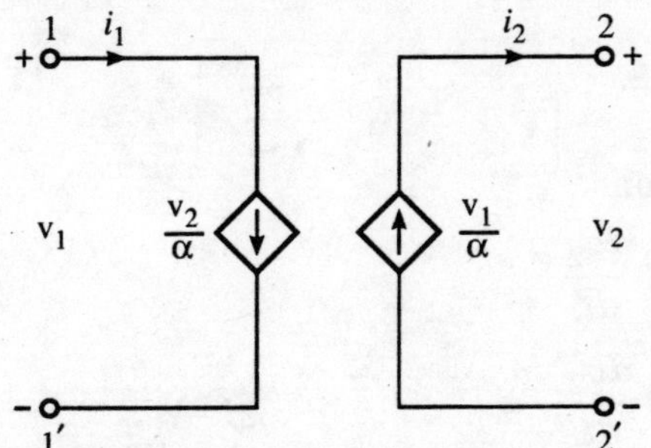

Fig 17.42 Equivaent circuit of a gyrator

The transmissin matrix of the gyrator is

$$\begin{bmatrix} A & B \\ C & D \end{bmatrix} = \begin{bmatrix} 0 & \alpha \\ \frac{1}{\alpha} & 0 \end{bmatrix}$$

therefore, $\quad A = 0 \,, D = 0 \,, B = \alpha \,, C = \frac{1}{\alpha}$

We see for the gyrator

$$AD - BC = 0 - \alpha \times \frac{1}{\alpha} = -1$$

Since for a reciprocal device $AD - BC = 1$, therefore, *the gyrator is a nonreciprocal device.*

In Eqs. (17.15.1) and (17.15.2), α is a constant. Thus, the gyrator is a linear time-invariant element.

The power delivered to the gyrator

$$= v_1 i_1 + v_2 i_2$$

$$= v_1 i_1 + (\alpha i_1)\left(-\frac{1}{\alpha} v_1\right) = 0$$

Thus, an ideal gyrator is a passive lossless device.

■ **EXAMPLE 17.24** Show that the circuit in Fig. 17.43 is equivalent to a single inductor. Find its inductance in terms of L and α.

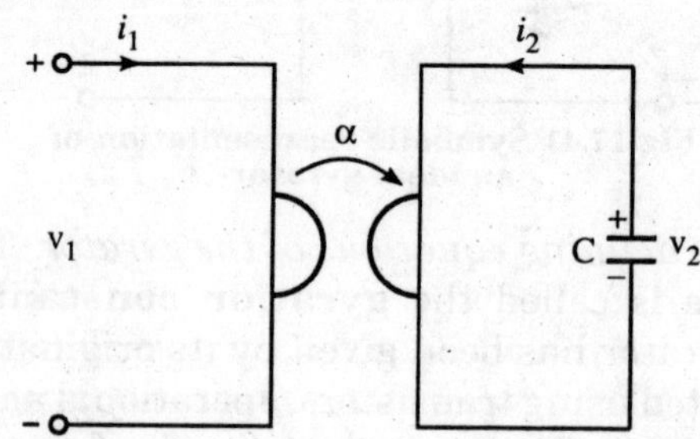

Fig 17.43 A gyrator circuit for simulating an inductor

SOLUTION :

We have $v_1 = -\alpha i_2$

According to polarity convention

$$i_2 = -C\frac{dv_2}{dt}$$

therefore, $$v_1 = \alpha C\frac{dv_2}{dt}$$

But $$v_2 = \alpha i_1$$

therefore, $$v_1 = \alpha C\frac{d}{dt}(\alpha i_1) = \alpha^2 C\frac{di_1}{dt}$$

$$= L\frac{di_1}{dt} \text{ (say)}$$

where $$L = \alpha^2 C$$

Thus, as for as the first port is concerned, the circuit given in Fig. 17.43 is equivalent to a linear time-invariant inductor with inductance

$$L = \alpha^2 C.$$

Since inductors are inherently bulky, the above property of gyrators can be applied to simulate inductors required in transistor and integrated circuits.

EXAMPLE 17.25 Show that the circuit in Fig. 17.44 is equivalent to a single capacitor. Find its capacitance in terms of L and α

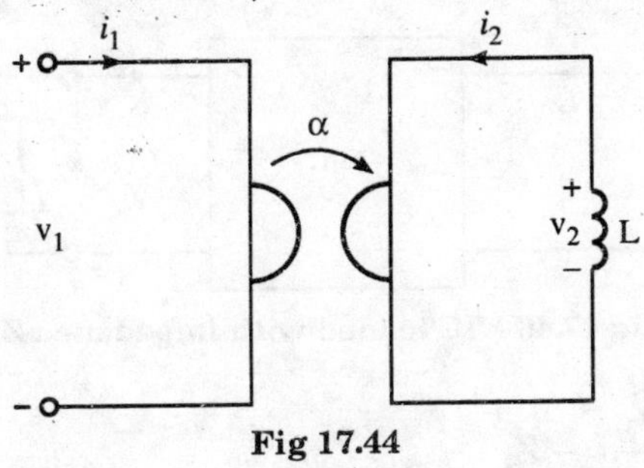

Fig 17.44

SOLUTION :

We have

$$v_2 = \alpha\, i_1 \,,\, v_1 = -\,\alpha i_2$$

According to polarity convention

$$-\,v_2 = L\frac{di_2}{dt}$$

$$-\,\alpha i_1 = L\frac{d}{dt}\left(-\frac{v_1}{\alpha}\right)$$

$$i_1 = \frac{L}{\alpha^2}\frac{dv_1}{dt} = C\frac{dv_1}{dt} \quad \text{(say)}$$

where $$C = \frac{L}{\alpha^2}$$

Thus the given circuit shows the use of a gyrator and inductor to simulate a capacitor.

17.16 NEGATIVE IMPEDANCE CONVERTER (NIC)

A negative impedance converter (NIC) is a 2-port element which converts a positive load impedance into a negative impedance at its input port. Thus, a negative impedance converter has the capability of generating negative R, L, or C simply by loading it with positive R, L, or C. This property is very useful in network theory; for example, losses in the positive resistances of a network can be cancelled by including resistively loaded negative impedance converters in the network. The introduction of NIC expands the number of network building blocks considerably, because it is possible to include $-R$, $-L$ and $-C$ elements in the network. The symbolic representation of NIC is given in Fig, 17.45.

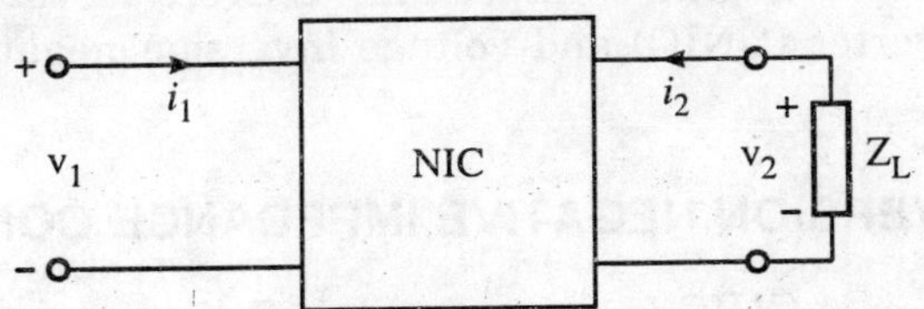

Fig 17.45 Symbol of NIC

Consider the NIC with a load impedance Z_L as shown in Fig. 17.46. An ideal NIC has the property that the impedance seen at the input port is equal to negative of the load impedance.

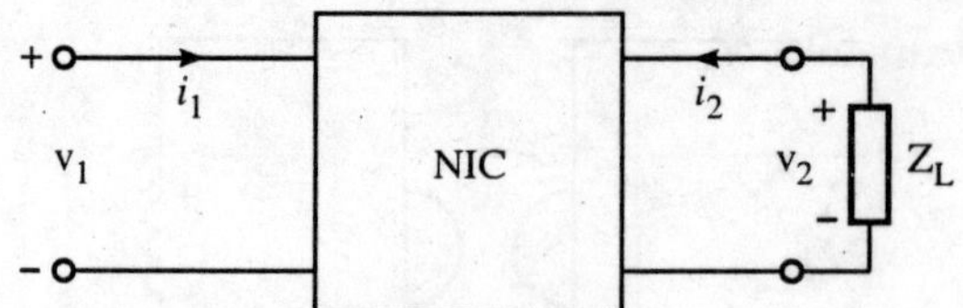

Fig 17.46 NIC loaded with impedance Z_L

$$Z_{in} = \frac{v_1}{i_1} = -Z_L \tag{17.16.1}$$

Since $Z_L = -\dfrac{v_2}{i_2}$ we can write

$$\frac{v_1}{i_1} = \frac{v_2}{i_2} \tag{17.16.2}$$

The following voltage-current relationships hold for the NIC

$$v_1 = kv_2 \tag{17.16.3}$$

$$i_1 = ki_2 \tag{17.16.4}$$

where k is a positive constant. These defining equations can be written in the form

$$v_1 = 0 \times i_1 + kv_2$$

$$i_2 = \frac{1}{k} i_1 + 0 \times v_2$$

In matrix form

$$\begin{bmatrix} v_1 \\ i_2 \end{bmatrix} = \begin{bmatrix} 0 & k \\ \frac{1}{k} & 0 \end{bmatrix} \begin{bmatrix} i_1 \\ v_2 \end{bmatrix} \tag{17.16.5}$$

Therefore, $h_{11} = 0$, $h_{22} = 0$

$h_{12} = k$, $h_{21} = 1/k$

For a reciprocal network we must have

$$h_{21} = -h_{12}$$

For NIC, $h_{21} \neq -h_{12}$, *hence NIC is a nonreciprocal device.*

There are two types of negative impedance converters, namely, current inversion negative impedance converter (CNIC) and voltage inversion negative impedance converter (VNIC).

17.17 CURRENT INVERSION NEGATIVE IMPEDANCE CONVERTER (CNIC)

The symbolic representation of CNIC is shown in Fig. 17.47. The defining equations of CNIC are

$$v_1 = kv_2 \tag{17.17.1}$$

$$i_1 = ki_2 \tag{17.17.2}$$

where k is a positive constant.

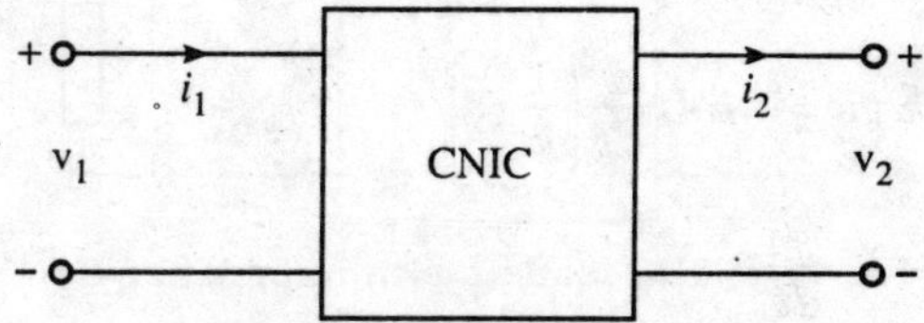

Fig 17.47 Symbol of CNIC

The basic property of CNIC is to invert the direction of current and keep the polarity of the voltage unchanged. Thus, the current after flowing through NIC is reversed in direction.

17.18 VOLTAGE INVERSION NEGATIVE IMPEDANCE CONVERTER (VNIC)

The symbolic representation of VNIC is shown in Fig. 17.48.

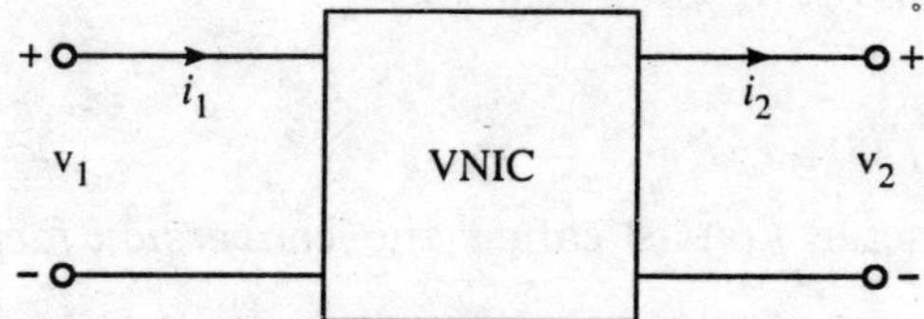

Fig 17.48 Symbol of VNIC

The defining equations of VNIC are

$$v_1 = -k\, v_2$$

$$i_1 = -k\, i_2 \tag{17.18.1}$$

where k is a positive constant. The basic property of VNIC is to invert the polarity of the input voltage. The current entering and leaving the VNIC have the same direction but the input and output voltages have opposite polarities. In the symbolic representation we always show that the upper terminals of the two ports are at higher potentials but in practice, the polarity of v_2 is opposite to that of v_1 in VNIC.

■ **EXAMPLE 17.26** Show that an inductor $+L$ connected at the output port of a INIC is equivalent to $-L$ at the input port.

SOLUTION :

Fig. 17.49 shows the connection of an inductor $+L$ at the output port. The defining equations of INIC are

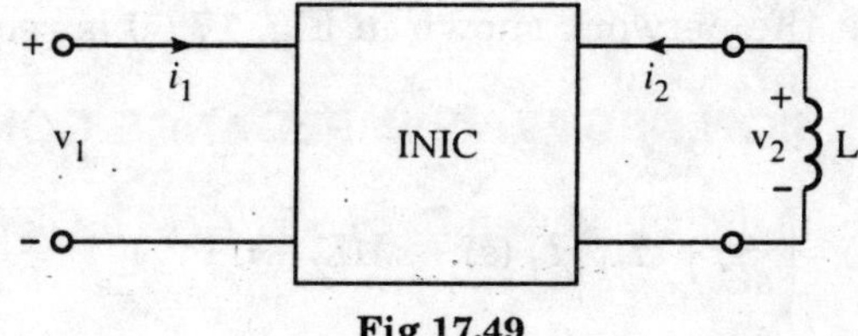

Fig 17.49

$$v_1 = -k\, v_2\,, \qquad i_1 = k\, i_2$$

In Fig. 17.49,

$$v_2 = L \frac{d}{dt}(-i_2)$$

$$v_1 = k\, v_2 = -k\, L \frac{d}{dt}\left(\frac{1}{k} i_1\right)$$

or
$$v_1 = -L \frac{di_1}{dt}$$

This relation shows that the equivalent inductance at input terminals is $-L$.

Thus INIC converts $+L$ connected at the output port into an inductance $-L$ at the input port. Similar conclusions would follow if the terminating elements were R or C.

17.19 GENERALIZED IMPEDANCE CONVERTER (GIC)

A generalized impedance converter (GIC) is a 2-port element having the property that if the output port is terminated by an impedance Z_{L2}, the impedance seen at its output port is proportional to Z_{L2}.

$$Z_1 = f(s)\, Z_{L2}$$

The proportionality factor $f(s)$ is called the *conversion factor*. It is a dimensionless quantity.

A special case of the GIC is the negative impedance converter (NIC). Here $f(s) = -1$ and hence

$$Z_1 = -Z_{L2}$$

$$Z_1 = \frac{v_1}{i_1}\,, \quad Z_{L2} = \frac{-v_2}{i_2}$$

We have two possibilities :

(*a*) VNIC

$$i_1 = -i_2$$

$$v_1 = -v_2$$

(*b*) CNIC

$$i_1 = i_2$$

$$v_1 = v_2$$

■ **EXAMPLE 17.27** Show that the network shown in Fig. 17.50 is reciprocal.

SOLUTION :

For the given network

$$V_1(s) = \left(\frac{1}{sC} + sL_1\right) I_1(s) + sMI_2(s)$$

$$V_2(s) = (R + sL_2)\, I_2(s) + sMI_1(s)$$

$$\begin{bmatrix} V_1(s) \\ V_2(s) \end{bmatrix} = \begin{bmatrix} \dfrac{1}{sC} + sL_1 & sM \\ sM & R + sL_2 \end{bmatrix} \begin{bmatrix} I_1(s) \\ I_2(s) \end{bmatrix}$$

Here $z_{12}(s) = sM$ and $z_{21}(s) = sM$.

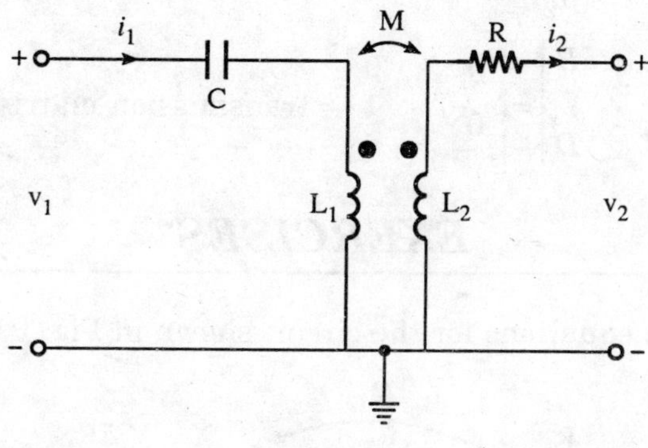

Fig 17.50

Since $z_{12}(s) = z_{21}(s)$, the given network is reciprocal.

EXAMPLE 17.28 Show that a cascade of two ideal gyrators is an ideal transformer.

SOLUTION : The transmision parameters of the two gyrators may be assumed as

$$\begin{bmatrix} A_1 & B_1 \\ C_1 & D_1 \end{bmatrix} = \begin{bmatrix} 0 & \alpha_1 \\ \dfrac{1}{\alpha_1} & 0 \end{bmatrix}$$

$$\begin{bmatrix} A_2 & B_2 \\ C_2 & D_2 \end{bmatrix} = \begin{bmatrix} 0 & \alpha_2 \\ \dfrac{1}{\alpha_2} & 0 \end{bmatrix}$$

The overall transmission parameters of the two gyrators in cascade are given by

$$\begin{bmatrix} A & B \\ C & D \end{bmatrix} = \begin{bmatrix} A_1 & B_1 \\ C_1 & D_1 \end{bmatrix} \begin{bmatrix} A_2 & B_2 \\ C_2 & D_2 \end{bmatrix}$$

$$= \begin{bmatrix} 0 & \alpha_1 \\ \dfrac{1}{\alpha_1} & 0 \end{bmatrix} \begin{bmatrix} 0 & \alpha_2 \\ \dfrac{1}{\alpha_2} & 0 \end{bmatrix}$$

$$= \begin{bmatrix} \dfrac{\alpha_1}{\alpha_2} & 0 \\ 0 & \dfrac{\alpha_2}{\alpha_1} \end{bmatrix}$$

Let $$\frac{\alpha_1}{\alpha_2} = n$$

$$\therefore \quad \begin{bmatrix} A & B \\ C & D \end{bmatrix} = \begin{bmatrix} n & 0 \\ 0 & \dfrac{1}{n} \end{bmatrix} = \text{transmission matrix of an ideal transformer}$$

EXERCISES

1. Write the mesh equations for the circuit shown in Fig. P17.1.

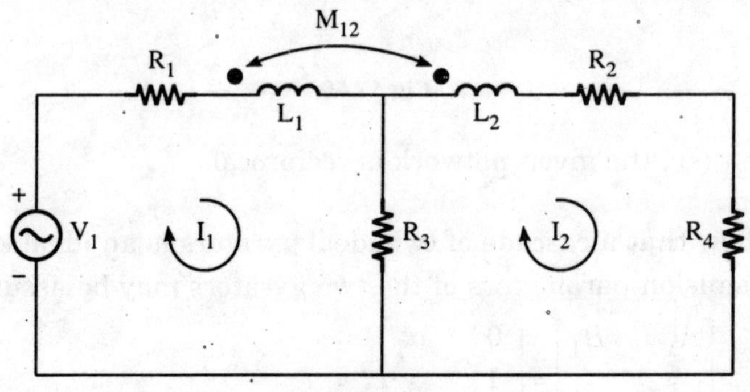

Fig P. 17.1

2 Find expressions for the input impedances of the circuits shown in Fig. P17.2.a and Fig. P17.2b.

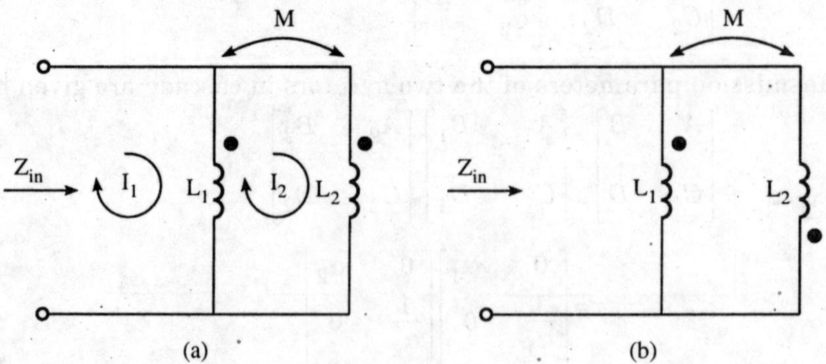

Fig P. 17.2

3. Write the mesh equations for the circuit shown in Fig. P17.3.

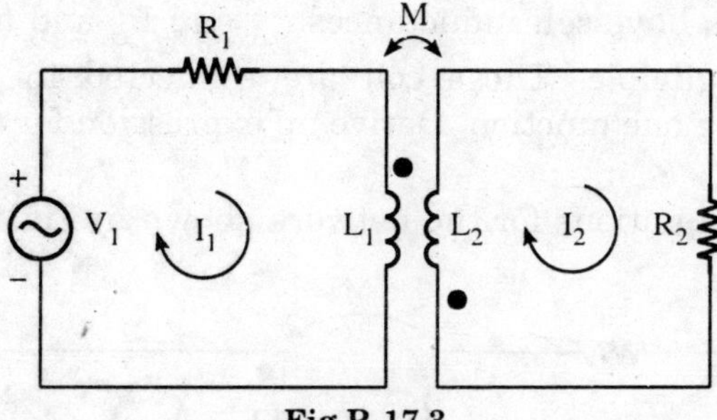

Fig P. 17.3

4. Find an expression for the input impedance of the circuit shown in Fig. P17.4.

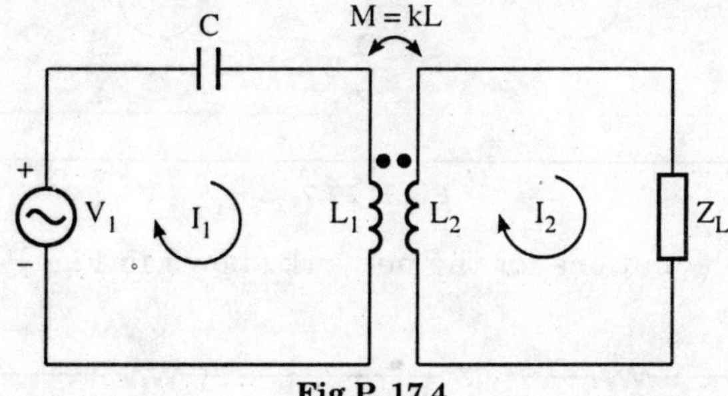

Fig P. 17.4

5. Write the mesh equations for the circuit of Fig. P17.5.

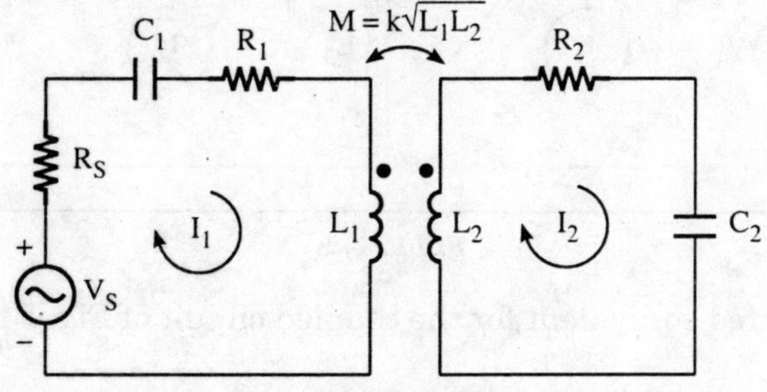

Fig P. 17.5

6. Three coils are mutually coupled as shown in Fig. P17.6. Write a set of equations describing $v_1\,(t)$, $v_2\,(t)$, and $v_3\,(t)$ in terms of the current and the inductances.

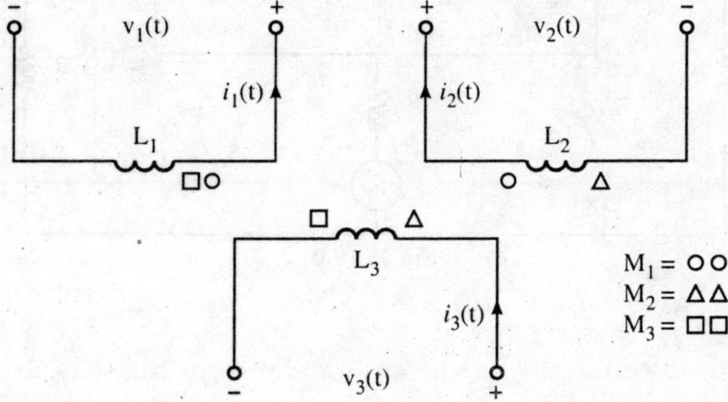

Fig P. 17.6

7. Two coupled coils have self inductances L_A and L_B and mutual inductance M. The resistance is negligible. These coil are connected in parallel with the dotted terminals forming one junction. Derive an expression for the total inductance of the circuit.
8. Write the mesh equations for the network shown in Fig. P17.7

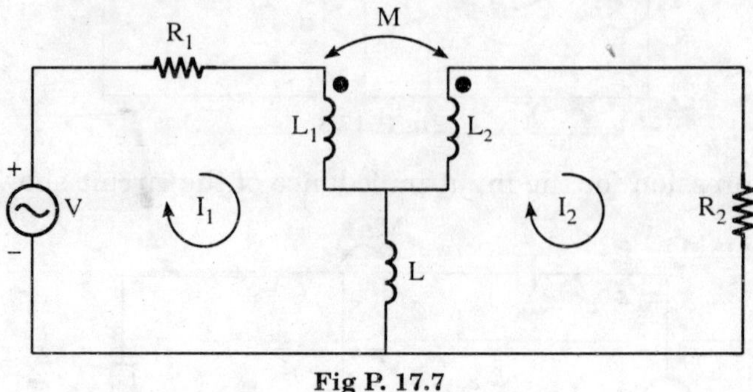

Fig P. 17.7

9. Write the mesh equations for the network shown in Fig. P17.8.

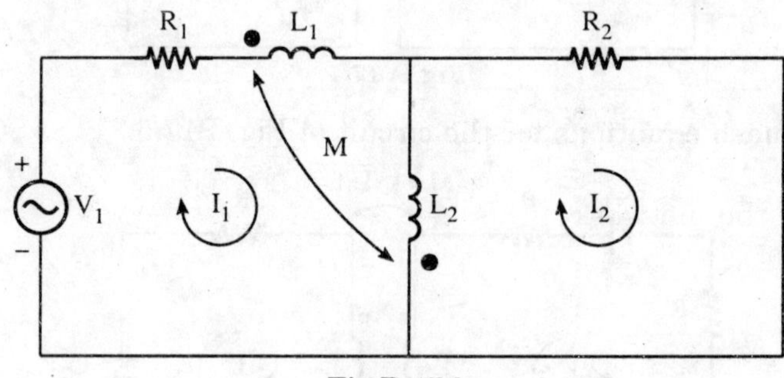

Fig P. 17.8

10. Obtain the dotted equivalent for the coupled circuit of Fig. P17.9 and write the mesh equations.

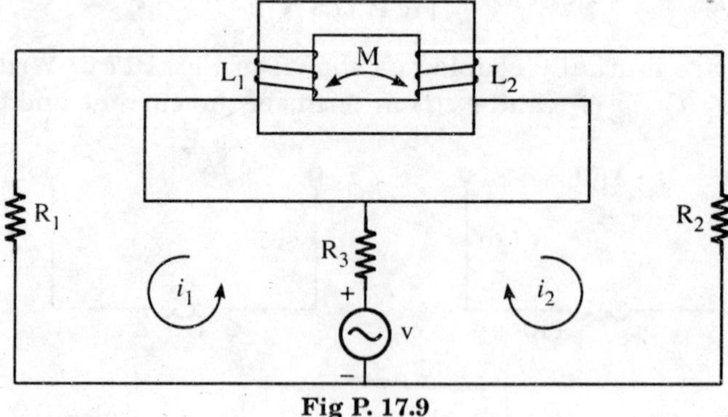

Fig P. 17.9

11. Write the mesh equations in matrix form for the network of Fig. P17.10.

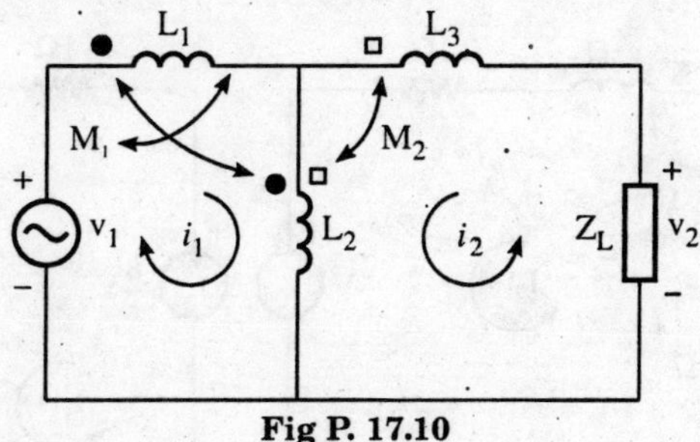

Fig P. 17.10

12. For the network shown in Fig. P17.11a, what inductance values should be chosen in order to make the circuit in Fig. P17.11b the T equivalent of Fig. P17.11a?

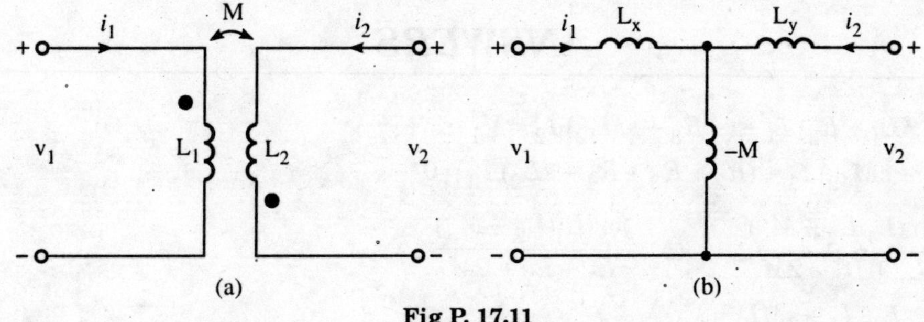

Fig P. 17.11

13. Show that the network in Fig. P17.12 is a π equivalent of the network in Fig. P17.11a.

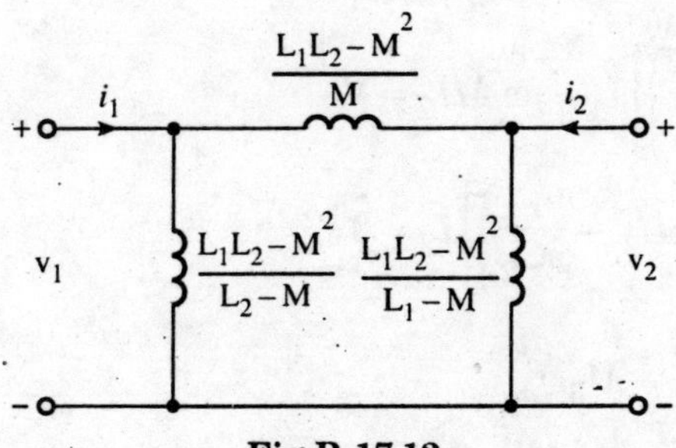

Fig P. 17.12

14. Write the mesh equations for the network of Fig. P17.13a and Fig. P17.13b.

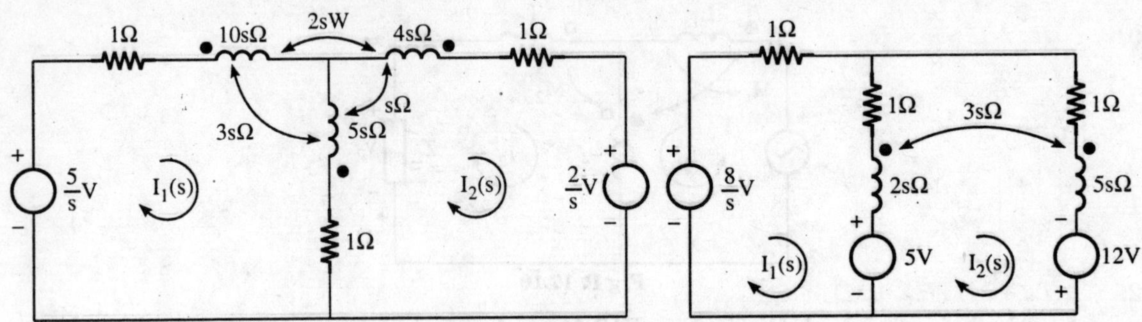

Fig P. 17.13

ANSWERS

1. $(R_1 + sL_1 + R_3)\, I_1 + (-R_3 + sM_{12})\, I_2 = V_1$

   $(-R_3 + sM_{12})\, I_1 + (R_2 + R_3 + R_4 + sL_2)\, I_2 = 0$

2. $(a)\ \dfrac{j\omega\,(L_1 L_2 - M^2)}{L_1 + L_2 - 2M}$, $(b)\ \dfrac{j\omega\,(L_1 L_2 - M^2)}{L_1 + L_2 + 2M}$

3. $(R_1 + sL_1)\, I_1 + sMI_2 = V_1$

   $sMI_1 + (R_2 + sL_2)\, I_2 = 0$

4. $j\left(\omega L - \dfrac{1}{\omega C}\right) + \dfrac{\omega^2 M^2}{j\omega L + Z_L}$

5. $\left[R_S + R_1 + j\left(\omega L_1 - \dfrac{1}{\omega C_1}\right)\right] I_1 - j\omega\, MI_2 = V_S$

   $-j\omega\, MI_1 + \left[R_2 + j\left(\omega L_2 - \dfrac{1}{\omega C_2}\right)\right] I_2 = 0$

6. $v_1(t) = L_1 \dfrac{di_1}{dt} + M_1 \dfrac{di_2}{dt} - M_3 \dfrac{di_3}{dt}$

   $v_2(t) = M_1 \dfrac{di_1}{dt} + L_2 \dfrac{di_2}{dt} - M_2 \dfrac{di_3}{dt}$

   $v_3(t) = -M_3 \dfrac{di_1}{dt} - M_2 \dfrac{di_2}{dt} + L_3 \dfrac{di_3}{dt}$

7. $L = \dfrac{M^2 - L_A L_B}{2M - L_A - L_B}$

8. $[R_1 + j\omega\,(L_1 + L)]\, I_1 + j\omega\,(M - L)\, I_2 = V$

   $j\omega\,(M - L)\, I_1 + [R_2 + j\omega\,(L_2 + L)]\, I_2 = 0$

9. $[R_1 = j\omega\,(L_1 + L_2 - 2M)]\, I_1 + j\omega\,(-\tfrac{1}{2} + M)\, I_2 = V_1$

$$j\omega(-L_2 + M) I_1 + (R_2 + j\omega L_2) I_2 = 0$$

10. $$(R_1 + R_3)\, i_1 + L_1 \frac{di_1}{dt} + R_3\, i_2 + M \frac{di_2}{dt} = v$$

$$(R_2 + R_3)\, i_2 + L_2 \frac{di_2}{dt} + R_3\, i_1 + M \frac{di_1}{dt} = v$$

11. $$\begin{bmatrix} v_1 \\ v_2 \end{bmatrix} = \begin{bmatrix} s\,(L_1 + L_2 + 2M_1) & s\,(L_2 + M_1 - M_2) \\ s\,(L_2 + M_1 - M_2) & s\,(L_2 + L_3 - 2M_2) \end{bmatrix} \begin{bmatrix} i_1 \\ i_2 \end{bmatrix}$$

12. $(L_1 + M)$, $(L_2 + M)$

14. (a) $$(2 + 9s)\, I_1(s) - (1 + 3s)\, I_2(s) = \frac{5}{s}$$

$$-(1 + 3s)\, I_1(s) + (2 + 7s)\, I_2(s) = -\frac{2}{s}$$

(b) $$(2 + 2s)\, I_1(s) - (1 - s)\, I_2(s) = \frac{8}{s} + 5$$

$$-(1 - s)\, I_1(s) + (2 + s)\, I_2(s) = 7$$

18

NETWORK FUNCTIONS

18.1 INTRODUCTION

Consider a network made up of resistances, inductances, capacitances, and possibly *dependent sources*. The applied signal $\boldsymbol{x}(t)$ and the response $\boldsymbol{y}(t)$ are in general related by a linear differential equation of the type

$$b_n \frac{d^n \boldsymbol{y}}{dt^n} + b_{n-1} \frac{d^{n-1} \boldsymbol{y}}{dt^{n-1}} + \ldots + b_1 \frac{dy}{dt} + b_0 \boldsymbol{y}$$

$$= a_m \frac{d^m \boldsymbol{x}}{dt^m} + a_{m-1} \frac{d^{m-1} \boldsymbol{x}}{dt^{m-1}} + \ldots + a_1 \frac{dx}{dt} + a_0 \boldsymbol{x} \quad (18.1.1)$$

where $a_0, a_1, \ldots, a_m$ and $b_0, b_1, \ldots, b_n$ are suitable coefficients, whose expressions depend on the elements comprising the network and are thus *real* and *time independent.*

With an applied signal of the complex exponential type

$$\boldsymbol{x}(t) = \mathbf{X}\, e^{st} \quad (18.1.2)$$

the response is also of the complex exponential type

$$\boldsymbol{y}(t) = \mathbf{Y}\, e^{st} \quad (18.1.3)$$

and the value of s in the expression for $\boldsymbol{y}(t)$ is the *same* as that given in the expression for $\boldsymbol{x}(t)$. Substituting $\boldsymbol{x}$ and $\boldsymbol{y}$ into Eq. (18.1.1) and using the relation

$$\frac{d^k (e^{st})}{dt^k} = s^k (e^{st}) \quad (18.1.4)$$

We get

$$(b_n s^n + b_{n-1} s^{n-1} + \ldots b_1 s + b_0)\, \mathbf{Y}\, e^{st}$$

$$= (a_m s^m + a_{m-1} s^{m-1} + \ldots + a_1 s + a_0)\, \mathbf{X}\, e^{st}$$

Cancelling e^{st} from both the sides we can write

$$D(s)\, \mathbf{Y} = N(s)\, \mathbf{X} \quad (18.1.5)$$

where $D(s)$ and $N(s)$ are *polynomials* in s with *real coefficients,*

$$D(s) = b_n s^n + b_{n-1} s^{n-1} + \ldots + b_1 s + b_0 \tag{18.1.6}$$

$$N(s) = a_m s^m + a_{m-1} s^{m-1} + \ldots + a_1 s + a_0 \tag{18.1.7}$$

and n and m are their degrees.

Let us define

$$\boldsymbol{H}(s) \stackrel{\Delta}{=} \frac{\mathbf{Y}}{\mathbf{X}} \tag{18.1.8}$$

The function $\boldsymbol{H}(s)$ is called the network function.

$$\boldsymbol{H}(s) = \frac{\mathbf{Y}}{\mathbf{X}} = \frac{N(s)}{D(s)} = \frac{a_m s^m + a_{m-1} s^{m-1} + .. + a_1 s + a_0}{b_n s^n + b_{n-1} s^{n-1} + \ldots + b_1 s + b_0} \tag{18.1.9}$$

Since the funciton $\boldsymbol{H}(s)$ is in the form of ratio of two polynomials with integer powers of s, it is a rational function of s.

Consider a linear, time-invariant network having an excitation $\boldsymbol{x}(t)$, which may be a current or a voltage, and a response $\boldsymbol{y}(t)$, which may also be either a current or a voltage. The **network function** $\boldsymbol{H}(s)$ is defined as the ratio of the Laplace transform of the response (effect) to the Laplace transform of the excitation (cause), provided that all the initial conditions, (ICs) of the network are set to zero.

$$\text{Network function} \stackrel{\Delta}{=} \frac{\pounds\,(\text{response with zero initial state})}{\pounds\,(\text{excitation})}$$

$$\boldsymbol{H}(s) \stackrel{\Delta}{=} \left.\frac{\pounds\,[y(t)]}{\pounds\,[x(t)]}\right|_{\text{all ICs} = 0}$$

$$\boxed{\mathbf{H}(s) = \left.\frac{\mathbf{Y}(s)}{\mathbf{X}(s)}\right|_{\text{all ICs} = 0}} \tag{18.1.10}$$

If a network has multiple independent sources, we can find the network function for each source and use superposition theorem to find the response to all sources.

In an electric network, the excitation (input) and the response (output) are each of two possible types, namely, current or voltage. Depending upon the nature of the excitation and the response we define two different sets of network functions, namely driving-point function and transfer functions.

18.2 DRIVING-POINT FUNCTIONS

When the excitation (input) and the response (output) are defined at the *same* pair of terminals (that is, the same port), these network functions are called driving point (*dp*) functions. Thus **driving-point functions** relate the voltage at a terminal or port to the current at the *same* pair of terminals or port. There are two types of driving-point functions : (*a*) **driving-point impedance function** and (*b*) **driving-point admittance function**. These two functions are given one common name, **immittance function.** For the input port, the **driving-point impedance function** is defined as

$$Z_{in}(s) \overset{\Delta}{=} \frac{V_{in}(s)}{I_{in}(s)} \tag{18.2.1}$$

The *driving-point admittance function* is defined as the reciprocal of the driving-point impedance function.

$$Y_{in}(s) = \frac{1}{Z_{in}(s)} = \frac{I_{in}(s)}{V_{in}(s)} \tag{18.2.2}$$

The output driving point functions are defined in a similar way.

18.3 TRANSFER FUNCTIONS

The transfer function is used to describe networks which have at least two ports. In general, the transfer function may have the following possible forms :

(*a*) If $Y(s)$ and $X(s)$ are voltage transforms [that is, $V_{out}(s)$ and $V_{in}(s)$ respectively], $H(s)$ is called a **voltage transfer function.**

(*b*) If $Y(s)$ and $X(s)$ are current transforms [that is, $I_{out}(s)$ and $I_{in}(s)$ respectively], $H(s)$ is called a **current transfer function.**

(*c*) If $Y(s)$ and $X(s)$ are voltage and current transforms [that is, $V_{out}(s)$ and $I_{in}(s)$ respectively], $H(s)$ is called a **transfer impedance function.**

(*d*) If $Y(s)$ and $X(s)$ are current and voltage transforms [that is, $I_{out}(s)$ and $V_{in}(s)$ respectively], $H(s)$ is called a **transfer admittance function**.

For a 2-port network, the input quantities are $V_1(s)$, $I_1(s)$ and the output quantities are $V_2(s)$ and $I_2(s)$. We can define four transfer functions as follows :

$$\frac{V_2(s)}{V_1(s)} = G_{21}(s) \text{ (voltage gain or voltage ratio transfer function)}$$

$$\frac{I_2(s)}{I_1(s)} = \alpha_{21}(s) \text{ (current gain or current ratio transfer function)}$$

$$\frac{V_2(s)}{I_1(s)} = Z_{21}(s) \text{ (transfer impedance function)}$$

$$\frac{I_2(s)}{V_1(s)} = Y_{21}(s) \text{ (transfer admittance function)}$$

The subscript notation in a transfer function has the following meaning.

The first subscript refers to the response and the second subscript, the input. In the above relations, the first subscript is for the numerator and the second for the denominator.

It should be noted that, for a one-port network, $Z(s) = 1/Y(s)$, but for a 2-port network $Z_{21}(s) \neq \dfrac{1}{Y_{21}(s)}$ and $G_{21}(s) \neq 1/\alpha_{21}(s)$.

18.4 COMPUTATION OF NETWORK FUNCTIONS USING MESH ANALYSIS

The relationships between currents and voltages in a network can be described by a set of mesh (loop) equations, The mesh equations for a network of m meshes without any dependent sources are

$$\begin{bmatrix} Z_{11} & Z_{12} & \cdots & Z_{1m} \\ Z_{21} & Z_{22} & \cdots & Z_{2m} \\ \cdot & & & \\ \cdot & & & \\ Z_{m1} & Z_{m2} & \cdots & Z_{mm} \end{bmatrix} \begin{bmatrix} I_1 \\ I_2 \\ \cdot \\ \cdot \\ I_m \end{bmatrix} = \begin{bmatrix} V_1 \\ V_2 \\ \cdot \\ \cdot \\ V_m \end{bmatrix} \tag{18.4.1}$$

or

$$[\mathbf{Z}_m]\,[\mathbf{I}_m] = [\mathbf{V}_m] \tag{18.4.2}$$

Here the voltages $V_1, V_2, \ldots, V_m$ are the applied voltage transforms (due to excitation and initial conditions), and $I_1, I_2, \ldots, I_m$ are the resulting current transforms. The matrix $[\mathbf{Z}_m]$ is the mesh impedance matrix. We can say that $V_1, V_2, \ldots, V_m$ are input variables and $I_1, I_2, \ldots, I_m$ are output variables. It should be noted that in Eq. (18.4.1) all quantities are functions of s. Solving for unknown currents $I_1, I_2, \ldots, I_m$ we get

$$I_1 = \frac{\Delta_{11}}{\Delta_m} V_1 + \frac{\Delta_{21}}{\Delta_m} V_2 + \ldots + \frac{\Delta_{k1}}{\Delta_m} V_k + \ldots + \frac{\Delta_{m1}}{\Delta_m} V_m$$

$$I_2 = \frac{\Delta_{12}}{\Delta_m} V_1 + \frac{\Delta_{22}}{\Delta_m} V_2 + \ldots + \frac{\Delta_{k2}}{\Delta_m} V_k + \ldots + \frac{\Delta_{m2}}{\Delta_m} V_m$$

$$\cdot$$

$$\cdot$$

$$I_l = \frac{\Delta_{1l}}{\Delta_m} V_1 + \frac{\Delta_{2l}}{\Delta_m} V_2 + \ldots + \frac{\Delta_{kl}}{\Delta_m} V_k + \ldots + \frac{\Delta_{ml}}{\Delta_m} V_m$$

$$\cdot$$

$$\cdot$$

$$I_m = \frac{\Delta_{1m}}{\Delta_m} V_1 + \frac{\Delta_{2m}}{\Delta_m} V_2 + \ldots + \frac{\Delta_{km}}{\Delta_m} V_k + \ldots + \frac{\Delta_{mm}}{\Delta_m} V_m \tag{18.4.3}$$

where Δ_m is the determinant of the mesh impedance matrix. That is,

$$\Delta_m = \begin{vmatrix} Z_{11} & Z_{12} & \cdots & Z_{1m} \\ Z_{21} & Z_{22} & \cdots & Z_{2m} \\ \cdot & & & \\ \cdot & & & \\ Z_{m1} & Z_{m2} & \cdots & Z_{mm} \end{vmatrix} \tag{18.4.4}$$

Δ_{ij} = are the various cofactors. That is,

$\Delta_{ij} = (-1)^{i+j} \times$ determinant obtained by deleting ith row and jth column of the determinant Δ_m.

Transfer admittance

Let us consider a special case when the network has zero initial conditions and only a single excitation V_k is applied, all other voltage sources being short circuited. Solving Eq. (18.4.3) for I_l we get

$$I_l = \frac{\Delta_{kl}}{\Delta_m} V_k \tag{18.4.5}$$

$$\frac{I_l}{V_k} = \frac{\Delta_{kl}}{\Delta_m} \tag{18.4.6}$$

The quantity I_l/V_k has the dimensions of admittance. It is denoted by Y_{lk} and is called transfer admittance. That is,

$$Y_{lk} \stackrel{\Delta}{=} \frac{I_l}{V_k} \tag{18.4.7}$$

$$I_l = Y_{lk} V_k \tag{18.4.8}$$

In Eq. (18.4.8) I_l is the loop (mesh) current, in the lth loop. Eq. (18.4.8) shows that admittance Y_{lk} is that function which transfers (through multiplication) the input V_k into the output I_l. From Eqs (18.4.6) and (18.4.7) we get

$$\boxed{Y_{lk} = \frac{I_l}{V_k} = \frac{\Delta_{kl}}{\Delta_m}} \tag{18.4.9}$$

The double subscript describes the properties of V_{lk}. The first subscript designates the output and the second subscript the input. In terms of Eq. (18.4.7), the first subscript stands for the numerator and the second for the denominator.

Driving-point admittance

For a m-loop network N excited by s single input voltage E_k, we can find m transfer admittances Y_{jk} ($j = 1, 2, \ldots, m$; $k = 1, 2, \ldots, m$) depending on the output current.

Consider the network N shown in Fig. 18.1. If E_k is the driving input voltage and I_k the

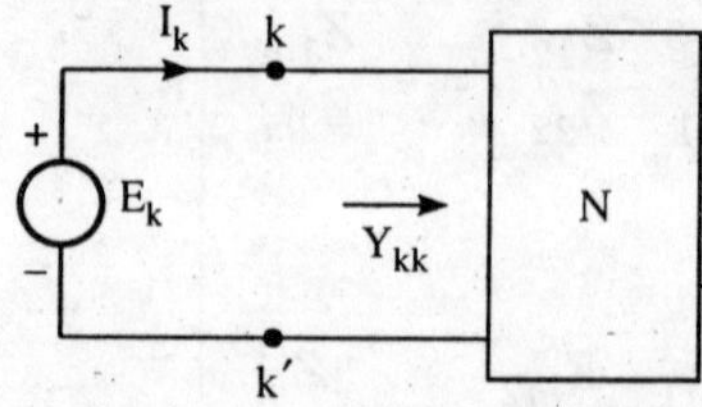

FIG. 18.1 A network driven by a voltage source E_k

current at the driving points $k - k'$, then the admittance Y_{kk} looking into the terminals $k - k'$ is defined as the *driving-point admittance* of the network. It is given by

$$\boxed{Y_{kk} \overset{\Delta}{=} \frac{I_k}{E_k}} \tag{18.4.10}$$

18.5 COMPUTATION OF NETWORK FUNCTIONS USING NODAL ANALYSIS

Network functions can also be determined by using nodal analysis. The current-voltage relationships in a network can also be described through a set of node equations. Let the number of nodes in the network be $(n + 1)$. We choose node $(n + 1)$ as the reference node and assume that the other n nodes are assumed to have unknown potentials $V_1, V_2, \ldots, V_n$. The current sources $I_1, I_2, \ldots, I_n$ are connected to each node except reference node. The nodal equations can, therefore, be written as

$$\begin{bmatrix} Y_{11} & Y_{12} & \cdots & Y_{1n} \\ Y_{21} & Y_{22} & \cdots & Y_{2n} \\ \cdot & & & \\ \cdot & & & \\ \cdot & & & \\ Y_{n1} & Y_{n2} & \cdots & Y_{nn} \end{bmatrix} \begin{bmatrix} V_1 \\ V_2 \\ \cdot \\ \cdot \\ \cdot \\ V_n \end{bmatrix} = \begin{bmatrix} I_1 \\ I_2 \\ \cdot \\ \cdot \\ \cdot \\ I_n \end{bmatrix} \tag{18.5.1}$$

It should be noted that in Eq. (18.5.1) all quantities are functions of s. In Eq. (18.5.1) $I_1, I_2, \ldots, I_n$ are the inputs and $V_1, V_2, \ldots, V_n$ the outputs. Eq. (18.5.1) can be written in a more compact form as

$$[\mathbf{Y}_n]\,[\mathbf{V}_n] = [\mathbf{I}_n] \tag{18.5.2}$$

Here $[\mathbf{Y}_n]$ is the *node-admittance matrix.*

The solutions of Eq. (18.5.1) for the unknown voltages $V_1, V_2, \ldots, V_n$ in terms of known admittances and current sources can be written as

$$V_1 = \frac{\Delta_{11}}{\Delta_n} I_1 + \frac{\Delta_{21}}{\Delta_n} I_2 + \ldots + \frac{\Delta_{n1}}{\Delta_n} I_n$$

$$V_2 = \frac{\Delta_{12}}{\Delta_n} I_1 + \frac{\Delta_{22}}{\Delta_n} I_2 + \ldots + \frac{\Delta_{n2}}{\Delta_n} I_n$$

$$V_q = \frac{\Delta_{1q}}{\Delta_n} I_1 + \frac{\Delta_{2q}}{\Delta_n} I_2 + \ldots + \frac{\Delta_{pq}}{\Delta_n} I_p + \ldots + \frac{\Delta_{nq}}{\Delta_n} I_n$$

$$V_n = \frac{\Delta_{1n}}{\Delta_n} I_1 + \frac{\Delta_{2n}}{\Delta_n} I_2 + \ldots + \frac{\Delta_{nn}}{\Delta_n} I_n \tag{18.5.3}$$

where Δ_n is the determinant of the nodal admittance matrix. That is,

$$\Delta_n = \begin{vmatrix} Y_{11} & Y_{12} & \dots & Y_{1n} \\ Y_{21} & Y_{22} & \dots & Y_{2n} \\ \cdot & & & \\ \cdot & & & \\ \cdot & & & \\ Y_{n1} & Y_{n2} & \dots & Y_{nn} \end{vmatrix} \tag{18.5.4}$$

Δ_{ij} are the various cofactors. That is,

$\Delta_{ij} = (-1) \times$ determinant obtained by deleting ith row and jth column of the determinant Δ_n

Transfer impedance

Nodal equations (18.5.1) and their solution given by Eq. (18.5.3) are valid for any linear network and for any type of current sources which drive the network. Now consider the case when the initial conditions are zero and the network is driven by only one current source, say I_p. Then solving Eq. (18.5.3) for any voltage V_q we get

$$V_q = \frac{\Delta_{pq}}{\Delta_n} I_p \quad q = 1, 2, \dots, n \tag{18.5.5}$$

$$\frac{V_q}{I_p} = \frac{\Delta_{pq}}{\Delta_n} \tag{18.5.6}$$

The quantity V_q/I_p has the dimensions of impedance. It is denoted by Z_{qp} and is called the **transfer impedance.** That is,

$$Z_{qp} \overset{\Delta}{=} \frac{V_q}{I_p} \tag{18.5.7}$$

$$V_q = Z_{qp} I_p \tag{18.5.8}$$

Eq. (18.5.8) shows that Z_{qp} is that function which transfers (through multiplication) the input I_p into the output V_q, where V_q is the transform of the voltage $v_q(t)$ at the qth node. From Eqs (18.5.6) and (18.5.7) we get,

$$Z_{qp} = \frac{V_q}{I_p} = \frac{\Delta_{pq}}{\Delta_n} \tag{18.5.9}$$

The first subscript denotes the output and the second subscript denotes the input. In terms of Eq. (18.5.7), the first subscript is for the numerator and the second for the denominator.

18.6 DRIVING-POINT IMPEDANCE

Let us consider the network N (Fig. 18.2) to be excited by a current source I_k at the terminals $k - k'$. Suppose that the driving input current I_k produces a voltage E_k across the driving points $k - k'$. The impedance $Z_{kk}(s)$ looking into the terminals $k - k'$ is defined as the *driving-point impedance* of the network. It is given by

$$Z_{kk}(s) = \frac{E_k(s)}{I_k(s)} \tag{18.6.1}$$

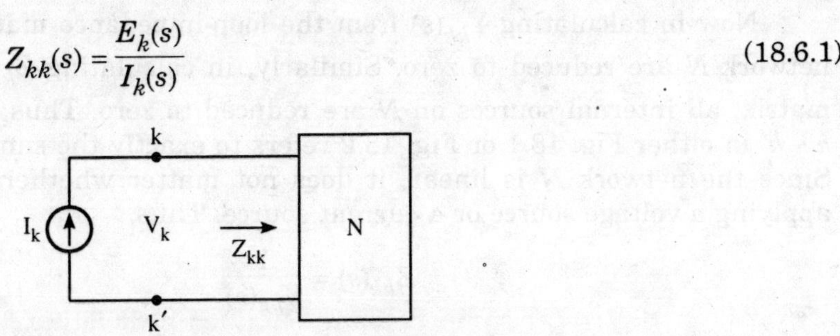

FIG. 18.2 A network driven by a current source

It is to be noted that the driving point and transfer immittances are not to be confused with the self and mutual immittances. The self and mutual immittances are actual immittances (impedances or admittances) forming the network. The driving point quantities are immittances looking into the entire network, and, in general, involve all the elements in the network. Similarly, the transfer immittances are given by the ratio of two determinants and can be a more general rational function.

A network of m meshes (loops) may have n nodes where $m \neq n$. On the loop basis, where the input is $E_k(s)$ and the output is $I_l(s)$, the transfer admittances is given by Eq. (18.4.9) as

$$Y_{lk}(s) = \frac{I_l(s)}{E_k(s)} = \frac{[\Delta_{kl}(s)]_{loop}}{\Delta_m(s)} \tag{18.6.2}$$

On the node basis, where the input is $I_k(s)$ and the output $E_l(s)$, the transfer impedance $Z_{lk}(s)$ is given by Eq.

$$Z_{lk}(s) = \frac{E_l(s)}{I_k(s)} = \frac{[\Delta_{kl}(s)]_{node}}{\Delta_n(s)} \tag{18.6.3}$$

Since the determinants given by Eqs (18.6.2) and (18.6.3) are, in general, not the same, the transfer quantities are *not* reciprocally related. That is,

$$Z_{lk}(s) \neq \frac{1}{Y_{lk}(s)} \quad (l \neq k) \tag{18.6.4}$$

Now we shall prove that the driving-point impedance (calculated from nodal-admittance matrix) and the driving-point admittance (calculated from the loop-impedance matrix) for the *same* port of the network are reciprocally related. That is

$$Z_{kk}(s) = \frac{1}{Y_{kk}(s)} \tag{18.6.5}$$

From Eq. (18.4.10),

$$Y_{kk}(s) = \frac{I_k(s)}{E_k(s)}$$

From Eq. (18.6.1)

$$Z_{kk}(s) = \frac{E_k(s)}{I_k(s)}$$

Now in calculating $Y_{kk}(s)$ from the loop-impedance matrix, all internal sources in the network N are reduced to zero. Similarly, in calculating $Z_{kk}(s)$ from the nodal-admittance matrix, all internal sources on N are reduced to zero. Thus, the measurements at the port $k - k'$ in either Fig. 18.1 or Fig. 18.2 refers to exactly the same conditions in the network N. Since the network N is linear, it does not matter whether the measurement is made by applying a voltage source or a current source. Thus,

$$Z_{kk}(s) = \frac{1}{Y_{kk}(s)}$$

18.7 PROCEDURE FOR FINDING NETWORK FUNCTIONS

A network function is found using *s-domain phasor techniques.* These techniques are identical to ω-domain techniques of ac circuits, except that we use s instead of $j\omega$.

The following procedure is used to derive and check the network function $\boldsymbol{H}(s)$.

1. Draw the circuit in phasor form, with each inductance L replaced by it generalized impedance sL and each capacitance C replaced by its generalized impedance $1/(sC)$.
2. Using standard phasor analysis techniques (Ohm's law, KCL, KVL, series/parallel impedance combinations, voltage and current dividers, the node and mesh methods, Thevenin/Norton conversions etc.), obtain the network function as the ratio $\boldsymbol{H}(s) = \dfrac{\boldsymbol{Y}(s)}{\boldsymbol{X}(s)}$
3. Check that all numerator terms have the same dimensions and that all denominator terms have the same dimensions. Also check that the ratio of the two sets has the same dimensions as the ratio $\boldsymbol{Y}(s)/\boldsymbol{X}(s)$. The dimensional verification can be made by observing that

$$[R] = [sL] = \left[\frac{1}{sC}\right] = \Omega$$

$$[sRC] = \left[\frac{sL}{R}\right] = [s^2LC] = \text{dimensionless}$$

   Here the physical dimensions of the quantity z is denoted by $[z]$.

4. Calculate $\boldsymbol{H}(0) = \lim_{s\to 0} \boldsymbol{H}(s)$. Verify this value physically by analyzing the circuit with all capacitances open circuited and all inductances short circuited.
5. Calculate $\boldsymbol{H}(\infty) = \lim_{s\to\infty} \boldsymbol{H}(s)$. Verify this value physically by analyzing the circuit with all capacitances short circuited and all inductances open circuited.

   The checks in steps 4 and 5 are based on the following facts :

$$Z_L = sL, \quad Z_C = \frac{1}{sC}$$

$$\therefore \quad \lim_{s\to 0} Z_L = 0$$

$$\lim_{s\to 0} Z_C = \infty$$

$$\lim_{s \to \infty} Z_L = \infty$$

$$\lim_{s \to \infty} Z_C = 0$$

In words, in the low-frequency or dc limit, an inductance act as a short circuit and a capacitance as an open circuit. In the high-frequency limit, an inductance acts as an open circuit and a capacitance as a short circuit.

EXAMPLE 18.1 Find the driving-point impedance function between ports 1 – 1′ of the network shown in Fig. 18.3.

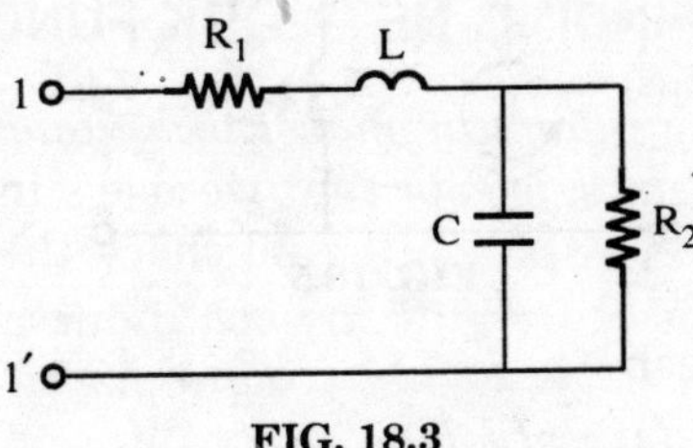

FIG. 18.3

SOLUTION : The driving-point impedance is given by

$$Z_{11}(s) = (R_1 + sL) + \left(R_2 \,||\, \frac{1}{sC}\right)$$

$$= R_1 + sL = \frac{R_2 \times \dfrac{1}{sC}}{R_2 + \dfrac{1}{sC}}$$

$$= R_1 + sL + \frac{R_2}{1 + R_2\,Cs}$$

EXAMPLE 18.2 Determine the condition under which the input impedance of the network of Fig. 18.4 will be equal to R.

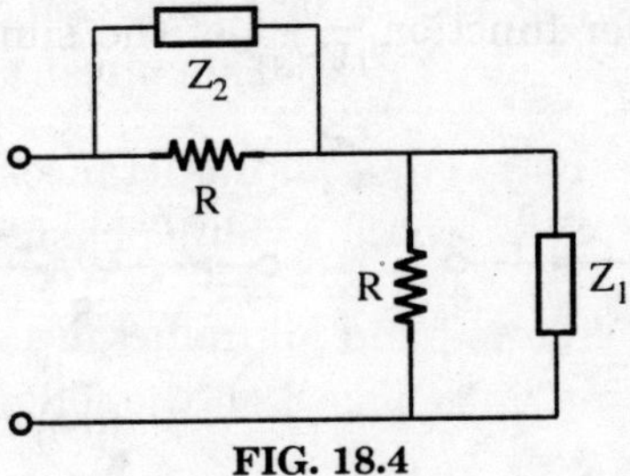

FIG. 18.4

SOLUTION : Input impedance

$$Z_{in} = (R \,||\, Z_2) + (R \,||\, Z_1)$$

$$R = \frac{R\,Z_2}{R + Z_2} + \frac{R\,Z_1}{R + Z_1}$$

$$(R + Z_2)(R + Z_1) = Z_2(R + Z_1) + Z_1(R + Z_2)$$

$$R^2 + R Z_1 + R Z_2 + Z_1 Z_2 = R Z_2 + Z_1 Z_2 + R Z_1 + Z_1 Z_2$$

$$R^2 = Z_1 Z_2$$

or $R = \sqrt{Z_1 Z_2}$, which is the required condition.

■ **EXAMPLE 18.3** Determine $Z_{11}(s)$, $Y_{11}(s)$, $G_{21}(s)$, and $Z_{21}(s)$ for the network of Fig. 18.5.

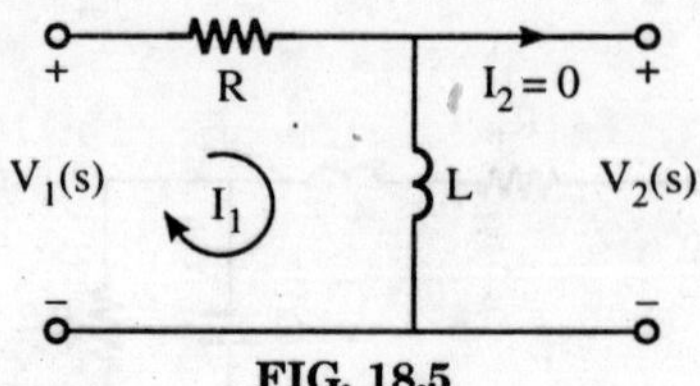

FIG. 18.5

SOLUTION : By KVL in the left mesh

$$V_1(s) = RI_1(s) + sLI_1(s)$$

$$Z_{11}(s) = \frac{V_1(s)}{I_1(s)} = R + sL$$

$$Y_{11}(s) = \frac{1}{Z_{11}(s)} = \frac{1}{R + sL}$$

Also, $$V_2(s) = (sL)\, I_1(s)$$

therefore, $$G_{21}(s) = \frac{V_2(s)}{V_1(s)} = \frac{sL\, I_1(s)}{(R + sL)\, I_1(s)} = \frac{sL}{R + sL}$$

$$Z_{21}(s) = \frac{V_2(s)}{I_1(s)} = \frac{sLI_1(s)}{I_1(s)} = sL$$

■ **EXAMPLE 18.4** Find the transfer function $\dfrac{V_0(s)}{V_i(s)}$ of the simple lag network shown in Fig. 18.6a.

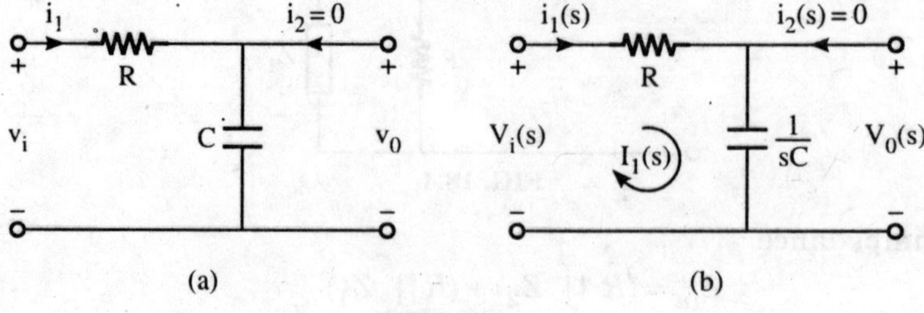

FIG. 18.6

SOLUTION : The transform network of Fig. 18.6a is shown in Fig. 18.6b. By KVL in the left-hand mesh of Fig. 18.6b,

$$\left(R + \frac{1}{sC}\right) I_1(s) = V_i(s)$$

Also,

$$\frac{1}{sC} I_1(s) = V_0(s)$$

therefore,

$$\frac{V_0(s)}{V_i(s)} = \frac{\frac{1}{sC}}{R + \frac{1}{sC}} = \frac{1}{RCs + 1}$$

■ **EXAMPLE 18.5** Detemine the transfer funcion $\frac{V_0(s)}{V_i(s)}$ of the *RC* network of the lag-lead compensator shown in Fig. 18.7a.

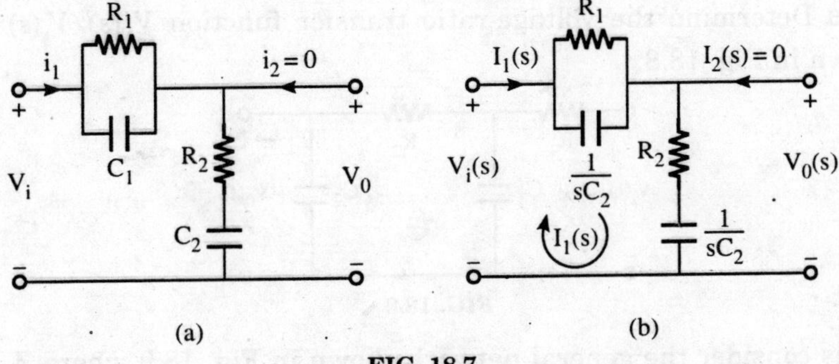

FIG. 18.7

SOLUTION: The transform network of Fig. 18.7a is shown in Fig. 18.7b. By KVL in the left-hand mesh of Fig. 18.7b,

$$V_i(s) = \left[\left(R_1 \,||\, \frac{1}{sC_1}\right) + \left(R_2 + \frac{1}{sC_2}\right)\right] I_1(s)$$

$$= \left[\frac{\frac{R_1}{sC_1}}{R_1 + \frac{1}{sC_1}} + \left(R_2 + \frac{1}{sC_2}\right)\right] I_1(s)$$

$$= \left[\frac{R_1}{R_1 C_1 s + 1} + \left(R_2 + \frac{1}{sC_2}\right)\right] I_1(s)$$

Also,

$$V_0(s) = \left(R_2 + \frac{1}{sC_2}\right) I_1(s)$$

therefore,

$$\frac{V_0(s)}{V_i(s)} = \left(R_2 + \frac{1}{sC_2}\right) \div \left(\frac{R_1}{R_1 C_1 s + 1} + R_2 + \frac{1}{sC_2}\right)$$

$$= \frac{R_2 C_2 s + 1}{sC_2} \div \frac{R_1 C_2 s + R_2 C_2 s + R_1 R_2 C_1 C_2 s^2 + 1 + R_1 C_1 s}{(R_1 C_1 s + 1)\, sC_2}$$

$$= \frac{(R_1 C_1 s + 1)(R_2 C_2 s + 1)}{R_1 R_2 C_1 C_2 s^2 + (R_1 C_1 + R_2 C_2 + R_1 C_2) s + 1}$$

$$= \frac{R_1 R_2 C_1 C_2 \left(s + \frac{1}{R_1 C_1}\right)\left(s + \frac{1}{R_2 C_2}\right)}{R_1 R_2 C_1 C_2 \left[s^2 + \left(\frac{1}{R_1 C_1} + \frac{1}{R_2 C_2} + \frac{1}{R_2 C_1}\right) s + \frac{1}{R_1 R_2 C_1 C_2}\right]}$$

$$= \frac{\left(s + \frac{1}{R_1 C_1}\right)\left(s + \frac{1}{R_2 C_2}\right)}{s^2 + \left(\frac{1}{R_1 C_1} + \frac{1}{R_2 C_2} + \frac{1}{R_2 C_1}\right) s + \frac{1}{R_1 R_2 C_1 C_2}}$$

■ **EXAMPLE 18.6** Determine the voltage-ratio transfer function $V_0(s)/V_i(s)$ for the ladder network shown in Fig. 18.8.

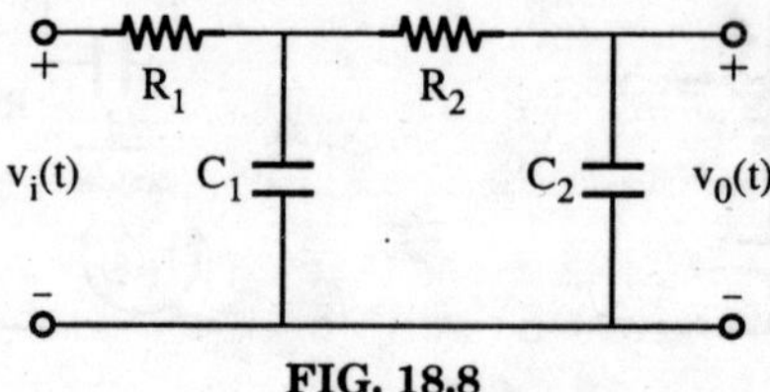

FIG. 18.8

SOLUTION : Let us consider the general network shown in Fig. 18.9 where Z_1, Z_2, Z_3 and Z_4 represent operational impedances.

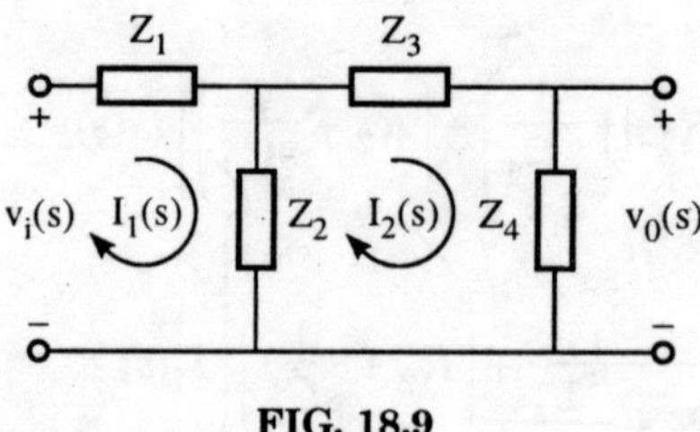

FIG. 18.9

By KVL

$$V_i(s) = (Z_1 + Z_2) I_1(s) - Z_2 I_2(s)$$

$$0 = -Z_2 I_1(s) + (Z_2 + Z_3 + Z_4) I_2(s)$$

Elimination of I_1 gives

$$\frac{V_i(s)}{I_2(s)} = [(Z_1 + Z_2)(Z_2 + Z_3 + Z_4) - Z_2^2]/Z_2$$

But $$V_0(s) = Z_4 I_2(s)$$

therefore,
$$\frac{V_0(s)}{V_i(s)} = \frac{Z_4 Z_2}{Z_1 Z_2 + Z_1 Z_3 + Z_1 Z_4 + Z_2 Z_3 + Z_2 Z_4}$$

Here
$$Z_1 = R_1, Z_3 = R_2, Z_2 = \frac{1}{sC_1}, Z_4 = \frac{1}{sC_2}$$

Substitution of these values gives
$$\frac{V_0(s)}{V_i(s)} = \frac{1}{R_1 R_2 C_1 C_2 s^2 + (R_1 C_1 + R_1 C_2 + R_2 C_2)\, s + 1}$$

EXAMPLE 18.7 For the network shown in Fig. 18.10a determine the voltage-ratio transfer function $V_2(s)/V_1(s)$.

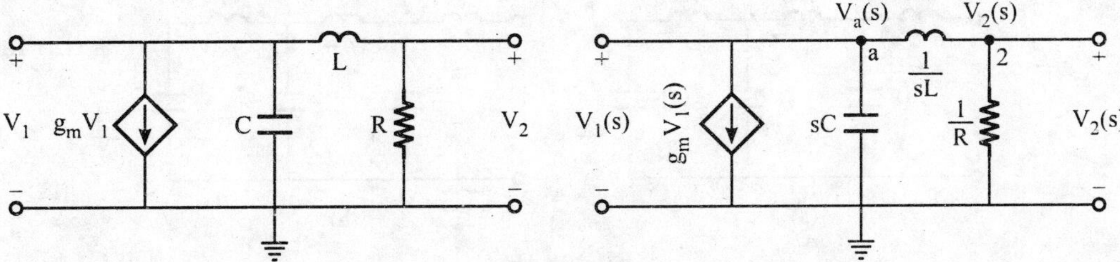

FIG. 18.10

SOLUTION : Let us express the admittance of each element in the s domain as shown in Fig. 18.10b. In this circuit, the voltages at nodes a and 2 with respect to ground are designated $V_a(s)$ and $V_2(s)$ respectively.

By KCL at node a,

sum of outgoing currents = sum of incoming currents

$$g_m V_1(s) + sCV_a(s) + \frac{1}{sL}[V_a(s) - V_2(s)] = 0$$

or
$$\left(sC + \frac{1}{sL}\right) V_a(s) - \frac{1}{sL} V_2(s) = -g_m V_1(s) \qquad \text{(E18.7.1)}$$

By KCL at node 2,

$$\frac{1}{R} V_2(s) = \frac{1}{sL}[V_a(s) - V_2(s)]$$

or
$$V_a(s) = sL\left(\frac{1}{sL} + \frac{1}{R}\right) V_2(s) \qquad \text{(E18.17.2)}$$

Substituting the value of $V_a(s)$ from Eq. (E18.7.2) in Eq. (E18.7.1) we get

$$sL\left(sC + \frac{1}{sL}\right)\left(\frac{1}{sL} + \frac{1}{R}\right) V_2(s) - \frac{1}{sL} V_2(s) = -g_m V_1(s)$$

or
$$(LCs^2 + RCs + 1)\, V_2(s) = -g_m R\, V_1(s)$$

therefore,
$$\frac{V_2(s)}{V_1(s)} = \frac{-g_m R}{LCs^2 + RCs + 1}$$

Alternatively Eqs. (E18.7.1) and (E18.7.2) can be written directly in the nodal matrix form as

$$\begin{bmatrix} sC + \frac{1}{sL} & -\frac{1}{sL} \\ -\frac{1}{sL} & \frac{1}{sL} + \frac{1}{R} \end{bmatrix} \begin{bmatrix} V_a(s) \\ V_2(s) \end{bmatrix} = \begin{bmatrix} -g_m V_1(s) \\ 0 \end{bmatrix}$$

■ EXAMPLE 18.8 Determine $Z_{11}(s)$ for the network shown in Fig. 18.11a.

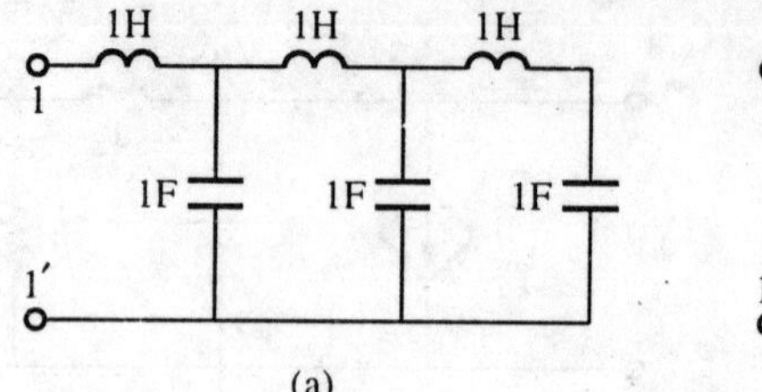

(a)

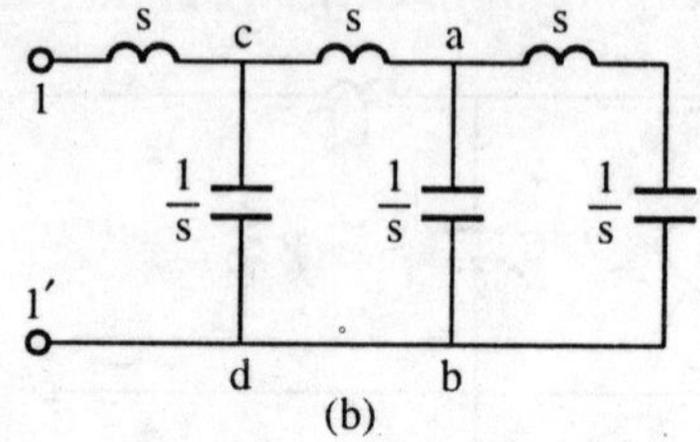

(b)

FIG. 18.11

SOLUTION : The s-domain network is shown in Fig. 18.11b.

$$Z_{ab}(s) = \left(s + \frac{1}{s}\right) || \frac{1}{s} = \frac{\left(s + \frac{1}{s}\right)\frac{1}{s}}{s + \frac{1}{s} + \frac{1}{s}} = \frac{s^2 + 1}{s^3 + 2s}$$

$$Z_{ab}(s) = s + \frac{s^2 + 1}{s^3 + 2s} = \frac{s^4 + 3s^2 + 1}{s^3 + 2s}$$

$$Z_{cd}(s) = Z_{cb}(s) || \frac{1}{s} = \frac{\left(\frac{s^4 + 3s^2 + 1}{s^3 + 2s}\right)\frac{1}{s}}{\frac{s^4 + 3s^2 + 1}{s^3 + 2s} + \frac{1}{s}}$$

$$= \frac{s^4 + 3s^2 + 1}{s^5 + 4s^3 + 3s}$$

$$Z_{11}(s) = s + Z_{cd}(s)$$

$$= s + \frac{s^4 + 3s^2 + 1}{s^5 + 4s^3 + 3s}$$

$$= \frac{s^6 + 5s^4 + 6s^2 + 1}{s^5 + 4^3 + 3s}$$

■ **EXAMPLE 18.9** Determine the driving-point functions for the network shown in Fig. 18.12a.

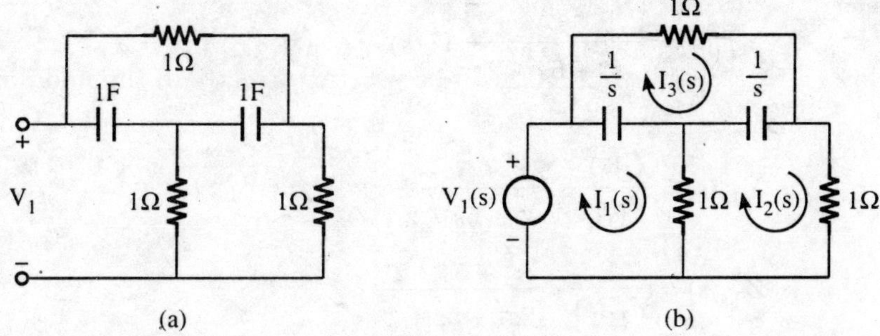

FIG. 18.12

SOLUTION : The s-domain network is shown in Fig. 18.12b. The mesh equations in matrix form can be written as follows :

$$\begin{bmatrix} 1+\frac{1}{s} & -1 & -\frac{1}{s} \\ -1 & 2+\frac{1}{s} & -\frac{1}{s} \\ -\frac{1}{s} & -\frac{1}{s} & \frac{2}{s}+1 \end{bmatrix} \begin{bmatrix} I_1(s) \\ I_2(s) \\ I_3(s) \end{bmatrix} = \begin{bmatrix} V_1(s) \\ 0 \\ 0 \end{bmatrix}$$

$$I_1(s) = \frac{\begin{vmatrix} V_1(s) & -1 & -\frac{1}{s} \\ 0 & 2+\frac{1}{s} & -\frac{1}{s} \\ 0 & -\frac{1}{s} & \frac{2}{s}+1 \end{vmatrix}}{\begin{vmatrix} 1+\frac{1}{s} & -1 & -\frac{1}{s} \\ -1 & 2+\frac{1}{s} & -\frac{1}{s} \\ -\frac{1}{s} & -\frac{1}{s} & \frac{2}{s}+1 \end{vmatrix}}$$

$$\begin{vmatrix} V_1(s) & -1 & -\frac{1}{s} \\ 0 & 2+\frac{1}{s} & -\frac{1}{s} \\ 0 & -\frac{1}{s} & \frac{2}{s}+1 \end{vmatrix} = V_1(s)\left[\left(2+\frac{1}{s}\right)\left(\frac{2}{s}+1\right)-\left(-\frac{1}{s}\right)\left(-\frac{1}{s}\right)\right] = \frac{2s^2+5s+1}{s^2}V_1(s)$$

$$\begin{vmatrix} 1+\frac{1}{s} & -1 & -\frac{1}{s} \\ -1 & 2+\frac{1}{s} & -\frac{1}{s} \\ -\frac{1}{s} & -\frac{1}{s} & \frac{2}{s}+1 \end{vmatrix} = \left(1+\frac{1}{s}\right)\left[\left(2+\frac{1}{s}\right)\left(1+\frac{2}{s}\right)-\frac{1}{s^2}\right]+\left[-1\left(1+\frac{2}{s}\right)-\frac{1}{s^2}\right]-\frac{1}{s}\left[\frac{1}{s}+\frac{1}{s}\left(2+\frac{1}{s}\right)\right]$$

$$= \frac{s^2+5s+2}{2}$$

$$\therefore \qquad I_1(s) = \frac{\left(\dfrac{2s^2 + 5s + 1}{s^2}\right)}{\dfrac{s^2 + 5s + 2}{s^2}} V_1(s)$$

$$Y_{11}(s) = \frac{I_1(s)}{V_1(s)} = \frac{2s^2 + 5s + 1}{s^2 + 5s + 2}$$

$$Z_{11}(s) = \frac{1}{Y_{11}(s)} = \frac{s^2 + 5s + 2}{2s^2 + 5s + 1}$$

18.8 POLES AND ZEROS OF NETWORK FUNCTION *H(s)*

The network function $H(s)$ of a lumped, linear network is the ratio of two polynomials of s with real coefficients. These polynomials are rational, that is, they can be factored. In the most general from $H(s)$ is given by

$$H(s) = \frac{N(s)}{D(s)} = \frac{a_m s^m + a_{m-1} s^{m-1} + \ldots + a_1 s + a_0}{b_n s^n + b_{n-1} s^{n-1} + \ldots + b_1 s + b_0} \qquad (18.8.1)$$

where $N(s)$ = numerator polynomial of degree m having coefficients $a_m, a_{m-1}, \ldots, a_0$

$D(s)$ = denominator polynomial of degree n having coefficients $b_n, b_{n-1}, \ldots, b_0$

Since a polynomial can be factored with a product of first-order polynomials, the network function $H(s)$ can therefore be written in the factored form as

$$H(s) = \frac{N(s)}{D(s)} = \frac{K(s - z_1)(s - z_2) \ldots (s - z_m)}{(s - p_1)(s - p_2) \ldots (s - p_n)} \qquad (18.8.2)$$

where $z_1, z_2, \ldots, z_m$ and $p_1, p_2, \ldots, p_n$ are called **complex frequencies.** The constant multiplier $K = a_m/b_n$ is called the **scale factor.** Since a_m and b_n are real, K is also real. Unless stated, otherwise, it is assumed that all common factors of the numerator and denominator have been cancelled.

A **zero** of the system function $H(s)$ is defined as the value of s which makes $H(s) = 0$. A **pole** of $H(s)$ is defined as the value of s which makes the denominator of $H(s)$ approach zero, or which makes $H(s)$ approach infinity. From Eq. (18.8.2) we conclude the following :

(*a*) When s has any of the values $z_1, z_2, \ldots, z_m$ the network function $H(s)$ becomes zero. Hence the complex frequencies $z_1, z_2, \ldots, z_m$ are called the **zeros** of the network function $H(s)$.

(*b*) When s has any of the values $p_1, p_2, \ldots, p_n$ the network function $H(s)$ becomes infinite. Hence the complex frequencies $p_1, p_2, \ldots, p_n$ are called the **poles** of the network function.

Thus, the zeros of a network function tell us what values of s will make the function zero, and the poles of a network function tell us what values of s will make the function infinite. At every other frequency, the network function has a finite non-zero value. For this reason, the poles and zeros *collectively* are called **critical frequencies,** or **critical values** of the network function $H(s)$. The poles and zeros are also known as the **roots** of the network function. They may be real, imaginary or complex. They depend only on the network, regardless of the applied signals, or its energy conditions of its reactive elements.

The roots of the numerator $N(s)$ give the ***finite zeros.*** The roots of the denominator $D(s)$ gives the ***finite poles.*** Eq. (18.8.2) shows that $H(s)$ has m finite zeros $z_1, z_2, ...z_m$ and n finite poles $p_1, p_2, \dots, p_n$. In case a network contains more than one pole at a particular point in the s-plane, we say that the function contains ***multiple poles.*** There may be ***multiple zeros*** also, A repeated pole appearing r times, is called either a ***rth order pole,*** or a ***pole of multiplicity r.*** Similarly, a zero appearing r times is called either a ***rth order zero*** or *a* ***zero of multiplicity r.*** The terms like double, triple, etc., are also used to describe multiplicity. In case the pole or zero is not repeated it is known as ***simple*** or ***distinct.*** Poles or zeros at $s = \infty$ are also assigned a degree (order).

We have seen from Eq. (18.8.2) that $H(s)$ has m finite zeros and n finite poles. There is also a zero at infinity if $n > m$, the order of this zero being equal to $(n - m)$. There is a pole at infinity if $m > n$, its order being equal to $(m - n)$, Thus,

total number of poles of H(s)

=number of finite poles of $H(s)$

+ number of poles of $H(s)$ at infinity

$= n + (m - n) = m =$ ***total number of zeros of H(s)***

Eq. (18.8.2) shows that any network function is completely specified if its zeros, poles, and scale factor are known. Hence poles and zeros are useful in describing the behaviour of any network.

Consider the function

$$H(s) = \frac{10\,(s+2)}{s\,(s+1)\,(s+3)^2}$$

Here the function $H(s)$ has four finite poles at $s = 0, -1, -3, -3$. There is only one finite zero at $s = -2$. Since the total number of zeros is equal to the total number of poles, there are three zeros at infinity. Thus, the function $H(s)$ has a total of four poles and four zeros in the entire s plane.

EXAMPLE 18.10 Find the critical frequencies for

$$H(s) = 5\,\frac{s+2}{s^2 + 6s + 25}$$

SOLUTION : Let us write the value of $H(s)$ in the factored form as

$$H(s) = 5\,\frac{s+2}{(s+3+j4)\,(s+3-j4)}$$

The standard form of $H(s)$ is given by

$$H(s) = k\,\frac{(s - z_1)}{(s - p_1)\,(s - p_2)}$$

Comparison of these equations shows that three is a zero at $s = -2$ and two poles, one at $s = -3 - j4$, and the other at $s = -3 + j4$.

18.9 POLE-ZERO PLOT

Since poles and zeros of a network function may be complex in general the quantity s is considered as a complex variable represented by $s = \sigma + j\omega$, where σ is the real part of s representing a variable called ***damping constant***. The quantity ω is the imaginary part of s representing a variable called **angular frequency**. The quantity s is considered as a ***complex frequency*** in a general sense.

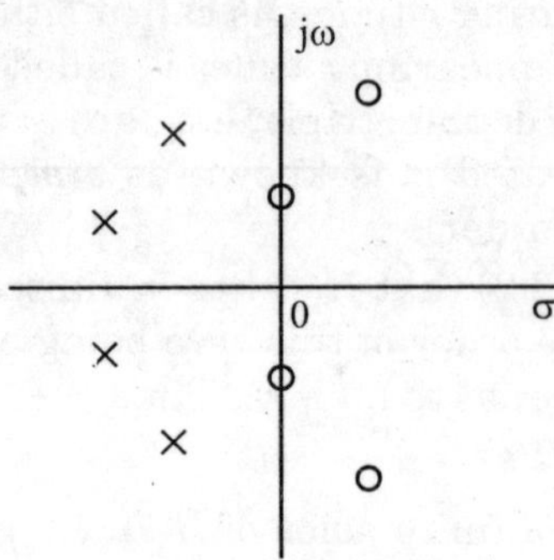

FIG. 18.13 Poles and zeros plotted in the complex s-plane

It is quite common to represent poles and zeros of a network function in a two-dimensional s-plane shown in Fig. 18.14. Such a representation is called a ***pole–zero plot*** or a ***pole–zero diagram.*** The horizontal real axis of the s-plane is called the **σ-axis** or ***damping axis,*** and the vertical (imaginary) axis is called the ***jω–axis*** or ***frequency axis.*** Zeros in the complex s-plane are drawn as little circles (o) and the poles are drawn as crosses (×). Thus, the symbol o represents a zero and the symbol × represents a pole in the s-plane. Fig. 18.14 shows the poles and zeros plotted in the complex s-plane. In pole-zero plots we are able to show only finite zeros and finite poles. The multiplicity (>1) is placed in paranthesis next to the zero or the pole.

In the s plane, the *left-hand half plane* includes all points to the left of, but not including, the $j\omega$-axis. Similarly, the *right-hand half plane* will include all points to the right of, but not including, the $j\omega$-axis. The *negative real axis* will include all points on the real axis to the left of, but not including, the origin. Similarly, the *positive real axis will include all points on the real axis to the right, but not including, the origin. The jω axis (including the origin) is considered as a separate domain.*

■ **EXAMPLE 18.11** Find the critical frequencies for

$$H(s) = \frac{(s+2)^2\,(s+1)}{(s+5+2j)\,(s+5-2j)\,(s+3)}$$

Draw the plot of the pole and zero locations.

SOLUTION : The given function has the following cirtical frequencies :.

$s = -1$ is simple zero

$s = -2$ is a second-order zero

$s = -5 + 2j$ is a simple pole

$s = -5 - 2j$ is a simple pole

$s = -3$ is a simple pole

The pole-zero plot is shown in Fig. 18.14. The multiplicity is placed in paranthesis next to zero at $s = -2$.

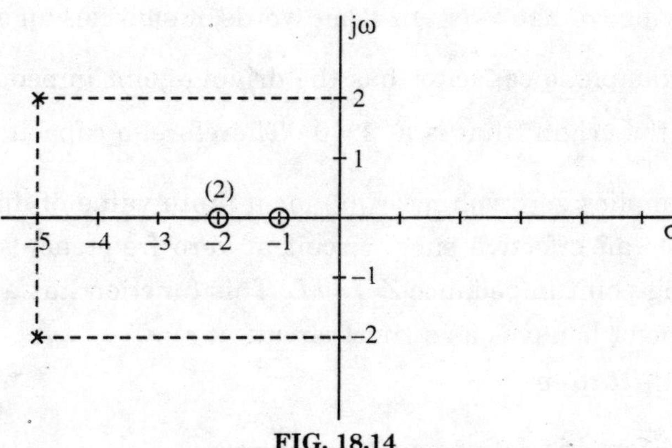

FIG. 18.14

18.10 RELATIONSHIP BETWEEN NETWORK FUNCTION AND UNIT IMPULSE FUNCTION

Consider a linear, time-invariant network having an excitation $x(t)$ and response $y(t)$. From the definition of the network function

$$Y(s) = H(s)\, X(s) \tag{18.10.1}$$

Consider the case when the excitation is a unit impulse source, that is,

$$x(t) = \delta(t) = u_0(t)$$

therefore,

$$X(s) = \pounds\, [x(t)] = \pounds\, u_0(t) = 1 \tag{18.10.2}$$

Substituting this reslut in Eq. (18.10.1) we obtain

$$Y(s)\Big|_{x(t) = u_0(t)} = H(s) \tag{18.10.3}$$

The inverse Laplace transforms of $H(s)$ is given by

$$\pounds^{-1} H(s) = h(t) = \pounds^{-1}[Y(s)]\Big|_{X(s) = 1} \tag{18.10.4}$$

The function $h(t)$ is called the ***impulse response*** or ***weighting*** functoin of the linear system. The ***impulse response*** *is defined as the zero-state response to a unit impulse function.*

Eq. (18.10.3) shows that **for any linear-time invariant network, the network function is the Laplace transform of the corresponding impulse *response*.** This is a very important result of practical significance. Eq. (18.10.4) implies that we can directly evaluate the impulse response, $h(t)$, of a network by finding the inverse Laplace transform of its network function. The impulse response provides a convenient method of determining the stability of the network.

18.11 SIGNIFICANCE OF POLES AND ZEROS IN NETWORK FUNCTIONS

Poles and zeros play significant roles in driving-point and transfer functions.

(a) Driving–point impedance

The driving-point impedance is defined as $Z_{11}(s) \triangleq \frac{V_1(s)}{I_1(s)}$ · A pole of $Z_{11}(s)$ implies zero current for a finite value of driving voltage $V_1(s)$. In other words, it signifies an open circuit at the pole frequency $s = p_i$. For example, a capacitor has the driving-point impedance $Z_{11}(s) = \frac{1}{sC}$ · This function has a pole at the origin (that is at $s = 0$). Therefore, a capacitor behaves as an open circuit at $s = 0$.

A zero of $Z_{11}(s)$ implies zero voltage $V_1(s)$ for a finite value of driving current $I_1(s)$. In other words, it signifies an effective short circuit at zero frequency $s = z_i$. For example an inductor has the driving-point impedance $Z_{11} = sL$. This function has a zero at origin (that is at $s = 0$). Thus, an inductor behaves as a short circuit at $s = 0$.

(b) Driving–point admittance

The driving point admittance is defined as $Y_{11} \triangleq \frac{I_1(s)}{V_1(s)}$ ·

A pole of $Z_{11}(s)$ implies zero voltage $V_1(s)$ for a finite value of driving current $I_1(s)$. Thus, it signifies a short circuit at pole frequency $s = p_i$. For example, an inductor has the driving point admittance $Y_{11}(s) = (1/sL)$. This function has a pole at the origin (that is, at $s = 0$). Thus, an inductor behaves as a short circuit at $s = 0$.

A zero of $Y_{11}(s)$ implies zero current for a finite value of driving voltage $V_1(s)$. In other words, a zero of $Y_{11}(s)$ implies an effective open circuit.

We have seen that an inductor has a zero at $s = 0$ and a pole at $s = \infty$. Similarly, a capacitor has a pole at $s = 0$ and a zero at $s = \infty$.

(c)Voltage transfer ratio

Consider the voltage transfer function

$$G_{21}(s) = \frac{V_2(s)}{V_1(s)}$$

therefore,

$$V_2(s) = G_{21}(s)\, V_1(s) \tag{18.11.1}$$

Usually the input voltage $V_1(s)$ is given and the output voltage $V_2(s)$ is to be determined. Eq. (18.11.1) shows that $V_2(s)$ can be determined if $G_{21}(s)$ and $V_1(s)$ are known. The expression $G_{21}(s)$ and $V_1(s)$ can be put in the form of a ratio of polynomials in s, for which partial fraction expansion can be obtained. Then the denominator of each term of the partial fractions gives a pole of either $G_{21}(s)$ or $V_1(s)$. If the transformed output $V_2(s)$ has no repeated poles, then we may write

$$V_2(s) = G_{21}(s)\, V_1(s)$$

$$= \sum_{i=1}^{m} \frac{k_i}{s - p_i} + \sum_{j=1}^{n} \frac{k_i}{s - p_j}$$

$$v_2(t) = \pounds^{-1}\left[\sum_{i=1}^{m} \frac{k_i}{s-p_i} + \sum_{j=1}^{n} \frac{k_j}{s-p_j}\right]$$

or
$$v_2(t) = \sum_{i=1}^{m} k_i e^{p_i t} + \sum_{j=1}^{n} k_j e^{p_j t}$$

where m and n are the number of poles of $G_{21}(s)$ and $V_1(s)$ respectively, and k_i's and k_j's are the residues. The frequencies s_i are the natural complex frequencies corresponding to *free oscillations* and s_j are the complex frequencies of the driving force corresponding to the *forced oscillations.* It is to be noted that the free oscillations depend entirely on the network function $G_{21}(s)$ while the forced oscillations depend entirely on the driving force $V_1(s)$.

Thus we have seen that the poles influence the nature of the network time response. The magnitude of the response depends on the coefficients k_i and k_j, which are influenced by the critical frequencies. Hence the zeros do not influence the nature of the response but they affect the magnitude of each part of the response.

(*d*) Other network functions

The significance of poles and zeros in other transfer functions, for example, $Z_{21}(s)$, $Y_{21}(s)$, and $\alpha_{21}(s)$ is similar to that of $G_{21}(s)$. In each of these transfer functions, poles determine the time-domain behavior of the response while zeros determine the magnitude of each term of the time response.

18.12 STABILITY

A system is said to be stable if its output (response) cannot be made to increase indefinitely by the application of a bounded input excitation. The concept of stability is very important in the system analysis and design. We shall define the stability concept in terms of the impulse response.The stability of a network function $F(s)$ can be conveniently determined by considering its response to an impulse function which is obtained by taking the inverse Laplace transform of the partial fraction expansion of the function. Let us take the following definitons. :

Stable System

A system is said to be stable if the impulse response approaches zero for sufficiently large time.

Unstable System

A system is said to be unstable if the impulse response grows without bound (that is, approaches infinity) for sufficiently large time.

Marginally Stable System

A system is said to be marginally stable if the impulse response approaches a constant non-zero value or a constant amplitude oscillation for sufficiently large time.

Almost all passive RLC networks are inherently stable. This is due to the fact that they do not contain energy sources that might inject additional energy into the network. The RLC networks can only dissipate energy.

18.13 EFFECT OF POLE POSITIONS ON STABILITY

Let us consider the relationship between pole positions and the corresponding impulse responses.

Poles on the Negative Real Axis

If the network has a simple pole on the negative real axis (Fig 18.15a), the corresponding impulse response for $t \geq 0$ is given by

$$f(t) = \pounds^{-1}(F)(s)$$

$$= \pounds^{-1}\frac{k_1}{s+\alpha}$$

$$= k_1 e^{-\alpha t}$$

When t is sufficiently large, $f(t)$ approaches zero and the system is *stable.* The time response is shown in Fig. 18.15b. If the network function has a multiple-order pole, the time response would be of the form

$$f(t) = k_1 t^r e^{-\alpha t}$$

This response is also *stable.*

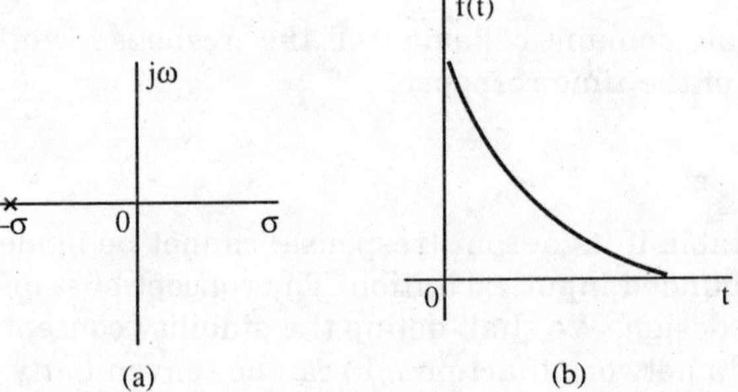

FIG. 18.15 Pole on negative real axis and the corresponding time response

Complex poles in the left-hand half plane

Suppose that $F(s)$ has a pair of first-order complex conjugate poles at $s = -\alpha \pm j\omega$ as shown in Fig. 11.16a. The time response due to this pair of poles is given by

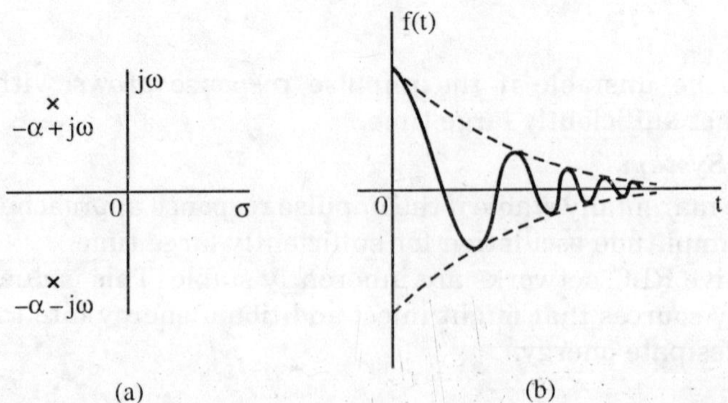

FIG. 18.16 Complex poles in the left-hand half plane and the corresponding time response

$$f(t) = £^{-1}\left(\frac{k_1}{s+\alpha-j\omega} + \frac{k_1}{s+\alpha+j\omega}\right)$$

$$= £^{-1}\,\frac{2k_1\,(s+\alpha)}{(s+\alpha)^2+\omega^2} = 2\,k_1\,e^{-\alpha t}\cos\omega t$$

When t is sufficiently large, $f(t)$ approaches zero and therefore the system is *stable.* The time response is shown in Fig. 18.16b.

If the pair of poles were multiple-order poles on the negative real axis, the time response would be of the form.

$$f(t) = 2\,k_1\,t^r\,e^{-\alpha t}\cos\omega t$$

Here also, $f(t)$ approaches zero for large t, and hence the system is stable. Thus, we have seen that both simple-order and multiple-order poles of a network function in the left-hand half of the s plane represent *stable* responses.

Pole on the positive real axis

If the network function has a simple pole on the positive real axis as shown in Fig. 18.17a, the time response is of the form

$$f(t) = k_1\,e^{\alpha t}$$

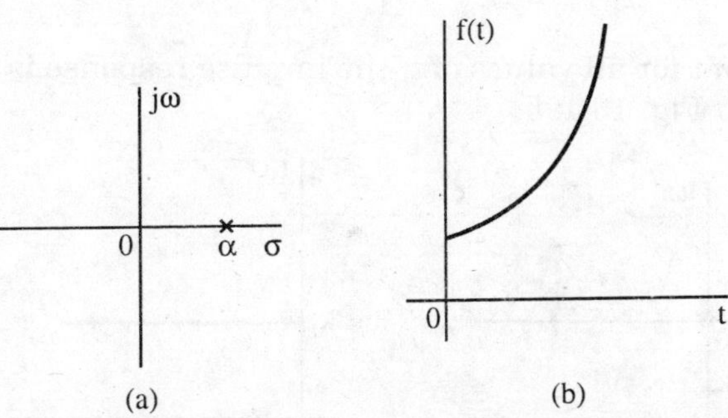

FIG. 18.17 Pole on positive real axis and the corresponding time response

This response increases exponentially with time (Fig. 18.17b) and therefore the response is *unstable.*

It can be shown that multiple-order poles on the positive real axis also represent *unstable* responses.

Complex poles in the right-hand half plane

Let us consider a pair of first-order complex poles in the right-hand half plane at $s = \alpha \pm j\omega$ as shown in Fig. 18.18a. The time response due to this pair of poles is given by

$$f(t) = £^{-1}\,F(s) = £^{-1}\left[\frac{k_1}{s-\alpha-j\omega} + \frac{k_1}{s-\alpha+j\omega}\right]$$

$$= £^{-1}\,\frac{2\,k_1\,(s-\alpha)}{(s-\alpha)^2+\omega^2} = 2\,k_1\,e^{\alpha t}\cos\omega t$$

It is seen that $f(t)$ is an exponentially increasing sinusoid and therefore the response in *unstable* (Fig. 18.18b).

It can be shown that multiple-order poles in the right-hand half plane also represent *unstable* responses. Thus, poles in the right-hand half plane always represent instability. It should be noted that right-hand half-poles can never occur in passive RLC networks.

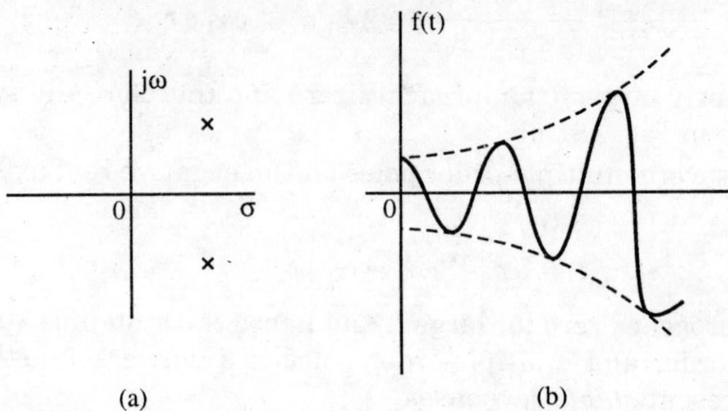

FIG. 18.18 Complex poles in right-hand half plane and the corresponding time-domain response representation

Pole at the origin

Consider a pole at the origin as shown in Fig. 18.19a. The impulse response is of the form

$$f(t) = £^{-1} \frac{k_1}{s} = k_1$$

Since $f(t)$ is a constant for all values of t, the impulse response is *marginally stable.* This time response is shown in Fig. 18.19b.

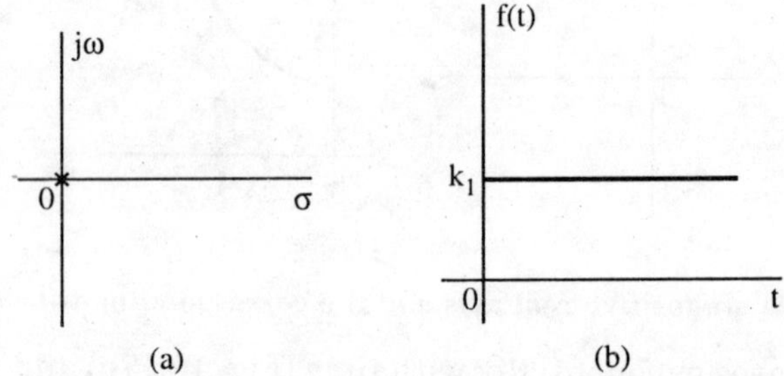

FIG. 18.19 Pole at the origin and the corresponding time response

If there is a double pole at the origin, the time response would be of the form

$$f(t) = £^{-1} \frac{k_2}{s^2} = k_2\, t$$

This relation shows that response is *unstable.* The response due to second-order poles at origin is shown if Fig. 18.20.

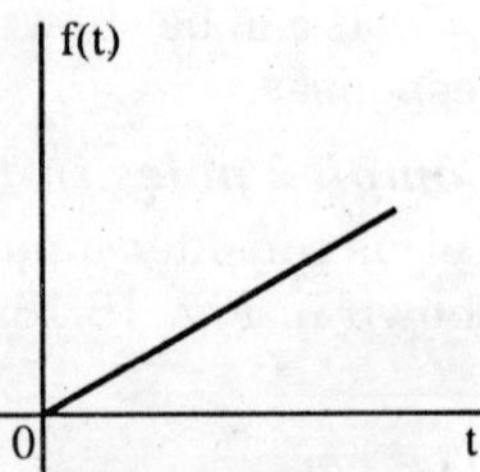

FIG. 18.20 Response due to second-order poles at the origin

Poles on jω axis

Consider a pair of first-order conjugate complex poles on the $j\omega$ axis as shown in Fig. 18.21a. The time response is given by

$$f(t) = \pounds^{-1} F(s) = \pounds^{-1}\left(\frac{k_1}{s+j\omega} + \frac{k_1}{s-j\omega}\right)$$

$$= \pounds^{-1} \frac{2k_1 s}{s^2+\omega^2} = 2\,k_1 \cos \omega t$$

This function is shown in Fig. 18.21b. Since the function $f(t)$ oscillates for all time, it is *marginally stable.*

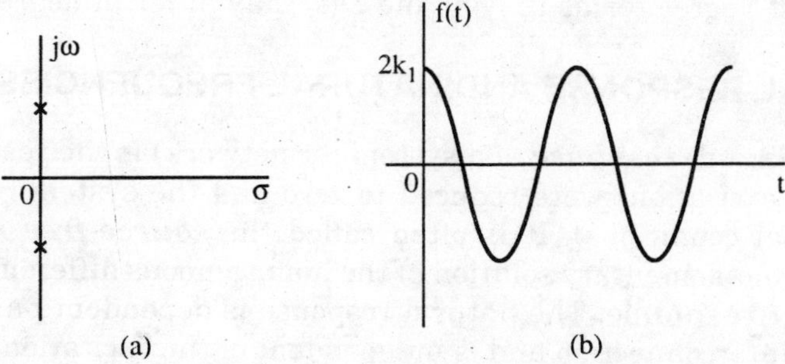

FIG. 18.21 Poles on imaginary axis and the corresponding time response

18.14 NECESSARY CONDITIONS FOR TRANSFER FUNCTIONS

The necessary conditions for a network function to be a transfer function with common factors in numerator polynomial $N(s)$ and denominator polynomial $D(s)$ cancelled are as follows :

1. The coefficients of different terms in $N(s)$ and $D(s)$ must be real, and those for $D(s)$ must be positive.
2. If the poles and zeros are complex or imaginary they must occur in complex conjugate pairs.
3. The real part of the pole must be negative or zero. If the real part is zero, then the pole must be simple. This includes the origin.
4. The polynomial $D(s)$ should not have any missing terms between those of the highest and the lowest degree, unless all even or all odd terms are missing.
5. The polynomial $N(s)$ may have missing terms between those of the highest and lowest degree Also, some of the coefficients of the powers of s in $N(s)$ may be negative.
6. The degree of $N(s)$ may be as small as zero, independent of the degree of $D(s)$.
7. For $G_{21}(s)$ and $\alpha_{21}(s)$, the maximum degree of $N(s)$ is the degree of $D(s)$.

For $Z_{21}(s)$ and $Y_{21}(s)$, the maximum degree of $N(s)$ equals the degree of $D(s)$ plus one.

18.15 NECESSARY CONDITIONS FOR DRIVING-POINT FUNCTIONS

The necessary conditions for a network function to be a driving point function with common factors in numerator polynomial $N(s)$ and denominator polynomial $D(s)$ cancelled are as follows:

1. The coefficients of different terms in $N(s)$ and $D(s)$ must be real and positive.
2. If the poles or zeros are complex or imaginary they must occur in complex conjugate pairs.

3. The real part of all poles and zeros must be either negative or zero.
4. If the real part of any pole or zero is equal to zero then that pole or zero must be simple.
5. The polynomials of $N(s)$ and $D(s)$ cannot have missing terms between those of highest and lowest degree of s, unless all even degree or all odd degree terms are missing.
6. The degree of $N(s)$ and $D(s)$ may differ by 0 or 1 only.
7. The lowest degree terms in $N(s)$ and $D(s)$ may differ in degree by, at most, 1.

18.16 NATURAL RESPONSE AND NATURAL FREQUENCIES

The ***natural*** (or **relaxed**) ***response*** of a system (or network) is the response that would take place if the source (excitation) were reduced to zero and the system responded to its initial stored energy (initial conditions). It is often called the ***source-free response* or *natural behavior*.** It is the complementary solution of the homogeneous differential equation. It is also called the ***transient* response.** The natural response is dependent only on the values of the elements and their interconnection and is independent of the excitation. The natural behavior of the system may also be described as the response to a *unit-impulse* excitation. This behavior is characterized by means of ***natural frequencies*** of oscillation (or natural modes, or free vibrations) of the system. Hence natural frequencies are defined as those to which a network responds when excited by an unit impulse function.

If an excitation (or forcing function) of frequency s_0 is applied to a network, the network is *forced* to react at that frequency. If the network is excited by an unit impulse function, it does not force any specific frequency into the network. It simply introduces some energy instantaneously and then leaves, allowing the network to react naturally. Thus, the natural response is the response of the network when it is excited by an impulse function. The resulting response singularities are defined as the natural frequencies.

The natural frequencies of a network are the zeros of the determinant of the matrix **Y**(s) formed by writing the node admittance equations for the transformed network.

The natural frequencies of a network are also the zeros of the determinant of the matrix **Z**(s) formed by writing the mesh impedance equations for the transformed network.

The above conclusions assume that none of the zeros of the determinants are cancelled by corresponding zeros in the cofactors. Such a case is called *degenerate* case.

18.17 OPEN-CIRCUIT NATURAL FREQUENCIES (OCNF)

Consider a linear network N excited at the kth pair of terminal by an ideal current source $I_k(s)$ as shown in Fig. 18.22. The voltage $V_k(s)$ across the terminals is given by

$$V_k(s) = Z_{kk}\,(s)\,I_k(s)$$

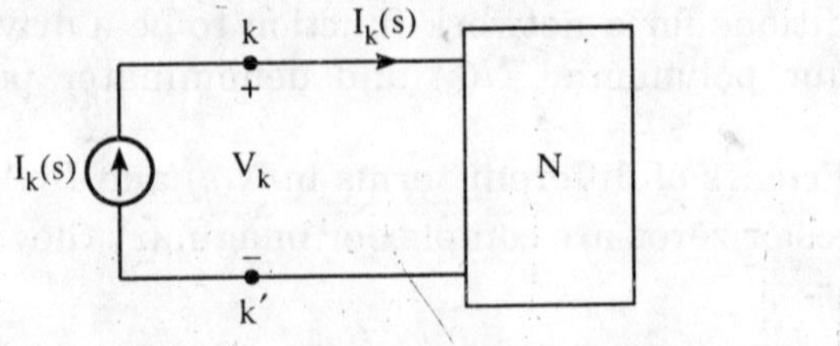

FIG. 18.22 Network *N* excited by an ideal current source

where $Z_{kk}(s)$ is the driving-point impedance. Let $I_k(s)$ become zero (the terminals k–k become open). Then $V_k(s)$ becomes zero unless $Z_{kk}(s)$ is infinite. In other words, for a voltage to exist at the open terminals with no excitation, it is necessary that the driving-point impedance $Z_{kk}(s)$ must be infinite.

But, $$Z_{kk}(s) = K\frac{(s-z_1)(s-z_2)\dots(s-z_m)}{(s-p_1)(s-p_2)\dots(s-p_n)}$$

Thus, the poles of $Z_{kk}(s)$, that is, $p_1, p_2, \dots, p_n$ are the natural frequencies of the network N with terminals $k - k'$ open. Therefore, these frequencies are called ***open-circuit natural frequencies (OCNF)*** of the network N.

18.18 SHORT-CIRCUIT NATURAL FREQUENCIES (SCNF)

Consider next the same network N excited by an ideal voltage source $E_k(s)$, as shown in Fig. 18.23. The current $I_k(s)$ is given by

$$I_k(s) = Y_{kk}(s)\,E_k(s) = \frac{1}{Z_{kk}(s)}\,E_k(s)$$

where $Y_{kk}(s)$ is the driving-point admittance. Now let us assume that $E_k(s)$ is zero (that is, the

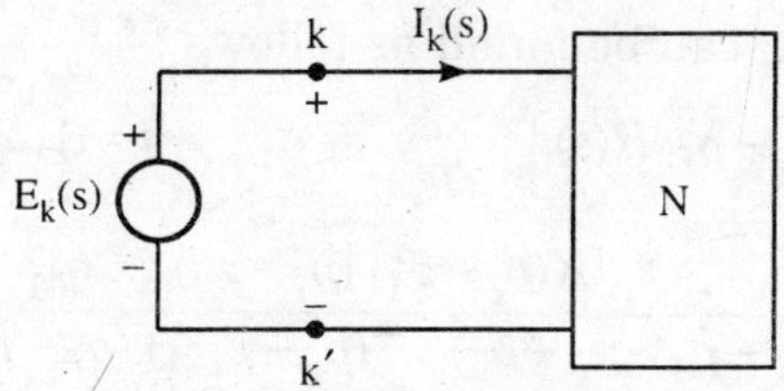

FIG. 18.23 Network *N* excited by an ideal voltage source

terminals $k - k'$ become shorted). Then $I_k(s)$ becomes zero unless $Y_{kk}(s)$ is infinite. In other words, for a current to exist at the shorted terminals with no excitation present, it is necessary that the driving-point admittance $Y_{kk}(s)$ there be infinite, or the driving-point impedance $Z_{kk}(s)$ there be zero. But

$$Y_{kk}(s) = \frac{1}{Z_{kk}(s)} = \frac{(s-p_1)(s-p_2)\dots(s-p_n)}{K(s-z_1)(s-z_2)\dots(s-z_m)}$$

For $Z_{kk}(s)$ to be zero,

$(s-z_1)(s-z_2)\dots(s-z_m)$ should be zero. Therefore the frequencies $z_1, z_2, \dots, z_m$ are called ***short–circuit natural frequencies*** (SCNF) of the network N. In other words, ***the SCNF are the zeros of $Z_{kk}(s)$***.

Thus, we make the following conclusions :

1. *The poles of an impedance function are the OCNF and the zeros of the impedance function are the SCNF of the network.*
2. *The zeros of an admittance function are the OCNF and the poles of the admittance function are the SCNF of the network.*

It is also clear from our previous discussion that the natural frequencies of a network are dependent upon the following :

(*a*) The type of excitation (voltage source or current source)

(*b*) The network configuration (that is, the values and the interconnection of the elements in the network).

18.19 TIME-DOMAIN RESPONSE FROM POLE-ZERO PLOT

The time-domain response can be obtained from the pole-zero plot of a network function. Consider a network function given by

$$H(s) = \frac{N(s)}{D(s)} = K\frac{(s - z_1)(s - z_2) \ldots (s - z_m)}{(s - p_1)(s - p_2) \ldots (s - p_n)} \tag{18.19.1}$$

where $z_1, z_2, \ldots, z_m$ are the zeros and $p_1, p_2, \ldots, p_n$ are the poles of the function $F(s)$.

Assume that the poles and zeros are distinct. The partial fraction expansion of the network function $H(s)$ can be written as

$$H(s) = \frac{A_1}{s - p_1} + \frac{A_2}{s - p_2} + \ldots + \frac{A_n}{s - p_n} \tag{18.19.2}$$

A particular coefficient A_i can be found as follows :

$$A_i = \left[(s - p_{i)}\, H(s)\right]_{s = p_i}, \qquad i = 1, 2, \ldots, n \tag{18.19.3}$$

or

$$A_i = \frac{K(p_i - z_1)(p_i - z_2) \ldots (p_1 - z_m)}{(p_i - p_1)(p_i - p_2) \ldots (p_i - p_{i-1})(p_i - p_{i+1}) \ldots (p_i - p_n)} \tag{18.19.4}$$

Here p_i, z_m and p_n are complex numbers. Eq. (18.19.4) contains factors like $(p_i - z_m)$ and $(p_i - p_n)$. The difference of two complex number is also a complex number.

Hence, the complex number $(p_i - z_i)$ is the directed line from z_1 to p_i and the complex number $(p_i - z_2)$ is the directed line from z_2 to p_i etc. Similarly, the complex number $(p_i - p_1)$ will be a directed line from p_1 to p_i, etc. as shown in Fig. 18.24. Each line is expressed in its polar form in magnitude and angle.

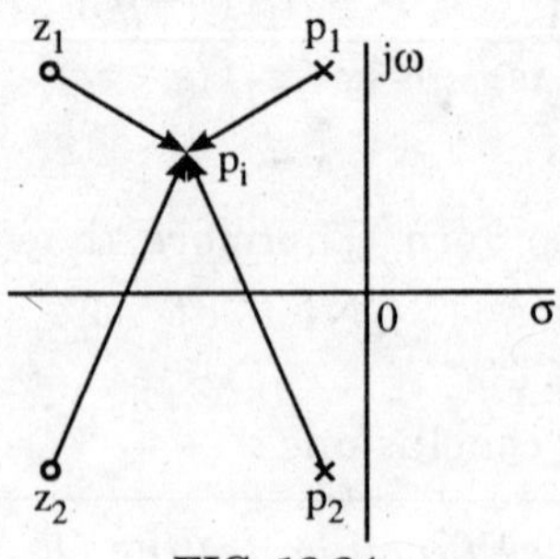

FIG. 18.24

Thus, let

$$p_i - z_1 = r_{1i} \angle \theta_{1i}$$

$$p_i - z_2 = r_{2i} \angle \theta_{2i}$$

$$\vdots$$

$$p_i - z_m = r_{mi} \angle \theta_{mi}$$

and

$$p_i - p_1 = R_{1i} \angle \phi_{1\iota}$$

$$p_i - p_2 = R_{2i} \angle \phi_{2\iota}$$

$$\vdots$$

$$p_i - p_n = R_{ni} \angle \phi_{v\iota}$$

therefore,

$$A_i = K \frac{r_{1i}\, r_{2i} \cdots r_{mi}}{R_{1i}\, R_{2i} \ldots R_{ni}} \angle (\theta_{i1} + \theta_{2i} + \ldots + \theta_{mi}) - (\phi_{1\iota} + \phi_{2\iota} + \ldots + \phi_{v\iota})$$

Similarly, the other constants $A_1, A_2, \ldots, A_{i-1}, A_{i+1}, \ldots A_n$ can be calculated. The time-domain response can be obtained by taking the inverse transform

$$h(t) = £^{-1} H(s)$$

$$= £^{-1}\left[\frac{A_1}{s-p_1} + \frac{A_2}{s-p_2} + \ldots + \frac{A_n}{s-p_n}\right] \qquad (18.19.6)$$

Eq. (18.19.4) shows that a particular coefficient A_i can be obtained from

$$A_i = K \frac{\text{product of all the directed lines from all zeros to } p_i}{\text{product of all the directed lines from other remaining poles to } p_i} \qquad (18.19.7)$$

This procedure can be summarised as follows :

1. Plot the poles and zeros of the network function on the s-plane.
2. Measure or calculate the distances from each of other fininte poles, and zeros to a given pole p_i.
3. Measure or calculate the angle from each of other finite poles or zeros to the given pole p_i.
4. Substitute the quantities in steps 2 and 3 in Eq. (18.19.5) to determine A_i.
5. Repeat the above procedure for all the finite poles.

It is to be noted that the above graphical method is applicable only for all non-repeated poles.

EXAMPLE 18.11

If $$F(s) = \frac{s\,(s+1)}{(s+4)\,(s^2+4s+8)}$$

find $f(t)$ using the pole-zero diagram of the function.

SOLUTION : Writing in the factored form

$$F(s) = \frac{s\,(s+1)}{(s+4)\,(s+2+j2)\,(s+2-j2)}$$

These are two zeros, one at $s = 0$ and the other at $s = -1$.

$$\therefore \qquad z_1 = 0 \text{ and } z_2 = -1$$

These are three poles $p_1 = -4$, $p_2 = -2 - j2$ and $p_3 = -2 + j2$.

The pole-zero plot of the function $F(s)$ is shown in Fig. 18.25.

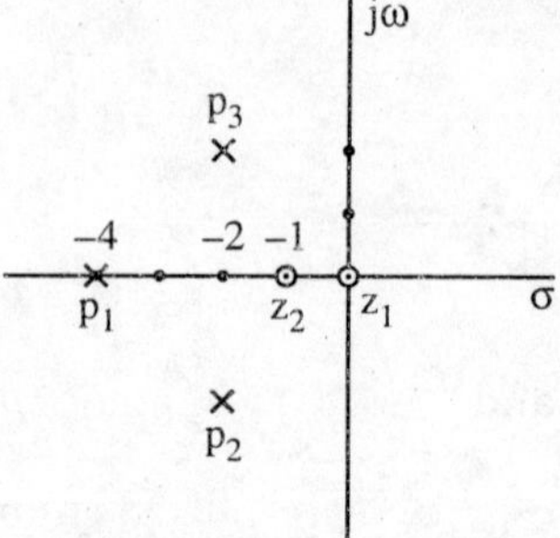

FIG. 18.25 Pole-zero plot of Example 18.11

therefore, $$F(s) = \frac{A_1}{s+4} + \frac{A_2}{s+2+j2} + \frac{A_3}{s+2-j2}$$

For the pole $p_1 = -4$

$$A_i = K\,\frac{(\text{product of all the phasors directed from all zeros towards } p_1)}{(\text{product of all the phasors directed from other remaining poles towards } p_1)} \tag{18.19.7}$$

Here $K = 1$

therefore, $$A_1 = \frac{(p_1 - z_1)\,(p_1 - z_2)}{(p_1 - p_2)\,(p_1 - p_3)}$$

$$= \frac{(-4-0)\,(-4+1)}{(-4+2+j2)\,(-4+2-j2)} = 1.5$$

For the pole $p_2 = -2 - j2$

$$A_2 = K\,\frac{(\text{product of all the phasors directed from all zeros towards } p_2)}{(\text{product of phasors directed from other poles } p_1 \text{ and } p_3)}$$

$$= \frac{(p_2 - z_1)\,(p_2 - z_2)}{(p_2 - p_1)\,(p_2 - p_3)}$$

$$= \frac{(-2-j2-0)\,(-2-j2+1)}{(-2-j2+4)\,(-2-j2+2-j2)} = \frac{1}{4}\,(-1-j2)$$

For the pole $p_3 = -2 + j2$

$$A_3 = \frac{(p_3 - z_1)\,(p_3 - z_2)}{(p_3 - p_1)\,(p_3 - p_2)}$$

$$= \frac{(-2+j2)\,(-2+j2+1)}{(-2+j2+4)\,(-2+j2+2+j2)} = \frac{1}{4}\,(-1+j2)$$

Since the poles p_2 and p_3 are conjugate points A_3 is the complex conjugate of A_2.

$$A_3 = A_2^* = \frac{1}{4}(-1-j2)^* = \frac{1}{4}(-1+j2)$$

$$F(s) = \frac{A_1}{s-p_1} + \frac{A_2}{s-p_2} + \frac{A_3}{s-p_3}$$

$$f(t) = \pounds^{-1}F(s)$$

$$= A_1 e^{p_1 t} + A_2 e^{p_2 t} + A_3 e^{p_3 t}$$

$$= 1.5\, e^{-4t} + \frac{1}{4}(-1-j2)\, e^{(-2-j2)t} + \frac{1}{4}(-1+j2)\, e^{(-2+j2)t}$$

$$= 1.5\, e^{-4t} + e^{-2t}\left[-\frac{1}{4}(e^{j2t} + e^{-j2t}) + \frac{1}{2} j\,(e^{j2t} - e^{-j2t})\right]$$

$$= 1.5\, e^{-4t} + e^{-2t}\left(-\frac{1}{2}\cos 2t + \sin 2t\right)$$

$$= 1.5\, e^{-4t} - \frac{1}{2} e^{-2t}(\cos 2t - 2\sin 2t)$$

■ **EXAMPLE 18.12** The transform voltage $V(s)$ of a network is given by

$$V(s) = \frac{3s}{(s+2)(s^2+2s+2)}$$

Plot its pole-zero diagram and hence obtain $v(t)$.

SOLUTION :

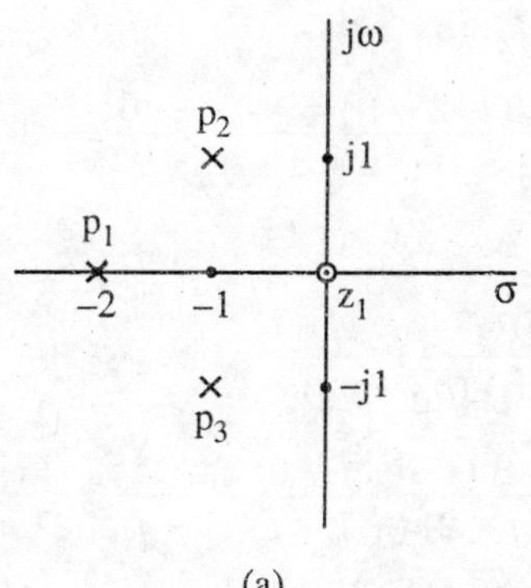

(a)

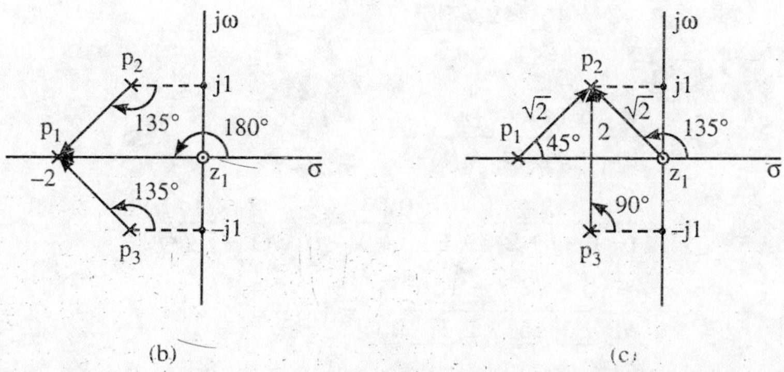

FIG. 18.26 (a) Pole-zero plot (b) Evaluation of A_1 (c)Evaluation of A_2

$$s^2 + 2s + 2 = 0$$

$$s = \frac{-2 \pm \sqrt{4 - 4 \times 1 \times 2}}{2 \times 1} = \frac{1}{2}(-2 \pm \sqrt{-4}) = -1 \pm j$$

therefore, $s_1 = -1 + j,\ s_2 = -1 - j$

We can write $V(s)$ in the factored form as

$$V(s) = \frac{3s}{(s+2)(s+1-j)(s+1+j)}$$

This function has only one zero, $z_1 = 0$ and three poles, $p_1 = -2, p_2 = -1 + j$ and $p_3 = -1 - j$.

The pole-zero plot of the function is shown in Fig. 18.26a.

We can write $V(s)$ in the partial fraction expansion as

$$V(s) = \frac{A_1}{s - p_1} + \frac{A_2}{s - p_2} + \frac{A_3}{s - p_3}$$

$$= \frac{A_1}{s+2} + \frac{A_2}{(s+1-j)} + \frac{A_3}{(s+1+j)}$$

Evaluation of A_1

In order to determine A_1 we draw phasors from the zero z_1 to p_1, from the pole p_2 to p_1 and from the pole p_3 to p_1. (Fig. 18.26b)

$$A_1 = \frac{K(p_1 - z_1)}{(p_1 - p_2)(p_1 - p_3)}$$

$$= \frac{3(-2-0)}{(-2+1-j)(-2+1+j)} = \frac{-6}{2} = -3$$

Evaluation of A_2

$$A_2 = \frac{K(p_2 - z_1)}{(p_2 - p_1)(p_2 - p_3)}$$

$$= \frac{3(-1+j-0)}{(-1+j+2)(-1+j+1+j)} = \frac{3(-1+j)}{(1+j)(j2)}$$

$$= \frac{3 \times \sqrt{2}\ \underline{/\ 135^\circ}}{\sqrt{2}\ \underline{/\ 45^\circ} \times 2\ \underline{/\ 90^\circ}} = 1.5$$

Evaluation of A_3

$$A_3 = \frac{K(p_3 - z_1)}{(p_3 - p_1)(p_3 - p_2)}$$

$$= \frac{3(-1-j-0)}{(-1-j+2)(-1-j+1-j)} = \frac{3(-1-j)}{(1-j)(-j2)}$$

$$= \frac{3 \times \sqrt{2}\ \underline{/\ -135^\circ}}{\sqrt{2}\ \underline{/\ -45^\circ} \times 2\ \underline{/\ -90^\circ}} = 1.5$$

It is to be noted that $A_3 = A_2^*$, therefore A_3 can be calculated directly by taking the complex conjugate of A_2.

Substituting the values of A_1, A_2 and A_3 in the partial fraction expansion of $V(s)$ we get

$$V(s) = \frac{-3}{s+2} + \frac{1.5}{s+1-j} + \frac{1.5}{s+1+j}$$

Taking the inverse Laplace transform we obtain

$$v(t) = -3\,e^{-2t} + 1.5\,e^{-(1-j)t} + 1.5\,e^{-(1+j)t}$$

$$= -\,3\,e^{-2t} + 3\,e^{-t}\,\frac{e^{jt} + e^{-jt}}{2}$$

$$= -3\,e^{-2t} + 3\,e^{-t}\cos t$$

$$= 3\,(e^{-t}\cos t - e^{-2t})$$

18.20 FREQUENCY RESPONSE FROM POLE-ZERO PLOT

The frequency response of a network function can be obtained from its pole-zero plot. Consider the function given by

$$H(s) = \frac{P(s)}{Q(s)} = K\,\frac{(s-z_1)\,(s-z_2)\,\ldots\,(s-z_m)}{(s-p_1)\,(s-p_2)\,\ldots\,(s-p_n)} \tag{18.20.1}$$

where $z_1, z_2, \ldots z_m$ are finite zeros, and $p_1, p_2, \ldots, p_n$ are finite poles. These can be simple or multiple. To calculate $H(s)$ for a specific $s = s_0$, we write Eq. (18.20.1) as

$$H(s_0) = K\,\frac{(s_0-z_1)\,(s_0-z_2)\,\ldots\,(s_0-z_m)}{(s_0-p_1)\,(s_0-p_2)\,\ldots\,(s_0-p_n)} \tag{18.20.2}$$

We consider the geometry of the term $(s_0 - z_1)$. Since s_0 and z_1 are complex numbers we can represent them as directed lines in the s-plane. In Fig. 18.27 s_0 is a directed line from the origin to the point s_0, and z_1 is a directed line form the origion to the point z_1. Hence the complex number $(s_0 - z_1)$ is the directed line *from* z_1 *to* s_0. Similarly, $(s_0 - z_2)$ will be the directed line *from* z_2 *to* s_0.

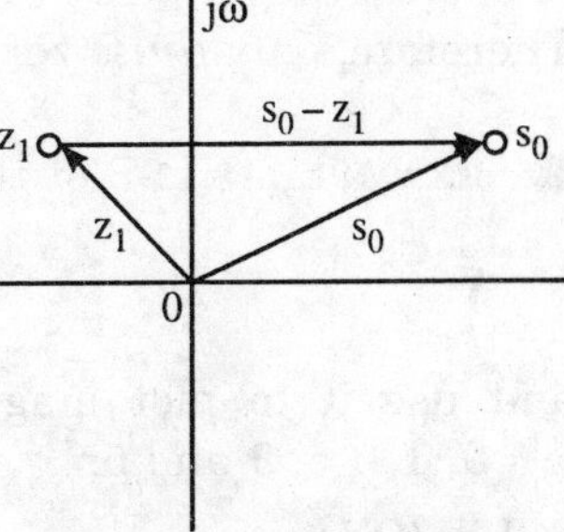

FIG. 18.27 Geometry of $(s_0 - z_1)$

Each will be called a *zero line* (from each zero to s_0). Similarly, the complex number $(s_0 - p_1)$ will be a directed line *from* p_1 *to* s_0, etc. Each of these will be called a *pole line.* (from each pole to s_0). Therefore, we can write Eq. (18.20.2) as

$$H(s_0) = K\,\frac{\text{product of all zeros lines to } s_0}{\text{product of all pole lines to } s_0} \tag{18.20.3}$$

Each zero line and pole line will be expressed in its polar form in magnitude and angle. For example,

$$s_0 - z_1 = M_{z1} \angle \theta_{z1}$$

$$s_0 - z_m = M_{zm} \angle \theta_{zm}$$

$$s_0 - p_1 = M_{p1} \angle \theta_{p1}$$

$$\vdots$$

$$s_0 - p_n = M_{pn} \angle \theta_{pn}$$

where M is the magnitude (length) of the line, θ is the angle measured in the anticlockwise direction from the horizontal axis. The subscripts on M and θ identify the respective zero and pole lines.

For the sinusoidal case we set $s = jw$.

Therefore the megnitude of the function is given by

$$|H(jw_0)| = \frac{M_{z1}\, M_{z2} \ldots M_{zm}}{M_{p1}\, M_{p2} \ldots M_{pn}}$$

while the phase function is

$$\theta(\omega_0) = (\theta_{z1} + \theta_{z2} + \ldots + \theta_{zm}) - (\theta_{p1} + \theta_{p2} + \ldots + \theta_{pn})$$

It is seen that a pole or a zero at $s = \infty$ does not appear explicitly as a factor in Eq. (18.20.1). Therefore, only *finite* zeros and poles are plotted in this graphical method.

EXAMPLE 18.13 Plot the poles and zeros of

$$H(s) = \frac{4s}{s^2 + 2s + s}$$

and use it to plot magnitude and phase characteristics for $\omega = 0, 1, 1.5, 3$ and 5.

SOLUTION :

$$s^2 + 2s + 2 = 0$$

$$s = \frac{-2 \pm \sqrt{(2)^2 - 4 \times 1 \times 2}}{2}$$

$$= \frac{1}{2}(-2 \pm \sqrt{-4}) = \frac{1}{2}(-2 \pm j^2)$$

therefore, $s_1 = -1 + j$, $s_2 = -1 - j$

Here the given function has only one zero at $s = 0$, *i.e.*, $z_1 = 0$.

This function has two poles $p_1 = -1 + j$ and $p_2 = -1 - j$

The pole-zero plot of the function is shown in Fig 18.28.

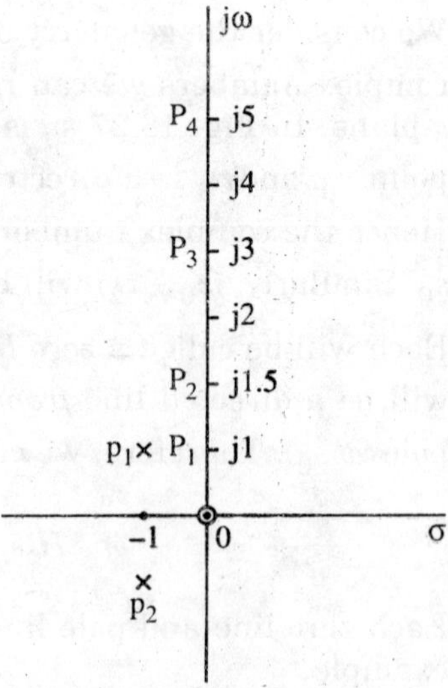

FIG. 18.28 Pole-zero plot of Example 18.13

The points O, P_1, P_2, P_3 and P_4 are corresponding to $\omega = 0, 1, 1.5, 3$ and 5 on the $j\omega$ axis. To calculate $H(s)$ for a specific $s = s_0$, we have from Eq. (18.20.3)

$$H(s_0) = K \frac{\text{product of all phasors from all zeros to } s_0}{\text{product of all phasors from all poles to } s_0}$$

Here $K = 4.$

therefore, $$H(s) = \frac{K(s_0 - z_1)}{(s_0 - p_1)(s_0 - p_2)}$$

(*a*) since $s = j\omega$, at $\omega = 0$ $\quad s_0 = j0$

therefore, $$H(s) = \frac{4(j0 - 0)}{[j0 - (-1 + j)][j0 - (-1 - j)]}$$

$$= 0\underline{/\ 0^\circ}$$

(*b*) At $\omega = 1$, $s_0 = j1$

$$H(s) = \frac{K(s_0 - z_1)}{(s_0 - p_1)(s_0 - p_2)}$$

therefore, $$H(s) = \frac{4(j1 - 0)}{[j1 - (-1 + j)][j1 - (-1 - j)]}$$

$$= \frac{j4}{1(1 + j2)} = \frac{4\underline{/\ 90^\circ}}{2.236\underline{/\ 63.4^\circ}}$$

$$= 1.8\underline{/\ 26.6^\circ}$$

(*c*) At $\omega = 1.5$, $s_0 = j1.5$

$$H(s) = \frac{K(s_0 - z_1)}{(s_0 - p_1)(s_0 - p_2)}$$

$$= \frac{4(j1.5 - 0)}{[j1.5 - (-1 + j)][j1.5 - (-1 - j)]}$$

$$= \frac{j6}{(1 + j0.5)(1 + j2.5)}$$

$$= \frac{6\underline{/\ 90^\circ}}{1.118\underline{/\ 26.6^\circ} \times 2.69\underline{/\ 68.2^\circ}}$$

$$= 1.995\underline{/\ -4.8^\circ}$$

(*d*) At $\omega = 3$, $s_0 = j3$

$$H(s) = \frac{K(s_0 - z_1)}{(s_0 - p_1)(s_0 - p_2)}$$

$$= \frac{4(j3 - 0)}{[j3 - (-1 + j)][j3 - (-1 - j)]}$$

$$= \frac{j12}{(1 + j2)(1 + j4)} = \frac{12\underline{/\ 90^\circ}}{(2.236\underline{/\ 63.4^\circ}) \times 4.123\underline{/\ 76^\circ}}$$

$$= 1.3 \underline{/ -49.4^\circ}$$

(*e*) At $\omega = 5$, $s_0 = j5$

$$H(s) = \frac{K(s_0 - z_1)}{(s_0 - p_1)(s_0 - p_2)}$$

$$= \frac{4(j5 - 0)}{[j5 - (-1 + j)]\ [j5 - (-1 - j)]}$$

$$= \frac{20 \underline{/ 90^\circ}}{(1 + j4)(1 + j6)}$$

$$= \frac{20 \underline{/ 90^\circ}}{(4.123 \underline{/ 76^\circ}) \times 6.08 \underline{/ 80.5^\circ}}$$

$$= 0.8 \underline{/ -66.5^\circ}$$

From these values, the magnitudes and phases can be plotted to a base of frequency.

EXAMPLE 18.14 For the network function

$$H(s) = 3\frac{s - 2}{s(s^2 + 4s + 5)}$$

find $H(j1)$.

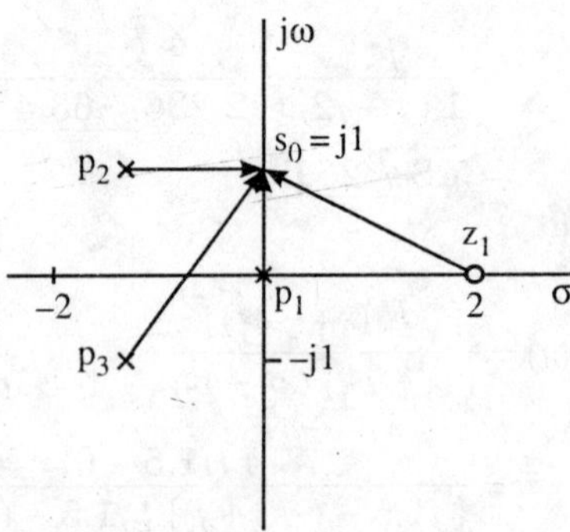

FIG. 18.29 Pole-zero plot of Example 18.14

SOLUTION : $s^2 + 4s + 5 = 0$

$$s = \frac{[-4 \pm \sqrt{(4)^2 - 4\times1\times5}]}{2} = -2 \pm j1$$

therefore, $$H(s) = \frac{3(s - 2)}{(s - 0)(s + 2 - j1)(s + 2 + j1)}$$

This function has only one zero $z_1 = 2$ and three poles $p_1 = 0$, $p_2 = -2 + j1$ and $p_3 = -2 - j1$. We have to find $H(j1)$, therefore $s_0 = j1$. The pole-zero plot of the function is shown in Fig 18.29. Here $s_0 = j1$. Since the given function has only one zero we draw the zero line from $z_1 = 2$ to $s_0 = j1$ and the pole lines from $p_1 = 0$ to $s_0 = j1$, from $p_2 = -2 + j1$ to $s_0 = j1$ and from $p_3 = -2 - j1$ to $s_0 = j1$.

We have here

$$s_0 - z_1 = j1 - 2 = 2.236 \underline{/ 153.5^\circ}$$

$$s_0 - p_1 = j1 - 0 = 1 \underline{/ 90^\circ}$$

$$s_0 - p_2 = j1 - (-2 + j1) = 2 + j0 = 2 \underline{/ 0^\circ}$$

$$s_0 - p_3 = j1 - (-2 - j1) = 2 + j2 = 2.828 \underline{/ 45^\circ}$$

The zero lines and pole lines may be measured directly if the plot is accurate, otherwise, they may be computed from the geometry of the plot. Hence

$$H(j1) = K \frac{\text{product of all zero lines to } s_0}{\text{product of all pole lines to } s_0}$$

$$= 3 \frac{2.236 \underline{/ 153.5^\circ}}{(1 \underline{/ 90^\circ})(2 \underline{/ 0^\circ}) \times 2.828 \underline{/ 45^\circ}}$$

$$= \frac{3 \times 2.236}{1 \times 2 \times 2.828} \underline{/ 153.5^\circ - 90^\circ - 0^\circ - 45^\circ}$$

$$= 1.186 \underline{/ 18.5^\circ}$$

18.21 RELATION BETWEEN *Q* AND POLE POSITIONS

Consider the network function

$$H(s) = \frac{Ks}{s^2 + a\,s + b} \tag{18.21.1}$$

where $a > 0, b > 0$ and $K > 0$.

Evaluating this for $s = j\omega$, we get

$$H(j\omega) = \frac{jK\omega}{b - \omega^2 + j\omega a}$$

or

$$H(j\omega) = \frac{K}{a + j[\omega - (b/\omega)]} \tag{18.21.2}$$

At resonance the denominator should be a minimum. This occurs when the imaginary part of the denomimator is zero. If ω_0 is the resonant frequency, we see that the following relation should be satisfied :

$$\omega_0 - \frac{b}{\omega_0} = 0$$

or

$$\boxed{\omega_0 = \sqrt{b}} \tag{18.21.3}$$

At resonance (that is, at $s = j\omega_0$), the network function is real and has the value.

$$\boxed{H(j\omega_0) = \frac{K}{a}} \tag{18.21.4}$$

Half-power frequencies

At half-power frequencies (that is the 3 dB frequencies), the magnitude of the network function $|H(j\omega)|$ is $\frac{1}{\sqrt{2}}$ times its value at resonance.

therefore,

$$|H(j\omega)| = \frac{1}{\sqrt{2}}|H(j\omega_0)| \tag{18.21.5}$$

From Eq (18.21.2),

$$|H(j\omega)| = \frac{K}{\sqrt{[\omega - (b/\omega)]^2 + a^2}} \tag{18.21.6}$$

Combination of Eqs. (18.21.5) and (18.21.6) gives

$$\frac{K}{\sqrt{[\omega - (b/\omega)]^2 + a^2}} = \frac{1}{\sqrt{2}} \cdot \frac{K}{a} \tag{18.21.7}$$

$$(\omega - \frac{b}{\omega})^2 + a^2 = 2a^2$$

$$(\omega - \frac{b}{\omega}) = \pm a \tag{18.21.8}$$

First choosing the positive sign of a on the right-hand side of Eq. (18.21.8),

$$\omega - \frac{b}{\omega} = a$$

$$\omega^2 - a\omega - b = 0$$

$$\omega = \frac{a}{2} \pm \frac{1}{2}\sqrt{a^2 + 4b} = \frac{a}{2} \pm \sqrt{\left(\frac{a}{2}\right)^2 + b} \tag{18.21.9}$$

If the negative sign of a in the R.H.S of Eq. (18.21.8) is considered

$$\omega - \frac{b}{\omega} = -a$$

$$\omega^2 + a\omega - b = 0$$

$$\omega = -\frac{a}{2} \pm \sqrt{\left(\frac{a}{2}\right)^2 + b} \tag{18.21.10}$$

The negative sign of the quantity under radical sign (that is, $\sqrt{\left(\frac{a}{2}\right)^2 + b}$) is not permissible. Hence we have upper-half-power frequency

$$\boxed{\omega_2 = \frac{a}{2} + \sqrt{\left(\frac{a}{2}\right)^2 + b}} \tag{18.21.11}$$

and lower half-power-frequency

$$\boxed{\omega_1 = -\frac{a}{2} + \sqrt{\left(\frac{a}{2}\right) + b}} \tag{18.21.12}$$

Bandwidth BW

$$\boxed{BW = \omega_2 - \omega_1 = a} \tag{18.21.13}$$

Quality factor Q_0

$$Q_0 = \frac{\omega_0}{BW} \tag{18.21.14}$$

From Eqs. (18.21.3), (18.21.13) and (18.21.14)

$$\boxed{Q_0 = \frac{\sqrt{b}}{a}} \tag{18.21.15a}$$

Hence we conclude that for a second order band pass network function having the form given in Eq. (18.21.1), the quality factor Q_0 is equal to the ratio of the square root of the zero-degree coefficient to the first degree coefficient. That is,

$$\boxed{Q_0 = \frac{\text{square root of the zero degree coefficient}}{\text{first degree coefficient}}} \tag{18.21.15b}$$

The denominator of the bandpass function in Eq. (18.21.1) can be written as

$$s^2 + as + b = s^2 + 2\xi\,\omega_0 s + \omega_0^2 \tag{18.21.16}$$

where $\omega_0 = \sqrt{b}$ is called the **natural frequency** and $\xi = \frac{a}{(2\sqrt{b})}$ is called the **damping ratio.**

The solution of the characteristic equation

$s^2 + 2\xi\,\omega_0 s + \omega_0^2 = 0$ gives two roots.

$$s_1 = -\xi\omega_0 + \omega_0\sqrt{\xi^2 - 1}$$

$$s_2 = -\xi\omega_0 - \omega_0\sqrt{\xi^2 - 1} \quad \text{and is}$$

For $\xi > 1$, the circuit has two distinct poles, on the negative real axis is and called *overdamped* . For $\xi < 1$, the circuit has, a pair of conjugate poles given by

$$s_1 = -\xi\omega_0 + j\omega_0\sqrt{1 - \xi^2} = p_1 \tag{18.21.17}$$

$$s_2 = -\xi\omega_0 - j\omega_0\sqrt{1 - \xi_2} = p_2 = p_1^* \tag{18.21.18}$$

Under the conditon $\xi < 1$, the circuit is said to be *underdamped.*

The poles p_1 and p_2 are shown in Fig. 18.30. From this figure,

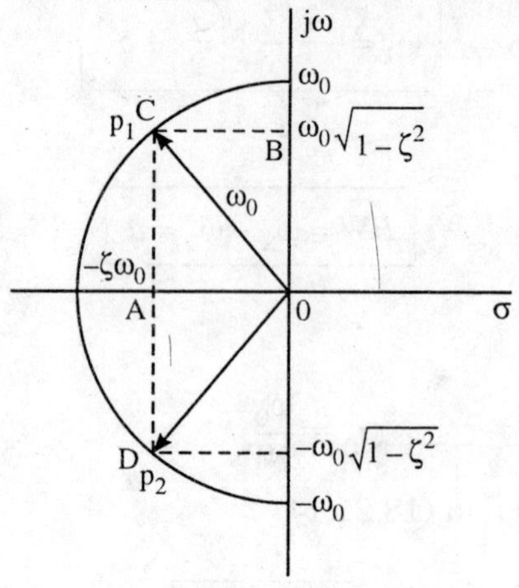

FIG. 18.30

$$OA = -\xi\,\omega_0$$

$$OB = \omega_0\sqrt{1-\xi^2}$$

$$OC^2 = OA^2 + OB^2 = \xi^2\omega_0^2 + \omega_0^2(1-\xi^2) = \omega_0^2$$

therefore, $$OC = \omega_0$$

Magnitude of the real part of the poles

$$OA = |-\xi\omega_0| = \left|\frac{a}{2\sqrt{b}} \times \sqrt{b}\right|$$

$$= \frac{a}{2} = \frac{1}{2}BW.$$

Bandwidth $BW = 2 \times$ magnitude of the real part of the poles (18.21.19)

Magnitude of each pole phasor

= distance of each pole from the origin

$= OC = OD = \omega_0$

Thus, the poles are positioned on a semicircle in the left half plane with radius ω_0.

Since $$Q_0 = \frac{\sqrt{b}}{a}$$

$$\boxed{Q_0 = \frac{\text{magnitude of each pole phasor}}{2 \times \text{magnitude of the real part of the poles}}} \quad (18.21.20)$$

The relations given in Eqs. (18.21.15) and (18.21.20) provide a simple means of calculating Q_0 in a network function which is characterized by a second-degree denominator polynomial.

For a given ω_0, the bandwidth is inversely proportional to Q_0. This means that closer the poles are to the $j\omega$ – axis, the narrower will be the bandwidth and larger will be the quality factor. Conversely, when the poles are far from the $j\omega$ – axis, Q_0 will be small and bandwidth will be large. This is shown in Fig. 18.31.

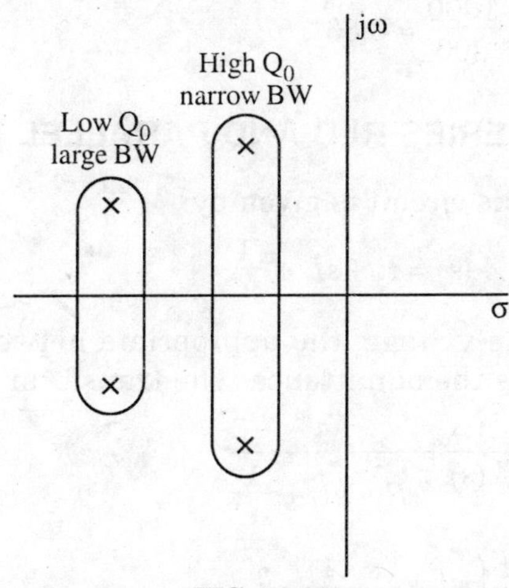

FIG. 18.31

■ **EXAMPLE 18.15** For the network function

$$H(s) = \frac{10s}{s^2 + 300s + 10^6}$$

determine the centre frequency, lower and upper half-power frequencies, the bandwidth, and the quality factor.

SOLUTION : The band-pass network function has the form

$$H(s) = \frac{Ks}{s^2 + as + b}$$

Comparing the given network function with the band-pass network function we have

$$a = 300, \qquad b = 10^6$$

∴ centre frequency = resonant frequency

$$\omega_0 = \sqrt{b} = \sqrt{10^6} = 10^3 \text{ rad/s}$$

Upper half-power frequency

$$\omega_2 = \frac{a}{2} + \sqrt{\left(\frac{a}{2}\right)^2 + b}$$

$$= \frac{300}{2} + \sqrt{\left(\frac{300}{2}\right)^2 + 10^6} = 1161.2 \text{ rad/s}$$

Lower half-power frequency

$$\omega_1 = -\frac{a}{2} + \sqrt{\left(\frac{a}{2}\right)^2 + b}$$

$$= -\frac{300}{2} + \sqrt{\left(\frac{300}{2}\right)^2 + 10^6} = 861.2 \text{ rad/s}$$

Bandwidth $BW = \omega_2 - \omega_1 = 300$ rad/s

Quality factor $Q_0 = \frac{\omega_0}{BW} = \frac{1000}{300} = 3.33$

18.22 RESONANCE IN SERIES RLC AND PARALLEL RLC CIRCUITS

The transform impedance of this circuit is given by

$$Z(s) = R + sL + \frac{1}{sC} \qquad (18.22.1)$$

Since the excitation is the voltage, the appropriate network function representing the response-to-excitation phasor is the admittance. The transform admittance is given by

$$Y(s) = \frac{1}{Z(s)} = \frac{1}{R + sL + \frac{1}{sC}}$$

or

$$Y(s) = \frac{1}{L}\left(\frac{s}{s^2 + \frac{R}{L}s + \frac{1}{LC}}\right)$$

The admittance function has a second-degree denominator polynomial $\left(s^2 + \frac{R}{L}s + \frac{1}{LC}\right)$. It is of the form $(s^2 + as + b)$.

By comparison,

$$a = \frac{R}{L},\ b = \frac{1}{LC}$$

$$\omega_0 = \sqrt{b} = \sqrt{\frac{1}{LC}}$$

$$\omega_1 = -\frac{a}{2} + \sqrt{\left(\frac{a}{2}\right)^2 + b}$$

$$\omega_2 = \frac{a}{2} + \sqrt{\left(\frac{a}{2}\right)^2 + b}$$

$$BW = a = \frac{R}{L}$$

$$Q_0 = \frac{\sqrt{b}}{a} = \frac{L}{R} \times \sqrt{\frac{1}{LC}} = \frac{1}{R}\sqrt{\frac{L}{C}}$$

(b) Parallel RLC Circuit

For a parallel RLC circuit, the transform admittance is

$$Y(s) = G + \frac{1}{sL} + sC$$

The transform impedance is

$$Z(s) = \frac{1}{Y(s)} = \frac{1}{G + \frac{1}{sL} + sC}$$

therefore,
$$Z(s) = \frac{s}{C\left[s^2 + \frac{G}{C}s + \frac{1}{LC}\right]}$$

The impedance function of a parallel RLC circuit has a second-degree denominator polynomial $\left(s^2 + \frac{G}{C}s + \frac{1}{LC}\right)$.

It is of the form $(s^2 + as + b)$.

By comparison

$$a = \frac{G}{C}, \; b = \frac{1}{LC}$$

therefore,

$$\omega_0 = \sqrt{b} = \sqrt{\frac{1}{LC}}$$

$$\omega_1 = -\frac{a}{2} + \sqrt{\left(\frac{a}{2}\right)^2 + b}$$

$$\omega_2 = \frac{a}{2} + \sqrt{\left(\frac{a}{2}\right)^2 + b}$$

$$BW = a = \frac{G}{C}$$

$$Q_0 = \frac{\sqrt{b}}{a} = \frac{C}{G} \times \sqrt{\frac{1}{LC}} = \frac{1}{G}\sqrt{\frac{C}{L}}$$

It is seen that the pole-zero diagrams can be drawn for series RLC and parallel RLC circuits. But it should be remembered that for series RLC circuit the pole-zero diagram is for admittance function and for parallel RLC circuit, the pole-zero diagram is drawn for impedance function.

EXAMPLE 18.16 The impedance of a series RLC circuit is given by

$$Z(s) = 20\,\frac{(s^2 + 2s + 101)}{s}$$

Determine the resonant frequency, bandwidth, quality factor, and the impedance at resonance.

SOLOUTION :

$$Z(s) = \frac{20}{s}(s^2 + 2s + 101)$$

$$Y(s) = \frac{1}{20} \frac{s}{s^2 + 2s + 101} \tag{E 18.16.1}$$

The admittance function has a second-degree denominator polynomial $(s^2 + 2s + 101)$. It is of the form $(s^2 + as + b)$. By comparison $a = 2$, $b = 101$

$\therefore$ resonant frequency $\omega_0 = \sqrt{b} = \sqrt{101} = 10.05$ rad/s

Bandwidth $= a = 2$ rad/s

Quality factor $\quad Q_0 = \frac{\sqrt{b}}{a} = \frac{\sqrt{101}}{2} = 5.025$

$$Y_0 = \frac{K}{a} = \frac{1}{20} \times \frac{1}{2} = \frac{1}{40} S$$

$\therefore \quad Z_0 = 40\Omega$

Alternatively

For a series RLC circuit

$$Y(s) = \frac{1}{L} \frac{s}{\left(s^2 + \frac{R}{L}s + \frac{1}{LC}\right)} \tag{E 18.16.2}$$

By comparing Eq. (E 18.16.1) with Eq. (E 18.16.2)

$$L = 20\text{H}, \quad \frac{R}{L} = 2$$

Impedance at resonance $Z_0 = R = 2L = 2 \times 20 = 40\Omega$

■ **EXAMPLE 18.17** The impedance of a parallel RLC circuit is given by

$$Z(s) = \frac{s}{20(s^2 + 2s + 101)}$$

Determine the resonant frequency, bandwidth, quality factor, and the impedance at resonance.

SOLUTION : The impedance function has a second-degree denominator polynomial $(s^2 + 2s + 101)$. It is of the form $(s^2 + as + b)$.

therefore, $\quad a = 2$, $b = 101$

Resonant frequency

$$\omega_0 = \sqrt{b} = \sqrt{101} = 10.05 \text{ rad/s}$$

Bandwidth $\quad = a = 2$ rad/s

Quality factor $\quad = \frac{\sqrt{b}}{a} = \frac{\sqrt{101}}{2} = 5.025$

Impedance at resonance $Z_0 = \frac{K}{a} = \frac{1}{20 \times 2} = \frac{1}{40}\,\Omega$

Alternatively

For a parallel *RLC* circuit

$$Z(s) = \frac{s}{C\left[s^2 + \frac{G}{C}s + \frac{1}{LC}\right]} \qquad \text{(E 18.17.1)}$$

Comparison of Eq. (E 18.17.1) with the given impedance function gives

$$C = 20F, \quad \frac{G}{C} = 2$$

therefore,

$$G = 2\,C = 2 \times 20 = 40 \text{ S}$$

Impedance at resonance

$$Z_0 = \frac{1}{G} = \frac{1}{40}\ \Omega$$

■ **EXAMPLE 18.18** In a 2 -port network the voltage gain is given by

$$\frac{V_2(s)}{V_1(s)} = \frac{s}{(s+2)(s+3)}$$

Determine the output voltage if the input is

(*a*) a unit impulse (*b*) a unit step (*c*) e^{-t}

SOLUTION :

$$\frac{V_2(s)}{V_1(s)} = \frac{s}{(s+2)(s+3)}$$

therefore,

$$V_2(s) = \frac{s\,V_1(s)}{(s+2)(s+3)}$$

(*a*) Impulse response

$$v_1(t) = u_0(t)$$

therefore,

$$V_1(s) = \pounds\, u_0(t) = 1$$

and

$$V_2(s) = \frac{s\,V_1(s)}{(s+2)(s+3)} = \frac{s}{(s+2)(s+3)}$$

Here

$$K = 1$$

For the function $V_2(s)$ there is one zero at $s = 0$, *i.e.*, $z_1 = 0$ two poles $p_1 = -2$ and $p_2 = -3$.

Let

$$V_2(s) = \frac{A_1}{s+2} + \frac{A_2}{s+3}$$

For pole $p_1 = -2$

$$A_1 = \frac{p_1 - z_1}{p_1 - p_2} = \frac{-2-0}{-2+3} = -2$$

For pole $p_2 = -3$

$$A_2 = \frac{p_2 - z_1}{p_2 - p_1}$$

$$= \frac{-3-0}{-3+2} = 3$$

therefore,
$$V_2(s) = \frac{A_1}{s+2} + \frac{A_2}{s+3}$$

$$= \frac{-2}{s+2} + \frac{3}{s+3}$$

therefore,
$$v_2(t) = £^{-1} V_2(s)$$

$$= £^{-1}\left[\frac{-2}{s+2} + \frac{3}{s+3}\right]$$

$$= -2e^{-2t} + 3e^{-3t}$$

(*b*) Step response

$$v_1(t) = u_{-1}(t)$$

$$V_1(s) = £u_{-1}(t) = \frac{1}{s}$$

therefore,
$$V_2(s) = \frac{s V_1(s)}{(s+2)(s+3)} = \frac{1}{(s+2)(s+3)}$$

or
$$V_1(s) = \frac{A_3}{s+2} + \frac{A_4}{s+3}$$

$$A_3 = \frac{1}{(p_1 - p_2)} = \frac{1}{-2-(-3)} = 1$$

$$A_4 = \frac{1}{p_2 - p_1} = \frac{1}{-3-(-2)} = -1$$

$$V_2(s) = \frac{1}{s+2} - \frac{1}{s+3}$$

$$v_2(t) = £^{-1}\left[\frac{1}{s+2} - \frac{1}{s+3}\right]$$

$$v_2(t) = e^{-2t} - e^{-3t}$$

(*c*) Exponential response

$$v_1(t) = e^{-t}$$

therefore,
$$V_1(s) = \frac{1}{s+1}$$

$$V_2(s) = \frac{s V_1(s)}{(s+2)(s+3)}$$

$$V_2(s) = \frac{s}{(s+1)(s+2)(s+3)}$$

$$= \frac{A_5}{s+1} + \frac{A_6}{s+2} + \frac{A_7}{s+3}$$

Here $\quad z_1 = 0\ ,\ p_1 = -1\ ,\ p_2 = -2\ ,\ p_3 = -3$

$$A_5 = \frac{p_1 - z_1}{(p_1 - p_2)(p_1 - p_3)}$$

$$= \frac{-1-0}{(-1+2)(-1+3)} = -\frac{1}{2}$$

$$A_6 = \frac{p_1 - z_1}{(p_2 - p_1)(p_2 - p_3)}$$

$$= \frac{-2-0}{(-2+1)(-2+3)} = 2$$

$$A_7 = \frac{p_3 - z_1}{(p_3 - p_1)(p_3 - p_2)}$$

$$= \frac{-3-0}{(-3+1)(-3+2)} = -\frac{3}{2}$$

therefore, $\quad V_1(s) = \frac{A_5}{s+1} + \frac{A_6}{s+2} + \frac{A_7}{s+3}$

$$= -\frac{1}{2}\ \frac{1}{s+1} + \frac{2}{s+2} - \frac{3}{2}\ \frac{1}{s+3}$$

$$v_1(t) = £^{-1}\ V_1(s)$$

$$= -\frac{1}{2} e^{-t} + 2\, e^{-2t} - \frac{3}{2} e^{-3t}$$

EXAMPLE 18.19 In a 2-port network the voltage gain is given by

$$\frac{V_2(s)}{V_1(s)} = \frac{s+2}{s^2+1}$$

Determine the output voltage if the input is

(*a*) a unit impulse

(*b*) a unit step

(*c*) e^{-t}

SOLUTION :

$$V_2(s) = \frac{s+2}{s^2+1} V_1(s)$$

(a) *Impulse response*

$$v_1(t) = u_0(t)$$

$$V_1(s) = £\, u_0(t) = 1$$

therefore, $$V_2(s) = \frac{s+2}{s^2+1} = \frac{s+2}{(s+j\,1)(s-j\,1)}$$

Here $$z_1 = -2\,,\ p_1 = -j1\,, p_2 = +j1$$

We can write

$$V_2(s) = \frac{A_1}{s+j\,1} + \frac{A_2}{s-j\,1}$$

$$A_1 = \frac{p_1 - z_1}{(p_1 - p_2)} - \frac{-j\,1-(-2)}{-j\,1-j\,1} = \frac{2-j\,1}{-j\,2}$$

$$= \frac{1}{2} + j\,1$$

$$A_2 = \frac{p_2 - z_1}{p_1 - p_2} = \frac{j\,1-(-2)}{j\,1-(-j\,1)} = \frac{2+j\,1}{j\,2}$$

$$= \frac{1}{2} - j\,1$$

therefore, $$V_2(s) = \left(\frac{1}{2} + j\,1\right)\cdot\frac{1}{s+j\,1} + \left(\frac{1}{2} - j\,1\right)\frac{1}{s-j\,1}$$

$$v_2(t) = \left(\frac{1}{2} + j\,1\right)e^{-jt} + \left(\frac{1}{2} - j\,1\right)e^{jt}$$

$$= \frac{1}{2}\left(e^{jt} + e^{-jt}\right) + j\left(e^{jt} - e^{-jt}\right)$$

$$= \cos t + 2\sin t$$

(b) *Step response*

$$v_1(t) = u_{-1}(t)$$

$$V_1(s) = \pounds\, u_{-1}(t) = \frac{1}{s}$$

$$V_2(s) = \frac{s+2}{s(s^2+1)} = \frac{s+2}{s(s+j\,1)(s-j\,1)}$$

Here $$z_1 = -2\,,\ p_1 = 0\,,\ p_2 = -j\,1\,,\ p_3 = +j\,1$$

Let $$V_2(s) = \frac{A_3}{s} + \frac{A_4}{s+j\,1} + \frac{A_5}{s-j\,1}$$

For pole p_1,

$$A_3 = \frac{p_1 - z_1}{(p_1 - p_2)(p_1 - p_3)}$$

$$= \frac{0-(-2)}{[0-(-j\,1)]\,(0-j\,1)} - \frac{2}{-j^2} = 2$$

For pole p_2,

$$A_4 = \frac{p_2 - z_1}{(p_2 - p_1)(p_2 - p_3)}$$

$$= \frac{-j\,1 - (-2)}{[-j\,1 - 0]\,[-j\,1 - j\,1]} = \frac{2 - j\,1}{-2}$$

$$= -1 + j\frac{1}{2}$$

For pole p_3,

$$A_5 = \frac{p_3 - z_1}{(p_3 - p_1)(p_3 - p_2)}$$

$$= \frac{j\,1 - (-2)}{(j\,1 - 0)\,[j\,1 - (-j\,1)]} = \frac{2 + j\,1}{j^2\,2} = \frac{2 + j\,1}{-2}$$

$$= -1 - j\frac{1}{2}$$

Alternatively

$$A_5 = A_4^* = \left(-1 + j\frac{1}{2}\right)^*$$

$$= -1 - j\frac{1}{2}$$

therefore,
$$V_2(s) = \frac{A_3}{s} + \frac{A_4}{s + j\,1} + \frac{A_5}{s - j\,1}$$

$$= 2 \times \frac{1}{s} + \left(-1 + j\frac{1}{2}\right) \cdot \frac{1}{s + j\,1} + \left(-1 - j\frac{1}{2}\right)\frac{1}{s - j\,1}$$

$$v_2(t) = £^{-1}\; V_2(s)$$

$$= 2\,u_{-1}(t) + \left(-1 + j\frac{1}{2}\right)e^{+jt} + \left(-1 - j\frac{1}{2}\right)e^{+jt}$$

$$= 2\,u_{-1}(t) - j\frac{1}{2}\left(e^{jt} - e^{-jt}\right) - \left(e^{jt} + e^{-jt}\right)$$

$$= 2\,u_{-1}(t) + \sin t - 2\cos t$$

(c) *Exponential response*

$$v_1(t) = e^{-t}$$

$$V_1(s) = \frac{1}{s + 1}$$

therefore,
$$V_2(s) = \frac{(s + 2)\,V_1(s)}{s^2 + 1} = \frac{s + 2}{(s + 1)(s^2 + 1)}$$

$$= \frac{s+2}{(s+1)(s+j\,1)(s-j\,1)}$$

Here $$z_1 = -2\,,\; p_1 = -1\,,\; p_2 = -j\,1\,,\; p_3 = j\,1$$

Let $$V_2(s) = \frac{A_6}{s+1} + \frac{A_7}{s+j\,1} + \frac{A_8}{s-j\,1}$$

For pole p_1

$$A_6 = \frac{(p_1 - z_1)}{(p_1 - p_2)(p_1 - p_3)}$$

or $$A_6 = \frac{-1+2}{(-1+j\,1)(-1-j\,1)} = \frac{1}{2}$$

For pole p_2

$$A_7 = \frac{(p_2 - z_1)}{(p_2 - p_1)(p_2 - p_3)}$$

$$= \frac{-j\,1+2}{(-j\,1+1)(-j\,1-j\,1)} = \frac{2-j\,1}{-2\,(1+j\,1)}$$

$$= \frac{(2-j\,1)(1-j\,1)}{-2\,(1+j)(1-j)} = \frac{-1+j\,3}{4}$$

For pole p_3

$$A_8 = \frac{(p_3 - z_1)}{(p_3 - p_1)(p_3 - p_2)}$$

$$= \frac{j\,1+2}{(j\,1+1)(j\,1+j\,1)} = \frac{-1-j\,3}{4}$$

Alternatively

$$A_8 = A_7^* = \frac{-1-j\,3}{4}$$

therefore, $$V_2(s) = \frac{A_6}{s+1} + \frac{A_7}{s+j\,1} + \frac{A_8}{s-j\,1} = \frac{1}{2\,(s+1)} + \frac{-1+j\,3}{4\,(s+j\,1)} + \frac{-1-j\,3}{4\,(s-j\,1)}$$

$$v_2(t) = \pounds^{-1}\, V_2(s)$$

$$= \frac{1}{2} e^{-t} + \frac{(-1+j\,3)\, e^{-jt}}{4} + \frac{(-1-j\,3)\, e^{+jt}}{4}$$

$$= \frac{1}{2} e^{-t} - \frac{1}{4}(e^{jt} + e^{-jt}) - j\,\frac{3}{4}(e^{jt} - e^{-jt})$$

$$= \frac{1}{2} e^{-t} - \frac{1}{2}\cos t + \frac{3}{2}\sin t$$

or $$v_2(t) = \frac{1}{2}\,[e^{-t} - \cos t + 3\sin t]$$

EXAMPLE 18.20 Find the driving-point impedance z_{11} at terminals 1 – 1′ in Fig. 18.32 from a consideration of the open-and short-circuit natural frequencies of the network.

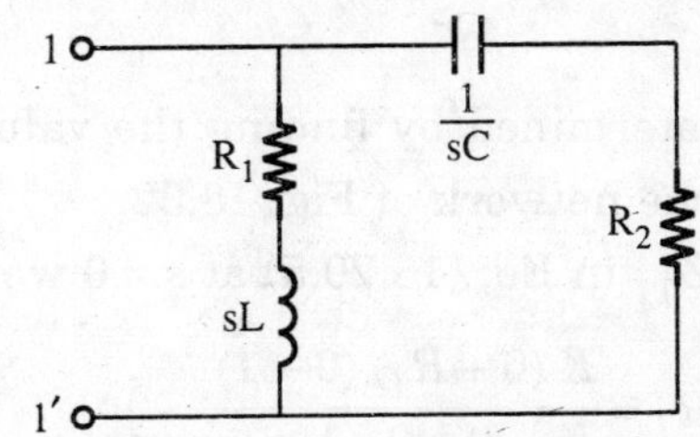

FIG. 18.32 Pertaining to Example 18.20

SOLUTION :

Determination of open-circuit natural frequencies (OCNF)

When terminals 1 – 1′are open, the driving-point impedance z_{11} is given by

$$z_{11} = (R_1 + sL)\,||\left(R_2 + \frac{1}{sC}\right) = \frac{(R_1 + sL)\left(R_2 + \frac{1}{sC}\right)}{R_1 + sL + R_2 + \frac{1}{sC}} \qquad \text{(E 18.20.1)}$$

Since the poles of z_{11} are the open-circuit natural frequencies, the denominator of Eq. (E 18.20.1) should be zero.

therefore, $$R_1 + sL + R_2 + \frac{1}{sC} = 0$$

or $$LCs^2 + (R_1 + R_2)\,Cs + 1 = 0 \qquad \text{(E18.20.2)}$$

The roots of Eq. (E 18.20.2) give the open circuit natural frequencies.

Determination of short-circuit natural frequencies (SCNF)

When terminals 1 – 1′ of Fig. 18.32 are shorted, the network takes the form shown in Fig. 18.33.

Since the zeros of Z_{11} are the natural frequencies of the network with terminals 1 – 1′ shorted, the zeros of Z_{11} are given by

$$sL + R_1 = 0 \qquad \text{(E 18.20.3)}$$

and $$R_2 + \frac{1}{sC} = 0$$

or $$CR_2 s + 1 = 0 \qquad \text{(E 18.20.4)}$$

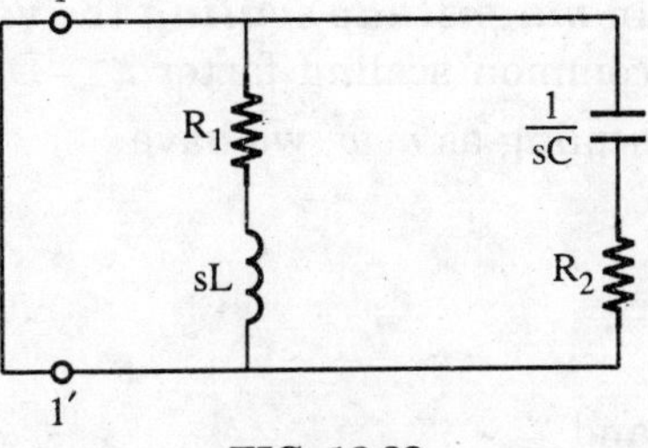

FIG. 18.33

Driving-point impedance

Since the poles, zeros and a constant K determine the network function, we can write

$$Z_{11} = \frac{K\,(s - z_1)\,(s - z_2)}{(s - p_1)\,(s - p_2)}$$

$$= \frac{K(sL + R_1)(CR_2 s + 1)}{LCs^2 + (R_1 + R_2)Cs + 1} \tag{E 18.20.5}$$

Determination of constant K

Here the constant K is to be determined by finding the value of Z_{11} at one frequency. For example, at $s = 0$, $z_{11} = R_1$ in the network of Fig. 18.32.

Substituting the value of Z_{11} in Eq. (18.20.5) at $s = 0$ we get

$$R_1 = \frac{K(0 + R_1)(0 + 1)}{0 + 0 + 1}$$

therefore, $K = 1$

Hence the driving-point impedance Z_{11} is obtained by putting $K = 1$ in Eq. (E 18.20.5).

$$Z_{11} = \frac{(sL + R_1)(CR_2 s + 1)}{LCs^2 + (R_1 + R_2)Cs + 1} \tag{E 18.20.6}$$

It is to be noted that all roots lie in the left half of the s-plane and the highest power of s in the numerator and denominator differ by no more than 1, and likewise for the lowest powers.

18.23 SCALING OR NORMALIZATION

The use of standard units (farads, hertz, henrys, ohms, seconds, radians per second, etc.) frequently gives very large or very small numbers in circuit analysis and design. Typically, such values may be for example, $R = 600\Omega$, $L = 0.8$ mH, $C = 3.5$ pF, $\omega = 2\pi \times 10^6$ rad/s.

Such numbers may often cause error in hand computations. In order to avoid this, it is desirable to scale all electrical quantities by some conveniently chosen factors. This process is called ***normalization* or *scaling.***

There are two types of scaling, *magnitude scaling* and *frequency scaling.*

18.24 MAGNITUDE SCALING OR IMPEDANCE SCALING

In ***magnitude scaling*** the values of all impedances at a given frequency are multiplied by a common scaling factor k_m. Denoting the original component values as *old*, and those after scaling as *new*, we have

$$R_{new} = k_m R_{old}$$

$$j\omega L_{new} = k_m (j\omega L_{old})$$

and

$$\frac{1}{j\omega C_{new}} = \frac{k_m}{j\omega C_{old}}$$

or

$$\boxed{\begin{aligned} R_{new} &= k_m R_{old} \\ L_{new} &= k_m L_{old} \\ C_{new} &= \frac{C_{old}}{k_m} \end{aligned}} \tag{18.24.1}$$

Magnitude scaling is also known as ***impedance scaling.*** It leaves the frequency response of a circuit unchanged. Magnitude scaling does not affect ω_0, Q and bandwidth. It only affects signal amplitudes and power dissipation because of resistance scaling.

18.25 FREQUENCY SCALING

In frequency scaling each element is changed so that its impedance at the new frequency is the same as at the original frequency. The frequency scaling factor is denoted by k_f. Suppose that

$$\omega_{new} = k_f \, \omega_{old}$$

Therefore, we have

$$R_{new} = R_{old}$$

since a resistor is not frequency dependent.

$$j\,\omega_{new} L_{new} = j\,\omega_{old} L_{old}$$

and

$$\frac{1}{j\,\omega_{new} C_{new}} = \frac{1}{j\,\omega_{old} C_{old}}$$

or

$$\boxed{\begin{aligned} R_{new} &= R_{old} \\ L_{new} &= \frac{L_{old}}{k_f} \\ C_{new} &= \frac{C_{old}}{k_f} \end{aligned}} \qquad (18.25.1)$$

Frequency scaling shifts the frequency response of a circuit up or down the frequency spectrum while leaving signal amplitudes and power dissipation unchanged. It leaves Q unchanged but it affects ω_0 and BW.

18.26 MAGNITUDE AND FREQUENCY SCALING

If we scale both magnitude and frequency, the total result is the combined effect of these scalings. A circuit can simultaneously be scaled in magnitude and frequency via

$$\boxed{\begin{aligned} R_{new} &= k_m R_{old} \\ L_{new} &= \left(\frac{k_m}{k_f}\right) L_{old} \\ C_{new} &= \frac{1}{k_m k_f} C_{old} \end{aligned}} \qquad (18.26.1)$$

These three equations are known as the ***element scaling equations***. Scaling results in simplification in calculations. Simultaneous scaling is widely used in the design of frequency-selective networks or filters. A scaled network can serve as a prototype (for example in design).

Its basic properties are unaffected by scaling. Thus, a large numbers of practical (unscaled) networks can be obtained from it by a simple unscaling operation.

■ **EXAMPLE 18.21** A series RLC circuit with

$$R_1 = 2\,,\ L_1 = 1 \text{ and } C_1 = \frac{1}{2} \text{ is to be scaled by } k_m = 500 \text{ and } k_f = 10^4.$$

Determine the new element values.

SOLUTION :

$$R_{new} = k_m\, R_{old} = 500 \times 2 = 1000$$

$$L_{new} = \frac{k_m}{k_f} L_{old}$$

$$= \frac{500}{10^4} \times 1 = 0.05$$

$$C_{new} = \frac{1}{k_m\, k_f} C_{old}$$

$$= \frac{1}{500 \times 10^4} \times \frac{1}{2} = 10^{-7}$$

■ **EXAMPLE 18.22** A parallel *RCL* circuit with

$$R_1 = 10\,,\ L_1 = \frac{1}{100} \text{ and } C_1 = \frac{1}{100} \text{ is scaled giving } R_2 = 10^4\,,\ L_2 = 10^{-3}.$$

Determine C_2.

SOLUTION :

$$R_{new} = k_m\, R_{old}$$

$$R_2 = k_m\, R_1$$

therefore,

$$k_m = \frac{R_2}{R_1} = \frac{10^4}{10} = 10^3$$

$$L_{new} = \frac{k_m}{k_f} L_{old}$$

$$L_2 = \frac{k_m}{k_f} L_1$$

$$10^{-3} = \frac{10^3}{k_f} \times \frac{1}{100}$$

$$k_f = \frac{10^3}{10^{-3} \times 100} = 10^4$$

$$C_{new} = \frac{1}{k_m\, k_f} C_{old}$$

$$C_2 = \frac{1}{k_m\, k_f} C_1$$

$$= \frac{1}{10^3 \times 10^4 \times 100} = 10^{-9}$$

EXAMPLE 18.23 The impedance of a parallel RLC network is given by

$$Z\,(s) = \frac{5s}{s^2 + 0.5\,s + 100}$$

(*a*) Determine the element values and Q_0.

(*b*) If the circuit is scaled so that $Z_{max} = 10^4$ occurs at $\omega = 10^8$, determine the new element values.

SOLUTION : For a parallel *RCL* circuit

$$Z(s) = \frac{s}{c\left(s^2 + \frac{G}{C} s + \frac{1}{LC}\right)}$$

For the given circuit

$$Z(s) = \frac{5\,s}{s^2 + 0.5\,s + 100}$$

This is of the form

$$Z(s) = \frac{Ks}{s^2 + as + b}$$

therefore, $K = 5\ ,\ a = 0.5\ ,\ b = 100$

Resonant frequency $\omega_0 = \sqrt{b} = \sqrt{100} = 10$ rad/s

Bandwidth $= a = 0.5$ rad/s

Quality factor $\quad Q_0 = \frac{\sqrt{b}}{a} = \frac{\sqrt{100}}{0.5} = 20$

$$R = Z_0 = \frac{K}{a} = \frac{5}{0.5} = 10\Omega$$

$$C = \frac{1}{5}$$

$$\omega_0^2 = \frac{1}{LC}$$

$$100 = \frac{1}{L} \times 5\ ,\ L = \frac{5}{100} = \frac{1}{20}$$

Scaling

$$k_m = \frac{10^4}{10} = 10^3$$

$$k_f = \frac{10^8}{10} = 10^7$$

The new values of elements are given by

$$R_{new} = R_{old} = 10^4$$

$$L_{new} = \frac{k_m}{k_f} L_{old} = \frac{10^3}{10^7} \times \frac{1}{20} = 5 \times 10^{-6}$$

$$C_{new} = \frac{C_{old}}{k_m \, k_f} = \frac{1}{5} \times \frac{1}{10^3 \times 10^7} = 20 \times 10^{-12}$$

Since $Q = \omega_0 \, CR$

$$Q_{new} = 10^8 \times 20 \times 10^{-12} \times 10^4 = 20$$

therefore, $R_{new} = 10^4 = 10 \text{ k}\Omega$

$$L_{new} = 5 \times 10^{-6} \, H = 5 \, \mu \, \text{H}$$

$$C_{new} = 20 \times 10^{-12} = 20 \text{pF}$$

$$Q_{new} = 20 = Q_{old}$$

EXAMPLE 18.24

(*a*) Obtain the system function $H(s) = \dfrac{V_c(s)}{I(s)}$ in terms of R, L and C for the network of Fig.18.34a.

If $R = 2$, $L = 1$ and $C = 1$, simplify $H(s)$

(*b*) If the elements are scaled up with the impedance scale factor of 10^3 and the frequency scale factor of 10^4, determine the new elements and the system function $H(s)$.

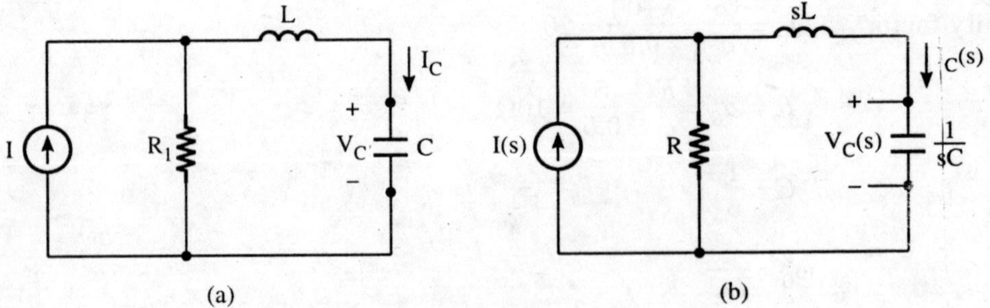

FIG. 18.34

SOLUTION :

(*a*) For the network of Fig. 18.34 *a*, the *s*-domain equivalent circuit is shown is Fig. 18.34b.

By the current - division rule,

current through the capacitor $I_C(s) = \dfrac{R}{sL + \dfrac{1}{sC} + R} I(s)$

Voltage across the capacitor

$$V_C(s) = \frac{1}{sC} \times I_C(s) = \frac{1}{sC} \times \frac{R\,I(s)}{sL + \frac{1}{sC} + R}$$

When $R = 2\,,\; L = 1\,,\; C = 1$

$$H(s) = \frac{2}{s^2 + 2s + 1}$$

(*b*) Here $k_m = 10^3\,,\; k_f = 10^4$

$$R_{new} = k_m\, R_{old} = 10^3 \times 2 = 2000\ \Omega$$

$$L_{new} = \frac{k_m}{k_f} L_{old} = \frac{10^3}{10^4} \times 1 = 0.1\ \text{H}$$

$$C_{new} = \frac{1}{k_m\, k_f} C_{old} = \frac{1}{10^3 \times 10^4} \times 1 = 0.1 \times 10^{-6}\ \text{F} = 0.1\ \mu\text{F}$$

$$H(s) = \frac{R}{s^2\, LC + sRC + 1}$$

$$= \frac{2000}{s^2 \times 0.1 \times 0.1 \times 10^{-6} + s \times 2000 \times 0.1 \times 10^{-6} + 1}$$

$$= \frac{2 \times 10^5}{s^2 + 2 \times 10^4\, s + 10^8}$$

■ **EXAMPLE 18.25** The impedance function of a parallel RLC circuit has poles located at $(-3 \pm j\,4)$ rad/s. If $L = 1$ H , determine R and C and the Q of the circuit.

SOLUTION :

$$\omega_0 = \sqrt{(-3)^2 + 4^2} = 5\ \text{rad/s}$$

$$\omega_0^2 = \frac{1}{LC}$$

$$C = \frac{1}{L\,\omega_0^2} = \frac{1}{1 \times 25} = 0.04\ F$$

Real part

$$-\frac{1}{2\,RC} = -3$$

$$R = \frac{1}{2 \times 3 \times C}$$

$$= \frac{1}{2 \times 3 \times 0.04} = = 4.166\ \Omega$$

$$Q_0 = \omega_0\, CR = 5 \times \frac{1}{6} = 0.833$$

EXERCISES

1. Determine $I(s)/V_i(s)$ for the network shown in Fig. P 18.1.

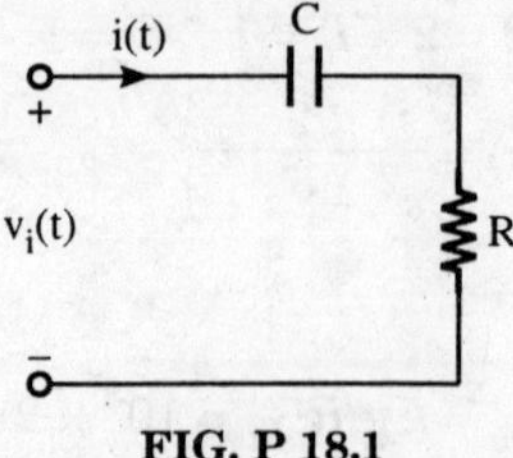

FIG. P 18.1

2. Determine $V_0(s)/I(s)$ for the network shown in Fig. P 18.2.

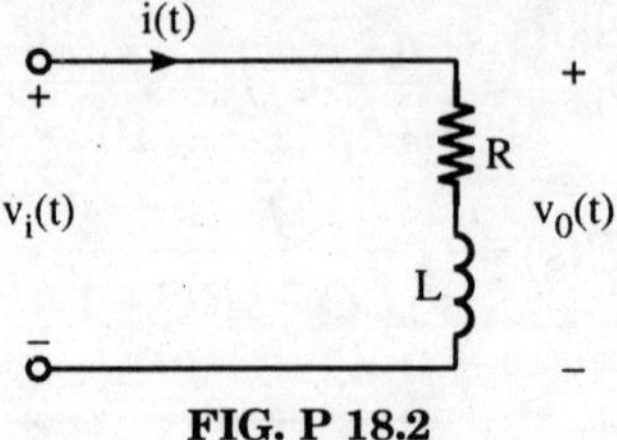

FIG. P 18.2

3. Determine $\dfrac{V_0(s)}{V_i(s)}$ for the network shown in Fig. P 18.3.

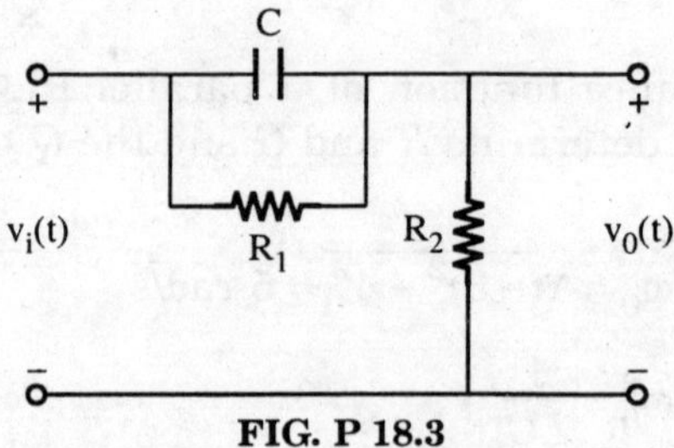

FIG. P 18.3

4. Find $\dfrac{V_0(s)}{I(s)}$ for the network shown in Fig. P 18.4.

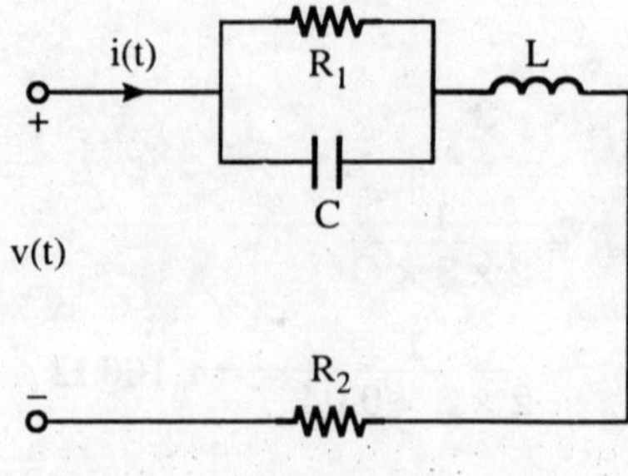

FIG. P 18.4

5. Determine $\dfrac{V_0(s)}{V_i(s)}$ of the RC network shown in Fig. P 18.5

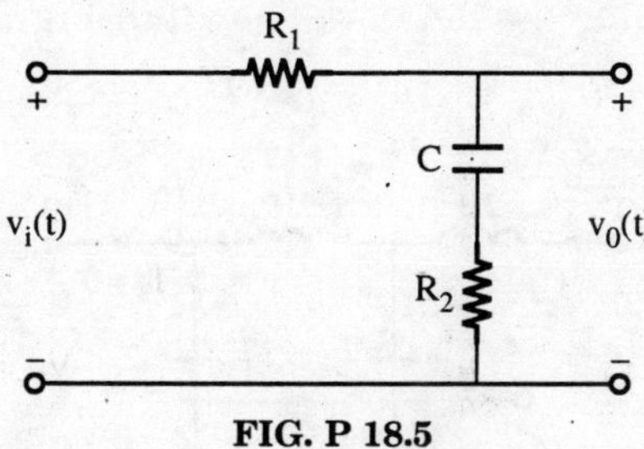

FIG. P 18.5

6. Find the input impedance Z_{in} of the circuit shown in Fig. P 18.6.

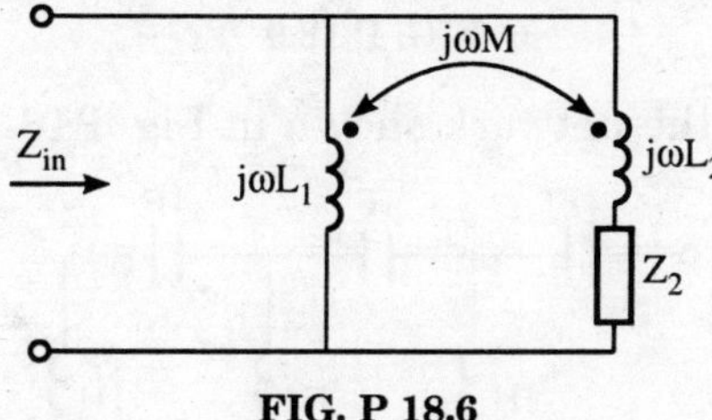

FIG. P 18.6

7. Find the input impedance Z_{in} of the circuit shown in Fig. P 18.7.

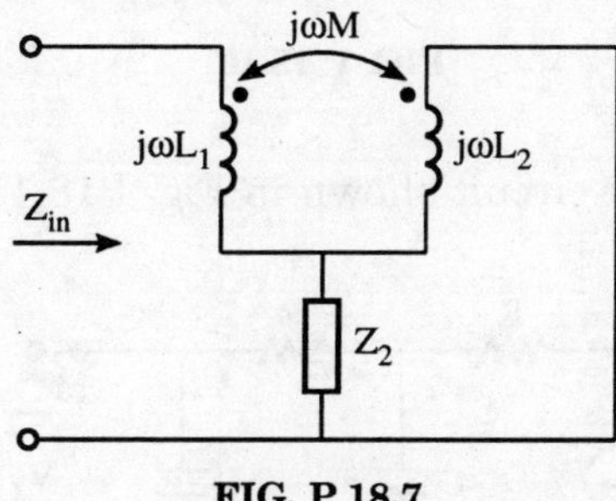

FIG. P 18.7

8. Determine the driving-point impedance of the ladder network shown in Fig. P18.8.

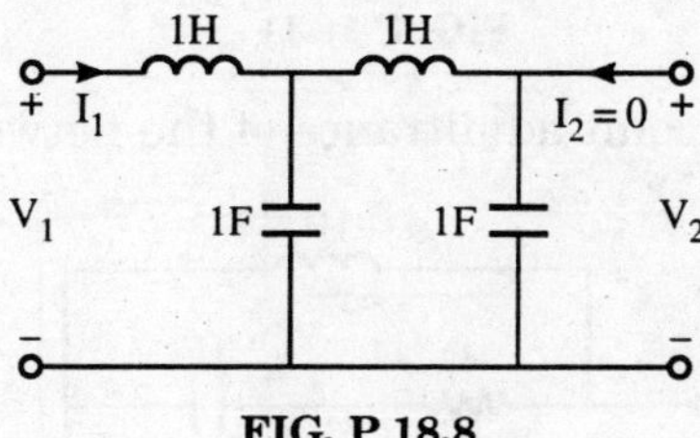

FIG. P 18.8

9. For the ladder network shown in Fig. P18.9, determine (*a*) open-circuit driving-point impedance at port 1, (*b*) open-circuit transfer impedance $\frac{V_2(s)}{I_1(s)}$, (*c*) open-circuit voltage ratio $\frac{V_2(s)}{V_1(s)}$.

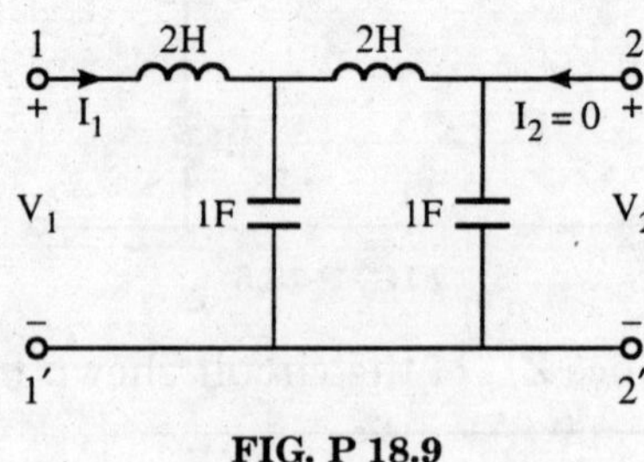

FIG. P 18.9

10. Determine $Y_{11}(s)$ for the network shown in Fig. P18.10.

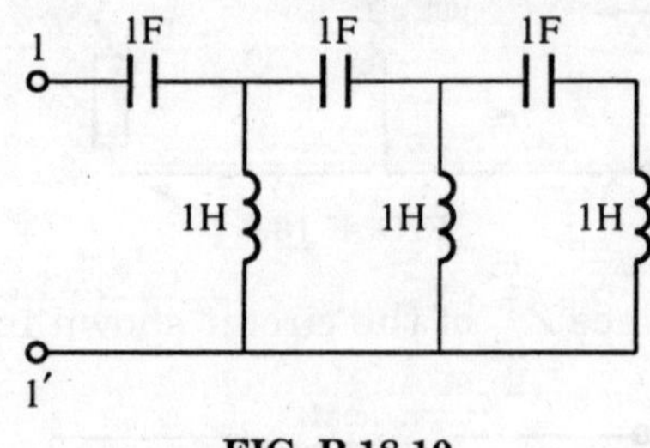

FIG. P 18.10

11. Determine $\frac{V_2(s)}{V_1(s)}$ for the circuit shown in Fig. P18.11.

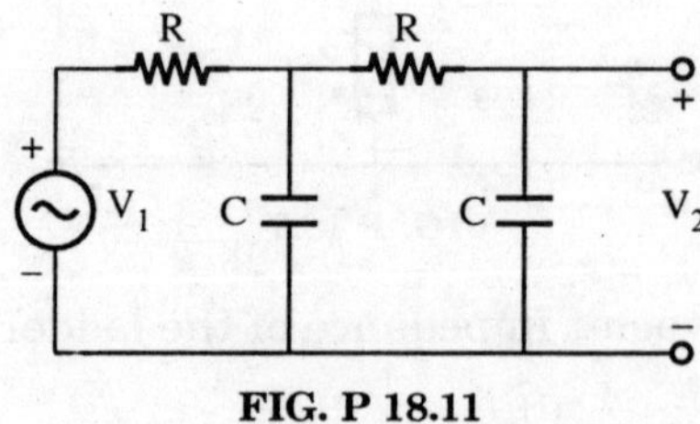

FIG. P 18.11

12. Determine the driving-point admittance of the network shown in Fig. P18.12.

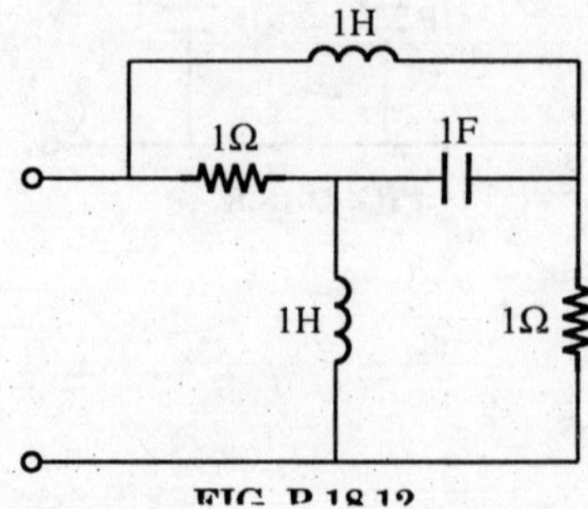

FIG. P 18.12

13. Find the voltage transfer ratio for the network shown in Fig. P18.13.

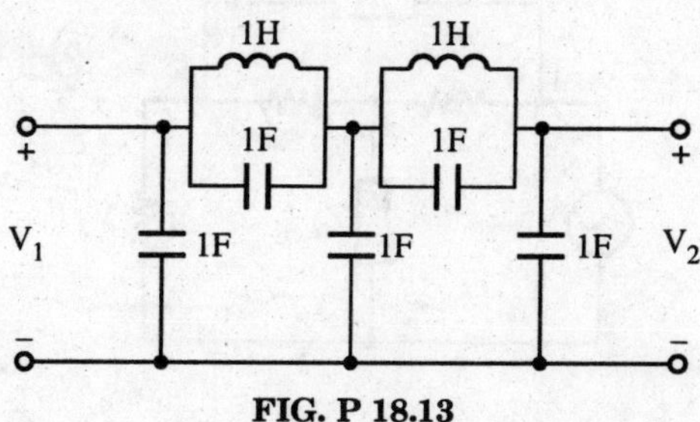

FIG. P 18.13

14. Show that the voltage transfer function of the network shown in Fig. P18.14 is given by

$$\frac{Z_1 Z_2 + Z_2' (Z_1 + Z_1' + Z_2)}{Z_1 Z_2' + (Z_1 + Z_2')(Z_1' + Z_2)}$$

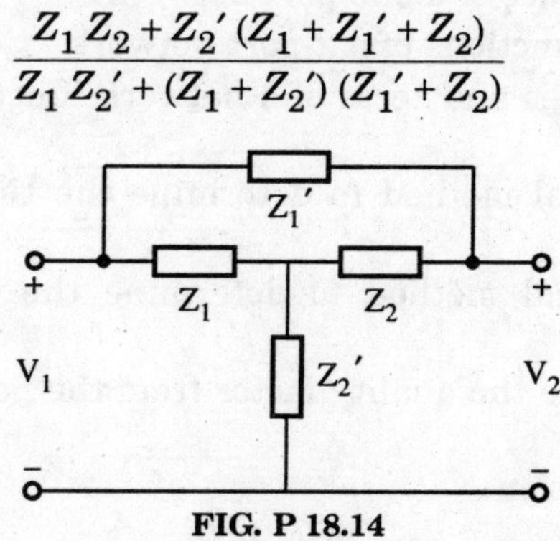

FIG. P 18.14

15. Show that the transfer function $\frac{V_0(s)}{V_i(s)}$ of the bridged-*T* circuit in Fig. P.18.15 is given by $\frac{Z_2}{Z_1 + Z_2}$

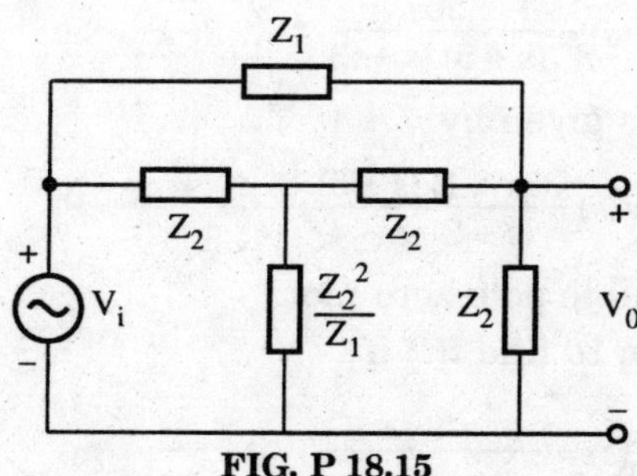

FIG. P 18.15

16. A bridged-*T* circuit is shown in Fig. P18.16. If $Z_1 Z_2 = R_2$, determine (*a*) $\frac{V_2(s)}{V_1(s)}$, (*b*) driving-point impedance at port 1.

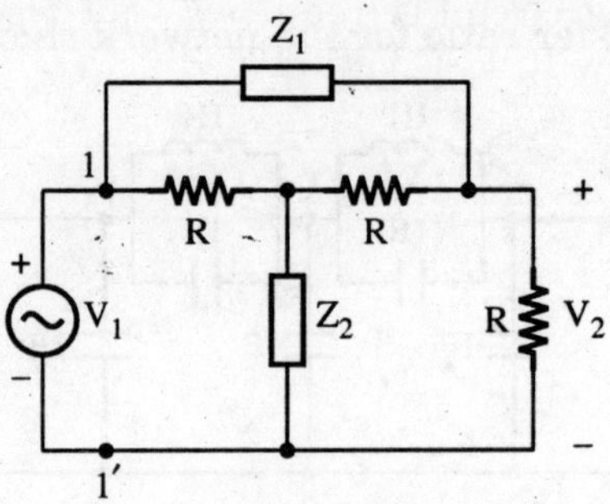

FIG. P 18.16

17. What is a network function ? Define the terms 'driving-point impedance' and driving-point admittance of a one-port network.
18. Define the network functions of a 2-port network.
19. What is meant by poles and zeros of a network function ? What is the significance of poles and zeros ?
20. Describe the graphical method to determine the time-domain response from the pole-zero plot.
21. Describe the graphical method to determine the frequency response from the pole-zero plot.
22. How do you determine the quality factor from the pole position of a given network function.
23. For the network function

$$H(s) = \frac{4\,(s+2)}{s^2 + 2s + 5}$$

graphically determine the magnitude of $H(s)$ for the case that (a) $s = j1$, (b) $s = j2$, and (c) $s = j3$

24. Use pole-zero diagram to find $i(t)$ if

$$I(s) = \frac{20s}{(s+5)\,(s+2)}$$

25. A voltage transform is given by

$$V(s) = \frac{(s+1)\,(s+3)}{(s+2)\,(s+4)}$$

Find $v(t)$ using a suitable pole-zero plot.

26. Use pole-zero diagram to find $i(t)$ if

$$I(s) = \frac{s^2 + 6s + 5}{s\,(s^2 + 4s + 5)}$$

27. For the network function

$$H(s) = \frac{10s}{s^2 + 30s + 10^6}$$

determine the resonant frequency, lower and upper half-power frequencies, the bandwidth, and the quality factor.

28. If a parallel RLC passive circuit is frequency scaled by 50 and then impedance scaled by 50, what are the new R, L, C values ?

ANSWERS

1. $$\frac{I(s)}{V_i(s)} = \frac{sC}{1 + sRC}$$

2. $$\frac{V_0(s)}{I(s)} = R + sL$$

3. $$\frac{V_0(s)}{V_i(s)} = \frac{sC + \frac{1}{R_1}}{sC + \frac{1}{R_1} + \frac{1}{R_2}}$$

4. $$\frac{V(s)}{I(s)} = \frac{L\,s^2 + \left(R_2 + \frac{L}{R_1\,C}\right)s + \frac{R_1 + R_2}{R_1\,C}}{s + \frac{1}{R_1\,C}}$$

5. $$\frac{V_0(s)}{V_i(s)} = \frac{R_2 + \frac{1}{sC}}{R_1 + R_2 + \frac{1}{sC}}$$

6. $$Z_{in} = j\omega L_1 = \frac{\omega^2\,(L_1 - M)^2}{j\omega(L_1 + L_2 - 2M) + Z_2}$$

7. $$Z_{in} = j\omega L_1 + Z_2 - \frac{(j\omega M + Z_2)^2}{j\omega L_2 + Z_2}$$

8. $$Z_{11}(s) = \frac{s^4 + 3s^2 + 1}{s^2 + 2s}$$

9. (a) $Z_{11}(s) = \dfrac{4s^4 + 6s^2 + 1}{2s^3 + 2s}$, (b) $\dfrac{V_2(s)}{I(s)} = \dfrac{1}{2s^3 + 2s}$, (c) $\dfrac{V_2(s)}{V_1(s)} = \dfrac{1}{4s^4 + 6s^2 + 1}$

10. $$Y_{11}(s) = \frac{3s^5 + 4s^3 + s}{s^6 + 6s^4 + 5s^2 + 1}$$

11. $$\frac{V_2(s)}{V_1(s)} = \frac{\left(\frac{1}{RC}\right)^2}{s^2 + \left(\frac{3}{RC}\right)s + \left(\frac{1}{RC}\right)^2}$$

12. $$Y_{11}(s) = \frac{2s^3 + 3s^2 + 2s + 1}{s^3 + 2s^2 + 3s + 2}$$

13. $$\frac{V_2(s)}{V_1(s)} = \frac{(s^2 + 1)^2}{5s^4 + 5s^2 + 1}$$

16. (*a*) $\dfrac{V_2(s)}{V_1(s)} = \dfrac{1}{1+\dfrac{Z}{R}}$, (*b*) $Z_{11}(s) = R$

23. (*a*) $2\ \underline{/\ 0^\circ}$, (*b*) $2.7\ \underline{/ -31^\circ}$, (*c*) $2\ \underline{/ -67.4^\circ}$

24. $i(t) = (-13.33\, e^{-2t} + 33.33\, e^{-5t})$ A

25. $v(t) = u_0(t) - 0.5\, e^{-2t} - 1.5\, e^{-4t}$

26. $i(t) = 1 + 2\, e^{-2t} \sin t$

27. $\omega_1 = 985.1$ rad/s, $\omega_2 = 1015.1$ rad/s
    $\omega_0 = 1000$ rad/s, $BW = 30$ rad/s, $Q_0 = 33.33$

28. $50\, R, L, \dfrac{C}{2500}$

19

FOURIER SERIES

19.1 PERIODIC FUNCTIONS

A periodic function is that which repeats itself after regular intervals of time. A function $f(t)$ is said to be *periodic*

if $$f(t) = f(t + T) \text{ for every } t \tag{19.1.1}$$

where T is called the *period* of the function. It follows that if n is an integer

$$f(t) = f(t \pm nT) \tag{19.1.2}$$

The smallest value of T that satisfies Eq. (19.1.2) is called the *period*. The functions $\sin \omega_0 t$ and $\cos \omega_0 t$ are most commonly used periodic functions. Their period T equals $2\pi/\omega_0$, where ω_0 is the angular frequency.

Fourier's theorem states that a periodic function can be expressed as a sum of sinusoids of appropriate amplitudes and frequencies. Therefore periodic nonsinusoidal inputs can be analysed with sinusoidal phasor techniques.

19.2 FOURIER SERIES

Suppose that the function $f(t)$ satisfies the following conditions :

1. It is a single-valued function.
2. The integral $\int_{t_1}^{t_1 + T} | f(t) | \, dt$ exists (that is, it is finite) for any arbitrary value of t_1.
3. If $f(t)$ is discontinuous there are finite number of discontinuities in any one period.
4. The function $f(t)$ has a finite number of maxima and minima in any one period.

These conditions are called **Dirichlet conditions**. A periodic function $f(t)$ of period T satisfying Dirichlet conditions can be expanded into the following series :

$$f(t) = a_0 + a_1 \cos \omega_0 t + a_2 \cos 2\omega_0 t + \ldots + a_n \cos n\omega_0 t + b_1 \sin \omega_0 t$$
$$+ b_2 \sin 2\omega_0 t + \ldots + b_n \sin n\omega_0 t$$

or

$$f(t) = a_0 + \sum_{n=1}^{\infty} (a_0 \cos n\omega_0 t + b_n \sin n\omega_0 t) \tag{19.2.1}$$

This series is known as **Fourier series.** Eq. (19.2.1) shows that a periodic function consists of sinusoidal components of angular frequencies 0, ω_0, $2\omega_0$, ..., $n\omega_0$ ..., and so. Thus, we see that the Fourier series representation is a description of $f(t)$ in the frequency domain. The component with zero angular frequency is called the **dc component**. The constants a_0, a_n, and b_n are called **Fourier series constants (coefficients)** of $f(t)$. These constants depend upon n and the nature of the waveform of $f(t)$. The sine and cosine terms in Eq. (19.2.1) are called **harmonics**. The number n represents the order of harmonics. Thus, when n = **1**, the component ($a_1 \cos \omega_0 t + b_1 \sin \omega_0 t$) is called the ***first harmonic*** or the **fundamental component** of the waveform. When $n = 2$, the component ($a_2 \cos 2\omega_0 t + b_2 \sin 2\omega_0 t$) is called the ***second harmonic component*** of the waveform. The term ($a_n \cos n\omega_0 t + b_n \sin n\omega_0 t$) is called the ***n*th harmonic component** of the waveform. The constant ω_0 is the radian frequency of the fundamental component. It is called the **fundamental frequency** and it is given by $\omega_0 = 2\pi/T$. All other frequencies are multiples of ω_0. The nth harmonic frequency is $n\omega_0$. Thus nth harmonic completes n cycles is one period T because its frequency is n times the fundamental frequency.

Eq. (19.2.1) is called the **trigonometric form of the Fourier series** for $f(t)$.

We have seen that a preiodic function $f(t)$ can be expanded into a Fourier series provided that $f(t)$ satisfies Dirichlet conditions which are sufficient conditions. These conditions are violated in certain cases. A train of equally spaced impulses is an example.

19.3 ALTERNATE FORMS OF TRIGONOMETRIC FOURIER SERIES

Eq. (19.2.1) can be written as

$$f(t) = a_0 + (a_1 \cos \omega_0 t + b_1 \sin \omega_0 t) + (a_2 \cos 2\omega_0 t + b_2 \sin 2\omega_0 t) + \ldots + (a_n \cos n\omega_0 t + b_n \sin n\omega_0 t)$$

We have

$$a_n \cos n\omega_0 t + b_n \sin n\omega_0 t = \gamma_n \sin (n\omega_0 t + \phi_n)$$

where

$$\gamma_n = \sqrt{(a_n^2 + b_n^2)}\ ,\ \phi_n = \tan^{-1} \frac{a_n}{b_n}$$

Also,

$$a_n \cos n\omega_0 t + b_n \sin n\omega_0 t = \gamma_n \cos (n\omega_0 + \theta_n)$$

where

$$\gamma_n = \sqrt{(a_n^2 + b_n^2)}\ ,\ \theta_n = \tan^{-1} \frac{b_n}{a_n}$$

Hence the trigonometric form of Fourier series can also be expressed in two other forms of sinusoids with phase angles as

$$f(t) = a_0 + \sum_{n=1}^{\infty} \gamma_n \sin (n\omega_0 t + \phi_n) \tag{19.3.1}$$

$$f(t) = a_0 + \sum_{n=1}^{\infty} \gamma_n \cos(n\omega_0 t - \theta_n) \tag{19.3.2}$$

These forms are useful to know the total percentage of any harmonic, since γ_n is the amplitude of the nth harmonic.

19.4 SOME USEFUL INTEGRALS

The following integrals are useful in determining Fourier series coefficients; If m and n are integers

$$\int_{t_1}^{t_1+T} \sin n\omega_0 t \, dt = 0, \quad \int_0^{2\pi} \sin n\theta \, d\theta = 0$$

$$\int_{t_1}^{t_1+T} \cos n\omega_0 t \, dt = 0, \quad \int_0^{2\pi} \cos n\theta \, d\theta = 0$$

$$\int_{t_1}^{t_1+T} \sin m\omega_0 t \cos n\omega_0 t = 0 \quad \text{for all } m \text{ and } n$$

$$\int_0^{2\pi} \sin m\theta \cos n\theta = 0 \quad \text{for all } m \text{ and } n$$

$$\int_{t^1}^{t_1+T} \sin m\omega_0 t \sin n\omega_0 t \, dt = \begin{cases} T/2 & m = n \neq 0 \\ 0 & m \neq n \\ 0 & m \text{ or } n = 0 \end{cases}$$

$$\int_0^{2\pi} \sin m\theta \sin n\theta \, d\theta = 0 \quad m \neq n$$

$$\int_0^{2\pi} \sin^2 n\theta \, d\theta = \pi$$

$$\int_{t_1}^{t_1+T} \cos m\omega_0 t \cos n\omega_0 t \, dt = \begin{cases} T/2, & m = n \neq 0 \\ 0 & m \neq n \\ T & m = n = 0 \end{cases}$$

$$\int_0^{2\pi} \cos^2 n\theta \, d\theta = \pi$$

$$\int_{t_1}^{t+T} e^{jn\omega_0 t} \, dt = \int_{t_1}^{t_1+T} (\cos n\omega_0 t + j \sin n\omega_0 t) \, dt = 0 \text{ for any } t_1 \text{ and for } n \neq 0$$

$$\int_{t_1}^{t_1+T} e^{j(n-m)\omega_0 t} \, dt = \begin{cases} 0 & n \neq m \\ T & n = m \end{cases}$$

These results can be proved easily.

19.5 EVALUATION OF FOURIER CONSTANTS

The process of determining the values of a_0, a_1, ..., a_n and b_1, b_2, ..., b_n is called ***Fourier analysis***.

Evaluation of a_0

Integrating both sides of Eq. (19.2.1) over one period $(t_1, t_1 + T)$ where t_1 is arbitrary. Thus

$$\int_{t_1}^{t_1+T} f(t)\,dt = a_0 \int_{t_1}^{t_1+T} 1\,dt + \int_{t_1}^{t_1+T} \sum_{n=1}^{\infty} (a_n \cos n\omega_0 t + b_n \sin n\omega_0 t)\,dt$$

Interchanging the order of integration and summation in the second term on the right-hand side, we get

$$\int_{t_1}^{t_1+T} f(t)\,dt = a_0 T + \sum_{n=1}^{\infty}\left[a_n \int_{t_1}^{t_1+T} \cos n\omega_0 t\,dt + b_n \int_{t_1}^{t_1+T} \sin n\omega_0 t\,dt\right]$$

But $\displaystyle\int_{t_1}^{t_1+T} \cos n\omega_0 t\;dt = 0,\ \int_{t_1}^{t_1+T} \sin n\omega_0\,dt = 0$

Hence $\displaystyle\int_{t_1}^{t_1+T} f(t)\,dt = a_0 T$

$$a_0 = \frac{1}{T}\int_{t_1}^{t_1+T} f(t)\,dt \tag{19.5.1}$$

or

$$a_0 = \text{average value of } f(t) \text{ over a period} \tag{19.5.2}$$

Evaluation of a_n

To evaluate a_n, the amplitude of $\cos n\omega_0 t$ component of $f(t)$, we multiply Eq. (19.2.1) by $\cos m\omega_0 t$ and integrate over one period $(t_1, t_1 + T)$ for any value of t_1.

$$\int_{t_1}^{t_1+T} f(t) \cos m\omega_0 t\,dt$$

$$= a_0 \int_{t_1}^{t_1+T} \cos m\omega_0 t\,dt + \int_{t_1}^{t_1+T} \sum_{n=1}^{\infty} (a_n \cos n\omega_0 t + b_n \sin n\omega_0 t) \cos m\omega_0 t\,dt \tag{19.5.3}$$

Changing the order of integration and summation on the right-hand side of Eq. (19.5.3), we get

$$\int_{t_1}^{t_1+T} f(t) \cos m\omega_0 t\,dt = a_0 + \int_{t_1}^{t_1+T} \cos m\omega_0\,dt$$

$$+ \sum_{n=1}^{\infty} a_n \int_{t_1}^{t_1+T} \cos m\omega_0 t \cos n\omega_0 t\,dt + \sum_{n=1}^{\infty} b_n \int_{t_1}^{t_1+T} \cos m\omega_0 t \sin n\omega_0 t\,dt \tag{19.5.4}$$

But $$\int_{t_1}^{t_1+T} \cos m\omega_0 t \, dt = 0$$

$$\int_{t_1}^{t_1+T} \cos m\omega_0 t \cos n\omega_0 t \, dt = \begin{cases} 0, & m \neq n \\ T/2, & m = n \end{cases}$$

$$\int_{t_1}^{t_1+T} \cos m\omega_0 t \sin n\omega_0 \, dt = 0$$

Therefore, every term on right-hand side of Eq. (19.5.3), except

$$a_n \int_{t_1}^{t_1+T} (\cos m\omega_0 t)(\cos n\omega_0 t) \, dt$$

(when $m = n$), is zero. Hence

$$\int_{t_1}^{t_1+T} f(t) \cos m\omega_0 t \, dt = a_n \int_{t_1}^{t_1+T} \cos n\omega_0 t \cos n\omega_0 t \, dt = 0$$

$$= a_n \frac{T}{2} \quad \text{(only when } m = n\text{)} \tag{19.5.5}$$

Putting $m = n$ on left-hand side of Eq. (19.5.5) gives

$$\boxed{a_n = \frac{2}{T} \int_{t_1}^{t_1+T} f(t) \cos n\omega_0 t \, dt, \qquad n = 1, 2, 3, \ldots} \tag{19.5.6}$$

Evaluation of b_n

To evaluate b_n, the amplitude of $\sin n\omega_0 t$ component of $f(t)$, we multiply Eq. (19.2.1) by $\sin m\omega_0 t$ and integrate over one interval $(t_1, t_1 + T)$ for any value of t_1.

$$\int_{t_1}^{t_1+T} f(t) \sin m\omega_0 t dt$$

$$= a_0 \int_{t_1}^{t_1+T} \sin m\omega_0 t \, dt + \int_{t_1}^{t_1+T} \sum_{n=1}^{\infty} (a_n \cos n\omega_0 t + b_n \sin n\omega_0 t) \sin m\omega_0 t \, dt \tag{19.5.7}$$

Changing the order of intergration and summation on the right-hand side of Eq. (19.5.7), we get,

$$\int_{t_1}^{t_1+T} f(t) \sin m\omega_0 t \, dt = a_0 \int_{t_1}^{t_1+T} \sin m\omega_0 t \, dt$$

$$+ \sum_{n=1}^{\infty} a_n \int_{t_1}^{t_1+T} \sin m\omega_0 t \cos n\omega_0 t \, dt$$

$$+\sum_{n=1}^{\infty} b_n \int_{t_1}^{t_1+T} \sin m\omega_0 t \sin n\omega_0 t \, dt \tag{19.5.8}$$

But
$$\int_{t_1}^{t_1+T} \sin m\omega_0 t \, dt = 0$$

$$\int_{t_1}^{t_1+T} \sin m\omega_0 t \sin n\omega_0 t \, dt = \begin{cases} 0, & m \neq n \\ T/2, & m = n \end{cases}$$

$$\int_{t_1}^{t_1+T} \sin m\omega_0 t \cos n\omega_0 t \, dt = 0$$

Therefore, every term on right-hand side of Eq. (19.5.7) except

$b_n \int_{t_1}^{t_1+T} \sin m\omega_0 t \sin n\omega_0 t \, dt$ (where $m = n$) is zero. Hence

$$\int_{t_1}^{t_1+T} f(t) \sin m\omega_0 t \, dt = b_n \int_{t_1}^{t_1+T} \sin m\omega_0 t \sin n\omega_0 t \, dt$$

$$= b_n \frac{T}{2}, \quad \text{when } m = n \tag{19.5.9}$$

Putting $m = n$ on left-hand side of Eq. (19.5.9) gives

$$b_n = \frac{2}{T}\int_{t_1}^{t_1+T} f(t) \sin n\omega_0 t \, dt, \qquad n = 1, 2, 3, \ldots \tag{19.5.10}$$

Note that $b_0 = 0$.

In summary, we can write

$$f(t) = a_0 + \sum_{n=1}^{\infty} (a_n \cos n\omega_0 t + b_n \sin n\omega_0 t), \;\; t_1 < t < t_1 + T$$

where
$$a_0 = \frac{1}{T}\int_{t_1}^{t_1+T} f(t) \, dt$$

$$a_n = \frac{2}{T}\int_{t_1}^{t_1+T} f(t) \cos n\omega_0 t \, dt$$

$$b_n = \frac{2}{T}\int_{t_1}^{t_1+T} f(t) \sin n\omega_0 t \, dt$$

19.5.1 Functions of Period 2π

Let $g(\theta)$ be a periodic function, with period 2π.

Thus, $g(\theta) = g(\theta + 2\pi)$

Define the variable

$$\theta \overset{\Delta}{=} \omega_0 t = \frac{2\pi}{T} t$$

and let
$$g(\theta) = f(t)$$

$$f(t) = a_0 + \sum_{n=1}^{\infty} a_n \cos \omega_0 t + \sum_{n=1}^{\infty} b_n \sin n\omega_0 t$$

$$g(\theta) = a_0 + \sum_{n=1}^{\infty} a_n \cos n\theta + \sum_{n=1}^{\infty} b_n \sin n\theta$$

Since
$$\theta = \frac{2\pi}{T} t, \; d\theta = \frac{2\pi}{T} dt, \; dt = \frac{T}{2\pi} d\theta$$

$$a_0 = \frac{1}{T} \int_0^T f(t)\, dt$$

The limits of integration will change from 0 to 2π instead of 0 to T.

Therefore,
$$a_0 = \frac{1}{T} \int_0^T f(t)\, dt = \frac{T}{2\pi} \times \frac{1}{T} \int_0^{2\pi} g(\theta)\, d\theta$$

or
$$a_0 = \frac{1}{2\pi} \int_0^{2\pi} g(\theta)\, d\theta$$

$$a_n = \frac{2}{T} \int_0^T f(t) \cos n\omega_0 t\, dt = \frac{1}{\pi} \int_0^{2\pi} g(\theta) \cos n\theta\, d\theta$$

$$b_n = \frac{1}{\pi} \int_0^{2\pi} g(\theta) \sin n\theta\, d\theta$$

19.6 LIMITS OF INTEGRATION

The general limits of integration for finding the Fourier series coefficients are t_1 and $(t_1 + T)$. The limits might be any values of t that are one fundamental cycle apart. Since t_1 is arbitrary it may be chosen according to convenience. Usually the limits of integration are chosen as follows :

$$0 \text{ to } T, \; -\frac{T}{2} \text{ to } \frac{T}{2}, \; -\frac{T}{4} \text{ to } \frac{3T}{4}$$

Thus, we may write

$$a_0 = \frac{1}{T} \int_{t_1}^{t_1 + T} f(t)\, dt = \frac{1}{T} \int_0^T dt = \frac{1}{T} \int_{-T/2}^{T/2} f(t)\, dt \tag{19.6.1}$$

$$a_n = \frac{2}{T}\int_{t_1}^{t_1+T} f(t)\cos n\omega_0 t\,dt = \frac{2}{T}\int_0^T f(t)\cos n\omega_0 t\,dt$$

$$= \frac{2}{T}\int_{-T/2}^{T/2} f(t)\cos n\omega_0 t\,dt \qquad (19.6.2)$$

$$b_n = \frac{2}{T}\int_{t_1}^{t_1+T} f(t)\sin n\omega_0 t\,dt = \frac{2}{T}\int_0^T f(t)\sin n\omega_0 t\,dt$$

$$= \frac{2}{T}\int_{-T/2}^{T/2} f(t)\sin n\omega_0 t\,dt \qquad (19.6.3)$$

19.7 EVEN AND ODD FUNCTIONS

A function $f(t)$ satisfying the condition $f(t) = f(-t)$ is said to be an *even function*. For example, the functions $\cos n\omega_0 t$. $|\sin t|$, t^n (for n even) are even functions.

A function $f(t)$ satisfying the condition $f(t) = -f(-t)$ is said to be an *odd function*. Examples of odd function are $\sin n\omega_0 t$, t^n (for n odd).

Let the symbols e and 0 denote even and odd.

If $f(t)$ is an even function, by definition

$$f_e(t) = f_e(-t) \qquad (19.7.1)$$

If $f(t)$ is an odd function, by definition

$$f_0(t) = -f_0(-t) \qquad (19.7.2)$$

Any function $f(t)$ can be written in terms of its even and odd parts such that

$$f(t) = f_e(t) + f_0(t) \qquad (19.7.3)$$

In Eq. (19.7.3), if we replace t by $-t$ we get,

$$f(-t) = f_e(-t) + f_0(-t)$$

or

$$f(-t) = f_e(t) - f_0(t) \qquad (19.7.4)$$

Solving Eqs. (19.7.3) and (19.7.4), we get

$$f_e(t) = \frac{1}{2}[f(t) + f(-t)] \qquad (19.7.5)$$

$$f_0(t) = \frac{1}{2}[f(t) - f(-t)] \qquad (19.7.6)$$

Thus, if $f(t)$ is given, we can always find its even and odd parts. Even and odd functions possess the following properties :

Sum of even functions = even function

Sum of odd functions = odd function

Even function × even function = even function

Odd function × odd function = even function

Even function × odd function = odd function

It is to be noted that if an even function is added to an odd function the result is neither even nor odd. Such a result is said to have an even and an odd part as given by Eq. (19.7.3).

For an even function $f_e(t)$,

$$\int_{-t_1}^{t_1} f_e(t)\,dt = 2\int_0^{t_1} f_e(t)\,dt \tag{19.7.7}$$

For an odd function $f_0(t)$,

$$\int_{-t_1}^{t_1} f_0(t)\,dt = 0 \tag{19.7.8}$$

In the above relations t_1 is arbitrary time. These results can be applied to evaluate the Fourier coefficients.

From Eq. (19.6.2)

$$a_n = \frac{2}{T}\int_{-T/2}^{T/2} f(t)\cos n\omega_0 t\,dt$$

But

$$f(t) = f_e(t) + f_0(t)$$

Therefore,

$$a_n = \frac{2}{T}\left[\int_{-T/2}^{T/2} f_e(t)\cos n\omega_0 t\,dt + \int_{-T/2}^{T/2} f_0(t)\cos n\omega_0 t\,dt\right] \tag{19.7.9}$$

Similarly, from Eq. (19.6.3)

$$b_n = \frac{2}{T}\left[\int_{-T/2}^{T/2} f_e(t)\sin n\omega_0 t\,dt + \int_{-T/2}^{T/2} f_0(t)\sin n\omega_0 t\,dt\right] \tag{19.7.10}$$

19.8 WAVEFORM SYMMETRIES

We have seen that the procedure of evaluating the Fourier coefficients is straight forward but the number of steps in evaluating them can be reduced by utilizing the symmetry properties of periodic waveforms. Waves with certain types of symmetry have some terms missing from their Fourier series. Usually, the symmetry is obvious by inspection. It can also be tested mathematically. The symmetry in waveforms can be classified as follows :

1. Even function symmetry.
2. Odd function symmetry.
3. Half-wave or mirror symmetry.
4. Quarter-wave symmetry

Even symmetry

A function $f(t)$ is said to possess even symmetry when

$$f(t) = f(-t)$$

In such a case the function is called *even function*. Geometrically, an even function is symmetrical not only with respect to the vertical axis passing through the origin $t = 0$ but also with respect to all vertical lines at $n\frac{T}{2}$ $(n = \pm 1, \pm 2, \ldots)$, since all periodic functions with period T also satisfy the condition $f(t) = f(t + T)$. Some of the waveforms possessing even symmetry are shown in Fig. 19.1.

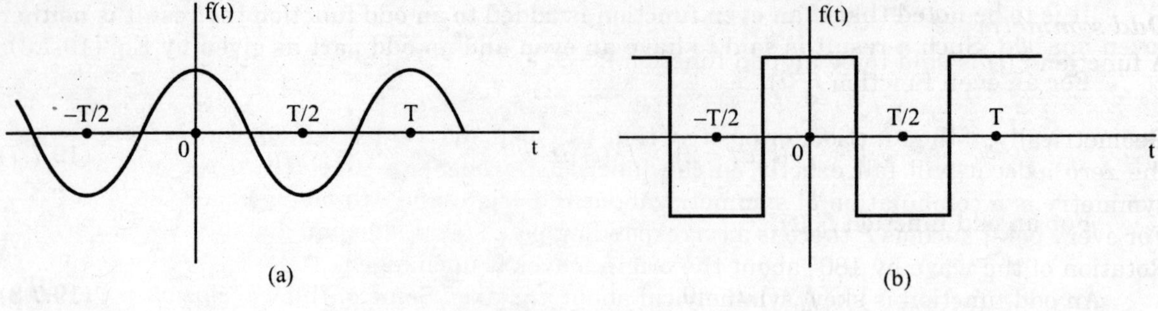

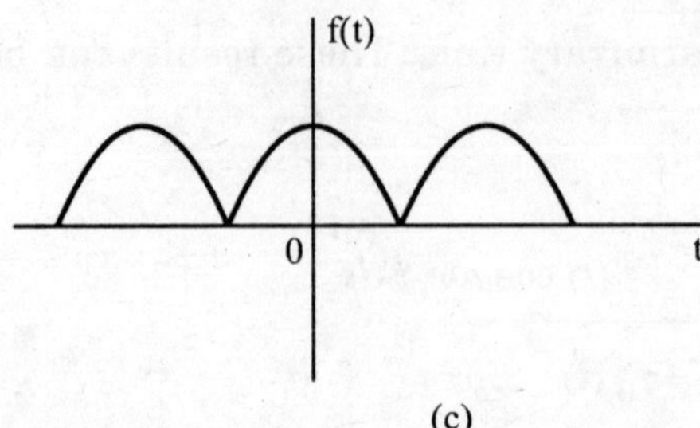

FIG. 19.1 Even waveforms possessing symmetry

For even functions, $f_0(t) = 0$ and therefore from Eq. (19.7.9).

$$a_n = \frac{2}{T}\int_{-T/2}^{T/2} f_e(t)\cos n\omega_0 t\, dt \tag{19.8.1}$$

Since $f_e(t)$ and $\cos n\omega_0 t$ are even, the product $f_e(t)\cos n\omega_0 t$ is also even. Therefore applying the condition given in Eq. (19.7.7) to Eq. (19.8.1), we obtain

$$a_n = \frac{4}{T}\int_{0}^{T/2} f_e(t)\cos n\omega_0 t\, dt \tag{19.8.2}$$

Similarly, from Eq. (19.7.10)

$$b_n = \frac{2}{T}\int_{-T/2}^{T/2} f_e(t)\sin n\omega_0 t\, dt$$

Since the product $f_e(t)\sin n\omega_0 t$ is odd, the application of property of even functions given in Eq. (19.7.8) yields.

$$b_n = 0 \tag{19.8.3}$$

Also, $$a_0 = \frac{1}{T}\int_{-T/2}^{T/2} f_e(t)\, dt$$

or $$a_0 = \frac{2}{T}\int_{0}^{T/2} f_e(t)\, dt \tag{19.8.4}$$

Equation (19.8.3) shows that the coefficients of all sine terms are zero. Thus, the Fourier series of an even function contains only cosine terms and the constant term (if $a_0 \neq 0$). It is to be noted that for an even function, we need only integrate over *half period*, that is it is sufficient to integrate from 0 to $\frac{T}{2}$ and then multiply the result by 2 in oder to evaluate a_n.

Odd symmetry

A function $f(t)$ is said to be an odd function if

$$f(t) = -f(-t)$$

Geometrically, if the function for positive time is folded about the origin, and then folded about the zero axis, it will fall exactly on the function for negative time. Thus, the odd function symmetry is a combination of symmetry about the origin and symmetry about the zero axis. For every point such as P there is a corresponding point P' at an equal distance beyond origin. Rotation of the wave by 180° about the origin leaves it unchanged.

An odd function is skew symmetrical about $f(t)$ axis. Some of the waveforms possessing odd symmetry are shown in Fig. 19.2.

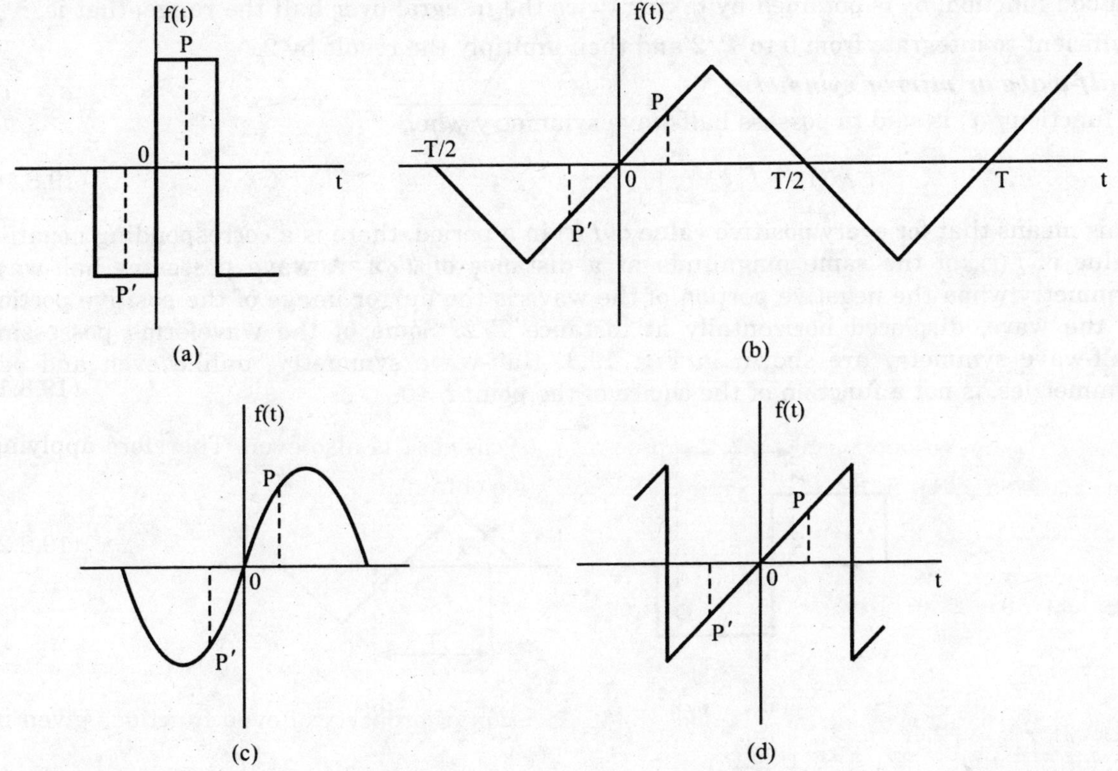

FIG. 19.2 Waveforms possessing odd symmetry

For odd functions, $f_e(t) = 0$ and therefore from Eq. (19.7.9)

$$a_n = \frac{2}{T}\int_{-T/2}^{T/2} f_0(t) \cos n\omega_0 t \, dt \tag{19.8.5}$$

Since $f_0(t)$ is odd and $\cos n\omega_0 t$ is even, the product $f_0(t) \cos n\omega_0 t$ is odd. Therefore applying the condition given for odd function in Eq. (19.7.8) to Eq. (19.8.5), we obtain

$$a_n = 0 \tag{19.8.6}$$

Similarly, from Eq. (19.7.10)

$$b_n = \frac{2}{T}\int_{-T/2}^{T/2} f_0\,(t)\,\sin n\omega_0 t\,dt \tag{19.8.7}$$

The product $f_0\,(t)\,\sin n\omega_0 t$ is even. Applying the condition given for even function in Eq. (19.7.7) to Eq. (19.8.7) yields

$$b_n = \frac{4}{T}\int_{0}^{T/2} f_0\,(t)\,\sin n\omega_0 t\,dt \tag{19.8.8}$$

Also,

$$a_0 = \frac{1}{T}\int_{-T/2}^{T/2} f_0\,(t)\,dt = 0 \tag{19.8.9}$$

Thus, the Fourier series of an odd function contains only sine terms. It is to be noted that for an odd function, b_n is obtained by taking twice the integral over half the range, that is, it is sufficient to integrate from 0 to $T/2$ and then multiply the result by 2.

Half-wave or mirror symmetry

A function $f\,(t)$ is said to possess half-wave symmetry when

$$f\,(t) = -f\left(t + \frac{T}{2}\right) \tag{19.8.10}$$

This means that for every positive value of $f\,(t)$ in a period, there is a corresponding negative value of $f\,(t)$, of the same magnitude at a distance of $T/2$. A wave possesses half-wave symmetry when the negative portion of the wave is the mirror image of the positive portion of the wave, displaced horizontally at distance $T/2$. Some of the waveforms possessing half-wave symmetry are shown in Fig. 19.3. Half-wave symmetry, unlike even and odd symmetries, is not a function of the choice of the point $t = 0$.

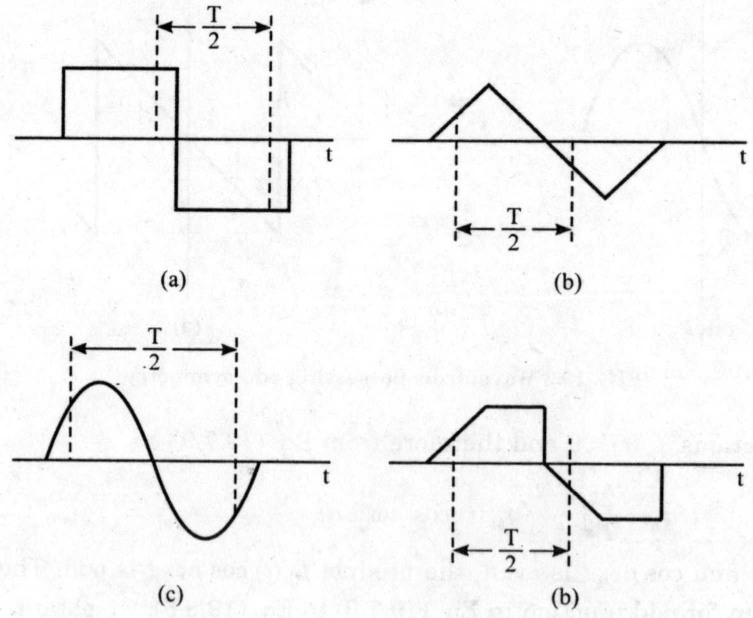

FIG. 19.3 Waveforms possessing half-wave symmetry

We have

$$a_n = \frac{2}{T}\int_{-T/2}^{T/2} f(t)\cos n\omega_0 t\, dt$$

$$= \frac{2}{T}\int_{-T/2}^{0} f(t)\cos n\omega_0 t\, dt + \frac{2}{T}\int_{0}^{T/2} f(t)\cos n\omega_0 t\, dt$$

$$= I_1 + I_2 \quad \text{(say)} \tag{19.8.11}$$

Let us substitute $\left(t + \dfrac{T}{2}\right)$ for t in the first integral of Eq. (19.8.11) so that the new limits of integration are 0 to $T/2$ instead of $-T/2$ to 0. The first integral in Eq. (19.8.11) becomes

$$I_1 = \frac{2}{T}\int_{0}^{T/2} f\left(t + \frac{T}{2}\right)\cos n\omega_0\left(t + \frac{T}{2}\right) d\left(t + \frac{T}{2}\right)$$

$$= \frac{2}{T}\int_{0}^{T/2} f\left(t + \frac{T}{2}\right)\left(\cos n\omega_0 t \cos n\omega_0 \frac{T}{2} - \sin n\omega_0 t \sin n\omega_0 \frac{T}{2}\right) dt$$

But $\qquad \omega_0 T = 2\pi, \; \sin \dfrac{n\omega_0 T}{2} = \sin n\pi = 0$

Also, for half-wave symmetry

$$f\left(t + \frac{T}{2}\right) = -f(t)$$

Therefore,

$$I_1 = \frac{2}{T}\int_{0}^{T/2} -f(t)\cos n\omega_0 t \cos n\pi\, dt$$

$$= -\frac{2\cos n\pi}{T}\int_{0}^{T/2} f(t)\cos n\omega_0 t\, dt$$

and $\qquad a_n = I_1 + I_2$

$$= \frac{2}{T}(1 - \cos n\pi)\int_{0}^{T/2} f(t)\cos n\omega_0 t\, dt \tag{19.8.12}$$

When n is even, $\cos n\pi = 1$ and $a_n = 0$.

When n is odd, $\cos n\pi = -1$ and

$$a_n = \frac{4}{T}\int_{0}^{T/2} f(t)\cos n\omega_0 t\, dt$$

Similarly,

$$b_n = 0, \text{ when } n \text{ is even}$$

$$b_n = \frac{4}{T}\int_{0}^{T/2} f(t)\sin n\omega_0 t, \text{ when } n \text{ is odd.}$$

Thus, *the Fourier series expansion of a periodic function with half-wave symmetry contains only odd harmonics*. A clear distinction should be made between odd functions and odd harmonics. For example, $b_2 \sin 2\omega_0 t$ is an even harmonic but an odd function.

Quarter-wave symmetry

A combination of half-wave symmetry with either even or odd symmetry results in a wave with quarter-wave symmetry. For example, the wave shown in Fig. 19.4 possesses half-wave symmetry as well as odd symmetry. The wave shown in Fig. 19.5 possesses both half-wave symmetry and even symmetry.

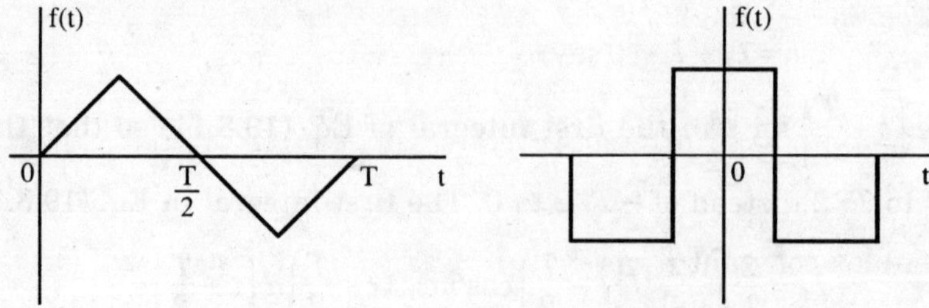

FIG. 19.4 A triangular wave possessing half-wave symmetry and odd symmetry

FIG. 19.5 A square wave possessing half-wave symmetry and even symmetry

The analytical test of quarter-wave symmetry is as follows :

Odd quarter-wave $\quad f(t) = -f\left(t + \dfrac{T}{2}\right)$ and $f(t) = -f(-t)$

Even quarter-wave $\quad f(t) = -f\left(t + \dfrac{T}{2}\right)$ and $f(t) = f(-t)$

Since each quarter cycle is the same in a wave with quarter-wave symmetry, it is sufficient to integrate over one quarter period (that is, from 0 to $T/4$) and then multiply the result by 4. Table 19.1 gives the Fourier coefficients of symmetrical periodic waveforms.

19.9 EQUALITY OF POSITIVE AND NEGATIVE AREAS

If the positive area of a waveform is equal to the negative area in one cycle, the average value of the function represented by the waveform is zero. Since a_0 is equal to the average value of the function over a period, therefore a_0 is equal to zero in such a case.

19.10 EXPONENTIAL FORM OF FOURIER SERIES

The *exponential* or *complex* **Fourier series representation** has two advantages. Firstly, it mathematically shortens the series representation and makes it in convenient and tidy form. Secondly, it generalizes the concept to the Fourier transform, which is useful to deal with nonperiodic functions.

The trigonometrical form of Fourier series is

$$f(t) = a_0 + \sum_{n=1}^{\infty} (a_n \cos n\omega_0 t + b_n \sin n\omega_0 t) \tag{19.10.1}$$

The sine and cosine terms can be expressed in exponential form as

$$\sin n\omega_0 t = \frac{1}{2j}\,(e^{jn\omega_0 t} - e^{-jn\omega_0 t}) \tag{19.10.2}$$

$$\cos n\omega_0 t = \frac{1}{2}\left(e^{jn\omega_0 t} + e^{-jn\omega_0 t}\right) \tag{19.10.3}$$

Substitution of these values in Eq. (19.10.1) gives

$$f(t) = a_0 + \sum_{n=1}^{\infty}\left[\frac{1}{2}\left(a_n\left(e^{jn\omega_0 t} + e^{-jn\omega_0 t}\right) + \frac{1}{2j}b_n\left(e^{jn\omega_0 t} - e^{-jn\omega_0 t}\right)\right]$$

$$= a_0 + \sum_{n=1}^{\infty}\left[\frac{1}{2}(a_n - jb_n)\,e^{jn\omega_0 t} + \frac{1}{2}(a_n + jb_n)\,e^{-jn\omega_0 t}\right] \tag{19.10.4}$$

Let us define a complex constant $\mathbf{c}_n$ as

$$\mathbf{c}_n \triangleq \frac{1}{2}(a_n - jb_n) \tag{19.10.5}$$

We have,

$$a_n = \frac{2}{T}\int_{t_1}^{t_1+T} f(t)\cos n\omega_0 t\,dt \tag{19.10.6}$$

$$b_n = \frac{2}{T}\int_{t_1}^{t_1+T} f(t)\sin n\omega_0 t\,dt \tag{19.10.7}$$

If we put $-n$ for n in Eqs. (10.10.6) and (10.10.7) we find that

$$a_{-n} = a_n,\quad b_{-n} = b_n$$

From Eq. (19.10.5)

$$\mathbf{c}_{-n} = \frac{1}{2}(a_n + jb_n) \tag{19.10.8}$$

Thus, we see that $\mathbf{c}_{-n}$ is the complex conjugate of $\mathbf{c}_n$, that is, $\mathbf{c}_{-n} = \mathbf{c}_n^*$ Substituting the values of a_n and b_n from Eq. (19.10.6) and (19.10.7) in Eq. (19.10.5), we obtain

$$\mathbf{c}_n = \frac{1}{2}\left[\frac{2}{T}\int_{t_1}^{t_1+T} f(t)\cos n\omega_0 t\,dt - j\,\frac{2}{T}\int_{t_1}^{t_1+T} f(t)\sin n\omega_0 t\,dt\right]$$

$$= \frac{1}{T}\int_{t_1}^{t_1+T} f(t)\left(\cos n\omega_0 t - j\sin n\omega_0 t\right)dt$$

$$= \frac{1}{T}\int_{t_1}^{t_1+T} f(t)\,e^{-jn\omega_0 t}\,\mathrm{dt} \tag{19.10.9}$$

If we put $n = 0$ in Eq. (19.10.9)

$$\mathbf{c}_0 = \frac{1}{T}\int_{t_1}^{t_1+T} f(t)\,dt = a_0 \tag{19.10.10}$$

Table 19.1

FOURIER COEFFICIENTS OF SYMMETRICAL PERIODIC WAVEFORMS

Type of symmetry	Conditions	Forms of the Fourier Series	Formulae for the Coefficients
Even function	$f(t) = f(-t)$	$f(t) = \sum_{n=0}^{\infty} a_n \cos n\omega_0 t$	$b_n = 0$; $a_n = \frac{4}{T}\int_0^{T/2} f(t) \cos n\omega_0 t\, dt$; $a_0 = \frac{2}{T}\int_0^{T/2} f(t)\, dt$
Odd function	$f(t) = -f(-t)$	$f(t) = \sum_{n=1}^{\infty} b_n \sin n\omega_0 t$	$a_n = 0$; $b_n = \frac{4}{T}\int_0^{T/2} f(t) \sin n\omega_0 t\, dt$
Half Wave	$f(t) = -f\left(t + \frac{T}{2}\right)$	$f(t) = \sum_{n=1}^{\infty} [a_{2n-1} \cos (2n-1)\omega_0 t + b_{2n-1} \sin (2n-1)\omega_0 t]$	$a_{2n-1} = \frac{4}{T}\int_0^{T/2} f(t) \cos (2n-1)\omega_0 t\, dt$ $b_{2n-1} = \frac{4}{T}\int_0^{T/2} f(t) \sin (2n-1)\omega_0 t\, dt$
Even Quarter Wave	$f(t) = f(-t)$ and $f(t) = -f\left(t + \frac{T}{2}\right)$	$f(t) = \sum_{n=1}^{\infty} a_{2n-1} \cos (2n-1)\omega_0 t$	$b_n = 0$; $a_{2n-1} = \frac{8}{T}\int_0^{T/4} f(t) \cos(2n-1)\omega_0 t\, dt$
Odd Quarter Wave	$f(t) = -f(-t)$ and $f(t) = -f\left(t + \frac{T}{2}\right)$	$f(t) = \sum_{n=1}^{\infty} b_{2n-1} \sin (2n-1)\omega_0 t$	$a_n = 0$; $b_{2n-1} = \frac{8}{T}\int_0^{T/4} f(t) \sin (2n-1)\omega_0 t\, dt$

Eq. (19.10.4) can now be written as

$$f(t) = \mathbf{c}_0 + \sum_{n=1}^{\infty} \mathbf{c}\, e^{jn\omega_0 t} + \sum_{n=1}^{\infty} \mathbf{c}_{-n}\, e^{-jn\omega_0 t}$$

$$= \sum_{n=0}^{\infty} \mathbf{c}_n\, e^{jn\omega_0 t} + \sum_{n=1}^{\infty} \mathbf{c}_{-n}\, e^{-jn\omega_0 t}$$

$$= \sum_{n=0}^{\infty} \mathbf{c}_n\, e^{jn\omega_0 t} + \sum_{n=-1}^{-\infty} \mathbf{c}_n\, e^{jn\omega_0 t}$$

Since the summation from -1 to $-\infty$ is the same as the summation from $-\infty$ to -1.

$$f(t) = \sum_{n=0}^{\infty} \mathbf{c}_n\, e^{jn\omega_0 t} + \sum_{-\infty}^{-1} \mathbf{c}_n\, e^{jn\omega_0 t}$$

$$f(t) = \sum_{n=-\infty}^{\infty} \mathbf{c}_n\, e^{jn\omega_0 t} \tag{19.10.11}$$

Thus n takes all integral values from $-\infty$ to ∞, including zero and where $\mathbf{c}_n$ is given by Eq. (19.10.9) that is,

$$\mathbf{c}_n = \frac{1}{T}\int_{t_1}^{t_1+T} f(t)\, e^{-jn\omega_0 t}\, dt \tag{19.10.12}$$

The series expressed by Eqs. (19.10.11) and (19.10.12) is called **exponential Fourier series.** This form is more compact than the trigonometric form. In the exponential form only one integration is required to find the Fourier coefficients. Exponential form of Fourier series also provides a convenient transition to Fourier integrals and Fourier transforms.

For the sake of convenience we shall define the following notation :

$$\mathbf{c}_n = \mathbf{F}(jn\omega_0) \overset{\Delta}{=} \frac{1}{2}(a_n - jb_n) \qquad n = 0, \pm 1, \pm, \ldots \tag{19.10.13}$$

The significance of this notation will be discussed shortly. With this notation Eqs. (19.10.11) and (19.10.12) can be written as

$$f(t) = \sum_{n=-\infty}^{\infty} \mathbf{F}(j\omega_0)\, e^{+jn\omega_0 t} \tag{19.10.14}$$

$$\mathbf{F}(jn\omega_0) = \frac{1}{T}\int_{t_1}^{t_1+T} f(t)\, e^{-jn\omega_0 t}\, dt \qquad \text{for any value of } t_1 \tag{19.10.15}$$

Eq. (19.10.15) shows that if we multiply a periodic function $f(t)$ by $e^{-jn\omega_0 t}$ and then integrate over one complete period. $(t_1, t_1 + T)$, we obtain a frequency representation $\mathbf{F}(jn\omega_0)$ of the function $f(t)$. This is known **direct Fourier series transform**.

Eq. (19.10.14) is called the ***inverse Fourier series transform.*** This transform is from the frequency domain to the time domain. Equation (19.10.15) is often called the **discrete Fourier transform** of $\boldsymbol{f(t)}$ and Eq. (19.10.14) is called the **inverse discrete transform of $\boldsymbol{F(jn\omega_0)}$**.

In Eq. (19.10.15), t_1 can have any value according to convenience. For $t_1 = 0$, the limits of integration are $(0, T)$. For $t_1 = -T/2$ the limits of integration are $(-T/2, T/2)$.

Real Functions

If $f(t)$ is a real function, then a_n and b_n are real. Since $\mathbf{c}_n$ is complex, it can be written as

$$\mathbf{c}_n = Re\,\mathbf{c}_n + j\,Im\,\mathbf{c}_n \tag{19.10.16}$$

where Re and Im denote real and imaginary respectively.

Also,
$$\mathbf{c}_n = \frac{1}{2}(a_n - jb_n)$$

Therefore
$$a_n = 2Re\,c_n,\; b_n = -2Im\,\mathbf{c}_n \qquad n > 0 \tag{19.10.17}$$

If $\mathbf{c}_n$ is real then $b_n = 0$ and the Fourier series will contain only cosine terms. If $\mathbf{c}_n$ is purely imaginary, then $a_n = 0$ and the series contains only sine terms.

19.11 FOURIER SERIES OF A SHIFTED FUNCTION (TIME-DISPLACEMENT THEOREM)

If the Fourier series expansion of a function $f(t)$ is known, the expansion of the shifted function $f(t - t_0)$ can be determined by replacing in the series the variable t by $(t - t_0)$,

That is,
$$f(t) = \sum_{n=-\infty}^{\infty} c_n\, e^{jn\omega_0 t}$$

becomes
$$f(t - t_0) = \sum_{n=-\infty}^{\infty} c_n\, e^{jn\omega_0 (t - t_0)}$$

$$= \sum_{n=-\infty}^{\infty} (c_n\, e^{jn\omega_0 t})\, e^{jn\omega_0 t_0}$$

Thus $(\mathbf{c}_n)_{\text{delayed function}} = (\mathbf{c}_n)_{\text{original function}} \times e^{-jn\omega_0 t_0}$

The factor $e^{-jn\omega_0 t_0}$ represents a phase shift of $(-n\omega_0 t_0)$ radians. Hence it is seen that a time shift (lag) of t_0 seconds produces no change in magnitude of the complex coefficients, but produces a phase shift of $(-n\omega_0 t_0)$ radians in those coefficients. This is called **time-displacement theorem**. If $f(t)$ has the Fourier series

$$f(t) = a_0 + \sum_{n=1}^{\infty} \gamma_n \cos(n\omega_0 t - \theta_n)$$

then
$$f(t - t_0) = a_0 + \sum_{n=1}^{\infty} \gamma_n \cos[n\omega_0 (t - t_0) - \theta_n]$$

Sometimes application of shifting theorem simplifies the evaluation of Fourier series coefficients of $f(t)$ by using known expansions.

19.12 SIGNIFICANCE OF NEGATIVE FREQUENCIES AND COMPLEX-VALUED FUNCTIONS

In the exponential Fourier series a periodic function is expressed as a sum of the complex exponential functions of frequencies 0, $\pm j\omega_0$, $\pm 2j\omega_0$, ..., etc. That is, a periodic function contains all frequencies (both positive and negative). The presence of negative frequencies and the use of complex-valued basis functions, namely, $e^{jn\omega_0 t}$ have no physical significance. The use of negative frequencies and complex valued basis functions is merely to provide a compact mathematical model of a periodic function which is most suitable for both theoretical and practical purposes.

19.13 DISCRETE FOURIER SPECTRUM

The representation of a periodic function by a Fourier series is equivalent to the resolution of the function in terms of its components of various frequencies. Eq. (19.10.11) gives the exponential Fourier series for a periodic function $f(t)$ of period T. This function contains components of frequencies 0, $\pm\,\omega_0$, $\pm\,2\omega_0$, ..., called the ***spectrum.*** This spectrum is not continuous but exists only at some discrete frequencies that are integer (both positive and negative) multiples of the fundamental freqnency ω_0. It is therefore called a ***discrete spectrum.*** Sometimes it is also known as a ***line spectrum.*** If we specify the periodic function $f(t)$, we can determine its spectrum, Conversely if we specify the spectrum, we can determine the corresponding periodic function $f(t)$. Thus, a periodic function $f(t)$ can be specified in two equivalent ways : (1) the time-domain representation, where $f(t)$ is expressed as a function of time, and (2) the frequency-domain representation, where the spectrum (that is, the amplitudes of various exponential components) is specified.

In general, $\mathbf{c}_n$ is complex it can be described by magnitude and phase. It can be written as

$$\mathbf{c}_n = |\,\mathbf{c}_n\,|\, e^{j\phi_n}$$

where ϕ_n is the phase of $\mathbf{c}_n$.

Similarly, $\qquad \mathbf{c}_{-n} = |\,\mathbf{c}_n\,|\, e^{-j\phi_n}$

Thus, the phase of $\qquad \mathbf{c}_{-n}$ is $-\,\phi_n$.

Also, $\qquad \mathbf{c}_{-n} = \mathbf{c}_n^*$

Therefore $\qquad |\,\mathbf{c}_n\,| = |\,\mathbf{c}_{-n}\,|$

The quantity $|\,\mathbf{c}_n\,|$ defines the nth harmonic component of the periodic function $f(t)$. The plot of $|\,\mathbf{c}_n\,|$ versus frequency is known as the **discrete amplitude spectrum of $f(t)$**. The plot of ϕ_n versus frequency is called the **discrete phase spectrum of $f(t)$.** The relation $|\,\mathbf{c}_n\,| = |\,\mathbf{c}_{-n}\,|$ shows that the amplitude spectrum is symmetrical (an even function of n) about the vertical axis passing through the origin. That is, the amplitude spectrum is a

two-sided spectrum. Since the phase of $\mathbf{c}_n$ is ϕ_n and the phase of $\mathbf{c}_{-n}$ is $-\phi_n$, therefore, the phase spectrum is antisymmetrical (an odd function of n) about the vertical passing through the origin.

The amplitude and phase spectra provide a convenient graphical description of the harmonic contents of a function $f(t)$. The amplitude spectrum also enables us to determine the number of terms sufficient to approximate $f(t)$.

■ **EXAMPLE 19.1** (*a*) Determine the Fourier series for the square wave voltage shown in Fig. 19.6a.

(*b*) Hence or otherwise, determine the Fourier series for the wave shown in Fig. 19.6b.

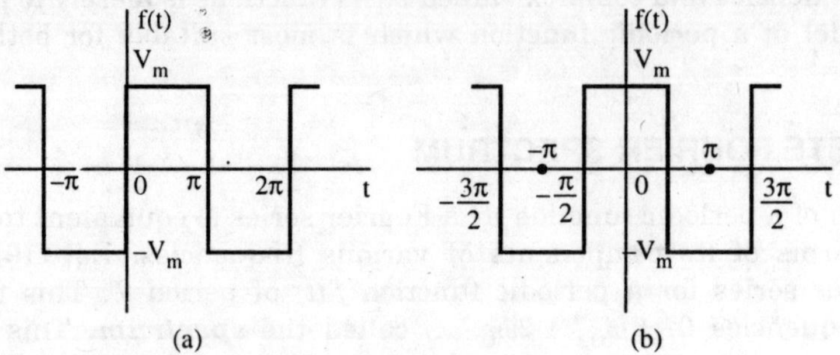

FIG. 19.6

SOLUTION : The wave may be defined over one period by

$$v = \begin{cases} V_m & 0 < t < \pi \\ -V_m & \pi < t < 2\pi \end{cases}$$

Here $T = 2\pi$, therefore $\omega_0 = \dfrac{2\pi}{T} = 1$

Determination of a_0

Since the positive half-cycle area is equal to the negative half-cycle area, a_0 is equal to zero.

Determination of a_n

$$a_n = \frac{1}{\pi}\int_0^{2\pi} f(t)\cos nt\,dt$$

$$= \frac{1}{\pi}\int_0^{\pi} f(t)\cos nt\,dt + \frac{1}{\pi}\int_{\pi}^{2\pi} f(t)\cos nt\,dt$$

$$= \frac{1}{\pi}\int_0^{\pi} V_m \cos nt\,dt + \frac{1}{\pi}\int_{\pi}^{2\pi} (-V_m)\cos nt\,dt$$

$$\frac{V_m}{\pi}\left[\frac{-\cos nt}{n}\right]_0^{\pi} + \frac{V_m}{\pi}\left[\frac{\cos nt}{n}\right]_{\pi}^{2\pi} = \frac{V_m}{\pi}\left[\frac{\sin nt}{n}\right]_0^{\pi} - \frac{V_m}{\pi}\left[\frac{\sin nt}{\pi}\right]_{\pi}^{2\pi}$$

$$= \frac{V_m}{n\pi}(\sin n\pi - \sin 0^\circ) - \frac{V_m}{n\pi}(\sin 2n\pi - \sin n\pi)$$

Since $\sin 0 = 0$, $\sin n\pi = 0$, $\sin 2n\pi = 0$, therefore

$$a_n = 0 \quad \text{for} \quad n = 1, 2, 3, \ldots$$

This shows that the Fourier series does not contain cosine terms.

Alternative method

Since the waveform possesses odd symmetry $a_n = 0$.

Determination of b_n

$$b_n = \frac{1}{\pi}\int_0^{2\pi} f(t) \sin nt \, dt$$

$$= \frac{1}{\pi}\int_0^{\pi} f(t) \sin nt \, dt + \frac{1}{\pi}\int_{\pi}^{2\pi} f(t) \sin nt \, dt$$

$$= \frac{1}{n}\int_0^{\pi} V_m \sin nt \, dt + \frac{1}{\pi}\int_{\pi}^{2\pi} (-V_m) \sin nt \, dt$$

$$= \frac{V}{\pi}\left[-\frac{\cos nt}{n}\right]_0^{\pi} + \frac{V_m}{\pi}\left[\frac{\cos nt}{n}\right]_{\pi}^{2\pi}$$

$$= -\frac{V_m}{n\pi}(\cos n\pi - \cos 0^\circ) + \frac{V_m}{n\pi}(\cos 2n\pi - \cos n\pi)$$

Since $\cos 0^\circ = 1, \cos 2n\pi = 1$

$$b_n = \frac{2V_m}{n\pi}(1 - \cos n\pi)$$

When n is even, $\cos n\pi = 1, b_n = 0$

When n is odd, $\cos n\pi = -1$

$$b_n = \frac{2V_m}{n\pi}(1 + 1) = \frac{4V_m}{n\pi} \quad \text{for} \quad n = 1, 3, 5, \ldots$$

Therefore, the Fourier series for the given square wave shown in Fig. 19.6a is

$$f(t) = \frac{4V_m}{\pi} \sum_{n = 1, 3, 5, \ldots,}^{\infty} \frac{1}{n} \sin nt$$

$$= \frac{4V_m}{\pi}\left(\sin t + \frac{1}{3}\sin 3t + \frac{1}{5}\sin 5t + \ldots\right)$$

(*b*) Here the vertical axis is shifted by $\pi/2$ radians. Replacing t by $(t + \pi/2)$ in the above relation, the Fourier series for the square wave shown in Fig. 19.6b is obtained as

$$f(t) = \frac{4V_m}{\pi}\left[\sin\left(t + \frac{\pi}{2}\right) + \frac{1}{3}\sin 3\left(t + \frac{\pi}{2}\right) + \frac{1}{5}\sin 5\left(t + \frac{\pi}{2}\right)\ldots\right]$$

$$= \frac{4V_m}{\pi}\left(\cos t - \frac{1}{3}\cos 3t + \frac{1}{5}\cos 5t - \ldots\right)$$

Alternative method

$$f(t) = \begin{cases} V_m & -\pi/2 < t < \pi/2 \\ -V_m & \pi/2 < t < 3\pi/2 \end{cases}$$

In one period,

positive area = negative area

Therefore $a_0 = 0$

$$a_n = \frac{1}{\pi}\int_{-\pi/2}^{3\pi/2} f(t) \cos nt\, dt$$

$$= \frac{1}{\pi}\int_{-\pi/2}^{\pi/2} f(t) \cos nt\, dt + \frac{1}{\pi}\int_{\pi/2}^{3\pi/2} f(t) \cos nt\, dt$$

$$= \frac{1}{\pi}\int_{-\pi/2}^{\pi/2} V_m \cos nt\, dt + \frac{1}{\pi}\int_{\pi/2}^{3\pi/2} (-V_m) \cos nt\, dt$$

$$= \frac{V_m}{\pi}\left[\frac{\sin nt}{n}\right]_{-\pi/2}^{\pi/2} - \frac{V_m}{\pi}\left[\frac{\sin nt}{n}\right]_{\pi/2}^{3\pi/2}$$

$$= \frac{V_m}{n\pi}\left[\sin\frac{n\pi}{2} - \sin\left(-\frac{n\pi}{2}\right) - \sin\frac{3n\pi}{2} + \sin\frac{n\pi}{2}\right]$$

$$= \frac{V_m}{n\pi}\left(3\sin\frac{n\pi}{2} - \sin\frac{3n\pi}{2}\right)$$

When n is even, $\sin\frac{n\pi}{2} = 0$ and $\sin\frac{3n\pi}{2} = 0$

Therefore, $a_n = 0$

When $n = 1,\ a_n = \frac{V_m}{\pi}\left(3\sin\frac{\pi}{2} - \sin\frac{3\pi}{2}\right) = \frac{4V_m}{\pi}$

When $n = 3,\ a_n = \frac{V_m}{3\pi}\left(3\sin\frac{3\pi}{2} - \sin\frac{9\pi}{2}\right) = \frac{-4V_m}{3\pi}$

When $n = 5,\ a_n = \frac{V_m}{5\pi}\left(3\sin\frac{5\pi}{2} - \sin\frac{15\pi}{2}\right) = \frac{4V_m}{5\pi}$

$$b_n = \frac{1}{\pi}\int_{-\pi/2}^{3\pi/2} f(t) \sin nt\, dt$$

$$= \frac{1}{\pi}\int_{-\pi/2}^{\pi/2} V_m \sin nt\, dt + \frac{1}{\pi}\int_{\pi/2}^{3\pi/2} (-V_m) \sin nt\, dt$$

$$= \frac{V_m}{\pi}\left[-\frac{1}{n}\cos nt\right]_{-\pi/2}^{\pi/2} + \frac{V_m}{\pi}\left[\frac{1}{n}\cos nt\right]_{\pi/2}^{3\pi/2}$$

$$= \frac{V_m}{n\pi}\left[-\left(\cos\frac{n\pi}{2} - \cos\frac{n\pi}{2}\right) + \left(\cos\frac{3n\pi}{2} - \cos\frac{n\pi}{2}\right)\right]$$

$$= 0, \quad n = 1, 2, 3, \ldots$$

Alternatively

Since the square wave of Fig. 19.6b possesses even function symmetry, the Fourier series has only cosine terms, that is, $b_n = 0$.

Hence the Fourier series for the square wave shown in Fig. 19.6b is

$$f(t) = \frac{4V_m}{\pi}\left(\cos t - \frac{1}{3}\cos 3t + \frac{1}{5}\cos 5t - \dots\right)$$

EXAMPLE 19.2 Determine the Fourier series for the sawtooth wave shown in Fig. 19.7.

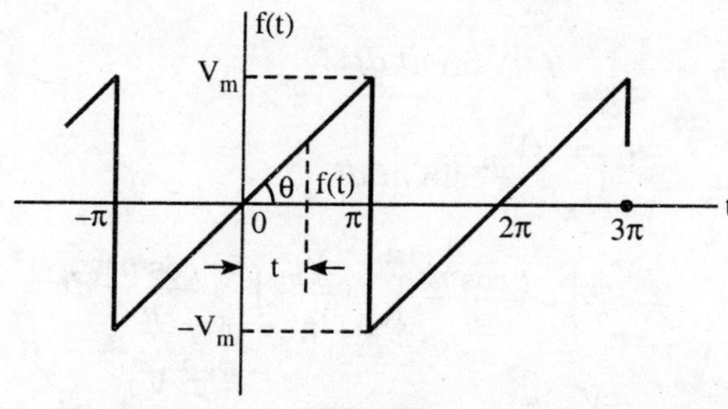

FIG. 19.7

SOLUTION : We have seen that the choice of the limits of integration depends upon the convenience. In the present case, if the limits are from 0 to 2π, two equations would be required to express the function. For simplicity we take the limits from $-\pi$ to π.

Here $T = 2\pi$, therefore $\omega_0 = \frac{2\pi}{T} = 1$

From Fig. 19.7,

$$\tan\theta = \frac{f(t)}{t} = \frac{V_m}{\pi}$$

Therefore,

$$f(t) = \frac{t}{\pi}V_m \text{ when } -\pi < t < \pi$$

Determination of a_0

In one cycle, positive area = negative area.

Therefore, $a_0 = 0$

Determination of a_n

$$a_n = \frac{1}{\pi}\int_{-\pi}^{\pi} f(t)\cos nt\,dt$$

$$= \frac{1}{\pi}\int_{-\pi}^{\pi} \frac{t}{\pi}V_m \cos nt\,dt$$

$$= \frac{V_m}{\pi^2}\left\{\left[\frac{t\sin nt}{n}\right]_{-\pi}^{\pi} - \int_{-\pi}^{\pi}\frac{\sin nt}{n}dt\right\}$$

$$= \frac{V_m}{n\pi^2}(\pi \sin n\pi - \pi \sin n\pi) + \frac{V_m}{n^2\pi^2}[\cos nt]_{-\pi}^{\pi}$$

$$= 0 + \frac{V_m}{n^2\pi^2}(\cos n\pi - \cos n\pi) = 0$$

Alternative method

Since the given waveform possesses odd-function symmetry, $a_n = 0$.

Determination of b_n

$$b_n = \frac{1}{\pi}\int_{-\pi}^{\pi} f(t) \sin nt \, dt$$

$$= \frac{1}{\pi}\int_{-\pi}^{\pi} \frac{tV_m}{\pi} \sin nt \, dt$$

$$= \frac{V_m}{\pi^2}\left[-\frac{t \cos nt}{n}\right]_{-\pi}^{\pi} + \frac{V_m}{\pi^2}\int_{-\pi}^{\pi} \frac{\cos nt}{n} dt$$

$$= \frac{-V_m}{n\pi^2}(\pi \cos n\pi + \pi \cos n\pi) + \frac{V_m}{n^2\pi^2}[\sin nt]_{-\pi}^{\pi}$$

$$= \frac{-V_m}{n\pi^2}(2\pi \cos n\pi) + \frac{V_m}{n^2\pi^2}(2 \sin n\pi)$$

But $\sin n\pi = 0$, therefore

$$b_n = \frac{-2V_m}{n\pi} \cos n\pi$$

When n is even, that is, $n = 2, 4, 6, \ldots$ $\cos n\pi = 1$

$$b_n = \frac{-2V_m}{n\pi}$$

$$b_2 = \frac{-2V_m}{2\pi}, \; b_4 = \frac{-2V_m}{4\pi}, \; b_6 = \frac{-2V_m}{6\pi}$$

When n is odd, that is, $n = 1, 3, 5, \ldots$

$$\cos n\pi = -1, \; b_n = \frac{2V_m}{n\pi}$$

$$b_1 = \frac{2V_m}{\pi}, \; b_3 = \frac{2V_m}{3\pi}, \; b_5 = \frac{2V_m}{5\pi}$$

Thus, the Fourier series for the given saw-tooth wave is

$$f(t) = \frac{2V_m}{\pi}\left(\sin t - \frac{1}{2}\sin 2t + \frac{1}{3}\sin 3t - \ldots\right)$$

EXAMPLE 19.3 Determine the Fourier series for the *half-wave rectified voltage* waveform shown in Fig. 19.8.

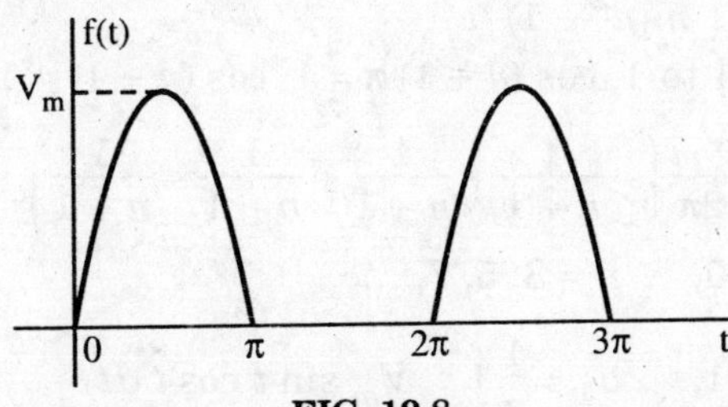

FIG. 19.8

SOLUTION :

Here $$T = 2\pi,\ \omega_0 = \frac{2\pi}{T} = 1$$

The given waveform is defined by

$$f(t) = \begin{cases} V_m \sin t & 0 < t < \pi \\ 0 & \pi < t < 2\pi \end{cases}$$

Determination of a_0

$$a_0 = \frac{1}{2\pi}\int_0^{2\pi} f(t)\, dt$$

$$= \frac{1}{2\pi}\int_0^{\pi} V_m \sin t\, dt + \frac{1}{2\pi}\int_{\pi}^{2\pi} 0\, dt$$

$$= \frac{V_m}{2\pi}[-\cos t]_0^{\pi} = \frac{-V_m}{2\pi}(\cos\pi - \cos 0^\circ) = \frac{V_m}{\pi}$$

Determination of a_n

$$a_n = \frac{1}{\pi}\int_0^{2\pi} f(t)\cos nt\, dt$$

$$= \frac{1}{\pi}\int_0^{\pi}(V_m \sin t)\cos nt\, dt + \frac{1}{\pi}\int_{\pi}^{2\pi}(0)\cos nt\, dt$$

$$= \frac{V_m}{2\pi}\int_0^{\pi}[\sin(n+1)t - \sin(n-1)t]\, dt + 0$$

$$= \frac{V_m}{2\pi}\left[-\frac{\cos(n+1)t}{n+1} + \frac{\cos(n-1)t}{n-1}\right]_0^{\pi}$$

$$= \frac{V_m}{2\pi}\left[-\frac{\cos(n+1)\pi}{n+1} + \frac{1}{n+1} + \frac{\cos(n-1)\pi}{n-1} - \frac{1}{n-1}\right]$$

When n is even, $\cos(n+1)\pi = -1$, $\cos(n-1)\pi = -1$

$$a_n = \frac{V_m}{2\pi}\left[\frac{1}{n+1} + \frac{1}{n+1} - \frac{1}{n-1} - \frac{1}{n-1}\right]$$

$$= -\frac{2V_m}{n\,(n^2 - 1)}, \qquad n = 2, 4, 6, \ldots$$

When n is odd and not equal to 1, $\cos (n + 1)\,\pi = 1$, $\cos (n - 1) = 1$

$$a_n = \frac{V_m}{2\pi}\left(-\frac{1}{n+1} + \frac{1}{n+1} + \frac{1}{n-1} - \frac{1}{n-1}\right)$$

$$= 0, \qquad n = 3, 5, 7, \ldots$$

When $\quad n = 1, \quad a_1 = \frac{1}{\pi}\int_0^{2\pi} V_m \sin t \cos t \, dt$

or $\quad a_1 = \frac{V_m}{2\pi}\int_0^{\pi} \sin 2t \, dt = 0$

Therefore $\quad a_n = 0 \;\text{ for }\; n = 1, 3, 5, \ldots$

Determination of b_n

$$b_n = \frac{1}{\pi}\int_0^{2\pi} f(t) \sin nt \, dt$$

$$= \frac{1}{\pi}\int_0^{\pi} V_m \sin t \sin nt \, dt + \int_{\pi}^{2\pi} (0) \sin nt \, dt$$

$$= \frac{V_m}{\pi}\int_0^{\pi} \sin t \sin nt \, dt$$

This integral is zero except for $n = 1$.

Therefore, $\quad b_n = 0 \quad \text{for} \quad n = 2, 3, 4, \ldots$

For $\quad n = 1, \quad b_n = b_1$

$$b_n = \frac{V_m}{\pi}\int_0^{\pi} \sin t \sin t \, dt = \frac{V_m}{\pi}\int_0^{\pi} \sin^2 t \, dt$$

$$= \frac{V_m}{\pi} \times \frac{\pi}{2} = \frac{V_m}{2}$$

Hence the Fourier series for the half-wave rectified voltage waveform is

$$f(t) = \frac{V_m}{\pi} + \frac{V_m}{2}\sin t - \frac{2V_m}{\pi}\sum_{n = 2, 4, 6, \ldots}^{\infty}\left(\frac{\cos nt}{n^2 - 1}\right)$$

or

$$f(t) = V_m\left(\frac{1}{\pi} + \frac{1}{2}\sin t - \frac{2}{3\pi}\cos 2t - \frac{2}{15\pi}\cos 4t - \frac{2}{35\pi}\cos 6t - \ldots\right)$$

EXAMPLE 19.4 Determine the Fourier series for the full-wave rectified voltage waveform shown in Fig. 19.9.

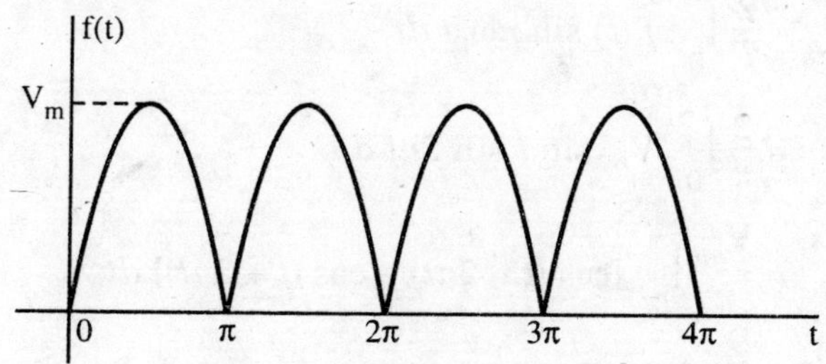

FIG. 19.9

SOLUTION : Here $T = \pi$ and $\omega_0 = \frac{2\pi}{T} = 2$

The function is given by

$$f(t) = |\ V_m \sin t\ |$$

Determination of a_0

$$a_0 = \frac{1}{T}\int_0^T f(t)\,dt$$

$$= \frac{1}{T}\int_0^\pi V_m \sin t\,dt$$

$$= \frac{V_m}{\pi}[-\cos t]_0^\pi = \frac{2V_m}{\pi}$$

Determination of a_n

$$a_n = \frac{2}{T}\int_0^T f(t)\cos n\omega_0 t\,dt$$

$$= \frac{2}{\pi}\int_0^\pi V_m \sin t \cos 2nt\,dt$$

$$= \frac{V_m}{\pi}\int_0^\pi [\sin(t + 2nt) + \sin(t - 2nt)]\,dt$$

$$= \frac{V_m}{\pi}\left[-\frac{\cos(1+2n)t}{1+2n} - \frac{\cos(1-2n)t}{1-2n}\right]_0^\pi$$

$$= -\frac{V_m}{\pi}\left[\frac{\cos(1+2n)\pi - 1}{1+2n} + \frac{\cos(1-2n)\pi - 1}{1-2n}\right]$$

$$\cos(1+2n)\pi = -1,\ \cos(1-2n)\pi = \cos(2n\pi - \pi) = -1$$

Therefore,

$$a_n = \frac{2V_m}{\pi}\left(\frac{1}{1+2n} + \frac{1}{1-2n}\right) = \frac{4V_m}{\pi}\frac{1}{1-4n^2}$$

Determination of b_n

$$b_n = \frac{2}{T}\int_0^T f(t) \sin n\omega_0 t \, dt$$

$$= \frac{2}{\pi}\int_0^{\pi} V_m \sin t \sin 2nt \, dt$$

$$= \frac{V_m}{\pi}\int_0^{\pi} [\cos (t - 2nt) - \cos (t + 2nt)] \, dt$$

$$= \frac{V_m}{\pi}\left[\frac{\sin (1 - 2n)\, t}{1 - 2n} - \frac{\sin (1 + 2n)\, t}{1 + 2n}\right]_0^{\pi}$$

$$= 0$$

Alternatively

Since the given function is even, it will have no sine terms in its Fourier series expansion, that is, $b_n = 0$.

Thus, the Fourier series is given by

$$f(t) = \frac{2V_m}{\pi} + \sum_{n=1}^{\infty} \frac{4V_m}{\pi} \frac{1}{(1 - 4n^2)} \cos 2nt$$

$$= \frac{2V_m}{\pi} - \frac{4V_m}{\pi}\left(\frac{1}{3}\cos 2t + \frac{1}{15}\cos 4t + \frac{1}{35}\cos 6t + \ldots\right)$$

Alternative Method

The given waveform may be treated as the sum of two waveforms 1 and 2 as shown in **Fig. 19.10.**

For waveform 1, from Example 19.3

$$f_1(t) = \frac{V_m}{\pi} + \frac{V_m}{2}\sin t - \frac{2V_m}{\pi} \sum_{n = 2, 4, 6, \ldots}^{\infty} \frac{1}{(n^2 - 1)} \cos nt$$

Waveform 2 lags behind waveform 1 by π radians. By shifting theorem, $f_2(t)$ can be found by putting $(t - \pi)$ for t in the above relation. That is,

$$f_2(t) = f_1(t - \pi)$$

$$= \frac{V_m}{\pi} + \frac{V_m}{2}\sin (t - \pi) - \frac{2V_m}{\pi} \sum_{n \text{ even}}^{\infty} \frac{\cos n(t - \pi)}{n^2 - 1}$$

$$= \frac{V_m}{\pi} - \frac{V_m}{2}\sin t - \frac{2V_m}{\pi} \sum_{n \text{ even}}^{\infty} \frac{\cos nt}{n^2 - 1}$$

For the given waveform, by superposition, we have

$$f(t) = f_1(t) + f_2(t)$$

$$= \frac{2V_m}{\pi} - \frac{4V_m}{\pi} \sum_{n \text{ even}}^{\infty} \left(\frac{1}{n^2 - 1} \right) \cos nt$$

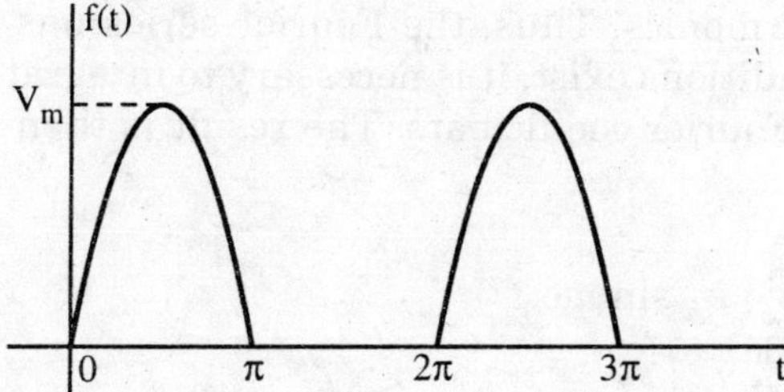

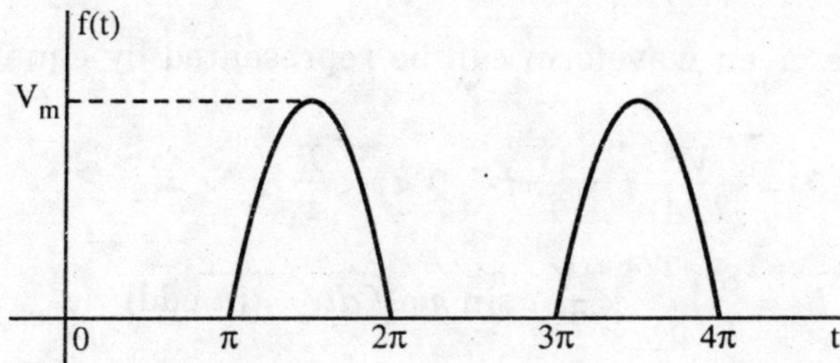

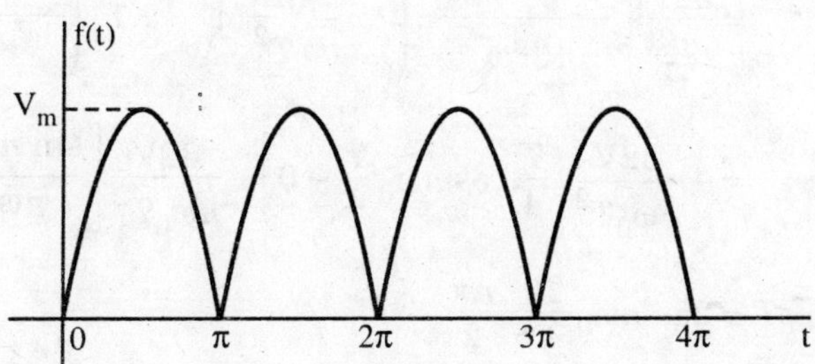

FIG. 19.10

■ **EXAMPLE 19.5** Represent the triangular waveform shown in Fig. 19.11 as a Fourier series.

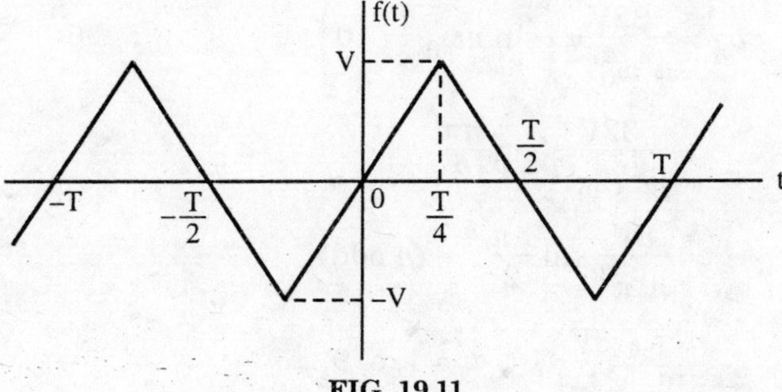

FIG. 19.11

SOLUTION : The given waveform possesses the following symmetries :

$f(t) = -f(-t)$ **odd function symmetry**

$$\left(t \pm \frac{T}{2}\right) = -f(t) \qquad \text{half-wave symmetry}$$

Since the function is odd, $a_0 = 0$ and $a_n = 0$. The Fourier series expansion with half-wave symmetry contains only odd harmonics. Thus, the Fourier series has only odd cosine terms. Further since two symmetry conditions exist, it is necessary to integrate over only one quarter period in the determination of Fourier coefficients. The result is then multiplied by 4. So we can write

$$f(t) = \sum_{n \text{ odd}}^{\infty} b_n \sin n\omega_0 t$$

where
$$b_n = 4 \times \frac{2}{T} \int_0^{T/4} f(t) \sin n\omega_0 t \, dt, \qquad n \text{ odd}$$

The quarter wave of the given waveform can be represented by equation of a straight line. Therefore

$$f(t) = \frac{V}{(T/4)} t = \frac{4V}{T} t \qquad 0 < t < \frac{T}{4}$$

Hence,
$$b_n = \frac{8}{T} \int_0^{T/4} \frac{4V}{T} t \sin n\omega_0 t \, dt, \qquad (n \text{ odd})$$

$$= \frac{32V}{T^2} \left[\frac{-t \cos n\omega_0 t}{n\omega_0} \right]_0^{T/4} - \frac{32V}{T^2} \int_0^{T/4} 1 \cdot \left(\frac{-\cos n\omega_0 t}{n\omega_0} \right) dt$$

$$= -\frac{32V}{n\omega_0 T^2} \left(\frac{T}{4} \cos n\omega_0 \frac{T}{4} - 0 \right) + \frac{32V}{n\omega_0 T^2} \left[\frac{\sin n\omega_0 t}{n\omega_0} \right]_0^{T/4}$$

$$\omega_0 T = 2\pi, \quad n\omega_0 \frac{T}{4} = \frac{n\pi}{2}$$

$$\cos \frac{n\omega_0 T}{4} = \cos \frac{n\pi}{2} = 0 \qquad \text{when } n \text{ is odd}$$

It follows that

$$b_n = \frac{32V}{n^2 \omega_0^2 T^2} \left(\sin n\omega_0 \frac{T}{4} - 0 \right)$$

$$= \frac{32V}{n^2 (2\pi)^2} \sin \frac{n\pi}{2}$$

$$= \frac{8V}{n^2 \pi^2} \sin \frac{n\pi}{2}, \qquad (n \text{ odd})$$

$$= \begin{cases} \dfrac{8V}{n^2 \pi^2}, & n = 1, 5, 9, \ldots \\ -\dfrac{8V}{n^2 \pi^2}, & n = 3, 7, 11, \ldots \end{cases}$$

Hence the Fourier series is

$$f(t) = \frac{8V}{\pi^2}\left(\sin \omega_0 t - \frac{1}{3^2}\sin 3\omega_0 t + \frac{1}{5^2}\sin 5\omega_0 t \ldots\right)$$

■ **EXAMPLE 19.6** Determine the Fourier series for the rectangular wave shown in Fig. 19.12.

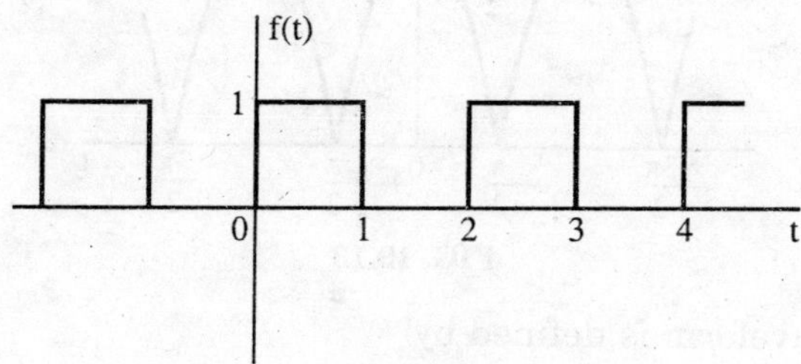

FIG. 19.12

SOLUTION : $T = 2,\ \omega_0 = \frac{2\pi}{T} = \frac{2\pi}{2} = \pi$

$$a_0 = \frac{1}{2}\int_0^2 f(t)\,dt = \frac{1}{2}\int_0^1 1\,dt + \frac{1}{2}\int_1^2 0\,dt = \frac{1}{2}$$

$$a_n = \frac{2}{2}\int_0^2 f(t)\cos n\pi t\,dt$$

$$= \int_0^1 1\cos n\pi t\,dt + \int_1^2 0\cos n\pi t\,dt$$

$$a_n = \frac{\sin n\pi}{n\pi} = 0$$

$$b_n = \int_0^1 \sin n\pi t\,dt$$

$$= -\frac{1}{n\pi}\cos n\pi t\Big|_0^1 = -\frac{1}{n\pi}(\cos n\pi - \cos 0^\circ) = -\frac{1}{n\pi}(\cos n\pi - 1)$$

$$b_n = \begin{cases} \dfrac{2}{n\pi} & \text{for } n \text{ odd} \\ 0 & \text{otherwise} \end{cases}$$

So the expansion for the rectangular wave is

$$f(t) = \frac{1}{2} + \frac{2}{\pi}\sin \pi t + \frac{2}{3\pi}\sin 3\pi t + \frac{2}{5\pi}\sin 5\pi t + \ldots$$

$$f(t) = \frac{1}{2} + \frac{2}{\pi}\sum_{n=1}^{\infty}\frac{1}{n}\sin n\pi t \quad \text{for } n \text{ odd}$$

■ **EXAMPLE 19.7** Determine the Fourier series for the wave shown Fig. 19.13.

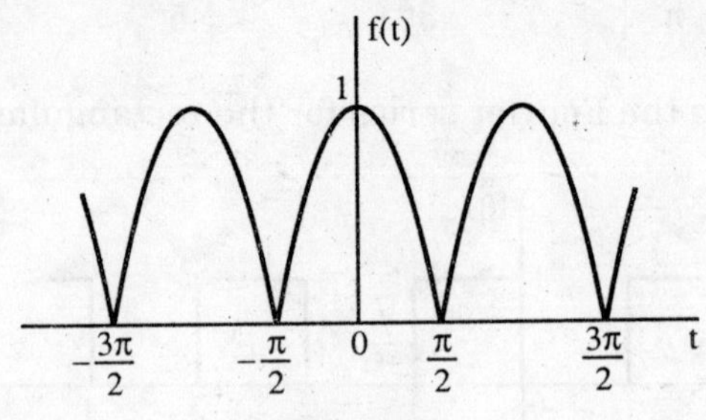

FIG. 19.13

SOLUTION : The given waveform is defined by

$$f(t) = |\cos t|$$

Here $$T = \pi, \omega_0 = \frac{2\pi}{\pi} = 2$$

Determination of a_0

$$a_0 = \frac{1}{T}\int_{-T/2}^{T/2} f(t)\, dt = \frac{1}{\pi}\int_{-\pi/2}^{\pi/2} \cos t\, dt$$

$$= \left[\frac{1}{\pi}\sin t\right]_{-\pi/2}^{\pi/2} = \frac{1}{\pi}\left[\sin\frac{\pi}{2} - \sin\left(-\frac{\pi}{2}\right)\right]$$

$$= \frac{1}{\pi}(1+1) = \frac{2}{\pi}$$

Determination of a_n

$$a_n = \frac{2}{T}\int_{-T/2}^{T/2} f(t)\cos n\omega_0 t\, dt$$

$$= \frac{2}{\pi}\int_{-\pi/2}^{\pi/2} \cos t \cos 2nt\, dt$$

$$= \frac{1}{\pi}\int_{-\pi/2}^{\pi/2} [\cos(2n-1)t + \cos(2n+1)t]\, dt$$

$$= \left[\frac{1}{\pi}\frac{\sin(2n-1)t}{(2n-1)}\right]_{-\pi/2}^{\pi/2} + \frac{1}{\pi}\left[\frac{\sin(2n+1)t}{(2n+1)}\right]_{-\pi/2}^{\pi/2}$$

$$= \frac{1}{\pi}\left[\frac{\sin(2n-1)\pi/2}{2n-1} - \frac{\sin\{(-2n-1)t\}\pi/2}{2n-1}\right] + \frac{1}{\pi}\left[\frac{\sin(2n+1)\pi/2}{2n+1} - \frac{\sin\{-(2n+1)t\}\pi/2}{2n+1}\right]$$

$$= \frac{2}{\pi}\frac{\sin(2n-1)\pi/2}{2n-1} + \frac{2}{\pi}\frac{\sin(2n+1)\pi/2}{2n+1}$$

We know that

$$\sin (2n - 1)\frac{\pi}{2} = (-1)^{n+1}$$

and $$\sin (2n + 1)\frac{\pi}{2} = (-1)^{n}$$

Therefore,

$$a_n = \frac{2}{n}\left[\frac{(-1)^{n+1}}{2n-1} - \frac{(-1)^n}{2n+1}\right]$$

Determination of b_n

$$b_n = \frac{2}{T}\int_{-T/2}^{T/2} f(t) \sin n\omega_0 t \, dt = \frac{2}{\pi}\int_{-\pi/2}^{\pi/2} \cos t \sin 2nt \, dt$$

Let $$\phi(t) = \cos t \sin 2nt$$

$\therefore$ $$\phi(-t) = \cos(-t)\sin(-2nt) = -\cos t \sin 2nt$$

This relation shows that $\phi(t)$ is an odd function. Therefore

$$\frac{1}{T}\int_{-T/2}^{T/2} \phi(t) \, dt = 0$$

and hence $$b_n = \frac{2}{\pi}\int_{-\pi/2}^{\pi/2} \cos t \sin 2\pi nt \, dt = 0$$

The Fourier series of the given function can be written as

$$f(t) = a_0 + \sum_{n=1}^{\infty} (a_n \cos 2nt + b_n \sin 2nt)$$

$$= \frac{2}{\pi} + \sum_{n=1}^{\infty} \frac{2}{\pi}\left[\frac{(-1)^{n+1}}{2n-1} + \frac{(-1)^n}{2n+1}\right]\cos 2nt$$

$$= \frac{2}{\pi}\left(1 + \frac{2}{3}\cos 2t - \frac{2}{15}\cos 4t + \frac{2}{35}\cos 6t \ldots\right)$$

■ **EXAMPLE 19.8** Use the results of Example 19.7 to find the exponential Fourier series and hence find the complex Fourier spectrum of the absolute value of a full-wave rectified cosine wave given by

$$f(t) = |\cos t| \quad \text{as shown in Fig. 19.13.}$$

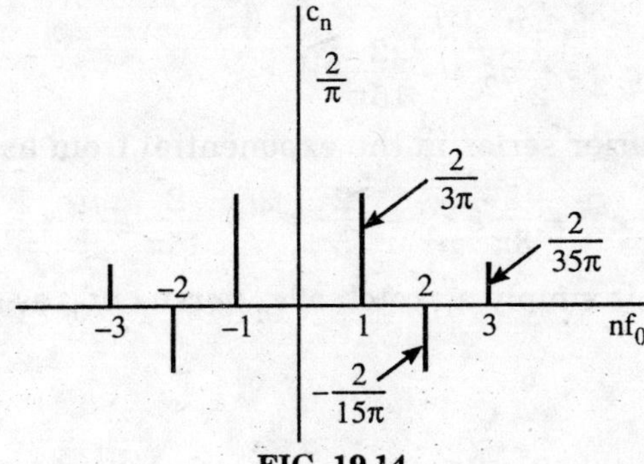

FIG. 19.14

SOLUTION : We have determined the trigonometric Fourier series expansion of $f(t) = |\cos t|$ as given below :

$$f(t) = \frac{2}{\pi} + \sum_{n=1}^{\infty} a_n \cos 2nt$$

where
$$a_n = \frac{2}{\pi}\left[\frac{(-1)^{n+1}}{2n-1} + \frac{(-1)^n}{2n+1}\right]$$

We know that

$$\cos 2nt = \frac{1}{2}(e^{j2nt} + e^{-j2nt})$$

$$f(t) = \frac{2}{\pi} + \sum_{n=1}^{\infty} \frac{a_n}{2} e^{j2nt} + \sum_{n=1}^{\infty} \frac{a_n}{2} e^{-j2nt}$$

If we change the variable, we can write

$$\sum_{n=1}^{\infty} \frac{a_n}{2} e^{-j2nt} \text{ as } \sum_{n=\infty}^{\infty} \frac{a_{-n}}{2} e^{+j2nt}$$

therefore,
$$f(t) = \frac{2}{\pi} + \sum_{n=1}^{\infty} \frac{a_n}{2} e^{j2nt} + \sum_{n=-\infty}^{-1} \frac{a_{-n}}{2} e^{+j2nt}$$

From Eq. (19.10.11), we have

$$f(t) = \sum_{n=-\infty}^{\infty} \mathbf{c}_n e^{jn\omega_0 t}$$

Comparing these equations we get

$$c_0 = a_0 = \frac{2}{\pi}$$

$$c_n = \frac{a_n}{2} \qquad (n > 0)$$

$$c_n = \frac{a_{-n}}{2} \qquad (n < 0)$$

therefore,
$$c_1 = c_{-1} = \frac{1}{2} a_1 = \frac{2}{3\pi}$$

$$c_2 = c_{-2} = \frac{1}{2} a_2 = -\frac{2}{15\pi}$$

Hence we can write the Fourier series in the exponential from as

$$f(t) = \frac{2}{\pi} + \frac{2}{3\pi} e^{j2t} + \frac{2}{3\pi} e^{-j2t} - \frac{2}{15\pi} e^{j4t} - \frac{2}{15\pi} e^{-j4t} + \ldots$$

The Fourier spectrum is simply a sketch of c_n versus nf_0, as shown in Fig. 19.14. for this example.

■ **EXAMPLE 19.9** Find the exponential Fourier series expansion of the square wave shown in Fig. 19.15.

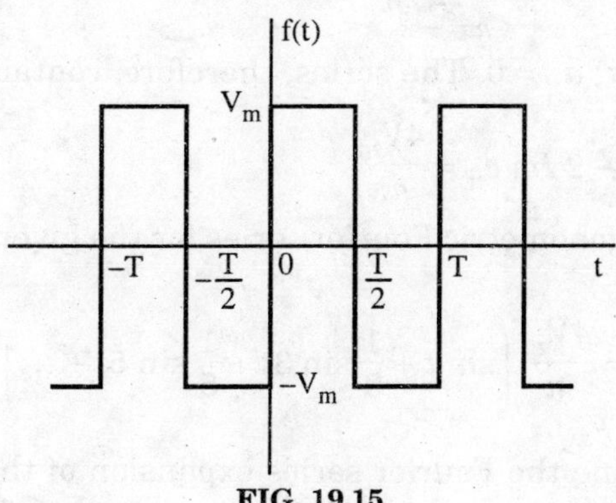

FIG. 19.15

SOLUTION :

Here
$$f(t) = \begin{cases} V_m & 0 < t < T/2 \\ -V_m & T/2 < t < T \end{cases}$$

$$c_n = \frac{1}{T}\int_0^T f(t)\, e^{-jn\omega_0 t}\, dt$$

$$c_n = \frac{1}{T}\int_0^{T/2} (V_m)\, e^{-jn\omega_0 t}\, dt + \frac{1}{T}\int_{T/2}^{T} (-V_m)\, e^{-jn\omega_0 t}\, dt$$

$$= \frac{V_m}{T}\left[\frac{e^{-jn\omega_0 t}}{-jn\omega_0}\right]_0^{T/2} - \frac{V_m}{T}\left[\frac{e^{-jn\omega_0 t}}{-jn\omega_0}\right]_{T/2}^{T}, \quad n \neq 0$$

$$= \frac{V_m}{jn\omega_0 T}\,[1 - e^{-jn\omega_0 T/2}] + \frac{V_m}{jn\omega_0 T}\,[e^{-jn\omega_0 T} - e^{-jn\omega_0 T/2}]$$

Now
$$e^{-jn\omega_0 T} = e^{-jn\,(2\pi)} = \cos 2n\pi - j \sin 2n\pi = 1$$

and
$$e^{-jn\omega_0 T/2} = e^{-jn\pi} = \cos n\pi - j \sin n\pi = (-1)^n$$

Therefore,
$$c_n = \frac{2V_m}{\pi j 2n\pi}\,[1 - (-1)^n], \quad n \neq 0$$

When n is odd,
$$c_n = \frac{2V_m}{jn\pi}$$

When n is even,
$$c_n = 0$$

$$c_0 = \frac{1}{T}\int_0^T f(t)\, dt = 0$$

Thus, the exponential series of the given square wave is,

$$f(t) = \frac{2V_m}{j\pi} \sum_{n=-\infty}^{\infty} \frac{1}{n}\, e^{jn\omega_0 t}, \qquad n \text{ odd}$$

$$= -j\frac{2V_m}{\pi}\sum_{n=-\infty}^{\infty}\frac{1}{n}e^{jn\omega_0 t}$$

Since c_n is purely imaginary, $a_n = 0$. The series, therefore, contains only sine terms.

Also, $$b_n = -2\,Im\,c_n = \frac{4V_m}{n\pi}$$

when n is odd. Thus, the trigonometric Fourier series for the given square wave can be written as

$$f(t) = \frac{4V_m}{\pi}\left(\sin t + \frac{1}{3}\sin 3t + \frac{1}{5}\sin 5t + \ldots\right)$$

■ **EXAMPLE 19.10** Determine the Fourier series expansion of the periodic function shown in Fig. 19.16.

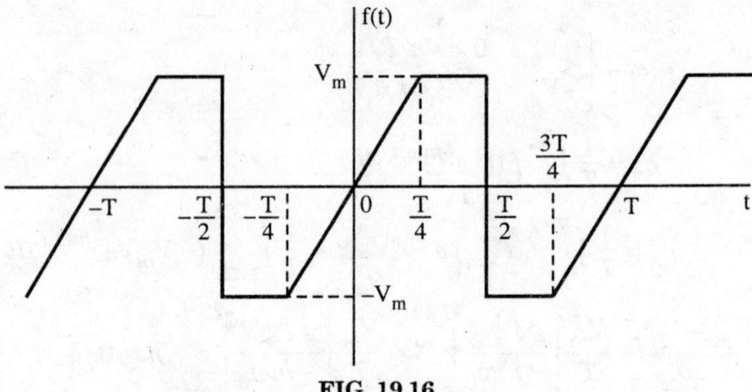

FIG. 19.16

SOLUTION : $$f(t) = \begin{cases} \dfrac{4V_m}{T}t & 0 < t < T/2 \\ V_m & T/4 < t < T/2 \end{cases}$$

By inspection it is found that the given function is odd. Therefore,

$$a_0 = 0\,,\quad a_n = 0,\quad n = 1, 2, 3, \ldots$$

$$b_n = 2\times\frac{2}{T}\int_0^{T/2} f(t)\sin n\omega_0 t\,dt$$

$$= \frac{4}{T}\int_0^{T/4}\left(\frac{4V_m}{T}t\right)\sin n\omega_0 t\,dt + \frac{4}{T}\int_{T/4}^{T/2} V_m \sin n\omega_0 t\,dt$$

$$= \frac{16V_m}{T^2}\left[-\frac{1}{n\omega_0}\cos n\omega_0 t + \frac{1}{n^2\omega_0^2}\sin n\omega_0 t\right]_0^{T/4} + \frac{4V_m}{T}\left[-\frac{\cos n\omega_0 t}{n\omega_0}\right]_{T/4}^{T/2}$$

Since $$\omega_0 = \frac{2\pi}{T}\,,\quad \omega_0\frac{T}{4} = \frac{\pi}{2}\,,\quad \omega_0\frac{T}{2} = \pi$$

$$\therefore \quad b_n = \frac{16V_m}{T}\left[-\frac{T\cos(n\pi/2)}{4n\omega_0} + \frac{\sin(n\pi/2)}{n^2\omega_0^2}\right] - \frac{4V_m}{T}\left[\frac{\cos n\pi}{n\omega_0} - \frac{\cos(n\pi/2)}{n\omega_0}\right]$$

$$= \frac{4V_m}{n^2\pi^2}\sin\frac{n\pi}{2} - \frac{2V_m}{n\pi}\cos n\pi$$

When n is even, $\sin\frac{n\pi}{2} = 0$, $\cos 2n\pi = 1$ and $b_n = -\frac{2V_m}{n\pi}$

When n is odd,

$$b_n = \frac{4V_m}{n^2\pi^2}(-1)^{\frac{n-1}{2}} + \frac{2V_m}{n\pi}$$

Therefore,

$$f(t) = \frac{4V_m}{\pi^2}\sum_{n=1,3,5,\ldots}^{\infty}\frac{(-1)^{\frac{n-1}{2}}}{n^2}\sin n\omega_0 t$$

$$-\frac{2V_m}{\pi}\sum_{n=1}^{\infty}\frac{1}{n}(-1)^n \sin n\omega_0 t$$

■ **EXAMPLE 19.11** Obtain the Fourier series in (a) trigonometric form, and (b) exponential form for the periodic train of pulses of period T, pulse height A, and width a shown in Fig. 19.17.

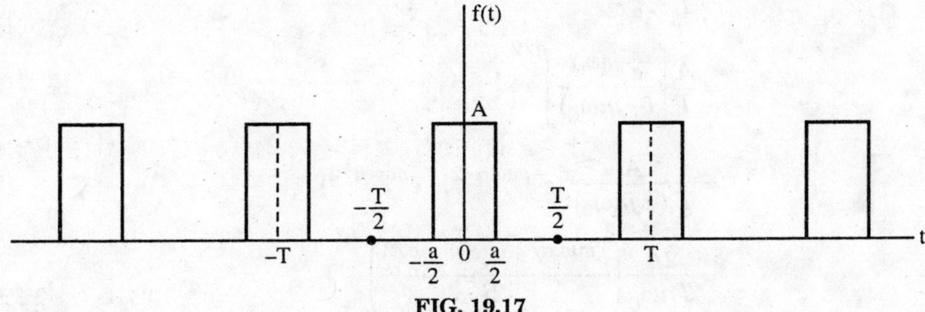

FIG. 19.17

SOLUTION :

Here
$$f(t) = \begin{cases} A, & -a/2 \le t \le a/2 \\ 0 & \text{elsewhere} \end{cases}$$

$$a_0 = \frac{1}{T}\int_{-T/2}^{T/2} f(t)\,dt = \frac{1}{T}\int_{-a/2}^{a/2} A\,dt = \frac{Aa}{T}$$

$$a_n = \frac{2}{T}\int_{-T/2}^{T/2} f(t)\cos n\omega_0 t\,dt$$

$$= \frac{2}{T}\int_{-a/2}^{a/2} A\cos n\omega_0 t\,dt = \frac{2A}{T}\left[\frac{\sin n\omega_0 t}{n\omega_0}\right]_{-a/2}^{a/2}$$

$$= \frac{2A}{n\omega_0 T}\left[\sin n\omega_0 \frac{a}{2} - \left(-\sin n\omega_0 \frac{a}{2}\right)\right]$$

$$= \frac{2A}{n\omega_0 T}\left(2 \sin n\omega_0 \frac{a}{2}\right)$$

This result can be put in the form $(\sin x)/x$.

Hence
$$a_n = \frac{2Aa}{T}\left(\frac{\sin n\omega_0 \frac{a}{2}}{n\omega_0 \frac{a}{2}}\right)$$

Since
$$\omega_0 = \frac{2\pi}{T}, \quad n\omega_0 \frac{a}{2} = \frac{n\pi a}{T}$$

Therefore,
$$a_n = \frac{2Aa}{T} \frac{\sin (n\pi a/T)}{(n\pi a/T)}$$

Since
$$a_n = r_n = 2c_n$$

Therefore,
$$c_n = \frac{1}{2} a_n = \frac{Aa}{T} \frac{\sin (n\pi a/T)}{(n\pi a/T)}$$

Alternatively

$$c_n = \frac{1}{T}\int_{-T/2}^{T/2} f(t)\, e^{-jn\omega_0 t}\, dt$$

$$= \frac{1}{T}\int_{-a/2}^{a/2} A\, e^{-jn\omega_0 t}\, dt$$

$$= \frac{A}{T}\left[\frac{e^{-jn\omega_0 t}}{(-jn\omega_0)}\right]_{-a/2}^{a/2}$$

$$= \frac{A}{T(-jn\omega_0)}\left(e^{-jn\omega_0 a/2} - e^{jn\omega_0 a/2}\right)$$

$$= \frac{2A}{Tn\omega_0}\left(\frac{e^{jn\omega_0 a/2} - e^{-jn\omega_0 a/2}}{2j}\right)$$

$$\frac{2A}{Tn\omega_0} \sin \omega_0 \frac{a}{2} = \frac{Aa}{2} \frac{\sin n\omega_0 \frac{a}{2}}{n\omega_0 \frac{a}{2}}$$

$$= \frac{Aa}{T} \frac{\sin (n\pi a/T)}{(n\pi a/T)}$$

The given function is even, that is,

$$f(t) = f(-t), \text{ therefore } b_n = 0$$

Also,
$$\theta_n = \tan^{-1} \frac{b_n}{a_n} = \tan^{-1} 0 = 0$$

Thus, the Fourier series of the given pulse train in trigonometric form is given by

$$f_T(t) = \frac{Aa}{T} + \frac{2Aa}{T} \sum_{n=1}^{\infty} \left(\frac{\sin n\pi \dfrac{A}{T}}{n\pi \dfrac{a}{T}} \right) \cos n\omega_0 t$$

In the exponential form, the Fourier series of the given pulse train is

$$f_T(t) = \sum_{n=-\infty}^{\infty} \frac{Aa}{T} \frac{\sin (n\pi a/T)}{(n\pi a/T)} e^{+jn\omega_0 t}$$

EXAMPLE 19.12 Obtain the Fourier series of the impulse train shown in Fig. 19.18.

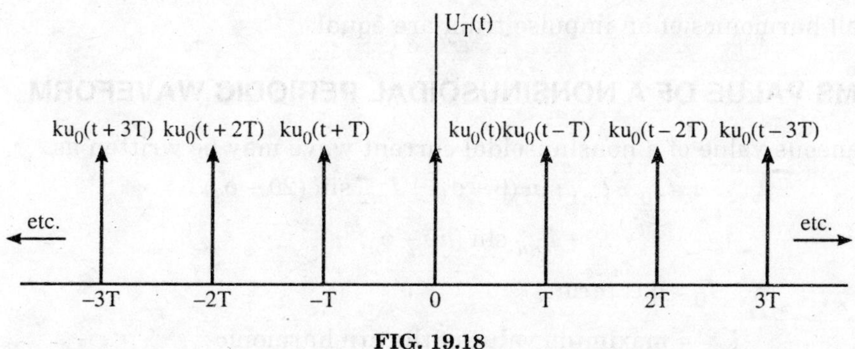

FIG. 19.18

SOLUTION: The given waveform consists of a periodic sequence of impulses of strength k. The separation between the impulses is T seconds. This periodic impulse is denoted by a special symbol $u_T(t)$.

$$u_T(t) = ku_0(t) + ku_0(t - T) + ku_0(t - 2T) + \ldots + ku_0(t - nT) + \ldots$$

$$+ ku_0(t + T) + ku_0(t + 2T) + \ldots + ku_0(t + nT) + \ldots$$

$$= \sum_{n=-\infty}^{\infty} ku_0(t - nT) \quad \text{(E19.12.1)}$$

Here $$f(t) = u_T(t) \quad \text{(E19.12.2)}$$

In the interval $(-T/2, T/2)$, the function $u_T(t)$ is simply $ku_0(t)$. The Fourier series for this periodic function is given by

$$u_T(t) = \sum_{n=-\infty}^{\infty} F(jn\omega_0) e^{jn\omega_0 t}$$

where $$F(jn\omega_0) = \frac{1}{T} \int_{-T/2}^{T/2} u_T(t) e^{-jn\omega_0 t} dt$$

$$= \frac{k}{T} \int_{-T/2}^{T/2} u_0(t) e^{-jn\omega_0 t} dt$$

Since the interval $(-T/2, T/2)$ contains the origin, the property of the impulse function gives

$$\int_{-T/2}^{T/2} u_0(t)\, e^{-jn\omega_0 t}\, dt = e^{-jn\omega_0 t}\Big|_{t=0} = e^0 = 1$$

$$\therefore \qquad F(jn\omega_0) = \frac{k}{T}$$

The Fourier series of the impulse train can be written in the forms

$$u_T(t) = \sum_{n=-\infty}^{\infty} \frac{k}{T} e^{jn\omega_0 t}$$

or $$u_T(t) = \frac{k}{T} + \frac{2k}{T}(\cos \omega_0 t + \cos 2\omega_0 t + \cos 3\omega_0 t + \ldots)$$

Thus, all harmonics of an impulse train are equal.

19.14 RMS VALUE OF A NONSINUSOIDAL PERIODIC WAVEFORM

The instantaneous value of a nonsinusoidal current wave may be written as

$$i = I_0 + I_{m1} \sin(\theta + \phi_1) + I_{m2} \sin(2\theta + \phi_2) + \ldots + I_{mn} \sin(n\theta + \phi_n) \tag{19.14.1}$$

where I_0 = d.c. term

I_{mn} = maximum value of the nth harmonic

ϕ_n = phase of nth harmonic

By definition, the r.m.s. value of an alternating current is given by

$$I = \sqrt{(\text{average value of } i^2)} = \sqrt{\left[\frac{1}{2\pi}\int_0^{2\pi} i^2\, d\theta\right]}$$

Let $I_{m1} \sin(\theta + \phi_1) = A_1$

$I_{m2} \sin(2\theta + \phi_2) = A_2$

... ...

$I_{mn} \sin(n\theta + \phi_n) = A_n$

Therefore,

$$i^2 = [I_0 + (A_1 + A_2 + \ldots + A_n)]^2 = I_0^2 + 2I_0(A_1 + A_2 + \ldots + A_n) + (A_1 + A_2 + \ldots + A_n)^2$$

$$\frac{1}{2\pi}\int_0^{2\pi} i^2\, d\theta = \frac{1}{2\pi}\int_0^{2\pi} I_0^2\, d\theta + \frac{1}{2\pi}\int_0^{2\pi} 2I_0(A_1 + A_2 + \ldots + A_n)\, d\theta + \frac{1}{2\pi}\int_0^{2\pi} (A_1 + A_2 + \ldots + A_n)^2\, d\theta \tag{19.14.2}$$

Let us evaluate the integrals on the right side separately.

$$\frac{1}{2\pi}\int_0^{2\pi} I_0^2\, d\theta = I_0^2$$

$$\frac{1}{2\pi}\int_0^{2\pi} 2I_0 (A_1 + A_2 \ldots + A_n)\, d\theta$$

$$= \frac{2I_0}{2\pi}\int_0^{2\pi} [I_{m1} \sin (\theta + \phi_1) + I_{m2} (2\theta + \phi_2) + \ldots + I_{mn} \sin (n\theta + \phi_n)]\, d\theta = 0,$$

since integral of any sine function over a period 0 to 2π is zero.

$$\frac{1}{2\pi}\int_0^{2\pi} (A_1 + A_2 + \ldots A_n)^2\, d\theta$$

$$= \frac{1}{2\pi}\int_0^{2\pi} [(A_1^2 + A_2^2 + \ldots + A_n^2) + (2A_1A_2 + 2A_2A_3 + \ldots)]\, d\theta$$

$$= \frac{1}{2\pi}\int_0^{2\pi} [I_{m1}^2 \sin^2 (\theta + \phi_1) + I_{m2}^2 \sin^2 (2\theta + \phi_2) + \ldots + I_{mn}^2 \sin^2 (n\theta + \phi_n)]\, d\theta$$

$$+ \frac{2}{2\pi}\int_0^{2\pi} [I_{m1} I_{m2} \sin (\theta + \phi_1) \sin (2\theta + \phi_2)$$

$$+ I_{m2} I_{m3} \sin (2\theta_2 + \phi_2) \sin (3\theta + \phi_3) + \ldots]\, d\theta$$

$$= \frac{1}{2\pi} (I_{m1}^2 \pi + I_{m2}^2 \pi + \ldots + I_{mn}^2 \pi) + 0$$

$$= \frac{1}{2} (I_{m1}^2 + I_{m2}^2 + \ldots + I_{mn}^2)$$

Substituting the values of these integrals in Eq. (19.14.2)

$$\frac{1}{2\pi}\int_0^{2\pi} i^2\, d\theta = I_0^2 + \frac{I_{m1}^2}{2} + \frac{I_{m2}^2}{2} + \ldots + \frac{I_{mn}^2}{2}$$

$$= I_0^2 + \left(\frac{I_{m1}}{\sqrt{2}}\right)^2 + \left(\frac{I_{m2}}{\sqrt{2}}\right)^2 + \ldots + \left(\frac{I_{mn}}{\sqrt{2}}\right)^2$$

$$= I_0^2 + I_1^2 + I_2^2 + \ldots + I_n^2$$

where $I_1, I_2, \ldots, I_n$ are the r.m.s. values of the harmonic components of the wave.

Therefore the r.m.s. value of the given nonsinusoidal wave is given by

$$I = \sqrt{\left(\frac{1}{2\pi}\int_0^{2\pi} i^2\, d\theta\right)}$$

$$I = \sqrt{(I_0^2 + I_1^2 + I_2^2 + \ldots + I_n^2)} \tag{19.14.3}$$

It is seen that the r.m.s. value is independent of the relative phase angles $\phi_1, \phi_2, \ldots, \phi_n$ of the various components.

Similarly, if

$$v = V_0 + \sum V_m \sin (n\omega t + \psi_n)$$

the effective value of the voltage is given by

$$V = \sqrt{(V_0^2 + V_1^2 + V_2^2 + \ldots + V_n^2)} \qquad (19.14.4)$$

19.15 POWER DUE TO NONSINUSOIDAL VOLTAGE AND CURRENT

Suppose that a complex voltage

$$v = V_0 + V_{m1} \sin(\theta + \psi_1) + V_{m2} \sin(2\theta + \psi_2) + \ldots$$

is applied to a circuit so that the current is

$$i = I_0 + I_{m1} \sin(\theta + \psi_1 - \phi_1) + I_{m2} \sin(2\theta + \psi_2 - \phi_2) + \ldots$$

Let

$$V_{m1} \sin(\theta + \psi_1) = A_1$$

$$V_{m2} \sin(2\theta + \psi_2) = A_2$$

$$\ldots \quad \ldots \quad \ldots \quad \ldots$$

$$V_{mn} \sin(n\theta + \psi_n) = A_n$$

and

$$I_{m1} \sin(\theta + \psi_1 - \phi_1) = B_1$$

$$I_{m2} \sin(2\theta + \psi_2 - \phi_2) = B_2$$

$$\ldots \quad \ldots \quad \ldots \quad \ldots$$

$$I_{mn} \sin(n\theta + \psi_n - \phi_n) = B_n$$

$$\begin{aligned} vi &= (V_0 + A_1 + A_2 + \ldots + A_n)(I_0 + B_1 + B_2 + \ldots + B_n) \\ &= V_0 I_0 + (A_1B_1 + A_2B_2 + \ldots + A_nB_n) \\ &\quad + (A_1B_2 + A_1B_3 + \ldots + A_1B_n + B_1A_2 + B_1A_3 + \ldots) \\ &\quad + V_0(B_1 + B_2 + \ldots + B_n) + I_0(A_1 + A_2 + \ldots + A_n) \end{aligned}$$

Thus, the product vi consists of

(*a*) product of constant terms namely, V_0I_0,

(*b*) product terms of like frequencies

(*c*) product terms of unlike frequencies

(*d*) product terms of the d.c. components and harmonic components

$$\text{Average power } P = \frac{1}{2\pi}\int_0^{2\pi} vi \, d\theta$$

We shall determine the average value of each type of product.

Average value of product of constant terms

$$P_0 = \frac{1}{2\pi}\int_0^{2\pi} V_0 I_0 \, d\theta = V_0 I_0$$

Average value of the product terms of like frequencies

$$\begin{aligned} F_{\text{like}} &= \frac{1}{2\pi}\int_0^{2\pi} (A_1B_1 + A_2B_2 + \ldots + A_nB_n)\, d\theta \\ &= P_1 + P_2 + \ldots + P_n \text{ (say)} \end{aligned}$$

$$P_n = \frac{1}{2\pi}\int_0^{2\pi} A_n B_n \, d\theta$$

$$= \frac{1}{2\pi}\int_0^{2\pi} V_{mn} I_{mn} \sin(n\theta + \psi_n) \sin(n\theta + \psi_n - \phi_n) \, d\theta$$

$$= \frac{V_{mn} I_{mn}}{2\pi}\int_0^{2\pi} \frac{1}{2}[\cos\phi_n - \cos(2n\theta + 2\psi_n - \phi_n)] \, d\theta$$

$$= \frac{V_{mn} I_{mn}}{4\pi}(2\pi \cos\phi_n - 0)$$

$$= \frac{V_{mn}}{\sqrt{2}} \times \frac{I_{mn}}{\sqrt{2}} \cos\phi_n$$

$$= V_n I_n \cos\phi_n$$

where V_n and I_n are the r.m.s. values.

Therefore, $P_1 = V_1 I_1 \cos\phi_1,\ P_2 = V_2 I_2 \cos\phi_2, \ldots$

and $P_{\text{like}} = V_1 I_1 \cos\phi_1 + V_2 I_2 \cos\phi_2 + \ldots + V_n I_n \cos\phi_n$

Average value of the product terms of unlike frequencies

$$P_{\text{unlike}} = \frac{1}{2\pi}\int_0^{2\pi} [(A_1B_2 + A_1B_3 + \ldots + A_1B_n) + (B_1A_2 + B_1A_3 + \ldots) + \ldots] \, d\theta$$

Consider the general terms for the product of two terms of unlike frequencies

$$\frac{1}{2\pi}\int_0^{2\pi} A_p B_q \, d\theta$$

$$= \frac{1}{2\pi}\int_0^{2\pi} V_{mp} I_{mq} \sin(p\theta + \psi_p) \sin(q\theta + \psi_q - \phi_q) \, d\theta$$

$$= \frac{V_{mp} I_{mq}}{4\pi}\int_0^{2\pi} \cos[(p - q)\theta + \psi_p - \psi_q + \phi_q]$$

$$- \cos[(p + q)\theta + \psi_p + \psi_q - \phi_q]\} \, d\theta = 0$$

Therefore, $F_{\text{unlike}} = 0$

Average value of the product terms of the d.c. components and harmonic components

Considering the general terms of the products

$$P_{0n} = \frac{1}{2\pi}\int_0^{2\pi} V_0 B_n \, d\theta$$

$$= \frac{1}{2\pi}\int_0^{2\pi} V_0 I_{mn} \sin(n\theta + \psi_n - \phi_n) \, d\theta = 0$$

and
$$P'_{0n} = \frac{1}{2\pi}\int_0^{2\pi} I_0 A_n \, d\theta$$

$$= \frac{1}{2\pi}\int_0^{2\pi} I_0 V_{mn} \sin(n\theta + \psi_n) \, d\theta = 0$$

The total power is given by

$$P = P_0 + P_{\text{like}} + P_{\text{unlike}} + P_{0n} + P'_{0n}$$

$$= V_0 I_0 + (V_1 I_1 \cos \phi_1 + V_2 I_2 \cos \phi_2 + \ldots + V_n I_n \cos \phi_n) + 0 + 0 + 0$$

$$P = V_0 I_0 + V_1 I_1 \cos \phi_1 + V_2 I_2 \cos \phi_2 + \ldots + V_n I_n \cos \phi_n \tag{19.15}$$

or

$$P = P_0 + P_1 + P_2 + \ldots + P_n$$

= sum of separate powers for each harmonic components alone

19.16 POWER FACTOR

The overall power factor of a single-phase circuit when harmonics are present is defined as

$$\text{power factor} \overset{\Delta}{=} \frac{\text{total power}}{\text{r.m.s. voltage} \times \text{r.m.s. current}} \tag{19.16.1}$$

Since harmonic currents have different phase angles with respect to their voltages, the power factor cannot be expressed in terms of a phase angle.

19.17 EFFECT OF HARMONICS IN SINGLE-PHASE CIRCUITS

Consider a nonsinusoidal voltage applied to a single-phase circuit containing linear circuit elements. The resulting current will also be nonsinusoidal. Each harmonic term is treated separately. The overall r.m.s. values and power consumption are calculated by the superposition principle. Let

$$v = V_{m1} \sin(\theta + \phi_1) + V_{m2} \sin(2\theta + \phi_2) + \ldots + V_{mn} \sin(n\theta + \phi_n) \tag{19.17.1}$$

Pure resistance *R*

Let the voltage given by Eq. (19.17.1) be applied to a pure resistance R. The expression for the current is

$$i = \frac{v}{R} = \frac{V_{m1}}{R} \sin(\theta + \phi_1) + \frac{V_{m2}}{R} \sin(2\theta + \phi_2) + \ldots + \frac{V_{mn}}{R} \sin(n\theta + \phi_n) \tag{19.17.2}$$

Eq. (19.17.2) shows that the harmonic contents in the current wave are the same as those in the voltage wave and the shapes of the two waves are identical.

Pure inductance *L*

Let the voltage given by Eq. (19.17.1) be applied to a pure inductance of L henrys. The reactance of the inductor at the fundamental frequency is ωL, and the reactance at the nth harmonic is $n\omega L$. For every harmonic term the current lags behind the voltage by $\pi/2$ radians. Hence the current is given by

$$i = \frac{V_{m1}}{\omega L}\sin\left(\theta + \phi_1 - \frac{\pi}{2}\right) + \frac{V_{m2}}{2\omega L}\sin(2\theta + \phi_2 - \pi/2)$$

$$+ \ldots + \frac{V_{mn}}{n\omega L}\sin(n\theta + \phi_n - \pi/2) \qquad (19.17.3)$$

For the voltage wave,

$$\frac{n\text{th harmonic}}{\text{fundamental}} = \frac{V_{mn}}{V_{m1}}$$

For the current wave,

$$\frac{n\text{th harmonic}}{\text{fundamental}} = \frac{V_{mn}/(n\omega L)}{V_{m1}/(\omega L)} = \frac{V_{mn}}{nV_{m1}}$$

Thus, the harmonic content in the current wave is smaller than that of the voltage wave in a circuit containing pure inductance. For example, for the nth harmonic, the harmonic content in the current wave is only $\frac{1}{n}$ of the corresponding harmonic content in the voltage wave.

Pure capacitance *C*

Let the voltage given by Eq. (19.17.1) be applied to a capacitor of capacitance C farads. The reactance of the capacitor at the fundamental frequency is $1/(\omega C)$. The reactance at the nth harmonic frequency is $1/(n\omega C)$. For every harmonic term, the current leads the voltage by $\pi/2$ radians. Hence the expression for the current is

$$i = V_{m1}\,\omega C \sin(\theta + \phi_1 + \pi/2) + 2V_{m2}\,\omega C \sin(2\theta + \phi_2 + \pi/2)$$

$$+ \ldots + nV_{mn}\,\omega C \sin(n\theta + \phi_n + \pi/2) \qquad (19.17.4)$$

For the voltage wave, $$\frac{n\text{th harmonic}}{\text{fundamental}} = \frac{V_{mn}}{V_{m1}}$$

For the current wave, $$\frac{n\text{th harmonic}}{\text{fundamental}} = \frac{nV_{mn}\,\omega C}{V_{m1}\,\omega C} = n\frac{V_{mn}}{V_{m1}}$$

Thus, the harmonic content of the current wave is larger than that of the voltage wave. For example, for the nth harmonic it is n times larger.

Harmonic resonance

When the voltage given by Eq. (19.17.1) is applied to a circuit containing inductance and capacitance, series or parallel resonance occurs at one of the harmonic frequencies of the applied voltage. This effect is called **selective or harmonic resonance**. In case of series selective reasonance, the current at resonant frequency is large producing large voltage drops in inductance and capacitance. When parallel selective resonance occurs, the supply current at resonant frequency selective resonance occurs, the supply current at resonant frequency is a minimum and the current through both the inductance and capacitance is very large.

19.18 RESPONSE TO COMPLEX WAVES

A periodic nonsinusoidal waveform can be written in terms of a d.c. term. a_0, and sinusoidal terms. The steady-state response of circuits to the waveform can be found by sinusoidal phasor techniques and superposition. Each term in a Fourier series can be considered as a sinusoidal source of a particular frequency (the constant term is considered a d.c. source). By superposition, we can find the response due to each of these sources separately and then add the responses to get the total response for the combination of sources. The sum of these responses in the time-domain form gives the Fourier series of the response.

If a voltage of complex waveshape is applied to a series RLC circuit, then to determine the overall response, it is necessary to determine the response due to each component separately, and then determine the resultant by superposition.

Let the complex voltage wave be expressed in the form

$$v = V_0 + V_{m1} \sin(\omega t + \phi_1) + V_{m2} \sin(2\omega t + \phi_2) + \ldots + V_{mn} \sin(n\omega t + \phi_n)$$

The impedance values of the given network due to various frequency voltages are first determined. Let the impedance to d.c. voltage be Z_0 and the impedances to other frequencies be

$$\mathbf{Z}_1 = Z_1 \,\underline{/\,\theta_1},\ \mathbf{Z}_2 = Z_2 \,\underline{/\,\theta_2},\ \ldots,\ \mathbf{Z}_n = Z_n \,\underline{/\,\theta_n}$$

The current is given by

$$i = \frac{V_0}{Z_0} + \frac{V_{m1}}{Z_1} \sin(\omega t + \phi_1 - \theta_1) + \frac{V_{m2}}{Z^2} \sin(2\omega t + \phi_2 - \theta_2)$$

$$+ \ldots + \frac{V_{mn}}{Z_n} \sin(n\omega t + \phi_n - \theta_n)$$

■ **EXAMPLE 19.13** A voltage wave is written as

$$v = 200 \sin(\omega t + 30°) - 80 \sin(3\omega t - 60°) + 50 \sin(5\omega t + 45°) \text{ (V)}$$

Determine the r.m.s value of the voltage.

SOLUTION : $V_{1m} = 200,\ V_{2m} = 80,\ V_{3m} = 50$

therefore, $V_1 = \dfrac{200}{\sqrt{2}},\ V_2 = \dfrac{80}{\sqrt{2}},\ V_3 = \dfrac{50}{\sqrt{2}}$

and $V = \sqrt{\left(\dfrac{200}{\sqrt{2}}\right)^2 + \left(\dfrac{80}{\sqrt{2}}\right)^2 + \left(\dfrac{50}{\sqrt{2}}\right)^2} = 156.36$ volts

■ **EXAMPLE 19.14** A complex wave, having an r.m.s. value of 200V has 15 per cent third harmonic and 8 per cent fifth harmonic. Find the r.m.s. value of fundamental and other harmonics.

SOLUTION : $V_3 = \dfrac{15}{100} V_1,\ V_5 = \dfrac{8}{100} V_1$

$$V = (V_1^2 + V_3^2 + V_5^2)^{1/2}$$

$$200 = \sqrt{[V_1^2 + (0.15V_1)^2 + (0.08V_1)^2]} = 1.014V_1$$

$$V_1 = \frac{200}{1.014} = 197.1 \text{ V}$$

$$V_3 = 0.15\, V_1 = 0.15 \times 197.1 = 29.57 \text{ V}$$

$$V_5 = 0.08\, V_1 = 0.08 \times 197.1 = 15.77 \text{ V}$$

■ **EXAMPLE 19.15** Find the sum of the following two current waves

$$i_a = 8 \sin (\omega t + 30^\circ) - 15 \sin (3\omega t - 10^\circ) \text{ (A)}$$

$$i_b = 12 \sin (\omega t - 60^\circ) + 5 \sin (3\omega t + 45^\circ) \text{ (A)}$$

SOLUTION : The r.m.s. value of the fundamental component of total current is given by

$$I_1 = \frac{8}{\sqrt{2}} \underline{/ 30^\circ} + \frac{12}{\sqrt{2}} \underline{/ -60^\circ}$$

$$= \frac{1}{\sqrt{2}} (6.928 + j4) + \frac{1}{\sqrt{2}} (6 - j10.392)$$

$$= \frac{1}{\sqrt{2}} (12.928 - j6.392) = \frac{14.42}{\sqrt{2}} \underline{/ -26.31^\circ} \text{ A}$$

The r.m.s. value of the third harmonic component of total current is given by

$$I_3 = -\frac{15}{\sqrt{2}} \underline{/ -10^\circ} + \frac{5}{\sqrt{2}} \underline{/ 45^\circ}$$

$$= \frac{1}{\sqrt{2}} (-14.772 + j2.604) + \frac{1}{\sqrt{2}} (3.535 + j3.535)$$

$$= \frac{1}{\sqrt{2}} (-11.237 + j6.139) = \frac{12.804}{\sqrt{2}} \underline{/ 151.35^\circ} \text{ A}$$

The total current is given by

$$i = I_{1m} \sin (\omega t - 26.31^\circ) + I_{3m} \sin (3\omega t + 151.35^\circ)$$

or

$$i = 14.42 \sin (\omega t - 26.31^\circ) + 12.804 \sin (3\omega t + 151.35^\circ) \text{ A}$$

■ **EXAMPLE 19.16** The voltage and current in a circuit are represented as follows :

$$v = 120 \sin (\omega t + 30^\circ) - 50 \sin (3\omega t + 60^\circ) + 30 \sin 5\omega t \text{ volts}$$

$$i = 20 \sin (\omega t - 30^\circ) + 16 \sin (3\omega t + 30^\circ) + 10 \cos (5\omega t - 60^\circ) \text{ amperes}$$

Determine the power consumed in the circuit and the overall power factor.

SOLUTION : The voltage and current can be written as

$$v = 120 \sin (\omega t + 30^\circ) + 50 \sin (3\omega t + 60^\circ + 180^\circ) + 30 \sin 5\omega t \text{ volts}$$

$$i = 20 \sin (\omega t - 30^\circ) + 16 \sin (3\omega t + 30^\circ) + 10 \sin (5\omega t - 60^\circ + 90^\circ) \text{ (A)}$$

Here the cosine term of the current wave is converted to the sine term.

$$P = V_1 I_1 \cos \phi_1 + V_3 I_3 \cos \phi_3 + V_5 I_5 \cos \phi_5$$

$$= \frac{120}{\sqrt{2}} \times \frac{20}{\sqrt{2}} \cos [30^\circ - (30^\circ)] + \frac{50}{\sqrt{2}} \times \frac{16}{\sqrt{2}} \cos (60^\circ + 180^\circ - 30^\circ)$$

$$+ \frac{30}{\sqrt{2}} \times \frac{10}{\sqrt{2}} \cos [0 - (-60^\circ + 90^\circ)]$$

$$= 600 - 346.4 + 129.9 = 383.5 \text{ W}$$

$$V = (V_1^2 + V_3^2 + V_5^2)^{1/2}$$

$$= \left[\left(\frac{120}{\sqrt{2}}\right)^2 + \left(\frac{50}{\sqrt{2}}\right)^2 + \left(\frac{30}{\sqrt{2}}\right)^2\right]^{1/2} = 94.32 \text{ volts}$$

$$I = \left[\left(\frac{20}{\sqrt{2}}\right)^2 + \left(\frac{16}{\sqrt{2}}\right)^2 + \left(\frac{10}{\sqrt{2}}\right)^2\right]^{1/2} = 19.439 \text{ amperes}$$

$$\text{Overall power factor} = \frac{\text{total power}}{\text{r.m.s. voltage} \times \text{r.m.s. current}}$$

$$= \frac{383.5}{94.32 \times 19.439} = 0.209$$

■ **EXAMPLE 19.17** The applied voltage and resulting current in a two-element series circuit are

$$v = 50 + 50 \sin 5000t + 30 \sin 10\,000t + 20 \sin 20\,000t \text{ (V)}$$

$$i = 11.2 \sin (5000t + 63.4^\circ) + 10.6 \sin (10\,000t + 45^\circ) + 8.97 \sin (20\,000t + 26.6^\circ) \text{ (A)}$$

Determine the network elements and the power supplied to the network.

SOLUTION : The voltage wave has a *d.c.* component, 50, but there is no corresponding *dc* component in current wave. This shows that one of the two elements in the circuit is a capacitor which blocks the *dc*. Since power is supplied to the circuit, the other element must be a resistor.

The r.m.s. value of the current

$$I = \sqrt{\left(\frac{11.2}{\sqrt{2}}\right)^2 + \left(\frac{10.6}{\sqrt{2}}\right)^2 + \left(\frac{8.97}{\sqrt{2}}\right)^2} = 12.61 \text{ A}$$

$$P = V_1 I_1 \cos \phi_1 + V_2 I_2 \cos \phi_2 + V_3 I_3 \cos \phi_3$$

$$= \frac{50}{\sqrt{2}} \times \frac{11.2}{\sqrt{2}} \cos 63.4^\circ + \frac{30}{\sqrt{2}} \times \frac{10.6}{\sqrt{2}} \cos 45^\circ + \frac{20}{\sqrt{2}} \times \frac{8.97}{\sqrt{2}} \cos 26.6^\circ = 318 \text{ W}$$

Also, $$P = I^2 R, R = \frac{P}{I^2} = \frac{318}{(12.61)^2} = 2\ \Omega$$

At $$\omega = 5\,000 \text{ rad/s},$$

$$|Z_1| = \frac{V_{1m}}{I_{1m}} = \frac{50}{11.2} = 4.464\ \Omega$$

Also, $$|Z_1|^2 = R^2 + X_c^2$$

$$19.929 = 2^2 + X_c^2$$

$$X_c = 3.991 = \frac{1}{\omega C}$$

$$C = \frac{1}{\omega \times 3.991} = \frac{1}{5000 \times 3.991} \text{ F} = 50\ \mu\text{F}$$

■ **EXAMPLE 19.18** A voltage $v = 200 \sin 314t + 50 \sin (942t + 45^\circ)$ volts is applied to a circuit consisting of a resistance of 20 Ω, an inductance 20 mH and a capacitance of 56.3 μF all connected in series. Find (*a*) the rms value of the applied voltage, (*b*) an expression for the instantaneous value of current, (*c*) the rms value of current, and (*d*) the power consumed in the circuit.

SOLUTION :

(*a*) The rms value of the voltage is

$$V = \sqrt{(V_1^2 + V_2^2)} = \sqrt{\left[\left(\frac{200}{\sqrt{2}}\right)^2 + \left(\frac{50}{\sqrt{2}}\right)^2\right]} = 145.8 \text{ V}$$

(*b*) For the *fundamental*, ω = 314 rad/s. Inductive reactance at fundamental frequency

$$X_{L1} = \omega L = 314 \times 20 \times 10^{-3} = 6.28\ \Omega$$

Capacitive reactance at fundamental frequency

$$X_{c1} = \frac{1}{\omega C} = \frac{1}{314 \times 56.3 \times 10^{-6}} = 56.6\ \Omega$$

Impedance at fundamental frequency

$$\mathbf{Z}_1 = R + j(X_{L1} - X_{C1})$$

$$= 20 + j(6.28 - 56.6) = 54 \angle -68.3^\circ\ \Omega$$

$$i_1 = \frac{V_{1m}}{Z_1} \sin(\omega t + \theta_1)$$

$$= \frac{200}{54} \sin(314t + 68.3^\circ)$$

or

$$i_1 = 3.7 \sin(314t + 68.3^\circ)$$

At third harmonic frequency

Inductive reactance, $X_{L3} = 3\omega L = 18.84\ \Omega$

Capacitive reactance, $X_{L3} = \frac{1}{3\omega C} = 18.87\ \Omega$

Impedance $\mathbf{Z}_3 = R + j(X_{L3} - X_{C3})$

$$= 20 + j(18.84 - 18.87) = 20 \angle 0^\circ$$

$$i_3 = \frac{V_{3m}}{Z_3} \sin(3\omega t + \theta_3)$$

$$= \frac{50}{20} \sin(942t + 45^\circ)$$

r

$$i_3 = 2.5 \sin(942t + 45^\circ)$$

The total current is equal to the sum of the harmonic currents. That is,

$$i = i_1 + i_3$$

$$i = 3.7 \sin(314t + 68.3^\circ) + 2.5 \sin(942t + 45^\circ) \text{ A}$$

(*c*) The rms value of current

$$I = \sqrt{\left[\left(\frac{3.7}{\sqrt{2}}\right)^2 + \left(\frac{2.5}{\sqrt{2}}\right)^2\right]} = 3.157 \text{ A}$$

(*d*) Power absorbed by the circuit

$$P = I^2R = (3.157)^2 \times 20 = 199.4 \text{ W}$$

Alternatively

$$P = P_1 + P_3 = V_1I_1 \cos \phi_1 + V_3 I_3 \cos \phi_3$$

$$= \frac{200}{\sqrt{2}} \times \frac{3.7}{\sqrt{2}} \cos 68.3^\circ + \frac{50}{\sqrt{2}} \times \frac{2.5}{\sqrt{2}} \cos 0^\circ$$

$$= 199.3 \text{ W}$$

■ **EXAMPLE 19.19** A three-element series circuit, with $R = 5\ \Omega$, $L = 5$ mH and $C = 20\ \mu$F, has an applied voltage

$$v = 200 \sin \omega t + 10 \sin 3\omega t \text{ (V)}$$

The frequency is adjusted so that the third harmonic is at resonance. Determine (*a*) the fundamental frequency ω, (*b*) expression for current, and (*c*) the rms value of current.

SOLUTION :

(*a*) Since the circuit is in resonance at third harmonic frequency,

$$3\omega L = \frac{1}{3\omega C}$$

or

$$\omega = \frac{1}{3\sqrt{LC}} = \frac{1}{3\,(5 \times 10^{-3} \times 20 \times 10^{-6})^{1/2}}$$

$$= 1\,054 \text{ rad/s}$$

(b) Impedance at fundamental frequency.

$$Z_1 = R + j\omega L - j\frac{1}{\omega C}$$

$$= 5 + j\left(1\,054 \times 5 \times 10^{-3} - \frac{1}{1054 \times 20 \times 10^{-6}}\right)$$

$$= 5 - j42.168 = 42.46 \underline{/ -83.2^\circ}\ \Omega$$

Impedance at third harmonic frequency

$$Z_3 = R + j3\omega L - j\frac{1}{3\omega C}$$

Since the circuit is in resonance at third harmonic frequency,

$$Z_3 = R = 5\ \Omega$$

$$I_1 = \frac{V_1}{Z_1} = \frac{200/\sqrt{2}}{42.46 \underline{/ -83.2^\circ}} = 3.33 \underline{/ 83.2^\circ} \text{ A}$$

$$I_3 = \frac{V_3}{Z_3} = \frac{10/\sqrt{2}}{5} = 1.414$$

The current wave is given by

$$i = I_{1m} \sin(\omega t + 83.2^\circ) + I_{3m} \sin 3\omega t$$

$$= 3.33 \times \sqrt{2} \sin(\omega t + 83.2^\circ) + 1.414 \times \sqrt{2} \sin 3\omega t$$

or $$i = 4.71 \sin(\omega t + 83.2^\circ) + 2 \sin 3\omega t$$

(*c*) The rms value of current

$$I = \sqrt{I_1^2 + I_3^2} = \sqrt{(3.33)^2 + (1.414)^2} = 3.617 \text{ A}$$

EXAMPLE 19.20 A voltage wave

$$v = 150 \sin \omega t + 50 \sin(3\omega t + 33^\circ) - 20 \sin(5\omega t - 30^\circ)$$

is applied to the circuit shown in Fig. 19.19. Find (*a*) expression for current wave, (*b*) rms value of current, and (*c*) total power consumed in the circuit. The reactances shown in Fig. 19.19 are for fundamental frequency.

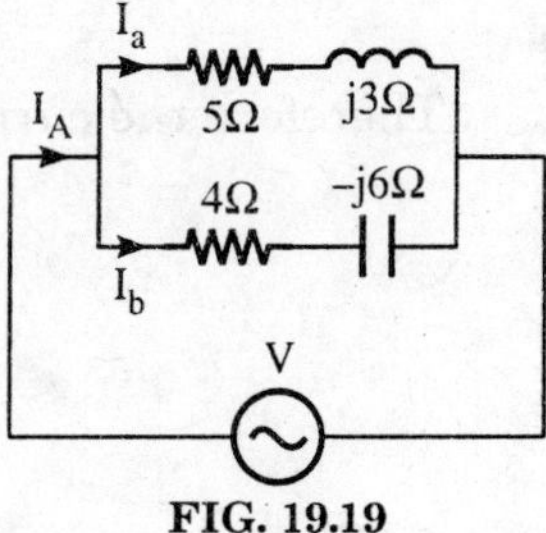

FIG. 19.19

SOLUTION : (*a*) For the fundamental frequency

$$V_1 = \frac{150}{\sqrt{2}} = 106 \text{ V}$$

$$\mathbf{I}_{a1} = \frac{106}{5 + j3} = 18.18 \underline{/ -30.96^\circ} \text{ A}$$

$$\mathbf{I}_{b1} = \frac{106}{4 - j6} = 14.7 \underline{/ 56.3^\circ} \text{ A}$$

$$\mathbf{I}_{A1} = \mathbf{I}_{a1} + \mathbf{I}_{b1}$$

$$= 18.18 \underline{/ -30.96^\circ} + 14.7 \underline{/ 56.3^\circ}$$

$$= 15.59 - j9.35 + 8.15 + j12.23$$

$$= 23.74 + j2.88 = 23.91 \underline{/ 6.91^\circ} \text{ A}$$

For the third harmonic, the inductive reactance of branch *a* is $j9\ \Omega$, and the capacitive reactance of branch *b* is $-j2\ \Omega$. Therefore,

$$V_3 = \frac{50}{\sqrt{2}} = 35.35 \text{ V}$$

$$\mathbf{I}_{a3} = \frac{35.35}{5 + j9} = 3.43 \underline{/ -60.94^\circ} \text{ A}$$

$$\mathbf{I}_{b3} = \frac{35.35}{4 - j2} = 7.9 \underline{/ 26.56^\circ} \text{ A}$$

$$\mathbf{I}_{A3} = \mathbf{I}_{a3} + \mathbf{I}_{b3} = 3.43 \underline{/ -60.94^\circ} + 7.9 \underline{/ 26.56^\circ}$$

$$= 1.666 - j2.998 + 7.066 + j3.532$$

$$= 8.732 + j0.534 = 8.748 \underline{/ 3.5^\circ} \text{ A}$$

For the fifth harmonic, the inductive reactance of branch *a* is $j5 \times 3\ \Omega$, and the capacitive reactance of branch *b* is $-j\frac{6}{5}\ \Omega$.

Also, $V_5 = \frac{20}{\sqrt{2}} = 14.14$ V. Therefore

$$\mathbf{I}_{a5} = \frac{14.14}{5 + j15} = 0.894 \underline{/ -71.56^\circ} \text{ A}$$

$$\mathbf{I}_{b5} = \frac{14.14}{4 - j1.2} = 3.386 \underline{/ 16.7^\circ} \text{ A}$$

$$\mathbf{I}_{A5} = \mathbf{I}_{a5} + \mathbf{I}_{b5} = 0.894 \underline{/ -71.56^\circ} + 3.386 \underline{/ 16.7^\circ}$$

$$= 0.283 + j0.848 + 3.243 + j0.973$$

$$= 3.526 + j0.125 = 3.528 \underline{/ 2.03^\circ} \text{ A}$$

Therefore the current wave can be written as

$$i = \sqrt{2} \times 23.91 \sin(\omega t + 6.91^\circ) + \sqrt{2} \times 8.748 \sin(3\omega t + 30^\circ + 3.5^\circ) - \sqrt{2} \times 3.528 \sin(5\omega t - 30^\circ + 2.03^\circ)$$

or

$$i = 33.81 \sin(\omega t + 6.91^\circ) + 12.37 \sin(3\omega t + 33.5^\circ) - 4.989 \sin(5\omega t - 27.97^\circ)$$

(*b*) The rms value of current is given by

$$I = \sqrt{(I_{A1}^2 + I_{A3}^2 + I_{A5}^2)} = \sqrt{[(23.91)^2 + (8.748)^2 + (3.528)^2]} = 25.7 \text{ A}$$

(*c*) $P = V_1 I_{A1} \cos\phi_1 + V_3 I_{A3} \cos\phi_3 + V_5 I_{A5} \cos\phi_5$

$$= 106 \times 23.91 \cos 6.91^\circ + 35.35 \times 8.748 \cos 3.5^\circ + (-14.14)(-3.528) \cos 2.03^\circ$$

$$= 2516 + 308.665 + 49.854 = 2874.5 \text{ W}$$

■ **EXAMPLE 19.21** A periodic voltage v_{in} shown in Fig. 19.20a is applied across a simple RC network shown in Fig. 19.20b. Determine the steady-state response v_{out} of the periodic input v_{in}. Write the first few terms of the Fourier series v_{out}. Plot the amplitude spectrum of v_{out}.

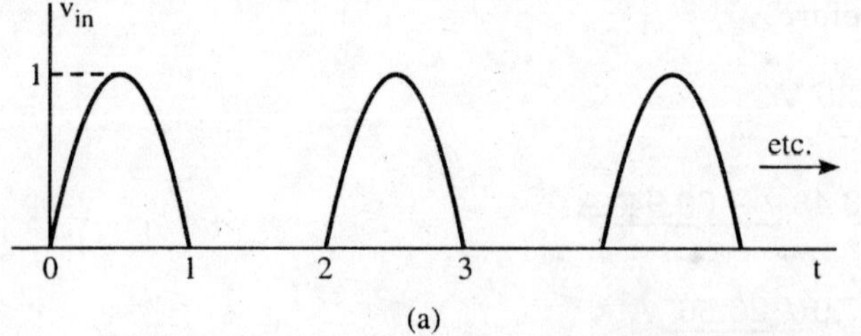

(a)

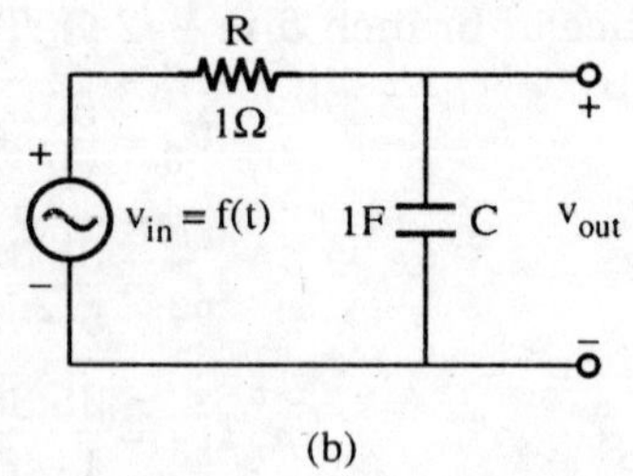

(b)

FIG. 19.20

SOLUTION : The given periodic function is half-wave rectified sine wave. For this function

$$T = 2s, \qquad \omega_0 = \frac{2\pi}{T} = \pi \text{ rad/s}$$

The given waveform can be defined as

$$v_{in}(t) = f(t) = \begin{cases} \sin \pi t, & 0 < t < 1 \\ 0, & 1 < t < 2 \end{cases} \tag{E19.21.1}$$

We have determined the Fourier series for the half-wave rectified voltage in Example 19.3. Hence we can write the first few terms of $v_{in}(t)$ as follows :

$$v_{in} = \frac{1}{\pi} + \frac{1}{2}\sin \pi t - \frac{2}{3\pi}\cos 2\pi t - \frac{2}{15\pi}\cos 4\pi t - \frac{2}{35\pi}\cos 6\pi t - \dots \tag{E19.21.2}$$

Let $$v_{in} = v_{i0} + v_{i1} + v_{i2} + v_{i4} + v_{i6} + \dots \tag{E19.21.3}$$

where $$v_{i0} = \frac{1}{\pi}, \quad v_{i1} = \frac{1}{2}\sin \pi t$$

$$v_{i2} = -\frac{2}{3\pi}\cos 2\pi t, \qquad v_{i4} = -\frac{2}{15\pi}\cos 4\pi t, \dots$$

There by the principle of superposition,

$$v_{out} = v_{00} + v_{01} + v_{02} + v_{04} + v_{06} + \dots \tag{E19.21.4}$$

where v_{00} is the output when the input is $v_{\cdot 0}$

v_{01} is the output when the input is v_{i1}

v_{02} is the output when the input is v_{i2}

v_{04} is the output when the input v_{i4}

ans so on.

From the circuit in Fig. 19.20b, by voltage division we have,

$$\mathbf{V}_{out} = \frac{1/(j\omega C)}{1/(j\omega C) + R}\mathbf{V}_{in} = \frac{1}{1 + j\omega RC}\mathbf{V}_{in}$$

$$= \frac{1}{1 + j\omega}\mathbf{V}_{in} = \frac{\mathbf{V}_{in}}{\sqrt{(1+\omega^2)}\ \underline{/\tan^{-1}\omega}} \tag{E19.21.5}$$

$$v_{i0} = \frac{1}{\pi}$$

$$\mathbf{V}_{i0} = \frac{1}{\pi} \quad (\omega = 0)$$

Substituting these values in Eq. (E19.21.5) we obtain

$$\mathbf{V}_{00} = \frac{\mathbf{V}_{i0}}{1 + j\omega} = \frac{1}{1 + j0}\left(\frac{1}{\pi}\right) = \frac{1}{\pi} \tag{E19.21.6}$$

therefore, $$v_{00} = \frac{1}{\pi} = 0.318$$

$$v_{i1} = \frac{1}{2}\sin \pi t$$

therefore, $$\mathbf{V}_{i1} = \frac{1}{2}\underline{/0^\circ} \quad (\omega = \pi)$$

The response in phasor to this input is given by

$$\mathbf{V}_{01} = \frac{\mathbf{V}_{i1}}{\sqrt{1+\omega^2}\ \underline{/\tan^{-1}\omega}} = \frac{\frac{1}{2}\underline{/0^\circ}}{\sqrt{(1+\pi^2)}\ \underline{/\tan^{-1}\pi}}$$

$$= \frac{1}{2\sqrt{(1+\pi^2)}} \underline{/ (0^\circ - \tan^{-1}\pi)} = 0.151 \underline{/ -72.3^\circ}$$

Therefore in the time domain

$$v_{01} = 0.151 \sin(\pi t - 72.3^\circ) \tag{E19.21.7}$$

Let us consider input

$$v_{\text{in}} = a_n \cos n\pi t$$

where n is even.

therefore, $\quad v_{\text{in}} = a_n \sin(n\pi t + 90^\circ)$

In the phasor form this input can be written as

$$\mathbf{V}_{in} = a_n \underline{/ 90^\circ} \qquad (\omega = n\pi)$$

Hence the output corresponding to this input in phasor form can be written from Eq. (E19.21.5) as

$$\mathbf{V}_{on} = \frac{a_n \underline{/ 90^\circ}}{\sqrt{1+(n\pi)^2} \underline{/ \tan^{-1}(n\pi)}}$$

$$= \frac{a_n}{\sqrt{1+(n\pi)^2}} \underline{/ 90^\circ - \tan^{-1}(n\pi)} \tag{E19.21.8}$$

We shall use Eq. (E19.21.8) to determine $\mathbf{V}_{02}$, $\mathbf{V}_{04}$, $\mathbf{V}_{06}$, etc.

$$\mathbf{V}_{02} = \frac{a_2}{\sqrt{1+(2\pi)^2}} \underline{/ 90^\circ - \tan^{-1}(2\pi)}$$

$$= \frac{2/(3\pi)}{\sqrt{1+(2\pi)^2}} \underline{/ 90^\circ - 80.95^\circ} = 0.0333 \underline{/ 9.05^\circ}$$

Therefore in the time domain

$$v_{02} = -0.0333 (\sin 2\pi t + 9.05^\circ)$$

$$\mathbf{V}_{04} = \frac{a_4}{\sqrt{1+(4\pi)^2}} \underline{/ 90^\circ - \tan^{-1}(4\pi)} = \frac{2/(15\pi)}{\sqrt{1+(4\pi)^2}} \underline{/ 90^\circ - 85.45^\circ}$$

$$= 0.00\,337 \underline{/ -4.54^\circ}$$

therefore, $\quad v_{04} = -0.0\,0337 \sin(4\pi t - 4.54^\circ)$

Hence the first few terms of the Fourier series of v_{out} can be written as

$$v_{\text{out}} = v_{00} + v_{01} + v_{02} + v_{04} + \ldots$$

$$= 0.318 + 0.151 \sin(\pi t - 72.3^\circ) - 0.0333 \sin(2\pi t + 9.05^\circ)$$

$$- 0.00337 \sin(4\pi t - 0.54^\circ) - \ldots$$

The amplitude spectrum of v_{out} is shown in Fig. 19.21.

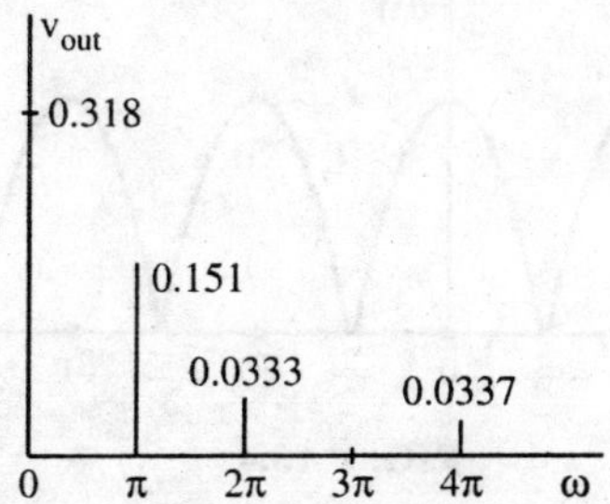

FIG. 19.21 Amplitude spectrum of v_{out} for Example 19.21

EXERCISES

1. Find the Fourier series for the voltage waveform shown in Fig. P.19.1.

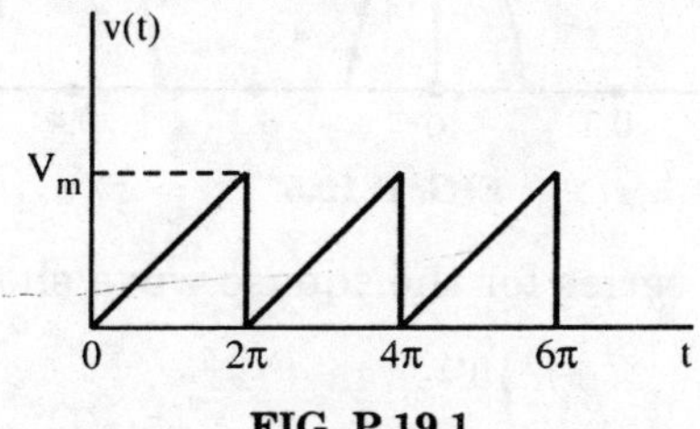

FIG. P 19.1

2. Find the Fourier series for the voltage waveform shown in Fig. P19.2.

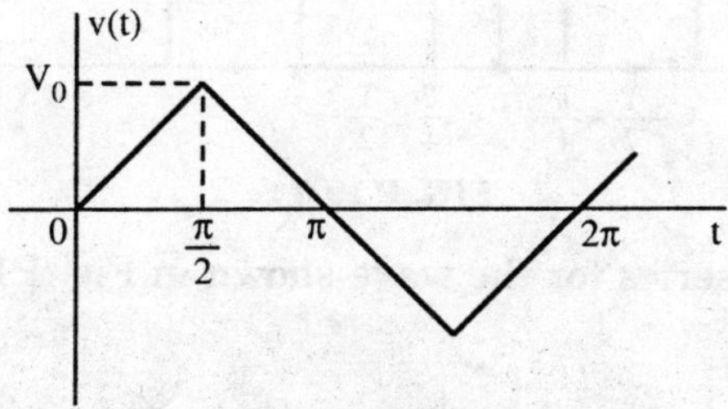

FIG. P 19.2

3. Find the Fourier series for the voltage waveform of a half-wave rectifier shown in Fig. P19.3.

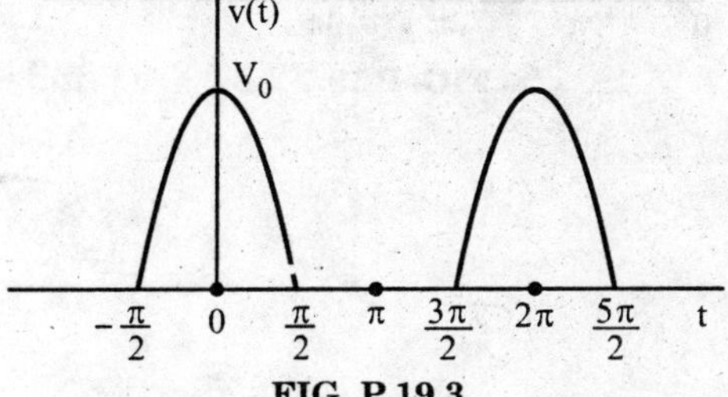

FIG. P 19.3

4. Find the Fourier series for the voltage waveform shown in Fig. P19.4.

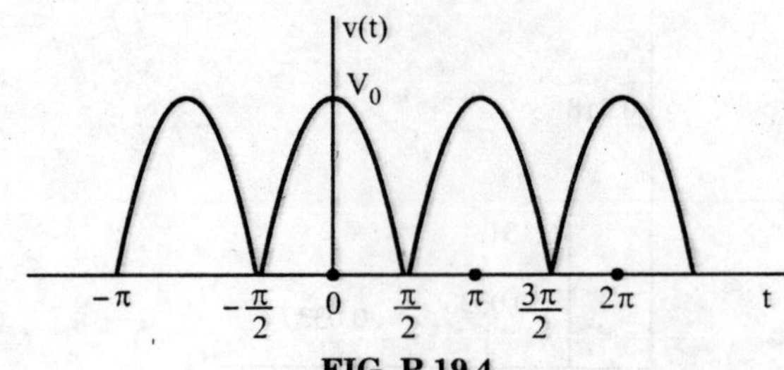

FIG. P 19.4

5. Find the Fourier series for the voltage waveform of a half-wave rectifier shown in Fig. P19.5.

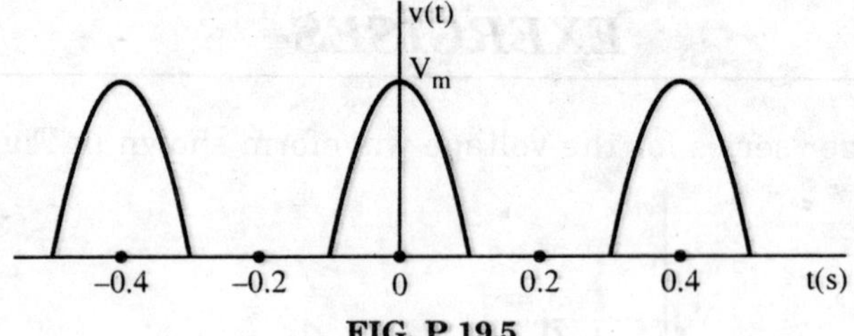

FIG. P 19.5

6. Determine the Fourier series for the square wave shown in Fig. P19.6.

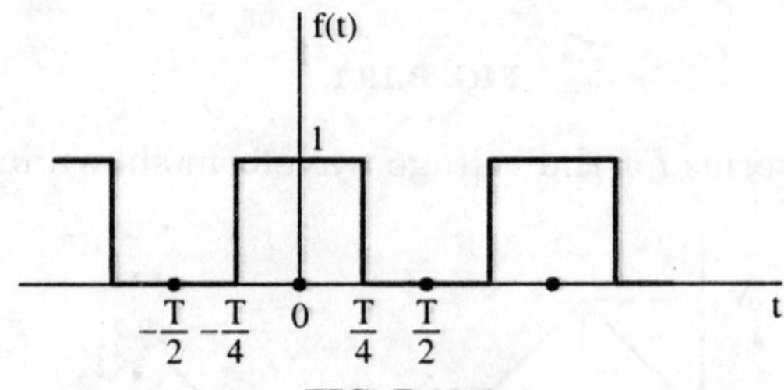

FIG. P 19.6

7. Obtain the Fourier series for the wave shown in Fig. P19.7.

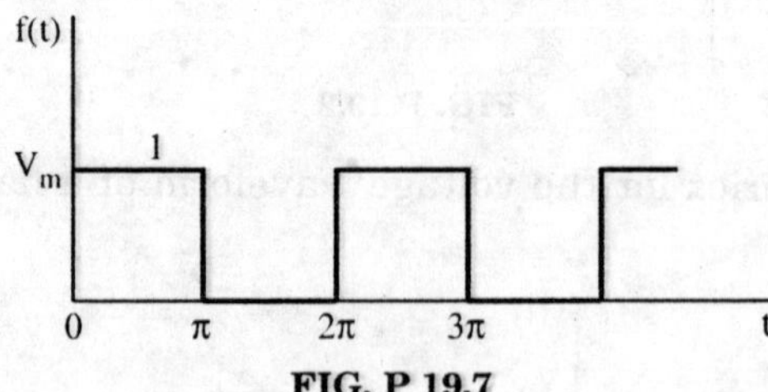

FIG. P 19.7

8. Determine the Fourier series for the sawtooth wave shown in Fig. P19.8.

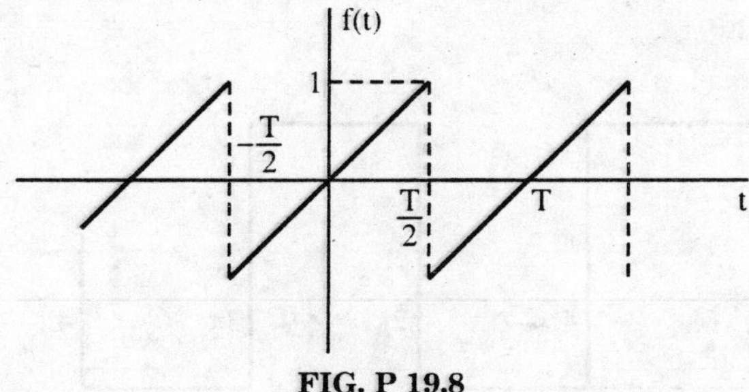

FIG. P 19.8

9. Find the Fourier series representation of the waveform shown in Fig. P19.9.

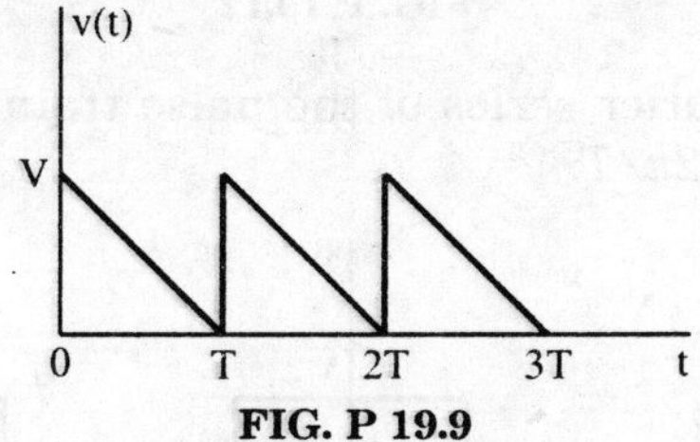

FIG. P 19.9

10. Find the trigonometric Fourier series for the waveform shown in Fig. P19.10.

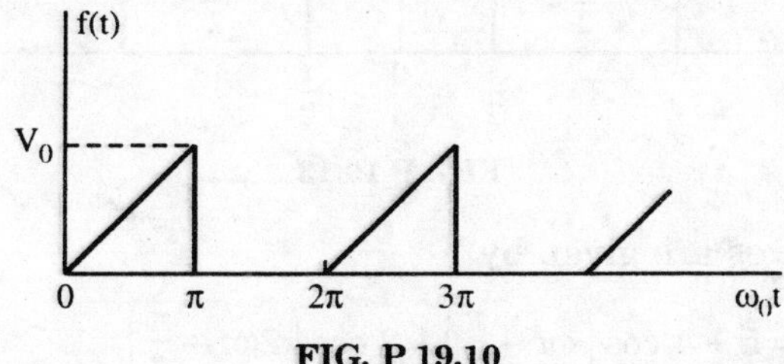

FIG. P 19.10

11. Determine the Fourier series for the waveform shown in Fig. P19.11.

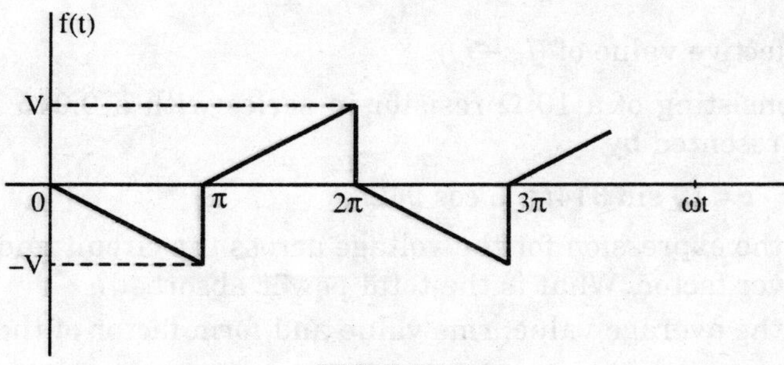

FIG. P 19.11

12. Find the Fourier series for the waveform shown in Fig. P19.12.

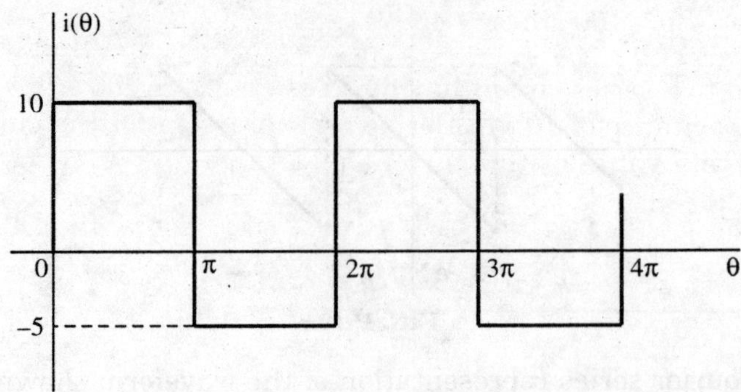

FIG. P 19.12

13. Find the complex Fourier series of the pulse train given in Fig. P19.13 where the period is 2π and $\omega_0 = 2\pi/T$.

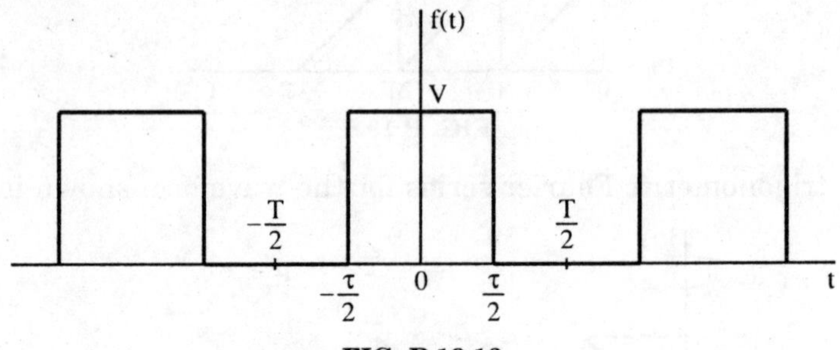

FIG. P 19.13

14. Two current waves are given by

$$i_a = 5 + 4\cos\left(\omega t - \frac{\pi}{3}\right) + 2\cos\left(2\omega t + \frac{\pi}{6}\right)$$

$$i_b = 3 - 5\cos\omega t - 3\cos\left(2\omega t + \frac{\pi}{4}\right)$$

Find the effective value of $(i_b - i_a)$.

15. A circuit consisting of a 10 Ω resistor in series with a 0.015 H inductor carries a current represented by

$$i = 10\sin 314t + 5\cos 942t \text{ A}$$

Determine the expression for the voltage across the circuit and calculate rms value and the power factor. What is the total power absorbed?

16. Determine the average value, rms value and form factor of the wave given by

$$v = 12.5\sin\omega_0 t + 5\sin\omega t \text{ (V)}$$

17. Find the average power in a resistor of 10 Ω if the current is given by

$$i = 10\sin\omega t + 5\sin 3\omega t + 2\sin 5\omega t \text{ (A)}$$

18. A voltage given by

$$v = 100 \sin \omega t + 40 \sin \left(3\omega t - \frac{\pi}{6}\right) + 10 \sin \left(5\omega t - \frac{\pi}{3}\right) \text{ volts}$$

is applied to a series circuit having a resistance of 100 Ω, an inductance of 40.6 mH and a capacitance of 10 μF. Derive an expression for the curent in the circuit. Also find the rms value of the current and the power consumed in the circuit. Take $\omega = 314$ rad/s.

ANSWERS

1. $v(t) = \frac{V_m}{2} - \frac{V_m}{\pi}\left(\sin \omega_0 t + \frac{1}{2}\sin 2\omega_0 t + \frac{1}{3}\sin 3\omega_0 t + \ldots\right)$

2. $v(t) = \frac{8V_0}{\pi^2}\left(\sin \omega_0 t - \frac{1}{9}\sin 3\omega_0 t + \frac{1}{25}\sin 5\omega_0 t - \ldots\right)$

3. $v(t) = \frac{V_0}{\pi}\left(1 + \frac{\pi}{2}\cos \omega_0 t + \frac{2}{3}\cos 2\omega_0 t - \frac{2}{15}\cos 4\omega_0 t + \frac{2}{35}\cos 6\omega_0 t - \ldots\right)$

4. $v(t) = \frac{2V_0}{\pi}\left(1 + \frac{2}{3}\cos \omega_0 t - \frac{2}{15}\cos 4\omega_0 t + \frac{2}{35}\cos 6\omega_0 t - \ldots\right)$

5. $v(t) = \frac{V_m}{\pi} + \frac{V_m}{2}\cos \omega_0 t + \frac{2V_m}{3\pi}\cos 10\pi t$

   $- \frac{2V_m}{15\pi}\cos 20\pi t + \frac{2V_m}{35\pi}\cos 30\pi t - \ldots$

6. $f(t) = \frac{1}{2} + \frac{2}{\pi}\left(\cos \omega_0 t - \frac{1}{3}\cos 3\omega_0 t + \frac{1}{5}\cos 5\omega_0 t - \ldots\right)$

7. $f(t) = \frac{V_m}{2} + \frac{2V_m}{\pi}\left(\sin t + \frac{1}{3}\sin 3t + \frac{1}{5}\sin 5t + \ldots\right)$

8. $f(t) = \frac{1}{\pi}\left(\sin \omega_0 t - \frac{1}{2}\sin 2\omega_0 t + \frac{1}{3}\sin 3\omega_0 t - \ldots\right)$

9. $v(t) = \frac{V}{2} + \frac{V}{\pi}\sin \frac{2\pi}{T}t + \frac{V}{2\pi}\sin \frac{4\pi}{T}t + \ldots$

10. $f(t) = \frac{V_0}{4} - \frac{2V_0}{\pi^2}\left(\cos \omega_0 t + \frac{1}{9}\cos 3\omega_0 t + \frac{1}{25}\cos 5\omega t + \ldots\right)$

    $+ \frac{V_0}{\pi}\left(\sin \omega_0 t - \frac{1}{2}\sin 2\omega_0 t + \frac{1}{3}\sin 3\omega_0 t - \ldots\right)$

11. $f(t) = \frac{4V}{\pi^2}\left(\cos \omega_0 t + \frac{1}{9}\cos 3\omega_0 t + \frac{1}{25}\cos 5\omega_0 t + \ldots\right)$

$$-\frac{2V}{\pi}\left(\sin \omega_0 t + \frac{1}{3}\sin 3\omega_0 t + \frac{1}{5}\sin 5\omega_0 t + \ldots\right)$$

12. $i(\theta) = 2.5 + \sum_{n=1}^{\infty} \frac{30}{n\pi} \sin n\theta \quad (n \text{ is odd})$

13. $f(t) = \sum_{n=-\infty}^{\infty} \frac{V_\tau}{T} \frac{\sin\left(\frac{n\pi\tau}{T}\right)}{\left(\frac{n\pi\tau}{T}\right)} e^{\frac{j2\pi nt}{T}}$

14. 6.85 A

15. $v = 110.5 \sin (314t + 25.2°) + 86.5 \cos (942t + 54.6°)$ V; 99 V; 0.8; 625 W

16. 8.594 V, 9.514, 1.107

17. 645 W

18. 0.329 A, 10.8 W

20

FOURIER TRANSFORM

20.1 INTRODUCTION

For a periodic waveform of period T, the amplitude and phase spectra are discrete. That is, they consist of lines spaced at intervals of $\Delta\omega = \omega_0 = 2\pi/T$. Every line erected at $n\omega_0$ for an integral value of n represents the amplitude of the nth harmonic. As T increases, $\Delta\omega$ become, smaller and the lines in the discrete spectrum become closer. In the limit as T tends to infinity, $\Delta\omega$ tends to zero and the lines become so dense that the discrete frequency spectrum becomes a continuous spectrum. This continuous spectrum exists for every value of ω and is a continuous function of ω. For a continuous spectrum n becomes meaningless and the angular frequency $n\omega_0$ can assume any value and it becomes a continuous variable ω.

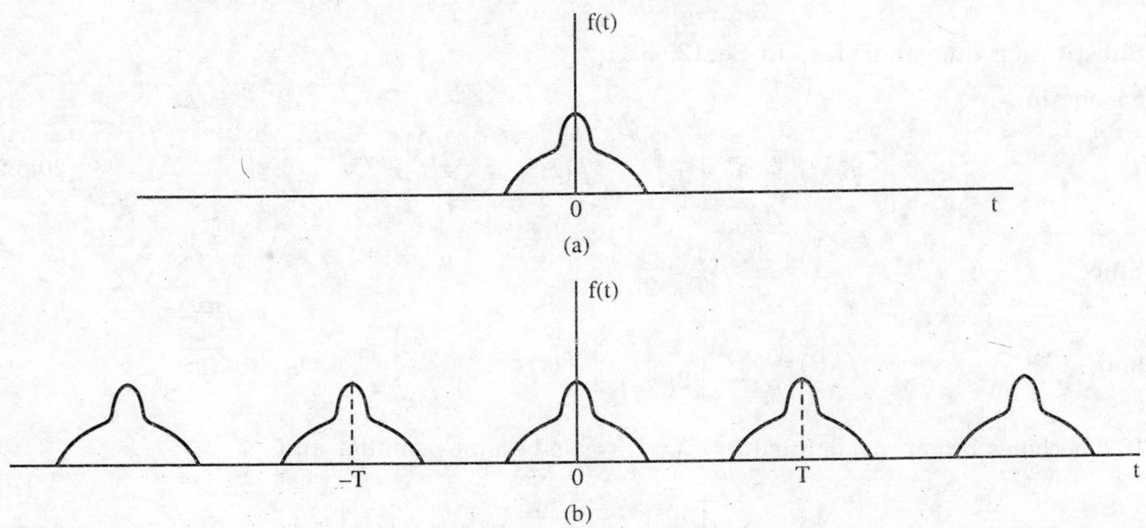

FIG. 20.1 (a) Arbitarily defined function of time $f(t)$ (b) Periodic function $f_T(t)$ based on $f(t)$

20.2 REPRESENTATION OF NONPERIODIC FUNCTIONS

We have used the Fourier series to represent a periodic function by a *discrete* sum of exponential functions. We desire to represent a nonperiodic (aperiodic) function t (shown in Fig. 20.1) in terms of exponential functions. In order to do this, we first construct a new periodic function $f_T(t)$ with period T in such a way that $f(t)$ defines one cycle of this periodic function, as shown in Fig. 20.1.

The period T is so large that there is no overlap of the pulses of the shape of $f(t)$. Since $f_T(t)$ is a periodic function, it can be represented in terms of the exponential form of the Fourier series. As T approaches infinity then the pulses in the periodic function $f_T(t)$ repeat after an infinite interval. In other words, as $T \to \infty$ we are left with $f(t)$ which never repeats and $f_T(t)$ and $f(t)$ become identical. That is,

$$\lim_{T \to \infty} f_T(t) = f(t)$$

Thus, the Fourier series representing $f_T(t)$ will also represent $f(t)$, if we let T become infinite in this series.

20.3 THE FOURIER INTEGRAL

The exponential Fourier series for $f_T(t)$ can be written as

$$f_T(t) = \sum_{n=-\infty}^{\infty} c_n\, e^{jn\omega_0 t} \tag{20.3.1}$$

where
$$\omega_0 = \frac{2\pi}{T}$$

and
$$c_n = \frac{1}{T}\int_{-T/2}^{T/2} f_T(t)\, e^{-jn\omega_0 t}\, dt = \frac{1}{T}\int_{-T/2}^{T/2} f(x)\, e^{-jn\omega_0 x}\, dx$$

Substituting the value of c_n in Eq. (20.3.1)

we obtain,

$$f_T(t) = \sum_{n=-\infty}^{\infty}\left[\frac{1}{T}\int_{-T/2}^{T/2} f(x)\, e^{-jn\omega_0 x}\, dx\right] e^{jn\omega_0 t} \tag{20.3.2}$$

Since
$$\omega_0 = \frac{2\pi}{T},\ \frac{1}{T} = \frac{\omega_0}{2\pi}$$

and
$$f_T(t) = \sum_{n=-\infty}^{\infty}\left[\frac{1}{2\pi}\int_{-T/2}^{T/2} f(x)\, e^{-jn\omega_0 x}\, dx\right] e^{jn\omega_0 t}\, \omega_0$$

If T becomes larger. $\Delta\omega$ defined by $\omega_0 = \frac{2\pi}{T} = \Delta\omega$ becomes smaller and

$$f_T(t) = \sum_{n=-\infty}^{\infty}\left[\frac{1}{2\pi}\int_{-T/2}^{T/2} f(x)\, e^{-jn\Delta\omega x}\, dx\right] e^{jn\Delta\omega t}\, \Delta\omega \tag{20.3.3}$$

As $T \mapsto \infty$, then $\Delta\omega \to d\omega$ and the discrete harmonics become a continuous frequency variable, say ω, that is, $n\omega_0 = n\Delta\omega \to \omega$.

Also, the discrete sum $\sum_{n=-\infty}^{\infty}$ becomes the continuous sum $\int_{-\infty}^{\infty}$.

Thus, in the limit. Eq. (20.3.3) becomes

$$f(t) = \int_{-\infty}^{\infty} \frac{1}{2\pi}\left[\int_{-\infty}^{\infty} f(x)\, e^{-j\omega x}\, dx\right] e^{j\omega t}\, d\omega \tag{20.3.4}$$

By changing the variable in Eq. (20.3.4)

We get,

$$f(t) = \frac{1}{2\pi}\int_{-\infty}^{\infty}\left[\int_{-\infty}^{\infty} f(t)\, e^{-j\omega t}\, dt\right] e^{j\omega t}\, d\omega$$

which has the form

$$f(t) = \frac{1}{2\pi}\int_{-\infty}^{\infty} F(j\omega)\, e^{j\omega t}\, d\omega \tag{20.3.5}$$

where

$$\boxed{F(j\omega) = \int_{-\infty}^{\infty} f(t)\, e^{-j\omega_0 t}\, dt} \tag{20.3.6}$$

The function $F(j\omega)$ is called the **Fourier transform of $f(t)$.** The Fourier transform of the function $f(t)$ is denoted by $\mathfrak{F}[f(t)]$. That is

$$\boxed{\mathfrak{F}[f(t)] = F(j\omega)} \tag{20.3.7}$$

The symbol $\mathfrak{F}$ is read as '**the direct Fourier transform of**'. **Eq. (20.3.5) states that, given the Fourier transform** $F(j\omega)$ **we can determine the original time function** $f(t)$. **Thus, the integral given by Eq. (20.3.5) is called the inverse Fourier transform of** $F(j\omega)$ denoted by $\mathfrak{F}^{-1}[F(j\omega)]$. That is,

$$\boxed{\mathfrak{F}^{-1}[F(j\omega)] = f(t)} \tag{20.3.8}$$

The symbol $\mathfrak{F}^{-1}$ is read as '**the inverse Fourier transform of**'. Eqs. (20.3.5) and (20.3.6) are known as the **Fourier transform pair.** Thus, a signal can be described by $f(t)$ in the time domain, or by $F(j\omega)$ in the frequency domain.

It should be noted that *Fourier transform pairs relate the time and frequency domains for nonperiodic (infinite period) functions while Fourier series pairs relate these two domains for periodic (finite period) functions.* A double-ended arrow is also often used to relate a time function to its Fourier transform. Thus we would write

$$f(t) \leftrightarrow F(j\omega)$$

or

$$F(j\omega) \leftrightarrow f(t)$$

20.4 GRAPHICAL REPRESENTATION OF THE FOURIER TRANSFORM

The function $F(j\omega)$ is in general complex. It can be represented graphically. Two sets of graphical representations are common. The plot of magnitude $|F(j\omega)|$ versus ω is called the ***amplitude spectrum*** of $f(t)$. The plot of the angle of $F(j\omega)$ versus frequency ω is called the ***phase spectrum*** of $f(t)$. These two spectra constitute the ***frequency spectrum*** of $f(t)$. Quite often the amplitude spectrum alone is called the frequency spectrum.

20.5 SOME PROPERTIES OF FOURIER TRANSFORMS

The Fourier integral is a special case of the Laplace transform. Therefore all Laplace transform theorems have a Fourier integral version. We shall now discuss some important properties of Fourier transforms.

20.5.1 Linearity Property

The operations of Fourier transforms are linear, that is, if

$$\mathfrak{F}[f_1(t)] = F_1(j\omega)$$

$$\mathfrak{F}[f_2(t)] = F_2(j\omega)$$

then

$$\mathfrak{F}[a_1 f_1(t) + a_2 f_2(t)] = a_1 F_1(j\omega) + a_1 F_2(j\omega)$$

Proof

$$\mathfrak{F}[a_1 f_1(t) + a_2 f_2(t) = \int_{-\infty}^{\infty} [a_1 f_1(t) + a_2 f_2(t)]\, e^{-j\omega t}\, dt$$

$$= a_1 \int_{-\infty}^{\infty} f_1(t)\, e^{-j\omega t}\, dt + a_2 \int_{-\infty}^{\infty} f_2(t)\, e^{-j\omega t}\, dt$$

$$= a_1 F_1(j\omega) + a_2 F_2(j\omega)$$

Here a_1 and a_2 are constants.

20.5.2 Time-Scaling Property

If $\mathfrak{F}[f(t)] = F(j\omega)$

then for a real constant a,

$$\mathfrak{F}[f(at)] = \frac{1}{|a|} F\left(\frac{j\omega}{a}\right)$$

Proof For a real positive constant a,

$$\mathfrak{F}[f(at)] = \int_{-\infty}^{\infty} f(at)\, e^{-j\omega t}\, dt$$

Put $at = x$, then

$$\mathfrak{F}[f(at)] = \frac{1}{a} \int_{-\infty}^{\infty} f(x)\, e^{-j\frac{\omega}{a}x}\, dx$$

therefore,

$$\mathfrak{F}[f(at)] = \frac{1}{a} F\left(\frac{j\omega}{a}\right) \qquad (20.5.2)$$

If a is negative (that is, $a < 0$) and at is replaced by x in Eq. (20.5.1), the lower limit of integration becomes $+\infty$ and the upper limit becomes $-\infty$ and

$$\mathfrak{F}[f(at)] = -\frac{1}{a} F\left(\frac{j\omega}{a}\right) \tag{20.5.3}$$

Combining the results of Eqs. (20.5.2) and (20.5.3) we get,

$$\boxed{\mathfrak{F}[f(at)] = \frac{1}{|a|} F\left(\frac{j\omega}{a}\right)} \tag{20.5.4}$$

The function $f(t)$ represents the function $f(t)$ compressed in time scale by a factor of a. The function $F\left(\frac{j\omega}{a}\right)$ represents the function $F(j\omega)$ expanded in the frequency scale by the same factor a. The expressions for scaling show that as the time function compresses the corresponding frequency spectum expands and vice versa.

20.5.3 Frequency-Scaling Property

From Eq. (20.5.4) it can be prove that

$$\boxed{\mathfrak{F}\left[\frac{1}{|a|} f\left(\frac{t}{a}\right)\right] = F(ja\omega)} \tag{20.5.5}$$

The expressions for scaling indicate that as time function spreads out, the corresponding frequency spectrum contracts and vice versa.

Reversal property

A special case of the scaling property is, when $a = -1$.

$$\boxed{\mathfrak{F}[f(-t)] = F(-j\omega)} \tag{20.5.6}$$

This result shows that if the time function is folded around the $t = 0$ axis, the corresponding frequency function is folded around the $\omega = 0$ axis.

20.5.4 Symmetry Property

If $\qquad \mathfrak{F}[f(t)] = F(j\omega)$

then $\qquad \mathfrak{F} F(jt) = 2\pi f(-\omega)$

Proof

We have

$$f(t) = \frac{1}{2\pi} \int_{-\infty}^{\infty} F(j\omega)\, e^{j\omega t}\, d\omega$$

$$2\pi f(t) = \int_{-\infty}^{\infty} F(j\omega)\, e^{j\omega t}\, d\omega \tag{20.5.7}$$

Replacing t by $-t$ in Eq. (20.5.7) gives

$$2\pi f(-t) = \int_{-\infty}^{\infty} F(j\omega)\, e^{-j\omega t}\, d\omega \qquad (20.5.8)$$

Now replacing t by ω and ω by t in Eq. (20.5.8) gives

$$2\pi f(-\omega) = \int_{-\infty}^{\infty} F(jt)\, e^{-j\omega t}\, dt \qquad (20.5.9)$$

The integral of Eq. (20.5.9) is the Fourier transform of $F(jt)$. Thus Eq. (20.5.9) becomes

$$\boxed{\mathfrak{F}[F(jt)] = 2\pi f(-\omega)} \qquad (20.5.10)$$

It can also be proved that

$$\boxed{\mathfrak{F}^{-1}[f(\omega)] = \frac{1}{2\pi} F(-jt)} \qquad (20.5.11)$$

The results given in Eqs (20.5.10) and (20.5.11) are known as the **symmetry property** of the Fourier transform.

20.5.5 Time-Shifting Property

If the time function $f(t)$ is shifted or delayed by t_0 then its Fourier transform $F(j\omega)$ is multiplied by the factor $e^{-j\omega t_0}$, that is, if

$$\mathfrak{F}[f(t)] = F(j\omega)$$

then

$$\mathfrak{F}[f(t - t_0)] = e^{-j\omega t_0} F(j\omega)$$

Proof

$$\mathfrak{F}[f(t - t_0)] = \int_{-\infty}^{\infty} f(t - t_0)\, e^{j\omega t}\, dt$$

Let $t - t_0 = x$, then

$$\mathfrak{F}[f(t - t_0)] = \int_{-\infty}^{\infty} f(x)\, e^{-j\omega (x + t_0)}\, dx$$

$$= e^{-j\omega t_0} \int_{-\infty}^{\infty} f(x)\, e^{-j\omega x}\, dx$$

therefore,

$$\boxed{\mathfrak{F}[f(t - t_0)] = e^{-j\omega t_0} F(j\omega)} \qquad (20.5.12)$$

This proves the required result.

Eq. (20.5.12) shows that *a shift of* t_0 *in the time domain in equivalent to multiplication* **by** $e^{-j\omega t_0}$ *in the frequency domain.*

In other words, *if a function is shifted in time domain by* t_0, *then its magnitude spectrum* $|F(j\omega)|$ *remains unchanged, but its phase spectrum is changed by* $(-\omega t_0)$.

20.5.6 Complex Translation

If $$\mathfrak{F}[f(t)] = F(j\omega)$$

then $$\mathfrak{F}[e^{-at} f(t)] = F(j\omega + a)$$

Proof

$$\mathfrak{F}[e^{-at} f(t)] = \int_{-\infty}^{\infty} e^{-at} f(t)\, e^{-j\omega t}\, dt$$

$$= \int_{-\infty}^{\infty} f(t)\, e^{-(j\omega + a)\,t}\, dt$$

$$= F(j\omega + a) \qquad (20.5.13)$$

That is, if $F(j\omega)$ is the Fourier transform of $f(t)$, then the Fourier transform of $e^{-at} f(t)$ can be obtained from $F(j\omega)$ by replacing $j\omega$ by $(j\omega + a)$. The constant a can be real or complex. This result is known as the **complex translation property** of the Fourier transform.

20.5.7. Frequency Shifting or Frequency Translation Property

If $$\mathfrak{F}[f(t)] = F(j\omega)$$

then $$\mathfrak{F}[f(t)\, e^{j\omega_0 t}] = F[j(\omega - \omega_0)]$$

Proof

$$\mathfrak{F}[f(t)\, e^{-j\omega_0 t}] = \int_{-\infty}^{\infty} f(t)\, e^{j\omega_0 t} \times e^{-j\omega t}\, dt$$

$$= \int_{-\infty}^{\infty} f(t)\, e^{-j(\omega - \omega_0)\,t}\, dt$$

$$= F[j(\omega - \omega_0)] \qquad (20.5.14)$$

This result shows that multiplication of a function $f(t)$ by a factor $e^{j\omega_0 t}$ is equivalent to shifting its Fourier transform (Fourier spectrum) in the positive direction by an amount ω_0. This property is known as **frequency-shifting or frequency translation property** of Fourier transforms. Note the duality between time-shifting and frequency shifting operations.

20.5.8. Modulaton Property

In communication systems the frequency spectrum of the signal $f(t)$ is shifted by multiplying the signal $f(t)$ by a sinusoidal signal. This process is known as **modulation**.

If $$\mathfrak{F} f(t) = F(j\omega)$$

then $$\mathfrak{F}[f(t) \cos \omega_0 t] = \frac{1}{2} F[j(\omega - \omega_0)] + \frac{1}{2} F[j(\omega + \omega_0)]$$

Proof

$$f(t) \cos \omega_0 t = \frac{1}{2} [f(t)\, e^{j\omega_0 t} + f(t)\, e^{-j\omega_0 t}]$$

Taking the Fourier transform of both sides we get

$$\mathfrak{F}[f(t) \cos \omega_0 t] = \mathfrak{F}\left[\frac{1}{2} f(t)\, e^{j\omega_0 t} + \frac{1}{2} f(t)\, e^{-j\omega_0 t}\right] \qquad (20.5.15)$$

Applying the frequency shift property on right-hand side of Eq. (20.5.15), we get

$$\Im[f(t)\cos\omega_0 t] = \frac{1}{2}F[j(\omega-\omega_0)] + \frac{1}{2}F[j(\omega+\omega_0)]$$

Thus, the result of multiplying a time function by a pure sinusoid is to shift the original transform both up and down by the frequency of the sinusoid and to cut the amplitude in half. This is known as **modulation theorem**. Similarly, we can prove that

$$\Im[f(t)\sin\omega_0 t] = \frac{1}{2j}F[j(\omega-\omega_0)] - \frac{1}{2j}F[j(\omega+\omega_0)]$$

Thus, the process of modulation translates (shifts) the frequency spectrum by $\pm\,\omega_0$. The signal $\cos\omega_0 t$ is called the **carrier**. The function $f(t)\cos\omega_0 t$ is called the **modulated signal with *envelope*** $f(t)$ and **carrier frequency** ω_0. Multiplication of $\cos\omega_0 t$ by $f(t)$ is equivalent to multiplying the amplitude of the carrier $\cos\omega_0 t$ in proportion to the modulating signal $f(t)$ Hence this type of moduation is known as **amplitude modulation** (AM).

20.5.9 Time Differentiation Property

If $\qquad \Im f(t) = F(j\omega)$

then $\qquad \Im\left[\frac{d}{dt}f(t)\right] = (j\omega)\,F(j\omega)$

and $\qquad \Im\left\{\frac{d^n}{dt^n}f(t)\right] = (j\omega)^n\,F(j\omega)$

or $\qquad \Im[f^n(t)] = (j\omega)^n\,\Im[f(t)]$

Proof

$$f(t) = \frac{1}{2\pi}\int_{-\infty}^{\infty} F(j\omega)\,e^{j\omega t}\,d\omega \tag{20.5.17}$$

$$\frac{d}{dt}f(t) = \frac{1}{2\pi}\frac{d}{dt}\left[\int_{-\infty}^{\infty} F(j\omega)\,e^{j\omega t}\,d\omega\right]$$

Changing the order of differentiation and integration, we get

$$\frac{d}{dt}f(t) = \frac{1}{2\pi}\int_{-\infty}^{\infty} j\omega\,F(j\omega)\,e^{j\omega t}\,d\omega$$

$$= (j\omega)\frac{1}{2\pi}\int_{-\infty}^{\infty} F(j\omega)\,e^{j\omega t}\,d\omega$$

$$\boxed{\Im\left[\frac{d}{dt}f(t)\right] = j\omega\,\Im[f(t)]}$$

or

$$\Im\left[\frac{d}{dt}f(t)\right] = (j\omega)\,F(j\omega) \tag{20.5.18}$$

Reapplying the result we get,

$$\Im[f''(t)] = \Im\left[\frac{d^2 f(t)}{dt^2}\right] = (j\omega)^2\,F(j\omega)$$

Differentiating Eq. (20.5.17) n times with respect to t and assuming that the order of differentiation and integration can be interchanged at each step, we get

$$\Im\left[\frac{d^n f(t)}{dt^n}\right] = (j\omega)^n F(j\omega) \tag{20.5.19}$$

20.5.10 Frequency Differentiation Property

If $\Im f(t) = F(j\omega)$

then $\frac{d}{d\omega} F(j\omega) = \Im[(-jt) f(t)]$

and $\frac{d^n}{d\omega^n} F(j\omega) = \Im[(-jt)^n F(t)]$

Proof

$$F(j\omega) = \int_{-\infty}^{\infty} f(t)\, e^{-j\omega t}\, dt \tag{20.5.20}$$

Differentiation Eq. (20.5.20) with respect to ω, we get

$$\frac{d}{d\omega} F(j\omega) = \frac{d}{d\omega} \int_{-\infty}^{\infty} f(t)\, e^{-j\omega t}\, dt$$

Changing the order of differentiation and integration, we get

$$\frac{d}{d\omega} F(j\omega) = \int_{-\infty}^{\infty} f(t)\,(-jt)\, e^{-j\omega t}\, dt$$

$$= \Im[(-jt) f(t)] \tag{20.5.21}$$

Differentiating Eq. (20.5.20) n times with respect to ω and assuming that the order of differenciation and integration can be interchanged at each step, we get

$$\frac{d^n}{d\omega^n} F(j\omega) = \Im[(-jt)^n f(t)] \tag{20.5.22}$$

or

$$(j)^n \frac{d^n}{d\omega^n} F(j\omega) = \Im[(t)^n f(t)] \tag{20.5.23}$$

20.5.11 Time-Convolution Theorem

If $\Im f_1(t) = F_1(j\omega)$

$\Im f_2(t) = F_2(j\omega)$

then $\Im[f_1(t) * f_2(t)] = F_1(j\omega)\, F_2(j\omega)$

where the symbol * denotes convolution.

Proof

$$f_1(t) * f_2(t) = \int_{-\infty}^{\infty} f_1(t-x) f_2(x)\, dx$$

$$\mathfrak{F}[f_1(t) * f_2(t)] = \int_{-\infty}^{\infty} \left[\int_{-\infty}^{\infty} f_1(t-x) f_2(x)\, dx \right] e^{j\omega t}\, dt$$

$$= \int_{-\infty}^{\infty} f_2(x) \left[\int_{-\infty}^{\infty} f_1(t-x)\, e^{-j\omega t}\, dt \right] dx$$

From the time-shifting property, the inner integral is equal to

$$F_1(j\omega)\, e^{-j\omega x}.$$

Hence $$\mathfrak{F}[f_1(t) * f_2(t)] = \int_{-\infty}^{\infty} f_2(x)\, e^{-j\omega x} F_1(j\omega)\, dx$$

Since $F_1(j\omega)$ is not a function of x, it can be taken out of the integral. Thus,

$$\mathfrak{F}[f_1(t) * f_2(t)] = F_1(j\omega) \int_{-\infty}^{\infty} f_2(x)\, e^{-j\omega x}\, dx$$

$$= F_1(j\omega)\, F_2(j\omega) \tag{20.5.24}$$

This property is known as **time-convolution theorem.**

Thus, *the convolution of two signals in the time domain is transformed into the multiplication of the their individual Fourier transforms in the frquency domain.*

20.5.11 Frequency Convolution Theorem

If $$\mathfrak{F}[f_1(t)] = F_1(j\omega)$$

$$\mathfrak{F}[f_2(t)] = F_2(j\omega)$$

then $$\mathfrak{F}[f_1(t) f_2(t)] = \frac{1}{2\pi} [F_1(j\omega) * F_2(j\omega)]$$

Proof

$$f_2(t) = \frac{1}{2\pi} \int_{-\infty}^{\infty} F_2(jy)\, e^{jyt}\, dy \tag{20.5.25}$$

Multiplying both the sides of Eq. (20.5.25), we get

$$f_1(t) f_2(t) = \frac{1}{2\pi} \int_{-\infty}^{\infty} F_2(jy) f_1(t)\, e^{jyt}\, dy \tag{20.5.26}$$

The right hand side of Eq. (20.5.26) is a function of time and its Fourier transform is equal to

$$\frac{1}{2\pi} \int_{-\infty}^{\infty} F_2(jy) F_1(j\omega - jy)\, dy$$

because $$\mathfrak{F}[f_1(t)\, e^{jyt}] = F_1(j\omega - jy)$$

Therefore, $$\mathfrak{F}[f_1(t) f_2(t)] = \frac{1}{2\pi} [F_1(j\omega) * F_2(j\omega)] \tag{20.5.27}$$

Also, $$\mathfrak{F}[f_1(t)\, f_2(t)] = \int_{-\infty}^{\infty} f_1(t)\, f_2(t)\, e^{-j\omega t}\, dt \tag{20.5.28}$$

Therefore, $$\int_{-\infty}^{\infty} f_1(t)\, f_2(t)\, e^{-j\omega t}\, dt$$

$$= \frac{1}{2\pi}\int_{-\infty}^{\infty} F_1(j\omega - jy)\, F_2(jy)\, dy \tag{20.5.29}$$

Eq. (20.5.27) shows that *the multiplication of two signals in the time domain is transformed into the convolution of their individual Fourier transforms in the frequency domain.*

20.5.12 Time-Integration Property

If $$\mathfrak{F}[f(t)] = F(j\omega)$$

$$\mathfrak{F}\left[\int_{-\infty}^{\infty} f(\tau)\, d\tau\right] = \pi F(0)\, u_0(j\omega) + \frac{1}{j\omega} F(j\omega)$$

Proof

Since $$\mathfrak{F}[f_1(t) * f_2(t)] = F_1(j\omega)\, F_2(j\omega)$$

$$\mathfrak{F}[f(t) * u_{-1}(t)] = F(j\omega)\, F[u_{-1}(t)]$$

$$= F(j\omega)\left[\pi u_0(j\omega) + \frac{1}{j\omega}\right] \tag{20.5.30}$$

But $$\int_{-\infty}^{\infty} f(\tau)\, d\tau = f(t) * u_{-1}(t) \tag{20.5.31}$$

Therefore, $$\mathfrak{F}\left[\int_{-\infty}^{\infty} f(\tau)\, d\tau\right] = F(j\omega)\left[\pi u_0(j\omega) + \frac{1}{j\omega}\right]$$

$$= \pi F(j\omega)\, u_0(j\omega) + \frac{1}{j\omega} F(j\omega) \tag{20.5.32}$$

Also, $$F(j\omega)\, u_0(j\omega) = F(0)\, u_0(j\omega) \tag{20.5.33}$$

This is because of the fact that $u_0(t) = 0$ for all values of t, except at the origin. The product $u_0(t)\,\phi(t)$ will also be zero for all t except at $t = 0$. At $t = 0$, $\phi(t) = \phi(0)$. Hence $u_0(t)\,\phi(t) = u_0(t)\,\phi(0)$ and Eq. (20.5.32) becomes

$$\mathfrak{F}\left[\int_{-\infty}^{\infty} f(\tau)\, d\tau\right] = \pi F(0)\, u_0(j\omega) + \frac{1}{j\omega} F(j\omega) \tag{20.5.34}$$

If $F(0) = 0$, then

$$\mathfrak{F}\left[\int_{-\infty}^{\infty} f(\tau)\, d\tau\right] = \frac{1}{j\omega} F(j\omega) \tag{20.5.35}$$

20.6 CONJUGATE FUNCTIONS

If $$\mathfrak{F}[f(t)] = F(j\omega)$$

then for a complex valued $f(t)$, we have

$$\mathfrak{F}[f^*(t)] = F^*(-j\omega) \tag{20.6.1}$$

where the asterisk * denotes the complex conjugate operation.

Proof

We have

$$f(t) = \frac{1}{2\pi}\int_{-\infty}^{\infty} F(j\omega)\, e^{j\omega t}\, d\omega$$

Taking the complex conjugate of both sides

$$f^*(t) = \frac{1}{2\pi}\int_{-\infty}^{\infty} F^*(j\omega)\, e^{-j\omega t}\, d\omega$$

Now replace ω by – ω,

$$f^*(t) = -\frac{1}{2\pi}\int_{\infty}^{-\infty} F(-j\omega)\, e^{j\omega t}\, d\omega$$

$$= \frac{1}{2\pi}\int_{-\infty}^{\infty} F(-j\omega)\, e^{j\omega t}\, d\omega$$

That is, $f^*(t)$ is the inverse Fourier transform of $F^*(-j\omega)$, which is the desired result.

20.7 REAL AND IMAGINARY PARTS OF A TIME FUNCTION

Let the function $f(t)$ be complex. We may express $f(t)$ in terms of its real and imaginary parts as

$$f(t) = Re\,[f(t)] + j\,Im\,[f(t)] \tag{20.7.1}$$

where *Re* denotes 'the real part of? and *Im* denotes 'imaginary part of'. The complex conjugate of $f(t)$ is written as

$$f^*(t) = Re\,[f(t)] - j\,Im\,[f(t)] \tag{20.7.2}$$

Addition of Eqs. (20.7.1) and (20.7.2) gives

$$Re\,[f(t)] = \frac{1}{2}\,[f(t) + f^*(i)] \tag{20.7.3}$$

Subtraction of Eqs. (20.7.1) and (20.7.2) gives

$$Im\,[f(t)] = \frac{1}{2}\,[f(t) - f^*(t)] \tag{20.7.4}$$

By applying the property of conjugates, we get

$$\mathfrak{F}[Re\,\{f(t)\}] = \frac{1}{2}\,[F(j\omega) + F^*(-j\omega)] \tag{20.7.5}$$

$$\mathfrak{F}[Im\,\{f(t)\}] = \frac{1}{2j}\,[F(j\omega) - F^*(-j\omega)] \tag{20.7.6}$$

From Eq. (20.7.6) it is clear that if the time function is real, we have

$$F(j\omega) = F^*(-j\omega) \tag{20.7.7}$$

that is, $F(j\omega)$ shows conjugate symmetry.

20.8 PARSEVAL'S THEOREM

$$\int_{-\infty}^{\infty} f_1(t) f_2^*(t)\, dt = \frac{1}{2\pi}\int_{-\infty}^{\infty} F_1(j\omega) F_2^*(j\omega)\, d\omega$$

Proof

We have $$\int_{-\infty}^{\infty} f_1(t) f_2(t) e^{-j\omega t}\, dt = \frac{1}{2\pi}\int_{-\infty}^{\infty} F_1(jy) F_2(j\omega - jy)\, dy \qquad (20.8.1)$$

Putting $\omega = 0$ in Eq. (20.8.1) gives

$$\int_{-\infty}^{\infty} f_1(t) f_2(t)\, dt = \frac{1}{2\pi}\int_{-\infty}^{\infty} F_1(jy) F_2(-jy)\, dy$$

or equivalently

$$\int_{-\infty}^{\infty} f_1(t) f_2(t)\, dt = \frac{1}{2\pi}\int_{-\infty}^{\infty} F_1(j\omega) F_2(-j\omega)\, d\omega \qquad (20.8.2)$$

Eq. (20.8.2) is true for any $f_1(t)$ and $f_2(t)$.

Let us replace $f_2(t)$ by $f_2^*(t)$. By the property of conjugate functions, the Fourier transform of $f_2(t)$ that is, $F_2(j\omega)$ should be replaced by $F_2^*(-j\omega)$. The function $F_2(-j\omega)$ then becomes $F_2^*(j\omega)$. Thus, Eq. (20.8.2) can be written as

$$\int_{-\infty}^{\infty} f_1(t) f_2^*(t)\, dt = \frac{1}{2\pi}\int_{-\infty}^{\infty} F_1(j\omega) F_2^*(j\omega)\, d\omega \qquad (20.8.3)$$

This is called **Parseval's theorem.**

We know that if $x = a + jb$

$$x^* = a - jb$$

and $$xx^* = (a + jb)(a - jb) = a^2 + b^2 = |x|^2$$

If $f_1(t) = f_2(t) = f(t)$ and $F_1(j\omega) = F_2(j\omega) = F(j\omega)$,

$$F(j\omega) F^*(j\omega) = |F(j\omega)|^2$$

and $$\int_{-\infty}^{\infty} |f(t)|^2\, dt = \frac{1}{2\pi}\int_{-\infty}^{\infty} |F(j\omega)|^2\, d\omega \qquad (20.8.4)$$

Eq. (24.9.4) is known as **Plancheral's theorem**. This theorem relates energy in the time domain to the energy in the frequency domain. A physical interpretation of this theorem can be obtained by considering $f(t)$ as the voltage across of current through a 1-Ω resistor. The intsantaneous power dissipated in the resistor is $f^2(t)$ watts. Therefore the total energy dissipated in the resistor taken over all time is equal to $\int_{-\infty}^{\infty} f^2(t)\, dt$. Eq. (20.8.4) shows that this energy is $\frac{1}{2\pi}$ times the area under the curve $|F(j\omega)|^2$. The quantity $|F(j\omega)|^2$, which is the square of the amplitude spectrum of $f(t)$, is sometimes known as the **energy density or energy spectrum** of $f(t)$.

20.9 FOURIER TRANSFORM OF THE UNIT IMPULSE FUNCTION u_0 (t)

The Fourier transform of the unit impulse function $u_0(t)$ is given by

$$\Im[u_0(t)] = \int_{-\infty}^{\infty} u_0(t)\, e^{-j\omega t}\, dt \qquad (20.9.1)$$

According to the sifting property of the impulse function,

$$\int_{-\infty}^{\infty} u_0(t)\, f(t)\, dt = f(0) \qquad (20.9.2)$$

Therefore,

$$\int_{-\infty}^{\infty} u_0(t)\, e^{-j\omega t}\, dt = e^{-j\omega t}\Big|_{t=0} = e^0 = 1$$

Hence, from Eq. (20.9.1)

$$\boxed{\Im[u_0(t)] = 1} \qquad (20.9.3)$$

Thus, the *Fourier transform of a unit impulse function is unity.* Fig. 20.2 shows an impulse function and its Fourier transform. The impulse function has a flat or constant Fourier spectrum. In other words, an impulse function contains all frequency components with the same relative amplitudes.

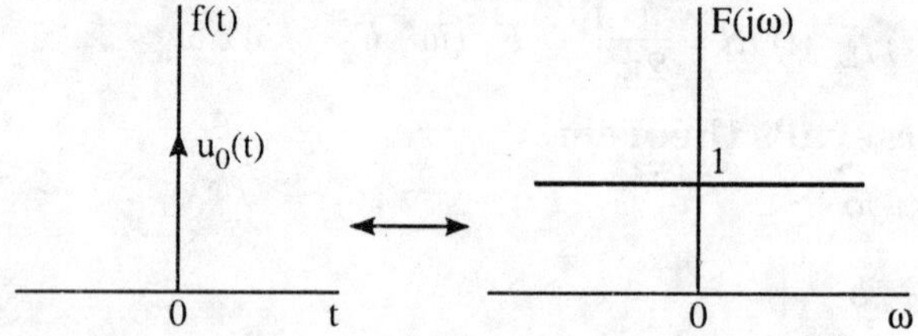

FIG. 20.2 An impulse function and its transform

$$\Im[u_0(t - t_0)] = e^{-j\omega t_0}$$

$$|\,\Im[u_0(t - t_0)]\,| = |\,e^{-j\omega t_0}\,| = |\,(\cos \omega t_0 - j \sin \omega t_0)\,|$$

$$= \sqrt{(\cos^2 \omega t_0 + \sin^2 \omega t_0)} = 1$$

or $$|\,\Im[u_0(t - t_0)]\,| = 1 \qquad (20.9.4)$$

Since the amplitude spectrum (and hence, energy density) of unit impulse function (delta function) is constant, *the energy content of an impulse futions is infinite.* This explains why *impulse functions cannot be physically realized.*

20.10 SAMPLING FUNCTION

The function of the form $\frac{\sin x}{x}$ is called the *sampling function.* It is denoted by Sa (x).

$$\boxed{\text{Sa}(x) \overset{\Delta}{=} \frac{\sin x}{x}} \qquad (20.10.1)$$

This function is plays an important role in communication theory. The sampling function is shown in Fig. 20.3.

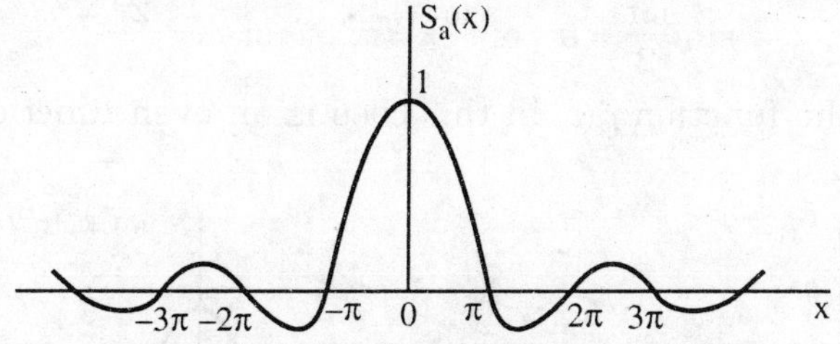

FIG. 20.3 The sampling function *Sa* (*x*)

It is an even function of x and has a maximum value of uniy at x =0. It oscillates through positive and negative values as the magnitude of x increases and approaches zero as x approaches plus or minus infinity. It has zeros at $x = \pm\,\pi, \pm\,2\pi, \pm\,3\pi, \ldots,$ etc.

20.11 FOURIER TRANSFORM OF A RECTANGULAR PULSE

Consider a rectangular pulse of width (duration) τ and height (amplitude) unity as shown in Fig. 20.4a. It can be defined mathematically as

$$f(t) = \begin{cases} 1 & -\tau/2 < t < \tau/2 \\ 0 & \text{elsewhere} \end{cases}$$

The Fourier transform of this rectangular pulse is given by

$$\mathfrak{F}[f(t)] = F(j\omega) = \int_{-\infty}^{\infty} f(t)\, e^{-j\omega t}\, dt$$

$$= \int_{-\tau/2}^{\tau/2} 1 \times e^{-j\omega t}\, dt = -\frac{1}{j\omega}\left[e^{-j\omega t}\right]_{-\tau/2}^{\tau/2}$$

$$= \frac{2}{\omega} \times \frac{e^{j\omega\tau/2} - e^{-j\omega\tau/2}}{2j}$$

$$\therefore \qquad F(j\omega) = \frac{2}{\omega} \sin\frac{\omega\tau}{2} \qquad (20.11.1)$$

This result can be put in the form $\frac{\sin x}{x}$ by multiplying the numerator and denominator by $\frac{\omega\tau}{2}$. Therefore

$$F(j\omega) = \frac{2}{\omega} \times \frac{\omega\tau}{2} \times \frac{\sin \omega\tau/2}{\frac{\omega\tau}{2}}$$

$$\boxed{F(j\omega) = \tau\,\frac{\sin\frac{\omega\tau}{2}}{\frac{\omega\tau}{2}} = \tau\,\text{Sa}\,\frac{\omega\tau}{2}} \qquad (20.11.2)$$

The Fourier transform of the rectangular pulse is shown in Fig. 20.4b. The first crossing of the ω axis occurs when

$$\sin\frac{\omega\tau}{2} = 0 \quad \text{or} \quad \frac{\omega\tau}{2} = \pi, \qquad \text{or } \omega = \frac{2\pi}{\tau}$$

It is to be noted that the function $f(t)$ in this case is an even function of t, and $F(j\omega)$ is an even function of ω.

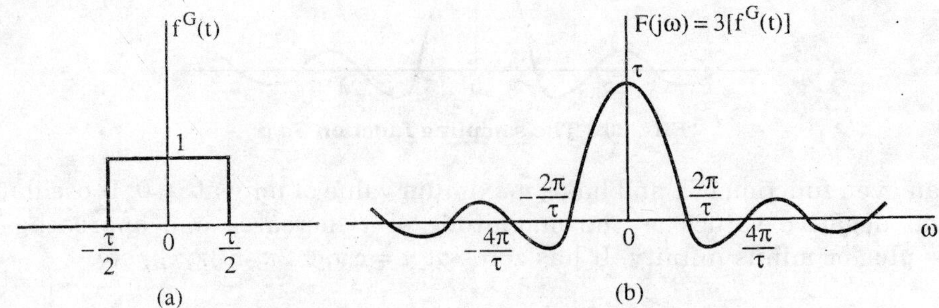

FIG. 20.4 Rectangular pulse and its Fourier transform

■ **EXAMPLE 20.1** Find the Fourier transform of a rectangular pulse shown in Fig. 20.5. Sketch the amplitude and phase spectra.

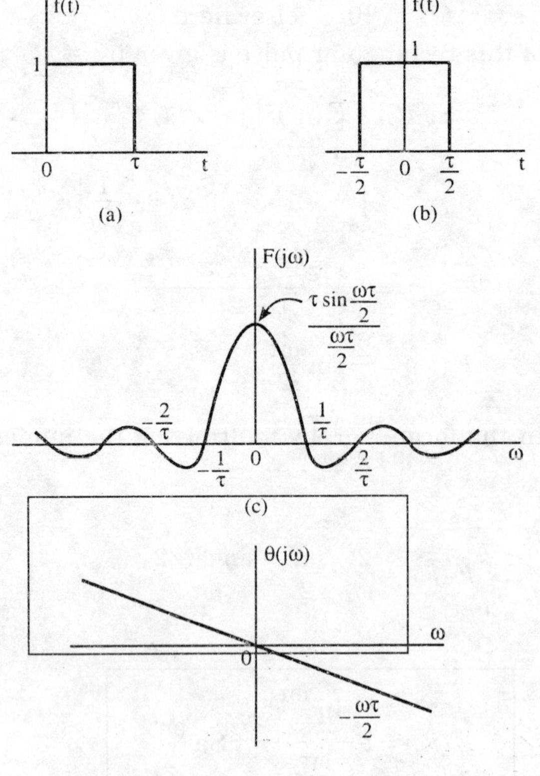

FIG. 20.5 (a) Rectangular pulse (b) Gate function (c) Ampitude spectrum (d) Phase spectrum

SOLUTION : We have

$$f(t) = \begin{cases} 1 & \text{for } 0 < t \le \tau \\ 0 & \text{otherwise} \end{cases}$$

$$\mathfrak{F}[f(t)] = F(j\omega) = \int_{-\infty}^{\infty} f(t)\, e^{-j\omega t}\, dt$$

$$= \int_0^{\tau} 1 \times e^{-j\omega t}\, dt = -\frac{1}{j\omega}\left[e^{-j\omega t}\right]_0^{\tau}$$

$$= \frac{1}{j\omega}(1 - e^{-j\omega\tau})$$

$$= \frac{1}{j\omega}\left\{1 - \frac{e^{-j\omega\tau/2}}{e^{j\omega\tau/2}}\right\}$$

$$= \frac{1}{j\omega}\, \frac{e^{j\omega\tau/2} - e^{-j\omega\tau/2}}{e^{j\omega\tau/2}}$$

$$= \frac{2}{\omega} \times \left(\frac{e^{j\omega\tau/2} - e^{-j\omega\tau/2}}{2j}\right) e^{-j\omega\tau/2}$$

$$= \frac{2}{\omega}\left(\frac{\omega\tau}{2}\right) \frac{\sin\frac{\omega\tau}{2}}{\left(\frac{\omega\tau}{2}\right)}\, e^{-j\omega\tau/2}$$

$$= \tau\, \frac{\sin\frac{\omega\tau}{2}}{\left(\frac{\omega\tau}{2}\right)}\, e^{-j\omega\frac{\tau}{2}}$$

Alternative method

The function shown in Fig. 20.5a is a gate function (shown in Fig. 20.5b) shifted by $\tau/2$ seconds. Let us denote the shifted function by $G_\tau(t - \tau/2)$

From Eq. 20.11.1

$$\mathfrak{F}[G_\tau(t)] = \tau\, \mathrm{Sa}\left(\frac{\omega\tau}{2}\right) = \tau\, \frac{\sin\frac{\omega\tau}{2}}{\frac{\tau}{2}}$$

By time-shifting property

$$\mathfrak{F}\left[G_\tau\left(t - \frac{\tau}{2}\right)\right] = G_\tau(t)\, e^{-j\omega\tau/2} = \tau\, \mathrm{Sa}\left(\frac{\omega\tau}{2}\right) e^{-j\omega\tau/2}$$

$$= \tau\, \frac{\sin\frac{\omega\tau}{2}}{\frac{\tau}{2}}\, e^{-j\omega\tau/2}$$

The amplitude and phase spectra are shown in Fig. 20.5c and 20.5d respectively.

20.12 FOURIER TRANSFORM OF A CONSTANT

Let the constant be unity. We are interested in finding the Fourier trasform $F(j\omega)$ of the function

$$f(t) = 1, \qquad -\infty < t < \infty$$

We are assuming a time function which is constant for *all t*. This function is shown in Fig. 20.6a. It is a d.c. signal.

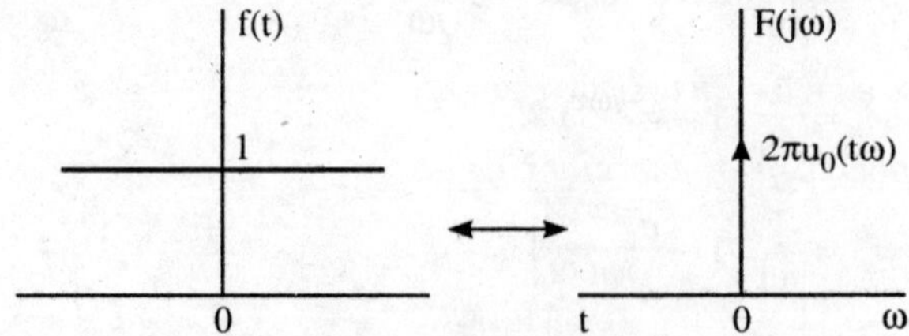

FIG. 20.6 A constant and its Fourier transform

$$\mathfrak{F}(j\omega) = \mathfrak{F}[1] = \int_{-\infty}^{\infty} 1 \times e^{-j\omega t}\, dt$$

This suggests that $F(j\omega)$ might be an impulse function.

$$\mathfrak{F}^{-1}[u_0(j\omega)] = \frac{1}{2\pi}\int_{-\infty}^{\infty} u_0(j\omega)\, e^{j\omega t}\, d\omega$$

According to the sifting (sampling) property of the impulse function

$$\int_{-\infty}^{\infty} u_0(j\omega)\, e^{j\omega t}\, d\omega = e^{j\omega t}\Big|_{t=0} = e^0 = 1$$

Therefore, $$\mathfrak{F}^{-1}[u_0(j\omega)] = \frac{1}{2\pi}$$

$$\mathfrak{F}[1] = 2\pi u_0(j\omega) \tag{20.12.1}$$

Thus the Fourier transform of constant 1 is an impulse of strength 2π occuring at $\omega = 0$. It is shown in Fig. 20.6b. This is an expected result since there is only one frequency present in a d.c. signal and this frequency is equal to zero.

The Fourier transform of a function

$$f(t) = A, \qquad -\infty < t < \infty$$

is given by

$$\mathfrak{F}[A] = 2\pi A\, u_0(j\omega) \tag{20.12.2}$$

This function and its Fourier transform are shown in Fig. 20.7.

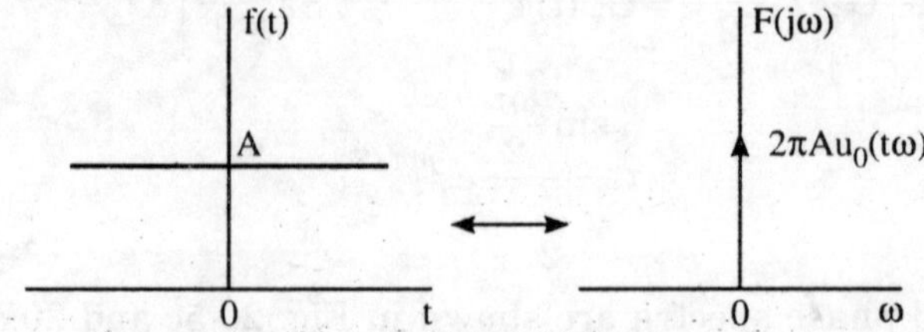

FIG. 20.7 A constant function *f*(*t*) and its Fourier transform

20.13 FOURIER TRANSFORM OF AN EXPONENTIAL FUNCTION

We have

$$\mathfrak{F}[1] = 2\pi u_0 (j\omega) \tag{20.13.1}$$

By frequency-shifting property

$$\mathfrak{F}[f(t)\, e^{j\omega_0 t}] = F(j\omega - j\omega_0) \tag{20.13.2}$$

Therefore,

$$\mathfrak{F}[1 \times e^{j\omega_0 t}] = 2\pi u_0 (j\omega - j\omega_0) \tag{20.13.3}$$

Thus, *the frequency spectrum of exponential* $e^{j\omega t}$ *is an impulse situated at* $\omega = \omega_0$.

Similarly,

$$\mathfrak{F}[1 \times e^{-j\omega_0 t}] = 2\pi u_0 (j\omega + j\omega_0) \tag{20.13.4}$$

20.13.1 Fourier Transforms of Sinusoidal Functions

$$\cos \omega_0 t = \frac{1}{2}(e^{j\omega_0 t} + e^{-j\omega_0 t})$$

$$\mathfrak{F}[\cos \omega_0 t] = \frac{1}{2}\mathfrak{F}[e^{j\omega_0 t}] + \frac{1}{2}\mathfrak{F}[e^{-j\omega_0 t}]$$

or

$$\mathfrak{F}[\cos \omega_0 t] = \pi u_0 (j\omega - j\omega_0) + \pi u_0 (j\omega + j\omega_0) \tag{20.13.5}$$

Thus, the magnitude spectrum of the cosine function $\cos \omega_0 t$ consists of a pair of impulses each of strength π occuring at $\omega = \omega_0$ and $\omega = -\omega_0$, as shown in Fig. 20.8.

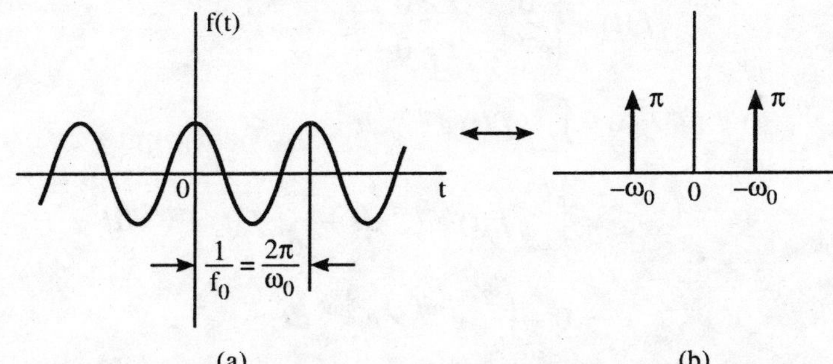

FIG. 20.8 (a) Cosine function cos $\omega_0 t$ (b) Magnitude spectrum of cos $\omega_0 t$

Similarly,

$$\boxed{\mathfrak{F}[\sin \omega_0 t] = j\pi\, [u_0 (j\omega + j\omega_0) - u_0 (j\omega - j\omega_0)]} \tag{20.13.6}$$

Since $j = e^{j\pi/2}$ and $-j = e^{-\pi/2}$, the Fourier transform of $\sin \omega_0 t$ can also be written as

$$\mathfrak{F}[\sin \omega_0 t] = \pi\, [u_0 (j\omega + j\omega_0)\, e^{j\pi/2} + u_0 (j\omega - j\omega_0)\, e^{-j\pi/2}] \tag{20.13.7}$$

Eq. (20.13.7) shows that the magnitude spectrum of the sine function also consists of a pair of impulse functions occuring at $\omega = \omega_0$ and $\omega = -\omega_0$ as shown in Fig. 20.9.

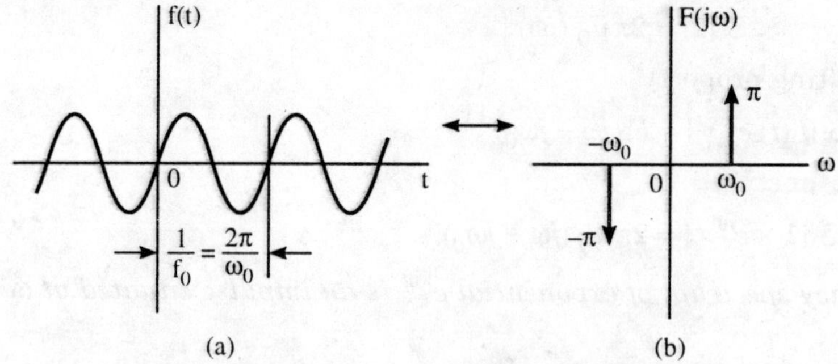

FIG. 20.9 (b) Sine function sin $\omega_0 t$ (b) Magnitude spectrum of sin $\omega_0 t$

Comparison of Eqs. (20.13.6) and (20.13.7) shows that there is a phase shift of $\frac{-\pi}{2}$ for sin $\omega_0 t$ in relation to cos $\omega_0 t$. This phase shift is indicated by the factors $e^{j\pi/2}$ and $e^{-j\pi/2}$ in Eq. (20.13.7).

20.14 FOURIER TRANSFORM OF A SINGLE-SIDED EXPONENTIAL SIGNAL $e^{-at}u_{-1}(t)$

$$f(t) = e^{-at}\, u_{-1}(t) \tag{20.14.1}$$

This function can also be written as

$$f(t) = \begin{cases} 0, & t < 0 \\ e^{-at}, & t > 0 \end{cases}$$

$$F(j\omega) = \int_{-\infty}^{\infty} f(t)\, e^{-j\omega t}\, dt$$

$$= \int_{-\infty}^{0} f(t)\, e^{-j\omega t}\, dt + \int_{0}^{\infty} f(t)\, e^{-j\omega t}\, dt$$

$$= 0 + \int_{0}^{\infty} e^{-at} \times e^{-j\omega t}\, dt$$

$$= \int_{0}^{\infty} e^{-(a+j\omega t)\,t}\, dt \tag{20.14.2}$$

$$= \frac{-1}{a+j\omega}\left[e^{-(a+j\omega)\,t}\right]_{0}^{\infty} = \frac{1}{a+j\omega}$$

$$\boxed{\mathfrak{F}\,[e^{-at}\, u_{-1}(t)] = \frac{1}{a+j\omega}} \tag{20.14.3}$$

It is to be noted that the integral in Eq. (20.14.2) converges only for $a > 0$. If $a < 0$, then the integral does not converge and the Fourier transform $F(j\omega)$ does not exist.

20.15 FOURIER TRANSFORM OF A UNIT STEP FUNCTION

The unit step function is given by

$$f(t) = u_{-1}(t) \qquad (20.15.1)$$

It is shown in Fig. 20.10

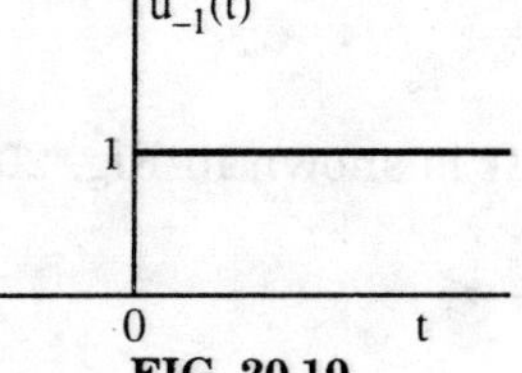

FIG. 20.10

This function violates Dirichlet conditions; yet its Fourier transform exists. It can be found by limiting process. Consider the single sided exponential function,

$$f_1(t) = e^{-at}\, u_{-1}(t) \qquad (20.15.2)$$

Now

$$\mathfrak{F}[f_1(t)] = \mathfrak{F}[e^{-at}\, u_{-1}(t)]$$

$$= \frac{1}{a + j\omega} = \frac{a - j\omega}{(a + j\omega)(a - j\omega)}$$

$$F_1(j\omega) = \frac{a}{a^2 + \omega^2} - j\frac{\omega}{a^2 + \omega^2} \qquad (20.15.3)$$

We shall examine the real and imaginary parts separately. The real part is shown in Fig. 20.11. The area under the curve is

$$\int_{-\infty}^{\infty} \frac{a}{a^2 + \omega^2}\, d\omega = \left[\tan^{-1}\frac{\omega}{a}\right]_{-\infty}^{\infty} = \frac{\pi}{2} - \left(-\frac{\pi}{2}\right) = \pi$$

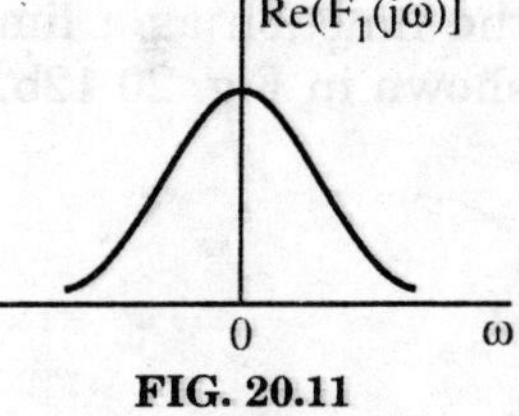

FIG. 20.11

As $a \to 0$, the height of the curve tends to infinity, and the width tends to zero, but the area under the curve remains constant throughout the limiting process. In the limit the real part of $F_1(j\omega)$ is, therefore, an impulse of value π. That is,

$$\lim_{a \to 0} Re\,[F_1(j\omega)] = \lim_{a \to 0} \frac{a}{a^2 + \omega^2} = \pi\, u_0(j\omega)$$

Now

$$\lim_{a \to 0} Im\,[F_1(j\omega)] = \lim_{a \to 0} \frac{-\omega}{a^2 + \omega^2} = -\frac{1}{\omega}$$

therefore,

$$\mathfrak{F}[u_{-1}(t)] = \lim_{a \to 0} \mathfrak{F}[e^{-at}\, u_{-1}(t)]$$

$$= \lim_{a \to 0} F_1(j\omega) = \lim_{a \to 0} \frac{a}{a^2 + \omega^2} - j \lim_{a \to 0} \frac{\omega}{a^2 + \omega^2}$$

$$= \pi\, u_0(j\omega) - \frac{j}{\omega}$$

or

$$\boxed{\mathfrak{F}[u_{-1}(t)] = \pi\, u_0(j\omega) + \frac{1}{j\omega}} \qquad (20.15.4)$$

20.16 FOURIER TRANSFORM OF A SIGNUM FUNCTION

The **signum function,** denoted by *sgn* (*t*), is defined as follows :

$$sgn\,(t) = \begin{cases} -1 & \text{for} \quad t < 0 \\ 1 & \text{for} \quad t > 0 \end{cases} \tag{20.16.1}$$

It is shown in Fig. 20.12a.

$$\mathfrak{F}[sgn\,(t)] = \int_{-\infty}^{\infty} sgn\,(t)\, e^{-j\omega t}\, dt$$

$$= \int_{-\infty}^{0} -1 \times e^{-j\omega t}\, dt + \int_{0}^{\infty} 1 \times e^{-j\omega t}\, dt$$

The Dirichlet's condition

$$\int_{-\infty}^{\infty} f\,(t)\, dt < \infty$$

should be satisfied. As this condition is not satisfied by the signum function, direct evaluation of the Fourier transform is not possible. This transform will be determined by first expressing the function as a limiting case of some other function. Therefore consider the function $f\,(t)$ shown in Fig. 20.12b, in which $a > 0$. In this case,

$$\mathfrak{F}[f\,(t)] = \int_{-\infty}^{\infty} f\,(t)\, e^{-j\omega t}\, dt$$

$$= \int_{-\infty}^{0} -e^{-at}\, e^{-j\omega t}\, dt + \int_{0}^{\infty} e^{-at}\, e^{-j\omega t}\, dt$$

$$= \frac{-1}{a - j\omega}\, e^{(a - j\omega)\,t} \Big|_{-\infty}^{0} + \frac{-1}{a + j\omega}\, e^{-(a + j\omega)\,t} \Big|_{0}^{\infty}$$

$$= \frac{-1}{a - j\omega} + \frac{1}{a + j\omega} = \frac{-2j\omega}{a^2 + \omega^2}$$

Since
$$\lim_{a \to 0} f\,(t) = sgn\,(t)$$

$$\mathfrak{F}[sgn\,(t)] = \lim_{a \to 0} \frac{-2j\omega}{a^2 + \omega^2} = \frac{-2j}{\omega} = \frac{2}{j\omega}$$

$$\boxed{\mathfrak{F}[sgn\,(t)] = \frac{2}{j\omega}}$$

The amplitude and phase spectra of the signum function are shown in Figs. 20.12c and 20.12d respectively.

The Fourier transform of unit step function may be found from the above result.

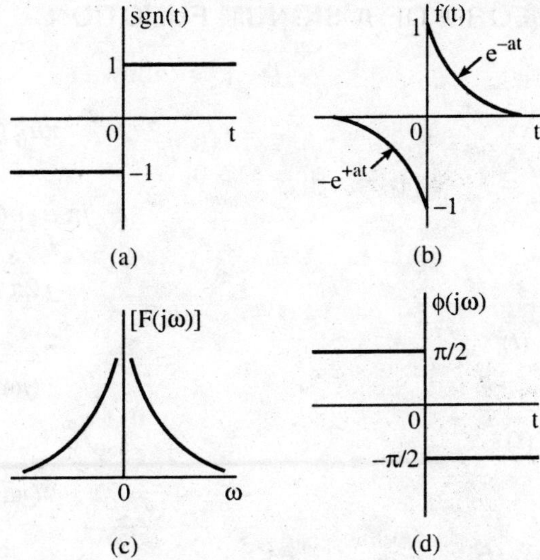

FIG. 20.12 (a) Signum function (b) Exponential function (c) Amplitude spectrum (d) Phase spectrum

$$u_{-1}(t) = \frac{1}{2} + \frac{1}{2}\, sgn\,(t)$$

$$\therefore \qquad \mathfrak{F}[u_{-1}(t)] = \mathfrak{F}\left[\frac{1}{2} + \frac{1}{2}\, sgn\,(t)\right]$$

$$= \mathfrak{F}\left[\frac{1}{2}\right] + \frac{1}{2}\,\mathfrak{F}[sgn\,(t)]$$

$$= \frac{1}{2} \times 2\pi\, u_0\,(j\omega) + \frac{1}{2}\left(\frac{2}{j\omega}\right) = \pi\, u_0\,(j\omega) + \frac{1}{j\omega}$$

Some common Fourier transforms are given in Table 20.1.

Table 20.1 Fourier Transforms

$f(t)$	$F(j\omega)$
$u_0(t)$	1
$u_0(t-\alpha)$	$e^{-j\omega\alpha}$
A	$2\pi A u_0(j\omega)$
$e^{j\omega_0 t}$	$2\pi u_0(j\omega - j\omega_0)$
$e^{-j\omega_0 t}$	$2\pi u_0(j\omega + j\omega_0)$
$\cos\omega_0 t$	$\pi u_0(j\omega - j\omega_0) + \pi u_0(j\omega + j\omega_0)$
$\sin\omega_0 t$	$j\pi\,[u_0(j\omega + j\omega_0) - u_0(j\omega - j\omega_0)]$
$e^{-\alpha t}\, u_{-1}(t)$	$\dfrac{1}{j\omega + \alpha}$

$sgn\,(t)$	$\frac{2}{j\omega}$
$u_{-1}(t)$	$\pi u_0(j\omega) + \frac{1}{j\omega}$
$tu_{-1}(t)$	$j\pi\, u_{+1}(j\omega) + \frac{1}{(j\omega)^2}$
t	$j\,2\pi\, u_{+1}(j\omega)$
$\cos \omega_0 t\, u_{-1}(t)$	$\frac{j\omega}{(j\omega)^2 + \omega_0^2}$
$\sin \omega_0 t\, u_{-1}(t)$	$\frac{\omega_0}{(j\omega)^2 + \omega_0^2}$
$e^{-\alpha t} \cos \omega_0 t\, u_{-1}(t)$	$\frac{j\omega + \alpha}{(j\omega + \alpha)^2 + \omega_0^2}$
$e^{-\alpha t} \sin \omega_0 t\, u_{-1}(t)$	$\frac{\omega_0}{(j\omega + \alpha)^2 + \omega_0^2}$
$a_1 f_1(t) + a_2 f_2(t)$	$a_1 F_1(j\omega) + a_2 F_2(j\omega)$
$\frac{d}{dt} f(t)$	$(j\omega)\, F(j\omega)$
$\frac{d^n f(t)}{dt^n}$	$(j\omega)^n F(j\omega)$
$f(at)$	$\frac{1}{\lvert a \rvert} F\left(\frac{j\omega}{a}\right)$
$f(t - t_0)$	$e^{-j\omega_0 t} F(j\omega)$
$e^{-\alpha t} f(t)$	$F(j\omega + \alpha)$
$e^{j\omega_0 t} f(t)$	$F[j(\omega - \omega_0)]$
$t\, e^{-\alpha t}\, u_{-1}(t)$	$\frac{1}{(j\omega + \alpha)^2}$
$u_{-1}(t + \alpha) - u_{-1}(t - \alpha)$	$2\alpha\left(\frac{\sin \alpha\omega}{\alpha\omega}\right)$
$\int_{-\infty}^{\infty} f_1(\tau) f_2(t - \tau)\, d\tau$	$F_1(j\omega)\, F_2(j\omega)$
$f_1(t)\, f_2(t)$	$\frac{1}{2\pi} \int_{-\infty}^{\infty} F_1(j\lambda)\, F_2(\overline{j\omega - \lambda})\, d\lambda$

■ **EXAMPLE 20.2** Find the Fourier transform of the function

$$f(t) = |t| \quad \text{for all } t$$

SOLUTION : If we multiply the signum function by t, we get the given function $f(t) = |t|$. That is,

$$|t| = t \, sgn\,(t)$$

By frequency-differentiation property,

if $$\mathfrak{F}[f(t)] = F(j\omega)$$

then $$\mathfrak{F}[t f(t)] = j \frac{d}{d\omega} F(j\omega)$$

therefore, $$\mathfrak{F}|t| = [t \, sgn\,(t)]$$

$$= j \frac{d}{d\omega}\left(\frac{2}{j\omega}\right) = -\frac{2}{\omega^2}$$

or $$\mathfrak{F}|t| = \frac{2}{(j\omega)^2}$$

20.17 FOURIER TRANSFORM OF A UNIT RAMP FUNCTION

The unit ramp function is given by

$$f(t) = u_{-2}(t) = t\, u_{-1}(t)$$

By frequency differentiation property,

if $$\mathfrak{F}[f(t)] = F(j\omega)$$

$$\mathfrak{F}[t f(t)] = j \frac{d}{d\omega} F(j\omega)$$

Therefore $$\mathfrak{F}[t\, u_{-1}(t)] = j \frac{d}{d\omega}\left[\pi u_0(j\omega) + \frac{1}{j\omega}\right]$$

$$= j\left[\pi u_{+1}(j\omega) - \frac{1}{j\omega^2}\right]$$

$$= j\pi u_{+1}(j\omega) - \frac{1}{\omega^2}$$

$$= j\pi u_{+1}(j\omega) + \frac{1}{(j\omega)^2} \qquad (20.17.1)$$

■ **EXAMPLE 20.3** Find the Fourier transfrom of the function

$$f(t) = t \quad \text{for all } t$$

SOLUTION : By frequency-differentiation property

if $$\mathfrak{F}[f_1(t)] = F_1(j\omega),$$

then $$\mathfrak{F}[t f_1(t)] = j \frac{d}{d\omega} F_1(j\omega)$$

If $$f_1(t) = 1,$$

then $$\mathfrak{F}[f_1(t)] = F_1(j\omega) = \mathfrak{F}[1] = 2\pi u_0(j\omega)$$

therefore, $$\mathfrak{F}[t f_1(t)] = \mathfrak{F}[t \times 1] = j \frac{d}{d\omega}[2\pi u_0(j\omega)]$$

or $$\mathfrak{F}[t] = j2\pi u_{+1}(j\omega) \tag{E20.3.1}$$

■ **EXAMPLE 20.4** Find the Fourier transform of the function

$$f(t) = e^{-\alpha t} \sin \omega_1 t \, u_{-1}(t)$$

SOLUTION :

$$\mathfrak{F}[e^{-\alpha t} \sin \omega_1 t \, u_{-1}(t)]$$

$$= \mathfrak{F}\left[e^{-\alpha t} \cdot \frac{e^{j\omega_1 t} - e^{-j\omega_1 t}}{2j} u_{-1}(t)\right]$$

$$= \frac{1}{2j} \mathfrak{F}[e^{-(\alpha - j\omega_1)t} u_{-1}(t) - e^{-(\alpha + j\omega_1)t} u_{-1}(t)]$$

$$= \frac{1}{2j}\left[\frac{1}{j\omega + (\alpha - j\omega_1)} - \frac{1}{j\omega + (\alpha + j\omega_1)}\right]$$

$$= \frac{1}{2j} \frac{(j\omega + \alpha + j\omega_1) - (j\omega + \alpha - j\omega_1)}{[(j\omega + \alpha) - j\omega_1]\,[(j\omega + \alpha) + j\omega_1]}$$

$$= \frac{\omega_1}{(j\omega + \alpha)^2 + \omega_1^2}$$

■ **EXAMPLE 20.5** Find the Fourier transform of the function

$$f(t) = e^{-\alpha t} \cos \omega_1 t \, u_{-1}(t)$$

SOLUTION :

$$\mathfrak{F}[e^{-\alpha t} \cos \omega_1 t u_{-1}(t)]$$

$$= \mathfrak{F}\left[e^{-\alpha t} \frac{e^{j\omega_1 t} + e^{-j\omega_1 t}}{2} u_{-1}(t)\right]$$

$$= \frac{1}{2} \mathfrak{F}[e^{-(\alpha - j\omega_1)t} u_{-1}(t) + e^{-(\alpha + j\omega_1)t} u_{-1}(t)]$$

$$= \frac{1}{2}\left[\frac{1}{j\omega + (\alpha - j\omega_1)} + \frac{1}{j\omega + (\alpha + j\omega_1)}\right]$$

$$= \frac{j\omega + \alpha}{(j\omega + \alpha)^2 + \omega_1^2}$$

20.18 FOURIER TRANSFORM OF A PERIODIC FUNCTION

By using the Fourier series, a periodic function $f_p(t)$ with period T can be expressed as a sum of complex exponentials.

$$f_p(t) = \sum_{n=\infty}^{\infty} \mathbf{c}_n e^{jn\omega_0 t} \tag{20.18.1}$$

where c_n is a complex Fourier coefficient defined by

$$\mathbf{c}_n \overset{\Delta}{=} \frac{1}{T}\int_{-T/2}^{T/2} f_p(t)\, e^{-jn\omega_0 t} dt \tag{20.18.2}$$

and ω_0 is the fundamental frequency given by

$$\omega_0 = \frac{2\pi}{T} \tag{20.18.3}$$

In a strict sense, the Fourier transform of periodic function does not exist since it violates the condition of absolute integrability. For any periodic function $f_p(t)$ we have

$$\int_{-\infty}^{\infty} | f_p(t) |\, dt = \infty \tag{20.18.4}$$

However, we shall show that the transform does exist in the limit. We assume that the periodic function exists only in the finite interval $(-T/2, T/2)$ and let T become infinite in the limit.

Taking the Fourier transform of both sides of Eq. (20.18.1), we obtain

$$\mathfrak{F}[f_p(t)] = \mathfrak{F} \sum_{n=-\infty}^{\infty} \mathbf{c}_n e^{jn\omega_0 t}$$

$$= \sum_{n=-\infty}^{\infty} [\mathbf{c}_n \mathfrak{F} e^{jn\omega_0 t}]$$

$$= \sum_{n=-\infty}^{\infty} [\mathbf{c}_n 2\pi\, u_0(j\omega - j\omega_0)]$$

$$= 2\pi \sum_{n=-\infty}^{\infty} \mathbf{c}_n u_0(j\omega - j\omega_0)$$

Thus the Fourier transform of a periodic function of time consists of a sum of impulse functions occuring at integral multiples of fundamental frequency and displaced in frequency by ω_0 (that is impulses at distance $\omega_0 = 2\pi/T$ apart). The value of each impulse is the same as 2π times the value of the corresponding coefficient $\mathbf{c}_n$ in the exponential Fourier series. In other words, *the frequency spectrum of a periodic function is not continuous but is discrete.*

A periodic function $f_p(t)$ and its Fourier transform are shown in Fig. 20.13.

Hence we have shown that *the Fourier series is a limiting case of the Fourier transform*.

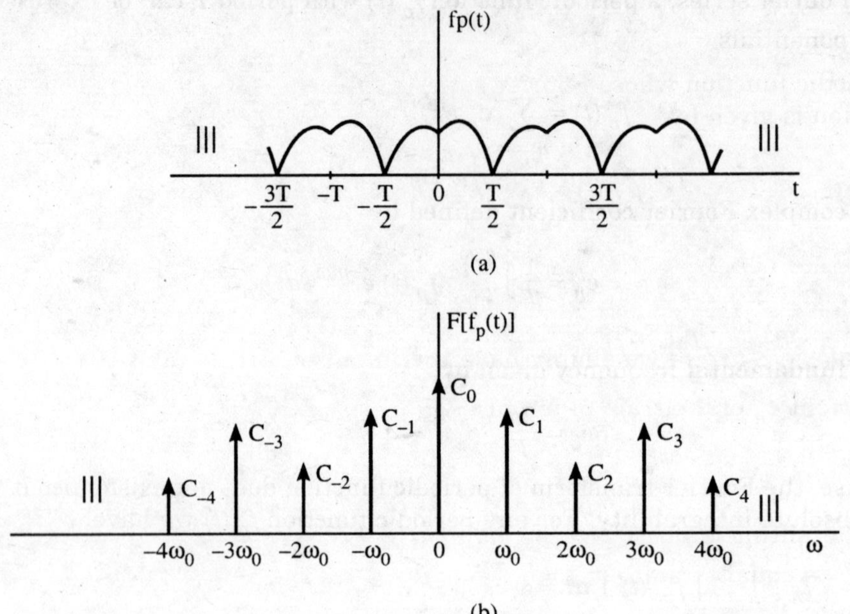

FIG. 20.13 (a) Periodic function of time (b) Fourier transform of the function

■ **EXAMPLE 20.6** Find the Fourier transform for an infinite train of unit impulses spaced T seconds apart as shown in Fig. 20.14a.

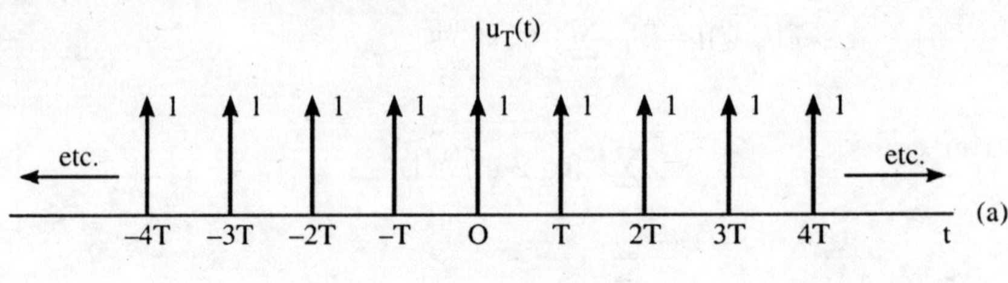

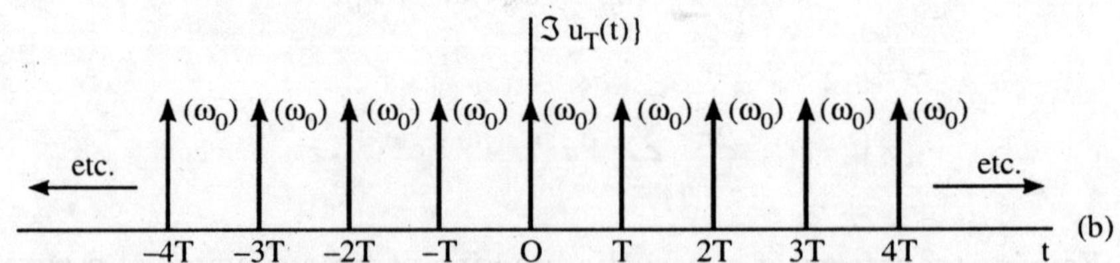

FIG. 20.14 (a) Train of regularly spaced impulses (b) Spectrum

SOLUTION : The train of regularly spaced impulses may be denoted by the symbol $u_T(t)$, where T denotes the period of the waveform. Thus, we can write,

$$u_T(t) = u_0(t) + u_0(t - T) + u_0(t - 2T) + u_0(t - nT) + \ldots$$

$$u_0(t + T) + u_0(t + 2T) + \ldots + u_0(t + nT) + \ldots$$

$$= \sum_{n=-\infty}^{\infty} u_0(t - nT) \tag{E20.6.1}$$

This is a periodic function whose Fourier series coefficients are all equal. The Fourier series for this function is given by

$$u_T(t) = \sum_{n=-\infty}^{\infty} \mathbf{c}_n \, e^{jn\omega_0 t} \tag{E20.6.2}$$

where

$$c_n = \frac{1}{T}\int_{-T/2}^{T/2} u_T(t)\, e^{-jn\omega_0 t}\, dt \tag{E20.6.3}$$

In the interval $-\frac{T}{2} \le t \le \frac{T}{2}$, we simply have the impulse $u_0(t)$ located at $t = 0$. The complex Fourier coefficient $\mathbf{c}_n$ of the train of impulses is therefore given by

$$\mathbf{c}_n = \frac{1}{T}\int_{-T/2}^{T/2} u_0(t)\, e^{-jn\omega_0 t}\, dt \tag{E20.6.4}$$

Using the sifting property of the impulse function, and noting that the exponential function $e^{-jn\omega_0 t}$ is equal to unity at $t = 0$, we obtain

$$\mathbf{c}_n = \frac{1}{T} \tag{E20.6.5}$$

With the help of Eqs. (E20.6.2) and (E20.6.5) we can express the function $u_T(t)$ as follows :

$$u_T(t) = \frac{1}{T} \sum_{n=-\infty}^{\infty} e^{jn\omega_0 t} \tag{E20.6.6}$$

Taking the Fourier transform of both sides of Eq. (E20.6.6), we get

$$\mathfrak{F}[u_T(t)] = \left[\frac{1}{T} \sum_{n=-\infty}^{\infty} e^{jn\omega_0 t}\right]$$

Since

$$\mathfrak{F}[e^{jn\omega_0 t}] = 2\pi\, u_0(j\omega - jn\omega_0)$$

$$\mathfrak{F}[u_T(t)] = \frac{2\pi}{T} \sum_{n=-\infty}^{\infty} u_0(j\omega - jn\omega_0) \tag{E20.6.7a}$$

$$= \omega_0 \sum_{n=-\infty}^{\infty} u_0(j\omega - jn\omega_0) \tag{E20.6.7b}$$

Relation (E20.6.7) is very important. It states that *the Fourier transform of a periodic train of impulse functions spaced T seconds apart consists of another set of impulses each of value* $\omega_0 \left(= \frac{2\pi}{T}\right)$. These impulses are regularly spaced by ω_0 radians along the frequency axis. Fig. 20.14a shows a train of regularly spaced impulses and Fig. 20.14b shows its Fourier transform.

It is to be noted that as the period T of the impulses in the time domain is increased, the impulses in the frequency domain decrease in value. In other words, the frequency spectrum becomes denser.

20.19 AREA UNDER f (t)

If $$\Im f(t) = F(j\omega)$$

then $$\int_{-\infty}^{\infty} f(t)\, dt = F(j0)$$

That is, *the area under a function $f(t)$ is equal to the value of its Fourier transform $F(j\omega)$ at $\omega = 0$.*

Proof

We have

$$F(j\omega) = \int_{-\infty}^{\infty} f(t)\, e^{-j\omega t}\, dt \tag{20.19.1}$$

Putting $\omega = 0$ on both the sides of Eq. (20.19.1) gives

$$F(j0) = \int_{-\infty}^{\infty} f(t)\, dt \tag{20.19.2}$$

The result in Eq. (20.19.2) can also be expressed as follows :

The height of the frequency spectrum at $\omega = 0$ is equal to the area under the $f(t)$ curve.

20.20 AREA UNDER $F(j\omega)$

If $$\Im f(t) = F(j\omega)$$

then $$f(0) = \frac{1}{2\pi}\int_{-\infty}^{\infty} F(j\omega)\, d\omega$$

That is, *the value of a function $f(t)$ at $t = 0$ is equal to $\frac{1}{2\pi}$ times the area under its Fourier transform $F(j\omega)$.*

Proof

We have

$$f(t) = \frac{1}{2\pi}\int_{-\infty}^{\infty} F(j\omega)\, e^{j\omega t}\, d\omega \tag{20.20.1}$$

Putting $t = 0$ on both sides of Eq. (20.20.1) gives

$$f(0) = \frac{1}{2\pi}\int_{-\infty}^{\infty} F(j\omega)\, d\omega \tag{20.20.2}$$

The result in Eq. (20.20.2) can also be expressed as follows :

The height of the time function $f(t)$ at $t = 0$ is equal to $\frac{1}{2\pi}$ times the area under the $F(j\omega)$ curve.

20.21 EVEN AND ODD FUNCTIONS

The Fourier transform $F(j\omega)$ of a function $f(t)$ is, in general, a complex function of ω because of the presence of $e^{-j\omega t}$ inside the integral. However, if $f(t)$ is even, or if $f(t)$ is odd certain simplifications result.

Direct Fourier Transform

Let the symbols e and o denote even and odd respectively.

We have

$$F(j\omega) = \int_{-\infty}^{\infty} F(t)\, e^{-j\omega t}\, dt$$

Since

$$f(t) = f_e(t) + f_0(t)$$

and

$$e^{-j\omega t} = \cos \omega t - j \sin \omega t$$

we can write

$$F(j\omega) = \int_{-\infty}^{\infty} [fe(t) + f_0(t)]\,(\cos \omega t - j \sin \omega t)\, dt$$

$$= \int_{-\infty}^{\infty} f_e(t) \cos \omega t\, dt + \int_{-\infty}^{\infty} f_0(t) \cos \omega t\, dt$$

$$-j \int_{-\infty}^{\infty} f_e(t) \sin \omega t\, dt - j \int_{-\infty}^{\infty} f_0(t) \sin \omega t\, dt \qquad (20.21.1)$$

The functions $f_e(t)$ and $\cos \omega t$ are even functions of t, and the functions $f_0(t)$ and $\sin \omega t$ are odd functions of t. Therefore, the product $f_e(t) \cos \omega t$ is an even function (since even times even is even). The product $f_0(t) \sin \omega t$ is also an even function (since odd times odd is even). For even functions, the integral from $-\infty$ to 0 is equal to the integral from 0 to ∞. Therefore, the integral from $-\infty$ to ∞ is equal to twice the integral from 0 to ∞. Thus,

$$\int_{-\infty}^{0} f_e(t) \cos \omega t\, dt = \int_{0}^{\infty} f_e(t) \cos \omega t\, dt$$

and

$$\int_{-\infty}^{\infty} f_e(t) \cos \omega t\, dt = 2 \int_{0}^{\infty} f_e(t) \cos \omega t\, dt \qquad (20.21.2)$$

Similarly

$$\int_{-\infty}^{\infty} f_0(t) \sin \omega t\, dt = 2 \int_{0}^{\infty} f_0(t) \sin \omega t\, dt \qquad (20.21.3)$$

The product $f_0(t) \cos \omega t$ is an odd function (since odd times even is odd). The product $f_0(t) \cos \omega t$ is also an odd function (since even times odd is odd). For odd functions, the integral from $-\infty$ to 0 is equal in magnitude to, but opposite in sign from, the integral from 0 to ∞. Therefore,

$$\int_{-\infty}^{0} f_0(t) \cos \omega\, dt = -\int_{0}^{\infty} f_0(t) \cos \omega t\, dt$$

$$\int_{-\infty}^{\infty} f_0(t)\cos\omega t\,dt = \int_{-\infty}^{0} f_0(t)\cos\omega t\,dt + \int_{0}^{\infty} f_0(t)\cos\omega t\,dt \quad (20.21.4)$$

Similarly,

$$\int_{-\infty}^{\infty} f_e(t)\sin\omega t\,dt = 0 \quad (20.21.5)$$

Therefore Eq. (20.21.1) can be written as

$$F(j\omega) = 2\int_{0}^{\infty} f_e(t)\cos\omega t\,dt - j2\int_{0}^{\infty} f_0(t)\sin\omega t\,dt \quad (20.21.6)$$

In Eq. (20.21.6), the integrations are with respect to t, but the results of integrations are in terms of ω. The first integral of Eq. (20.21.6) is an even part of $F(j\omega)$. Let the even and odd parts of $F(j\omega)$ be denoted by $F_e(j\omega)$ and $F_0(j\omega)$ respectively. Therefore,

$$F_e(j\omega) = 2\int_{0}^{\infty} f_e(t)\cos\omega t\,dt \quad (20.21.7)$$

and

$$F_0(j\omega) = -j2\int_{0}^{\infty} f_0(t)\sin\omega t\,dt \quad (20.21.8)$$

The function $F(j\omega)$ can be written as

$$F(j\omega) = F_e(j\omega) + F_0(j\omega) \quad (20.21.9)$$

The function $F(j\omega)$ is, in general, a complex function of ω. But if $f(t)$ is an even function of time, Eq. (20.21.7) shows that $F(j\omega)$ will be real and even in ω, whereas if $f(t)$ is an odd function of time, Eq. (20.21.8) shows that $F(j\omega)$ will be imaginary and odd in ω.

Inverse Fourier Transform

$$f(t) = \frac{1}{2\pi}\int_{-\infty}^{\infty} F(j\omega)\,e^{j\omega t}\,d\omega \quad (20.21.10)$$

$$= \frac{1}{2\pi}\int_{-\infty}^{\infty} [F_e(j\omega) + F_0(j\omega)]\,(\cos\omega t + j\sin\omega t)\,dt \quad (20.21.11)$$

Eq. (20.21.11) is split into four terms analogous to Eq. (20.21.1). With similar arguments, Eq. (20.21.11) can be written as

$$f(t) = \frac{1}{2\pi}\left[2\int_{0}^{\infty} F_e(j\omega)\cos\omega t\,d\omega + 2j\int_{0}^{\infty} F_0(j\omega)\sin\omega t\,d\omega\right] \quad (20.21.12)$$

These integrals are with respect to ω and the results of integration are in terms of t. Since $\cos\omega t$ and $\sin\omega t$ are even and odd functions of time respectively, Eq. (20.21.12) can be written as

$$f_e(t) = \frac{1}{\pi}\int_{0}^{\infty} F_e(j\omega)\cos\omega t\,d\omega \quad (20.21.13)$$

$$f_0(t) = \frac{j}{\pi}\int_{0}^{\infty} F_0(j\omega)\sin\omega t\,d\omega \quad (20.21.14)$$

Hence we can write

$$f(t) = f_e(t) + f_0(t) \quad (20.21.15)$$

EXAMPLE 20.7 Find the Fourier transform of the function $f_1(t)$ shown in Fig. 20.15a where

$$f_1(t) = \begin{cases} e^{at}, & t < 0 \\ e^{-at}, & t > 0 \end{cases}$$

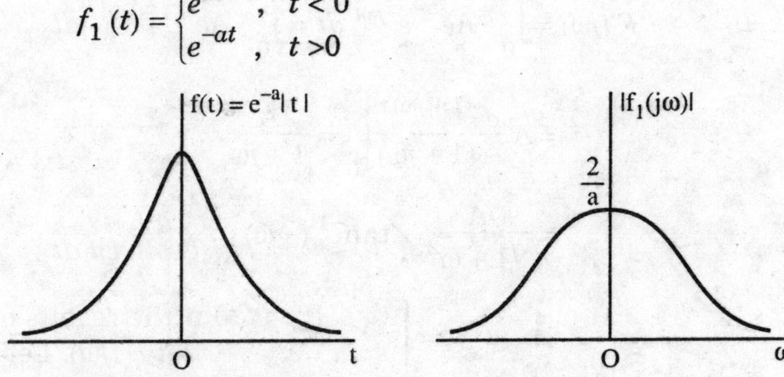

FIG. 20.15 (a) Function $f_1(t)$ (b) Amplitude spectrum

SOLUTION :

$$F_1(j\omega) = \int_{-\infty}^{\infty} f_1(t)\, e^{-j\omega t}\, dt$$

$$= \int_{-\infty}^{0} e^{at} e^{-j\omega t}\, dt + \int_{0}^{\infty} e^{-at} e^{-j\omega t}\, dt$$

$$= \int_{-\infty}^{0} e^{(a - j\omega)t}\, dt + \int_{0}^{\infty} e^{-(a + j\omega)t}\, dt$$

$$= \frac{1}{(a - j\omega)} e^{(a - j\omega)t}\Bigg|_{-\infty}^{0} + \frac{-1}{(a + j\omega)} e^{(a + j\omega)t}\Bigg|_{0}^{\infty}$$

$$= \frac{1}{(a - j\omega)}(e^0 - 0) - \frac{1}{(a + j\omega)}(0 - e^0)$$

$$= \frac{1}{(a - j\omega)} + \frac{1}{a + j\omega} = \frac{2a}{a^2 + \omega^2}$$

The amplitude spectrum is shown in Fig. 20.15b. For the given function $\theta(j\omega) = 0$

EXAMPLE 20.8 Determine the Fourier transform of the exciting voltage

$$v(t) = \begin{cases} Ae^{-t}, & t \geq 0 \\ 0, & t < 0 \end{cases}$$

and sketch the amplitude and phase spectra. If the above voltage is applied across a series RC circuit, find $i(t)$.

SOLUTION :

$$F(j\omega) = \int_{-\infty}^{\infty} f(t)\, e^{-j\omega t}\, dt$$

$$= \int_{-\infty}^{0} f(t)\, e^{-j\omega t}\, dt + \int_{0}^{\infty} f(t)\, e^{-j\omega t}\, dt$$

Since $f(t) = 0 \quad \text{for } t < 0,$

$$F(j\omega) = \int_0^\infty Ae^{-t}\, e^{-j\omega t}\, dt = \int_0^\infty Ae^{-(1+j\omega)t}\, dt$$

$$= A \left. \frac{e^{-(1+j\omega)t}}{-(1+j\omega)} \right|_0^\infty = \frac{A}{1+j\omega}$$

$$= \frac{A}{\sqrt{1+\omega^2}} \angle \tan^{-1}(-\omega)$$

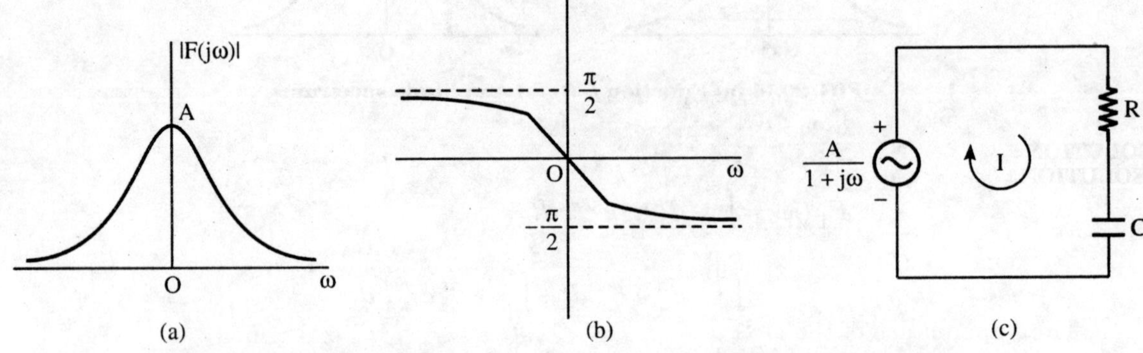

FIG. 20.16 (a) Amplitude spectrum (b) Phase spectrum (c) Series RC circuit

Fig. 20.16a shows the amplitude spectrum and Fig. 20.16b shows the phase spectrum of $F(j\omega)$.

The series RC circuit is shown in Fig. 20.16c.

The impedance of the circuit

$$Z(j\omega) = R + \frac{1}{j\omega C} = \frac{1 + j\omega CR}{j\omega C}$$

therefore,

$$Y(j\omega) = \frac{1}{Z(j\omega)} = \frac{j\omega C}{1 + j\omega CR}$$

$$= \frac{\omega C}{\sqrt{(1 + \omega^2 C^2 R^2)}} \angle (90^\circ - \tan^{-1} \omega CR)$$

$$I(j\omega) = V(j\omega)\, Y(j\omega)$$

$$= \left[\frac{V}{\sqrt{1+\omega^2}} \angle \tan^{-1}(-\omega) \right] \times \left[\frac{\omega C}{\sqrt{(1+\omega^2 C^2 R^2)}} \angle (90^\circ - \tan^{-1} \omega CR) \right]$$

$$= \frac{V\omega C}{\sqrt{(1+\omega^2)(1+\omega^2 C^2 R^2)}} \angle \phi_1$$

where

$$\phi_1 = 90^\circ - \tan^{-1} \omega CR - \tan^{-1} \omega$$

EXAMPLE 20.9 The voltage source in the circuit of Fig. 20.17a is given by

$$v(t) = 10\, sgn\,(t)$$

Use Fourier-transform method to determine $v_c(t)$ and sketch $v_c(t)$ versus t curve.

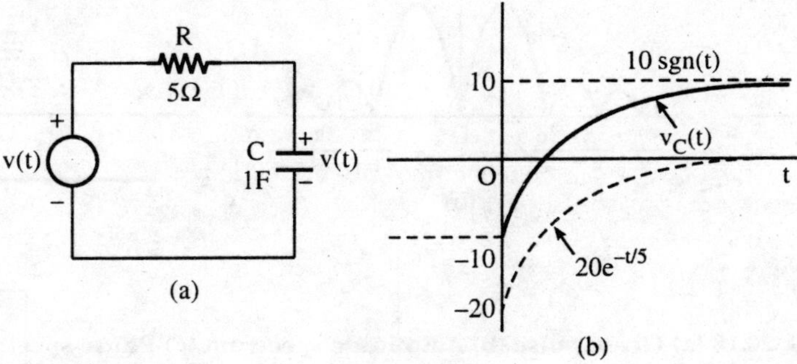

FIG. 20.17

SOLUTION :

$$v(t) = 10\, sgn\,(t)$$

$$V(j\omega) = 10\left(\frac{2}{j\omega}\right) = \frac{20}{j\omega}$$

The impendance of the circuit

$$Z(j\omega) = R + \frac{1}{j\omega C} = \frac{1 + j\omega CR}{j\omega C}$$

$$Y(j\omega) = \frac{1}{Z(j\omega)} = \frac{j\omega C}{1 + j\omega CR} = \frac{j\omega}{1 + j\omega 5}$$

$$I(j\omega) = V(j\omega)\, Y(j\omega) = \frac{20}{j\omega} \times \frac{j\omega}{1 + j\omega 5} = \frac{20}{1 + j\omega 5}$$

$$V_c(j\omega) = \left(\frac{1}{j\omega C}\right) I(j\omega) = \frac{20}{j\omega\,(1 + j\omega 5)}$$

Let
$$\frac{20}{j\omega\,(1 + j\omega 5)} = \frac{k_1}{j\omega} + \frac{k_2}{1 + j\omega 5}$$

$$k_1 = \left.\frac{20}{j\omega 5}\right|_{\omega = 0} = 20$$

$$k_2 = \left.\frac{20}{j\omega}\right|_{\omega = -\frac{1}{j5}} = \frac{20}{-1/5} = -100$$

$\therefore$
$$V_c(j\omega) = \frac{20}{j\omega} - \frac{100}{1 + j\omega 5} = 10\left(\frac{2}{j\omega}\right) - 20\,\frac{1}{(1/5) + j\omega}$$

The inverse Fourier transform gives

$$v_c(t) = 10\, sgn\,(t) - 20\, e^{-t/5}$$

The required curve in shown in 20.17b.

EXAMPLE 20.10 Find the magnitude and phase spectrum of the Fourier transform of the pulse shown in Fig. 20.18a.

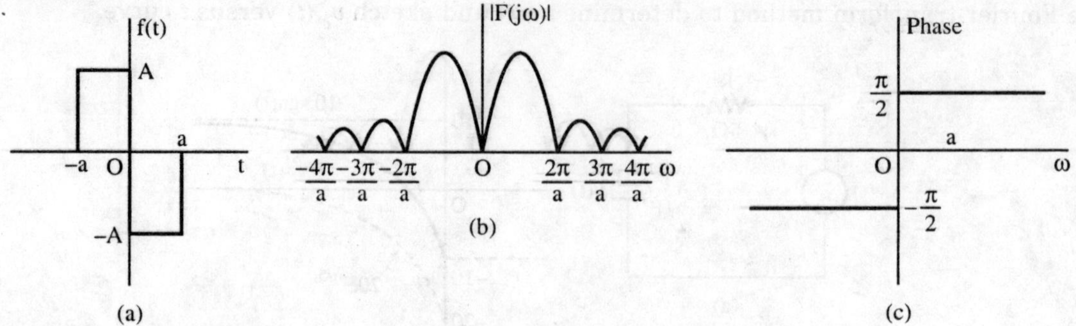

FIG. 20.18 (a) Given pulse (b) Amplitude spectrum (c) Phase spectrum

SOLUTION :

Here
$$\begin{aligned} f(t) &= A && \text{for } -a \le t \le 0 \\ &= -A && \text{for } 0 \le t \le a \\ &= 0 && \text{for all other values of } t \end{aligned}$$

$$F(j\omega) = \int_{-\infty}^{\infty} f(t)\, e^{-j\omega t}\, dt$$

$$= \int_{-\infty}^{0} f(t)\, e^{-j\omega t}\, dt + \int_{0}^{\infty} f(t)\, e^{-j\omega t}\, dt$$

$$= \int_{-a}^{0} f(t)\, e^{j\omega t}\, dt + \int_{0}^{a} f(t)\, e^{-j\omega t}\, dt$$

$$= \int_{-a}^{0} Ae^{-j\omega t}\, dt + \int_{0}^{a} -Ae^{-j\omega t}\, dt$$

If we replace t by $-t$ in the first integral

$$F(j\omega) = \int_{0}^{a} Ae^{j\omega t}\, dt - \int_{0}^{a} Ae^{-j\omega t}\, dt$$

$$= A\int_{0}^{a} (e^{+j\omega t} - e^{-j\omega t})\, dt$$

$$= A\int_{0}^{a} 2j \sin \omega t\, dt = 2Aj\left[\frac{\cos \omega t}{-\omega}\right]_0^a$$

$$= -2Aj\left(\frac{\cos \omega a}{\omega} - \frac{\cos 0^\circ}{\omega}\right) = \frac{j2A}{\omega}(1 - \cos \omega a)$$

$$|F(j\omega)| = \frac{2A}{\omega}(1 - \cos \omega a)$$

Therefore the magnitude $|F(j\omega)|$ is zero when $1 - \cos \omega a = 0$.

$$\therefore \qquad \omega a = 2n\pi, \text{ or } \omega = n\frac{2\pi}{a}$$

The amplitude spectrum is shown in Fig. 20.18b.

$$\tan \alpha_n = \frac{\text{imaginary part of } F(j\omega)}{\text{real part of } F(j\omega)}$$

$$= \frac{\frac{2A}{\omega}(1 - \cos \omega a)}{0}$$

$$= \alpha \qquad \text{when } \omega \text{ is positive}$$

$$= -\alpha \qquad \text{when } \omega \text{ is negative}$$

$$\alpha_n = \tan^{-1} \alpha \qquad \text{for } \omega \text{ positive} = +\frac{\pi}{2}$$

$$\alpha_n = \tan^{-1} (-\alpha) \qquad \text{for } \omega \text{ negative} = -\frac{\pi}{2}$$

The phase spectrum is shown in Fig. 20.18c.

■ **EXAMPLE 20.11** Find the amplitude spectrum of the Fourier transform of a single triangular pulse shown in Fig. 20.19a.

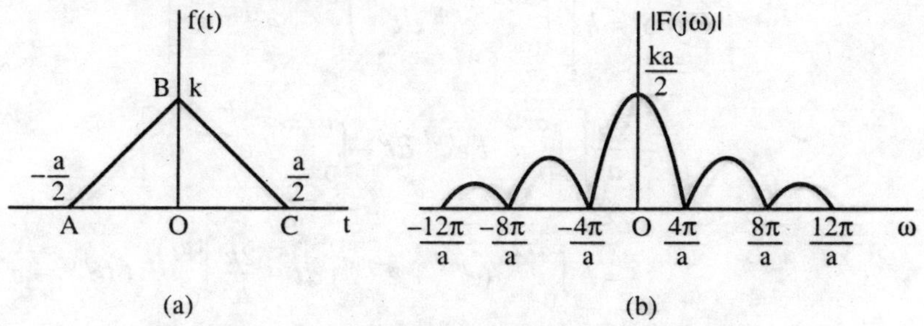

FIG. 20.19 (a) Given pulse (b) Amplitude spectrum

SOLUTION : The general equation of a straight line is

$$y = mx + c$$

The equation of line AB is given by

$$f(t) = \frac{k}{a/2} t + k$$

or
$$f(t) = k\left(1 + \frac{2}{a} t\right)$$

The equation of line BC is given by

$$f(t) = -\frac{k}{a/2} t + k$$

or
$$f(t) = k\left(1 - \frac{2}{a}\right)$$

Hence we can write

$$f(t) = k\left(1 + \frac{2}{a} t\right) \qquad \text{for} \qquad -\frac{a}{2} \le t \le 0$$

$$f(t) = k\left(1 - \frac{2}{a}t\right) \quad \text{for} \quad 0 \le t \le \frac{a}{2}$$

$$F(j\omega) = \int_{-\infty}^{\infty} f(t)\, e^{-j\omega t}\, dt$$

$$= \int_{-a/2}^{a/2} f(t)\, e^{-j\omega t}\, dt$$

$$= \int_{-a/2}^{0} f(t)\, e^{-j\omega t}\, dt + \int_{0}^{a/2} f(t)\, e^{-j\omega t}\, dt$$

$$= \int_{-a/2}^{0} k\left(1 + \frac{2}{a}t\right) e^{-j\omega t}\, dt + \int_{0}^{a/2} k\left(1 - \frac{2}{a}t\right) e^{-j\omega t}\, dt$$

$$= k\int_{-a/2}^{0} e^{-j\omega t}\, dt + k\int_{0}^{a/2} e^{-j\omega t}\, dt$$

$$+ \frac{2k}{a}\int_{-a/2}^{0} t\, e^{-j\omega t}\, dt - \frac{2k}{a}\int_{0}^{a/2} t\, e^{-j\omega t}\, dt$$

Replacing t by $-t$ in first and third integrals we get

$$F(j\omega) = k\left[\int_{0}^{a/2} e^{j\omega t}\, dt + \int_{0}^{a/2} e^{-j\omega t}\, dt\right]$$

$$-\frac{2k}{a}\left[\int_{0}^{a/2} t\, e^{j\omega t}\, dt + \int_{0}^{a/2} t\, e^{-j\omega t}\, dt\right]$$

$$= k\int_{0}^{a/2} (e^{j\omega t} + e^{-j\omega t})\, dt - \frac{2k}{a}\int_{0}^{a/2} t\,(e^{j\omega t} + e^{-j\omega t})\, dt$$

$$= k\int_{0}^{a/2} 2\cos\omega t\, dt - \frac{2k}{a}\int_{0}^{a/2} 2t\cos\omega t\, dt$$

$$= 2k\,\frac{\sin\frac{\omega a}{2}}{\omega} - \frac{4k}{a}\left[\frac{a}{2}\,\frac{\sin\frac{\omega a}{2}}{\omega} - \int_{0}^{a/2} \frac{\sin\omega t}{\omega}\, dt\right]$$

$$= 2k\,\frac{\sin\frac{\omega a}{2}}{\omega} - \frac{2k}{\omega}\sin\frac{\omega a}{2} - \frac{4k}{a\omega}\left[\frac{\cos\omega t}{\omega}\right]_{0}^{a/2}$$

$$= \frac{-4k}{a\omega^2}\left(\cos\frac{\omega a}{2} - 1\right)$$

$$= \frac{8k}{a\omega^2}\sin^2\frac{\omega a}{4}$$

$$= \frac{8k}{a\omega^2}\left(\frac{\omega a}{4}\right)^2 \frac{\sin^2 \frac{\omega a}{4}}{\left(\frac{\omega a}{4}\right)^2}$$

$$= \frac{ka}{2} \frac{\sin^2 \beta}{\beta^2} \qquad \text{where } \beta = \frac{\omega a}{4}$$

The amplitude spectrum is shown in Fig. 20.19b.

EXAMPLE 20.12 Determine the Fourier transform of the function

$$f(t) = t\, e^{-\alpha t}\, u_{-1}(t)$$

SOLUTION :

$$f(t) = t\, e^{-\alpha t}\, u_{-1}(t)$$

$$\frac{d}{dt} f(t) = \frac{d}{dt}[t\, e^{-\alpha t}\, u_{-1}(t)]$$

$$= t \frac{d}{dt}[e^{-\alpha t}\, u_{-1}(t)] + e^{-\alpha t}\, u_{-1}(t) \frac{dt}{dt}$$

$$= t\left[e^{-\alpha t} \frac{d}{dt} u_{-1}(t) + u_{-1}(t) \frac{d}{dt} e^{-\alpha t}\right] + e^{-\alpha t}\, u_{-1}(t)$$

$$= t\,[e^{-\alpha t}\, u_0(t) - \alpha\, u_{-1}(t)\, e^{-\alpha t}] + e^{-\alpha t}\, u_{-1}(t)$$

$$= te^{-\alpha t}\, u_0(t) - \alpha\, te^{-\alpha t}\, u_{-1}(t) + e^{-\alpha t}\, u_{-1}(t)$$

$$= 0 \times u_0(t) - \alpha f(t) + e^{-\alpha t}\, u_{-1}(t)$$

Taking the Fourier transform, since

$$\lim_{t \to -\infty} f(t) = \lim_{t \to \infty} f(t) = 0, \text{ we obtain}$$

$$j\omega\, F(j\omega) = -\,\alpha F(j\omega) + \frac{1}{j\omega + \alpha}$$

$$(j\omega + \alpha)\, F(j\omega) = \frac{1}{j\omega + \alpha}$$

$$\because \qquad F(j\omega) = \frac{1}{(j\omega + \alpha)^2}$$

That is, $\mathfrak{F}[t\, e^{-\alpha t}\, u_{-1}(t)] = \dfrac{1}{(j\omega + \alpha)^2}$

EXAMPLE 20.13 A voltage $v(t) = e^{-\alpha t} u_{-1}(t)$ is applied across a 1-Ω resistor. Determine the total energy absorbed by the resistor.

SOLUTION : The total energy absorbed by the resistor

$$= \int_{-\infty}^{\infty} v^2(t)\,dt = \int_{-\infty}^{\infty} [e^{-\alpha t} u_{-1}(t)]^2\,dt$$

$$= \int_{-\infty}^{\infty} e^{-2\alpha t} u_{-1}(t)\,dt = \int_{0}^{\infty} e^{-2\alpha t}\,dt$$

$$= \frac{-1}{2\alpha} e^{-2\alpha} \Big|_0^{\infty} = -\frac{1}{2\alpha}(0-1) = \frac{1}{2\alpha} \text{ joules}$$

20.22 RELATIONSHIPS AMONG THE FOURIER SERIES, THE FOURIER TRANSFORM AND THE LAPLACE TRANSFORM

Table 20.2 shows the similarities and differences among the one-sided Laplace transform, the two-sided Laplace transform the Fourier transform, and the Fourier series transform.

Table 20.2

Function	Direct transform	Inverse transform
One-side Laplace tranform	$F(s) = \int_{0}^{\infty} f(t)\,e^{-st}\,dt$	$f(t) = \frac{1}{2\pi j}\int_{c-j\infty}^{c+j\infty} F(s)\,e^{+st}\,ds$
Two-side Laplace transform	$F(s) = \int_{-\infty}^{\infty} f(t)\,e^{-st}\,dt$	$f(t) = \frac{1}{2\pi j}\int_{c-j\infty}^{c+j\infty} F(s)\,e^{+st}\,ds$
Fourier transform	$F(j\omega) = \int_{-\infty}^{\infty} f(t)\,e^{-j\omega t}\,dt$	$f(t) = \frac{1}{2\pi}\int_{-\infty}^{\infty} F(j\omega)\,e^{+j\omega t}\,dt$
Fourier-series transform	$F(jn\omega_0) = \frac{1}{T}\int_{t_1}^{t_1+T} f(t)\,e^{-jn\omega_0 t}\,dt$	$f(t) = \sum_{n=-\infty}^{\infty} F(jn\omega_0)\,e^{+jn\omega_0 t}$

EXERCISES

1. Show that

$$\int_{-\infty}^{\infty} x(t)\,u_0(at - t_0)\,dt = \frac{1}{a}\,x\left(\frac{t_0}{a}\right)$$

2. Show that

$$\mathfrak{F}[u_{-1}(t+a) - u_{-1}(t-a)] = 2a\left(\frac{\sin a\omega}{a\omega}\right)$$

3. Find the Fourier transform of the exponential function

$$v(t) = 2e^{-2t}$$

4. **Find the Fourier transform of the function shown in Fig. P.20.1.**

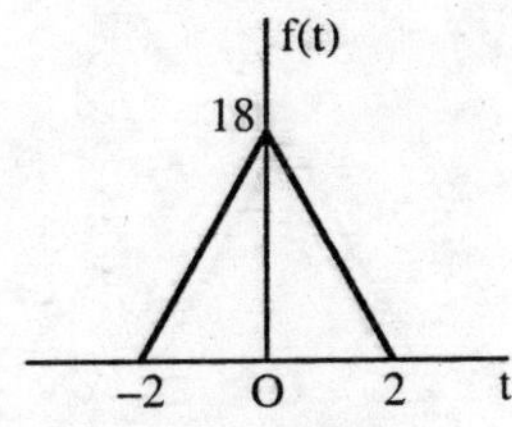

FIG. P 20.1

ANSWERS

3. $V(j\omega) = \dfrac{2}{2 + j\omega}$

4. $F(j\omega) = 36\left(\dfrac{\sin \omega}{\omega}\right)^2$

21

PASSIVE NETWORK SYNTHESIS

21.1 INTRODUCTION

Network analysis is the determination of response for a given excitation of the network. In **network synthesis**, the excitation and response are given and the network element types, element values and topology is to be determined. Thus, the problem of synthesis deals with the design and fabrication of a network that satisfies the prescribed specifications. A problem in network analysis usually has a unique solution, while a problem in network synthesis generally has several equivalent solutions and in some cases, may have no solution at all. In other words, a synthesis problem does not have a unique solution, and for a given excitation-response pair, it is possible to synthesize many different networks.

The excitation and response may be specified either in the time domain or frequency domain.

Instead of excitation and response, usually the driving-point immittance (that is driving-point impedance $Z(s)$ or driving-point admittance $Y(s)$) is given and the network elements are determined from this immittance. Such a synthesis is called **driving-point synthesis.**

In this chapter we shall study some aspects of designing a linear time-invariant network with given driving-point impedance or admittance function with passive elements only. Such a synthesis is called **passive network synthesis**. We shall consider the synthesis of driving-point immittance functions only.

21.2 SOME PROPERTIES OF THE DRIVING-POINT IMMITTANCE FUNCTION OF A PASSIVE NETWORK; CONDITIONS FOR PHYSICAL REALIZABILITY

The driving-point immittance function $F(s)$ of a passive network must have the following properties. Any function which does not have these properties cannot be a driving-point immittance function of a passive network. These properties may be regarded as the conditions for physical realizability for passive networks.

Property 1

$F(s)$ is representable as a ratio of two polynomials in s with real coefficients.

Corollary A. The real and imaginary components of $F(j\omega)$ are, respectively, even and odd functions of the angular frequency ω, namely

$$\text{Re } F(-j\omega) = \text{Re } F(j\omega)$$

and

$$\text{Im } F(-j\omega) = -\text{ Im } F(j\omega)$$

Corrollary B. The poles and zeros of F(s) which are not real appear in complex conjugate pairs.

Property 2.

The poles and zeros of F(s), if not on the jω axis, are in the left-hand half of the s-plane.

Property 3.

The poles and zeros of F(s) on the jω axis are simple.

This property also rules out multiple zeros and multiple poles at s = 0 and s = ∞, which may be considered to be on the jω axis. This leads to

Corrollary C. The highest and lowest powers of N(s) and D(s) cannot differ by more than 1 in the following equation

$$F(s) = \frac{N(s)}{D(s)} = \frac{a_m s^m + a_{m-1} s^{m-1} + \ldots + a_1 s + a_0}{b_n s^n + b_{n-1} s^{n-1} + \ldots + b_1 s + b_0} \tag{21.2.1}$$

or

$$F(s) = \frac{N(s)}{D(s)} = \frac{K(s - z_1)(s - z_2) \ldots (s - z_m)}{(s - p_1)(s - p_2) \ldots (s - p_n)} \tag{21.2.2}$$

where N(s) = numerator polynomial of degree m having coefficients $a_m, a_{m-1}, \ldots, a_0$

D(s) = denominator polynomial of degree n having coefficients $b_n, b_{n-1}, \ldots b_0$

and $K = a_m / b_n$

Property 4.

F(s) as a driving-point immittance function, when expressed in the form of Eq. (21.2.1), has only positive coefficients in its numerator and denominator.

Property 5.

The real component of $F(j\omega)$ is never negative, that is $\text{Re } F(j\omega) \geq 0$ for any value of ω.

21.3 HURWITZ POLYNOMIALS

A polynomial having all its zeros in the left half of the *s*-plane is called a **Hurwitz polynomial.**

A polynomial *P*(*s*) is Hurwitz when the following conditions are satisfied :

(i) *P*(*s*) is real when *s* is real.

(ii) The roots of *P*(*s*) have real parts which are zero or negative.

Properties of Hurwitz Polynomials

Let *P*(*s*) be a real polynomial of degree *n* given by

$$P(s) = a_n s^n + a_{n-1} s^{n-1} + \ldots + a_2 s^2 + a_1 s + a_0$$

A Hurwitz polynomial has the following properties :

1. *The coefficients of s terms must be positive.*
   That is $a_1, a_2, \ldots, a_n$ are positive.
2. *There must not be any power of s missing between the highest degree and lowest degree of the polynomial unless the polynomial is completely even or completely odd.*
3. *Both the even and odd parts of the Hurwitz polynomial have roots on the jω axis only.*

Let
$$P(s) = M(s) + N(s)$$

where $M(s)$ = even part of $P(s)$ and $N(s)$ = odd part of $P(s)$
then $M(s)$ and $N(s)$ both have roots on $j\omega$ axis only.

4. The continued fraction expansion of the ratio of even to odd parts (or the ratio of odd to even parts) of a Hurwitz polynomial gives all positive quotient terms.

21.4 PROCEDURE FOR OBTAINING THE CONTINUED-FRACTION EXPANSION

Property 4 is a necessary and sufficient condition for a polynomial to be Hurwitz. The continued fraction expansion is obtained by synthetic division. For this purpose the polynomial $P(s)$ is written as

$$P(s) = M(s) + N(s)$$

Suppose that
$$F(s) = \frac{M(s)}{N(s)}$$

where $M(s)$ is exactly one degree higher than $N(s)$. Now divide $M(s)$ by $N(s)$ once. The result is

$$\frac{M(s)}{N(s)} = \alpha_1 s + \frac{M_1(s)}{N(s)}$$

where $\alpha_1 > 0$ and $M_1(s)$ is one degree lower than $N(s)$. Invert the remainder and divide once again,

$$\frac{N_1(s)}{M_1(s)} = \alpha_2 s + \frac{M_2(s)}{M_1(s)}$$

Repeat the operation, each time inverting the remainder and dividing once. This process is finite, since the degree of every $M_k(s)$, $k = 1, 2, 3, \ldots$ is reduced after each step. The final result is the continued fraction,

$$F(s) = \frac{M(s)}{N(s)} \text{ or } \frac{N(s)}{M(s)}$$

$$= \alpha_1(s) + \cfrac{1}{\alpha_2(s) + \cfrac{1}{\alpha_3(s) + \cfrac{1}{\alpha_4(s) + \cfrac{1}{\ddots + \cfrac{1}{\alpha_n(s)}}}}} \quad (21.4.1)$$

where $\alpha_1, \alpha_2, \ldots, \alpha_n$ are the quotients in each step of division $\frac{M(s)}{N(s)}$ or $\frac{N(s)}{M(s)}$.

This is known as **"divide once, invert, divide"** procedure.

These quotients must be positive when the polynomial $P(s) = M(s) + N(s)$ is Hurwitz, because a negative quotient denotes a zero of $P(s)$ in the right half of the s plane.

21.5 PROCEDURE FOR TESTING HURWITZ POLYNOMIAL

Two steps are involved in testing a given polynomial for Hurwitz.

Step 1 Inspection Test

This test can be performed by a simple inspection of the polynomial $P(s)$. The following conditions are **necessary but not sufficient** for a polynomial to be strict Hurwitz :

1. All a_i coefficients of the polynomial must be real and positive.
2. No powers of s can be missing in the polynomial between the highest degree and lowest degree of s, unless the polynomial is completely even or completely odd.

It is to be noted that if **all** a_i coefficients in a given $P(s)$ are **negative**, then $P(s)$ can be multiplied by -1 so as to make all a_i coefficients positive. For example,

$$P(s) = -2s^2 - 4s - 8 \text{ can be written as } P(s) = 2s^2 + 4s + 8$$

which satisfies condition 1 above. Conditions 1 and 2 provide a simple test by inspection. The given polynomial must pass these before any further test. If the polynomial fails to satisfy any of the two inspection tests, there is no need of further test because the given polynomial is not strict Hurwitz. However, if the polynomial passes this inspection test, we cannot draw any definite conclusion as yet, since this test is necessary, but not sufficient, condition.

Step 2. Confirmatory Test

After the inspection test, the given polynomial must be subjected to a conclusive test which will establish whether the polynomial $P(s)$ is strict Hurwitz or not.

The necessary and sufficient conditions for $P(s)$ to be a Hurwitz polynomial is that the quotients $\alpha_1, \alpha_2, \ldots, \alpha_n$ in Eq. (21.4.1) in the continued-fraction expansion be real and positive.

Sometimes continued fraction expansion $\frac{M(s)}{N(s)}$ or $\frac{N(s)}{M(s)}$ may terminate prematurely if $M(s)$ and $N(s)$ contain a common factor (the last divisor). Such a factor shows that $F(s)$ is not Hurwitz. In such a case $F(s)$ can be written a

$$F(s) = W(s)\, F_1(s)$$

and the polynomial $F(s)$ is Hurwitz if both $W(s)$ and $F_1(s)$ are Hurwitz. The imaginary axis zeros of the function $F(s)$ are contributed by the common factor $W(s)$.

If the polynomial is either only even or only odd, it is not possible to obtain the continued fraction expansion of $\frac{M(s)}{N(s)}$ or $\frac{N(s)}{M(s)}$. In such cases the ratio of the polynomial $P(s)$ and its derivative $P'(s)$ is found. That is, if $P(s)$ is completely even or completely odd then $F(s)$ is constructed as

$$F(s) = \frac{P(s)}{P'(s)} = \frac{P(s)}{\frac{d}{ds}P(s)}$$

■ **EXAMPLE 21.1** Show that the function

$$Y(s) = \frac{2s^2 - 4s + 5}{s^2 + 7s + 6}$$

is not physically realizable with a passive network.

SOLUTION : Since the coefficient of one term in the numerator is negative, the given function is not physically realizable with a passive network.

■ **EXAMPLE 21.2** Show that the driving-point impedance function

$$Z(s) = \frac{2s^3 + 2s + 1}{s^5 + 2s^3 + s}$$

is not physically realizable with a passive network.

SOLUTION : The difference between the highest degrees of numerator and denominator is more than 1. Hence this function is not physically realizable as a passive network.

■ **EXAMPLE 21.3** Test whether the following polynomials are Hurwitz :

(a) $s^4 + 4s^3 + 3s + 2$

(b) $s^6 + s^5 + 3s^4 - 2s^3 + 3s^2 + 4s + 3$

(c) $s^4 + 3s^2 + 2$

(d) $s^2 + 5s + 9$

(e) $(s + 1)(s^2 + 2s + 8)(s^2 + 3s + 2)$

(f) $s^5 + 7s^4 + 5s^3 + s^2 + 2s$

SOLUTION : (a) As the s^2 term is missing, the given polynomial is not Hurwitz.

(b) Not Hurwitz because of the negative coefficient of the cubic term.

(c) Not Hurwitz because it has missing terms.

(d) The given polynomial is Hurwitz because it is a quadratic polynomial with no missing terms and all its coefficients are positive.It is to be noted that a quadratic polynomial with positive coefficients is always Hurwitz, since the roots of the quadratic polynomial lie on left half of s-plane.

(e) Since all the factors are Hurwitz, the given polynomial is Hurwitz.

(f) Not Hurwitz because a zero at the origin is present. That is, a root of this polynomial lies at the origin. This is not permitted according to the condition of the Hurwitz polynomial.

■ **EXAMPLE 21.4** Test whether the polynomial

$$P(s) = s^2 + 2s + 5 \text{ is Hurwitz.}$$

SOLUTION :

$$P(s) = s^2 + 2s + 5$$

The roots of the quadratic $s^2 + 2s + 5$ are

$$s_1, s_2 = \frac{-2 \pm \sqrt{4 - 4 \times 5}}{2} = -1 \pm j2$$

Since the roots s_1 and s_2 have negative real parts, the polynomial $(s^2 + 2s + 5)$ is Hurwitz.

Actually a quadratic polynomial with real positive coefficients is always Hurwitz, because the roots of the quadratic polynomial must lie on the left half of the s plane. If any coefficient a_i is negative, then the quadratic polynomial is **not** Hurwitz, since at least one root lies on the right half of the s- plane.

■ **EXAMPLE 21.5** If $M(s) + N(s)$ is a Hurwitz polynomial, where $M(s)$ is the even and $N(s)$ is the odd part, prove that $M(s) - N(s)$ has all its zeros in the right half plane.

SOLUTION : Let $P(s) = M(s) + N(s)$

Therefore $P(s)$ has all its zeros in the left half plane. Since the polynomial $P(-s)$ has zeros that are negatives of those of $P(s)$, all its zeros are in the right half plane. But

$$P(-s) = M(-s) + N(-s) = M(s) - N(s)$$

Therefore $M(s) - N(s)$ has all its zeros in the right half plane.

■ **EXAMPLE 21.6** Test if the polynomial

$$P(s) = s^4 + s^3 + 5s^2 + 3s + 4 \text{ is Hurwitz.}$$

SOLUTION : Since all coefficients of $P(s)$ are positive, $P(s)$ is real for s real.

The even and odd parts of $P(s)$are

$$M(s) = s^4 + 5s^2 + 4$$

$$N(s) = s^3 + 3s$$

The continued fraction of $F(s) = \dfrac{M(s)}{N(s)}$ is given by

$$\begin{array}{l}
s^3 + 3s \;)\; s^4 + 5s^2 + 4 \;(\; s \\
\qquad\quad\;\; s^4 + 3s^2 \\
\hline
\qquad 2s^2 + 4 \;)\; s^3 + 3s \;(\; \frac{1}{2}s \\
\qquad\qquad\qquad s^3 + 2s \\
\hline
\qquad\qquad\qquad\quad s \;)\; 2s^2 + 4 \;(\; 2s \\
\qquad\qquad\qquad\qquad\quad 2s^2 \\
\hline
\qquad\qquad\qquad\qquad\qquad 4 \;)\; s \;(\; \frac{1}{4}s \\
\qquad\qquad\qquad\qquad\qquad\quad\;\; s \\
\hline
\qquad\qquad\qquad\qquad\qquad\quad\;\; 0
\end{array}$$

Thus, the continued fraction expansion of $F(s)$ is

$$F(s) = \frac{M(s)}{N(s)} = s + \cfrac{1}{\frac{1}{2}s + \cfrac{1}{2s + \cfrac{1}{1/4}}}$$

Since all the quotient terms of the continued fraction expansion are positive, $P(s)$ is Hurwitz.

■ **EXAMPLE 21.7** Find the limits of m so that the polynomial $P(s) = 2s^4 + s^3 + ms^2 + s + 2$ may be Hurwitz.

SOLUTION :

Since the coefficients of s terms must be positive, $m > 0$.

$$P(s) = 2s^4 + s^3 + ms^2 + s + 2$$

$$M(s) = 2s^4 + ms^2 + 2$$

$$N(s) = s^3 + s$$

The continued-fraction expansion is given by

$$\begin{array}{l} s^3 + s \;)\; 2s^4 + ms^2 + 2 \;(\; 2s \\ \qquad\quad 2s^4 + 2s^2 \\ \hline (m-2)s^2 + 2 \;)\; s^3 + s \;\left(\; \dfrac{s}{m-2} \right. \\ \qquad\qquad\qquad s^3 + \dfrac{2s}{m-2} \\ \hline \qquad\qquad\qquad \left(1 - \dfrac{2}{m-2}\right)s \end{array}$$

If the function is Hurwitz then the quotients of the continued fraction should be positive.

therefore, $$m - 2 > 0$$

and $$1 - \frac{2}{m-2} > 0$$

or $$1 > \frac{2}{m-2}$$

that is, $m - 2 > 2$ or $m > 4$

Thus, the range of m is that $m > 4$.

■ **EXAMPLE 21.8** Test whether the polynomial

$P(s) = s^3 + 2s^2 + 3s + 6$ is Hurwitz.

SOLUTION : Since all coefficients of $P(s)$ are positive, $P(s)$ is real for s real.

$$P(s) = s^3 + 2s^2 + 3s + 6$$

Even and odd parts of $P(s)$ are

$$M(s) = 2s^2 + 6$$

$$N(s) = s^3 + 3s$$

Since degree of $N(s)$ is higher than that of $M(s)$ we shall determine the continued fraction expansion of $\frac{N(s)}{M(s)}$. This expansion is given by

$$2s^2 + 6\ )\ s^3 + 3s\ (\ \frac{1}{2}s$$

$$\underline{s^3 + 3s}$$

$$0$$

It is seen that the division has terminated abruptly by a common factor $(s^3 + 3s)$.

Thus, $$P(s) = (s^3 + 3s)\left(1 + \frac{2}{s}\right) = W(s)\,P_1(s)$$

It is seen that $P_1(s) = \left(1 + \frac{2}{s}\right)$ is Hurwitz. Let us check whether $W(s) = (s^3 + 3s)$ is Hurwitz or not.

$$W(s) = s^3 + 3s = s(s^2 + 3)$$
$$= s\,(s + j\sqrt{3})\,(s - j\sqrt{3})$$

Since all roots of $W(s)$ have zero real parts, $W(s)$ is Hurwitz.

Alternative Method

Since the polynomial $W(s) = s^3 + 3s$ contains only odd terms, it is not possible to obtain the contained fraction expansion.

$$W(s) = s^3 + 3s$$

The derivative of $W(s)$ is

$$W'(s) = 3s^2 + 3$$

Let us find the continued fraction expansion of $\frac{W(s)}{W'(s)}$ as follows :

$$\frac{W(s)}{W'(s)} = \frac{s^3 + 3s}{3s^2 + 3}$$

$$3s^2 + 3\ )\ s^3 + 3s\ (\ \frac{1}{3}s$$

$$\underline{s^3 + s}$$

$$2s\ )\ 3s^2 + 3\ (\ \frac{3}{2}s$$

$$\underline{3s^2}$$

$$3\ )\ 2s\ (\ \frac{2}{3}s$$

$$\underline{2s}$$

$$0$$

Since all the quotients of the continued fraction expansion of $\frac{W(s)}{W'(s)}$ are positive, $W(s)$ is Hurwitz. Since both the factors $(s^3 + 3s)$ and $\left(1 + \frac{2}{s}\right)$ of $P(s)$ are Hurwitz, $P(s)$ is Hurwitz.

■ **EXAMPLE 21.9** Test whether the polynomial $P(s) = s^3 + s^2 + 2s + 2$ is Hurwitz.

SOLUTION :

$$P(s) = s^3 + s^2 + 2s + 2$$

Therefore, $M(s) = s^2 + 2$ [even part of $P(s)$]

$N(s) = s^3 + 2s$ [odd part of $P(s)$]

Since degree of $N(s)$ is higher than that of $M(s)$,

Therefore,
$$\frac{N(s)}{M(s)} = \frac{s^3 + 2s}{s^2 + 2}$$

We shall determine the continued fraction expansion of $\frac{N(s)}{M(s)}$

$$s^2 + 2 \,)\, s^3 + 2s \,(\, s$$
$$\underline{s^3 + 2s}$$
$$0$$

It is seen that the division has terminated abruptly by a common factor $(s^2 + 2s)$. The polynomial can be written as

$$P(s) = (s^2 + 2s)\left(1 + \frac{1}{s}\right)$$

It is seen that both the factors $(s^2 + 2s)$ and $\left(1 + \frac{1}{s}\right)$ are Hurwitz. Therefore, the given polynomial $P(s)$ is Hurwitz.

■ **EXAMPLE 21.10** Test whether the following polynomial is Hurwitz or not.

$$P(s) = s^4 + 3s^2 + 2$$

SOLUTION : Since the given polynomial contains only even terms, it is not possible to obtain the continued fraction expansion.

$$P(s) = s^4 + 3s^2 + 2$$

The derivative of $P(s)$ is

$$P'(s) = 4s^3 + 6s$$

Let us find the continued fraction expansion of $\frac{P(s)}{P'(s)}$

$$\frac{P(s)}{P'(s)} = \frac{s^4 + 3s^2 + 2}{4s^3 + 6s}$$

$$
\begin{array}{l}
4s^3+6s\ )\ s^4+3s^2+2\ (\ \frac{1}{4}s \\
\qquad\qquad s^4+\frac{3}{2}s^2 \\
\qquad\qquad \overline{\frac{3}{2}s^2+2\ )\ 4s^3+6s\ (\ \frac{8}{3}s} \\
\qquad\qquad\qquad\qquad 4s^3+\frac{16}{3}s \\
\qquad\qquad\qquad\qquad \overline{\frac{2}{3}s\ \ )\ \frac{3}{2}s^2+2\ (\ \frac{9}{4}s} \\
\qquad\qquad\qquad\qquad\qquad\qquad \frac{3}{2}s^2 \\
\qquad\qquad\qquad\qquad\qquad\qquad \overline{2\ \ )\ \frac{2}{3}s\ (\ \frac{1}{3}s} \\
\qquad\qquad\qquad\qquad\qquad\qquad\qquad\quad \frac{2}{3}s \\
\qquad\qquad\qquad\qquad\qquad\qquad\qquad\quad \overline{0}
\end{array}
$$

Since all the quotients of the continued fraction expansion are positive, the given polynomial $P(s)$ is Hurwitz.

■ **EXAMPLE 21.11** Check whether the polynomial

$$P(s) = s^3 + 4s^2 + 2s + 8$$

is Hurwitz or not.

SOLUTION :

$$P(s) = s^3 + 4s^2 + 2s + 8$$

$$M(s) = 4s^2 + 8 \qquad \text{[even part of } P(s)]$$

$$N(s) = s^3 + 2s \qquad \text{[odd part of } P(s)]$$

Since the degree of $N(s)$ is more than the degree of $M(s)$, hence instead of $\frac{M(s)}{N(s)}$ we will find the continued fraction expansion of $\frac{N(s)}{M(s)}$.

$$\frac{N(s)}{M(s)} = \frac{s^3 + 2s}{4s^2 + 8}$$

The continued fraction expansion is given by

$$
\begin{array}{l}
4s^2+8\ )\ s^3+2s\ (\ \frac{s}{4} \\
\qquad\qquad s^3+2s \\
\qquad\qquad \overline{\qquad 0 \qquad}
\end{array}
$$

It is found that the division has terminated abruptly by a common factor $(s^2 + 2s)$. Therefore, the polynomial can be written as

$$P(s) = (s^2 + 2s)\left(1 + \frac{4}{s}\right)$$

Since both the factors $(s^2 + 2s)$ and $\left(1 + \frac{4}{s}\right)$ are Hurwitz, the given polynomial is Hurwitz.

■ **EXAMPLE 21.12** Check whether the given polynomial

$$P(s) = s^4 + s^3 + 2s^2 + 4s + 1$$

is Hurwitz or not.

SOLUTION : Here all the coefficients a_i are positive.

Even and odd parts of $P(s)$ are

$$M(s) = s^4 + 2s^2 + 1$$

$$N(s) = s^3 + 4s$$

The continued fraction expansion of $\frac{M(s)}{N(s)}$ is as follows :

$$s^3 + 4s\ )\ s^4 + 2s^2 + 1\ (\ s$$

$$\underline{s^4 + 4s^2}$$

$$-2s^2 + 1\ )\ s^2 + 4s\ (\ -\frac{1}{2}s$$

$$\underline{s^2 - \frac{s}{2}}$$

$$\frac{9}{2}s$$

There is no need to proceed with further division as one of the quotient terms $\left(-\frac{s}{2}\right)$ is negative. Hence the given polynomial is **not** Hurwitz.

21.6 POSITIVE REAL FUNCTIONS (PRF)

A positive real function is defined as follows :

A function $F(s)$ with $s = \sigma + j\omega$ is said to be positive real if it satisfies the following conditions :

1. $F(s)$ is real when s is real (that is, $\omega = 0$)
2. The real part of $F(s)$ is greater than or equal to zero when the real part of s is greater than or equal to zero. That is,

$$\text{Re}\,[F(s)] \geq 0, \text{ for Re}\,[s] \geq 0$$

or

$$\text{Re}\,[F(\sigma + j\omega)] \geq 0 \text{ when } \sigma \geq 0$$

Positive real functions are abbreviated as P.R.F. Such functions were first defined by Otto Brune in 1931 and are sometimes called **Brune's functions**. These functions are important because they represent physically realizable passive driving point immittances. Thus, **positive realness is the condition for physical realizability of a function $F(s)$ with a passive linear network.**

The conditions stated in the definition of a positive real function are equivalent to the conditions given as follows :

A rational function $F(s)$ is positive real if and only if

(a) $F(s)$ has no poles in the right half of the s plane.

(b) The poles of $F(s)$ on the $j\omega$ axis are simple with real and positive residues.

(c) Re $[F(j\omega)] \geq 0$ for all ω

A simplification of (c) is possible.

Let $$F(s) = \frac{N^r(s)}{D^r(s)}$$

where $N^r(s)$ = numerator polynomial

$D^r(s)$ = denominator polynomial

To find Re $[F(j\omega)]$ we separate even and odd parts of $F(s)$.

Let $M_1(s)$ = even part of $N^r(s)$

$N_1(s)$ = odd part of $N^r(s)$

$M_2(s)$ = even part of $D^r(s)$

$N_2(s)$ = odd part of $D^r(s)$

$$\therefore \quad F(s) = \frac{N^r(s)}{D^r(s)} = \frac{M_1(s) + N_1(s)}{M_2(s) + N_2(s)} \tag{21.6.1}$$

Multiplying and dividing the numerator and denominator of Eq. (21.6.1) by $[M_2(s) - N_2(s)]$ we get

$$F(s) = \frac{M_1(s) + N_1(s)}{M_2(s) + N_2(s)} \times \frac{M_2(s) - N_2(s)}{M_2(s) - N_2(s)}$$

$$= \frac{M_1(s)\,M_2(s) - N_1(s)\,N_2(s)}{[M_2(s)]^2 - [N_2(s)]^2} + \frac{N_1(s)\,M_2(s) - M_1(s)\,N_2(s)}{[M_2(s)]^2 - [N_2(s)]^2}$$

The product of two even or two odd functions is an even function while the product of an even and an odd function is an odd function. We see that the products $M_1(s)\,M_2(s)$ and $N_1(s)\,N_2(s)$ are even functions, while $M_1(s)\,N_2(s)$ and $M_2(s)\,N_1(s)$ are odd functions. Therefore, even part of $F(s)$ is

$$Ev\ [F(s)] = \frac{M_1(s)\,M_2(s) - N_1(s)\,N_2(s)}{[M_2(s)]^2 - [N_2(s)]^2} \tag{21.6.2}$$

and odd part of $F(s)$ is

$$\text{Odd}\ [F(s)] = \frac{M_2(s)\,N_1(s) - M_1(s)\,N_2(s)}{[M_2(s)]^2 - [N_2(s)]^2} \tag{21.6.3}$$

If we substitute $s = j\omega$, we see that the even part of any polynomial is real, while the odd part of the polynomial is imaginary, so that if $F(j\omega)$ is written as

$$F(j\omega) = \text{Re}\,[F(j\omega)] + j\,\text{Im}\,[F(j\omega)] \tag{21.6.4}$$

Therefore

$$Ev\,[F(s)]\,|_{s=j\omega} = \text{Re}\,[F(j\omega)] \tag{21.6.5}$$

and

$$\text{Odd}\,[F(s)]\,|_{s=j\omega} = \text{Odd}\,[F(j\omega)] \tag{21.6.6}$$

When $s = j\omega$, the denominator of Eq. (21.6.2) is always positive because

$$[M_2\,(j\omega)]^2 - [N_2\,(j\omega)]^2 = [M_2\,(\omega)]^2 + [N_2\,(\omega)]^2 \geq 0 \tag{21.6.7}$$

That is, there is an extra j or imaginary term in $N_2\,(j\omega)$ which, when squared, gives –1, so that the denominator of Re $[F(j\omega)$ is the sum of two squared numbers and is always positive.

Therefore it is sufficient to test the numerator of Eq. (21.6.2)

This numerator is

$$M_1\,(j\omega)\,M_2\,(j\omega) - N_1\,(j\omega)\,N_2\,(j\omega) \triangleq A(\omega) \tag{21.6.8}$$

We have to check whether $A(\omega) \geq 0$

To check if $A(\omega) \geq 0$, it is necessary to factorize it and find its roots. The criterion for $A(\omega) \geq 0$ is that $A(\omega)$ has no positive real roots of odd multiplicity.

21.7 PROPERTIES OF POSITIVE REAL FUNCTIONS

A number of properties of positive real function (PRF) can be deduced from its definition. Some useful properties of positive real function are as follows :

1. All the coefficients of the numerator and denominator polynomials are real and positive.
2. If $F(s)$ is positive real then its reciprocal $1/F(s)$ is also positive real.
   This property implies that if a driving-point impedance function $Z(s)$ is positive real, then its reciprocal $[1/Z(s)]$, the driving-point admittance function $[Y(s)]$ is also positive real.
3. The sum of positive real functions is also a positive real function.

This means that if two impedances are connected in series, the sum of the impedances is positive real. Similarly, the resultant of two admittances in parallel is positive real. It is to be noted that the difference of two positive real function is not necessarily positive real. For example,

$$F(s) = \left(s - \frac{1}{s}\right) \text{ is not a positive real function.}$$

4. The poles and zeros of a positive zeros of a positive real function have either negative or zero real parts. They cannot have positive real parts. That is, they cannot lie on the right half of the s-plane.
5. Those poles of $F(s)$ or $1/F(s)$ which lie on the $j\omega$ axis are simple and their residues are real and positive.
6. The poles and zeros of a positive real function are real or occur in conjugate pairs.
7. The highest powers of the numerator and denominator polynomials differ at most by unity.
   This condition prohibits multiple poles and zeros at $s = \infty$.

8. The lowest powers of the numerator or denominator polynomials also may differ at most by unity.
This condition prevents the possibility of multiple poles or zeros at the origin (that at $s = 0$).
9. The numerator and denominator polynomials do not have missing terms between those of the highest and lowest degrees unless all odd or all even terms are missing.
10. For $F(s)$ to be positive real the numerator and denominator should be Hurwitz polynomials.

Thus, if a given function $F(s)$ fails to satisfy any of these conditions, there is no need for further tests-the function is not positive real. If, however, a given function satisfies all these conditions, we should check to see whether it satisfies condition (c) of positive real function or not.

21.8 EQUIVALENT SETS OF CONDITIONS FOR PHYSICAL REALIZABILTY OF PASSIVE NETWORKS

There are three equivalent ways of expressing the conditions for the physical realizability of passive networks. These equivalent sets of conditions are as follows :

1. Conditions (or properties) in Section 21.2.
2. Conditions involving Hurwitz polynomials in Section 21.3.
3. Conditions for positive realness as defined in Section 21.6.

The choice of one of these three sets of conditions to test a given driving-point immittance function $F(s)$ is a matter of personal preference. However, the conditions in Section 21.2 are explicitly expressed and, therefore, quite convenient to use. If a function $F(s)$ violates **any** of the five conditions in Section 21.2, it is non realizable. It is, therefore, not necessary to continue checking it against the remaining four conditions.

EXAMPLE 21.13 Check the following functions for positive reality :

(a) $\dfrac{s^2 + 7s + 12}{s^2 + s - 2}$

(b) $\dfrac{1}{s^2 + 3s + 2}$

(c) $\dfrac{s^2}{s^3 + 3s + 2}$

SOLUTION : (a) $F(s) = \dfrac{s^2 + 7s + 12}{s^2 + s - 2}$

$$= \frac{(s+4)(s+3)}{(s+2)(s-1)}$$

Since there is a right-hand pole at $s = 1$, $F(s)$ is not a prf.

(b) $$F(s) = \frac{1}{s^2 + 3s + 2}$$

Since the degrees of numerator and denominator differ by 2, $F(s)$ is not positive real.

(c) $$F(s) = \frac{s^2}{s^3 + 3s + 2}$$

$$\frac{1}{F(s)} = \frac{s^3 + 3s + 2}{s^2}$$

$$= s + \frac{3}{s} + \frac{2}{s^2}$$

Since the reciprocal of $F(s)$ has a double-ordered pole at $s = 0$, the function $F(s)$ is not positive real.

EXAMPLE 21.14 Test whether the function

$$F(s) = \frac{as}{s^2 + b}$$

is positive real or not if $ab \geq 0$ is positive real.

SOLUTION :

$$F(s) = \frac{as}{s^2 + b}$$

$$= \frac{1}{\frac{s}{a} + \frac{b}{as}}$$

It is seen that $\frac{s}{a}$ and $\frac{b}{as}$ are positive real. Since the sum of two positive real functions is also a positive real function, the denominator $D(s) = \frac{s}{a} + \frac{b}{as}$ is also a positive real function.

Now $$F(s) = \frac{1}{D(s)}$$

Since the reciprocal of a positive real function is also a positive real function, the given function $F(s)$ is positive real.

EXAMPLE 21.15 Check the positive realness of the network function

$$Z(s) = \frac{s^3 + 2s}{s^2 + 1}$$

SOLUTION: The given function has a pair of poles at $s = \pm j1$. Since the degree of the numerator is 1 higher than that of the denominator there is a pole at $s \to \infty$. All the $j\omega$-axis poles are simple. Hence $Z(s)$ is positive real if its residues at all poles are real and positive. The partial fraction expansion of $Z(s)$ is given by

$$Z(s) = k_1 s + \frac{k_2}{s - j1} + \frac{k_3}{s + j1} = \frac{s^3 + 2s}{s^2 + 1}$$

where

$$k_1 = \left[\frac{Z(s)}{s}\right]_{s \to \infty} = \frac{s^2 + 2}{s^2 + 1}\Big|_{s \to \infty}$$

$$= \frac{1+\dfrac{2}{s^2}}{1+\dfrac{1}{s^2}}\Bigg|_{s\to\infty} = 1$$

$$k_2 = [(s-j1)\,Z(s)]_{s=j1} = \left[\frac{s^3+2s}{s+j1}\right]_{s=j1} = \frac{1}{2}$$

$$k_3 = [(s+j1)\,Z(s)]_{s=-j1} = \left[\frac{s^3+2s}{s-j1}\right]_{s=-j1} = \frac{1}{2}$$

Since all the residues are real and positive, $Z(s)$ is positive real.

■ **EXAMPLE 21.16** **Check the positive realness of the network function**

$$Z(s) = \frac{s+4}{s^2+2s+1}$$

SOLUTION :

$$Z(s) = \frac{s+4}{s^2+2s+1}$$

$$= \frac{s+4}{(s+1)^2}$$

The partial-fraction expansion of $Z(s)$ can be wrtten as

$$Z(s) = \frac{1}{s+1} + \frac{3}{(s+1)^2}$$

Since $Z(s)$ has multiple poles at $s = -1$, it is not a positive real function.

■ **EXAMPLE 21.17** **Check the positive realness of the function**

$$Z(s) = \frac{(s+2)\,(s+4)}{(s+1)\,(s+3)}$$

SOLUTION :

$$Z(s) = \frac{(s+2)\,(s+4)}{(s+1)\,(s+3)}$$

$$= \frac{s^2+6s+8}{s^2+4s+3} = 1 + \frac{2s+5}{(s^2+4s+3)}$$

$$= 1 + \frac{2s+5}{(s+1)\,(s+3)}$$

$$= 1 + \frac{k_1}{s+1} + \frac{k_2}{s+3}$$

$$k_1 = \frac{2s+5}{s+3}\Bigg|_{s=-1} = \frac{-2+5}{-1+3} = \frac{3}{2}$$

$$k_2 = \frac{2s+5}{s+1}\Bigg|_{s=-3} = \frac{-2\times 3+5}{-3+1} = \frac{1}{2}$$

$$\therefore \qquad Z(s) = 1 + \frac{3/2}{s+1} + \frac{1/2}{s+3}$$

Since the residues at the poles at $s = -1$ and $s = -3$ are $\frac{3}{2}$ and $\frac{1}{2}$ and are positive, $Z(s)$ is a positive real function.

EXAMPLE 21.18 Check the positive realness of the function

$$Z(s) = \frac{s^2 + s + 6}{s^2 + s + 1}$$

SOLUTION :

$$Z(s) = \frac{s^2 + s + 6}{s^2 + s + 1}$$

$$= 1 + \frac{5}{s^2 + s + 1}$$

$$= 1 + \frac{k_1}{s + \frac{1}{2} + j\frac{\sqrt{3}}{2}} + \frac{k_2}{s + \frac{1}{2} - j\frac{\sqrt{3}}{2}}$$

$$k_1 = \frac{5}{s + \frac{1}{2} - j\frac{\sqrt{3}}{2}} \Bigg|_{s = -\frac{1}{2} - j\frac{\sqrt{3}}{2}}$$

$$= \frac{5}{-\frac{1}{2} - j\frac{\sqrt{3}}{2} + \frac{1}{2} - j\frac{\sqrt{3}}{2}} = \frac{5}{-j\sqrt{3}} = +j\frac{5}{\sqrt{3}}$$

$$k_2 = \frac{5}{s + \frac{1}{2} + j\frac{\sqrt{3}}{2}} \Bigg|_{s = -\frac{1}{2} + j\frac{\sqrt{3}}{2}}$$

$$= \frac{5}{-\frac{1}{2} + j\frac{\sqrt{3}}{2} + \frac{1}{2} + j\sqrt{3}} = \frac{5}{j\sqrt{3}} = -j\frac{5}{\sqrt{3}}$$

$$Z(s) = 1 + \frac{+j\,5/\sqrt{3}}{s + \frac{1}{2} + j\frac{\sqrt{3}}{2}} + \frac{-j\,5/\sqrt{3}}{s + \frac{1}{2} - j\frac{\sqrt{3}}{2}}$$

It is seen that the residue at pole $s = \left(-\frac{1}{2} + j\frac{\sqrt{3}}{2}\right)$ is $-j\frac{5}{\sqrt{3}}$, that is negative.

Hence the given $Z(s)$ is not a positive real function.

EXAMPLE 21.19 Determine whether the function $Z(s) = \frac{2s + 3}{s + 1}$ is positive real or not.

SOLUTION :

$$Z(s) = \frac{2s + 3}{s + 1}$$

By inspection, all the coefficients of s are real, hence $Z(s)$ is real if s is real. The function

$Z(s)$ has a pole $s = -1$ which lies in the left half of the s plane. Thus, the given function satisfies all the **necessary** conditions of PRF. We shall check the sufficient condition that

Re $[Z(j\omega)] \geq 0$ for all ω we have

$$Z(j\omega) = \frac{2j\omega + 3}{j\omega + 1}$$

$$= \frac{(2j\omega + 3)(-j\omega + 1)}{(j\omega + 1)(-j\omega + 1)}$$

$$= \frac{(3 + 2\omega^2) + j(2\omega - 3\omega)}{\omega^2 + 1}$$

$$\text{Re}\,[Z(j\omega)] = \frac{2\omega^2 + 3}{\omega^2 + 1}$$

It is seen that for all values of ω, Re $[Z(j\omega)] \geq 0$.

Hence the sufficient condition is also satisfied. Thus, $Z(s)$ is PRF.

■ **EXAMPLE 21.20** Show that the following function

$$P(s) = \frac{3s^2 + 5}{s(s^2 + 1)}$$

is not a positive real function.

SOLUTION : The given functions has a pair of poles at $s = \pm j1$. The partial fraction expansion can be written as

$$P(s) = \frac{k_1}{s} + \frac{k_2 s + k_3}{s^2 + 1}$$

$$= \frac{k_1(s^2 + 1) + k_2 s^2 + k_3 s}{s(s^2 + 1)}$$

$$P(s) = \frac{(k_1 + k_2)s^2 + k_3 s + k_1}{s(s^2 + 1)}$$

Also, $$P(s) = \frac{3s^2 + 5}{s(s^2 + 1)}$$

Comparison of the above relations gives

$$k_1 + k_2 = 3$$

$$k_3 = 0, \; k_1 = 5$$

therefore, $$k_2 = 3 - k_1 = 3 - 5 = -2$$

Thus, the partial fraction expansion is given by

$$P(s) = \frac{5}{s} + \frac{-2}{s^2 + 1}$$

Since the coefficients of the poles at $s = \pm j1$ are negative, $P(s)$ is not a positive real function.

■ **EXAMPLE 21.21** Find the conditions for

$$F(s) = \frac{s+a}{s^2+bs+c}$$

to be positive real.

SOLUTION :
$$F(s) = \frac{s+a}{s^2+bs+c} \quad \text{(E 21.21.1)}$$

(a) First Condition

The poles and zeros of $F(s)$ should be in the left half or on the $j\omega$ axis of the s plane. Therefore the coefficients a, b and c must be greater than or equal to zero.

(b) Second Condition

"The poles of $F(s)$ on the $j\omega$ axis are simple with real and positive residues."

$F(s)$ will have poles on the $j\omega$ axis if $b = 0$. Therefore $F(s)$ can be written as

$$F(s) = \frac{s+a}{s^2+c} \quad \text{(E 21.21.2)}$$

$$F(s) = \frac{s}{s^2+c} + \frac{a}{s^2+c} \quad \text{(E 21.21.3)}$$

The second term in Eq. (E 21.21.3) violates the condition of positive real function that the highest powers of numerator and denominator polynomials must differ at most by unity. Thus, if $b = 0$ and a should also be zero.

(c) Third condition

$$\text{Re } [F(j\omega)] \geq 0 \text{ for all } \omega$$

Alternatively,

therefore, $M_1(j\omega)\, M_2(j\omega) - N_1\,(j\omega)\, N_2(j\omega) \geq 0$

For Eq. (E21.21.1)

$M_1(s)$ = even part of numerator = a

$M_2(s)$ = even part of denominator = $s^2 + c$

$N_1(s)$ = odd part of numerator = s

$N_2(s)$ = odd part of denominator = bs

therefore, $M_1(s)\, M_2(s) - N_1(s)\, N_2(s) = a(s^2 + c) - s(bs)$

Putting $s + j\omega$

$$M_1(j\omega)\, M_2(j\omega) - N_1(j\omega)\, N_2\,(j\omega) = A(\omega)$$

therefore, $A(\omega) = a\; [(j\omega)^2 + c] - j\omega\; [b\; (j\omega)]$

$$= a(-\omega^2 + c) + b\omega^2$$

$$= (b - a)\,\omega^2 + ac$$

For $F(s)$ to be positive real $A(\omega) \geq 0$.

To check if $A(\omega) \geq 0$, it is necessary to factorize it and find its roots. The criterion for $A(\omega) \geq 0$ is that $A(\omega)$ has no positive real roots of odd multiplicity. In order to ensure that $A(\omega)$ does not have positive real roots, $b \geq a$

Therefore, if $b = 0$, then $a > 0$

Hence the conditions for $F(s)$ to be positive real are

(i) $a, b, c \geq 0$

(ii) $b \geq a$

EXAMPLE 21.22 Assuming the coefficients a_0, a_1, b_0 and b_1 to be all real, positive constants, determine the condition for the function

$$Z(s) = \frac{s^2 + a_1 s + a_0}{s^2 + b_1 s + b_0}$$

to be positive real.

SOLUTION :

$$Z(s) = \frac{s^2 + a_1 s + a_0}{s^2 + b_1 s + b_0} \qquad \text{...(E 21.21.1)}$$

Since a_0, a_1, b_0 and b_1 are all real and positive constants, the numerator and denominator polynomials are Hurwitz polynomials.

If $b_1 \neq 0$ then $F(s)$ has no poles on $j\omega$ axis. We may, therefore, ignore the second condition of PRF. We shall check the third requirement of PRF. That is

$$\text{Re}\,[Z(j\omega)] \geq 0 \text{ for all } \omega$$

or $\quad A(\omega) = M_1(s)\,M_2(s) - N_1(s)\,N_2(s)\ |_{s = j\omega} \geq 0$ for all ω.

From Eq. (E 21.22.1)

$$M_1(s) = s^2 + a_0$$

$$M_2(s) = s^2 + b_0$$

$$N_1(s) = a_1 s$$

$$N_2(s) = b_1 s$$

$$A(\omega) = (s^2 + a_0)\,(s^2 + b_0) - a_1 b_1 s^2\ |_{s = j\omega}$$

$$= s^4 + [(a_0 + b_0) - a_1 b_1]\,s^2 + a_0 b_0\ |_{s = j\omega}$$

or

$$A(\omega) = \omega^4 - [(a_0 + b_0) - a_1 b_1]\,\omega^2 + a_0 b_0 \qquad \text{(E 21.22.2)}$$

To check if $A(\omega) \geq 0$, it is necessary to factorize $A(\omega)$ and find its roots. The criterion for $A(\omega) \geq 0$ is that $A(\omega)$ has no positive real roots of odd multiplicity. The roots of $A(\omega)$ are

$$w_{1,2}^2 = \frac{(a_0 + b_0) - a_1 b_1}{2} \pm \frac{1}{2}\sqrt{[(a_0 + b_0) - a_1 b_1]^2 - 4a_0 b_0} \quad \text{(E 21.22.3)}$$

There are two posisiblities in which $A(\omega)$ does not have simple real roots.

Case 1. When the quantity under radical sign of Eq. (21.22.3) is zero giving double real roots or negative giving complex roots. That is,

$$[(a_0 + b_0) - a_1 b_1]^2 - 4a_0 b_0 \le 0 \quad \text{(E 21.22.4)}$$

or $$[(a_0 + b_0) - a_1 b_1]^2 \le 4a_0 b_0 \quad \text{(E 21.22.5)}$$

If $$(a_0 + b_0) - a_1 b_1 \ge 0$$

then $$(a_0 + b_0) - a_1 b_1 \le 2\sqrt{a_0 b_0}$$

or $$a_1 b_1 \ge (\sqrt{a_0} - \sqrt{b_0})^2$$

If $$(a_0 + b_0) - a_1 b_1 \le 0$$

then $$a_1 b_1 - (a_0 + b_0) \le 2\sqrt{a_0 b_0}$$

or $$a_1 b_1 \ge (\sqrt{a_0} - \sqrt{b_0})^2$$

Case 2. The second situation in which $A(\omega)$ does not have simple real roots is when $\omega_{1,2}^2$ in Eq. (E 21.22.3) in negative so that the roots are imaginary. This happens when

$$[(a_0 + b_0) - a_1 b_1]^2 - 4a_0 b_0 > 0$$

and $$(a_0 + b_0) - a_1 b_1 < 0$$

The above equations give

$$a_1 b_1 - (a_0 + b_0) > 2\sqrt{a_0 b_0} > (a_0 + b_0) - a_1 b_1$$

or $$a_1 b_1 > (\sqrt{a_0} - \sqrt{b_0})^2$$

Hence the function $Z(s)$ is positive real when

$$a_1 b_1 \ge (\sqrt{a_0} - \sqrt{b_0})^2$$

■ **EXAMPLE 21.23** Check the function

$$Z(s) = \frac{s^2 + 2s + 25}{s + 4}$$

for positive reality.

SOLUTION : Since all the coefficients of s are positive and real $Z(s)$ is real if s is real. The function $Z(s)$ has a pole at $s = -4$ which lies in the left half of the s plane. Thus, the given function satisfies necessary conditions of PRF. We shall check whether the sufficient condition Re $[Z(j\omega)] \ge 0$ for all ω is also satisfied.

$$Z(j\omega) = \frac{(j\omega)^2 + 2\,(j\omega) + 25}{(j\omega) + 4}$$

$$= \frac{-\omega^2 + j(2\omega) + 25}{j\omega + 4}$$

$$= \frac{[-\omega^2 + j(2\omega) + 25]\,(-j\omega + 4)}{(j\omega + 4)\,(-j\omega + 4)}$$

$$= \frac{(-2\omega^2 + 100) + j(\omega^2 - 17\omega)}{\omega^2 + 16}$$

$$\text{Re}\,[Z(j\omega)] = \frac{-2\omega^2 + 100}{\omega^2 + 16}$$

It is observed that for values of ω^2 greater than 50, Re $[Z(j\omega)] < 0$. Since sufficient condition is not satisfied, $Z(s)$ is **not** positive real.

Alternative Method

$$Z(s) = \frac{s^2 + 2s + 25}{s + 4}$$

By inspection

$$M_1(s) = s^2 + 25$$

$$N_1(s) = 2s$$

$$M_2(s) = 4$$

$$N_2(s) = s$$

$$A(\omega) = M_1(j\omega)\, M_2(j\omega) - N_1(j\omega)\, N_2(j\omega)$$

$$= (-\omega^2 + 25)\,(4) - (2j\omega)\,(j\omega)$$

$$= -2\omega^2 + 100$$

Thus, the same condition is reached as before.

■ **EXAMPLE 21.24** Test the following functions for positive realness :

1. $F(s) = \dfrac{1}{s^2 + 1}$

2. $F(s) = \dfrac{s^3 + 3s^2 - 2s + 1}{s^3 + s^2 + s + 6}$

3. $F(s) = \dfrac{s^3 + 3s + 2}{s^4 + 3s^3 + s^2}$

4. $F(s) = \dfrac{s^4 + 4s^3 + 3s^2 + 1}{s^4 + 2s^2 + 1}$

5. $Z(s) = \dfrac{s + 1}{s + 2}$

6. $Z(s) = \dfrac{(s+1)(s+2)}{(s+3)(s+4)}$

SOLUTION :

1. The difference between highest and lowest degrees of numerator and denominator polynomials is more than 1. Hence the function is not positive real.

2. The coefficient of one term in the numerator is negative. Hence $F(s)$ is not positive real.

3. Since $F(s)$ has multiple poles at the origin, the function is not positive real.

4.
$$F(s) = \frac{s^4 + 4s^3 + 3s^2 + 1}{s^4 + 2s^2 + 1}$$
$$= \frac{s^4 + 4s^3 + 3s^2 + 1}{(s^2+1)^2}$$

This function has multiple poles at $s = \pm j1$. Therefore, the function is not positive real.

5. The given function satisfies all the necessary conditions of P.R.F. But we have to check the sufficient condition that

$$\text{Re}\,[F(j\omega)] \geq 0 \text{ for all } \omega.$$

We have

$$F(s) = \frac{s+1}{s+2}$$
$$F(j\omega) = \frac{1+j\omega}{2+j\omega} = \frac{(1+j\omega)(2-j\omega)}{(2+j\omega)(2-j\omega)}$$
$$= \frac{2+\omega^2}{4+\omega^2}$$

Since $F(j\omega)$ is positive for all ω, the given function si positive real.

6. The given function satisfies all the necessary conditions of PRF. We shall check the sufficient condition that

$$\text{Re}\,[F(j\omega)] \geq 0 \text{ for all } \omega$$

We have

$$F(j\omega) = \frac{(1+j\omega)(2+j\omega)}{(3+j\omega)(4+j\omega)}$$
$$= \frac{(1+j\omega)(2+j\omega)(3-j\omega)(4-j\omega)}{(3+j\omega)(4+j\omega)(3-j\omega)(4-j\omega)}$$

therefore,
$$\text{Re}\,[F(j\omega)] = \frac{\omega^4 + 7\omega^2 + 24}{(9+\omega^2)(16+\omega^2)}$$

which is positive real for all positive ω. Hence $F(s)$ is PRF.

21.9 BASIC METHODS OF SYNTHESIS

The basic method for realizing a positive real function is based upon breaking the given function into several simpler positive real functions. These individual functions are then synthesized by inspection. The overall network is realized as a combination of these simpler building blocks.

The given positive real function can be broken into smaller positive real functions by one or more following methods :

(a) Removal of a pole at infinity

(b) Removal of the poles from the imaginary axis.

(c) Removal of a pole at the origin.

(d) Removal of a constant.

(a) Removal of a Pole at Infinity

If the numerator polynomial of $Z(s)$ is 1 degree higher than its denominator, then $Z(s) \to \infty$ as $s \to \infty$. In this case $Z(s)$ is said to have a pole at infinity. The general form of $Z(s)$ in this case is

$$Z(s) = \frac{a_{n+1} s^{n+1} + a_n s^n + \ldots + a_1 s + a_0}{b_n s^n + b_{n-1} s^{n-1} + \ldots + b_1 s + b_0} \quad (21.9.1)$$

The first step in synthesizing $Z(s)$ is to "extract" the pole at infinity. This can be done by dividing the numerator of $Z(s)$ by its denominator. The result is

$$Z(s) = sL_1 + \frac{a' s^n + a'_{n-1} s^{n-1} + \ldots + a'_1 s + a'_0}{b_n s^n + b_{n-1} s^{n-1} + \ldots + b_1 s + b_0} = sL_1 + Z_1(s) \quad (21.9.2)$$

where L_1 is a real positive constant given by

$$L_1 = \frac{a_{n+1}}{b_n} \quad (21.9.3)$$

and $$Z_1(s) = \frac{a' s^n + a'_{n-1} s^{n-1} + \ldots + a'_1 s + a'_0}{b_n s^n + b_{n-1} s^{n-1} + \ldots + b_1 s + b_0} \quad (21.9.4)$$

Thus, $Z(s)$ can be considered as the driving-point impedance of an inductor with inductance L_1 in series with another network whose driving-point impedance is $Z_1(s)$ as shown in Fig. 21.1.

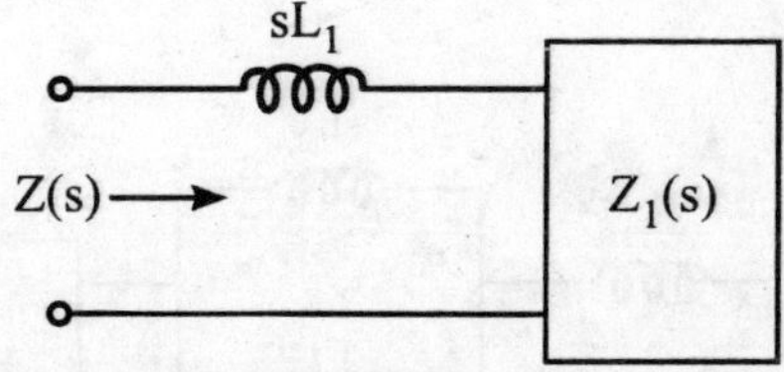

Fig. 21.1 Removal of a pole at infinity

We should prove that $Z_1(s)$ is also positive real, otherwise, it cannot be synthesized as the driving-point impedance of a linear time-invariant network with passive elements. For this purpose, we check the necessary and sufficient conditions of Theorem 2 of Section 21.6 :

(a) The poles of $Z_1(s)$ are the same as the poles of $Z(s)$. Since $Z(s)$ is positive real, all the poles of $Z_1(s)$ are in the left-half plane.

(b) Removal of a pole at infinity will not affect the residues of the poles on the $j\omega$ axis. This can be seen by a partial fraction expansion of $Z(s)$. Thus, $j\omega$-axis poles of $Z_1(s)$ are simple and have real and positive residues.

(c) The real part of $Z_1\,(j\omega)$ can be obtained form

$$\text{Re}\,[Z(j\omega)] = \text{Re}\,[j\omega L_1] + \text{Re}\,[Z_1\,(j\omega)]$$

and since $\text{Re}\,[j\omega L_1] = 0$, we get

$$\text{Re}\,[Z_1\,(j\omega)] = \text{Re}\,[Z\,(j\omega)] \geq 0 \text{ for } 0 \leq \omega \leq \infty \tag{21.9.5}$$

Hence, $Z_1(s)$ is positive real, and it can be realized by linear time-invariant passive elements.

The next subnetwork that can be extracted from $Z_1(s)$ may be obtained by removing poles of $Z_1(s)$ which lie on the $j\omega$ axis.

(b) Removal of the poles from the Imaginary Axis.

If $Z_1(s)$ contains a pair of poles on the $j\omega$ axis (say at $\omega = \pm j\omega_1$), then using partial fraction expansion $Z_1(s)$ can be written as

$$Z_1(s) = \frac{k_1}{s + j\omega_1} + \frac{k_2}{s - j\omega_1} + Z_2(s) \tag{21.9.6}$$

where k_1 and k_2 are the residues at the poles $s = -j\omega$ and $s = j\omega_1$ respectively. Since $Z_1(s)$ is positive real, the residues at both the imaginary-axis poles are real and equal or $k_1 = k_2$ and Eq. (21.9.6) becomes

$$Z_1(s) = \frac{2k_1 s}{s^2 + \omega_1^2} + Z_2(s) \tag{21.9.7}$$

The first term in Fig. (21.9.7) is the driving-point impedance of a parallel combination of an inductor L_2 and a capacitor C_2 where

$$L_2 = \frac{2k_1}{\omega_1^2} \quad \text{and} \quad C_2 = \frac{1}{2k_1}$$

This is shown in Fig. 21.2.

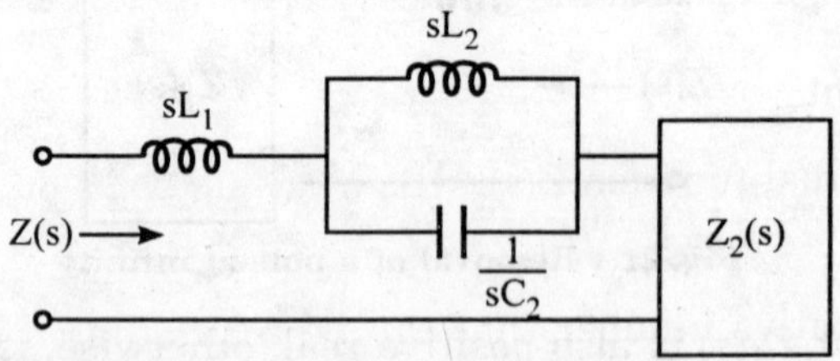

Fig. 21.2 Removal of a pair of poles from the $j\omega$ axis.

By the same method used in the previous case, it can be shown that $Z_2(s)$ given in Eq. (21.9.7) is positive real.

(c) Removal of a Pole at the Origin

If the given positive real function has a pole at the origin, it can be written as

$$Z(s) = \frac{a_n s^n + a_{n-1} s^{n-1} + \ldots + a_1 s + a_0}{s(b_n s^n + b_{n-1} s^{n-1} + \ldots + b_1 s + b_0)} \tag{21.9.8}$$

Using a partial fraction expansion, Eq. (21.9.8) can be written as

$$Z(s) = \frac{1}{sC_1} + Z_1(s) \tag{21.9.9}$$

where

$$C_1 = b_0/a_0$$

and

$$Z_1(s) = \frac{a'_n s^n + a'_{n-1} s^{n-1} + \ldots + a'_1 s + a'_0}{b_n s^n + b_{n-1} s^{n-1} + \ldots + b_1 s + b_0} \tag{21.9.10}$$

This means that the removal of a pole at the origin of a driving-point impedance function corresponds to the removal of a capacitor a, shown in Fig. 21.3.

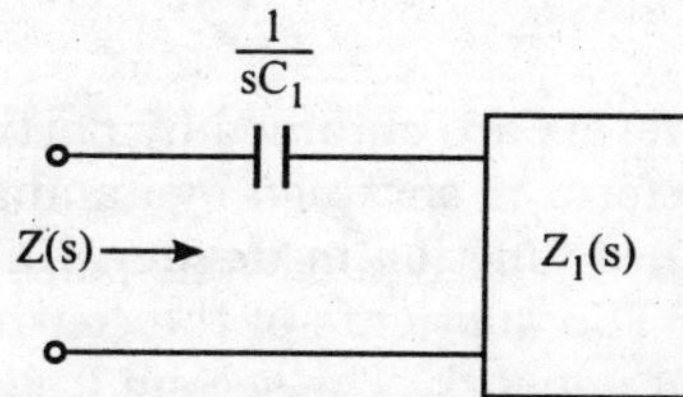

Fig. 21.3. Removal of a pole at the origin of a driving-point impedance function.

On the other hand, if the driving-point function is an admittance function, removal of a pole at the origin results in the removal of an inductor as shown in Fig. 21.4.

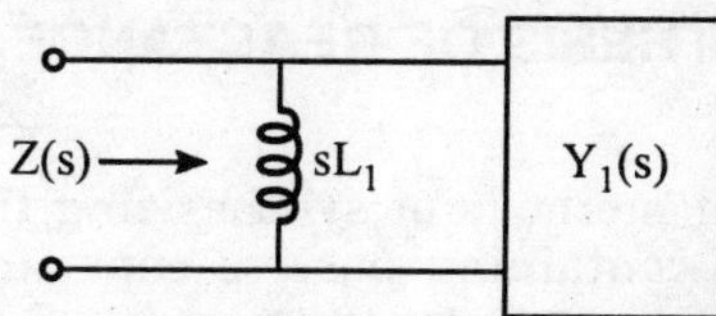

Fig. 21.4. Removal of a pole at the origin of an admittance driving-point function.

(d) Removal of a Constant

Consider a positive real function $Z(s)$, which has no poles on the $j\omega$ axis. Let

$$Z(s) = R + Z_1(s)$$

where R is a real and positive number that can be realized as a resistor with resistance R, and $Z_1(s)$ is a positive real function. This function imposes the following constraint on the value of resistance R.

$$R \le [\min [\text{Re } Z(j\omega)] \tag{21.9.11}$$

That is, the value of resistance R must be less than or equal to the minimum value of the real part of $Z(j\omega)$ for all positive ω. The removal of a constant from $Z(s)$ is shown in Fig. 21.5.

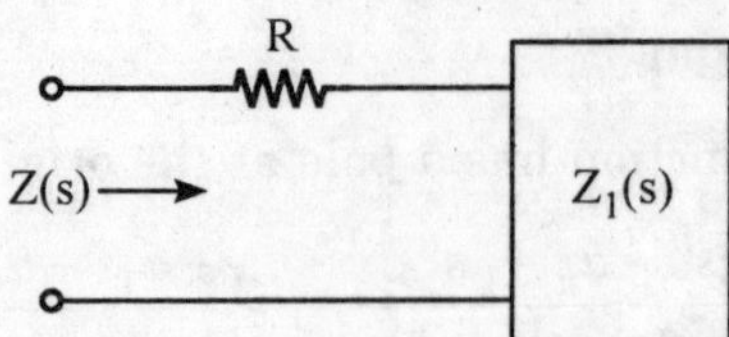

Fig. 21.5. Removal of a constant

Thus, a constant can be removed as a series resistor.

21.10 SYNTHESIS OF ONE-PORT NETWORKS WITH TWO KINDS OF ELEMENTS

Now we shall introduce the design of one ports which contain only two types of elements L and C, or R and C, or R and L. First we will discuss the properties of a particular type of one port network, and then we will synthesize it.

There are a number of methods available for synthesizing (or realizing) a one-port network. We shall consider the four basic forms only.

(a) First Foster form

(b) Second Foster form

(c) First Cauer form

(d) Second Cauer form

The first and second Foster forms are obtained by partial-fraction expansion of $Z(s)$ and $Y(s)$ respectively. The first Cauer form is obtained by arranging the numerator and denominator polynomials of the immittance function in **descending powers of *s*** and then finding the continued fraction expansion. The quotients of the continued fraction expansion give the elements of the first Cauer form of network. The second Cauer form is obtained by arranging the numerator and denominator polynomials of the immittance function in **ascending powers of *s*** and the finding the continued-fraction expansion. The quotients of the continued-fraction expansion give the elements of the second Cauer form of network.

21.11 DRIVING-POINT SYNTHESIS OF REACTANCE FUNCTIONS OR LC FUNCTIONS

In this section we discuss several methods of synthesizing the driving point impedance or admittance function of networks containing passive capacitors and inductors only. These networks are commonly known as **reactance networks**. Such networks are lossless. The driving-point impedance and admittance of such networks are among a special class of positive real functions called the reactance functions.

The **reactance function** is defined as follows :

The rational function $Z(s)$ is a reactance function if and only if

(a) All poles of $Z(s)$ are simple and lie on $j\omega$ axis with real and positive residues.

(b) Real $[Z(j\omega)] = 0$ for all $\omega \geq 0$

If $Z(s)$ is positive real and $Z(s) + Z(-s) = 0$ then $Z(s)$ is called a **reactance function** or a **Foster function**.

A reactance function must be the ratio of an even to an odd polynomial or vice versa, since it is an odd function.

Since $Z(s)$ is positive real it cannot have any right-half plane poles or zeros. Similarly, $Z(-s)$ cannot have any left-half plane poles or zeros. But since $Z(s) = -Z(-s)$ so that $Z(s)$ cannot have either right or left half plane poles or zeros. Thus, **the poles and zeros of a reactance function lie entirely on the imaginary axis.**

Since $Z(s)$ is the ratio of an odd to an even polynomial or vice versa and its poles and zeros lie on the imaginary axis, the difference in degree of numerator and denominator is not greater than one. Therefore, we can write an expression for the driving point impedance of an LC network either as

$$Z(s) = k\frac{(s^2+\omega_1^2)(s^2+\omega_3^2)+\dots}{s(s^2+\omega_2^2)(s^2+\omega_4^2)+\dots} \quad (21.11.1)$$

or as
$$Z(s) = \frac{ks\,(s^2+\omega_1^2)(s^2+\omega_3^2)\dots}{(s^2+\omega_2^2)(s^2+\omega_4^2)\dots} \quad (21.11.2)$$

where k is a scale factor. The zeros occur at $s = \pm j\omega_1, \pm j\omega_3, \dots$ and the poles at $s = \pm j\omega_2, \pm j\omega_4, \dots$ The origin is either a pole or a zero and there is either a pole or a zero at infinity depending on whether the numerator or the denominator has a higher degree.

Now since $Z(s)$ must be positive real, and since it poles are on the imaginary axis they must be simple with real positive residues. Therefore $Z(s)$ may be expanded in partial fraction form as

$$Z(s) = k_\infty s + \frac{k_0}{s} + \frac{k_2}{s+j\omega_2} + \frac{k'_2}{s-j\omega_2} + \frac{k_4}{s+j\omega_4} + \frac{k'_4}{s-j\omega_4} + \dots \quad (21.11.3)$$

where all the k are real and positive. Equation (21.11.3) can be written as

$$Z(s) = k_\infty s + \frac{k_0}{s} + \frac{s(k_2+k'_2)+j\omega_2\,(k'_2-k_2)}{s^2+\omega_2^2} + \frac{s(k_4+k'_4)+j\omega_4\,(k'_4-k_4)}{s^2+\omega_4^2} + \dots \quad (21.11.4)$$

But $Z(s)$ is real for s real it is required that $k_2 = k'_2$, $k_4 = k'_4$, …

Therefore Eq. (21.11.4) can be written as

$$Z(s) = k_\infty s + \frac{k_0}{s} + \frac{2k_2 s}{s^2+\omega_2^2} + \frac{2k_4\, s}{s^2+\omega_4^2} + \dots \quad (21.11.5)$$

If there is no pole at infinity $k_\infty = 0$ and if there is no pole at the origin $k_0 = 0$.

Similarly, in general we may write,

$$Y(s) = \alpha_\infty s + \frac{\alpha_0}{s} + \frac{2\alpha_1 s}{s^2+\omega_1^2} + \frac{2\alpha_3 s}{s^2+\omega_3^2} + \dots \quad (21.11.6)$$

with all α positive, since the poles are simple. $Y(s)$ is positive real. Again if $Y(s)$ has no pole at infinity α_∞ is zero and if no pole at the origin α_0 is zero.

Since $Z(s)$ is positive real there exists a passive network for which $Z(s)$ is either the driving point impedance function or driving-point admittance function. Moreover, since $Z(s)$ is a reactance function, it can be synthesized using LC elements only.

21.12 THEOREM

Z_{LC} or Y_{LC} is the ratio of odd to even or even to odd polynomials.

From Eq. (21.6.2), even part of $F(s)$ is given by

$$Ev\ [F(s)] = \frac{M_1(s)\, M_2(s) - N_1(s)\, N_2(s)}{[M_2(s)]^2 - [N_2(s)]^2} \tag{21.12.1}$$

For LC network, power dissipation is zero. The average power is given by

$$P = \frac{1}{2} \text{ Re } [Z(j\omega)]\ I^2 = 0$$

From Eq. (21.6.5)

$$Ev\ [F(s)]\ |_{s = j\omega} = \text{Re } [F(j\omega)]$$

That is,

$$0 = Ev\ [F(j\omega)] = \frac{M_1(j\omega)\, M_2(j\omega) - N_1(j\omega)\, N_2(j\omega)}{[M_2(j\omega)]^2 - [N_2(j\omega)]^2}$$

or
$$M_1(j\omega)\, M_2(j\omega) - N_1(j\omega)\, N_2(j\omega) = 0 \tag{21.12.2}$$

That is, either

$$M_1(j\omega) = N_2(j\omega) = 0$$

when $F(s)$ becomes $\dfrac{N_1(s)}{M_2(s)}$

or
$$M_2(j\omega) = N_1(j\omega) = 0$$

when $F(s)$ becomes $\dfrac{M_1(s)}{N_2(s)}$.

Hence, **the driving-point immittance function of a *LC* network is either a ratio of odd to even or even to odd.**

21.13 PROPERTIES OF A REACTANCE (or LC) FUNCTION

1. A reactance function must be the ratio of an odd to an even polynomial or vice versa.
2. Its poles and zeros must lie on the imaginary ($j\omega$), axis, with the difference in degree of numerator and denominator not greater than one. But since one is even and the other odd they must differ in degree by one.
3. The origin is either a pole or a zero and there is either a pole or a zero at infinity depending on whether the numerator or the denominator has the higher degree.
4. The poles and zeros of a reactance function **alternate** on the $j\omega$ axis that is, in Eq. (21.11.1) we must have $0 \le \omega_1 < \omega_2 < \omega_3 < \omega_4 \cdots$

The alternation of poles and zeros is also expressed by stating that the poles and zeros are **interlaced** or that they **mutually separate each other**. This is frequently referred to as the **separation property** of the poles and zeros.

5. The constant multiplier k is real positive.

21.14 SERIES AND PARALLEL COMBINATION OF LC ELEMENTS

The driving-point impedance or driving-point admittance may be found by combining impedance or admittance expressions for the elements in the network. An inductance L has impedance of Ls and admittance of $\frac{1}{Ls}$. A capacitor C has impedance of $\frac{1}{Cs}$ and admittance Cs.

A given network may be split up into either series or parallel combination of L and C elements.

Series Combination of LC Elements

Let the entire network be formed by series combination of L and C elements as shown in Fig. 21.6.

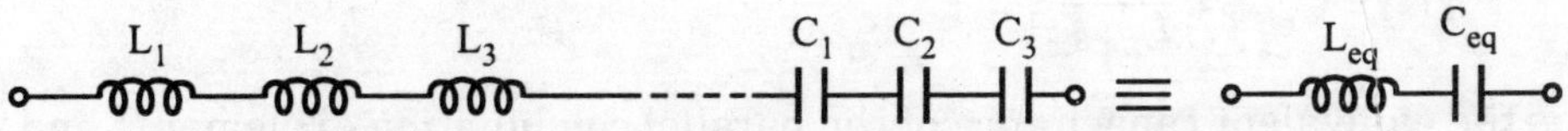

Fig. 21.6. Series combination of inductances and capacitances.

The total impedance of the series combination is given by

$$Z(s) = Z_1(s) + Z_2(s) + \ldots + Z_n(s) \tag{21.14.1}$$

$$= (sL_1 + sL_2 + \ldots sL_3) + \left(\frac{1}{C_1 s} + \frac{1}{C_2 s} + \ldots + \frac{1}{C_n s}\right)$$

$$= s(L_1 + L_2 + \ldots + L_n) + \left(\frac{1}{C_1} + \frac{1}{C_2} + \ldots + \frac{1}{C_n}\right)\frac{1}{s}$$

$$Z(s) = L_{eq}\, s + \frac{1}{C_{eq}\, s} \tag{21.14.2}$$

where L_{eq} is the equivalent inductance and C_{eq} is the equivalent capacitance of the series combination of elements. Hence the series combination of L and C elements may be represented by L_{eq} in series with C_{eq} as shown in Fig. 21.6.

Equation (21.14.2) may be written as

$$Z(s) = L_{eq} \frac{s^2 + \frac{1}{L_{eq}\, C_{eq}}}{s} \tag{21.14.3}$$

Parallel Combination of LC Elements

Let the entire network be formed by parallel combination of L and C elements as shown in Fig. 21.7.

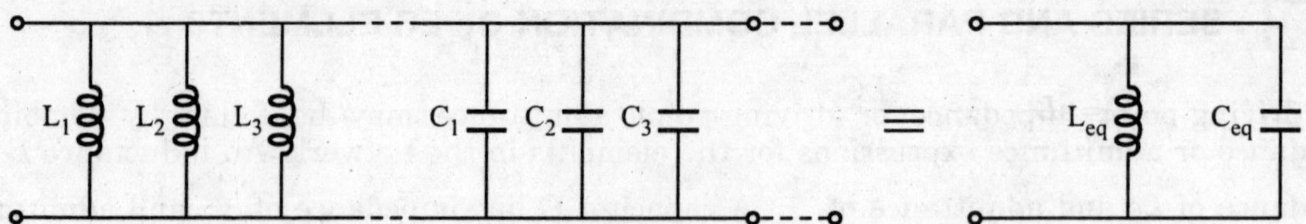

Fig. 21.7 Parallel combination of inductances and capacitances.

The total admittance of the parallel combination is given by

$$Y(s) = Y_1(s) + Y_2(s) + \dots + Y_n(s) \tag{21.14.4}$$

$$= (C_1 + C_2 + C_3 + \dots +)\, s + \left(\frac{1}{L_1} + \frac{1}{L_2} + \frac{1}{L_3} + \dots + \frac{1}{L_n}\right)\frac{1}{s}$$

$$Y(s) = C_{eq}\, s + \frac{1}{L_{eq}\, s} \tag{21.14.5}$$

where C_{eq} is the equivalent capacitance of the parallel combination of elements and L_{eq} is the equivalent inductance of the parallel combination of elements. Hence the parallel combination of L and C elements may be represented by L_{eq} in parallel with C_{eq} as shown in Fig. 21.7.

Equation (21.14.5) may be written as

$$Y(s) = C_{eq} \frac{s^2 + \dfrac{1}{L_{eq}\, C_{eq}}}{s} \tag{21.14.6}$$

It is seen that Eqs. (21.14.3) and (21.14.6) are similar in form. Only the multiplying constants are different.

21.15 SYNTHESIS BY INSPECTION

Simple driving-point functions can be synthesized by inspection. Such functions can be directly recognized as the sum of the impedances (or admittances) of resistors, capacitors and inductors. These are shown in Fig. 21.8. More complex network function can be realized as a combination of these simple building blocks.

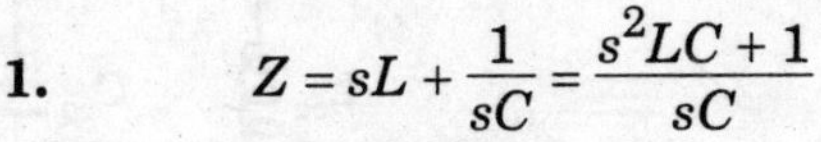

1. $Z = sL + \frac{1}{sC} = \frac{s^2LC + 1}{sC}$

2. $Z = sL + R$

3. $Z = R + \frac{1}{sC} = \frac{sCR + 1}{sC}$

4. $Z = R + sL + \frac{1}{sC} = \frac{s^2LC + sCR + 1}{C}$

5. $Z = \frac{1}{sC + \frac{1}{sL}} = \frac{sL}{s^2LC + 1}$

6. $Z = \frac{1}{\frac{1}{sL} + \frac{1}{R}} = \frac{sRL}{R + sL}$

7. $Z = \frac{1}{sC + \frac{1}{R}} = \frac{R}{sCR + 1}$

8. $Z = \frac{1}{sC + \frac{1}{R} + \frac{1}{sL}} = \frac{1}{C\left(s^2 + s\frac{1}{RC} + \frac{1}{LC}\right)}$

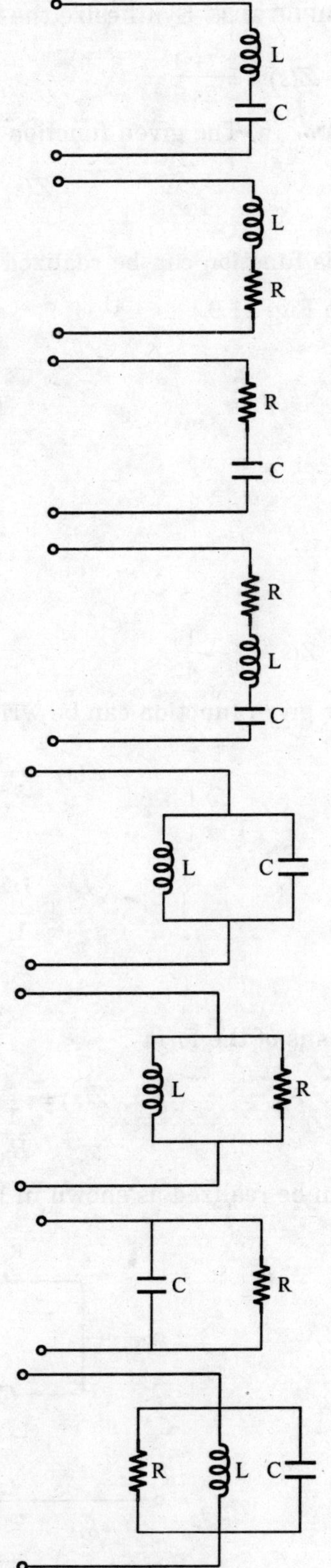

Fig. 21.8 Some simple functions that can be synthesized by inspection.

■ **EXAMPLE 21.25** Synthesize the function

(a) $Z(s) = \dfrac{s^2+1}{2s}$ (b) $Z(s) = \dfrac{s+1}{s+4}$

SOLUTION : (a) The given function may be written as

$$Z(s) = \frac{s}{2} + \frac{1}{2s}$$

This function can be realized as an inductor of $\frac{1}{2}$ H in series with a capacitor of $2F$ as shown in Fig. 21.9.

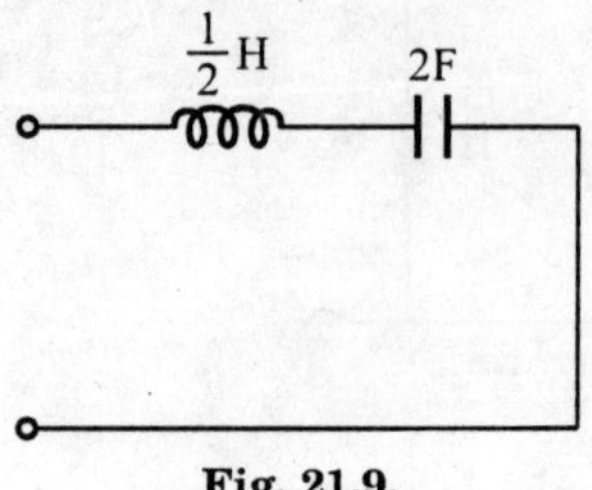

Fig. 21.9.

(b) $Z(s) = \dfrac{s+1}{s+4}$

The given function can be written as

$$Z(s) = \frac{s}{s+4} + \frac{1}{s+4} = \frac{1}{1+\frac{4}{s}} + \frac{1}{s+\frac{1}{\frac{1}{4}}}$$

$$= \frac{1}{\frac{1}{1} + \frac{1}{s\left(\frac{1}{4}\right)}} + \frac{1}{s\cdot 1 + \frac{1}{1/4}}$$

This is of the form

$$Z(s) = \frac{1}{\frac{1}{R_1} + \frac{1}{sL_1}} + \frac{1}{sC_2 + \frac{1}{R_2}}$$

which can be realized as shown in Fig. 21.10.

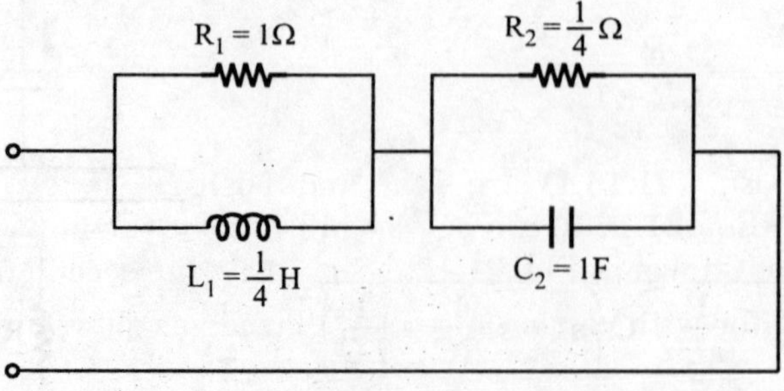

Fig. 21.10.

Alternative Method

The corresponding admittance function can be written as

$$Y(s) = \frac{1}{Z(s)} = \frac{s+4}{s+1} = 1 + \frac{3}{s+1}$$

$$= 1 + \frac{1}{\frac{s}{3} + \frac{1}{3}}$$

which can be realized as shown in Fig. 21.11.

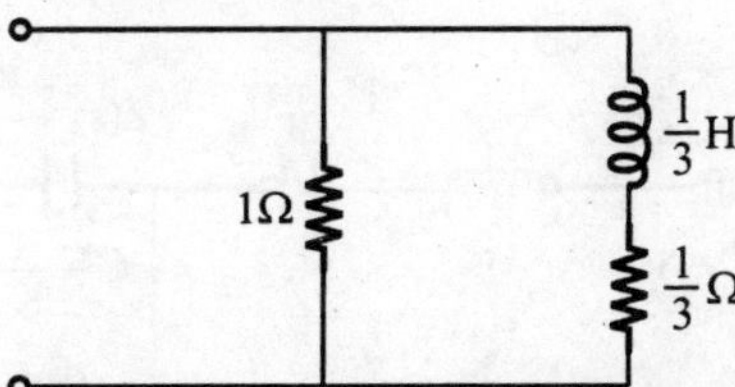

Fig. 21.11. It is seen that this realization is superior to the circuit of Fig. 21.10 in that it requires one less component.

21.16 FOSTER'S SYNTHESIS OF LC DRIVING-POINT IMMITTANCES

(a) First Foster Form

The partial fraction expansion of $Z(s)$ is writteni n the form

$$Z(s) = k_\infty s + \frac{k_0}{s} + \frac{2k_2 s}{s^2 + \omega_2^2} + \frac{2k_4 s}{s^2 + \omega_4^2} + \dots + \frac{2k_m s}{s^2 + \omega_m^2} \qquad (21.16.1)$$

where

$$k_\infty = \frac{Z(s)}{s} \Big|_{s \to \infty}$$

$$k_0 = s\, Z(s) \Big|_{s = 0}$$

$$k_2 = (s + j\omega_2)\, Z(s) \Big|_{s = -j\omega_2}$$

$$= \frac{(s^2 + \omega_2^2)\, Z(s)}{2s} \Big|_{s^2 = -\omega_2^2}$$

$$k_4 = (s + j\omega_4)\, Z(s) \Big|_{s = -j\omega_4}$$

$$= \frac{(s^2 + \omega_4^2)\, Z(s)}{2s} \Big|_{s^2 = -\omega_4^2}$$

and so on.

All the constants of Eq. (21.16.1) are real and positive. If $Z(s)$ is a driving-point impedance, the first term in Eq. (21.16.1) can be recognized to be the impedance of an inductor with inductance k_∞ henrys as shown in Fig. 21.12a. Similarly, the second term in Eq. (21.16.1) is the impedance of a capacitor with capacitance $(1/k_0)$ farads as shown in Fig. (21.12b).

The third term in Eq. (21.16.1) may be considered as the impedance of a parallel connection of a capacitor with capacitance $(1/2k_2)$ farads and an inductor with inductance $(2k_2/\omega_2^2)$ henrys. The fourth term may also be considered in the same manner. Thus, each term subsequent to second term in Eq. (21.16.1) may be considered as the impedance of a parallel connection of a capacitor with capacitance $\left(\dfrac{1}{2k_i}\right)$ farads and an inductor with inductance $\left(\dfrac{2k_i}{\omega_i^2}\right)$ henrys where $i = 2, 4, 6, \ldots, m$ as shown in Fig. 21.12c.

$Z(s) = K_\infty s$, $L = K_\infty$

(a)

$Z(s) = \dfrac{K_o}{s}$, $C = \dfrac{1}{K_o}$

(b)

$L = \dfrac{2K_i}{\omega_i^2}$, $C = \dfrac{1}{2K_i}$, $Z(s) = \dfrac{2K_i s}{s^2 + \omega_i^2}$

(c)

Fig. 21.12

The realization of the final network is obtained by a series connection of these impedances. The resulting network is known as **first order form** (or **Foster series network**) as shown in Fig. 21.13a.

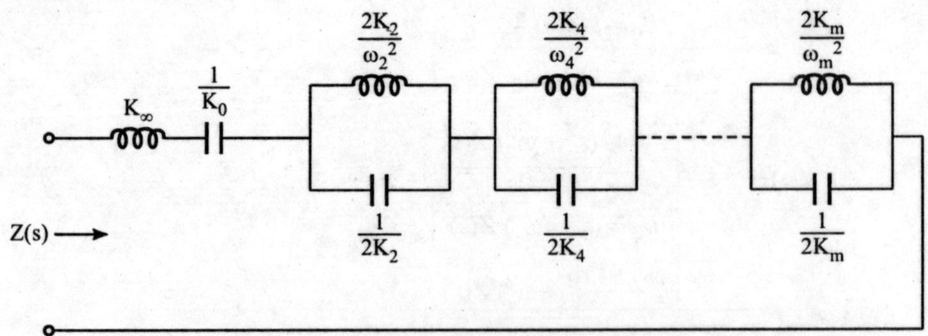

Fig. 21.13 First Foster form realization of a reactance function

(b) Second Foster Form

The second Foster form is the dual of the first Foster form for realization of a reactance function. The second form is realized from the admittance function $Y(s)$. To obtain the second Foster realization of the impedance function $Z(s)$ we find the reciprocal of $Z(s)$ to obtain the corresponding admittance function $Y(s)$. It is clear that $Y(s)$ is also a reactance function. The partial fraction expansion of $Y(s)$ can be written as

$$Y(s) = \frac{1}{Z(s)} = k'_\infty s + \frac{k'_0}{s} + \frac{2k'_2 s}{s^2 + \omega_2^2} + \frac{2k'_4 s}{s^2 + \omega_4^2} + \ldots + \frac{2k'_m s}{s^2 + \omega_m^2} \quad (21.16.2)$$

where all k' are real and positive constants in Eq. (21.16.2). The first term on the right-hand side is the admittance of a capacitor with capacitance k'_∞ farads (Fig. 21.14a). The second term is the admittance of an inductor with inductance $1/k'_0$ henrys (Fig. 21.14b). Each subsequent term represents the series connection of an inductance $\left(\frac{1}{2k'_i}\right)$ henrys and a capacitor with capacitance $\left(\frac{2k'_i}{\omega_i^2}\right)$ farads where $i = 2, 4, 6, \ldots, m$ (Fig. 21.14c)

$C = K'_\infty$

$Y(s) = K'_\infty s$

(a)

$L = \frac{1}{K'_o}$

$Y(s) = \frac{K'_o}{s}$

(b)

$L = \frac{1}{2K'_i}$ $\quad C = \frac{2K'_i}{\omega_i^2}$

$Y(s) = \frac{2K'_i s}{s^2 + \omega_i^2}$

(c)

Fig. 21.14

The final network is obtained by connecting these admittances in parallel as shown in Fig. 21.15.

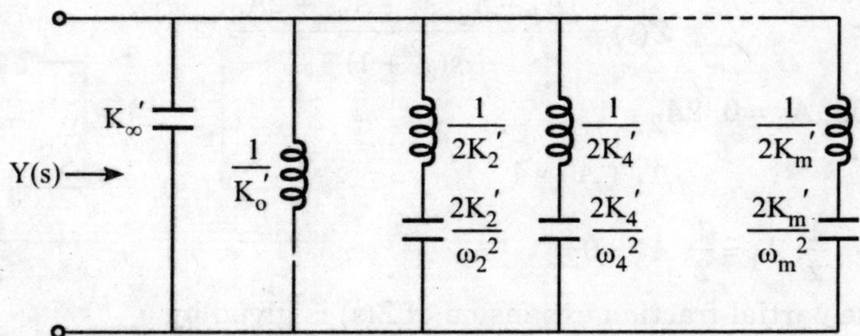

Fig. 21.15. Second Foster form realization of a reactace function $Z(s)$

Foster forms are said to be **canonical** in that they are capable of realizing **every** reactance function. They are said to be **minimal** in that they use minimum possible number of circuit elements. The total number of elements is equal to the degree of the highest power of s in the immittance function.

EXAMPLE 21.26 : (a) Show that the function given by

$$Z(s) = \frac{s^2 + 1}{s(s^2 + 2)}$$

is a reactance function.

(b) Obtain the first and second Foster realizations of $Z(s)$.

SOLUTION : (a) The poles of $Z(s)$ are at $s = 0$ and $s = \pm j\sqrt{2}$.

$$[\text{Residue at } s = 0] = \frac{1}{2}$$

$$[\text{Residue at } s = +j\sqrt{2}]$$

$$= [\text{Residue at } s = -j\sqrt{2}] = \frac{1}{4}$$

$$\text{Re}\,[Z(j\omega)] = \text{Re}\left[\frac{1-\omega^2}{j\omega\,(2-\omega^2)}\right]$$

$$= \text{Re}\left[\frac{-j\,(1-\omega^2)}{\omega(2-\omega^2)}\right] = 0$$

Since all the poles of $Z(s)$ are on the $j\omega$ axis with real and positive residues and Re $[Z(j\omega)] = 0$ for all values of ω, therefore, $Z(s)$ is a reactance function.

(b) First Foster Form

$$Z(s) = \frac{s^2 + 1}{s(s^2 + 2)}$$

The partial-fraction expansion of $Z(s)$ is given by

$$Z(s) = \frac{A_1}{s} + \frac{A_2 s + A_3}{s^2 + 1}$$

$$Z(s) = \frac{(A_1 + A_2)\,s^2 + A_3 s + 2A_2}{s(s^2 + 1)}$$

By comparison, $A_3 = 0$, $2A_2 = 1$

$$A_1 + A_2 = 1$$

therefore, $A_2 = \frac{1}{2}$, $A_1 = \frac{1}{2}$, $A_3 = 0$

Therefore, the partial-fraction expansion of $Z(s)$ is given by

$$Z(s) = \frac{1/2}{s} + \frac{\left(\frac{1}{2}\right)s}{s^2 + 2}$$

$$= \frac{1}{2s} + \frac{1}{\frac{s^2}{\frac{s}{2}} + \frac{2}{\frac{s}{2}}}$$

$$= \frac{1}{2s} + \frac{1}{2s + \frac{4}{s}} = Z_1(s) + Z_2(s) \text{ (say)}$$

$$Y_1(s) = \frac{1}{Z_1(s)}$$

$$Z_2(s) = \frac{1}{Y_2(s)} = \frac{1}{Y_2 + Y'_2} = \frac{1}{sC + \frac{1}{sL}}$$

$$Y_2(s) = sC = 2s, \; C = 2F$$

$$Y'_2(s) = \frac{1}{sL} = \frac{4}{s}, \; \frac{1}{L} = 4, \; L = \frac{1}{4}H$$

Thus, $Z(s)$ is realized by the series connection of two networks. One is a capacitor C_1 of $2F$ and the other is a parallel circuit consisting of a capacitor $C_2 = 2F$ and an inductor $L_2 = \frac{1}{4}H$ as shown in Fig. 21.16a.

(c) Second Foster Form

In order to obtain second Foster form, we find the reciprocal fo $Z(s)$ to obtain the corresponding admittance function $Y(s)$.

$$Y(s) = \frac{1}{Z(s)} = \frac{s(s^2+2)}{s^2+1}$$

$$\frac{Y(s)}{s} = \frac{s^2+2}{s^2+1} = \frac{s^2+1+1}{s^2+1} = 1 + \frac{1}{s^2+1}$$

$$Y(s) = s + \frac{s}{s^2+1}$$

$$Y(s) = s \times 1 + \frac{1}{s \times 1 + \frac{1}{s \times 1}} \qquad \text{(E 21.16.1)}$$

$$Y(s) = sC_1 + \frac{1}{sC_2 + \frac{1}{sL_2}} \qquad \text{(E 21.26.2)}$$

Comparison of Eqs. (E 21.26.1) and (E 21.26.2) shows that $Y(s)$ is realized by the parallel connection of two networks. One is a capacitor of capacitance $C_1 = 1F$ and the other is an inductor of inductance $L_2 = 1H$ in series with a capacitor of capacitance $C_2 = 1F$ as shown in Fig. 21.16b.

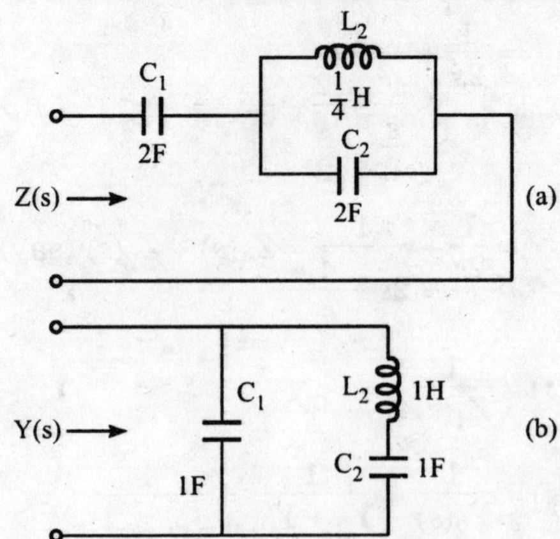

Fig. 21.16 (a) First Foster form, (b) Second Foster form realization of $Z(s) = \dfrac{(s^2+1)}{s(s^2+2)}$

EXAMPLE 21.27 : Find the two Foster realizations of

$$Z(s) = \frac{2s^3 + 8s}{s^2 + 1}$$

SOLUTION : (a) First Foster Realization

We expand $Z(s)$ into partial fractions.

$$s^2 + 1 \overline{)\, 2s^3 + 8s \,}(2s$$
$$\underline{2s^3 + 2s}$$
$$6s$$

therefore,
$$Z(s) = 2s + \frac{6s}{s^2+1}$$

$$Z(s) = 2s + \frac{1}{\dfrac{1}{6}s + \dfrac{1}{6s}} \tag{E 21.27.1}$$

$$= Z_1(s) + Z_2(s) \qquad \text{(say)}$$

This is of the form

$$Z(s) = sL_1 + \frac{1}{sC_2 + \dfrac{1}{sL_2}}$$

By comparison

$$L_1 = 2H,\ L_2 = 6H,\ C_2 = \frac{1}{6}F$$

Equation (E 21.27.1) shows that the resulting network is realized by the series combination of two impedances $Z_1(s)$ and $Z_2(s)$. The impedance $Z_1(s)$ has an inductor of $L_1 = 2H$. The impedance $Z_2(s)$ is a parallel combination of inductance of $L_2 = 6H$ and capacitance of $C_2 = \frac{1}{6} F$. The resulting network is shown in Fig. 21.17a.

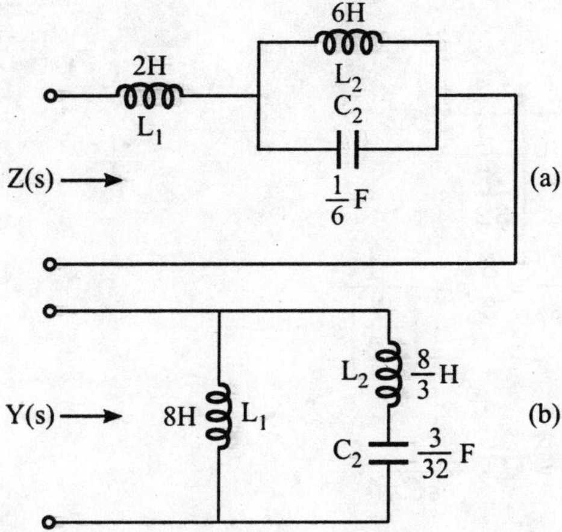

Fig. 21.17 (a) First Foster realization and (b) Second Foster realization of $Z(s) = \frac{2s^3 + 8s}{s^2 + 1}$

(b) Second Foster Realization

For second Foster realization we find the reciprocal of $Z(s)$ to obtain the corresponding admittance function $Y(s)$.

$$Y(s) = \frac{1}{Z(s)} = \frac{s^2 + 1}{2s^3 + 8s} = \frac{s^2 + 1}{2s(s^2 + 4)}$$

The partial-fraction expansion of $Y(s)$ is given by

$$\frac{s^2 + 1}{2s(s^2 + 4)} = \frac{A}{s} + \frac{Bs + D}{s^2 + 4}$$

$$= \frac{(A + B)s^2 + Ds + 4A}{s(s^2 + 4)}$$

therefore, $2(A + B) = 1,\ 8A = 1,\ D = 0$

$$A = \frac{1}{8},\ B = \frac{3}{8}$$

therefore, $Y(s) = \frac{1}{8s} + \frac{\frac{3}{8}s}{s^2 + 4}$

or
$$Y(s) = \frac{1}{8s} + \frac{1}{\frac{8}{3}s + \frac{1}{\left(\frac{3}{32}\right)s}}$$

$$= Y_1(s) + Y_2(s) \quad \text{(say)}$$

therefore, $Y_1(s) = \dfrac{1}{8s}$, $Z_1(s) = \dfrac{1}{Y_1(s)} = 8s = sL_1$

$$L_1 = 8H$$

$$Y_2(s) = \cfrac{1}{\cfrac{8}{3}s + \cfrac{1}{\left(\cfrac{3}{32}\right)s}}$$

$$Z_2(s) = \frac{1}{Y_2(s)} = \frac{8}{3}s + \cfrac{1}{\left(\cfrac{3}{32}\right)s}$$

This is of the form

$$Z_2(s) = sL_2 + \frac{1}{sC_2}$$

$$\therefore \quad L_2 = \frac{8}{3}H,\ C_2 = \frac{3}{32}F$$

Therefore $Y(s)$ is realized a parallel combination of two networks. One network has an inductance of $L_1 = 8H$ and the other network is a series combination of an inductance of $L_2 = \dfrac{8}{3}H$ and capacitance $C_2 = \dfrac{3}{32}F$. The resulting network is shown in Fig. 21.17b.

■ **EXAMPLE 21.28:** Obtain the first and second Foster forms of the reactance function

$$Z(s) = \frac{s(s^2+9)}{(s^2+4)(s^2+16)}$$

SOLUTION : (a) First Foster Realization

$$\frac{Z(s)}{s} = \frac{s^2+9}{(s^2+4)(s^2+16)}$$

The partial-fraction expansion of $\dfrac{Z(s)}{s}$ is given by

$$\frac{Z(s)}{s} = \frac{A}{s^2+4} + \frac{B}{s^2+16}$$

$$= \frac{(A+B)s^2 + 16A + 4B}{(s^2+4)(s^2+16)}$$

By comparison of the two expressions of $\dfrac{Z(s)}{s}$ we get

$$A + B = 1$$

$$16A + 4B = 9$$

By solving these equations we get

$$A = \frac{5}{12}, B = \frac{7}{12}$$

therefore,
$$\frac{Z(s)}{s} = \frac{\frac{5}{12}}{s^2+4} + \frac{\frac{7}{12}}{s^2+16}$$

$$Z(s) = \frac{\frac{5}{12}s}{s^2+4} + \frac{\frac{7}{12}s}{s^2+16}$$

$$= Z_1(s) + Z_2(s) \quad \text{(say)} \qquad \text{(E 21.28.1)}$$

$$Z_1(s) = \frac{\frac{5}{12}s}{s^2+4}$$

$$= \frac{1}{\frac{s^2}{(5/12)\,s} + \frac{4}{(5/12)\,s}}$$

$$= \frac{1}{s\left(\frac{12}{5}\right) + \frac{1}{s\left(\frac{5}{48}\right)}}$$

$Z_1(s)$ is of the form

$$Z_1(s) = \frac{1}{sC_1 + \frac{1}{sL_1}}$$

Thus, $Z_1(s)$ is realized by the parallel combination of two circuits. One is capacitor of capacitance $C_1 = \frac{12}{5}$ F and the other is an inductor of inductance $L_1 = \frac{5}{48}$ H

$$Z_2(s) = \frac{\frac{7}{12}s}{s^2+16}$$

$$= \frac{1}{\frac{s^2}{\left(\frac{7}{12}\right)s} + \frac{16}{\left(\frac{7}{12}\right)s}} = \frac{1}{s\left(\frac{12}{7}\right) + \frac{1}{s\left(\frac{7}{192}\right)}}$$

$Z_2(s)$ is of the form

$$Z_2(s) = \frac{1}{sC_2 + \frac{1}{sL_2}}$$

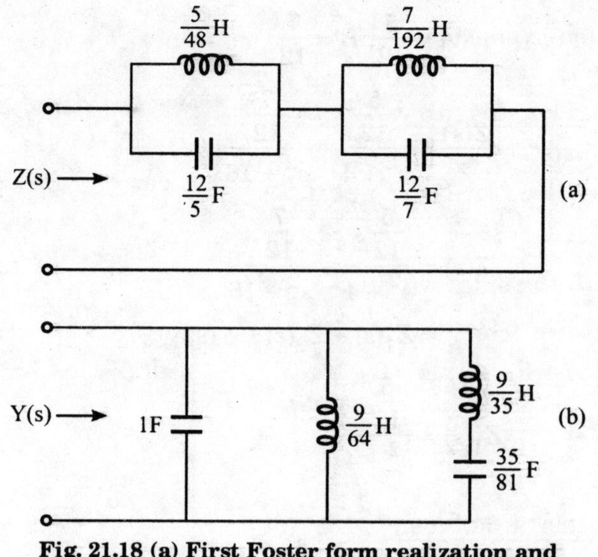

Fig. 21.18 (a) First Foster form realization and (b) Second Foster form realization of $Z(s) = \dfrac{s(s^2+9)}{(s^2+4)(s^2+16)}$

Thus, $Z_2(s)$ is realized by the parallel combination of two circuits. One is a capacitor of capacitance $C_2 = \dfrac{7}{12}\,F$ and the other is an inductor of inductance $L_2 = \dfrac{7}{192}\,H$.

Equation (E 21.28.1) shows that $Z(s)$ is a series combination of $Z_1(s)$ and $Z_2(s)$. Thus, $Z(s)$ is realized as a series connection of two parallel LC networks as shown in Fig. 21.18a.

(b) Second Foster Realization

In order to obtain second Foster form, we find the reciprocal of $Z(s)$ to obtain the corresponding admittance function $Y(s)$.

$$Y(s) = \frac{1}{Z(s)} = \frac{(s^2+4)(s^2+16)}{s(s^2+9)}$$

$$= \frac{s^4 + 20s^2 + 64}{s^3 + 9s}$$

$$s^3 + 9s \,\big)\, s^4 + 20s^2 + 64 \,\big(\, s$$
$$\underline{s^4 + 9s^2}$$
$$11s^2 + 64$$

$$Y(s) = s + \frac{11s^2 + 64}{s^3 + 9s}$$

$$= s + \frac{11s^2 + 64}{s(s^2+9)}$$

The partial fraction expansion of $\dfrac{11s^2+64}{s(s^2+9)}$ is given by

$$\frac{11s^2+64}{s(s^2+9)}=\frac{A}{s}+\frac{Bs+D}{s^2+9}$$

$$=\frac{(A+B)s^2+Ds+9A}{s(s^2+9)}$$

By comparing both the sides we get

$$9A=64$$

$$A+B=11,\ D=0$$

therefore, $A=\dfrac{64}{9},\ B=11-\dfrac{64}{9}=\dfrac{35}{9}$

therefore, $\dfrac{11s^2+64}{s(s^2+9)}=\dfrac{64/9}{s}+\dfrac{\frac{35}{9}s}{s^2+9}$

therefore, $$Y(s)=s+\frac{64/9}{s}+\frac{\frac{35}{9}s}{s^2+9} \qquad \text{(E 21.28.2)}$$

$$=Y_1(s)+Y_2(s)+Y_3(s) \qquad \text{(say)}$$

Equation (E 21.28.2) shows that $Y(s)$ is a parallel combination of $Y_1(s)$, $Y_2(s)$ and $Y_3(s)$. The admittance $Y_1(s)$ is a capacitor of $1F$ with admittance s. The term $\dfrac{64}{9s}$ is produced by the admittance of an inductor $L=\left(\dfrac{9}{64}\right)H$.

$$Y_3(s)=\frac{\frac{35}{9}s}{s^2+9}$$

$$=\frac{1}{\dfrac{s^2}{\left(\frac{35}{9}\right)s}+\dfrac{9}{\left(\frac{35}{9}\right)s}}$$

$$=\frac{1}{\dfrac{9}{35}s+\dfrac{1}{\frac{35}{81}s}}$$

$$Z_3(s)=\frac{1}{Y_3(s)}=\frac{9}{35}s+\frac{1}{\left(\frac{35}{81}\right)s}$$

Therefore, branch 3 of the network is a series LC circuit with $L=\dfrac{9}{35}H$ and $C=\dfrac{35}{81}F$.

The resulting network is realized as a parallel combination of three admittances $Y_1(s)$, $Y_2(s)$ and $Y_3(s)$. It is shown in Fig. 21.18b.

■ **EXAMPLE 21.29 :** Synthesize the first and second Foster forms of the *LC* driving-point impedance

$$Z(s) = \frac{2(s^2+1)(s^2+9)}{s(s^2+4)}$$

SOLUTION :

(a) First Foster Form

$$Z(s) = \frac{2(s^2+1)(s^2+9)}{s(s^2+4)}$$

$$= \frac{2s^4+20s^2+18}{s^3+4s}$$

$$s^3+4s\,\big)\; 2s^4+20s^2+18 \;\big(\; 2s$$
$$\underline{2s^4+8s^2}$$
$$12s^2+18$$

therefore, $$Z(s) = 2s + \frac{12s^2+18}{s(s^2+4)}$$

The partial-fraction expansion of $\frac{12s^2+18}{s(s^2+4)}$ may be written as

$$\frac{12s^2+18}{s(s^2+4)} = \frac{A}{s} + \frac{Bs+D}{s^2+4}$$

$$= \frac{A(s^2+4)+Bs^2+Ds}{s(s^2+4)}$$

or $$\frac{12s^2+18}{s(s^2+4)} = \frac{(A+B)s^2+Ds+4A}{s(s^2+4)}$$

Comparison of the two sides of the above equation gives

$$A+B = 12,\; D = 0,\; 4A = 18$$

therefore, $$A = \frac{9}{2},\; B = 12 - \frac{9}{2} = \frac{15}{2},\; D = 0$$

therefore, $$Z(s) = 2s + \frac{9/2}{s} + \frac{\frac{15}{2}s}{s^2+4}$$

$$= 2s + \frac{1}{\frac{2}{9}s} + \frac{1}{\frac{s^2}{\frac{15}{2}s} + \frac{4}{\frac{15}{2}s}}$$

$$= 2s + \frac{1}{\frac{2}{9}s} + \frac{1}{\frac{2}{15}s + \frac{1}{\frac{15}{8}s}}$$

$Z(s)$ may be written in the form

$$Z(s) = sL_1 + \frac{1}{sC_1} + \frac{1}{C_2 s + \frac{1}{L_2 s}}$$

where $L_1 = 2H$, $C_1 = \frac{2}{9}F$, $C_2 = \frac{2}{15}F$, $L_2 = \frac{15}{8}H$

The resulting network is realized as shown in Fig. 21.19a.

(b) Second Foster Form

$$Y(s) = \frac{1}{Z(s)} = \frac{s(s^2+4)}{2(s^2+1)(s^2+9)}$$

The partial fraction expansion of $Y(s)$ may be written as

$$Y(s) = \frac{s(s^2+4)}{2(s^2+1)(s^2+9)} = \frac{1}{2}\left[\frac{A_1 s + B_1}{s^2+1} + \frac{A_2 s + B_2}{s^2+9}\right]$$

$$= \frac{[(A_1+A_2)s^3 + (B_1+B_2)s^2 + (9A_1+A_2)s + (9B_1+B_2)]}{2(s^2+1)(s^2+9)}$$

By comparison

$$A_1 + A_2 = 1,\ B_1 + B_2 = 0$$

$$9A_1 + A_2 = 4,\ 9B_1 + B_2 = 0$$

therefore, $B_1 = 0, B_2 = 0$

$$A_1 = \frac{3}{8}, A_2 = \frac{5}{8}$$

$$Y(s) = \frac{1}{2}\left[\frac{\frac{3}{8}s}{s^2+1} + \frac{\frac{5}{8}s}{s^2+9}\right]$$

$$Y(s) = \frac{\frac{3}{16}s}{s^2+1} + \frac{\frac{5}{16}s}{s^2+9}$$

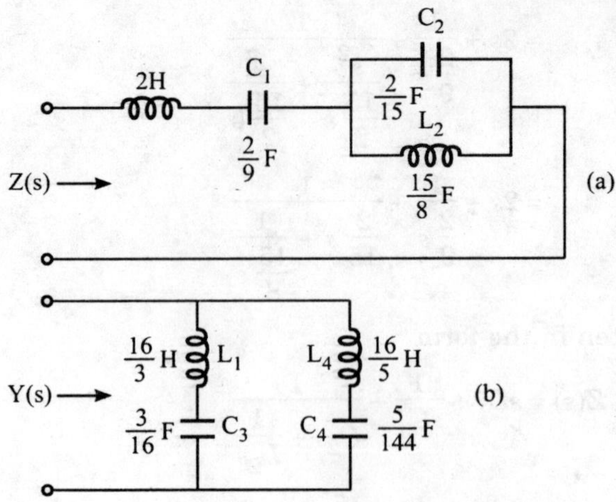

Fig. 21.19 (a) First Foster Form (b) Second Foster Form of $Z(s) = \dfrac{2(s^2+1)(s^2+9)}{s(s^2+4)}$

or

$$Y(s) = \frac{1}{\frac{16}{3}s + \frac{1}{\frac{3}{16}s}} + \frac{1}{\frac{16}{5}s + \frac{1}{\frac{5}{144}s}}$$

$Y(s)$ may be written in the form

$$Y(s) = \frac{1}{sL_3 + \frac{1}{sC_3}} + \frac{1}{sL_4 + \frac{1}{sC_4}}$$

therefore, $L_3 = \dfrac{16}{3}\,H,\ C_3 = \dfrac{3}{16}\,F$

$$L_4 = \frac{16}{5}\,H,\ C_4 = \frac{5}{144}\,F$$

Therefore the resulting network is shown in Fig. 21.19b.

21.17 CAUER'S SYNTHESIS METHOD FOR LC NETWORKS

Another form of synthesis for *LC* driving-point impedances is the Cauer or ladder form. There are two possibilities and both are canonic and minimal. One is based on extraction of poles at the origin and the other is based on extraction of poles at infinity.

(a) First Cauer's Form

This method is based on removal of poles at infinity.

Consider the reactance function

$$Z(s) = \frac{Ks\,(s^2+\omega_1^2)\,(s^2+\omega_3^2)\,\ldots}{(s^2+\omega_2^2)\,(s^2+\omega_4^2)\,\ldots} \qquad (21.17.1)$$

In this equation the numerator polynomial is one degree higher than the denominator polynomial, that is, $Z(s)$ has a pole at infinity. Therefore, $Z(s)$ can be written as

$$Z(s) = sL_1 + Z_1(s) \tag{21.17.2}$$

where $L_1 = K$ and $Z_1(s)$ is another reactance function whose denominator polynomial is one degree higher than its numerator polynomial. $Z_1(s)$ can be inverted and written as

$$\frac{1}{Z_1(s)} = C_2 s + Y_2(s) \tag{21.17.3}$$

where $Y_2(s)$ is also a reactance function whose pole is at infinity. This process is continued until the remainader is simply $\frac{K_i}{s}$ or $K_i s$. Therefore, $Z(s)$ can be expressed as

$$Z(s) = L_1 s + \cfrac{1}{C_2 s + \cfrac{1}{L_3 s + \cfrac{1}{C_4\, s + .}}} \tag{21.17.4}$$

It is seen that in Eq. (21.17.4) $L_1 s$ is the impedance of an inductor with inductance L_1 henrys, $C_2 s$ is the admittance of a capacitor with capacitance C_2 farads, and so on. The resulting network can be synthesized as shown in Fig. 21.20a. This network is known as the **First Cauer's form**. It is a ladder network with inductors in series arms and capacitors in shunt arms.

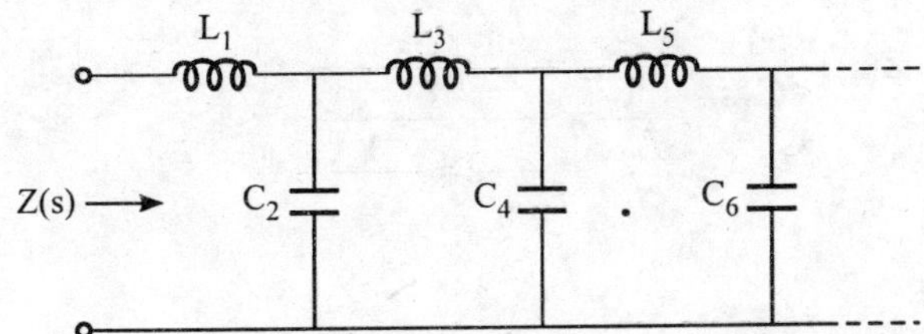

Fig. 21.20a. First Cauer realization of a reactance function

Equation (21.17.4) is known as **continued fraction expansion** of $Z(s)$.

The Cauer (ladder) first form of network realization is obtained by **repeated removal of poles at infinity**. This corresponds to arranging the numerator and denominator polynomials of the driving-point impedance in **descending powers of *s*** and then finding the continued-fraction expansion.

The following steps are used to obtain the continued fraction expansion of $Z(s)$:

1. Write numerator and denominator in descending powers of s.
2. Divide the denominator into the numerator to find the quotient $L_1 s$.
3. Divide the remainder into the previous divisor to find the second quotient $C_2 s$ and a new remainder.
4. Divide that remainder into the immediately preceding divisor and continue the process until a zero remainder is obtained.

The quotients are the element values of the network and are alternately inductance and capacitance.

If the original function is an impedance, the first quotient is an inductance.
If the original function is an admittance the first quotient is a capacitance.

The process of **division-inversion-division** generates the coefficients of the continued fraction expansion of a function. When $Z(s)$ is a reactance function the coefficients are positive. This type of expansion provides a very convenient means for checking whether a function is in fact a reactance function without the necessity of determining the poles and zeros which may be difficult if the degree is high. The method can also be used to check whether a polynomial is Hurwitz since, if it is, then the ratio of its even to its odd part must be a reactance function and we have merely to carry out a continued fraction expansion of this and check whether the coefficients are positive.

(b) Second Cauer's Form

This method of synthesis is based on the removal of successive poles at the origin. Consider the reactance function

$$Z(s) = \frac{a_0 + a_1 s + a_2 s^2 + \dots + a_n s^n}{b_1 s + b_2 s^2 + b_3 s^3 + \dots + b_{n+1} s^{n+1}} = \frac{N(s)}{D(s)} \tag{21.17.5}$$

It is to be noted that the numerator and denominator polynomials are arranged an **ascending order**.

Now divide the numerator by the denominator. Invert the remainder and divide the numerator by the denominator again. Continue the process until a zero remainder is obtained. The continued fraction expansion takes the form

$$Z(s) = \frac{1}{C_1 s} + \cfrac{1}{\cfrac{1}{L_2 s} + \cfrac{1}{\cfrac{1}{C_3} s + \cfrac{1}{\cfrac{1}{L_4} s + \vdots}}} \tag{21.17.6}$$

where C_i and L_i are fixed constants obtained in the process of continued fraction expansion. It is seen from Eq. (21.17.6) that the resulting network can be synthesized as shown in Fig. 21.20b.

This network is known as the **second Cauer form**. It is a **ladder network** with capacitors in series arms and inductors in shunt arms.

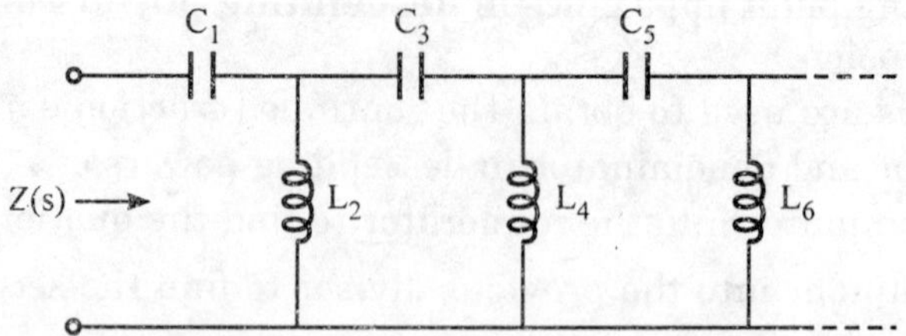

Fig. 21.20b Second Cauer realization of a reactance function.

It is seen that driving-point synthesis technique, proposed by Cauer, is based on the continued fraction expansion of the given function.

The second Cauer form of network realization is obtained by repeated removal of poles at the origin. This corresponds to arranging the numerator and denominator polynomial of the driving-point impedance function in **ascending powers of** s and then finding the continued-fraction expansion.

Both the Foster and Cauer forms give minimum number of elements for a specified LC driving-point function. These relations are sometimes known as **canonical forms.**

EXAMPLE 21.30 : Find the Cauer realization of

$$Z(s) = \frac{(s^2+2)(s^2+4)}{s(s^2+3)}$$

SOLUTION :

(a) First Cauer Form

The given impedance function has a pole at infinity. We write the numerator and denominator in ascending powers of s to give

$$Z(s) = \frac{s^4+6s^2+8}{s^3+3s}$$

The continued-fraction expansion is

$$s^3+3s\ )\ s^4+6s^2+8\ (\ s \rightarrow Z_1(s) = sL_1,\ L_1 = 1H$$

$$\underline{s^4+3s^2}$$

$$3s^2+8\ )\ s^3+3s\ (\ \frac{s}{3} \rightarrow Y_2(s) = sC_2,\ C_2 = \frac{1}{3}F$$

$$s^3+\frac{8}{3}s$$

$$\overline{\qquad\qquad} \frac{s}{3}\ )\ 3s^2+8\ (\ 9s \rightarrow Z_3(s) = sL_3,\ L_3 = 9H$$

$$\underline{3s^2}$$

$$8\ )\ \frac{s}{3}\ (\ \frac{s}{24} \rightarrow Y_4(s) = sC_4,\ C_4 = \frac{1}{24}F$$

$$\frac{s}{3}$$

$$\overline{\ 0\ }$$

The resulting network is shown in Fig. 2.1.21a.

(b) Second Cauer Form

In the second Cauer form, we extract poles at the origin. We write the numerator and denominator in ascending powers of s and starting with $Z(s)$ or $Y(s)$ (whichever has the pole at the origin) we divide the numerator by the denominator and find the continued fraction expansion. The given function $Z(s)$ has a pole at the origin.

$$Z(s) = \frac{8+6s^2+s^4}{3s+s^3}$$

The continued-fraction expansion is

$$3s + s^3\ \big)\ 8 + 6s^2 + s^4\ \big(\ \frac{8}{3s} \rightarrow Z_1(s) = \frac{1}{sC_1},\ C_1 = \frac{3}{8}\ F$$

$$\underline{8 + \frac{8}{3}s^2}$$

$$\frac{10}{3}s^2 + s^4\ \big)\ 3s + s^3\ \big(\ \frac{9}{10s} \rightarrow Y_2(s) = \frac{1}{sL_2},\ L_2 = \frac{10}{9}\ H$$

$$\underline{3s + \frac{9}{10}s^3}$$

$$\frac{s^3}{10}\ \big)\ 10s^2 + s^4\ \big(\ \frac{100}{3s} \rightarrow Z_3(s) = \frac{1}{sC_3},\ C_3 = \frac{3}{100}\ F$$

$$\underline{\frac{10}{3}s^2}$$

$$s^4\ \big)\ \frac{s^3}{10}\ \big(\ \frac{1}{10s} \rightarrow Y_4(s) = \frac{1}{sL_4},\ L_4 = 10H$$

$$\underline{\frac{s^3}{10}}$$

$$0$$

The synthesized network in second Cauer form is shown in Fig. 21.21b.

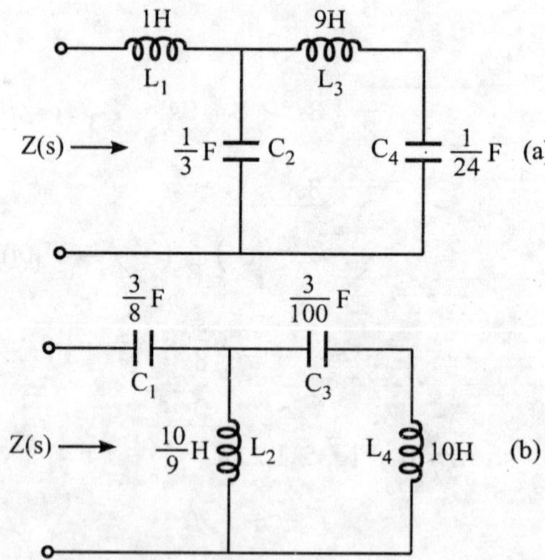

Fig. 21.21 (a) First Cauer form and (b) Second Cauer form realization of $Z(s) = \dfrac{(s^2+2)(s^2+4)}{s(s^2+3)}$

■ **EXAMPLE 21.31 :** Find the Cauer 1 and Cauer 2 realizations of $Z(s) = \dfrac{2s^3 + 8s}{s^2 + 1}$

SOLUTION : (a) Cauer 1 Realisation

The continued fraction expansion of $Z(s)$ is given by

$$s^2 + 1\;)\;2s^3 + 8s\;(\;2s \rightarrow Z_1(s) = sL_1,\; L_1 = 2H$$

$$\underline{2s^3 + 2s}$$

$$6s\;)\;s^2 + 1\;(\;\frac{1}{6}s \rightarrow Y_2(s) = sC_2,\; C_2 = \frac{1}{6}F$$

$$\underline{s^2}$$

$$1\;)\;6s\;(\;6s \rightarrow Z_3(s) = sL_3,\; L_3 = 6H$$

$$\underline{6s}$$

$$0$$

The first form of Cauer realization is shown in Fig. 21.22a.
It is identical with the first Foster form of Fig. 21.17a

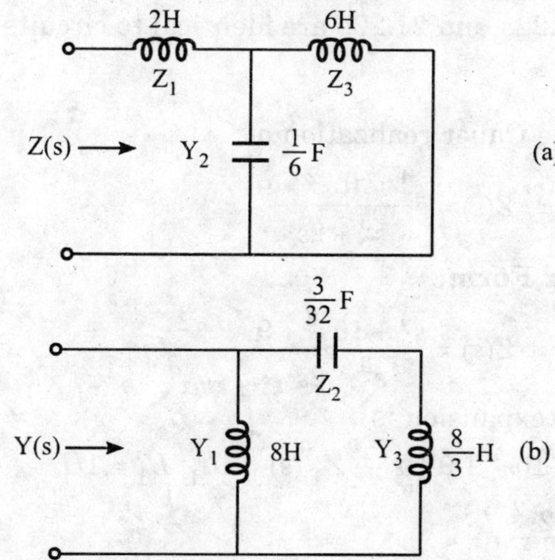

Fig. 21.22 (a) Cauer 1 and (b) Cauer 2 realizations of Fig. 21.22 (a) $Z(s) = \dfrac{2s^3 + 8s}{s^2 + 1}$

(b) Cauer 2 Realization

The numerator and denominator polynomials are arranged in ascending powers of s

$$Z(s) = \frac{8s + 2s^3}{1 + s^2}$$

$$Y(s) = \frac{1 + s^2}{8s + 2s^3}$$

The continued-fraction expansion of $Y(s)$ is given by

$$8s + 2s^3 \;)\; 1 + s^2 \;\left(\; \frac{1}{8s} \to Y_1(s) = \frac{1}{sL_1},\; L_1 = 8H\right.$$

$$\frac{1 + \frac{1}{4}s^2}{\frac{3}{4}s^2} \;)\; 8s + 2s^3 \;\left(\; \frac{32}{3s} \to Z_2(s) = \frac{1}{sC_2},\; C_2 = \frac{3}{32}F\right.$$

$$\frac{8s}{2s^3} \;)\; \frac{3}{4}s^2 \;\left(\; \frac{3}{8s} \to Y_3(s) = \frac{1}{sL_3},\; L_3 = \frac{8}{3}H\right.$$

$$\frac{\frac{3}{4}s^2}{0}$$

Therefore the synthesized network in the second Cauer form is shown in Fig. 21.22. It is identical with the second Foster form of Fig. 21.17b.

IMPORTANT NOTE

For degree $N \le 3$, the Foster and Cauer expansions lead to the same circuits. For $N > 3$, the networks obtained by Cauer and Foster realizations are different. Hence the circuits of Fig. 21.22a and 21.22b are identical to circuits of Fig. 21.17a and 21.17b respectively.

EXAMPLE 21.32 : Find the Cauer realization of

$$Z(s) = \frac{s^4 + 10s^2 + 9}{s^3 + 2s}$$

SOLUTION : (a) First Cauer Form

$$Z(s) = \frac{s^4 + 10s^2 + 9}{s^3 + 2s}$$

The continued fraction expansion is

$$s^3 + 2s \;)\; s^4 + 10s^2 + 9 \;(\; s \to Z_1(s) = sL_1,\; L_1 = 1H$$

$$\frac{s^4 + 2s^2}{8s^2 + 9} \;)\; s^3 + 2s \;\left(\; \frac{s}{8} \to Y_2(s) = sC_2,\; C_2 = \frac{1}{8}F\right.$$

$$\frac{s^3 + \frac{9s}{8}}{\frac{7}{8}s} \;)\; 8s^2 + 9 \;\left(\; \frac{64}{7}s \to Z_3(s) = sL_3,\; L_3 = \frac{64}{7}H\right.$$

$$\frac{8s^2}{9} \;)\; \frac{7}{8}s \;\left(\; \frac{7}{72}s \to Y_4(s) + sC_4,\; C_4 = \frac{7}{72}F\right.$$

$$\frac{\frac{7}{8}s}{0}$$

The network is Cauer's first form is shown in Fig. 21.23a.

(b) Second Cauer Form

$$Z(s) = \frac{9 + 10s^2 + s^4}{2s + s^3}$$

The continued-fraction expansion is

$$2s + s^3 \;)\; 9 + 10s^2 + s^4 \;(\; \frac{9}{2s} \to Z_1(s) + \frac{1}{sC_1},\; C_1 = \frac{2}{9} F$$

$$9 + \frac{9}{2}s^2$$

$$\frac{11}{2}s^2 + s^4 \;)\; 2s + s^3 \;(\; \frac{4}{11s} \to Y_2(s) = \frac{1}{sL_2},\; L_2 = \frac{11}{4} H$$

$$2s + \frac{4}{11}s^3$$

$$\frac{7}{11}s^3 \;)\; \frac{11s^2}{2} + s^4 \;(\; \frac{121}{14s} \to Z_3(s) = \frac{1}{sC_3},\; C_3 = \frac{14}{121} F$$

$$\frac{11s^2}{2}$$

$$s^4 \;)\; \frac{7}{11}s^3 \;(\; \frac{7}{11s} \to Y_4(s) = \frac{1}{sL_4},\; L_4 = \frac{11}{7} H$$

$$\frac{7}{11}s^3$$

$$0$$

Cauer 2 realization is shown in 21.23b.

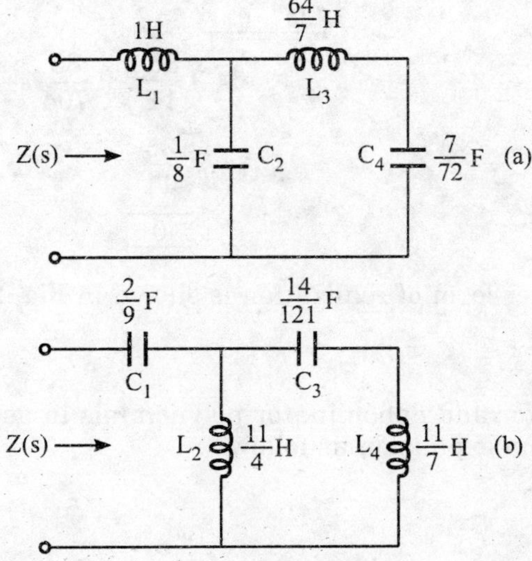

Fig. 21.23 (a) Cauer 1 and (b) Cauer 2 realizations of $Z(s) = \dfrac{s^4 + 10s^2 + 9}{s^3 + 2s}$

EXAMPLE 21.33 : Realize the following function using the Cauer forms with

$$Z(s) = \frac{s(s^2+9)}{(s^2+4)(s^2+16)}$$

SOLUTION :

$$Z(s) = \frac{s^3+9s}{s^4+20s^2+64}$$

(a) First Cauer Form

The Cauer first form of network realization is obtained by repeated removal of poles at infinity. It is seen that it is $Y(s)$ which possesses the pole at infinity and its residue is 1. Therefore we begin with writing $Y(s)$.

$$Y(s) = \frac{s^4+20s^2+64}{s^3+9s}$$

For Cauer first-form realization the numerator and denominator polynomials should be arranged in descending powers of s. Then we find the continued-fraction expansion as follows:

$$s^3+9s\ )\ s^4+20s^2+64\ (\ s \rightarrow Y_1(s) = sC_1,\ C_1 = 1F$$

$$\underline{s^4+9s^2}$$

$$11s^2+64\ )\ s^3+9s\ (\ \frac{1}{11}s \rightarrow Z_2(s) = sL_2,\ L_2 = \frac{1}{11}H$$

$$\underline{s^3+\frac{64}{11}s}$$

$$\frac{35}{11}s\ )\ 11s^2+64\ (\ \frac{121}{35}s \rightarrow Y_3(s) = sC_3,\ C_3 = \frac{121}{35}F$$

$$\underline{11s^2}$$

$$64\ )\ \frac{35}{11}s\ (\ \frac{35}{704}s \rightarrow Z_4(s) = sL_4,\ L_4 = \frac{35}{704}H$$

$$\underline{\frac{35}{11}s}$$

$$\underline{0}$$

Thus, the first Cauer form of realization is shown in Fig. 21.24a.

(b) Second Cauer form

We arrange the numerator and denominator polynomials in ascending powers of s. Then we find the continued fraction expansion as follows :

$$9s + s^3\ )\ 64 + 20s^2 + s^4\ \left(\ \frac{64}{9s} \rightarrow Y_1(s) = \frac{1}{sL_1},\ L_1 = \frac{9}{64}\ H\right.$$

$$64 + \frac{64}{9}s^2$$

$$\frac{116s^2}{9} + s^4\ )\ 9s + s^3\ \left(\ \frac{81}{116s} \rightarrow Z_2(s) = \frac{1}{sC_2},\ C_2 = \frac{116}{81}\ F\right.$$

$$9s + \frac{81}{116}s^3$$

$$\frac{35}{116}s^3\ )\ \frac{116}{9}s^2 + s^4\ \left(\ \frac{(116)^2}{315s} \rightarrow Y_3(s) = \frac{1}{sL_3},\ L_3 = \frac{315}{(116)^2}\ H\right.$$

$$\frac{116}{9}s^2$$

$$s^4\ )\ \frac{35}{116}s^3\ \left(\ \frac{35}{116s} \rightarrow Z_4(s) = \frac{1}{sC_4},\ C_4 = \frac{116}{35}\ F\right.$$

$$\frac{35}{116}s^3$$

$$0$$

Thus, the second Cauer realization is shown in Fig. 21.24b.

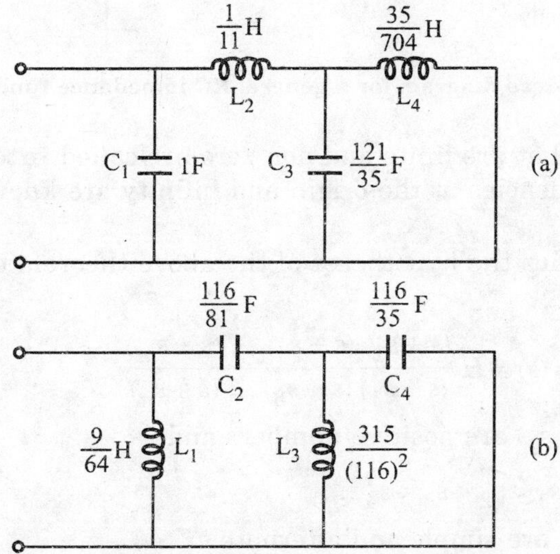

Fig. 21.24(a) First Caver form and (b) Second Caver form realizations of $Z(s) = \dfrac{s(s^2 + 9)}{(s^2 + 4)(s^2 + 16)}$

21.18 RC DRIVING-POINT IMPEDANCE FUNCTION $Z_{RC}(s)$

This class of networks is of considerable practical as well as theoretical importance. From a practical viewpoint the difficulty of constructing inductors without loss means that the realization of networks involving inductors is seldom very exact, while the resistors and capacitors can be made with a high degree of purity.

The *RC* driving-point impedance function $Z_{RC}(s)$ is defined by the following theorem :

Theorem

A rational function is an *RC* driving-point function if and only if :

1. All the zeros and poles are simple and lie on the negative real axis or at the origin in the *s*-plane.
2. The poles and zeros alternate on the negative real axis.
3. The critical frequency of smallest magnitude, that is, the frequency at or nearest the origin, is a **pole**.
4. The constant multiplier is positive.

As a consequence of conditions 2 and 3, since the number of poles is always equal to the number of zeros, it is clear that the critical frequency of largest magnitude is a zero. In other words the highest critical frequency at infinity or nearest to infinity is a **zero**.

The pole-zero diagram for a general *RC* impedance function $Z_{RC}(s)$ is shown in Fig. 21.25.

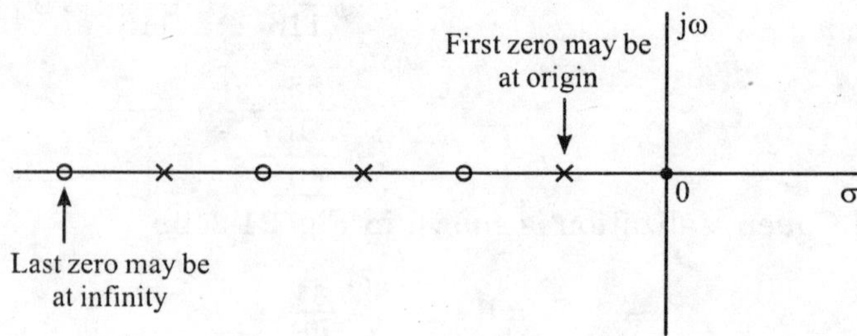

Fig. 21.25 Pole-zero diagram for a general *RC* impedance function $Z_{RC}(s)$.

The zeros and poles that are finite and non zero are called **internal critical frequencies**, whereas critical frequencies at the origin and infinity are known as **external critical frequencies**.

A function that satisfies the hypotheses of the above theorem either is a constant or is given in factored form by

$$Z_{RC}(s) = H\frac{(s+s_2)(s+s_4)\ldots(s+s_m)}{(s+s_1)(s+s_3)\ldots(s+s_n)} \quad (21.18.1)$$

where H and the s_i (except s_1) are positive numbers and

$$0 \le s_1 < s_2 < s_3 < s_4 \ldots,$$

that is, the poles and zeros are simple and alternate.

The subscripts m and n are finite, and either $m = n - 1$ or $m = n + 1$.

A pole at the origin is permitted, but a pole at infinity or a zero at the origin is **not** permitted. If $m > n$, the numerator and denominator are of equal degree and thus the number of finite zeros is equal to the number of poles. If $m < n$, there is a simple zero at infinity.

21.19 PROPERTIES OF RC DRIVING-POINT IMPEDANCE $Z_{RC}(s)$

A rational function is *RC* driving-point impedance $Z_{RC}(s)$ if and only if

1. All poles and zeros are simple and are located on the negative real axis or at the origin of the *s*-plane.
2. Poles and zeros alternate on the negative real axis.
3. The smallest critical frequency at the origin or nearest to the origini is a **pole**.
4. The highest critical frequency at infinity or nearest to infinity is a **zero**.
5. The residues of poles of $Z_{RC}(s)$ are real and positive.
6. The slope of the impedance curve is **negative** that is, $\dfrac{dZ(\sigma)}{d\sigma} \le 0$ for all σ.
7. The impedance at $s = 0$ is always greater than the impedance at $s = \infty$. That is,

$$Z(0) > Z(\infty).$$

A pole-zero diagram for a general *RC* impedance function is shown in Fig. 21.25.

The zeros and poles that are finite and non zero are called **internal** critical frequencies, whereas critical frequencies at the origin and infinity are designated as **external**.

The alternation of the poles and zeros is also expressed by stating that the poles and zeros are **interlaced** or that they **mutually separate each other**; this is frequently referred to as the **separation property** of the poles and zeros. A pole at the origin is permitted, but a pole at infinity or a zero at the origin is not.

21.20 PROPERTIES OF DRIVING-POINT ADMITTANCE FUNCTION $Y_{RC}(s)$

The driving-point admittance function of an *RC* network, that is, $Y_{RC}(s)$ has the following properties :

1. All poles and zeros are simple and are located on the negative real axis or at the origin of the *s*-plane.
2. Poles and zeros alternate on the negative real axis.
3. The smallest critical frequency at the origin or nearest to the origin a **zero**.
4. The highest critical frequency at infinity or nearest to infinity is a **pole**.
5. The residues of the poles of $Y_{RC}(s)$ are real and negative but the residues of $\dfrac{Y_{RC}(s)}{s}$ are real and positive.
6. The slope of the admittance curve is **positive**, that is

$$\frac{dY(\sigma)}{d\sigma} \ge 0 \text{ for all } \sigma$$

7. The admittance at $s = 0$ is always less than the admittance at $s = \infty$. That is, $Y(0) < Y(\infty)$.

Note that for synthesis of *RL* and *RC* networks

(a) expand $\dfrac{Z(s)}{s}$ instead of $Z(s)$ for *RL* driving-point impedance.

(b) expand $\dfrac{Y(s)}{s}$ instead of $Y(s)$ for *RC* driving-point admittance.

21.21 IMPEDANCE OF A SERIES RC NETWORK

(a) The impedance of a **series RC network** (Fig. 21.26a) is given by

$$Z_{RC}(s) = R + \frac{1}{sC} = \frac{sRC + 1}{sC}$$

$$= \frac{s + \dfrac{1}{RC}}{\dfrac{1}{R}s}$$

(b) The admittance of a **series RC network** is given by

$$Y_{RC}(s) = \frac{\dfrac{1}{R}s}{s + \dfrac{1}{RC}}$$

(c) The admittance of a **parallel** *RC* network (Fig. 21.26b) is

$$Y_{RC}(s) = \frac{1}{R} + sC$$

$$= \frac{1 + sRC}{R}$$

$$= \frac{s + \dfrac{1}{RC}}{\dfrac{1}{C}}$$

(d) The impedance of a **parallel** *RC* network is

$$Z_{RC}(s) = \frac{\dfrac{1}{C}}{s + \dfrac{1}{RC}}$$

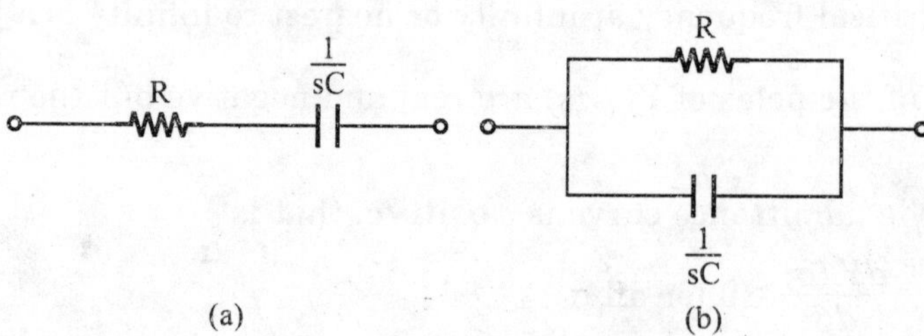

Fig. 21.26

21.22 TESTING OF RC DRIVING-POINT IMPEDANCE FUNCTIONS

In order to identify an *RC* impedance function we require first five properties of $Z_{RC}(s)$. If a given function fails to satisfy any of these properties, the given function is **not** an *RC* impedance function. If, however, a given function satisfies all these properties, the given function is an *RC* impedance function.

EXAMPLE 21.34 : Test whether the following functions represent the *RC* driving-point impedance functions :

(a) $Z_1(s) = \dfrac{(s+3)(s+7)}{(s+1)(s-4)}$

(b) $Z_2(s) = \dfrac{s(s+1)}{(s+2)(s+4)}$

(c) $Z_3(s) = \dfrac{(s+1)(s+5)}{(s+2)(s+3)}$

(d) $Z_4(s) = \dfrac{(s+2)(s+5)}{s+1}$

(e) $Z_5(s) = \dfrac{(s+1)(s+3)}{s(s+5)}$

(f) $Z_6(s) = \dfrac{(s+1)(s+7)(s+10)}{s(s+2)(s+9)}$

(g) $Z_7(s) = \dfrac{s(s+1)(s+6)}{(s+1)(s+4)}$

SOLUTION : (a) The poles and zeros of $Z_1(s)$ are :

Poles at $s = -1$ and $s = 4$

Zeros at $s = -3$ and $s = -7$

Since the pole at $s = 4$ does not lie on the negative real axis, $Z_1(s)$ is not an *RC* impedance function.

(b) The poles and zeros of $Z_2(s)$ are :

Poles at $s = -2$ and $s = -4$

Zeros at $s = 0$, $s = -1$

Since the poles and zeros do not alternate along the negative real axis, $Z_2(s)$ is not an *RC* impedance function.

(c) The poles and zeros of $Z_3(s)$ are :

Poles at $s = -2$, and $s = -3$

Zeros at $s = -1$, and $s = -5$

Since the singularity nearest to the origin is a zero, $Z_3(s)$ is not an *RC* impedance function.

(d) The poles and zeros of $Z_4(s)$ are :

Pole at $s = -1$

Zeros at $s = -2$ and $s = -5$

The poles and zeros do not alternate and that singularity at $s = \infty$ is a pole. Since $Z_4(s)$ violates properties 2 and 4, it is not an *RC* impedance function.

(e) The poles and zeros of $Z_5(s)$ are :

Poles at $s = 0$ and $s = -5$

Zeros at $s = -1$ and $s = -3$

Since the poles and zeros do not alternate, $Z_5(s)$ is not an *RC* impedance function.

(f) The singularity at the origin is a pole. Hence $Z_6(s)$ is not an *RC* impedance function.

(g) The function $Z_7(s)$ satisfies all the required condition and therefore, it represents an *RC* impedance function.

21.23 METHODS OF SYNTHESIS OF RC NETWORKS

RC networks can also be synthesized in four different ways namely

(a) First Foster form

(b) Second Foster form

(c) First Cauer form

(d) Second Cauer form

(a) First Foster Form of RC Networks

In order to obtain first Foster form, we expand $Z_{RC}(s)$ by partial fraction expansion. The general partial-fraction expansion is given by

$$Z_{RC}(s) = \frac{k_0}{s} + k_\infty + \sum_{i=1}^{n} \frac{k_i}{s+\sigma_i} \tag{21.23.1a}$$

or

$$Z_{RC}(s) = \frac{k_0}{s} + k_\infty + \frac{k_1}{s+\sigma_1} + \frac{k_2}{s+\sigma_2} + \ldots + \frac{k_n}{s+\sigma_n} \tag{21.23.1b}$$

where

$$k_0 = s\, Z_{RC}(s)\,\big|_{s=0}$$

$$k_\infty = \lim_{s\to\infty} Z_{RC}(s)$$

$$k_i = (s+\sigma_i)\, Z_{RC}(s)\,\big|_{s=-\sigma_i}$$

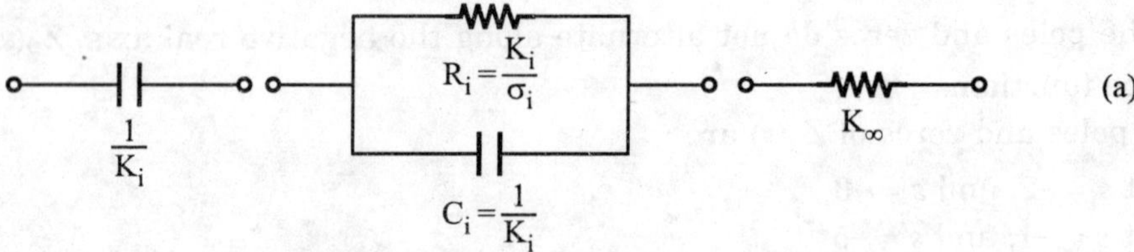

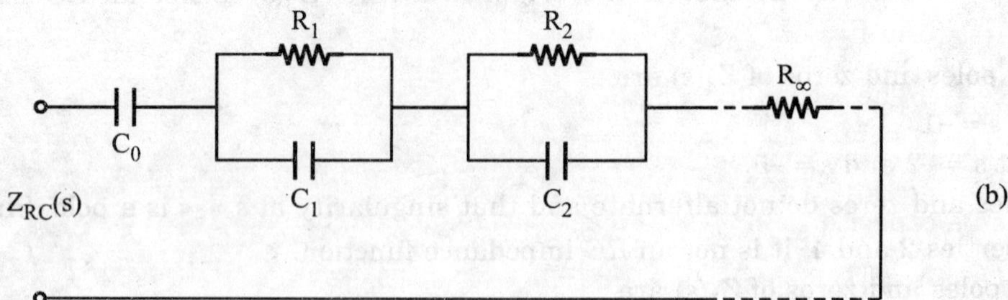

Fig. 21.27 (a) Building blocks of impedance $Z_{RC}(s)$, (b) First Foster form of *RC* network

The first term $\frac{k_0}{s}$ in Eq. (21.23.1) is the impedance of a capacitor having capacitance $\frac{1}{k}$. The second term k_∞, is the impedance of a resistor having the resistance k_∞. Each of the remaining terms represents the impedance of a parallel connection of a resistance R_i (equal to $\frac{k_i}{\sigma_i}$) and a capacitance $C_i = \frac{1}{k_i}$. These simple building blocks of $Z_{RC}(s)$ are shown in Fig. 21.27a. The final network is obtained by a series connection of all these impedances as shown in Fig. $N_1(b)$.

(b) Second Foster Form of RC Networks

It has been stated in the properties of *RC* admittance function that the residues at the poles of $Y_{RC}(s)$ are real and negative while the residues at the poles of $\frac{Y_{RC}(s)}{s}$ are real and positive. Therefore the proper function to expand to obtain second Foster form is $\frac{Y_{RC}(s)}{s}$. This expansion is of the form

$$\frac{Y_{RC}(s)}{s} = \frac{k'_0}{s} + k'_\infty + \sum_{i=1}^{n'} \frac{k'_i}{s+\sigma'_i} \qquad (21.23.2)$$

where,

$$k'_0 = Y_{RC}(s)\Big|_{s=0}$$

$$k'_\infty = \lim_{s\to\infty} \frac{Y_{RC}(s)}{s}$$

$$k'_i = \frac{Y_{RC}(s)}{s}(s+\sigma'_i)\Big|_{s=-\sigma'_i}$$

From Eq. (21.23.2)

$$Y_{RC}(s) = k'_0 + k'_\infty s + \sum_{i=1}^{n} \frac{k'_i s}{s+\sigma'_i} \qquad (21.23.3)$$

The first term in Eq. (21.23.3) is the admittance of a resistor having a conductance k'_0 or resistance $\frac{1}{k'_0}$. The second term is the admittance of a capacitor having capacitance k'_∞ and each subsequent term represents the admittance of a series, connection of a resistance $R_i\left(=\frac{1}{k'_i}\right)$ and a capacitance $C_i\left(=\frac{k'_i}{\sigma'_i}\right)$. These elemental admittances are shown in Fig. 21.28a. The final network is realized by a parallel connection of these elemental admittances as shown in Fig. 21.28b.

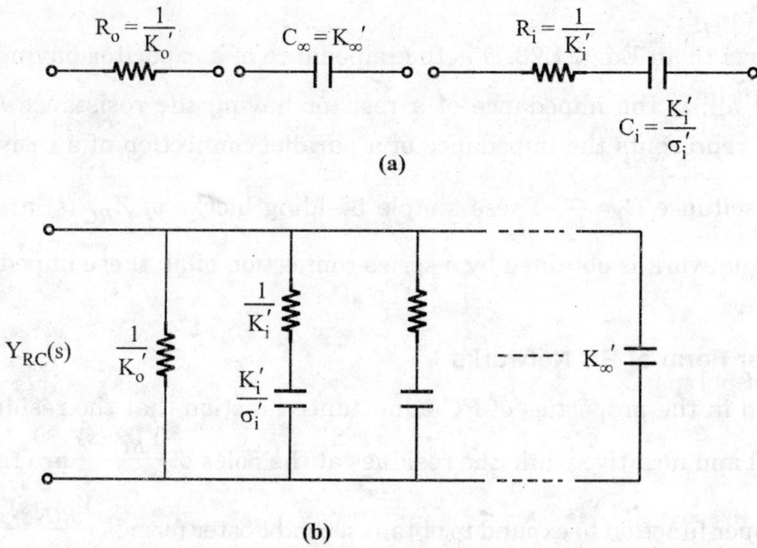

Fig. 21.28 (a) Elemental admittances of $Y_{RC}(s)$ (b) Second Foster form of *RC* admittance

(c) First Cauer form of RC Networks

The first Cauer form of *RC* driving-point impedance function is obtained by expanding it in continued fraction about **infinity**. For this purpose, we arrange the numerator and denominator polynomials in **descending powers** of *s* and then find the continued fraction expansion. The quotients of the continued-fraction expansion give the elements of a ladder network.

The first Cauer form of *RC* impedance is shown in Fig. 21.29a.

(d) Second Cauer Form of RC Networks

The second Caver form is obtained by expanding the *RC* impedance function in continued fraction about the **origin**. For this purpose, we arrange the numerator and denominator polynomials in **ascending powers** of *s*, and then find the continued-fraction expansion. The quotients of the expansion represent the elements of a ladder network. The second Cauer form of *RC* impedance function has capacitors in series branches and resistors are in shunt branches as shown in Fig. 21.29.

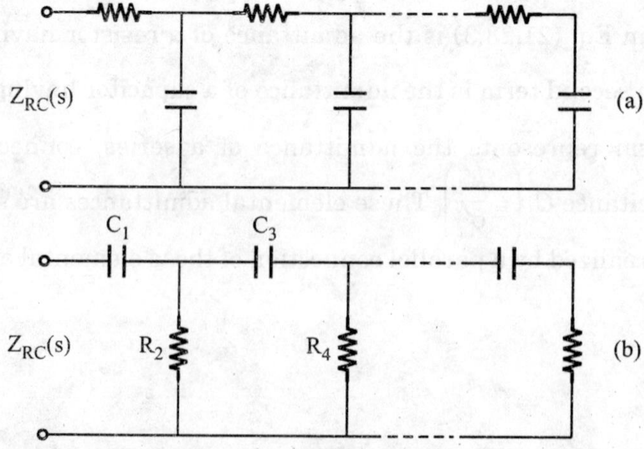

Fig. 21.29 (a) First Cauer form of *RC* impedance (b) Second Caver form of *RC* impedance

In the continued fraction expansion for the two Cauer forms of RC networks (as in case of LC networks), if the given function $F(s)$ represents an impedance $Z_{RC}(s)$, then the first quoteint is also an impedance. Similarly, if the given function $F(s)$ represents an admittance $Y_{RC}(s)$, then the first quotient is also an admittance.

It is to be noted that in each of the four networks (two Foster form and two Cauer forms) for RC impedance of RL admittance) functions, the number of elements in the same and is equal to the sum of the degree of the numerator and the degree of the denominator polynomials in $F(s)$. Therefore if

$$F(s) = \frac{P(s)}{Q(s)}$$

and m = degree of polynomial $P(s)$

n = degree of polynomial $Q(s)$

then the numbers of elements in the network = $m + n$

Important Note

An RC impedance, $Z_{RC}(s)$, can also be realized as an RL admittance, $Y_{RL}(s)$. All the properties of RL admittances are the same as the properties of RC impedance. It is, therefore, important to specify whether a function is to be realized as an RC impedance or an RL admittance.

EXAMPLE 21.35 : Show that $\dfrac{s^2 + 7s + 6}{s + 2}$ is an RC admittance. Synthesize it in first Cauer form and second Foster form.

SOLUTION : (a) $Y(s) = \dfrac{s^2 + 7s + 6}{s + 2}$

The continued-fraction expansion with the numerator and denominator polynomials arranged in descending powers of s is

$$s + 2 \,)\, s^2 + 7s + 6 \,(\, s \rightarrow Y_1(s) = sC_1,\ C_1 = 1F$$

$$\underline{s^2 + 2s}$$

$$5s + 6 \,)\, s + 2 \,(\, \frac{1}{5} \rightarrow Z_2(s) = R_2,\ R_2 = \frac{1}{5}\,\Omega$$

$$\underline{s + \frac{6}{5}}$$

$$\frac{4}{5} \,)\, 5s + 6 \,(\, \frac{25s}{4} \rightarrow Y_3(s) = sC_3,\ C_3 = \frac{25}{4}\,F$$

$$\underline{5s}$$

$$6 \,)\, \frac{4}{5} \,(\, \frac{2}{15} \rightarrow Z_4(s) = R_4,\ R_4 = \frac{2}{15}\,\Omega$$

$$\underline{\frac{4}{5}}$$

$$\underline{0}$$

It is seen that quotient, $s, \frac{1}{5}, \frac{25s}{4}, \frac{2}{15}$ are all positive. Hence the given function in an *RC* admittance function. The first form of Cauer realization is shown in Fig. 21.30a

(b) Second Foster Realization

$$Y(s) = \frac{s^2 + 7s + 6}{s + 2}$$

$$\frac{Y(s)}{s} = \frac{s^2 + 7s + 6}{s(s + 2)} = 1 + \frac{5s + 6}{s(s + 2)} = 1 + \frac{3}{s} + \frac{2}{s + 2}$$

$$Y(s) = s + 3 + \frac{2s}{s + 2} = s + 3 + \frac{1}{\frac{1}{2} + \frac{1}{s}}$$

$$= Y_1(s) + Y_2(s) + Y_3(s)$$

$$Y_1(s) = s = sC_1,\ C_1 = 1F$$

$$Y_2(s) = 3 = \frac{1}{R_2},\ R_2 = \frac{1}{3}\ \Omega$$

$$Y_3(s) = \frac{1}{Z_3(s)}$$

$$Z_3(s) = \frac{1}{2} + \frac{1}{s} = R_3 + \frac{1}{sC_3}$$

$$R_3 = \frac{1}{2}\ \Omega,\ C_3 = 1F$$

Thus, $Y(s)$ is a parallel combination of $Y_1(s)$, $Y_2(s)$ and $Y_3(s)$. That is, the second form of Foster network is a parallel combination of C_1, R_2 and (series combination of R_3 and C_3) as shown in Fig. 21.30b.

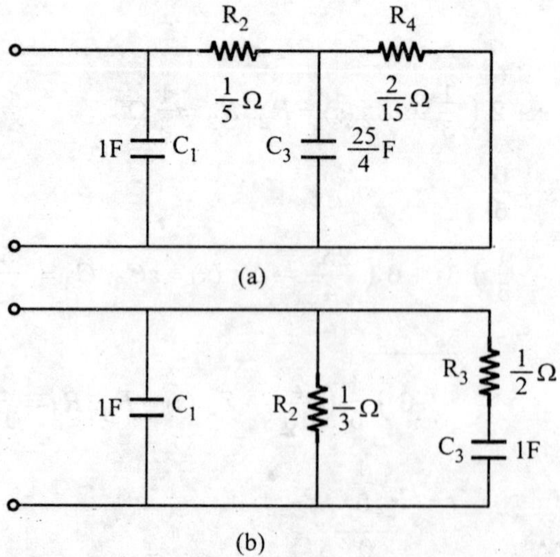

Fig. 21.30 (a) First Cauer realization and (b) Second Foster realization of $Y(s) = \frac{s^2 + 7s + 6}{s + 2}$

■ **EXAMPLE 21.36 :** Synthesize the RC driving-point impedance

$$Z_{RC}(s) = \frac{(s+1)(s+3)}{s(s+2)}$$

to get the Cauer first and second forms of network realizations.

SOLUTION : $Z_{RC}(s) = \dfrac{(s+1)(s+3)}{s(s+2)} = \dfrac{s^2+4s+3}{s^2+2s}$

(a) First Cauer Form

The continued fraction expansion of $Z_{RC}(s)$ is given by

$$s^2+2s\ \big)\ s^2+4s+3\ \big(\ 1 \rightarrow Z_1(s),\ R_1 = 1\ \Omega$$

$$\underline{s^2+2s}$$

$$2s+3\ \big)\ s^2+2s\ \big(\ \frac{1}{2}s \rightarrow Y_2(s) = sC_2,\ C_2 = \frac{1}{2}F$$

$$\underline{s^2+\frac{3}{2}s}$$

$$\frac{s}{2}\ \big)\ 2s+3\ \big(\ 4 \rightarrow Z_3(s) = R_3\ ,\ R_3 = 4\ \Omega$$

$$\underline{2s}$$

$$3\ \big)\ \frac{s}{2}\ \big(\ \frac{s}{6} \rightarrow Y_4(s) = sC_4,\ C_4 = \frac{1}{6}F$$

$$\underline{\frac{s}{2}}$$

$$\underline{0}$$

Therefore the first form of the Cauer network is shown in Fig. 21.31a.

(b) Second Cauer Form

The numerator and denominator polynomials are arranged in **ascending powers** of s.

$$Z_{RC}(s) = \frac{3+4s+s^2}{2s+s^2}$$

The continued fraction expansion is

$$2s+s^2\ \big)\ 3+4s+s^2\ \big(\ \frac{3}{2s} \rightarrow Z_1(s) = \frac{1}{sC_1},\ C_1 = \frac{2}{3}F$$

$$\underline{3+\frac{3}{2}s}$$

$$\frac{5}{2}s+s^2\ \big)\ 2s+s^2\ \big(\ \frac{4}{5} \rightarrow Y_2(s) = \frac{1}{R_2},\ R_2 = \frac{5}{4}\ \Omega$$

$$\begin{array}{l} 2s + \frac{4}{5}s^2 \\ \hline \frac{s^2}{5} \Big) \frac{5}{2}s + s^2 \Big(\frac{25}{2s} \rightarrow Z_3(s) = \frac{1}{sC_3},\ C_3 = \frac{2}{25}F \\ \quad \frac{5}{2}s \\ \hline \quad s^2 \Big) \frac{s^2}{5} \Big(\frac{1}{5} \rightarrow Y_4(s) = \frac{1}{R_4},\ R_4 = 5\ \Omega \\ \quad\quad \frac{s^2}{5} \\ \hline \quad\quad 0 \end{array}$$

The synthesized network in the second Cauer form is shown in Fig. 21.31b.

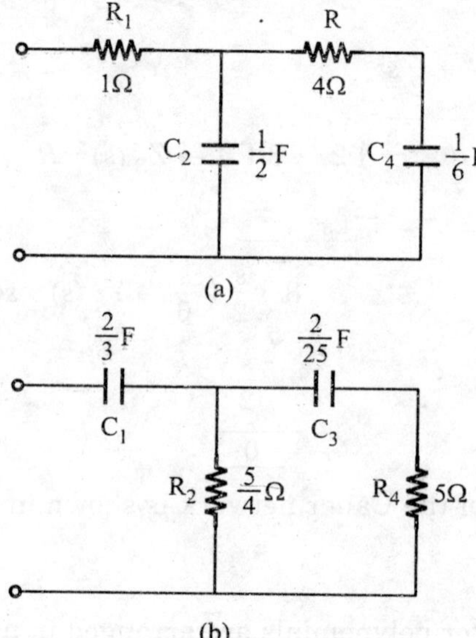

Fig. 21.31(a) First Cauer form realization (b) Second Cauer form realization of $Z_{RC}(s) = \dfrac{(s+1)(s+3)}{s(s+3)}$

■ **EXAMPLE 21.37 :** Find the Foster's first and second forms and Cauer's first and second forms of the RC driving-point impedance

$$Z_{RC}(s) = \frac{(s+2)(s+5)}{(s+1)(s+3)}$$

SOLUTION :

(a) First Foster Network

$$Z_{RC}(s) = \frac{(s+2)(s+5)}{(s+1)(s+3)}$$

$$= \frac{s^2 + 7s + 10}{s^2 + 4s + 3} = \frac{(s^2 + 4s + 3) + (3s + 7)}{s^2 + 4s + 3}$$

$$= 1 + \frac{3s + 7}{s^2 + 4s + 3} = 1 + \frac{3s + 7}{(s + 1)(s + 3)}$$

The partial-fraction expansion of $Z_{RC}(s)$ is given by

$$Z_{RC}(s) = 1 + \frac{k_1}{s + 1} + \frac{k_2}{s + 3}$$

$$k_1 = \frac{3s + 7}{s + 3}\Big|_{s=-1} = \frac{-3 + 7}{-1 + 3} = 2$$

$$k_2 = \frac{3s + 7}{s + 1}\Big|_{s=-3} = \frac{3(-3) + 7}{-3 + 1} = 1$$

$$Z_{RC}(s) = 1 + \frac{2}{s + 1} + \frac{1}{s + 3}$$

$$Z_{RC}(s) = 1 + \frac{1}{\frac{1}{2}s + \frac{1}{2}} + \frac{1}{s + \frac{1}{\frac{1}{3}}}$$

Each term is realized directly. The first term is a 1 Ω resistor. The second term represents a capacitor of $\frac{1}{2}F$ and a resistor of 2 Ω connected in parallel. The third term represents a parallel combination of 1*F* capacitance and $\frac{1}{3}$ Ω resistance. These partial networks are connected in series to obtain the resulting network shown in Fig. 21.32a.

(b) Second Foster Network

In order to obtain second Foster realization, we find the reciprocal of $Z_{RC}(s)$ to obtain the corresponding admittance function $Y_{RC}(s)$.

$$Y_{RC}(s) = \frac{1}{Z_{RC}(s)} = \frac{(s + 1)(s + 3)}{(s + 2)(s + 5)}$$

$$\frac{Y_{RC}(s)}{s} = \frac{(s + 1)(s + 3)}{s(s + 2)(s + 5)}$$

The partial-fraction expansion of $\frac{Y_{RC}(s)}{s}$ is given by

$$\frac{Y_{RC}(s)}{s} = \frac{k_1}{s} + \frac{k_2}{s + 2} + \frac{k_3}{s + 5}$$

$$k_1 = \frac{(s + 1)(s + 3)}{(s + 2)(s + 5)}\Big|_{s=0} = \frac{1 \times 3}{2 \times 5} = \frac{3}{10}$$

$$k_2 = \frac{(s + 1)(s + 3)}{s(s + 5)}\Big|_{s=-2} = \frac{(-2 + 1)(-2 + 3)}{(-2)(-2 + 5)} = \frac{1}{6}$$

$$k_3 = \frac{(s+1)(s+3)}{s(s+2)}\Big|_{s=-5} = \frac{(-5+1)(-5+3)}{(-5)(-5+2)} = \frac{8}{15}$$

therefore, $$\frac{Y_{RC}(s)}{s} = \frac{\frac{3}{10}}{s} + \frac{\frac{1}{6}}{s+2} + \frac{\frac{8}{15}}{s+5}$$

$$Y_{RC}(s) = \frac{3}{10} + \frac{\frac{1}{6}s}{s+2} + \frac{\frac{8}{15}s}{s+5} = \frac{1}{\frac{10}{3}} + \frac{1}{\frac{s}{\frac{1}{6}s} + \frac{2}{\frac{1}{6}s}} + \frac{1}{\frac{s}{\frac{8}{15}s} + \frac{5}{\frac{8}{15}s}}$$

$$= \frac{1}{\frac{10}{3}} + \frac{1}{6 + \frac{1}{12s}} + \frac{1}{\frac{15}{8} + \frac{1}{\frac{8}{75}s}}$$

The resulting network is realized as shown in Fig. 21.32b.

(c) First Cauer Network

To obtain the first Cauer form, we arrange the numerator and denominator polynomials in **descending powers** of s and then find the continued fraction expansion.

$$Z_{RC}(s) = \frac{(s+2)(s+5)}{(s+1)(s+3)} = \frac{s^2+7s+10}{s^2+4s+3}$$

The continued-fraction expansion is

$s^2 + 4s + 3 \;)\; s^2 + 7s + 10 \;(\; 1 \rightarrow Z_1(s) = R_1,\ R_1 = 1\ \Omega$

$\quad s^2 + 4s + 3$

$\quad\quad 3s + 7 \;)\; s^2 + 4s + 3 \;(\; \frac{s}{3} \rightarrow Y_2(s) = sC_2,\ C_2 = \frac{1}{3}F$

$\quad\quad s^2 + \frac{7}{3}s$

$\quad\quad\quad \frac{5s}{3} + 3 \;)\; 3s + 7 \;(\; \frac{9}{5} \rightarrow Z_3(s) = R_3,\ R_3 = \frac{9}{5}\ \Omega$

$\quad\quad\quad 3s + \frac{27}{5}$

$\quad\quad\quad\quad \frac{8}{5} \;)\; \frac{5s}{3} + 3 \;(\; \frac{25}{24}s \rightarrow Y_4(s) = sC_4,\ C_4 = \frac{25}{24}F$

$\quad\quad\quad\quad \frac{5s}{3}$

$\quad\quad\quad\quad\quad 3 \;)\; \frac{8}{5} \;(\; \frac{8}{15} \rightarrow Z_4(s) = R_5,\ R_5 = \frac{8}{15}\ \Omega$

$\quad\quad\quad\quad\quad \frac{8}{5}$

$\quad\quad\quad\quad\quad 0$

Therefore, the synthesized network is shown in Fig. 21.32c

(d) Second Cauer Network

The numerator and denominator polynomials are arranged in **ascending powers** of s.

$$Z_{RC}(s) = \frac{10 + 7s + s^2}{3 + 4s + s^2}$$

or

$$Y_{RC}(s) = \frac{3 + 4s + s^2}{10 + 7s + s^2}$$

The continued-fraction expansion of $Y_{RC}(s)$ is as follows :

$$10 + 7s + s^2 \,\big)\, 3 + 4s + s^2 \,\big(\, \frac{3}{10} \rightarrow Y_1(s) = \frac{1}{R_1},\; R_1 = \frac{10}{3}\,\Omega$$

$$\underline{3 + \frac{21}{10}s + \frac{3}{10}s^2}$$

$$\frac{19}{10}s + \frac{7}{10}s^2 \,\big)\, 10 + 7s + s^2 \,\big(\, \frac{100}{19s} \rightarrow Z_2(s) = \frac{1}{sC_2},\; C_2 = \frac{19}{100}\,F$$

$$\underline{10 + \frac{70}{19}s}$$

$$\frac{63}{19}s + s^2 \,\big)\, \frac{19}{10}s + \frac{7}{10}s^2 \,\big(\, \frac{361}{630} \rightarrow Y_3(s) = \frac{1}{R_3},\; R_3 = \frac{630}{361}\,\Omega$$

$$\underline{\frac{19}{10}s + \frac{361}{630}s^2}$$

$$\frac{8}{63}s^2 \,\big)\, \frac{63}{19}s + s^2 \,\big(\, \frac{3969}{152s} \rightarrow Z_4(s) = \frac{1}{sC_4},\; C_4 = \frac{152}{3969}\,F$$

$$\underline{\frac{63}{19}s}$$

$$s^2 \,\big)\, \frac{8}{63}s^2 \,\big(\, \frac{8}{63} \rightarrow Y_5(s) = \frac{1}{R_5},\; R_5 = \frac{63}{8}\,\Omega$$

$$\underline{\frac{8}{63}s^2}$$

$$0$$

The resulting network is shown in Fig. 21.32d.

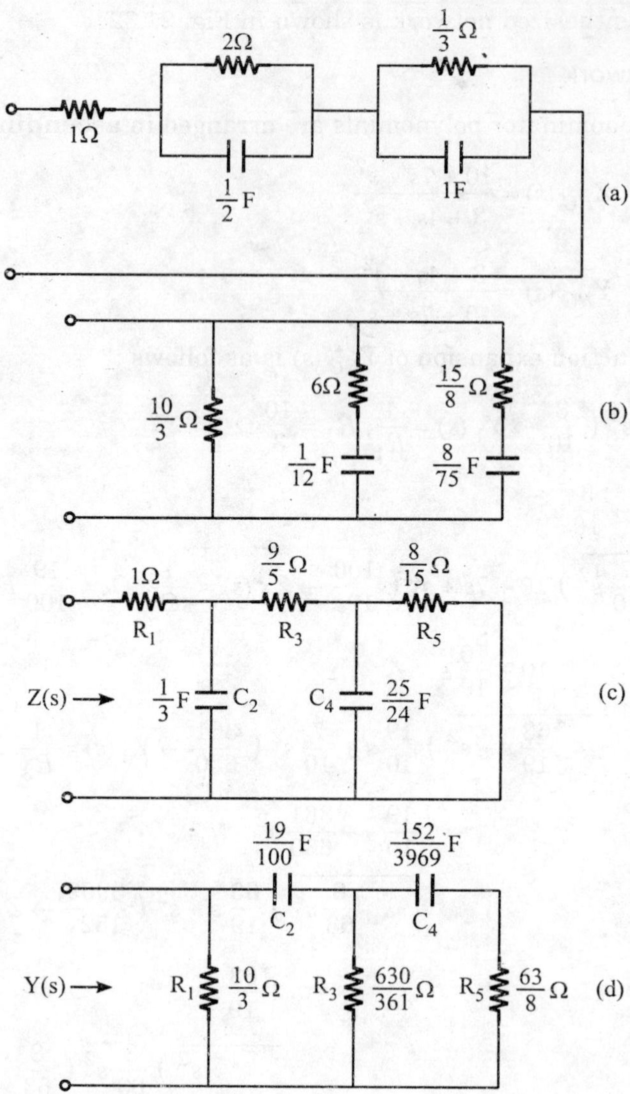

Fig. 21.32 (a) First Foster form realization, (b) Second Foster form realization, (c) First Cauer form realization, (d) Second Cauer form realization of $Z_{RC}(s) = \frac{(s+2)(s+5)}{(s+1)(s+3)}$

EXAMPLE 21.38 : An impedance function is given by

$$Z(s) = \frac{(s+1)(s+4)}{s(s+2)(s+5)}$$

Find the *RC* representation of

(a) First Foster form

(b) Second Foster form

(c) First Cauer form

(d) Second Cauer form

SOLUTION :

(a) First Foster Form

The partial-fraction expansion of $Z(s)$ is given by

$$Z(s) = \frac{(s+1)(s+4)}{s(s+2)(s+5)} = \frac{k_1}{s} + \frac{k_2}{s+2} + \frac{k_3}{s+5}$$

$$k_1 = sZ(s)\,|_{s=0} = \frac{1 \times 4}{2 \times 5} = \frac{2}{5}$$

$$k_2 = (s+2)\,Z(s)\,|_{s=-2} = \frac{(-2+1)(-2+4)}{-2(-2+5)} = \frac{1}{3}$$

$$k_3 = (s+5)\,Z(s)\,|_{s=-5} = \frac{(-5+1)(-5+4)}{-5(-5+2)} = \frac{4}{15}$$

therefore, $$Z(s) = \frac{\frac{2}{5}}{s} + \frac{\frac{1}{3}}{s+2} + \frac{\frac{4}{15}}{s+5}$$

$$Z(s) = \frac{1}{\frac{5}{2}s} + \frac{1}{3s + \frac{1}{\frac{1}{6}}} + \frac{1}{\frac{15}{4}s + \frac{1}{\frac{4}{75}}}$$

The resulting network is shown in Fig. 21.33a.

(b) Second Foster Form

$$Y(s) = \frac{1}{Z(s)} = \frac{s(s+2)(s+5)}{(s+1)(s+4)}$$

$$\frac{Y(s)}{s} = \frac{(s+2)(s+5)}{(s+1)(s+4)} = \frac{s^2+7s+10}{s^2+5s+4}$$

$$s^2+5s+4\ \big)\ s^2+7s+10\ \big(\ 1$$
$$\underline{s^2+5s+4}$$
$$2s+6$$

therefore, $$\frac{Y(s)}{s} = 1 + \frac{2s+6}{s^2+5s+4}$$

$$= 1 + \frac{2s+6}{(s+1)(s+4)}$$

By partial-fraction expansion

$$\frac{2s+6}{(s+1)(s+4)} = \frac{k_1}{s+1} + \frac{k_2}{s+4}$$

therefore, $$k_1 = \frac{2s+6}{s+4}\,|_{s=-1} = \frac{-2+6}{-1+4} = \frac{4}{3}$$

$$k_2 = \frac{2s+6}{s+1}\Big|_{s=-4} = \frac{-8+6}{-4+1} = \frac{2}{3}$$

therefore,

$$\frac{Y(s)}{s} = 1 + \frac{\frac{4}{3}}{s+1} + \frac{\frac{2}{3}}{s+4}$$

$$Y(s) = s + \frac{\frac{4}{3}s}{s+1} + \frac{\frac{2}{3}s}{s+4}$$

$$= s + \frac{1}{\frac{3}{4} + \frac{1}{\frac{4}{3}s}} + \frac{1}{\frac{3}{2} + \frac{1}{\frac{1}{6}s}}$$

The resulting network is shown in Fig. 21.33b.

(c) First Cauer Form

$$Z(s) = \frac{(s+1)(s+4)}{s(s+2)(s+5)}$$

$$= \frac{s^2+5s+4}{s^3+7s^2+10s}$$

$$Y(s) = \frac{1}{Z(s)} = \frac{s^3+7s^2+10s}{s^2+5s+4}$$

The continued-fraction expansion is

$$s^2+5s+4 \;\big)\; s^3+7s^2+10s \;\big(\; s \rightarrow Y_1(s) = sC_1,\ C_1 = 1F$$
$$\underline{s^3+5s^2+4s}$$

$$2s^2+6s \;\big)\; s^2+5s+4 \;\big(\; \frac{1}{2} \rightarrow Z_2(s) = R_2,\ R_2 = \frac{1}{2}\,\Omega$$
$$\underline{s^2+3s}$$

$$2s+4 \;\big)\; 2s^2+6s \;\big(\; s \rightarrow Y_3(s) = sC_3,\ C_1 = 1F$$
$$\underline{2s^2+4s}$$

$$2s \;\big)\; 2s+4 \;\big(\; 1 \rightarrow Z_4(s) = R_4,\ R_4 = 1\Omega$$
$$\underline{2s}$$

$$4 \;\big)\; 2s \;\big(\; \frac{1}{2}s \rightarrow Y_5(s) = sC_5,\ C_5 = \frac{1}{2}F$$
$$\underline{2s}$$
$$0$$

The synthesized network is shown in Fig. 21.33c.

(d) Second Cauer Form

$$Z(s)=\frac{s^2+5s+4}{s^3+7s^2+10s}$$

The numerator and denominator are arranged in **ascending powers** of s.

$$Z(s)=\frac{4+5s+s^2}{10s+7s^2+s^3}$$

The continued-fraction expansion is given by

$$10s+7s^2+s^3\ \big)\ 4+5s+s^2\ \Big(\ \frac{4}{10s}=\frac{2}{5s}\rightarrow Z_1(s)=\frac{1}{sC_1},\ C_1=\frac{5}{2}\,F$$

$$4+\frac{14}{5}s+\frac{2}{5}s^2$$

$$\frac{11}{5}s+\frac{3}{5}s^2\ \big)\ 10s+7s^2+s^3\ \Big(\ \frac{50}{11}\rightarrow Y_2(s)=\frac{1}{R_2},\ R_2=\frac{11}{50}\,\Omega$$

$$10s+\frac{30}{11}s$$

$$\frac{47}{11}s^2+s^3\ \big)\ \frac{11}{5}s+\frac{3}{5}s^2\ \Big(\ \frac{121}{235}s\rightarrow Z_3(s)=\frac{1}{sC_3},\ C_3=\frac{235}{121}\,F$$

$$\frac{11}{5}s+\frac{121}{235}s^2$$

$$\frac{20}{235}s^2\ \big)\ \frac{47}{11}s^2+s^3\ \Big(\ \frac{2209}{44}\rightarrow Y_4(s),=\frac{1}{R_4},\ R_4=\frac{44}{2209}\,\Omega$$

$$\frac{47}{11}s^2$$

$$s^3\ \big)\ \frac{20}{235}s^2\ \Big(\ \frac{235}{20s}\rightarrow Z_5(s),=\frac{1}{sC_5},\ C_5=\frac{20}{235}=\frac{4}{47}\,F$$

$$\frac{20}{235}s^2$$

$$0$$

The synthesized network is shown in Fig. 21.33 d.

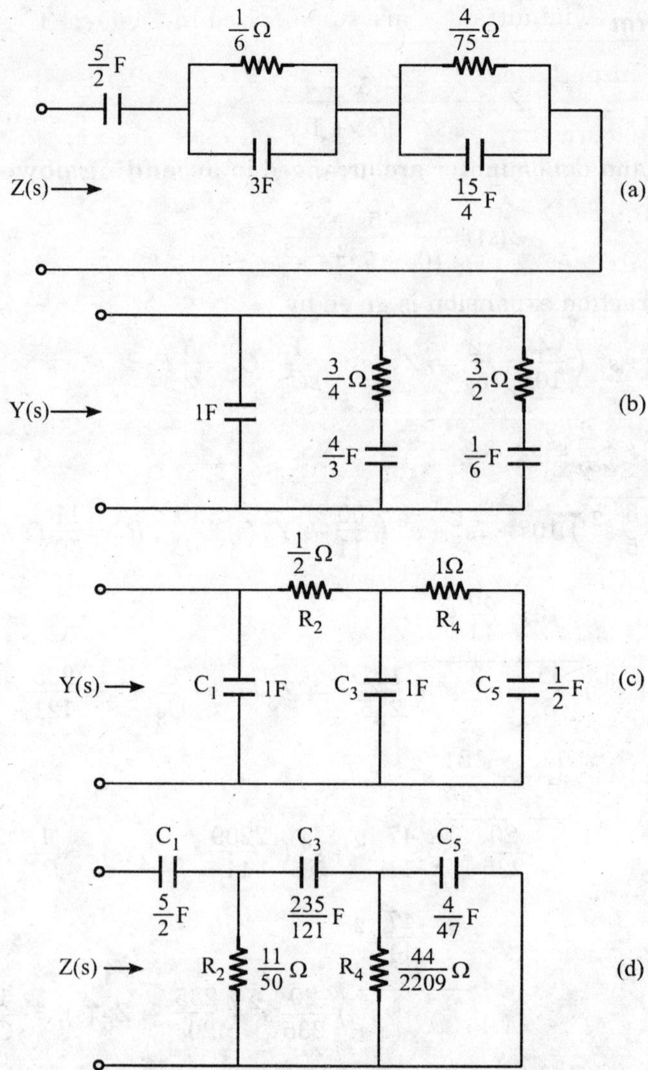

Fig. 21.33 (a) First Foster Form, (b) Second Foster form, (c) First Cauer form, (d) Second Cauer form of the function $Z(s) = \dfrac{(s+1)(s+4)}{s(s+2)(s+5)}$

21.24 RL DRIVING-POINT IMMITTANCES

RL networks are of much less importance than *LC* or *RC* networks. They have little practical importance. The properties of an *RL* impedance are the same as those of *RC* admittance and the properties of an *RL* admittance are the same as those of an *RC* impedance. Accordingly, the *RL* immittances have the partial fraction expansions

$$Z_{RL}(s) = K_0 + K_\infty s + \sum_{i=1}^{n} \frac{K_i s}{s + \sigma_i} \tag{21.24.1}$$

and

$$Y_{RL}(s) = \frac{K'_0}{s} + K'_\infty + \sum_{i=1}^{n'} \frac{K'_i}{s + \sigma'_i} \tag{21.24.2}$$

where all constants are positive.

The properties of *RL* immittances are summarized in Table 21.1.

***RL* Impedance**	***RL* admittance**
1. All poles are zeros are simple and lie on the $-\sigma$ (real negative) axis of the s plane	1. All poles are zeros are simple and lie on the $-\sigma$ (real negative) axis of the s plane.
2. Poles are zeros alternate on $-\sigma$ axis.	2. Poles are zero alternate on $-\sigma$ axis.
3. The first critical frequency on the $-\sigma$ axis is a **zero**, it may be at $s = 0$.	3. The first critical frequency on the $-\sigma$ axis is a **pole**, it may be $s = 0$.
4. The last critical frequency is a **pole**, which may be at infinity.	4. The last critical frequency is a **zero**, which may be at infinity.
5. The residues at the poles of $Z(s)$ are real and negative; the residues of $\frac{Z(s)}{s}$ are real and positive.	5. The residues at the poles of $Y(s)$ are real and positive.

IMPORTANT NOTE

An *RL* impedance, $Z(s)$, can also be realized as an *RC* admittances, $Y_{RC}(s)$. All the properties of *RC* admittance are the same as the properties of *RL* impedances. It is, therefore, important to specify whether a function is to be realized as an *RL* impedance or an *RC* admittance.

■ **EXAMPLE 21.39 :** Synthesize the *RL* driving-point impedance

$$Z_{RL}(s) = \frac{2(s+1)(s+3)}{(s+2)(s+6)}$$

SOLUTION : It is to be noted that in the synthesis of *RL* driving-point impedance we expand $\frac{Z_{RL}(s)}{s}$ by partial fraction method.

$$\frac{Z_{RL}(s)}{s} = \frac{2(s+1)(s+3)}{s(s+2)(s+6)}$$

$$= \frac{A_1}{s} + \frac{A_2}{s+2} + \frac{A_3}{s+6}$$

$$A_1 = \frac{2(s+1)(s+3)}{(s+2)(s+6)}\Big|_{s=0} = \frac{2(1)(3)}{2(6)} = \frac{1}{2}$$

$$A_2 = \frac{2(s+1)(s+3)}{s(s+6)}\Big|_{s=-2} = \frac{2(-2+1)(-2+3)}{(-2)(-2+6)} = \frac{1}{4}$$

$$A_3 = \frac{2(s+1)(s+3)}{s(s+2)}\Big|_{s=-6} = \frac{2(-6+1)(-6+3)}{(-6)(-6+2)} = \frac{5}{4}$$

therefore, $$\frac{Z_{RL}(s)}{s} = \frac{\frac{1}{2}}{s} + \frac{\frac{1}{4}}{s+2} + \frac{\frac{5}{4}}{s+6}$$

$$Z_{RL}(s) = \frac{1}{2} + \frac{\frac{s}{4}}{s+2} + \frac{\frac{5}{4}s}{s+6}$$

$$= \frac{1}{2} + \frac{1}{\frac{s}{\frac{s}{4}} + \frac{2}{\frac{s}{4}}} + \frac{1}{\frac{s}{\frac{5}{4}s} + \frac{6}{\frac{5}{4}s}}$$

$$= \frac{1}{2} + \frac{1}{\frac{1}{\frac{1}{4}} + \frac{1}{\frac{1}{8}s}} + \frac{1}{\frac{1}{\frac{5}{4}} + \frac{1}{\frac{5}{24}s}}$$

$$= R + \frac{1}{\frac{1}{R_1} + \frac{1}{sL_1}} + \frac{1}{\frac{1}{R_2} + \frac{1}{sL_2}} \text{ (say)}$$

therefore, $R = \frac{1}{2}\Omega,\ R_1 = \frac{1}{4}\ \Omega,\ L_1 = \frac{1}{8}\ H,\ R_2 = \frac{5}{4}\ \Omega,\ L_2 = \frac{5}{24}\ H$

Thus, the network synthesized in first Foster form is shown in Fig. 21.34.

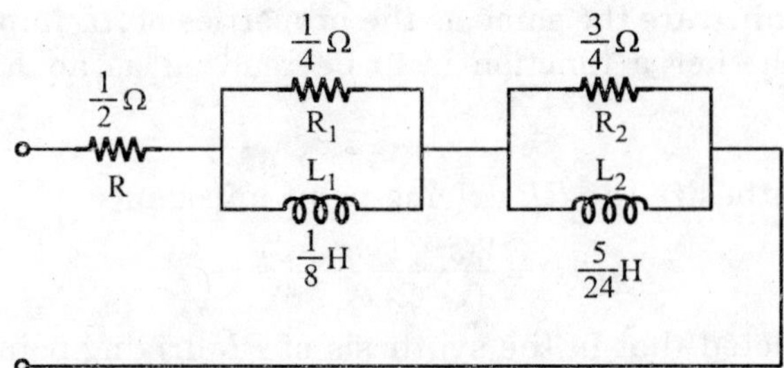

Fig. 21.34 First Foster form realization of $Z_{RL}(s) = \frac{2(s+1)(s+3)}{(s+2)(s+6)}$

■ **EXAMPLE 21.40 :** Show that $\frac{s(3s+8)}{(s+1)(s+3)}$ represents an *RL* immittance. Realize the impedance in Foster form.

SOLUTION :

$$F(s) = \frac{s(3s+8)}{(s+1)(s+3)} = \frac{3s(s+8/3)}{(s+1)(s+3)}$$

The given function has zeros at $s = 0$ and $s = -\frac{8}{3}$.

It has pole. at $s = -1$ and $s = -3$. The smallest critical frequency is at the origin. The poles and zeros alternate on the negative real axis. Hence $F(s)$ is an *RC* admittance or an *RL* impedance.

therefore, $F(s) = Z_{RL} = Y_{RC} = \frac{s(3s+8)}{(s+1)(s+3)} = \frac{3s\left(s+\frac{8}{3}\right)}{(s+1)(s+3)}$

$$\frac{Y_{RC}}{s} = \frac{3\left(s+\frac{8}{3}\right)}{(s+1)(s+3)}$$

$$= \frac{\frac{5}{2}}{s+1} + \frac{\frac{1}{2}}{s+3}$$

$$Y_{RC} = \frac{\frac{5}{2}s}{s+1} + \frac{\frac{1}{2}s}{s+3} = Z_{RL}$$

$$Z_{RL} = \frac{1}{\frac{2}{5}+\frac{2}{5s}} + \frac{1}{2+\frac{6}{s}} = Z_1(s) + Z_2(s)$$

$$Y_1(s) = \frac{1}{Z_1(s)} = \frac{2}{5} + \frac{2}{5s} = \frac{1}{R_1} + \frac{1}{sL_1}$$

$$Y_2(s) = \frac{1}{Z_2(s)} = 2 + \frac{6}{s} = \frac{1}{R_2} + \frac{1}{sL_2}$$

Thus, $Z_1(s)$ is a parallel combination of R_1 and L_1.

$$R_1 = \frac{5}{2}\ \Omega,\ L_1 = \frac{5}{2}\ H$$

$Z_2(s)$ is a parallel combination of R_2 and L_2.

$$R_2 = \frac{1}{2}\ \Omega,\ L_2 = \frac{1}{6}\ H.$$

$Z_1(s)$ and $Z_2(s)$ are in series. Therefore, the Foster form of realization is shown in Fig. 21.35.

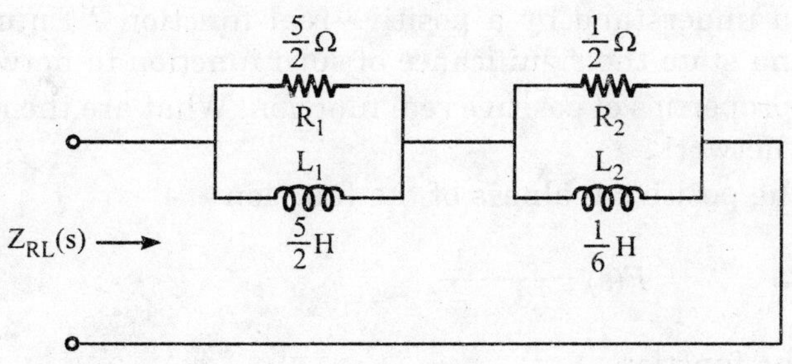

Fig. 21.35 Foster realization of $Z_{RL} = \frac{s(3s+8)}{(s+1)(s+3)}$

EXERCISES

1. Show that the function

$$Z(s) = \frac{2s^2 + s + 4}{(s^2 + 1)(s + 2)}$$

is not physically realizable with a passive network.

2. Show that the driving-point admittance function

$$Y(s) = \frac{6s^2 + 5s + 6}{2s^2 + 4s + 4}$$

is physically realizable with a passive network.

3. Show that the driving-point impedance function

$$Z(s) = \frac{3s^2 + 2s + 2}{(s^2 + 1)(s + 2)}$$

is physically realizable with a passive network.

4. Test whether the following polynomials are Hurwitz :

(a) $s^4 + 3s^3 + 2s + 1$

(b) $s^4 - 3s^3 + 2s^2 + s + 1$

(c) $s^4 + s^2 + s + 1$

(d) $s^3 + 6s^2 + 2s + 8$

(e) $s^4 + 3s^2 + 2$

(f) $s^4 + 7s^3 + 6s^2 + 21s + 8$

(g) $s^4 + 2s^3 + 6s^2 + 6s + 8$

(h) $s^4 + 5s^3 + 5s^2 + 4s + 10$

5. What do you understand by a positive real function ? Enumerarte its important properties and state the significance of such function in network synthesis.

6. Discuss the properties of positive real function. What are the realizability conditions of a passive network.

7. Determine the positive realness of the function

$$F(s) = \frac{s^2 + 1}{s^3 + 4s}$$

8. Show that the function

$$F(s) = \frac{2s^2 + 9}{s(s^2 + 3)}$$

is not p.r.f.

9. Show that the function

$$F(s) = \frac{ks}{s + \alpha}$$

is p.r.f. if k and α are real and positive quantities.

10. Determine the positive realness of the functions

(a) $Z(s) = \dfrac{2s^2 + 5s + 4}{s + 2}$

(b) $Z(s) = \dfrac{2s^2 + 2s + 2}{s + 2}$

(c) $F(s) = \dfrac{s^2 + 7s + 12}{s^2 + s - 2}$

(d) $Z(s) = \dfrac{2s + 3}{s + 1}$

(e) $Z(s) = \dfrac{s^2 + 2s + 25}{s + 4}$

(f) $Z(s) = \dfrac{s^2 + 2s}{s^2 + 1}$

11. Show that the function

$$Y(s) = \frac{2s^2 - 4s + 5}{s^2 + 7s + 6}$$

is not physically realizable with a passive network.

12. Obtain the first and second Cauer realizations of the reactance function

$$Z(s) = \frac{(s^2 + 1)(s^2 + 3)}{s(s^2 + 2)}$$

13. Realize the ladder network corresponding to the continued fraction expansion about zero of the impedance

$$Z(s) = \frac{s^2 + 1}{s(s^2 + 2)}$$

14. Obtain the first and second Foster forms of the function

$$Z(s) = \frac{s(s^2 + 2)}{(s^2 + 1)(s^2 + 3)}$$

15. Synthesize the function

$$Z(s) = \frac{s + 2}{(s + 1)(s + 3)}$$

in first and second Foster forms.

ANSWERS

4. (a) Not Hurwitz
(b) Not Hurwitz
(c) Not Hurwitz
(d) Hurwitz

(e) Not Hurwitz
(f) Hurwitz
(g) Hurwitz
(h) Not Hurwitz

7. Positive real

10. (a) PR (b) Not PR (c) Not PR (d) PR (e) Not PR (f) PR

12.

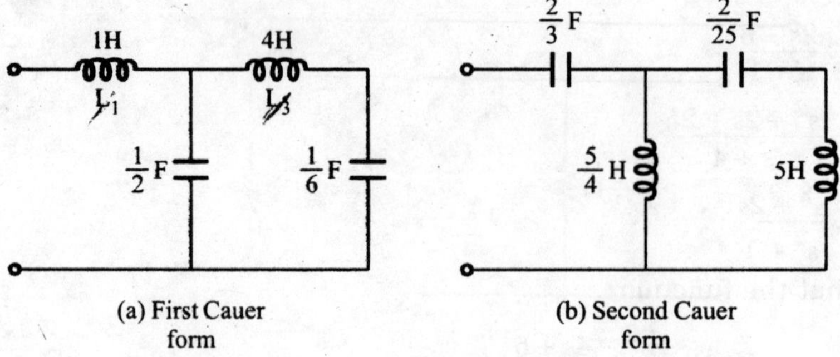

Fig. 21.36

13.

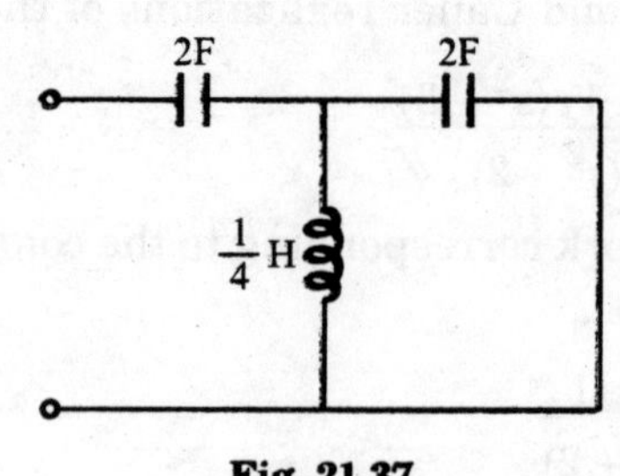

Fig. 21.37

14.

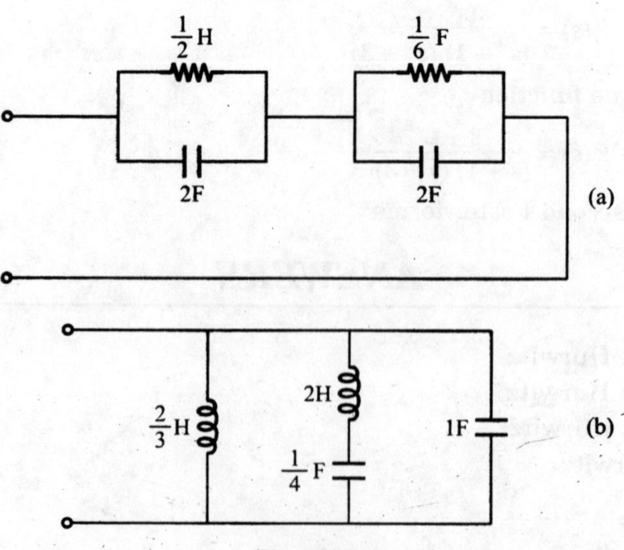

Fig. 21.38

15.

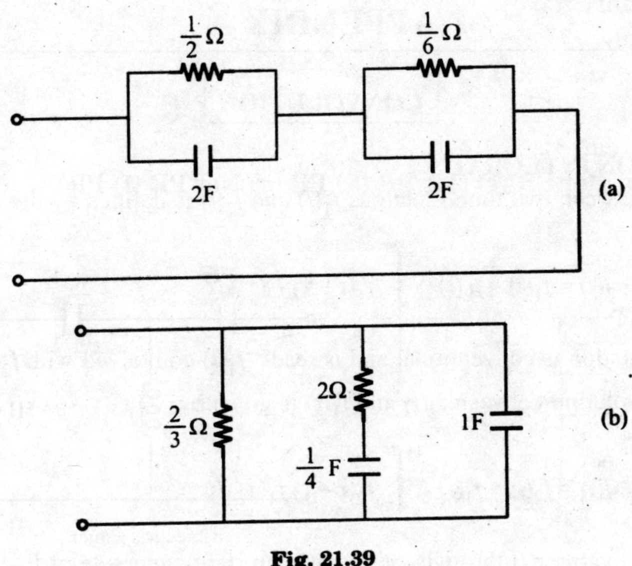

$\frac{1}{2}\Omega$
$\frac{1}{6}\Omega$
2F
2F
(a)
2Ω
$\frac{2}{3}\Omega$
1F
$\frac{1}{4}$ F
(b)

Fig. 21.39

APPENDIX A

CONVOLUTION

A.1 CONVOLUTION INTEGRAL

The convolution between two time functions $f_1(t)$ and $f_2(t)$ is defined by the integral

$$f(t) = f_1(t) * f_2(t) \triangleq \int_0^t f_1(t-\tau) f_2(\tau)\, d\tau \tag{A.1.1}$$

The asterisk (*) notation is conventional and is read "$f_1(t)$ convolved with $f_2(t)$".

Similarly, the convolution between $f_2(t)$ and $f_1(t)$ is given by

$$f(t) = f_2(t) * f_1(t) \triangleq \int_0^t f_2(t-\tau) f_1(\tau)\, d\tau \tag{A.1.2}$$

The process of evaluating $f(t)$ through the integral on right-hand side of Eq. (A.1.1) or Eq. (A.1.2.) is referred to as **taking the convolution** of the functions $f_1(t)$ and $f_2(t)$. The integrals of Eq. (A.1.1) or Eq. (A.1.2) are called the **convolution integrals.**

A.2 BILATERAL CONVOLUTION

In our preceding discussion, we have considered the time functions defined for positive time. In general, we consider the time functions defined for positive as well as negative time. The **bilateral convolution** of two functions $f_1(t)$ and $f_2(t)$ is the integral

$$f_1(t) * f_2(t) = \int_{-\infty}^{\infty} f_1(t-\tau) f_2(\tau)\, d\tau = \int_{-\infty}^{\infty} f_2(t-\tau) f_1(\tau)\, d\tau \tag{A.2.1}$$

defined for every t positive and negative.

The convolution integral applies to linear, time-invariant networks and requires only that $f_1(t)$ and $f_2(t)$ are Laplace transformable.

It should be noted that t is taken as a parameter (constant) in the integration.

If we put $t - \tau = x$, then

$$f_1(t) * f_2(t) = \int_{-\infty}^{\infty} f_1(x) f_2(t-x)\, dx$$

Therefore,

$$f_1(t) * f_2(t) = f_2(t) * f_1(t) \tag{A.2.3}$$

Thus, the order of convolution is unimportant.

It is possible to take the convolution of more than two functions, for example

$$f_1(t) * f_2(t) * f_3(t).$$

In this case, the convolution $f_1(t) * f_2(t)$ is obtained. The convolution of this function with $f_3(t)$ gives the desired result. Again, the order may be changed.

$$f_1(t) * f_2(t) * f_3(t) = \left[f_1(t) * f_2(t)\right] * f_3(t) = f_1(t) * \left[f_2(t) * f_3(t)\right] \quad \text{(A.2.4)}$$

A.3 CONVOLUTION THEOREM

Statement

The convolution theorem states that the Laplace transform of the convolution of two time functions is equal to the product of their individual transforms.

If

$$\pounds f_1(t) = F_1(s)$$

$$\pounds f_2(t) = F_2(s)$$

the convolution theorem can be expressed as

$$\pounds f(t) = \pounds [f_1(t) * f_2(t)] = \pounds [f_2(t) * f_1(t)] = F_1(s)\, F_2(s) \quad \text{(A.3.1)}$$

Proof

$$F_1(s) = \int_0^{\infty} f_1(\tau)\, e^{-s\tau}\, d\tau \quad \text{(A.3.2)}$$

Multiplying both the sides by $F_2(s)$, we obtain

$$F_1(s)\, F_2(s) = F_2(s) \int_0^{\infty} f_1(\tau)\, e^{-s\tau}\, d\tau$$

Since the integration is with respect to τ, $F_2(s)$ can be moved under the integral sign. Thus,

$$F_1(s)\, F_2(s) = \int_0^{\infty} f_1(\tau)\, e^{-s\tau}\, F_2(s)\, d\tau \quad \text{(A.3.3)}$$

By Shifting theorem

$$F_2(s)\, e^{-s\tau} = \pounds\, [f_2(t-\tau)\, u_{-1}\, (t-\tau)]$$

$$= \int_0^{\infty} f_2(t-\tau)\, u_{-1}\, (t-\tau)\, dt \quad \text{(A.3.4)}$$

Therefore, Eq. (A.3.3) can be written as

$$F_1(s)\, F_2(s) = \int_0^{\infty} \left[\int_0^{\infty} f_2(t-\tau)\, u_{-1}\, (t-\tau)\, e^{-st}\, dt \right] f_1(\tau)\, d\tau$$

Since $f_1(t)$ and $f_2(t)$ are both Laplace transformable, the order of integration can be interchanged and

$$F_1(s)\, F_2(s) = \int_0^{\infty} \left[\int_0^{\infty} f_2(t-\tau)\, f_1(\tau)\, u_{-1}\, (t-\tau)\, d\tau \right] e^{-st}\, dt \quad \text{(A.3.5)}$$

Since $u_{-1}\,(t-\tau) = 0$ when $\tau < t$

and $u_{-1}\,(t-\tau) = 1$ when $\tau \geq t$,

therefore, we can write

$$\int_0^\infty f_2(t-\tau) f_1(\tau)\, u_{-1}(t-\tau)\, d\tau = \int_0^t f_2(t-\tau) f_1(\tau)\, d\tau \tag{A.3.6}$$

Thus, Eq. (A.3.5) becomes,

$$F_1(s) F_2(s) = \int_0^\infty \left[\int_0^t f_2(t-\tau) f_1(\tau)\, d\tau \right] e^{-st}\, d\tau$$

$$= £\left[\int_0^t f_2(t-\tau) f_1(\tau)\, d\tau \right] \tag{A.3.7}$$

$$F_1(s) F_2(s) = £\left[f_2(t) * f_1(t) \right]$$

In the above treatment, the functions $f_1(t)$ and $f_2(t)$ can be interchanged, and we can write

$$F_1(s) F_2(s) = £\left[\int_0^t f_2(t-\tau) f_2(\tau)\, d\tau \right] = £\left[f_1(t) * f_2(t) \right] \tag{A.3.8}$$

Since $F_1(s) F_2(s) = F_2(s) F_1(s)$, we have

$$£ f(t) = £\,[f_1(t) * f_2(t)] = £\,[f_2(t) * f_1(t)] = F_1(s) F_2(s) \tag{A.3.9}$$

Equation (A.3.9) gives the result we wanted to prove.

From eq. (A.3.9)

$$f_1(t) * f_2(t) = £^{-1}\,[F_1(s) F_2(s)] \tag{A.3.10}$$

Equation shows that the convolution theorem provides a convenient and simple method of finding the inverse Laplace transform of a product of Laplace transforms.

EXAMPLE A.1

If $f_1(t) = e^{-\alpha t}$ and $f_2(t) = e^{-\beta t}$

Find $f_1(t) * f_2(t)$ and $f_2(t) * f_1(t)$

and hence show that

$$f_1(t) * f_2(t) = f_2(t) * f_1(t)$$

SOLUTION :

$$f_1(t) * f_2(t) = \int_0^t f_1(t-\tau) f_2(\tau)\, d\tau$$

$$= \int_0^t e^{-\alpha(t-\tau)}\, e^{-\beta\tau}\, d\tau$$

$$= \int_0^t e^{-\alpha t} e^{(\alpha-\beta)\tau}\, d\tau$$

$$= e^{-\alpha t} \int_0^t e^{(\alpha-\beta)\tau}\, d\tau$$

Case 1.
When $\alpha = \beta$

$$f_1(t) * f_2(t) = e^{-\alpha t} \int_0^t e^0\, d\tau = t\, e^{-\alpha t}$$

Case 2.
When $\alpha \neq \beta$

$$f_1(t) * f_2(t) = e^{-\alpha t} \left(\frac{1}{\alpha-\beta}\right) \left[e^{(\alpha-\beta)\tau}\right]_0^t$$

$$= \frac{e^{-\alpha t}}{\alpha-\beta} (e^{-\beta t} - e^{-\alpha t})$$

$$f_2(t) * f_1(t) = \int_0^t f_2(t-\tau) f_1(\tau)\, d\tau$$

$$= \int_0^t e^{-\beta(t-\tau)} e^{-\alpha\tau}\, d\tau$$

$$= e^{-\beta t} \int_0^t e^{(\beta-\alpha)\tau}\, d\tau$$

When $\alpha = \beta$

$$f_2(t) * f_1(t) = e^{-\alpha t} \int_0^t e^0\, d\tau = t e^{-\alpha t}$$

When $\alpha \neq \beta$

$$f_2(t) * f_1(t) = \frac{e^{-\beta t}}{\beta-\alpha} \left[e^{(\beta-\alpha)\tau}\right]_0^t$$

$$= \frac{e^{-\beta t}}{\beta-\alpha} \left[e^{(\beta-\alpha)t} - e^0\right]$$

$$= \frac{1}{\beta-\alpha} (e^{-\alpha t} - e^{-\beta t})$$

Hence we see that

$$f_1(t) * f_2(t) = f_2(t) * f_1(t)$$

A.4 CONVOLUTION OF A FUNCTION WITH A UNIT IMPULSE FUNCTION

The convolution of any function $f(t)$ with the unit impulse function $u_0(t)$ is the function $f(t)$ itself.

That is, $$f(t) * u_0(t) = f(t) \tag{A.4.1}$$

Proof

$$f(t) * u_0(t) = \int_{-\infty}^{\infty} f(t-\tau)\, u_0(\tau)\, d\tau$$

$$= \int_{-\infty}^{\infty} f(\tau)\, u_0(t-\tau)\, d\tau$$

$\therefore$ $f(t) * u_0(t) = f(t)$ by Sampling property

Alternative Method

We have

$$£\,[f(t) * u_0(t)] = F(s)\, £\, u_0(t) \tag{A.4.2}$$

Since $£\, u_0(t) = 1$, Eq. (A.4.2) becomes

$$£\,[f(t) * u_0(t)] = F(s) \cdot 1$$

$$f(t) * u_0(t) = £^{-1}\,[F(s)] = f(t) \tag{A.4.3}$$

A simple extension of Eq. (A.4.1) gives

$$f(t) * u_0(t-\tau) = f(t-\tau) \tag{A.4.4}$$

$$f(t-t_1) * u_0(t-\tau_2) = f(t-t_1-t_2) \tag{A.4.5}$$

$$u_0(t-t_1) * u_0(t_1-t_2) = u_0\,(t-t_1-t_2) \tag{A.4.6}$$

INDEX

E

F

G

H

S